Quantum Chemistry

Sixth Edition

IRA N. LEVINE
Chemistry Department
Brooklyn College
City University of New York

Pearson Education International

Assistant Editor: *Carol DuPont*
Senior Editor: *Kent Porter Hamman*
Editor in Chief, Science: *Nicole Folchetti*
Marketing Manager: *Liz Averbeck*
Managing Editor, Production: *Gina M. Cheselka*
Art Director: *Jayne Conte*
Cover Designer: *Bruce Kenselaar*
Senior Operations Supervisor: *Alan Fischer*
Production Supervision/Composition: *Prepare, Inc.*

Pearson Prentice Hall
Pearson Education, Inc.
Upper Saddle River, NJ 07458

Printed in the United States of America
10 9 8 7 6 5 4 3 2 1

ISBN-13 978-0-13-235850-7
ISBN-10 0-13-235850-6

Pearson Education Ltd., *London*
Pearson Education Singapore, Pte. Ltd
Pearson Education, Canada, Ltd
Pearson Education—Japan
Pearson Education Australia Pty, Limited
Pearson Education North Asia, Ltd
Pearson Educación de Mexico, S.A. de C.V.
Pearson Education Malaysia, Pte. Ltd.
Pearson Education, Upper Saddle River, New Jersey

To my quantum chemistry students: Vincent Adams, Margaret Adamson, Emanuel Akinfeleye, Ricardo Alkins, Byongjae An, Salvatore Atzeni, Abe Auerbach, Andrew Auerbach, Nikolay Azar, Joseph Barbuto, David Baron, Christie Basseth, Sene Bauman, Laurance Beaton, Howard Becker, Michael Beitchman, Anna Berne, Kamal Bharucha, Susan Bienenfeld, Mark Blackman, Toby Block, Allen Bloom, Gina Bolnet, Demetrios Boyce, Diza Braksmayer, Steve Braunstein, Paul Brumer, Jean Brun, Margaret Buckley, Lynn Caporale, Richard Carter, Julianne Caton-Williams, Shih-ching Chang, Ching-hong Chen, Hongbin Chen, Huifen Chen, Kangmin Chen, Kangping Chen, Guang-Yu Cheng, Yu-Chi Cheng, El-hadi Cherchar, Jeonghwan Cho, Ting-Yi Chu, Kyu Suk Chung, Joseph Cincotta, Robert Curran, Joseph D'Amore, Ronald Davy, Jody Delsol, Aly Dominique, Xiao-Hong Dong, Barry DuRon, Azaria Eisenberg, Myron Elgart, Musa Elmagadam, Anna Eng, Stephen Engel, Quianping Fang, Nicola Farina, Larry Filler, Seymour Fishman, Donald Franceschetti, Mark Freilich, Michael Freshwater, Tobi Eisenstein Fried, Joel Friedman, Kenneth Friedman, Aryeh Frimer, Mark Froimowitz, Irina Gaberman, Paul Gallant, Hong Gan, Mark Gold, Stephen Goldman, Neil Goodman, Roy Goodman, Isaac Gorbaty, Aleksander Gorbenko, Steven Greenberg, Walter Greigg, Michael Gross, Zhijie Gu, Judy Guiseppi-Henry, Lin Guo, Hasan Hajomar, Runyu Han, Sheila Handler, Noyes Harrigan, Jun He, Warren Hirsch, Richard Hom, Kuo-zong Hong, Mohammed Hossain, Fu-juan Hsu, Bo Hu, Jong-chin Hwan, Leonard Itzkowitz, Colin John, Mark Johnson, Kirby Juengst, Abraham Karkowsky, Spiros Kassomenakis, Abdelahad Khajo, Mohammed Khan, Michael Kittay, Colette Knight, Barry Kohn, Yasemin Kopkalli, Malgorzata Kulcyk-Stanko, David Kurnit, Athanasios Ladas, Alan Lambowitz, Eirini Lampiri, Bentley Lane, Yedidyah Langsam, Noah Lansner, Surin Laosooksathit, Chi-Yin Lee, Chiu Hong Lee, Stephen Lemont, Elliot Lerner, Jiang Li, Zheng Li, Israel Liebersohn, Joel Liebman, Steven Lipp, Huiyu Liu, Letian Liu, James Liubicich, John Lobo, Rachel Loftoa, Wei Luo, Dennis Lynch, Michelle Maison, Mohammad Malik, Pietro Mangiaracina, Louis Maresca, Allen Marks, Tom McDonough, Keisha McMillan, Antonio Mennito, Leonid Metlitsky, Ira Michaels, Tziril Miller, Bin Mo, Qi Mo, Paul Mogolesko, Alim Monir, Safrudin Mustopa, Irving Nadler, Stuart Nagourney, Kwazi Ndlovu, Harold Nelson, Wen-Hui Pan, Padmanabhan Parakat, Frank Pecci, Albert Pierre-Louis, Paloma Pimenta, Eli Pines, Jerry Polesuk, Arlene Gallanter Pollin, James Pollin, Lahanda Punyasena, Cynthia Racer, Munira Rampersaud, Caleen Ramsook, Robert Richman, Richard Rigg, Bruce Rosenberg, Martin Rosenberg, Robert Rundberg, Edward Sachs, Mohamed Salem, Mahendra Sawh, David Schaeffer, Gary Schneier, Neil Schweid, Judith Rosenkranz Selwyn, Gunnar Senum, Simone Shaker, Steven Shaya, Allen Sheffron, Wu-mian Shen, Yuan Shi, Lawrence Shore, Alvin Silverstein, Barry Siskind, Jerome Solomon, De Zai Song, Henry Sperling, Joseph Springer, Charles Stimler, Helen Sussman, Sybil Tobierre, Dana McGowan Tormey, David Trauber, Choi Han Tsang, King-hung Tse, Michele Tujague, Irina Vasilkin, Natalya Voluschuk, Sammy Wainhaus, Alan Waldman, Huai Zhen Wang, Zheng Wang, Robert Washington, Janet Weaver, William Wihlborg, Peter Williamsen, Frederic Wills, Shiming Wo, Guohua Wu, Jinan Wu, Xiaowen Wu, Ming Min Xia, Wei-Guo Xia, Xiaoming Ye, Ching-Chun Yiu, Wen Young, Xue-yi Yuan, Ken Zaner, Juin-tao Zhang, Hannian Zhao, Li Li Zhou, Shan Zhou, Yun Zhou.

Contents

Preface

This book is intended for first-year graduate and advanced undergraduate courses in quantum chemistry.

The following improvements were made in the sixth edition:

- Exercises with answers were added to most of the in-chapter examples.
- The problems were revised and are now classified by section.
- Section 1.9 containing calculus formulas for review was added.
- Chapter 13 was shortened by moving three of its sections to Chapter 14.
- Chapter 15 was shortened by moving discussion of correlation methods to a new Chapter 16.
- Some of the mathematical derivations have been moved from the text to homework problems with step-by-step hints; examples are the derivations of the He $1/r_{12}$ integral (Section 9.3) and the Numerov-method formula (Section 4.4).
- References to online computer simulations of several quantum-mechanical systems were added.
- The discussion of NMR spectroscopy in Section 10.9 was expanded.
- Mention of Gaussian units was dropped.

New material in the sixth edition includes the following:

- Mayer bond orders (Section 15.6)
- The RI (density-fitting) approximation (Sections 15.16 and 16.2)
- The MP2-R12 and MP2-F12 methods (Section 16.2)
- Extrapolation to the complete-basis-set limit (Sections 15.5 and 16.3)
- The SCS-MP2 and SOS-MP2 methods (Section 16.2)
- Meta-GGA functionals (Section 16.4)
- Time-dependent DFT (Section 16.4)
- The G4 , W1, W2, W3, and W4 methods (Section 16.5)
- The PM5, PM6, RM1, PDDG/PM3, and SCC-DFTB methods (Section 17.4)

A solutions manual for the problems in the book is available.

The expanding role of quantum chemistry makes it highly desirable for students in all areas of chemistry to understand modern methods of electronic structure calculation, and this book has been written with this goal in mind.

I have tried to make explanations clear and complete, without glossing over difficult or subtle points. Derivations are given with enough detail to make them easy to follow, and I avoid resorting to the frustrating phrase "it can be shown that" wherever possible. The aim is to give students a solid understanding of the physical and mathematical aspects of quantum mechanics and molecular electronic structure. The book is designed to be useful to students in all branches of chemistry, not just future quantum chemists. However, the presentation is such that those who do go on in quantum chemistry will have a good foundation and will not be hampered by misconceptions.

An obstacle faced by many chemistry students in learning quantum mechanics is their unfamiliarity with much of the required mathematics. In this text I have included detailed treatments of operators, differential equations, simultaneous linear equations, and other needed topics. Rather than putting all the mathematics in an introductory chapter or a series of appendices, I have integrated the mathematics with the physics and chemistry. Immediate application of the mathematics to solving a quantum-mechanical problem will make the mathematics more meaningful to students than would separate study of the mathematics. I have also kept in mind the limited physics background of many chemistry students by reviewing topics in physics.

This book has benefited from the reviews and suggestions of Leland Allen, N. Colin Baird, Steven Bernasek, James Bolton, W. David Chandler, Donald Chesnut, R. James Cross, David Farrelly, Melvyn Feinberg, Gordon A. Gallup, David Goldberg, Tracy Hamilton, John Head, Warren Hehre, Hans Jaffé, Miklos Kertesz, Neil Kestner, Harry King, Peter Kollman, Mel Levy, Errol Lewars, Joel Liebman, Frank Meeks, Robert Metzger, Pedro Muiño, William Palke, Sharon Palmer, Gary Pfeiffer, Russell Pitzer, Kenneth Sando, Harrison Shull, James J. P. Stewart, Richard Stratt, Arieh Warshel, John S. Winn, and Michael Zerner. The following people provided reviews for the sixth edition: Gary DeBoer, Douglas Doren, Daniel Gerrity, Robert Griffin, Sharon Hammes-Schiffer, James Harrison, Robert Hinde, Anna Krylov, Tien-Sung Tom Lin, Ryan McLaughlin, Charles Millner, John H. Moore, Kirk Peterson, Oleg Prezhdo, Frank Rioux, Fu-Ming Tao, Ronald Terry, Alexander Van Hook, and Peter Weber. I wish to thank these people and several anonymous reviewers.

I would appreciate receiving any suggestions that readers may have for improving the book.

Ira N. Levine
INLevine@brooklyn.cuny.edu

Quantum Chemistry

CHAPTER 1

The Schrödinger Equation

1.1 QUANTUM CHEMISTRY

In the late seventeenth century, Isaac Newton discovered **classical mechanics**, the laws of motion of macroscopic objects. In the early twentieth century, physicists found that classical mechanics does not correctly describe the behavior of very small particles such as the electrons and nuclei of atoms and molecules. The behavior of such particles is described by a set of laws called **quantum mechanics**.

Quantum chemistry applies quantum mechanics to problems in chemistry. The influence of quantum chemistry is evident in all branches of chemistry. Physical chemists use quantum mechanics to calculate (with the aid of statistical mechanics) thermodynamic properties (for example, entropy, heat capacity) of gases; to interpret molecular spectra, thereby allowing experimental determination of molecular properties (for example, molecular geometries, dipole moments, barriers to internal rotation, energy differences between conformational isomers); to calculate molecular properties theoretically; to calculate properties of transition states in chemical reactions, thereby allowing estimation of rate constants; to understand intermolecular forces; and to deal with bonding in solids.

Organic chemists use quantum mechanics to estimate the relative stabilities of molecules, to calculate properties of reaction intermediates, to investigate the mechanisms of chemical reactions, and to analyze nuclear-magnetic-resonance spectra.

Analytical chemists use spectroscopic methods extensively. The frequencies and intensities of lines in a spectrum can be properly understood and interpreted only through the use of quantum mechanics.

Inorganic chemists use ligand field theory, an approximate quantum-mechanical method, to predict and explain the properties of transition-metal complex ions.

Although the large size of biologically important molecules makes quantum-mechanical calculations on them extremely hard, biochemists are beginning to benefit from quantum-mechanical studies of conformations of biological molecules, enzyme–substrate binding, and solvation of biological molecules.

Quantum mechanics determines the properties of nanomaterials (objects with at least one dimension in the range 1 to 100 nm), and calculational methods to deal with nanomaterials are being developed (E. K. Wilson, *Chem. Eng. News,* April 28, 2003, p. 27). When one or more dimensions of a material fall below 100 nm (and especially below 20 nm), dramatic changes in the optical, electronic, chemical, and other properties from those of the bulk material can occur. A semiconductor or metal object with one dimension in the 1 to 100 nm range is called a *quantum well*; one with two dimensions in this range is a *quantum wire*; and one with all three dimensions in this range is a *quantum dot*. The word *quantum* in these names indicates the key role played by quantum mechanics in determining the properties of such materials. Many people have speculated that nanoscience and nanotechnology will bring about the "next industrial revolution," while others have criticized this field as, so far, containing more hype than achievement.

The rapid increase in computer speed and the development of new methods (such as density functional theory—Section 16.4) of doing molecular calculations have made quantum chemistry a practical tool in all areas of chemistry. Nowadays, several companies sell quantum-chemistry software for doing molecular quantum-chemistry calculations. These programs are designed to be used by all kinds of chemists, not just quantum chemists. Because of the rapidly expanding role of quantum chemistry and related theoretical and computational methods, the American Chemical Society began publication of the new periodical the *Journal of Chemical Theory and Computation* in 2005.

"Quantum mechanics ... underlies nearly all of modern science and technology. It governs the behavior of transistors and integrated circuits ... and is ... the basis of modern chemistry and biology" (Stephen Hawking, *A Brief History of Time*, 1988, Bantam, chap. 4).

1.2 HISTORICAL BACKGROUND OF QUANTUM MECHANICS

The development of quantum mechanics began in 1900 with Planck's study of the light emitted by heated solids, so we start by discussing the nature of light.

In 1801, Thomas Young gave convincing experimental evidence for the wave nature of light by observing diffraction and interference when light went through two adjacent pinholes. (*Diffraction* is the bending of a wave around an obstacle. *Interference* is the combining of two waves of the same frequency to give a wave whose disturbance at each point in space is the algebraic or vector sum of the disturbances at that point resulting from each interfering wave. See any first-year physics text.)

In 1864, James Clerk Maxwell published four equations, known as Maxwell's equations, which unified the laws of electricity and magnetism. Maxwell's equations predicted that an accelerated electric charge would radiate energy in the form of electromagnetic waves consisting of oscillating electric and magnetic fields. The speed predicted by Maxwell's equations for these waves turned out to be the same as the experimentally measured speed of light. Maxwell concluded that light is an electromagnetic wave.

In 1888, Heinrich Hertz detected radio waves produced by accelerated electric charges in a spark, as predicted by Maxwell's equations. This convinced physicists that light is indeed an electromagnetic wave.

All electromagnetic waves travel at speed $c = 2.998 \times 10^8$ m/s in vacuum. The frequency ν and wavelength λ of an electromagnetic wave are related by

$$\boxed{\lambda\nu = c} \qquad \textbf{(1.1)}$$

(Equations that are enclosed in a box should be memorized.) Various conventional labels are applied to electromagnetic waves depending on their frequency. In order of increasing frequency are radio waves, microwaves, infrared radiation, visible light, ultraviolet radiation, X-rays, and gamma rays. We shall use the term **light** to denote any kind of electromagnetic radiation. Wavelengths of visible and ultraviolet radiation were formerly given in **angstroms** (Å) and are now given in **nanometers** (nm):

$$\boxed{1\text{ nm} = 10^{-9}\text{ m}, \qquad 1\text{ Å} = 10^{-10}\text{ m} = 0.1\text{ nm}} \tag{1.2}$$

In the late 1800s, physicists measured the intensity of light at various frequencies emitted by a heated blackbody at a fixed temperature. A *blackbody* is an object that absorbs all light falling on it. A good approximation to a blackbody is a cavity with a tiny hole. When physicists used statistical mechanics and the electromagnetic-wave model of light to predict the intensity-versus-frequency curve for emitted blackbody radiation, they found a result in complete disagreement with the high-frequency portion of the experimental curves.

In 1900, Max Planck developed a theory that gave excellent agreement with the observed blackbody-radiation curves. Planck assumed the radiation emitters and absorbers in the blackbody to be harmonically oscillating electric charges ("resonators") in equilibrium with electromagnetic radiation in a cavity. Planck assumed that the total energy of those resonators whose frequency is ν consisted of N indivisible "energy elements," each of magnitude $h\nu$, where N is an integer and h (**Planck's constant**) was a new constant in physics. The value $h = 6.6 \times 10^{-34}$ J·s led to curves that agreed with the experimental blackbody curves. Planck distributed these energy elements among the resonators. In effect, this restricted the energy of each resonator to be a whole-number multiple of $h\nu$ (although Planck did not explicitly say this). Thus the energy of each resonator was **quantized**, meaning that only certain discrete values were allowed for a resonator energy. Planck's work is usually considered to mark the beginning of quantum mechanics. However, historians of physics have debated whether Planck in 1900 viewed energy quantization as a description of physical reality or as merely a mathematical approximation that allowed him to obtain the correct blackbody radiation formula. [See C. A. Gearhart, *Phys. Perspect.*, **4**, 170 (2002); S. G. Brush, *Am. J. Phys.*, **70**, 119 (2002).]

The concept of energy quantization is in direct contradiction to all previous ideas of physics. According to Newtonian mechanics, the energy of a material body can vary continuously. However, only with the hypothesis of quantized energy does one obtain the correct blackbody-radiation curves.

The second application of energy quantization was to the photoelectric effect. In the *photoelectric effect,* light shining on a metal causes emission of electrons. The energy of a wave is proportional to its intensity and is not related to its frequency, so the electromagnetic-wave picture of light leads one to expect that the kinetic energy of an emitted photoelectron would increase as the light intensity increases but would not change as the light frequency changes. Instead, one observes that the kinetic energy of an emitted electron is independent of the light's intensity but increases as the light's frequency increases.

In 1905, Einstein showed that these observations could be explained by regarding light as composed of particlelike entities (called **photons**), with each photon having an energy

$$\boxed{E_{\text{photon}} = h\nu} \tag{1.3}$$

When an electron in the metal absorbs a photon, part of the absorbed photon energy is used to overcome the forces holding the electron in the metal, and the remainder appears as kinetic energy of the electron after it has left the metal. Conservation of energy gives $h\nu = \Phi + T$, where Φ is the minimum energy needed by an electron to escape the metal (the metal's *work function*), and T is the maximum kinetic energy of an emitted electron. An increase in the light's frequency ν increases the photon energy and hence increases the kinetic energy of the emitted electron. An increase in light intensity at fixed frequency increases the rate at which photons strike the metal and hence increases the rate of emission of electrons, but does not change the kinetic energy of each emitted electron.

The photoelectric effect shows that light can exhibit particlelike behavior in addition to the wavelike behavior it shows in diffraction experiments.

In 1907, Einstein applied energy quantization to the vibrations of atoms in a solid element, assuming that each atom's vibrational energy in each direction (x, y, z) is restricted to be an integer times $h\nu_{\text{vib}}$, where the vibrational frequency ν_{vib} is characteristic of the element. Using statistical mechanics, Einstein derived an expression for the constant-volume heat capacity C_V of the solid. Einstein's equation agreed fairly well with known C_V-versus-temperature data for diamond.

Now let us consider the structure of matter.

In the late nineteenth century, investigations of electric discharge tubes and natural radioactivity showed that atoms and molecules are composed of charged particles. Electrons have a negative charge. The proton has a positive charge equal in magnitude but opposite in sign to the electron charge and is 1836 times as heavy as the electron. The third constituent of atoms, the neutron (discovered in 1932), is uncharged and slightly heavier than the proton.

Starting in 1909, Rutherford, Geiger, and Marsden repeatedly passed a beam of alpha particles through a thin metal foil and observed the deflections of the particles by allowing them to fall on a fluorescent screen. Alpha particles are positively charged helium nuclei obtained from natural radioactive decay. Most of the alpha particles passed through the foil essentially undeflected, but, surprisingly, a few underwent large deflections, some being deflected backward. To get large deflections, one needs a very close approach between the charges, so that the Coulombic repulsive force is great. If the positive charge were spread throughout the atom (as J. J. Thomson had proposed in 1904), once the high-energy alpha particle penetrated the atom, the repulsive force would fall off, becoming zero at the center of the atom, according to classical electrostatics. Hence Rutherford concluded that such large deflections could occur only if the positive charge were concentrated in a tiny, heavy nucleus.

An atom contains a tiny (10^{-13} to 10^{-12} cm radius), heavy nucleus consisting of neutrons and Z protons, where Z is the atomic number. Outside the nucleus there are Z electrons. The charged particles interact according to Coulomb's law. (The nucleons are held together in the nucleus by strong, short-range nuclear forces, which will not concern us.) The radius of an atom is about one angstrom, as shown, for example, by results from the kinetic theory of gases. Molecules have more than one nucleus.

The chemical properties of atoms and molecules are determined by their electronic structure, and so the question arises as to the nature of the motions and energies of the electrons. Since the nucleus is much more massive than the electron, we expect the motion of the nucleus to be slight compared with the electrons' motions.

In 1911, Rutherford proposed his planetary model of the atom in which the electrons revolved about the nucleus in various orbits, just as the planets revolve about the sun. However, there is a fundamental difficulty with this model. According to classical electromagnetic theory, an accelerated charged particle radiates energy in the form of electromagnetic (light) waves. An electron circling the nucleus at constant speed is being accelerated, since the direction of its velocity vector is continually changing. Hence the electrons in the Rutherford model should continually lose energy by radiation and therefore would spiral toward the nucleus. Thus, according to classical (nineteenth-century) physics, the Rutherford atom is unstable and would collapse.

A possible way out of this difficulty was proposed by Niels Bohr in 1913, when he applied the concept of quantization of energy to the hydrogen atom. Bohr assumed that the energy of the electron in a hydrogen atom was quantized, with the electron constrained to move only on one of a number of allowed circles. When an electron makes a transition from one Bohr orbit to another, a photon of light whose frequency ν satisfies

$$\boxed{E_{\text{upper}} - E_{\text{lower}} = h\nu} \qquad \textbf{(1.4)}$$

is absorbed or emitted, where E_{upper} and E_{lower} are the energies of the upper and lower states (conservation of energy). With the assumption that an electron making a transition from a free (ionized) state to one of the bound orbits emits a photon whose frequency is an integral multiple of one-half the classical frequency of revolution of the electron in the bound orbit, Bohr used classical mechanics to derive a formula for the hydrogen-atom energy levels. Using (1.4), he got agreement with the observed hydrogen spectrum. However, attempts to fit the helium spectrum using the Bohr theory failed. Moreover, the theory could not account for chemical bonds in molecules.

The failure of the Bohr model arises from the use of classical mechanics to describe the electronic motions in atoms. The evidence of atomic spectra, which show discrete frequencies, indicates that only certain energies of motion are allowed; the electronic energy is quantized. However, classical mechanics allows a continuous range of energies. Quantization does occur in wave motion; for example, the fundamental and overtone frequencies of a violin string. Hence Louis de Broglie suggested in 1923 that the motion of electrons might have a wave aspect; that an electron of mass m and speed v would have a wavelength

$$\lambda = \frac{h}{mv} = \frac{h}{p} \qquad (1.5)$$

associated with it, where p is the linear momentum. De Broglie arrived at Eq. (1.5) by reasoning in analogy with photons. The energy of a photon can be expressed, according to Einstein's special theory of relativity, as $E = pc$, where c is the speed of light and p is the photon's momentum. Using $E_{\text{photon}} = h\nu$, we get $pc = h\nu = hc/\lambda$ and $\lambda = h/p$ for a photon traveling at speed c. Equation (1.5) is the corresponding equation for an electron.

In 1927, Davisson and Germer experimentally confirmed de Broglie's hypothesis by reflecting electrons from metals and observing diffraction effects. In 1932, Stern observed the same effects with helium atoms and hydrogen molecules, thus verifying that the wave effects are not peculiar to electrons, but result from some general law of motion for microscopic particles.

Thus electrons behave in some respects like particles and in other respects like waves. We are faced with the apparently contradictory "wave–particle duality" of matter (and of light). How can an electron be both a particle, which is a localized entity, and a wave, which is nonlocalized? The answer is that an electron is neither a wave nor a particle, but something else. An accurate pictorial description of an electron's behavior is impossible using the wave or particle concept of classical physics. The concepts of classical physics have been developed from experience in the macroscopic world and do not properly describe the microscopic world. Evolution has shaped the human brain to allow it to understand and deal effectively with macroscopic phenomena. The human nervous system was not developed to deal with phenomena at the atomic and molecular level, so it is not surprising if we cannot fully understand such phenomena.

Although both photons and electrons show an apparent duality, they are not the same kinds of entities. Photons always travel at speed c and have zero rest mass; electrons always have $v < c$ and a nonzero rest mass. Photons must always be treated relativistically, but electrons whose speed is much less than c can be treated nonrelativistically.

1.3 THE UNCERTAINTY PRINCIPLE

Let us consider what effect the wave–particle duality has on attempts to measure simultaneously the x coordinate and the x component of linear momentum of a microscopic particle. We start with a beam of particles with momentum p, traveling in the y direction, and we let the beam fall on a narrow slit. Behind this slit is a photographic plate. See Fig. 1.1.

Particles that pass through the slit of width w have an uncertainty w in their x coordinate at the time of going through the slit. Calling this spread in x values Δx, we have $\Delta x = w$.

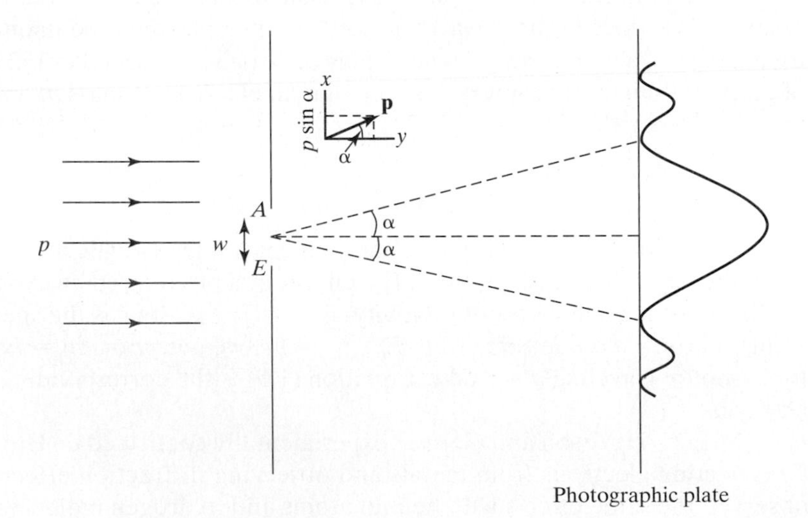

FIGURE 1.1 Diffraction of electrons by a slit.

Since microscopic particles have wave properties, they are diffracted by the slit producing (as would a light beam) a diffraction pattern on the plate. The height of the graph in Fig. 1.1 is a measure of the number of particles reaching a given point. The diffraction pattern shows that when the particles were diffracted by the slit, their direction of motion was changed so that part of their momentum was transferred to the x direction. The x component of momentum p_x equals the projection of the momentum vector $\mathbf{p}$ in the x direction. A particle deflected upward by an angle α has $p_x = p \sin \alpha$. A particle deflected downward by α has $p_x = -p \sin \alpha$. Since most of the particles undergo deflections in the range $-\alpha$ to α, where α is the angle to the first minimum in the diffraction pattern, we shall take one-half the spread of momentum values in the central diffraction peak as a measure of the uncertainty Δp_x in the x component of momentum: $\Delta p_x = p \sin \alpha$.

Hence at the slit, where the measurement is made,

$$\Delta x \, \Delta p_x = pw \sin \alpha \tag{1.6}$$

The angle α at which the first diffraction minimum occurs is readily calculated. The condition for the first minimum is that the difference in the distances traveled by particles passing through the slit at its upper edge and particles passing through the center of the slit be equal to $\frac{1}{2}\lambda$, where λ is the wavelength of the associated wave. Waves originating from the top of the slit are then exactly out of phase with waves originating from the center of the slit, and they cancel each other. Waves originating from a point in the slit at a distance d below the slit midpoint cancel with waves originating at a distance d below the top of the slit. Drawing AC in Fig. 1.2 so that $AD = CD$, we have the difference in path length as BC. The distance from the slit to the screen is large compared with the slit width. Hence AD and BD are nearly parallel. This makes the angle ACB essentially a right angle, and so angle $BAC = \alpha$. The path difference BC is then $\frac{1}{2}w \sin \alpha$. Setting BC equal to $\frac{1}{2}\lambda$, we have $w \sin \alpha = \lambda$, and Eq. (1.6) becomes $\Delta x \, \Delta p_x = p\lambda$. The wavelength λ is given by the de Broglie relation $\lambda = h/p$, so $\Delta x \, \Delta p_x = h$. Since the uncertainties have not been precisely defined, the equality sign is not really justified. Instead we write

$$\Delta x \, \Delta p_x \approx h \tag{1.7}$$

indicating that the product of the uncertainties in x and p_x is of the order of magnitude of Planck's constant.

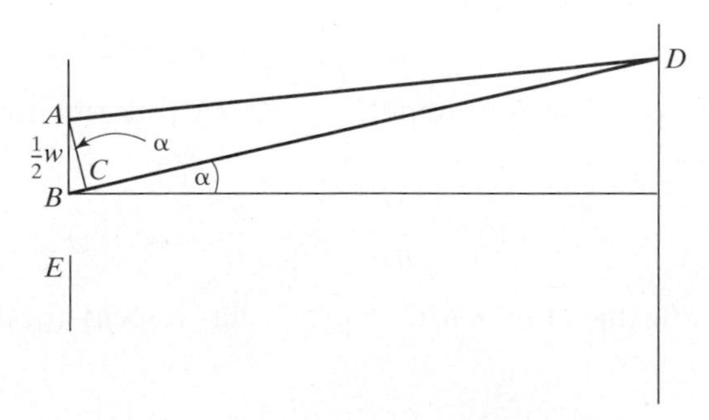

FIGURE 1.2 Calculation of first diffraction minimum.

Although we have demonstrated (1.7) for only one experimental setup, its validity is general. No matter what attempts are made, the wave–particle duality of microscopic "particles" imposes a limit on our ability to measure simultaneously the position and momentum of such particles. The more precisely we determine the position, the less accurate is our determination of momentum. (In Fig. 1.1, $\sin \alpha = \lambda/w$, so narrowing the slit increases the spread of the diffraction pattern.) This limitation is the **uncertainty principle**, discovered in 1927 by Werner Heisenberg.

Because of the wave–particle duality, the act of measurement introduces an uncontrollable disturbance in the system being measured. We started with particles having a precise value of p_x (zero). By imposing the slit, we measured the x coordinate of the particles to an accuracy w, but this measurement introduced an uncertainty into the p_x values of the particles. The measurement changed the state of the system.

1.4 THE TIME-DEPENDENT SCHRÖDINGER EQUATION

Classical mechanics applies only to macroscopic particles. For microscopic "particles" we require a new form of mechanics, called **quantum mechanics**. We now consider some of the contrasts between classical and quantum mechanics. For simplicity a one-particle, one-dimensional system will be discussed.

In classical mechanics the motion of a particle is governed by Newton's second law:

$$\boxed{F = ma = m\frac{d^2x}{dt^2}} \tag{1.8}$$

where F is the force acting on the particle, m is its mass, and t is the time; a is the acceleration, given by $a = dv/dt = (d/dt)(dx/dt) = d^2x/dt^2$, where v is the velocity. Equation (1.8) contains the second derivative of the coordinate x with respect to time. To solve it, we must carry out two integrations. This introduces two arbitrary constants c_1 and c_2 into the solution, and

$$x = g(t, c_1, c_2) \tag{1.9}$$

where g is some function of time. We now ask: What information must we possess at a given time t_0 to be able to predict the future motion of the particle? If we know that at t_0 the particle is at point x_0, we have

$$x_0 = g(t_0, c_1, c_2) \tag{1.10}$$

Since we have two constants to determine, more information is needed. Differentiating (1.9), we have

$$\frac{dx}{dt} = v = \frac{d}{dt}g(t, c_1, c_2)$$

If we also know that at time t_0 the particle has velocity v_0, then we have the additional relation

$$v_0 = \frac{d}{dt}g(t, c_1, c_2)\bigg|_{t=t_0} \tag{1.11}$$

We may then use (1.10) and (1.11) to solve for c_1 and c_2 in terms of x_0 and v_0. Knowing c_1 and c_2, we can use Eq. (1.9) to predict the exact future motion of the particle.

As an example of Eqs. (1.8) to (1.11), consider the vertical motion of a particle in the earth's gravitational field. Let the x axis point upward. The force on the particle is downward and is $F = -mg$, where g is the gravitational acceleration constant. Newton's second law (1.8) is $-mg = m\, d^2x/dt^2$, so $d^2x/dt^2 = -g$. A single integration gives $dx/dt = -gt + c_1$. The arbitrary constant c_1 can be found if we know that at time t_0 the particle had velocity v_0. Since $v = dx/dt$, we have $v_0 = -gt_0 + c_1$ and $c_1 = v_0 + gt_0$. Therefore, $dx/dt = -gt + gt_0 + v_0$. Integration gives $x = -\frac{1}{2}gt^2 + (gt_0 + v_0)t + c_2$. If we know that at time t_0 the particle had position x_0, then $x_0 = -\frac{1}{2}gt_0^2 + (gt_0 + v_0)t_0 + c_2$ and $c_2 = x_0 - \frac{1}{2}gt_0^2 - v_0t_0$. The expression for x as a function of time becomes $x = -\frac{1}{2}gt^2 + (gt_0 + v_0)t + x_0 - \frac{1}{2}gt_0^2 - v_0t_0$ or $x = x_0 - \frac{1}{2}g(t - t_0)^2 + v_0(t - t_0)$. Knowing x_0 and v_0 at time t_0, we can predict the future position of the particle.

The classical-mechanical potential energy V of a particle moving in one dimension is defined to satisfy

$$\boxed{\partial V(x,t)/\partial x = -F(x,t)} \qquad \textbf{(1.12)}$$

For example, for a particle moving in the earth's gravitational field, $\partial V/\partial x = -F = mg$ and integration gives $V = mgx + c$, where c is an arbitrary constant. We are free to set the zero level of potential energy wherever we please. Choosing $c = 0$, we have $V = mgx$ as the potential-energy function.

The word **state** in classical mechanics means a specification of the position and velocity of each particle of the system at some instant of time, plus specification of the forces acting on the particles. According to Newton's second law, given the state of a system at any time, its future state and future motions are exactly determined, as shown by Eqs. (1.9)–(1.11). The impressive success of Newton's laws in explaining planetary motions led many philosophers to use Newton's laws as an argument for philosophical determinism. The mathematician and astronomer Laplace (1749–1827) assumed that the universe consisted of nothing but particles that obeyed Newton's laws. Therefore, given the state of the universe at some instant, the future motion of everything in the universe was completely determined. A super-being able to know the state of the universe at any instant could, in principle, calculate all future motions.

Although classical mechanics is deterministic, many classical-mechanical systems (for example, a pendulum oscillating under the influence of gravity, friction, and a periodically varying driving force) show chaotic behavior for certain ranges of the systems' parameters. In a chaotic system, the motion is extraordinarily sensitive to the initial values of the particles' positions and velocities and to the forces acting, and two initial states that differ by an experimentally undetectable amount will eventually lead to very different future behavior of the system. Thus, because the accuracy with which one can measure the initial state is limited, prediction of the long-term behavior of a chaotic classical-mechanical system is, in practice, impossible, even though the system obeys deterministic equations. Computer calculations of solar-system planetary orbits over tens of millions of years indicate that the motions of the planets are chaotic [*Science,* **257**, 33 (1992); I. Peterson, *Newton's Clock: Chaos in the Solar System,* Freeman, 1993; J. J. Lissauer, *Rev. Mod. Phys.*, **71**, 835 (1999)].

Given exact knowledge of the present state of a classical-mechanical system, we can predict its future state. However, the Heisenberg uncertainty principle shows that we cannot determine simultaneously the exact position and velocity of a microscopic particle, so the very knowledge required by classical mechanics for predicting the future motions of a system cannot be obtained. We must be content in quantum mechanics with something less than complete prediction of the exact future motion.

Our approach to quantum mechanics will be to *postulate* the basic principles and then use these postulates to deduce experimentally testable consequences such as the energy levels of atoms. To describe the **state** of a system in quantum mechanics, we postulate the existence of a function of the particles' coordinates called the **state function** or **wave function** Ψ. Since the state will, in general, change with time, Ψ is also a function of time. For a one-particle, one-dimensional system, we have $\Psi = \Psi(x, t)$. The wave function contains all possible information about a system, so instead of speaking of "the state described by the wave function Ψ," we simply say "the state Ψ." Newton's second law tells us how to find the future state of a classical-mechanical system from knowledge of its present state. To find the future state of a quantum-mechanical system from knowledge of its present state, we want an equation that tells us how the wave function changes with time. For a one-particle, one-dimensional system, this equation is postulated to be

$$-\frac{\hbar}{i}\frac{\partial \Psi(x,t)}{\partial t} = -\frac{\hbar^2}{2m}\frac{\partial^2 \Psi(x,t)}{\partial x^2} + V(x,t)\Psi(x,t) \tag{1.13}$$

where the constant $\hbar$ (**h-bar**) is defined as

$$\boxed{\hbar \equiv \frac{h}{2\pi}} \tag{1.14}$$

The concept of the wave function and the equation governing its change with time were discovered in 1926 by the Austrian physicist Erwin Schrödinger (1887–1961). In this equation, known as the **time-dependent Schrödinger equation** (or the **Schrödinger wave equation**), $i = \sqrt{-1}$, m is the mass of the particle, and $V(x, t)$ is the potential-energy function of the system.

The time-dependent Schrödinger equation contains the first derivative of the wave function with respect to time and allows us to calculate the future wave function (state) at any time, if we know the wave function at time t_0.

The wave function contains all the information we can possibly know about the system it describes. What information does Ψ give us about the result of a measurement of the x coordinate of the particle? We cannot expect Ψ to involve the definite specification of position that the state of a classical-mechanical system does. The correct answer to this question was provided by Max Born shortly after Schrödinger discovered the Schrödinger equation. Born postulated that for a one-particle, one-dimensional system,

$$\boxed{|\Psi(x,t)|^2\, dx} \tag{1.15}$$

gives the *probability* at time t of finding the particle in the region of the x axis lying between x and $x + dx$. In (1.15) the bars denote the absolute value and dx is an infinitesimal length on the x axis. The function $|\Psi(x, t)|^2$ is the **probability density** for finding the particle at various places on the x axis. (Probability is reviewed in Section 1.6.) For example, suppose that at some particular time t_0 the particle is in a state characterized by the wave function ae^{-bx^2}, where a and b are real constants. If we measure the particle's position at time t_0, we might get any value of x, because the probability density $a^2e^{-2bx^2}$ is nonzero everywhere. Values of x in the region around $x = 0$ are more likely to be found than other values, since $|\Psi|^2$ is a maximum at the origin in this case.

To relate $|\Psi|^2$ to experimental measurements, we would take many identical noninteracting systems, each of which was in the same state Ψ. Then the particle's position in each system is measured. If we had n systems and made n measurements, and if dn_x denotes the number of measurements for which we found the particle between x and $x + dx$, then dn_x/n is the probability for finding the particle between x and $x + dx$. Thus

$$\frac{dn_x}{n} = |\Psi|^2\, dx$$

and a graph of $(1/n)dn_x/dx$ versus x gives the probability density $|\Psi|^2$ as a function of x. It might be thought that we could find the probability-density function by taking one system that was in the state Ψ and repeatedly measuring the particle's position. This procedure will not do because the process of measurement generally changes the state of a system. We saw an example of this in the discussion of the uncertainty principle (Section 1.3).

Quantum mechanics is *statistical* in nature. Knowing the state, we cannot predict the result of a position measurement with certainty; we can only predict the *probabilities* of various possible results. The Bohr theory of the hydrogen atom specified the precise path of the electron and is therefore not a correct quantum-mechanical picture.

Quantum mechanics does not say that an electron is distributed over a large region of space as a wave is distributed. Rather, it is the probability patterns (wave functions) used to describe the electron's motion that behave like waves and satisfy a wave equation.

How the wave function gives us information on other properties besides the position is discussed in later chapters.

The postulates of thermodynamics (the first, second, and third laws of thermodynamics) are stated in terms of macroscopic experience and hence are fairly readily understood. The postulates of quantum mechanics are stated in terms of the microscopic world and appear quite abstract. You should not expect to fully understand the postulates of quantum mechanics at first reading. As we treat various examples, understanding of the postulates will increase.

It may bother the reader that we wrote down the Schrödinger equation without any attempt to prove its plausibility. By using analogies between geometrical optics and classical mechanics on the one hand, and wave optics and quantum mechanics on the other hand, one can show the plausibility of the Schrödinger equation. Geometrical optics is an approximation to wave optics, valid when the wavelength of the light is much less than the size of the apparatus. (Recall its use in treating lenses and mirrors.) Likewise, classical mechanics is an approximation to wave mechanics, valid when the particle's wavelength is much less than the size of the apparatus. One can make a plausible guess as to how to get the proper equation for quantum mechanics from classical mechanics based on the known relation between the equations of geometrical and wave optics. Since many chemists are not particularly familiar with optics, these arguments have been omitted. In any case, such analogies can only make the Schrödinger equation seem *plausible.* They cannot be used to *derive* or *prove* this equation. The Schrödinger equation is a *postulate* of the theory, to be tested by agreement of its predictions with experiment. (Details of the reasoning that led Schrödinger to his equation are given in *Jammer,* Section 5.3. A reference with the author's name italicized is listed in the Bibliography.)

Quantum mechanics provides the law of motion for microscopic particles. Experimentally, macroscopic objects obey classical mechanics. Hence for quantum mechanics to be a valid theory, it should reduce to classical mechanics as we make the transition from microscopic to macroscopic particles. Quantum effects are associated with the de Broglie wavelength $\lambda = h/mv$. Since h is very small, the de Broglie wavelength of macroscopic objects is essentially zero. Thus, in the limit $\lambda \rightarrow 0$, we expect the time-dependent Schrödinger equation to reduce to Newton's second law. We can prove this to be so (see Prob. 7.59).

A similar situation holds in the relation between special relativity and classical mechanics. In the limit $v/c \rightarrow 0$, where c is the velocity of light, special relativity reduces to classical mechanics. The form of quantum mechanics that we will develop will be nonrelativistic. A complete integration of relativity with quantum mechanics has not been achieved.

Historically, quantum mechanics was first formulated in 1925 by Heisenberg, Born, and Jordan using matrices, several months before Schrödinger's 1926 formulation using differential equations. Schrödinger proved that the Heisenberg formulation (called **matrix mechanics**) is equivalent to the Schrödinger formulation (called **wave mechanics**). In 1926, Dirac and Jordan, working independently, formulated quantum mechanics in an abstract version called *transformation theory* that is a generalization of matrix mechanics and wave mechanics (see *Dirac*). In 1948, Feynman devised the *path integral* formulation of quantum mechanics [R. P. Feynman, *Rev. Mod. Phys.,* **20,** 367 (1948); R. P. Feynman and A. R. Hibbs, *Quantum Mechanics and Path Integrals,* McGraw-Hill, 1965].

1.5 THE TIME-INDEPENDENT SCHRÖDINGER EQUATION

The time-dependent Schrödinger equation (1.13) is formidable looking. Fortunately, many applications of quantum mechanics to chemistry do not use this equation. Instead, the simpler time-independent Schrödinger equation is used. We now derive the time-independent from the time-dependent Schrödinger equation for the one-particle, one-dimensional case.

We begin by restricting ourselves to the special case where the potential energy V is not a function of time but depends only on x. This will be true if the system experiences no time-dependent external forces. The time-dependent Schrödinger equation reads

$$-\frac{\hbar}{i}\frac{\partial\Psi(x,t)}{\partial t} = -\frac{\hbar^2}{2m}\frac{\partial^2\Psi(x,t)}{\partial x^2} + V(x)\Psi(x,t) \tag{1.16}$$

We now restrict ourselves to looking for those solutions of (1.16) that can be written as the product of a function of time and a function of x:

$$\boxed{\Psi(x,t) = f(t)\psi(x)} \tag{1.17}$$

Capital psi is used for the time-dependent wave function and lowercase psi for the factor that depends only on the coordinate x. States corresponding to wave functions of the form (1.17) possess certain properties (to be discussed shortly) that make them of great interest. [Not all solutions of (1.16) have the form (1.17); see Prob. 3.48.] Taking partial derivatives of (1.17), we have

$$\frac{\partial\Psi(x,t)}{\partial t} = \frac{df(t)}{dt}\psi(x), \qquad \frac{\partial^2\Psi(x,t)}{\partial x^2} = f(t)\frac{d^2\psi(x)}{dx^2}$$

Substitution into (1.16) gives

$$-\frac{\hbar}{i}\frac{df(t)}{dt}\psi(x) = -\frac{\hbar^2}{2m}f(t)\frac{d^2\psi(x)}{dx^2} + V(x)f(t)\psi(x)$$

$$-\frac{\hbar}{i}\frac{1}{f(t)}\frac{df(t)}{dt} = -\frac{\hbar^2}{2m}\frac{1}{\psi(x)}\frac{d^2\psi(x)}{dx^2} + V(x) \tag{1.18}$$

where we divided by $f\psi$. In general, we expect the quantity to which each side of (1.18) is equal to be a certain function of x and t. However, the right side of (1.18) does not depend on t, so the function to which each side of (1.18) is equal must be independent of t. The left side of (1.18) is independent of x, so this function must also be independent of x. Since the function is independent of both variables, x and t, it must be a constant. We call this constant E.

Equating the left side of (1.18) to E, we get

$$\frac{df(t)}{f(t)} = -\frac{iE}{\hbar}dt$$

Integrating both sides of this equation with respect to t, we have

$$\ln f(t) = -iEt/\hbar + C$$

where C is an arbitrary constant of integration. Hence

$$f(t) = e^C e^{-iEt/\hbar} = Ae^{-iEt/\hbar}$$

where the arbitrary constant A has replaced e^C. Since A can be included as a factor in the function $\psi(x)$ that multiplies $f(t)$ in (1.17), A can be omitted from $f(t)$. Thus

$$f(t) = e^{-iEt/\hbar}$$

Equating the right side of (1.18) to E, we have

$$-\frac{\hbar^2}{2m}\frac{d^2\psi(x)}{dx^2} + V(x)\psi(x) = E\psi(x) \tag{1.19}$$

Equation (1.19) is the **time-independent Schrödinger equation** for a single particle of mass m moving in one dimension.

What is the significance of the constant E? Since E occurs as $[E - V(x)]$ in (1.19), E has the same dimensions as V, so E has the dimensions of energy. In fact, we postulate that E is the energy of the system. (This is a special case of a more general postulate to be discussed in a later chapter.) Thus, for cases where the potential energy is a function of x only, there exist wave functions of the form

$$\Psi(x, t) = e^{-iEt/\hbar}\psi(x) \tag{1.20}$$

and these wave functions correspond to states of constant energy E. Much of our attention in the next few chapters will be devoted to finding the solutions of (1.19) for various systems.

The wave function in (1.20) is complex, but the quantity that is experimentally observable is the probability density $|\Psi(x, t)|^2$. The square of the absolute value of a complex quantity is given by the product of the quantity with its complex conjugate, the complex conjugate being formed by replacing i with $-i$ wherever it occurs. (See Section 1.7.) Thus

$$|\Psi|^2 = \Psi^*\Psi \tag{1.21}$$

where the star denotes the complex conjugate. For the wave function (1.20),

$$\begin{aligned} |\Psi(x,t)|^2 &= [e^{-iEt/\hbar}\psi(x)]^* e^{-iEt/\hbar}\psi(x) \\ &= e^{iEt/\hbar}\psi^*(x)e^{-iEt/\hbar}\psi(x) \\ &= e^0\psi^*(x)\psi(x) = \psi^*(x)\psi(x) \end{aligned}$$

$$|\Psi(x,t)|^2 = |\psi(x)|^2 \tag{1.22}$$

In deriving (1.22), we assumed that E is a real number, so $E = E^*$. This fact will be proved in Section 7.2.

Hence for states of the form (1.20), the probability density is given by $|\psi(x)|^2$ and does not change with time. Such states are called **stationary states**. Since the physically significant quantity is $|\Psi(x, t)|^2$, and since for stationary states $|\Psi(x, t)|^2 = |\psi(x)|^2$, the function $\psi(x)$ is often called the **wave function**, although the complete wave function of a stationary state is obtained by multiplying $\psi(x)$ by $e^{-iEt/\hbar}$. The term *stationary state* should not mislead the reader into thinking that a particle in a stationary statc is at rest. What is stationary is the probability density $|\Psi|^2$, not the particle itself.

We will be concerned mostly with states of constant energy (stationary states) and hence will usually deal with the time-independent Schrödinger equation (1.19). For simplicity we will refer to this equation as "the Schrödinger equation." Note that the Schrödinger equation contains *two* unknowns, the allowed energies E and the allowed wave functions ψ. To solve for two unknowns, we need to impose additional conditions (called boundary conditions) on ψ besides requiring that it satisfy (1.19). The boundary conditions determine the allowed energies, since it turns out that only certain values of E allow ψ to satisfy the boundary conditions. This will become clearer when we discuss specific examples in later chapters.

1.6 PROBABILITY

Probability plays a fundamental role in quantum mechanics. This section reviews the mathematics of probability.

There has been much controversy about the proper definition of probability. One definition is the following: If an experiment has n equally probable outcomes, m of which are favorable to the occurrence of a certain event A, then the probability that A occurs is m/n. Note that this definition is circular, since it specifies equally *probable* outcomes when *probability* is what we are trying to define. It is simply assumed that we can recognize equally probable outcomes. An alternative definition is based on actually performing the experiment many times. Suppose that we perform the experiment N times and that in M of these trials the event A occurs. The probability of A occurring is then defined as

$$\lim_{N \to \infty} \frac{M}{N}$$

Thus, if we toss a coin repeatedly, the fraction of heads will approach 1/2 as we increase the number of tosses.

For example, suppose we ask for the probability of drawing a heart when a card is picked at random from a standard 52-card deck containing 13 hearts. There are 52 cards and hence 52 equally probable outcomes. There are 13 hearts and hence 13 favorable outcomes. Therefore, $m/n = 13/52 = 1/4$. The probability for drawing a heart is 1/4.

Sometimes we ask for the probability of two related events both occurring. For example, we may ask for the probability of drawing two hearts from a 52-card deck, assuming we do not replace the first card after it is drawn. There are 52 possible outcomes of the first draw, and for each of these possibilities there are 51 possible second draws. We have $52 \cdot 51$ possible outcomes. Since there are 13 hearts, there are $13 \cdot 12$ different ways to draw two hearts. The desired probability is $(13 \cdot 12)/(52 \cdot 51) = 1/17$. This calculation illustrates the theorem: The probability that two events A and B both occur is the probability that A occurs, multiplied by the conditional probability that B then occurs, calculated with the assumption that A occurred. Thus, if A is the probability of drawing a heart on the first draw, the probability of A is 13/52. The probability of drawing a heart on the second draw, given that the first draw yielded a heart, is 12/51 since there remain 12 hearts in the deck. The probability of drawing two hearts is then $(13/52)(12/51) = 1/17$, as found previously.

In quantum mechanics we must deal with probabilities involving a continuous variable, for example, the x coordinate. It does not make much sense to talk about the probability of a particle being found *at* a particular point such as $x = 0.5000\ldots$, since there are an infinite number of points on the x axis, and for any finite number of measurements we make, the probability of getting *exactly* $0.5000\ldots$ is vanishingly small.

Instead we talk of the probability of finding the particle in a tiny interval of the x axis lying between x and $x + dx$, dx being an infinitesimal element of length. This probability will naturally be proportional to the length of the interval, dx, and will vary for different regions of the x axis. Hence the probability that the particle will be found between x and $x + dx$ is equal to $g(x)\,dx$, where $g(x)$ is some function that tells how the probability varies over the x axis. The function $g(x)$ is called the **probability density**, since it is a probability per unit length. Since probabilities are real, nonnegative numbers, $g(x)$ must be a real function that is everywhere nonnegative. The wave function Ψ can take on negative and complex values and is not a probability density. Quantum mechanics postulates that the probability density is $|\Psi|^2$ [Eq. (1.15)].

What is the probability that the particle lies in some finite region of space $a \le x \le b$? To find this probability, we sum up the probabilities $|\Psi|^2\,dx$ of finding the particle in all the infinitesimal regions lying between a and b. This is just the definition of the definite integral

$$\int_a^b |\Psi|^2\,dx = \Pr(a \le x \le b) \tag{1.23}$$

where Pr denotes a probability. A probability of 1 represents certainty. Since it is certain that the particle is somewhere on the x axis, we have the requirement

$$\int_{-\infty}^{\infty} |\Psi|^2\,dx = 1 \tag{1.24}$$

When Ψ satisfies (1.24), it is said to be **normalized**. For a stationary state, $|\Psi|^2 = |\psi|^2$ and $\int_{-\infty}^{\infty} |\psi|^2\,dx = 1$.

EXAMPLE A one-particle, one-dimensional system has $\Psi = a^{-1/2}e^{-|x|/a}$ at $t = 0$, where $a = 1.0000$ nm. At $t = 0$, the particle's position is measured. (a) Find the probability that the measured value lies between $x = 1.5000$ nm and $x = 1.5001$ nm. (b) Find the probability that the measured value is between $x = 0$ and $x = 2$ nm. (c) Verify that Ψ is normalized.

(a) In this tiny interval, x changes by only 0.0001 nm, and Ψ goes from $e^{-1.5000}\text{ nm}^{-1/2} = 0.22313\text{ nm}^{-1/2}$ to $e^{-1.5001}\text{ nm}^{-1/2} = 0.22311\text{ nm}^{-1/2}$, so Ψ is nearly constant in this interval, and it is a very good approximation to consider this interval as infinitesimal. The desired probability is given by (1.15) as

$$|\Psi|^2\,dx = a^{-1}e^{-2|x|/a}\,dx = (1\text{ nm})^{-1}e^{-2(1.5\text{ nm})/(1\text{ nm})}(0.0001\text{ nm}) = 4.979 \times 10^{-6}$$

(See also Prob. 1.12.)

(b) Use of Eq. (1.23) and $|x| = x$ for $x \ge 0$ gives

$$\Pr(0 \le x \le 2\text{ nm}) = \int_0^{2\text{ nm}} |\Psi|^2\,dx = a^{-1}\int_0^{2\text{ nm}} e^{-2x/a}\,dx$$
$$= -\tfrac{1}{2}e^{-2x/a}\Big|_0^{2\text{ nm}} = -\tfrac{1}{2}(e^{-4} - 1) = 0.4908$$

(c) Use of $\int_{-\infty}^{\infty} f(x)\,dx = \int_{-\infty}^{0} f(x)\,dx + \int_{0}^{\infty} f(x)\,dx$, $|x| = -x$ for $x \le 0$, and $|x| = x$ for $x \ge 0$, gives

$$\int_{-\infty}^{\infty} |\Psi|^2\,dx = a^{-1}\int_{-\infty}^{0} e^{2x/a}\,dx + a^{-1}\int_{0}^{\infty} e^{-2x/a}\,dx$$

$$= a^{-1}\left(\tfrac{1}{2}ae^{2x/a}\big|_{-\infty}^{0}\right) + a^{-1}\left(-\tfrac{1}{2}ae^{-2x/a}\big|_{0}^{\infty}\right) = \tfrac{1}{2} + \tfrac{1}{2} = 1$$

EXERCISE For a system whose state function at the time of a position measurement is $\Psi = (32a^3/\pi)^{1/4}xe^{-ax^2}$, where $a = 1.0000\text{ nm}^{-2}$, find the probability that the particle is found between $x = 1.2000$ nm and 1.2001 nm. Treat the interval as infinitesimal. (*Answer:* 0.0000258.)

1.7 COMPLEX NUMBERS

We have seen that the wave function can be complex, so we now review some properties of complex numbers.

A **complex number** z is a number of the form

$$z = x + iy, \quad \text{where } i = \sqrt{-1} \tag{1.25}$$

and where x and y are **real numbers** (numbers that do not involve the square root of a negative quantity). If $y = 0$ in (1.25), then z is a real number. If $y \neq 0$, then z is an **imaginary number**. If $x = 0$ and $y \neq 0$, then z is a **pure imaginary number**. For example, 6.83 is a real number, $5.4 - 3i$ is an imaginary number, and $0.60i$ is a pure imaginary number. Real and pure imaginary numbers are special cases of complex numbers. In (1.25), x and y are called the real and imaginary parts of z, respectively: $x = \text{Re}(z)$; $y = \text{Im}(z)$.

The complex number z can be represented as a point in the **complex plane** (Fig. 1.3), where the real part of z is plotted on the horizontal axis and the imaginary part on the vertical axis. This diagram immediately suggests defining two quantities that

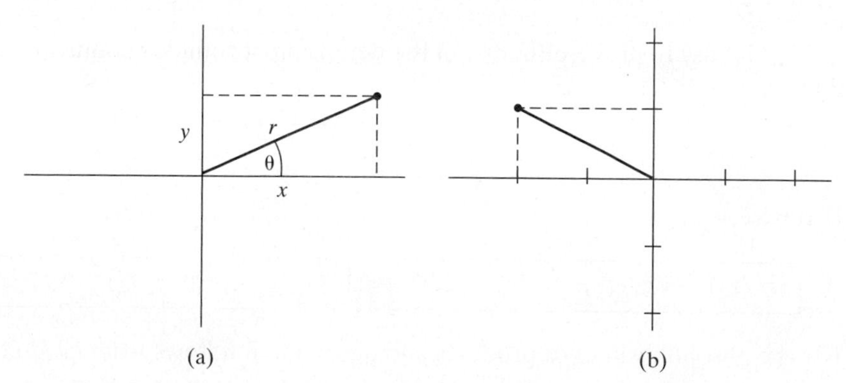

FIGURE 1.3 (a) Plot of a complex number $z = x + iy$. (b) Plot of the number $-2 + i$.

characterize the complex number z: the distance r of the point z from the origin is called the **absolute value** or modulus of z and is denoted by $|z|$; the angle θ that the radius vector to the point z makes with the positive horizontal axis is called the **phase** or argument of z. We have

$$|z| = r = (x^2 + y^2)^{1/2}, \qquad \tan\theta = y/x \tag{1.26}$$

$$x = r\cos\theta, \qquad y = r\sin\theta$$

So we may write $z = x + iy$ as

$$z = r\cos\theta + ir\sin\theta = re^{i\theta} \tag{1.27}$$

since (Prob. 4.3)

$$\boxed{e^{i\theta} = \cos\theta + i\sin\theta} \tag{1.28}$$

The angle θ in these equations is in radians.

If $z = x + iy$, the **complex conjugate** z^* of the complex number z is defined as

$$\boxed{z^* \equiv x - iy = re^{-i\theta}} \tag{1.29}$$

If z is a real number, its imaginary part is zero. Thus z is real if and only if $z = z^*$. Taking the complex conjugate twice, we get z back again, $(z^*)^* = z$. Forming the product of z and its complex conjugate and using $i^2 = -1$, we have

$$zz^* = (x + iy)(x - iy) = x^2 + iyx - iyx - i^2y^2$$

$$\boxed{zz^* = x^2 + y^2 = r^2 = |z|^2} \tag{1.30}$$

For the product and quotient of two complex numbers $z_1 = r_1e^{i\theta_1}$ and $z_2 = r_2e^{i\theta_2}$, we have

$$z_1z_2 = r_1r_2e^{i(\theta_1+\theta_2)}, \qquad \frac{z_1}{z_2} = \frac{r_1}{r_2}e^{i(\theta_1-\theta_2)} \tag{1.31}$$

It is easy to prove, either from the definition of complex conjugate or from (1.31), that

$$\boxed{(z_1z_2)^* = z_1^*z_2^*} \tag{1.32}$$

Likewise,

$$\boxed{(z_1/z_2)^* = z_1^*/z_2^*, \qquad (z_1 + z_2)^* = z_1^* + z_2^*, \qquad (z_1 - z_2)^* = z_1^* - z_2^*} \tag{1.33}$$

For the absolute values of products and quotients, it follows from (1.31) that

$$|z_1z_2| = |z_1||z_2|, \qquad \left|\frac{z_1}{z_2}\right| = \frac{|z_1|}{|z_2|} \tag{1.34}$$

Therefore, if ψ is a complex wave function, we have

$$|\psi^2| = |\psi|^2 = \psi^*\psi \tag{1.35}$$

We now obtain a formula for the nth roots of the number 1. We may take the phase of the number 1 to be 0 or 2π or 4π, and so on. Hence $1 = e^{i2\pi k}$, where k is any integer, zero, negative, or positive. Now consider the number ω, where $\omega \equiv e^{i2\pi k/n}$, n being a positive integer. Using (1.31) n times, we see that $\omega^n = e^{i2\pi k} = 1$. Thus ω is an nth root of unity. There are n different complex nth roots of unity, and taking n successive values of the integer k gives us all of them:

$$\omega = e^{i2\pi k/n}, \qquad k = 0, 1, 2, \ldots, n - 1 \tag{1.36}$$

Any other value of k besides those in (1.36) gives a number whose phase differs by an integral multiple of 2π from one of the numbers in (1.36) and hence is not a different root. For $n = 2$ in (1.36), we get the two square roots of 1; for $n = 3$, the three cube roots of 1; and so on.

1.8 UNITS

This book uses SI units. In the International System (SI), the units of length, mass, and time are the meter (m), kilogram (kg), and second (s). Force is measured in newtons (N) and energy in joules (J). Coulomb's law for the magnitude of the force between two charges Q_1 and Q_2 separated by a distance r in vacuum is written in SI units as $F = Q_1Q_2/4\pi\varepsilon_0 r^2$, where the charges Q_1 and Q_2 are in coulombs (C) and ε_0 is a constant (called the permittivity of vacuum or the electric constant) whose value is $8.854 \times 10^{-12}\ \mathrm{C^2\ N^{-1}\ m^{-2}}$.

In this book, Coulomb's law is sometimes written as

$$\boxed{F = Q_1'Q_2'/r^2} \tag{1.37}$$

where Q_1' and Q_2' are abbreviations for $Q_1/(4\pi\varepsilon_0)^{1/2}$ and $Q_2/(4\pi\varepsilon_0)^{1/2}$:

$$\boxed{Q' \equiv Q/(4\pi\varepsilon_0)^{1/2}} \tag{1.38}$$

1.9 CALCULUS

Calculus is heavily used in quantum chemistry, and the following formulas (in which c is a constant and f and g are functions of x) should be memorized.

$$\boxed{\frac{dc}{dx} = 0, \qquad \frac{d(cf)}{dx} = c\frac{df}{dx}, \qquad \frac{dx^n}{dx} = nx^{n-1} \qquad \frac{de^{cx}}{dx} = ce^{cx}}$$

$$\boxed{\frac{d(\sin cx)}{dx} = c\cos cx, \qquad \frac{d(\cos cx)}{dx} = -c\sin cx, \qquad \frac{d\ln cx}{dx} = \frac{1}{x}}$$

$$\boxed{\frac{d(f+g)}{dx} = \frac{df}{dx} + \frac{dg}{dx}, \qquad \frac{d(fg)}{dx} = f\frac{dg}{dx} + g\frac{df}{dx}}$$

$$\frac{d(f/g)}{dx} = \frac{d(fg^{-1})}{dx} = -fg^{-2}\frac{dg}{dx} + g^{-1}\frac{df}{dx}$$

$$\frac{d}{dx}f(g(x)) = \frac{df}{dg}\frac{dg}{dx}$$

An example of the last formula is $d[\sin(cx^2)]/dx = 2cx\cos(cx^2)$. Here, $g(x) = cx^2$ and $f = \sin$.

$$\int cf(x)\,dx = c\int f(x)\,dx, \qquad \int [f(x) + g(x)]\,dx = \int f(x)\,dx + \int g(x)\,dx$$

$$\int dx = x, \qquad \int x^n\,dx = \frac{x^{n+1}}{n+1} \quad \text{for } n \neq -1, \qquad \int \frac{1}{x}\,dx = \ln x$$

$$\int e^{cx}\,dx = \frac{e^{cx}}{c}, \qquad \int \sin cx\,dx = -\frac{\cos cx}{c}, \qquad \int \cos cx\,dx = \frac{\sin cx}{c}$$

$$\int_b^c f(x)\,dx = g(c) - g(b) \quad \text{where } \frac{dg}{dx} = f(x)$$

1.10 SUMMARY

The state of a quantum-mechanical system is described by a state function or wave function Ψ, which is a function of the coordinates of the particles of the system and of the time. The state function changes with time according to the time-dependent Schrödinger equation, which for a one-particle, one-dimensional system is Eq. (1.13). For such a system, the quantity $|\Psi(x,t)|^2\,dx$ gives the probability that a measurement of the particle's position at time t will find it between x and $x + dx$. The state function is normalized according to $\int_{-\infty}^{\infty}|\Psi|^2 dx = 1$. If the system's potential-energy function does not depend on t, then the system can exist in one of a number of stationary states of fixed energy. For a stationary state of a one-particle, one-dimensional system, $\Psi(x,t) = e^{-iEt/\hbar}\psi(x)$, where the time-independent wave function $\psi(x)$ is a solution of the time-independent Schrödinger equation (1.19).

PROBLEMS

Answers to numerical problems are given at the end of the book.

Sec.	1.2	1.3	1.4	1.5	1.6	1.7	1.8	general
Probs.	1.1–1.4	1.5	1.6–1.7	1.8–1.9	1.10–1.17	1.18–1.27	1.28–1.29	1.30

1.1 True or false? (a) All photons have the same energy. (b) As the frequency of light increases, its wavelength decreases. (c) If violet light with $\lambda = 400$ nm does not cause the photoelectric effect in a certain metal, then it is certain that red light with $\lambda = 700$ nm will not cause the photoelectric effect in that metal. (d) The de Broglie wavelength of an H_2 molecule traveling at the root-mean-square speed v_{rms} of molecules in H_2 gas at temperature T is greater than that of an O_2 molecule traveling at the v_{rms} of molecules in O_2 gas at the same T. (Recall that the kinetic theory of gases gives $\frac{1}{2}mv_{rms}^2 = \frac{3}{2}kT$, where k is Boltzmann's constant.)

1.2 (a) Calculate the energy of one photon of infrared radiation whose wavelength is 1064 nm. (b) An Nd:YAG laser emits a pulse of 1064-nm radiation of average power 5×10^6 W and duration 2×10^{-8} s. Find the number of photons emitted in this pulse. (Recall that 1 W = 1 J/s.)

1.3 The work function of very pure Na is 2.75 eV, where 1 eV = 1.602×10^{-19} J. (a) Calculate the maximum kinetic energy of photoelectrons emitted from Na exposed to 200 nm ultraviolet radiation. (b) Calculate the longest wavelength that will cause the photoelectric effect in pure Na. (c) The work function of sodium that has not been very carefully purified is substantially less than 2.75 eV, because of adsorbed sulfur and other substances derived from atmospheric gases. When impure Na is exposed to 200-nm radiation, will the maximum photoelectron kinetic energy be less than or greater than that for pure Na exposed to 200-nm radiation?

1.4 Calculate the de Broglie wavelength of an electron moving at 1/137th the speed of light. (At this speed, the relativistic correction to the mass is negligible.)

1.5 When J. J. Thomson investigated electrons in cathode-ray tubes, he observed the behavior expected for particles obeying classical mechanics. (a) Electrons are accelerated through a potential difference of 1000 volts and passed through a collimating slit of width 0.100 cm. [These electrons each have a kinetic energy of 1000 electronvolts (eV), where 1 eV = 1.602×10^{-19} J.] Calculate the diffraction angle α in Fig. 1.1. (b) What slit width is needed to give $\alpha = 1.00°$ for 1000-volt electrons?

1.6. In classical mechanics, the kinetic energy of a particle is defined as $T \equiv \frac{1}{2}mv^2$. Use results from Section 1.4 to show that, for a particle moving vertically in the earth's gravitational field (with g assumed constant), $T + V = \frac{1}{2}mv_0^2 + mgx_0$, so $T + V$ is constant.

1.7 A certain one-particle, one-dimensional system has $\Psi = ae^{-ibt}e^{-bmx^2/\hbar}$, where a and b are constants and m is the particle's mass. Find the potential-energy function V for this system. *Hint:* Use the time-dependent Schrödinger equation.

1.8 True or false? (a) For all quantum-mechanical states, $|\Psi(x, t)|^2 = |\psi(x)|^2$. (b) For all quantum-mechanical states, $\Psi(x, t)$ is the product of a function of x and a function of t.

1.9 A certain one-particle, one-dimensional system has the potential energy $V = 2c^2\hbar^2x^2/m$ and is in a stationary state with $\psi(x) = bxe^{-cx^2}$, where b is a constant, $c = 2.00\ \text{nm}^{-2}$, and $m = 1.00 \times 10^{-27}$ g. Find the particle's energy.

1.10 At a certain instant of time, a one-particle, one-dimensional system has $\Psi = (2/b^3)^{1/2}xe^{-|x|/b}$, where $b = 3.000$ nm. If a measurement of x is made at this time in the system, find the probability that the result (a) lies between 0.9000 nm and 0.9001 nm (treat this interval as infinitesimal); (b) lies between 0 and 2 nm (use the table of integrals in the Appendix, if necessary). (c) For what value of x is the probability density a minimum? (There is no need to use calculus to answer this.) (d) Verify that Ψ is normalized.

1.11 A one-particle, one-dimensional system has the state function

$$\Psi = (\sin at)(2/\pi c^2)^{1/4}e^{-x^2/c^2} + (\cos at)(32/\pi c^6)^{1/4}xe^{-x^2/c^2}$$

where a is a constant and $c = 2.000$ Å. If the particle's position is measured at $t = 0$, estimate the probability that the result will lie between 2.000 Å and 2.001 Å.

1.12 Use Eq. (1.23) to find the answer to part (a) of the example at the end of Section 1.6 and compare it with the approximate answer found in the example.

1.13 Which of the following functions meet *all* the requirements of a probability-density function (a and b are positive constants)? (a) e^{iax}; (b) xe^{-bx^2}; (c) e^{-bx^2}.

1.14 (a) Frank and Phyllis Eisenberg have two children; they have at least one female child. What is the probability that both their children are girls? (b) Bob and Barbara Shrodinger have two children. The older child is a girl. What is the probability the younger child is a girl? (Assume the odds of giving birth to a boy or girl are equal.)

1.15 If the peak in the mass spectrum of C_2F_6 at mass number 138 is 100 units high, calculate the heights of the peaks at mass numbers 139 and 140. Isotopic abundances: ^{12}C, 98.89%; ^{13}C, 1.11%; ^{19}F, 100%.

1.16 In bridge, each of the four players (A, B, C, D) receives 13 cards. Suppose A and C have 11 of the 13 spades between them. What is the probability that the remaining two spades are distributed so that B and D have one spade apiece?

1.17 What important probability-density function occurs in (a) the kinetic theory of gases? (b) the analysis of random errors of measurement?

1.18 Classify each of the following as a real number or an imaginary number: (a) -17; (b) $2 + i$; (c) $\sqrt{7}$; (d) $\sqrt{-1}$; (e) $\sqrt{-6}$; (f) $2/3$; (g) π; (h) i^2; (i) $(a + bi)(a - bi)$, where a and b are real numbers.

1.19 Plot these points in the complex plane: (a) 3; (b) $-i$; (c) $-2 + 3i$.

1.20 Show that $1/i = -i$.

1.21 Simplify (a) i^2; (b) i^3; (c) i^4; (d) i^*i; (e) $(1 + 5i)(2 - 3i)$; (f) $(1 - 3i)/(4 + 2i)$. *Hint*: In (f), multiply numerator and denominator by the complex conjugate of the denominator.

1.22 Find the complex conjugate of (a) -4; (b) $-2i$; (c) $6 + 3i$; (d) $2e^{-i\pi/5}$.

1.23 Find the absolute value and the phase of (a) i; (b) $2e^{i\pi/3}$; (c) $-2e^{i\pi/3}$; (d) $1 - 2i$.

1.24 Where in the complex plane are all points whose absolute value is 5 located? Where are all points with phase $\pi/4$ located?

1.25 Write each of the following in the form $re^{i\theta}$: (a) i; (b) -1; (c) $1 - 2i$; (d) $-1 - i$.

1.26 (a) Find the cube roots of 1. (b) Explain why the n nth roots of 1 when plotted in the complex plane lie on a circle of radius 1 and are separated by an angle $2\pi/n$ from one another.

1.27 Verify that

$$\sin \theta = \frac{e^{i\theta} - e^{-i\theta}}{2i}, \qquad \cos \theta = \frac{e^{i\theta} + e^{-i\theta}}{2}$$

1.28 Express each of the following units in terms of fundamental SI units (m, kg, s): (a) newton; (b) joule.

1.29 Calculate the force on an alpha particle passing a gold atomic nucleus at a distance of 0.00300 Å.

1.30 True or false? (a) A probability density can never be negative. (b) The state function Ψ can never be negative. (c) The state function Ψ must be a real function. (d) If $z = z^*$, then z must be a real number. (e) $\int_{-\infty}^{\infty} \Psi \, dx = 1$ for a one-particle, one-dimensional system. (f) Thc product of a number and its complex conjugate is always a real number.

CHAPTER 2

The Particle in a Box

The stationary-state wave functions and energy levels of a one-particle, one-dimensional system are found by solving the time-independent Schrödinger equation (1.19). In this chapter, we solve the time-independent Schrödinger equation for a very simple system, a particle in a one-dimensional box (Section 2.2). Because the Schrödinger equation is a differential equation, we first review the mathematics of differential equations.

2.1 DIFFERENTIAL EQUATIONS

This section considers only **ordinary** differential equations, which are those with only one independent variable. [A **partial** differential equation has more than one independent variable. An example is the time-dependent Schrödinger equation (1.16), in which t and x are the independent variables.] An ordinary differential equation is a relation involving an independent variable x, a dependent variable $y(x)$, and the first, second, ..., nth derivatives of y (y', y'', ..., $y^{(n)}$). An example is

$$y''' + 2x(y')^2 + \sin x \cos y = 3e^x \tag{2.1}$$

The **order** of a differential equation is the order of the highest derivative in the equation. Thus, (2.1) is of third order.

A special kind of differential equation is the **linear differential equation**, which has the form

$$A_n(x)y^{(n)} + A_{n-1}(x)y^{(n-1)} + \cdots + A_1(x)y' + A_0(x)y = g(x) \tag{2.2}$$

where the A's and g (some of which may be zero) are functions of x only. In the nth-order linear differential equation (2.2), y and its derivatives appear to the first power. A differential equation that cannot be put in the form (2.2) is *nonlinear*. If $g(x) = 0$ in (2.2), the linear differential equation is **homogeneous**; otherwise it is **inhomogeneous**. The one-dimensional Schrödinger equation (1.19) is a linear homogenous differential equation of second order.

By dividing by the coefficient of y'', we can put any linear homogeneous second-order differential equation into the form

$$y'' + P(x)y' + Q(x)y = 0 \tag{2.3}$$

Suppose y_1 and y_2 are two independent functions, each of which satisfies (2.3). By independent, we mean that y_2 is not simply a multiple of y_1. Then the general solution of the linear homogeneous differential equation (2.3) is

$$y = c_1y_1 + c_2y_2 \tag{2.4}$$

where c_1 and c_2 are arbitrary constants. This is readily verified by substituting (2.4) into the left side of (2.3):

$$c_1 y_1'' + c_2 y_2'' + P(x)c_1 y_1' + P(x)c_2 y_2' + Q(x)c_1 y_1 + Q(x)c_2 y_2$$
$$= c_1[y_1'' + P(x)y_1' + Q(x)y_1] + c_2[y_2'' + P(x)y_2' + Q(x)y_2]$$
$$= c_1 \cdot 0 + c_2 \cdot 0 = 0 \tag{2.5}$$

where the fact that y_1 and y_2 satisfy (2.3) has been used.

The general solution of a differential equation of nth order usually has n arbitrary constants. To fix these constants, we may have **boundary conditions**, which are conditions that specify the value of y or various of its derivatives at a point or points. Thus, if y represents the displacement of a vibrating string held fixed at two points, we know y must be zero at these points.

An important case is the linear homogeneous second-order differential equation with *constant coefficients:*

$$y'' + py' + qy = 0 \tag{2.6}$$

where p and q are constants. To solve (2.6), let us tentatively assume a solution of the form $y = e^{sx}$. We are looking for a function whose derivatives when multiplied by constants will cancel the original function. The exponential function repeats itself when differentiated and is thus the correct choice. Substitution in (2.6) gives

$$s^2 e^{sx} + pse^{sx} + qe^{sx} = 0$$

$$\boxed{s^2 + ps + q = 0} \tag{2.7}$$

Equation (2.7) is called the **auxiliary equation**. It is a quadratic equation with two roots s_1 and s_2 that, provided s_1 and s_2 are not equal, give two independent solutions to (2.6). Thus, the general solution of (2.6) is

$$\boxed{y = c_1 e^{s_1 x} + c_2 e^{s_2 x}} \tag{2.8}$$

For example, for $y'' + 6y' - 7y = 0$, the auxiliary equation is $s^2 + 6s - 7 = 0$. The quadratic formula gives $s_1 = 1$, $s_2 = -7$, so the general solution is $c_1 e^x + c_2 e^{-7x}$.

2.2 PARTICLE IN A ONE-DIMENSIONAL BOX

Having found the solution of one kind of differential equation, let us look at a case where we can use this solution to solve the time-independent Schrödinger equation. We consider a particle in a one-dimensional box. By this we mean a particle subjected to a potential-energy function that is infinite everywhere along the x axis except for a line segment of length l, where the potential energy is zero. Such a system may seem physically unreal, but this model can be applied with some success to certain conjugated molecules; see Prob. 2.15. We put the origin at the left end of the line segment (Fig. 2.1).

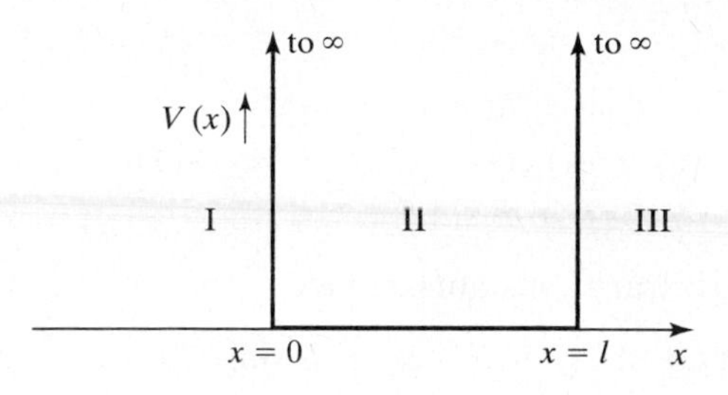

FIGURE 2.1 Potential energy function $V(x)$ for the particle in a one-dimensional box.

We have three regions to consider. In regions I and III, the potential energy V equals infinity and the time-independent Schrödinger equation (1.19) is

$$-\frac{\hbar^2}{2m}\frac{d^2\psi}{dx^2} = (E - \infty)\psi$$

Neglecting E in comparison with ∞, we have

$$\frac{d^2\psi}{dx^2} = \infty\psi, \qquad \psi = \frac{1}{\infty}\frac{d^2\psi}{dx^2}$$

and we conclude that ψ is zero outside the box:

$$\psi_{\text{I}} = 0, \qquad \psi_{\text{III}} = 0 \tag{2.9}$$

For region II, x between zero and l, the potential energy V is zero, and the Schrödinger equation (1.19) becomes

$$\frac{d^2\psi_{\text{II}}}{dx^2} + \frac{2m}{\hbar^2}E\psi_{\text{II}} = 0 \tag{2.10}$$

where m is the mass of the particle and E is its energy. We recognize (2.10) as a linear homogeneous second-order differential equation with constant coefficients. The auxiliary equation (2.7) gives

$$s^2 + 2mE\hbar^{-2} = 0$$

$$s = \pm(-2mE)^{1/2}\hbar^{-1} \tag{2.11}$$

$$s = \pm i(2mE)^{1/2}/\hbar \tag{2.12}$$

where $i = \sqrt{-1}$. Using (2.8), we have

$$\psi_{\text{II}} = c_1 e^{i(2mE)^{1/2}x/\hbar} + c_2 e^{-i(2mE)^{1/2}x/\hbar} \tag{2.13}$$

Temporarily, let

$$\theta \equiv (2mE)^{1/2}x/\hbar$$

$$\psi_{\text{II}} = c_1 e^{i\theta} + c_2 e^{-i\theta}$$

We have $e^{i\theta} = \cos\theta + i\sin\theta$ [Eq. (1.28)] and $e^{-i\theta} = \cos(-\theta) + i\sin(-\theta) = \cos\theta - i\sin\theta$, since

$$\boxed{\cos(-\theta) = \cos\theta \quad \text{and} \quad \sin(-\theta) = -\sin\theta} \tag{2.14}$$

Therefore,

$$\begin{aligned}\psi_{\mathrm{II}} &= c_1 \cos\theta + ic_1 \sin\theta + c_2 \cos\theta - ic_2 \sin\theta \\ &= (c_1 + c_2)\cos\theta + (ic_1 - ic_2)\sin\theta \\ &= A\cos\theta + B\sin\theta\end{aligned}$$

where A and B are new arbitrary constants. Hence,

$$\psi_{\mathrm{II}} = A\cos[\hbar^{-1}(2mE)^{1/2}x] + B\sin[\hbar^{-1}(2mE)^{1/2}x] \tag{2.15}$$

Now we find A and B by applying boundary conditions. It seems reasonable to postulate that the wave function will be continuous; that is, it will make no sudden jumps in value (see Fig. 3.4). If ψ is to be continuous at the point $x = 0$, then ψ_{I} and ψ_{II} must approach the same value at $x = 0$:

$$\begin{aligned}\lim_{x\to 0}\psi_{\mathrm{I}} &= \lim_{x\to 0}\psi_{\mathrm{II}} \\ 0 &= \lim_{x\to 0}\{A\cos[\hbar^{-1}(2mE)^{1/2}x] + B\sin[\hbar^{-1}(2mE)^{1/2}x]\} \\ 0 &= A\end{aligned}$$

since

$$\boxed{\sin 0 = 0 \quad \text{and} \quad \cos 0 = 1} \tag{2.16}$$

With $A = 0$, Eq. (2.15) becomes

$$\psi_{\mathrm{II}} = B\sin[(2\pi/h)(2mE)^{1/2}x] \tag{2.17}$$

Applying the continuity condition at $x = l$, we get

$$B\sin[(2\pi/h)(2mE)^{1/2}l] = 0 \tag{2.18}$$

B cannot be zero, because this would make the wave function zero everywhere—we would have an empty box. Therefore,

$$\sin[(2\pi/h)(2mE)^{1/2}l] = 0$$

The zeros of the sine function occur at $0, \pm\pi, \pm 2\pi, \pm 3\pi, \ldots = \pm n\pi$. Hence,

$$(2\pi/h)(2mE)^{1/2}l = \pm n\pi \tag{2.19}$$

The value $n = 0$ is a special case. From (2.19), $n = 0$ corresponds to $E = 0$. For $E = 0$, the roots (2.12) of the auxiliary equation are equal and (2.13) is not the complete solution of the Schrödinger equation. To find the complete solution, we return to (2.10), which for $E = 0$ reads $d^2\psi_{\mathrm{II}}/dx^2 = 0$. Integration gives $d\psi_{\mathrm{II}}/dx = c$ and $\psi_{\mathrm{II}} = cx + d$, where c and d are constants. The boundary condition that $\psi_{\mathrm{II}} = 0$ at $x = 0$ gives $d = 0$, and the condition that $\psi_{\mathrm{II}} = 0$ at $x = l$ then gives $c = 0$. Thus, $\psi_{\mathrm{II}} = 0$ for $E = 0$, and therefore $E = 0$ is not an allowed energy value. Hence, $n = 0$ is not allowed.

Solving (2.19) for E, we have

$$\boxed{E = \frac{n^2h^2}{8ml^2}, \qquad n = 1, 2, 3, \ldots} \tag{2.20}$$

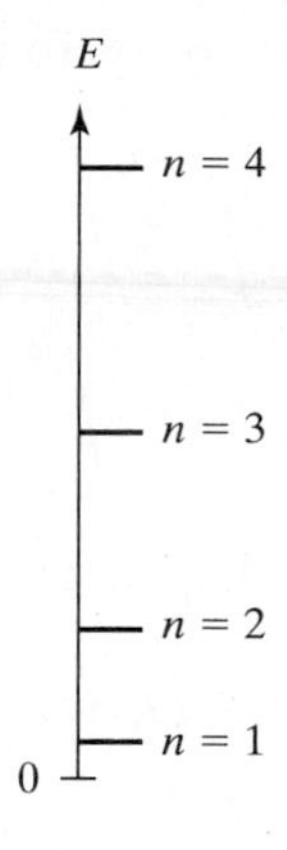

FIGURE 2.2 Lowest four energy levels for the particle in a one-dimensional box.

Only the energy values (2.20) allow ψ to satisfy the boundary condition of continuity at $x = l$. Application of a boundary condition has forced us to the conclusion that the values of the energy are quantized (Fig. 2.2). This is in striking contrast to the classical result that the particle in the box can have any nonnegative energy. Note that there is a minimum value, greater than zero, for the energy of the particle. The state of lowest energy is called the **ground state**. States with energies higher than the ground-state energy are **excited states**.

EXAMPLE A particle of mass 2.00×10^{-26} g is in a one-dimensional box of length 4.00 nm. Find the frequency and wavelength of the photon emitted when this particle goes from the $n = 3$ to the $n = 2$ level.

By conservation of energy, the energy $h\nu$ of the emitted photon equals the energy difference between the two stationary states [Eq. (1.4); see also Section 9.10]:

$$h\nu = E_{\text{upper}} - E_{\text{lower}} = n_u^2 h^2/8ml^2 - n_l^2 h^2/8ml^2$$

$$\nu = \frac{(n_u^2 - n_l^2)h}{8ml^2} = \frac{(3^2 - 2^2)(6.626 \times 10^{-34}\text{ J s})}{8(2.00 \times 10^{-29}\text{ kg})(4.00 \times 10^{-9}\text{ m})^2} = 1.29 \times 10^{12}\text{ s}^{-1}$$

where u and l stand for upper and lower. Use of $\lambda\nu = c$ gives $\lambda = 2.32 \times 10^{-4}$ m. (A common student error is to set $h\nu$ equal to the energy of one of the states instead of the energy *difference* between states.)

EXERCISE For an electron in a certain one-dimensional box, the longest-wavelength transition occurs at 400 nm. Find the length of the box. (*Answer:* 0.603 nm.)

Substitution of (2.19) into (2.17) gives for the wave function

$$\psi_{\text{II}} = B \sin\left(\frac{n\pi x}{l}\right), \qquad n = 1, 2, 3, \ldots \tag{2.21}$$

The use of the negative sign in front of $n\pi$ does not give us another independent solution. Since $\sin(-\theta) = -\sin\theta$, we would simply get a constant, -1, times the solution with the plus sign.

The constant B in Eq. (2.21) is still arbitrary. To fix its value, we use the normalization requirement, Eqs. (1.24) and (1.22):

$$\int_{-\infty}^{\infty} |\Psi|^2\,dx = \int_{-\infty}^{\infty} |\psi|^2\,dx = 1$$

$$\int_{-\infty}^{0} |\psi_{\mathrm{I}}|^2\,dx + \int_{0}^{l} |\psi_{\mathrm{II}}|^2\,dx + \int_{l}^{\infty} |\psi_{\mathrm{III}}|^2\,dx = 1$$

$$|B|^2 \int_0^l \sin^2\!\left(\frac{n\pi x}{l}\right) dx = 1 = |B|^2 \frac{l}{2} \qquad (2.22)$$

where the integral was evaluated by using Eq. (A.2) in the Appendix. We have

$$|B| = (2/l)^{1/2}$$

Note that only the absolute value of B has been determined. B could be $-(2/l)^{1/2}$ as well as $(2/l)^{1/2}$. Moreover, B need not be a real number. We could use any complex number with absolute value $(2/l)^{1/2}$. All we can say is that $B = (2/l)^{1/2}e^{i\alpha}$, where α is the phase of B and could be any value in the range 0 to 2π (Section 1.7). Choosing the phase to be zero, we write as the stationary-state wave functions for the particle in a box

$$\boxed{\psi_{\mathrm{II}} = \left(\frac{2}{l}\right)^{1/2} \sin\!\left(\frac{n\pi x}{l}\right), \qquad n = 1, 2, 3, \ldots} \qquad \mathbf{(2.23)}$$

Graphs of the wave functions and the probability densities are shown in Figs. 2.3 and 2.4.

The number n in the energies (2.20) and the wave functions (2.23) is called a **quantum number**. Each different value of the quantum number n gives a different wave function and a different state.

The wave function is zero at certain points; these points are called **nodes**. For each increase of one in the value of the quantum number n, ψ has one more node. The existence of nodes in ψ and $|\psi|^2$ may seem surprising. Thus, for $n = 2$, Fig. 2.4 says that there is zero probability of finding the particle in the center of the box at $x = l/2$. How can the particle get from one side of the box to the other without at any time being found in the center? This apparent paradox arises from trying to understand the motion of microscopic particles using our everyday experience of the motions of macroscopic particles. However, as stated in Chapter 1, electrons and other microscopic "particles" cannot be fully and correctly described in terms of concepts of classical physics drawn from the macroscopic world.

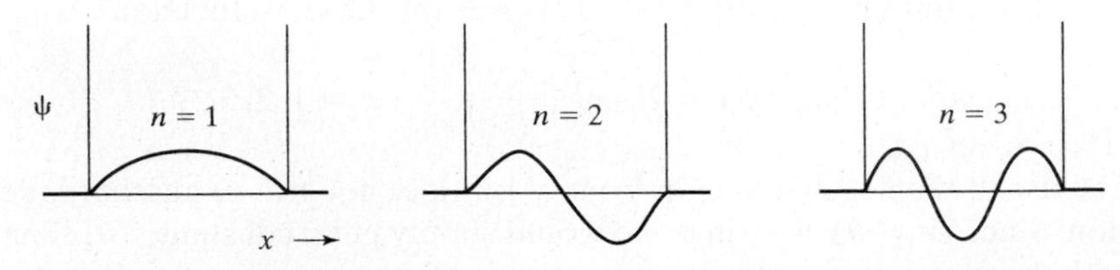

FIGURE 2.3 Graphs of ψ for the three lowest-energy particle-in-a-box states.

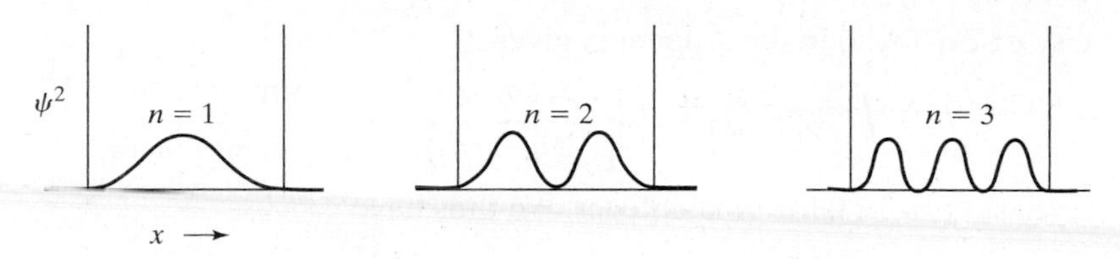

FIGURE 2.4 Graphs of $|\psi|^2$ for the lowest particle-in-a-box states.

Figure 2.4 shows that the probability of finding the particle at various places in the box is quite different from the classical result. Classically, a particle of fixed energy in a box bounces back and forth elastically between the two walls, moving at constant speed. Thus it is equally likely to be found at any point in the box. Quantum mechanically, we find a maximum in probability at the center of the box for the lowest energy level. As we go to higher energy levels with more nodes, the maxima and minima of probability come closer together, and the variations in probability along the length of the box ultimately become undetectable. For high quantum numbers, we approach the classical result of uniform probability density.

This result, that in the limit of large quantum numbers quantum mechanics goes over into classical mechanics, is known as the *Bohr correspondence principle.* Since Newtonian mechanics is valid for macroscopic bodies (moving at speeds much less than the speed of light), we expect nonrelativistic quantum mechanics to give the same answer as classical mechanics for macroscopic bodies. Because of the extremely small size of Planck's constant, quantization of energy is unobservable for macroscopic bodies. Since the mass of the particle and the length of the box squared appear in the denominator of Eq. (2.20), a macroscopic object in a macroscopic box having a macroscopic energy of motion would have a huge value for n, and hence, according to the correspondence principle, would show classical behavior.

We have a whole set of wave functions, each corresponding to a different value of the energy and characterized by the quantum number n, which may have integral values from 1 up. Let the subscript i denote a particular wave function with the value n_i for its quantum number:

$$\psi_i = \left(\frac{2}{l}\right)^{1/2} \sin\left(\frac{n_i \pi x}{l}\right), \quad 0 < x < l$$

$$\psi_i = 0 \quad \text{elsewhere}$$

Since the wave function has been normalized, we have

$$\int_{-\infty}^{\infty} \psi_i^* \psi_j \, dx = 1 \quad \text{if } i = j \tag{2.24}$$

We now ask for the value of this integral when we use wave functions corresponding to *different* energy levels:

$$\int_{-\infty}^{\infty} \psi_i^* \psi_j \, dx = \int_0^l \left(\frac{2}{l}\right)^{1/2} \sin\left(\frac{n_i \pi x}{l}\right) \left(\frac{2}{l}\right)^{1/2} \sin\left(\frac{n_j \pi x}{l}\right) dx, \qquad n_i \neq n_j$$

Use of Eq. (A.5) in the Appendix gives

$$\int_{-\infty}^{\infty} \psi_i^* \psi_j \, dx = \frac{2}{l}\left[\frac{\sin[(n_i - n_j)\pi]}{2(n_i - n_j)\pi/l} - \frac{\sin[(n_i + n_j)\pi]}{2(n_i + n_j)\pi/l}\right] = 0 \quad (2.25)$$

since $\sin m\pi = 0$ for m an integer. We thus have

$$\int_{-\infty}^{\infty} \psi_i^* \psi_j \, dx = 0, \quad i \neq j \quad (2.26)$$

When (2.26) holds, we say that the functions ψ_i and ψ_j are **orthogonal** to each other for $i \neq j$. We can combine (2.24) and (2.26) by writing

$$\int_{-\infty}^{\infty} \psi_i^* \psi_j \, dx = \delta_{ij} \quad (2.27)$$

The symbol δ_{ij} is called the **Kronecker delta** (after a mathematician). It equals 1 when the two indexes i and j are equal, and equals 0 when i and j are unequal:

$$\boxed{\delta_{ij} \equiv \begin{cases} 0 & \text{for } i \neq j \\ 1 & \text{for } i = j \end{cases}} \quad \textbf{(2.28)}$$

The property (2.27) of the wave functions is called **orthonormality**. We proved orthonormality only for the particle-in-a-box wave functions. We shall prove it more generally in Section 7.2.

A more rigorous way to look at the particle in a box with infinite walls is to first treat the particle in a box with a finite jump in potential energy at the walls and then take the limit as the jump in V becomes infinite. The results, when the limit is taken, will be the same as (2.20) and (2.23) (see Prob. 2.20).

Some online computer simulations of the particle in a box are www.chem.uci.edu/undergrad/applets/dwell/dwell.htm (shows the effects on the wave functions and energy levels when a barrier of variable height and width is introduced into the middle of the box); www.williams.edu/Chemistry/dbingemann/Chem153/particle.html (shows quantization by plotting the solution to the Schrödinger equation as the energy is varied and as the box length is varied); http://falstad.com/qm1d/ (shows both time-independent and time-dependent states; see Prob. 7.47).

2.3 THE FREE PARTICLE IN ONE DIMENSION

By a free particle, we mean a particle subject to no forces whatever. For a free particle, integration of (1.12) shows the potential energy remains constant no matter what the value of x is. Since the choice of the zero level of energy is arbitrary, we may set $V(x) = 0$. The Schrödinger equation (1.19) becomes

$$\frac{d^2\psi}{dx^2} + \frac{2m}{\hbar^2} E\psi = 0 \quad (2.29)$$

Equation (2.29) is the same as Eq. (2.10) (except for the boundary conditions). Therefore, the general solution of (2.29) is (2.13):

$$\psi = c_1 e^{i(2mE)^{1/2}x/\hbar} + c_2 e^{-i(2mE)^{1/2}x/\hbar} \quad (2.30)$$

What boundary condition might we impose? It seems reasonable to postulate (since $\psi^*\psi\, dx$ represents a probability) that ψ will remain finite as x goes to $\pm\infty$. If the energy E is less than zero, then this boundary condition will be violated, since for $E < 0$ we have

$$i(2mE)^{1/2} = i(-2m|E|)^{1/2} = i \cdot i \cdot (2m|E|)^{1/2} = -(2m|E|)^{1/2}$$

and therefore the first term in (2.30) will become infinite as x approaches minus infinity. Similarly, if E is negative, the second term in (2.30) becomes infinite as x approaches plus infinity. Thus the boundary condition requires

$$E \geq 0 \tag{2.31}$$

for the free particle. The wave function is oscillatory and is a linear combination of a sine and a cosine term [Eq. (2.15)]. For the free particle, the energy is not quantized; all nonnegative energies are allowed. Since we set $V = 0$, the energy E is in this case all kinetic energy. If we try to evaluate the arbitrary constants c_1 and c_2 by normalization, we will find that the integral $\int_{-\infty}^{\infty} \psi^*(x)\psi(x)\, dx$ is infinite. In other words, the free-particle wave function is not normalizable in the usual sense. This is to be expected on physical grounds, because there is no reason for the probability of finding the free particle to approach zero as x goes to $\pm\infty$.

The free-particle problem is an unreal situation because we could not actually have a particle that had no interaction with any other particle in the universe.

2.4 PARTICLE IN A RECTANGULAR WELL

Consider a particle in a one-dimensional box with walls of finite height (Fig. 2.5a). The potential-energy function is $V = V_0$ for $x < 0$, $V = 0$ for $0 \leq x \leq l$, and $V = V_0$ for $x > l$. There are two cases to examine, depending on whether the particle's energy E is less than or greater than V_0.

We first consider $E < V_0$. The Schrödinger equation (1.19) in regions I and III is $d^2\psi/dx^2 + (2m/\hbar^2)(E - V_0)\psi = 0$. This is a linear homogenous differential equation with constant coefficients, and the auxiliary equation (2.7) is $s^2 + (2m/\hbar^2)(E - V_0) = 0$ with roots $s = \pm(2m/\hbar^2)^{1/2}(V_0 - E)^{1/2}$. Therefore,

$$\psi_{\mathrm{I}} = C \exp[(2m/\hbar^2)^{1/2}(V_0 - E)^{1/2}x] + D \exp[-(2m/\hbar^2)^{1/2}(V_0 - E)^{1/2}x]$$

$$\psi_{\mathrm{III}} = F \exp[(2m/\hbar^2)^{1/2}(V_0 - E)^{1/2}x] + G \exp[-(2m/\hbar^2)^{1/2}(V_0 - E)^{1/2}x]$$

where C, D, F, and G are constants.

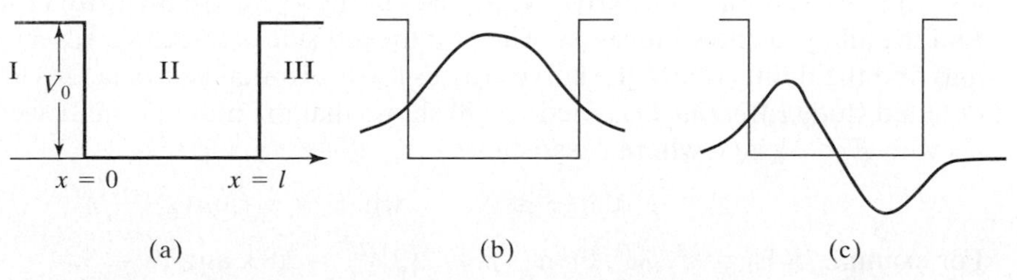

FIGURE 2.5 (a) Potential energy for a particle in a one-dimensional rectangular well. (b) The ground-state wave function for this potential. (c) The first excited-state wave function.

As in Section 2.3, we must prevent ψ_I from becoming infinite as $x \to -\infty$. Since we are assuming $E < V_0$, the quantity $(V_0 - E)^{1/2}$ is a real, positive number, and to keep ψ_I finite as $x \to -\infty$, we must have $D = 0$. Similarly, to keep ψ_III finite as $x \to +\infty$, we must have $F = 0$. Therefore,

$$\psi_\text{I} = C \exp[(2m/\hbar^2)^{1/2}(V_0 - E)^{1/2}x], \qquad \psi_\text{III} = G \exp[-(2m/\hbar^2)^{1/2}(V_0 - E)^{1/2}x]$$

In region II, $V = 0$, the Schrödinger equation is (2.10) and its solution is (2.15):

$$\psi_\text{II} = A \cos[(2m/\hbar^2)^{1/2}E^{1/2}x] + B \sin[(2m/\hbar^2)^{1/2}E^{1/2}x] \tag{2.32}$$

To complete the problem, we must apply the boundary conditions. As with the particle in a box with infinite walls, we require the wave function to be continuous at $x = 0$ and at $x = l$; so $\psi_\text{I}(0) = \psi_\text{II}(0)$ and $\psi_\text{II}(l) = \psi_\text{III}(l)$. The wave function has four arbitrary constants, so more than these two boundary conditions are needed. As well as requiring ψ to be continuous, we shall require that its derivative $d\psi/dx$ be continuous everywhere. To justify this requirement, we note that if $d\psi/dx$ changed discontinuously at a point then its derivative (its instantaneous rate of change) $d^2\psi/dx^2$ would become infinite at that point. However, for the particle in a rectangular well, the Schrödinger equation $d^2\psi/dx^2 = (2m/\hbar^2)(V - E)\psi$ does not contain anything infinite on the right side, so $d^2\psi/dx^2$ cannot become infinite. [For a more rigorous argument, see D. Branson, *Am. J. Phys.*, **47**, 1000 (1979).] Therefore, $d\psi_\text{I}/dx = d\psi_\text{II}/dx$ at $x = 0$ and $d\psi_\text{II}/dx = d\psi_\text{III}/dx$ at $x = l$.

From $\psi_\text{I}(0) = \psi_\text{II}(0)$, we get $C = A$. From $\psi_\text{I}'(0) = \psi_\text{II}'(0)$, we get (Prob. 2.19a) $B = (V_0 - E)^{1/2}A/E^{1/2}$. From $\psi_\text{II}(l) = \psi_\text{III}(l)$, we get a complicated equation that allows G to be found in terms of A. The constant A is found by normalization.

Taking $\psi_\text{II}'(l) = \psi_\text{III}'(l)$, dividing it by $\psi_\text{II}(l) = \psi_\text{III}(l)$, and expressing B in terms of A, we get the following equation for the energy levels (Prob. 2.19b):

$$(2E - V_0) \sin[(2mE)^{1/2}l/\hbar] = 2(V_0E - E^2)^{1/2} \cos[(2mE)^{1/2}l/\hbar] \tag{2.33}$$

[Although $E = 0$ satisfies (2.33), it is not an allowed energy value, since it gives $\psi = 0$ (Prob. 2.28).] Defining the dimensionless constants ε and b as

$$\varepsilon \equiv E/V_0 \quad \text{and} \quad b \equiv (2mV_0)^{1/2}l/\hbar \tag{2.34}$$

we divide (2.33) by V_0 to get

$$(2\varepsilon - 1)\sin(b\varepsilon^{1/2}) - 2(\varepsilon - \varepsilon^2)^{1/2}\cos(b\varepsilon^{1/2}) = 0 \tag{2.35}$$

Only the particular values of E that satisfy (2.33) give a wave function that is continuous and has a continuous derivative, so the energy levels are quantized for $E < V_0$. To find the allowed energy levels, we can plot the left side of (2.35) versus ε for $0 < \varepsilon < 1$ and find the points where the curve crosses the horizontal axis (see also Prob. 4.30c). A detailed study (*Merzbacher,* Section 6.8) shows that the number of allowed energy levels with $E < V_0$ is N, where N satisfies

$$N - 1 < b/\pi \le N, \qquad \text{where } b \equiv (2mV_0)^{1/2}l/\hbar$$

For example, if $V_0 = h^2/ml^2$, then $b/\pi = 2(2^{1/2}) = 2.83$, and $N = 3$.

Figure 2.5 shows ψ for the lowest two energy levels. The wave function is oscillatory inside the box and dies off exponentially outside the box. It turns out that the number of nodes increases by one for each higher level.

So far we have considered only states with $E < V_0$. For $E > V_0$, the quantity $(V_0 - E)^{1/2}$ is imaginary, and instead of dying off to zero as x goes to $\pm\infty$, ψ_{I} and ψ_{III} oscillate (similar to the free-particle ψ). We no longer have any reason to set D in ψ_{I} and F in ψ_{III} equal to zero, and with these additional constants available to satisfy the boundary conditions on ψ and ψ', one finds that E need not be restricted to obtain properly behaved wave functions. Therefore, all energies above V_0 are allowed.

A state in which $\psi \to 0$ as $x \to \infty$ and as $x \to -\infty$ is called a **bound state**. For a bound state, significant probability for finding the particle exists in only a finite region of space. For an **unbound state**, ψ does not go to zero as $x \to \pm\infty$ and is not normalizable. For the particle in a rectangular well, states with $E < V_0$ are bound and states with $E > V_0$ are unbound. For the particle in a box with infinitely high walls, all states are bound. For the free particle, all states are unbound.

For an online simulation of the particle in a well, go to www.falstad.com/qm1d/ and choose Finite Well in the Setup box. You can vary the well width and depth and see the effect on the energy levels and wave functions.

2.5 TUNNELING

For the particle in a rectangular well (Section 2.4), Fig. 2.5 and the equations for ψ_{I} and ψ_{III} show that for the bound states there is a nonzero probability of finding the particle in regions I and III, where its total energy E is less than its potential energy $V = V_0$. Classically, this behavior is not allowed. The classical equations $E = T + V$ and $T \geq 0$, where T is the kinetic energy, mean that E cannot be less than V in classical mechanics.

Consider a particle in a one-dimensional box with walls of finite height *and* finite thickness (Fig. 2.6). Classically, the particle cannot escape from the box unless its energy is greater than the potential-energy barrier V_0. However, a quantum-mechanical treatment (which is omitted) shows that there is a finite probability for a particle of total energy less than V_0 to be found outside the box.

The term **tunneling** denotes the penetration of a particle into a classically forbidden region (as in Fig. 2.5) or the passage of a particle through a potential-energy barrier whose height exceeds the particle's energy. Since tunneling is a quantum effect, its probability of occurrence is greater the less classical is the behavior of the particle. Therefore, tunneling is most prevalent with particles of small mass. (Note that the greater the mass m, the more rapidly the functions ψ_{I} and ψ_{III} of Section 2.4 die away to zero.) Electrons tunnel quite readily. Hydrogen atoms and ions tunnel more readily than heavier atoms.

The emission of alpha particles from a radioactive nucleus involves tunneling of the alpha particles through the potential-energy barrier produced by the short-range attractive nuclear forces and the Coulombic repulsive force between the daughter nucleus and the alpha particle. The NH_3 molecule is pyramidal. There is a potential-energy barrier to

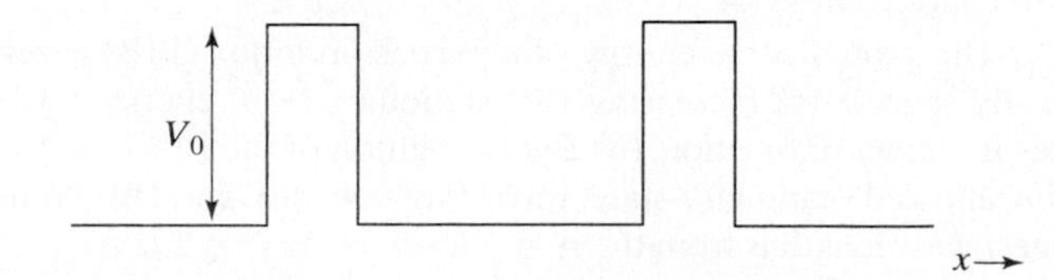

FIGURE 2.6 Potential energy for a particle in a one-dimensional box of finite height and thickness.

inversion of the molecule, with the potential-energy maximum occurring at the planar configuration. The hydrogen atoms can tunnel through this barrier, thereby inverting the molecule. In CH_3CH_3 there is a barrier to internal rotation, with a potential-energy maximum at the eclipsed position of the hydrogens. The hydrogens can tunnel through this barrier from one staggered position to the next. Tunneling of electrons is important in oxidation–reduction reactions and in electrode processes. Tunneling usually contributes significantly to the rate of chemical reactions that involve transfer of hydrogen atoms. See R. P. Bell, *The Tunnel Effect in Chemistry,* Chapman & Hall, 1980.

Tunneling occurs in some enzyme-catalyzed reactions, but the extent to which it contributes to the rate enhancement provided by the enzyme is the subject of controversy (S. Borman, *Chem. Eng. News*, Feb. 23, 2004, p. 35).

The scanning tunneling microscope, invented in 1981, uses the tunneling of electrons through the space between the extremely fine tip of a metal wire and the surface of an electrically conducting solid to produce images of individual atoms on the solid's surface. A small voltage is applied between the solid and the wire, and as the tip is moved across the surface at a height of a few angstroms, the tip height is adjusted to keep the current flow constant. A plot of tip height versus position yields an image of the surface.

2.6 SUMMARY

The general solution to the linear, homogeneous, second-order, constant-coefficients differential equation $y''(x) + py'(x) + qy(x) = 0$ is $y = c_1e^{s_1x} + c_2e^{s_2x}$, where s_1 and s_2 are the solutions to the auxiliary equation $s^2 + ps + q = 0$.

For a particle in a one-dimensional box (potential energy $V = 0$ for $0 \le x \le l$ and $V = \infty$ elsewhere), the stationary-state wave functions and energies are $\psi = (2/l)^{1/2} \sin(n\pi x/l)$ for $0 \le x \le l$, $\psi = 0$ elsewhere, and $E = n^2h^2/8ml^2$, where $n = 1, 2, 3, \ldots$.

PROBLEMS

Sec.	2.1	2.2	2.3	2.4	general
Probs.	2.1–2.3	2.4–2.17	2.18	2.19–2.28	2.29–2.30

2.1 (a) Solve $y''(x) + y'(x) - 6y(x) = 0$. (b) Evaluate the arbitrary constants in the solution if the boundary conditions are $y = 0$ at $x = 0$ and $y' = 1$ at $x = 0$.

2.2 (a) For the case of equal roots of the auxiliary equation, $s_1 = s_2 = s$, we have found only one independent solution of the linear homogeneous second-order differential equation: e^{sx}. Verify that xe^{sx} is the second solution in this case. (b) Solve $y''(x) - 2y'(x) + y(x) = 0$.

2.3 For each of these choices of F in Newton's second law (1.8), classify (1.8) as a linear or nonlinear differential equation (a, b, c, and k are constants). (a) $F = c$; (b) $F = -kx$; (c) $F = -ax^3$; (d) $F = b \sin ax$; (e) $F = a - kx$. [Classical-mechanical systems that show chaotic behavior (Section 1.4) obey nonlinear differential equations, but not all nonlinear differential equations give rise to chaotic behavior.]

2.4 True or false? (a) The ground-state energy of a particle in a box (PIB) is zero. (b) The PIB energy levels are equally spaced. (c) Increasing PIB stationary-state energy corresponds to increasing number of nodes in the wave function. (d) Every solution of the PIB time-independent Schrödinger equation is an allowed stationary-state wave function. (e) The PIB transition that absorbs the photon of longest wavelength is from the $n = 1$ level to the $n = 2$ level.

2.5 For a particle in a one-dimensional box of length l, consider an infinitesimal interval of length dl that lies within the box. For each of the following stationary states, state which location(s) of this interval gives a maximum probability and which gives a minimum probability to find the particle in the interval. (a) $n = 1$; (b) $n = 2$; (c) $n = 3$.

2.6 Consider a particle with quantum number n moving in a one-dimensional box of length l. (a) Find the probability of finding the particle in the left quarter of the box. (b) For what value of n is this probability a maximum? (c) What is the limit of this probability for $n \rightarrow \infty$? (d) What principle is illustrated in (c)?

2.7 Consider an electron in a one-dimensional box of length 2.000 Å with the left end of the box at $x = 0$. (a) Suppose we have one million of these systems, each in the $n = 1$ state, and we measure the x coordinate of the electron in each system. About how many times will the electron be found between 0.600 Å and 0.601 Å? Consider the interval to be infinitesimal. *Hint:* Check whether your calculator is set to degrees or radians. (b) Suppose we have a large number of these systems, each in the $n = 1$ state, and we measure the x coordinate of the electron in each system and find the electron between 0.700 Å and 0.701 Å in 126 of the measurements. In about how many measurements will the electron be found between 1.000 Å and 1.001 Å?

2.8 (a) Sketch rough graphs of ψ and of ψ^2 for the $n = 4$ and $n = 5$ particle-in-a-box states. (b) Use calculus to find the slope of the ψ^2 curve for $n = 4$ at $x = \frac{1}{2}l$ and check that your curve was drawn with the correct slope.

2.9 An extremely crude picture of an electron in an atom or molecule treats it as a particle in a one-dimensional box whose length is on the order of the size of atoms and molecules. (a) For an electron in a one-dimensional box of length 1.0 Å, calculate the separation between the two lowest energy levels. (b) Calculate the wavelength of a photon corresponding to a transition between these two levels. (c) In what portion of the electromagnetic spectrum is this wavelength?

2.10 For a macroscopic object of mass 1.0 g moving with speed 1.0 cm/s in a one-dimensional box of length 1.0 cm, find the quantum number n.

2.11 When a particle of mass 9.1×10^{-28} g in a certain one-dimensional box goes from the $n = 5$ level to the $n = 2$ level, it emits a photon of frequency $6.0 \times 10^{14}\ s^{-1}$. Find the length of the box.

2.12 When an electron in a certain excited energy level in a one-dimensional box of length 2.00 Å makes a transition to the ground state, a photon of wavelength 8.79 nm is emitted. Find the quantum number of the initial state.

2.13 The $n = 1$ to $n = 2$ absorption frequency for a certain particle in a certain one-dimensional box is $6.0 \times 10^{12}\ s^{-1}$. Find the $n = 2$ to $n = 3$ absorption frequency for this system.

2.14 An electron in a stationary state of a one-dimensional box of length 0.300 nm emits a photon of frequency $5.05 \times 10^{15}\ s^{-1}$. Find the initial and final quantum numbers for this transition.

2.15 A crude treatment of the pi electrons of a conjugated molecule regards these electrons as moving in the particle-in-a-box potential of Fig. 2.1, where the box length is somewhat more than the length of the conjugated chain. The Pauli exclusion principle (Chapter 10) allows no more than two electrons to occupy each box level. (These two have opposite spins.) For butadiene, $CH_2{=}CHCH{=}CH_2$, take the box length as 7.0 Å and use this model to estimate the wavelength of light absorbed when a pi electron is excited from the highest-occupied to the lowest-vacant box level. The experimental value is 217 nm.

2.16 For the particle in a one-dimensional box of length l, we could have put the coordinate origin at the center of the box. Find the wave functions and energy levels for this choice of origin.

2.17 The particle-in-a-box time-independent Schrödinger equation contains the constants h and m, and the boundary conditions involve the box length l. We therefore expect the stationary-state energies to be a function of h, m, and l; that is, $E = f(h, m, l)$. [We found $E = (n^2/8)(h^2/ml^2)$.] Prove that the only values of a, b, and c that give the product $h^a m^b l^c$ the dimensions of energy are $a = 2$, $b = -1$, $c = -2$.

2.18 Write down the time-dependent wave function for a free particle with energy E.

2.19 (a) For the particle in a rectangular well (Section 2.4), verify that $B = (V_0 - E)^{1/2}A/E^{1/2}$. (b) Verify Eq. (2.33).

2.20 For the particle in a rectangular well (Section 2.4), show that in the limit $V_0 \rightarrow \infty$ (a) Eq. (2.33) gives $E = n^2h^2/8ml^2$ as in Eq. (2.20); (b) the wave function goes to Eqs. (2.9) and (2.21).

2.21 For an electron in a 15.0-eV-deep one-dimensional rectangular well of width 2.00 Å, calculate the number of bound states. Use (6.107).

2.22 Find the allowed bound-state energy levels for the system of Prob. 2.21 by using a programmable calculator or computer to calculate the left side of (2.35) for ε going from 0 to 1 in small steps.

2.23 Sketch ψ for the next-lowest bound level in Fig. 2.5.

2.24 For a particle in a one-dimensional rectangular well, (a) must there be at least one bound state? (b) is ψ'' continuous at $x = 0$?

2.25 For an electron in a certain rectangular well with a depth of 20.0 eV, the lowest energy level lies 3.00 eV above the bottom of the well. Find the width of this well. *Hint:* Use $\tan\theta = \sin\theta/\cos\theta$.

2.26 For an electron in a certain rectangular well with a depth of 2.00 aJ, there are three bound energy levels. Find the minimum and maximum possible widths of this well. One attojoule (aJ) equals 10^{-18} J.

2.27. For a particle in a rectangular well of depth V_0 and width l, state whether the number of bound-state energy levels increases, decreases, or remains the same (a) as V_0 increases at fixed l; (b) as l increases at fixed V_0.

2.28 For the case $E = 0$ of a particle in a rectangular well of depth V_0, the solution to the Schrödinger equation inside the well is $\psi_{\text{II}} = ax + b$ [see the discussion after Eq. (2.19)]. (a) Use the boundary conditions to find four equations relating the constants a and b to C and G in the equation preceding (2.32). (b) Show that if $C > 0$ (or if $C < 0$), the equations in (a) lead to the contradiction that G is both less than and greater than zero. Hence, $C = 0$. (c) Show that $C = 0$ gives $\psi = 0$ everywhere. Hence, $E = 0$ is not allowed.

2.29 The energy of most stars results from the fusion of hydrogen nuclei to helium nuclei. The temperature of the interior of the sun (a typical star) is 15×10^6 K. At this temperature, virtually no nuclei have enough kinetic energy to overcome the electrostatic repulsion between nuclei and approach each other closely enough to undergo fusion. Therefore, when Eddington proposed in 1920 that nuclear fusion is the source of stellar energy, his idea was rejected. Explain why fusion does occur in stars, despite the above-mentioned apparent difficulty.

2.30 True or false? (a) The particle-in-a-box ground state has quantum number $n = 0$. (b) The particle-in-a-box stationary-state wave functions are discontinuous at certain points. (c) The first derivative of each particle-in-a-box stationary-state wave function is discontinuous at certain points. (d) The maximum probability density for every particle-in-a-box stationary state is at the center of the box. (e) For the particle-in-a-box $n = 2$ stationary state, the probability of finding the particle in the left quarter of the box equals the probability of finding it in the right quarter of the box. (Answer this and the next question without evaluating any integrals.) (f) For the $n = 1$ particle-in-a-box stationary state, the probability of finding the particle in the left third of the box equals the probability of finding it in the middle third. (g) The wavelength of the particle-in-a-box absorption transition from quantum number n to $n + 1$ decreases as the value of the quantum number n increases. (h) The probability density for the particle-in-a-box stationary states is constant along the length of the box.

CHAPTER 3

Operators

3.1 OPERATORS

We now develop the theory of quantum mechanics in a more general fashion than previously. We begin by writing the one-particle, one-dimensional time-independent Schrödinger equation (1.19) in the form

$$\left[-\frac{\hbar^2}{2m}\frac{d^2}{dx^2} + V(x)\right]\psi(x) = E\psi(x) \tag{3.1}$$

The entity in brackets in (3.1) is an *operator.* Equation (3.1) suggests that we have an energy operator, which, operating on the wave function, gives us the wave function back again, but multiplied by an allowed value of the energy. We therefore discuss operators.

An **operator** is a rule that transforms a given function into another function. For example, let $\hat{D}$ be the operator that differentiates a function with respect to x. We use a circumflex to denote an operator. Provided $f(x)$ is differentiable, the result of operating on $f(x)$ with $\hat{D}$ is $\hat{D}f(x) = f'(x)$. For example, $\hat{D}(x^2 + 3e^{2x}) = 2x + 6e^{2x}$. If $\hat{3}$ is the operator that multiplies a function by 3, then $\hat{3}(x^2 + 3e^x) = 3x^2 + 9e^x$. If cos is the operator that takes the cosine of a function, then application of cos to the function $x^2 + 1$ gives $\cos(x^2 + 1)$. If the operator $\hat{A}$ transforms the function $f(x)$ into the function $g(x)$, we write $\hat{A}f(x) = g(x)$.

We define the **sum** and the **difference** of two operators $\hat{A}$ and $\hat{B}$ by

$$\boxed{(\hat{A} + \hat{B})f(x) \equiv \hat{A}f(x) + \hat{B}f(x)} \tag{3.2}$$

$$(\hat{A} - \hat{B})f(x) \equiv \hat{A}f(x) - \hat{B}f(x)$$

For example, if $\hat{D} \equiv d/dx$, then

$$(\hat{D} + \hat{3})(x^3 - 5) \equiv \hat{D}(x^3 - 5) + 3(x^3 - 5) = 3x^2 + (3x^3 - 15) = 3x^3 + 3x^2 - 15$$

The **product** of two operators $\hat{A}$ and $\hat{B}$ is defined by

$$\boxed{\hat{A}\hat{B}f(x) \equiv \hat{A}[\hat{B}f(x)]} \tag{3.3}$$

In other words, we first operate on $f(x)$ with the operator on the right of the operator product, and then we take the resulting function and operate on it with the operator on the left of the operator product. For example, $\hat{3}\hat{D}f(x) = \hat{3}[\hat{D}f(x)] = \hat{3}f'(x) = 3f'(x)$.

The operators $\hat{A}\hat{B}$ and $\hat{B}\hat{A}$ may not have the same effect. Consider, for example, the operators d/dx and $\hat{x}$:

$$\hat{D}\hat{x}f(x) = \frac{d}{dx}[xf(x)] = f(x) + xf'(x) = (\hat{1} + \hat{x}\hat{D})f(x) \tag{3.4}$$

$$\hat{x}\hat{D}f(x) = \hat{x}\left[\frac{d}{dx}f(x)\right] = xf'(x)$$

Thus $\hat{A}\hat{B}$ and $\hat{B}\hat{A}$ are different operators in this case.

We can develop an *operator algebra* as follows. Two operators $\hat{A}$ and $\hat{B}$ are said to be **equal** if $\hat{A}f = \hat{B}f$ for all functions *f*. Equal operators produce the same result when they operate on a given function. For example, (3.4) shows that

$$\hat{D}\hat{x} = 1 + \hat{x}\hat{D} \tag{3.5}$$

The operator $\hat{1}$ (multiplication by 1) is the *unit operator*. The operator $\hat{0}$ (multiplication by 0) is the *null operator*. We usually omit the circumflex over operators that are simply multiplication by a constant. We can transfer operators from one side of an operator equation to the other (Prob. 3.7). Thus (3.5) is equivalent to $\hat{D}\hat{x} - \hat{x}\hat{D} - 1 = 0$, where circumflexes over the null and unit operators were omitted.

Operators obey the associative law of multiplication:

$$\hat{A}(\hat{B}\hat{C}) = (\hat{A}\hat{B})\hat{C} \tag{3.6}$$

The proof of (3.6) is outlined in Prob. 3.11. As an example, let $\hat{A} = d/dx$, $\hat{B} = \hat{x}$, and $\hat{C} = 3$. Using (3.5), we have

$$(\hat{A}\hat{B}) = \hat{D}\hat{x} = 1 + \hat{x}\hat{D}, \qquad [(\hat{A}\hat{B})\hat{C}]f = (1 + \hat{x}\hat{D})3f = 3f + 3xf'$$

$$(\hat{B}\hat{C}) = 3\hat{x}, \qquad [\hat{A}(\hat{B}\hat{C})]f = \hat{D}(3xf) = 3f + 3xf'$$

A major difference between operator algebra and ordinary algebra is that numbers obey the commutative law of multiplication, but operators do not necessarily do so; *ab* = *ba* if *a* and *b* are numbers, but $\hat{A}\hat{B}$ and $\hat{B}\hat{A}$ are not necessarily equal operators. We define the **commutator** $[\hat{A}, \hat{B}]$ of the operators $\hat{A}$ and $\hat{B}$ as the operator $\hat{A}\hat{B} - \hat{B}\hat{A}$:

$$\boxed{[\hat{A}, \hat{B}] \equiv \hat{A}\hat{B} - \hat{B}\hat{A}} \tag{3.7}$$

If $\hat{A}\hat{B} = \hat{B}\hat{A}$, then $[\hat{A}, \hat{B}] = 0$, and we say that $\hat{A}$ and $\hat{B}$ **commute**. If $\hat{A}\hat{B} \neq \hat{B}\hat{A}$, then $\hat{A}$ and $\hat{B}$ do not commute. Note that $[\hat{A}, \hat{B}]f = \hat{A}\hat{B}f - \hat{B}\hat{A}f$. Since the order in which we apply the operators 3 and d/dx makes no difference, we have

$$\left[\hat{3}, \frac{d}{dx}\right] = \hat{3}\frac{d}{dx} - \frac{d}{dx}\hat{3} = 0$$

From Eq. (3.5) we have

$$\left[\frac{d}{dx}, \hat{x}\right] = \hat{D}\hat{x} - \hat{x}\hat{D} = 1 \tag{3.8}$$

The operators d/dx and $\hat{x}$ do not commute.

EXAMPLE Find $[z^3, d/dz]$.

To find $[z^3, d/dz]$, we apply this operator to an arbitrary function $g(z)$. Using the commutator definition (3.7) and the definitions of the difference and product of two operators, we have

$$[z^3, d/dz]g = [z^3(d/dz) - (d/dz)z^3]g = z^3(d/dz)g - (d/dz)(z^3g)$$
$$= z^3g' - 3z^2g - z^3g' = -3z^2g$$

Deleting the arbitrary function g, we get the operator equation $[z^3, d/dz] = -3z^2$.

EXERCISE Find $[d/dx, 5x^2 + 3x + 4]$. (*Answer:* $10x + 3$.)

The **square** of an operator is defined as the product of the operator with itself: $\hat{A}^2 = \hat{A}\hat{A}$. Let us find the square of the differentiation operator:

$$\hat{D}^2 f(x) = \hat{D}(\hat{D}f) = \hat{D}f' = f''$$
$$\hat{D}^2 = d^2/dx^2$$

As another example, the square of the operator that takes the complex conjugate of a function is equal to the unit operator, since taking the complex conjugate twice gives the original function. The operator $\hat{A}^n$ ($n = 1, 2, 3, \dots$) is defined to mean applying the operator $\hat{A}$ n times in succession.

It turns out that the operators occurring in quantum mechanics are linear. $\hat{A}$ is a **linear operator** if and only if it has the following two properties:

$$\boxed{\hat{A}[f(x) + g(x)] = \hat{A}f(x) + \hat{A}g(x)} \quad \textbf{(3.9)}$$

$$\boxed{\hat{A}[cf(x)] = c\hat{A}f(x)} \quad \textbf{(3.10)}$$

where f and g are arbitrary functions and c is an arbitrary constant (not necessarily real). Examples of linear operators include $\hat{x}^2$, d/dx, and d^2/dx^2. Some nonlinear operators are cos and $(\ )^2$, where $(\ )^2$ squares the function it acts on.

EXAMPLE Is d/dx a linear operator? Is $\sqrt{\ }$ a linear operator?

We have

$$(d/dx)[f(x) + g(x)] = df/dx + dg/dx = (d/dx)f(x) + (d/dx)g(x)$$
$$(d/dx)[cf(x)] = c\, df(x)/dx$$

so d/dx obeys (3.9) and (3.10) and is a linear operator. However,

$$\sqrt{f(x) + g(x)} \neq \sqrt{f(x)} + \sqrt{g(x)}$$

so $\sqrt{\ }$ does not obey (3.9) and is nonlinear.

EXERCISE Is the operator $x^2 \cdot$ (multiplication by x^2) linear? (*Answer:* Yes.)

Useful identities in linear-operator manipulations are

$$(\hat{A} + \hat{B})\hat{C} = \hat{A}\hat{C} + \hat{B}\hat{C} \tag{3.11}$$

$$\hat{A}(\hat{B} + \hat{C}) = \hat{A}\hat{B} + \hat{A}\hat{C} \tag{3.12}$$

EXAMPLE Prove the distributive law (3.11) for linear operators.

A good way to begin a proof is to first write down what is given and what is to be proved. We are given that $\hat{A}$, $\hat{B}$, and $\hat{C}$ are linear operators. We must prove that $(\hat{A} + \hat{B})\hat{C} = \hat{A}\hat{C} + \hat{B}\hat{C}$.

To prove that the operator $(\hat{A} + \hat{B})\hat{C}$ is equal to the operator $\hat{A}\hat{C} + \hat{B}\hat{C}$, we must prove that these two operators give the same result when applied to an arbitrary function f. Thus we must prove that

$$[(\hat{A} + \hat{B})\hat{C}]f = (\hat{A}\hat{C} + \hat{B}\hat{C})f$$

We start with $[(\hat{A} + \hat{B})\hat{C}]f$. This expression involves the product of the two operators $\hat{A} + \hat{B}$ and $\hat{C}$. The operator-product definition (3.3) with $\hat{A}$ replaced by $\hat{A} + \hat{B}$ and $\hat{B}$ replaced by $\hat{C}$ gives $[(\hat{A} + \hat{B})\hat{C}]f = (\hat{A} + \hat{B})(\hat{C}f)$. The entity $\hat{C}f$ is a function, and use of the definition (3.2) of the sum $\hat{A} + \hat{B}$ of the two operators $\hat{A}$ and $\hat{B}$ gives $(\hat{A} + \hat{B})(\hat{C}f) = \hat{A}(\hat{C}f) + \hat{B}(\hat{C}f)$. Thus

$$[(\hat{A} + \hat{B})\hat{C}]f = (\hat{A} + \hat{B})(\hat{C}f) = \hat{A}(\hat{C}f) + \hat{B}(\hat{C}f)$$

Use of the operator-product definition (3.3) gives $\hat{A}(\hat{C}f) = \hat{A}\hat{C}f$ and $\hat{B}(\hat{C}f) = \hat{B}\hat{C}f$. Hence

$$[(\hat{A} + \hat{B})\hat{C}]f = \hat{A}\hat{C}f + \hat{B}\hat{C}f \tag{3.13}$$

Use of the operator-sum definition (3.2) with $\hat{A}$ replaced by $\hat{A}\hat{C}$ and $\hat{B}$ replaced by $\hat{B}\hat{C}$ gives $(\hat{A}\hat{C} + \hat{B}\hat{C})f = \hat{A}\hat{C}f + \hat{B}\hat{C}f$, so (3.13) becomes

$$[(\hat{A} + \hat{B})\hat{C}]f = (\hat{A}\hat{C} + \hat{B}\hat{C})f$$

which is what we wanted to prove. Hence $(\hat{A} + \hat{B})\hat{C} = \hat{A}\hat{C} + \hat{B}\hat{C}$.

Note that we did not need to use the linearity of $\hat{A}$, $\hat{B}$, and $\hat{C}$. Hence (3.11) holds for any operators. However, (3.12) holds only if $\hat{A}$ is linear (see Prob. 3.18).

EXAMPLE Find the square of the operator $d/dx + \hat{x}$.

To find the effect of $(d/dx + \hat{x})^2$, we apply this operator to an arbitrary function $f(x)$. Letting $\hat{D} \equiv d/dx$, we have

$$\begin{aligned}(\hat{D} + \hat{x})^2 f(x) &= (\hat{D} + \hat{x})[(\hat{D} + x)f] = (\hat{D} + \hat{x})(f' + xf)\\ &= f'' + f + xf' + xf' + x^2 f = (\hat{D}^2 + 2\hat{x}\hat{D} + \hat{x}^2 + 1)f(x)\\ (\hat{D} + \hat{x})^2 &= \hat{D}^2 + 2\hat{x}\hat{D} + \hat{x}^2 + 1\end{aligned}$$

Let us repeat this calculation, using only operator equations:

$$\begin{aligned}(\hat{D} + \hat{x})^2 &= (\hat{D} + \hat{x})(\hat{D} + \hat{x}) = \hat{D}(\hat{D} + \hat{x}) + \hat{x}(\hat{D} + \hat{x})\\ &= \hat{D}^2 + \hat{D}\hat{x} + \hat{x}\hat{D} + \hat{x}^2 = \hat{D}^2 + \hat{x}\hat{D} + 1 + \hat{x}\hat{D} + \hat{x}^2\\ &= \hat{D}^2 + 2x\hat{D} + x^2 + 1\end{aligned}$$

where (3.11), (3.12), and (3.5) have been used and the circumflex over the operator "multiplication by x" has been omitted. *Until you have become thoroughly experienced with operators, it is safest when doing operator manipulations always to let the operator operate on an arbitrary function f and then delete f at the end.*

EXERCISE Find $(d^2/dx^2 + x)^2$. (*Answer:* $d^4/dx^4 + 2xd^2/dx^2 + 2d/dx + x^2$.)

3.2 EIGENFUNCTIONS AND EIGENVALUES

Suppose that the effect of operating on some function $f(x)$ with the linear operator $\hat{A}$ is simply to multiply $f(x)$ by a certain constant k. We then say that $f(x)$ is an **eigenfunction** of $\hat{A}$ with **eigenvalue** k. As part of the definition, we shall require that the eigenfunction $f(x)$ is not identically zero. By this we mean that, although $f(x)$ may vanish at various points, it is not everywhere zero. We have

$$\boxed{\hat{A}f(x) = kf(x)} \tag{3.14}$$

(*Eigen* is a German word meaning *characteristic*.) As an example of (3.14), e^{2x} is an eigenfunction of the operator d/dx with eigenvalue 2:

$$(d/dx)e^{2x} = 2e^{2x}$$

However, $\sin 2x$ is not an eigenfunction of d/dx, since $(d/dx)(\sin 2x) = 2\cos 2x$, which is not a constant times $\sin 2x$.

EXAMPLE If $f(x)$ is an eigenfunction of the linear operator $\hat{A}$ and c is any constant, prove that $cf(x)$ is an eigenfunction of $\hat{A}$ with the same eigenvalue as $f(x)$.

A good way to see how to do a proof is to carry out the following steps:

1. Write down the given information and translate this information from words into equations.
2. Write down what is to be proved in the form of an equation or equations.
3. (a) Manipulate the given equations of step 1 so as to transform them to the desired equations of step 2. (b) Alternatively, start with one side of the equation that we want to prove and use the given equations of step 1 to manipulate this side until it is transformed into the other side of the equation to be proved.

We are given three pieces of information: f is an eigenfunction of $\hat{A}$; $\hat{A}$ is a linear operator; c is a constant. Translating these statements into equations, we have [see Eqs. (3.14), (3.9), and (3.10)]

$$\hat{A}f = kf \tag{3.15}$$

$$\hat{A}(f + g) = \hat{A}f + \hat{A}g \quad \text{and} \quad \hat{A}(bf) = b\hat{A}f \tag{3.16}$$

$$c = \text{a constant}$$

where k and b are constants and f and g are functions.

We want to prove that cf is an eigenfunction of $\hat{A}$ with the same eigenvalue as f, which, written as an equation, is

$$\hat{A}(cf) = k(cf)$$

Using the strategy of step 3(b), we start with the left side $\hat{A}(cf)$ of this last equation and try to show that it equals $k(cf)$. Using the second equation in the linearity definition (3.16), we have $\hat{A}(cf) = c\hat{A}f$. Using the eigenvalue equation (3.15), we have $c\hat{A}f = ckf$. Hence

$$\hat{A}(cf) = c\hat{A}f = ckf = k(cf)$$

which completes the proof.

EXAMPLE (a) Find the eigenfunctions and eigenvalues of the operator d/dx. (b) If we impose the boundary condition that the eigenfunctions remain finite as $x \to \pm\infty$, find the eigenvalues.

(a) Equation (3.14) with $\hat{A} = d/dx$ becomes

$$df(x)/dx = kf(x) \tag{3.17}$$

$$df/f = k\,dx$$

Integration gives

$$\ln f = kx + \text{constant}$$

$$f = e^{\text{constant}}e^{kx}$$

$$f = ce^{kx} \tag{3.18}$$

The eigenfunctions of d/dx are given by (3.18). The eigenvalues are k, which can be any number whatever and (3.17) will still be satisfied. The eigenfunctions contain an arbitrary multiplicative constant c. This is true for the eigenfunctions of any linear operator, as was proved in the previous example. Each different value of k in (3.18) gives a different eigenfunction. However, eigenfunctions with the same value of k but different values of c are not independent of each other.

(b) Since k can be complex, we write it as $k = a + ib$, where a and b are real numbers. We then have $f(x) = ce^{ax}e^{ibx}$. The factor e^{ax} goes to infinity as x goes to infinity if a is positive; it goes to infinity as x goes to minus infinity if a is negative. Hence the boundary conditions require that $a = 0$, and the eigenvalues are $k = ib$.

In the first example in Section 3.1, we found that $[z^3, d/dz]g(z) = -3z^2g(z)$ for *every* function g, and we concluded that $[z^3, d/dz] = -3z^2$. In contrast, the eigenvalue equation $\hat{A}f(x) = kf(x)$ [Eq. (3.14)] does not hold for every function $f(x)$ and we cannot conclude from this equation that $\hat{A} = k$. Thus the fact that $(d/dx)e^{2x} = 2e^{2x}$ does not mean that the operator d/dx equals multiplication by 2.

3.3 OPERATORS AND QUANTUM MECHANICS

We now examine the relationship between operators and quantum mechanics. Comparing Eq. (3.1) with (3.14), we see that the Schrödinger equation is an eigenvalue problem. The values of the energy E are the eigenvalues. The eigenfunctions are the wave functions ψ. The operator whose eigenfunctions and eigenvalues are desired is $-(\hbar^2/2m)\, d^2/dx^2 + V(x)$. This operator is called the **Hamiltonian operator** for the system.

Sir William Rowan Hamilton (1805–1865) devised an alternative form of Newton's equations of motion involving a function H, the Hamiltonian function for the system. For a system where the potential energy is a function of the coordinates only, the total energy remains constant with time; that is, E is conserved. We shall restrict ourselves to such conservative systems. For conservative systems, the classical-mechanical **Hamiltonian function** turns out to be simply the total energy expressed in terms of coordinates and conjugate momenta. For Cartesian coordinates x, y, z, the **conjugate momenta** are the components of linear momentum in the x, y, and z directions: p_x, p_y, and p_z:

$$\boxed{p_x \equiv mv_x, \quad p_y \equiv mv_y, \quad p_z \equiv mv_z} \tag{3.19}$$

where v_x, v_y, and v_z are the components of the particle's velocity in the x, y, and z directions.

Let us find the classical-mechanical Hamiltonian function for a particle of mass m moving in one dimension and subject to a potential energy $V(x)$. The Hamiltonian function is equal to the energy, which is composed of kinetic and potential energies. The familiar form of the kinetic energy, $\frac{1}{2}mv_x^2$, will not do, however, since we must express the Hamiltonian as a function of coordinates and momenta, not velocities. Since $v_x = p_x/m$, the form of the kinetic energy we want is $p_x^2/2m$. The Hamiltonian function is

$$H = \frac{p_x^2}{2m} + V(x) \tag{3.20}$$

The time-independent Schrödinger equation (3.1) indicates that, corresponding to the Hamiltonian function (3.20), we have a quantum-mechanical operator

$$-\frac{\hbar^2}{2m}\frac{d^2}{dx^2} + V(x)$$

whose eigenvalues are the possible values of the system's energy. This correspondence between physical quantities in classical mechanics and operators in quantum mechanics is general. It is a fundamental postulate of quantum mechanics that *every physical property* (for example, the energy, the x coordinate, the momentum) *has a corresponding quantum-mechanical operator*. We further postulate that the operator corresponding to the property B is found by writing the classical mechanical expression for B as a function of Cartesian coordinates and corresponding momenta and then making the following replacements. Each Cartesian coordinate q is replaced by the operator multiplication by that coordinate:

$$\hat{q} = q\,\cdot$$

Each Cartesian component of linear momentum p_q is replaced by the operator

$$\hat{p}_q = \frac{\hbar}{i}\frac{\partial}{\partial q} = -i\hbar\frac{\partial}{\partial q}$$

where $i = \sqrt{-1}$ and $\partial/\partial q$ is the operator for the partial derivative with respect to the coordinate q. Note that $1/i = i/i^2 = i/(-1) = -i$.

Consider some examples. The operator corresponding to the x coordinate is multiplication by x:

$$\boxed{\hat{x} = x \cdot} \tag{3.21}$$

Also,

$$\boxed{\hat{y} = y \cdot \quad \text{and} \quad \hat{z} = z \cdot} \tag{3.22}$$

The operators for the components of linear momentum are

$$\boxed{\hat{p}_x = \frac{\hbar}{i}\frac{\partial}{\partial x}, \quad \hat{p}_y = \frac{\hbar}{i}\frac{\partial}{\partial y}, \quad \hat{p}_z = \frac{\hbar}{i}\frac{\partial}{\partial z}} \tag{3.23}$$

The operator corresponding to p_x^2 is

$$\hat{p}_x^2 = \left(\frac{\hbar}{i}\frac{\partial}{\partial x}\right)^2 = \frac{\hbar}{i}\frac{\partial}{\partial x}\frac{\hbar}{i}\frac{\partial}{\partial x} = -\hbar^2\frac{\partial^2}{\partial x^2} \tag{3.24}$$

with similar expressions for $\hat{p}_y^2$ and $\hat{p}_z^2$.

Now consider the potential-energy and kinetic-energy operators in one dimension. Suppose we had a system with the potential-energy function $V(x) = ax^2$, where a is a constant. Replacing x with $x \cdot$, we see that the potential-energy operator is simply multiplication by ax^2; that is, $\hat{V}(x) = ax^2 \cdot$. In general, we have for any potential-energy function

$$\boxed{\hat{V}(x) = V(x) \cdot} \tag{3.25}$$

The classical-mechanical expression for the kinetic energy T in (3.20) is

$$\boxed{T = p_x^2/2m} \tag{3.26}$$

Replacing p_x by the corresponding operator (3.23), we have

$$\hat{T} = -\frac{\hbar^2}{2m}\frac{\partial^2}{\partial x^2} = -\frac{\hbar^2}{2m}\frac{d^2}{dx^2} \tag{3.27}$$

where (3.24) has been used, and the partial derivative becomes an ordinary derivative in one dimension. The classical-mechanical Hamiltonian (3.20) is

$$H = T + V = p_x^2/2m + V(x) \tag{3.28}$$

The corresponding quantum-mechanical Hamiltonian (or energy) operator is

$$\boxed{\hat{H} = \hat{T} + \hat{V} = -\frac{\hbar^2}{2m}\frac{d^2}{dx^2} + V(x)} \tag{3.29}$$

which agrees with the operator in the Schrödinger equation (3.1). Note that all these operators are linear.

How are the quantum-mechanical operators related to the corresponding properties of a system? Each such operator has its own set of eigenfunctions and eigenvalues. Let $\hat{B}$ be the quantum-mechanical operator that corresponds to the physical property B. Letting f_i and b_i symbolize the eigenfunctions and eigenvalues of $\hat{B}$, we have [Eq. (3.14)]

$$\hat{B}f_i = b_i f_i, \quad i = 1, 2, 3, \ldots \tag{3.30}$$

The operator $\hat{B}$ has many eigenfunctions and eigenvalues, and the subscript i is used to indicate this. $\hat{B}$ is usually a differential operator, and (3.30) is a differential equation whose solutions give the eigenfunctions and eigenvalues. Quantum mechanics postulates that (no matter what the state function of the system happens to be) *a measurement of the property B must yield one of the eigenvalues b_i of the operator $\hat{B}$*. For example, the only values that can be found for the energy of a system are the eigenvalues of the energy (Hamiltonian) operator $\hat{H}$. Using ψ_i to symbolize the eigenfunctions of $\hat{H}$, we have as the eigenvalue equation (3.30)

$$\boxed{\hat{H}\psi_i = E_i\psi_i} \tag{3.31}$$

Using the Hamiltonian (3.29) in (3.31), we obtain for a one-dimensional, one-particle system

$$\left[-\frac{\hbar^2}{2m}\frac{d^2}{dx^2} + V(x)\right]\psi_i = E_i\psi_i \tag{3.32}$$

which is the time-independent Schrödinger equation, (3.1). Thus our postulates about operators are consistent with our previous work. We shall later further justify the choice (3.23) for the momentum operator by showing that in the limiting transition to classical mechanics this choice yields $p_x = m(dx/dt)$, as it should. (See Prob. 7.59.)

In Chapter 1 we postulated that the state of a quantum-mechanical system is specified by a state function $\Psi(x, t)$, which contains all the information we can know about the system. How does Ψ give us information about the property B? We postulate that *if Ψ is an eigenfunction of $\hat{B}$ with eigenvalue b_k, then a measurement of B is certain to yield the value b_k*. Consider, for example, the energy. The eigenfunctions of the energy operator are the solutions $\psi(x)$ of the time-independent Schrödinger equation (3.32). Suppose the system is in a stationary state with state function [Eq. (1.20)]

$$\Psi(x, t) = e^{-iEt/\hbar}\psi(x) \tag{3.33}$$

Is $\Psi(x, t)$ an eigenfunction of the energy operator $\hat{H}$? We have

$$\hat{H}\Psi(x, t) = \hat{H}e^{-iEt/\hbar}\psi(x)$$

$\hat{H}$ contains no derivatives with respect to time and therefore does not affect the exponential factor $e^{-iEt/\hbar}$. We have

$$\hat{H}\Psi(x, t) = e^{-iEt/\hbar}\hat{H}\psi(x) = Ee^{-iEt/\hbar}\psi(x) = E\Psi(x, t)$$

$$\hat{H}\Psi = E\Psi \tag{3.34}$$

where (3.31) was used. Hence, for a stationary state, $\Psi(x, t)$ is an eigenfunction of $\hat{H}$, and we are certain to obtain the value E when we measure the energy.

As an example of another property, consider momentum. The eigenfunctions g of $\hat{p}_x$ are found by solving

$$\hat{p}_x g = kg$$
$$\frac{\hbar}{i}\frac{dg}{dx} = kg \tag{3.35}$$

We find (Prob. 3.30)

$$g = Ae^{ikx/\hbar} \tag{3.36}$$

where A is an arbitrary constant. To keep g finite for large $|x|$, the eigenvalues k must be real. Thus the eigenvalues of $\hat{p}_x$ are all the real numbers

$$-\infty < k < \infty \tag{3.37}$$

which is reasonable. Any measurement of p_x must yield one of the eigenvalues (3.37) of $\hat{p}_x$. Each different value of k in (3.36) gives a different eigenfunction g. It might seem surprising that the operator for the physical property momentum involves the imaginary number i. Actually, the presence of i in $\hat{p}_x$ ensures that the eigenvalues k are real. Recall that the eigenvalues of d/dx are imaginary (Section 3.2).

Comparing the free-particle wave function (2.30) with the eigenfunctions (3.36) of $\hat{p}_x$, we note the following physical interpretation: The first term in (2.30) corresponds to positive momentum and represents motion in the $+x$ direction, while the second term in (2.30) corresponds to negative momentum and represents motion in the $-x$ direction.

Now consider the momentum of a particle in a box. The state function for a particle in a stationary state in a one-dimensional box is [Eqs. (3.33), (2.20), and (2.23)]

$$\Psi(x, t) = e^{-iEt/\hbar}\left(\frac{2}{l}\right)^{1/2}\sin\left(\frac{n\pi x}{l}\right) \tag{3.38}$$

where $E = n^2h^2/8ml^2$. Does the particle have a definite value of p_x? That is, is $\Psi(x, t)$ an eigenfunction of $\hat{p}_x$? Looking at the eigenfunctions (3.36) of $\hat{p}_x$, we see that there is no numerical value of the real constant k that will make the exponential function in (3.36) become a sine function, as in (3.38). Hence Ψ is not an eigenfunction of $\hat{p}_x$. We can verify this directly; we have

$$\hat{p}_x\Psi = \frac{\hbar}{i}\frac{\partial}{\partial x}e^{-iEt/\hbar}\left(\frac{2}{l}\right)^{1/2}\sin\left(\frac{n\pi x}{l}\right) = \frac{n\pi\hbar}{il}e^{-iEt/\hbar}\left(\frac{2}{l}\right)^{1/2}\cos\left(\frac{n\pi x}{l}\right)$$

Since $\hat{p}_x\Psi \neq \text{constant} \cdot \Psi$, the state function Ψ is not an eigenfunction of $\hat{p}_x$.

Note that the system's state function Ψ need not be an eigenfunction f_i of the operator $\hat{B}$ in (3.30) that corresponds to the physical property B of the system. Thus, the particle-in-a-box stationary-state wave functions are not eigenfunctions of $\hat{p}_x$. Despite this, we still must get one of the eigenvalues (3.37) of $\hat{p}_x$ when we measure p_x for a particle-in-a-box stationary state.

Are the particle-in-a-box stationary-state wave functions eigenfunctions of $\hat{p}_x^2$? We have [Eq. (3.24)]

$$\hat{p}_x^2\Psi = -\hbar^2\frac{\partial^2}{\partial x^2}e^{-iEt/\hbar}\left(\frac{2}{l}\right)^{1/2}\sin\left(\frac{n\pi x}{l}\right) = \frac{n^2\pi^2\hbar^2}{l^2}e^{-iEt/\hbar}\left(\frac{2}{l}\right)^{1/2}\sin\left(\frac{n\pi x}{l}\right)$$

$$\hat{p}_x^2\Psi = \frac{n^2h^2}{4l^2}\Psi \tag{3.39}$$

Hence a measurement of p_x^2 will always give the result $n^2h^2/4l^2$ when the particle is in the stationary state with quantum number n. This should come as no surprise: The potential energy in the box is zero, and the Hamiltonian operator is

$$\hat{H} = \hat{T} + \hat{V} = \hat{T} = \hat{p}_x^2/2m$$

We then have [Eq. (3.34)]

$$\hat{H}\Psi = E\Psi = \frac{\hat{p}_x^2}{2m}\Psi$$

$$\hat{p}_x^2\Psi = 2mE\Psi = 2m\frac{n^2h^2}{8ml^2}\Psi = \frac{n^2h^2}{4l^2}\Psi \tag{3.40}$$

in agreement with (3.39). The only possible value for p_x^2 is

$$p_x^2 = n^2h^2/4l^2 \tag{3.41}$$

Equation (3.41) suggests that a measurement of p_x would necessarily yield one of the two values $\pm\frac{1}{2}nh/l$, corresponding to the particle moving to the right or to the left in the box. This plausible suggestion is not accurate. An analysis using the methods of Chapter 7 shows that there is a high probability that the measured value will be close to one of the two values $\pm\frac{1}{2}nh/l$, but that any value consistent with (3.37) can result from a measurement of p_x for the particle in a box; see Prob. 7.41.

We postulated that a measurement of the property B must give a result that is one of the eigenvalues of the operator $\hat{B}$. If the state function Ψ happens to be an eigenfunction of $\hat{B}$ with eigenvalue b, we are certain to get b when we measure B. Suppose, however, that Ψ is not one of the eigenfunctions of $\hat{B}$. What then? We still assert that *we will get one of the eigenvalues of* $\hat{B}$ *when we measure B, but we cannot predict which eigenvalue will be obtained.* We shall see in Chapter 7 that the probabilities for obtaining the various eigenvalues of $\hat{B}$ can be predicted.

EXAMPLE The energy of a particle of mass m in a one-dimensional box of length l is measured. What are the possible values that can result from the measurement if at the time the measurement begins, the particle's state function is (a) $\Psi = (30/l^5)^{1/2}x(l - x)$ for $0 \le x \le l$; (b) $\Psi = (2/l)^{1/2}\sin(3\pi x/l)$ for $0 \le x \le l$?

(a) The possible outcomes of a measurement of the property E are the eigenvalues of the system's energy (Hamiltonian) operator $\hat{H}$. Therefore, the measured value must be one of the numbers $n^2h^2/8ml^2$, where $n = 1, 2, 3, \ldots$. Since Ψ is not one of the eigenfunctions $(2/l)^{1/2}\sin(n\pi x/l)$ [Eq. (2.23)] of $\hat{H}$, we cannot predict which one of these eigenvalues will be obtained for this nonstationary state. (b) Since Ψ is an eigenfunction of $\hat{H}$ with eigenvalue $3^2h^2/8ml^2$ [Eq. (2.20)], the measurement must give $9h^2/8ml^2$.

3.4 THE THREE-DIMENSIONAL, MANY-PARTICLE SCHRÖDINGER EQUATION

Up to now we have restricted ourselves to one-dimensional, one-particle systems. The operator formalism developed in the last section allows us to extend our work to three-dimensional, many-particle systems. The time-dependent Schrödinger equation for the time development of the state function is postulated to have the form of Eq. (1.13):

$$\boxed{i\hbar \frac{\partial \Psi}{\partial t} = \hat{H}\Psi} \tag{3.42}$$

The time-independent Schrödinger equation for the energy eigenfunctions and eigenvalues is

$$\boxed{\hat{H}\psi = E\psi} \tag{3.43}$$

which is obtained from (3.42) by taking the potential energy as independent of time and applying the separation-of-variables procedure used to obtain (1.19) from (1.13).

For a one-particle, three-dimensional system, the classical-mechanical Hamiltonian is

$$H = T + V = \frac{1}{2m}(p_x^2 + p_y^2 + p_z^2) + V(x, y, z) \tag{3.44}$$

Introducing the quantum-mechanical operators [Eq. (3.24)], we have for the Hamiltonian operator

$$\hat{H} = -\frac{\hbar^2}{2m}\left(\frac{\partial^2}{\partial x^2} + \frac{\partial^2}{\partial y^2} + \frac{\partial^2}{\partial z^2}\right) + V(x, y, z) \tag{3.45}$$

The operator in parentheses in (3.45) is called the **Laplacian operator** ∇^2 (read as "del squared"):

$$\boxed{\nabla^2 \equiv \frac{\partial^2}{\partial x^2} + \frac{\partial^2}{\partial y^2} + \frac{\partial^2}{\partial z^2}} \tag{3.46}$$

The one-particle, three-dimensional, time-independent Schrödinger equation is then

$$-\frac{\hbar^2}{2m}\nabla^2\psi + V\psi = E\psi \tag{3.47}$$

Now consider a three-dimensional system with n particles. Let particle i have mass m_i and coordinates (x_i, y_i, z_i), where $i = 1, 2, 3, \ldots, n$. The kinetic energy is the sum of the kinetic energies of the individual particles:

$$T = \frac{1}{2m_1}(p_{x_1}^2 + p_{y_1}^2 + p_{z_1}^2) + \frac{1}{2m_2}(p_{x_2}^2 + p_{y_2}^2 + p_{z_2}^2) + \cdots + \frac{1}{2m_n}(p_{x_n}^2 + p_{y_n}^2 + p_{z_n}^2)$$

where p_{x_i} is the x component of the linear momentum of particle i, and so on. The kinetic-energy operator is

$$\hat{T} = -\frac{\hbar^2}{2m_1}\left(\frac{\partial^2}{\partial x_1^2} + \frac{\partial^2}{\partial y_1^2} + \frac{\partial^2}{\partial z_1^2}\right) - \cdots - \frac{\hbar^2}{2m_n}\left(\frac{\partial^2}{\partial x_n^2} + \frac{\partial^2}{\partial y_n^2} + \frac{\partial^2}{\partial z_n^2}\right)$$

$$\boxed{\hat{T} = -\sum_{i=1}^{n} \frac{\hbar^2}{2m_i} \nabla_i^2} \tag{3.48}$$

$$\boxed{\nabla_i^2 \equiv \frac{\partial^2}{\partial x_i^2} + \frac{\partial^2}{\partial y_i^2} + \frac{\partial^2}{\partial z_i^2}} \tag{3.49}$$

We shall usually restrict ourselves to cases where the potential energy depends only on the $3n$ coordinates:

$$V = V(x_1, y_1, z_1, \ldots, x_n, y_n, z_n)$$

The Hamiltonian operator for an n-particle, three-dimensional system is then

$$\boxed{\hat{H} = -\sum_{i=1}^{n} \frac{\hbar^2}{2m_i} \nabla_i^2 + V(x_1, \ldots, z_n)} \tag{3.50}$$

and the time-independent Schrödinger equation is

$$\left[-\sum_{i=1}^{n} \frac{\hbar^2}{2m_i} \nabla_i^2 + V(x_1, \ldots, z_n)\right]\psi = E\psi \tag{3.51}$$

where the time-independent wave function is a function of the $3n$ coordinates of the n particles:

$$\psi = \psi(x_1, y_1, z_1, \ldots, x_n, y_n, z_n) \tag{3.52}$$

The Schrödinger equation (3.51) is a linear partial differential equation.

As an example, consider a system of two particles interacting so that the potential energy is inversely proportional to the distance between them, with c being the proportionality constant. The Schrödinger equation (3.51) becomes

$$\left[-\frac{\hbar^2}{2m_1}\left(\frac{\partial^2}{\partial x_1^2} + \frac{\partial^2}{\partial y_1^2} + \frac{\partial^2}{\partial z_1^2}\right) - \frac{\hbar^2}{2m_2}\left(\frac{\partial^2}{\partial x_2^2} + \frac{\partial^2}{\partial y_2^2} + \frac{\partial^2}{\partial z_2^2}\right) + \frac{c}{[(x_1 - x_2)^2 + (y_1 - y_2)^2 + (z_1 - z_2)^2]^{1/2}}\right]\psi = E\psi \tag{3.53}$$

$$\psi = \psi(x_1, y_1, z_1, x_2, y_2, z_2)$$

Although (3.53) looks formidable, we shall solve it in Chapter 6.

For a one-particle, one-dimensional system, the Born postulate [Eq. (1.15)] states that $|\Psi(x', t)|^2\, dx$ is the probability of observing the particle between x' and $x' + dx$

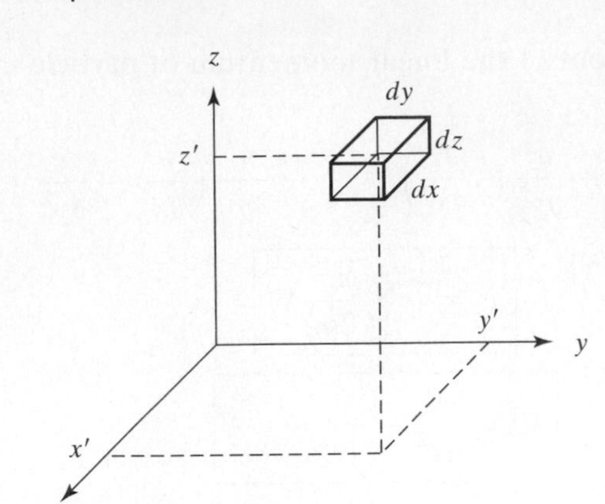

FIGURE 3.1 An infinitesimal box-shaped region located at x', y', z'.

at time t, where x' is a particular value of x. We extend this postulate as follows. *For a three-dimensional, one-particle system, the quantity*

$$\boxed{|\Psi(x', y', z', t)|^2\, dx\, dy\, dz} \tag{3.54}$$

is the probability of finding the particle in the infinitesimal region of space with the x coordinate lying between x' and $x' + dx$, the y coordinate lying between y' and $y' + dy$, and the z coordinate between z' and $z' + dz$ (Fig. 3.1). Since the total probability of finding the particle is 1, the normalization condition is

$$\int_{-\infty}^{\infty}\int_{-\infty}^{\infty}\int_{-\infty}^{\infty} |\Psi(x, y, z, t)|^2\, dx\, dy\, dz = 1 \tag{3.55}$$

For a three-dimensional, n-particle system, we postulate that

$$\boxed{|\Psi(x_1', y_1', z_1', x_2', y_2', z_2', \ldots, x_n', y_n', z_n', t)|^2\, dx_1\, dy_1\, dz_1\, dx_2\, dy_2\, dz_2 \cdots dx_n\, dy_n\, dz_n} \tag{3.56}$$

is the probability at time t of simultaneously finding particle 1 in the infinitesimal rectangular box-shaped region at (x_1', y_1', z_1') with edges dx_1, dy_1, dz_1, particle 2 in the infinitesimal box-shaped region at (x_2', y_2', z_2') with edges $dx_2, dy_2, dz_2, \ldots,$ and particle n in the infinitesimal box-shaped region at (x_n', y_n', z_n') with edges dx_n, dy_n, dz_n. The total probability of finding all the particles is 1, and the normalization condition is

$$\int_{-\infty}^{\infty}\int_{-\infty}^{\infty}\int_{-\infty}^{\infty}\cdots\int_{-\infty}^{\infty}\int_{-\infty}^{\infty}\int_{-\infty}^{\infty} |\Psi|^2\, dx_1\, dy_1\, dz_1 \cdots dx_n\, dy_n\, dz_n = 1 \tag{3.57}$$

It is customary in quantum mechanics to denote integration over the full range of all the coordinates of a system by $\int d\tau$. A shorthand way of writing (3.55) or (3.57) is

$$\boxed{\int |\Psi|^2\, d\tau = 1} \tag{3.58}$$

Although (3.58) may look like an indefinite integral, it is understood to be a definite integral. The integration variables and their ranges are understood from the context.

For a stationary state, $|\Psi|^2 = |\psi|^2$, and

$$\boxed{\int |\psi|^2 \, d\tau = 1} \qquad \textbf{(3.59)}$$

3.5 THE PARTICLE IN A THREE-DIMENSIONAL BOX

For the present, we confine ourselves to one-particle problems. In this section we consider the three-dimensional case of the problem solved in Section 2.2, the particle in a box.

There are many possible shapes for a three-dimensional box. The box we consider is a rectangular parallelepiped with edges of length *a, b,* and *c*. We choose our coordinate system so that one corner of the box lies at the origin and the box lies in the first octant of space (Fig. 3.2). Within the box, the potential energy is zero. Outside the box, it is infinite:

$$V(x, y, z) = 0 \quad \text{in the region} \begin{cases} 0 < x < a \\ 0 < y < b \\ 0 < z < c \end{cases} \tag{3.60}$$

$$V = \infty \quad \text{elsewhere}$$

Since the probability for the particle to have infinite energy is zero, the wave function must be zero outside the box. Within the box, the potential-energy operator is zero and the Schrödinger equation (3.47) is

$$-\frac{\hbar^2}{2m}\left(\frac{\partial^2\psi}{\partial x^2} + \frac{\partial^2\psi}{\partial y^2} + \frac{\partial^2\psi}{\partial z^2}\right) = E\psi \tag{3.61}$$

To solve (3.61), we assume that the solution can be written as the product of a function of x alone times a function of y alone times a function of z alone:

$$\psi(x, y, z) = f(x)g(y)h(z) \tag{3.62}$$

It might be thought that this assumption throws away solutions that are not of the form (3.62). However, it can be shown that, if we can find solutions of the form (3.62) that

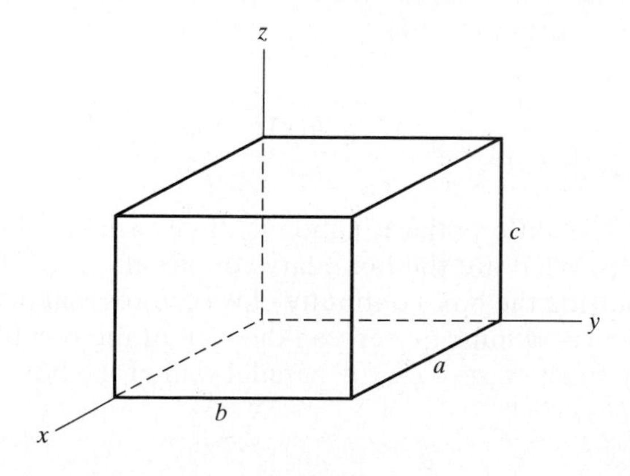

FIGURE 3.2 Inside the box-shaped region, $V = 0$.

satisfy the boundary conditions, then there are no other solutions of the Schrödinger equation that will satisfy the boundary conditions. (For a proof, see G. F. D. Duff and D. Naylor, *Differential Equations of Applied Mathematics*, Wiley, 1966, pp. 257–258.) The method we are using to solve (3.62) is called **separation of variables**.

From (3.62), we find

$$\frac{\partial^2\psi}{\partial x^2} = f''(x)g(y)h(z), \quad \frac{\partial^2\psi}{\partial y^2} = f(x)g''(y)h(z), \quad \frac{\partial^2\psi}{\partial z^2} = f(x)g(y)h''(z) \tag{3.63}$$

Substitution of (3.62) and (3.63) into (3.61) gives

$$-(\hbar^2/2m)f''gh - (\hbar^2/2m)fg''h - (\hbar^2/2m)fgh'' - Efgh = 0 \tag{3.64}$$

Division of this equation by fgh gives

$$-\frac{\hbar^2 f''}{2mf} - \frac{\hbar^2 g''}{2mg} - \frac{\hbar^2 h''}{2mh} - E = 0 \tag{3.65}$$

$$-\frac{\hbar^2 f''(x)}{2mf(x)} = \frac{\hbar^2 g''(y)}{2mg(y)} + \frac{\hbar^2 h''(z)}{2mh(z)} + E \tag{3.66}$$

Let us define E_x as equal to the left side of (3.66):

$$E_x \equiv -\hbar^2 f''(x)/2mf(x) \tag{3.67}$$

The definition (3.67) shows that E_x is independent of y and z. Equation (3.66) shows that E_x equals $\hbar^2 g''(y)/2mg(y) + \hbar^2 h''(z)/2mh(z) + E$; therefore, E_x must be independent of x. Being independent of x, y, and z, the quantity E_x must be a constant.

Similar to (3.67), we define E_y and E_z by

$$E_y \equiv -\hbar^2 g''(y)/2mg(y), \qquad E_z \equiv -\hbar^2 h''(z)/2mh(z) \tag{3.68}$$

Since x, y, and z occur symmetrically in (3.65), the same reasoning that showed E_x to be a constant shows that E_y and E_z are constants. Substitution of the definitions (3.67) and (3.68) into (3.65) gives

$$E_x + E_y + E_z = E \tag{3.69}$$

Equations (3.67) and (3.68) are

$$\frac{d^2 f(x)}{dx^2} + \frac{2m}{\hbar^2}E_x f(x) = 0 \tag{3.70}$$

$$\frac{d^2 g(y)}{dy^2} + \frac{2m}{\hbar^2}E_y g(y) = 0, \qquad \frac{d^2 h(z)}{dz^2} + \frac{2m}{\hbar^2}E_z h(z) = 0 \tag{3.71}$$

We have converted the partial differential equation in three variables into three ordinary differential equations. What are the boundary conditions on (3.70)? Since the wave function vanishes outside the box, continuity of ψ requires that it vanish on the walls of the box. In particular, ψ must be zero on the wall of the box lying in the yz plane, where $x = 0$, and it must be zero on the parallel wall of the box, where $x = a$. Therefore, $f(0) = 0$ and $f(a) = 0$.

Now compare Eq. (3.70) with the Schrödinger equation [Eq. (2.10)] for a particle in a one-dimensional box. The equations are the same in form, with E_x in (3.70) corresponding to E in (2.10). Are the boundary conditions the same? Yes, except that we have $x = a$ instead of $x = l$ as the second point where the independent variable vanishes. Thus we can use the work in Section 2.2 to write as the solution [see Eqs. (2.23) and (2.20)]

$$f(x) = \left(\frac{2}{a}\right)^{1/2} \sin\left(\frac{n_x \pi x}{a}\right)$$

$$E_x = \frac{n_x^2 h^2}{8ma^2}, \quad n_x = 1, 2, 3 \ldots$$

The same reasoning applied to the y and z equations gives

$$g(y) = \left(\frac{2}{b}\right)^{1/2} \sin\left(\frac{n_y \pi y}{b}\right), \qquad h(z) = \left(\frac{2}{c}\right)^{1/2} \sin\left(\frac{n_z \pi z}{c}\right)$$

$$E_y = \frac{n_y^2 h^2}{8mb^2}, \quad n_y = 1, 2, 3 \ldots \quad \text{and} \quad E_z = \frac{n_z^2 h^2}{8mc^2}, \quad n_z = 1, 2, 3 \ldots$$

From (3.69), the energy is

$$E = \frac{h^2}{8m}\left(\frac{n_x^2}{a^2} + \frac{n_y^2}{b^2} + \frac{n_z^2}{c^2}\right) \tag{3.72}$$

From (3.62), the wave function inside the box is

$$\psi(x, y, z) = \left(\frac{8}{abc}\right)^{1/2} \sin\left(\frac{n_x \pi x}{a}\right) \sin\left(\frac{n_y \pi y}{b}\right) \sin\left(\frac{n_z \pi z}{c}\right) \tag{3.73}$$

The wave function has three quantum numbers, n_x, n_y, n_z. We can attribute this to the three-dimensional nature of the problem. The three quantum numbers vary independently of one another.

Since the x, y, and z factors in the wave function are each independently normalized, the wave function is normalized:

$$\int_{-\infty}^{\infty}\int_{-\infty}^{\infty}\int_{-\infty}^{\infty} |\psi|^2 \, dx\, dy\, dz = \int_0^a |f(x)|^2 \, dx \int_0^b |g(y)|^2 \, dy \int_0^c |h(z)|^2 \, dz = 1$$

where we used (Prob. 3.38)

$$\boxed{\iiint F(x)G(y)H(z)\, dx\, dy\, dz = \int F(x)\, dx \int G(y)\, dy \int H(z)\, dz} \tag{3.74}$$

What are the dimensions of $\psi(x, y, z)$ in (3.73)? For a one-particle, three-dimensional system, $|\psi|^2 \, dx\, dy\, dz$ is a probability, and probabilities are dimensionless. Since the dimensions of $dx\, dy\, dz$ are length^3, $\psi(x, y, z)$ must have dimensions of length$^{-3/2}$ to make $|\psi|^2 \, dx\, dy\, dz$ dimensionless.

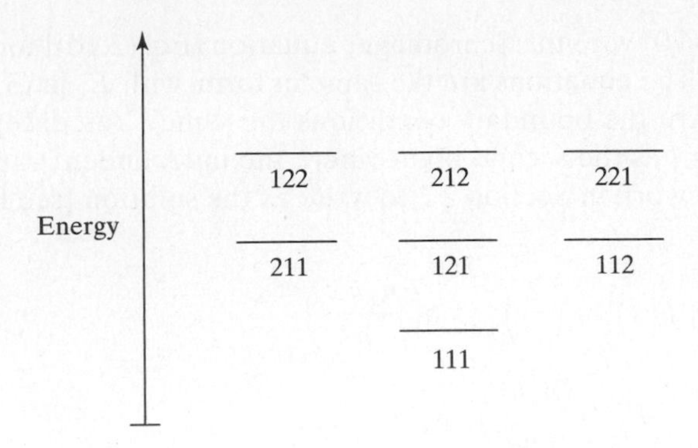

FIGURE 3.3 Energies of the lowest few states of a particle in a cubic box.

Suppose that $a = b = c$. We then have a cube. The energy levels are then

$$E = (h^2/8ma^2)(n_x^2 + n_y^2 + n_z^2) \tag{3.75}$$

Let us tabulate some of the allowed energies of a particle confined to a cube with infinitely strong walls:

$n_x n_y n_z$	111	211	121	112	122	212	221	113	131	311	222
$E(8ma^2/h^2)$	3	6	6	6	9	9	9	11	11	11	12

Observe that states with different quantum numbers may have the same energy (Fig. 3.3). For example, the states ψ_{211}, ψ_{121}, and ψ_{112} (where the subscripts give the quantum numbers) all have the same energy. However, Eq. (3.73) shows that these three sets of quantum numbers give three different, independent wave functions and therefore do represent different states of the system. When two or more independent wave functions correspond to states with the same energy eigenvalue, the eigenvalue is said to be **degenerate**. The **degree of degeneracy** (or, simply, the **degeneracy**) of an energy level is the number of states that have that energy. Thus the second-lowest energy level of the particle in a cube is threefold degenerate. We got the degeneracy when we made the edges of the box equal. Degeneracy is usually related to the symmetry of the system. Note that the wave functions ψ_{211}, ψ_{121}, and ψ_{112} can be transformed into one another by rotating the cubic box. Usually, the bound-state energy levels in one-dimensional problems are nondegenerate.

In the statistical-mechanical evaluation of the molecular partition function of an ideal gas, the translational energy levels of each gas molecule are taken to be the levels of a particle in a three-dimensional rectangular box; see *Levine, Physical Chemistry,* Sections 22.6 and 22.7.

3.6 DEGENERACY

For an n-fold degenerate energy level, there are n independent wave functions $\psi_1, \psi_2, \ldots, \psi_n$, each having the same energy eigenvalue w:

$$\hat{H}\psi_1 = w\psi_1, \quad \hat{H}\psi_2 = w\psi_2, \quad \ldots, \quad \hat{H}\psi_n = w\psi_n \tag{3.76}$$

We wish to prove the following important theorem: *Every linear combination*

$$\phi \equiv c_1\psi_1 + c_2\psi_2 + \cdots + c_n\psi_n \tag{3.77}$$

of the wave functions of a degenerate level with energy eigenvalue w is an eigenfunction of the Hamiltonian operator with eigenvalue w. [A **linear combination** of the functions $\psi_1, \psi_2, \ldots, \psi_n$ is defined as a function of the form (3.77) where the c's are constants (some of which might be zero).] To prove this theorem, we must show that $\hat{H}\phi = w\phi$ or

$$\hat{H}(c_1\psi_1 + c_2\psi_2 + \cdots + c_n\psi_n) = w(c_1\psi_1 + c_2\psi_2 + \cdots + c_n\psi_n) \tag{3.78}$$

Since $\hat{H}$ is a linear operator, we can apply Eq. (3.9) $n - 1$ times to the left side of (3.78) to get

$$\hat{H}(c_1\psi_1 + c_2\psi_2 + \cdots + c_n\psi_n) = \hat{H}(c_1\psi_1) + \hat{H}(c_2\psi_2) + \cdots + \hat{H}(c_n\psi_n)$$

Use of Eqs. (3.10) and (3.76) gives

$$\hat{H}(c_1\psi_1 + c_2\psi_2 + \cdots + c_n\psi_n) = c_1\hat{H}\psi_1 + c_2\hat{H}\psi_2 + \cdots + c_n\hat{H}\psi_n$$

$$= c_1 w\psi_1 + c_2 w\psi_2 + \cdots + c_n w\psi_n$$

$$\hat{H}(c_1\psi_1 + c_2\psi_2 + \cdots + c_n\psi_n) = w(c_1\psi_1 + c_2\psi_2 + \cdots + c_n\psi_n)$$

which completes the proof.

As an example, the stationary-state wave functions ψ_{211}, ψ_{121}, and ψ_{112} for the particle in a cubic box are degenerate, and the linear combination $c_1\psi_{211} + c_2\psi_{121} + c_3\psi_{112}$ is an eigenfunction of the particle-in-a-cubic-box Hamiltonian with eigenvalue $6h^2/8ma^2$, the same eigenvalue as for each of ψ_{211}, ψ_{121}, and ψ_{112}.

Note that the linear combination $c_1\psi_1 + c_2\psi_2$ is *not* an eigenfunction of $\hat{H}$ if ψ_1 and ψ_2 correspond to different energy eigenvalues ($\hat{H}\psi_1 = E_1\psi_1$ and $\hat{H}\psi_2 = E_2\psi_2$ with $E_1 \neq E_2$).

Since any linear combination of the wave functions corresponding to a degenerate energy level is an eigenfunction of $\hat{H}$ with the same eigenvalue, we can construct an infinite number of different wave functions for any degenerate energy level. Actually, we are only interested in eigenfunctions that are linearly independent. The n functions $f_1, \ldots, f_n$ are said to be **linearly independent** if the equation $c_1f_1 + \cdots + c_nf_n = 0$ can only be satisfied with all the constants $c_1, \ldots, c_n$ equal to zero. This means that *no member of a set of linearly independent functions can be expressed as a linear combination of the remaining members.* For example, the functions $f_1 = 3x$, $f_2 = 5x^2 - x$, $f_3 = x^2$ are not linearly independent, since $f_2 = 5f_3 - \frac{1}{3}f_1$. The functions $g_1 = 1$, $g_2 = x$, $g_3 = x^2$ are linearly independent, since none of them can be written as a linear combination of the other two.

The **degree of degeneracy** of an energy level is equal to the number of linearly independent wave functions corresponding to that value of the energy. The one-dimensional free-particle wave functions (2.30) are linear combinations of two linearly independent functions that are each eigenfunctions with the same energy eigenvalue E, so each such energy eigenvalue (except $E = 0$) is doubly degenerate (meaning that the degree of degeneracy is two).

3.7 AVERAGE VALUES

It was pointed out in Section 3.3 that, when the state function Ψ is not an eigenfunction of the operator $\hat{B}$, a measurement of B will give one of a number of possible values (the eigenvalues of $\hat{B}$). We now consider the average value of the property B for a system whose state is Ψ.

To determine the average value of B experimentally, we take many identical, noninteracting systems each in the same state Ψ and we measure B in each system. The **average value** of B, symbolized by $\langle B \rangle$, is defined as the arithmetic mean of the observed values $b_1, b_2, \ldots, b_N$:

$$\langle B \rangle = \frac{\sum_{j=1}^{N} b_j}{N} \tag{3.79}$$

where N, the number of systems, is extremely large.

Instead of summing over the observed values of B, we can sum over all the possible values of B, multiplying each possible value by the number of times it is observed, to get the equivalent expression

$$\langle B \rangle = \frac{\sum_{b} n_b b}{N} \tag{3.80}$$

where n_b is the number of times the value b is observed. An example will make this clear. Suppose a class of nine students takes a quiz that has five questions and that the students receive these grades: 0, 20, 20, 60, 60, 80, 80, 80, 100. Calculating the average grade according to (3.79), we have

$$\frac{1}{N}\sum_{j=1}^{N} b_j = \frac{0 + 20 + 20 + 60 + 60 + 80 + 80 + 80 + 100}{9} = 56$$

To calculate the average grade according to (3.80), we sum over the possible grades: 0, 20, 40, 60, 80, 100. We have

$$\frac{1}{N}\sum_{b} n_b b = \frac{1(0) + 2(20) + 0(40) + 2(60) + 3(80) + 1(100)}{9} = 56$$

Equation (3.80) can be written as

$$\langle B \rangle = \sum_{b}\left(\frac{n_b}{N}\right) b$$

Since N is very large, n_b/N is the probability P_b of observing the value b, and

$$\langle B \rangle = \sum_{b} P_b b \tag{3.81}$$

Now consider the average value of the x coordinate for a one-particle, one-dimensional system in the state $\Psi(x, t)$. The x coordinate takes on a continuous range of values, and the probability of observing the particle between x and $x + dx$ is $|\Psi|^2\,dx$. The summation over the infinitesimal probabilities is equivalent to an integration over the full range of x, and (3.81) becomes

$$\langle x \rangle = \int_{-\infty}^{\infty} x|\Psi(x,t)|^2\,dx \tag{3.82}$$

For the one-particle, three-dimensional case, the probability of finding the particle in the volume element at point (x, y, z) with edges dx, dy, dz is

$$|\Psi(x,y,z,t)|^2\,dx\,dy\,dz \tag{3.83}$$

If we want the probability that the particle is between x and $x + dx$, we must integrate (3.83) over all possible values of y and z, since the particle can have any values for its y and z coordinates while its x coordinate lies between x and $x + dx$. Hence, in the three-dimensional case (3.82) becomes

$$\langle x \rangle = \int_{-\infty}^{\infty}\left[\int_{-\infty}^{\infty}\int_{-\infty}^{\infty}|\Psi(x,y,z,t)|^2\,dy\,dz\right]x\,dx$$

$$\langle x \rangle = \int_{-\infty}^{\infty}\int_{-\infty}^{\infty}\int_{-\infty}^{\infty}|\Psi(x,y,z,t)|^2 x\,dx\,dy\,dz \tag{3.84}$$

Now consider the average value of some physical property $B(x, y, z)$ that is a function of the particle's coordinates. An example is the potential energy $V(x, y, z)$. The same reasoning that gave Eq. (3.84) yields

$$\langle B(x,y,z) \rangle = \int_{-\infty}^{\infty}\int_{-\infty}^{\infty}\int_{-\infty}^{\infty}|\Psi(x,y,z,t)|^2 B(x,y,z)\,dx\,dy\,dz \tag{3.85}$$

$$\langle B(x,y,z) \rangle = \int_{-\infty}^{\infty}\int_{-\infty}^{\infty}\int_{-\infty}^{\infty}\Psi^* B\Psi\,dx\,dy\,dz \tag{3.86}$$

The form (3.86) might seem like a bit of whimsy, since it is no different from (3.85). In a moment we shall see its significance.

In general, the property B depends on both coordinates *and* momenta:

$$B = B(x, y, z, p_x, p_y, p_z)$$

for the one-particle, three-dimensional case. How do we find the average value of B? We *postulate* that $\langle B \rangle$ for a system in state Ψ is

$$\langle B \rangle = \int_{-\infty}^{\infty}\int_{-\infty}^{\infty}\int_{-\infty}^{\infty}\Psi^* B\left(x, y, z, \frac{\hbar}{i}\frac{\partial}{\partial x}, \frac{\hbar}{i}\frac{\partial}{\partial y}, \frac{\hbar}{i}\frac{\partial}{\partial z}\right)\Psi\,dx\,dy\,dz$$

$$\langle B \rangle = \int_{-\infty}^{\infty}\int_{-\infty}^{\infty}\int_{-\infty}^{\infty}\Psi^* \hat{B}\Psi\,dx\,dy\,dz \tag{3.87}$$

where $\hat{B}$ is the quantum-mechanical operator for the property B. [Later we shall provide some justification for this postulate by using (3.87) to show that the time-dependent Schrödinger equation reduces to Newton's second law in the transition from quantum to classical mechanics; see Prob. 7.59.] For the n-particle case, we postulate that

$$\boxed{\langle B\rangle = \int \Psi^* \hat{B} \Psi \, d\tau} \qquad \textbf{(3.88)}$$

where $\int d\tau$ denotes a definite integral over the full range of the $3n$ coordinates. The state function in (3.88) must be normalized, since we took $\Psi^*\Psi$ as the probability density. It is important to have the operator properly sandwiched between Ψ^* and Ψ. The quantities $\hat{B}\Psi^*\Psi$ and $\Psi^*\Psi\hat{B}$ are not the same as $\Psi^*\hat{B}\Psi$, unless B is a function of coordinates only. In $\int \Psi^*\hat{B}\Psi\, d\tau$, one first operates on Ψ with $\hat{B}$ to produce a new function $\hat{B}\Psi$, which is then multiplied by Ψ^*; one then integrates over all space to produce a number, which is $\langle B\rangle$.

For a stationary state, we have [Eq. (1.20)]

$$\Psi^*\hat{B}\Psi = e^{iEt/\hbar}\psi^*\hat{B}e^{-iEt/\hbar}\psi = e^0\psi^*\hat{B}\psi = \psi^*\hat{B}\psi$$

since $\hat{B}$ contains no time derivatives and does not affect the time factor in Ψ. Hence, for a stationary state,

$$\langle B\rangle = \int \psi^*\hat{B}\psi\, d\tau \qquad (3.89)$$

Thus, if $\hat{B}$ is time-independent, then $\langle B\rangle$ is time-independent in a stationary state.

Consider the special case where Ψ is an eigenfunction of $\hat{B}$. When $\hat{B}\Psi = k\Psi$, Eq. (3.88) becomes

$$\langle B\rangle = \int \Psi^*\hat{B}\Psi\, d\tau = \int \Psi^* k\Psi\, d\tau = k\int \Psi^*\Psi\, d\tau = k$$

since Ψ is normalized. This result is reasonable, since when $\hat{B}\Psi = k\Psi$, k is the only possible value we can find for B when we make a measurement (Section 3.3).

The following properties of average values are easily proved from Eq. (3.88) (see Prob. 3.46):

$$\boxed{\langle A + B\rangle = \langle A\rangle + \langle B\rangle} \quad \boxed{\langle cB\rangle = c\langle B\rangle} \qquad \textbf{(3.90)}$$

where A and B are any two properties and c is a constant. However, the average value of a product need not equal the product of the average values: $\langle AB\rangle \neq \langle A\rangle\langle B\rangle$.

The term **expectation value** is often used instead of average value. The expectation value is not necessarily one of the possible values we might observe.

EXAMPLE Find $\langle x\rangle$ and $\langle p_x\rangle$ for the ground stationary state of a particle in a three-dimensional box.

Substitution of the stationary-state wave function $\psi = f(x)g(y)h(z)$ [Eq. (3.62)] into the average-value postulate (3.89) gives

$$\langle x\rangle = \int \psi^*\hat{x}\psi\, d\tau = \int_0^c\int_0^b\int_0^a f^*g^*h^*xfgh\, dx\, dy\, dz$$

since $\psi = 0$ outside the box. Use of (3.74) gives

$$\langle x \rangle = \int_0^a x|f(x)|^2\,dx \int_0^b |g(y)|^2\,dy \int_0^c |h(z)|^2\,dz = \int_0^a x|f(x)|^2\,dx$$

since $g(y)$ and $h(z)$ are each normalized. For the ground state, $n_x = 1$ and $f(x) = (2/a)^{1/2}\sin(\pi x/a)$. So

$$\langle x \rangle = \frac{2}{a}\int_0^a x\sin^2\!\left(\frac{\pi x}{a}\right)dx = \frac{a}{2} \tag{3.91}$$

where the Appendix integral (A.3) was used. A glance at Fig. 2.4 shows that this result is reasonable.

Also,

$$\langle p_x \rangle = \int \psi^*\hat{p}_x\psi\,d\tau = \int_0^c\int_0^b\int_0^a f^*g^*h^*\frac{\hbar}{i}\frac{\partial}{\partial x}[f(x)g(y)h(z)]\,dx\,dy\,dz$$

$$\langle p_x \rangle = \frac{\hbar}{i}\int_0^a f^*(x)f'(x)\,dx \int_0^b |g(y)|^2\,dy \int_0^c |h(z)|^2\,dz$$

$$\langle p_x \rangle = \frac{\hbar}{i}\int_0^a f(x)f'(x)\,dx = \frac{\hbar}{2i}f^2(x)\bigg|_0^a = 0 \tag{3.92}$$

where the boundary conditions $f(0) = 0$ and $f(a) = 0$ were used. The result (3.92) is reasonable since the particle is equally likely to be headed in the $+x$ or $-x$ direction.

EXERCISE Find $\langle p_x^2 \rangle$ for the ground state of a particle in a three-dimensional box. (*Answer:* $h^2/4l^2$.)

3.8 REQUIREMENTS FOR AN ACCEPTABLE WAVE FUNCTION

In solving the particle in a box, we required ψ to be continuous. We now discuss other requirements the wave function must satisfy.

Since $|\psi|^2\,d\tau$ is a probability, we want to be able to normalize the wave function by choosing a suitable **normalization constant** N as a multiplier of the wave function. If ψ is unnormalized and $N\psi$ is normalized, the normalization condition (3.59) gives

$$1 = \int |N\psi|^2\,d\tau = |N|^2 \int |\psi|^2\,d\tau$$

$$|N| = 1\bigg/\left(\int |\psi|^2\,d\tau\right)^{1/2} \tag{3.93}$$

The definite integral $\int |\psi|^2\,d\tau$ will equal zero only if the function ψ is zero everywhere. However, ψ cannot be zero everywhere (this would mean no particles were present), so this integral is never zero. If $\int |\psi|^2\,d\tau$ is infinite, then the magnitude $|N|$ of the normalization constant is zero and ψ cannot be normalized. We can normalize ψ if and only if $\int |\psi|^2\,d\tau$ has a finite, rather than infinite, value. If the integral over all space $\int |\psi|^2\,d\tau$ is finite, ψ is said to be **quadratically integrable**. Thus we generally demand that ψ be quadratically integrable. The important exception is a particle that is not bound. Thus the

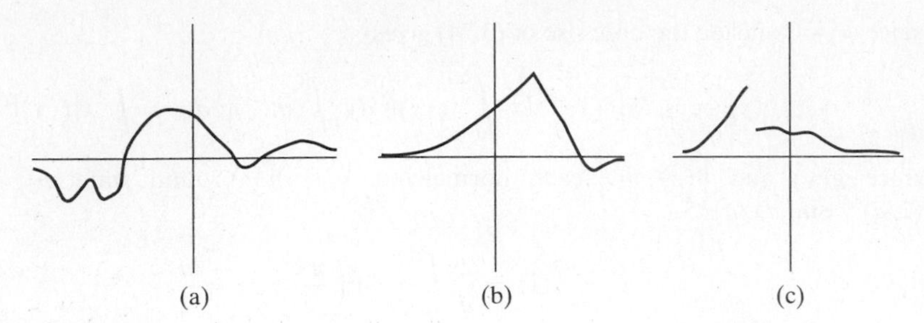

FIGURE 3.4 Function (a) is continuous and its first derivative is continuous. Function (b) is continuous, but its first derivative has a discontinuity. Function (c) is discontinuous.

wave functions for the unbound states of the particle in a well (Section 2.4) and for a free particle are not quadratically integrable.

Since $\psi^*\psi$ is the probability density, it must be single-valued. It would be embarrassing if our theory gave two different values for the probability of finding a particle at a certain point. If we demand that ψ be single-valued, then surely $\psi^*\psi$ will be single-valued. It is possible to have ψ multivalued [for example, $\psi(q) = -1, +1, i$] and still have $\psi^*\psi$ single-valued. We will, however, demand single-valuedness for ψ.

In addition to demanding that ψ be continuous, we usually also require that all the partial derivatives $\partial\psi/\partial x$, $\partial\psi/\partial y$, and so on, be continuous. (See Fig. 3.4.) Referring back to Section 2.2, however, we note that for the particle in a box, $d\psi/dx$ is discontinuous at the walls of the box; ψ and $d\psi/dx$ are zero everywhere outside the box; but from Eq. (2.23) we see that $d\psi/dx$ does not become zero at the walls. The discontinuity in ψ' is due to the infinite jump in potential energy at the walls of the box. For a box with walls of finite height, ψ' is continuous at the walls (Section 2.4).

In line with the requirement of quadratic integrability, it is sometimes stated that the wave function must be finite everywhere, including infinity. However, this is usually a much stronger requirement than quadratic integrability, and, in fact, it turns out that some of the relativistic wave functions for the hydrogen atom are infinite at the origin, but are quadratically integrable. Occasionally, one encounters nonrelativistic wave functions that are infinite at the origin [L. D. Landau and E. M. Lifshitz, *Quantum Mechanics*, 3rd ed. (1977), Section 35]. Thus the fundamental requirement is quadratic integrability, rather than finiteness.

We require that the eigenfunctions of any operator representing a physical quantity meet the above requirements. A function meeting these requirements is said to be **well-behaved**.

3.9 SUMMARY

An operator is a rule that transforms one function into another function. The sum and product of operators are defined by $(\hat{A} + \hat{B})f(x) \equiv \hat{A}f(x) + \hat{B}f(x)$ and $\hat{A}\hat{B}f(x) \equiv \hat{A}[\hat{B}f(x)]$. The commutator of two operators is $[\hat{A}, \hat{B}] \equiv \hat{A}\hat{B} - \hat{B}\hat{A}$. The operators in quantum mechanics are linear, meaning that they satisfy

$\hat{A}[f(x) + g(x)] = \hat{A}f(x) + \hat{A}g(x)$ and $\hat{A}[cf(x)] = c\hat{A}f(x)$. The eigenfunctions F_i and eigenvalues b_i of the operator $\hat{B}$ obey $\hat{B}F_i = b_iF_i$.

So far, the following quantum-mechanical postulates have been introduced:

(a) The state of a system is described by a function Ψ (the state function or wave function) of the particles' coordinates and the time. Ψ is single-valued, continuous, and (except for unbound states) quadratically integrable.

(b) To each physical property B of a system, there corresponds an operator $\hat{B}$. This operator is found by taking the classical-mechanical expression for the property in terms of Cartesian coordinates and momenta and replacing each coordinate x by $x\cdot$ and each momentum component p_x by $(\hbar/i)\partial/\partial x$.

(c) The only possible values that can result from measurements of the property B are the eigenvalues b_i of the equation $\hat{B}g_i = b_ig_i$ where the eigenfunctions g_i are required to be well-behaved.

(d) The average value of the property B is given by $\langle B\rangle = \int \Psi^*\hat{B}\Psi\, d\tau$, where Ψ is the system's state function.

(e) The state function of an undisturbed system changes with time according to $-(\hbar/i)(\partial\Psi/\partial t) = \hat{H}\Psi$, where $\hat{H}$ is the Hamiltonian operator (the energy operator) of the system.

(f) For a three-dimensional n-particle system, the quantity (3.56) is the probability of finding the system's particles in the infinitesimal regions of space listed after (3.56).

The Hamiltonian operator for an n-particle, three-dimensional system is $\hat{H} = -\sum_{i=1}^{n}(\hbar^2/2m_i)\nabla_i^2 + V$, where $\nabla_i^2 \equiv \partial^2/\partial x_i^2 + \partial^2/\partial y_i^2 + \partial^2/\partial z_i^2$. The time-independent Schrödinger equation is $\hat{H}\psi_i = E_i\psi_i$, where the index i labels the different stationary states.

The stationary-state wave functions and energy levels of a particle in a three-dimensional rectangular box were found by the use of separation of variables.

The degree of degeneracy of an energy level is the number of linearly independent wave functions that correspond to that energy value. Any linear combination of wave functions of a degenerate level with energy w is an eigenfunction of $\hat{H}$ with eigenvalue w.

PROBLEMS

Sec.	3.1	3.2	3.3	3.4	3.5
Probs.	3.1–3.24	3.25–3.28	3.29–3.30	3.31–3.34	3.35–3.40

Sec.	3.6	3.7	3.8	general
Probs.	3.41–3.44	3.45–3.46	3.47	3.48–3.50

3.1 If $g = \hat{A}f$, find g for each of these choices of $\hat{A}$ and f. (a) $\hat{A} = d/dx$ and $f = \cos(x^2 + 1)$; (b) $\hat{A} = \hat{5}$ and $f = \sin x$; (c) $\hat{A} = (\ \ )^2$ and $f = \sin x$; (d) $\hat{A} = \exp$ and $f = \ln x$; (e) $\hat{A} = d^2/dx^2$ and $f = \ln(3x)$; (f) $\hat{A} = d^2/dx^2 + 3x\, d/dx$ and $f = 4x^3$.

3.2 State whether each of the following entities is an operator or a function: (a) $\hat{A}\hat{B}$; (b) $\hat{A}f(x)$; (c) $\hat{B}\hat{A}f(x)$; (d) $[\hat{B}, \hat{A}]$; (e) $f(x)\hat{A}$; (f) $f(x)\hat{A}\hat{B}g(x)$.

3.3 If $\hat{A}f(x) = 3x^2 f(x) + 2x\, df/dx$, where f is an arbitrary function, give an expression for $\hat{A}$.

3.4 Give three different operators $\hat{A}$ that satisfy $\hat{A}e^x = e^x$.

3.5 For each of the following pairs of functions f and g, give two possible forms for $\hat{A}$ that will allow the relation $g(x) = \hat{A}f(x)$ to be satisfied. (a) $f = 2x^4$, $g = 8x^3$; (b) $f = 2x$, $g = x^2$; (c) $f = x^4$, $g = x^2$.

3.6 Prove that $\hat{A} + \hat{B} = \hat{B} + \hat{A}$.

3.7 Prove that if $\hat{A} + \hat{B} = \hat{C}$, then $\hat{A} = \hat{C} - \hat{B}$.

3.8 If $\hat{A} = d^2/dx^2$ and $\hat{B} = x^2\cdot$, find (a) $\hat{A}\hat{B}x^3$; (b) $\hat{B}\hat{A}x^3$; (c) $\hat{A}\hat{B}f(x)$; (d) $\hat{B}\hat{A}f(x)$.

3.9 Let $\hat{A} = x^3$ and $\hat{B} = d/dx$. Find $\hat{A}\hat{B}$ and $\hat{B}\hat{A}$.

3.10 What do you suppose is meant by the zeroth power of an operator?

3.11 By repeated application of the definition of the product of two operators, show that

$$[(\hat{A}\hat{B})\hat{C}]f = \hat{A}[\hat{B}(\hat{C}f)]$$
$$[\hat{A}(\hat{B}\hat{C})]f = \hat{A}[\hat{B}(\hat{C}f)]$$

3.12 (a) Show that $(\hat{A} + \hat{B})^2 = (\hat{B} + \hat{A})^2$ for any two operators (linear or nonlinear). (b) Under what conditions is $(\hat{A} + \hat{B})^2$ equal to $\hat{A}^2 + 2\hat{A}\hat{B} + \hat{B}^2$?

3.13 Verify the commutator identity $[\hat{A}, \hat{B}] = -[\hat{B}, \hat{A}]$.

3.14 Evaluate (a) $[\sin z, d/dz]$; (b) $[d^2/dx^2, ax^2 + bx + c]$, where a, b, and c are constants; (c) $[d/dx, d^2/dx^2]$.

3.15 Classify each of these operators as linear or nonlinear: (a) $3x^2\, d^2/dx^2$; (b) $(\ \ )^2$; (c) $\int dx$; (d) exp; (e) $\sum_{x=1}^{n}$.

3.16 Write the linear differential equation (2.2) in the form $\hat{B}y(x) = g(x)$, where $\hat{B}$ is a certain linear operator.

3.17 Prove that the product of two linear operators is a linear operator.

3.18 Prove that $\hat{A}(\hat{B} + \hat{C}) = \hat{A}\hat{B} + \hat{A}\hat{C}$ for linear operators.

3.19 (a) If $\hat{A}$ is linear, show that

$$\hat{A}(bf + cg) = b\hat{A}f + c\hat{A}g \tag{3.94}$$

where b and c are arbitrary constants and f and g are arbitrary functions. (b) If (3.94) is true, show that $\hat{A}$ is linear.

3.20 (a) Give an example of an operator that satisfies Eq. (3.9) but does not satisfy (3.10). (b) Give an example of an operator that satisfies (3.10) but not (3.9).

3.21 Which of the following equations are true for all operators $\hat{A}$ and $\hat{B}$ and all functions f and g? (a) $(\hat{A} + \hat{B})f = \hat{A}f + \hat{B}f$; (b) $\hat{A}(f + g) = \hat{A}f + \hat{A}g$; (c) $(\hat{A}f)/f = \hat{A}$, provided $f \neq 0$; (d) $\hat{A}\hat{B}f = \hat{B}\hat{A}f$; (e) $\hat{A}f = f\hat{A}$.

3.22 The *Laplace transform operator* $\hat{L}$ is defined by

$$\hat{L}f(x) = \int_0^\infty e^{-px} f(x)\, dx$$

(a) Is $\hat{L}$ linear? (b) Evaluate $\hat{L}(1)$. (c) Evaluate $\hat{L}e^{ax}$, assuming that $p > a$.

3.23 We define the *translation operator* $\hat{T}_h$ by $\hat{T}_h f(x) = f(x + h)$. (a) Is $\hat{T}_h$ a linear operator? (b) Evaluate $(\hat{T}_1^2 - 3\hat{T}_1 + 2)x^2$.

3.24 The operator $e^{\hat{A}}$ is defined by the equation

$$e^{\hat{A}} = \hat{1} + \hat{A} + \frac{\hat{A}^2}{2!} + \frac{\hat{A}^3}{3!} + \cdots = \sum_{k=0}^{\infty} \frac{\hat{A}^k}{k!}$$

Show that $e^{\hat{D}} = \hat{T}_1$, where $\hat{D} = d/dx$ and $\hat{T}_1$ is defined in Prob. 3.23.

3.25 Which of the following functions are eigenfunctions of d^2/dx^2? (a) e^x; (b) x^2; (c) $\sin x$; (d) $3 \cos x$; (e) $\sin x + \cos x$. Give the eigenvalue for each eigenfunction.

3.26 Find the eigenfunctions of $-(\hbar^2/2m)\, d^2/dx^2$. If the eigenfunctions are to remain finite for $x \to \pm\infty$, what are the allowed eigenvalues?

3.27 Find the eigenfunctions and eigenvalues of $\int dx$.

3.28 Find the eigenfunctions and eigenvalues of $d^2/dx^2 + 2\, d/dx$.

3.29 Give the quantum-mechanical operators for the following physical quantities: (a) p_y^3; (b) $xp_y - yp_x$; (c) $(xp_y)^2$.

3.30 Fill in the details leading to (3.36) and (3.37) as the eigenfunctions and eigenvalues of $\hat{p}_x$.

3.31 Evaluate the commutators (a) $[\hat{x}, \hat{p}_x]$; (b) $[\hat{x}, \hat{p}_x^2]$; (c) $[\hat{x}, \hat{p}_y]$; (d) $[\hat{x}, \hat{V}(x, y, z)]$; (e) $[\hat{x}, \hat{H}]$, where the Hamiltonian operator is given by Eq. (3.45); (f) $[\hat{x}\hat{y}\hat{z}, \hat{p}_x^2]$.

3.32 Write the kinetic-energy operator $\hat{T}$ for a two-particle, three-dimensional system.

3.33 Write the expression for the probability of finding particle number 1 with its x coordinate between 0 and 2 for (a) a one-particle, one-dimensional system; (b) a one-particle, three-dimensional system; (c) a two-particle, three-dimensional system.

3.34 If ψ is a normalized wave function, what are its SI units for (a) the one-particle, one-dimensional case; (b) the one-particle, three-dimensional case; (c) the n-particle, three-dimensional case?

3.35 An electron in a three-dimensional rectangular box with dimensions of 5.00 Å, 3.00 Å, and 6.00 Å makes a radiative transition from the lowest-lying excited state to the ground state. Calculate the frequency of the photon emitted.

3.36 An electron is in the ground state in a three-dimensional box with V given by (3.60), with $a = 1.00$ nm, $b = 2.00$ nm, and $c = 5.00$ nm. Find the probability that a measurement of the electron's position will find it in the region defined by (a) $0 \le x \le 0.40$ nm, $1.50 \text{ nm} \le y \le 2.00$ nm, $2.00 \text{ nm} \le z \le 3.00$ nm; (b) $0 \le x \le 0.40$ nm, $0 \le y \le 2.00$ nm, $0 \le z \le 5.00$ nm. (c) What is the probability that a measurement of the position of this electron will find it with its x coordinate between 0 and 0.40 nm?

3.37 The stationary-state wave functions of a particle in a three-dimensional rectangular box are eigenfunctions of which of these operators? (a) $\hat{p}_x$; (b) $\hat{p}_x^2$; (c) $\hat{p}_z^2$; (d) $\hat{x}$. For each of these operators that $\psi_{n_x n_y n_z}$ is an eigenfunction of, state in terms of n_x, n_y, and n_z what value will be observed when the corresponding property is measured.

3.38 Prove the multiple-integral identity (3.74).

3.39 Explain how degeneracy can occur for a particle in a rectangular box with $a \ne b \ne c$.

3.40 Solve the one-particle, three-dimensional, time-independent Schrödinger equation for the free particle.

3.41 Which of the following combinations of particle-in-a-cubic-box stationary-state wave functions are eigenfunctions of the particle-in-a-cubic-box Hamiltonian operator? (a) $2^{-1/2}(\psi_{138} - \psi_{381})$; (b) $2^{-1/2}(\psi_{212} + \psi_{131})$; (c) $\frac{1}{2}\psi_{151} - \frac{1}{2}\psi_{333} + 2^{-1/2}\psi_{511}$.

3.42 The terms *state* and *energy level* are not synonymous in quantum mechanics. For the particle in a cubic box, consider the energy range $E < 15h^2/8ma^2$. (a) How many states lie in this range? (b) How many energy levels lie in this range?

3.43 For the particle in a cubic box, what is the degree of degeneracy of the energy levels with the following values of $8ma^2E/h^2$? (a) 12; (b) 14; (c) 27.

3.44 Which of the following are sets of linearly independent functions? (a) x, x^2, x^6; (b) 8, $x, x^2, 3x^2 - 1$; (c) $\sin x, \cos x$; (d) $\sin z, \cos z, \tan z$; (e) $\sin x, \cos x, e^{ix}$; (f) $\sin^2 x, \cos^2 x, 1$; (g) $\sin^2 x, \cos^2 y, 1$.

3.45 For the particle confined to a box with dimensions a, b, and c, find the following values for the state with quantum numbers n_x, n_y, n_z. (a) $\langle x \rangle$; (b) $\langle y \rangle, \langle z \rangle$. Use symmetry considerations and the answer to part a. (c) $\langle p_x \rangle$; (d) $\langle x^2 \rangle$. Is $\langle x^2 \rangle = \langle x \rangle^2$? Is $\langle xy \rangle = \langle x \rangle \langle y \rangle$?

3.46 Derive the average-value relations in (3.90) from (3.88).

3.47 Which of the following functions, when multiplied by a normalization constant, would be acceptable one-dimensional wave functions for a bound particle? (a and b are positive constants and x goes from $-\infty$ to ∞.) (a) e^{-ax}; (b) e^{-bx^2}; (c) xe^{-bx^2}; (d) ie^{-bx^2}; (e) $f(x) = e^{-bx^2}$ for $x < 0$, $f(x) = 2e^{-bx^2}$ for $x \geq 0$.

3.48 Show that, if Ψ_1 and Ψ_2 satisfy the time-dependent Schrödinger equation, then $c_1\Psi_1 + c_2\Psi_2$ satisfies this equation, where c_1 and c_2 are constants.

3.49 (a) Write a computer program that will find all sets of positive integers n_x, n_y, n_z for which $n_x^2 + n_y^2 + n_z^2 \leq 60$, will print the sets in order of increasing $n_x^2 + n_y^2 + n_z^2$ and will print the $n_x^2 + n_y^2 + n_z^2$ values. (b) What is the degeneracy of the particle-in-a-cubic-box level with $n_x^2 + n_y^2 + n_z^2 = 54$?

3.50 True or false? (a) If g is an eigenfunction of the linear operator $\hat{B}$, then cg is an eigenfunction of $\hat{B}$, where c is an arbitrary constant. (b) If we measure the property B when the system's state function is not an eigenfunction of $\hat{B}$, then we can get a result that is not an eigenvalue of $\hat{B}$. (c) If f_1 and f_2 are eigenfunctions of $\hat{B}$, then $c_1f_1 + c_2f_2$ must be an eigenfunction of $\hat{B}$, where c_1 and c_2 are constants. (d) The state function Ψ must be an eigenfunction of each operator $\hat{B}$ that represents a physical property of the system. (e) A linear combination of two solutions to the time-independent Schrödinger equation must be a solution of this equation. (f) The system's state function Ψ must be an eigenfunction of $\hat{H}$. (g) $5x$ is an eigenvalue of the position operator $\hat{x}$. (h) $5x$ is an eigenfunction of $\hat{x}$. (i) For a stationary state, Ψ is an eigenfunction of $\hat{H}$. (j) For a stationary state, Ψ equals a function of time multiplied by a function of the particles' coordinates. (k) For a stationary state, the probability density is independent of time. (l) In the equation $\langle B \rangle = \int \Psi^* \hat{B} \Psi \, d\tau$, the integral is an indefinite integral. (m) If f is an eigenfunction of the linear operator $\hat{A}$ with eigenvalue a, then f is an eigenfunction of $\hat{A}^2$ with eigenvalue a^2. (n) $\langle B \rangle = \int \hat{B}|\Psi|^2 \, d\tau$. (o) Every function of x, y, and z has the form $f(x)g(y)h(z)$.

CHAPTER 4

The Harmonic Oscillator

We shall see in Section 13.2 that the energy of a gas-phase molecule can be approximated as the sum of translational, rotational, vibrational, and electronic energies. Calculation of electronic energies is considered in Chapters 13 to 18. The translational energy is the kinetic energy of motion of the molecule as a whole in the space of the container holding the substance. The translational energy levels can be taken as those of a particle in a three-dimensional box (Section 3.5). The rotational energy levels of a diatomic (two-atom) molecule can be approximated by those of a rigid, two-particle rotor, which is discussed in Section 6.4. The lowest few vibrational energy levels of a diatomic molecule can be approximated by the levels of a harmonic oscillator. The Schrödinger equation for the harmonic oscillator is solved in this chapter in Section 4.2. As a preliminary, Section 4.1 discusses the power-series method of solving differential equations, which is used to solve the Schrödinger equation for the harmonic oscillator. Section 4.3 discusses molecular vibration. Section 4.4 presents a numerical method that finds eigenvalues and eigenfunctions for the one-dimensional Schrödinger equation.

4.1 POWER-SERIES SOLUTION OF DIFFERENTIAL EQUATIONS

So far we have considered only cases where the potential energy $V(x)$ is a constant. This makes the Schrödinger equation a second-order linear homogeneous differential equation with *constant* coefficients, which we know how to solve. For cases in which V varies with x, a useful approach is to try a power-series solution of the Schrödinger equation.

To illustrate the method, consider the differential equation

$$y''(x) + c^2 y(x) = 0 \tag{4.1}$$

where $c^2 > 0$. Of course, this differential equation has *constant* coefficients, but we can solve it with the power-series method if we want. Let us first find the solution by using the auxiliary equation, which is $s^2 + c^2 = 0$. We find $s = \pm ic$. Recalling the work in Section 2.2 [Eqs. (2.10) and (4.1) are the same], we get trigonometric solutions when the roots of the auxiliary equation are pure imaginary:

$$y = A\cos cx + B\sin cx \tag{4.2}$$

where A and B are the constants of integration. A different form of (4.2) is

$$y = D\sin(cx + e) \tag{4.3}$$

where D and e are arbitrary constants. Using the formula for the sine of the sum of two angles, we can show that (4.3) is equivalent to (4.2).

Now let us solve (4.1) using the power-series method. We start by assuming that the solution can be expanded in a Taylor series (see Prob. 4.1) about $x = 0$; that is, we assume that

$$y(x) = \sum_{n=0}^{\infty} a_n x^n = a_0 + a_1 x + a_2 x^2 + a_3 x^3 + \cdots \tag{4.4}$$

where the a's are constant coefficients to be determined so as to satisfy (4.1). Differentiating (4.4), we have

$$y'(x) = a_1 + 2a_2 x + 3a_3 x^2 + \cdots = \sum_{n=1}^{\infty} n a_n x^{n-1} \tag{4.5}$$

where we assumed that term-by-term differentiation is valid for the series. (This is not always true for infinite series.) For y'', we have

$$y''(x) = 2a_2 + 3(2)a_3 x + \cdots = \sum_{n=2}^{\infty} n(n-1)a_n x^{n-2} \tag{4.6}$$

Substituting (4.4) and (4.6) into (4.1), we get

$$\sum_{n=2}^{\infty} n(n-1)a_n x^{n-2} + \sum_{n=0}^{\infty} c^2 a_n x^n = 0 \tag{4.7}$$

We want to combine the two sums in (4.7). Provided certain conditions are met, we can add two infinite series term by term to get their sum:

$$\sum_{j=0}^{\infty} b_j x^j + \sum_{j=0}^{\infty} c_j x^j = \sum_{j=0}^{\infty} (b_j + c_j) x^j \tag{4.8}$$

To apply (4.8) to the two sums in (4.7), we want the limits in each sum to be the same and the powers of x to be the same. We therefore change the summation index in the first sum in (4.7), defining k as $k \equiv n - 2$. The limits $n = 2$ to ∞ correspond to $k = 0$ to ∞, and use of $n = k + 2$ gives

$$\sum_{n=2}^{\infty} n(n-1)a_n x^{n-2} = \sum_{k=0}^{\infty} (k+2)(k+1)a_{k+2} x^k = \sum_{n=0}^{\infty} (n+2)(n+1)a_{n+2} x^n \tag{4.9}$$

The last equality in (4.9) is valid because the summation index is a **dummy variable**; it makes no difference what letter we use to denote this variable. For example, the sums $\sum_{i=1}^{3} c_i x^i$ and $\sum_{m=1}^{3} c_m x^m$ are equal because only the dummy variables in the two sums differ. This equality is easy to see if we write out the sums:

$$\sum_{i=1}^{3} c_i x^i = c_1 x + c_2 x^2 + c_3 x^3 \quad \text{and} \quad \sum_{m=1}^{3} c_m x^m = c_1 x + c_2 x^2 + c_3 x^3$$

In the last equality in (4.9), we simply changed the symbol denoting the summation index from k to n.

The integration variable in a definite integral is also a dummy variable, since the value of a definite integral is unaffected by what letter we use for this variable:

$$\int_a^b f(x)\, dx = \int_a^b f(t)\, dt \tag{4.10}$$

Using (4.9) in (4.7), we find, after applying (4.8), that

$$\sum_{n=0}^{\infty}[(n+2)(n+1)a_{n+2}+c^2a_n]x^n = 0 \tag{4.11}$$

If (4.11) is to be true for all values of x, then the coefficient of each power of x must vanish. To see this, consider the equation

$$\sum_{j=0}^{\infty} b_j x^j = 0 \tag{4.12}$$

Putting $x = 0$ in (4.12) shows that $b_0 = 0$. Taking the first derivative of (4.12) with respect to x and then putting $x = 0$ shows that $b_1 = 0$. Taking the nth derivative and putting $x = 0$ gives $b_n = 0$. Thus, from (4.11), we have

$$(n+2)(n+1)a_{n+2}+c^2a_n = 0 \tag{4.13}$$

$$a_{n+2} = -\frac{c^2}{(n+1)(n+2)}a_n \tag{4.14}$$

Equation (4.14) is a **recursion relation**. If we know the value of a_0, we can use (4.14) to find $a_2, a_4, a_6, \ldots$. If we know a_1, we can find $a_3, a_5, a_7, \ldots$. Since there is no restriction on the values of a_0 and a_1, they are arbitrary constants, which we denote by A and Bc:

$$a_0 = A, \qquad a_1 = Bc \tag{4.15}$$

Using (4.14), we find for the coefficients

$$a_0 = A, \quad a_2 = -\frac{c^2A}{1\cdot 2}, \quad a_4 = \frac{c^4A}{4\cdot 3\cdot 2\cdot 1}, \quad a_6 = -\frac{c^6A}{6!}, \ldots \tag{4.16}$$

$$a_{2k} = (-1)^k\frac{c^{2k}A}{(2k)!}, \qquad k = 0, 1, 2, 3, \ldots \tag{4.17}$$

$$a_1 = Bc, \quad a_3 = -\frac{c^3B}{2\cdot 3}, \quad a_5 = \frac{c^5B}{5\cdot 4\cdot 3\cdot 2}, \quad a_7 = -\frac{c^7B}{7!}, \ldots$$

$$a_{2k+1} = (-1)^k\frac{c^{2k+1}B}{(2k+1)!}, \qquad k = 0, 1, 2, \ldots \tag{4.18}$$

From (4.4), (4.17), and (4.18), we have

$$y = \sum_{n=0}^{\infty} a_n x^n = \sum_{n=0,2,4,\ldots}^{\infty} a_n x^n + \sum_{n=1,3,5,\ldots}^{\infty} a_n x^n \tag{4.19}$$

$$y = A\sum_{k=0}^{\infty}(-1)^k\frac{c^{2k}x^{2k}}{(2k)!} + B\sum_{k=0}^{\infty}(-1)^k\frac{c^{2k+1}x^{2k+1}}{(2k+1)!} \tag{4.20}$$

The two series in (4.20) are the Taylor series for $\cos cx$ and $\sin cx$ (Prob. 4.2). Hence, in agreement with (4.2), we have

$$y = A\cos cx + B\sin cx \tag{4.21}$$

4.2 THE ONE-DIMENSIONAL HARMONIC OSCILLATOR

In this section we will increase our quantum-mechanical repertoire by solving the Schrödinger equation for the one-dimensional harmonic oscillator. This system is important as a model for molecular vibrations.

Classical-Mechanical Treatment. Before looking at the wave mechanics of the harmonic oscillator, we review the classical treatment. We have a single particle of mass m attracted toward the origin by a force proportional to the particle's displacement from the origin:

$$F_x = -kx \tag{4.22}$$

The proportionality constant k is called the **force constant**. F_x is the x component of the force on the particle. This is also the total force in this one-dimensional problem. Equation (4.22) is obeyed by a particle attached to a spring, provided the spring is not stretched greatly from its equilibrium position.

Newton's second law, $F = ma$, gives

$$-kx = m\frac{d^2x}{dt^2} \tag{4.23}$$

where t is the time. Equation (4.23) is the same as Eq. (4.1) with $c^2 = k/m$; hence the solution is [Eq. (4.3) with $c = (k/m)^{1/2}$]

$$x = A\sin(2\pi\nu t + b) \tag{4.24}$$

where A (the **amplitude** of the vibration) and b are the integration constants, and the **vibration frequency** ν is

$$\boxed{\nu = \frac{1}{2\pi}\left(\frac{k}{m}\right)^{1/2}} \tag{4.25}$$

Since the sine function has maximum and minimum values of 1 and -1, respectively, x in (4.24) oscillates between A and $-A$. The sine function repeats itself every 2π radians, and the time needed for one complete oscillation (called the **period**) is the time it takes for the argument of the sine function to increase by 2π. At time $t + 1/\nu$, the argument of the sine function is $2\pi\nu(t + 1/\nu) + b = 2\pi\nu t + 2\pi + b$, which is 2π greater than the argument at time t, so the period is $1/\nu$. The reciprocal of the period is the number of vibrations per unit time (the vibrational frequency), and so the frequency is ν.

Now consider the energy. The potential energy V is related to the components of force in the three-dimensional case by

$$\boxed{F_x = -\frac{\partial V}{\partial x}, \qquad F_y = -\frac{\partial V}{\partial y}, \qquad F_z = -\frac{\partial V}{\partial z}} \tag{4.26}$$

Equation (4.26) is the definition of potential energy. Since this is a one-dimensional problem, we have [Eq. (1.12)]

$$F_x = -\frac{dV}{dx} = -kx \tag{4.27}$$

Integration of (4.27) gives $V = \int kx\,dx = \frac{1}{2}kx^2 + C$, where C is a constant. The potential energy always has an arbitrary additive constant. Choosing $C = 0$, we have [Eq. (4.25)]

$$\boxed{V = \tfrac{1}{2}kx^2} \tag{4.28}$$

$$V = 2\pi^2\nu^2 mx^2 \tag{4.29}$$

The graph of $V(x)$ is a parabola (Fig. 4.4). The kinetic energy T is

$$T = \tfrac{1}{2}m(dx/dt)^2 \tag{4.30}$$

and can be found by differentiating (4.24) with respect to x. Adding T and V, one finds for the total energy (Prob. 4.5)

$$E = T + V = \tfrac{1}{2}kA^2 = 2\pi^2\nu^2 mA^2 \tag{4.31}$$

where the identity $\sin^2\theta + \cos^2\theta = 1$ was used.

Quantum-Mechanical Treatment. The harmonic-oscillator Hamiltonian operator is [Eqs. (3.27) and (4.29)]

$$\hat{H} = \hat{T} + \hat{V} = -\frac{\hbar^2}{2m}\frac{d^2}{dx^2} + 2\pi^2\nu^2 mx^2 = -\frac{\hbar^2}{2m}\left(\frac{d^2}{dx^2} - \alpha^2 x^2\right) \tag{4.32}$$

where, to save time in writing, α was defined as

$$\alpha \equiv 2\pi\nu m/\hbar \tag{4.33}$$

The Schrödinger equation $\hat{H}\psi = E\psi$ reads, after multiplication by $2m/\hbar^2$,

$$\frac{d^2\psi}{dx^2} + (2mE\hbar^{-2} - \alpha^2x^2)\psi = 0 \tag{4.34}$$

We might now attempt a power-series solution of (4.34). If we do now try a power series for ψ of the form (4.4), we will find that it leads to a three-term recursion relation, which is harder to deal with than a two-term recursion relation like Eq. (4.14). We therefore modify the form of (4.34) so as to get a two-term recursion relation when we try a series solution. A substitution that will achieve this purpose is (see Prob. 4.21) $f(x) \equiv e^{\alpha x^2/2}\psi(x)$. Thus

$$\psi = e^{-\alpha x^2/2}f(x) \tag{4.35}$$

This equation is simply the definition of a new function $f(x)$ that replaces $\psi(x)$ as the unknown function to be solved for. (We can make any substitution we please in a differential equation.) Differentiating (4.35) twice, we have

$$\psi'' = e^{-\alpha x^2/2}(f'' - 2\alpha xf' - \alpha f + \alpha^2x^2f) \tag{4.36}$$

Substituting (4.35) and (4.36) into (4.34), we find

$$f''(x) - 2\alpha xf'(x) + (2mE\hbar^{-2} - \alpha)f(x) = 0 \tag{4.37}$$

Now we try a series solution for $f(x)$:

$$f(x) = \sum_{n=0}^{\infty} c_n x^n \tag{4.38}$$

Assuming the validity of term-by-term differentiation of (4.38), we get

$$f'(x) = \sum_{n=1}^{\infty} nc_n x^{n-1} = \sum_{n=0}^{\infty} nc_n x^{n-1} \tag{4.39}$$

[The first term in the second sum in (4.39) is zero.] Also,

$$f''(x) = \sum_{n=2}^{\infty} n(n-1)c_n x^{n-2} = \sum_{j=0}^{\infty} (j+2)(j+1)c_{j+2}x^j = \sum_{n=0}^{\infty} (n+2)(n+1)c_{n+2}x^n$$

where we made the substitution $j = n - 2$ and then changed the summation index from j to n. [See Eq. (4.9).] Substitution into (4.37) gives

$$\sum_{n=0}^{\infty} (n+2)(n+1)c_{n+2}x^n - 2\alpha \sum_{n=0}^{\infty} nc_n x^n + (2mE\hbar^{-2} - \alpha)\sum_{n=0}^{\infty} c_n x^n = 0$$

$$\sum_{n=0}^{\infty} [(n+2)(n+1)c_{n+2} - 2\alpha nc_n + (2mE\hbar^{-2} - \alpha)c_n]x^n = 0 \tag{4.40}$$

Setting the coefficient of x^n equal to zero [for the same reason as in Eq. (4.11)], we have

$$c_{n+2} = \frac{\alpha + 2\alpha n - 2mE\hbar^{-2}}{(n+1)(n+2)} c_n \tag{4.41}$$

which is the desired two-term recursion relation. Equation (4.41) has the same form as (4.14), in that knowing c_n we can calculate c_{n+2}. We thus have two arbitrary constants: c_0 and c_1. If we set c_1 equal to zero, then we will have as a solution a power series containing only even powers of x, multiplied by the exponential factor:

$$\psi = e^{-\alpha x^2/2} f(x) = e^{-\alpha x^2/2} \sum_{n=0,2,4,\ldots}^{\infty} c_n x^n = e^{-\alpha x^2/2} \sum_{l=0}^{\infty} c_{2l} x^{2l} \tag{4.42}$$

If we set c_0 equal to zero, we get another independent solution:

$$\psi = e^{-\alpha x^2/2} \sum_{n=1,3,\ldots}^{\infty} c_n x^n = e^{-\alpha x^2/2} \sum_{l=0}^{\infty} c_{2l+1} x^{2l+1} \tag{4.43}$$

The general solution of the Schrödinger equation is a linear combination of these two independent solutions [recall Eq. (2.4)]:

$$\psi = Ae^{-\alpha x^2/2} \sum_{l=0}^{\infty} c_{2l+1} x^{2l+1} + Be^{-\alpha x^2/2} \sum_{l=0}^{\infty} c_{2l} x^{2l} \tag{4.44}$$

where A and B are arbitrary constants.

We now must see if the boundary conditions on the wave function lead to any restrictions on the solution. To see how the two infinite series behave for large x, we examine the ratio of successive coefficients in each series. The ratio of the coefficient of x^{2l+2} to that of x^{2l} in the second series is [set $n = 2l$ in (4.41)]

$$\frac{c_{2l+2}}{c_{2l}} = \frac{\alpha + 4\alpha l - 2mE\hbar^{-2}}{(2l + 1)(2l + 2)}$$

Assuming that for large values of x the later terms in the series are the dominant ones, we look at this ratio for large values of l:

$$\frac{c_{2l+2}}{c_{2l}} \sim \frac{4\alpha l}{(2l)(2l)} = \frac{\alpha}{l} \qquad \text{for } l \text{ large} \tag{4.45}$$

Setting $n = 2l + 1$ in (4.41), we find that for large l the ratio of successive coefficients in the first series is also α/l. Now consider the power-series expansion for the function $e^{\alpha x^2}$. Using (Prob. 4.3)

$$e^z = \sum_{n=0}^{\infty} \frac{z^n}{n!} = 1 + z + \frac{z^2}{2!} + \cdots \tag{4.46}$$

we get

$$e^{\alpha x^2} = 1 + \alpha x^2 + \cdots + \frac{\alpha^l x^{2l}}{l!} + \frac{\alpha^{l+1} x^{2l+2}}{(l + 1)!} + \cdots$$

The ratio of the coefficients of x^{2l+2} and x^{2l} in this series is

$$\frac{\alpha^{l+1}}{(l + 1)!} \div \frac{\alpha^l}{l!} = \frac{\alpha}{l + 1} \sim \frac{\alpha}{l} \qquad \text{for large } l$$

Thus the ratio of successive coefficients in each of the infinite series in the solution (4.44) is the same as in the series for $e^{\alpha x^2}$ for large l. We conclude that, for large x, each series behaves as $e^{\alpha x^2}$. [This is not a rigorous proof. A proper mathematical derivation is given in H. A. Buchdahl, *Am. J. Phys.*, **42,** 47 (1974); see also M. Bowen and J. Coster, *Am. J. Phys.*, **48,** 307 (1980).]

If each series behaves as $e^{\alpha x^2}$, then (4.44) shows that ψ will behave as $e^{\alpha x^2/2}$ for large x. The wave function will become infinite as x goes to infinity and will not be quadratically integrable. If we could somehow break off the series after a finite number of terms, then the factor $e^{-\alpha x^2/2}$ would ensure that ψ went to zero as x became infinite. (Using l'Hôpital's rule, it is easy to show that $x^p e^{-\alpha x^2/2}$ goes to zero as $x \to \infty$, where p is any finite power.) To have one of the series break off after a finite number of terms, the coefficient of c_n in the recursion relation (4.41) must become zero for some value of n, say for $n = v$. This makes $c_{v+2}, c_{v+4}, \ldots$ all equal to zero, and one of the series in (4.44) will have a finite number of terms. In the recursion relation (4.41), there is one quantity whose value is not yet fixed, but can be adjusted to make the coefficient of c_v vanish. This quantity is the energy E. Setting the coefficient of c_v equal to zero in (4.41) and using (4.33) for α, we get

$$\alpha + 2\alpha v - 2mE\hbar^{-2} = 0$$

$$2mE\hbar^{-2} = (2v + 1)2\pi\nu m\hbar^{-1}$$

$$\boxed{E = \left(v + \tfrac{1}{2}\right)h\nu, \qquad v = 0, 1, 2, \ldots} \tag{4.47}$$

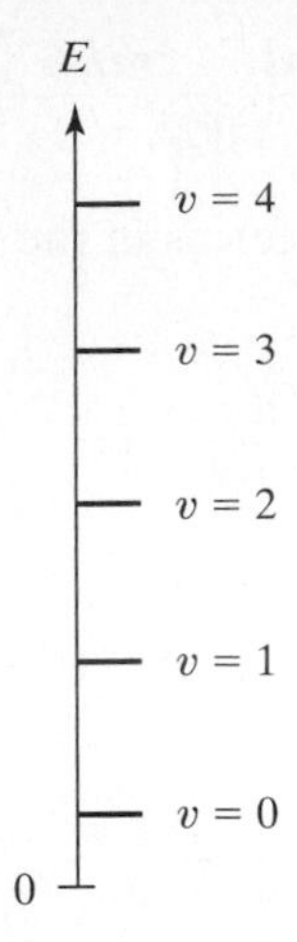

FIGURE 4.1 Lowest five energy levels for the harmonic oscillator.

The harmonic-oscillator stationary-state energy levels (4.47) are equally spaced (Fig. 4.1). Do not confuse the quantum number v (vee) with the vibrational frequency ν (nu).

Substitution of (4.47) into the recursion relation (4.41) gives

$$c_{n+2} = \frac{2\alpha(n - v)}{(n + 1)(n + 2)} c_n \tag{4.48}$$

By quantizing the energy according to (4.47), we have made one of the series break off after a finite number of terms. To get rid of the other infinite series in (4.44), we must set the arbitrary constant that multiplies it equal to zero. This leaves us with a wave function that is $e^{-\alpha x^2/2}$ times a finite power series containing only even or only odd powers of x, depending on whether v is even or odd, respectively. The highest power of x in this power series is x^v, since we chose E to make $c_{v+2}, c_{v+4}, \ldots$ all vanish. The wave functions (4.44) are thus

$$\psi_v = \begin{cases} e^{-\alpha x^2/2}(c_0 + c_2 x^2 + \cdots + c_v x^v) & \text{for } v \text{ even} \\ e^{-\alpha x^2/2}(c_1 x + c_3 x^3 + \cdots + c_v x^v) & \text{for } v \text{ odd} \end{cases} \tag{4.49}$$

where the arbitrary constants A and B in (4.44) can be absorbed into c_1 and c_0, respectively, and can therefore be omitted. The coefficients after c_0 and c_1 are found from the recursion relation (4.48). Since the quantum number v occurs in the recursion relation, we get a different set of coefficients c_i for each different v. For example, c_2 in ψ_4 differs from c_2 in ψ_2.

As in the particle in a box, it is the boundary conditions that force us to quantize the energy. For values of E that differ from (4.47), ψ is not quadratically integrable. For example, Fig. 4.2 plots ψ of Eq. (4.42) for the values $E/h\nu = 0.499$, 0.500, and 0.501, where the recursion relation (4.41) is used to calculate the coefficients c_n (see also Prob. 4.22).

The harmonic-oscillator ground-state energy is nonzero. This energy, $\frac{1}{2}h\nu$, is called the **zero-point energy**. This would be the vibrational energy of a harmonic oscillator in a collection of harmonic oscillators at a temperature of absolute zero. The zero-point

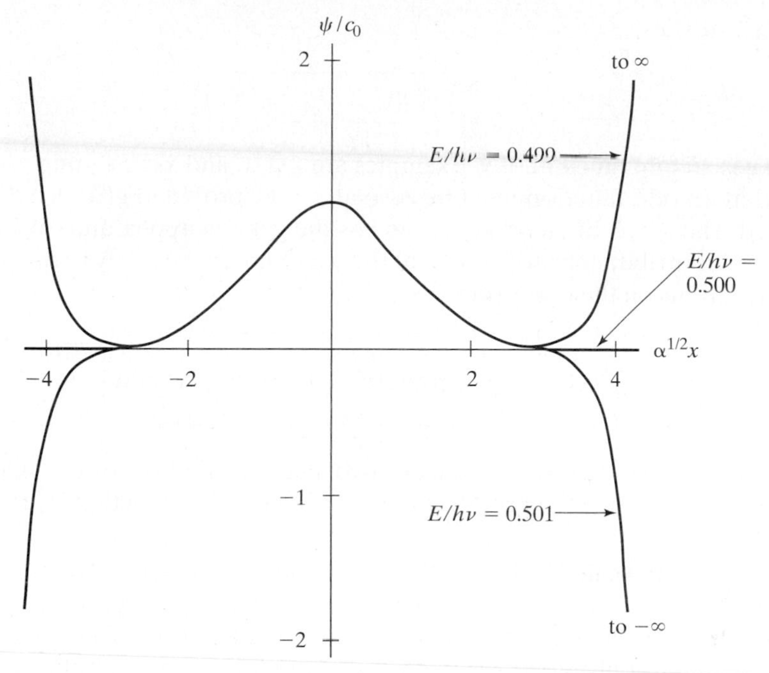

FIGURE 4.2 Plots of the harmonic-oscillator-Schrödinger-equation solution containing only even powers of x for $E = 0.499h\nu$, $E = 0.500h\nu$, and $E = 0.501h\nu$. In the region around $x = 0$, the three curves nearly coincide. For $|\alpha^{1/2}x| > 3$, the $E/h\nu = 0.500$ curve nearly coincides with the x axis.

energy can be understood from the uncertainty principle. If the lowest state had an energy of zero, both its potential and kinetic energies (which are nonnegative) would have to be zero. Zero kinetic energy would mean that the momentum was exactly zero, and Δp_x would be zero. Zero potential energy would mean that the particle was always located at the origin, and Δx would be zero. But we cannot have both Δx and Δp_x equal to zero. Hence the need for a nonzero ground-state energy. Similar ideas apply for the particle in a box. The definition of the zero-point energy (ZPE) is $E_{\text{ZPE}} = E_{gs} - V_{\text{min}}$, where E_{gs} and V_{min} are the ground-state energy and the minimum value of the potential-energy function.

Even and Odd Functions. Before considering the wave functions in detail, we define even and odd functions. If $f(x)$ satisfies

$$\boxed{f(-x) = f(x)} \tag{4.50}$$

then f is an **even function** of x. Thus x^2 and e^{-bx^2} are both even functions of x since $(-x)^2 = x^2$ and $e^{-b(-x)^2} = e^{-bx^2}$. The graph of an even function is symmetric about the y axis (for example, see Fig. 4.3a); hence

$$\boxed{\int_{-a}^{+a} f(x)\,dx = 2\int_0^a f(x)\,dx \qquad \text{for } f(x) \text{ even}} \tag{4.51}$$

If $g(x)$ satisfies

$$\boxed{g(-x) = -g(x)} \tag{4.52}$$

then g is an **odd function** of x. Examples are x, $1/x$, and xe^{x^2}. Setting $x = 0$ in (4.52), we see that an odd function must be zero at $x = 0$, provided $g(0)$ is defined and single-valued. The graph of an odd function has the general appearance of Fig. 4.3b. Because positive contributions on one side of the y axis are canceled by corresponding negative contributions on the other side, we have

$$\boxed{\int_{-a}^{+a} g(x)\,dx = 0 \qquad \text{for } g(x) \text{ odd}} \tag{4.53}$$

It is easy to show that the product of two even functions or of two odd functions is an even function, while the product of an even and an odd function is an odd function.

The Harmonic-Oscillator Wave Functions. The exponential factor $e^{-\alpha x^2/2}$ in (4.49) is an even function of x. If v is an even number, the polynomial factor contains only even powers of x, which makes ψ_v an even function. If v is odd, the polynomial factor contains only odd powers of x, and ψ_v, being the product of an even function and an odd function, is an odd function. Each harmonic-oscillator stationary state ψ is either an even or odd function according to whether the quantum number v is even or odd. In Section 7.5, we shall see that, when the potential energy V is an even function, the wave functions of nondegenerate levels must be either even or odd functions.

We now find the explicit forms of the wave functions of the lowest three levels. For the $v = 0$ ground state, Eq. (4.49) gives

$$\psi_0 = c_0 e^{-\alpha x^2/2} \tag{4.54}$$

where the subscript on ψ gives the value of v. We fix c_0 by normalization:

$$1 = \int_{-\infty}^{\infty} |c_0|^2 e^{-\alpha x^2}\,dx = 2|c_0|^2 \int_0^{\infty} e^{-\alpha x^2}\,dx$$

where Eq. (4.51) has been used. Using the integral (A.9) in the Appendix, we find $|c_0| = (\alpha/\pi)^{1/4}$. Therefore,

$$\psi_0 = (\alpha/\pi)^{1/4} e^{-\alpha x^2/2} \tag{4.55}$$

if we choose the phase of the normalization constant to be zero. The wave function (4.55) is a Gaussian function (Fig. 4.3a).

For the $v = 1$ state, Eq. (4.49) gives

$$\psi_1 = c_1 x e^{-\alpha x^2/2} \tag{4.56}$$

After normalization using the integral in Eq. (A.10), we have

$$\psi_1 = (4\alpha^3/\pi)^{1/4} x e^{-\alpha x^2/2} \tag{4.57}$$

ψ_1 is graphed in Fig. 4.3b.

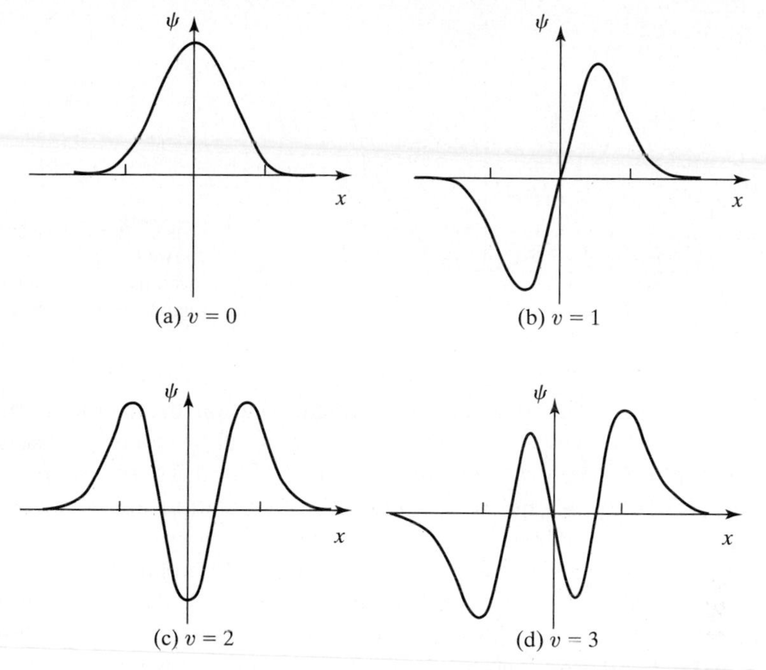

FIGURE 4.3 Harmonic-oscillator wave functions. The same scale is used for all graphs. The points marked on the x axes are for $\alpha^{1/2}x = \pm 2$.

For $v = 2$, Eq. (4.49) gives

$$\psi_2 = (c_0 + c_2x^2)e^{-\alpha x^2/2}$$

The recursion relation (4.48) with $v = 2$, gives

$$c_2 = \frac{2\alpha(-2)}{1 \cdot 2}c_0 = -2\alpha c_0$$

Hence

$$\psi_2 = c_0(1 - 2\alpha x^2)e^{-\alpha x^2/2} \tag{4.58}$$

Evaluating c_0 by normalization, we find (Prob. 4.11)

$$\psi_2 = (\alpha/4\pi)^{1/4}(2\alpha x^2 - 1)e^{-\alpha x^2/2} \tag{4.59}$$

Note that c_0 in ψ_2 is not the same as c_0 in ψ_0.

The number of nodes in the wave function equals the quantum number v. It can be proved (see *Messiah*, pages 109–110) that *for the bound stationary states of a one-dimensional problem, the number of nodes interior to the boundary points is zero for the ground-state ψ and increases by one for each successive excited state*. The boundary points for the harmonic oscillator are $\pm\infty$.

The polynomial factors in the harmonic-oscillator wave functions are well known in mathematics and are called *Hermite polynomials*, after a French mathematician. (See Prob. 4.20.)

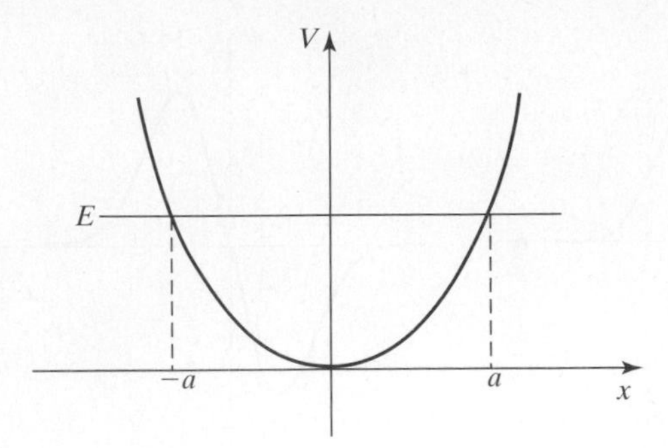

FIGURE 4.4 The classically allowed ($-a \le x \le a$) and forbidden ($x < -a$ and $x > a$) regions for the harmonic oscillator.

According to the quantum-mechanical solution, there is some probability of finding the particle at any point on the x axis (except at the nodes). Classically, $E = T + V$ and the kinetic energy T cannot be negative: $T \ge 0$. Therefore, $E - V = T \ge 0$ and $V \le E$. The potential energy V is a function of position, and a classical particle is confined to the region of space where $V \le E$; that is, where the potential energy does not exceed the total energy. In Fig. 4.4, the horizontal line labeled E gives the energy of a harmonic oscillator, and the parabolic curve gives the potential energy $\frac{1}{2}kx^2$. For the regions $x < -a$ and $x > a$, we have $V > E$, and these regions are **classically forbidden**. The **classically allowed region** $-a \le x \le a$ in Fig. 4.4 is where $V \le E$.

In quantum mechanics, the stationary-state wave functions are not eigenfunctions of $\hat{T}$ or $\hat{V}$ and we cannot assign definite values to T or V for a stationary state. Instead of the classical equations $E = T + V$ and $T \ge 0$, we have in quantum mechanics that $E = \langle T \rangle + \langle V \rangle$ (Prob. 6.33) and $\langle T \rangle \ge 0$ (Prob. 7.7), so $\langle V \rangle \le E$ in quantum mechanics, but we cannot write $V \le E$, and a particle has some probability to be found in classically forbidden regions where $V > E$.

It might seem that, by saying the particle can be found outside the classically allowed region, we are allowing it to have negative kinetic energy. Actually, there is no paradox in the quantum-mechanical view. To verify that the particle is in the classically forbidden region, we must measure its position. This measurement changes the state of the system (Sections 1.3 and 1.4). The interaction of the oscillator with the measuring apparatus transfers enough energy to the oscillator for it to be in the classically forbidden region. An accurate measurement of x introduces a large uncertainty in the momentum and hence in the kinetic energy. Penetration of classically forbidden regions was previously discussed in Sections 2.4 and 2.5.

A harmonic-oscillator stationary state has $E = \left(v + \frac{1}{2}\right)h\nu$ and $V = \frac{1}{2}kx^2 = 2\pi^2\nu^2 mx^2$, so the classically allowed region where $V \le E$ is where $2\pi^2\nu^2 mx^2 \le \left(v + \frac{1}{2}\right)h\nu$, which gives $x^2 \le \left(v + \frac{1}{2}\right)h/2\pi^2\nu m = (2v + 1)/\alpha$, where $\alpha \equiv 2\pi\nu m/\hbar$ [Eq. (4.33)]. Therefore, the classically allowed region for the harmonic oscillator is where $-(2v + 1)^{1/2} \le \alpha^{1/2}x \le (2v + 1)^{1/2}$.

Note from Fig. 4.3 that ψ *oscillates in the classically allowed region and decreases exponentially to zero in the classically forbidden region.* We previously saw this behavior for the particle in a rectangular well (Section 2.4).

Figure 4.3 shows that, as we go to higher-energy states of the harmonic oscillator, ψ and $|\psi|^2$ tend to have maxima farther and farther from the origin. Since $V = \frac{1}{2}kx^2$

increases as we go farther from the origin, the average potential energy $\langle V \rangle = \int_{-\infty}^{\infty} |\psi|^2 V\,dx$ increases as the quantum number increases. The average kinetic energy is $\langle T \rangle = -(\hbar^2/2m)\int_{-\infty}^{\infty} \psi^* \psi''\,dx$. Integration by parts gives (Prob. 7.7b) $\langle T \rangle = (\hbar^2/2m)\int_{-\infty}^{\infty} |d\psi/dx|^2\,dx$. The higher number of nodes in states with higher quantum number produces a faster rate of change of ψ, so $\langle T \rangle$ increases as the quantum number increases.

Some online simulations of the quantum harmonic oscillator are available at www.phy.davidson.edu/StuHome/cabell_f/energy.html (shows energy levels and wave functions and shows how the wave function diverges when the energy is changed from an allowed value); www.falstad.com/qm1d/ (choose harmonic oscillator from the drop-down menu at the top; double click on one of the small circles at the bottom to show a stationary state; shows energies, wave functions, probability densities; m and k can be varied).

4.3 VIBRATION OF MOLECULES

We shall see in Section 13.1 that to an excellent approximation one can treat separately the motions of the electrons and the motions of the nuclei of a molecule. (This is due to the much heavier mass of the nuclei.) One first imagines the nuclei to be held stationary and solves a Schrödinger equation for the electronic energy U. (U also includes the energy of nuclear repulsion.) For a diatomic (two-atom) molecule, the electronic energy U depends on the distance R between the nuclei, $U = U(R)$, and the U versus R curve has the typical appearance of Fig. 13.1.

After finding $U(R)$, one solves a Schrödinger equation for nuclear motion, using $U(R)$ as the potential energy for nuclear motion. For a diatomic molecule, the nuclear Schrödinger equation is a two-particle equation. We shall see in Section 6.3 that, when the potential energy of a two-particle system depends only on the distance between the particles, the energy of the system is the sum of (a) the kinetic energy of translational motion of the entire system through space and (b) the energy of internal motion of the particles relative to each other. The classical expression for the two-particle internal-motion energy turns out to be the sum of the potential energy of interaction between the particles and the kinetic energy of a hypothetical particle whose mass is $m_1 m_2/(m_1 + m_2)$ (where m_1 and m_2 are the masses of the two particles) and whose coordinates are the coordinates of one particle relative to the other. The quantity $m_1 m_2/(m_1 + m_2)$ is called the **reduced mass** μ.

The internal motion of a diatomic molecule consists of **vibration**, corresponding to a change in the distance R between the two nuclei, and **rotation**, corresponding to a change in the spatial orientation of the line joining the nuclei. To a good approximation, one can usually treat the vibrational and rotational motions separately. The rotational energy levels are found in Section 6.4. Here we consider the vibrational levels.

The Schrödinger equation for the vibration of a diatomic molecule has a kinetic-energy operator for the hypothetical particle of mass $\mu = m_1 m_2/(m_1 + m_2)$ and a potential-energy term given by $U(R)$. If we place the origin to coincide with the minimum point of the U curve in Fig. 13.1 and take the zero of potential energy at the energy of this minimum point, then the lower portion of the $U(R)$ curve will nearly coincide with the potential-energy curve of a harmonic oscillator with the appropriate

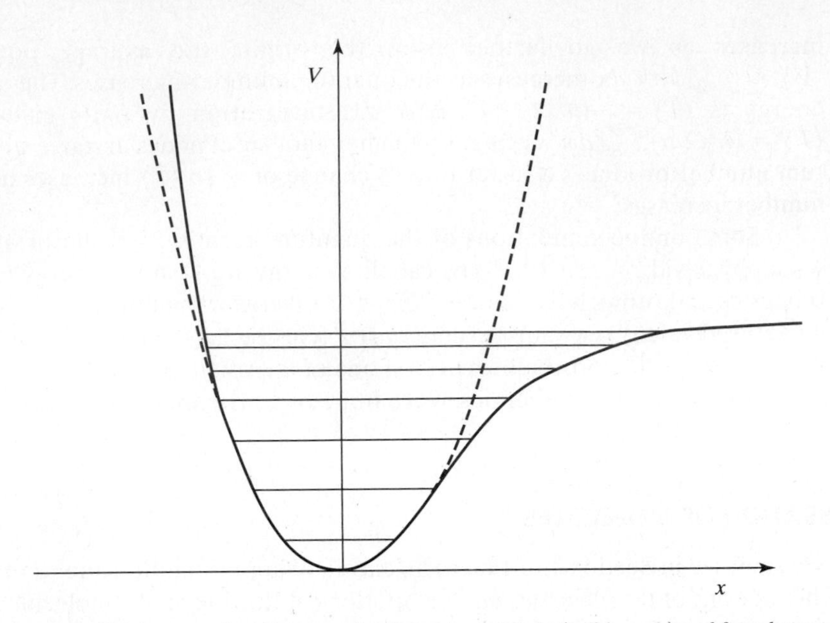

FIGURE 4.5 Potential energy for vibration of a diatomic molecule (solid curve) and for a harmonic oscillator (dashed curve). Also shown are the bound-state vibrational energy levels for the diatomic molecule. In contrast to the harmonic oscillator, a diatomic molecule has only a finite number of bound vibrational levels

force constant k (see Fig. 4.5 and Prob. 4.27). The minimum in the $U(R)$ curve occurs at the **equilibrium distance** R_e between the nuclei. In Fig. 4.5, x is the deviation of the internuclear distance from its equilibrium value: $x \equiv R - R_e$.

The harmonic-oscillator force constant k in Eq. (4.28) is obtained as $k = d^2V/dx^2$, and the harmonic-oscillator curve essentially coincides with the $U(R)$ curve at $R = R_e$, so the molecular force constant is $k = d^2U/dR^2|_{R=R_e}$ (see also Prob. 4.27). *Differences in nuclear mass have virtually no effect on the electronic-energy curve $U(R)$, so different isotopic species of the same molecule have essentially the same force constant k.*

We expect, therefore, that a reasonable approximation to the vibrational energy levels E_{vib} of a diatomic molecule would be the harmonic-oscillator vibrational energy levels; Eqs. (4.47) and (4.25) give

$$\boxed{E_{\text{vib}} \approx \left(v + \tfrac{1}{2}\right)h\nu_e, \qquad v = 0, 1, 2, \ldots} \tag{4.60}$$

$$\boxed{\nu_e = \frac{1}{2\pi}\left(\frac{k}{\mu}\right)^{1/2}, \qquad \mu = \frac{m_1 m_2}{m_1 + m_2}, \qquad k = \left.\frac{d^2U}{dR^2}\right|_{R=R_e}} \tag{4.61}$$

ν_e is called the **equilibrium** (or **harmonic**) **vibrational frequency**. This approximation is best for the lower vibrational levels. As v increases, the nuclei spend more time in regions far from their equilibrium separation. For such regions the potential energy deviates substantially from that of a harmonic oscillator and the harmonic-oscillator approximation is poor. Instead of being equally spaced, one finds that the vibrational levels of a diatomic molecule come closer and closer together as v increases (Fig. 4.5). Eventually, the vibrational energy is large enough to dissociate the diatomic molecule into atoms that are not bound to each other. Unlike the harmonic oscillator, a diatomic molecule has only a finite number of bound-state vibrational levels. A more accurate expression for the molecular vibrational energy that allows for the anharmonicity of the vibration is

$$E_{\text{vib}} = \left(v + \tfrac{1}{2}\right)h\nu_e - \left(v + \tfrac{1}{2}\right)^2 h\nu_e x_e \tag{4.62}$$

where the *anharmonicity constant* $\nu_e x_e$ is positive in nearly all cases.

Using the time-dependent Schrödinger equation, one finds (Section 9.10) that the most probable vibrational transitions when a diatomic molecule is exposed to electromagnetic radiation are those where v changes by ± 1. Furthermore, for absorption or emission of electromagnetic radiation to occur, the vibration must change the molecule's dipole moment. Hence homonuclear diatomics (such as H_2 or N_2) cannot undergo transitions between vibrational levels by absorption or emission of radiation. (Such transitions can occur during intermolecular collisions.) The relation $E_{\text{upper}} - E_{\text{lower}} = h\nu$, the approximate equation (4.60), and the **selection rule** $\Delta v = 1$ for absorption of radiation show that a heteronuclear diatomic molecule whose vibrational frequency is ν_e will most strongly absorb light of frequency ν_{light} given approximately by

$$\nu_{\text{light}} = (E_2 - E_1)/h \approx \left[\left(v_2 + \tfrac{1}{2}\right)h\nu_e - \left(v_1 + \tfrac{1}{2}\right)h\nu_e\right]/h = (v_2 - v_1)\nu_e = \nu_e \tag{4.63}$$

The values of k and μ in (4.61) for diatomic molecules are such that ν_{light} usually falls in the infrared region of the spectrum. Transitions with $\Delta v = 2, 3, \ldots$ also occur, but these (called *overtones*) are much weaker than the $\Delta v = 1$ absorption.

Use of the more accurate equation (4.62) gives (Prob. 4.26)

$$\nu_{\text{light}} = \nu_e - 2\nu_e x_e(v_1 + 1) \tag{4.64}$$

where v_1 is the quantum number of the lower level and $\Delta v = 1$.

The relative population of two molecular energy levels is given by the **Boltzmann distribution law** (see any physical chemistry text) as

$$\boxed{\frac{N_i}{N_j} = \frac{g_i}{g_j}e^{-(E_i - E_j)/kT}} \tag{4.65}$$

where energy levels i and j have energies E_i and E_j and degeneracies g_i and g_j and are populated by N_i and N_j molecules, and where k is Boltzmann's constant and T the absolute temperature. For a nondegenerate level, $g_i = 1$.

The magnitude of $\nu = (1/2\pi)(k/\mu)^{1/2}$ is such that for light diatomics (for example, H_2, HCl, CO) only the $v = 0$ vibrational level is significantly populated at room temperature. For heavy diatomics (for example, I_2), there is significant room-temperature population of one or more excited vibrational levels.

The vibrational absorption spectrum of a polar diatomic molecule consists of a $v = 0 \rightarrow 1$ band, much weaker overtone bands ($v = 0 \rightarrow 2, 0 \rightarrow 3, \ldots$), and, if $v > 0$ levels are significantly populated, *hot bands* such as $v = 1 \rightarrow 2, 2 \rightarrow 3$. Each band corresponding to a particular vibrational transition consists of several closely spaced lines.

Each such line corresponds to a different change in rotational state simultaneous with the change in vibrational state. Each line is the result of a vibration–rotation transition.

The SI unit for spectroscopic frequencies is the **hertz** (Hz), defined by $1\text{ Hz} \equiv 1\text{ s}^{-1}$. Multiples such as the megahertz (MHz) equal to 10^6 Hz and the gigahertz (GHz) equal to 10^9 Hz are often used. Infrared absorption lines are usually specified by giving their **wavenumber** $\tilde{\nu}$, defined as

$$\boxed{\tilde{\nu} \equiv 1/\lambda = \nu/c} \tag{4.66}$$

where λ is the wavelength in vacuum.

In the harmonic-oscillator approximation, the quantum-mechanical energy levels of a polyatomic molecule turn out to be $E_{\text{vib}} = \sum_i \left(v_i + \frac{1}{2}\right)h\nu_i$, where the ν_i's are the frequencies of the normal modes of vibration of the molecule and v_i is the vibrational quantum number of the ith normal mode. Each v_i takes on the values $0, 1, 2, \ldots$ independently of the values of the other vibrational quantum numbers. A linear molecule with n atoms has $3n - 5$ normal modes; a nonlinear molecule has $3n - 6$ normal modes. (See *Levine, Molecular Spectroscopy,* Chapter 6 for details.)

To calculate the reduced mass μ in (4.61), one needs the masses of isotopic species. Some relative isotopic masses are listed in Table A.3 in the Appendix.

EXAMPLE The strongest infrared band of $^{12}C^{16}O$ occurs at $\tilde{\nu} = 2143\text{ cm}^{-1}$. Find the force constant of $^{12}C^{16}O$. State any approximation made.

The strongest infrared band corresponds to the $v = 0 \rightarrow 1$ transition. We approximate the molecular vibration as that of a harmonic oscillator. From (4.63), the equilibrium molecular vibrational frequency is approximately

$$\nu_e \approx \nu_{\text{light}} = \tilde{\nu}c = (2143\text{ cm}^{-1})(2.9979 \times 10^{10}\text{ cm/s}) = 6.424 \times 10^{13}\text{s}^{-1}$$

To relate k to ν_e in (4.61), we need the reduced mass $\mu = m_1m_2/(m_1 + m_2)$. One mole of ^{12}C has a mass of 12 g and contains Avogadro's number of atoms. Hence the mass of one atom of ^{12}C is $(12\text{ g})/(6.02214 \times 10^{23})$. The reduced mass and force constant are

$$\mu = \frac{12(15.9949)\text{ g}}{27.9949}\,\frac{1}{6.02214 \times 10^{23}} = 1.1385 \times 10^{-23}\text{ g}$$

$$k = 4\pi^2\nu_e^2\mu = 4\pi^2(6.424 \times 10^{13}\text{ s}^{-1})^2(1.1385 \times 10^{-26}\text{ kg}) = 1855\text{ N/m}$$

EXERCISE (a) Find the approximate zero-point energy of $^{12}C^{16}O$. (*Answer:* 2.1×10^{-20} J.) (b) Estimate ν_e of $^{13}C^{16}O$. (*Answer:* $6.28 \times 10^{13}\text{ s}^{-1}$.)

4.4 NUMERICAL SOLUTION OF THE ONE-DIMENSIONAL TIME-INDEPENDENT SCHRÖDINGER EQUATION

The Numerov Method. We solved the Schrödinger equation exactly for the particle in a box and the harmonic oscillator. For many potential-energy functions $V(x)$, the one-particle, one-dimensional Schrödinger equation cannot be solved exactly. This section presents a numerical method (the *Numerov method*) for computer solution of the one-particle, one-dimensional Schrödinger equation that allows one to get accurate bound-state eigenvalues and eigenfunctions for an arbitrary $V(x)$.

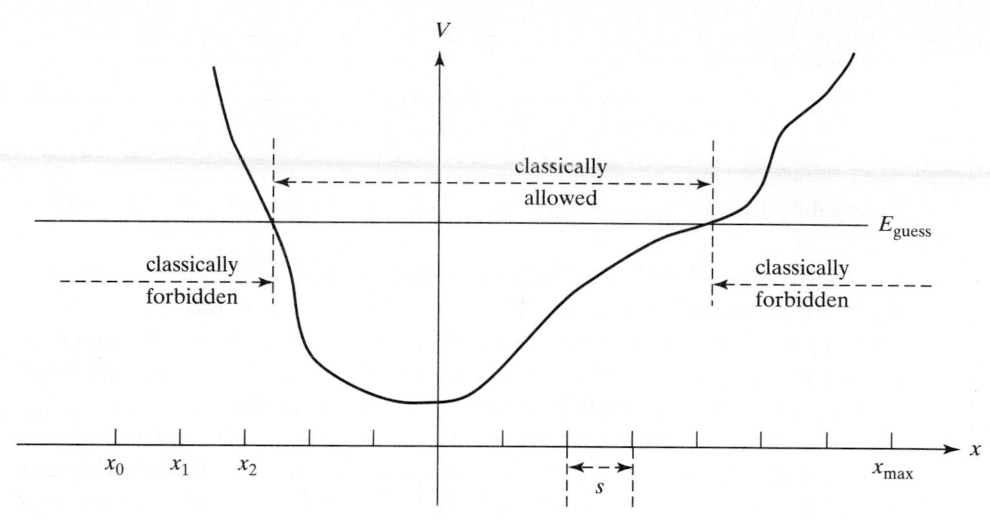

FIGURE 4.6 V versus x for a one-particle, one-dimensional system.

To solve the Schrödinger equation numerically, we deal with a portion of the x axis that includes the classically allowed region and that extends somewhat into the classically forbidden region at each end of the classically allowed region. We divide this portion of the x axis into small intervals, each of length s (Fig. 4.6). The points x_0 and x_{max} are the endpoints of this portion, and x_n is the endpoint of the nth interval. Let ψ_{n-1}, ψ_n, and ψ_{n+1} denote the values of ψ at the points $x_n - s$, x_n, and $x_n + s$, respectively (these are the endpoints of adjacent intervals)

$$\psi_{n-1} \equiv \psi(x_n - s), \qquad \psi_n \equiv \psi(x_n), \qquad \psi_{n+1} \equiv \psi(x_n + s) \tag{4.67}$$

Don't be confused by the notation. The subscripts $n - 1$, n, and $n + 1$ do not label different states but rather indicate values of one particular wave function ψ at points on the x axis separated by the interval s. The n subscript means ψ is evaluated at the point x_n. We write the Schrödinger equation $-(\hbar^2/2m)\psi'' + V\psi = E\psi$ as

$$\psi'' = G\psi, \quad \text{where } G \equiv m\hbar^{-2}[2V(x) - 2E] \tag{4.68}$$

By expanding $\psi(x_n + s)$ and $\psi(x_n - s)$ in Taylor series involving powers of s, adding these two expansions to eliminate odd powers of s, using the Schrödinger equation to express ψ'' and $\psi^{(iv)}$ in terms of ψ, and neglecting terms in s^6 and higher powers of s (an approximation that will be accurate if s is small), one finds that (Prob. 4.42)

$$\psi_{n+1} \approx \frac{2\psi_n - \psi_{n-1} + 5G_n\psi_n s^2/6 + G_{n-1}\psi_{n-1}s^2/12}{1 - G_{n+1}s^2/12} \tag{4.69}$$

where $G_n \equiv G(x_n) \equiv m\hbar^{-2}[2V(x_n) - 2E]$ [Eqs. (4.68) and (4.67)]. Equation (4.69) allows us to calculate ψ_{n+1}, the value of ψ at point $x_n + s$, if we know ψ_n and ψ_{n-1}, the values of ψ at the preceding two points x_n and $x_n - s$.

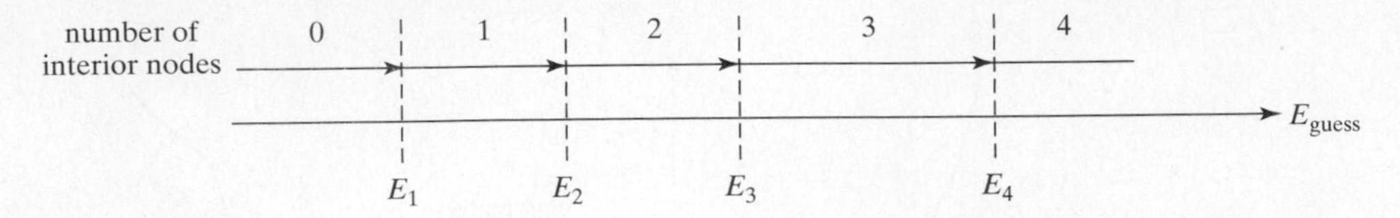

FIGURE 4.7 The number of nodes in a Numerov-method solution as a function of the energy E_{guess}.

How do we use (4.69) to solve the Schrödinger equation? We first guess a value E_{guess} for an energy eigenvalue. We start at a point x_0 well into the left-hand classically forbidden region (Fig. 4.6), where ψ will be very small, and we approximate ψ as zero at this point: $\psi_0 \equiv \psi(x_0) = 0$. Also, we pick a point x_{max} well into the right-hand classically forbidden region, where ψ will be very small and we shall demand that $\psi(x_{\text{max}}) = 0$. We pick a small value for the interval s between successive points and we take ψ at $x_0 + s$ as some small number, say, 0.0001: $\psi_1 \equiv \psi(x_1) \equiv \psi(x_0 + s) = 0.0001$. The value of ψ_1 will not make any difference in the eigenvalues found. If 0.001 were used instead of 0.0001 for ψ_1, Eq. (4.69) shows that this would simply multiply all values of ψ at subsequent points by 10 (Prob. 4.40). This would not affect the eigenvalues [see the example after Eq. (3.14)]. The wave function can be normalized after each eigenvalue is found.

Having chosen values for ψ_0 and ψ_1, we then use (4.69) with $n = 1$ to calculate $\psi_2 \equiv \psi(x_2) \equiv \psi(x_1 + s)$, where the G values are calculated using E_{guess}. Next, (4.69) with $n = 2$ gives ψ_3; and so on. We continue until we reach x_{max}. If E_{guess} is not equal to or very close to an eigenvalue, ψ will not be quadratically integrable and $|\psi(x_{\text{max}})|$ will be very large. If $\psi(x_{\text{max}})$ is not found to be close to zero, we start again at x_0 and repeat the process using a new E_{guess}. The process is repeated until we find an E_{guess} that makes $\psi(x_{\text{max}})$ very close to zero. E_{guess} is then essentially equal to an eigenvalue. The systematic way to locate the eigenvalues is to count the nodes in the ψ produced by E_{guess}. Recall (Section 4.2) that in a one-dimensional problem, the number of interior nodes is 0 for the ground state, 1 for the first excited state, etc. Let $E_1, E_2, E_3, \ldots$ denote the energies of the ground state, the first excited state, the second excited state, etc., respectively. If ψ_{guess} contains no nodes between x_0 and x_{max}, then E_{guess} is less than or equal to E_1; if ψ_{guess} contains one interior node, then E_{guess} is between E_1 and E_2; etc. (Fig. 4.7). Examples are given later.

Dimensionless Variables. The Numerov method requires that we guess values of E. What should be the order of magnitude of our guesses: 10^{-20} J, 10^{-15} J, ...? To answer this question, we reformulate the Schrödinger equation using dimensionless variables, taking the harmonic oscillator as the example.

The harmonic oscillator has $V = \frac{1}{2}kx^2$ and the harmonic-oscillator Schrödinger equation contains the three constants k, m, and $\hbar$. We seek to find a dimensionless reduced energy E_r and a dimensionless reduced x coordinate x_r that are defined by

$$E_r \equiv E/A, \qquad x_r \equiv x/B \tag{4.70}$$

where the constant A is a combination of k, m, and $\hbar$ that has dimensions of energy, and B is a combination with dimensions of length. The dimensions of energy are mass $\times$ length^2 $\times$ time^{-2}, which we write as

$$[E] = \text{ML}^2\text{T}^{-2} \tag{4.71}$$

where the brackets around E denote its dimensions, and M, L, and T stand for the dimensions mass, length, and time, respectively. The equation $V = \frac{1}{2}kx^2$ shows that k has dimensions of energy $\times$ length^{-2}, and (4.71) gives $[k] = \text{MT}^{-2}$. The dimensions of $\hbar$ are energy $\times$ time. Thus

$$[m] = \text{M}, \qquad [k] = \text{MT}^{-2}, \qquad [\hbar] = \text{ML}^2\text{T}^{-1} \tag{4.72}$$

The dimensions of A and B in (4.70) are energy and length, respectively, so

$$[A] = \text{ML}^2\text{T}^{-2}, \qquad [B] = \text{L} \tag{4.73}$$

Let $A = m^a k^b \hbar^c$, where a, b, and c are powers that are determined by the requirement that the dimensions of A must be ML^2T^{-2}. We have

$$[A] = [m^a k^b \hbar^c] = \text{M}^a(\text{MT}^{-2})^b(\text{ML}^2\text{T}^{-1})^c = \text{M}^{a+b+c}\text{L}^{2c}\text{T}^{-2b-c} \tag{4.74}$$

Equating the exponents of each of M, L, and T in (4.73) and (4.74), we have

$$a + b + c = 1, \qquad 2c = 2, \qquad -2b - c = -2$$

Solving these equations, we get $c = 1, b = \frac{1}{2}, a = -\frac{1}{2}$. Therefore,

$$A = m^{-1/2}k^{1/2}\hbar \tag{4.75}$$

Let $B = m^d k^e \hbar^f$. The same dimensional-analysis procedure that gave (4.75) gives (Prob. 4.43)

$$B = m^{-1/4}k^{-1/4}\hbar^{1/2} \tag{4.76}$$

From (4.70), (4.75), and (4.76), the reduced variables for the harmonic oscillator are

$$E_r = E/m^{-1/2}k^{1/2}\hbar, \qquad x_r = x/m^{-1/4}k^{-1/4}\hbar^{1/2} \tag{4.77}$$

Using $k^{1/2} = 2\pi\nu m^{1/2}$ [Eq. (4.25)] to eliminate k from (4.77) and recalling the definition $\alpha \equiv 2\pi\nu m/\hbar$ [Eq. (4.33)], we have the alternative expressions

$$E_r = E/h\nu, \qquad x_r = \alpha^{1/2}x \tag{4.78}$$

Similar to the equation $E_r \equiv E/A$ [Eq. (4.70)], we define the reduced potential energy function V_r as

$$V_r \equiv V/A \tag{4.79}$$

Since $|\psi(x)|^2\,dx$ is a probability, and probabilities are dimensionless, the normalized $\psi(x)$ must have the dimensions of length$^{-1/2}$. We therefore define a reduced normalized wave function ψ_r that is dimensionless. From (4.73), B has dimensions of length, so $B^{-1/2}$ has units of length$^{-1/2}$. Therefore,

$$\psi_r = \psi/B^{-1/2} \tag{4.80}$$

ψ_r satisfies $\int_{-\infty}^{\infty}|\psi_r|^2\,dx_r = 1$; this follows from (4.70), (4.80), and $\int_{-\infty}^{\infty}|\psi|^2\,dx = 1$.

We now rewrite the Schrödinger equation in terms of the reduced variables x_r, ψ_r, V_r, and E_r. We have

$$\frac{d^2\psi}{dx^2} = \frac{d^2}{dx^2}B^{-1/2}\psi_r = B^{-1/2}\frac{d}{dx}\frac{d\psi_r}{dx} = B^{-1/2}\frac{d}{dx}\frac{d\psi_r}{dx_r}\frac{dx_r}{dx} = B^{-1/2}\frac{d(d\psi_r/dx_r)}{dx_r}\frac{dx_r}{dx}\frac{dx_r}{dx}$$

$$\frac{d^2\psi}{dx^2} = B^{-5/2}\frac{d^2\psi_r}{dx_r^2} \tag{4.81}$$

since $dx_r/dx = B^{-1}$ [Eq. (4.70)]. Substitution of (4.70), (4.79), and (4.81) into the Schrödinger equation $-(\hbar^2/2m)(d^2\psi/dx^2) + V\psi = E\psi$ gives

$$-\frac{\hbar^2}{2m}B^{-5/2}\frac{d^2\psi_r}{dx_r^2} + AV_rB^{-1/2}\psi_r = AE_rB^{-1/2}\psi_r$$

$$-\frac{\hbar^2}{2m}\frac{1}{AB^2}\frac{d^2\psi_r}{dx_r^2} + V_r\psi_r = E_r\psi_r \tag{4.82}$$

From (4.75) and (4.76), we get $AB^2 = \hbar^2/m$, so $\hbar^2/mAB^2 = 1$ for the harmonic oscillator.

More generally, let V contain a single parameter c that is not dimensionless. For example, we might have $V = cx^4$ or $V = cx^2(1 + 0.05m^{1/2}c^{1/2}\hbar^{-1}x^2)$. (Note that $m^{1/2}c^{1/2}\hbar^{-1}x^2$ is dimensionless, as it must be, since 1 is dimensionless.) The quantity AB^2 in (4.82) must have the form $AB^2 = \hbar^r m^s c^t$, where r, s, and t are certain powers. Since the term $V_r\psi_r$ in (4.82) is dimensionless, the first term is dimensionless. Therefore, $\hbar^2/mAB^2$ is dimensionless and AB^2 has the same dimensions as $\hbar^2/m$; so $r = 2$, $s = -1$, and $t = 0$. With $AB^2 = \hbar^2/m$, Eq. (4.82) gives as the dimensionless Schrödinger equation

$$\frac{d^2\psi_r}{dx_r^2} = (2V_r - 2E_r)\psi_r \tag{4.83}$$

$$\psi_r'' = G_r\psi_r, \quad \text{where} \quad G_r \equiv 2V_r - 2E_r \tag{4.84}$$

For the harmonic oscillator, $V_r \equiv V/A = \frac{1}{2}kx^2/m^{-1/2}k^{1/2}\hbar = \frac{1}{2}x_r^2$ [Eqs. (4.75) and (4.77)]:

$$V_r = \tfrac{1}{2}x_r^2 \tag{4.85}$$

Having reduced the harmonic-oscillator Schrödinger equation to the form (4.83) involving only dimensionless quantities, we can expect that the lowest energy eigenvalues will be of the order of magnitude 1.

The reduced harmonic-oscillator Schrödinger equation (4.84) has the same form as (4.68), so we can use the Numerov formula (4.69) with ψ, G, and s replaced by ψ_r, G_r, and s_r, respectively, where, similar to (4.70), $s_r \equiv s/B$.

Once numerical values of the reduced energy E_r have been found, the energies E are found from (4.77) or (4.78).

Choice of $x_{r,0}$, $x_{r,\max}$, and s_r. We now need to choose initial and final values of x_r and the value of the interval s_r between adjacent points. Suppose we want to find all the harmonic-oscillator eigenvalues and eigenfunctions with $E_r \le 5$. We start the solution in the left-hand classically forbidden region, so we first locate the classically forbidden regions for $E_r = 5$. The boundaries between the classically allowed and forbidden regions are where $E_r = V_r$. From (4.85), $V_r = \frac{1}{2}x_r^2$. Thus $E_r = V_r$ becomes $5 = \frac{1}{2}x_r^2$ and the classically allowed region for $E_r = 5$ is from $x_r = -(10)^{1/2} = -3.16$ to $+3.16$. For $E_r < 5$, the classically allowed region is smaller. We want to start the solution at a point well into the left-hand classically forbidden region, where ψ is very small and to end the solution well into the right-hand classically forbidden region. The left-hand classically forbidden region ends at $x_r = -3.16$ for $E_r = 5$, and a reasonable choice is to start at $x_r = -5$. [Starting too far into the classically forbidden region can sometimes lead to trouble (see the following), so some trial-and-error might be needed in picking the starting point.] Since V is symmetrical, we shall end the solution at $x_r = 5$.

For reasonable accuracy, one usually needs a minimum of 100 points, so we shall take $s_r = 0.1$ to give us 100 points. As is evident from the derivation of the Numerov method, s_r must be small. A reasonable rule might be to have s_r no greater than 0.1.

If, as is often true, $V \to \infty$ as $x \to \pm\infty$, then starting too far into the classically forbidden region can make the denominator $1 - G_{n+1}s^2/12$ in the Numerov formula (4.69) negative. We have $G_r = 2V_r - 2E_r$, and if we start at a point x_0 where V_r is extremely large, G_r at that point might be large enough to make the Numerov denominator negative. The method will then fail to work. We are taking ψ_0 as zero and ψ_1 as a positive number. The Numerov formula (4.69) shows that if the denominator is negative, then ψ_2 will be negative, and we will have produced a spurious node in ψ between x_1 and x_2. To avoid this problem, we can either decrease the step size s_r or decrease $x_{r,\max} - x_{r,0}$ (see Prob. 4.45).

Computer Program for the Numerov Method. Table 4.1 contains a BASIC computer program that applies the Numerov method to the harmonic-oscillator Schrödinger equation. The # character in the names of variables makes these variables double precision. M is the number of intervals between $x_{r,0}$ and $x_{r,\max}$ and equals $(x_{r,\max} - x_{r,0})/s$. Lines 55 and 75 contain two times the potential-energy function, which must be modified if the problem is not the harmonic oscillator. If there is a node between two successive values of x_r, then the ψ_r values at these two points will have opposite signs (see Prob. 4.44) and statement 90 will increase the nodes counter NN by 1.

TABLE 4.1 BASIC Program for Numerov Solution of the One-Dimensional Schrödinger Equation

```
10   DIM X#(1500), G#(1500), P#(1500)
15   INPUT "Xr0"; X#(0): INPUT "Sr"; S#
25   INPUT "number of intervals"; M
30   INPUT "Er (enter 1E15 to quit)"; E#
33   IF E#>1E14 THEN STOP
35   NN=0: P#(0)=0: P#(1)=0.0001
50   X#(1)=X#(0)+S#
55   G#(0)=X#(0)^2-2*E#: G#(1)=X#(1)^2-2*E#
65   SS#=S#^2/12
70   FOR I=1 TO M-1
75   X#(I+1)=X#(I)+S#: G#(I+1)=X#(I+1)^2-2*E#
85   P#(I+1)=(-P#(I-1)+2*P#(I)+10*G#(I)*P#(I)*SS#+G#(I-1)*P#(I-1)*SS#)/(1-G#(I+1)*SS#)
90   IF P#(I+1)*P#(I)<0 THEN NN=NN+1
95   NEXT I
100  PRINT " Er="; E#; " Nodes="; NN; " Psir(XM)="; P#(M)
105  GOTO 30
```

For example, suppose we want the harmonic-oscillator ground-state energy. The program of Table 4.1, with $s_r = 0.1$, $x_{r,0} = -5$, and M = 100 gives the following results. The guess $E_r = 0$ gives a wave function with zero nodes (NN = 0), telling us (Fig. 4.7) that the ground-state energy $E_{r,1}$ is above 0. If we now guess 0.9 for E_r, we get a function with one node, so (Fig. 4.7) 0.9 is between $E_{r,1}$ and $E_{r,2}$. Hence the ground state E_r is between 0 and 0.9. Averaging these, we try 0.45. This value gives a function with no nodes and so 0.45 is below $E_{r,1}$. Averaging 0.45 and 0.9, we get 0.675, which is found to give one node and so is too high. Averaging 0.675 and 0.45, we try 0.5625, which gives one node and is too high. We next try 0.50625. And so on. The program's results show that as we get closer to the true $E_{r,1}$, $\psi_r(5)$ comes closer to zero.

Use of a Spreadsheet to Solve the One-Dimensional Schrödinger Equation. An alternative to a Numerov-method computer program is a spreadsheet.

The following directions for the Excel 2003 spreadsheet apply the Numerov method to solve the harmonic-oscillator Schrödinger equation. (Excel 2007, which uses tabs instead of menus, can be used with modified directions.)

The columns in the spreadsheet are labeled A, B, C, . . . and the rows are labeled 1, 2, 3, . . . (see Fig. 4.8 later in this section). A cell's location is specified by its row and column. For example, the cell at the upper left is cell A1. To enter something into a cell, you first **select** that cell, either by moving the mouse pointer over the desired cell and then clicking the (left) mouse button, or by using the arrow keys to move from the currently selected cell (which has a heavy outline) to the desired cell. After a cell has been selected, type the entry for that cell and press Enter or one of the four arrow keys.

To begin, enter a title in cell A1. Then enter `Er=` in cell A3. We shall enter our guesses for E_r in cell B3. We shall look first for the ground-state (lowest) eigenvalue, pretending that we don't know the answer. The minimum value of $V(x)$ for the harmonic oscillator is zero, so E_r cannot be negative. We shall take zero as our initial guess for E_r, so enter 0 in cell B3. Enter `sr=` in cell C3.

Enter 0.1 (the s_r value chosen earlier) in cell D3. Enter xr in cell A5, Gr in cell B5, and psir in C5. (These entries are labels for the data columns that we shall construct.) Enter −5 (the starting value for x_r) in cell A7. Enter `=A7+$D$3` in cell A8. The equal sign at the beginning of the entry tells the spreadsheet that a **formula** is being entered. This formula tells the spreadsheet to add the numbers in cells A7 and D3. The reason for the $ signs in D3 will be explained shortly. When you type a formula, you will see it displayed in the formula bar above the spreadsheet. When you press Enter, the value −4.9 is displayed in cell A8. This is the sum of cells A7 and D3. (If you see a different value in A8 or get an error message, you probably mistyped the formula. To correct the formula, select cell A8 and click in the formula displayed in the formula bar to make the correction.)

Select cell A8 and then on the Edit menu choose the Copy command. This copies the contents of cell A8 to a storage area called the Clipboard. Now sclect all cells from A9 through A107 by dragging the mouse. From the Edit menu choose the Paste command. This pastes the cell A8 formula (which was stored on the Clipboard) into each of cells A9 through A107. To see how this works, click on cell A9. You will see the formula `=A8+$D$3` in the formula bar. Note that when the cell A8 formula `=A7+$D$3` was copied one cell down to cell A9, the A7 in the formula was changed to A8. However, the $ signs prevented cell D3 in the formula from being changed when it was copied. In a spreadsheet formula, a cell address without $ signs is called a **relative** reference, whereas a cell address with $ signs is called an **absolute** reference. When a relative reference is copied to the next row in a column, the row number is increased by one; when it is copied two rows below the original row, the row number is increased by two; etc. Click in some of the other cells in column A to see their formulas. The net result of this copy-and-paste procedure is to fill the column-A cells with numbers from -5 to 5 in increments of 0.1. (Spreadsheets have faster ways to accomplish this than using Copy and Paste.)

We next fill in the G_r column. From Eqs. (4.84) and (4.85), $G_r = x_r^2 - 2E_r$ for the harmonic oscillator. We therefore enter `=A7^2-2*$B$3` in cell B7 (which will contain the value of G_r at $x_r = -5$). Cell A7 contains the $x_r = -5$ value and the ^ symbol denotes exponentiation. The * denotes multiplication. Cell B3 contains the E_r value. Next, the rest of the G_r values are calculated. Select cell B7. Then choose Copy from the Edit menu. Now select cells B8 through B107. Then choose Paste from the Edit menu. This fills the cells with the appropriate G_r values. (Click on cell B8 or B9 and see how its formula compares with that of B7.)

We now go to the ψ_r values. Enter 0 in cell C7. Cell C7 contains the value of ψ_r at $x_r = -5$. Since this point is well into the classically forbidden region, ψ_r will be very small here, and we can approximate it as zero. Cell C8 contains the value of ψ_r at $x_r = -4.9$. This value will be very small and we can enter any small number in C8 without affecting the eigenvalues. Enter `1E-4` in cell C8 (where E denotes the power of 10). Now that we have values in cells C7 and C8 for ψ_r at the first two points -5.0 and -4.9 [points x_{n-1} and x_n in (4.69)], we use (4.69) to calculate ψ_r at $x_r = -4.8$ (point x_{n+1}). Therefore, enter the Eq. (4.69) formula

```
=(2*C8-C7+5*B8*C8*$D$3^2/6+B7*C7*$D$3^2/12)/(1-B9*$D$3^2/12)
```

in cell C9. After the Enter key is pressed, the value 0.000224 for ψ_r at $x_r = -4.8$ appears in cell C9. Select cell C9. Choose Copy from the Edit menu. Select C10 through C107. Choose Paste from the Edit menu. Cells C10 through C107 will now be filled with their appropriate ψ_r values. As a further check that you entered the C9 formula correctly, verify that cell C10 contains 0.000401.

Since the number of nodes tells us which eigenvalues our energy guess is between (Fig. 4.7), we want to count and display the number of nodes in ψ_r. To do this, enter into cell D9 the formula `=IF(C9*C8<0,1,0)`. This formula enters the number 1 into D9 if C9 times C8 is negative and enters 0 into D9 if C9 times C8 is not negative. If there is a node between the x_r values in A8 and A9, then the ψ_r values in C8 and C9 will have opposite signs, and the value 1 will be entered into D9. Use Copy and Paste to copy the cell D9 formula into cells D10 through D107. Enter `nodes=` into cell E2. Enter `=SUM(D9:D107)` into F2. This formula gives the sum of the values in D9 through D107 and thus gives the number of interior nodes in ψ_r.

Next, we graph ψ_r versus x_r. Select cells A7 through A107 and C7 through C107 by clicking on Go To on the Edit menu, typing `A7:A107,C7:C107` in the Reference box, and clicking OK. From the Insert menu, choose Chart. Choose the Chart type as XY (Scatter). Choose the Chart-subtype as data points connected by smoothed lines. Click Next and make sure that Series in columns is chosen. Click Next, click the Gridlines tab and uncheck any checked boxes; click the Legend tab and uncheck Show

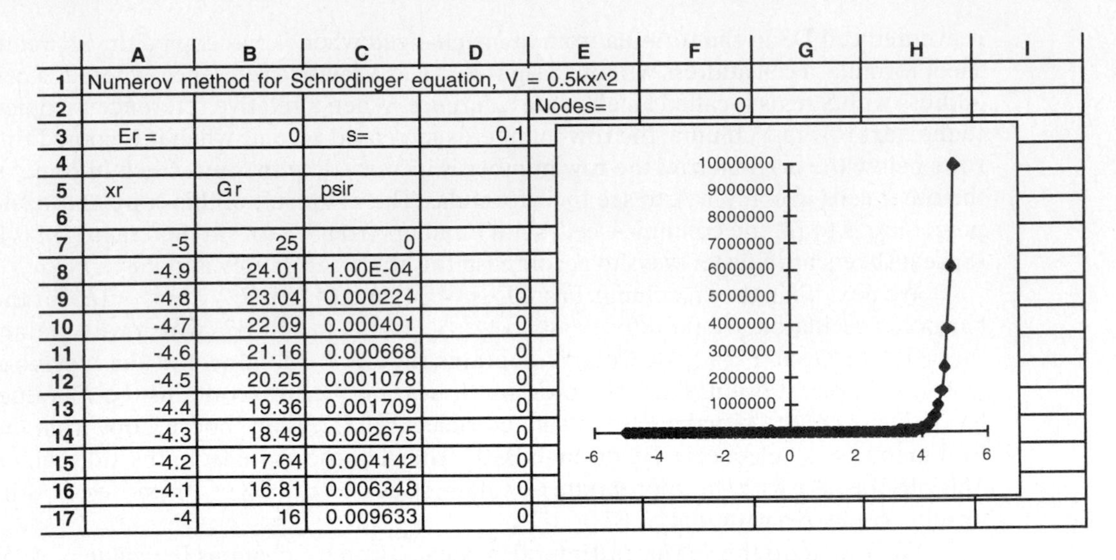

	A	B	C	D	E	F	G	H	I
1	Numerov method for Schrodinger equation, V = 0.5kx^2								
2					Nodes=	0			
3	Er =	0	s=	0.1					
4									
5	xr	Gr	psir						
6									
7	-5	25	0						
8	-4.9	24.01	1.00E-04						
9	-4.8	23.04	0.000224	0					
10	-4.7	22.09	0.000401	0					
11	-4.6	21.16	0.000668	0					
12	-4.5	20.25	0.001078	0					
13	-4.4	19.36	0.001709	0					
14	-4.3	18.49	0.002675	0					
15	-4.2	17.64	0.004142	0					
16	-4.1	16.81	0.006348	0					
17	-4	16	0.009633	0					

FIGURE 4.8 Spreadsheet for Numerov-method solution of the harmonic oscillator.

legend. Click Next and check that Place Chart As Object in Sheet1 is chosen. Click Finish. The graph appears, and you can position the pointer in the graph margin and drag the graph to any desired location. The spreadsheet looks like Fig. 4.8.

Since $x_r = 5$ is well into the right-hand classically forbidden region, ψ_r should be very close to zero at this point. However, the graph shows that for our choice $E_r = 0$, the wave function ψ_r is very large at $x_r = 5$. Therefore, $E_r = 0$ does not give a well-behaved ψ_r, and we must try a different E_r. Cell F2 has a zero, so this ψ_r has no nodes. Therefore (Fig. 4.7), the guess $E_r = 0$ is less than the true ground-state energy. Let us try $E_r = 2$. Select cell B3 and enter 2 into it. After you press Enter, the spreadsheet will recalculate every column B and column C cell whose value depends on E_r (all column B and C cells except C7 and C8 change) and will then replot the graph. The graph for $E_r = 2$ goes to a very large positive ψ_r value at $x_r = 5$. Also, cell F2 tells us that ψ_r for $E_r = 2$ contains two nodes. These are not readily visible on the graph, but the column-C data show that ψ_r changes sign between the x_r values -0.4 and -0.3 and between 1.2 and 1.3. We are looking for the ground-state eigenfunction, which does not contain a node, or rather, in our approximation, will contain nodes at -5 and 5. The presence of the two interior nodes shows (Fig. 4.7) that the value 2 for E_r is not only too high for the ground state, but is higher than E_r for the first excited state (whose wave function contains one interior node). We therefore need to try a lower value of E_r.

Before doing so, let us change the graph scale, so as to make the nodes more visible. In Excel 2003, double click on the y axis of the graph. In the Format Axis dialog box that appears, click on the Scale tab. Replace the original numbers in the Minimum and Maximum boxes with -10 and 10, respectively. Then click on OK. The graph will be redrawn with -10 and 10 as the minimum and maximum y-axis values, making the two nodes easily visible.

We now change E_r to a smaller value, say 1.2. Enter 1.2 in cell B3. We get a ψ_r that goes to a large negative value on the right and that has only one node. The presence of one node tells us that we are now below the energy of the second-lowest state and above the ground-state energy (see Fig. 4.7). We have bracketed the ground-state energy to lie between 0 and 1.2. Let us average these two numbers and try 0.6 as the energy. When we enter 0.6 into cell B3, we get a function with one node, so we are still above the ground-state energy. Since the maximum on the graph is now off scale, it's a good idea to change the graph scale and reset the y maximum and minimum values to 25 and -25.

We have found the lowest eigenvalue to be between 0 and 0.6. Averaging these values, we enter 0.3 into B3. This gives a function that has no nodes, so 0.3 is below the ground-state reduced energy, and E_r is between 0.3 and 0.6. Averaging these, we enter 0.45 into B3. We get a function that has no nodes, so we are still below the correct eigenvalue. However, if we rescale the y axis suitably (taking, for example, -15 and 30 as the minimum and maximum values), we see a function that for values of x_r less than 0.2 resembles closely what we expect for the ground state, so we are getting warm. The eigenvalue is now known to be between 0.45 and 0.60. Averaging these, we enter 0.525 into B3. We get a function with one node, so we are too high. Averaging 0.45 and 0.525, we try 0.4875, which we find to be too low.

Continuing in this manner, we get the successive values 0.50625 (too high), 0.496875 (too low), . . . , 0.4999995943548 (low), 0.4999996301176 (high). Thus we have found 0.4999996 as the lowest eigenvalue. Since $E_r = E/h\nu$, we should have gotten 0.5.

Suppose we want the second-lowest eigenvalue. We previously found this eigenvalue to be below 2.0, so it lies between 0.5 and 2.0. Averaging these numbers, we enter 1.25 into B3. We get a function that has the desired one node but that goes to a large negative value at the right, rather than returning to the x axis at the right. Therefore, 1.25 is too low (Fig. 4.7). Averaging 1.25 and 2.0, we next try 1.625. This gives a function with two nodes, so we are too high. Continuing, we get the successive values 1.4375 (low), 1.53125 (high), 1.484375 (low), 1.5078125 (high), etc.

To test the accuracy of the eigenvalues found, we can repeat the calculation with s_r half as large and see if the new eigenvalues differ significantly from those found with the larger s_r. Also, we can start further into the classically forbidden region.

Finding eigenvalues as we have done by trial-and-error is instructive and fun the first few times, but if you have a lot of eigenvalues to find, you can use a faster method. Most spreadsheets have a built-in program that can adjust the value in one cell so as to yield a desired value in a second cell. To see how this works, enter 0 into cell B3. From the Tools menu, choose Solver. (If Solver is missing from the Tools menu, choose Add-Ins on the Tools menu, click the box for Solver Add-In, and click OK. If Solver Add-In is missing from Add-Ins, click Browse and find the Solver.xla file.) The ψ_r value at $x_r = 5$ is in cell C107, and we want to make this value zero. Therefore, in the Set Target Cell box of the Solver, enter `$C$107`. Click on Value of and enter 0. We want to adjust the energy so as to satisfy the boundary condition at $x_r = 5$, so click in the By Changing Cells box and enter `$B$3`. Then click on Solve. The solution found by Excel has 0.4999996089 . . . in cell B3 (the formula bar will show the full value if you select cell B3). Cell C107 will have the value -2×10^{-7}. This value is not precisely zero, because Excel allows a certain margin of error. (You can have fun by watching Excel trying different solutions by using the Show Iteration Results option. The Solver uses either the quasi-Newton or the conjugate-gradient method, both of which are discussed in Section 15.10.)

To find higher eigenvalues automatically, start with a value in B3 that is well above the previously found eigenvalue. If the program converges to the previous eigenvalue, start with a still-higher value in B3. You can check which eigenvalue you have found by counting the nodes in the wave function. If the Solver fails to find the desired eigenvalue, use trial-and-error to find an approximate value and then use the Solver starting from the approximate eigenvalue.

The wave function we have found is unnormalized. The normalization constant is given by Eq. (3.93) as $N = [\int_{-\infty}^{\infty} |\psi_r|^2 \, dx_r]^{-1/2}$. We have $\int_{-\infty}^{\infty} |\psi_r|^2 \, dx_r \approx \int_{-5}^{5} \psi_r^2 \, dx_r \approx \sum_{i=1}^{100} \psi_{r,i}^2 s_r$, where the $\psi_{r,i}$ values are in column B. Enter npsir in E5. In cell D109, enter the formula `=SUMSQ(C8:C107)*$D$3`. The SUMSQ function adds the squares of a series of numbers. Enter `=C7/$D$109^0.5` in E7. Copy and paste E7 into E8 through E107. Column E will contain the normalized ψ_r values if E_{guess} is equal to an eigenvalue.

Excel is widely used to do statistical and scientific calculations, but studies of Excel 97, Excel 2000, and Excel 2002 (Excel XP) found many significant errors in Excel [B. D. McCullough and B. Wilson, *Comput. Statist. Data Anal.*, **31**, 27 (1999); **40**, 713 (2002)]. For example, McCullough and Wilson found that "*the Solver has a marked tendency to stop at a point that is not a solution and declare that it has found a solution.*" Therefore one should always verify the correctness of the Solver's solution. Studies of Excel 2003 found that some of the problems present in earlier versions of Excel were fixed, but some problems remain [McCullough and Wilson, *Comput. Statist. Data Anal.*, **49**, 1244 (2005)]. An excellent survey of deficiencies of Excel 2007 and earlier versions is www.daheiser.info/excel/frontpage.html.

Use of Mathcad to Solve the One-Dimensional Schrödinger Equation. Several programs classified as *computer algebra systems* do a wide variety of mathematical procedures, including symbolic integration and differentiation, numerical integration, algebraic manipulations, solving systems of equations, graphing, and matrix computations. Examples of such computer algebra systems are Maple, *Mathematica*, MATLAB, Mathcad, Derive, and LiveMath Maker. The Numerov procedure can be performed using these programs. One nice feature of Mathcad is its ability to produce animations ("movies"). With Mathcad one can create a movie showing how ψ_r changes as E_r goes through an eigenvalue.

Summary of Numerov-Method Steps. Problems 4.29–4.37 apply the Numerov method to several other one-dimensional problems. In solving these problems, you need to (a) find combinations of the constants in the problem that will give a dimensionless reduced energy and length [Eq. (4.70)]; (b) convert the Schrödinger equation to dimensionless form and find what $G_r(x_r)$ in (4.84) is; (c) decide on a maximum E_r value, below which you will find all eigenvalues; (d) locate the boundaries between the classically allowed and forbidden regions for this maximum E_r value and choose $x_{r,0}$ and $x_{r,\max}$ values in the classically forbidden regions (for the particle in a box with infinitely high walls, use $x_0 = 0$ and $x_{\max} = l$); (e) decide on a value for the interval s_r.

4.5 SUMMARY

The one-dimensional harmonic oscillator has $V = \frac{1}{2}kx^2$. Its stationary-state energies are $E = \left(v + \frac{1}{2}\right)h\nu$, where the vibrational frequency is $\nu = (1/2\pi)(k/m)^{1/2}$ and the quantum number v is $v = 0, 1, 2, \ldots$. The eigenfunctions are even or odd functions and are given by (4.49). An even function satisfies $f(-x) = f(x)$. An odd function satisfies $f(-x) = -f(x)$. If f is even, then $\int_{-a}^{a} f(x)\,dx = 2\int_0^a f(x)\,dx$. If f is odd, then $\int_{-a}^{a} f(x)\,dx = 0$. The vibrational energy of a diatomic molecule can be roughly approximated by the harmonic-oscillator energies with $\nu = (1/2\pi)(k/\mu)^{1/2}$, where the reduced mass is $\mu = m_1m_2/(m_1 + m_2)$.

The Numerov method is a numerical method that allows one to find bound-state energies and wave functions for the one-particle, one-dimensional Schrödinger equation.

PROBLEMS

Sec.	4.1	4.2	4.3	4.4	general
Probs.	4.1–4.3	4.4–4.22	4.23–4.28	4.29–4.47	4.48–4.52

4.1 Provided certain conditions are met, we can expand the function $f(x)$ in an infinite power series about the point $x = a$:

$$f(x) = \sum_{n=0}^{\infty} c_n(x - a)^n \tag{4.86}$$

Differentiate (4.86) m times, and then set $x = a$ to show that $c_n = f^{(n)}(a)/n!$, thus giving the familiar **Taylor series**:

$$\boxed{f(x) = \sum_{n=0}^{\infty} \frac{f^{(n)}(a)}{n!}(x - a)^n} \tag{4.87}$$

4.2 (a) Use (4.87) to derive the first few terms in the Taylor-series expansion about $x = 0$ for the function $\sin x$ and infer the general formula. (b) Differentiate the Taylor series in (a) to obtain the Taylor series for $\cos x$.

4.3 (a) Find the Taylor-series expansion about $x = 0$ for e^x. (b) Use the Taylor series (about $x = 0$) of $\sin x$, $\cos x$, and e^x to verify that $e^{i\theta} = \cos\theta + i\sin\theta$ [Eq. (1.28)].

4.4 True or false? (a) All harmonic-oscillator wave functions with v an odd integer must have a node at the origin. (b) The $v = 10$ harmonic-oscillator wave function has 10 interior nodes. (c) The $v = 1$ harmonic-oscillator wave function must be negative for $x < 0$. (d) The harmonic-oscillator energy levels are equally spaced. (e) The one-dimensional harmonic-oscillator energy levels are nondegenerate.

4.5 Derive (4.31) for E of a classical oscillator.

4.6 (a) Find the recursion relation for the coefficients c_n in the power-series solution of $(1 - x^2)y''(x) - 2xy'(x) + 3y(x) = 0$. (b) Express c_4 in terms of c_0 and c_5 in terms of c_1.

4.7 Which of the following are even functions? are odd functions? (a) $\sin x$; (b) $\cos x$; (c) $\tan x$; (d) e^x; (e) 13; (f) $x\cos x$; (g) $2 - 2x$; (h) $(3 + x)(3 - x)$.

4.8 Prove the statements made after Eq. (4.53) about products of even and odd functions.

4.9 (a) If $f(x)$ is an even function that is everywhere differentiable, prove that $f'(x)$ is an odd function. Do not assume that $f(x)$ can be expanded in a Taylor series. (b) Prove that the derivative of an everywhere-differentiable odd function is an even function. (c) If $f(x)$ is an even function that is differentiable at the origin, find $f'(0)$.

4.10 For the ground state of the one-dimensional harmonic oscillator, find the average value of the kinetic energy and of the potential energy; verify that $\langle T \rangle = \langle V \rangle$ in this case.

4.11 Verify the normalization factors for the $v = 1$ and $v = 2$ harmonic-oscillator wave functions.

4.12 Use the recursion relation (4.48) to find the $v = 3$ normalized harmonic-oscillator wave function.

4.13 Find ψ/c_0 for the $v = 4$ harmonic-oscillator wave function.

4.14 For the $v = 1$ harmonic-oscillator state, find the most likely position(s) of the particle.

4.15 Draw rough graphs of ψ and of ψ^2 for the $v = 5$ state of the one-dimensional harmonic oscillator without finding the explicit formula for ψ.

4.16 Find $\langle x \rangle$ for the harmonic-oscillator state with quantum number v.

4.17 Point out the similarities and differences between the one-dimensional particle-in-a-box and the harmonic-oscillator wave functions and energies.

4.18 Find the eigenvalues and eigenfunctions of $\hat{H}$ for a one-dimensional system with $V(x) = \infty$ for $x < 0, V(x) = \frac{1}{2}kx^2$ for $x \geq 0$.

4.19 (a) The three-dimensional harmonic oscillator has the potential-energy function

$$V = \tfrac{1}{2}k_x x^2 + \tfrac{1}{2}k_y y^2 + \tfrac{1}{2}k_z z^2$$

where the k's are three force constants. Find the energy eigenvalues by solving the Schrödinger equation. (b) If $k_x = k_y = k_z$, find the degree of degeneracy of each of the four lowest energy levels.

4.20 The *Hermite polynomials* are defined by

$$H_n(z) = (-1)^n e^{z^2} \frac{d^n e^{-z^2}}{dz^n}$$

(a) Verify that

$$H_0 = 1, \qquad H_1 = 2z, \qquad H_2 = 4z^2 - 2, \qquad H_3 = 8z^3 - 12z$$

(b) The Hermite polynomials obey the relation (*Pauling and Wilson*, pages 77–79)

$$zH_n(z) = nH_{n-1}(z) + \tfrac{1}{2}H_{n+1}(z)$$

Verify this identity for $n = 0, 1,$ and 2. (c) The normalized harmonic-oscillator wave functions can be written as (*Pauling and Wilson*, pages 79–80)

$$\psi_v(x) = (2^v v!)^{-1/2}(\alpha/\pi)^{1/4} e^{-\alpha x^2/2} H_v(\alpha^{1/2}x) \tag{4.88}$$

Verify (4.88) for the three lowest states.

4.21 When a second-order linear homogeneous differential equation is written in the form (2.3), any point at which $P(x)$ or $Q(x)$ becomes infinite is called a *singular point* or *singularity*. In solving a differential equation by the power-series method, one can often find the proper substitution to give a two-term recursion relation by examining the differential equation near its singularities. For the harmonic-oscillator Schrödinger equation (4.34), the singularities are at $x = \pm\infty$. To check whether $x = \infty$ is a singular point, one substitutes $z = 1/x$ and examines the coefficients at $z = 0$. Verify that $\exp(-\alpha x^2/2)$ is an approximate solution of (4.34) for very large $|x|$.

4.22 (a) Write a computer program that uses the recursion relation (4.41) to calculate ψ/c_0 of (4.42) versus $\alpha^{1/2}x$ for $\alpha^{1/2}x$ values from 0 to 6 in increments of 0.5 for specified values of $mE\hbar^{-2}/\alpha = E/h\nu$. Include a test to stop adding terms in the infinite series when the last calculated term is small enough. (b) Run the program for $E/h\nu = 0.499$, 0.5, and 0.501 to verify Fig. 4.2.

4.23 (a) The infrared absorption spectrum of $^1H^{35}Cl$ has its strongest band at 8.65×10^{13} Hz. Calculate the force constant of the bond in this molecule. (b) Find the approximate zero-point vibrational energy of $^1H^{35}Cl$. (c) Predict the frequency of the strongest infrared band of $^2H^{35}Cl$.

4.24 The $v = 0 \rightarrow 1$ and $v = 0 \rightarrow 2$ bands of $^1H^{35}Cl$ occur at 2885.98 cm^{-1} and 5667.98 cm^{-1}. (a) Calculate ν_e/c and $\nu_e x_e/c$ for this molecule. (b) Predict the wavenumber of the $v = 0 \rightarrow 3$ band of $^1H^{35}Cl$.

4.25 (a) The $v = 0 \rightarrow 1$ band of LiH occurs at 1359 cm^{-1}. Calculate the ratio of the $v = 1$ to $v = 0$ populations at 25°C and at 200°C. (b) Do the same as in (a) for ICl, whose strongest infrared band occurs at 381 cm^{-1}.

4.26 (a) Verify (4.64). (b) Find the corresponding equation for the $v = 0 \rightarrow v_2$ transition.

4.27 Show that if one expands $U(R)$ in Fig. 4.5 in a Taylor series about $R = R_e$ and neglects terms containing $(R - R_e)^3$ and higher powers (these terms are small for R near R_e) then one obtains a harmonic-oscillator potential with $k = d^2U/dR^2|_{R=R_e}$.

4.28 The *Morse function* $U(R) = D_e[1 - e^{-a(R-R_e)}]^2$ is often used to approximate the $U(R)$ curve of a diatomic molecule, where the molecule's equilibrium dissociation energy D_e is $D_e = U(\infty) - U(R_e)$. (a) Verify that this equation for D_e is satisfied by the Morse function. (b) Show that $a = (k_e/2D_e)^{1/2}$.

For Probs. 4.29–4.37, use either a program similar to that in Table 4.1, a spreadsheet, or a computer-algebra system such as Mathcad.

4.29 Use the Numerov method to find the lowest three stationary-state energies for a particle in a one-dimensional box of length l with walls of infinite height.

4.30 (a) Use the Numerov method to find all the bound-state eigenvalues for a particle in a rectangular well (Section 2.4) of length l with $V_0 = 20\hbar^2/ml^2$. Note that V is different in different regions and $\psi \neq 0$ at the walls. (b) Repeat (a) for $V_0 = 50\hbar^2/ml^2$. (c) Check your results using the automatic solver in your spreadsheet program or Mathcad to find the roots of Eq. (2.35). (Before doing this, find approximate values from a spreadsheet graph.)

4.31 Use the Numerov method to find the lowest three energy eigenvalues for a one-particle system with $V = cx^4$, where c is a constant.

4.32 Use the Numerov method to find the lowest three energy eigenvalues for a one-particle system with $V = ax^8$, where a is a constant.

4.33 Use the Numerov method to find the lowest four energy eigenvalues for a one-particle system with $V = \infty$ for $x \leq 0$ and $V = bx$ for $x > 0$, where b is a positive constant. (For $b = mg$, this is a particle in a gravitational field.)

4.34 Consider a one-particle system with $V = -31.5(\hbar^2/ma^2)/(e^{x/a} + e^{-x/a})^2$, where a is a positive constant. (a) Find V_r. (b) Use a spreadsheet, graphing calculator, or Mathcad to graph V_r versus x_r. (See Prob. 4.38.) (c) Use the Numerov method to find all bound-state eigenvalues less than -0.1. [The exponential function is written EXP(A7) in Excel.]

4.35 Consider a one-particle system with $V = \frac{1}{4}b^2/c - bx^2 + cx^4$, where b and c are positive constants. If we use $\hbar$, m, and b to find A and B in $E_r = E/A$ and $x_r = x/B$, we will get the same results as for the harmonic oscillator, except that k is replaced by b. Thus, Eq. (4.76) gives $B = m^{-1/4}b^{-1/4}\hbar^{1/2}$. The equation for V in this problem shows that $[bx^2] = [cx^4]$, so $[c] = [b]/\mathrm{L}^2$ and we write $c = ab/B^2 = ab/m^{-1/2}b^{-1/2}\hbar$, where a is a dimensionless constant. (a) Verify that $V_r = 1/4a - x_r^2 + ax_r^4$. (b) Use a spreadsheet or graphing calculator to plot V_r versus x_r for $a = 0.05$. (The form of V_r roughly resembles the potential energy for the inversion of the NH_3 molecule.) (c) For $a = 0.05$, use the Numerov method to find all eigenvalues with $E_r < 10$. *Hint:* Some of the eigenvalues lie very close together. (See also Prob. 4.36.)

4.36 A one-dimensional double-well potential has $V = \infty$ for $x < -\frac{1}{2}l$, $V = 0$ for $-\frac{1}{2}l \le x \le -\frac{1}{4}l$, $V = V_0$ for $-\frac{1}{4}l < x < \frac{1}{4}l$, $V = 0$ for $\frac{1}{4}l \le x \le \frac{1}{2}l$, and $V = \infty$ for $x > \frac{1}{2}l$, where l and V_0 are positive constants. Sketch V. Use the Numerov method to find the lowest four eigenvalues and the corresponding unnormalized eigenfunctions for the following values of $V_0/(\hbar^2/ml^2)$: (a) 1; (b) 100; (c) 1000. Compare the wave functions and energies for (a) with those of a particle in a box of length l, and those for (c) with those of a particle in a box of length $\frac{1}{4}l$. *Hints:* In (b) and (c), some of the eigenvalues lie very close together. In (c), the eigenvalues need to be located to many decimal places to get decent-looking eigenfunctions. The eigenfunctions must be either even or odd functions.

4.37 (a) For the harmonic-oscillator Numerov example, we went from –5 to 5 in steps of 0.1 and found 0.4999996 as the lowest eigenvalue. For this choice of $x_{r,0}$ and s_r, find all eigenvalues with $E_r < 6$; then find the eigenvalue that lies between 11 and 12 and explain why the result is not accurate. Then change $x_{r,0}$ or s_r or both to get an accurate value for this eigenvalue. (b) Find the harmonic-oscillator eigenvalues with $E_r < 6$ if we go from –5 to 5 in steps of 0.5. (c) Find the harmonic-oscillator eigenvalues with $E_r < 6$ if we go from –3 to 3 in steps of 0.1.

4.38 Spreadsheets contain pitfalls for the unwary. (a) If cell A1 contains the value 5, what would you expect the formula `=-A1^2+A1^2` to give? (Note the minus sign.) (b) Enter 5 in cell A1, enter `=-A1^2+A1^2` in cell A2, and enter `=+A1^2-A1^2` in A3. What results do you get? Repeat this with as many different spreadsheet programs as are available to you. Do the different spreadsheets agree with one another?

4.39 Spreadsheet formulas can be written more elegantly if cells containing parameters are given names. Find out how to name cells in the spreadsheet program you are using. (a) In the harmonic-oscillator example, name cells B3 and D3 as Er and sr, respectively. Then replace $B$3 and $D$3 by Er and sr in all the formulas in the spreadsheet. (b) Why is x2 not allowed as the name for a spreadsheet cell?

4.40 Use (4.69) to show that if one multiplies ψ_1 in the Numerov method by a constant c, then $\psi_2, \psi_3, \ldots$ are all multiplied by c, so the entire wave function is multiplied by c, which does not affect the eigenvalues we find.

4.41 Use the normalized Numerov-method harmonic-oscillator wave functions found by going from –5 to 5 in steps of 0.1 to estimate the probability of being in the classically forbidden region for the $v = 0$ and $v = 1$ states.

4.42 In the Taylor series (4.87) of Prob. 4.1, let the point $x = a$ be called x_n (that is, $x_n \equiv a$) and let $s \equiv x - a = x - x_n$, so $x = x_n + s$. (a) Use this notation to write (4.87) as $f(x_n + s)$ equal to a power series in s and evaluate all terms through s^5. (b) In the result of part (a), change s to $-s$ to find a series for $f(x_n - s)$. Then add the two series and neglect terms in s^6 and higher powers of s to show that

$$f(x_n + s) + f(x_n - s) \approx 2f(x_n) + f''(x_n)s^2 + \tfrac{1}{12}f^{(\mathrm{iv})}(x_n)s^4$$

$$f_{n+1} \approx -f_{n-1} + 2f_n + f_n''s^2 + \tfrac{1}{12}f_n^{(\mathrm{iv})}s^4 \tag{4.89}$$

$$\psi_{n+1} \approx -\psi_{n-1} + 2\psi_n + \psi_n''s^2 + \tfrac{1}{12}\psi_n^{(\mathrm{iv})}s^4 \tag{4.90}$$

where the notation of (4.67) with ψ replaced by f was used and then f was replaced by ψ. (c) Replace f in (4.89) by ψ'', multiply the resulting equation by s^2, neglect the s^6 term, solve for $\psi_n^{(iv)}s^4$, and use $\psi'' = G\psi$ [Eq. (4.68)] to show that

$$\psi_n^{(iv)}s^4 \approx G_{n+1}\psi_{n+1}s^2 + G_{n-1}\psi_{n-1}s^2 - 2G_n\psi_n s^2 \tag{4.91}$$

Substitute (4.91) and $\psi_n'' = G_n\psi_n$ into (4.90) and solve for ψ_{n+1} to show that Eq. (4.69) holds.

4.43 Use dimensional analysis to verify (4.76) for B.

4.44 In applying the Numerov method to count the nodes in ψ_r, we assumed that ψ changes sign as it goes through a node. However, there are functions that do not have opposite signs on each side of a node. For example, the functions $y = x^2$ and $y = x^4$ are positive on both sides of the node at $x = 0$. For a function y that is positive at points just to the left of $x = a$, is zero at $x = a$, and is positive just to the right of $x = a$, the definition $y' = \lim_{\Delta x \to 0} \Delta y/\Delta x$ shows that the derivative y' is negative just to the left of $x = a$ and is positive just to the right of $x = a$. Therefore (assuming y' is a continuous function), y' is zero at $x = a$. (An exception is a function such as the V-shaped function $y = |x|$, whose derivative is discontinuous at $x = a$. But such a function is ruled out by the requirement that ψ' be continuous.) (a) Use the Schrödinger equation to show that if $\psi(x) = 0$ at $x = a$, then $\psi'' = 0$ at $x = a$ (provided $V(a) \neq \infty$). (b) Differentiate the Schrödinger equation to show that if both ψ and ψ' are zero at $x = a$, then $\psi'''(a) = 0$ [provided $V'(a) \neq \infty$]. Then show that all higher derivatives of ψ are zero at $x = a$ if both ψ and ψ' are zero at $x = a$ (and no derivatives of V are infinite at $x = a$). If ψ and all its derivatives are zero at $x = a$, the Taylor series (4.87) shows that ψ is zero everywhere. But a zero function is not allowed as a wave function. Therefore, ψ and ψ' cannot both be zero at a point, and the wave function must have opposite signs on the two sides of a node.

4.45 Suppose $V = cx^8$, where c is a positive constant, and we want all eigenvalues with $E_r < 10$. (a) Show that $V_r = x_r^8$ and that for $E_r = 10$ the boundaries of the classically allowed region are at $x_r = \pm 1.33$. (b) Set up a spreadsheet and verify that if we take $x_{r,0} = -3$, $x_{r,\max} = 3$, and $s_r = 0.05$, ψ undergoes spurious oscillations for $|x_r| > 2.65$. (c) Verify that $1 - G_r s_r^2/12 \approx 0$ for $|x_r| = 2.65$, so $1 - G_r s_r^2/12$ is negative for $|x_r| > 2.65$. (d) Use your spreadsheet to verify that the spurious oscillations are eliminated if we take either $x_{r,0} = -2.5$, $x_{r,\max} = 2.5$, and $s_r = 0.05$; or $x_{r,0} = -3$, $x_{r,\max} = 3$, and $s_r = 0.02$.

4.46 Modify the program of Table 4.1 to find the normalized wave function.

4.47 Rewrite the program of Table 4.1 to eliminate all array variables.

4.48 Spreadsheets and computer-algebra systems can easily be used to solve equations of the form $f(x) = 0$. For example, suppose we want to solve $e^x = 2 - x^2$. A computer-made graph shows that the function $e^x - 2 + x^2$ equals zero at only two points, one positive and one negative. In Excel, we enter an initial guess of 0 for x in cell A1 and `=exp(A1)-2+A1^2` in cell A3. Then from the Tools menu, choose Solver. In the Solver Parameters Box, we enter A3 next to Set Target Cell, click Value of and enter 0 after Equal to, and enter A1 after By Changing Cells; then click Options and change Precision to 1E-14. Click on OK and then click on Solve. Excel gives us the result 0.537274449173857, with 2.39×10^{-15} in cell A3. To find the negative root, start with -1 in cell A1 and use the Solver. (a) Use a spreadsheet or Mathcad to solve Prob. 2.22 for the bound-state energies of a particle in a well. (b) For the double-well potential of Prob. 4.36, application of the procedures used in Section 2.4 shows that the allowed bound-state energy levels satisfy

$$[(V_{0r} - E_r)/E_r]^{1/2} \tan[(E_r/8)^{1/2}] = -(\tanh\{[(V_{0r} - E_r)/8]^{1/2}\})^p \tag{4.92}$$

where $V_{0r} \equiv V_0/(\hbar^2/ml^2)$, $E_r \equiv E/(\hbar^2/ml^2)$, and where $p = -1$ for the even wave functions and $p = 1$ for the odd wave functions. The hyperbolic tangent function, defined by $\tanh z \equiv (e^z - e^{-z})/(e^z + e^{-z})$, can be produced by typing TANH in Excel. Use (4.92) and a spreadsheet or Mathcad to find the lowest four double-well energies for each of the V_{0r} values of Prob. 4.36.

4.49 (a) Show that if k_i and f_i are eigenvalues and eigenfunctions of the linear operator $\hat{A}$ then ck_i and f_i are eigenvalues and eigenfunctions of $c\hat{A}$. (b) Give an operator whose eigenvalues are $\frac{1}{2}, \frac{3}{2}, \frac{5}{2}, \ldots$. (c) Give an operator whose eigenvalues are $1, 2, 3, \ldots$.

4.50 (a) A certain system in a certain stationary state has $\psi = Ne^{-\alpha x^4}$ (N is the normalization constant.) Find the system's potential-energy function $V(x)$ and its energy E. *Hint:* The zero level of energy is arbitrary, so choose $V(0) = 0$. (b) Sketch $V(x)$. (c) Is this the ground-state ψ? Explain.

4.51 Show that adding a constant C to the potential energy leaves the stationary-state wave functions unchanged and simply adds C to the energy eigenvalues.

4.52 True or false? (a) In the classically forbidden region, $E > V$ for a stationary state. (b) If the harmonic-oscillator wave function ψ_v is an even function, then ψ_{v+1} is an odd function. (c) For harmonic-oscillator wave functions, $\int_{-\infty}^{\infty} \psi_v^*(x)\psi_{v+1}(x)\,dx = 0$. (d) At a node in a bound stationary-state wave function, ψ'' must be zero provided V is not infinite at the node. (e) The spacing between adjacent vibrational levels of a diatomic molecule remains constant as the vibrational energy increases. (f) For the $v = 25$ harmonic-oscillator wave function, the sign of ψ in the right-hand classically forbidden region is opposite the sign in the left-hand classically forbidden region.

C H A P T E R 5

Angular Momentum

5.1 SIMULTANEOUS SPECIFICATION OF SEVERAL PROPERTIES

In this chapter we discuss angular momentum, and in the next chapter we show that for the stationary states of the hydrogen atom the magnitude of the electron's angular momentum is constant. As a preliminary, we consider what criterion we can use to decide which properties of a system can be simultaneously assigned definite values.

In Section 3.3 we postulated that if the state function Ψ is an eigenfunction of the operator $\hat{A}$ with eigenvalue s, then a measurement of the physical property A is certain to yield the result s. If Ψ is simultaneously an eigenfunction of the two operators $\hat{A}$ and $\hat{B}$, that is, if $\hat{A}\Psi = s\Psi$ and $\hat{B}\Psi = t\Psi$, then we can simultaneously assign definite values to the physical quantities A and B. When will it be possible for Ψ to be simultaneously an eigenfunction of two different operators? In Chapter 7, we shall prove the following two theorems. First, a necessary condition for the existence of a complete set of simultaneous eigenfunctions of two operators is that the operators commute with each other. (The word *complete* is used here in a certain technical sense, which we won't worry about until Chapter 7.) Conversely, if $\hat{A}$ and $\hat{B}$ are two commuting operators that correspond to physical quantities, then there exists a complete set of functions that are eigenfunctions of both $\hat{A}$ and $\hat{B}$. Thus, if $[\hat{A}, \hat{B}] = 0$, then Ψ can be an eigenfunction of both $\hat{A}$ and $\hat{B}$.

Recall that the commutator of $\hat{A}$ and $\hat{B}$ is $[\hat{A}, \hat{B}] \equiv \hat{A}\hat{B} - \hat{B}\hat{A}$ [Eq. (3.7)]. The following identities are helpful in evaluating commutators. These identities are proved by writing out the commutators in detail (Prob. 5.2):

$$[\hat{A}, \hat{B}] = -[\hat{B}, \hat{A}] \tag{5.1}$$

$$[\hat{A}, \hat{A}^n] = 0, \qquad n = 1, 2, 3, \ldots \tag{5.2}$$

$$[k\hat{A}, \hat{B}] = [\hat{A}, k\hat{B}] = k[\hat{A}, \hat{B}] \tag{5.3}$$

$$[\hat{A}, \hat{B} + \hat{C}] = [\hat{A}, \hat{B}] + [\hat{A}, \hat{C}] \qquad [\hat{A} + \hat{B}, \hat{C}] = [\hat{A}, \hat{C}] + [\hat{B}, \hat{C}] \tag{5.4}$$

$$[\hat{A}, \hat{B}\hat{C}] = [\hat{A}, \hat{B}]\hat{C} + \hat{B}[\hat{A}, \hat{C}] \qquad [\hat{A}\hat{B}, \hat{C}] = [\hat{A}, \hat{C}]\hat{B} + \hat{A}[\hat{B}, \hat{C}] \tag{5.5}$$

where k is a constant and the operators are assumed to be linear.

EXAMPLE Starting from $[\partial/\partial x, x] = 1$ [Eq. (3.8)], use the commutator identities (5.1)–(5.5) to find (a) $[\hat{x}, \hat{p}_x]$; (b) $[\hat{x}, \hat{p}_x^2]$, (c) $[\hat{x}, \hat{H}]$ for a one-particle, three-dimensional system.

(a) Use of (5.3), (5.1), and $[\partial/\partial x, x] = 1$ gives

$$[\hat{x}, \hat{p}_x] = \left[x, \frac{\hbar}{i}\frac{\partial}{\partial x}\right] = \frac{\hbar}{i}\left[x, \frac{\partial}{\partial x}\right] = -\frac{\hbar}{i}\left[\frac{\partial}{\partial x}, x\right] = -\frac{\hbar}{i}$$

$$[\hat{x}, \hat{p}_x] = i\hbar \tag{5.6}$$

(b) Use of (5.5) and (5.6) gives

$$[\hat{x}, \hat{p}_x^2] = [\hat{x}, \hat{p}_x]\hat{p}_x + \hat{p}_x[\hat{x}, \hat{p}_x] = i\hbar \cdot \frac{\hbar}{i}\frac{\partial}{\partial x} + \frac{\hbar}{i}\frac{\partial}{\partial x} \cdot i\hbar$$

$$[\hat{x}, \hat{p}_x^2] = 2\hbar^2\frac{\partial}{\partial x} \tag{5.7}$$

(c) Use of (5.4), (5.3), and (5.7) gives

$$\begin{aligned}[\hat{x}, \hat{H}] &= [\hat{x}, \hat{T} + \hat{V}] = [\hat{x}, \hat{T}] + [\hat{x}, \hat{V}(x, y, z)] = [\hat{x}, \hat{T}] \\ &= [x, (1/2m)(\hat{p}_x^2 + \hat{p}_y^2 + \hat{p}_z^2)] \\ &= (1/2m)[\hat{x}, \hat{p}_x^2] + (1/2m)[\hat{x}, \hat{p}_y^2] + (1/2m)[\hat{x}, \hat{p}_z^2] \\ &= \frac{1}{2m}2\hbar^2\frac{\partial}{\partial x} + 0 + 0\end{aligned}$$

$$[\hat{x}, \hat{H}] = \frac{\hbar^2}{m}\frac{\partial}{\partial x} = \frac{i\hbar}{m}\hat{p}_x \tag{5.8}$$

EXERCISE Show that for a one-particle, three-dimensional system,

$$[\hat{p}_x, \hat{H}] = -i\hbar\, \partial V(x, y, z)/\partial x \tag{5.9}$$

The above commutators have important physical consequences. Since $[\hat{x}, \hat{p}_x] \neq 0$, we cannot expect the state function to be simultaneously an eigenfunction of $\hat{x}$ and of $\hat{p}_x$. Hence we cannot simultaneously assign definite values to x and p_x, in agreement with the uncertainty principle. Since $\hat{x}$ and $\hat{H}$ do not commute, we cannot expect to assign definite values to the energy and the x coordinate at the same time. A stationary state (which has a definite energy) shows a spread of possible values for x, the probabilities for observing various values of x being given by the Born postulate.

For a state function Ψ that is not an eigenfunction of $\hat{A}$, we get various possible outcomes when we measure A in identical systems. We want some measure of the spread or dispersion in the set of observed values A_i. If $\langle A \rangle$ is the average of these values, then the deviation of each measurement from the average is $A_i - \langle A \rangle$. If we averaged all the deviations, we would get zero, since positive and negative deviations would cancel. Hence to make all deviations positive, we square them. The average of the squares of the deviations is called the *variance* of A, symbolized in statistics by σ_A^2 and in quantum mechanics by $(\Delta A)^2$:

$$(\Delta A)^2 \equiv \sigma_A^2 \equiv \langle (A - \langle A \rangle)^2 \rangle = \int \Psi^*(\hat{A} - \langle A \rangle)^2 \Psi \, d\tau \tag{5.10}$$

where the average-value expression (3.88) was used. The definition (5.10) is equivalent to (Prob. 5.7)

$$(\Delta A)^2 = \langle A^2 \rangle - \langle A \rangle^2 \tag{5.11}$$

The positive square root of the variance is called the **standard deviation**, σ_A or ΔA. The standard deviation is the most commonly used measure of spread, and we shall take it as the measure of the "uncertainty" in the property A.

For the product of the standard deviations of two properties of a quantum-mechanical system whose state function is Ψ, one can show that (Prob. 7.60)

$$\Delta A \, \Delta B \geq \frac{1}{2}\left| \int \Psi^*[\hat{A}, \hat{B}]\Psi \, d\tau \right| \tag{5.12}$$

If $\hat{A}$ and $\hat{B}$ commute, then the integral in (5.12) is zero, and ΔA and ΔB may both be zero, in agreement with the previous discussion.

As an example of (5.12), we find, using (5.6) and $|z_1 z_2| = |z_1||z_2|$ [Eq. (1.34)],

$$\Delta x \, \Delta p_x \geq \frac{1}{2}\left| \int \Psi^*[\hat{x}, \hat{p}_x]\Psi \, d\tau \right| = \frac{1}{2}\left| \int \Psi^* i\hbar \Psi \, d\tau \right| = \frac{1}{2}\hbar |i| \left| \int \Psi^* \Psi \, d\tau \right|$$

$$\Delta x \, \Delta p_x \geq \tfrac{1}{2}\hbar \tag{5.13}$$

Equation (5.13) is the quantitative statement of the **Heisenberg uncertainty principle**.

EXAMPLE Equations (3.91), (3.92), (3.39), the equation following (3.89), and Prob. 3.45 give for the ground state of the particle in a three-dimensional box

$$\langle x \rangle = a/2, \qquad \langle x^2 \rangle = a^2(1/3 - 1/2\pi^2), \qquad \langle p_x \rangle = 0, \qquad \langle p_x^2 \rangle = h^2/4a^2$$

Use these results to check that the uncertainty principle (5.13) is obeyed.

We have

$$(\Delta x)^2 = \langle x^2 \rangle - \langle x \rangle^2 = a^2(1/3 - 1/2\pi^2) - a^2/4 = a^2(\pi^2 - 6)/12\pi^2$$

$$\Delta x = a(\pi^2 - 6)^{1/2}/(12)^{1/2}\pi$$

$$(\Delta p_x)^2 = \langle p_x^2 \rangle - \langle p_x \rangle^2 = h^2/4a^2, \qquad \Delta p_x = h/2a$$

$$\Delta x \, \Delta p_x = \frac{h}{2\pi}\left(\frac{\pi^2 - 6}{12}\right)^{1/2} = 0.568\hbar > \tfrac{1}{2}\hbar$$

There is also an uncertainty relation involving energy and time:

$$\Delta E \, \Delta t \geq \tfrac{1}{2}\hbar \tag{5.14}$$

Some texts state that (5.14) is derived from (5.12) by taking $i\hbar \, \partial/\partial t$ as the energy operator and multiplication by t as the time operator. However, the energy operator is the Hamiltonian $\hat{H}$, and not $i\hbar \, \partial/\partial t$. Moreover, time is not an observable but is a parameter in quantum mechanics. Hence there is no quantum-mechanical time operator. (The noun **observable** in quantum mechanics means a physically measurable property of a system.)

Equation (5.14) must be derived by a special treatment, which we omit. (See *Ballentine,* Section 12.3.) The derivation of (5.14) shows that Δt is to be interpreted as the lifetime of the state whose energy is uncertain by ΔE. It is often stated that Δt in (5.14) is the duration of the energy measurement. However, Aharonov and Bohm have shown that "energy can be measured reproducibly in an arbitrarily short time" [Y. Aharonov and D. Bohm, *Phys. Rev.,* **122**, 1649 (1961); **134**, B1417 (1964); see also S. Massar and S. Popescu, *Phys. Rev. A*, **71**, 042106 (2005); P. Busch, *The Time–Energy Uncertainty Relation*, arxiv.org/abs/quant-ph/0105049].

Now consider the possibility of simultaneously assigning definite values to *three* physical quantities: *A, B,* and *C.* Suppose

$$[\hat{A}, \hat{B}] = 0 \tag{5.15}$$

$$[\hat{A}, \hat{C}] = 0 \tag{5.16}$$

Is this enough to ensure that there exist simultaneous eigenfunctions of all three operators? Equation (5.15) ensures that we can construct a common set of eigenfunctions for $\hat{A}$ and $\hat{B}$; Eq. (5.16) ensures that we can construct a common set of eigenfunctions for $\hat{A}$ and $\hat{C}$. If these two sets of eigenfunctions are the same, then we will have a common set of eigenfunctions for all three operators. Hence we ask: Is the set of eigenfunctions of the linear operator $\hat{A}$ uniquely determined (apart from arbitrary multiplicative constants)? The answer is, in general, no. If there is more than one independent eigenfunction corresponding to an eigenvalue of $\hat{A}$ (that is, degeneracy), then any linear combination of the eigenfunctions of the degenerate eigenvalue is an eigenfunction of $\hat{A}$ (Section 3.6). It might well be that the proper linear combinations needed to give eigenfunctions of $\hat{B}$ would differ from the linear combinations that give eigenfunctions of $\hat{C}$. It turns out that, to have a common complete set of eigenfunctions of all three operators, we require that $[\hat{B}, \hat{C}] = 0$, in addition to (5.15) and (5.16). *To have a complete set of functions that are simultaneous eigenfunctions of several operators, each operator must commute with every other operator.*

5.2 VECTORS

In the next section we shall solve the eigenvalue problem for angular momentum, which is a vector property. We therefore first review vectors.

Physical properties (for example, mass, length, energy) that are completely specified by their magnitude are called **scalars**. Physical properties (for example, force, velocity, momentum) that require specification of both magnitude and direction are called **vectors**. A vector is represented by a directed line segment whose length and direction give the magnitude and direction of the property.

The sum of two vectors **A** and **B** is defined by the following procedure: Slide the first vector so that its tail touches the head of the second vector, keeping the direction of the first vector fixed. Then draw a new vector from the tail of the second vector to the head of the first vector. See Fig. 5.1. The product of a vector and a scalar, $c\mathbf{A}$, is defined as a vector of length $|c|$ times the length of **A** with the same direction as **A** if c is positive, or the opposite direction to **A** if c is negative.

To obtain an algebraic (as well as geometric) way of representing vectors, we set up Cartesian coordinates in space. We draw a vector of unit length directed along the

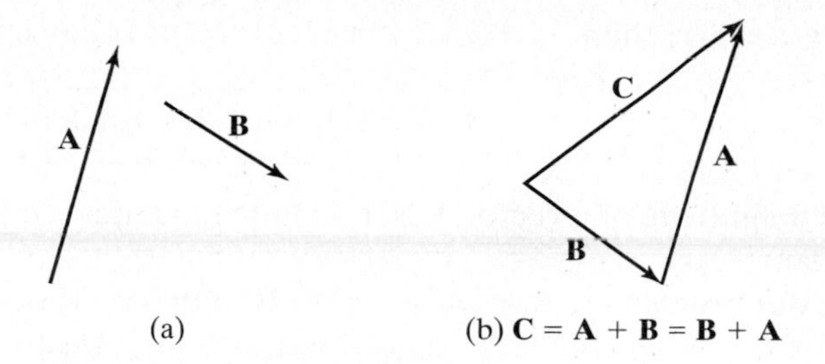

FIGURE 5.1 Addition of two vectors.

positive x axis and call it **i**. (No connection with $i = \sqrt{-1}$.) Unit vectors in the positive y and z directions are called **j** and **k** (Fig. 5.2). To represent any vector **A** in terms of the three unit vectors, we first slide **A** so that its tail is at the origin, preserving its direction during this process. We then find the projections of **A** on the x, y, and z axes: A_x, A_y, and A_z. From the definition of vector addition, it follows that (Fig. 5.2)

$$\boxed{\mathbf{A} = A_x\mathbf{i} + A_y\mathbf{j} + A_z\mathbf{k}} \tag{5.17}$$

To specify **A**, it is sufficient to specify its three components: (A_x, A_y, A_z). We can therefore define a vector in three-dimensional space as an ordered set of three numbers.

Two vectors **A** and **B** are equal if and only if all the corresponding components are equal: $A_x = B_x$, $A_y = B_y$, $A_z = B_z$. Hence a vector equation is equivalent to three scalar equations.

To add two vectors analytically, we add corresponding components:

$$\mathbf{A} + \mathbf{B} = A_x\mathbf{i} + A_y\mathbf{j} + A_z\mathbf{k} + B_x\mathbf{i} + B_y\mathbf{j} + B_z\mathbf{k}$$

$$\boxed{\mathbf{A} + \mathbf{B} = (A_x + B_x)\mathbf{i} + (A_y + B_y)\mathbf{j} + (A_z + B_z)\mathbf{k}} \tag{5.18}$$

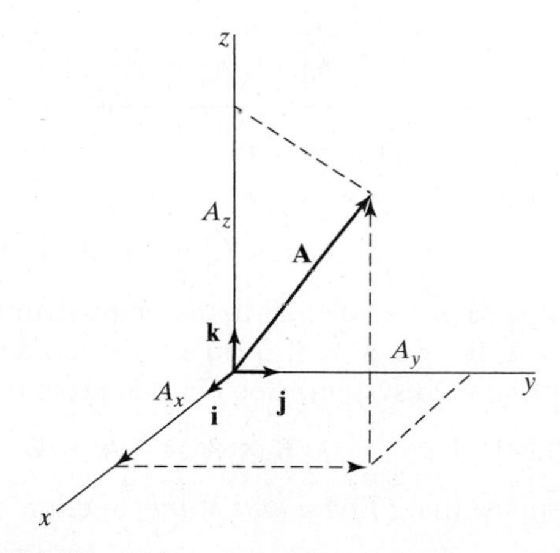

FIGURE 5.2 Unit vectors **i**, **j**, **k**, and components of **A**.

Also, if c is a scalar, then

$$\boxed{c\mathbf{A} = cA_x\mathbf{i} + cA_y\mathbf{j} + cA_z\mathbf{k}} \tag{5.19}$$

The **magnitude** of a vector **A** is its length and is denoted by A or $|\mathbf{A}|$. The quantity A is a scalar.

The **dot product** or *scalar product* $\mathbf{A} \cdot \mathbf{B}$ of two vectors is defined by

$$\boxed{\mathbf{A} \cdot \mathbf{B} = |\mathbf{A}||\mathbf{B}| \cos\theta = \mathbf{B} \cdot \mathbf{A}} \tag{5.20}$$

where θ is the angle between the vectors. The dot product, being the product of three scalars, is a scalar. Note that $|\mathbf{A}| \cos\theta$ is the projection of **A** on **B**. From the definition of vector addition, it follows that the projection of the vector $\mathbf{A} + \mathbf{B}$ on some vector **C** is the sum of the projections of **A** and of **B** on **C**. Hence

$$(\mathbf{A} + \mathbf{B}) \cdot \mathbf{C} = \mathbf{A} \cdot \mathbf{C} + \mathbf{B} \cdot \mathbf{C} \tag{5.21}$$

Since the three unit vectors **i**, **j**, and **k** are each of unit length and are mutually perpendicular, we have

$$\mathbf{i} \cdot \mathbf{i} = \mathbf{j} \cdot \mathbf{j} = \mathbf{k} \cdot \mathbf{k} = \cos 0 = 1, \qquad \mathbf{i} \cdot \mathbf{j} = \mathbf{j} \cdot \mathbf{k} = \mathbf{k} \cdot \mathbf{i} = \cos(\pi/2) = 0 \tag{5.22}$$

Using (5.22) and the distributive law (5.21), we have

$$\mathbf{A} \cdot \mathbf{B} = (A_x\mathbf{i} + A_y\mathbf{j} + A_z\mathbf{k}) \cdot (B_x\mathbf{i} + B_y\mathbf{j} + B_z\mathbf{k})$$

$$\boxed{\mathbf{A} \cdot \mathbf{B} = A_xB_x + A_yB_y + A_zB_z} \tag{5.23}$$

where six of the nine terms in the dot product are zero.

Equation (5.20) gives

$$\boxed{\mathbf{A} \cdot \mathbf{A} = |\mathbf{A}|^2} \tag{5.24}$$

Using (5.23), we therefore have

$$\boxed{|\mathbf{A}| = (A_x^2 + A_y^2 + A_z^2)^{1/2}} \tag{5.25}$$

For three-dimensional vectors, there is another type of product. The **cross product** or *vector product* $\mathbf{A} \times \mathbf{B}$ is a vector whose magnitude is

$$|\mathbf{A} \times \mathbf{B}| = |\mathbf{A}||\mathbf{B}| \sin\theta \tag{5.26}$$

whose line segment is perpendicular to the plane defined by **A** and **B**, and whose direction is such that **A**, **B**, and $\mathbf{A} \times \mathbf{B}$ form a right-handed system (just as the x, y, and z axes form a right-handed system). See Fig. 5.3. From the definition it follows that

$$\mathbf{B} \times \mathbf{A} = -\mathbf{A} \times \mathbf{B}$$

Also, it can be shown that (*Taylor and Mann,* Section 10.2)

$$\mathbf{A} \times (\mathbf{B} + \mathbf{C}) = \mathbf{A} \times \mathbf{B} + \mathbf{A} \times \mathbf{C} \tag{5.27}$$

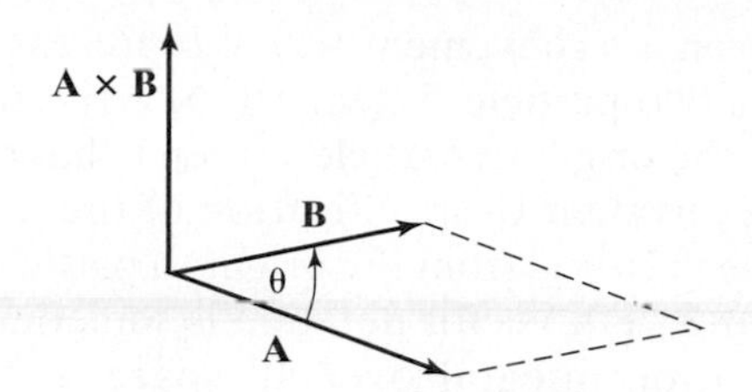

FIGURE 5.3 Cross product of two vectors.

For the three unit vectors, we have

$$\mathbf{i} \times \mathbf{i} = \mathbf{j} \times \mathbf{j} = \mathbf{k} \times \mathbf{k} = \sin 0 = 0$$

$$\mathbf{i} \times \mathbf{j} = \mathbf{k}, \quad \mathbf{j} \times \mathbf{i} = -\mathbf{k}, \quad \mathbf{j} \times \mathbf{k} = \mathbf{i}, \quad \mathbf{k} \times \mathbf{j} = -\mathbf{i}, \quad \mathbf{k} \times \mathbf{i} = \mathbf{j}, \quad \mathbf{i} \times \mathbf{k} = -\mathbf{j}$$

Using these equations and the distributive property (5.27), we find

$$\mathbf{A} \times \mathbf{B} = (A_x\mathbf{i} + A_y\mathbf{j} + A_z\mathbf{k}) \times (B_x\mathbf{i} + B_y\mathbf{j} + B_z\mathbf{k})$$

$$\mathbf{A} \times \mathbf{B} = (A_yB_z - A_zB_y)\mathbf{i} + (A_zB_x - A_xB_z)\mathbf{j} + (A_xB_y - A_yB_x)\mathbf{k}$$

As a memory aid, we can express the cross product as a determinant (see Section 8.3):

$$\mathbf{A} \times \mathbf{B} = \begin{vmatrix} \mathbf{i} & \mathbf{j} & \mathbf{k} \\ A_x & A_y & A_z \\ B_x & B_y & B_z \end{vmatrix} = \mathbf{i}\begin{vmatrix} A_y & A_z \\ B_y & B_z \end{vmatrix} - \mathbf{j}\begin{vmatrix} A_x & A_z \\ B_x & B_z \end{vmatrix} + \mathbf{k}\begin{vmatrix} A_x & A_y \\ B_x & B_y \end{vmatrix} \quad (5.28)$$

We define the vector operator **del** as

$$\boxed{\nabla \equiv \mathbf{i}\frac{\partial}{\partial x} + \mathbf{j}\frac{\partial}{\partial y} + \mathbf{k}\frac{\partial}{\partial z}} \quad \mathbf{(5.29)}$$

From Eq. (3.23), the operator for the linear-momentum vector is $\hat{\mathbf{p}} = -i\hbar\nabla$.

The **gradient** of a function $g(x, y, z)$ is defined as the result of operating on g with del:

$$\boxed{\mathbf{grad}\; g(x, y, z) \equiv \nabla g(x, y, z) \equiv \mathbf{i}\frac{\partial g}{\partial x} + \mathbf{j}\frac{\partial g}{\partial y} + \mathbf{k}\frac{\partial g}{\partial z}} \quad \mathbf{(5.30)}$$

The gradient of a scalar function is a vector function. The vector $\nabla g(x, y, z)$ represents the spatial rate of change of the function g; the x component of ∇g is the rate of change of g with respect to x, and so on. It can be shown that the vector ∇g points in the direction in which the rate of change of g is greatest. From Eq. (4.26), the relation between force and potential energy is

$$\mathbf{F} = -\nabla V(x, y, z) = -\mathbf{i}\frac{\partial V}{\partial x} - \mathbf{j}\frac{\partial V}{\partial y} - \mathbf{k}\frac{\partial V}{\partial z} \quad (5.31)$$

Suppose that the components of the vector **A** are each functions of some parameter t; $A_x = A_x(t)$, $A_y = A_y(t)$, $A_z = A_z(t)$. We define the derivative of **A** with respect to t as

$$\frac{d\mathbf{A}}{dt} = \mathbf{i}\frac{dA_x}{dt} + \mathbf{j}\frac{dA_y}{dt} + \mathbf{k}\frac{dA_z}{dt} \quad (5.32)$$

Vector notation is a convenient way to represent the variables of a function. The wave function of a two-particle system can be written as $\psi(x_1, y_1, z_1, x_2, y_2, z_2)$. If $\mathbf{r}_1$ is the vector from the origin to particle 1, then $\mathbf{r}_1$ has components x_1, y_1, z_1 and specification of $\mathbf{r}_1$ is equivalent to specification of the three coordinates x_1, y_1, z_1. The same is true for the vector $\mathbf{r}_2$ from the origin to particle 2. Therefore, we can write the wave function as $\psi(\mathbf{r}_1, \mathbf{r}_2)$. Vector notation is sometimes used in integrals. For example, the normalization integral over all space in Eq. (3.57) is often written as $\int \cdots \int |\Psi(\mathbf{r}_1, \ldots, \mathbf{r}_n, t)|^2 \, d\mathbf{r}_1 \cdots d\mathbf{r}_n$.

Vectors in *n*-Dimensional Space. The definition of a vector can be generalized to more than three dimensions. A vector $\mathbf{A}$ in three-dimensional space can be defined by its magnitude $|\mathbf{A}|$ and its direction, or can be defined by its three components (A_x, A_y, A_z) in a Cartesian coordinate system. Therefore, we can define a three-dimensional vector as a set of three real numbers (A_x, A_y, A_z) in a particular order. A vector $\mathbf{B}$ in an n-dimensional real vector "space" (sometimes called a hyperspace) is defined as an ordered set of n real numbers $(B_1, B_2, \ldots, B_n)$, where $B_1, B_2, \ldots, B_n$ are the **components** of $\mathbf{B}$. Don't be concerned that you can't visualize vectors in an n-dimensional space.

The variables of a function are often denoted using n-dimensional vector notation. For example, instead of writing the wave function of a two-particle system as $\psi(\mathbf{r}_1, \mathbf{r}_2)$, we can define a six-dimensional vector $\mathbf{q}$ whose components are $q_1 = x_1$, $q_2 = y_1, q_3 = z_1, q_4 = x_2, q_5 = y_2, q_6 = z_2$ and write the wave function as $\psi(\mathbf{q})$. For an n-particle system, we can define $\mathbf{q}$ to have $3n$ components and write the wave function as $\psi(\mathbf{q})$ and the normalization integral over all space as $\int |\psi(\mathbf{q})|^2 \, d\mathbf{q}$.

The theory of searching for the equilibrium geometry of a molecule uses n-dimensional vectors (Section 15.10). The rest of Section 5.2 is relevant to Section 15.10 and need not be read until you study Section 15.10.

Two n-dimensional vectors are **equal** if all their corresponding components are equal; $\mathbf{B} = \mathbf{C}$ if and only if $B_1 = C_1, B_2 = C_2, \ldots, B_n = C_n$. Thus, in n-dimensional space, a vector equation is equivalent to n scalar equations. The **sum** of two n-dimensional vectors $\mathbf{B}$ and $\mathbf{D}$ is defined as the vector $(B_1 + D_1, B_2 + D_2, \ldots, B_n + D_n)$. The difference is defined similarly. The vector $k\mathbf{B}$ is defined as the vector $(kB_1, kB_2, \ldots, kB_n)$, where k is a scalar. In three-dimensional space the vectors $k\mathbf{A}$, where $k > 0$, all lie in the same direction. In n-dimensional space the vectors $k\mathbf{B}$ all lie in the same direction. Just as the numbers (A_x, A_y, A_z) define a point in three-dimensional space, the numbers $(B_1, B_2, \ldots, B_n)$ define a **point** in n-dimensional space.

The **length** (or **magnitude** or **Euclidean norm**) $|\mathbf{B}|$ (sometimes denoted $\|\mathbf{B}\|$) of an n-dimensional real vector is defined as

$$|\mathbf{B}| \equiv (\mathbf{B} \cdot \mathbf{B})^{1/2} = (B_1^2 + B_2^2 + \cdots + B_n^2)^{1/2}$$

A vector whose length is 1 is said to be **normalized**.

The **inner product** (or **scalar product**) $\mathbf{B} \cdot \mathbf{G}$ of two real n-dimensional vectors $\mathbf{B}$ and $\mathbf{G}$ is defined as the scalar

$$\mathbf{B} \cdot \mathbf{G} \equiv B_1 G_1 + B_2 G_2 + \cdots + B_n G_n$$

If $\mathbf{B} \cdot \mathbf{G} = 0$, the vectors **B** and **G** are said to be **orthogonal**. The cosine of the angle θ between two n-dimensional vectors **B** and **C** is defined by analogy to (5.20) as $\cos\theta \equiv \mathbf{B} \cdot \mathbf{C}/|\mathbf{B}||\mathbf{C}|$. One can show that this definition makes $\cos\theta$ lie in the range -1 to 1.

In three-dimensional space, the unit vectors $\mathbf{i} = (1, 0, 0)$, $\mathbf{j} = (0, 1, 0)$, $\mathbf{k} = (0, 0, 1)$ are mutually perpendicular. Also, any vector can be written as a linear combination of these three vectors [Eq. (5.17)]. In an n-dimensional real vector space, the unit vectors $\mathbf{e}_1 \equiv (1, 0, 0, \ldots, 0)$, $\mathbf{e}_2 \equiv (0, 1, 0, \ldots, 0), \ldots, \mathbf{e}_n \equiv (0, 0, 0, \ldots, 1)$ are mutually orthogonal. Since the n-dimensional vector **B** equals $B_1\mathbf{e}_1 + B_2\mathbf{e}_2 + \cdots + B_n\mathbf{e}_n$, any n-dimensional real vector can be written as a linear combination of the n unit vectors $\mathbf{e}_1, \mathbf{e}_2, \ldots, \mathbf{e}_n$. This set of n vectors is therefore said to be a **basis** for the n-dimensional real vector space. Since the vectors $\mathbf{e}_1, \mathbf{e}_2, \ldots, \mathbf{e}_n$ are orthogonal and normalized, they are an **orthonormal** basis for the real vector space. The scalar product $\mathbf{B} \cdot \mathbf{e}_i$ gives the component of **B** in the direction of the basis vector $\mathbf{e}_i$. A vector space has many possible basis sets. Any set of n linearly independent real vectors can serve as a basis for the n-dimensional real vector space.

A three-dimensional vector can be specified by its three components or by its length and its direction. The direction can be specified by giving the three angles the vector makes with the positive halves of the *x, y,* and *z* axes. These angles are the **direction angles** of the vector and lie in the range 0 to 180°. However, the direction angle with the *z* axis is fixed once the other two direction angles have been given, so only two direction angles are independent. Thus a three-dimensional vector can be specified by its length and two direction angles. Similarly, in n-dimensional space, the direction angles between a vector and each unit vector $\mathbf{e}_1, \mathbf{e}_2, \ldots, \mathbf{e}_n$ can be found from the above formula for the cosine of the angle between two vectors. An n-dimensional vector can thus be specified by its length and $n - 1$ direction angles.

The gradient of a function of three variables is defined by (5.30). The **gradient** ∇f of a function $f(q_1, q_2, \ldots, q_n)$ of n variables is defined as the n-dimensional vector whose components are the first partial derivatives of *f*:

$$\nabla f \equiv (\partial f/\partial q_1)\mathbf{e}_1 + (\partial f/\partial q_2)\mathbf{e}_2 + \cdots + (\partial f/\partial q_n)\mathbf{e}_n$$

We have considered real, n-dimensional vector spaces. Dirac's formulation of quantum mechanics uses a complex, infinite-dimensional vector space, discussion of which is omitted.

5.3 ANGULAR MOMENTUM OF A ONE-PARTICLE SYSTEM

In Section 3.3 we found the eigenfunctions and eigenvalues for the linear-momentum operator $\hat{p}_x$. In this section we consider the same problem for the angular momentum of a particle. Angular momentum is important in the quantum mechanics of atomic structure. We begin by reviewing the classical mechanics of angular momentum.

Classical Mechanics of One-Particle Angular Momentum. Consider a moving particle of mass *m*. We set up a Cartesian coordinate system that is fixed in space. Let **r** be the vector from the origin to the instantaneous position of the particle. We have

$$\mathbf{r} = \mathbf{i}x + \mathbf{j}y + \mathbf{k}z \tag{5.33}$$

where x, y, and z are the particle's coordinates at a given instant. These coordinates are functions of time, and defining the velocity vector **v** as the time derivative of the position vector, we have [Eq. (5.32)]

$$\mathbf{v} \equiv \frac{d\mathbf{r}}{dt} = \mathbf{i}\frac{dx}{dt} + \mathbf{j}\frac{dy}{dt} + \mathbf{k}\frac{dz}{dt} \tag{5.34}$$

$$v_x = dx/dt, \quad v_y = dy/dt, \quad v_z = dz/dt$$

We define the particle's **linear momentum** vector **p** by

$$\boxed{\mathbf{p} \equiv m\mathbf{v}} \tag{5.35}$$

$$p_x = mv_x, \quad p_y = mv_y, \quad p_z = mv_z \tag{5.36}$$

The particle's **angular momentum L** with respect to the coordinate origin is defined in classical mechanics as

$$\boxed{\mathbf{L} \equiv \mathbf{r} \times \mathbf{p}} \tag{5.37}$$

$$L = \begin{vmatrix} \mathbf{i} & \mathbf{j} & \mathbf{k} \\ x & y & z \\ p_x & p_y & p_z \end{vmatrix} \tag{5.38}$$

$$L_x = yp_z - zp_y, \quad L_y = zp_x - xp_z, \quad L_z = xp_y - yp_x \tag{5.39}$$

where (5.28) was used. L_x, L_y, and L_z are the components of **L** along the x, y, and z axes. The angular-momentum vector **L** is perpendicular to the plane defined by the particle's position vector **r** and its velocity **v** (Fig. 5.4).

The *torque* $\boldsymbol{\tau}$ acting on a particle is defined as the cross product of **r** and the force **F** acting on the particle: $\boldsymbol{\tau} \equiv \mathbf{r} \times \mathbf{F}$. One can show that (*Halliday and Resnick,* Section 12-3) $\boldsymbol{\tau} = d\mathbf{L}/dt$. When no torque acts on a particle, the rate of change of its angular momentum is zero; that is, its angular momentum is constant (or conserved). For a planet orbiting the sun, the gravitational force is radially directed. Since the cross product of two parallel vectors is zero, there is no torque on the planet and its angular momentum is conserved.

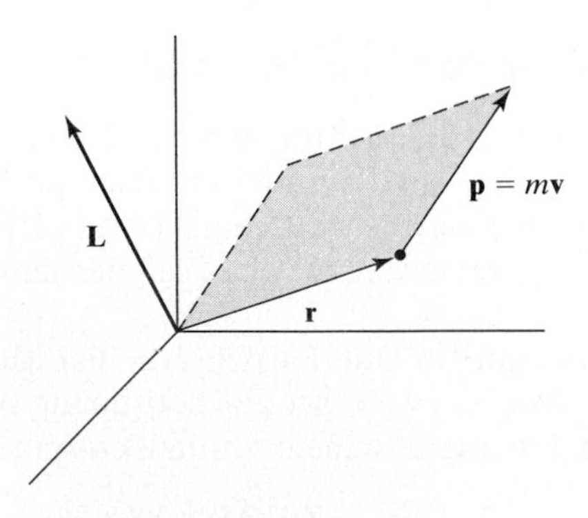

FIGURE 5.4 $\mathbf{L} = \mathbf{r} \times \mathbf{p}$.

One-Particle Orbital-Angular-Momentum Operators. Now for the quantum-mechanical treatment. In quantum mechanics, there are two kinds of angular momenta: **Orbital angular momentum** results from the motion of a particle through space, and is the analog of the classical-mechanical quantity **L**; **spin angular momentum** (Chapter 10) is an intrinsic property of many microscopic particles and has no classical-mechanical analog. We are now considering only orbital angular momentum. We get the quantum-mechanical operators for the components of orbital angular momentum of a particle by replacing the coordinates and momenta in the classical equations (5.39) by their corresponding operators [Eqs. (3.21)–(3.23)]. We find

$$\hat{L}_x = -i\hbar\left(y\frac{\partial}{\partial z} - z\frac{\partial}{\partial y}\right) \tag{5.40}$$

$$\hat{L}_y = -i\hbar\left(z\frac{\partial}{\partial x} - x\frac{\partial}{\partial z}\right) \tag{5.41}$$

$$\hat{L}_z = -i\hbar\left(x\frac{\partial}{\partial y} - y\frac{\partial}{\partial x}\right) \tag{5.42}$$

(Since $\hat{y}\,\hat{p}_z = \hat{p}_z\hat{y}$, and so on, we do not run into any problems of noncommutativity in constructing these operators.) Using

$$\hat{L}^2 = |\hat{\mathbf{L}}|^2 = \hat{\mathbf{L}}\cdot\hat{\mathbf{L}} = \hat{L}_x^2 + \hat{L}_y^2 + \hat{L}_z^2 \tag{5.43}$$

we can construct the operator for the square of the angular-momentum magnitude from the operators in (5.40)–(5.42).

Since the commutation relations determine which physical quantities can be simultaneously assigned definite values, we investigate these relations for angular momentum. Operating on some function $f(x, y, z)$ with $\hat{L}_y$, we have

$$\hat{L}_y f = -i\hbar\left(z\frac{\partial f}{\partial x} - x\frac{\partial f}{\partial z}\right)$$

Operating on this last equation with $\hat{L}_x$, we get

$$\hat{L}_x\hat{L}_y f = -\hbar^2\left(y\frac{\partial f}{\partial x} + yz\frac{\partial^2 f}{\partial z\,\partial x} - yx\frac{\partial^2 f}{\partial z^2} - z^2\frac{\partial^2 f}{\partial y\,\partial x} + zx\frac{\partial^2 f}{\partial y\,\partial z}\right) \tag{5.44}$$

Similarly,

$$\hat{L}_x f = -i\hbar\left(y\frac{\partial f}{\partial z} - z\frac{\partial f}{\partial y}\right)$$

$$\hat{L}_y\hat{L}_x f = -\hbar^2\left(zy\frac{\partial^2 f}{\partial x\,\partial z} - z^2\frac{\partial^2 f}{\partial x\,\partial y} - xy\frac{\partial^2 f}{\partial z^2} + x\frac{\partial f}{\partial y} + xz\frac{\partial^2 f}{\partial z\,\partial y}\right) \tag{5.45}$$

Subtracting (5.45) from (5.44), we have

$$\hat{L}_x\hat{L}_y f - \hat{L}_y\hat{L}_x f = -\hbar^2\left(y\frac{\partial f}{\partial x} - x\frac{\partial f}{\partial y}\right)$$

$$[\hat{L}_x, \hat{L}_y] = i\hbar\hat{L}_z \tag{5.46}$$

where we used relations such as

$$\boxed{\frac{\partial^2 f}{\partial z\,\partial x} = \frac{\partial^2 f}{\partial x\,\partial z}} \tag{5.47}$$

which are true for well-behaved functions. We could use the same procedure to find $[\hat{L}_y, \hat{L}_z]$ and $[\hat{L}_z, \hat{L}_x]$, but we can save time by noting a certain kind of symmetry in (5.40)–(5.42). By a *cyclic permutation* of x, y, and z, we mean replacing x by y, replacing y by z, and replacing z by x. If we carry out a cyclic permutation in $\hat{L}_x$, we get $\hat{L}_y$; a cyclic permutation in $\hat{L}_y$ gives $\hat{L}_z$; and $\hat{L}_z$ is transformed into $\hat{L}_x$ by a cyclic permutation. Hence, by carrying out two successive cyclic permutations on (5.46), we get

$$[\hat{L}_y, \hat{L}_z] = i\hbar\hat{L}_x, \quad [\hat{L}_z, \hat{L}_x] = i\hbar\hat{L}_y \tag{5.48}$$

Now we evaluate the commutators of $\hat{L}^2$ with each of its components, using commutator identities of Section 5.1.

$$\begin{aligned}
[\hat{L}^2, \hat{L}_x] &= [\hat{L}_x^2 + \hat{L}_y^2 + \hat{L}_z^2, \hat{L}_x] \\
&= [\hat{L}_x^2, \hat{L}_x] + [\hat{L}_y^2, \hat{L}_x] + [\hat{L}_z^2, \hat{L}_x] \\
&= [\hat{L}_y^2, \hat{L}_x] + [\hat{L}_z^2, \hat{L}_x] \\
&= [\hat{L}_y, \hat{L}_x]\hat{L}_y + \hat{L}_y[\hat{L}_y, \hat{L}_x] + [\hat{L}_z, \hat{L}_x]\hat{L}_z + \hat{L}_z[\hat{L}_z, \hat{L}_x] \\
&= -i\hbar\hat{L}_z\hat{L}_y - i\hbar\hat{L}_y\hat{L}_z + i\hbar\hat{L}_y\hat{L}_z + i\hbar\hat{L}_z\hat{L}_y
\end{aligned}$$

$$\boxed{[\hat{L}^2, \hat{L}_x] = 0} \tag{5.49}$$

Since a cyclic permutation of x, y, and z leaves $\hat{L}^2 = \hat{L}_x^2 + \hat{L}_y^2 + \hat{L}_z^2$ unchanged, if we carry out two such permutations on (5.49), we get

$$\boxed{[\hat{L}^2, \hat{L}_y] = 0, \quad [\hat{L}^2, \hat{L}_z] = 0} \tag{5.50}$$

To which of the quantities L^2, L_x, L_y, L_z can we assign definite values simultaneously? Because $\hat{L}^2$ commutes with each of its components, we can specify an exact value for L^2 and any *one* component. However, no two components of $\hat{\mathbf{L}}$ commute with each other, so we cannot specify more than one component simultaneously. (There is one exception to this statement, which will be discussed shortly.) It is traditional to take L_z as the component of angular momentum that will be specified along with L^2. Note that in specifying $L^2 = |\mathbf{L}|^2$ we are not specifying the vector $\mathbf{L}$, only its magnitude. A complete specification of $\mathbf{L}$ requires simultaneous specification of each of its three components, which we usually cannot do. In classical mechanics when angular momentum is conserved, each of its three components has a definite value. In quantum mechanics when angular momentum is conserved, only its magnitude and one of its components are specifiable.

We could now try to find the eigenvalues and common eigenfunctions of $\hat{L}^2$ and $\hat{L}_z$ by using the forms for these operators in Cartesian coordinates. However, we would find that the partial differential equations obtained would not be separable. For this reason we carry out a transformation to **spherical coordinates** (Fig. 5.5). The coordinate r is the distance from the origin to the point (x, y, z). The angle θ is the angle the vector $\mathbf{r}$ makes with the positive z axis. The angle that the projection of $\mathbf{r}$ in

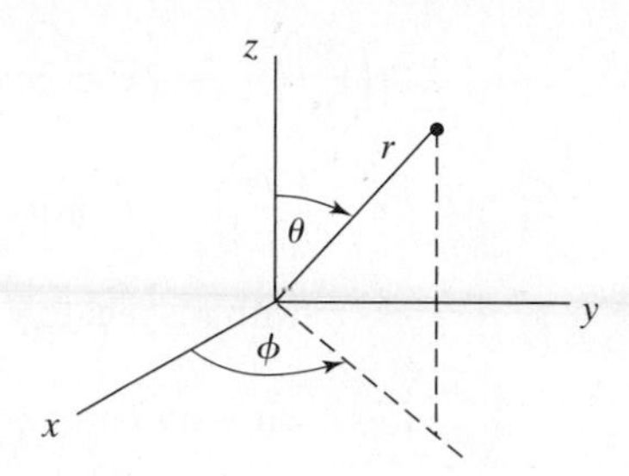

FIGURE 5.5 Spherical coordinates.

the xy plane makes with the positive x axis is ϕ. (Mathematics texts often interchange θ and ϕ.) A little trigonometry gives

$$x = r\sin\theta\cos\phi, \qquad y = r\sin\theta\sin\phi, \qquad z = r\cos\theta \tag{5.51}$$

$$r^2 = x^2 + y^2 + z^2, \qquad \cos\theta = \frac{z}{(x^2 + y^2 + z^2)^{1/2}}, \qquad \tan\phi = y/x \tag{5.52}$$

To transform the angular-momentum operators to spherical coordinates, we must transform $\partial/\partial x$, $\partial/\partial y$, and $\partial/\partial z$ into these coordinates. [This transformation may be skimmed if desired. Begin reading again after Eq. (5.64).]

To perform this transformation, we use the *chain rule.* Suppose we have a function of r, θ, and ϕ: $f(r, \theta, \phi)$. If we change the independent variables by substituting

$$r = r(x, y, z), \quad \theta = \theta(x, y, z), \quad \phi = \phi(x, y, z)$$

into f, we transform it into a function of x, y, and z:

$$f[r(x, y, z), \theta(x, y, z), \phi(x, y, z)] = g(x, y, z)$$

For example, suppose that $f(r, \theta, \phi) = 3r\cos\theta + 2\tan^2\phi$. Using (5.52), we have $g(x, y, z) = 3z + 2y^2x^{-2}$.

The chain rule tells us how the partial derivatives of $g(x, y, z)$ are related to those of $f(r, \theta, \phi)$. In fact,

$$\left(\frac{\partial g}{\partial x}\right)_{y,z} = \left(\frac{\partial f}{\partial r}\right)_{\theta,\phi}\left(\frac{\partial r}{\partial x}\right)_{y,z} + \left(\frac{\partial f}{\partial \theta}\right)_{r,\phi}\left(\frac{\partial \theta}{\partial x}\right)_{y,z} + \left(\frac{\partial f}{\partial \phi}\right)_{r,\theta}\left(\frac{\partial \phi}{\partial x}\right)_{y,z} \tag{5.53}$$

$$\left(\frac{\partial g}{\partial y}\right)_{x,z} = \left(\frac{\partial f}{\partial r}\right)_{\theta,\phi}\left(\frac{\partial r}{\partial y}\right)_{x,z} + \left(\frac{\partial f}{\partial \theta}\right)_{r,\phi}\left(\frac{\partial \theta}{\partial y}\right)_{x,z} + \left(\frac{\partial f}{\partial \phi}\right)_{r,\theta}\left(\frac{\partial \phi}{\partial y}\right)_{x,z} \tag{5.54}$$

$$\left(\frac{\partial g}{\partial z}\right)_{x,y} = \left(\frac{\partial f}{\partial r}\right)_{\theta,\phi}\left(\frac{\partial r}{\partial z}\right)_{x,y} + \left(\frac{\partial f}{\partial \theta}\right)_{r,\phi}\left(\frac{\partial \theta}{\partial z}\right)_{x,y} + \left(\frac{\partial f}{\partial \phi}\right)_{r,\theta}\left(\frac{\partial \phi}{\partial z}\right)_{x,y} \tag{5.55}$$

To convert these equations to operator equations, we delete f and g to give

$$\frac{\partial}{\partial x} = \left(\frac{\partial r}{\partial x}\right)_{y,z}\frac{\partial}{\partial r} + \left(\frac{\partial \theta}{\partial x}\right)_{y,z}\frac{\partial}{\partial \theta} + \left(\frac{\partial \phi}{\partial x}\right)_{y,z}\frac{\partial}{\partial \phi} \tag{5.56}$$

with similar equations for $\partial/\partial y$ and $\partial/\partial z$. The task now is to evaluate the partial derivatives such as $(\partial r/\partial x)_{y,z}$. Taking the partial derivative of the first equation in (5.52) with respect to x at constant y and z, we have

$$2r\left(\frac{\partial r}{\partial x}\right)_{y,z} = 2x = 2r\sin\theta\cos\phi$$

$$\left(\frac{\partial r}{\partial x}\right)_{y,z} = \sin\theta\cos\phi \tag{5.57}$$

Differentiating $r^2 = x^2 + y^2 + z^2$ with respect to y and with respect to z, we find

$$\left(\frac{\partial r}{\partial y}\right)_{x,z} = \sin\theta\sin\phi, \qquad \left(\frac{\partial r}{\partial z}\right)_{x,y} = \cos\theta \tag{5.58}$$

From the second equation in (5.52), we find

$$-\sin\theta\left(\frac{\partial \theta}{\partial x}\right)_{y,z} = -\frac{xz}{r^3}$$

$$\left(\frac{\partial \theta}{\partial x}\right)_{y,z} = \frac{\cos\theta\cos\phi}{r} \tag{5.59}$$

Also,

$$\left(\frac{\partial \theta}{\partial y}\right)_{x,z} = \frac{\cos\theta\sin\phi}{r}, \qquad \left(\frac{\partial \theta}{\partial z}\right)_{x,y} = -\frac{\sin\theta}{r} \tag{5.60}$$

From $\tan\phi = y/x$, we find

$$\left(\frac{\partial \phi}{\partial x}\right)_{y,z} = -\frac{\sin\phi}{r\sin\theta}, \qquad \left(\frac{\partial \phi}{\partial y}\right)_{x,z} = \frac{\cos\phi}{r\sin\theta}, \qquad \left(\frac{\partial \phi}{\partial z}\right)_{x,y} = 0 \tag{5.61}$$

Substituting (5.57), (5.59), and (5.61) into (5.56), we find

$$\frac{\partial}{\partial x} = \sin\theta\cos\phi\frac{\partial}{\partial r} + \frac{\cos\theta\cos\phi}{r}\frac{\partial}{\partial\theta} - \frac{\sin\phi}{r\sin\theta}\frac{\partial}{\partial\phi} \tag{5.62}$$

Similarly,

$$\frac{\partial}{\partial y} = \sin\theta\sin\phi\frac{\partial}{\partial r} + \frac{\cos\theta\sin\phi}{r}\frac{\partial}{\partial\theta} + \frac{\cos\phi}{r\sin\theta}\frac{\partial}{\partial\phi} \tag{5.63}$$

$$\frac{\partial}{\partial z} = \cos\theta\frac{\partial}{\partial r} - \frac{\sin\theta}{r}\frac{\partial}{\partial\theta} \tag{5.64}$$

At last, we are ready to express the angular-momentum components in spherical coordinates. Substituting (5.51), (5.63), and (5.64) into (5.40), we have

$$\hat{L}_x = -i\hbar\left[r\sin\theta\sin\phi\left(\cos\theta\frac{\partial}{\partial r} - \frac{\sin\theta}{r}\frac{\partial}{\partial\theta}\right)\right.$$
$$\left. - r\cos\theta\left(\sin\theta\sin\phi\frac{\partial}{\partial r} + \frac{\cos\theta\sin\phi}{r}\frac{\partial}{\partial\theta} + \frac{\cos\phi}{r\sin\theta}\frac{\partial}{\partial\phi}\right)\right]$$

$$\hat{L}_x = i\hbar\left(\sin\phi\frac{\partial}{\partial\theta} + \cot\theta\cos\phi\frac{\partial}{\partial\phi}\right) \tag{5.65}$$

Also, we find

$$\hat{L}_y = -i\hbar\left(\cos\phi\frac{\partial}{\partial\theta} - \cot\theta\sin\phi\frac{\partial}{\partial\phi}\right) \tag{5.66}$$

$$\hat{L}_z = -i\hbar\frac{\partial}{\partial\phi} \tag{5.67}$$

By squaring each of $\hat{L}_x$, $\hat{L}_y$, and $\hat{L}_z$ and then adding their squares, we can construct $\hat{L}^2 = \hat{L}_x^2 + \hat{L}_y^2 + \hat{L}_z^2$ [Eq. (5.43)]. The result is (Prob. 5.16)

$$\hat{L}^2 = -\hbar^2\left(\frac{\partial^2}{\partial\theta^2} + \cot\theta\frac{\partial}{\partial\theta} + \frac{1}{\sin^2\theta}\frac{\partial^2}{\partial\phi^2}\right) \tag{5.68}$$

Although the angular-momentum operators depend on all three Cartesian coordinates, x, y, and z, they involve only the two spherical coordinates θ and ϕ.

One-Particle Orbital-Angular-Momentum Eigenfunctions and Eigenvalues. We now find the common eigenfunctions of $\hat{L}^2$ and $\hat{L}_z$, which we denote by Y. Since these operators involve only θ and ϕ, Y is a function of these two coordinates: $Y = Y(\theta, \phi)$. (Of course, since the operators are linear, we can multiply Y by an arbitrary function of r and still have an eigenfunction of $\hat{L}^2$ and $\hat{L}_z$.) We must solve

$$\hat{L}_z Y(\theta, \phi) = bY(\theta, \phi) \tag{5.69}$$

$$\hat{L}^2 Y(\theta, \phi) = cY(\theta, \phi) \tag{5.70}$$

where b and c are the eigenvalues of $\hat{L}_z$ and $\hat{L}^2$.

Using the $\hat{L}_z$ operator, we have

$$-i\hbar\frac{\partial}{\partial\phi}Y(\theta, \phi) = bY(\theta, \phi) \tag{5.71}$$

Since the operator in (5.71) does not involve θ, we try a separation of variables, writing

$$Y(\theta, \phi) = S(\theta)T(\phi) \tag{5.72}$$

Equation (5.71) becomes

$$-i\hbar\frac{\partial}{\partial\phi}[S(\theta)T(\phi)] = bS(\theta)T(\phi)$$

$$-i\hbar S(\theta)\frac{dT(\phi)}{d\phi} = bS(\theta)T(\phi)$$

$$\frac{dT(\phi)}{T(\phi)} = \frac{ib}{\hbar}\,d\phi$$

$$T(\phi) = Ae^{ib\phi/\hbar} \tag{5.73}$$

where A is an arbitrary constant.

Is T suitable as an eigenfunction? The answer is no, since it is not, in general, a single-valued function. If we add 2π to ϕ, we will still be at the same point in space, and hence we want no change in T when this is done. For T to be single-valued, we have the restriction

$$T(\phi + 2\pi) = T(\phi)$$

$$Ae^{ib\phi/\hbar}e^{ib2\pi/\hbar} = Ae^{ib\phi/\hbar}$$

$$e^{ib2\pi/\hbar} = 1 \tag{5.74}$$

To satisfy $e^{i\alpha} = \cos\alpha + i\sin\alpha = 1$, we must have $\alpha = 2\pi m$, where

$$m = 0, \pm1, \pm2, \pm\cdots$$

Therefore, (5.74) gives

$$2\pi b/\hbar = 2\pi m$$

$$b = m\hbar, \qquad m = \dots -2, -1, 0, 1, 2, \dots \tag{5.75}$$

and (5.73) becomes

$$T(\phi) = Ae^{im\phi}, \quad m = 0, \pm 1, \pm 2, \dots \tag{5.76}$$

The eigenvalues for the z component of angular momentum are quantized.

We fix A by normalizing T. First let us consider normalizing some function F of r, θ, and ϕ. The ranges of the independent variables are (see Fig. 5.5)

$$\boxed{0 \le r \le \infty, \qquad 0 \le \theta \le \pi, \qquad 0 \le \phi \le 2\pi} \tag{5.77}$$

The infinitesimal volume element in spherical coordinates is (*Taylor and Mann*, Section 13.9)

$$\boxed{d\tau = r^2 \sin\theta \, dr \, d\theta \, d\phi} \tag{5.78}$$

The quantity (5.78) is the volume of an infinitesimal region of space for which the spherical coordinates lie in the ranges r to $r + dr$, θ to $\theta + d\theta$, and ϕ to $\phi + d\phi$. The normalization condition for F in spherical coordinates is therefore

$$\int_0^\infty \left[\int_0^\pi \left[\int_0^{2\pi} |F(r, \theta, \phi)|^2 \, d\phi \right] \sin\theta \, d\theta \right] r^2 \, dr = 1 \tag{5.79}$$

If F happens to have the form

$$F(r, \theta, \phi) = R(r)S(\theta)T(\phi)$$

then use of the integral identity (3.74) gives for (5.79)

$$\int_0^\infty |R(r)|^2 r^2 \, dr \int_0^\pi |S(\theta)|^2 \sin\theta \, d\theta \int_0^{2\pi} |T(\phi)|^2 \, d\phi = 1$$

and it is convenient to normalize each factor of F separately:

$$\int_0^\infty |R|^2 r^2 \, dr = 1, \qquad \int_0^\pi |S|^2 \sin\theta \, d\theta = 1, \qquad \int_0^{2\pi} |T|^2 \, d\phi = 1 \tag{5.80}$$

Therefore,

$$\int_0^{2\pi} (Ae^{im\phi})^* Ae^{im\phi} \, d\phi = 1 = |A|^2 \int_0^{2\pi} d\phi$$

$$|A| = (2\pi)^{-1/2}$$

$$T(\phi) = \frac{1}{\sqrt{2\pi}} e^{im\phi}, \qquad m = 0, \pm 1, \pm 2, \dots \tag{5.81}$$

We now solve $\hat{L}^2 Y = cY$ [Eq. (5.70)] for the eigenvalues c of $\hat{L}^2$. Using (5.68) for $\hat{L}^2$, (5.72) for Y, and (5.81), we have

$$-\hbar^2\left(\frac{\partial^2}{\partial\theta^2} + \cot\theta\frac{\partial}{\partial\theta} + \frac{1}{\sin^2\theta}\frac{\partial^2}{\partial\phi^2}\right)\left(S(\theta)\frac{1}{\sqrt{2\pi}}e^{im\phi}\right) = cS(\theta)\frac{1}{\sqrt{2\pi}}e^{im\phi}$$

$$\frac{d^2S}{d\theta^2} + \cot\theta\frac{dS}{d\theta} - \frac{m^2}{\sin^2\theta}S = -\frac{c}{\hbar^2}S \tag{5.82}$$

To solve (5.82), we carry out some tedious manipulations, which may be skimmed if desired. Begin reading again at Eq. (5.91). First, for convenience, we change the independent variable by making the substitution

$$w = \cos\theta \tag{5.83}$$

This transforms S into some new function of w:

$$S(\theta) = G(w) \tag{5.84}$$

The chain rule gives

$$\frac{dS}{d\theta} = \frac{dG}{dw}\frac{dw}{d\theta} = -\sin\theta\frac{dG}{dw} = -(1 - w^2)^{1/2}\frac{dG}{dw} \tag{5.85}$$

Similarly, we find (Prob. 5.24)

$$\frac{d^2S}{d\theta^2} = (1 - w^2)\frac{d^2G}{dw^2} - w\frac{dG}{dw} \tag{5.86}$$

Using (5.86), (5.85), and $\cot\theta = \cos\theta/\sin\theta = w/(1 - w^2)^{1/2}$, we find that (5.82) becomes

$$(1 - w^2)\frac{d^2G}{dw^2} - 2w\frac{dG}{dw} + \left[\frac{c}{\hbar^2} - \frac{m^2}{1 - w^2}\right]G(w) = 0 \tag{5.87}$$

The range of w is $-1 \le w \le 1$.

To get a two-term recursion relation when we try a power-series solution, we make the following change of dependent variable:

$$G(w) = (1 - w^2)^{|m|/2}H(w) \tag{5.88}$$

Differentiating (5.88), we evaluate G' and G'', and (5.87) becomes, after we divide by $(1 - w^2)^{|m|/2}$,

$$(1 - w^2)H'' - 2(|m| + 1)wH' + [c\hbar^{-2} - |m|(|m| + 1)]H = 0 \tag{5.89}$$

We now try a power series for H:

$$H(w) = \sum_{j=0}^{\infty} a_j w^j \tag{5.90}$$

Differentiating [compare Eqs. (4.38)–(4.40)], we have

$$H'(w) = \sum_{j=0}^{\infty} ja_j w^{j-1}$$

$$H''(w) = \sum_{j=0}^{\infty} j(j-1)a_j w^{j-2} = \sum_{j=0}^{\infty} (j+2)(j+1)a_{j+2}w^j$$

Substitution of these power series into (5.89) yields, after combining sums,

$$\sum_{j=0}^{\infty}\left[(j+2)(j+1)a_{j+2} + \left(-j^2 - j - 2|m|j + \frac{c}{\hbar^2} - |m|^2 - |m|\right)a_j\right]w^j = 0$$

Setting the coefficient of w^j equal to zero, we get the recursion relation

$$a_{j+2} = \frac{[(j + |m|)(j + |m| + 1) - c/\hbar^2]}{(j + 1)(j + 2)} a_j \tag{5.91}$$

Just as in the harmonic-oscillator case, the general solution of (5.89) is an arbitrary linear combination of a series of even powers (whose coefficients are determined by a_0) and a series of odd powers (whose coefficients are determined by a_1). It can be shown that the infinite series defined by the recursion relation (5.91) does not give well-behaved eigenfunctions. [Many texts point out that the infinite series diverges at $w = \pm 1$. However, this is not sufficient cause to reject the infinite series, since the eigenfunctions might be quadratically integrable, even though infinite at two points. For a careful discussion, see M. Whippman, *Am. J. Phys.*, **34**, 656 (1966).] Hence, as in the harmonic-oscillator case, we must cause one of the series to break off, its last term being $a_k w^k$. We eliminate the other series by setting a_0 or a_1 equal to zero, depending on whether k is odd or even.

Setting the coefficient of a_k in (5.91) equal to zero, we have

$$c = \hbar^2(k + |m|)(k + |m| + 1), \qquad k = 0, 1, 2, \ldots \tag{5.92}$$

Since $|m|$ takes on the values $0, 1, 2, \ldots$, the quantity $k + |m|$ takes on the values $0, 1, 2, \ldots$. We therefore define the quantum number l as

$$l \equiv k + |m| \tag{5.93}$$

and the eigenvalues for the square of the magnitude of angular momentum are

$$c = l(l + 1)\hbar^2, \qquad l = 0, 1, 2, \ldots \tag{5.94}$$

The magnitude of the orbital angular momentum of a particle is

$$|\mathbf{L}| = [l(l + 1)]^{1/2}\hbar \tag{5.95}$$

From (5.93), it follows that $|m| \le l$. The possible values for m are thus

$$m = -l, -l + 1, -l + 2, \ldots, -1, 0, 1, \ldots, l - 2, l - 1, l \tag{5.96}$$

Let us examine the angular-momentum eigenfunctions. From (5.83), (5.84), (5.88), (5.90), and (5.93), the theta factor in the eigenfunctions is

$$S_{l,m}(\theta) = \sin^{|m|}\theta \sum_{\substack{j=1,3,\ldots \\ \text{or } j=0,2,\ldots}}^{l-|m|} a_j \cos^j \theta \tag{5.97}$$

where the sum is over even or odd values of j, depending on whether $l - |m|$ is even or odd. The coefficients a_j satisfy the recursion relation (5.91), which, using (5.94), becomes

$$a_{j+2} = \frac{[(j + |m|)(j + |m| + 1) - l(l + 1)]}{(j + 1)(j + 2)} a_j \tag{5.98}$$

The $\hat{L}^2$ and $\hat{L}_z$ eigenfunctions are given by Eqs. (5.72) and (5.81) as

$$\boxed{Y_l^m(\theta, \phi) = S_{l,m}(\theta)T(\phi) = \frac{1}{\sqrt{2\pi}} S_{l,m}(\theta)e^{im\phi}} \tag{5.99}$$

EXAMPLE Find $Y_l^m(\theta, \phi)$ and the $\hat{L}^2$ and $\hat{L}_z$ eigenvalues for (a) $l = 0$; (b) $l = 1$.

(a) For $l = 0$, Eq. (5.96) gives $m = 0$, and (5.97) becomes

$$S_{0,0}(\theta) = a_0 \tag{5.100}$$

The normalization condition (5.80) gives

$$\int_0^{\pi} |a_0^2| \sin\theta \, d\theta = 1 = 2|a_0^2|$$

$$|a_0| = 2^{-1/2}$$

Equation (5.99) gives

$$Y_0^0(\theta, \phi) = \frac{1}{\sqrt{4\pi}} \tag{5.101}$$

[Obviously, (5.101) is an eigenfunction of $\hat{L}^2$, $\hat{L}_x$, $\hat{L}_y$, and $\hat{L}_z$, Eqs. (5.65)–(5.68).] For $l = 0$, there is no angular dependence in the eigenfunction; we say that the eigenfunctions are **spherically symmetric** for $l = 0$.

For $l = 0$ and $m = 0$, Eqs. (5.69), (5.70), (5.75), and (5.94) give the $\hat{L}^2$ eigenvalue as $c = 0$ and the $\hat{L}_z$ eigenvalue as $b = 0$.

(b) For $l = 1$, the possible values for m in (5.96) are $-1, 0$, and 1. For $|m| = 1$, (5.97) gives

$$S_{1,\pm 1}(\theta) = a_0 \sin\theta \tag{5.102}$$

a_0 in (5.102) is not necessarily the same as a_0 in (5.100). Normalization gives

$$1 = |a_0^2| \int_0^{\pi} \sin^2\theta \sin\theta \, d\theta = |a_0^2| \int_{-1}^{1} (1 - w^2) \, dw$$

$$|a_0| = \sqrt{3}/2$$

where the substitution $w = \cos\theta$ was made. Thus $S_{1,\pm 1} = (3^{1/2}/2)\sin\theta$ and (5.99) gives

$$Y_1^1 = (3/8\pi)^{1/2} \sin\theta \, e^{i\phi}, \qquad Y_1^{-1} = (3/8\pi)^{1/2} \sin\theta \, e^{-i\phi} \tag{5.103}$$

For $l = 1$ and $m = 0$, we find (see the following exercise) $S_{1,0} = (3/2)^{1/2}\cos\theta$ and $Y_1^0 = (3/4\pi)^{1/2}\cos\theta$.

For $l = 1$, (5.94) gives the $\hat{L}^2$ eigenvalue as $2\hbar^2$; for $m = -1, 0$, and 1, (5.75) gives the $\hat{L}_z$ eigenvalues as $-\hbar$, 0, and $\hbar$, respectively.

EXERCISE Verify the expressions for $S_{1,0}$ and Y_1^0.

The functions $S_{l,m}(\theta)$ are well known in mathematics and are *associated Legendre functions* multiplied by a normalization constant. The associated Legendre functions are defined in Prob. 5.31. Table 5.1 gives the $S_{l,m}(\theta)$ functions for $l \le 3$.

The angular-momentum eigenfunctions (5.99) are called **spherical harmonics** (or surface harmonics).

In summary, the one-particle orbital angular-momentum eigenfunctions and eigenvalues are [Eqs. (5.69), (5.70), (5.75), and (5.94)]

$$\boxed{\hat{L}^2 Y_l^m(\theta, \phi) = l(l+1)\hbar^2 Y_l^m(\theta, \phi), \quad l = 0, 1, 2, \ldots} \tag{5.104}$$

TABLE 5.1 $S_{l,m}(\theta)$

$l = 0$:	$S_{0,0} = \frac{1}{2}\sqrt{2}$
$l = 1$:	$S_{1,0} = \frac{1}{2}\sqrt{6}\cos\theta$
	$S_{1,\pm1} = \frac{1}{2}\sqrt{3}\sin\theta$
$l = 2$:	$S_{2,0} = \frac{1}{4}\sqrt{10}(3\cos^2\theta - 1)$
	$S_{2,\pm1} = \frac{1}{2}\sqrt{15}\sin\theta\cos\theta$
	$S_{2,\pm2} = \frac{1}{4}\sqrt{15}\sin^2\theta$
$l = 3$:	$S_{3,0} = \frac{3}{4}\sqrt{14}\left(\frac{5}{3}\cos^3\theta - \cos\theta\right)$
	$S_{3,\pm1} = \frac{1}{8}\sqrt{42}\sin\theta(5\cos^2\theta - 1)$
	$S_{3,\pm2} = \frac{1}{4}\sqrt{105}\sin^2\theta\cos\theta$
	$S_{3,\pm3} = \frac{1}{8}\sqrt{70}\sin^3\theta$

$$\boxed{\hat{L}_z Y_l^m(\theta, \phi) = m\hbar Y_l^m(\theta, \phi), \quad m = -l, -l + 1, \ldots, l - 1, l} \tag{5.105}$$

where the eigenfunctions are given by (5.99). Often the symbol m_l is used instead of m for the L_z quantum number.

Since $l \geq |m|$, the magnitude $[l(l + 1)]^{1/2}\hbar$ of the orbital angular momentum **L** is greater than the magnitude $|m|\hbar$ of its z component L_z, except for $l = 0$. If it were possible to have the angular-momentum magnitude equal to its z component, this would mean that the x and y components were zero, and we would have specified all three components of **L**. However, since the components of angular momentum do not commute with each other, we cannot do this. The one exception is when l is zero. In this case, $|\mathbf{L}|^2 = L_x^2 + L_y^2 + L_z^2$ has zero for its eigenvalue, and it must be true that all three components L_x, L_y, and L_z have zero eigenvalues. From Eq. (5.12), the uncertainties in angular-momentum components satisfy

$$\Delta L_x \, \Delta L_y \geq \frac{1}{2}\left|\int \Psi^*[\hat{L}_x, \hat{L}_y]\Psi \, d\tau\right| = \frac{\hbar}{2}\left|\int \Psi^*\hat{L}_z\Psi \, d\tau\right| \tag{5.106}$$

and two similar equations obtained by cyclic permutation. When the eigenvalues of $\hat{L}_z$, $\hat{L}_x$, and $\hat{L}_y$ are zero, $\hat{L}_x\Psi = 0$, $\hat{L}_y\Psi = 0$, $\hat{L}_z\Psi = 0$, the right-hand sides of (5.106) and the two similar equations are zero, and having $\Delta L_x = \Delta L_y = \Delta L_z = 0$ is permitted. But what about the statement in Section 5.1 that to have simultaneous eigenfunctions of two operators the operators must commute? The answer is that this theorem refers to the possibility of having a complete set of eigenfunctions of one operator be eigenfunctions of the other operator. Thus, even though $\hat{L}_x$ and $\hat{L}_z$ do not commute, it is possible to have *some* of the eigenfunctions of $\hat{L}_z$ (those with $l = 0 = m$) be eigenfunctions of $\hat{L}_x$. However, it is impossible to have *all* the $\hat{L}_z$ eigenfunctions also be eigenfunctions of $\hat{L}_x$.

Since we cannot specify L_x and L_y, the vector **L** can lie anywhere on the surface of a cone whose axis is the z axis, whose altitude is $m\hbar$, and whose slant height is $\sqrt{l(l + 1)}\,\hbar$ (Fig. 5.6). The possible orientations of **L** with respect to the z axis for the case $l = 1$ are shown in Fig. 5.7. For each eigenvalue of $\hat{L}^2$, there are $2l + 1$ different

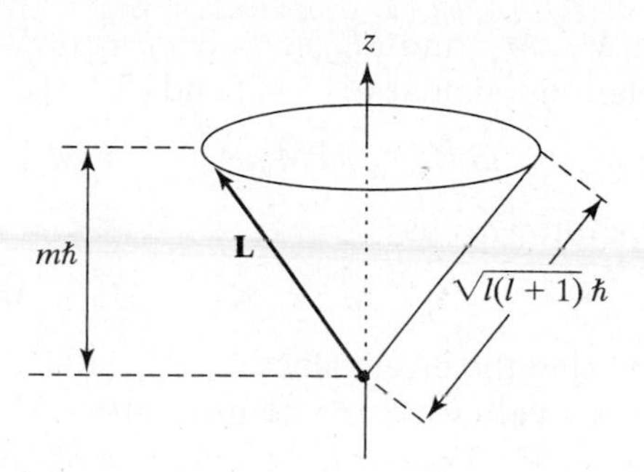

FIGURE 5.6 Orientation of **L**.

eigenfunctions Y_l^m, corresponding to the $2l + 1$ values of m. We say that the $\hat{L}^2$ eigenvalues are $(2l + 1)$-fold degenerate. The term **degeneracy** is applicable to the eigenvalues of any operator, not just the Hamiltonian.

Of course, there is nothing special about the z axis; all directions of space are equivalent. If we had chosen to specify L^2 and L_x (rather than L_z), we would have gotten the same eigenvalues for L_x as we found for L_z. However, it is easier to solve the $\hat{L}_z$ eigenvalue equation because $\hat{L}_z$ has a simple form in spherical coordinates, which involve the angle of rotation ϕ about the z axis.

5.4 THE LADDER-OPERATOR METHOD FOR ANGULAR MOMENTUM

We found the eigenvalues of $\hat{L}^2$ and $\hat{L}_z$ by expressing these orbital angular-momentum operators as differential operators and solving the resulting differential equations. We now show that these eigenvalues can be found using only the operator commutation relations. The work in this section applies to any operators that satisfy the angular-momentum commutation relations. In particular, it applies to spin angular momentum (Chapter 10) as well as orbital angular momentum.

We used the letter L for orbital angular momentum. Here we will use the letter M to indicate that we are dealing with any kind of angular momentum. We have three

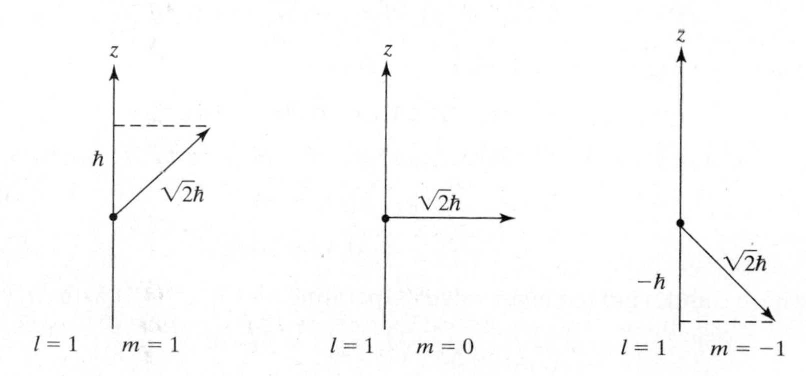

FIGURE 5.7 Orientations of **L** for $l = 1$.

linear operators $\hat{M}_x$, $\hat{M}_y$, and $\hat{M}_z$, and all we know about them is that they obey the commutation relations [similar to (5.46) and (5.48)]

$$[\hat{M}_x, \hat{M}_y] = i\hbar\hat{M}_z, \qquad [\hat{M}_y, \hat{M}_z] = i\hbar\hat{M}_x, \qquad [\hat{M}_z, \hat{M}_x] = i\hbar\hat{M}_y \tag{5.107}$$

We define the operator $\hat{M}^2$ as

$$\hat{M}^2 = \hat{M}_x^2 + \hat{M}_y^2 + \hat{M}_z^2 \tag{5.108}$$

Our problem is to find the eigenvalues of $\hat{M}^2$ and $\hat{M}_z$.

We begin by evaluating the commutators of $\hat{M}^2$ with its components, using Eqs. (5.107) and (5.108). The work is identical with that used to derive Eqs. (5.49) and (5.50), and we have

$$[\hat{M}^2, \hat{M}_x] = [\hat{M}^2, \hat{M}_y] = [\hat{M}^2, \hat{M}_z] = 0 \tag{5.109}$$

Hence we can have simultaneous eigenfunctions of $\hat{M}^2$ and $\hat{M}_z$.

Next we define two new operators, the **raising operator** $\hat{M}_+$ and the **lowering operator** $\hat{M}_-$:

$$\hat{M}_+ \equiv \hat{M}_x + i\hat{M}_y \tag{5.110}$$

$$\hat{M}_- \equiv \hat{M}_x - i\hat{M}_y \tag{5.111}$$

These are examples of **ladder operators**. The reason for the terminology will become clear shortly. We have

$$\hat{M}_+\hat{M}_- = (\hat{M}_x + i\hat{M}_y)(\hat{M}_x - i\hat{M}_y) = \hat{M}_x(\hat{M}_x - i\hat{M}_y) + i\hat{M}_y(\hat{M}_x - i\hat{M}_y)$$

$$= \hat{M}_x^2 - i\hat{M}_x\hat{M}_y + i\hat{M}_y\hat{M}_x + \hat{M}_y^2 = \hat{M}^2 - \hat{M}_z^2 + i[\hat{M}_y, \hat{M}_x]$$

$$\hat{M}_+\hat{M}_- = \hat{M}^2 - \hat{M}_z^2 + \hbar\hat{M}_z \tag{5.112}$$

Similarly, we find

$$\hat{M}_-\hat{M}_+ = \hat{M}^2 - \hat{M}_z^2 - \hbar\hat{M}_z \tag{5.113}$$

For the commutators of these operators with $\hat{M}_z$, we have

$$[\hat{M}_+, \hat{M}_z] = [\hat{M}_x + i\hat{M}_y, \hat{M}_z] = [\hat{M}_x, \hat{M}_z] + i[\hat{M}_y, \hat{M}_z] = -i\hbar\hat{M}_y - \hbar\hat{M}_x$$

$$[\hat{M}_+, \hat{M}_z] = -\hbar\hat{M}_+$$

$$\hat{M}_+\hat{M}_z = \hat{M}_z\hat{M}_+ - \hbar\hat{M}_+ \tag{5.114}$$

where (5.107) was used. Similarly, we find

$$\hat{M}_-\hat{M}_z = \hat{M}_z\hat{M}_- + \hbar\hat{M}_- \tag{5.115}$$

Using Y for the common eigenfunctions of $\hat{M}^2$ and $\hat{M}_z$, we have

$$\hat{M}^2Y = cY \tag{5.116}$$

$$\hat{M}_zY = bY \tag{5.117}$$

where c and b are the eigenvalues. Operating on Eq. (5.117) with $\hat{M}_+$, we get

$$\hat{M}_+\hat{M}_zY = \hat{M}_+bY$$

Using Eq. (5.114) and the fact that $\hat{M}_+$ is linear, we have

$$(\hat{M}_z\hat{M}_+ - \hbar\hat{M}_+)Y = b\hat{M}_+Y$$

$$\hat{M}_z(\hat{M}_+Y) = (b + \hbar)(\hat{M}_+Y) \tag{5.118}$$

This last equation says that the function $\hat{M}_+Y$ is an eigenfunction of $\hat{M}_z$ with eigenvalue $b + \hbar$. In other words, operating on the eigenfunction Y with the raising operator $\hat{M}_+$ converts Y into another eigenfunction of $\hat{M}_z$ with eigenvalue $\hbar$ higher than the eigenvalue of Y. If we now apply the raising operator to (5.118) and use (5.114) again, we find similarly

$$\hat{M}_z(\hat{M}_+^2Y) = (b + 2\hbar)(\hat{M}_+^2Y)$$

Repeated application of the raising operator gives

$$\hat{M}_z(\hat{M}_+^kY) = (b + k\hbar)(\hat{M}_+^kY), \quad k = 0, 1, 2, \ldots \tag{5.119}$$

If we operate on (5.117) with the lowering operator and apply (5.115), we find in the same manner

$$\hat{M}_z(\hat{M}_-Y) = (b - \hbar)(\hat{M}_-Y) \tag{5.120}$$

$$\hat{M}_z(\hat{M}_-^kY) = (b - k\hbar)(\hat{M}_-^kY) \tag{5.121}$$

Thus by using the raising and lowering operators on the eigenfunction with the eigenvalue b, we generate a ladder of eigenvalues, the difference from step to step being $\hbar$:

$$\cdots \quad b - 2\hbar, \quad b - \hbar, \quad b, \quad b + \hbar, \quad b + 2\hbar, \quad \cdots$$

The functions $\hat{M}_\pm^kY$ are eigenfunctions of $\hat{M}_z$ with eigenvalues $b \pm k\hbar$ [Eqs. (5.119) and (5.121)]. We now show that these functions are also eigenfunctions of $\hat{M}^2$, all with the *same* eigenvalue c:

$$\hat{M}_z\hat{M}_\pm^kY = (b + k\hbar)\hat{M}_\pm^kY \tag{5.122}$$

$$\hat{M}^2\hat{M}_\pm^kY = c\hat{M}_\pm^kY, \qquad k = 0, 1, 2, \ldots \tag{5.123}$$

To prove (5.123), we first show that $\hat{M}^2$ commutes with $\hat{M}_+$ and $\hat{M}_-$:

$$[\hat{M}^2, \hat{M}_\pm] = [\hat{M}^2, \hat{M}_x \pm i\hat{M}_y] = [\hat{M}^2, \hat{M}_x] \pm i[\hat{M}^2, \hat{M}_y] = 0 \pm 0 = 0$$

We also have

$$[\hat{M}^2, \hat{M}_\pm^2] = [\hat{M}^2, \hat{M}_\pm]\hat{M}_\pm + \hat{M}_\pm[\hat{M}^2, \hat{M}_\pm] = 0 + 0 = 0$$

and it follows by induction that

$$[\hat{M}^2, \hat{M}_\pm^k] = 0 \quad \text{or} \quad \hat{M}^2\hat{M}_\pm^k = \hat{M}_\pm^k\hat{M}^2, \qquad k = 0, 1, 2, \ldots \tag{5.124}$$

If we operate on (5.116) with $\hat{M}_\pm^k$ and use (5.124), we get

$$\hat{M}_\pm^k\hat{M}^2Y = \hat{M}_\pm^kcY$$

$$\hat{M}^2(\hat{M}_\pm^kY) = c(\hat{M}_\pm^kY) \tag{5.125}$$

which is what we wanted to prove.

Next we show that the set of eigenvalues of $\hat{M}_z$ generated using the ladder operators must be bounded. For the particular eigenfunction Y with $\hat{M}_z$ eigenvalue b, we have

$$\hat{M}_zY = bY$$

and for the set of eigenfunctions and eigenvalues generated by the ladder operators, we have

$$\hat{M}_z Y_k = b_k Y_k \tag{5.126}$$

where

$$Y_k = \hat{M}_{\pm}^k Y \tag{5.127}$$

$$b_k = b \pm k\hbar \tag{5.128}$$

(Application of $\hat{M}_+$ or $\hat{M}_-$ destroys the normalization of Y, so Y_k is not normalized. For the normalization constant, see Prob. 10.28.)

Operating on (5.126) with $\hat{M}_z$, we have

$$\hat{M}_z^2 Y_k = b_k \hat{M}_z Y_k$$

$$\hat{M}_z^2 Y_k = b_k^2 Y_k \tag{5.129}$$

Now subtract (5.129) from (5.123), and use (5.127) and (5.108):

$$\hat{M}^2 Y_k - \hat{M}_z^2 Y_k = cY_k - b_k^2 Y_k$$

$$(\hat{M}_x^2 + \hat{M}_y^2) Y_k = (c - b_k^2) Y_k \tag{5.130}$$

The operator $\hat{M}_x^2 + \hat{M}_y^2$ corresponds to a nonnegative physical quantity and hence has nonnegative eigenvalues. (This is proved in Prob. 7.11.) Therefore, (5.130) implies that $c - b_k^2 \geq 0$ and $c^{1/2} \geq |b_k|$. Thus

$$c^{1/2} \geq b_k \geq -c^{1/2}, \qquad k = 0, \pm 1, \pm 2, \ldots \tag{5.131}$$

Since c remains constant as k varies, (5.131) shows that the set of eigenvalues b_k is bounded above and below. Let $b_{\max}$ and $b_{\min}$ denote the maximum and minimum values of b_k. $Y_{\max}$ and $Y_{\min}$ are the corresponding eigenfunctions:

$$\hat{M}_z Y_{\max} = b_{\max} Y_{\max} \tag{5.132}$$

$$\hat{M}_z Y_{\min} = b_{\min} Y_{\min} \tag{5.133}$$

Now operate on (5.132) with the raising operator and use (5.114):

$$\hat{M}_+ \hat{M}_z Y_{\max} = b_{\max} \hat{M}_+ Y_{\max}$$

$$\hat{M}_z(\hat{M}_+ Y_{\max}) = (b_{\max} + \hbar)(\hat{M}_+ Y_{\max}) \tag{5.134}$$

This last equation seems to contradict the statement that $b_{\max}$ is the largest eigenvalue of $\hat{M}_z$, since it says that $\hat{M}_+ Y_{\max}$ is an eigenfunction of $\hat{M}_z$ with eigenvalue $b_{\max} + \hbar$. The only way out of this contradiction is to have $\hat{M}_+ Y_{\max}$ vanish. (We always reject zero as an eigenfunction on physical grounds.) Thus

$$\hat{M}_+ Y_{\max} = 0 \tag{5.135}$$

Operating on (5.135) with the lowering operator and using (5.113), (5.132), and (5.116), we have

$$0 = \hat{M}_- \hat{M}_+ Y_{\max} = (\hat{M}^2 - \hat{M}_z^2 - \hbar \hat{M}_z) Y_{\max} = (c - b_{\max}^2 - \hbar b_{\max}) Y_{\max}$$

$$c - b_{\text{max}}^2 - \hbar b_{\text{max}} = 0$$

$$c = b_{\text{max}}^2 + \hbar b_{\text{max}} \tag{5.136}$$

A similar argument shows that

$$\hat{M}_- Y_{\text{min}} = 0 \tag{5.137}$$

and by applying the raising operator to this equation and using (5.112), we find

$$c = b_{\text{min}}^2 - \hbar b_{\text{min}}$$

Subtracting this last equation from (5.136), we have

$$b_{\text{max}}^2 + \hbar b_{\text{max}} + (\hbar b_{\text{min}} - b_{\text{min}}^2) = 0$$

This is a quadratic equation in the unknown b_{max}, and using the usual formula (it still works in quantum mechanics), we find

$$b_{\text{max}} = -b_{\text{min}}, \qquad b_{\text{max}} = b_{\text{min}} - \hbar$$

The second root is rejected, since it says that b_{max} is less than b_{min}. So

$$b_{\text{min}} = -b_{\text{max}} \tag{5.138}$$

Moreover, (5.128) says that b_{max} and b_{min} differ by an integral multiple of $\hbar$:

$$b_{\text{max}} - b_{\text{min}} = n\hbar, \qquad n = 0, 1, 2, \ldots \tag{5.139}$$

Substituting (5.138) in (5.139), we have for the $\hat{M}_z$ eigenvalues

$$b_{\text{max}} = \tfrac{1}{2}n\hbar$$

$$b_{\text{max}} = j\hbar, \qquad j = 0, \tfrac{1}{2}, 1, \tfrac{3}{2}, 2, \ldots \tag{5.140}$$

$$b_{\text{min}} = -j\hbar$$

$$b = -j\hbar, (-j + 1)\hbar, (-j + 2)\hbar, \ldots, (j - 2)\hbar, (j - 1)\hbar, j\hbar \tag{5.141}$$

and from (5.136) we find as the $\hat{M}^2$ eigenvalues

$$c = j(j + 1)\hbar^2, \quad j = 0, \tfrac{1}{2}, 1, \tfrac{3}{2}, \ldots \tag{5.142}$$

Thus

$$\hat{M}^2 Y = j(j + 1)\hbar^2 Y, \quad j = 0, \tfrac{1}{2}, 1, \tfrac{3}{2}, 2, \ldots \tag{5.143}$$

$$\hat{M}_z Y = m_j \hbar Y, \quad m_j = -j, -j + 1, \ldots, j - 1, j \tag{5.144}$$

We have found the eigenvalues of $\hat{M}^2$ and $\hat{M}_z$ using just the commutation relations. However, comparison of (5.143) and (5.144) with (5.104) and (5.105) shows that in addition to integral values for the angular-momentum quantum number ($l = 0, 1, 2, \ldots$) we now also have the possibility for half-integral values $\left(j = 0, \frac{1}{2}, 1, \frac{3}{2}, \ldots\right)$. This perhaps suggests that there might be another kind of angular momentum besides orbital angular momentum. In Chapter 10 we shall see that spin angular momentum can have half-integral, as well as integral, quantum numbers. For orbital angular momentum, the boundary condition of single-valuedness of the $T(\phi)$ eigenfunctions [see the equation following (5.73)] eliminates the half-integral values of the angular-momentum quantum numbers.

[Not everyone accepts single-valuedness as a valid boundary condition on wave functions, and many other reasons have been given for rejecting half-integral orbital-angular-momentum quantum numbers; see C. G. Gray, *Am. J. Phys.*, **37**, 559 (1969); M. L. Whippman, *Am. J. Phys.*, **34**, 656 (1966).]

The ladder-operator method can be used to solve other eigenvalue problems; see Prob. 5.33.

5.5 SUMMARY

For a complete set of eigenfunctions to be simultaneously eigenfunctions of several operators, each operator must commute with every other operator.

The standard deviation ΔA measures the uncertainty in a quantum-mechanical property A, where $(\Delta A)^2 = \langle A^2 \rangle - \langle A \rangle^2$. For the properties x and p_x, we have $\Delta x \, \Delta p_x \geq \frac{1}{2}\hbar$.

The vector $\mathbf{B}$ can be written as $\mathbf{B} = B_x\mathbf{i} + B_y\mathbf{j} + B_z\mathbf{k}$, where $\mathbf{i}, \mathbf{j}$, and $\mathbf{k}$ are unit vectors along the x, y, and z axes, and B_x, B_y, and B_z are the components of $\mathbf{B}$. The magnitude of $\mathbf{B}$ is $|\mathbf{B}| = (B_x^2 + B_y^2 + B_z^2)^{1/2}$. The dot product of two vectors that make an angle θ with each other is $\mathbf{B} \cdot \mathbf{A} = B_xA_x + B_yA_y + B_zA_z = |\mathbf{B}||\mathbf{A}| \cos\theta$. The cross product is given by (5.28).

The classical-mechanical definition of orbital angular momentum is $\mathbf{L} \equiv \mathbf{r} \times \mathbf{p}$. The operator $\hat{L}^2$ commutes with $\hat{L}_x$, $\hat{L}_y$, and $\hat{L}_z$ but $\hat{L}_x$, $\hat{L}_y$, and $\hat{L}_z$ do not commute with one another. When expressed in spherical coordinates, the operators $\hat{L}^2$, $\hat{L}_x$, $\hat{L}_y$, and $\hat{L}_z$ depend only on the angles θ (the angle between the z axis and $\mathbf{r}$) and ϕ (the angle between the projection of $\mathbf{r}$ in the xy plane and the x axis) and not on the radial coordinate r.

The ranges of the spherical coordinates are 0 to π for θ, 0 to 2π for ϕ, and 0 to ∞ for r. The infinitesimal volume element in spherical coordinates is $d\tau = r^2 \sin\theta \, dr \, d\theta \, d\phi$.

The common eigenfunctions and eigenvalues of $\hat{L}_z$ and $\hat{L}^2$ are given by $\hat{L}_zY_l^m = m\hbar Y_l^m$ and $\hat{L}^2Y_l^m = l(l+1)\hbar^2Y_l^m$, where the functions $Y_l^m(\theta, \phi)$ are spherical harmonics and the angular-momentum quantum numbers are $l = 0, 1, 2, \ldots$ and $m = -l, -l+1, \ldots, l-1, l$.

For operators $\hat{M}_x$, $\hat{M}_y$, $\hat{M}_z$, and $\hat{M}^2 = \hat{M}_x^2 + \hat{M}_y^2 + \hat{M}_z^2$ that obey the angular-momentum commutation relations, use of the ladder operators $\hat{M}_+ = \hat{M}_x + i\hat{M}_y$ and $\hat{M}_- = \hat{M}_x - i\hat{M}_y$ gives the possible $\hat{M}^2$ eigenvalues as $j(j+1)\hbar^2$, where j can be integral or half-integral, and gives the $\hat{M}_z$ eigenvalues as $m_j\hbar$, where m_j ranges from $-j$ to j in integral steps.

PROBLEMS

Sec.	5.1	5.2	5.3	5.4	general
Probs.	5.1–5.8	5.9–5.14	5.15–5.31	5.32–5.33	5.34

5.1 State whether the two operators in each of the following pairs commute with each other. (a) $\hat{x}^2$ and $\partial/\partial x$; (b) $\hat{y}^2$ and $\partial/\partial z$; (c) $\hat{x}^2$ and $\hat{z}^2$; (d) $\partial/\partial x$ and $\partial^2/\partial x^2$; (e) $\partial/\partial x$ and $\partial/\partial y$.

5.2 Verify the commutator identities (5.1)–(5.5).

5.3 Find $[\hat{x}, \hat{p}_x^3]$ starting from (5.7) for $[\hat{x}, \hat{p}_x^2]$.

5.4 For the ground state of the one-dimensional harmonic oscillator, compute the standard deviations Δx and Δp_x and check that the uncertainty principle is obeyed. Use the results of Prob. 4.10 to save time.

5.5 At a certain instant of time, a particle in a one-dimensional box of length l (Fig. 2.1) is in a nonstationary state with $\Psi = (105/l^7)^{1/2}x^2(l - x)$ inside the box. For this state, find Δx and Δp_x and verify that the uncertainty principle $\Delta x\, \Delta p_x \geq \frac{1}{2}\hbar$ is obeyed.

5.6 Show that the standard deviation ΔA is 0 when Ψ is an eigenfunction of $\hat{A}$.

5.7 Derive $(\Delta A)^2 = \langle A^2 \rangle - \langle A \rangle^2$ [Eq. (5.11)].

5.8 Let w be the variable defined as the number of heads that show when two coins are tossed simultaneously. Find $\langle w \rangle$ and σ_w. [*Hint:* Use (5.11).]

5.9 Classify each of these as a scalar or vector. (a) $3\mathbf{B}$; (b) $\mathbf{C} \times \mathbf{B}$; (c) $\mathbf{C} \cdot \mathbf{B}$; (d) $|\mathbf{B}|$; (e) velocity; (f) potential energy.

5.10 Let $\mathbf{A}$ have the components $(3, -2, 6)$; let $\mathbf{B}$ have the components $(-1, 4, 4)$. Find $|\mathbf{A}|$, $|\mathbf{B}|$, $\mathbf{A} + \mathbf{B}$, $\mathbf{A} - \mathbf{B}$, $\mathbf{A} \cdot \mathbf{B}$, $\mathbf{A} \times \mathbf{B}$. Find the angle between $\mathbf{A}$ and $\mathbf{B}$.

5.11 Use the vector dot product to find the obtuse angle between two diagonals of a cube. What is the chemical significance of this angle?

5.12 (a) Use the vector dot product to show that in $HCBr_3$, $\cos(\angle \text{BrCBr}) = 1 - 1.5 \sin^2(\angle \text{HCBr})$. (b) In $HCBr_3$, $\angle \text{HCBr} = 107.2°$. Find $\angle \text{BrCBr}$ in $HCBr_3$.

5.13 Let $f = 2x^2 - 5xyz + z^2 - 1$. Find **grad** f. Find $\nabla^2 f$.

5.14 The *divergence* of a vector function $\mathbf{A}$ is a scalar function defined by

$$\text{div}\,\mathbf{A} \equiv \nabla \cdot \mathbf{A} = \left(\mathbf{i}\frac{\partial}{\partial x} + \mathbf{j}\frac{\partial}{\partial y} + \mathbf{k}\frac{\partial}{\partial z}\right) \cdot (A_x\mathbf{i} + A_y\mathbf{j} + A_z\mathbf{k})$$

$$\text{div}\,\mathbf{A} \equiv \frac{\partial A_x}{\partial x} + \frac{\partial A_y}{\partial y} + \frac{\partial A_z}{\partial z}$$

(a) Verify that div [**grad** $g(x, y, z)$] $\equiv \nabla \cdot \nabla g = \partial^2 g/\partial x^2 + \partial^2 g/\partial y^2 + \partial^2 g/\partial z^2$. This is the origin of the notation ∇^2 for $\partial^2/\partial x^2 + \partial^2/\partial y^2 + \partial^2/\partial z^2$. (b) Find $\nabla \cdot \mathbf{r}$, where $\mathbf{r} = \mathbf{i}x + \mathbf{j}y + \mathbf{k}z$.

5.15 State whether the two operators in each of the following pairs commute with each other. (a) $\hat{L}_x$ and $\hat{L}_z$; (b) $\hat{L}_z$ and $\hat{L}^2$; (c) $\hat{L}_x$ and $\hat{L}^2$; (d) $\hat{L}_x^2$ and $\hat{L}^2$.

5.16 Derive Eq. (5.68) for $\hat{L}^2$ from Eqs. (5.65)–(5.67).

5.17 Find $[\hat{L}_x^2, \hat{L}_y]$.

5.18 Find the spherical coordinates for points with the following (x, y, z) coordinates: (a) $(1, 2, 0)$; (b) $(-1, 0, 3)$; (c) $(3, 1, -2)$; (d) $(-1, -1, -1)$.

5.19 Find the (x, y, z) coordinates of the points with the following spherical coordinates: (a) $r = 1, \theta = \pi/2, \phi = \pi$; (b) $r = 2, \theta = \pi/4, \phi = 0$.

5.20 Give the shape of a surface on which (a) r is constant; (b) θ is constant; (c) ϕ is constant.

5.21 By integrating the spherical-coordinates differential volume element $d\tau$ over appropriate limits, verify the formula $\frac{4}{3}\pi R^3$ for the volume of a sphere of radius R.

5.22 Calculate the possible angles between $\mathbf{L}$ and the z axis for $l = 2$.

5.23 Consider the following incorrect derivation. Differentiation of x in (5.51) gives $\partial x/\partial r = \sin\theta \cos\phi$. Then, since $\partial r/\partial x = 1/(\partial x/\partial r)$, we have $(\partial r/\partial x)_{y,z} = 1/(\sin\theta \cos\phi)$ (?). But this result disagrees with (5.57). Find the error in this reasoning.

5.24 Substitute $d/d\theta = -(1 - w^2)^{1/2}(d/dw)$ [which follows from (5.85)] and Eq. (5.85) into $d^2S/d\theta^2 = (d/d\theta)(dS/d\theta)$ to verify Eq. (5.86).

5.25 Derive the formula for $S_{2,0}$ (Table 5.1) in two ways: (a) by using (5.146) in Prob. 5.31; (b) by using the recursion relation and normalization.

5.26 Use the recursion relation (5.98) and normalization to find (a) Y_3^0; (b) Y_3^1.

5.27 Complete this equation: $\hat{L}_z^3 Y_l^m = ?$

5.28 Show that the spherical harmonics are eigenfunctions of the operator $\hat{L}_x^2 + \hat{L}_y^2$. (The proof is short.) What are the eigenvalues?

5.29 (a) If we measure L_z of a particle that has angular-momentum quantum number $l = 2$, what are the possible outcomes of the measurement? (b) If we measure L_z of a particle whose state function is an eigenfunction of $\hat{L}^2$ with eigenvalue $12\hbar^2$, what are the possible outcomes of the measurement?

5.30 At a certain instant of time t', a particle has the state function $\Psi = Ne^{-ar^2}Y_2^1(\theta, \phi)$, where N and a are constants. (a) If L^2 of this particle were to be measured at time t', what would be the outcome? Give a numerical answer. (b) If L_z of this particle were to be measured at t', what would be the outcome? Give a numerical answer.

5.31 The *associated Legendre functions* $P_l^{|m|}(w)$ are defined by

$$P_l^{|m|}(w) \equiv \frac{1}{2^l l!}(1 - w^2)^{|m|/2}\frac{d^{l+|m|}}{dw^{l+|m|}}(w^2 - 1)^l, \qquad l = 0, 1, 2, \ldots \tag{5.145}$$

Verify that $P_0^0(w) = 1$, $P_1^0(w) = w$, $P_1^1(w) = (1 - w^2)^{1/2}$ and find $P_2^0(w)$, $P_2^1(w)$, and $P_2^2(w)$. It can be shown that (*Pauling and Wilson*, page 129)

$$S_{l,m}(\theta) = \left[\frac{2l + 1}{2}\frac{(l - |m|)!}{(l + |m|)!}\right]^{1/2} P_l^{|m|}(\cos\theta) \tag{5.146}$$

Equations (5.146) and (5.145) give the explicit formula for the normalized theta factor in the angular-momentum eigenfunctions. The normalized eigenfunctions of $\hat{L}^2$ and $\hat{L}_z$ (the **spherical harmonics**) are given by Eqs. (5.146) and (5.99) as

$$Y_l^m(\theta, \phi) = \left[\frac{2l + 1}{4\pi}\frac{(l - |m|)!}{(l + |m|)!}\right]^{1/2} P_l^{|m|}(\cos\theta)e^{im\phi} \tag{5.147}$$

The phase of the normalization constant of the spherical harmonics is arbitrary (see Section 1.7). Many texts use a different phase convention than in (5.147). Thus Y_l^m differs from text to text by a minus sign.

5.32 Apply the lowering operator $\hat{L}_-$ three times in succession to $Y_1^1(\theta, \phi)$ and verify that we obtain functions that are proportional to Y_1^0, Y_1^{-1}, and zero.

5.33 The one-dimensional harmonic-oscillator Hamiltonian is

$$\hat{H} = \frac{\hat{p}_x^2}{2m} + 2\pi^2\nu^2 m\hat{x}^2$$

The raising and lowering operators for this problem are defined as

$$\hat{A}_+ \equiv (2m)^{-1/2}(\hat{p}_x + 2\pi i\nu m\hat{x}), \quad \hat{A}_- \equiv (2m)^{-1/2}(\hat{p}_x - 2\pi i\nu m\hat{x})$$

Show that

$$\hat{A}_+\hat{A}_- = \hat{H} - \tfrac{1}{2}h\nu, \quad \hat{A}_-\hat{A}_+ = \hat{H} + \tfrac{1}{2}h\nu, \quad [\hat{A}_+, \hat{A}_-] = -h\nu$$
$$[\hat{H}, \hat{A}_+] = h\nu\hat{A}_+, \quad [\hat{H}, \hat{A}_-] = -h\nu\hat{A}_-$$

Show that $\hat{A}_+$ and $\hat{A}_-$ are indeed ladder operators and that the eigenvalues are spaced at intervals of $h\nu$. Since both the kinetic energy and the potential energy are nonnegative, we expect the energy eigenvalues to be nonnegative. Hence there must be a state of minimum energy. Operate on the wave function for this state first with $\hat{A}_-$ and then with $\hat{A}_+$ and show that the lowest-energy eigenvalue is $\frac{1}{2}h\nu$. Finally, conclude that

$$E = \left(n + \tfrac{1}{2}\right)h\nu, \quad n = 0, 1, 2, \ldots$$

(See also Prob. 7.62.)

5.34 True or false? (a) The $\hat{L}^2$ eigenvalues are degenerate except for $l = 0$. (b) Since $\hat{L}^2 Y_l^m = l(l+1)\hbar^2 Y_l^m$, it follows that $\hat{L}^2 = l(l+1)\hbar^2$. (c) $\hat{L}^2$ commutes with $\hat{L}_x$. (d) Y_0^0 is a constant. (e) Y_0^0 is an eigenfunction of $\hat{L}^2$, $\hat{L}_x$, $\hat{L}_y$, and $\hat{L}_z$. (f) If $\hat{A}$ and $\hat{B}$ do not commute, it is impossible for an eigenfunction of $\hat{A}$ to be also an eigenfunction of $\hat{B}$.

CHAPTER 6

The Hydrogen Atom

6.1 THE ONE-PARTICLE CENTRAL-FORCE PROBLEM

Before studying the hydrogen atom, we shall consider the more general problem of a single particle moving under a central force. The results of this section will apply to any central-force problem; examples are the hydrogen atom (Section 6.5) and the isotropic three-dimensional harmonic oscillator (Prob. 6.3).

A **central force** is one derived from a potential-energy function that is spherically symmetric, which means that it is a function only of the distance of the particle from the origin: $V = V(r)$. The relation between force and potential energy is given by (5.31) as

$$\mathbf{F} = -\nabla V(x, y, z) = -\mathbf{i}(\partial V/\partial x) - \mathbf{j}(\partial V/\partial y) - \mathbf{k}(\partial V/\partial z) \tag{6.1}$$

The partial derivatives in (6.1) can be found by the chain rule [Eqs. (5.53)–(5.55)]. Since V in this case is a function of r only, we have $(\partial V/\partial \theta)_{r,\phi} = 0$ and $(\partial V/\partial \phi)_{r,\theta} = 0$. Therefore,

$$\left(\frac{\partial V}{\partial x}\right)_{y,z} = \frac{dV}{dr}\left(\frac{\partial r}{\partial x}\right)_{y,z} = \frac{x}{r}\frac{dV}{dr} \tag{6.2}$$

$$\left(\frac{\partial V}{\partial y}\right)_{x,z} = \frac{y}{r}\frac{dV}{dr}, \qquad \left(\frac{\partial V}{\partial z}\right)_{x,y} = \frac{z}{r}\frac{dV}{dr} \tag{6.3}$$

where Eqs. (5.57) and (5.58) have been used. Equation (6.1) becomes

$$\mathbf{F} = -\frac{1}{r}\frac{dV}{dr}(x\mathbf{i} + y\mathbf{j} + z\mathbf{k}) = -\frac{dV(r)}{dr}\frac{\mathbf{r}}{r} \tag{6.4}$$

where (5.33) for $\mathbf{r}$ was used. The quantity $\mathbf{r}/r$ in (6.4) is a unit vector in the radial direction. A central force is radially directed.

Now we consider the quantum mechanics of a single particle subject to a central force. The Hamiltonian operator is

$$\hat{H} = \hat{T} + \hat{V} = -(\hbar^2/2m)\nabla^2 + V(r) \tag{6.5}$$

where $\nabla^2 \equiv \partial^2/\partial x^2 + \partial^2/\partial y^2 + \partial^2/\partial z^2$ [Eq. (3.46)]. Since V is spherically symmetric, we shall work in spherical coordinates. Hence we want to transform the Laplacian operator to these coordinates. We already have the forms of the operators $\partial/\partial x$, $\partial/\partial y$, and $\partial/\partial z$ in these coordinates [Eqs. (5.62)–(5.64)], and by squaring each of these operators and then adding their squares, we get the Laplacian. This calculation is left as an exercise. The result is (Prob. 6.4)

$$\nabla^2 = \frac{\partial^2}{\partial r^2} + \frac{2}{r}\frac{\partial}{\partial r} + \frac{1}{r^2}\frac{\partial^2}{\partial \theta^2} + \frac{1}{r^2}\cot\theta\frac{\partial}{\partial \theta} + \frac{1}{r^2\sin^2\theta}\frac{\partial^2}{\partial \phi^2} \tag{6.6}$$

Looking back to (5.68), which gives the operator $\hat{L}^2$ for the square of the magnitude of the orbital angular momentum of a single particle, we see that

$$\nabla^2 = \frac{\partial^2}{\partial r^2} + \frac{2}{r}\frac{\partial}{\partial r} - \frac{1}{r^2\hbar^2}\hat{L}^2 \tag{6.7}$$

The Hamiltonian (6.5) becomes

$$\hat{H} = -\frac{\hbar^2}{2m}\left(\frac{\partial^2}{\partial r^2} + \frac{2}{r}\frac{\partial}{\partial r}\right) + \frac{1}{2mr^2}\hat{L}^2 + V(r) \tag{6.8}$$

In classical mechanics a particle subject to a central force has its angular momentum conserved (Section 5.3). In quantum mechanics we might ask whether we can have states with definite values for both the energy and the angular momentum. To have the set of eigenfunctions of $\hat{H}$ also be eigenfunctions of $\hat{L}^2$, the commutator $[\hat{H}, \hat{L}^2]$ must vanish. We have

$$[\hat{H}, \hat{L}^2] = [\hat{T}, \hat{L}^2] + [\hat{V}, \hat{L}^2]$$

$$[\hat{T}, \hat{L}^2] = \left[-\frac{\hbar^2}{2m}\left(\frac{\partial^2}{\partial r^2} + \frac{2}{r}\frac{\partial}{\partial r}\right) + \frac{1}{2mr^2}\hat{L}^2, \hat{L}^2\right]$$

$$[\hat{T}, \hat{L}^2] = -\frac{\hbar^2}{2m}\left[\frac{\partial^2}{\partial r^2} + \frac{2}{r}\frac{\partial}{\partial r}, \hat{L}^2\right] + \frac{1}{2m}\left[\frac{1}{r^2}\hat{L}^2, \hat{L}^2\right] \tag{6.9}$$

Recall that $\hat{L}^2$ involves only θ and ϕ and not r [Eq. (5.68)]. Hence it commutes with any operator that involves only r. [To reach this conclusion, we must use relations like (5.47) with x and z replaced by r and θ.] Thus the first commutator in (6.9) is zero. Moreover, since any operator commutes with itself, the second commutator in (6.9) is zero. Therefore, $[\hat{T}, \hat{L}^2] = 0$. Also, since $\hat{L}^2$ does not involve r and V is a function of r only, we have $[\hat{V}, \hat{L}^2] = 0$. Therefore,

$$[\hat{H}, \hat{L}^2] = 0 \qquad \text{if } V = V(r) \tag{6.10}$$

$\hat{H}$ commutes with $\hat{L}^2$ when the potential-energy function is independent of θ and ϕ.

Now consider the operator $\hat{L}_z = -i\hbar\,\partial/\partial\phi$ [Eq. (5.67)]. Since $\hat{L}_z$ does not involve r and since it commutes with $\hat{L}^2$ [Eq. (5.50)], it follows that $\hat{L}_z$ commutes with the Hamiltonian (6.8):

$$[\hat{H}, \hat{L}_z] = 0 \qquad \text{if } V = V(r) \tag{6.11}$$

We can therefore have a set of simultaneous eigenfunctions of $\hat{H}$, $\hat{L}^2$, and $\hat{L}_z$ for the central-force problem. Let ψ denote these common eigenfunctions:

$$\boxed{\hat{H}\psi = E\psi} \tag{6.12}$$

$$\boxed{\hat{L}^2\psi = l(l+1)\hbar^2\psi, \quad l = 0, 1, 2, \ldots} \tag{6.13}$$

$$\boxed{\hat{L}_z\psi = m\hbar\psi, \quad m = -l, -l+1, \ldots, l} \tag{6.14}$$

where Eqs. (5.104) and (5.105) have been used.

Using (6.8) and (6.13), we have for the Schrödinger equation (6.12)

$$-\frac{\hbar^2}{2m}\left(\frac{\partial^2\psi}{\partial r^2} + \frac{2}{r}\frac{\partial\psi}{\partial r}\right) + \frac{1}{2mr^2}\hat{L}^2\psi + V(r)\psi = E\psi$$

$$-\frac{\hbar}{2m}\left(\frac{\partial^2\psi}{\partial r^2} + \frac{2}{r}\frac{\partial\psi}{\partial r}\right) + \frac{l(l+1)\hbar^2}{2mr^2}\psi + V(r)\psi = E\psi \tag{6.15}$$

The eigenfunctions of $\hat{L}^2$ are the spherical harmonics $Y_l^m(\theta, \phi)$, and since $\hat{L}^2$ does not involve r, we can multiply Y_l^m by an arbitrary function of r and still have eigenfunctions of $\hat{L}^2$ and $\hat{L}_z$. Therefore,

$$\boxed{\psi = R(r)Y_l^m(\theta, \phi)} \tag{6.16}$$

Using (6.16) in (6.15), we then divide both sides by Y_l^m to get an ordinary differential equation for the unknown function $R(r)$:

$$-\frac{\hbar^2}{2m}\left(R'' + \frac{2}{r}R'\right) + \frac{l(l+1)\hbar^2}{2mr^2}R + V(r)R = ER(r) \tag{6.17}$$

We have shown that, *for any one-particle problem with a spherically symmetric potential-energy function $V(r)$, the stationary-state wave functions are $\psi = R(r)Y_l^m(\theta, \phi)$, where the radial factor $R(r)$ satisfies (6.17).* By using a specific form for $V(r)$ in (6.17), we can solve it for a particular problem.

6.2 NONINTERACTING PARTICLES AND SEPARATION OF VARIABLES

Up to this point, we have solved only one-particle quantum-mechanical problems. The hydrogen atom is a two-particle system, and as a preliminary to dealing with the H atom, we first consider a simpler case, that of two noninteracting particles.

Suppose that a system is composed of the noninteracting particles 1 and 2. Let q_1 and q_2 symbolize the coordinates (x_1, y_1, z_1) and (x_2, y_2, z_2) of particles 1 and 2. Because the particles exert no forces on each other, the classical-mechanical energy of the system is the sum of the energies of the two particles: $E = E_1 + E_2 = T_1 + V_1 + T_2 + V_2$, and the classical Hamiltonian is the sum of Hamiltonians for each particle: $H = H_1 + H_2$. Therefore, the Hamiltonian operator is

$$\hat{H} = \hat{H}_1 + \hat{H}_2$$

where $\hat{H}_1$ involves only the coordinates q_1 and the momentum operators $\hat{p}_1$ that correspond to q_1. The Schrödinger equation for the system is

$$(\hat{H}_1 + \hat{H}_2)\psi(q_1, q_2) = E\psi(q_1, q_2) \tag{6.18}$$

We try a solution of (6.18) by separation of variables, setting

$$\psi(q_1, q_2) = G_1(q_1)G_2(q_2) \tag{6.19}$$

We have

$$\hat{H}_1 G_1(q_1)G_2(q_2) + \hat{H}_2 G_1(q_1)G_2(q_2) = EG_1(q_1)G_2(q_2) \tag{6.20}$$

Since $\hat{H}_1$ involves only the coordinate and momentum operators of particle 1, we have $\hat{H}_1[G_1(q_1)G_2(q_2)] = G_2(q_2)\hat{H}_1 G_1(q_1)$, since, as far as $\hat{H}_1$ is concerned, G_2 is a constant. Using this equation and a similar equation for $\hat{H}_2$, we find that Eq. (6.20) becomes

$$G_2(q_2)\hat{H}_1 G_1(q_1) + G_1(q_1)\hat{H}_2 G_2(q_2) = EG_1(q_1)G_2(q_2) \tag{6.21}$$

$$\frac{\hat{H}_1 G_1(q_1)}{G_1(q_1)} + \frac{\hat{H}_2 G_2(q_2)}{G_2(q_2)} = E \tag{6.22}$$

Now, by the same arguments used in connection with Eq. (3.65), we conclude that each term on the left in (6.22) must be a constant. Using E_1 and E_2 to denote these constants, we have

$$\frac{\hat{H}_1 G_1(q_1)}{G_1(q_1)} = E_1, \qquad \frac{\hat{H}_2 G_2(q_2)}{G_2(q_2)} = E_2$$

$$E = E_1 + E_2 \tag{6.23}$$

Thus, when the system is composed of two noninteracting particles, we can reduce the two-particle problem to two separate one-particle problems by solving

$$\hat{H}_1 G_1(q_1) = E_1 G_1(q_1), \qquad \hat{H}_2 G_2(q_2) = E_2 G_2(q_2) \tag{6.24}$$

which are separate Schrödinger equations for each particle.

Generalizing this result to n noninteracting particles, we have

$$\hat{H} = \hat{H}_1 + \hat{H}_2 + \cdots + \hat{H}_n$$

$$\boxed{\psi(q_1, q_2, \ldots, q_n) = G_1(q_1)G_2(q_2)\cdots G_n(q_n)} \tag{6.25}$$

$$\boxed{E = E_1 + E_2 + \cdots + E_n} \tag{6.26}$$

$$\boxed{\hat{H}_i G_i = E_i G_i, \quad i = 1, 2, \ldots, n} \tag{6.27}$$

For a system of noninteracting particles, the energy is the sum of the individual energies of each particle and the wave function is the product of wave functions for each particle; the wave function G_i of particle i is found by solving a Schrödinger equation for particle i using the Hamiltonian $\hat{H}_i$.

These results also apply to a single particle whose Hamiltonian is the sum of separate terms for each coordinate:

$$\hat{H} = \hat{H}_x(\hat{x}, \hat{p}_x) + \hat{H}_y(\hat{y}, \hat{p}_y) + \hat{H}_z(\hat{z}, \hat{p}_z)$$

In this case, we conclude that the wave functions and energies are

$$\psi(x, y, z) = F(x)G(y)K(z), \quad E = E_x + E_y + E_z$$

$$\hat{H}_x F(x) = E_x F(x), \quad \hat{H}_y G(y) = E_y G(y), \quad \hat{H}_z K(z) = E_z K(z)$$

Examples include the particle in a three-dimensional box (Section 3.5), the three-dimensional free particle (Prob. 3.40), and the three-dimensional harmonic oscillator (Prob. 4.19).

6.3 REDUCTION OF THE TWO-PARTICLE PROBLEM TO TWO ONE-PARTICLE PROBLEMS

The hydrogen atom contains two particles, the proton and the electron. For a system of two particles 1 and 2 with coordinates (x_1, y_1, z_1) and (x_2, y_2, z_2), the potential energy of interaction between the particles is usually a function of only the relative coordinates $x_2 - x_1$, $y_2 - y_1$, and $z_2 - z_1$ of the particles. In this case the two-particle problem can be simplified to two separate one-particle problems, as we now prove.

Consider the classical-mechanical treatment of two interacting particles of masses m_1 and m_2. We specify their positions by the radius vectors $\mathbf{r}_1$ and $\mathbf{r}_2$ drawn from the origin of a Cartesian coordinate system (Fig. 6.1). Particles 1 and 2 have coordinates (x_1, y_1, z_1) and (x_2, y_2, z_2). We draw the vector $\mathbf{r} = \mathbf{r}_2 - \mathbf{r}_1$ from particle 1 to 2 and denote the components of $\mathbf{r}$ by x, y, and z:

$$\boxed{x = x_2 - x_1, \quad y = y_2 - y_1, \quad z = z_2 - z_1} \qquad \textbf{(6.28)}$$

The coordinates x, y, and z are called the **relative** or **internal coordinates**.

We now draw the vector $\mathbf{R}$ from the origin to the system's center of mass, point C, and denote the coordinates of C by X, Y, and Z:

$$\mathbf{R} = \mathbf{i}X + \mathbf{j}Y + \mathbf{k}Z \qquad (6.29)$$

The definition of the center of mass of this two-particle system gives

$$X = \frac{m_1 x_1 + m_2 x_2}{m_1 + m_2}, \quad Y = \frac{m_1 y_1 + m_2 y_2}{m_1 + m_2}, \quad Z = \frac{m_1 z_1 + m_2 z_2}{m_1 + m_2} \qquad (6.30)$$

These three equations are equivalent to the vector equation

$$\mathbf{R} = \frac{m_1 \mathbf{r}_1 + m_2 \mathbf{r}_2}{m_1 + m_2} \qquad (6.31)$$

We also have

$$\mathbf{r} = \mathbf{r}_2 - \mathbf{r}_1 \qquad (6.32)$$

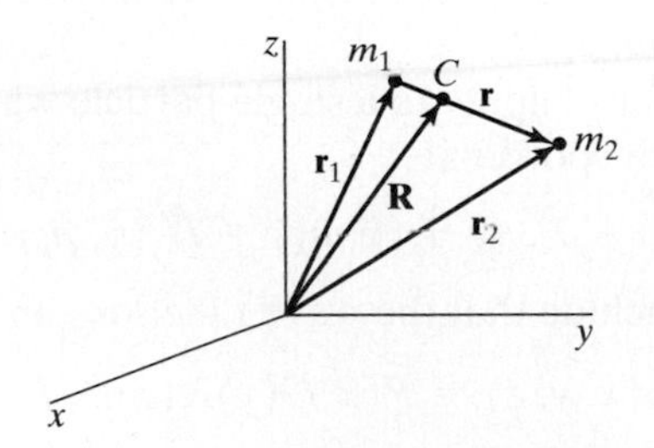

FIGURE 6.1 A two-particle system with center of mass at C.

We regard (6.31) and (6.32) as simultaneous equations in the two unknowns $\mathbf{r}_1$ and $\mathbf{r}_2$ and solve for them to get

$$\mathbf{r}_1 = \mathbf{R} - \frac{m_2}{m_1 + m_2}\mathbf{r}, \qquad \mathbf{r}_2 = \mathbf{R} + \frac{m_1}{m_1 + m_2}\mathbf{r} \tag{6.33}$$

Equations (6.31) and (6.32) represent a transformation of coordinates from $x_1, y_1, z_1, x_2, y_2, z_2$ to X, Y, Z, x, y, z. Consider what happens to the Hamiltonian under this transformation. Let an overhead dot indicate differentiation with respect to time. The velocity of particle 1 is [Eq. (5.34)] $\mathbf{v}_1 = d\mathbf{r}_1/dt = \dot{\mathbf{r}}_1$. The kinetic energy is the sum of the kinetic energies of the two particles:

$$T = \tfrac{1}{2}m_1|\dot{\mathbf{r}}_1|^2 + \tfrac{1}{2}m_2|\dot{\mathbf{r}}_2|^2 \tag{6.34}$$

Introducing the time derivatives of Eqs. (6.33) into (6.34), we have

$$T = \frac{1}{2}m_1\left(\dot{\mathbf{R}} - \frac{m_2}{m_1 + m_2}\dot{\mathbf{r}}\right)\cdot\left(\dot{\mathbf{R}} - \frac{m_2}{m_1 + m_2}\dot{\mathbf{r}}\right)$$

$$+ \frac{1}{2}m_2\left(\dot{\mathbf{R}} + \frac{m_1}{m_1 + m_2}\dot{\mathbf{r}}\right)\cdot\left(\dot{\mathbf{R}} + \frac{m_1}{m_1 + m_2}\dot{\mathbf{r}}\right)$$

where $|\mathbf{A}|^2 = \mathbf{A}\cdot\mathbf{A}$ [Eq. (5.24)] has been used. Using the distributive law for the dot products, we find, after simplifying,

$$T = \frac{1}{2}(m_1 + m_2)|\dot{\mathbf{R}}|^2 + \frac{1}{2}\frac{m_1 m_2}{m_1 + m_2}|\dot{\mathbf{r}}|^2 \tag{6.35}$$

Let M be the total mass of the system:

$$M \equiv m_1 + m_2 \tag{6.36}$$

We define the **reduced mass** μ of the two-particle system as

$$\boxed{\mu \equiv \frac{m_1 m_2}{m_1 + m_2}} \tag{6.37}$$

Then

$$T = \tfrac{1}{2}M|\dot{\mathbf{R}}|^2 + \tfrac{1}{2}\mu|\dot{\mathbf{r}}|^2 \tag{6.38}$$

The first term in (6.38) is the kinetic energy due to translational motion of the whole system of mass M. **Translational motion** is motion in which each particle undergoes the same displacement. The quantity $\frac{1}{2}M|\dot{\mathbf{R}}|^2$ would be the kinetic energy of a hypothetical particle of mass M located at the center of mass. The second term in (6.38) is the kinetic energy of internal (relative) motion of the two particles. This internal motion is of two types. The distance r between the two particles can change (vibration), and the direction of the $\mathbf{r}$ vector can change (rotation). Note that $|\dot{\mathbf{r}}| = |d\mathbf{r}/dt| \neq d|\mathbf{r}|/dt$.

Corresponding to the original coordinates $x_1, y_1, z_1, x_2, y_2, z_2$, we had six linear momenta:

$$p_{x_1} = m_1\dot{x}_1, \quad \ldots, \quad p_{z_2} = m_2\dot{z}_2 \tag{6.39}$$

Comparing Eqs. (6.34) and (6.38), we define the six linear momenta for the new coordinates X, Y, Z, x, y, z as

$$p_X \equiv M\dot{X}, \qquad p_Y \equiv M\dot{Y}, \qquad p_Z \equiv M\dot{Z}$$
$$p_x \equiv \mu\dot{x}, \qquad p_y \equiv \mu\dot{y}, \qquad p_z \equiv \mu\dot{z}$$

We define two new momentum vectors as

$$\mathbf{p}_M \equiv \mathbf{i}M\dot{X} + \mathbf{j}M\dot{Y} + \mathbf{k}M\dot{Z} \quad \text{and} \quad \mathbf{p}_\mu \equiv \mathbf{i}\mu\dot{x} + \mathbf{j}\mu\dot{y} + \mathbf{k}\mu\dot{z}$$

Introducing these momenta into (6.38), we have

$$T = \frac{|\mathbf{p}_M|^2}{2M} + \frac{|\mathbf{p}_\mu|^2}{2\mu} \tag{6.40}$$

Now consider the potential energy. We make the restriction that V is a function *only* of the relative coordinates x, y, and z of the two particles:

$$V = V(x, y, z) \tag{6.41}$$

An example of (6.41) is two charged particles interacting according to Coulomb's law [see Eq. (3.53)]. With this restriction on V, the Hamiltonian function is

$$H = \frac{p_M^2}{2M} + \left[\frac{p_\mu^2}{2\mu} + V(x, y, z)\right] \tag{6.42}$$

Now suppose we had a system composed of a particle of mass M subject to no forces and a particle of mass μ subject to the potential-energy function $V(x, y, z)$, and further suppose that there was no interaction between these particles. If (X, Y, Z) are the coordinates of the particle of mass M, and (x, y, z) are the coordinates of the particle of mass μ, what is the Hamiltonian of this hypothetical system? Clearly, it is identical with (6.42).

The Hamiltonian (6.42) can be viewed as the sum of the Hamiltonians $p_M^2/2M$ and $[p_\mu^2/2\mu + V(x, y, z)]$ of two hypothetical noninteracting particles with masses M and μ. Therefore, the results of Section 6.2 show that the system's quantum-mechanical energy is the sum of energies of the two hypothetical particles [Eq. (6.23)]: $E = E_M + E_\mu$. From Eqs. (6.24) and (6.42), the translational energy E_M is found by solving the Schrödinger equation $(\hat{p}_M^2/2M)\psi_M = E_M\psi_M$. This is the Schrödinger equation for a free particle of mass M, so its possible eigenvalues are all nonnegative numbers: $E_M \geq 0$ [Eq. (2.31)]. From (6.24) and (6.42), the energy E_μ is found by solving the Schrödinger equation

$$\left[\frac{\hat{p}_\mu^2}{2\mu} + V(x, y, z)\right]\psi_\mu(x, y, z) = E_\mu\psi_\mu(x, y, z) \tag{6.43}$$

We have thus separated the problem of two particles interacting according to a potential-energy function $V(x, y, z)$ that depends on only the relative coordinates x, y, z into two separate one-particle problems: (1) the translational motion of the entire system of mass M, which simply adds a nonnegative constant energy E_M to the system's energy, and (2) the relative or internal motion, which is dealt with by solving the Schrödinger equation (6.43) for a hypothetical particle of mass μ whose coordinates are the relative coordinates x, y, z and that moves subject to the potential energy $V(x, y, z)$.

For example, for the hydrogen atom, which is composed of an electron (e) and a proton (p), the atom's total energy is $E = E_M + E_\mu$, where E_M is the translational energy of motion through space of the entire atom of mass $M = m_e + m_p$, and where E_μ is found by solving (6.43) with $\mu = m_e m_p/(m_e + m_p)$ and V being the Coulomb's law potential energy of interaction of the electron and proton; see Section 6.5.

6.4 THE TWO-PARTICLE RIGID ROTOR

Before solving the Schrödinger equation for the hydrogen atom, we will first deal with the two-particle rigid rotor. This is a two-particle system with the particles held at a fixed distance from each other by a rigid massless rod of length d. For this problem, the vector $\mathbf{r}$ in Fig. 6.1 has the constant magnitude $|\mathbf{r}| = d$. Therefore (see Section 6.3), the kinetic energy of internal motion is wholly rotational energy. The energy of the rotor is entirely kinetic, and

$$V = 0 \tag{6.44}$$

Equation (6.44) is a special case of Eq. (6.41), and we may therefore use the results of the last section to separate off the translational motion of the system as a whole. We will concern ourselves only with the rotational energy. The Hamiltonian operator for the rotation is given by the terms in brackets in (6.43) as

$$\hat{H} = \frac{\hat{p}_\mu^2}{2\mu} = -\frac{\hbar^2}{2\mu}\nabla^2, \qquad \mu = \frac{m_1 m_2}{m_1 + m_2} \tag{6.45}$$

where m_1 and m_2 are the masses of the two particles. The coordinates of the fictitious particle of mass μ are the relative coordinates of m_1 and m_2 [Eq. (6.28)].

Instead of the relative Cartesian coordinates x, y, z, it will prove more fruitful to use the relative spherical coordinates r, θ, ϕ. The r coordinate is equal to the magnitude of the $\mathbf{r}$ vector in Fig. 6.1, and since m_1 and m_2 are constrained to remain a fixed distance apart, we have $r = d$. Thus the problem is equivalent to a particle of mass μ constrained to move on the surface of a sphere of radius d. Because the radial coordinate is constant, the wave function will be a function of θ and ϕ only. Hence the first two terms of the Laplacian operator in (6.8) will give zero when operating on the wave function and may be omitted. Looking at things in a slightly different way, we note that the operators in (6.8) that involve r derivatives correspond to the kinetic energy of radial motion, and since there is no radial motion, the r derivatives are omitted from $\hat{H}$.

Since $V = 0$ is a special case of $V = V(r)$, the results of Section 6.1 tell us that the eigenfunctions are given by (6.16) with the r factor omitted:

$$\psi = Y_J^m(\theta, \phi) \tag{6.46}$$

where J rather than l is used for the rotational angular-momentum quantum number.

The Hamiltonian operator is given by Eq. (6.8) with the r derivatives omitted and $V(r) = 0$. Thus

$$\hat{H} = (2\mu d^2)^{-1}\hat{L}^2$$

Use of (6.13) gives

$$\hat{H}\psi = E\psi$$

$$(2\mu d^2)^{-1}\hat{L}^2 Y_J^m(\theta, \phi) = EY_J^m(\theta, \phi)$$

$$(2\mu d^2)^{-1}J(J + 1)\hbar^2 Y_J^m(\theta, \phi) = EY_J^m(\theta, \phi)$$

$$E = \frac{J(J + 1)\hbar^2}{2\mu d^2}, \qquad J = 0, 1, 2 \cdots \tag{6.47}$$

The **moment of inertia** I of a system of n particles about some particular axis in space as defined as

$$I \equiv \sum_{i=1}^{n} m_i \rho_i^2 \tag{6.48}$$

where m_i is the mass of the ith particle and ρ_i is the perpendicular distance from this particle to the axis. The value of I depends on the choice of axis. For the two-particle rigid rotor, we choose our axis to be a line that passes through the center of mass and is perpendicular to the line joining m_1 and m_2 (Fig. 6.2). If we place the rotor so that the center of mass, point C, lies at the origin of a Cartesian coordinate system and the line joining m_1 and m_2 lies on the x axis, then C will have the coordinates $(0, 0, 0)$, m_1 will have the coordinates $(-\rho_1, 0, 0)$, and m_2 will have the coordinates $(\rho_2, 0, 0)$. Using these coordinates in (6.30), we find

$$m_1\rho_1 = m_2\rho_2 \tag{6.49}$$

The moment of inertia of the rotor about the axis we have chosen is

$$I = m_1\rho_1^2 + m_2\rho_2^2 \tag{6.50}$$

Using (6.49), we transform Eq. (6.50) to (see Prob. 6.12)

$$\boxed{I = \mu d^2} \tag{6.51}$$

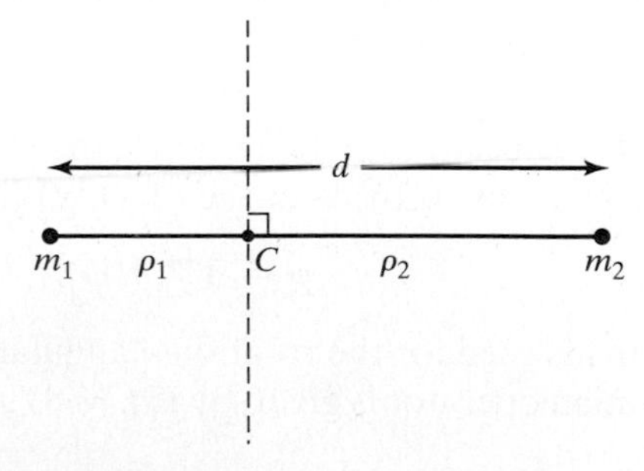

FIGURE 6.2 Axis (dashed line) for calculating the moment of inertia of a two-particle rigid rotor. C is the center of mass.

where $\mu \equiv m_1 m_2/(m_1 + m_2)$ is the reduced mass of the system and $d \equiv \rho_1 + \rho_2$ is the distance between m_1 and m_2. The allowed energy levels (6.47) of the two-particle rigid rotor are

$$\boxed{E = \frac{J(J+1)\hbar^2}{2I}, \quad J = 0, 1, 2, \ldots} \tag{6.52}$$

The lowest level is $E = 0$, so there is no zero-point rotational energy. Having zero rotational energy and therefore zero angular momentum for the rotor does not violate the uncertainty principle; recall the discussion following Eq. (5.105). Note that E increases as $J^2 + J$, so the spacing between adjacent rotational levels increases as J increases.

Are the rotor energy levels (6.52) degenerate? The energy depends on J only, but the wave function (6.46) depends on J and m, where $m\hbar$ is the z component of the rotor's angular momentum. For each value of J, there are $2J + 1$ values of m, ranging from $-J$ to J. Hence the levels are $(2J + 1)$-fold degenerate. The states of a degenerate level have different orientations of the angular-momentum vector of the rotor about a space-fixed axis.

The angles θ and ϕ in the wave function (6.46) are relative coordinates of the two point masses. If we set up a Cartesian coordinate system with the origin at the rotor's center of mass, θ and ϕ will be as shown in Fig. 6.3. This coordinate system undergoes the same translational motion as the rotor's center of mass but does not rotate in space.

The rotational angular momentum $[J(J+1)\hbar^2]^{1/2}$ is the angular momentum of the two particles with respect to an origin at the system's center of mass C.

The rotational levels of a diatomic molecule can be well approximated by the two-particle rigid-rotor energies (6.52). It is found (*Levine, Molecular Spectroscopy*, Section 4.4) that when a diatomic molecule absorbs or emits radiation, the allowed pure-rotational transitions are

$$\Delta J = \pm 1 \tag{6.53}$$

In addition, a molecule must have a nonzero dipole moment in order to show a pure-rotational spectrum. A *pure-rotational transition* is one where only the rotational

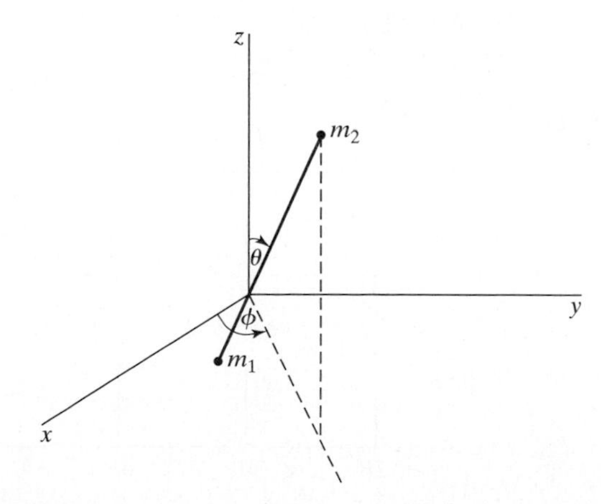

FIGURE 6.3 Coordinate system for the two-particle rigid rotor.

quantum number changes. [Vibration–rotation transitions (Section 4.3) involve changes in both vibrational and rotational quantum numbers.] The spacing between adjacent low-lying rotational levels is significantly less than that between adjacent vibrational levels, and the pure-rotational spectrum falls in the microwave (or the far-infrared) region. The frequencies of the pure-rotational spectral lines of a diatomic molecule are then (approximately)

$$\nu = \frac{E_{J+1} - E_J}{h} = \frac{[(J+1)(J+2) - J(J+1)]h}{8\pi^2 I} = 2(J+1)B \qquad (6.54)$$

$$B \equiv h/8\pi^2 I, \qquad J = 0, 1, 2, \ldots \qquad (6.55)$$

B is called the **rotational constant** of the molecule.

The spacings between the diatomic rotational levels (6.52) for low and moderate values of J are generally less than or of the same order of magnitude as kT at room temperature, so the Boltzmann distribution law (4.65) shows that many rotational levels are significantly populated at room temperature. Absorption of radiation by diatomic molecules having $J = 0$ (the $J = 0 \rightarrow 1$ transition) gives a line at the frequency $2B$; absorption by molecules having $J = 1$ (the $J = 1 \rightarrow 2$ transition) gives a line at $4B$; absorption by $J = 2$ molecules gives a line at $6B$; and so on. See Fig. 6.4.

Measurement of the rotational absorption frequencies allows B to be found. From B, we get the molecule's moment of inertia I, and from I we get the bond distance d. The value of d found is an average over the $v = 0$ vibrational motion. Because of the asymmetry of the potential-energy curve in Figs. 4.5 and 13.1, d is very slightly longer than the equilibrium bond length R_e in Fig. 13.1.

As noted in Section 4.3, isotopic species such as $^1H^{35}Cl$ and $^1H^{37}Cl$ have virtually the same electronic energy curve $U(R)$ and so have virtually the same equilibrium bond distance. However, the different isotopic masses produce different moments of inertia and hence different rotational absorption frequencies.

For more discussion of nuclear motion in diatomic molecules, see Section 13.2. For the rotational energies of polyatomic molecules, see *Levine, Molecular Spectroscopy,* Chapter 5.

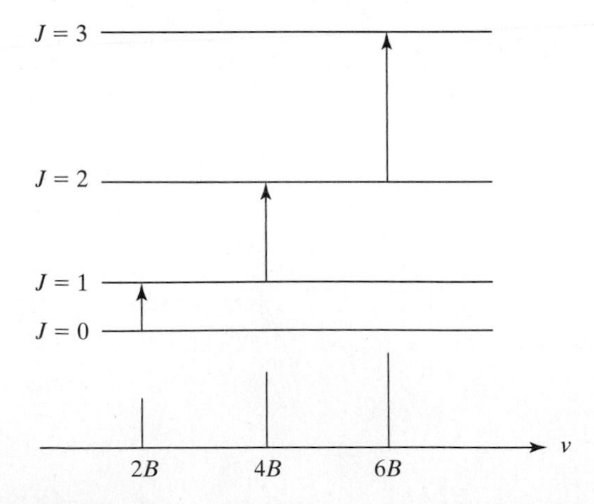

FIGURE 6.4 Two-particle rigid-rotor absorption transitions.

EXAMPLE The lowest-frequency pure-rotational absorption line of $^{12}C^{32}S$ occurs at 48991.0 MHz. Find the bond distance in $^{12}C^{32}S$.

The lowest-frequency rotational absorption is the $J = 0 \rightarrow 1$ line. Equations (1.4), (6.52), and (6.51) give

$$h\nu = E_{\text{upper}} - E_{\text{lower}} = \frac{1(2)\hbar^2}{2\mu d^2} - \frac{0(1)\hbar^2}{2\mu d^2}$$

which gives $d = (h/4\pi^2\nu\mu)^{1/2}$. Table A.3 in the Appendix gives

$$\mu = \frac{m_1 m_2}{m_1 + m_2} = \frac{12(31.97207)}{12 + 31.97207} \frac{1}{6.02214 \times 10^{23}} \text{ g} = 1.44885 \times 10^{-23} \text{ g}$$

The SI unit of mass is the kilogram, and

$$d = \frac{1}{2\pi}\left(\frac{h}{\nu_{0\rightarrow 1}\mu}\right)^{1/2} = \frac{1}{2\pi}\left[\frac{6.62607 \times 10^{-34} \text{ J s}}{(48991.0 \times 10^6 \text{ s}^{-1})(1.44885 \times 10^{-26} \text{ kg})}\right]^{1/2}$$

$$= 1.5377 \times 10^{-10} \text{ m} = 1.5377 \text{ Å}$$

EXERCISE The $J = 1$ to $J = 2$ pure rotational transition of $^{12}C^{16}O$ occurs at 230.538 GHz (1 GHz $= 10^9$ Hz.) Find the bond distance in this molecule. (*Answer:* 1.1309×10^{-10} m.)

6.5 THE HYDROGEN ATOM

The hydrogen atom consists of a proton and an electron. If e symbolizes the charge on the proton ($e = +1.6 \times 10^{-19}$ C), then the electron's charge is $-e$.

A few scientists have speculated that the proton and electron charges might not be exactly equal in magnitude. Experiments show that the magnitudes of the electron and proton charges are equal to within one part in 10^{21}. See H. F. Dylla and J. G. King, *Phys. Rev. A*, **7**, 1224 (1973); M. Marinelli and G. Morpurgo, *Phys. Lett. B*, **137**, 439 (1984).

We shall assume the electron and proton to be point masses whose interaction is given by Coulomb's law. In discussing atoms and molecules, we shall usually be considering isolated systems, ignoring interatomic and intermolecular interactions.

Instead of treating just the hydrogen atom, we consider a slightly more general problem: the **hydrogenlike atom**, which consists of one electron and a nucleus of charge Ze. For $Z = 1$, we have the hydrogen atom; for $Z = 2$, the He^+ ion; for $Z = 3$, the Li^{2+} ion; and so on. The hydrogenlike atom is the most important system in quantum chemistry. An exact solution of the Schrödinger equation for atoms with more than one electron cannot be obtained because of the interelectronic repulsions. If, as a crude first approximation, we ignore these repulsions, then the electrons can be treated independently. (See Section 6.2.) The atomic wave function will be approximated by a product of one-electron functions, which will be hydrogenlike wave functions. A one-electron wave function (whether or not it is hydrogenlike) is called an **orbital**. (More precisely, an **orbital** is a one-electron spatial wave function, where the word *spatial* means that the wave function depends on the electron's three spatial coordinates x, y, and z or r, θ, and ϕ. We shall see in Chapter 10 that the existence of electron spin adds a fourth coordinate to a one-electron wave function, giving what is called a spin-orbital.) An orbital for an electron in an atom is called an **atomic orbital**. We shall use atomic orbitals to construct approximate wave functions for atoms with many electrons (Chapter 11). Orbitals are also used to construct approximate wave functions for molecules.

For the hydrogenlike atom, let (x, y, z) be the coordinates of the electron relative to the nucleus, and let $\mathbf{r} = \mathbf{i}x + \mathbf{j}y + \mathbf{k}z$. The Coulomb's law force on the electron in the hydrogenlike atom is [see Eq. (1.37)]

$$\mathbf{F} = -\frac{Ze'^2}{r^2}\frac{\mathbf{r}}{r} \tag{6.56}$$

where $\mathbf{r}/r$ is a unit vector in the $\mathbf{r}$ direction, and $e' \equiv e/(4\pi\varepsilon_0)^{1/2}$ [Eq. (1.38)]. The minus sign indicates an attractive force.

The possibility of small deviations from Coulomb's law has been considered. Experiments have shown that if the Coulomb's-law force is written as being proportional to r^{-2+s}, then $|s| < 10^{-16}$. It can be shown that a deviation from Coulomb's law would imply a nonzero rest mass for the photon; see A. S. Goldhaber and M. M. Nieto, *Rev. Mod. Phys.*, **43**, 277 (1971). There is little or no evidence for a nonzero photon rest mass, and data indicate that any such mass must be less than 10^{-49} g; L.-C. Tuo, J. Luo, and G. T. Gillies, *Rep. Progr. Phys.*, **68**, 77 (2005).

The force in (6.56) is central, and comparison with Eq. (6.4) gives $dV(r)/dr = Ze'^2/r^2$. Integration gives

$$V = Ze'^2 \int \frac{dr}{r^2} = -\frac{Ze'^2}{r} \tag{6.57}$$

where the integration constant has been taken as 0 to make $V = 0$ at infinite separation between the charges. For any two charges Q_1 and Q_2 separated by distance r_{12}, Eq. (6.57) becomes

$$\boxed{V = \frac{Q_1'Q_2'}{r_{12}}} \tag{6.58}$$

Since the potential energy of this two-particle system depends only on the relative coordinates of the particles, we can apply the results of Section 6.3 to reduce the problem to two one-particle problems. The translational motion of the atom as a whole simply adds some constant to the total energy, and we shall not concern ourselves with it. To deal with the internal motion of the system, we introduce a fictitious particle of mass

$$\mu = \frac{m_e m_N}{m_e + m_N} \tag{6.59}$$

where m_e and m_N are the electronic and nuclear masses. The particle of reduced mass μ moves subject to the potential-energy function (6.57), and its coordinates (r, θ, ϕ) are the spherical coordinates of one particle relative to the other (Fig. 6.5).

The Hamiltonian for the internal motion is [Eq. (6.43)]

$$\hat{H} = -\frac{\hbar^2}{2\mu}\nabla^2 - \frac{Ze'^2}{r} \tag{6.60}$$

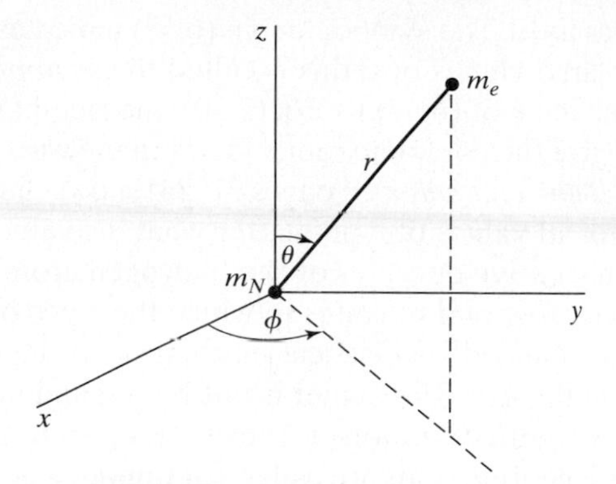

FIGURE 6.5 Relative spherical coordinates.

Since V is a function of the r coordinate only, we have a one-particle central-force problem, and we may apply the results of Section 6.1. Using Eqs. (6.16) and (6.17), we have for the wave function

$$\psi(r, \theta, \phi) = R(r)Y_l^m(\theta, \phi), \qquad l = 0, 1, 2, \ldots, \qquad |m| \le l \tag{6.61}$$

where Y_l^m is a spherical harmonic, and the radial function $R(r)$ satisfies

$$-\frac{\hbar^2}{2\mu}\left(R'' + \frac{2}{r}R'\right) + \frac{l(l+1)\hbar^2}{2\mu r^2}R - \frac{Ze'^2}{r}R = ER(r) \tag{6.62}$$

To save time in writing, we define the constant a as

$$a \equiv \hbar^2/\mu e'^2 = 4\pi\varepsilon_0\hbar^2/\mu e^2 \tag{6.63}$$

and (6.62) becomes

$$R'' + \frac{2}{r}R' + \left[\frac{2E}{ae'^2} + \frac{2Z}{ar} - \frac{l(l+1)}{r^2}\right]R = 0 \tag{6.64}$$

Solution of the Radial Equation. We could now try a power-series solution of (6.64), but we would get a three-term rather than a two-term recursion relation. We therefore seek a substitution that will lead to a two-term recursion relation. It turns out that the proper substitution can be found by examining the behavior of the solution for large values of r. For large r, (6.64) becomes

$$R'' + \frac{2E}{ae'^2}R = 0, \qquad r \text{ large} \tag{6.65}$$

which may be solved using the auxiliary equation (2.7). The solutions are

$$\exp[\pm(-2E/ae'^2)^{1/2}r] \tag{6.66}$$

Suppose that E is positive. The quantity under the square-root sign in (6.66) is negative, and the factor multiplying r is imaginary:

$$R(r) \sim e^{\pm i\sqrt{2\mu E}r/\hbar}, \qquad E \ge 0 \tag{6.67}$$

where (6.63) was used. The symbol $\sim$ in (6.67) indicates that we are giving the behavior of $R(r)$ for large values of r; this is called the *asymptotic* behavior of the function. Note the resemblance of (6.67) to Eq. (2.30), the free-particle wave function. Equation (6.67) does not give the complete radial factor in the wave function for positive energies. Further study (*Bethe and Salpeter,* pages 21–24) shows that the radial function for $E \geq 0$ remains finite for all values of r, no matter what the value of E. Thus, just as for the free particle, all nonnegative energies of the hydrogen atom are allowed. Physically, these eigenfunctions correspond to states in which the electron is not bound to the nucleus; that is, the atom is ionized. (A classical-mechanical analogy is a comet moving in a hyperbolic orbit about the sun. The comet is not bound and makes but one visit to the solar system.) Since we get continuous rather than discrete allowed values for $E \geq 0$, the positive-energy eigenfunctions are called **continuum eigenfunctions**. The angular part of a continuum wave function is a spherical harmonic. Like the free-particle wave functions, the continuum eigenfunctions are not normalizable in the usual sense.

We now consider the *bound* states of the hydrogen atom, with $E < 0$. (For a **bound state,** $\psi \to 0$ as $x \to \pm\infty$.) In this case, the quantity in parentheses in (6.66) is positive. Since we want the wave functions to remain finite as r goes to infinity, we prefer the minus sign in (6.66), and in order to get a two-term recursion relation, we make the substitution

$$R(r) = e^{-Cr}K(r) \tag{6.68}$$

$$C \equiv \left(-\frac{2E}{ae'^2}\right)^{1/2} \tag{6.69}$$

where e in (6.68) stands for the base of natural logarithms, and not the proton charge. Use of the substitution (6.68) will guarantee nothing about the behavior of the wave function for large r. The differential equation we obtain from this substitution will still have two linearly independent solutions. We can make any substitution we please in a differential equation; in fact, we could make the substitution $R(r) = e^{+Cr}J(r)$ and still wind up with the correct eigenfunctions and eigenvalues. The relation between J and K would naturally be $J(r) = e^{-2Cr}K(r)$.

Proceeding with (6.68), we evaluate R' and R'', substitute into (6.64), multiply by r^2e^{Cr}, and use (6.69) to get the following differential equation for $K(r)$:

$$r^2K'' + (2r - 2Cr^2)K' + [(2Za^{-1} - 2C)r - l(l+1)]K = 0 \tag{6.70}$$

We could now substitute a power series of the form

$$K = \sum_{k=0}^{\infty} c_k r^k \tag{6.71}$$

for K. If we did we would find that, in general, the first few coefficients in (6.71) are zero. If c_s is the first nonzero coefficient, (6.71) can be written as

$$K = \sum_{k=s}^{\infty} c_k r^k, \qquad c_s \neq 0 \tag{6.72}$$

Letting $j \equiv k - s$, and then defining b_j as $b_j \equiv c_{j+s}$, we have

$$K = \sum_{j=0}^{\infty} c_{j+s} r^{j+s} = r^s \sum_{j=0}^{\infty} b_j r^j, \qquad b_0 \neq 0 \tag{6.73}$$

(Although the various substitutions we are making might seem arbitrary, they are standard procedure in solving differential equations by power series.) The integer s is evaluated by substitution into the differential equation. Equation (6.73) is

$$K(r) = r^s M(r) \tag{6.74}$$

$$M(r) = \sum_{j=0}^{\infty} b_j r^j, \qquad b_0 \neq 0 \tag{6.75}$$

Evaluating K' and K'' from (6.74) and substituting into (6.70), we get

$$r^2 M'' + [(2s+2)r - 2Cr^2]M' + [s^2 + s + (2Za^{-1} - 2C - 2Cs)r - l(l+1)]M = 0 \tag{6.76}$$

To find s, we look at (6.76) for $r = 0$. From (6.75), we have

$$M(0) = b_0, \qquad M'(0) = b_1, \qquad M''(0) = 2b_2 \tag{6.77}$$

Using (6.77) in (6.76), we find for $r = 0$

$$b_0(s^2 + s - l^2 - l) = 0 \tag{6.78}$$

Since b_0 is not zero, the terms in parentheses must vanish: $s^2 + s - l^2 - l = 0$. This is a quadratic equation in the unknown s, with the roots

$$s = l, \qquad s = -l - 1 \tag{6.79}$$

These roots correspond to the two linearly independent solutions of the differential equation. Let us examine them from the standpoint of proper behavior of the wave function. From Eqs. (6.68), (6.74), and (6.75), we have

$$R(r) = e^{-Cr} r^s \sum_{j=0}^{\infty} b_j r^j \tag{6.80}$$

Since $e^{-Cr} = 1 - Cr + \dots$, the function $R(r)$ behaves for small r as $b_0 r^s$. For the root $s = l$, $R(r)$ behaves properly at the origin. However, for $s = -l - 1$, $R(r)$ is proportional to

$$\frac{1}{r^{l+1}} \tag{6.81}$$

for small r. Since $l = 0, 1, 2, \dots$, the root $s = -l - 1$ makes the radial factor in the wave function infinite at the origin. Many texts take this as sufficient reason for rejecting this root. However, this is not a good argument, since for the *relativistic* hydrogen atom, the $l = 0$ eigenfunctions are infinite at $r = 0$. Let us therefore look at (6.81) from the standpoint of quadratic integrability, since we certainly require the bound-state eigenfunctions to be normalizable.

The normalization integral [Eq. (5.80)] for the radial functions that behave like (6.81) looks like

$$\int_0 |R|^2 r^2\, dr \approx \int_0 \frac{1}{r^{2l}}\, dr \tag{6.82}$$

for small r. The behavior of the integral at the lower limit of integration is

$$\left.\frac{1}{r^{2l-1}}\right|_{r=0} \tag{6.83}$$

For $l = 1, 2, 3, \ldots$, (6.83) is infinite, and the normalization integral is infinite. Hence we must reject the root $s = -l - 1$ for $l \geq 1$. However, for $l = 0$, (6.83) is finite, and there is no trouble with quadratic integrability. Thus there is a quadratically integrable solution to the radial equation that behaves as r^{-1} for small r.

Further study of this solution shows that it corresponds to an energy value that the experimental hydrogen-atom spectrum shows does not exist. Thus the r^{-1} solution must be rejected, but there is some dispute over the reason for doing so. One view is that the $1/r$ solution satisfies the Schrödinger equation everywhere in space except at the origin and hence must be rejected [*Dirac,* page 156; B. H. Armstrong and E. A. Power, *Am. J. Phys.,* **31**, 262 (1963)]. A second view is that the $1/r$ solution must be rejected because the Hamiltonian operator is not Hermitian with respect to it (*Merzbacher,* Section 10.5). (In Chapter 7 we shall define Hermitian operators and show that quantum-mechanical operators are required to be Hermitian.)

Taking the first root in (6.79), we have for the radial factor (6.80)

$$R(r) = e^{-Cr} r^l M(r) \tag{6.84}$$

With $s = l$, Eq. (6.76) becomes

$$rM'' + (2l + 2 - 2Cr)M' + (2Za^{-1} - 2C - 2Cl)M = 0 \tag{6.85}$$

From (6.75), we have

$$M(r) = \sum_{j=0}^{\infty} b_j r^j \tag{6.86}$$

$$M' = \sum_{j=0}^{\infty} j b_j r^{j-1} = \sum_{j=1}^{\infty} j b_j r^{j-1} = \sum_{k=0}^{\infty} (k + 1) b_{k+1} r^k = \sum_{j=0}^{\infty} (j + 1) b_{j+1} r^j$$

$$M'' = \sum_{j=0}^{\infty} j(j - 1) b_j r^{j-2} = \sum_{j=1}^{\infty} j(j - 1) b_j r^{j-2} = \sum_{k=0}^{\infty} (k + 1) k b_{k+1} r^{k-1}$$

$$M'' = \sum_{j=0}^{\infty} (j + 1) j b_{j+1} r^{j-1} \tag{6.87}$$

Substituting these expressions in (6.85) and combining sums, we get

$$\sum_{j=0}^{\infty} \left[j(j + 1) b_{j+1} + 2(l + 1)(j + 1) b_{j+1} + \left(\frac{2Z}{a} - 2C - 2Cl - 2Cj \right) b_j \right] r^j = 0$$

Setting the coefficient of r^j equal to zero, we get the recursion relation

$$b_{j+1} = \frac{2C + 2Cl + 2Cj - 2Za^{-1}}{j(j + 1) + 2(l + 1)(j + 1)} b_j \tag{6.88}$$

We now must examine the behavior of the infinite series (6.86) for large r. The result of the same procedure used to examine the harmonic-oscillator power series in (4.44) suggests that for large r the infinite series (6.86) behaves like e^{2Cr}. (See Prob. 6.18.) For large r, the radial function (6.84) behaves like

$$R(r) \sim e^{-Cr} r^l e^{2Cr} = r^l e^{Cr} \tag{6.89}$$

Therefore, $R(r)$ will become infinite as r goes to infinity and will not be quadratically integrable. The only way to avoid this "infinity catastrophe" (as in the harmonic-oscillator case) is to have the series terminate after a finite number of terms, in which case the e^{-Cr} factor will ensure that the wave function goes to zero as r goes to infinity. Let the last term in the series be $b_k r^k$. Then, to have $b_{k+1}, b_{k+2}, \ldots$ all vanish, the fraction multiplying b_j in the recursion relation (6.88) must vanish when $j = k$. We have

$$2C(k + l + 1) = 2Za^{-1}, \qquad k = 0, 1, 2, \ldots \tag{6.90}$$

k and l are integers, and we now define a new integer n by

$$n \equiv k + l + 1, \qquad n = 1, 2, 3, \ldots \tag{6.91}$$

From (6.91) the quantum number l must satisfy

$$l \leq n - 1 \tag{6.92}$$

Hence l ranges from 0 to $n - 1$.

Energy Levels. Use of (6.91) in (6.90) gives

$$Cn = Za^{-1} \tag{6.93}$$

Substituting $C \equiv (-2E/ae'^2)^{1/2}$ [Eq. (6.69)] into (6.93) and solving for E, we get

$$E = -\frac{Z^2}{n^2}\frac{e'^2}{2a} = -\frac{Z^2\mu e'^4}{2n^2\hbar^2} = -\frac{Z^2\mu e^4}{8\varepsilon_0^2 n^2 h^2} \tag{6.94}$$

where $a \equiv \hbar^2/\mu e'^2$ [Eq. (6.63)]. These are the bound-state energy levels of the hydrogenlike atom, and they are discrete. Figure 6.6 shows the potential-energy curve [Eq. (6.57)] and some of the allowed energy levels for the hydrogen atom ($Z = 1$). The crosshatching indicates that all positive energies are allowed.

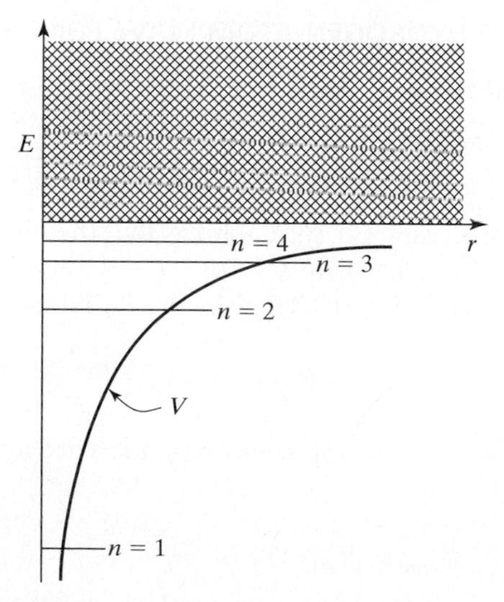

FIGURE 6.6 Energy levels of the hydrogen atom.

It turns out that all changes in n are allowed in light absorption and emission. The wavenumbers [Eq. (4.66)] of H-atom spectral lines are then

$$\tilde{\nu} \equiv \frac{1}{\lambda} = \frac{\nu}{c} = \frac{E_2 - E_1}{hc} = \frac{e'^2}{2ahc}\left(\frac{1}{n_1^2} - \frac{1}{n_2^2}\right) \equiv R_{\mathrm{H}}\left(\frac{1}{n_1^2} - \frac{1}{n_2^2}\right) \quad (6.95)$$

where $R_{\mathrm{H}} = 109677.6 \text{ cm}^{-1}$ is the *Rydberg constant* for hydrogen.

Degeneracy. Are the hydrogen-atom energy levels degenerate? For the bound states, the energy (6.94) depends only on n. However, the wave function (6.61) depends on all three quantum numbers n, l, and m, whose allowed values are [Eqs. (6.91), (6.92), (5.104), and (5.105)]

$$\boxed{n = 1, 2, 3, \ldots} \quad \mathbf{(6.96)}$$

$$\boxed{l = 0, 1, 2, \ldots, n - 1} \quad \mathbf{(6.97)}$$

$$\boxed{m = -l, -l + 1, \ldots, 0, \ldots, l - 1, l} \quad \mathbf{(6.98)}$$

Hydrogen-atom states with different values of l or m, but the same value of n, have the same energy. The energy levels are degenerate, except for $n = 1$, where l and m must both be 0. For a given value of n, we can have n different values of l. For each of these values of l, we can have $2l + 1$ values of m. The degree of degeneracy of an H-atom bound-state level is found to equal n^2 (spin considerations being omitted); see Prob. 6.14. For the continuum levels, it turns out that for a given energy there is no restriction on the maximum value of l; hence these levels are infinity-fold degenerate.

The radial equation for the hydrogen atom can also be solved by the use of ladder operators (also known as *factorization*); see Z. W. Salsburg, *Am. J. Phys.*, **33**, 36 (1965).

6.6 THE BOUND-STATE HYDROGEN-ATOM WAVE FUNCTIONS

The Radial Factor. Using (6.93), we have for the recursion relation (6.88)

$$b_{j+1} = \frac{2Z}{na}\frac{j + l + 1 - n}{(j + 1)(j + 2l + 2)}b_j \quad (6.99)$$

The discussion preceding Eq. (6.91) shows that the highest power of r in the polynomial $M(r) = \Sigma_j b_j r^j$ [Eq. (6.86)] is $k = n - l - 1$; hence use of $C = Z/na$ [Eq. (6.93)] in $R(r) = e^{-Cr}r^l M(r)$ [Eq. (6.84)] gives the radial factor in the hydrogen-atom ψ as

$$R_{nl}(r) = r^l e^{-Zr/na} \sum_{j=0}^{n-l-1} b_j r^j \quad (6.100)$$

where $a \equiv \hbar^2/\mu e'^2$ [Eq. (6.63)]. The complete hydrogenlike bound-state wave functions are [Eq. (6.61)]

$$\psi_{nlm} = R_{nl}(r)Y_l^m(\theta, \phi) = R_{nl}(r)S_{lm}(\theta)\frac{1}{\sqrt{2\pi}}e^{im\phi} \quad (6.101)$$

where the first few theta functions are given in Table 5.1.

How many nodes does $R(r)$ have? The radial function is zero at $r = \infty$, at $r = 0$ for $l \neq 0$, and at values of r that make $M(r)$ vanish. $M(r)$ is a polynomial of degree $n - l - 1$, and it can be shown that the roots of $M(r) = 0$ are all real and positive. Thus, aside from the origin and infinity, there are $n - l - 1$ nodes in $R(r)$. The nodes of the spherical harmonics are discussed in Prob. 6.39.

Ground-State Wave Function and Energy. For the ground state of the hydrogenlike atom, we have $n = 1$, $l = 0$, and $m = 0$. The radial factor (6.100) is

$$R_{10}(r) = b_0 e^{-Zr/a} \tag{6.102}$$

The constant b_0 is determined by normalization [Eq. (5.80)]:

$$|b_0|^2 \int_0^\infty e^{-2Zr/a} r^2 \, dr = 1$$

Using the Appendix integral (A.8), we find

$$R_{10}(r) = 2\left(\frac{Z}{a}\right)^{3/2} e^{-Zr/a} \tag{6.103}$$

Multiplying by $Y_0^0 = 1/(4\pi)^{1/2}$, we have as the ground-state wave function

$$\psi_{100} = \frac{1}{\pi^{1/2}}\left(\frac{Z}{a}\right)^{3/2} e^{-Zr/a} \tag{6.104}$$

The hydrogen-atom energies and wave functions involve the reduced mass, given by (6.59) as

$$\mu_{\text{H}} = \frac{m_e m_p}{m_e + m_p} = \frac{m_e}{1 + m_e/m_p} = \frac{m_e}{1 + 0.000544617} = 0.9994557 m_e \tag{6.105}$$

where m_p is the proton mass and m_e/m_p was found from Table A.1. The reduced mass is very close to the electron mass. Because of this, some texts use the electron mass instead of the reduced mass in the H atom Schrödinger equation. This corresponds to assuming that the proton mass is infinite compared with the electron mass in (6.105) and that all the internal motion is motion of the electron. The error introduced by using the electron mass for the reduced mass is about 1 part in 2000 for the hydrogen atom. For heavier atoms, the error introduced by assuming an infinitely heavy nucleus is even less than this. Also, for many-electron atoms, the form of the correction for nuclear motion is quite complicated. For these reasons we shall assume in the future an infinitely heavy nucleus and simply use the electron mass in writing the Schrödinger equation for atoms.

If we replace the reduced mass of the hydrogen atom by the electron mass, the quantity a defined by (6.63) becomes

$$a_0 \equiv \frac{\hbar^2}{m_e e'^2} = 0.52918 \text{ Å} \tag{6.106}$$

where the subscript zero indicates use of the electron mass instead of the reduced mass. a_0 is called the **Bohr radius**, since it was the radius of the circle in which the electron moved in the ground state of the hydrogen atom in the Bohr theory. Of course, since the ground-state wave function (6.104) is nonzero for all finite values of r, there is some probability of finding the electron at any distance from the nucleus. The electron is certainly not confined to a circle.

A convenient unit for electronic energies is the **electronvolt** (eV), defined as the kinetic energy acquired by an electron accelerated through a potential difference of 1 volt (V). Potential difference is defined as energy per unit charge. Since $e = 1.6021765 \times 10^{-19}$ C and 1 V C = 1 J, we have

$$1 \text{ eV} = 1.6021765 \times 10^{-19} \text{ J} \tag{6.107}$$

EXAMPLE Calculate the ground-state energy of the hydrogen atom using SI units and convert the result to electronvolts.

The H atom ground-state energy is given by (6.94) with $n = 1$ and $Z = 1$ as $E = -\mu e^4/8h^2\varepsilon_0^2$. Use of (6.105) for μ gives

$$E = -\frac{0.9994557(9.10938 \times 10^{-31} \text{ kg})(1.6021765 \times 10^{-19} \text{ C})^4}{8(6.62607 \times 10^{-34} \text{ J s})^2(8.8541878 \times 10^{-12} \text{ C}^2/\text{N-m}^2)^2} \frac{Z^2}{n^2}$$

$$E = -(2.17868 \times 10^{-18} \text{ J})(Z^2/n^2)[(1 \text{ eV})/(1.6021765 \times 10^{-19} \text{ J})]$$

$$E = -(13.598 \text{ eV})(Z^2/n^2) = -13.598 \text{ eV} \tag{6.108}$$

a number worth remembering. The minimum energy needed to ionize a ground-state hydrogen atom is 13.598 eV.

EXERCISE Find the $n = 2$ energy of Li^{2+} in eV; do the minimum amount of calculation needed. (*Answer:* -30.60 eV.)

EXAMPLE Find $\langle T \rangle$ for the hydrogen-atom ground state.

Equations (3.89) for $\langle T \rangle$ and (6.7) for $\nabla^2\psi$ give

$$\langle T \rangle = \int \psi^* \hat{T}\psi \, d\tau = -\frac{\hbar^2}{2\mu}\int \psi^* \nabla^2\psi \, d\tau$$

$$\nabla^2\psi = \frac{\partial^2\psi}{\partial r^2} + \frac{2}{r}\frac{\partial\psi}{\partial r} - \frac{1}{r^2\hbar^2}\hat{L}^2\psi = \frac{\partial^2\psi}{\partial r^2} + \frac{2}{r}\frac{\partial\psi}{\partial r}$$

since $\hat{L}^2\psi = l(l+1)\hbar^2\psi$ and $l = 0$ for an s state. From (6.104) with $Z = 1$, we have $\psi = \pi^{-1/2}a^{-3/2}e^{-r/a}$, so $\partial\psi/\partial r = -\pi^{-1/2}a^{-5/2}e^{-r/a}$ and $\partial^2\psi/\partial r^2 = \pi^{-1/2}a^{-7/2}e^{-r/a}$. Using $d\tau = r^2 \sin\theta \, dr \, d\theta \, d\phi$ [Eq. (5.78)], we have

$$\langle T \rangle = -\frac{\hbar^2}{2\mu}\frac{1}{\pi a^4}\int_0^{2\pi}\int_0^{\pi}\int_0^{\infty}\left(\frac{1}{a}e^{-2r/a} - \frac{2}{r}e^{-2r/a}\right)r^2 \sin\theta \, dr \, d\theta \, d\phi$$

$$= -\frac{\hbar^2}{2\mu\pi a^4}\int_0^{2\pi} d\phi \int_0^{\pi} \sin\theta \, d\theta \int_0^{\infty}\left(\frac{r^2}{a}e^{-2r/a} - 2re^{-2r/a}\right) dr = \frac{\hbar^2}{2\mu a^2} = \frac{e'^2}{2a}$$

where Appendix integral A.8 and $a = \hbar^2/\mu e'^2$ were used. From (6.94), $e'^2/2a$ is minus the ground-state H-atom energy, and (6.108) gives $\langle T \rangle = 13.598$ eV. (See also Sec. 14.4.)

EXERCISE Find $\langle T \rangle$ for the hydrogen-atom $2p_0$ state using (6.113). (*Answer:* $e'^2/8a = 3.40$ eV.)

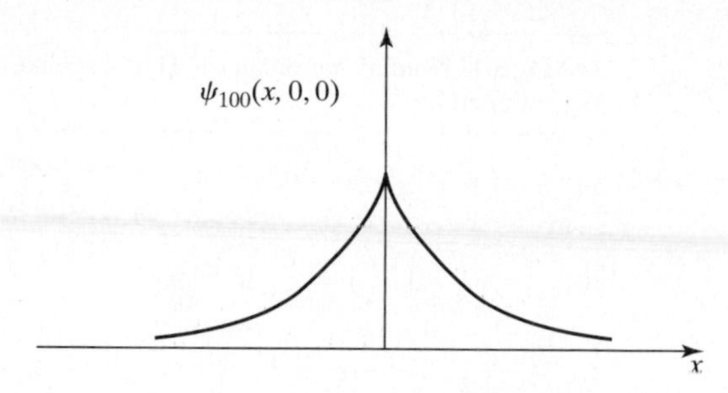

FIGURE 6.7 Cusp in the hydrogen-atom ground-state wave function.

Let us examine a significant property of the ground-state wave function (6.104). We have $r = (x^2 + y^2 + z^2)^{1/2}$. For points on the x axis, where $y = 0$ and $z = 0$, we have $r = (x^2)^{1/2} = |x|$, and

$$\psi_{100}(x, 0, 0) = \pi^{-1/2}(Z/a)^{3/2}e^{-Z|x|/a} \tag{6.109}$$

Figure 6.7 shows how (6.109) varies along the x axis. Although ψ_{100} is continuous at the origin, the slope of the tangent to the curve is positive at the left of the origin but negative at its right. Thus $\partial\psi/\partial x$ is discontinuous at the origin. We say that the wave function has a *cusp* at the origin. The cusp is present because the potential energy $V = -Ze'^2/r$ becomes infinite at the origin. Recall the discontinuous slope of the particle-in-a-box wave functions at the walls of the box.

We denoted the hydrogen-atom bound-state wave functions by three subscripts that give the values of n, l, and m. In an alternative notation, the value of l is indicated by a letter:

letter	s	p	d	f	g	h	i	k	$\cdots$
l	0	1	2	3	4	5	6	7	$\cdots$

(6.110)

The letters s, p, d, f are of spectroscopic origin, standing for sharp, principal, diffuse, fundamental. After these we go alphabetically, except that j is omitted. Preceding the code letter for l, we write the value of n. Thus the ground-state wave function ψ_{100} is called ψ_{1s} or, more simply, $1s$.

Wave Functions for $n = 2$. For $n = 2$, we have the states ψ_{200}, ψ_{21-1}, ψ_{210}, and ψ_{211}. We denote ψ_{200} as ψ_{2s} or simply as $2s$. To distinguish the three $2p$ functions, we use a subscript giving the m value and denote them as $2p_1$, $2p_0$, and $2p_{-1}$. The radial factor in the wave function depends on n and l, but not on m, as can be seen from (6.100). Each of the three $2p$ wave functions thus has the same radial factor. The $2s$ and $2p$ radial factors may be found in the usual way from (6.100) and (6.99), followed by normalization. The results are given in Table 6.1. Note that the exponential factor in the $n = 2$ radial functions is not the same as in the R_{1s} function. The complete wave

TABLE 6.1 Radial Factors in the Hydrogenlike-Atom Wave Functions

$$R_{1s} = 2\left(\frac{Z}{a}\right)^{3/2} e^{-Zr/a}$$

$$R_{2s} = \frac{1}{\sqrt{2}}\left(\frac{Z}{a}\right)^{3/2}\left(1 - \frac{Zr}{2a}\right)e^{-Zr/2a}$$

$$R_{2p} = \frac{1}{2\sqrt{6}}\left(\frac{Z}{a}\right)^{5/2} re^{-Zr/2a}$$

$$R_{3s} = \frac{2}{3\sqrt{3}}\left(\frac{Z}{a}\right)^{3/2}\left(1 - \frac{2Zr}{3a} + \frac{2Z^2r^2}{27a^2}\right)e^{-Zr/3a}$$

$$R_{3p} = \frac{8}{27\sqrt{6}}\left(\frac{Z}{a}\right)^{3/2}\left(\frac{Zr}{a} - \frac{Z^2r^2}{6a^2}\right)e^{-Zr/3a}$$

$$R_{3d} = \frac{4}{81\sqrt{30}}\left(\frac{Z}{a}\right)^{7/2} r^2e^{-Zr/3a}$$

function is found by multiplying the radial factor by the appropriate spherical harmonic. Using (6.101), Table 6.1, and Table 5.1, we have

$$2s = \frac{1}{\pi^{1/2}}\left(\frac{Z}{2a}\right)^{3/2}\left(1 - \frac{Zr}{2a}\right)e^{-Zr/2a} \tag{6.111}$$

$$2p_{-1} = \frac{1}{8\pi^{1/2}}\left(\frac{Z}{a}\right)^{5/2} re^{-Zr/2a}\sin\theta\, e^{-i\phi} \tag{6.112}$$

$$2p_0 = \frac{1}{\pi^{1/2}}\left(\frac{Z}{2a}\right)^{5/2} re^{-Zr/2a}\cos\theta \tag{6.113}$$

$$2p_1 = \frac{1}{8\pi^{1/2}}\left(\frac{Z}{a}\right)^{5/2} re^{-Zr/2a}\sin\theta\, e^{i\phi} \tag{6.114}$$

Table 6.1 lists some of the normalized radial factors in the hydrogenlike wave functions. Figure 6.8 graphs some of the radial functions. The r^l factor makes the radial functions zero at $r = 0$, except for s states.

The Radial Distribution Function. The probability of finding the electron in the region of space where its coordinates lie in the ranges r to $r + dr$, θ to $\theta + d\theta$, and ϕ to $\phi + d\phi$ is [Eq. (5.78)]

$$|\psi|^2\, d\tau = [R_{nl}(r)]^2|Y_l^m(\theta, \phi)|^2r^2 \sin\theta\, dr\, d\theta\, d\phi \tag{6.115}$$

We now ask: What is the probability of the electron having its radial coordinate between r and $r + dr$ with no restriction on the values of θ and ϕ? We are asking for the probability of finding the electron in a thin spherical shell centered at the origin, of inner radius r and outer radius $r + dr$. We must thus add up the infinitesimal

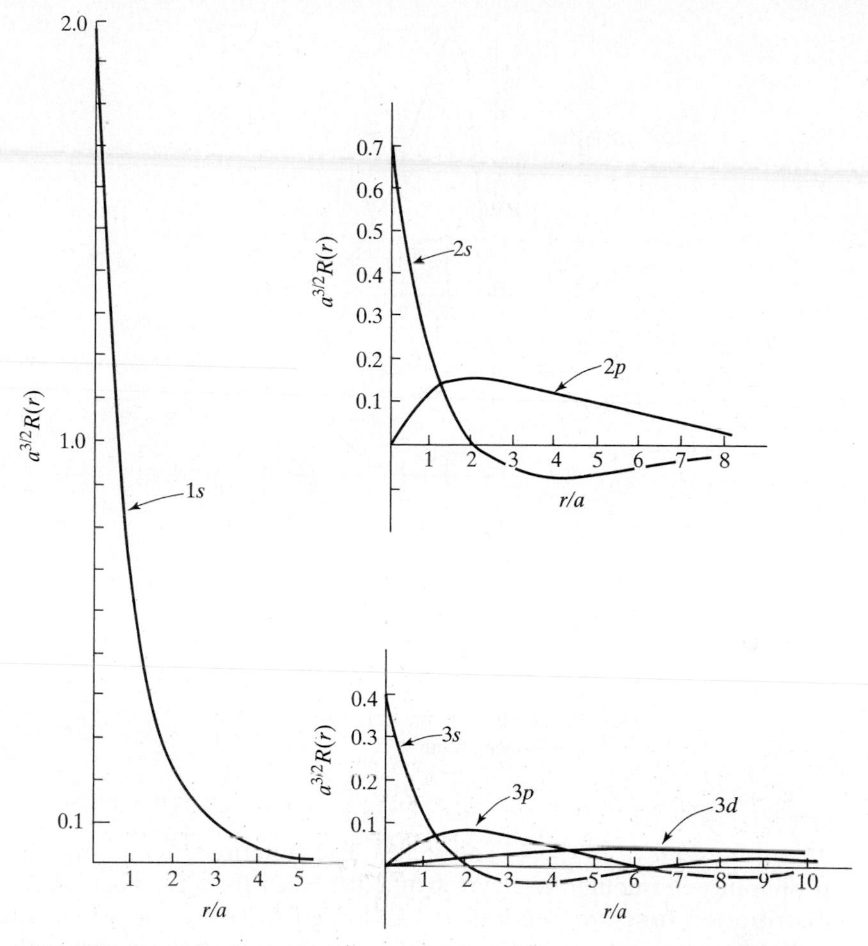

FIGURE 6.8 Graphs of the radial factor $R_{nl}(r)$ in the hydrogen-atom ($Z = 1$) wave functions. The same scale is used in all graphs. (In some texts, these functions are not properly drawn to scale.)

probabilities (6.115) for all possible values of θ and ϕ, keeping r fixed. This amounts to integrating (6.115) over θ and ϕ. Hence the probability of finding the electron between r and $r + dr$ is

$$[R_{nl}(r)]^2 r^2\,dr \int_0^{2\pi}\int_0^{\pi} |Y_l^m(\theta,\phi)|^2 \sin\theta\,d\theta\,d\phi = [R_{nl}(r)]^2 r^2\,dr \tag{6.116}$$

since the spherical harmonics are normalized:

$$\boxed{\int_0^{2\pi}\int_0^{\pi} |Y_l^m(\theta,\phi)|^2 \sin\theta\,d\theta\,d\phi = 1} \tag{6.117}$$

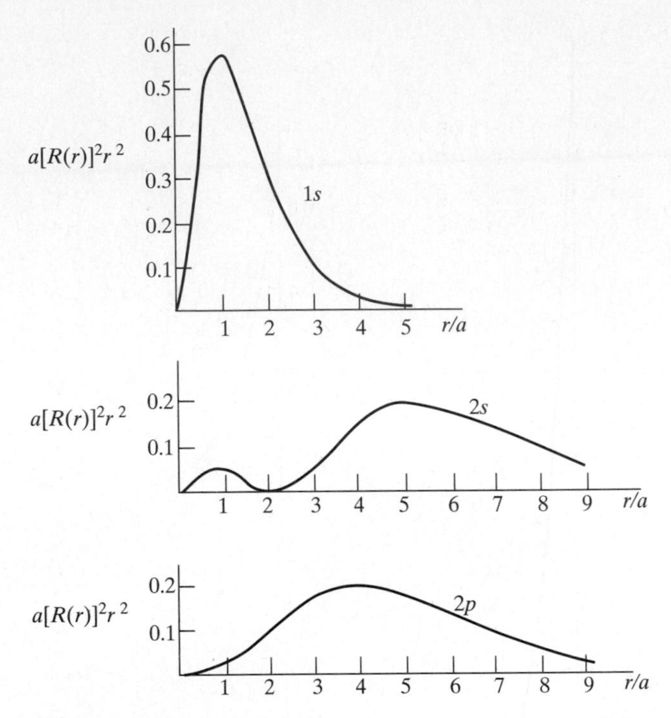

FIGURE 6.9 Plots of the radial distribution function $[R_{nl}(r)]^2r^2$ for the hydrogen atom.

as can be seen from (5.72) and (5.80). The function $R^2(r)r^2$, which determines the probability of finding the electron at a distance r from the nucleus, is called the **radial distribution function**; see Fig. 6.9. Although $R_{1s}(r)$ is not zero at the origin, the 1s radial distribution function is zero at $r = 0$ because of the r^2 factor; the volume of the thin spherical shell becomes zero as r goes to zero. The maximum in the radial distribution function for the 1s state of hydrogen is at $r = a$.

EXAMPLE Find the probability that the electron in the ground-state H atom is less than a distance a from the nucleus.

We want the probability that the radial coordinate lies between 0 and a. This is found by taking the infinitesimal probability (6.116) of being between r and $r + dr$ and summing it over the range from 0 to a. This sum of infinitesimal quantities is the definite integral

$$\int_0^a R_{nl}^2 r^2\,dr = \frac{4}{a^3}\int_0^a e^{-2r/a}r^2\,dr = \frac{4}{a^3}e^{-2r/a}\left(-\frac{r^2a}{2} - \frac{2ra^2}{4} - \frac{2a^3}{8}\right)\Bigg|_0^a$$

$$= 4[e^{-2}(-5/4) - (-1/4)] = 0.323$$

where R_{10} was taken from Table 6.1 and the Appendix integral A.7 was used.

EXERCISE Find the probability that the electron in a $2p_1$ H atom is less than a distance a from the nucleus. Use a table of integrals or the website integrals.wolfram.com. (*Answer:* 0.00366.)

Real Hydrogenlike Functions. The factor $e^{im\phi}$ makes the spherical harmonics complex, except when $m = 0$. Instead of working with complex wave functions such as (6.112) and (6.114), chemists often use real hydrogenlike wave functions formed by taking linear combinations of the complex functions. The justification for this procedure is given by the theorem of Section 3.6: Any linear combination of eigenfunctions of a degenerate energy level is an eigenfunction of the Hamiltonian with the same eigenvalue. Since the energy of the hydrogen atom does not depend on m, the $2p_1$ and $2p_{-1}$ states belong to a degenerate energy level. Any linear combination of them is an eigenfunction of the Hamiltonian with the same energy eigenvalue.

One way to combine these two functions to obtain a real function is

$$2p_x \equiv \frac{1}{\sqrt{2}}(2p_{-1} + 2p_1) = \frac{1}{4\sqrt{2\pi}}\left(\frac{Z}{a}\right)^{5/2} re^{-Zr/2a}\sin\theta\cos\phi \tag{6.118}$$

where we used (6.112), (6.114), and $e^{\pm i\phi} = \cos\phi \pm i\sin\phi$. The $1/\sqrt{2}$ factor normalizes $2p_x$:

$$\int |2p_x|^2\,d\tau = \frac{1}{2}\left(\int |2p_{-1}|^2\,d\tau + \int |2p_1|^2\,d\tau + \int (2p_{-1})^*2p_1\,d\tau + \int (2p_1)^*2p_{-1}\,d\tau\right)$$
$$= \tfrac{1}{2}(1 + 1 + 0 + 0) = 1$$

Here we used the fact that $2p_1$ and $2p_{-1}$ are normalized and are orthogonal to each other, since

$$\int_0^{2\pi} (e^{-i\phi})^* e^{i\phi}\,d\phi = \int_0^{2\pi} e^{2i\phi}\,d\phi = 0$$

The designation $2p_x$ for (6.118) becomes clearer if we note that (5.51) gives

$$2p_x = \frac{1}{4\sqrt{2\pi}}\left(\frac{Z}{a}\right)^{5/2} xe^{-Zr/2a} \tag{6.119}$$

A second way of combining the functions is

$$2p_y \equiv \frac{1}{i\sqrt{2}}(2p_1 - 2p_{-1}) = \frac{1}{4\sqrt{2\pi}}\left(\frac{Z}{a}\right)^{5/2} r\sin\theta\sin\phi\, e^{-Zr/2a} \tag{6.120}$$

$$2p_y = \frac{1}{4\sqrt{2\pi}}\left(\frac{Z}{a}\right)^{5/2} ye^{-Zr/2a} \tag{6.121}$$

The function $2p_0$ is real and is often denoted by

$$2p_0 = 2p_z = \frac{1}{\sqrt{\pi}}\left(\frac{Z}{2a}\right)^{5/2} ze^{-Zr/2a} \tag{6.122}$$

where capital Z stands for the number of protons in the nucleus, and small z is the z coordinate of the electron. The functions $2p_x$, $2p_y$, and $2p_z$ are mutually orthogonal (Prob. 6.40). Note that $2p_z$ is zero in the xy plane, positive above this plane, and negative below it.

The functions $2p_{-1}$ and $2p_1$ are eigenfunctions of $\hat{L}^2$ with the *same* eigenvalue: $2\hbar^2$. The reasoning of Section 3.6 shows that the linear combinations (6.118) and (6.120) are also eigenfunctions of $\hat{L}^2$ with eigenvalue $2\hbar^2$. However, $2p_{-1}$ and $2p_1$ are eigenfunctions of $\hat{L}_z$ with *different* eigenvalues: $-\hbar$ and $+\hbar$. Therefore, $2p_x$ and $2p_y$ are not eigenfunctions of $\hat{L}_z$.

We can extend this procedure to construct real wave functions for higher states. Since m ranges from $-l$ to $+l$, for each complex function containing the factor $e^{-i|m|\phi}$, there is a function with the same value of n and l but having the factor $e^{+i|m|\phi}$. Addition and subtraction of these functions gives two real functions, one with the factor $\cos(|m|\phi)$, the other with the factor $\sin(|m|\phi)$. Table 6.2 lists these real wave functions for the hydrogenlike atom. The subscripts on these functions come from similar considerations as for the $2p_x$, $2p_y$, and $2p_z$ functions. For example, the $3d_{xy}$ function is proportional to xy (Prob. 6.35).

TABLE 6.2 Real Hydrogenlike Wave Functions

$$1s = \frac{1}{\pi^{1/2}}\left(\frac{Z}{a}\right)^{3/2} e^{-Zr/a}$$

$$2s = \frac{1}{4(2\pi)^{1/2}}\left(\frac{Z}{a}\right)^{3/2}\left(2 - \frac{Zr}{a}\right)e^{-Zr/2a}$$

$$2p_z = \frac{1}{4(2\pi)^{1/2}}\left(\frac{Z}{a}\right)^{5/2} re^{-Zr/2a}\cos\theta$$

$$2p_x = \frac{1}{4(2\pi)^{1/2}}\left(\frac{Z}{a}\right)^{5/2} re^{-Zr/2a}\sin\theta\cos\phi$$

$$2p_y = \frac{1}{4(2\pi)^{1/2}}\left(\frac{Z}{a}\right)^{5/2} re^{-Zr/2a}\sin\theta\sin\phi$$

$$3s = \frac{1}{81(3\pi)^{1/2}}\left(\frac{Z}{a}\right)^{3/2}\left(27 - 18\frac{Zr}{a} + 2\frac{Z^2r^2}{a^2}\right)e^{-Zr/3a}$$

$$3p_z = \frac{2^{1/2}}{81\pi^{1/2}}\left(\frac{Z}{a}\right)^{5/2}\left(6 - \frac{Zr}{a}\right)re^{-Zr/3a}\cos\theta$$

$$3p_x = \frac{2^{1/2}}{81\pi^{1/2}}\left(\frac{Z}{a}\right)^{5/2}\left(6 - \frac{Zr}{a}\right)re^{-Zr/3a}\sin\theta\cos\phi$$

$$3p_y = \frac{2^{1/2}}{81\pi^{1/2}}\left(\frac{Z}{a}\right)^{5/2}\left(6 - \frac{Zr}{a}\right)re^{-Zr/3a}\sin\theta\sin\phi$$

$$3d_{z^2} = \frac{1}{81(6\pi)^{1/2}}\left(\frac{Z}{a}\right)^{7/2} r^2e^{-Zr/3a}(3\cos^2\theta - 1)$$

$$3d_{xz} = \frac{2^{1/2}}{81\pi^{1/2}}\left(\frac{Z}{a}\right)^{7/2} r^2e^{-Zr/3a}\sin\theta\cos\theta\cos\phi$$

$$3d_{yz} = \frac{2^{1/2}}{81\pi^{1/2}}\left(\frac{Z}{a}\right)^{7/2} r^2e^{-Zr/3a}\sin\theta\cos\theta\sin\phi$$

$$3d_{x^2-y^2} = \frac{1}{81(2\pi)^{1/2}}\left(\frac{Z}{a}\right)^{7/2} r^2e^{-Zr/3a}\sin^2\theta\cos 2\phi$$

$$3d_{xy} = \frac{1}{81(2\pi)^{1/2}}\left(\frac{Z}{a}\right)^{7/2} r^2e^{-Zr/3a}\sin^2\theta\sin 2\phi$$

The real hydrogenlike functions are derived from the complex functions by replacing $e^{im\phi}/(2\pi)^{1/2}$ with $\pi^{-1/2}\sin(|m|\phi)$ or $\pi^{-1/2}\cos(|m|\phi)$ for $m \neq 0$; for $m = 0$ the ϕ factor is $1/(2\pi)^{1/2}$ for both real and complex functions.

In dealing with molecules, the real hydrogenlike orbitals are more useful than the complex ones. For example, we shall see in Section 15.5 that the real atomic orbitals $2p_x$, $2p_y$, and $2p_z$ of the oxygen atom have the proper symmetry to be used in constructing a wave function for the H_2O molecule, whereas the complex $2p$ orbitals do not.

6.7 HYDROGENLIKE ORBITALS

The hydrogenlike wave functions are one-electron spatial wave functions and so are hydrogenlike orbitals (Section 6.5). These functions have been derived for a one-electron atom, and we cannot expect to use them to get a truly accurate representation of the wave function of a many-electron atom. The use of the orbital concept to approximate many-electron atomic wave functions is discussed in Chapter 11. For now we restrict ourselves to one-electron atoms.

There are two fundamentally different ways of depicting orbitals: One way is to draw graphs of the functions; a second way is to draw contour surfaces of constant probability density.

First consider drawing graphs. To graph the variation of ψ as a function of the three independent variables r, θ, and ϕ, we need four dimensions. The three-dimensional nature of our world prevents us from drawing such a graph. Instead, we draw graphs of the factors in ψ. Graphing $R(r)$ versus r, we get the curves of Fig. 6.8, which contain no information on the angular variation of ψ.

Now consider graphs of $S(\theta)$. We have (Table 5.1)

$$S_{0,0} = 1/\sqrt{2}, \qquad S_{1,0} = \tfrac{1}{2}\sqrt{6}\cos\theta$$

We can graph these functions using two-dimensional Cartesian coordinates, plotting S on the vertical axis and θ on the horizontal axis. $S_{0,0}$ gives a horizontal straight line, and $S_{1,0}$ gives a cosine curve. More commonly, S is graphed using plane polar coordinates. The variable θ is the angle with the positive z axis, and $S(\theta)$ is the distance from the origin to the point on the graph. For $S_{0,0}$, we get a circle; for $S_{1,0}$ we obtain two tangent circles (Fig. 6.10). The negative sign on the lower circle of the graph of $S_{1,0}$ indicates that

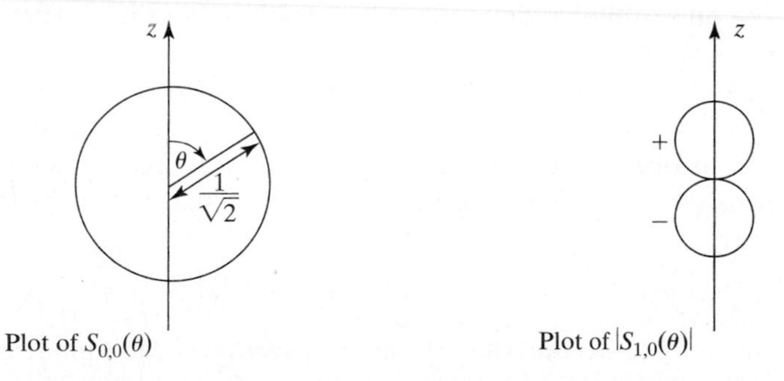

FIGURE 6.10 Polar graphs of the θ factors in the s and p_z hydrogen-atom wave functions.

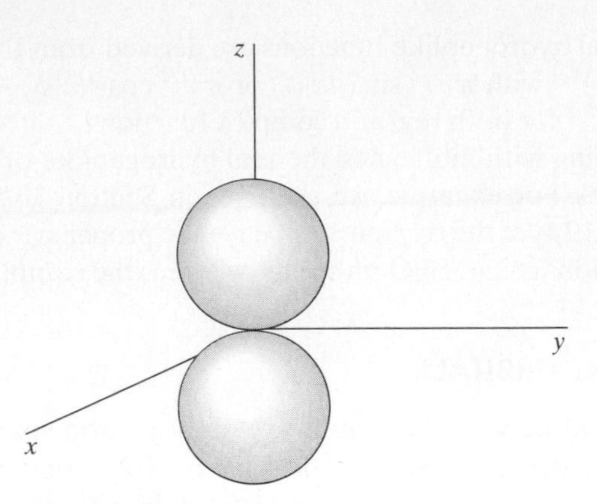

FIGURE 6.11 Graph of $|Y_1^0(\theta, \phi)|$, the angular factor in a p_z wave function.

$S_{1,0}$ is negative for $\frac{1}{2}\pi < \theta \le \pi$. Strictly speaking, in graphing $\cos\theta$ we only get the upper circle, which is traced out twice; to get two tangent circles, we must graph $|\cos\theta|$.

Instead of graphing the angular factors separately, we can draw a single graph that plots $|S(\theta)T(\phi)|$ as a function of θ and ϕ. We will use spherical coordinates, and the distance from the origin to a point on the graph will be $|S(\theta)T(\phi)|$. For an s state, ST is independent of the angles, and we get a sphere of radius $1/(4\pi)^{1/2}$ as the graph. For a p_z state, $ST = \frac{1}{2}(3/\pi)^{1/2}\cos\theta$, and the graph of $|ST|$ consists of two spheres with centers on the z axis and tangent at the origin (Fig. 6.11). No doubt Fig. 6.11 is familiar. Some texts say this gives the shape of a p_z orbital, which is wrong; Fig. 6.11 is simply a *graph* of the *angular factor* in a p_z wave function. Graphs of the p_x and p_y angular factors give tangent spheres lying on the x and y axes, respectively. If we graph S^2T^2 in spherical coordinates, we get surfaces with the familiar figure-eight cross sections; to repeat, these are graphs and not orbital shapes.

Now consider drawing contour surfaces of constant probability density. We shall draw surfaces in space, on each of which the value of $|\psi|^2$, the probability density, is constant. Naturally, if $|\psi|^2$ is constant on a given surface, $|\psi|$ is also constant on that surface; the contour surfaces for $|\psi|^2$ and for $|\psi|$ are identical.

For an s orbital, ψ depends only on r, and a contour surface is a surface of constant r, that is, a sphere centered at the origin. To pin down the size of an orbital, we take a contour surface within which the probability of finding the electron is, say, 95%; thus we want $\int_V |\psi|^2\, d\tau = 0.95$, where V is the volume enclosed by the orbital contour surface.

Let us obtain the cross section of the $2p_y$ hydrogenlike orbital in the yz plane. In this plane, $\phi = \pi/2$ (Fig. 6.5), and $\sin\phi = 1$; hence Table 6.2 gives for this orbital in the yz plane

$$|2p_y| = k^{5/2}\pi^{-1/2}re^{-kr}|\sin\theta| \tag{6.123}$$

where $k = Z/2a$. To find the orbital cross section, we use plane polar coordinates to plot (6.123) for a fixed value of ψ; r is the distance from the origin, and θ is the angle with the

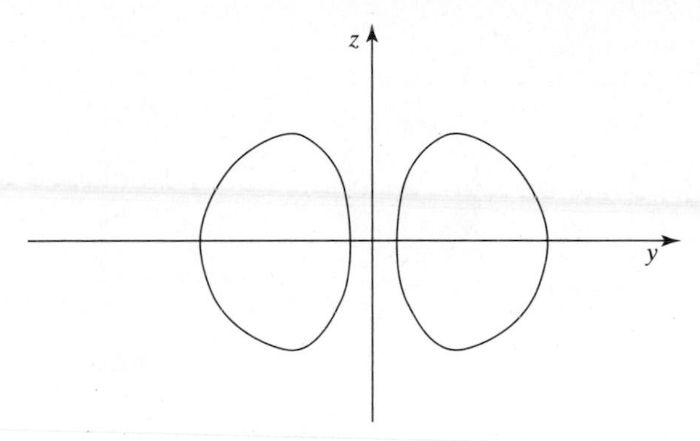

FIGURE 6.12 Contour of a $2p_y$ orbital.

z axis. The result for a typical contour (Prob. 6.42) is shown in Fig. 6.12. Since $ye^{-kr} = y\exp[-k(x^2 + y^2 + z^2)^{1/2}]$, we see that the $2p_y$ orbital is a function of y and $(x^2 + z^2)$. Hence, on a circle centered on the y axis and parallel to the xz plane, $2p_y$ is constant. Thus a three-dimensional contour surface may be developed by rotating the cross section in Fig. 6.12 about the y axis, giving a pair of distorted ellipsoids. The shape of a real $2p$ orbital is two separated, distorted ellipsoids, and not two tangent spheres.

Now consider the shape of the two complex orbitals $2p_{\pm 1}$. We have

$$2p_{\pm 1} = k^{5/2}\pi^{-1/2}re^{-kr}\sin\theta\, e^{\pm i\phi}$$

$$|2p_{\pm 1}| = k^{5/2}\pi^{-1/2}e^{-kr}r|\sin\theta| \tag{6.124}$$

and these two orbitals have the same shape. Since the right sides of (6.124) and (6.123) are identical, we conclude that Fig. 6.12 also gives the cross section of the $2p_{\pm 1}$ orbitals in the yz plane. Since [Eq. (5.51)]

$$e^{-kr}r|\sin\theta| = \exp[-k(x^2 + y^2 + z^2)^{1/2}](x^2 + y^2)^{1/2}$$

we see that $2p_{\pm 1}$ is a function of z and $x^2 + y^2$; so we get the three-dimensional orbital shape by rotating Fig. 6.12 about the z axis. This gives a doughnut-shaped surface.

Some hydrogenlike orbital surfaces are shown in Fig. 6.13. The $2s$ orbital has a spherical node, which is not visible; the $3s$ orbital has two such nodes. The $3p_z$ orbital has a spherical node (indicated by a dashed line) and a nodal plane (the xy plane). The $3d_{z^2}$ orbital has two nodal cones. The $3d_{x^2-y^2}$ orbital has two nodal planes. Note that the view shown is not the same for the various orbitals. The relative signs of the wave functions are indicated. The other three real $3d$ orbitals in Table 6.2 have the same shape as the $3d_{x^2-y^2}$ orbital but have different orientations. The $3d_{xy}$ orbital has its lobes lying between the x and y axes and is obtained by rotating the $3d_{x^2-y^2}$ orbital by 45° about the z axis. The $3d_{yz}$ and $3d_{xz}$ orbitals have their lobes between the y and z axes and between the x and z axes, respectively. (Online three-dimensional views of the real hydrogenlike orbitals are at www.falstad.com/qmatom; these can be rotated using a mouse.)

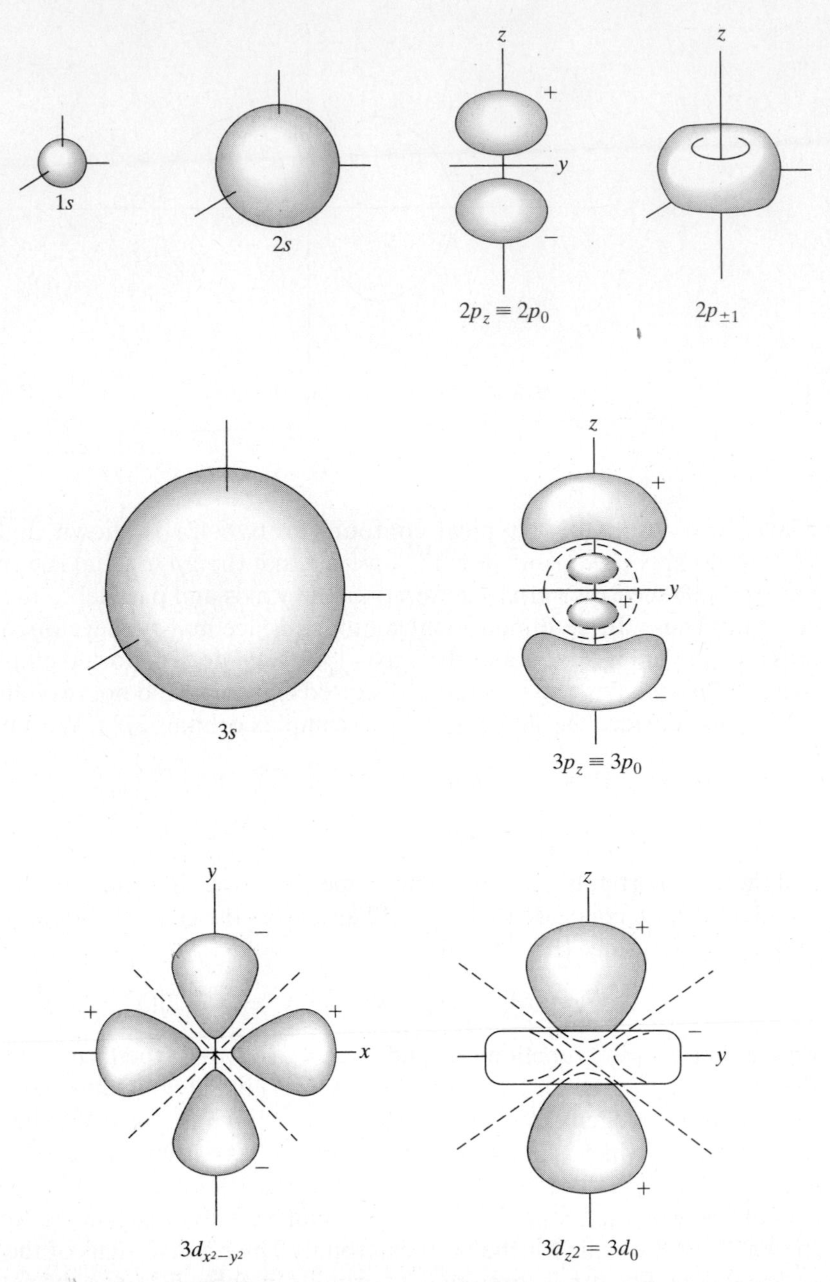

FIGURE 6.13 Shapes of some hydrogen-atom orbitals.

Figure 6.14 represents the probability density in the yz plane for various orbitals; the number of dots in a given region is proportional to the value of $|\psi|^2$ in that region. Rotation of these diagrams about the vertical (z) axis gives the three-dimensional probability density. The 2s orbital has a constant for its angular factor and hence has no

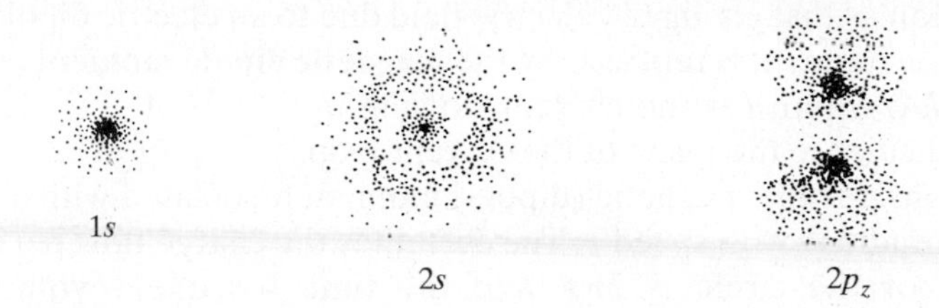

FIGURE 6.14 Probability densities for some hydrogen-atom states. [For accurate stereo plots, see D. T. Cromer, *J. Chem. Educ.*, **45**, 626 (1968).]

angular nodes; for this orbital, $n - l - 1 = 1$, indicating one radial node. The sphere on which $\psi_{2s} = 0$ is evident in Fig. 6.14.

Schrödinger's original interpretation of $|\psi|^2$ was that the electron is "smeared out" into a charge cloud. If we consider an electron passing from one medium to another, we find that $|\psi|^2$ is nonzero in both mediums. According to the charge-cloud interpretation, this would mean that part of the electron was reflected and part transmitted. However, experimentally one never detects a fraction of an electron; electrons behave as indivisible entities. This difficulty is removed by the Born interpretation, according to which the values of $|\psi|^2$ in the two mediums give the *probabilities* for reflection and transmission. The orbital shapes we have drawn give the regions of space in which the total probability of finding the electron is 95%.

6.8 THE ZEEMAN EFFECT

In 1896, Zeeman observed that application of an external magnetic field caused a splitting of atomic spectral lines. We shall consider this *Zeeman effect* for the hydrogen atom. We begin by reviewing magnetism.

Magnetic fields arise from moving electric charges. A charge Q with velocity $\mathbf{v}$ gives rise to a magnetic field $\mathbf{B}$ at point P in space, such that

$$\mathbf{B} = \frac{\mu_0}{4\pi}\frac{Q\mathbf{v} \times \mathbf{r}}{r^3} \tag{6.125}$$

where $\mathbf{r}$ is the vector from Q to point P and where μ_0 (the *permeability of vacuum*) is defined as $4\pi \times 10^{-7}$ N C^{-2} s^2. [Equation (6.125) is valid only for a nonaccelerated charge moving with a speed much less than the speed of light.] The vector $\mathbf{B}$ is called the **magnetic induction** or **magnetic flux density**. (It was formerly believed that the vector $\mathbf{H}$ was the fundamental magnetic field vector, so $\mathbf{H}$ was called the *magnetic field strength*. It is now known that $\mathbf{B}$ is the fundamental magnetic vector.) Equation (6.125) is in SI units with Q in coulombs and $\mathbf{B}$ in teslas (T), where 1 T = 1 N C^{-1} m^{-1} s. Only SI units will be used in this section.

Two electric charges $+Q$ and $-Q$ separated by a small distance b constitute an electric dipole. The **electric dipole moment** is defined as a vector from $-Q$ to $+Q$ with magnitude Qb. For a small planar loop of electric current, it turns out that the magnetic field generated by the moving charges of the current is given by the same mathematical expression as that giving the electric field due to an electric dipole, except that the electric dipole moment is replaced by the **magnetic dipole moment m**; **m** is a vector of magnitude IA, where I is the current flowing in a loop of area A; the direction of **m** is perpendicular to the plane of the current loop.

Consider the magnetic (dipole) moment associated with a charge Q moving in a circle of radius r with speed v. The current is the charge flow per unit time. The circumference of the circle is $2\pi r$, and the time for one revolution is $2\pi r/v$. Hence $I = Qv/2\pi r$. The magnitude of **m** is

$$|\mathbf{m}| = IA = (Qv/2\pi r)\pi r^2 = Qvr/2 = Qrp/2m \tag{6.126}$$

where m is the mass of the charged particle and p is its linear momentum. Since the radius vector **r** is perpendicular to **p**, we have

$$\mathbf{m}_L = \frac{Q\mathbf{r} \times \mathbf{p}}{2m} = \frac{Q}{2m}\mathbf{L} \tag{6.127}$$

where the definition of orbital angular momentum **L** was used and the subscript on **m** indicates that it arises from the orbital motion of the particle. Although we derived (6.127) for the special case of circular motion, its validity is general. For an electron, $Q = -e$, and the magnetic moment due to its orbital motion is

$$\mathbf{m}_L = -\frac{e}{2m_e}\mathbf{L} \tag{6.128}$$

The magnitude of **L** is given by (5.95), and the magnitude of the orbital magnetic moment of an electron with orbital-angular-momentum quantum number l is

$$|\mathbf{m}_L| = \frac{e\hbar}{2m_e}[l(l + 1)]^{1/2} = \beta_e[l(l + 1)]^{1/2} \tag{6.129}$$

The constant $e\hbar/2m_e$ is called the *Bohr magneton* β_e:

$$\beta_e = e\hbar/2m_e = 9.274 \times 10^{-24}\ \text{J/T} \tag{6.130}$$

Now consider applying an external magnetic field to the hydrogen atom. The energy of interaction between a magnetic dipole **m** and an external magnetic field **B** is [*Halliday and Resnick,* Eq. (33-12)]

$$E_B = -\mathbf{m} \cdot \mathbf{B} \tag{6.131}$$

Using Eq. (6.128), we have

$$E_B = \frac{e}{2m_e}\mathbf{L} \cdot \mathbf{B} \tag{6.132}$$

We take the z axis along the direction of the applied field: $\mathbf{B} = B\mathbf{k}$, where **k** is a unit vector in the z direction. We have

$$E_B = \frac{e}{2m_e}B(L_x\mathbf{i} + L_y\mathbf{j} + L_z\mathbf{k}) \cdot \mathbf{k} = \frac{e}{2m_e}BL_z = \frac{\beta_e}{\hbar}BL_z$$

where L_z is the z component of orbital angular momentum. We now replace L_z by the operator $\hat{L}_z$ to give the following additional term in the Hamiltonian operator, resulting from the external magnetic field:

$$\hat{H}_B = \beta_e B\hbar^{-1}\hat{L}_z \tag{6.133}$$

The Schrödinger equation for the hydrogen atom in a magnetic field is

$$(\hat{H} + \hat{H}_B)\psi = E\psi \tag{6.134}$$

where $\hat{H}$ is the hydrogen-atom Hamiltonian in the absence of an external field. We readily verify that the solutions of Eq. (6.134) are the complex hydrogenlike wave functions (6.61):

$$(\hat{H} + \hat{H}_B)R(r)Y_l^m(\theta, \phi) = \hat{H}RY_l^m + \beta_e\hbar^{-1}B\hat{L}_zRY_l^m = \left(-\frac{Z^2}{n^2}\frac{e'^2}{2a} + \beta_e Bm\right)RY_l^m \tag{6.135}$$

where Eqs. (6.94) and (5.105) were used. Thus there is an additional term $\beta_e Bm$ in the energy, and the external magnetic field removes the m degeneracy. For obvious reasons, m is often called the *magnetic quantum number.* Actually, the observed energy shifts do *not* match the predictions of Eq. (6.135) because of the existence of electron spin magnetic moment (Chapter 10 and Section 11.7).

In Chapter 5 we found that in quantum mechanics $\mathbf{L}$ lies on the surface of a cone. A *classical*-mechanical treatment (*Halliday and Resnick,* Sections 13-2 and 37-7) of the motion of $\mathbf{L}$ in an applied magnetic field shows that the field exerts a torque on $\mathbf{m}_L$, causing $\mathbf{L}$ to revolve about the direction of $\mathbf{B}$ at a constant frequency given by $|\mathbf{m}_L|B/2\pi|\mathbf{L}|$, while maintaining a constant angle with $\mathbf{B}$. This gyroscopic motion is called *precession.* In quantum mechanics, a complete specification of $\mathbf{L}$ is impossible; however, one finds that $\langle \mathbf{L} \rangle$ precesses about the field direction (*Dicke and Wittke,* Section 12-3).

6.9 NUMERICAL SOLUTION OF THE RADIAL SCHRÖDINGER EQUATION

For a one-particle central-force problem, the wave function is given by (6.16) as $\psi = R(r)Y_l^m(\theta, \phi)$ and the radial factor $R(r)$ is found by solving the radial equation (6.17). The Numerov method of Section 4.4 applies to differential equations of the form $\psi'' = G(x)\psi(x)$ [Eq. (4.68)], so we need to eliminate the first derivative R' in (6.17). Let us define $F(r)$ by $F(r) \equiv rR(r)$, so

$$R(r) = r^{-1}F(r) \tag{6.136}$$

Then $R' = -r^{-2}F + r^{-1}F'$ and $R'' = 2r^{-3}F - 2r^{-2}F' + r^{-1}F''$. Substitution in (6.17) transforms the radial equation to

$$-\frac{\hbar^2}{2m}F''(r) + \left[V(r) + \frac{l(l+1)\hbar^2}{2mr^2}\right]F(r) = EF(r) \tag{6.137}$$

$$F''(r) = G(r)F(r), \quad \text{where } G(r) \equiv \frac{m}{\hbar^2}(2V - 2E) + \frac{l(l+1)}{r^2} \tag{6.138}$$

which has the form needed for the Numerov method. In solving (6.137) numerically, one deals separately with each value of l. Equation (6.137) resembles the one-dimensional Schrödinger equation $-(\hbar^2/2m)\psi''(x) + V(x)\psi(x) = E\psi(x)$, except that r (whose range is 0 to ∞) replaces x (whose range is $-\infty$ to ∞), $F(r) \equiv rR(r)$ replaces ψ, and $V(r) + l(l+1)\hbar^2/2mr^2$ replaces $V(x)$. We can expect that for each value of l, the lowest-energy solution will have 0 interior nodes (that is, nodes with $0 < r < \infty$), the next lowest will have 1 interior node, etc.

Recall from the discussion after (6.81) that if $R(r)$ behaves as $1/r^b$ near the origin, then if $b > 1$, $R(r)$ is not quadratically integrable; also, the value $b = 1$ is not allowed, as noted after (6.83). Hence $F(r) \equiv rR(r)$ must be zero at $r = 0$.

For $l \neq 0$, $G(r)$ in (6.138) is infinite at $r = 0$, which upsets most computers. To avoid this problem, one starts the solution at an extremely small value of r (for example, 10^{-15} for the dimensionless r_r) and approximates $F(r)$ as zero at this point.

As an example, we shall use the Numerov method to solve for the lowest bound-state H-atom energies. Here, $V = -e'^2/r$, and the radial equation (6.62) contains the three constants e', μ, and $\hbar$, where $e' \equiv e/(4\pi\varepsilon_0)^{1/2}$ has SI units of m $N^{1/2}$ (see Table A.1 of the Appendix) and hence has the dimensions $[e'] = L^{3/2}M^{1/2}T^{-1}$. Following the procedure used to derive Eq. (4.77), we find the H-atom reduced energy and reduced radial coordinate to be (Prob. 6.45)

$$E_r = E/\mu e'^4\hbar^{-2}, \qquad r_r = r/B = r/\hbar^2\mu^{-1}e'^{-2} \tag{6.139}$$

and use of (6.139) and (4.80) and (4.81) with ψ replaced by F and $B = \hbar^2\mu^{-1}e'^{-2}$ transforms (6.137) for the H atom to (Prob. 6.45)

$$F_r'' = G_r F_r, \qquad \text{where } G_r = l(l+1)/r_r^2 - 2/r_r - 2E_r \tag{6.140}$$

and where $F_r = F/B^{-1/2}$.

The bound-state H-atom energies are all less than zero. Suppose we want to find the H-atom bound-state eigenvalues with $E_r \leq -0.04$. Equating this energy to V_r, we have (Prob. 6.45) $-0.04 = -1/r_r$ and the classically allowed region for this energy value extends from $r_r = 0$ to $r_r = 25$. Going two units into the classically forbidden region, we take $r_{r,\max} = 27$ and require that $F_r(27) = 0$. We shall take $s_r = 0.1$, giving 270 points from 0 to 27 (more precisely, from 10^{-15} to $27 + 10^{-15}$).

G_r in (6.140) contains the parameter l, so the program of Table 4.1 has to be modified to input the value of l. When setting up a spreadsheet, enter the l value in some cell and refer to this cell when you type the formula for cell B7 (Fig. 4.8) that defines G_r. Start column A at $r_r = 1 \times 10^{-15}$. Column C of the spreadsheet will contain F_r values instead of ψ_r values, and F_r will differ negligibly from zero at $r_r = 1 \times 10^{-15}$, and will be taken as zero at this point.

With these choices, we find (Prob. 6.46a) the lowest three H-atom eigenvalues for $l = 0$ to be $E_r = -0.4970, -0.1246$, and -0.05499; the lowest two $l = 1$ eigenvalues found are -0.1250 and -0.05526. The true values [Eqs. (6.94) and (6.139)] are -0.5000, -0.1250, and -0.05555. The mediocre accuracy can be attributed mainly to the rapid variation of $G(r)$ near $r = 0$. If s_r is taken as 0.025 instead of 0.1 (giving 1080 points), the $l = 0$ eigenvalues are improved to -0.4998, -0.12497, and -0.05510. See also Prob. 6.46b.

6.10 SUMMARY

For a one-particle system with potential energy a function of r only [$V = V(r)$, a central-force problem], the stationary-state wave functions have the form $\psi = R(r)Y_l^m(\theta, \phi)$, where $R(r)$ satisfies the radial equation (6.17) and Y_l^m are the spherical harmonics.

For a system of two noninteracting particles 1 and 2, the Hamiltonian operator is $\hat{H} = \hat{H}_1 + \hat{H}_2$, and the stationary-state wave functions and energies satisfy $\psi = \psi_1(q_1)\psi_2(q_2)$, $E = E_1 + E_2$, where $\hat{H}_1\psi_1 = E_1\psi_1$ and $\hat{H}_2\psi_2 = E_2\psi_2$; q_1 and q_2 stand for the coordinates of particles 1 and 2.

For a system of two interacting particles 1 and 2 with Hamiltonian $\hat{H} = \hat{T}_1 + \hat{T}_2 + \hat{V}$, where V is a function of only the relative coordinates x, y, z of the particles, the energy is the sum of the energies of two hypothetical particles: $E = E_M + E_\mu$. One hypothetical particle has mass $M = m_1 + m_2$; its coordinates are the coordinates of the center of mass and its energy E_M is that of a free particle. The second particle has mass $\mu \equiv m_1m_2/(m_1 + m_2)$; its coordinates are the relative coordinates $x, y, z,$ and its energy E_μ is found by solving the Schrödinger equation for internal motion: $[(-\hbar^2/2\mu)\nabla^2 + V]\psi(x, y, z) = E_\mu\psi(x, y, z)$.

The two-particle rigid rotor consists of particles of masses m_1 and m_2 separated by a fixed distance d. Its energy is the sum of the energy E_M of translation and the energy E_μ of rotation. Its stationary-state rotational wave functions are $\psi = Y_J^m(\theta, \phi)$, where θ and ϕ give the orientation of the rotor axis with respect to an origin at the rotor's center of mass, and the quantum numbers are $J = 0, 1, 2, \ldots,$ and $m = -J, -J + 1, \ldots, J - 1, J$. The rotational energy levels are $E_\mu = J(J + 1)\hbar^2/2I$, where $I = \mu d^2$, with $\mu = m_1m_2/(m_1 + m_2)$. The selection rule for spectroscopic transitions is $\Delta J = \pm 1$.

The hydrogenlike atom has $V = -Ze'^2/r$. With the translational energy separated off, the internal motion is a central-force problem and $\psi = R(r)Y_l^m(\theta, \phi)$. The continuum states have $E \geq 0$ and correspond to an ionized atom. The bound states have the allowed energies $E = -(Z^2/n^2)(e'^2/2a)$, where $a \equiv \hbar^2/\mu e'^2$; the bound-state radial wave function is (6.101). The bound-state quantum numbers are $n = 1, 2, 3, \ldots$; $l = 0, 1, 2, \ldots, n - 1$; $m = -l, -l + 1, \ldots, l - 1, l$.

A one-electron spatial wave function is called an orbital. The shape of an orbital is defined by a contour surface of constant $|\psi|$ that encloses a specified amount of probability.

The Numerov method can be used to numerically solve the radial Schrödinger equation for a one-particle system with a spherically symmetric potential energy.

PROBLEMS

Sec.	6.1	6.2	6.3	6.4	6.5	6.6
Probs.	6.1–6.4	6.5–6.6	6.7	6.8–6.12	6.13–6.19	6.20–6.40

Sec.	6.7	6.8	6.9	general
Probs.	6.41–6.43	6.44	6.45–6.49	6.50–6.54

6.1 True or false? (a) For a one-particle problem with $V = br^3$, where b is a positive constant, the stationary-state wave functions have the form $\psi = f(r)Y_l^m(\theta, \phi)$. (b) Every one-particle Hamiltonian operator commutes with $\hat{L}^2$ and with $\hat{L}_z$.

6.2 The particle in a spherical box has $V = 0$ for $r \le b$ and $V = \infty$ for $r > b$. For this system: (a) Explain why $\psi = R(r)f(\theta, \phi)$, where $R(r)$ satisfies (6.17). What is the function $f(\theta, \phi)$? (b) Solve (6.17) for $R(r)$ for the $l = 0$ states. *Hints:* The substitution $R(r) = g(r)/r$ reduces (6.17) to an easily solved equation. Use the boundary condition that ψ is finite at $r = 0$ [see the discussion after Eq. (6.83)] and use a second boundary condition. Show that for the $l = 0$ states, $\psi = N[\sin(n\pi r/b)]/r$ and $E = n^2h^2/8mb^2$ with $n = 1, 2, 3, \ldots$. (For $l \ne 0$, the energy-level formula is more complicated.)

6.3 If the three force constants in Prob. 4.19 are all equal, we have a three-dimensional isotropic harmonic oscillator. (a) State why the wave functions for this case can be written as $\psi = f(r)G(\theta, \phi)$. (b) What is the function G? (c) Write a differential equation satisfied by $f(r)$. (d) Use the results found in Prob. 4.19 to show that the ground-state wave function does have the form $f(r)G(\theta, \phi)$ and verify that the ground-state $f(r)$ satisfies the differential equation in (c).

6.4 Verify Eq. (6.6) for the Laplacian in spherical coordinates. (This is a long, tedious problem, and you probably have better things to spend your time on.)

6.5 True or false? (a) For a system of n noninteracting particles, each stationary state wave function has the form $\psi = \psi(q_1) + \psi(q_2) + \cdots + \psi(q_n)$. (b) The energy of a system of noninteracting particles is the sum of the energies of the individual particles, where the energy of each particle is found by solving a one-particle Schrödinger equation.

6.6 For a system of two noninteracting particles of mass 9.0×10^{-26} g and 5.0×10^{-26} g in a one-dimensional box of length 1.00×10^{-8} cm, calculate the energies of the six lowest stationary states.

6.7 True or false? (a) The reduced mass of a two-particle system is always less than m_1 and less than m_2. (b) When we solve a two-particle system (whose potential-energy V is a function of only the relative coordinates of the two particles) by dealing with two separate one-particle systems, V is part of the Hamiltonian operator of the fictitious particle with mass equal to the reduced mass.

6.8 True or false? (a) The degeneracy of the $J = 4$ two-particle rigid-rotor energy level is 9. (b) The spacings between successive two-particle-rigid-rotor energy levels remain constant as J increases. (c) The spacings between successive two-particle-rigid-rotor absorption frequencies remain constant as the J of the lower level increases. (d) The molecules $^1H^{35}Cl$ and $^1H^{37}Cl$ have essentially the same equilibrium bond length. (e) The H_2 molecule does not have a pure-rotational absorption spectrum.

6.9 The lowest observed microwave absorption frequency of $^{12}C^{16}O$ is 115271 MHz. (a) Compute the bond distance in $^{12}C^{16}O$. (b) Predict the next two lowest microwave absorption frequencies of $^{12}C^{16}O$. (c) Predict the lowest microwave absorption frequency of $^{13}C^{16}O$. (d) For $^{12}C^{16}O$ at 25°C, calculate the ratio of the $J = 1$ population to the $J = 0$ population. Repeat for the $J = 2$ to $J = 0$ ratio. Don't forget degeneracy.

6.10 The $J = 2$ to 3 rotational transition in a certain diatomic molecule occurs at 126.4 GHz, where 1 GHz $\equiv 10^9$ Hz. Find the frequency of the $J = 5$ to 6 absorption in this molecule.

6.11 The $J = 7$ to 8 rotational transition in gas-phase $^{23}Na^{35}Cl$ occurs at 104189.7 MHz. The relative atomic mass of ^{23}Na is 22.989770. Find the bond distance in $^{23}Na^{35}Cl$.

6.12 Verify Eq. (6.51) for I of a two-particle rotor. Begin by multiplying and dividing the right side of (6.50) by $m_1m_2/(m_1 + m_2)$. Then use (6.49).

6.13 Calculate the ratio of the electrical and gravitational forces between a proton and an electron. Is neglect of the gravitational force justified?

6.14 (a) Explain why the degree of degeneracy of an H-atom energy level is given by $\sum_{l=0}^{n-1}(2l + 1)$. (b) Break this sum into two sums. Evaluate the first sum using the fact that

$\sum_{j=1}^{k} j = \frac{1}{2}k(k+1)$. Show that the degree of degeneracy of the H-atom levels is n^2 (spin omitted). (c) Prove that $\sum_{j=1}^{k} j = \frac{1}{2}k(k+1)$ by adding corresponding terms of the two series $1, 2, 3, \ldots, k$ and $k, k-1, k-2, \ldots, 1$.

6.15 (a) Calculate the wavelength and frequency for the spectral line that arises from an $n = 6$ to $n = 3$ transition in the hydrogen atom. (b) Repeat the calculations for He^+; neglect the change in reduced mass from H to He^+.

6.16 Assign each of the following observed vacuum wavelengths to a transition between two hydrogen-atom levels:

6564.7 Å, 4862.7 Å, 4341.7 Å, 4102.9 Å (Balmer series)

Predict the wavelengths of the next two lines in this series and the wavelength of the series limit. (Balmer was a Swiss mathematician who, in 1885, came up with an empirical formula that fitted lines of the hydrogen spectrum.)

6.17 Each hydrogen-atom line of Prob. 6.16 shows a very weak nearby satellite line. Two of the satellites occur at the vacuum wavelengths 6562.9 Å and 4861.4 Å. (a) Explain their origin. (The person who first answered this question got a Nobel Prize.) (b) Calculate the other two satellite wavelengths.

6.18 Verify that for large values of j, the ratio b_{j+1}/b_j in (6.88) is the same as the ratio of the coefficient of r^{j+1} to that of r^j in the power series for e^{2Cr}.

6.19 For the particle in a box with infinitely high walls and for the harmonic oscillator, there are no continuum eigenfunctions, whereas for the hydrogen atom we do have continuum functions. Explain this in terms of the nature of the potential-energy function for each problem.

6.20 The positron has charge $+e$ and mass equal to the electron mass. Calculate in electronvolts the ground-state energy of positronium—an "atom" that consists of a positron and an electron.

6.21 For the ground state of the hydrogenlike atom, show that $\langle r \rangle = 3a/2Z$.

6.22 Find $\langle r \rangle$ for the $2p_0$ state of the hydrogenlike atom.

6.23 Find $\langle r^2 \rangle$ for the $2p_1$ state of the hydrogenlike atom.

6.24 For a hydrogenlike atom in a stationary state with quantum numbers n, l, and m, prove that $\langle r \rangle = \int_0^\infty r^3 |R_{nl}|^2 \, dr$.

6.25 Derive the 2*s* and 2*p* radial hydrogenlike functions.

6.26 For which hydrogen-atom states is ψ nonzero at the nucleus?

6.27 What is the value of the angular-momentum quantum number l for a t orbital?

6.28 If we were to ignore the interelectronic repulsion in helium, what would be its ground-state energy and wave function? (See Section 6.2.) Compute the percent error in the energy; the experimental He ground-state energy is -79.0 eV.

6.29 For the ground state of the hydrogenlike atom, find the most probable value of r.

6.30 Where is the probability density a maximum for the hydrogen-atom ground state?

6.31 (a) For the hydrogen-atom ground state, find the probability of finding the electron farther than $2a$ from the nucleus. (b) For the H-atom ground state, find the probability of finding the electron in the classically forbidden region.

6.32 A stationary-state wave function is an eigenfunction of the Hamiltonian operator $\hat{H} = \hat{T} + \hat{V}$. Students sometimes erroneously believe that ψ is an eigenfunction of $\hat{T}$ and of $\hat{V}$. For the ground state of the hydrogen atom, verify directly that ψ is not an eigenfunction of $\hat{T}$ or of $\hat{V}$, but is an eigenfunction of $\hat{T} + \hat{V}$. Can you think of a problem we solved where ψ is an eigenfunction of $\hat{T}$ and of $\hat{V}$?

6.33 Show that $\langle T\rangle + \langle V\rangle = E$ for a stationary state.

6.34 For the hydrogen-atom ground state, (a) find $\langle V\rangle$; (b) use the results of (a) and Prob. 6.33 to find $\langle T\rangle$; then find $\langle T\rangle/\langle V\rangle$; (c) use $\langle T\rangle$ to calculate the root-mean-square speed $\langle v^2\rangle^{1/2}$ of the electron; then find the numerical value of $\langle v^2\rangle^{1/2}/c$, where c is the speed of light.

6.35 (a) The $3d_{xy}$ function is defined as $3d_{xy} \equiv (3d_2 - 3d_{-2})/2^{1/2}i$. Use Tables 6.1 and 5.1, Eq. (6.101), and an identity for $\sin 2\phi$ to show that $3d_{xy}$ is proportional to xy. (b) Express the other real $3d$ functions of Table 6.2 as linear combinations of the complex functions $3d_2, 3d_1, \ldots, 3d_{-2}$. (c) Use a trigonometric identity to show that $3d_{x^2-y^2}$ contains the factor $x^2 - y^2$.

6.36 The hydrogenlike wave functions $2p_1, 2p_0$, and $2p_{-1}$ can be characterized as those $2p$ functions that are eigenfunctions of $\hat{L}_z$. What operators can we use to characterize the functions $2p_x, 2p_y$, and $2p_z$, and what are the corresponding eigenvalues?

6.37 Given that $\hat{A}f = af$ and $\hat{A}g = bg$, where f and g are functions and a and b are constants, under what condition(s) is the linear combination $c_1 f + c_2 g$ an eigenfunction of the linear operator $\hat{A}$?

6.38 State which of the three operators $\hat{L}^2$, $\hat{L}_z$, and the H-atom $\hat{H}$ each of the following functions is an eigenfunction of: (a) $2p_z$; (b) $2p_x$; (c) $2p_1$.

6.39 For the *real* hydrogenlike functions: (a) What is the shape of the $n - l - 1$ nodal surfaces for which the radial factor is zero? (b) The nodal surfaces for which the ϕ factor vanishes are of the form ϕ = constant; thus they are planes perpendicular to the xy plane. How many such planes are there? (Values of ϕ that differ by π are considered to be part of the same plane.) (c) It can be shown that there are $l - m$ surfaces on which the θ factor vanishes. What is the shape of these surfaces? (d) How many nodal surfaces are there for the real hydrogenlike wave functions?

6.40 Verify the orthogonality of the $2p_x, 2p_y$, and $2p_z$ functions.

6.41 Find the radius of the sphere defining the $1s$ hydrogen orbital using the 95% probability definition.

6.42 Show that the maximum value for $2p_y$ [Eq. (6.123)] is $k^{3/2}\pi^{-1/2}e^{-1}$. Use Eq. (6.123) to plot the $2p_y$ contour for which $\psi = 0.316\psi_{\max}$.

6.43 Sketch rough contours of constant $|\psi|$ for each of the following states of a particle in a two-dimensional square box: $n_x n_y = 11; 12; 21; 22$. What are you reminded of?

6.44 (Answer this question based on Sec. 6.8, which omits the effects of electron spin.) How many energy levels is the $n = 2$ H-atom energy level split into when an external magnetic field is applied? Give the degeneracy of each of these levels.

6.45 (a) Verify the equations (6.139) for the H-atom dimensionless E_r and r_r. (b) Verify (6.140) for F_r. (c) Verify that $V_r = -1/r_r$ for the H atom.

For Probs. 6.46–6.49, use a modified version of the program in Table 4.1 or a spreadsheet or Mathcad.

6.46 (a) Verify the $l = 0$ and $l = 1$ Numerov-method H-atom energies given in Section 6.9 for 270 points and for 1080 points, with r_r going from 10^{-15} to 27. (b) For 1080 points and r_r going to 27, the $n = 3$, $l = 0$ Numerov E_r of -0.05510 is still substantially in error. Use the Numerov method to improve this energy significantly without decreasing s_r.

6.47 Use the Numerov method to calculate the lowest four $l = 0$ energy eigenvalues and the lowest four $l = 1$ eigenvalues of the three-dimensional isotropic harmonic oscillator, which has $V = \frac{1}{2}kr^2$. Compare with the exact results (Prob. 4.19).

6.48 Use the Numerov method to calculate and plot the reduced radial function $R_r(r_r)$ for the lowest $l = 0$ H-atom state. Explain why the value of R_r at $r_r = 10^{-15}$ calculated from $R_r = F_r/r_r$ [the equation corresponding to (6.136)] is wrong for this state.

6.49 For the particle in a spherical box (Prob. 6.2), use the Numerov method to find the lowest three $l = 0$ energy eigenvalues and the lowest three $l = 1$ eigenvalues. Compare your $l = 0$ results with the exact results of Prob. 6.2.

6.50 For each of the following systems, give the expression for $d\tau$ and give the limits of each variable in the equation $\int |\psi|^2 \, d\tau = 1$. (a) The particle in a one-dimensional box of length l. (b) The one-dimensional harmonic oscillator. (c) A one-particle, three-dimensional system where Cartesian coordinates are used. (d) Internal motion in the hydrogen atom, using spherical coordinates.

6.51 Find the $n = 2$ to $n = 1$ energy-level population ratio for a gas of hydrogen atoms at (a) 25°C; (b) 1000 K; (c) 10000 K.

6.52 Name a quantum-mechanical system for which the spacing between adjacent bound-state energy levels (a) remains constant as E increases; (b) increases as E increases; (c) decreases as E increases.

6.53 (a) Name two quantum-mechanical systems that have an infinite number of bound-state energy levels. (b) Name a quantum-mechanical system that has a finite number of bound-state energy levels. (c) Name a quantum-mechanical system that has no zero-point energy.

6.54 True or false? (a) The value zero is never allowed for an eigenvalue. (b) The function $f = 0$ is never allowed as an eigenfunction. (c) The symbol e stands for the charge on an electron. (d) In the equation $\langle B \rangle = \int \psi^* \hat{B} \psi \, d\tau$, where $d\tau = r^2 \sin\theta \, dr \, d\theta \, d\phi$, $\hat{B}$ operates on ψ only and does not operate on $r^2 \sin\theta$. (e) In the $n = 1$ state of an H atom, the electron moves on a circular orbit whose radius is the Bohr radius. (f) The electron probability density at the nucleus is zero for all H-atom states. (g) In the ground-state of an H atom, the electron is restricted to move on the surface of a sphere. (h) For the ground-state H atom, the electron probability density is greater than zero at all locations in the atom.

CHAPTER 7

Theorems of Quantum Mechanics

7.1 INTRODUCTION

The Schrödinger equation for the one-electron atom (Chapter 6) is exactly solvable. However, because of the interelectronic-repulsion terms in the Hamiltonian, the Schrödinger equation for many-electron atoms and molecules cannot be solved exactly. Hence we must seek approximate methods of solution. The two main approximation methods, the variation method and perturbation theory, will be presented in Chapters 8 and 9. To derive these methods, we must develop further the theory of quantum mechanics, which is what is done in this chapter.

Before starting, we introduce some notation. The definite integral over all space of an operator sandwiched between two functions occurs often, and various abbreviations are used:

$$\boxed{\int f_m^* \hat{A} f_n \, d\tau \equiv \langle f_m | \hat{A} | f_n \rangle \equiv \langle m | \hat{A} | n \rangle} \tag{7.1}$$

where f_m and f_n are two functions. If it is clear what functions are meant, we can use just the indexes, as indicated in (7.1). The above notation, introduced by Dirac, is called **bracket notation**. Another notation is

$$\boxed{\int f_m^* \hat{A} f_n \, d\tau \equiv A_{mn}} \tag{7.2}$$

The notations A_{mn} and $\langle m | \hat{A} | n \rangle$ imply that we use the complex conjugate of the function whose letter appears first. The definite integral $\langle m | \hat{A} | n \rangle$ is called a **matrix element** of the operator $\hat{A}$. Matrices are rectangular arrays of numbers and obey certain rules of combination (see Section 7.10).

For the definite integral over all space between two functions, we write

$$\boxed{\int f_m^* f_n \, d\tau \equiv \langle f_m | f_n \rangle \equiv (f_m, f_n) \equiv \langle m | n \rangle} \tag{7.3}$$

Note that

$$\langle f | \hat{B} | g \rangle = \langle f | \hat{B} g \rangle$$

where f and g are functions. Since $[\int f_m^* f_n \, d\tau]^* = \int f_n^* f_m \, d\tau$, we have the identity

$$\boxed{\langle m|n\rangle^* = \langle n|m\rangle} \qquad \textbf{(7.4)}$$

Since we take the complex conjugate of f_m in (7.1), it follows that

$$\boxed{\langle cf|\hat{B}|g\rangle = c^*\langle f|\hat{B}|g\rangle \quad \text{and} \quad \langle f|\hat{B}|cg\rangle = c\langle f|\hat{B}|g\rangle} \qquad \textbf{(7.5)}$$

if $\hat{B}$ is a linear operator.

7.2 HERMITIAN OPERATORS

The quantum-mechanical operators that represent physical quantities are linear (Section 3.1). These operators must meet an additional requirement, which we now discuss.

Definition of Hermitian Operators. Let $\hat{A}$ be the linear operator representing the physical property A. The average value of A is [Eq. (3.88)]

$$\langle A\rangle = \int \Psi^* \hat{A} \Psi \, d\tau$$

where Ψ is the state function of the system. Since the average value of a physical quantity must be a real number, we demand that

$$\langle A\rangle = \langle A\rangle^*$$

$$\int \Psi^* \hat{A} \Psi \, d\tau = \left[\int \Psi^* \hat{A} \Psi \, d\tau\right]^* = \int (\Psi^*)^* (\hat{A}\Psi)^* \, d\tau$$

$$\int \Psi^* \hat{A} \Psi \, d\tau = \int \Psi (\hat{A}\Psi)^* \, d\tau \qquad (7.6)$$

Equation (7.6) must hold for every function Ψ that can represent a possible state of the system; that is, it must hold for all well-behaved functions Ψ. A linear operator that satisfies (7.6) for all well-behaved functions is called a **Hermitian operator** (after the mathematician Charles Hermite).

Many texts define a Hermitian operator as a linear operator that satisfies

$$\int f^* \hat{A} g \, d\tau = \int g (\hat{A} f)^* \, d\tau \qquad (7.7)$$

for all well-behaved functions f and g. Note especially that on the left side of (7.7) $\hat{A}$ operates on g, but on the right side $\hat{A}$ operates on f. For the special case $f = g$, (7.7) reduces to (7.6). Equation (7.7) is apparently a more stringent requirement than (7.6), but we shall prove that (7.7) is a consequence of (7.6). Hence the two definitions of a Hermitian operator are equivalent.

We begin the proof by setting $\Psi = f + cg$ in (7.6), where c is an arbitrary parameter; this gives

$$\int (f + cg)^*\hat{A}(f + cg)\,d\tau = \int (f + cg)[\hat{A}(f + cg)]^*\,d\tau$$

$$\int (f^* + c^*g^*)\hat{A}f\,d\tau + \int (f^* + c^*g^*)\hat{A}cg\,d\tau$$

$$= \int (f + cg)(\hat{A}f)^*\,d\tau + \int (f + cg)(\hat{A}cg)^*\,d\tau$$

$$\int f^*\hat{A}f\,d\tau + c^*\int g^*\hat{A}f\,d\tau + c\int f^*\hat{A}g\,d\tau + c^*c\int g^*\hat{A}g\,d\tau$$

$$= \int f(\hat{A}f)^*\,d\tau + c\int g(\hat{A}f)^*\,d\tau + c^*\int f(\hat{A}g)^*\,d\tau + cc^*\int g(\hat{A}g)^*\,d\tau$$

By virtue of (7.6), the first terms on each side of this last equation are equal to each other; likewise, the last terms on each side are equal. Hence

$$c^*\int g^*\hat{A}f\,d\tau + c\int f^*\hat{A}g\,d\tau = c\int g(\hat{A}f)^*\,d\tau + c^*\int f(\hat{A}g)^*\,d\tau \tag{7.8}$$

Setting $c = 1$ in (7.8), we have

$$\int g^*\hat{A}f\,d\tau + \int f^*\hat{A}g\,d\tau = \int g(\hat{A}f)^*\,d\tau + \int f(\hat{A}g)^*\,d\tau \tag{7.9}$$

Setting $c = i$ in (7.8), we have, after dividing by i,

$$-\int g^*\hat{A}f\,d\tau + \int f^*\hat{A}g\,d\tau = \int g(\hat{A}f)^*\,d\tau - \int f(\hat{A}g)^*\,d\tau \tag{7.10}$$

We now add (7.9) and (7.10) to get (7.7). This completes the proof.

Therefore, *a Hermitian operator $\hat{A}$ is a linear operator that satisfies*

$$\boxed{\int f_m^*\hat{A}f_n\,d\tau = \int f_n(\hat{A}f_m)^*\,d\tau} \tag{7.11}$$

where f_m and f_n are arbitrary well-behaved functions and the integrals are definite integrals over all space. Using the bracket and matrix-element notations, we write

$$\boxed{\langle f_m|\hat{A}|f_n\rangle = \langle f_n|\hat{A}|f_m\rangle^*} \tag{7.12}$$

$$\langle m|\hat{A}|n\rangle = \langle n|\hat{A}|m\rangle^* \tag{7.13}$$

$$\boxed{A_{mn} = (A_{nm})^*} \tag{7.14}$$

The two sides of (7.12) differ by having the functions interchanged and the complex conjugate taken.

Examples of Hermitian Operators. Let us show that some of the operators we have been using are indeed Hermitian. For simplicity, we shall work in one dimension. To prove that an operator is Hermitian, it suffices to show that it satisfies (7.6) for all well-behaved functions. However, we shall make things a bit harder by proving that (7.11) is satisfied.

First consider the one-particle, one-dimensional potential-energy operator. The right side of (7.11) is

$$\int_{-\infty}^{\infty} f_n(x)[V(x)f_m(x)]^* \, dx \tag{7.15}$$

We have $V^* = V$, since the potential energy is a real function. Moreover, the order of the factors in (7.15) does not matter. Therefore,

$$\int_{-\infty}^{\infty} f_n(Vf_m)^* \, dx = \int_{-\infty}^{\infty} f_n V^* f_m^* \, dx = \int_{-\infty}^{\infty} f_m^* V f_n \, dx$$

which proves that V is Hermitian.

The operator for the x component of linear momentum is $\hat{p}_x = -i\hbar \, d/dx$ [Eq. (3.23)]. For this operator, the left side of (7.11) is

$$-i\hbar \int_{-\infty}^{\infty} f_m^*(x) \frac{df_n(x)}{dx} \, dx$$

Now we use the formula for integration by parts:

$$\int_a^b u(x) \frac{dv(x)}{dx} \, dx = u(x)v(x) \Big|_a^b - \int_a^b v(x) \frac{du(x)}{dx} \, dx \tag{7.16}$$

Let

$$u(x) \equiv -i\hbar f_m^*(x), \qquad v(x) \equiv f_n(x)$$

Then

$$-i\hbar \int_{-\infty}^{\infty} f_m^* \frac{df_n}{dx} \, dx = -i\hbar f_m^* f_n \Big|_{-\infty}^{\infty} + i\hbar \int_{-\infty}^{\infty} f_n(x) \frac{df_m^*(x)}{dx} \, dx \tag{7.17}$$

Because f_m and f_n are well-behaved functions, they vanish at $x = \pm\infty$. (If they didn't vanish at infinity, they wouldn't be quadratically integrable.) Therefore, (7.17) becomes

$$\int_{-\infty}^{\infty} f_m^* \left(-i\hbar \frac{df_n}{dx} \right) dx = \int_{-\infty}^{\infty} f_n \left(-i\hbar \frac{df_m}{dx} \right)^* dx$$

which is the same as (7.11) and proves that $\hat{p}_x$ is Hermitian. The proof that the kinetic-energy operator is Hermitian is left to the reader. The sum of two Hermitian operators can be shown to be Hermitian. Hence the Hamiltonian operator $\hat{H} = \hat{T} + \hat{V}$ is Hermitian.

Theorems about Hermitian Operators. We now prove some important theorems about the eigenvalues and eigenfunctions of Hermitian operators.

Since the eigenvalues of the operator $\hat{A}$ corresponding to the physical quantity A are the possible results of a measurement of A (Section 3.3), these eigenvalues should all be real numbers. We now prove that the eigenvalues of a Hermitian operator are real numbers.

We are given that $\hat{A}$ is Hermitian. Translating these words into an equation, we have [Eq. (7.11)]

$$\int f_m^* \hat{A} f_n \, d\tau = \int f_n (\hat{A} f_m)^* \, d\tau \tag{7.18}$$

for all well-behaved functions f_m and f_n. We want to prove that every eigenvalue of $\hat{A}$ is a real number. Translating this into equations, we want to show that $a_i = a_i^*$, where the eigenvalues a_i satisfy $\hat{A} g_i = a_i g_i$; the functions g_i are the eigenfunctions.

To introduce the eigenvalues a_i into (7.18), we write (7.18) for the special case where $f_m = g_i$ and $f_n = g_i$:

$$\int g_i^* \hat{A} g_i \, d\tau = \int g_i (\hat{A} g_i)^* \, d\tau$$

Use of $\hat{A} g_i = a_i g_i$ gives

$$a_i \int g_i^* g_i \, d\tau = \int g_i (a_i g_i)^* \, d\tau = a_i^* \int g_i g_i^* \, d\tau$$

$$(a_i - a_i^*) \int |g_i|^2 \, d\tau = 0 \tag{7.19}$$

Since the integrand $|g_i|^2$ is never negative, the only way the integral in (7.19) could be zero would be if g_i were zero for all values of the coordinates. However, we always reject $g_i = 0$ as an eigenfunction on physical grounds. Hence the integral in (7.19) cannot be zero. Therefore, $(a_i - a_i^*) = 0$, and $a_i = a_i^*$. We have proved:

THEOREM 1. The eigenvalues of a Hermitian operator are real numbers.

To help become familiar with bracket notation, we shall repeat the proof of Theorem 1 using bracket notation. We begin by setting $m = i$ and $n = i$ in (7.13) to get $\langle i|\hat{A}|i\rangle = \langle i|\hat{A}|i\rangle^*$. Choosing the function with index i to be an eigenfunction of $\hat{A}$ and using the eigenvalue equation $\hat{A} g_i = a_i g_i$, we have $\langle i|a_i|i\rangle = \langle i|a_i|i\rangle^*$. Hence $a_i\langle i|i\rangle = a_i^*\langle i|i\rangle^* = a_i^*\langle i|i\rangle$ and $(a_i - a_i^*)\langle i|i\rangle = 0$. So $a_i = a_i^*$, where (7.4) with $m = n$ was used.

We showed that two different particle-in-a-box energy eigenfunctions ψ_i and ψ_j are orthogonal, meaning that $\int_{-\infty}^{\infty} \psi_i^* \psi_j \, dx = 0$ for $i \neq j$ [Eq. (2.26)]. Two functions f_1 and f_2 of the same set of coordinates are said to be **orthogonal** if

$$\boxed{\int f_1^* f_2 \, d\tau = 0} \tag{7.20}$$

where the integral is a definite integral over the full range of the coordinates. We now prove the general theorem that *the eigenfunctions of a Hermitian operator are, or can be chosen to be, mutually orthogonal.* Given that

$$\hat{B} F = sF, \qquad \hat{B} G = tG \tag{7.21}$$

where F and G are two linearly independent eigenfunctions of the Hermitian operator $\hat{B}$, we want to prove that

$$\int F^*G\,d\tau \equiv \langle F|G\rangle = 0$$

We begin with Eq. (7.12), which expresses the Hermitian nature of $\hat{B}$:

$$\langle F|\hat{B}|G\rangle = \langle G|\hat{B}|F\rangle^*$$

Using (7.21), we have

$$\langle F|t|G\rangle = \langle G|s|F\rangle^*$$

$$t\langle F|G\rangle = s^*\langle G|F\rangle^*$$

Since eigenvalues of Hermitian operators are real (Theorem 1), we have $s^* = s$. Use of $\langle G|F\rangle^* = \langle F|G\rangle$ [Eq. (7.4)] gives

$$t\langle F|G\rangle = s\langle F|G\rangle$$

$$(t - s)\langle F|G\rangle = 0$$

If $s \neq t$, then

$$\langle F|G\rangle = 0 \tag{7.22}$$

We have proved that two eigenfunctions of a Hermitian operator that correspond to *different* eigenvalues are orthogonal. The question now is: Can we have two independent eigenfunctions that have the *same* eigenvalue? The answer is yes. In the case of *degeneracy,* we have the same eigenvalue for more than one independent eigenfunction. Therefore, we can only be certain that two independent eigenfunctions of a Hermitian operator are orthogonal to each other if they do not correspond to a degenerate eigenvalue. We now show that in the case of degeneracy we may *construct* eigenfunctions that will be orthogonal to one another. We shall use the theorem proved in Section 3.6, that any linear combination of eigenfunctions corresponding to a degenerate eigenvalue is an eigenfunction with the same eigenvalue. Let us therefore suppose that F and G are independent eigenfunctions that have the same eigenvalue:

$$\hat{B}F = sF, \qquad \hat{B}G = sG$$

We take linear combinations of F and G to form two new eigenfunctions g_1 and g_2 that will be orthogonal to each other. We choose

$$g_1 \equiv F, \qquad g_2 \equiv G + cF$$

where the constant c will be chosen to ensure orthogonality. We want

$$\int g_1^* g_2\,d\tau = 0$$

$$\int F^*(G + cF)\,d\tau = \int F^*G\,d\tau + c\int F^*F\,d\tau = 0$$

Hence choosing

$$c = -\int F^*G\,d\tau \bigg/ \int F^*F\,d\tau \tag{7.23}$$

we have two orthogonal eigenfunctions g_1 and g_2 corresponding to the degenerate eigenvalue. This procedure (called **Schmidt** or **Gram–Schmidt orthogonalization**) can be extended to the case of n-fold degeneracy to give n linearly independent orthogonal eigenfunctions corresponding to the degenerate eigenvalue.

Thus, although there is no guarantee that the eigenfunctions of a degenerate eigenvalue are orthogonal, we can always *choose* them to be orthogonal, if we desire, by using the Schmidt (or some other) orthogonalization method. In fact, unless stated otherwise, we shall always assume that we have chosen the eigenfunctions to be orthogonal:

$$\int g_i^* g_k\,d\tau = 0, \qquad i \neq k \tag{7.24}$$

where g_i and g_k are independent eigenfunctions of a Hermitian operator. We have proved:

THEOREM 2. Two eigenfunctions of a Hermitian operator $\hat{B}$ that correspond to different eigenvalues are orthogonal; eigenfunctions of $\hat{B}$ that belong to a degenerate eigenvalue can always be chosen to be orthogonal.

An eigenfunction can usually be multiplied by a constant to normalize it, and we shall assume, unless stated otherwise, that all eigenfunctions are normalized:

$$\int g_i^* g_i\,d\tau = 1 \tag{7.25}$$

The exception is where the eigenvalues form a continuum, rather than a discrete set of values. In this case, the eigenfunctions are not quadratically integrable. Examples are the linear-momentum eigenfunctions, the free-particle energy eigenfunctions, and the hydrogen-atom continuum energy eigenfunctions.

Using the Kronecker delta, defined by $\delta_{ik} \equiv 1$ if $i = k$ and $\delta_{ik} \equiv 0$ if $i \neq k$ [Eq. (2.28)], we can combine (7.24) and (7.25) into one equation:

$$\boxed{\int g_i^* g_k\,d\tau = \langle i|k\rangle = \delta_{ik}} \tag{7.26}$$

where g_i and g_k are eigenfunctions of some Hermitian operator.

As an example, consider the spherical harmonics. We shall prove that

$$\int_0^{2\pi}\int_0^{\pi} [Y_l^m(\theta,\phi)]^* Y_{l'}^{m'}(\theta,\phi)\sin\theta\,d\theta\,d\phi = \delta_{l,l'}\delta_{m,m'} \tag{7.27}$$

where the $\sin\theta$ factor comes from the volume element in spherical coordinates, (5.78). The spherical harmonics are eigenfunctions of the Hermitian operator $\hat{L}^2$ [Eq. (5.104)]. Since eigenfunctions of a Hermitian operator belonging to different eigenvalues are orthogonal, we conclude that the integral in (7.27) is zero unless $l = l'$. Similarly, since the Y_l^m functions are eigenfunctions of $\hat{L}_z$ [Eq. (5.105)], we conclude that the integral in (7.27) is zero unless $m = m'$. Also, the multiplicative constant in Y_l^m [Eq. (5.147) of Prob. 5.31] has been chosen so that the spherical harmonics are normalized [Eq. (6.117)]. Therefore (7.27) is valid.

A proof of the uncertainty principle is outlined in Prob. 7.60.

7.3 EXPANSION IN TERMS OF EIGENFUNCTIONS

In the previous section, we proved the orthogonality of the eigenfunctions of a Hermitian operator. We now discuss another important property of these functions; this property allows us to expand an arbitrary well-behaved function in terms of these eigenfunctions.

We have used the Taylor-series expansion (Prob. 4.1) of a function as a linear combination of the nonnegative integral powers of $(x - a)$. Can we expand a function as a linear combination of some other set of functions besides $1, (x - a), (x - a)^2, \ldots$? The answer is yes, as was first shown by Fourier in 1807. A Fourier series is an expansion of a function as a linear combination of an infinite number of sine and cosine functions. We shall not go into detail about Fourier series, but shall simply look at one example.

Expansion of a Function Using Particle-in-a-Box Wave Functions. Let us consider expanding a function in terms of the particle-in-a-box stationary-state wave functions, which are [Eq. (2.23)]

$$\psi_n = \left(\frac{2}{l}\right)^{1/2} \sin\left(\frac{n\pi x}{l}\right), \qquad n = 1, 2, 3, \ldots \tag{7.28}$$

for x between 0 and l. What are our chances for representing an arbitrary function $f(x)$ in the interval $0 \le x \le l$ by a series of the form

$$f(x) = \sum_{n=1}^{\infty} a_n \psi_n = \left(\frac{2}{l}\right)^{1/2} \sum_{n=1}^{\infty} a_n \sin\left(\frac{n\pi x}{l}\right), \qquad 0 \le x \le l \tag{7.29}$$

where the a_n's are constants. Substitution of $x = 0$ and $x = l$ in (7.29) gives the restrictions that $f(0) = 0$ and $f(l) = 0$. In other words, $f(x)$ must satisfy the same boundary conditions as the ψ_n functions. We shall also assume that $f(x)$ is finite, single-valued, and continuous, but not necessarily differentiable. With these assumptions it can be shown that the expansion (7.29) is valid. We shall not prove (7.29) but simply illustrate its use to represent a function.

Before we can apply (7.29) to a specific $f(x)$, we must derive an expression for the expansion coefficients a_n. We start by multiplying (7.29) by ψ_m^*:

$$\psi_m^* f(x) = \sum_{n=1}^{\infty} a_n \psi_m^* \psi_n = \left(\frac{2}{l}\right) \sum_{n=1}^{\infty} a_n \sin\left(\frac{n\pi x}{l}\right) \sin\left(\frac{m\pi x}{l}\right) \tag{7.30}$$

Now we integrate this equation from 0 to l. Assuming the validity of interchanging the integration and the infinite summation, we have

$$\int_0^l \psi_m^* f(x)\, dx = \sum_{n=1}^{\infty} a_n \int_0^l \psi_m^* \psi_n\, dx = \sum_{n=1}^{\infty} a_n \left(\frac{2}{l}\right) \int_0^l \sin\left(\frac{n\pi x}{l}\right) \sin\left(\frac{m\pi x}{l}\right) dx$$

We proved the orthonormality of the particle-in-a-box wave functions [Eq. (2.27)]. Therefore, the last equation becomes

$$\int_0^l \psi_m^* f(x)\, dx = \sum_{n=1}^{\infty} a_n \delta_{mn} \tag{7.31}$$

The type of sum in (7.31) occurs often. Writing it in detail, we have

$$\sum_{n=1}^{\infty} a_n \delta_{mn} = a_1 \delta_{m,1} + a_2 \delta_{m,2} + \cdots + a_m \delta_{m,m} + a_{m+1} \delta_{m,m+1} + \cdots$$

$$= 0 + 0 + \cdots + a_m + 0 + \cdots$$

$$\sum_{n=1}^{\infty} a_n \delta_{mn} = a_m \tag{7.32}$$

Thus, since δ_{mn} is zero except when the summation index n is equal to m, all terms but one vanish, and (7.31) becomes

$$a_m = \int_0^l \psi_m^* f(x)\, dx \tag{7.33}$$

which is the desired expression for the expansion coefficients.

Changing m to n in (7.33) and substituting it into (7.29), we have

$$f(x) = \sum_{n=1}^{\infty} \left[\int_0^l \psi_n^* f(x)\, dx \right] \psi_n(x) \tag{7.34}$$

This is the desired expression for the expansion of an arbitrary well-behaved function $f(x)$ $(0 \le x \le l)$ as a linear combination of the particle-in-a-box wave functions ψ_n. Note that the definite integral $\int_0^l \psi_n^* f(x)\, dx$ is a number and not a function of x.

We now use (7.29) to represent a specific function, the function of Fig. 7.1, which is defined by

$$\begin{aligned} f(x) &= x && \text{for } 0 \le x \le \tfrac{1}{2}l \\ f(x) &= l - x && \text{for } \tfrac{1}{2}l \le x \le l \end{aligned} \tag{7.35}$$

To find the expansion coefficients a_n, we substitute (7.28) and (7.35) into (7.33):

$$a_n = \int_0^l \psi_n^* f(x)\, dx = \left(\frac{2}{l}\right)^{1/2} \int_0^l \sin\left(\frac{n\pi x}{l}\right) f(x)\, dx$$

$$a_n = \left(\frac{2}{l}\right)^{1/2} \int_0^{l/2} x \sin\left(\frac{n\pi x}{l}\right) dx + \left(\frac{2}{l}\right)^{1/2} \int_{l/2}^{l} (l - x) \sin\left(\frac{n\pi x}{l}\right) dx$$

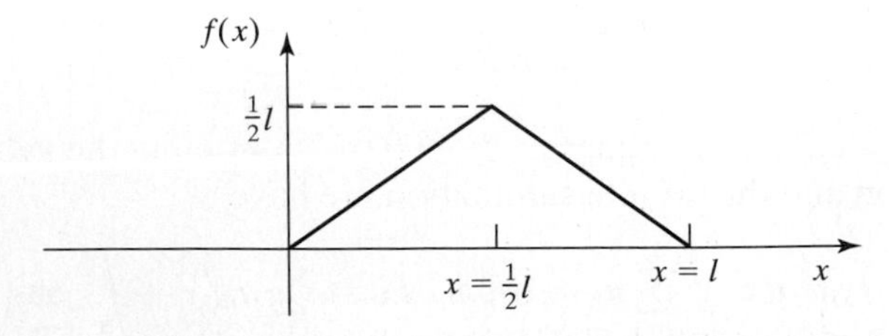

FIGURE 7.1 Function to be expanded in terms of particle-in-a-box functions.

Using the Appendix integral (A.1), we find

$$a_n = \frac{(2l)^{3/2}}{n^2\pi^2} \sin\left(\frac{n\pi}{2}\right) \tag{7.36}$$

Using (7.36) in the expansion (7.29), we have [note that $\sin(n\pi/2)$ equals zero for n even and equals $+1$ or -1 for n odd]

$$f(x) = \frac{4l}{\pi^2}\left[\sin\left(\frac{\pi x}{l}\right) - \frac{1}{3^2}\sin\left(\frac{3\pi x}{l}\right) + \frac{1}{5^2}\sin\left(\frac{5\pi x}{l}\right) - \cdots\right]$$

$$f(x) = \frac{4l}{\pi^2}\sum_{n=1}^{\infty}(-1)^{n+1}\frac{1}{(2n-1)^2}\sin\left[(2n-1)\frac{\pi x}{l}\right] \tag{7.37}$$

where $f(x)$ is given by (7.35). Let us check (7.37) at $x = \frac{1}{2}l$. We have

$$f\left(\frac{l}{2}\right) = \frac{4l}{\pi^2}\left(1 + \frac{1}{3^2} + \frac{1}{5^2} + \frac{1}{7^2} + \cdots\right) \tag{7.38}$$

Let us tabulate the right side of (7.38) as a function of the number of terms we take in the infinite series:

Number of terms	1	2	3	4	5	20	100
Right side of (7.38)	$0.405l$	$0.450l$	$0.467l$	$0.475l$	$0.480l$	$0.495l$	$0.499l$

If we take an infinite number of terms, the series should sum to $\frac{1}{2}l$, which is the value of $f\left(\frac{1}{2}l\right)$. Assuming the validity of the series, we have the interesting result that the infinite sum in parentheses in (7.38) equals $\pi^2/8$. Figure 7.2 plots $f(x) - \sum_{n=1}^{k} a_n\psi_n$ [where

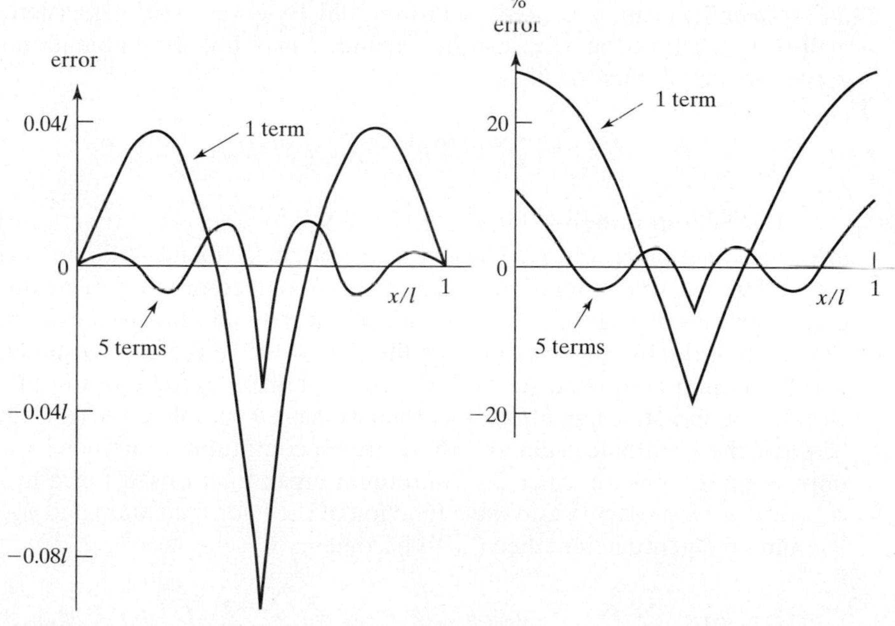

FIGURE 7.2 Plots of (a) the error and (b) the percent error in the expansion of the function of Fig. 7.1 in terms of particle-in-a-box wave functions when 1 and 5 terms are taken in the expansion.

f, a_n, and ψ_n are given by (7.35), (7.36), and (7.28)] for k values of 1 and 5. As k, the number of terms in the expansion, increases, the series comes closer to $f(x)$, and the difference between f and the series goes to zero.

Expansion of a Function in Terms of Eigenfunctions. We have seen an example of the expansion of a function in terms of a set of functions—the particle-in-a-box energy eigenfunctions. Many different sets of functions can be used to expand an arbitrary function. A set of functions $g_1, g_2, \ldots, g_i, \ldots$ is said to be a **complete set** if any well-behaved function f that obeys the same boundary conditions as the g_i's can be expanded as a linear combination of the g_i's according to

$$\boxed{f = \sum_i a_i g_i} \qquad \textbf{(7.39)}$$

where the a_i's are constants. Of course, it is understood that f and the g_i's are all functions of the same set of variables. The limits have been omitted from the sum in (7.39). It is understood that this sum goes over all members of the complete set. By virtue of theorems of Fourier analysis (which we have not proved), the particle-in-a-box energy eigenfunctions can be shown to be a complete set.

We now *postulate* that *the set of eigenfunctions of every Hermitian operator that represents a physical quantity is a complete set.* (Completeness of the eigenfunctions can be proved in many cases, but must be postulated in the general case.) Thus, every well-behaved function that satisfies the same boundary conditions as the set of eigenfunctions can be expanded according to (7.39). Equation (7.29) is an example of (7.39).

The harmonic-oscillator wave functions are given by a Hermite polynomial H_v times an exponential factor [Eq. (4.88) of Prob. 4.20b]. By virtue of the expansion postulate, any well-behaved function $f(x)$ can be expanded as a linear combination of harmonic-oscillator energy eigenfunctions:

$$f(x) = \sum_{n=0}^{\infty} a_n (2^n n!)^{-1/2} (\alpha/\pi)^{1/4} H_n(\alpha^{1/2}x) e^{-\alpha x^2/2}$$

How about using the hydrogen-atom bound-state wave functions to expand an arbitrary function $f(r, \theta, \phi)$? The answer is that these functions do *not* form a complete set, and we cannot expand f using them. To have a complete set, we must use *all* the eigenfunctions of a particular Hermitian operator. In addition to the bound-state eigenfunctions of the hydrogen-atom Hamiltonian, we have the continuum eigenfunctions, corresponding to ionized states. If the continuum eigenfunctions are included along with the bound-state eigenfunctions, then we have a complete set. (For the particle in a box and the harmonic oscillator, there are no continuum functions.) Equation (7.39) implies an integration over the continuum eigenfunctions, if there are any. Thus, if $\psi_{nlm}(r, \theta, \phi)$ is a bound-state wave function of the hydrogen atom and $\psi_{Elm}(r, \theta, \phi)$ is a continuum eigenfunction, then (7.39) becomes

$$f(r, \theta, \phi) = \sum_{n=1}^{\infty} \sum_{l=0}^{n-1} \sum_{m=-l}^{l} a_{nlm} \psi_{nlm}(r, \theta, \phi) + \sum_{l=0}^{\infty} \sum_{m=-l}^{l} \int_0^{\infty} a_{lm}(E) \psi_{Elm}(r, \theta, \phi)\, dE$$

As another example, consider the eigenfunctions of $\hat{p}_x$ [Eq. (3.36)]:

$$g_k = e^{ikx/\hbar}, \qquad -\infty < k < \infty$$

Here the eigenvalues are all continuous, and the eigenfunction expansion (7.39) of an arbitrary function f becomes

$$f(x) = \int_{-\infty}^{\infty} a(k)e^{ikx/\hbar}\, dk$$

The reader with a good mathematical background may recognize this integral as very nearly the Fourier transform of $a(k)$.

Let us evaluate the expansion coefficients in $f = \sum_i a_i g_i$ [Eq. (7.39)], where the g_i functions are the complete set of eigenfunctions of a Hermitian operator. The procedure is the same as used to derive (7.33). We multiply $f = \sum_i a_i g_i$ by g_k^* and integrate over all space:

$$g_k^* f = \sum_i a_i g_k^* g_i$$

$$\int g_k^* f\, d\tau = \sum_i a_i \int g_k^* g_i\, d\tau = \sum_i a_i \delta_{ik} = a_k$$

$$a_k = \int g_k^* f\, d\tau \tag{7.40}$$

where we used the orthonormality of the eigenfunctions of a Hermitian operator: $\int g_k^* g_i\, d\tau = \delta_{ik}$ [Eq. (7.26)]. The procedure that led to (7.40) will be used often and is worth remembering. Substitution of (7.40) for a_i in $f = \sum_i a_i g_i$ gives

$$f = \sum_i \left[\int g_i^* f\, d\tau\right] g_i = \sum_i \langle g_i | f \rangle g_i \tag{7.41}$$

EXAMPLE Let $F(x) = x(l - x)$ for $0 \le x \le l$ and $F(x) = 0$ elsewhere. Expand F in terms of the particle-in-a-box energy eigenfunctions $\psi_n = (2/l)^{1/2} \sin(n\pi x/l)$ for $0 \le x \le l$.

We begin by noting that $F(0) = 0$ and $F(l) = 0$, so F obeys the same boundary conditions as the ψ_n's and can be expanded using the ψ_n's. The expansion is $F = \sum_{n=1}^{\infty} a_n \psi_n$, where $a_n = \int \psi_n^* F\, d\tau$ [Eqs. (7.39) and (7.40)]. Thus

$$a_n = \int \psi_n^* F\, d\tau = \left(\frac{2}{l}\right)^{1/2} \int_0^l \left(\sin \frac{n\pi x}{l}\right) x(l - x)\, dx = \frac{2^{3/2} l^{5/2}}{n^3 \pi^3}[1 - (-1)^n]$$

where details of the integral evaluation are left as a problem (Prob. 7.17). The expansion $F = \sum_{n=1}^{\infty} a_n \psi_n$ is

$$x(l - x) = \frac{4l^2}{\pi^3} \sum_{n=1}^{\infty} \frac{1 - (-1)^n}{n^3} \sin \frac{n\pi x}{l}, \qquad \text{for } 0 \le x \le l$$

EXERCISE Let $G(x) = 1$ for $0 \le x \le l$ and $G(x) = 0$ elsewhere. Expand G in terms of the particle-in-a-box energy eigenfunctions. Since G is not zero at 0 and at l, the expansion will not represent G at these points but will represent G elsewhere. Use the first seven nonzero terms of the expansion to calculate G at $x = \frac{1}{4}l$. Repeat this using the first 70 nonzero terms (use a programmable calculator). (*Answers:* (0.1219, 0.9977.)

A useful theorem is the following:

THEOREM 3. Let the functions $g_1, g_2, \ldots$ be the complete set of eigenfunctions of the Hermitian operator $\hat{A}$, and let the function F be an eigenfunction of $\hat{A}$ with eigenvalue k (that is, $\hat{A}F = kF$). Then if F is expanded as $F = \sum_i a_i g_i$, the only nonzero coefficients a_i are those for which g_i has the eigenvalue k. (Because of degeneracy, several g_i's may have the same eigenvalue k.)

Thus in the expansion of F, we include only those eigenfunctions that have the same eigenvalue as F. The proof of Theorem 3 follows at once from $a_k = \int g_k^* F\, d\tau$ [Eq. (7.40)]; if F and g_k correspond to different eigenvalues of the Hermitian operator $\hat{A}$, they will be orthogonal [Eq. (7.22)] and a_k will vanish.

We shall occasionally use a notation (called *ket* notation) in which the function f is denoted by the symbol $|f\rangle$. There doesn't seem to be any point to this notation, but in advanced formulations of quantum mechanics, it takes on a special significance. In ket notation, Eq. (7.41) reads

$$|f\rangle = \sum_i |g_i\rangle\langle g_i|f\rangle = \sum_i |i\rangle\langle i|f\rangle \tag{7.42}$$

Ket notation is conveniently used to specify eigenfunctions by listing their eigenvalues. For example, the hydrogen-atom wave function with quantum numbers n, l, m is denoted by $\psi_{nlm} = |nlm\rangle$.

The contents of Sections 7.2 and 7.3 can be summarized by the statement that *the eigenfunctions of a Hermitian operator form a complete, orthonormal set, and the eigenvalues are real.*

7.4 EIGENFUNCTIONS OF COMMUTING OPERATORS

If the state function Ψ is simultaneously an eigenfunction of the two operators $\hat{A}$ and $\hat{B}$ with eigenvalues a_j and b_j, respectively, then a measurement of the physical property A will yield the result a_j and a measurement of B will yield b_j. Hence the two properties A and B have definite values when Ψ is simultaneously an eigenfunction of $\hat{A}$ and $\hat{B}$.

In Section 5.1, some statements were made about simultaneous eigenfunctions of two operators. We now prove these statements.

First, we show that if there exists a common *complete set* of eigenfunctions for two linear operators then these operators commute. Let $\hat{A}$ and $\hat{B}$ denote two linear operators that have a common complete set of eigenfunctions $g_1, g_2, \ldots$:

$$\hat{A}g_i = a_i g_i, \qquad \hat{B}g_i = b_i g_i \tag{7.43}$$

where a_i and b_i are the eigenvalues. We must prove that

$$[\hat{A}, \hat{B}] = \hat{0} \tag{7.44}$$

Equation (7.44) is an operator equation. For two operators to be equal, the results of operating with either of them on an arbitrary well-behaved function f must be the same. Hence we must show that

$$(\hat{A}\hat{B} - \hat{B}\hat{A})f = \hat{0}f = 0$$

where f is an arbitrary function. We begin the proof by expanding f (assuming that it obeys the proper boundary conditions) in terms of the complete set of eigenfunctions g_i:

$$f = \sum_i c_i g_i$$

Operating on each side of this last equation with $\hat{A}\hat{B} - \hat{B}\hat{A}$, we have

$$(\hat{A}\hat{B} - \hat{B}\hat{A})f = (\hat{A}\hat{B} - \hat{B}\hat{A}) \sum_i c_i g_i$$

Since the products $\hat{A}\hat{B}$ and $\hat{B}\hat{A}$ are linear operators (Prob. 3.17), we have

$$(\hat{A}\hat{B} - \hat{B}\hat{A})f = \sum_i c_i(\hat{A}\hat{B} - \hat{B}\hat{A})g_i = \sum_i c_i[\hat{A}(\hat{B}g_i) - \hat{B}(\hat{A}g_i)]$$

where the definitions of the sum and the product of operators were used. Use of the eigenvalue equations (7.43) gives

$$(\hat{A}\hat{B} - \hat{B}\hat{A})f = \sum_i c_i[\hat{A}(b_i g_i) - \hat{B}(a_i g_i)] = \sum_i c_i(b_i a_i g_i - a_i b_i g_i) = 0$$

This completes the proof of:

THEOREM 4. If the linear operators $\hat{A}$ and $\hat{B}$ have a common complete set of eigenfunctions, then $\hat{A}$ and $\hat{B}$ commute.

It is sometimes erroneously stated that if a common eigenfunction of $\hat{A}$ and $\hat{B}$ exists, then they commute. An example that shows this statement to be false is the fact that the spherical harmonic Y_0^0 is an eigenfunction of both $\hat{L}_z$ and $\hat{L}_x$ even though these two operators do not commute (Section 5.3). It is instructive to examine the so-called proof that is given for this erroneous statement. Let g be the common eigenfunction: $\hat{A}g = ag$ and $\hat{B}g = bg$. We have

$$\hat{A}\hat{B}g = \hat{A}bg = abg \quad \text{and} \quad \hat{B}\hat{A}g = \hat{B}ag = bag = abg$$

$$\hat{A}\hat{B}g = \hat{B}\hat{A}g \tag{7.45}$$

The "proof" is completed by canceling g from each side of (7.45) to get

$$\hat{A}\hat{B} = \hat{B}\hat{A}(?) \tag{7.46}$$

It is in going from (7.45) to (7.46) that the error occurs. Just because the two operators $\hat{A}\hat{B}$ and $\hat{B}\hat{A}$ give the same result when acting on the single function g is no reason to conclude that $\hat{A}\hat{B} = \hat{B}\hat{A}$. (For example, d/dx and d^2/dx^2 give the same result when operating on e^x, but d/dx is certainly not equal to d^2/dx^2.) The two operators must give the same result when acting on *every* well-behaved function before we can conclude that they are equal. Thus, even though $\hat{A}$ and $\hat{B}$ do not commute, one or more common eigenfunctions of $\hat{A}$ and $\hat{B}$ might exist. However, we cannot have a common *complete* set of eigenfunctions of two noncommuting operators, as we proved earlier in this section.

We have shown that, if there exists a common complete set of eigenfunctions of the linear operators $\hat{A}$ and $\hat{B}$, then they commute. We now prove the following:

THEOREM 5. If the Hermitian operators $\hat{A}$ and $\hat{B}$ commute, we can select a common complete set of eigenfunctions for them.

The proof is as follows. Let the functions g_i and the numbers a_i be the eigenfunctions and eigenvalues of $\hat{A}$:

$$\hat{A}g_i = a_i g_i$$

Operating on both sides of this equation with $\hat{B}$, we have

$$\hat{B}\hat{A}g_i = \hat{B}(a_i g_i)$$

Since $\hat{A}$ and $\hat{B}$ commute and since $\hat{B}$ is linear, we have

$$\hat{A}(\hat{B}g_i) = a_i(\hat{B}g_i) \tag{7.47}$$

This equation states that the function $\hat{B}g_i$ is an eigenfunction of the operator $\hat{A}$ with the same eigenvalue a_i as the eigenfunction g_i. Suppose the eigenvalues of $\hat{A}$ are nondegenerate, so that for any given eigenvalue a_i one and only one linearly independent eigenfunction exists. If this is so, then the two eigenfunctions g_i and $\hat{B}g_i$, which correspond to the same eigenvalue a_i, must be linearly dependent; that is, one function must be simply a multiple of the other:

$$\hat{B}g_i = k_i g_i \tag{7.48}$$

where k_i is a constant. This equation states that the functions g_i are eigenfunctions of $\hat{B}$, which is what we wanted to prove. In Section 7.3, we postulated that the eigenfunctions of any operator that represents a physical quantity form a complete set. Hence the g_i's form a complete set.

We have just proved the desired theorem for the nondegenerate case, but what about the degenerate case? Let the eigenvalue a_i be n-fold degenerate. We know from Eq. (7.47) that $\hat{B}g_i$ is an eigenfunction of $\hat{A}$ with eigenvalue a_i. Hence, Theorem 3 of Section 7.3 tells us that, if the function $\hat{B}g_i$ is expanded in terms of the complete set of eigenfunctions of $\hat{A}$, then all the expansion coefficients will be zero except those for which the $\hat{A}$ eigenfunction has the eigenvalue a_i. In other words, $\hat{B}g_i$ must be a linear combination of the n linearly independent $\hat{A}$ eigenfunctions that correspond to the eigenvalue a_i:

$$\hat{B}g_i = \sum_{k=1}^{n} c_k g_k, \quad \text{where } \hat{A}g_k = a_i g_k \quad \text{for } k = 1 \text{ to } n \tag{7.49}$$

where $g_1, \ldots, g_n$ denote those $\hat{A}$ eigenfunctions that have the degenerate eigenvalue a_i. Equation (7.49) shows that g_i is not necessarily an eigenfunction of $\hat{B}$. However, by taking suitable linear combinations of the n linearly independent $\hat{A}$ eigenfunctions corresponding to the degenerate eigenvalue a_i, one can construct a new set of n linearly independent eigenfunctions of $\hat{A}$ that will also be eigenfunctions of $\hat{B}$. Proof of this statement is given in *Merzbacher,* Section 8.5.

Thus, when $\hat{A}$ and $\hat{B}$ commute, it is always possible to *select* a common complete set of eigenfunctions for them. For example, consider the hydrogen atom, where the operators $\hat{L}_z$ and $\hat{H}$ were shown to commute. If we desired, we could take the phi factor in the eigenfunctions of $\hat{H}$ as $\sin m\phi$ and $\cos m\phi$ (Section 6.6). If we did this, we would not have eigenfunctions of $\hat{L}_z$, except for $m = 0$. However, the linear combinations

$$R(r)S(\theta)(\cos m\phi + i \sin m\phi) = RSe^{im\phi}, \qquad m = -l, \ldots, l$$

give us eigenfunctions of $\hat{L}_z$ that are still eigenfunctions of $\hat{H}$ by virtue of the theorem in Section 3.6.

Extension of the above proofs to the case of more than two operators shows that for a set of Hermitian operators $\hat{A}, \hat{B}, \hat{C}, \ldots$ there exists a common complete set of eigenfunctions if and only if every operator commutes with every other operator.

A useful theorem that is related to Theorem 5 is:

THEOREM 6. If g_i and g_j are eigenfunctions of the Hermitian operator $\hat{A}$ with different eigenvalues (that is, if $\hat{A}g_i = a_i g_i$ and $\hat{A}g_j = a_j g_j$ with $a_i \neq a_j$), and if $\hat{B}$ is a linear operator that commutes with $\hat{A}$, then

$$\langle g_j|\hat{B}|g_i\rangle = 0 \qquad \text{for } a_j \neq a_i \tag{7.50}$$

The proof of (7.50) is as follows. If the eigenvalue a_i is nondegenerate, then, because $\hat{B}$ commutes with $\hat{A}$, Theorem 5 shows that g_i is also an eigenfunction of $\hat{B}$, with $\hat{B}g_i = k_i g_i$ [Eq. (7.48)]. We have $\langle g_j|\hat{B}|g_i\rangle = \langle g_j|\hat{B}g_i\rangle = k_i\langle g_j|g_i\rangle = 0$, since g_j and g_i are eigenfunctions of the Hermitian operator $\hat{A}$ and correspond to different eigenvalues ($a_j \neq a_i$) and so are orthogonal. If a_i is degenerate, then Eq. (7.49) shows that $\hat{B}g_i$ is a linear combination of the n eigenfunctions of the degenerate eigenvalue a_i, according to $\hat{B}g_i = \sum_{k=1}^{n} c_k g_k$; we then have $\langle g_j|\hat{B}g_i\rangle = \langle g_j|\sum_{k=1}^{n} c_k g_k\rangle = \sum_{k=1}^{n} c_k\langle g_j|g_k\rangle = 0$ because $\langle g_j|g_k\rangle = 0$. The integrals $\langle g_j|g_k\rangle$ are zero because g_j corresponds to the eigenvalue a_j, and each g_k corresponds to the eigenvalue a_i, which differs from a_j.

7.5 PARITY

Certain quantum-mechanical operators have no classical analog. An example is the parity operator. Recall that the harmonic oscillator wave functions are either even or odd. We shall show how this property is related to the parity operator.

The **parity operator** $\hat{\Pi}$ is defined in terms of its effect on an arbitrary function f:

$$\boxed{\hat{\Pi}f(x, y, z) = f(-x, -y, -z)} \tag{7.51}$$

The parity operator replaces each Cartesian coordinate with its negative. For example, $\hat{\Pi}(x^2 - ze^{ay}) = x^2 + ze^{-ay}$.

As with any quantum-mechanical operator, we are interested in the eigenvalues c_i and the eigenfunctions g_i of the parity operator:

$$\hat{\Pi}g_i = c_i g_i \tag{7.52}$$

The key to the problem is to calculate the square of $\hat{\Pi}$:

$$\hat{\Pi}^2 f(x, y, z) = \hat{\Pi}[\hat{\Pi}f(x, y, z)] = \hat{\Pi}[f(-x, -y, -z)] = f(x, y, z)$$

Since f is arbitrary, we conclude that $\hat{\Pi}^2$ equals the unit operator:

$$\hat{\Pi}^2 = \hat{1} \tag{7.53}$$

We now operate on (7.52) with $\hat{\Pi}$ to get $\hat{\Pi}\hat{\Pi}g_i = \hat{\Pi}c_i g_i$. Since $\hat{\Pi}$ is linear (Prob. 7.25), we have

$$\hat{\Pi}^2 g_i = c_i \hat{\Pi} g_i$$

$$\hat{\Pi}^2 g_i = c_i^2 g_i \tag{7.54}$$

where the eigenvalue equation (7.52) was used. Since $\hat{\Pi}^2$ is the unit operator, the left side of (7.54) is simply g_i, and

$$g_i = c_i^2 g_i$$

The function g_i cannot be zero everywhere (zero is always rejected as an eigenfunction on physical grounds); hence we can divide by g_i to get $c_i^2 = 1$ and

$$c_i = \pm 1 \tag{7.55}$$

The eigenvalues of $\hat{\Pi}$ are $+1$ and -1. Note that this derivation applies to any operator whose square is the unit operator.

What are the eigenfunctions g_i? The eigenvalue equation (7.52) reads

$$\hat{\Pi} g_i(x, y, z) = \pm g_i(x, y, z)$$

$$g_i(-x, -y, -z) = \pm g_i(x, y, z)$$

If the eigenvalue is $+1$, then $g_i(-x, -y, -z) = g_i(x, y, z)$ and g_i is an even function. If the eigenvalue is -1, then g_i is odd: $g_i(-x, -y, -z) = -g_i(x, y, z)$. Hence, *the eigenfunctions of the parity operator* $\hat{\Pi}$ *are all possible well-behaved even and odd functions.*

When the parity operator commutes with the Hamiltonian operator $\hat{H}$, we can select a common set of eigenfunctions for these operators, as proved in Section 7.4. The eigenfunctions of $\hat{H}$ are the stationary-state wave functions ψ_i. Hence when

$$[\hat{\Pi}, \hat{H}] = 0 \tag{7.56}$$

the wave functions ψ_i can be chosen to be eigenfunctions of $\hat{\Pi}$. We just proved that the eigenfunctions of $\hat{\Pi}$ are either even or odd. Hence, when (7.56) holds, each wave function can be chosen to be either even or odd. Let us find out when the parity and Hamiltonian operators commute.

We have, for a one-particle system,

$$[\hat{H}, \hat{\Pi}] = [\hat{T}, \hat{\Pi}] + [\hat{V}, \hat{\Pi}] = -\frac{\hbar^2}{2m}\left(\left[\frac{\partial^2}{\partial x^2}, \hat{\Pi}\right] + \left[\frac{\partial^2}{\partial y^2}, \hat{\Pi}\right] + \left[\frac{\partial^2}{\partial z^2}, \hat{\Pi}\right]\right) + [\hat{V}, \hat{\Pi}]$$

Since

$$\hat{\Pi}\left[\frac{\partial^2}{\partial x^2} f(x, y, z)\right] = \frac{\partial}{\partial(-x)}\frac{\partial}{\partial(-x)} f(-x, -y, -z) = \frac{\partial^2}{\partial x^2} f(-x, -y, -z)$$

$$= \frac{\partial^2}{\partial x^2} \hat{\Pi} f(x, y, z)$$

where f is any function, we conclude that

$$\left[\frac{\partial^2}{\partial x^2}, \hat{\Pi}\right] = 0$$

Similar equations hold for the y and z coordinates, and $[\hat{H}, \hat{\Pi}]$ becomes

$$[\hat{H}, \hat{\Pi}] = [\hat{V}, \hat{\Pi}] \tag{7.57}$$

Now

$$\hat{\Pi}[V(x, y, z)f(x, y, z)] = V(-x, -y, -z)f(-x, -y, -z) \tag{7.58}$$

If the potential energy is an even function, that is, if $V(-x, -y, -z) = V(x, y, z)$, then (7.58) becomes

$$\hat{\Pi}[V(x, y, z)f(x, y, z)] = V(x, y, z)f(-x, -y, -z) = V(x, y, z)\hat{\Pi} f(x, y, z)$$

so $[\hat{V}, \hat{\Pi}] = 0$. Hence, when the potential energy is an even function, the parity operator commutes with the Hamiltonian:

$$[\hat{H}, \hat{\Pi}] = 0 \qquad \text{if } V \text{ is even} \tag{7.59}$$

These results are easily extended to the n-particle case. For an n-particle system, the parity operator is defined by

$$\hat{\Pi} f(x_1, y_1, z_1, \ldots, x_n, y_n, z_n) = f(-x_1, -y_1, -z_1, \ldots, -x_n, -y_n, -z_n) \tag{7.60}$$

It is easy to see that (7.56) holds when

$$V(x_1, y_1, z_1, \ldots, x_n, y_n, z_n) = V(-x_1, -y_1, -z_1, \ldots, -x_n, -y_n, -z_n) \tag{7.61}$$

If V satisfies this equation, V is said to be an **even function** of the $3n$ coordinates. In summary, we have:

THEOREM 7. When the potential energy V is an even function, we can choose the stationary-state wave functions so that each ψ_i is either an even function or an odd function.

A function that is either even or odd is said to be of *definite parity.*

If the energy levels are all nondegenerate (as is usually true in one-dimensional problems), then only one independent wave function corresponds to each energy eigenvalue and there is no element of choice (apart from an arbitrary multiplicative constant) in the wave functions. Thus, for the nondegenerate case, the stationary-state wave functions must be of definite parity when V is an even function. For example, the one-dimensional harmonic oscillator has $V = \frac{1}{2}kx^2$, which is an even function, and the wave functions have definite parity.

The hydrogen-atom potential-energy function is even, and the hydrogenlike orbitals can be chosen to have definite parity (Probs. 7.21 and 7.28).

For the degenerate case, we have an element of choice in the wave functions, since an arbitrary linear combination of the functions corresponding to the degenerate level is an eigenfunction of $\hat{H}$. For a degenerate energy level, by taking appropriate linear combinations we can choose wave functions that are of definite parity, but there is no necessity that they be of definite parity.

Parity aids in evaluating integrals. We showed that $\int_{-\infty}^{\infty} f(x)\, dx = 0$ when $f(x)$ is an odd function [Eq. (4.53)]. Let us extend this result to the $3n$-dimensional case. An **odd function** of $3n$ variables satisfies

$$g(-x_1, -y_1, -z_1, \ldots, -x_n, -y_n, -z_n) = -g(x_1, y_1, z_1, \ldots, x_n, y_n, z_n) \tag{7.62}$$

If g is an odd function of the $3n$ variables, then

$$\int_{-\infty}^{\infty} \cdots \int_{-\infty}^{\infty} g(x_1, \ldots, z_n)\, dx_1 \cdots dz_n = 0 \tag{7.63}$$

where the integration is over the $3n$ coordinates. This equation holds because the contribution to the integral from the value of g at $(x_1, y_1, z_1, \ldots, x_n, y_n, z_n)$ is canceled by the contribution from $(-x_1, -y_1, -z_1, \ldots, -x_n, -y_n, -z_n)$. Equation (7.63) also holds when the integrand is an odd function of some (but not necessarily all) of the variables. See Prob. 7.29.

7.6 MEASUREMENT AND THE SUPERPOSITION OF STATES

Quantum mechanics can be regarded as a scheme for calculating the probabilities of the various possible outcomes of a measurement. For example, if we know the state function $\Psi(x, t)$, then the probability that a measurement at time t of the particle's position yields a value between x and $x + dx$ is given by $|\Psi(x, t)|^2\, dx$. We now consider measurement of the general property B. Our aim is to find out how to use Ψ to calculate the probabilities for each possible result of a measurement of B. The results of this section, which tell us what information is contained in the state function Ψ, lie at the heart of quantum mechanics.

We shall deal with an n-particle system and use q to symbolize the $3n$ coordinates. We have postulated that the eigenvalues b_i of the operator $\hat{B}$ are the only possible results of a measurement of the property B. Using g_i for the eigenfunctions of $\hat{B}$, we have

$$\hat{B}g_i(q) = b_i g_i(q) \tag{7.64}$$

We postulated in Section 7.3 that the eigenfunctions of any Hermitian operator that represents a physically observable property form a complete set. Since the g_i's form a complete set, we can expand the state function Ψ as

$$\Psi(q, t) = \sum_i c_i(t) g_i(q) \tag{7.65}$$

To allow for the change of Ψ with time, the expansion coefficients c_i vary with time.

Since $|\Psi|^2$ is a probability density, we require that

$$\int \Psi^* \Psi\, d\tau = 1 \tag{7.66}$$

Substituting (7.65) into the normalization condition and using (1.33) and (1.32), we get

$$1 = \int \sum_i c_i^* g_i^* \sum_i c_i g_i\, d\tau = \int \sum_i c_i^* g_i^* \sum_k c_k g_k\, d\tau = \int \sum_i \sum_k c_i^* c_k g_i^* g_k\, d\tau \tag{7.67}$$

Since the summation indexes in the two sums in (7.67) need not have the same value, different symbols must be used for these two dummy indexes. For example, consider the following product of two sums:

$$\sum_{i=1}^{2} s_i \sum_{i=1}^{2} t_i = (s_1 + s_2)(t_1 + t_2) = s_1 t_1 + s_1 t_2 + s_2 t_1 + s_2 t_2$$

If we carelessly write

$$\sum_{i=1}^{2} s_i \sum_{i=1}^{2} t_i \overset{\text{(wrong)}}{=} \sum_{i=1}^{2}\sum_{i=1}^{2} s_i t_i = \sum_{i=1}^{2} (s_1t_1 + s_2t_2) = 2(s_1t_1 + s_2t_2)$$

we get the wrong answer. The correct way to write the product is

$$\sum_{i=1}^{2} s_i \sum_{i=1}^{2} t_i = \sum_{i=1}^{2} s_i \sum_{k=1}^{2} t_k = \sum_{i=1}^{2}\sum_{k=1}^{2} s_i t_k = \sum_{i=1}^{2} (s_it_1 + s_it_2) = s_1t_1 + s_1t_2 + s_2t_1 + s_2t_2$$

which gives the right answer.

Assuming the validity of interchanging the infinite summation and the integration in (7.67), we have

$$\sum_i \sum_k c_i^* c_k \int g_i^* g_k \, d\tau = 1$$

Since $\hat{B}$ is Hermitian, its eigenfunctions g_i are orthonormal [Eq. (7.26)]; hence

$$\sum_i \sum_k c_i^* c_k \delta_{ik} = 1$$

$$\sum_i |c_i|^2 = 1 \tag{7.68}$$

We shall point out the significance of (7.68) shortly.

Recall the postulate (Section 3.7) that, if Ψ is the normalized state function of a system, then the average value of the property B is

$$\langle B \rangle = \int \Psi^*(q,t)\hat{B}\Psi(q,t)\, d\tau$$

Using the expansion (7.65) in the average-value expression, we have

$$\langle B \rangle = \int \sum_i c_i^* g_i^* \hat{B} \sum_k c_k g_k \, d\tau = \sum_i \sum_k c_i^* c_k \int g_i^* \hat{B} g_k \, d\tau$$

where the linearity of $\hat{B}$ was used. Use of $\hat{B}g_k = b_k g_k$ [Eq. (7.64)] gives

$$\langle B \rangle = \sum_i \sum_k c_i^* c_k b_k \int g_i^* g_k \, d\tau = \sum_i \sum_k c_i^* c_k b_k \delta_{ik}$$

$$\langle B \rangle = \sum_i |c_i|^2 b_i \tag{7.69}$$

How do we interpret (7.69)? We postulated in Section 3.3 that the eigenvalues of an operator are the only possible numbers we can get when we measure the property that the operator represents. In any measurement of B, we get one of the values b_i (assuming there is no experimental error). Now recall Eq. (3.81):

$$\langle B \rangle = \sum_{b_i} P_{b_i} b_i \tag{7.70}$$

where P_{b_i} is the probability of getting b_i in a measurement of B. The sum in (7.70) goes over the different eigenvalues b_i, whereas the sum in (7.69) goes over the different eigenfunctions g_i, since the expansion (7.65) is over the g_i's . If there is only one independent eigenfunction for each eigenvalue, then a sum over eigenfunctions is the same as a sum over eigenvalues, and comparison of (7.70) and (7.69) shows that, when there is no degeneracy in the $\hat{B}$ eigenvalues, $|c_i|^2$ is the probability of getting the value b_i in a measurement of the property B. Note that the $|c_i|^2$ values sum to 1, as probabilities should [Eq. (7.68)]. Suppose the eigenvalue b_i is degenerate. From (7.70), P_{b_i} is given by the quantity that multiplies b_i. With degeneracy, more than one term in (7.69) contains b_i, so the probability P_{b_i} of getting b_i in a measurement is found by adding the $|c_i|^2$ values for those eigenfunctions that have the same eigenvalue b_i. We have proved the following:

THEOREM 8. If b_m is a nondegenerate eigenvalue of the operator $\hat{B}$ and g_m is the corresponding normalized eigenfunction ($\hat{B}g_m = b_m g_m$), then, when the property B is measured in a quantum-mechanical system whose state function at the time of the measurement is Ψ, the probability of getting the result b_m is given by $|c_m|^2$, where c_m is the coefficient of g_m in the expansion $\Psi = \sum_i c_i g_i$. If the eigenvalue b_m is degenerate, the probability of obtaining b_m when B is measured is found by adding the $|c_i|^2$ values for those eigenfunctions whose eigenvalue is b_m.

When can the result of a measurement of B be predicted with certainty? We can do this if all the coefficients in the expansion $\Psi = \sum_i^n c_i g_i$ are zero, except one: $c_i = 0$ for all $i \neq k$ and $c_k \neq 0$. For this case, Eq. (7.68) gives $|c_k|^2 = 1$ and we are certain to find the result b_k. In this case, the state function $\Psi = \sum_i c_i g_i$ is given by $\Psi = g_k$. When Ψ is an eigenfunction of $\hat{B}$ with eigenvalue b_k, we are certain to get the value b_k when we measure B.

We can thus view the expansion $\Psi = \sum_i c_i g_i$ [Eq. (7.65)] as expressing the general state Ψ as a *superposition* of the eigenstates g_i of the operator $\hat{B}$. Each eigenstate g_i corresponds to the value b_i for the property B. The degree to which any eigenfunction g_i occurs in the expansion of Ψ, as measured by $|c_i|^2$, determines the probability of getting the value b_i in a measurement of B.

How do we calculate the expansion coefficients c_i so that we can get the probabilities $|c_i|^2$? We multiply $\Psi = \sum_i c_i g_i$ by g_j^*, integrate over all space, and use the orthonormality of the eigenfunctions of the Hermitian operator $\hat{B}$ to get

$$\int g_j^* \Psi \, d\tau = \sum_i c_i \int g_j^* g_i \, d\tau = \sum_i c_i \delta_{ij}$$

$$c_j = \int g_j^* \Psi \, d\tau = \langle g_j | \Psi \rangle \tag{7.71}$$

The probability of finding the nondegenerate eigenvalue b_j in a measurement of B is

$$|c_j|^2 = \left| \int g_j^* \Psi \, d\tau \right|^2 = |\langle g_j | \Psi \rangle|^2 \tag{7.72}$$

where $\hat{B}g_j = b_j g_j$. The quantity $\langle g_j | \Psi \rangle$ is called a **probability amplitude**.

Thus, if we know the state of the system as determined by the state function Ψ, we can use (7.72) to predict the probabilities of the various possible outcomes of a measurement of any property B. Determination of the eigenfunctions g_j and eigenvalues b_j of $\hat{B}$ is a mathematical problem.

To determine experimentally the probability of finding g_j when B is measured, we take a very large number n of identical, noninteracting systems, each in the same state Ψ, and measure B in each system. If n_j of the measurements yield b_j, then $P_{b_j} = n_j/n = |\langle g_j|\Psi\rangle|^2$.

We can restate the first part of Theorem 8 as follows:

THEOREM 9. If the property B is measured in a quantum-mechanical system whose state function at the time of the measurement is Ψ, then the probability of observing the nondegenerate $\hat{B}$ eigenvalue b_j is $|\langle g_j|\Psi\rangle|^2$, where g_j is the normalized eigenfunction corresponding to the eigenvalue b_j.

The integral $\langle g_j|\Psi\rangle = \int g_j^*\Psi\, d\tau$ will have a substantial absolute value if the normalized functions g_j and Ψ resemble each other closely and so have similar magnitudes in each region of space. If g_j and Ψ do not resemble each other, then in regions where g_j is large Ψ will be small (and vice versa), so the product $g_j^*\Psi$ will always be small and the absolute value of the integral $\int g_j^*\Psi\, d\tau$ will be small; the probability $|\langle g_j|\Psi\rangle|^2$ of getting b_j will then be small.

EXAMPLE Suppose that we measure L_z of the electron in a hydrogen atom whose state at the time the measurement begins is the $2p_x$ state. Give the possible outcomes of the measurement and give the probability of each outcome.

From (6.118),

$$\Psi = 2p_x = 2^{-1/2}(2p_1) + 2^{-1/2}(2p_{-1})$$

This equation is the expansion of Ψ as a linear combination of $\hat{L}_z$ eigenfunctions. The only nonzero coefficients are for $2p_1$ and $2p_{-1}$, which are eigenfunctions of $\hat{L}_z$ with eigenvalues $\hbar$ and $-\hbar$, respectively. (Recall that the 1 and -1 subscripts give the m quantum number and that the $\hat{L}_z$ eigenvalues are $m\hbar$.) Using Theorem 8, we take the squares of the absolute values of the coefficients in the expansion of Ψ to get the probabilities. Hence the probability for getting $\hbar$ when L_z is measured is $|2^{-1/2}|^2 = 0.5$, and the probability for getting $-\hbar$ is $|2^{-1/2}|^2 = 0.5$.

EXERCISE Write down a hydrogen-atom wave function for which the probability of getting the $n = 2$ energy if E is measured is 1, the probability of getting $2\hbar^2$ if L^2 is measured is 1, and there are equal probabilities for getting $-\hbar$, 0, and $\hbar$ if L_z is measured. Is there only one possible answer? (*Partial answer:* No.)

EXAMPLE Suppose that the energy E is measured for a particle in a box of length l and that at the time of the measurement the particle is in the nonstationary state $\Psi = 30^{1/2}l^{-5/2}x(l - x)$ for $0 \le x \le l$. Give the possible outcomes of the measurement and give the probability of each possible outcome.

The possible outcomes are given by postulate (c) of Section 3.9 as the eigenvalues of the energy operator $\hat{H}$. The eigenvalues of the particle-in-a-box Hamiltonian are $E = n^2h^2/8ml^2$ $(n = 1, 2, 3, \dots)$ and these are nondegenerate. The probabilities are found by expanding Ψ in terms of the eigenfunctions ψ_n of $\hat{H}$; $\Psi = \sum_{n=1}^{\infty} c_n\psi_n$, where $\psi_n = (2/l)^{1/2}\sin(n\pi x/l)$. In the example after Eq. (7.41), the function $x(l - x)$ was expanded in terms of the particle-in-a-box energy eigenfunctions. The state function $30^{1/2}l^{-5/2}x(l - x)$ equals $x(l - x)$ multiplied by the normalization constant $30^{1/2}l^{-5/2}$. Hence the expansion coefficients c_n are found by multiplying the a_n coefficients in the earlier example by $30^{1/2}l^{-5/2}$ to get

$$c_n = \frac{(240)^{1/2}}{n^3\pi^3}[1 - (-1)^n]$$

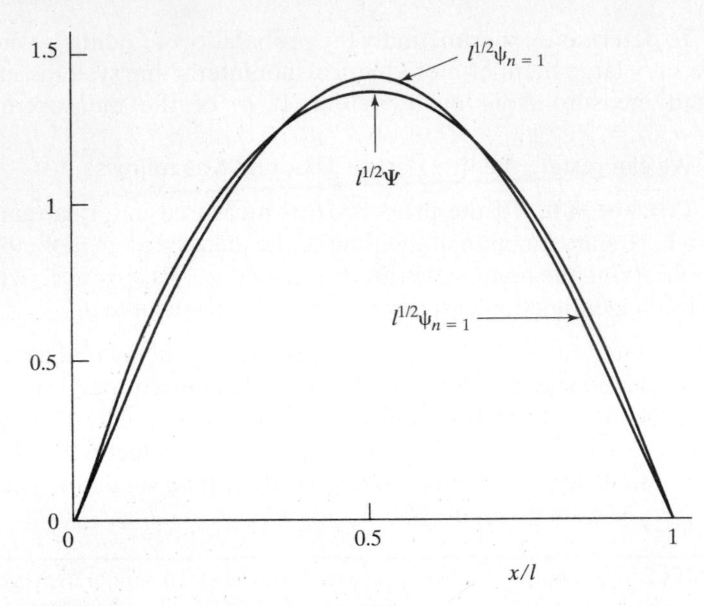

FIGURE 7.3 Plots of $\Psi = (30)^{1/2}l^{-5/2}x(l - x)$ and the $n = 1$ particle-in-a-box wave function.

The probability P_{E_n} of observing the value $E_n = n^2h^2/8ml^2$ equals $|c_n|^2$:

$$P_{E_n} = \frac{240}{n^6\pi^6}[1 - (-1)^n]^2 \tag{7.73}$$

The first few probabilities are

n	1	2	3	4	5
E_n	$h^2/8ml^2$	$4h^2/8ml^2$	$9h^2/8ml^2$	$16h^2/8ml^2$	$25h^2/8ml^2$
P_{E_n}	0.99855	0	0.001370	0	0.000064

The very high probability of finding the $n = 1$ energy is related to the fact that the parabolic state function $30^{1/2}l^{-5/2}x(l - x)$ closely resembles the $n = 1$ particle-in-a-box wave function $(2/l)^{1/2}\sin(\pi x/l)$ (Fig. 7.3). The zero probabilities for $n = 2, 4, 6, \ldots$ are due to the fact that, if the origin is put at the center of the box, the state function $\Psi = 30^{1/2}l^{-5/2}x(l - x)$ is an even function, whereas the $n = 2, 4, 6, \ldots$ functions are odd functions (Fig. 2.3) and so cannot contribute to the expansion of Ψ; the integral $\langle g_n|\Psi\rangle$ vanishes when the integrand is an odd function.

If the property B has a continuous range of eigenvalues (for example, position; Section 7.7), the summation in the expansion (7.65) of Ψ is replaced by an integration over the values of b:

$$\Psi = \int c_b g_b(q)\, db \tag{7.74}$$

and $|\langle g_b(q)|\Psi\rangle|^2$ is interpreted as a probability density; that is, the probability of finding a value of B between b and $b + db$ for a system in the state Ψ is

$$|\langle g_b(q)|\Psi(q,t)\rangle|^2\, db \tag{7.75}$$

7.7 POSITION EIGENFUNCTIONS

We derived the eigenfunctions of the linear-momentum and angular-momentum operators. We now ask: What are the eigenfunctions of the position operator?

We have

$$\hat{x} = x \cdot$$

Denoting the position eigenfunctions by $g_a(x)$, we write

$$xg_a(x) = ag_a(x) \tag{7.76}$$

where a symbolizes the possible eigenvalues. It follows that

$$(x - a)g_a(x) = 0 \tag{7.77}$$

We conclude from (7.77) that

$$g_a(x) = 0 \quad \text{for } x \neq a \tag{7.78}$$

Moreover, since an eigenfunction that is zero everywhere is unacceptable, we have

$$g_a(x) \neq 0 \quad \text{for } x = a \tag{7.79}$$

These conclusions make sense. If the state function is an eigenfunction of $\hat{x}$ with eigenvalue a, $\Psi = g_a(x)$, we know (Section 7.6) that a measurement of x is certain to give the value a. This can be true only if the probability density $|\Psi|^2$ is zero for $x \neq a$, in agreement with (7.78).

Before considering further properties of $g_a(x)$, we define the *Heaviside step function* $H(x)$ by (see Fig. 7.4)

$$\begin{aligned} H(x) &= 1 \quad \text{for } x > 0 \\ H(x) &= 0 \quad \text{for } x < 0 \\ H(x) &= \tfrac{1}{2} \quad \text{for } x = 0 \end{aligned} \tag{7.80}$$

We next define the **Dirac delta function** $\delta(x)$ as the derivative of the Heaviside step function:

$$\delta(x) = dH(x)/dx \tag{7.81}$$

From (7.80) and (7.81), we have at once (see also Fig. 7.4)

$$\delta(x) = 0 \quad \text{for } x \neq 0 \tag{7.82}$$

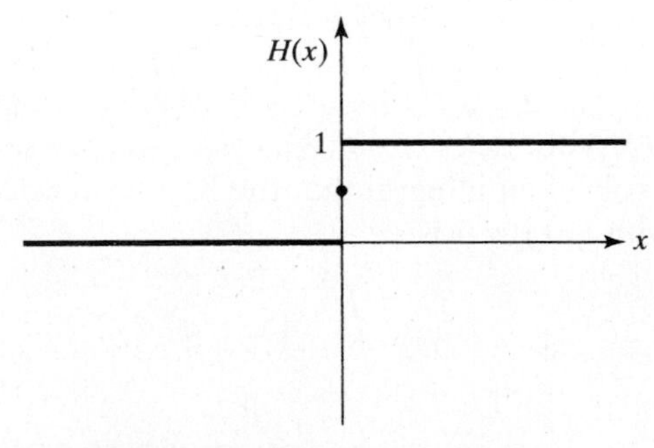

FIGURE 7.4 The Heaviside step function.

Since $H(x)$ makes a sudden jump at $x = 0$, its derivative is infinite at the origin:

$$\delta(x) = \infty \quad \text{for } x = 0 \tag{7.83}$$

We can generalize these equations slightly by setting $x = t - a$ and then changing the symbol t to x. Equations (7.80) to (7.83) become

$$H(x - a) = 1, \quad x > a \tag{7.84}$$

$$H(x - a) = 0, \quad x < a \tag{7.85}$$

$$H(x - a) = \tfrac{1}{2}, \quad x = a \tag{7.86}$$

$$\delta(x - a) = dH(x - a)/dx \tag{7.87}$$

$$\delta(x - a) = 0, \quad x \neq a \quad \text{and} \quad \delta(x - a) = \infty, \quad x = a \tag{7.88}$$

Now consider the following integral:

$$\int_{-\infty}^{\infty} f(x)\delta(x - a)\, dx$$

We evaluate it using integration by parts:

$$\int u\, dv = uv - \int v\, du$$

$$u = f(x), \quad dv = \delta(x - a)\, dx$$

Using (7.87), we have

$$du = f'(x)\, dx, \quad v = H(x - a)$$

$$\int_{-\infty}^{\infty} f(x)\delta(x - a)\, dx = f(x)H(x - a)\Big|_{-\infty}^{\infty} - \int_{-\infty}^{\infty} H(x - a)f'(x)\, dx$$

$$\int_{-\infty}^{\infty} f(x)\delta(x - a)\, dx = f(\infty) - \int_{-\infty}^{\infty} H(x - a)f'(x)\, dx \tag{7.89}$$

where (7.84) and (7.85) were used. Since $H(x - a)$ vanishes for $x < a$, (7.89) becomes

$$\int_{-\infty}^{\infty} f(x)\delta(x - a)\, dx = f(\infty) - \int_{a}^{\infty} H(x - a)f'(x)\, dx$$

$$= f(\infty) - \int_{a}^{\infty} f'(x)\, dx = f(\infty) - f(x)\Big|_{a}^{\infty}$$

$$\int_{-\infty}^{\infty} f(x)\delta(x - a)\, dx = f(a) \tag{7.90}$$

Comparing (7.90) with the equation $\sum_k c_k \delta_{ik} = c_i$, we see that the Dirac delta function plays the same role in an integral that the Kronecker delta plays in a sum. The special case of (7.90) with $f(x) = 1$ is

$$\int_{-\infty}^{\infty} \delta(x - a)\, dx = 1$$

The properties (7.88) of the Dirac delta function agree with the properties (7.78) and (7.79) of the position eigenfunctions $g_a(x)$. We therefore tentatively set

$$g_a(x) = \delta(x - a) \tag{7.91}$$

To verify (7.91), we now show it to be in accord with the Born postulate that $|\Psi(a, t)|^2\, da$ is the probability of observing a value of x between a and $a + da$. According to (7.75), this probability is given by

$$|\langle g_a(x)|\Psi(x, t)\rangle|^2\, da = \left|\int_{-\infty}^{\infty} g_a^*(x)\Psi(x, t)\, dx\right|^2 da \tag{7.92}$$

Using (7.91) and then (7.90), we have for (7.92)

$$\left|\int_{-\infty}^{\infty} \delta(x - a)\Psi(x, t)\, dx\right|^2 da = |\Psi(a, t)|^2\, da$$

which completes the proof.

Since the quantity a in $\delta(x - a)$ can have any real value, the eigenvalues of $\hat{x}$ form a continuum: $-\infty < a < \infty$. As usual for continuum eigenfunctions, $\delta(x - a)$ is not quadratically integrable (Prob. 7.43).

Summarizing, the eigenfunctions and eigenvalues of position are

$$\hat{x}\delta(x - a) = a\delta(x - a) \tag{7.93}$$

where a is any real number.

The delta function is badly behaved, and consequently the manipulations we performed are lacking in rigor and would make a mathematician shudder. However, one can formulate things rigorously by considering the delta function to be the limiting case of a function that becomes successively more peaked at the origin (Fig. 7.5).

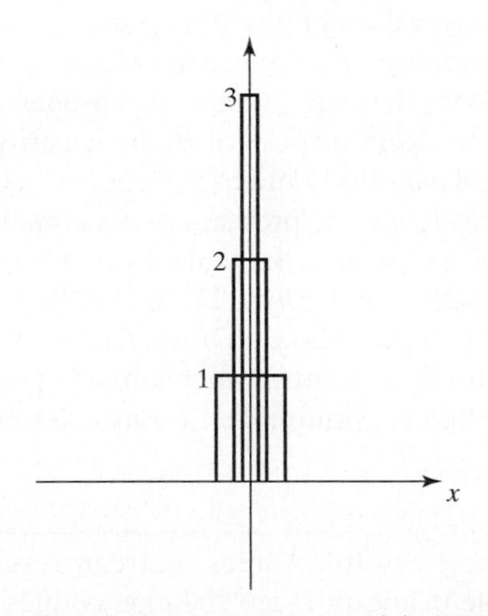

FIGURE 7.5 Functions that approximate $\delta(x)$ with successively increasing accuracy. The area under each curve is 1.

7.8 THE POSTULATES OF QUANTUM MECHANICS

This section summarizes the postulates of quantum mechanics introduced in previous chapters.

Postulate 1. The state of a system is described by a function Ψ of the coordinates of the particles and the time. This function, called the state function or wave function, contains all the information that can be determined about the system. We further postulate that Ψ is single-valued, continuous, and quadratically integrable. For continuum states, the quadratic integrability requirement is omitted.

The designation "wave function" for Ψ is perhaps not the best choice. A physical wave moving in three-dimensional space is a function of the three spatial coordinates and the time. However, for an n-particle system, the function Ψ is a function of $3n$ spatial coordinates and the time. Hence, for a many-particle system, we cannot interpret Ψ as any sort of physical wave. The state function is best thought of as a function from which we can calculate various properties of the system. The nature of the information that Ψ contains is the subject of Postulate 5 and its consequences.

Postulate 2. To every physically observable property there corresponds a linear Hermitian operator. To find this operator, write down the classical-mechanical expression for the observable in terms of Cartesian coordinates and corresponding linear-momentum components, and then replace each coordinate x by the operator $x\cdot$ and each momentum component p_x by the operator $-i\hbar\partial/\partial x$.

We saw in Section 7.2 that the restriction to Hermitian operators arises from the requirement that average values of physical quantities be real numbers. The requirement of linearity is closely connected to the superposition of states discussed in Section 7.6. In our derivation of (7.69) for the average value of a property B for a state that was expanded as a superposition of the eigenfunctions of $\hat{B}$, the linearity of $\hat{B}$ played a key role.

When the classical quantity contains a product of a Cartesian coordinate and its conjugate momentum, we run into the problem of noncommutativity in constructing the correct quantum-mechanical operator. Several different rules have been proposed to handle this case. See J. R. Shewell, *Am. J. Phys.*, **27**, 16 (1959); E. H. Kerner and W. G. Sutcliffe, *J. Math. Phys.*, **11**, 391 (1970); A. de Souza Dutra, *J. Phys. A: Math. Gen.*, **39**, 203 (2006).

The process of finding quantum-mechanical operators in non-Cartesian coordinates is complicated. See K. Simon, *Am. J. Phys.*, **33**, 60 (1965); G. R. Gruber, *Found. Phys.*, **1**, 227 (1971).

Postulate 3. The only possible values that can result from measurements of the physically observable property B are the eigenvalues b_i in the equation $\hat{B}g_i = b_i g_i$, where $\hat{B}$ is the operator corresponding to the property B. The eigenfunctions g_i are required to be well behaved.

Our main concern is with the energy levels of atoms and molecules. These are given by the eigenvalues of the energy operator, the Hamiltonian $\hat{H}$. The eigenvalue equation for $\hat{H}$, $\hat{H}\psi = E\psi$, is the time-independent Schrödinger equation. However, finding the possible values of any property involves solving an eigenvalue equation.

Postulate 4. If $\hat{B}$ is a linear Hermitian operator that represents a physically observable property, then the eigenfunctions g_i of $\hat{B}$ form a complete set.

There are Hermitian operators whose eigenfunctions do not form a complete set (see *Griffiths*, pp. 99, 106; *Messiah*, p. 188; *Ballentine*, Sec. 1.3.) The completeness requirement is essential to developing the theory of quantum mechanics, so it necessary to postulate that all Hermitian operators that correspond to observable properties have a complete set of eigenfunctions. Postulate 4 allows us to expand the wave function for any state as a superposition of the orthonormal eigenfunctions of any quantum-mechanical operator:

$$\Psi = \sum_i c_i g_i = \sum_i |g_i\rangle\langle g_i|\Psi\rangle \tag{7.94}$$

Postulate 5. If $\Psi(q, t)$ is the normalized state function of a system at time t, then the average value of a physical observable B at time t is

$$\boxed{\langle B\rangle = \int \Psi^* \hat{B}\Psi\, d\tau} \tag{7.95}$$

The definition of the quantum-mechanical average value is given in Section 3.7 and should not be confused with the time average used in classical mechanics.

From Postulates 4 and 5, we showed in Section 7.6 that the probability of observing the nondegenerate eigenvalue b_i in a measurement of B is $P_{b_i} = |\int g_i^* \Psi\, d\tau|^2 = |\langle g_i|\Psi\rangle|^2$, where $\hat{B}g_i = b_i g_i$. If the state function happens to be one of the eigenfunctions of $\hat{B}$, that is, if $\Psi = g_k$, then P_{b_i} becomes $P_{b_i} = |\int g_i^* g_k\, d\tau|^2 = |\delta_{ik}|^2 = \delta_{ik}$, where the orthonormality of the eigenfunctions of the Hermitian operator $\hat{B}$ was used. We are certain to observe the value b_k when $\Psi = g_k$.

Postulate 6. The time development of the state of an undisturbed quantum-mechanical system is given by the Schrödinger time-dependent equation

$$\boxed{-\frac{\hbar}{i}\frac{\partial \Psi}{\partial t} = \hat{H}\Psi} \tag{7.96}$$

where $\hat{H}$ is the Hamiltonian (that is, energy) operator of the system.

The time-dependent Schrödinger equation is a first-order differential equation in the time, so that, just as in classical mechanics, the present state of an undisturbed system determines the future state. However, unlike knowledge of the state in classical mechanics, knowledge of the state in quantum mechanics involves knowledge of only the *probabilities* for various possible outcomes of a measurement. Thus, suppose we have several identical noninteracting systems, each having the same state function $\Psi(t_0)$ at time t_0. If we leave each system undisturbed, then the state function for each system will change in accord with (7.96). Since each system has the same Hamiltonian, each system will have the same state function $\Psi(t_1)$ at any future time t_1. However, suppose that at time t_2 we measure property B in each system. Although each system has the same state function $\Psi(t_2)$ at the instant the measurement begins, we will not get the same result for each system. Rather, we will get a spread of possible values b_i, where b_i are the eigenvalues of $\hat{B}$. The relative number of times we get each b_i can be calculated from the quantities $|c_i|^2$, where $\Psi(t_2) = \sum_i c_i g_i$, with the g_i's being the eigenfunctions of $\hat{B}$.

If the Hamiltonian is independent of time, we have the possibility of states of definite energy E. For such states the state function must satisfy

$$\hat{H}\Psi = E\Psi \tag{7.97}$$

and the time-dependent Schrödinger equation becomes

$$-\frac{\hbar}{i}\frac{\partial \Psi}{\partial t} = E\Psi$$

which integrates to $\Psi = Ae^{-iEt/\hbar}$, where A, the integration "constant," is independent of time. The function Ψ depends on the coordinates and the time, so A is some function of the coordinates, which we designate as $\psi(q)$. We have

$$\Psi(q, t) = e^{-iEt/\hbar}\psi(q) \tag{7.98}$$

for a state of constant energy. The function $\psi(q)$ satisfies the time-independent Schrödinger equation

$$\hat{H}\psi(q) = E\psi(q)$$

which follows from (7.97) and (7.98). The factor $e^{-iEt/\hbar}$ simply indicates a change in the phase of the wave function $\Psi(q, t)$ with time and has no direct physical significance. Hence we generally refer to $\psi(q)$ as "the wave function." The Hamiltonian operator plays a unique role in quantum mechanics in that it occurs in the fundamental dynamical equation, the time-dependent Schrödinger equation. The eigenstates of $\hat{H}$ (known as stationary states) have the special property that the probability density $|\Psi|^2$ is independent of time.

The time-dependent Schrödinger equation (7.96) is $(i\hbar\partial/\partial t - \hat{H})\Psi = 0$. Because the operator $i\hbar\partial/\partial t - \hat{H}$ is linear, any linear combination of solutions of the time-dependent Schrödinger equation (7.96) is a solution of (7.96). For example, if the Hamiltonian $\hat{H}$ is independent of time, then there exist stationary-state solutions $\Psi_n = e^{-iE_n t/\hbar}\psi_n(q)$ [Eq. (7.98)] of the time-dependent Schrödinger equation. Any linear combination

$$\Psi = \sum_n c_n \Psi_n = \sum_n c_n e^{-iE_n t/\hbar}\psi_n(q) \tag{7.99}$$

where the c_n's are time-independent constants, is a solution of the time-dependent Schrödinger equation, although it is not an eigenfunction of $\hat{H}$. Because of the completeness of the eigenfunctions ψ_n, any state function can be written in the form (7.99) if $\hat{H}$ is independent of time. (See also Section 9.9.) The state function (7.99) represents a state that does not have a definite energy. Rather, when we measure the energy, the probability of getting E_n is $|c_n e^{-iE_n t/\hbar}|^2 = |c_n|^2$.

To find the constants c_n in (7.99), we write (7.99) at time t_0 as $\Psi(q, t_0) = \sum_n c_n e^{-iE_n t_0/\hbar}\psi_n(q)$. Multiplication by $\psi_j^*(q)$ followed by integration over all space gives

$$\langle\psi_j(q)|\Psi(q,t_0)\rangle = \sum_n c_n e^{-iE_n t_0/\hbar}\langle\psi_j|\psi_n\rangle = \sum_n c_n e^{-iE_n t_0/\hbar}\delta_{jn} = c_j e^{-iE_j t_0/\hbar}$$

so $c_j = \langle\psi_j|\Psi(q,t_0)\rangle e^{iE_j t_0/\hbar}$ and (7.99) becomes

$$\Psi(q,t) = \sum_j \langle\psi_j(q)|\Psi(q,t_0)\rangle e^{\,iE_j(t-t_0)/\hbar}\psi_j(q) \quad \text{if } \hat{H} \text{ ind. of } t \tag{7.100}$$

where $\hat{H}\psi_j = E_j\psi_j$ and the ψ_j's have been chosen to be orthonormal. Equation (7.100) tells us how to find Ψ at time t from Ψ at an initial time t_0 and is the general solution of the time-dependent Schrödinger equation when $\hat{H}$ is independent of t. [Equation (7.100) can also be derived directly from the time-dependent Schrödinger equation; see Prob. 7.46.]

EXAMPLE A particle in a one-dimensional box of length l has a time-independent Hamiltonian and has the state function $\Psi = 2^{-1/2}\psi_1 + 2^{-1/2}\psi_2$ at time $t = 0$, where ψ_1 and ψ_2 are particle-in-a-box time-independent energy eigenfunctions [Eq. (2.23)] with $n = 1$ and $n = 2$, respectively. (a) Find the probability density as a function of time. (b) Show that $|\Psi|^2$ oscillates with a period $T = 8ml^2/3h$. (c) Use a spreadsheet or Mathcad to plot $l|\Psi|^2$ versus x/l at each of the times $jT/8$, where $j = 0, 1, 2, \ldots, 8$.

(a) Since $\hat{H}$ is independent of time, Ψ at any future time will be given by (7.99) with $c_1 = 2^{-1/2}$, $c_2 = 2^{-1/2}$, and all other c's equal to zero. Therefore,

$$\Psi = \frac{1}{\sqrt{2}}e^{-iE_1t/\hbar}\left(\frac{2}{l}\right)^{1/2}\sin\frac{\pi x}{l} + \frac{1}{\sqrt{2}}e^{-iE_2t/\hbar}\left(\frac{2}{l}\right)^{1/2}\sin\frac{2\pi x}{l} = \frac{1}{\sqrt{2}}e^{-iE_1t/\hbar}\psi_1 + \frac{1}{\sqrt{2}}e^{-iE_2t/\hbar}\psi_2$$

We find for the probability density (Prob. 7.47)

$$\Psi^*\Psi = \tfrac{1}{2}\psi_1^2 + \tfrac{1}{2}\psi_2^2 + \psi_1\psi_2\cos[(E_2 - E_1)t/\hbar] \tag{7.101}$$

(b) The time-dependent part of $|\Psi|^2$ is the cosine factor in (7.101). The period T is the time it takes for the cosine to increase by 2π, so $(E_2 - E_1)T/\hbar = 2\pi$ and $T = 2\pi\hbar/(E_2 - E_1) = 8ml^2/3h$, since $E_n = n^2h^2/8ml^2$. (c) Using (7.101), the expressions for ψ_1 and ψ_2, and $T = 2\pi\hbar/(E_2 - E_1)$, we have

$$l|\Psi|^2 = \sin^2(\pi x_r) + \sin^2(2\pi x_r) + 2\sin(\pi x_r)\sin(2\pi x_r)\cos(2\pi t/T) \tag{7.102}$$

where $x_r \equiv x/l$. With $t = jT/8$, the graphs are easily plotted for each j value. The plots show that the probability-density maximum oscillates between the left and right sides of the box. Using Mathcad, one can produce a movie of $|\Psi|^2$ as time changes (Prob. 7.47). (An online resource that allows one to follow $|\Psi|^2$ as a function of time for systems such as the particle in a box or the harmonic oscillator for any chosen initial mixture of stationary states is at www.falstad.com/qm1d; one chooses the mixture by clicking on the small circles at the bottom and dragging on each rotating arrow within a circle.)

Equation (7.99) with the c_n's being constant is the general solution of the time-dependent Schrödinger equation when $\hat{H}$ is independent of time. For a system acted on by an external time-dependent force, the Hamiltonian contains a time-dependent part: $\hat{H} = \hat{H}^0 + \hat{H}'(t)$, where $\hat{H}^0$ is the Hamiltonian of the system in the absence of the external force and $\hat{H}'(t)$ is the time-dependent potential energy of interaction of the system with the external force. In this case, we can use the stationary-state time-independent eigenfunctions of $\hat{H}^0$ to expand Ψ in an equation like (7.99), except that now the c_n's depend on t. An example is an atom or molecule exposed to the time-dependent electric field of electromagnetic radiation (light); see Section 9.9.

What determines whether a system is in a stationary state such as (7.98) or a nonstationary state such as (7.99)? The answer is that the history of the system determines its present state. For example, if we take a system that is in a stationary state and expose it to radiation, the time-dependent Schrödinger equation shows that the radiation causes the state to change to a nonstationary state; see Section 9.9.

You might be wondering about the absence from the list of postulates of the Born postulate that $|\Psi(x,t)|^2\,dx$ is the probability of finding the particle between x and $x + dx$. This postulate is a consequence of Postulate 5, as we now show. Equation (3.81) is $\langle B \rangle = \sum_b P_b b$, where P_b is the probability of observing the value b in a measurement of the property B that takes on discrete values. The corresponding equation for the continuous variable x is $\langle x \rangle = \int_{-\infty}^{\infty} P(x)x\,dx$, where $P(x)$ is the probability density for observing various values of x. According to Postulate 5, we have $\langle x \rangle = \int_{-\infty}^{\infty} \Psi^* \hat{x} \Psi\,dx = \int_{-\infty}^{\infty} |\Psi|^2 x\,dx$. Comparison of these two expressions for $\langle x \rangle$ shows that $|\Psi|^2$ is the probability density $P(x)$.

Chapter 10 gives two further quantum-mechanical postulates that deal with spin and the spin–statistics theorem.

7.9 MEASUREMENT AND THE INTERPRETATION OF QUANTUM MECHANICS

In quantum mechanics, the state function of a system changes in two ways. [See E. P. Wigner, *Am. J. Phys.*, **31**, 6 (1963).] First, there is the continuous, causal change with time given by the time-dependent Schrödinger equation (7.96). Second, there is the sudden, discontinuous, probabilistic change that occurs when a measurement is made on the system. This kind of change cannot be predicted with certainty, since the result of a measurement cannot be predicted with certainty; only the probabilities (7.72) are predictable. The sudden change in Ψ caused by a measurement is called the *reduction* (or *collapse*) *of the wave function. A measurement of the property B that yields the result b_k changes the state function to g_k, the eigenfunction of $\hat{B}$ whose eigenvalue is b_k.* (If b_k is degenerate, Ψ is changed to a linear combination of the eigenfunctions corresponding to b_k.) The probability of finding the nondegenerate eigenvalue b_k is given by Eq. (7.72) and Theorem 9 as $|\langle g_k|\Psi\rangle|^2$, so the quantity $|\langle g_k|\Psi\rangle|^2$ is the probability the system will make a transition from the state Ψ to the state g_k when B is measured.

Consider an example. Suppose that at time t we measure a particle's position. Let $\Psi(x, t_-)$ be the state function of the particle the instant before the measurement is

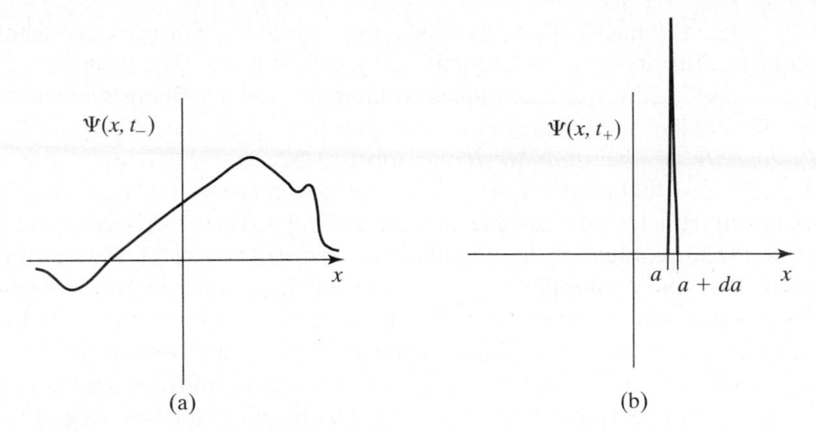

FIGURE 7.6 Reduction of the wave function caused by a measurement of position.

made (Fig. 7.6a). We further suppose that the result of the measurement is that the particle is found to be in the small region of space

$$a < x < a + da \tag{7.103}$$

We ask: What is the state function $\Psi(x, t_+)$ the instant after the measurement? To answer this question, suppose we were to make a second measurement of position at time t_+. Since t_+ differs from the time t of the first measurement by an infinitesimal amount, we must still find that the particle is confined to the region (7.103). If the particle moved a finite distance in an infinitesimal amount of time, it would have infinite velocity, which is unacceptable. Since $|\Psi(x, t_+)|^2$ is the probability density for finding various values of x, we conclude that $\Psi(x, t_+)$ must be zero outside the region (7.103) and must look something like Fig. 7.6b. Thus the position measurement at time t has reduced Ψ from a function that is spread out over all space to one that is localized in the region (7.103). The change from $\Psi(x, t_-)$ to $\Psi(x, t_+)$ is a probabilistic change.

The measurement process is one of the most controversial areas in quantum mechanics. Just how and at what stage in the measurement process reduction occurs is unclear. Some physicists take the reduction of Ψ as an additional quantum-mechanical postulate, while others claim it is a theorem derivable from the other postulates. Some physicists reject the idea of reduction [see M. Jammer, *The Philosophy of Quantum Mechanics,* Wiley, 1974, Section 11.4; L. E. Ballentine, *Am. J. Phys.,* **55**, 785 (1987)]. Ballentine advocates Einstein's statistical-ensemble interpretation of quantum mechanics, in which the wave function does not describe the state of a single system (as in the orthodox interpretation) but gives a statistical description of a collection of a large number of systems each prepared in the same way (an ensemble). In this interpretation, the need for reduction of the wave function does not occur. [See L. E. Ballentine, *Am. J. Phys.,* **40**, 1763 (1972); *Rev. Mod. Phys.,* **42**, 358 (1970).] There are many serious problems with the statistical-ensemble interpretation [see *Whitaker,* pp. 213–217; D. Home and M. A. B. Whitaker, *Phys. Rep.,* **210**, 223 (1992); Prob. 10.5], and this interpretation has been largely rejected.

"For the majority of physicists the problem of finding a consistent and plausible quantum theory of measurement is still unsolved. ... The immense diversity of opinion ... concerning quantum measurements ... [is] a reflection of the fundamental disagreement as to the interpretation of quantum mechanics as a whole." (M. Jammer, *The Philosophy of Quantum Mechanics,* pp. 519, 521.)

The probabilistic nature of quantum mechanics has disturbed many physicists, including Einstein, de Broglie, and Schrödinger. These physicists and others have suggested that quantum mechanics may not furnish a complete description of physical reality. Rather, the probabilistic laws of quantum mechanics might be simply a reflection of deterministic laws that operate at a subquantum-mechanical level and that involve "hidden variables." An analogy given by the physicist Bohm is the Brownian motion of a dust particle in air. The particle undergoes random fluctuations of position, and its motion is not completely determined by its position and velocity. Of course, Brownian motion is a result of collisions with the gas molecules and is determined by variables existing on the level of molecular motion. Analogously, the motions of electrons might be determined by hidden variables existing on a subquantum-mechanical level. The orthodox interpretation (often called the Copenhagen interpretation) of quantum mechanics, which was developed by Heisenberg and Bohr, denies the existence of hidden variables and asserts that the laws of quantum mechanics provide a complete description of physical reality. (Hidden-variables theories are discussed in F. J. Belinfante, *A Survey of Hidden-Variables Theories,* Pergamon, 1973.)

In 1964, J. S. Bell proved that, in certain experiments involving measurements on two widely separated particles that originally were in the same region of space, any possible local hidden-variable theory must make predictions that differ from those that quantum mechanics makes (see *Ballentine,* Chapter 20). In a *local theory,* two systems very far from each other act independently of each other. The results of such experiments agree with quantum-mechanical predictions, thus providing very strong evidence against all deterministic, local hidden-variable theories but do not rule out nonlocal hidden-variable theories. These experiments are described in A. Aspect, in *The Wave-Particle Dualism,* S. Diner et al. (eds.), Reidel, 1984, pp. 377–390; A. Shimony, *Scientific American,* Jan. 1988, p. 46; A. Zeilinger, *Rev. Mod. Phys*, **71**, S288 (1999).

Further analysis by Bell and others shows that the results of these experiments and the predictions of quantum mechanics are incompatible with a view of the world in which both realism and locality hold. Realism (also called objectivity) is the doctrine that external reality exists and has definite properties independent of whether or not this reality is observed by us. Locality excludes instantaneous action-at-a-distance and asserts that any influence from one system to another must travel at a speed that does not exceed the speed of light. Clauser and Shimony stated that quantum mechanics leads to the "philosophically startling" conclusion that we must either "totally abandon the realistic philosophy of most working scientists, or dramatically revise our concept of space–time" to permit "some kind of action-at a distance" [J. F. Clauser and A. Shimony, *Rep. Prog. Phys.*, **41**, 1881 (1978); see also B. d'Espagnat, *Scientific American,* Nov. 1979, p. 158; A. Aspect, *Nature*, **446**, 866 (2007); S. Gröblacher et al., *Nature*, **446**, 871 (2007)].

Quantum theory predicts and experiments confirm that when measurements are made on two particles that once interacted but now are separated by an unlimited distance the results obtained in the measurement on one particle depend on the results obtained from the measurement on the second particle and depend on which property of the second particle is measured. Such instantaneous "spooky actions at a distance" (Einstein's phrase) have led one physicist to remark that "quantum mechanics is magic" (D. Greenberger, quoted in N. D. Mermin, *Physics Today,* April 1985, p. 38).

The relation between quantum mechanics and the mind has been the subject of much speculation. Wigner argued that the reduction of the wave function occurs when the result of a measurement enters the consciousness of an observer and thus "the being with consciousness must have a different role in quantum mechanics than the inanimate measuring device." He believed it likely that conscious beings obey different laws of nature than inanimate objects and proposed that scientists look for unusual effects of consciousness acting on matter. [E. P. Wigner, "Remarks on the Mind–Body Question," in *The Scientist Speculates,* I. J. Good, ed., Capricorn, 1965, p. 284; *Proc. Amer. Phil. Soc.,* **113**, 95 (1969); *Found. Phys.,* **1**, 35 (1970).]

In 1952, David Bohm (following a suggestion made by de Broglie in 1927 that the wave function might act as a pilot wave guiding the motion of the particle) devised a nonlocal deterministic hidden-variable theory that predicts the same experimental results as quantum mechanics [D. Bohm, *Phys. Rev.,* **85**, 166, 180 (1952)]. In Bohm's theory, a particle at any instant of time possesses both a definite position and a definite momentum (although these quantities are not observable), and it travels on a definite path. The particle also possesses a wave function Ψ whose time development obeys the time-dependent Schrödinger equation. In Bohm's theory, the wave function is a real physical entity that determines the motion of the particle. If we are given a particle at a particular position with a particular wave function at a particular time t, Bohm's theory postulates a certain equation that allows us to calculate the velocity of the particle at that time from its wave function and position; knowing the position and velocity at t, we can find the position at time $t + dt$ and can use the time-dependent Schrödinger equation to find the wave function at $t + dt$; then we calculate the velocity at $t + dt$ from the position and wave function at $t + dt$; and so on. Hence the path can be calculated from the initial position and wave function (assuming we know the potential energy). In Bohm's theory, the particle's position turns out to obey an equation like Newton's second law $m\, d^2x/dt^2 = -\partial V/\partial x$ [Eqs. (1.8) and (1.12)], except that the potential energy V is replaced by $V + Q$, where Q is a quantum potential that is calculated in a certain way from the wave function. In Bohm's theory, collapse of the wave function does not occur; rather, the interaction of the system with the measuring apparatus follows the equations of Bohm's theory, but this interaction leads to the system evolving after the measurement in the manner that would occur if the wave function had been collapsed.

Bohm's work was largely ignored for many years, but interest in Bohm's theory has increased. For more on Bohm's theory, see D. Z Albert, *Quantum Mechanics and Experience,* Harvard Univ. Press, 1992, Chapter 7; *Whitaker,* Chapter 7; P. Holland, *The Quantum Theory of Motion,* Cambridge Univ. Press, 1993; D. Bohm and B. J. Hiley, *The Undivided Universe,* Routledge, 1992; D. Z Albert, *Scientific American,* May 1994, p. 58; S. Goldstein, "Bohmian Mechanics," plato.stanford.edu/entries/qm-bohm/. One of the main characters in Rebecca Goldstein's novel *Properties of Light* (Houghton Mifflin, 2000) is modeled in part on David Bohm.

Although the experimental predictions of quantum mechanics are not arguable, its conceptual interpretation is still the subject of heated debate. Excellent bibliographies with commentary on this subject are B. S. DeWitt and R. N. Graham, *Am. J. Phys.*, **39**, 724 (1971); L. E. Ballentine, *Am. J. Phys.*, **55**, 785 (1987). See also, B. d'Espagnat, *Conceptual Foundations of Quantum Mechanics,* 2nd ed., Benjamin, 1976; M. Jammer, *The Philosophy of Quantum Mechanics,* Wiley, 1974; *Whitaker,* Chapter 8; P. Yam, *Scientific American,* June 1997, p. 124. An online bibliography by A. Cabello on the foundations of quantum mechanics lists 12 different interpretations of quantum mechanics (arxiv.org/abs/quant-ph/0012089).

7.10 MATRICES

Matrix algebra is a key mathematical tool in doing modern-day quantum-mechanical calculations on molecules. Matrices also furnish a convenient way to formulate much of the theory of quantum mechanics. Matrix methods will be used in some later chapters, but this book is written so that the material on matrices can be omitted if time does not allow this material to be covered.

A **matrix** is a rectangular array of numbers. The numbers that compose a matrix are called the **matrix elements**. Let the matrix $\mathbf{A}$ have m rows and n columns, and let a_{ij} $(i = 1, 2, \dots, m$ and $j = 1, 2, \dots, n)$ denote the element in row i and column j. Then

$$\mathbf{A} = \begin{pmatrix} a_{11} & a_{12} & \cdots & a_{1n} \\ a_{21} & a_{22} & \cdots & a_{2n} \\ \cdot & \cdot & \cdots & \cdot \\ a_{m1} & a_{m2} & \cdots & a_{mn} \end{pmatrix}$$

$\mathbf{A}$ is said to be an m by n matrix. Do not confuse $\mathbf{A}$ with a determinant (Section 8.3); a matrix need not be square and is not equal to a single number.

A **row matrix** (also called a **row vector**) is a matrix having only one row. A **column matrix** or **column vector** has only one column.

Two matrices $\mathbf{R}$ and $\mathbf{S}$ are **equal** if they have the same number of rows, the same number of columns, and have corresponding elements equal. If $\mathbf{R} = \mathbf{S}$, then $r_{ij} = s_{ij}$ for $i = 1, \dots, m$ and $j = 1, \dots, n$, where m and n are the dimensions of $\mathbf{R}$ and $\mathbf{S}$. A matrix equation is thus equivalent to mn scalar equations.

The **sum** of two matrices $\mathbf{A}$ and $\mathbf{B}$ is defined as the matrix formed by adding corresponding elements of $\mathbf{A}$ and $\mathbf{B}$; the sum is defined only if $\mathbf{A}$ and $\mathbf{B}$ have the same dimensions. If $\mathbf{C} = \mathbf{A} + \mathbf{B}$, then we have the mn scalar equations $c_{ij} = a_{ij} + b_{ij}$ for $i = 1, \dots, m$ and $j = 1, \dots, n$.

$$\boxed{\text{If} \quad \mathbf{C} = \mathbf{A} + \mathbf{B}, \quad \text{then} \quad c_{ij} = a_{ij} + b_{ij}} \tag{7.104}$$

The product of the scalar k and the matrix $\mathbf{A}$ is defined as the matrix formed by multiplying every element of $\mathbf{A}$ by k.

$$\boxed{\text{If} \quad \mathbf{D} = k\mathbf{A}, \quad \text{then} \quad d_{ij} = ka_{ij}} \tag{7.105}$$

If $\mathbf{A}$ is an m by n matrix and $\mathbf{B}$ is an n by p matrix, the **matrix product** $\mathbf{C} = \mathbf{AB}$ is defined to be the m by p matrix whose elements are

$$c_{ij} \equiv a_{i1}b_{1j} + a_{i2}b_{2j} + \cdots + a_{in}b_{nj} = \sum_{k=1}^{n} a_{ik}b_{kj} \tag{7.106}$$

To calculate c_{ij}, we take row i of $\mathbf{A}$ (this row's elements are $a_{i1}, a_{i2}, \ldots, a_{in}$), multiply each element of this row by the corresponding element in column j of $\mathbf{B}$ (this column's elements are $b_{1j}, b_{2j}, \ldots, b_{nj}$), and add the n products. For example, suppose

$$\mathbf{A} = \begin{pmatrix} -1 & 3 & \frac{1}{2} \\ 0 & 4 & 1 \end{pmatrix} \quad \text{and} \quad \mathbf{B} = \begin{pmatrix} 1 & 0 & -2 \\ 2 & 5 & 6 \\ -8 & 3 & 10 \end{pmatrix}$$

The number of columns of $\mathbf{A}$ equals the number of rows of $\mathbf{B}$, so the matrix product $\mathbf{AB}$ is defined. $\mathbf{AB}$ is the product of the 2 by 3 matrix $\mathbf{A}$ and the 3 by 3 matrix $\mathbf{B}$, so $\mathbf{C} \equiv \mathbf{AB}$ is a 2 by 3 matrix. The element c_{21} is found from the second row of $\mathbf{A}$ and the first column of $\mathbf{B}$ as follows: $c_{21} = 0(1) + 4(2) + 1(-8) = 0$. Calculation of the remaining elements gives

$$\mathbf{C} = \begin{pmatrix} 1 & 16\frac{1}{2} & 25 \\ 0 & 23 & 34 \end{pmatrix}$$

Matrix multiplication is not commutative; the products $\mathbf{AB}$ and $\mathbf{BA}$ need not be equal. (In the preceding example, the product $\mathbf{BA}$ happens to be undefined.) Matrix multiplication can be shown to be **associative**, meaning that $\mathbf{A}(\mathbf{BC}) = (\mathbf{AB})\mathbf{C}$ and can be shown to be **distributive**, meaning that $\mathbf{A}(\mathbf{B} + \mathbf{C}) = \mathbf{AB} + \mathbf{AC}$ and $(\mathbf{B} + \mathbf{C})\mathbf{D} = \mathbf{BD} + \mathbf{CD}$.

A matrix with equal numbers of rows and columns is a **square matrix**. The **order** of a square matrix equals the number of rows.

If $\mathbf{A}$ is a square matrix, its square, cube, ... are defined by $\mathbf{A}^2 \equiv \mathbf{AA}$, $\mathbf{A}^3 \equiv \mathbf{AAA}, \ldots$.

The elements $a_{11}, a_{22}, \ldots, a_{nn}$ of a square matrix of order n lie on its **principal diagonal**. A **diagonal matrix** is a square matrix having zero as the value of each element not on the principal diagonal.

The **trace** of a square matrix is the sum of the elements on the principal diagonal. If $\mathbf{A}$ is a square matrix of order n, its trace is $\text{Tr}\,\mathbf{A} = \sum_{i=1}^{n} a_{ii}$.

A diagonal matrix whose diagonal elements are each equal to 1 is called a **unit matrix** or an **identity matrix**. The (i, j)th element of a unit matrix is the Kronecker delta δ_{ij}; $(\mathbf{I})_{ij} = \delta_{ij}$, where $\mathbf{I}$ is a unit matrix. For example, the unit matrix of order 3 is

$$\begin{pmatrix} 1 & 0 & 0 \\ 0 & 1 & 0 \\ 0 & 0 & 1 \end{pmatrix}$$

Let $\mathbf{B}$ be a square matrix of the same order as a unit matrix $\mathbf{I}$. The (i, j)th element of the product $\mathbf{IB}$ is given by (7.106) as $(\mathbf{IB})_{ij} = \sum_k (\mathbf{I})_{ik}b_{kj} = \sum_k \delta_{ik}b_{kj} = b_{ij}$. Since the (i, j)th elements of $\mathbf{IB}$ and $\mathbf{B}$ are equal for all i and j, we have $\mathbf{IB} = \mathbf{B}$. Similarly, we find $\mathbf{BI} = \mathbf{B}$. Multiplication by a unit matrix has no effect.

A matrix all of whose elements are zero is called a **zero matrix**, symbolized by **0**. A **nonzero matrix** has at least one element not equal to zero. These definitions apply to row vectors and column vectors.

Most matrices in quantum chemistry are either square matrices or row or column matrices.

Matrices and Quantum Mechanics. In Section 7.1, the integral $\int f_i^* \hat{A} f_j \, d\tau$ was called a matrix element of $\hat{A}$. We now justify this name by showing that such integrals obey the rules of matrix algebra.

Let the functions $f_1, f_2, \ldots$ be a complete, orthonormal set and let the symbol $\{f_i\}$ denote this complete set. The numbers $A_{ij} \equiv \langle f_i|\hat{A}|f_j\rangle \equiv \int f_i^* \hat{A} f_j \, d\tau$ are called **matrix elements of the linear operator $\hat{A}$ in the basis** $\{f_i\}$. The square matrix

$$\mathbf{A} = \begin{pmatrix} A_{11} & A_{12} & \cdots \\ A_{21} & A_{22} & \cdots \\ \cdots & \cdots & \cdots \end{pmatrix} = \begin{pmatrix} \langle f_1|\hat{A}|f_1\rangle & \langle f_1|\hat{A}|f_2\rangle & \cdots \\ \langle f_2|\hat{A}|f_1\rangle & \langle f_2|\hat{A}|f_2\rangle & \cdots \\ \cdots & \cdots & \cdots \end{pmatrix} \tag{7.107}$$

is called the **matrix representative** or **matrix representation** of the linear operator $\hat{A}$ in the $\{f_i\}$ **basis**. Since $\{f_i\}$ usually consists of an infinite number of functions, **A** is an infinite-order matrix.

Consider the addition of matrix-element integrals. Suppose $\hat{C} = \hat{A} + \hat{B}$. A typical matrix element of $\hat{C}$ in the $\{f_i\}$ basis is

$$C_{ij} = \langle f_i|\hat{C}|f_j\rangle = \langle f_i|\hat{A} + \hat{B}|f_j\rangle = \int f_i^*(\hat{A} + \hat{B}) f_j \, d\tau$$

$$= \int f_i^* \hat{A} f_j \, d\tau + \int f_i^* \hat{B} f_j \, d\tau = A_{ij} + B_{ij}$$

Thus, if $\hat{C} = \hat{A} + \hat{B}$, then $C_{ij} = A_{ij} + B_{ij}$, which is the rule (7.104) for matrix addition. Hence, if $\hat{C} = \hat{A} + \hat{B}$, then $\mathbf{C} = \mathbf{A} + \mathbf{B}$, where **A**, **B**, and **C** are the matrix representatives of the operators $\hat{A}, \hat{B}, \hat{C}$.

Similarly, if $\hat{D} = k\hat{C}$, then we find (Prob. 7.52) $D_{ij} = kC_{ij}$, which is the rule for multiplication of a matrix by a scalar.

Finally, suppose that $\hat{R} = \hat{S}\hat{T}$. We have

$$R_{ij} = \int f_i^* \hat{R} f_j \, d\tau = \int f_i^* \hat{S}\hat{T} f_j \, d\tau \tag{7.108}$$

The function $\hat{T} f_j$ can be expanded in terms of the complete orthonormal set $\{f_i\}$ as [Eq. (7.41)]:

$$\hat{T} f_j = \sum_k c_k f_k = \sum_k \langle f_k|\hat{T} f_j\rangle f_k = \sum_k \langle f_k|\hat{T}|f_j\rangle f_k = \sum_k T_{kj} f_k$$

and R_{ij} becomes

$$R_{ij} = \int f_i^* \hat{S} \sum_k T_{kj} f_k \, d\tau = \sum_k \int f_i^* \hat{S} f_k \, d\tau \, T_{kj} = \sum_k S_{ik} T_{kj} \tag{7.109}$$

The equation $R_{ij} = \Sigma_k S_{ik}T_{kj}$ is the rule (7.106) for matrix multiplication. Hence, if $\hat{R} = \hat{S}\hat{T}$, then $\mathbf{R} = \mathbf{ST}$.

We have proved that *the matrix representatives of linear operators in a complete orthonormal basis set obey the same equations that the operators obey.* Combining Eqs. (7.108) and (7.109), we have the useful sum rule

$$\sum_k \langle i|\hat{S}|k\rangle \langle k|\hat{T}|j\rangle = \langle i|\hat{S}\hat{T}|j\rangle \tag{7.110}$$

Suppose the basis set $\{f_i\}$ is chosen to be the complete, orthonormal set of eigenfunctions g_i of $\hat{A}$, where $\hat{A}g_i = a_i g_i$. Then the matrix element A_{ij} is

$$A_{ij} = \langle g_i|\hat{A}|g_j\rangle = \langle g_i|\hat{A}g_j\rangle = \langle g_i|a_j g_j\rangle = a_j\langle g_i|g_j\rangle = a_j\delta_{ij}$$

The matrix that represents $\hat{A}$ in the basis of orthonormal $\hat{A}$ eigenfunctions is thus a diagonal matrix whose diagonal elements are the eigenvalues of $\hat{A}$. Conversely, one can prove (Prob. 7.53) that, when the matrix representative of $\hat{A}$ using a complete orthonormal set is a diagonal matrix, then the basis functions are the eigenfunctions of $\hat{A}$ and the diagonal matrix elements are the eigenvalues of $\hat{A}$.

We have used the complete, orthonormal basis $\{f_i\}$ to represent the operator $\hat{A}$ by the matrix $\mathbf{A}$ of (7.107). The basis $\{f_i\}$ can also be used to represent an arbitrary function u, as follows. We expand u in terms of the complete set $\{f_i\}$, according to $u = \Sigma_i u_i f_i$, where the expansion coefficients u_i are numbers (not functions) given by Eq. (7.40) as $u_i = \langle f_i|u\rangle$. The set of expansion coefficients $u_1, u_2, \ldots$ is formed into a column matrix (column vector), which we call $\mathbf{u}$, and $\mathbf{u}$ is said to be the **representative** of the function u in the $\{f_i\}$ basis. If $\hat{A}u = w$, where w is another function, then we can show (Prob. 7.54) that $\mathbf{Au} = \mathbf{w}$, where $\mathbf{A}$, $\mathbf{u}$, and $\mathbf{w}$ are the matrix representatives of $\hat{A}$, u, and w in the $\{f_i\}$ basis. Thus, the effect of the linear operator $\hat{A}$ on an arbitrary function u can be found if the matrix representative $\mathbf{A}$ of $\hat{A}$ is known. Hence, knowing the matrix representative $\mathbf{A}$ is equivalent to knowing what the operator $\hat{A}$ is.

7.11 SUMMARY

A linear quantum-mechanical operator $\hat{A}$ that represents a physical quantity must be Hermitian, meaning that it satisfies $\int f^*\hat{A}u\,d\tau = \int u(\hat{A}f)^*\,d\tau$ for all well-behaved functions f and u. The eigenvalues of Hermitian operators are real numbers. For a Hermitian operator, eigenfunctions that correspond to different eigenvalues are orthogonal, and eigenfunctions that correspond to a degenerate eigenvalue can be chosen to be orthogonal.

We postulated that the eigenfunctions g_i of any Hermitian operator that represents a physical quantity form a complete set, meaning that any well-behaved function f can be expanded as $f = \Sigma_k c_k g_k$, where the g_k's are orthonormal and where $c_k = \int g_k^* f\,d\tau$.

If two quantum-mechanical operators commute, they have a complete set of eigenfunctions in common. Two quantum-mechanical operators that have a common complete set of eigenfunctions must commute.

If the potential energy V is an even function, each stationary-state wave function can be chosen to be either even or odd.

If the property B is measured in a system whose state function is Ψ, the probability that the nondegenerate eigenvalue b_i is found is given by $|\langle g_i|\Psi\rangle|^2$, where $\hat{B}g_i = b_i g_i$.

The postulates of quantum mechanics are summarized in Section 7.8.

Matrices are rectangular arrays of numbers and obey the rules (7.104)–(7.106) for addition, scalar multiplication, and matrix multiplication. If $\{f_i\}$ is a complete, orthonormal set of functions, then the matrix **A** with elements $A_{mn} \equiv \langle f_m|\hat{A}|f_n\rangle$ represents the linear operator $\hat{A}$ in the $\{f_i\}$ basis. Also, the column matrix **u** with elements u_i equal to the coefficients in the expansion $u = \Sigma_i u_i f_i$ represents the function u in the $\{f_i\}$ basis. The matrix representatives of operators and functions in a given basis obey the same relations as the operators and functions. For example, if $\hat{C} = \hat{A} + \hat{B}$, $\hat{R} = \hat{S}\hat{T}$, and $w = \hat{A}u$, then $\mathbf{C} = \mathbf{A} + \mathbf{B}$, $\mathbf{R} = \mathbf{ST}$, and $\mathbf{w} = \mathbf{Au}$.

PROBLEMS

Sec.	7.1	7.2	7.3	7.4	7.5
Probs.	7.1–7.3	7.4–7.16	7.17–7.18	7.19	7.20–7.29

Sec.	7.6	7.7	7.8	7.10	general
Probs.	7.30–7.41	7.42–7.45	7.46–7.48	7.49–7.55	7.56–7.63

7.1 True or false? (If a relation is sometimes true and sometimes false, answer "false".) Here, f and g are functions, c is a constant, and $\hat{B}$ is a linear operator. (a) $\langle f|\hat{B}|g\rangle = \langle f|\hat{B}g\rangle$; (b) $\langle f|\hat{B}|cg\rangle = c\langle f|\hat{B}|g\rangle$; (c) $\langle cf|\hat{B}|g\rangle = c\langle f|\hat{B}|g\rangle$.

7.2 If c is a constant, under what condition is $\langle cf_m|\hat{A}|f_n\rangle$ equal to $\langle f_m|\hat{A}|cf_n\rangle$?

7.3 Verify the equation after (7.3). Verify the equations in (7.5).

7.4 What operator is shown to be Hermitian by the equation $\langle m|n\rangle = \langle n|m\rangle^*$?

7.5 If $\hat{B}$ is Hermitian, prove that $\langle f|\hat{B}|g\rangle = \langle \hat{B}f|g\rangle$.

7.6 Let $\hat{A}$ and $\hat{B}$ be Hermitian operators and let c be a constant. (a) Show that $c\hat{A}$ is Hermitian if c is a real number and that $c\hat{A}$ is not Hermitian if c is not real. (b) Show that $\hat{A} + \hat{B}$ is Hermitian.

7.7 (a) Show that d^2/dx^2 and $\hat{T}_x$ are Hermitian, where $\hat{T}_x \equiv -(\hbar^2/2m)\, d^2/dx^2$. (See Prob. 7.6a.) (b) Show that $\langle T_x\rangle = (\hbar^2/2m) \int |\partial\Psi/\partial x|^2\, d\tau$. (c) For a one-particle system, does $\langle T\rangle$ equal $\langle T_x\rangle + \langle T_y\rangle + \langle T_z\rangle$? (d) Show that $\langle T\rangle \geq 0$ for a one-particle system.

7.8 Which of the following operators are Hermitian: d/dx, id/dx, $4d^2/dx^2$, id^2/dx^2?

7.9 Which of the following operators meet the requirements for a quantum-mechanical operator that is to represent a physical quantity: (a) $(\quad)^{1/2}$, (b) d/dx; (c) d^2/dx^2; (d) id/dx?

7.10 Verify that $\hat{L}_z$ is Hermitian using spherical coordinates.

7.11 Let $\hat{A}$ be a Hermitian operator. Show that $\langle A^2\rangle = \int |\hat{A}\psi|^2\, d\tau$ and therefore $\langle A^2\rangle \geq 0$. [This result can be used to derive Eq. (5.131) more rigorously than in the text. Thus, since $\hat{M}^2 - \hat{M}_z^2 = \hat{M}_x^2 + \hat{M}_y^2$ we have $\langle M^2\rangle - \langle M_z^2\rangle = \langle M_x^2\rangle + \langle M_y^2\rangle$. Now $\langle M^2\rangle = c$, $\langle M_z^2\rangle = b_k^2$, and by the theorem of this exercise $\langle M_x^2\rangle \geq 0$, $\langle M_y^2\rangle \geq 0$. Hence $c - b_k^2 \geq 0$.]

7.12 (a) If $\hat{A}$ and $\hat{B}$ are Hermitian operators, prove that their product $\hat{A}\hat{B}$ is Hermitian if and only if $\hat{A}$ and $\hat{B}$ commute. (b) If $\hat{A}$ and $\hat{B}$ are Hermitian, prove that $\hat{A}\hat{B} + \hat{B}\hat{A}$ is Hermitian. The operator $\hat{A}\hat{B} + \hat{B}\hat{A}$ is called the **anticommutator** of $\hat{A}$ and $\hat{B}$. (c) Is $\hat{x}\hat{p}_x$ Hermitian? (d) Is $\frac{1}{2}(\hat{x}\hat{p}_x + \hat{p}_x\hat{x})$ Hermitian?

7.13 A linear operator $\hat{C}$ is **anti-Hermitian** if $\langle f|\hat{C}|g\rangle = -\langle g|\hat{C}|f\rangle^*$. (a) Show that d/dx is anti-Hermitian. (b) If $\hat{A}$ and $\hat{B}$ are Hermitian, prove that their commutator $\hat{A}\hat{B} - \hat{B}\hat{A}$ is anti-Hermitian.

7.14 Explain why each of the following integrals must be zero, where the functions are hydrogenlike wave functions: (a) $\langle 2p_1|\hat{L}_z|3p_{-1}\rangle$; (b) $\langle 3p_0|\hat{L}_z|3p_0\rangle$.

7.15 Evaluate $\langle f_m|\hat{H}|f_n\rangle$ if (a) $\hat{H}$ is the harmonic-oscillator Hamiltonian operator and f_m and f_n are harmonic-oscillator stationary-state wave functions with vibrational quantum numbers m and n; (b) $\hat{H}$ is the particle-in-a-box $\hat{H}$ and f_m and f_n are particle-in-a-box energy eigenfunctions with quantum numbers m and n.

7.16 (a) Show that the hydrogenlike wave functions $2p_x$ and $2p_1$ are not orthogonal. (b) Use the Schmidt procedure to construct linear combinations of $2p_x$ and $2p_1$ that will be orthogonal. Then normalize these functions. Which of the operators $\hat{H}, \hat{L}^2, \hat{L}_z$ are your linear combinations eigenfunctions of?

7.17 (a) Fill in the details in the evaluation of the expansion coefficients a_n in the example in Section 7.3. (b) Write the $x(l - x)$ expansion of this example for $x = \frac{1}{2}l$ and use the first five nonzero terms to approximate π^3. (c) Calculate the percent error in this $x(l - x)$ expansion at $x = l/4$ when the first 1, 3, and 5 nonzero terms are taken.

7.18 (a) Expand the function $f(x) = -1$ for $0 \le x \le \frac{1}{2}l$ and $f(x) = 1$ for $\frac{1}{2}l < x \le l$ in terms of the particle-in-a-box wave functions. (Since f is discontinuous at $\frac{1}{2}l$, the expansion cannot be expected to represent f at this point. Since f does not obey the particle-in-a-box boundary conditions of being zero at $x = 0$ and $x = l$, the expansion will not represent f at these points. The expansion will represent f at other points.) (b) Calculate the percent error at $x = l/4$ when the first 1, 3, 5, and 7 nonzero terms are taken in this expansion.

7.19 True or false? (a) If two Hermitian operators do not commute, then they cannot possess any common eigenfunctions. (b) If two Hermitian operators commute, then every eigenfunction of one must be an eigenfunction of the other. (c) The integral $\langle 2p_x|\hat{L}_z|3p_x\rangle$ involving hydrogenlike orbitals must equal zero.

7.20 Let $\hat{\Pi}$ be the parity operator and m be a positive integer. What is $\hat{\Pi}^m$ if m is even? if m is odd?

7.21 For the hydrogenlike atom, $V = -Ze'^2(x^2 + y^2 + z^2)^{-1/2}$ and the potential energy is an even function of the coordinates. (a) What is the parity of the $2s$ wave function? (b) What is the parity of $2p_x$? (c) Consider $2s + 2p_x$. Is it an eigenfunction of $\hat{H}$? Does it have definite parity?

7.22 Let $\hat{\Pi}$ be the parity operator and let $\psi_i(x)$ be a normalized harmonic-oscillator wave function. We define the matrix elements Π_{ij} as $\Pi_{ij} = \int_{-\infty}^{\infty} \psi_i^* \hat{\Pi} \psi_j \, dx$. Show that $\Pi_{ij} = 0$ for $i \ne j, \Pi_{ii} = \pm 1$.

7.23 Use parity to find which of the following integrals must be zero: (a) $\langle 2s|x|2p_x\rangle$; (b) $\langle 2s|x^2|2p_x\rangle$; (c) $\langle 2p_y|x|2p_x\rangle$. The functions in these integrals are hydrogenlike wave functions.

7.24 Let $\hat{R}$ be a linear operator such that $\hat{R}^n = \hat{1}$, where n is a positive integer and no lower power of $\hat{R}$ equals $\hat{1}$. Find the eigenvalues of $\hat{R}$.

7.25 (a) Show that the parity operator is linear. (b) Show that the parity operator is Hermitian; a proof in one dimension is sufficient.

7.26 Since the parity operator is Hermitian (Prob. 7.25), two eigenfunctions of $\hat{\Pi}$ that correspond to different eigenvalues must be orthogonal. Show directly that this is so.

7.27 Consider the integral $\langle v_2|x|v_1\rangle$, where the functions are one-dimensional harmonic-oscillator wave functions with quantum numbers v_2 and v_1. Under what conditions do parity considerations allow us to conclude that this integral must be zero? Might the integral be zero in other cases as well? (This integral is important in discussing radiative transitions.)

7.28 (a) In Cartesian coordinates the parity operator $\hat{\Pi}$ corresponds to the transformation of variables $x \to -x$, $y \to -y$, $z \to -z$. Show that in spherical coordinates the parity operator corresponds to the transformation $r \to r, \theta \to \pi - \theta, \phi \to \phi + \pi$. (b) Show that $\hat{\Pi}e^{im\phi} = (-1)^m e^{im\phi}$. (c) Use Eq. (5.97) to show that

$$\hat{\Pi}S_{l,m}(\theta) = (-1)^{l-|m|}S_{l,m}(\theta) = (-1)^{l-m}S_{l,m}(\theta)$$

(d) Combine the results of (b) and (c) to conclude that the spherical harmonic $Y_l^m(\theta, \phi)$ is an even function if l is even and is an odd function if l is odd.

7.29 Let f be an odd function of only some of its variables, meaning that

$$f(-q_1, -q_2, \ldots, -q_k, q_{k+1}, q_{k+2}, \ldots, q_m) = -f(q_1, q_2, \ldots, q_k, q_{k+1}, q_{k+2}, \ldots, q_m)$$

where $1 \le k \le m$. Explain why

$$\int_{-\infty}^{\infty} \cdots \int_{-\infty}^{\infty} f(q_1, \ldots, q_m)\, dq_1 \cdots dq_m = 0$$

Hint: Write the integral as

$$\int_{-\infty}^{\infty} \cdots \int_{-\infty}^{\infty}\left[\int_{-\infty}^{\infty} \cdots \int_{-\infty}^{\infty} f(q_1, \ldots, q_k, q_{k+1}, \ldots, q_m)\, dq_1 \cdots dq_k\right] dq_{k+1} \cdots dq_m$$

7.30 For a hydrogen atom in a p state, the possible outcomes of a measurement of L_z are $-\hbar$, 0, and $\hbar$. For each of the following wave functions, give the probabilities of each of these three results: (a) $2p_z$; (b) $2p_y$; (c) $2p_1$. Then find $\langle L_z\rangle$ for each of these three wave functions.

7.31 Suppose that at time t' a hydrogen atom is in a nonstationary state with

$$\Psi = 6^{-1/2}(2p_1) - 2^{-1/2}i(2p_0) - 3^{-1/2}(3d_1)$$

If L_z is measured at t', give the possible outcomes and give the probability of each possible outcome.

7.32 If L^2 is measured in a hydrogen atom whose state function is that in Prob. 7.31, give the possible outcomes and their probabilities.

7.33 If E is measured in a hydrogen atom whose state function is that in Prob. 7.31, give the possible outcomes and their probabilities.

7.34 Find $\langle L_z\rangle$ for the state of Prob. 7.31.

7.35 A measurement yields $2\hbar^2$ for the square of the magnitude of a particle's orbital angular momentum. If L_x is now measured, what are the possible outcomes?

7.36 Suppose that a particle in a box of length l is in a nonstationary state with $\Psi = 0$ for $x < 0$ and for $x > l$, and

$$\Psi = \tfrac{1}{2}e^{-ih^2t/8ml^2\hbar}\left(\frac{2}{l}\right)^{1/2}\sin\frac{\pi x}{l} + \frac{1}{2}\sqrt{3}e^{i\pi}e^{-ih^2t/2ml^2\hbar}\left(\frac{2}{l}\right)^{1/2}\sin\frac{2\pi x}{l}$$

for $0 \le x \le l$. If the energy is measured at time t, give the possible outcomes and their probabilities.

7.37 Suppose that a particle in a box of length l is in the nonstationary state $\Psi = (105/l^7)^{1/2}x^2(l - x)$ for $0 \le x \le l$ (and $\Psi = 0$ elsewhere) at the time the energy is measured. Give the possible outcomes and their probabilities. Calculate the probability for $n = 1, 2,$ and 3.

7.38 Suppose that the electron in a hydrogen atom has the state function $\Psi = (27/\pi a^3)^{1/2} e^{-3r/a}$ (where $a = \hbar^2/\mu e'^2$) at the time its energy is measured. Find the probability that the energy value $-e'^2/2a$ is found.

7.39 (a) Combine Eqs. (2.30) and (3.33) to write down Ψ for a free particle in one dimension. (b) Show that this Ψ is a linear combination of two eigenfunctions of $\hat{p}_x$. What are the eigenvalues for these eigenfunctions? (c) If p_x of a free particle in one dimension is measured, give the possible outcomes and give their probabilities.

7.40 Use the results of the last example in Section 7.6 to evaluate the sum $\sum_{m=0}^{\infty}[1/(2m + 1)^6]$.

7.41 (a) Show that, for a particle in a one-dimensional box (Fig. 2.1) of length l, the probability of observing a value of p_x between p and $p + dp$ is

$$\frac{4|N|^2 s^2}{l(s^2 - b^2)^2}[1 - (-1)^n \cos bl]\, dp \tag{7.111}$$

where $s \equiv n\pi l^{-1}$ and $b \equiv p\hbar^{-1}$. The constant N is to be chosen so that the integral from minus infinity to infinity of (7.111) is unity. Do not evaluate N. (b) Evaluate (7.111) in terms of $|N|$ for $p = \pm nh/2l$. [At these values of p_x, the denominator of (7.111) is zero and the probability reaches a large but finite value.]

7.42 Find (a) $\int_{-\infty}^{\infty} \delta(x)\, dx$; (b) $\int_{-\infty}^{-1} \delta(x)\, dx$; (c) $\int_{-1}^{1} \delta(x)\, dx$; (d) $\int_{1}^{2} f(x)\delta(x - 3)\, dx$.

7.43 Show that $\int_{-\infty}^{\infty} |\delta(x - a)|^2\, dx = \infty$.

7.44 What is the value of $\int_{0}^{\infty} f(x)\delta(x)\, dx$?

7.45 The functions in Fig. 7.5 approximate the Dirac delta function. Draw graphs of the corresponding functions that approximate the Heaviside step function with successively increasing accuracy.

7.46 Quantum mechanics postulates that the present state of an undisturbed system determines its future state. Consider the special case of a system with a time-independent Hamiltonian $\hat{H}$. Suppose it is known that at time t_0 the state function is $\Psi(x, t_0)$. Derive Eq. (7.100) by substituting the expansion (7.65) with $g_i = \psi_i$ into the time-dependent Schrödinger equation (7.96); multiply the result by ψ_m^*, integrate over all space, and solve for c_m.

7.47 For the time-dependent particle-in-a-box states with Ψ at $t = 0$ equal to $2^{-1/2}\psi_1 + 2^{-1/2}\psi_2$ in the example after Eq. (7.100), (a) calculate the period T for an electron in a box of length 2.00 Å; (b) verify (7.101) for $|\Psi|^2$; (c) use a spreadsheet or program like Mathcad to plot $l|\Psi|^2$ in (7.102) for $t = jT/8$ with $j = 0, 1, \ldots, 8$; (d) use Mathcad to produce an animation of $|\Psi|^2$ as time changes.

7.48 Consider the particle-in-a-box time-dependent states with Ψ at $t = 0$ given by $2^{-1/2}\psi_1 + 2^{-1/2}\psi_n$, where $n \neq 1$. (a) Find the equations that correspond to (7.101) and (7.102) for $|\Psi|^2$ and $l|\Psi|^2$; (b) produce animations of $|\Psi|^2$ as time changes for $n = 3$, $n = 4$, and $n = 5$. For which of the values $n = 2, 3, 4, 5$ does $|\Psi|^2$ remain symmetrical about the box midpoint as t changes?

7.49 For the matrices

$$\mathbf{A} = \begin{pmatrix} 2 & 1 \\ 0 & -3 \end{pmatrix}, \quad \mathbf{B} = \begin{pmatrix} 1 & -1 \\ 4 & 4 \end{pmatrix}$$

find (a) **AB**; (b) **BA**; (c) **A** + **B**; (d) 3**A**; (e) **A** − 4**B**.

7.50 Calculate the matrix products **CD** and **DC**, where

$$\mathbf{C} = \begin{pmatrix} 5 \\ 0 \\ -1 \end{pmatrix} \quad \text{and} \quad \mathbf{D} = (i \;\; 2 \;\; 1)$$

7.51 Find the matrix representative of the unit operator $\hat{1}$ in a complete, orthonormal basis.

7.52 If $\hat{D} = k\hat{C}$, show that $D_{ij} = kC_{ij}$.

7.53 If $\{f_i\}$ is a complete, orthonormal basis such that $\langle f_i|\hat{A}|f_j\rangle = a_i\delta_{ij}$ for all i and j (that is, the matrix representative of $\hat{A}$ is diagonal), prove that the functions $\{f_i\}$ are eigenfunctions of $\hat{A}$ and the a_i's are the eigenvalues of $\hat{A}$. *Hint:* Expand $\hat{A}f_j$ in terms of the set $\{f_i\}$.

7.54 (a) If $\hat{A}$ is a linear operator, $\{f_i\}$ is a complete, orthonormal basis, and u is an arbitrary function, show that $\hat{A}u = \sum_j(\sum_i\langle f_j|\hat{A}|f_i\rangle\langle f_i|u\rangle)f_j$. *Hint:* Expand u in terms of the set $\{f_i\}$, apply $\hat{A}$ to this expansion, and then expand $\hat{A}f_i$. (b) If $\hat{A}u = w$, with $w = \sum_j w_j f_j$ and $u = \sum_i u_i f_i$ (where the expansion coefficients w_j and u_i are numbers), show that the equation in (a) says that $w_j = \sum_i A_{ji}u_i$, and hence that $\mathbf{w} = \mathbf{Au}$, where $\mathbf{w}$, $\mathbf{A}$, and $\mathbf{u}$ are the representatives of w, $\hat{A}$, and u.

7.55 Write down the $l = 2$ portion of the matrix representative of $\hat{L}_z$ in the $\{Y_l^m\}$ basis.

7.56 Calculate the uncertainty ΔL_z for these hydrogen-atom stationary states: (a) $2p_z$; (b) $2p_x$.

7.57 For the system of Prob. 7.36, (a) use orthonormality [Eq. (2.27)] to verify that Ψ is normalized; (b) find $\langle E\rangle$ at time t; (c) find $\langle x\rangle$ at time t, and find the maximum and minimum values of $\langle x\rangle$ considered as a function of t.

7.58 Consider an operator $\hat{A}$ that contains the time as a parameter. We are interested in how the average value of the property A changes with time, and we have

$$\frac{d\langle A\rangle}{dt} = \frac{d}{dt}\int \Psi^*\hat{A}\Psi\,d\tau$$

The definite integral on the right side of this equation is a function of the parameter t, and it is generally a valid mathematical operation to calculate its derivative with respect to t by differentiating the integrand with respect to t:

$$\frac{d\langle A\rangle}{dt} = \int \frac{\partial}{\partial t}(\Psi^*\hat{A}\Psi)\,d\tau = \int \frac{\partial\Psi^*}{\partial t}\hat{A}\Psi\,d\tau + \int \Psi^*\frac{\partial\hat{A}}{\partial t}\Psi\,d\tau + \int \Psi^*\hat{A}\frac{\partial\Psi}{\partial t}\,d\tau$$

Use the time-dependent Schrödinger equation, its complex conjugate, and the Hermitian property of $\hat{H}$ to show that

$$\frac{d\langle A\rangle}{dt} = \int \Psi^*\frac{\partial\hat{A}}{\partial t}\Psi\,d\tau + \frac{i}{\hbar}\int \Psi^*(\hat{H}\hat{A} - \hat{A}\hat{H})\Psi\,d\tau$$

$$\frac{d\langle A\rangle}{dt} = \left\langle\frac{\partial\hat{A}}{\partial t}\right\rangle + \frac{i}{\hbar}\langle[\hat{H}\hat{A} - \hat{A}\hat{H}]\rangle \tag{7.112}$$

7.59 Use (7.112) and (5.8) to show that

$$\frac{d}{dt}\langle x\rangle = \frac{\langle p_x\rangle}{m} = \frac{1}{m}\int \Psi^*\frac{\hbar}{i}\frac{\partial\Psi}{\partial x}\,d\tau \tag{7.113}$$

From (7.113) it follows that

$$\frac{d^2\langle x\rangle}{dt^2} = \frac{1}{m}\frac{d}{dt}\langle p_x\rangle$$

Use (7.112), (5.9), and (4.26) to show that

$$\langle F_x\rangle = m\frac{d^2\langle x\rangle}{dt^2} \tag{7.114}$$

If we consider a classical-mechanical particle, its wave function will be large only in a very small region corresponding to its position, and we may then drop the averages in (7.114) to obtain Newton's second law. Thus classical mechanics is a special case of quantum mechanics. Equation (7.114) is known as *Ehrenfest's theorem,* after the physicist who derived it in 1927.

7.60 Prove the uncertainty principle (5.12) for any two Hermitian operators $\hat{A}$ and $\hat{B}$ as follows. (a) Define the function u as $u \equiv f - cg$, where $c \equiv \langle g|f\rangle/\langle g|g\rangle$ and f and g are any two quadratically integrable functions. Consider the integral $\langle u|u\rangle$. Since $u^*u = |u|^2 \geq 0$ everywhere, the integrand is never negative, so we have $\langle u|u\rangle \geq 0$. Substitute the definition of u into this last inequality, multiply the result by the positive quantity $\langle g|g\rangle$, and show that this leads to $\langle f|f\rangle\langle g|g\rangle - \langle f|g\rangle\langle g|f\rangle \geq 0$ or $\langle f|f\rangle\langle g|g\rangle \geq |\langle f|g\rangle|^2$, a result known as the *Schwarz inequality.* (b) Let $f \equiv (\hat{A} - \langle A\rangle)\Psi$ and $g \equiv (\hat{B} - \langle B\rangle)\Psi$. Use the Hermitian property of $\hat{A} - \langle A\rangle$ and $\hat{B} - \langle B\rangle$ to show that $(\Delta A)^2 = \langle f|f\rangle$ and $(\Delta B)^2 = \langle g|g\rangle$, where the uncertainties are defined by Eq. (5.11). (c) For the complex number $z \equiv x + iy$, where x and y are real numbers, we have $|z|^2 = x^2 + y^2 \geq y^2$. Verify that $y = (z - z^*)/2i$. Therefore, $|z|^2 \geq [(z - z^*)/2i]^2$. Let $z \equiv \langle f|g\rangle$. Then $|\langle f|g\rangle|^2 \geq [(\langle f|g\rangle - \langle g|f\rangle)/2i]^2$. Combine this result with the results of parts (a) and (b) to show that $(\Delta A)^2(\Delta B)^2 \geq -\frac{1}{4}(\langle f|g\rangle - \langle g|f\rangle)^2$, where f and g are defined in (b). (d) Use the definitions of f and g and the normalization of Ψ to verify that $\langle f|g\rangle = \langle g|f\rangle^* = \langle\Psi|\hat{A}\hat{B}|\Psi\rangle - \langle A\rangle\langle B\rangle$. From the definitions of f and g, interchanging f and g interchanges A and B, so it follows that $\langle g|f\rangle = \langle\Psi|\hat{B}\hat{A}|\Psi\rangle - \langle B\rangle\langle A\rangle$. Verify that the result of part (c) becomes $(\Delta A)^2(\Delta B)^2 \geq -\frac{1}{4}\langle\Psi|[\hat{A}, \hat{B}]|\Psi\rangle^2$. (e) Use the results of Prob. 7.13 to show that $\langle\Psi|[\hat{A}, \hat{B}]|\Psi\rangle = -\langle\Psi|[\hat{A}, \hat{B}]|\Psi\rangle^*$. Then substitute this result for one of the integrals on the right of the inequality in (d) to show that $(\Delta A)^2(\Delta B)^2 \geq \frac{1}{4}|\langle\Psi|[\hat{A}, \hat{B}]|\Psi\rangle|^2$, which gives the uncertainty principle (5.12).

7.61 Write a computer program that uses the expansion (7.37) to calculate the function (7.35) at x values of $0, 0.1l, 0.2l, \ldots, l$; have the program do the calculations for 5, 10, 15, and 20 terms taken in the expansion.

7.62 In finding eigenvalues by solving differential equations, quantization occurs only when we apply the condition that the eigenfunctions be well behaved. Ladder operators were used in Section 5.4 to find the eigenvalues of $\hat{M}^2$ and $\hat{M}_z$ and in Prob. 5.33 to find the harmonic-oscillator energy eigenvalues. Where was the condition that the eigenfunctions be well-behaved used in the derivations in Section 5.4 and Prob. 5.33? *Hint*: See Prob. 7.11.

7.63 True or false? (a) The state function is always equal to a function of time multiplied by a function of the coordinates. (b) In both classical and quantum mechanics, knowledge of the present state of an isolated system allows its future state to be calculated. (c) The state function is always an eigenfunction of the Hamiltonian. (d) Every linear combination of eigenfunctions of the Hamiltonian is an eigenfunction of the Hamiltonian operator. (e) If the state function is not an eigenfunction of the operator $\hat{A}$, then a measurement of the property A might give a value that is not one of the eigenvalues of $\hat{A}$. (f) The probability density is independent of time for a stationary state. (g) If two Hermitian operators do not commute, then they cannot possess any common eigenfunctions. (h) If two Hermitian operators commute, then every eigenfunction of one must be an eigenfunction of the other. (i) Two linearly independent eigenfunctions of the same Hermitian operator are always orthogonal to each other. (j) If the operator $\hat{B}$ corresponds to a physical property of a quantum-mechanical system, the state function Ψ must be an eigenfunction of $\hat{B}$. (k) Every linear combination of solutions of the time-dependent Schrödinger equation is a solution of this equation. (l) The normalized state function Ψ is dimensionless (that is, has no units). (m) All eigenfunctions of Hermitian operators must be real functions. (n) The quantities $\langle f_m|\hat{A}|f_n\rangle$ and $\langle f_m|f_n\rangle$ are numbers. (o) When Ψ is an eigenfunction of $\hat{B}$ with eigenvalue b_k, we are certain to observe the value b_k when the property B is measured. (p) The relation $\langle f|\hat{B}|g\rangle = \langle g|\hat{B}|f\rangle^*$ is valid only when f and g are eigenfunctions of the Hermitian operator $\hat{B}$.

CHAPTER 8

The Variation Method

8.1 THE VARIATION THEOREM

We now begin the study of the approximation methods needed to deal with the time-independent Schrödinger equation for systems (such as atoms or molecules) that contain interacting particles. This chapter deals with the variation method, which allows us to approximate the ground-state energy of a system without solving the Schrödinger equation. The variation method is based on the following theorem:

THE VARIATION THEOREM. Given a system whose Hamiltonian operator $\hat{H}$ is time independent and whose lowest-energy eigenvalue is E_1, if ϕ is any normalized, well-behaved function of the coordinates of the system's particles that satisfies the boundary conditions of the problem, then

$$\boxed{\int \phi^* \hat{H} \phi \, d\tau \geq E_1, \qquad \phi \text{ normalized}} \tag{8.1}$$

The variation theorem allows us to calculate an upper bound for the system's ground-state energy.

To prove (8.1), we expand ϕ in terms of the complete, orthonormal set of eigenfunctions of $\hat{H}$, the stationary-state eigenfunctions ψ_k:

$$\phi = \sum_k a_k \psi_k \tag{8.2}$$

where

$$\hat{H}\psi_k = E_k \psi_k \tag{8.3}$$

Note that the expansion (8.2) requires that ϕ obey the same boundary conditions as the ψ_k's. Substitution of (8.2) into the left side of (8.1) gives

$$\int \phi^* \hat{H} \phi \, d\tau = \int \sum_k a_k^* \psi_k^* \hat{H} \sum_j a_j \psi_j \, d\tau = \int \sum_k a_k^* \psi_k^* \sum_j a_j \hat{H} \psi_j \, d\tau$$

Using the eigenvalue equation (8.3) and assuming the validity of interchanging the integration and the infinite summations, we get

$$\int \phi^* \hat{H} \phi \, d\tau = \int \sum_k a_k^* \psi_k^* \sum_j a_j E_j \psi_j \, d\tau = \sum_k \sum_j a_k^* a_j E_j \int \psi_k^* \psi_j \, d\tau$$

$$= \sum_k \sum_j a_k^* a_j E_j \delta_{kj}$$

where the orthonormality of the eigenfunctions ψ_k was used. We perform the sum over j, and, as usual, the Kronecker delta makes all terms zero except the one with $j = k$, giving

$$\int \phi^* \hat{H} \phi \, d\tau = \sum_k a_k^* a_k E_k = \sum_k |a_k|^2 E_k \tag{8.4}$$

Since E_1 is the lowest-energy eigenvalue of $\hat{H}$, we have $E_k \geq E_1$. Since $|a_k|^2$ is never negative, we can multiply the inequality $E_k \geq E_1$ by $|a_k|^2$ without changing the direction of the inequality sign to get $|a_k|^2 E_k \geq |a_k|^2 E_1$. Therefore, $\sum_k |a_k|^2 E_k \geq \sum_k |a_k|^2 E_1$, and use of (8.4) gives

$$\int \phi^* \hat{H} \phi \, d\tau = \sum_k |a_k|^2 E_k \geq \sum_k |a_k|^2 E_1 = E_1 \sum_k |a_k|^2 \tag{8.5}$$

Because ϕ is normalized, we have $\int \phi^* \phi \, d\tau = 1$. Substitution of the expansion (8.2) into the normalization condition gives

$$1 = \int \phi^* \phi \, d\tau = \int \sum_k a_k^* \psi_k^* \sum_j a_j \psi_j \, d\tau = \sum_k \sum_j a_k^* a_j \int \psi_k^* \psi_j \, d\tau = \sum_k \sum_j a_k^* a_j \delta_{kj}$$

$$1 = \sum_k |a_k|^2 \tag{8.6}$$

[Note that in deriving Eqs. (8.4) and (8.6) we essentially repeated the derivations of Eqs. (7.69) and (7.68), respectively.]

Use of (8.6) in (8.5) gives the variation theorem (8.1):

$$\int \phi^* \hat{H} \phi \, d\tau \geq E_1, \qquad \phi \text{ normalized} \tag{8.7}$$

Suppose we have a function ϕ that is not normalized. To apply the variation theorem, we multiply ϕ by a normalization constant N so that $N\phi$ is normalized. Replacing ϕ by $N\phi$ in (8.7), we have

$$|N|^2 \int \phi^* \hat{H} \phi \, d\tau \geq E_1 \tag{8.8}$$

N is determined by $\int (N\phi)^* N\phi \, d\tau = |N|^2 \int \phi^* \phi \, d\tau = 1$; so $|N|^2 = 1/\int \phi^* \phi \, d\tau$ and (8.8) becomes

$$\boxed{\frac{\int \phi^* \hat{H} \phi \, d\tau}{\int \phi^* \phi \, d\tau} \geq E_1} \tag{8.9}$$

where ϕ is any well-behaved function (not necessarily normalized) that satisfies the boundary conditions of the problem.

The function ϕ is called a **trial variation function**, and the integral in (8.1) [or the ratio of integrals in (8.9)] is called the **variational integral**. To arrive at a good approximation to the ground-state energy E_1, we try many trial variation functions and look for the one that gives the lowest value of the variational integral. From (8.1), the lower the value of the variational integral, the better the approximation we have to E_1. One way to disprove quantum mechanics would be to find a trial variation function that made the variational integral less than E_1 for some system where E_1 is known.

Let ψ_1 be the true ground-state wave function:

$$\hat{H}\psi_1 = E_1\psi_1 \tag{8.10}$$

If we happened to be lucky enough to hit upon a variation function that was equal to ψ_1, then, using (8.10) in (8.1), we see that the variational integral will be equal to E_1. Thus the ground-state wave function gives the minimum value of the variational integral. We therefore expect that the lower the value of the variational integral, the closer the trial variational function will approach the true ground-state wave function. However, it turns out that the variational integral approaches E_1 a lot faster than the trial variation function approaches ψ_1, and it is possible to get a rather good approximation to E_1 using a rather poor ϕ.

In practice, one usually puts several parameters into the trial function ϕ and then varies these parameters so as to minimize the variational integral. Successful use of the variation method depends on the ability to make a good choice for the trial function.

Let us look at some examples of the variation method. Although the real utility of the method is for problems to which we do not know the true solutions, we will consider problems that are exactly solvable so that we can judge the accuracy of our results.

EXAMPLE Devise a trial variation function for the particle in a one-dimensional box of length l.

The wave function is zero outside the box and the boundary conditions require that $\psi = 0$ at $x = 0$ and at $x = l$. The variation function ϕ must meet these boundary conditions of being zero at the ends of the box. As noted after Eq. (4.59), the ground-state ψ has no nodes interior to the boundary points, so it is desirable that ϕ have no interior nodes. A simple function that has these properties is the parabolic function

$$\phi = x(l - x) \quad \text{for } 0 \le x \le l \tag{8.11}$$

and $\phi = 0$ outside the box. Since we have not normalized ϕ, we use Eq. (8.9). Inside the box the Hamiltonian is $-(\hbar^2/2m)\,d^2/dx^2$. For the numerator and denominator of (8.9), we have

$$\int \phi^* \hat{H} \phi \, d\tau = -\frac{\hbar^2}{2m}\int_0^l (lx - x^2)\frac{d^2}{dx^2}(lx - x^2)\,dx = -\frac{\hbar^2}{m}\int_0^l (x^2 - lx)\,dx = \frac{\hbar^2 l^3}{6m} \tag{8.12}$$

$$\int \phi^* \phi \, d\tau = \int_0^l x^2(l - x)^2\,dx = \frac{l^5}{30} \tag{8.13}$$

Substituting in the variation theorem (8.9), we get

$$E_1 \le \frac{5h^2}{4\pi^2 ml^2} = 0.1266515\frac{h^2}{ml^2}$$

From Eq. (2.20), $E_1 = h^2/8ml^2 = 0.125h^2/ml^2$, and the energy error is 1.3%.

Since $\int |\phi|^2\,d\tau = l^5/30$, the normalized form of (8.11) is $(30/l^5)^{1/2}x(l - x)$. Figure 7.3 shows that this function rather closely resembles the true ground-state particle-in-a-box wave function.

EXERCISE A one-particle, one-dimensional system has the potential energy function $V = V_0$ for $0 \le x \le l$ and $V = \infty$ elsewhere. (a) Use the variation function $\phi = \sin(\pi x/l)$ for $0 \le x \le l$ and $\phi = 0$ elsewhere to estimate the ground-state energy of this system. (b) Explain why the result of (a) is the exact ground-state energy. *Hint*: see one of the Chapter 4 problems. (*Answer:* (a) $V_0 + h^2/8ml^2$.)

The preceding example did not have a parameter in the trial function. The next example does.

EXAMPLE For the one-dimensional harmonic oscillator, devise a variation function with a parameter and find the optimum value of the parameter. Estimate the ground-state energy.

The variation function ϕ must be quadratically integrable and so must go to zero as x goes to $\pm\infty$. The function e^{-x} has the proper behavior at $+\infty$ but becomes infinite at $-\infty$. The function e^{-x^2} has the proper behavior at $\pm\infty$. However, it is dimensionally unsatisfactory, since the power to which we raise e must be dimensionless. This can be seen from the Taylor series $e^z = 1 + z + z^2/2! + \cdots$ [Eq. (4.46)]. Since all the terms in this series must have the same dimensions, z must have the same dimensions as 1; that is, z in e^z must be dimensionless. Hence we modify e^{-x^2} to e^{-cx^2}, where c has units of length^{-2}. We shall take c as a variational parameter. The true ground-state ψ must have no nodes. Also, since $V = \frac{1}{2}kx^2$ is an even function, the ground-state ψ must have definite parity and must be an even function, since an odd function has a node at the origin. The trial function e^{-cx^2} has the desired properties of having no nodes and of being an even function.

Use of (4.32) for $\hat{H}$ and Appendix integrals gives (Prob. 8.3)

$$\int \phi^* \hat{H} \phi \, d\tau = -\frac{\hbar^2}{2m} \int_{-\infty}^{\infty} e^{-cx^2} \frac{d^2 e^{-cx^2}}{dx^2} \, dx + 2\pi^2 \nu^2 m \int_{-\infty}^{\infty} x^2 e^{-2cx^2} \, dx$$

$$= (\hbar^2/m)(\pi/8)^{1/2} c^{1/2} + \nu^2 m (\pi^5/8)^{1/2} c^{-3/2}$$

$$\int \phi^* \phi \, d\tau = \int_{-\infty}^{\infty} e^{-2cx^2} \, dx = 2 \int_0^{\infty} e^{-2cx^2} \, dx = (\pi/2)^{1/2} c^{-1/2}$$

The variational integral W is

$$W \equiv \frac{\int \phi^* \hat{H} \phi \, d\tau}{\int \phi^* \phi \, d\tau} = (\hbar^2/2m)c + (\pi^2/2)\nu^2 m c^{-1} \tag{8.14}$$

We now vary c to minimize the variational integral (8.14). A necessary condition that W be minimized is that

$$dW/dc = 0 = (\hbar^2/2m) - (\pi^2/2)\nu^2 m c^{-2}$$

$$c = \pm \pi \nu m/\hbar \tag{8.15}$$

The negative root $c = -\pi\nu m/\hbar$ is rejected, since it would make $\phi = e^{-cx^2}$ not quadratically integrable. Substitution of $c = \pi\nu m/\hbar$ into (8.14) gives $W = \frac{1}{2}h\nu$. This is the exact ground-state harmonic-oscillator energy. With $c = \pi\nu m/\hbar$ the variation function ϕ is the same (except for being unnormalized) as the harmonic-oscillator ground-state wave function (4.55) and (4.33).

For the normalized harmonic-oscillator variation function $\phi = (2c/\pi)^{1/4} e^{-cx^2}$, a large value of c makes ϕ fall off very rapidly from its maximum value at $x = 0$. This makes the probability density large only near $x = 0$. The potential energy $V = \frac{1}{2}kx^2$ is low near $x = 0$, so a large c

means a low $\langle V \rangle = \langle \phi | V | \phi \rangle$. [Note also that $\langle V \rangle$ equals the second term on the right side of (8.14).] However, because a large c makes ϕ fall off very rapidly from its maximum, it makes $|d\phi/dx|$ large in the region near $x = 0$. From Prob. 7.7b, a large $|d\phi/dx|$ means a large value of $\langle T \rangle$ [which equals the first term on the right side of (8.14)]. The optimum value of c minimizes the sum $\langle T \rangle + \langle V \rangle = W$. In atoms and molecules, the true wave function is a compromise between the tendency to minimize $\langle V \rangle$ by confining the electrons to regions of low V (near the nuclei) and the tendency to minimize $\langle T \rangle$ by allowing the electron probability density to spread out over a large region.

EXERCISE Consider a one-particle, one-dimensional system with $V = 0$ for $-\frac{1}{2}l \le x \le \frac{1}{2}l$ and $V = b\hbar^2/ml^2$ elsewhere (where b is a positive constant) (Fig. 2.5 with $V_0 = b\hbar^2/ml^2$ and the origin shifted). (a) For the variation function $\phi = (x - c)(x + c) = x^2 - c^2$ for $-c \le x \le c$ and $\phi = 0$ elsewhere, where the variational parameter c satisfies $c > \frac{1}{2}l$, one finds that the variational integral W is given by

$$W = \frac{\hbar^2}{ml^2}\left[\frac{5l^2}{4c^2} + b\left(1 - \frac{15l}{16c} + \frac{5l^3}{32c^3} - \frac{3l^5}{256c^5}\right)\right]$$

Sketch ϕ and V on the same plot. Find the equation satisfied by the value of c that minimizes W. (b) Find the optimum c and W for $V_0 = 20\hbar^2/ml^2$ and compare with the true ground-state energy $2.814\hbar^2/ml^2$ (Prob. 4.30c). (*Hint:* You may want to use the Solver in a spreadsheet or a programmable calculator to find c/l.) (*Answer:* (a) $48t^4 - 24t^2 - 128t^3/b + 3 = 0$, where $t \equiv c/l$. (b) $c = 0.6715l$, $W = 3.454\hbar^2/ml^2$.)

8.2 EXTENSION OF THE VARIATION METHOD

The variation method as presented in the last section gives information about only the *ground*-state energy and wave function. We now discuss extension of the variation method to excited states. (See also Section 8.5.)

Consider how we might extend the variation method to estimate the energy of the first excited state. We number the stationary states of the system 1, 2, 3, ... in order of increasing energy:

$$E_1 \le E_2 \le E_3 \le \cdots$$

We showed that for a normalized variational function ϕ [Eqs. (8.4) and (8.6)]

$$\int \phi^* \hat{H} \phi \, d\tau = \sum_{k=1}^{\infty} |a_k|^2 E_k \quad \text{and} \quad \int \phi^* \phi \, d\tau = \sum_{k=1}^{\infty} |a_k|^2 = 1$$

where the a_k's are the expansion coefficients in $\phi = \sum_k a_k \psi_k$ [Eq. (8.2)]. We have $a_k = \langle \psi_k | \phi \rangle$ [Eq. (7.40)]. Let us restrict ourselves to normalized functions ϕ that are orthogonal to the true ground-state wave function ψ_1. Then we have $a_1 = \langle \psi_1 | \phi \rangle = 0$ and

$$\int \phi^* \hat{H} \phi \, d\tau = \sum_{k=2}^{\infty} |a_k|^2 E_k \quad \text{and} \quad \int \phi^* \phi \, d\tau = \sum_{k=2}^{\infty} |a_k|^2 = 1 \tag{8.16}$$

For $k \ge 2$, we have $E_k \ge E_2$ and $|a_k|^2 E_k \ge |a_k|^2 E_2$. Hence

$$\sum_{k=2}^{\infty} |a_k|^2 E_k \ge \sum_{k=2}^{\infty} |a_k|^2 E_2 = E_2 \sum_{k=2}^{\infty} |a_k|^2 = E_2 \tag{8.17}$$

Combining (8.16) and (8.17), we have the desired result:

$$\int \phi^* \hat{H} \phi \, d\tau \geq E_2 \quad \text{if} \quad \int \psi_1^* \phi \, d\tau = 0 \quad \text{and} \quad \int \phi^* \phi \, d\tau = 1 \tag{8.18}$$

The inequality (8.18) allows us to get an upper bound to the energy E_2 of the first excited state. However, the restriction $\langle \psi_1 | \phi \rangle = 0$ makes this method troublesome to apply.

For certain systems, it is possible to be sure that $\langle \psi_1 | \phi \rangle = 0$ even though we do not know the true ground-state wave function. An example is a one-dimensional problem for which V is an even function of x. In this case the ground-state wave function is always an even function, while the first excited-state wave function is odd. (All the wave functions must be of definite parity. The ground-state wave function is nodeless, and, since an odd function vanishes at the origin, the ground-state wave function must be even. The first excited-state wave function has one node and must be odd.) Therefore, for odd trial functions, it must be true that $\langle \psi_1 | \phi \rangle = 0$; the even function ψ_1 times the odd function ϕ gives an odd integrand whose integral from $-\infty$ to ∞ is zero.

Another example is a particle moving in a central field (Section 6.1). The form of the potential energy might be such that we could not solve for the radial factor $R(r)$ in the eigenfunction. However, the angular factor in ψ is a spherical harmonic [Eq. (6.16)] and spherical harmonics with different values of l are orthogonal. Thus we can get an upper bound to the energy of the lowest state with any given angular momentum l by using the factor Y_l^m in the trial function. This result depends on the extension of (8.18) to higher excited states:

$$\frac{\int \phi^* \hat{H} \phi \, d\tau}{\int \phi^* \phi \, d\tau} \geq E_{k+1} \quad \text{if} \quad \int \psi_1^* \phi \, d\tau = \int \psi_2^* \phi \, d\tau = \cdots = \int \psi_k^* \phi \, d\tau = 0 \tag{8.19}$$

8.3 DETERMINANTS

Section 8.5 discusses a kind of variation function that gives rise to an equation involving a determinant. Therefore, we now discuss determinants.

A **determinant** is a square array of n^2 quantities (called **elements**); the value of the determinant is calculated from its elements in a manner to be given shortly. The number n is the **order** of the determinant. Using a_{ij} to represent a typical element, we write the nth-order determinant as

$$\det(a_{ij}) = \begin{vmatrix} a_{11} & a_{12} & a_{13} & \cdots & a_{1n} \\ a_{21} & a_{22} & a_{23} & \cdots & a_{2n} \\ . & . & . & \cdots & . \\ . & . & . & \cdots & . \\ . & . & . & \cdots & . \\ a_{n1} & a_{n2} & a_{n3} & \cdots & a_{nn} \end{vmatrix} \tag{8.20}$$

The vertical lines in (8.20) have nothing to do with absolute value. Before considering how the value of the nth-order determinant is defined, we consider determinants of first, second, and third orders.

A first-order determinant has one element, and its value is simply the value of that element. Thus

$$|a_{11}| = a_{11} \tag{8.21}$$

where the vertical lines indicate a determinant and not absolute value.

A second-order determinant has four elements, and its value is defined by

$$\boxed{\begin{vmatrix} a_{11} & a_{12} \\ a_{21} & a_{22} \end{vmatrix} = a_{11}a_{22} - a_{12}a_{21}} \tag{8.22}$$

The value of a third-order determinant is defined by

$$\boxed{\begin{vmatrix} a_{11} & a_{12} & a_{13} \\ a_{21} & a_{22} & a_{23} \\ a_{31} & a_{32} & a_{33} \end{vmatrix} = a_{11}\begin{vmatrix} a_{22} & a_{23} \\ a_{32} & a_{33} \end{vmatrix} - a_{12}\begin{vmatrix} a_{21} & a_{23} \\ a_{31} & a_{33} \end{vmatrix} + a_{13}\begin{vmatrix} a_{21} & a_{22} \\ a_{31} & a_{32} \end{vmatrix}} \tag{8.23}$$

$$= a_{11}a_{22}a_{33} - a_{11}a_{32}a_{23} - a_{12}a_{21}a_{33} + a_{12}a_{31}a_{23} + a_{13}a_{21}a_{32} - a_{13}a_{31}a_{22} \tag{8.24}$$

A third-order determinant is evaluated by writing down the elements of the top row with alternating plus and minus signs and then multiplying each element by a certain second-order determinant; the second-order determinant that multiplies a given element is found by crossing out the row and column of the third-order determinant in which that element appears. The $(n - 1)$-order determinant obtained by striking out the ith row and the jth column of the nth-order determinant is called the **minor** of the element a_{ij}. We define the **cofactor** of a_{ij} as the minor of a_{ij} times the factor $(-1)^{i+j}$. Thus (8.23) states that a third-order determinant is evaluated by multiplying each element of the top row by its cofactor and adding up the three products. [Note that (8.22) conforms to this evaluation by means of cofactors, since the cofactor of a_{11} in (8.22) is a_{22}, and the cofactor of a_{12} is $-a_{21}$.] A numerical example is

$$\begin{vmatrix} 5 & 10 & 2 \\ 0.1 & 3 & 1 \\ 0 & 4 & 4 \end{vmatrix} = 5\begin{vmatrix} 3 & 1 \\ 4 & 4 \end{vmatrix} - 10\begin{vmatrix} 0.1 & 1 \\ 0 & 4 \end{vmatrix} + 2\begin{vmatrix} 0.1 & 3 \\ 0 & 4 \end{vmatrix}$$

$$= 5(8) - 10(0.4) + 2(0.4) = 36.8$$

Denoting the minor of a_{ij} by M_{ij} and the cofactor of a_{ij} by C_{ij}, we have

$$C_{ij} = (-1)^{i+j}M_{ij} \tag{8.25}$$

The expansion (8.23) of the third-order determinant can be written as

$$\det(a_{ij}) = \begin{vmatrix} a_{11} & a_{12} & a_{13} \\ a_{21} & a_{22} & a_{23} \\ a_{31} & a_{32} & a_{33} \end{vmatrix} = a_{11}C_{11} + a_{12}C_{12} + a_{13}C_{13} \tag{8.26}$$

A third-order determinant can be expanded using the elements of any row and the corresponding cofactors. For example, using the second row to expand the third-order determinant, we have

$$\det(a_{ij}) = a_{21}C_{21} + a_{22}C_{22} + a_{23}C_{23} \tag{8.27}$$

$$\det(a_{ij}) = -a_{21}\begin{vmatrix} a_{12} & a_{13} \\ a_{32} & a_{33} \end{vmatrix} + a_{22}\begin{vmatrix} a_{11} & a_{13} \\ a_{31} & a_{33} \end{vmatrix} - a_{23}\begin{vmatrix} a_{11} & a_{12} \\ a_{31} & a_{32} \end{vmatrix} \tag{8.28}$$

and expansion of the second-order determinants shows that (8.28) is equal to (8.24). We may also use the elements of any column and the corresponding cofactors to expand the determinant, as can be readily verified. Thus for the third-order determinant, we can write

$$\det(a_{ij}) = a_{k1}C_{k1} + a_{k2}C_{k2} + a_{k3}C_{k3} = \sum_{l=1}^{3} a_{kl}C_{kl}, \qquad k = 1 \text{ or } 2 \text{ or } 3$$

$$\det(a_{ij}) = a_{1k}C_{1k} + a_{2k}C_{2k} + a_{3k}C_{3k} = \sum_{l=1}^{3} a_{lk}C_{lk}, \qquad k = 1 \text{ or } 2 \text{ or } 3$$

The first expansion uses one of the rows; the second uses one of the columns.

We define determinants of higher order by an analogous row (or column) expansion. For an nth-order determinant,

$$\det(a_{ij}) = \sum_{l=1}^{n} a_{kl}C_{kl} = \sum_{l=1}^{n} a_{lk}C_{lk}, \quad k = 1 \text{ or } 2 \text{ or} \ldots \text{or } n \tag{8.29}$$

Some theorems on determinants are as follows (for proofs, see *Sokolnikoff and Redheffer*, pp. 702–707):

I. If every element of a row (or column) of a determinant is zero, the value of the determinant is zero.

II. Interchanging any two rows (or columns) multiplies the value of a determinant by -1.

III. If any two rows (or columns) of a determinant are identical, the determinant has the value zero.

IV. Multiplication of each element of any one row (or any one column) by some constant k multiplies the value of the determinant by k.

V. Addition to each element of one row of the same constant multiple of the corresponding element of another row leaves the value of the determinant unchanged. This theorem also applies to the addition of a multiple of one column to another column.

VI. The interchange of all corresponding rows and columns leaves the value of the determinant unchanged. (This interchange means that column 1 becomes row 1, column 2 becomes row 2, etc.)

EXAMPLE Use Theorem V to evaluate

$$B = \begin{vmatrix} 1 & 2 & 3 & 4 \\ 4 & 1 & 2 & 3 \\ 3 & 4 & 1 & 2 \\ 2 & 3 & 4 & 1 \end{vmatrix} \tag{8.30}$$

Addition of -2 times the elements of row one to the corresponding elements of row four changes row four to $2 + (-2)1 = 0$, $3 + (-2)(2) = -1$, $4 + (-2)3 = -2$, $1 + (-2)4 = -7$. Then, addition of -3 times row one to row three and -4 times row one to row two gives

$$B = \begin{vmatrix} 1 & 2 & 3 & 4 \\ 0 & -7 & -10 & -13 \\ 0 & -2 & -8 & -10 \\ 0 & -1 & -2 & -7 \end{vmatrix} = 1 \begin{vmatrix} -7 & -10 & -13 \\ -2 & -8 & -10 \\ -1 & -2 & -7 \end{vmatrix} \tag{8.31}$$

where we expanded B in terms of elements of the first column. Subtracting twice row three from row two and seven times row three from row one, we have

$$B = \begin{vmatrix} 0 & 4 & 36 \\ 0 & -4 & 4 \\ -1 & -2 & -7 \end{vmatrix} = (-1) \begin{vmatrix} 4 & 36 \\ -4 & 4 \end{vmatrix} = -(16 + 144) = -160 \tag{8.32}$$

The diagonal of a determinant that runs from the top left to the lower right is the **principal diagonal**. A **diagonal determinant** is a determinant all of whose elements are zero except those on the principal diagonal. For a diagonal determinant,

$$\begin{vmatrix} a_{11} & 0 & 0 & \cdots & 0 \\ 0 & a_{22} & 0 & \cdots & 0 \\ 0 & 0 & a_{33} & \cdots & 0 \\ . & . & . & \cdots & . \\ 0 & 0 & 0 & \cdots & a_{nn} \end{vmatrix} = a_{11} \begin{vmatrix} a_{22} & 0 & \cdots & 0 \\ 0 & a_{33} & \cdots & 0 \\ . & . & \cdots & . \\ 0 & 0 & \cdots & a_{nn} \end{vmatrix} = a_{11}a_{22} \begin{vmatrix} a_{33} & 0 & \cdots & 0 \\ 0 & a_{44} & \cdots & 0 \\ . & . & \cdots & . \\ 0 & 0 & \cdots & a_{nn} \end{vmatrix}$$

$$= \cdots = a_{11}a_{22}a_{33}\cdots a_{nn} \tag{8.33}$$

A diagonal determinant is equal to the product of its diagonal elements.

A determinant whose only nonzero elements occur in square blocks centered about the principal diagonal is in **block-diagonal form**. If we regard each square block as a determinant, then a block-diagonal determinant is equal to the product of the blocks. For example,

$$\begin{vmatrix} a & b & 0 & 0 & 0 & 0 \\ c & d & 0 & 0 & 0 & 0 \\ 0 & 0 & e & 0 & 0 & 0 \\ 0 & 0 & 0 & f & g & h \\ 0 & 0 & 0 & i & j & k \\ 0 & 0 & 0 & l & m & n \end{vmatrix} = \begin{vmatrix} a & b \\ c & d \end{vmatrix} (e) \begin{vmatrix} f & g & h \\ i & j & k \\ l & m & n \end{vmatrix} \tag{8.34}$$

The dashed lines outline the blocks. Equation (8.34) is readily proved by expanding the left side in terms of elements of the top row, and expanding several subsequent determinants using their top rows (Prob. 8.20).

8.4 SIMULTANEOUS LINEAR EQUATIONS

To deal with the kind of variation function discussed in the next section, we need to know about simultaneous linear equations.

Consider the following system of n linear equations in n unknowns:

$$\begin{aligned} a_{11}x_1 + a_{12}x_2 + \cdots + a_{1n}x_n &= b_1 \\ a_{21}x_1 + a_{22}x_2 + \cdots + a_{2n}x_n &= b_2 \\ \cdot \quad \cdot \quad \cdot \quad \cdot \quad \cdots \quad \cdot \quad \cdot & \quad \cdot \quad \cdot \\ a_{n1}x_1 + a_{n2}x_2 + \cdots + a_{nn}x_n &= b_n \end{aligned} \tag{8.35}$$

where the a's and b's are known constants and $x_1, x_2, \ldots, x_n$ are the unknowns. If at least one of the b's is not zero, we have a system of **inhomogeneous** linear equations. Such a system can be solved by Cramer's rule. (For a proof of Cramer's rule, see *Sokolnikoff and Redheffer*, p. 708.) Let $\det(a_{ij})$ be the determinant of the coefficients of the unknowns in (8.35). *Cramer's rule* states that x_k $(k = 1, 2, \ldots, n)$ is given by

$$x_k = \frac{\begin{vmatrix} a_{11} & a_{12} & \cdots & a_{1,k-1} & b_1 & a_{1,k+1} & \cdots & a_{1n} \\ a_{21} & a_{22} & \cdots & a_{2,k-1} & b_2 & a_{2,k+1} & \cdots & a_{2n} \\ \cdot & \cdot & \cdots & \cdot & \cdot & \cdot & \cdots & \cdot \\ a_{n1} & a_{n2} & \cdots & a_{n,k-1} & b_n & a_{n,k+1} & \cdots & a_{nn} \end{vmatrix}}{\det(a_{ij})}, \quad k = 1, 2, \ldots, n \tag{8.36}$$

where $\det(a_{ij})$ is given by (8.20) and the numerator is the determinant obtained by replacing the kth column of $\det(a_{ij})$ with the elements $b_1, b_2, \ldots, b_n$. Although Cramer's rule is of theoretical significance, it should not be used for numerical calculations, since successive elimination of unknowns is much more efficient.

A widely used successive-elimination procedure is **Gaussian elimination**, which proceeds as follows: Divide the first equation in (8.35) by the coefficient a_{11} of x_1, thereby making the coefficient of x_1 equal to 1 in this equation. Then subtract a_{21} times the first equation from the second equation, subtract a_{31} times the first equation from the third equation, . . . , and subtract a_{n1} times the first equation from the nth equation. This eliminates x_1 from all equations but the first. Now divide the second equation by the coefficient of x_2; then subtract appropriate multiples of the second equation from the 3rd, 4th, . . . , nth equations, so as to eliminate x_2 from all equations but the first and second. Continue in this manner. Ultimately, equation n will contain only x_n, equation $n - 1$ only x_{n-1} and x_n, and so on. The value of x_n found from equation n is substituted into equation $n - 1$ to give x_{n-1}; the values of x_n and x_{n-1} are substituted into equation $n - 2$ to give x_{n-2}; and so on. If at any stage a coefficient we want to divide by happens to be zero, the equation with the zero coefficient is exchanged with a later equation that has a nonzero coefficient in the desired position. (The Gaussian elimination procedure also gives an efficient way to evaluate a determinant; see Prob. 8.24.)

A related method is **Gauss–Jordan elimination**, which proceeds the same way as Gaussian elimination, except that instead of eliminating x_2 from equations $3, 4, \ldots, n$, we eliminate x_2 from equations $1, 3, 4, \ldots, n$, by subtracting appropriate multiples of the second equation from equations $1, 3, 4, \ldots, n$; instead of eliminating x_3 from equations $4, 5, \ldots, n$, we eliminate x_3 from equations $1, 2, 4, 5, \ldots, n$; and so on. At the end of Gauss–Jordan elimination, equation 1 contains only x_1, equation 2 contains only $x_2, \ldots$, equation n contains only x_n. Gauss–Jordan elimination requires more computation than Gaussian elimination.

If all the b's in (8.35) are zero, we have a system of **linear homogeneous equations**:

$$\begin{aligned} a_{11}x_1 + a_{12}x_2 + \cdots + a_{1n}x_n &= 0 \\ a_{21}x_1 + a_{22}x_2 + \cdots + a_{2n}x_n &= 0 \\ \cdots\cdots\cdots\cdots\cdots\cdots\cdots & \\ a_{n1}x_1 + a_{n2}x_2 + \cdots + a_{nn}x_n &= 0 \end{aligned} \tag{8.37}$$

One obvious solution of (8.37) is $x_1 = x_2 = \cdots = x_n = 0$, which is called the **trivial solution**. If the determinant of the coefficients in (8.37) is not equal to zero, $\det(a_{ij}) \neq 0$, then we can use Cramer's rule (8.36) to solve for the unknowns, and we find $x_k = 0$, $k = 1, 2, \ldots, n$, since the determinant in the numerator of (8.36) has a column all of whose elements are zero. Thus, when $\det(a_{ij}) \neq 0$, the only solution is the trivial solution, which is of no interest. For there to be a nontrivial solution of a system of n linear homogeneous equations in n unknowns, the determinant of the coefficients must be zero. Also, this condition can be shown to be sufficient to ensure the existence of a nontrivial solution (see T. L. Wade, *The Algebra of Vectors and Matrices*, Addison-Wesley, 1951, p. 146). We thus have the extremely important theorem:

> *A system of n linear homogeneous equations in n unknowns has a nontrivial solution if and only if the determinant of the coefficients is zero.*

Suppose that $\det(a_{ij}) = 0$, so that (8.37) has a nontrivial solution. How do we find it? With $\det(a_{ij}) = 0$, Cramer's rule (8.36) gives $x_k = 0/0$, $k = 1, \ldots, n$, which is indeterminate. Thus Cramer's rule is of no immediate help. We also observe that, if $x_1 = d_1, x_2 = d_2, \ldots, x_n = d_n$ is a solution of (8.37), then so is $x_1 = cd_1$, $x_2 = cd_2, \ldots, x_n = cd_n$, where c is an arbitrary constant. This is easily seen, since

$$a_{11}cd_1 + a_{12}cd_2 + \cdots + a_{1n}cd_n = c(a_{11}d_1 + a_{12}d_2 + \cdots + a_{1n}d_n) = c \cdot 0 = 0$$

and so on. Therefore, the solution to the linear homogeneous system of equations will contain an arbitrary constant, and we cannot determine a unique value for each unknown. To solve (8.37), we therefore assign an arbitrary value to any one of the unknowns, say x_n; we set $x_n = c$, where c is an arbitrary constant. Having assigned a value to x_n, we transfer the last term in each of the equations of (8.37) to the right side to get

$$\begin{aligned} a_{11}x_1 + a_{12}x_2 + \cdots + a_{1,n-1}x_{n-1} &= -a_{1,n}c \\ a_{21}x_1 + a_{22}x_2 + \cdots + a_{2,n-1}x_{n-1} &= -a_{2,n}c \\ \cdots\cdots\cdots\cdots\cdots\cdots\cdots & \\ a_{n-1,1}x_1 + a_{n-1,2}x_2 + \cdots + a_{n-1,n-1}x_{n-1} &= -a_{n-1,n}c \\ a_{n1}x_1 + a_{n2}x_2 + \cdots + a_{n,n-1}x_{n-1} &= -a_{nn}c \end{aligned} \tag{8.38}$$

We now have n equations in $n - 1$ unknowns, which is one more equation than we need. We therefore discard any one of the equations of (8.38), say the last one. This gives a system of $n - 1$ linear *in*homogeneous equations in $n - 1$ unknowns. We could then apply Cramer's rule (8.36) to solve for $x_1, x_2, \ldots, x_{n-1}$. Since the constants on the right side of the equations in (8.38) all contain the factor c, Theorem IV in Section 8.3 shows that all the unknowns contain this arbitrary constant as a factor. The form of the solution is therefore

$$x_1 = ce_1, \quad x_2 = ce_2, \quad \ldots, \quad x_{n-1} = ce_{n-1}, \quad x_n = c \tag{8.39}$$

where $e_1, \ldots, e_{n-1}$ are numbers and c is an arbitrary constant.

EXAMPLE Solve

$$\begin{aligned} 3x_1 + 4x_2 + x_3 &= 0 \\ x_1 + 3x_2 - 2x_3 &= 0 \\ x_1 - 2x_2 + 5x_3 &= 0 \end{aligned}$$

This is a set of linear homogeneous equations, and we begin by evaluating the determinant of the coefficients. We find (see the Exercise)

$$\begin{vmatrix} 3 & 4 & 1 \\ 1 & 3 & -2 \\ 1 & -2 & 5 \end{vmatrix} = 0$$

Therefore, a nontrivial solution exists. We set x_3 equal to an arbitrary constant c ($x_3 = c$) and discard the third equation to give

$$\begin{aligned} 3x_1 + 4x_2 &= -c \\ x_1 + 3x_2 &= 2c \end{aligned}$$

Subtracting 3 times the second equation from the first, we get $-5x_2 = -7c$, so $x_2 = \frac{7}{5}c$. Substitution into $x_1 + 3x_2 = 2c$ gives $x_1 + \frac{21}{5}c = 2c$, so $x_1 = -\frac{11}{5}c$. Hence the general solution is $x_1 = -\frac{11}{5}c$, $x_2 = \frac{7}{5}c$, $x_3 = c$. For those allergic to fractions, we define a new arbitrary constant s as $s \equiv \frac{1}{5}c$ and write $x_1 = -11s$, $x_2 = 7s$, $x_3 = 5s$.

EXERCISE (a) Verify that the coefficient determinant in this example is zero. (b) Verify that the third equation in this example can be obtained by adding a certain constant times the second equation to the first equation.

The procedure just outlined fails if the determinant of the inhomogeneous system of $n - 1$ equations in $n - 1$ unknowns [(8.38) with the last equation omitted] happens to be zero. Cramer's rule then has a zero in the denominator and is of no use. We could try to get around this difficulty by initially assigning the arbitrary value to another of the unknowns rather than to x_n. We could also try discarding some other equation of (8.38), rather than the last one. What we are looking for is a nonvanishing determinant of order $n - 1$ formed from the determinant of the coefficients of the system (8.37) by striking out a row and a column. If such a determinant exists, then by the procedure given, with the right choice of the equation to be discarded and the right choice of the unknown to be assigned an arbitrary value, we can solve the system and will get solutions of the form (8.39). If no such determinant exists, we must assign arbitrary values to two of the unknowns and attempt to proceed from there. Thus the solution to (8.37) might contain two (or even more) arbitrary constants.

An efficient way to solve a system of linear homogeneous equations is to do Gauss–Jordan elimination on the equations. If only the trivial solution exists, the final set of equations obtained will be $x_1 = 0, x_2 = 0, \ldots, x_n = 0$. If a nontrivial solution exists, at least one equation will be reduced to the form $0 = 0$; if m equations of the form $0 = 0$ are obtained, we assign arbitrary constants to m of the unknowns and express the remaining unknowns in terms of these m unknowns.

EXAMPLE Use Gauss–Jordan elimination to solve the set of equations in the preceding example.

In doing Gaussian or Gauss–Jordan elimination on a set of n inhomogeneous or homogeneous equations, we can eliminate needless writing by omitting the variables $x_1, \ldots, x_n$ and writing down only the n-row, $(n + 1)$-column array of coefficients and constant terms (including any zero coefficients); we then produce the next array by operating on the numbers of each row as if that row were the equation it represents.

To eliminate one set of divisions, we interchange the first and second equations so that we start with $a_{11} = 1$. Detaching the coefficients and proceeding with Gauss–Jordan elimination, we have

$$\begin{matrix} 1 & 3 & -2 & 0 \\ 3 & 4 & 1 & 0 \\ 1 & -2 & 5 & 0 \end{matrix} \quad \rightarrow \quad \begin{matrix} 1 & 3 & -2 & 0 \\ 0 & -5 & 7 & 0 \\ 0 & -5 & 7 & 0 \end{matrix} \quad \rightarrow \quad \begin{matrix} 1 & 3 & -2 & 0 \\ 0 & 1 & -\frac{7}{5} & 0 \\ 0 & -5 & 7 & 0 \end{matrix} \quad \rightarrow \quad \begin{matrix} 1 & 0 & \frac{11}{5} & 0 \\ 0 & 1 & -\frac{7}{5} & 0 \\ 0 & 0 & 0 & 0 \end{matrix}$$

The first array is the original set of equations with the first and second equations interchanged. To eliminate x_1 from the second and third equations, we subtract 3 times row one from row two and 1 times row one from row three, thereby producing the second array. Division of row two by -5 produces the third array. To eliminate x_2 from the first and third equations, we subtract 3 times row two from row one and -5 times row two from row three, thereby producing the fourth array. Because the fourth array has the x_3 coefficient in row three equal to zero, we cannot use row three to eliminate x_3 from rows one and two (as would be the last step in the Gauss–Jordan algorithm). Discarding the last equation, which reads $0 = 0$, we assign $x_3 = k$, where k is an arbitrary constant. The first and second equations in the last array read $x_1 + \frac{11}{5}x_3 = 0$ and $x_2 - \frac{7}{5}x_3 = 0$, or $x_1 = -\frac{11}{5}x_3$, $x_2 = \frac{7}{5}x_3$. The general solution is $x_1 = -\frac{11}{5}k$, $x_2 = \frac{7}{5}k$, $x_3 = k$.

8.5 LINEAR VARIATION FUNCTIONS

A special kind of variation function widely used in the study of molecules is the linear variation function. A **linear variation function** is a linear combination of n linearly independent functions $f_1, f_2, \ldots, f_n$:

$$\phi = c_1 f_1 + c_2 f_2 + \cdots + c_n f_n = \sum_{j=1}^{n} c_j f_j \tag{8.40}$$

where ϕ is the trial variation function and the coefficients c_j are parameters to be determined by minimizing the variational integral. The functions f_j (which are called **basis functions**) must satisfy the boundary conditions of the problem. We shall restrict ourselves to *real* ϕ so that the c_j's and f_j's are all real. In (8.40), the functions f_j are known functions.

We now apply the variation theorem (8.9). For the real linear variation function, we have

$$\int \phi^* \phi \, d\tau = \int \sum_{j=1}^{n} c_j f_j \sum_{k=1}^{n} c_k f_k \, d\tau = \sum_{j=1}^{n} \sum_{k=1}^{n} c_j c_k \int f_j f_k \, d\tau \equiv \sum_{j=1}^{n} \sum_{k=1}^{n} c_j c_k S_{jk} \tag{8.41}$$

where we defined the **overlap integral** S_{jk} as

$$\boxed{S_{jk} \equiv \int f_j^* f_k \, d\tau} \tag{8.42}$$

Note that S_{jk} is not necessarily equal to δ_{jk}, since there is no reason to suppose that the functions f_j are mutually orthogonal. They are not necessarily the eigenfunctions of any operator. The numerator in (8.9) is

$$\int \phi^* \hat{H} \phi \, d\tau = \int \sum_{j=1}^{n} c_j f_j \hat{H} \sum_{k=1}^{n} c_k f_k \, d\tau$$

$$= \sum_{j=1}^{n} \sum_{k=1}^{n} c_j c_k \int f_j \hat{H} f_k \, d\tau \equiv \sum_{j=1}^{n} \sum_{k=1}^{n} c_j c_k H_{jk} \tag{8.43}$$

where we defined H_{jk} as

$$\boxed{H_{jk} \equiv \int f_j^* \hat{H} f_k \, d\tau} \tag{8.44}$$

The variational integral W is

$$W \equiv \frac{\int \phi^* \hat{H} \phi \, d\tau}{\int \phi^* \phi \, d\tau} = \frac{\sum_{j=1}^{n} \sum_{k=1}^{n} c_j c_k H_{jk}}{\sum_{j=1}^{n} \sum_{k=1}^{n} c_j c_k S_{jk}} \tag{8.45}$$

$$W \sum_{j=1}^{n} \sum_{k=1}^{n} c_j c_k S_{jk} = \sum_{j=1}^{n} \sum_{k=1}^{n} c_j c_k H_{jk} \tag{8.46}$$

We now minimize W so as to approach as closely as we can to E_1 ($W \geq E_1$). The variational integral W is a function of the n independent variables $c_1, c_2, \ldots, c_n$:

$$W = W(c_1, c_2, \ldots, c_n)$$

A necessary condition for a minimum in a function W of several variables is that its partial derivatives with respect to each of the variables must be zero at the minimum:

$$\frac{\partial W}{\partial c_i} = 0, \qquad i = 1, 2, \ldots, n \tag{8.47}$$

We now differentiate (8.46) partially with respect to each c_i to obtain n equations:

$$\frac{\partial W}{\partial c_i} \sum_{j=1}^{n} \sum_{k=1}^{n} c_j c_k S_{jk} + W \frac{\partial}{\partial c_i} \sum_{j=1}^{n} \sum_{k=1}^{n} c_j c_k S_{jk} = \frac{\partial}{\partial c_i} \sum_{j=1}^{n} \sum_{k=1}^{n} c_j c_k H_{jk}, \quad i = 1, 2, \ldots, n \tag{8.48}$$

Now

$$\frac{\partial}{\partial c_i}\sum_{j=1}^{n}\sum_{k=1}^{n} c_j c_k S_{jk} = \sum_{j=1}^{n}\sum_{k=1}^{n}\left[\frac{\partial}{\partial c_i}(c_j c_k)\right]S_{jk} = \sum_{j=1}^{n}\sum_{k=1}^{n}\left(c_k\frac{\partial c_j}{\partial c_i} + c_j\frac{\partial c_k}{\partial c_i}\right)S_{jk}$$

The c_j's are independent variables, and therefore

$$\frac{\partial c_j}{\partial c_i} = 0 \quad \text{if } j \neq i, \qquad \frac{\partial c_j}{\partial c_i} = 1 \quad \text{if } j = i$$

$$\frac{\partial c_j}{\partial c_i} = \delta_{ij} \tag{8.49}$$

We then have

$$\frac{\partial}{\partial c_i}\sum_{j=1}^{n}\sum_{k=1}^{n} c_j c_k S_{jk} = \sum_{k=1}^{n}\sum_{j=1}^{n} c_k \delta_{ij} S_{jk} + \sum_{j=1}^{n}\sum_{k=1}^{n} c_j \delta_{ik} S_{jk} = \sum_{k=1}^{n} c_k S_{ik} + \sum_{j=1}^{n} c_j S_{ji}$$

where we evaluated one of the sums in each double summation using Eq. (7.32). Use of (7.4) gives

$$S_{ji} = S_{ij}^* = S_{ij} \tag{8.50}$$

where the last equality follows because we are dealing with real functions. Hence,

$$\frac{\partial}{\partial c_i}\sum_{j=1}^{n}\sum_{k=1}^{n} c_j c_k S_{jk} = \sum_{k=1}^{n} c_k S_{ik} + \sum_{j=1}^{n} c_j S_{ij} = \sum_{k=1}^{n} c_k S_{ik} + \sum_{k=1}^{n} c_k S_{ik} = 2\sum_{k=1}^{n} c_k S_{ik} \tag{8.51}$$

where the fact that j is a dummy variable was used.

By replacing S_{jk} by H_{jk} in each of these manipulations, we get

$$\frac{\partial}{\partial c_i}\sum_{j=1}^{n}\sum_{k=1}^{n} c_j c_k H_{jk} = 2\sum_{k=1}^{n} c_k H_{ik} \tag{8.52}$$

This result depends on the fact that

$$H_{ji} = H_{ij}^* = H_{ij} \tag{8.53}$$

which is true because $\hat{H}$ is Hermitian, and f_i, f_j, and $\hat{H}$ are real.

Substitution of Eqs. (8.47), (8.51), and (8.52) into (8.48) gives

$$2W\sum_{k=1}^{n} c_k S_{ik} = 2\sum_{k=1}^{n} c_k H_{ik}, \qquad i = 1, 2, \ldots, n$$

$$\sum_{k=1}^{n}[(H_{ik} - S_{ik}W)c_k] = 0, \qquad i = 1, 2, \ldots, n \tag{8.54}$$

Equation (8.54) is a set of n simultaneous, linear, homogeneous equations in the n unknowns $c_1, c_2, \ldots, c_n$ [the coefficients in the linear variation function (8.40)]. For example, for $n = 2$, (8.54) gives

$$\begin{aligned}(H_{11} - S_{11}W)c_1 + (H_{12} - S_{12}W)c_2 &= 0\\ (H_{21} - S_{21}W)c_1 + (H_{22} - S_{22}W)c_2 &= 0\end{aligned} \tag{8.55}$$

For the general case of n functions $f_1, \ldots, f_n$, (8.54) is

$$\begin{array}{|l|}\hline (H_{11} - S_{11}W)c_1 + (H_{12} - S_{12}W)c_2 + \cdots + (H_{1n} - S_{1n}W)c_n = 0 \\ (H_{21} - S_{21}W)c_1 + (H_{22} - S_{22}W)c_2 + \cdots + (H_{2n} - S_{2n}W)c_n = 0 \\ \cdots\cdots\cdots\cdots\cdots\cdots\cdots\cdots\cdots\cdots\cdots\cdots\cdots\cdots\cdots \\ (H_{n1} - S_{n1}W)c_1 + (H_{n2} - S_{n2}W)c_2 + \cdots + (H_{nn} - S_{nn}W)c_n = 0 \\ \hline \end{array} \quad \textbf{(8.56)}$$

From the theorem of Section 8.4, for there to be a solution to the linear homogeneous equations (8.56) besides the trivial solution $0 = c_1 = c_2 = \cdots = c_n$ (which would make the variation function ϕ zero), the determinant of the coefficients must vanish. For $n = 2$, we have

$$\begin{vmatrix} H_{11} - S_{11}W & H_{12} - S_{12}W \\ H_{21} - S_{21}W & H_{22} - S_{22}W \end{vmatrix} = 0 \quad (8.57)$$

and for the general case

$$\boxed{\det(H_{ij} - S_{ij}W) = 0} \quad \textbf{(8.58)}$$

$$\begin{vmatrix} H_{11} - S_{11}W & H_{12} - S_{12}W & \cdots & H_{1n} - S_{1n}W \\ H_{21} - S_{21}W & H_{22} - S_{22}W & \cdots & H_{2n} - S_{2n}W \\ \cdot & \cdot & \cdots & \cdot \\ \cdot & \cdot & \cdots & \cdot \\ \cdot & \cdot & \cdots & \cdot \\ H_{n1} - S_{n1}W & H_{n2} - S_{n2}W & \cdots & H_{nn} - S_{nn}W \end{vmatrix} = 0 \quad (8.59)$$

Expansion of the determinant in (8.59) gives an algebraic equation of degree n in the unknown W. This algebraic equation has n roots, which can be shown to be real. Arranging these roots in order of increasing value, we denote them as

$$W_1 \leq W_2 \leq \cdots \leq W_n \quad (8.60)$$

If we number the bound states of the system in order of increasing energy, we have

$$E_1 \leq E_2 \leq \cdots \leq E_n \leq E_{n+1} \leq \cdots \quad (8.61)$$

where the E's denote the true energies of various states. From the variation theorem, we know that $E_1 \leq W_1$. Moreover, it can be proved that [J. K. L. MacDonald, *Phys. Rev.*, **43**, 830 (1933); R. H. Young, *Int. J. Quantum Chem.*, **6**, 596 (1972); see Prob. 8.39]

$$E_1 \leq W_1, \quad E_2 \leq W_2, \quad E_3 \leq W_3, \ldots, \quad E_n \leq W_n \quad (8.62)$$

Thus, the linear variation method provides upper bounds to the energies of the lowest n bound states of the system. We use the roots $W_1, W_2, \ldots, W_n$ as approximations to the energies of the lowest states. If approximations to the energies of more states are wanted, we add more functions f_k to the trial function ϕ. The addition of more functions f_k can be shown to increase (or cause no change in) the accuracy of the previously calculated energies. If the functions f_k in $\phi = \sum_k c_k f_k$ form a complete set, then we will obtain the true wave functions of the system. Unfortunately, we usually need an infinite number of functions to have a complete set.

Quantum chemists may use dozens, hundreds, thousands, or even millions of terms in linear variation functions so as to get accurate results for molecules. Obviously, a computer is essential for this work. The most efficient way to solve (8.59) (which is called the **secular equation**) and the associated linear equations (8.56) is by matrix methods (Section 8.6).

To obtain an approximation to the wave function of the ground state, we take the lowest root W_1 of the secular equation and substitute it in the set of equations (8.56); we then solve this set of equations for the coefficients $c_1^{(1)}, c_2^{(1)}, \ldots, c_n^{(1)}$, where the superscript $^{(1)}$ was added to indicate that these coefficients correspond to W_1. [As noted in the previous section, we can determine only the ratios of the coefficients. We solve for $c_2^{(1)}, \ldots, c_n^{(1)}$ in terms of $c_1^{(1)}$, and then determine $c_1^{(1)}$ by normalization.] Having found the $c_k^{(1)}$'s, we take $\phi_1 = \sum_k c_k^{(1)} f_k$ as an approximate ground-state wave function. Use of higher roots of (8.59) in (8.56) gives approximations to excited-state wave functions. These approximate wave functions can be shown to be orthogonal (Prob. 8.39).

Solution of (8.59) and (8.56) is simplified by having as many of the integrals equal to zero as possible. We can make some of the off-diagonal H_{ij}'s vanish by choosing the functions f_k as eigenfunctions of some operator $\hat{A}$ that commutes with $\hat{H}$. If f_i and f_j correspond to different eigenvalues of $\hat{A}$, then H_{ij} vanishes (Theorem 6 of Section 7.4). If the functions f_k are orthonormal, the off-diagonal S_{ij}'s vanish ($S_{ij} = \delta_{ij}$). If the initially chosen f_k's are not orthogonal, we can use the Schmidt (or some other) procedure to find n linear combinations of these f_k's that are orthogonal and then use the orthogonalized functions.

Equations (8.56) and (8.59) are also valid when the restriction that the variation function be real is removed (Prob. 8.38).

EXAMPLE Add functions to the function $x(l - x)$ of the first example of Section 8.1 to form a linear variation function for the particle in a one-dimensional box of length l and find approximate energies and wave functions for the lowest four states.

In the trial function $\phi = \sum_{k=1}^{n} c_k f_k$, we take $f_1 = x(l - x)$. Since we want approximations to the lowest four states, n must be at least 4. There are an infinite number of possible well-behaved functions that could be used for f_2, f_3, and f_4. The function $x^2(l - x)^2$ obeys the boundary conditions of vanishing at $x = 0$ and $x = l$ and leads to simple integrals, so we take $f_2 = x^2(l - x)^2$.

If the origin is placed at the center of the box, the potential energy (Fig. 2.1) is an even function, and, as noted in Section 8.2, the wave functions alternate between being even and odd functions (see also Fig. 2.3). (Throughout this example, the terms even and odd will refer to having the origin at the box's center.) The functions $f_1 = x(l - x)$ and $f_2 = x^2(l - x)^2$ are both even functions (see Prob. 8.33). If we were to take $\phi = c_1 x(l - x) + c_2 x^2(l - x)^2$, we would end up with upper bounds to the energies of the lowest two states with even wave functions (the $n = 1$ and $n = 3$ states) and would get approximate wave functions for these two states. Since we also want to approximate the $n = 2$ and $n = 4$ states, we shall add in two functions that are odd. An odd function must vanish at the origin [as noted after Eq. (4.52)], so we need functions that vanish at the box midpoint $x = \frac{1}{2}l$, as well as at $x = 0$ and l. A simple function with these properties is $f_3 = x(l - x)(\frac{1}{2}l - x)$. To get f_4, we shall multiply f_2 by $(\frac{1}{2}l - x)$. Thus we take $\phi = \sum_{k=1}^{4} c_k f_k$, with

$$f_1 = x(l - x), f_2 = x^2(l - x)^2, f_3 = x(l - x)(\tfrac{1}{2}l - x), f_4 = x^2(l - x)^2(\tfrac{1}{2}l - x) \quad (8.63)$$

Note that f_1, f_2, f_3, and f_4 are linearly independent, as assumed in (8.40).

Because f_1 and f_2 are even, while f_3 and f_4 are odd, many integrals will vanish. Thus

$$S_{13} = S_{31} = 0, \quad S_{14} = S_{41} = 0, \quad S_{23} = S_{32} = 0, \quad S_{24} = S_{42} = 0 \quad (8.64)$$

because the integrand in each of these overlap integrals is an odd function with respect to the origin at the box center. The functions f_1, f_2, f_3, f_4 are eigenfunctions of the parity operator $\hat{\Pi}$ (Section 7.5) with the even functions f_1 and f_2 having parity eigenvalue +1 and f_3 and f_4 having eigenvalue −1. The operator $\hat{\Pi}$ commutes with $\hat{H}$ (since V is an even function), so by Theorem 6 of Section 7.4, H_{ij} vanishes if f_i is an odd function and f_j is even, or vice versa. Thus

$$H_{13} = H_{31} = 0, \quad H_{14} = H_{41} = 0, \quad H_{23} = H_{32} = 0, \quad H_{24} = H_{42} = 0 \quad (8.65)$$

From (8.64) and (8.65), the $n = 4$ secular equation (8.59) becomes

$$\begin{vmatrix} H_{11} - S_{11}W & H_{12} - S_{12}W & 0 & 0 \\ H_{21} - S_{21}W & H_{22} - S_{22}W & 0 & 0 \\ 0 & 0 & H_{33} - S_{33}W & H_{34} - S_{34}W \\ 0 & 0 & H_{43} - S_{43}W & H_{44} - S_{44}W \end{vmatrix} = 0 \quad (8.66)$$

The secular determinant is in block-diagonal form and so is equal to the product of its blocks [Eq. (8.34)]:

$$\begin{vmatrix} H_{11} - S_{11}W & H_{12} - S_{12}W \\ H_{21} - S_{21}W & H_{22} - S_{22}W \end{vmatrix} \times \begin{vmatrix} H_{33} - S_{33}W & H_{34} - S_{34}W \\ H_{43} - S_{43}W & H_{44} - S_{44}W \end{vmatrix} = 0$$

The four roots of this equation are found from the equations

$$\begin{vmatrix} H_{11} - S_{11}W & H_{12} - S_{12}W \\ H_{21} - S_{21}W & H_{22} - S_{22}W \end{vmatrix} = 0 \quad (8.67)$$

$$\begin{vmatrix} H_{33} - S_{33}W & H_{34} - S_{34}W \\ H_{43} - S_{43}W & H_{44} - S_{44}W \end{vmatrix} = 0 \quad (8.68)$$

Let the roots of (8.67) (which are approximations to the $n = 1$ and $n = 3$ energies) be W_1 and W_3 and let the roots of (8.68) be W_2 and W_4. After solving the secular equation for the W's, we substitute them one at a time into the set of equations (8.56) to find the coefficients c_k in the variation function. From the secular equation (8.66), the set of equations (8.56) with the root W_1 is

$$\left.\begin{aligned} (H_{11} - S_{11}W_1)c_1^{(1)} + (H_{12} - S_{12}W_1)c_2^{(1)} &= 0 \\ (H_{21} - S_{21}W_1)c_1^{(1)} + (H_{22} - S_{22}W_1)c_2^{(1)} &= 0 \end{aligned}\right\} \quad (8.69a)$$

$$\left.\begin{aligned} (H_{33} - S_{33}W_1)c_3^{(1)} + (H_{34} - S_{34}W_1)c_4^{(1)} &= 0 \\ (H_{43} - S_{43}W_1)c_3^{(1)} + (H_{44} - S_{44}W_1)c_4^{(1)} &= 0 \end{aligned}\right\} \quad (8.69b)$$

Because W_1 is a root of (8.67), the set of equations (8.69a) has the determinant of its coefficients [which is the determinant in (8.67)] equal to zero. Hence (8.69a) has a nontrivial solution for $c_1^{(1)}$ and $c_2^{(1)}$. However, W_1 is not a root of (8.68), so the determinant of the coefficients of the set of equations (8.69b) is nonzero. Hence, (8.69b) has only the trivial solution $c_3^{(1)} = c_4^{(1)} = 0$. The trial function ϕ_1 corresponding to the root W_1 thus has the form $\phi_1 = \sum_{k=1}^{4} c_k^{(1)} f_k = c_1^{(1)} f_1 + c_2^{(1)} f_2$.

The same reasoning shows that ϕ_3 is a linear combination of f_1 and f_2, while ϕ_2 and ϕ_4 are each linear combinations of f_3 and f_4:

$$\begin{aligned} \phi_1 &= c_1^{(1)} f_1 + c_2^{(1)} f_2, \quad \phi_3 = c_1^{(3)} f_1 + c_2^{(3)} f_2 \\ \phi_2 &= c_3^{(2)} f_3 + c_4^{(2)} f_4, \quad \phi_4 = c_3^{(4)} f_3 + c_4^{(4)} f_4 \end{aligned} \tag{8.70}$$

The even wave functions ψ_1 and ψ_3 are approximated by linear combinations of the even functions f_1 and f_2; the odd functions ψ_2 and ψ_4 are approximated by linear combinations of the odd functions f_3 and f_4.

When the secular equation is in block-diagonal form, it factors into two or more smaller secular equations, and the set of simultaneous equations (8.56) breaks up into two or more smaller sets of equations.

We now must evaluate the H_{ij} and S_{ij} integrals so as to solve (8.67) and (8.68) for W_1, W_2, W_3, and W_4. We have

$$H_{11} = \langle f_1|\hat{H}|f_1\rangle = \int_0^l x(l-x)\left(\frac{-\hbar^2}{2m}\right)\frac{d^2}{dx^2}[x(l-x)]\,dx = \frac{\hbar^2 l^3}{6m}$$

$$S_{11} = \langle f_1|f_1\rangle = \int_0^l x^2(l-x)^2\,dx = \frac{l^5}{30}$$

where Eqs. (8.12) and (8.13) were used. Evaluation of the remaining integrals using (8.63), (8.50), and (8.53) gives (Prob. 8.34)

$$\begin{aligned} &H_{12} = H_{21} = \langle f_2|\hat{H}|f_1\rangle = \hbar^2 l^5/30m, \quad H_{22} = \hbar^2 l^7/105m \\ &H_{33} = \hbar^2 l^5/40m, \quad H_{44} = \hbar^2 l^9/1260m, \quad H_{34} = H_{43} = \hbar^2 l^7/280m \\ &S_{12} = S_{21} = \langle f_1|f_2\rangle = l^7/140, \quad S_{22} = l^9/630 \\ &S_{33} = l^7/840, \quad S_{44} = l^{11}/27720, \quad S_{34} = S_{43} = l^9/5040 \end{aligned}$$

Equation (8.67) becomes

$$\begin{vmatrix} \dfrac{\hbar^2 l^3}{6m} - \dfrac{l^5}{30}W & \dfrac{\hbar^2 l^5}{30m} - \dfrac{l^7}{140}W \\ \dfrac{\hbar^2 l^5}{30m} - \dfrac{l^7}{140}W & \dfrac{\hbar^2 l^7}{105m} - \dfrac{l^9}{630}W \end{vmatrix} = 0 \tag{8.71}$$

Using Theorem IV of Section 8.3, we eliminate the fractions by multiplying row 1 of the determinant by $420m/l^3$, row 2 by $1260m/l^5$, and the right side of (8.71) by both factors, to get

$$\begin{vmatrix} 70\hbar^2 - 14ml^2W & 14\hbar^2 l^2 - 3ml^4W \\ 42\hbar^2 - 9ml^2W & 12\hbar^2 l^2 - 2ml^4W \end{vmatrix} = 0$$

$$m^2 l^4 W^2 - 56ml^2\hbar^2 W + 252\hbar^4 = 0$$

$$W = (\hbar^2/ml^2)\left(28 \pm \sqrt{532}\right) = 0.1250018h^2/ml^2, \quad 1.293495h^2/ml^2 \tag{8.72}$$

Substitution of the integrals into (8.68) leads to the roots (Prob. 8.35)

$$W = (\hbar^2/ml^2)\left(60 \pm \sqrt{1620}\right) = 0.5002930h^2/ml^2, \quad 2.5393425h^2/ml^2 \tag{8.73}$$

The approximate values $(ml^2/h^2)W = 0.1250018, 0.5002930, 1.293495$, and 2.5393425 may be compared with the exact values [Eq. (2.20)] $(ml^2/h^2)E = 0.125, 0.5, 1.125$, and 2 for the four lowest states. The percent errors are 0.0014%, 0.059%, 15.0%, and 27.0% for $n = 1, 2, 3$, and 4, respectively. We did great for $n = 1$ and 2; lousy for $n = 3$ and 4.

We now find the approximate wave functions corresponding to these W's. Substitution of $W_1 = 0.1250018h^2/ml^2$ into the set of equations (8.69a) corresponding to (8.72) gives (after division by h^2)

$$0.023095c_1^{(1)} - 0.020381c_2^{(1)}l^2 = 0$$
$$-0.061144c_1^{(1)} + 0.053960c_2^{(1)}l^2 = 0 \quad (8.74)$$

where, for example, the first coefficient is found from

$$70\hbar^2 - 14ml^2W = 70h^2/4\pi^2 - 14(0.1250018)h^2 = 0.023095h^2$$

To solve the homogeneous equations (8.74), we follow the procedure given near the end of Section 8.4. We discard the second equation of (8.74), transfer the $c_2^{(1)}$ term to the right side, and solve for the coefficient ratio; we get

$$c_1^{(1)} = k, \quad c_2^{(1)} = 1.133k/l^2$$

where k is a constant. We find k from the normalization condition:

$$\langle \phi_1 | \phi_1 \rangle = 1 = \langle kf_1 + 1.133kf_2/l^2 | kf_1 + 1.133kf_2/l^2 \rangle$$
$$= k^2(\langle f_1|f_1\rangle + 2.266\langle f_1|f_2\rangle/l^2 + 1.284\langle f_2|f_2\rangle/l^4)$$
$$= k^2(S_{11} + 2.266S_{12}/l^2 + 1.284S_{22}/l^4) = 0.05156k^2l^5$$

where the previously found values of the overlap integrals were used. Hence $k = 4.404/l^{5/2}$ and

$$\phi_1 = c_1^{(1)}f_1 + c_2^{(1)}f_2 = 4.404f_1/l^{5/2} + 4.990f_2/l^{9/2}$$
$$\phi_1 = l^{-1/2}[4.404(x/l)(1 - x/l) + 4.990(x/l)^2(1 - x/l)^2] \quad (8.75)$$

where (8.63) was used.

Using W_2, W_3, and W_4 in turn in (8.56), we find the following normalized linear variation functions (Prob. 8.37), where $X \equiv x/l$:

$$\phi_2 = l^{-1/2}\left[16.78X(1 - X)\left(\tfrac{1}{2} - X\right) + 71.85X^2(1 - X)^2\left(\tfrac{1}{2} - X\right)\right]$$
$$\phi_3 = l^{-1/2}[28.65X(1 - X) - 132.7X^2(1 - X)^2] \quad (8.76)$$
$$\phi_4 = l^{-1/2}\left[98.99X(1 - X)\left(\tfrac{1}{2} - X\right) - 572.3X^2(1 - X)^2\left(\tfrac{1}{2} - X\right)\right]$$

8.6 MATRICES, EIGENVALUES, AND EIGENVECTORS

Matrices (Section 7.10) were introduced in 1857 by the mathematician and lawyer Arthur Cayley as a shorthand way of dealing with simultaneous linear equations and linear transformations from one set of variables to another. The set of linear inhomogeneous equations (8.35) can be written as the matrix equation

$$\begin{pmatrix} a_{11} & a_{12} & \cdots & a_{1n} \\ a_{21} & a_{22} & \cdots & a_{2n} \\ \vdots & \vdots & \ddots & \vdots \\ a_{n1} & a_{n2} & \cdots & a_{nn} \end{pmatrix} \begin{pmatrix} x_1 \\ x_2 \\ \vdots \\ x_n \end{pmatrix} = \begin{pmatrix} b_1 \\ b_2 \\ \vdots \\ b_n \end{pmatrix} \quad (8.77)$$

$$\mathbf{Ax} = \mathbf{b} \quad (8.78)$$

where $\mathbf{A}$ is the coefficient matrix and $\mathbf{x}$ and $\mathbf{b}$ are column matrices. The equivalence of (8.35) and (8.77) is readily verified using the matrix-multiplication rule (7.106).

The **determinant** of a square matrix **A** is the determinant whose elements are the same as the elements of **A**. If det $\mathbf{A} \neq 0$, the matrix **A** is said to be *nonsingular.*

The **inverse** of a square matrix **A** of order n is the square matrix whose product with **A** is the unit matrix of order n. Denoting the inverse by $\mathbf{A}^{-1}$, we have

$$\boxed{\mathbf{A}\mathbf{A}^{-1} = \mathbf{A}^{-1}\mathbf{A} = \mathbf{I}} \tag{8.79}$$

One can prove that $\mathbf{A}^{-1}$ exists if and only if det $\mathbf{A} \neq 0$. (For efficient methods of computing $\mathbf{A}^{-1}$, see *Press et al.*, Section 2.3; *Shoup*, Section 3.3; Prob. 8.50. Many spreadsheets have a built-in capability to find the inverse of a matrix.)

If det $\mathbf{A} \neq 0$ for the coefficient matrix **A** in (8.77), then we can multiply each side of (8.78) by $\mathbf{A}^{-1}$ on the left to get $\mathbf{A}^{-1}(\mathbf{A}\mathbf{x}) = \mathbf{A}^{-1}\mathbf{b}$. Since matrix multiplication is associative (Section 7.10), we have $\mathbf{A}^{-1}(\mathbf{A}\mathbf{x}) = (\mathbf{A}^{-1}\mathbf{A})\mathbf{x} = \mathbf{I}\mathbf{x} = \mathbf{x}$. Thus, left multiplication of (8.78) by $\mathbf{A}^{-1}$ gives $\mathbf{x} = \mathbf{A}^{-1}\mathbf{b}$ as the solution for the unknowns in a set of linear inhomogeneous equations.

The linear variation method is widely used to find approximate molecular wave functions, and matrix algebra gives the most computationally efficient method to solve the equations of the linear variation method. If the functions $f_1, \ldots, f_n$ in the linear variation function $\phi = \sum_{k=1}^{n} c_k f_k$ are made to be orthonormal, then $S_{ij} \equiv \int f_i^* f_j \, d\tau = \delta_{ij}$, and the homogeneous set of equations (8.56) for the coefficients c_k that minimize the variational integral becomes

$$\begin{aligned} H_{11}c_1 + H_{12}c_2 + \cdots + H_{1n}c_n &= Wc_1 \\ H_{21}c_1 + H_{22}c_2 + \cdots + H_{2n}c_n &= Wc_2 \\ \vdots \qquad\quad \vdots \qquad\quad \ddots \qquad \vdots \quad &\quad\ \vdots \\ H_{n1}c_1 + H_{n2}c_2 + \cdots + H_{nn}c_n &= Wc_n \end{aligned} \tag{8.80a}$$

$$\begin{pmatrix} H_{11} & H_{12} & \cdots & H_{1n} \\ H_{21} & H_{22} & \cdots & H_{2n} \\ \vdots & \vdots & \ddots & \vdots \\ H_{n1} & H_{n2} & \cdots & H_{nn} \end{pmatrix} \begin{pmatrix} c_1 \\ c_2 \\ \vdots \\ c_n \end{pmatrix} = W \begin{pmatrix} c_1 \\ c_2 \\ \vdots \\ c_n \end{pmatrix} \tag{8.80b}$$

$$\mathbf{H}\mathbf{c} = W\mathbf{c} \tag{8.80c}$$

where **H** is the square matrix whose elements are $H_{ij} = \langle f_i | \hat{H} | f_j \rangle$ and **c** is the column vector of coefficients $c_1, \ldots, c_n$. In (8.80c), **H** is a known matrix and **c** and W are unknowns to be solved for.

If

$$\boxed{\mathbf{A}\mathbf{c} = \lambda\mathbf{c}} \tag{8.81}$$

where **A** is a square matrix, **c** is a column vector with at least one nonzero element, and λ is a scalar, then **c** is said to be an **eigenvector** (or characteristic vector) of **A** and λ is an **eigenvalue** (or characteristic value) of **A**.

Comparison of (8.81) with (8.80c) shows that solving the linear variation problem with $S_{ij} = \delta_{ij}$ amounts to finding the eigenvalues and eigenvectors of the matrix **H**.

The matrix eigenvalue equation $\mathbf{Hc} = W\mathbf{c}$ is equivalent to the set of homogeneous equations (8.56), which has a nontrivial solution for the c's if and only if $\det(H_{ij} - \delta_{ij}W) = 0$ [Eq. (8.58) with $S_{ij} = \delta_{ij}$]. For a general square matrix $\mathbf{A}$ of order n, the corresponding equation satisfied by the eigenvalues is

$$\boxed{\det(A_{ij} - \delta_{ij}\lambda) = 0} \qquad \mathbf{(8.82)}$$

Equation (8.82) is called the **characteristic equation** of matrix $\mathbf{A}$. When the nth-order determinant in (8.82) is expanded, it gives a polynomial in λ (called the **characteristic polynomial**) whose highest power is λ^n. The characteristic polynomial has n roots for λ (some of which may be equal to each other and some of which may be imaginary), so a square matrix of order n has n eigenvalues.

The matrix equation (8.80c) for $\mathbf{H}$ corresponds to (8.81) for $\mathbf{A}$. The elements of the eigenvectors of $\mathbf{A}$ satisfy the following set of equations that corresponds to (8.80a):

$$\begin{array}{ccccccccc} A_{11}c_1 & + & A_{12}c_2 & + & \cdots & + & A_{1n}c_n & = & \lambda c_1 \\ \vdots & & \vdots & & \ddots & & \vdots & & \vdots \\ A_{n1}c_1 & + & A_{n2}c_2 & + & \cdots & + & A_{nn}c_n & = & \lambda c_n \end{array} \qquad (8.83)$$

For each different eigenvalue, we have a different set of equations (8.83) and a different set of numbers $c_1, c_2, \ldots, c_n$, giving a different eigenvector.

If all the eigenvalues of a matrix are different, one can show that solving (8.83) leads to n linearly independent eigenvectors (see *Strang*, Section 5.2), where **linear independence** means that no eigenvector can be written as a linear combination of the other eigenvectors. If some eigenvalues are equal, then the matrix may have fewer than n linearly independent eigenvectors. The matrices that occur in quantum mechanics are usually Hermitian (this term is defined later in this section), and a Hermitian matrix of order n always has n linearly independent eigenvectors even if some of its eigenvalues are equal (see *Strang*, Section 5.6 for the proof).

If $\mathbf{A}$ is a diagonal matrix ($a_{ij} = 0$ for $i \neq j$), then the determinant in (8.82) is diagonal. A diagonal determinant equals the product of its diagonal elements [Eq. (8.33)], so the characteristic equation for a diagonal matrix is

$$(a_{11} - \lambda)(a_{12} - \lambda)\cdots(a_{nn} - \lambda) = 0$$

The roots of this equation are $\lambda_1 = a_{11}, \lambda_2 = a_{22}, \ldots, \lambda_n = a_{nn}$. *The eigenvalues of a diagonal matrix are equal to its diagonal elements*. (For the eigenvectors, see Prob. 8.45.)

If $\mathbf{c}$ is an eigenvector of $\mathbf{A}$, then clearly $\mathbf{d} \equiv k\mathbf{c}$ is also an eigenvector of $\mathbf{A}$, where k is any constant. If k is chosen so that

$$\boxed{\sum_{i=1}^{n} |d_i|^2 = 1} \qquad \mathbf{(8.84)}$$

then the column vector $\mathbf{d}$ is said to be **normalized**. Two column vectors $\mathbf{b}$ and $\mathbf{c}$ that each have n elements are said to be **orthogonal** if

$$\boxed{\sum_{i=1}^{n} b_i^* c_i = 0} \qquad \mathbf{(8.85)}$$

Let us denote the n eigenvalues and the corresponding eigenvectors of $\mathbf{H}$ in the variation-method equations (8.80) by $W_1, W_2, \ldots, W_n$ and $\mathbf{c}^{(1)}, \mathbf{c}^{(2)}, \ldots, \mathbf{c}^{(n)}$, so that

$$\mathbf{H}\mathbf{c}^{(i)} = W_i\mathbf{c}^{(i)} \quad \text{for } i = 1, 2, \ldots, n \tag{8.86}$$

where $\mathbf{c}^{(i)}$ is a column vector whose elements are $c_1^{(i)}, \ldots, c_n^{(i)}$ and the basis functions f_i are orthonormal. Furthermore, let $\mathbf{C}$ be the square matrix whose columns are the eigenvectors of $\mathbf{H}$, and let $\mathbf{W}$ be the diagonal matrix whose diagonal elements are the eigenvalues of $\mathbf{H}$:

$$\mathbf{C} = \begin{pmatrix} c_1^{(1)} & c_1^{(2)} & \cdots & c_1^{(n)} \\ c_2^{(1)} & c_2^{(2)} & \cdots & c_2^{(n)} \\ \vdots & \vdots & \ddots & \vdots \\ c_n^{(1)} & c_n^{(2)} & \cdots & c_n^{(n)} \end{pmatrix}, \quad \mathbf{W} = \begin{pmatrix} W_1 & 0 & \cdots & 0 \\ 0 & W_2 & \cdots & 0 \\ \vdots & \vdots & \ddots & \vdots \\ 0 & 0 & \cdots & W_n \end{pmatrix} \tag{8.87}$$

The set of n eigenvalue equations (8.86) can be written as the single equation:

$$\mathbf{HC} = \mathbf{CW} \tag{8.88}$$

To verify the matrix equation (8.88), we show that each element $(\mathbf{HC})_{ij}$ of the matrix $\mathbf{HC}$ equals the corresponding element $(\mathbf{CW})_{ij}$ of $\mathbf{CW}$. The matrix-multiplication rule (7.106) gives $(\mathbf{HC})_{ij} = \sum_k H_{ik}(\mathbf{C})_{kj} = \sum_k H_{ik}c_k^{(j)}$. Consider the eigenvalue equation $\mathbf{HC}^{(j)} = W_j\mathbf{c}^{(j)}$ [Eq. (8.86)]. $\mathbf{Hc}^{(j)}$ and $W_j\mathbf{c}^{(j)}$ are column matrices. Using (7.106) to equate the elements in row i of each of these column matrices, we have $\sum_k H_{ik}c_k^{(j)} = W_jc_i^{(j)}$. Then

$$(\mathbf{HC})_{ij} = \sum_k H_{ik}c_k^{(j)} = W_jc_i^{(j)}, \quad (\mathbf{CW})_{ij} = \sum_k (\mathbf{C})_{ik}(\mathbf{W})_{kj} = \sum_k c_i^{(k)}\delta_{kj}W_k = c_i^{(j)}W_j$$

Hence $(\mathbf{HC})_{ij} = (\mathbf{CW})_{ij}$ and (8.88) is proved.

Provided $\mathbf{C}$ has an inverse (see below), we can multiply each side of (8.88) by $\mathbf{C}^{-1}$ on the left to get $\mathbf{C}^{-1}\mathbf{HC} = \mathbf{C}^{-1}(\mathbf{CW})$. [Since matrix multiplication is not commutative, when we multiply each side of $\mathbf{HC} = \mathbf{CW}$ by $\mathbf{C}^{-1}$, we must put the factor $\mathbf{C}^{-1}$ on the left of $\mathbf{HC}$ and on the left of $\mathbf{CW}$ (or on the right of $\mathbf{HC}$ and the right of $\mathbf{CW}$).] We have $\mathbf{C}^{-1}\mathbf{HC} = \mathbf{C}^{-1}(\mathbf{CW}) = (\mathbf{C}^{-1}\mathbf{C})\mathbf{W} = \mathbf{IW} = \mathbf{W}$:

$$\mathbf{C}^{-1}\mathbf{HC} = \mathbf{W} \tag{8.89}$$

To simplify (8.89), we must learn more about matrices.

A square matrix $\mathbf{B}$ is a **symmetric matrix** if all its elements satisfy $b_{ij} = b_{ji}$. The elements of a symmetric matrix are symmetric about the principal diagonal; for example, $b_{12} = b_{21}$. A square matrix $\mathbf{D}$ is a **Hermitian matrix** if all its elements satisfy $d_{ij} = d_{ji}^*$. For example, if

$$\mathbf{M} = \begin{pmatrix} 2 & 5 & 0 \\ 5 & i & 2i \\ 0 & 2i & 4 \end{pmatrix}, \quad \mathbf{N} = \begin{pmatrix} 6 & 1+2i & 8 \\ 1-2i & -1 & -i \\ 8 & i & 0 \end{pmatrix} \tag{8.90}$$

then $\mathbf{M}$ is symmetric and $\mathbf{N}$ is Hermitian. (Note that the diagonal elements of a Hermitian matrix must be real; $d_{ii} = d_{ii}^*$.) A **real matrix** is one whose elements are all real numbers. A real Hermitian matrix is a symmetric matrix.

The **transpose** $\mathbf{A}^{\mathrm{T}}$ (often written $\widetilde{\mathbf{A}}$) of the matrix $\mathbf{A}$ is the matrix formed by interchanging rows and columns of $\mathbf{A}$ so that column 1 becomes row 1, column 2 becomes row 2, and so on. The elements a_{ij}^{T} of $\mathbf{A}^{\mathrm{T}}$ are related to the elements of $\mathbf{A}$ by $a_{ij}^{\mathrm{T}} = a_{ji}$. For a square matrix, the transpose is found by reflecting the elements about the principal diagonal. A symmetric matrix is equal to its transpose. Thus, for the matrix $\mathbf{M}$ in (8.90), we have $\mathbf{M}^{\mathrm{T}} = \mathbf{M}$.

The **complex conjugate** $\mathbf{A}^*$ of $\mathbf{A}$ is the matrix formed by taking the complex conjugate of each element of $\mathbf{A}$. The **conjugate transpose** $\mathbf{A}^\dagger$ of the matrix $\mathbf{A}$ is formed by taking the transpose of $\mathbf{A}^*$; thus $\mathbf{A}^\dagger = (\mathbf{A}^*)^{\mathrm{T}}$ and

$$a_{ij}^\dagger = (a_{ji})^* \tag{8.91}$$

(Physicists call $\mathbf{A}^\dagger$ the *adjoint* of $\mathbf{A}$, a name that is used by mathematicians to refer to an entirely different matrix.) An example is

$$\mathbf{B} = \begin{pmatrix} 2 & 3+i \\ 0 & 4i \end{pmatrix}, \quad \mathbf{B}^{\mathrm{T}} = \begin{pmatrix} 2 & 0 \\ 3+i & 4i \end{pmatrix}, \quad \mathbf{B}^\dagger = \begin{pmatrix} 2 & 0 \\ 3-i & -4i \end{pmatrix}$$

An **orthogonal matrix** is a square matrix whose inverse is equal to its transpose:

$$\mathbf{A}^{-1} = \mathbf{A}^{\mathrm{T}} \quad \text{if } \mathbf{A} \text{ is orthogonal} \tag{8.92}$$

A **unitary matrix** is one whose inverse is equal to its conjugate transpose:

$$\mathbf{U}^{-1} = \mathbf{U}^\dagger \quad \text{if } \mathbf{U} \text{ is unitary} \tag{8.93}$$

From the definition (8.93), we have $\mathbf{U}^\dagger\mathbf{U} = \mathbf{I}$ if $\mathbf{U}$ is unitary. By equating $(\mathbf{U}^\dagger\mathbf{U})_{ij}$ to $(\mathbf{I})_{ij}$, we find (Prob. 8.42)

$$\sum_k u_{ki}^* u_{kj} = \delta_{ij} \tag{8.94}$$

for columns i and j of a unitary matrix. Thus the columns of a unitary matrix (viewed as column vectors) are orthogonal and normalized (orthonormal), as defined by (8.85) and (8.84). Conversely, if (8.94) is true for all columns, then $\mathbf{U}$ is a unitary matrix. If $\mathbf{U}$ is unitary and real, then $\mathbf{U}^\dagger = \mathbf{U}^{\mathrm{T}}$, and $\mathbf{U}$ is an orthogonal matrix.

One can prove that two eigenvectors of a Hermitian matrix $\mathbf{H}$ that correspond to different eigenvalues are orthogonal (see *Strang*, Section 5.5). For eigenvectors of $\mathbf{H}$ that correspond to the same eigenvalue, one can take linear combinations of them that will be orthogonal eigenvectors of $\mathbf{H}$. Moreover, the elements of an eigenvector can be multiplied by a constant to normalize the eigenvector. Hence, *the eigenvectors of a Hermitian matrix can be chosen to be orthonormal.* If the eigenvectors are chosen to be orthonormal, then the eigenvector matrix $\mathbf{C}$ in (8.87) is a unitary matrix, and $\mathbf{C}^{-1} = \mathbf{C}^\dagger$; Eq. (8.89) then becomes

$$\mathbf{C}^\dagger\mathbf{H}\mathbf{C} = \mathbf{W} \quad \text{if } \mathbf{H} \text{ is Hermitian} \tag{8.95}$$

For the common case that $\mathbf{H}$ is real as well as Hermitian (that is, $\mathbf{H}$ is real and symmetric), the c's in (8.80a) are real (since W and the H_{ij}'s are real) and $\mathbf{C}$ is real as well as unitary; that is, $\mathbf{C}$ is orthogonal, with $\mathbf{C}^{-1} = \mathbf{C}^{\mathrm{T}}$; Eq. (8.95) becomes

$$\mathbf{C}^{\mathrm{T}}\mathbf{H}\mathbf{C} = \mathbf{W} \quad \text{if } \mathbf{H} \text{ is real and symmetric} \tag{8.96}$$

The eigenvalues of a Hermitian matrix can be proven to be real numbers (*Strang*, Section 5.5).

EXAMPLE Find the eigenvalues and normalized eigenvectors of the Hermitian matrix

$$\mathbf{A} = \begin{pmatrix} 3 & 2i \\ -2i & 0 \end{pmatrix}$$

by solving algebraic equations. Then verify that $\mathbf{C}^{\dagger}\mathbf{A}\mathbf{C}$ is diagonal, where $\mathbf{C}$ is the eigenvector matrix.

The characteristic equation (8.82) for the eigenvalues λ is $\det(a_{ij} - \delta_{ij}\lambda) = 0$, which becomes

$$\begin{vmatrix} 3 - \lambda & 2i \\ -2i & -\lambda \end{vmatrix} = 0$$

$$\lambda^2 - 3\lambda - 4 = 0$$

$$\lambda_1 = 4, \quad \lambda_2 = -1$$

A useful theorem in checking eigenvalue calculations is the following (*Strang*, Exercise 5.1.9): The sum of the diagonal elements of a square matrix $\mathbf{A}$ of order n is equal to the sum of the eigenvalues λ_i of $\mathbf{A}$; that is, $\sum_{i=1}^{n} a_{ii} = \sum_{i=1}^{n} \lambda_i$. In this example, $\sum_i a_{ii} = 3 + 0 = 3$,which equals the sum $4 - 1 = 3$ of the eigenvalues.

For the root $\lambda_1 = 4$, the set of simultaneous equations (8.83) is

$$(3 - \lambda_1)c_1^{(1)} + 2ic_2^{(1)} = 0$$

$$-2ic_1^{(1)} - \lambda_1 c_2^{(1)} = 0$$

or

$$-c_1^{(1)} + 2ic_2^{(1)} = 0$$

$$-2ic_1^{(1)} - 4c_2^{(1)} = 0$$

Discarding either one of these equations, we find

$$c_1^{(1)} = 2ic_2^{(1)}$$

Normalization gives

$$1 = |c_1^{(1)}|^2 + |c_2^{(1)}|^2 = 4|c_2^{(1)}|^2 + |c_2^{(1)}|^2$$

$$|c_2^{(1)}| = 1/\sqrt{5}\,, \quad c_2^{(1)} = 1/\sqrt{5}$$

$$c_1^{(1)} = 2ic_2^{(1)} = 2i/\sqrt{5}$$

where the phase of $c_2^{(1)}$ was chosen to be zero.

Similarly, we find for $\lambda_2 = -1$ (Prob. 8.48)

$$c_1^{(2)} = -i/\sqrt{5}, \quad c_2^{(2)} = 2/\sqrt{5}$$

The normalized eigenvectors are then

$$\mathbf{c}^{(1)} = \begin{pmatrix} 2i/\sqrt{5} \\ 1/\sqrt{5} \end{pmatrix}, \qquad \mathbf{c}^{(2)} = \begin{pmatrix} -i/\sqrt{5} \\ 2/\sqrt{5} \end{pmatrix}$$

Because the eigenvalues λ_1 and λ_2 of the Hermitian matrix **A** differ, $\mathbf{c}^{(1)}$ and $\mathbf{c}^{(2)}$ are orthogonal (as the reader should verify). Also, $\mathbf{c}^{(1)}$ and $\mathbf{c}^{(2)}$ are normalized. Therefore, **C** is unitary and $\mathbf{C}^{-1} = \mathbf{C}^{\dagger}$. Forming **C** and its conjugate transpose, we have

$$\mathbf{C}^{-1}\mathbf{A}\mathbf{C} = \mathbf{C}^{\dagger}\mathbf{A}\mathbf{C} = \begin{pmatrix} -2i/\sqrt{5} & 1/\sqrt{5} \\ i/\sqrt{5} & 2/\sqrt{5} \end{pmatrix}\begin{pmatrix} 3 & 2i \\ -2i & 0 \end{pmatrix}\begin{pmatrix} 2i/\sqrt{5} & -i/\sqrt{5} \\ 1/\sqrt{5} & 2/\sqrt{5} \end{pmatrix}$$

$$= \begin{pmatrix} -2i/\sqrt{5} & 1/\sqrt{5} \\ i/\sqrt{5} & 2/\sqrt{5} \end{pmatrix}\begin{pmatrix} 8i/\sqrt{5} & i/\sqrt{5} \\ 4/\sqrt{5} & -2/\sqrt{5} \end{pmatrix} = \begin{pmatrix} 4 & 0 \\ 0 & -1 \end{pmatrix}$$

which is the diagonal matrix of eigenvectors.

We have shown that if **H** is a real symmetric matrix with eigenvalues W_i and orthonormal eigenvectors $\mathbf{c}^{(i)}$ (that is, if $\mathbf{H}\mathbf{c}^{(i)} = W_i\mathbf{c}^{(i)}$ for $i = 1, 2, \ldots, n$), then $\mathbf{C}^{\mathrm{T}}\mathbf{H}\mathbf{C} = \mathbf{W}$ [Eq. (8.96)], where **C** is the real orthogonal matrix whose columns are the eigenvectors $\mathbf{c}^{(i)}$ and **W** is the diagonal matrix of eigenvalues W_i. The converse of this theorem is readily proved; that is, if **H** is a real symmetric matrix, **B** is a real orthogonal matrix, and $\mathbf{B}^{\mathrm{T}}\mathbf{H}\mathbf{B}$ equals a diagonal matrix $\mathbf{\Lambda}$, then the columns of **B** are the eigenvectors of **H** and the diagonal elements of $\mathbf{\Lambda}$ are the eigenvalues of **H**.

To find the eigenvalues and eigenvectors of a Hermitian matrix of order n, we can use either of the following procedures: (1) Solve the characteristic equation $\det(H_{ij} - \delta_{ij}W) = 0$ [Eq. (8.82)] for the eigenvalues $W_1, \ldots, W_n$. Then substitute each W_k into the set of algebraic equations (8.80a) and solve for the elements $c_1^{(k)}, \ldots, c_n^{(k)}$ of the kth eigenvector. (2) Search for a unitary matrix **C** such that $\mathbf{C}^{\dagger}\mathbf{H}\mathbf{C}$ is a diagonal matrix. The diagonal elements of $\mathbf{C}^{\dagger}\mathbf{H}\mathbf{C}$ are the eigenvalues of **H**, and the columns of **C** are the orthonormal eigenvectors of **H**. For the large matrices that occur in quantum chemistry, procedure (2) (called **matrix diagonalization**) is computationally much faster than (1).

One reason that expanding the characteristic determinant and solving the characteristic equation is not a good way to find the eigenvalues of large matrices is that, for large matrices, a very small change in a coefficient in the characteristic polynomial may produce a large change in the eigenvalues (see Prob. 8.53). Hence we might have to calculate the coefficients in the characteristic polynomial to hundreds or thousands of decimal places in order to get eigenvalues accurate to a few decimal places. Although it is true that for certain matrices a tiny change in the value of an element of that matrix might produce large changes in the eigenvalues, one can prove that for Hermitian matrices, a small change in a matrix element always produces only small changes in the eigenvalues. Hence method (2) of the preceding paragraph is the correct way to get accurate eigenvalues.

A systematic way to diagonalize a real symmetric matrix **H** is as follows. Construct an orthogonal matrix $\mathbf{O}_1$ such that the matrix $\mathbf{H}_1 \equiv \mathbf{O}_1^{\mathrm{T}}\mathbf{H}\mathbf{O}_1$ has zero in place of the off-diagonal elements H_{12} and H_{21} of **H**. (Because **H** is symmetric, we have $H_{12} = H_{21}$. Also, the transformed matrices $\mathbf{H}_1, \mathbf{H}_2, \ldots$ are symmetric.) Then construct an orthogonal matrix $\mathbf{O}_2$ such that $\mathbf{H}_2 \equiv \mathbf{O}_2^{\mathrm{T}}\mathbf{H}_1\mathbf{O}_2 = \mathbf{O}_2^{\mathrm{T}}\mathbf{O}_1^{\mathrm{T}}\mathbf{H}\mathbf{O}_1\mathbf{O}_2$ has zeros in place of the elements $(\mathbf{H}_1)_{13}$ and $(\mathbf{H}_1)_{31}$ of $\mathbf{H}_1$; and so on. Unfortunately, when a given pair of off-diagonal elements is made zero in a step, some off-diagonal elements made zero in a previous step are likely to become nonzero, so one has to go back and recycle through the off-diagonal elements over and over again. Generally an infinite number of steps are required to make all off-diagonal elements equal to zero. In practice, one skips a step if the absolute value of the off-diagonal elements to be zeroed in that step is less than some tiny number, and one stops the procedure when the absolute values of all off-diagonal elements are less than some tiny number. The eigenvalues are then the diagonal elements of the transformed matrix $\cdots \mathbf{O}_3^{\mathrm{T}}\mathbf{O}_2^{\mathrm{T}}\mathbf{O}_1^{\mathrm{T}}\mathbf{H}\mathbf{O}_1\mathbf{O}_2\mathbf{O}_3\cdots$, and the eigenvector matrix is the product $\mathbf{O}_1\mathbf{O}_2\mathbf{O}_3\cdots$. This method (the *cyclic Jacobi method*) is not very efficient for large matrices when run on a serial computer but is efficient on a parallel computer.

More-efficient approaches to diagonalize real, symmetric matrices than the Jacobi method begin by carrying out a series of orthogonal transformations to reduce the original matrix $\mathbf{H}$ to a symmetric tridiagonal matrix $\mathbf{T}$. A *tridiagonal matrix* is one whose elements are all zero except for those on the principal diagonal (elements t_{ii}) and those on the diagonals immediately above and immediately below the principal diagonal (elements $t_{i-1,i}$ and $t_{i+1,i}$, respectively). The relation between $\mathbf{T}$ and $\mathbf{H}$ is $\mathbf{T} = \mathbf{O}^{\mathrm{T}}\mathbf{H}\mathbf{O}$, where $\mathbf{O}$ is a real orthogonal matrix that is the product of the orthogonal matrices used in the individual steps of going from $\mathbf{H}$ to $\mathbf{T}$. Two efficient methods of transforming $\mathbf{H}$ to tridiagonal form are due to Givens and to Householder. An efficient method to find the eigenvalues of a symmetric tridiagonal matrix is the $\mathbf{QR}$ method. Here, $\mathbf{T}$ is expressed as the product of an orthogonal matrix $\mathbf{Q}$ and an upper triangular matrix $\mathbf{R}$ (one whose elements below the principal diagonal are all zero). A series of iterative steps yields matrices converging to a diagonal matrix whose diagonal elements are the eigenvalues of $\mathbf{T}$, which equal the eigenvalues of $\mathbf{H}$ (Prob. 8.54). With certain refinements, the $\mathbf{QR}$ method is a very efficient way to find eigenvalues and eigenvectors (see *Strang*, Sections 5.3 and 7.3 for details).

Details of matrix diagonalization procedures and computer programs are given in *Press et al.*, Chapter 11; *Acton,* Chapters 8 and 13; *Shoup,* Chapter 4.

A major compilation of procedures and computer programs for scientific and engineering calculations is *Press et al*; the text of older editions of this book is available free on the Internet at www.nr.com/. For comments on older editions of this book, see amath.colorado.edu/computing/Fortran/numrec.html.

Programs for mathematical and scientific calculations can be found at www.netlib.org and at gams.nist.gov. Downloadable free personal-computer mathematical software and demonstration software for such commercial programs as Mathcad and Maple can be found at archives.math.utk.edu.

The procedure for using matrix algebra to solve the linear variation equations when nonorthonormal basis functions are used is outlined in Prob. 8.56.

The Excel spreadsheet can be used to find eigenvalues and eigenvectors; see Prob. 8.52.

Computer algebra systems such as Mathcad and some electronic calculators have built-in commands to easily find eigenvalues and eigenvectors.

The methods for finding matrix eigenvalues and eigenvectors discussed in this section are useful for matrices of order up to 10^3. Special methods are used to find the lowest few eigenvalues and corresponding eigenvectors of matrices of order up to 10^9 that occur in certain quantum-chemistry calculations (see Section 16.1).

As noted after Eq. (8.81), for the linear variation function $\sum_{i=1}^{n} c_i f_i$ with orthonormal basis functions f_i, the eigenvalues of the matrix $\mathbf{H}$ formed from the matrix elements $\langle f_j|\hat{H}|f_k\rangle$ are the roots of the secular equation and the eigenvector corresponding to the eigenvalue W_m gives the coefficients in the variation function that corresponds to

W_m. Problems 8.59 to 8.64 apply the linear variation method to problems such as the double well and the harmonic oscillator using particle-in-a-box wave functions as basis functions and using a computer algebra program such as Mathcad to find the eigenvalues and eigenvectors of the **H** matrix.

We have discussed matrix diagonalization in the context of the linear variation method. However, *finding the eigenvalues a_k and eigenfunctions g_k of any Hermitian operator $\hat{A}$ ($\hat{A}g_k = a_k g_k$) can be formulated as a matrix-diagonalization problem.* If we choose a complete, orthonormal basis set $\{f_i\}$ and expand the eigenfunctions as $g_k = \sum_i c_i^{(k)} f_i$, then (Prob. 8.58) the eigenvalues of the matrix **A** whose elements are $a_{ij} = \langle f_i|\hat{A}|f_j\rangle$ are the eigenvalues of the operator $\hat{A}$, and the elements $c_i^{(k)}$ of the eigenvectors $\mathbf{c}^{(k)}$ of **A** give the coefficients in the expansions of the eigenfunctions g_k.

The material of this section further emphasizes the correspondence between linear operators and matrices and the correspondence between functions and column vectors (Section 7.10).

8.7 SUMMARY

The variation theorem states that for a system with time-independent Hamiltonian $\hat{H}$ and ground-state energy E_1, we have $\int \phi^* \hat{H} \phi \, d\tau / \int \phi^* \phi \, d\tau \geq E_1$, where ϕ is any well-behaved function that obeys the boundary conditions of the problem. The variation theorem allows us to obtain approximations for the ground-state energy and wave function.

The mathematics of determinants was reviewed in Section 8.3. A set of n linear homogeneous equations in n unknowns has a nontrivial solution if and only if the determinant of the coefficients is equal to zero.

For the linear variation function $\phi = \sum_{i=1}^{n} c_i f_i$, variation of the coefficients c_i to minimize the variational integral W leads to the secular equation $\det(H_{ij} - S_{ij}W) = 0$, whose roots $W_1, \ldots, W_n$ are upper bounds for the n lowest bound-state energy eigenvalues; here, $H_{ij} \equiv \langle f_i|\hat{H}|f_j\rangle$ and $S_{ij} \equiv \langle f_i|f_j\rangle$. Substitution of $W_1, \ldots, W_n$ one at a time into the simultaneous homogeneous equations (8.56) allows the coefficients c_i that correspond to each W to be found.

The eigenvalues λ_i and eigenvectors $\mathbf{c}_i$ of a square matrix **A** satisfy $\mathbf{A}\mathbf{c}_i = \lambda_i \mathbf{c}_i$. For a Hermitian matrix **H**, the eigenvector matrix **C** (whose columns are the eigenvectors of **H**) is unitary (meaning that its inverse equals its conjugate transpose $\mathbf{C}^\dagger$), and $\mathbf{C}^\dagger\mathbf{H}\mathbf{C}$ equals the diagonal matrix whose diagonal elements are the eigenvalues of **H**. For a real symmetric matrix, the eigenvector matrix is orthogonal. For the linear variation function $\phi = \sum_i c_i f_i$ formed from orthonormal functions f_i, each set of optimized coefficients c_i is an eigenvector of the matrix whose elements are $\langle f_i|\hat{H}|f_j\rangle$ and the corresponding values of the variational integral are the eigenvalues of this matrix.

PROBLEMS

Sec.	8.1	8.2	8.3	8.4	8.5	8.6	general
Probs.	8.1–8.15	8.16–8.17	8.18–8.21	8.22–8.27	8.28–8.39	8.40–8.55	8.56–8.65

8.1 Calculations on the Li atom with three well-behaved variation functions gave the following values of the variational integral: −203.2 eV, −192.0 eV, and −201.2 eV. Therefore, the true ground-state energy of Li must be ($\leq$ or $\geq$) than ? eV. (Choose one of the two inequality signs and replace the ? with a number.)

8.2 (a) Consider a one-particle, one-dimensional system with potential energy

$$V = V_0 \quad \text{for } \tfrac{1}{4}l \leq x \leq \tfrac{3}{4}l, \qquad V = 0 \quad \text{for } 0 \leq x \leq \tfrac{1}{4}l \text{ and } \tfrac{3}{4}l \leq x \leq l$$

and $V = \infty$ elsewhere (where V_0 is a constant). Plot V versus x. Use the trial variation function $\phi_1 = (2/l)^{1/2}\sin(\pi x/l)$ for $0 \leq x \leq l$ to estimate the ground-state energy for $V_0 = \hbar^2/ml^2$ and compare with the true ground-state energy $E = 5.750345\hbar^2/ml^2$. To save time in evaluating integrals, note that $\langle\phi_1|\hat{H}|\phi_1\rangle = \langle\phi_1|\hat{T}|\phi_1\rangle + \langle\phi_1|\hat{V}|\phi_1\rangle$ and explain why $\langle\phi_1|\hat{T}|\phi_1\rangle$ equals the particle-in-a-box ground-state energy $h^2/8ml^2$. (b) For this system, use the variation function $\phi_2 = x(l - x)$. To save time, note that $\langle\phi_2|\hat{T}|\phi_2\rangle$ is given by Eq. (8.12). (Why?)

8.3 Verify the result for $\langle\phi|\hat{H}|\phi\rangle$ in the last example of Section 8.1.

8.4 If the normalized variation function $\phi = (3/l^3)^{1/2}x$ for $0 \leq x \leq l$ is applied to the particle-in-a-one-dimensional-box problem, one finds that the variational integral equals zero, which is *less* than the true ground-state energy. What is wrong?

8.5 For a particle in a three-dimensional box with sides of length a, b, c, write down the variation function that is the three-dimensional extension of the variation function $\phi = x(l - x)$ used in Section 8.1 for the particle in a one-dimensional box. Use the integrals in Eqs. (8.12) and (8.13) to evaluate the variational integral for the three-dimensional case. Find the percent error in the ground-state energy.

8.6 (a) A particle in a spherical box of radius b has $V = 0$ for $0 \leq r \leq b$ and $V = \infty$ for $r > b$. Use the trial function $\phi = b - r$ for $0 \leq r \leq b$ and $\phi = 0$ for $r > b$ to estimate the ground-state energy and compare with the true value $h^2/8mb^2$ (Prob. 6.2). (b) Devise another simple variation function that obeys the boundary conditions for this problem and work out the percent error in the ground-state energy given by your function.

8.7 Application of the variation function $\phi = e^{-cx^2}$ (where c is a variational parameter) to a problem with $V = af(x)$, where a is a positive constant and $f(x)$ is a certain function of x, gives the variational integral as $W = c\hbar^2/2m + 15a/64c^3$. Find the minimum value of W (in terms of a and physical constants) for this variation function.

8.8 Consider the variation functions $\phi_1 = af + bg$ and $\phi_2 = f + cg$, where f and g in ϕ_1 are the same functions as f and g in ϕ_2, and a, b, and c are parameters whose values are chosen to minimize the variational integral W. Explain why ϕ_1 and ϕ_2 will give the same result for W when applied to the same problem.

8.9 Apply the variation function $\phi = e^{-cr}$ to the hydrogen atom. Choose the parameter c (which is real) to minimize the variational integral, and calculate the percent error in the ground-state energy.

8.10 A one-dimensional quartic oscillator has $V = cx^4$, where c is a constant. Devise a variation function with a parameter for this problem, and find the optimum value of the parameter to minimize the variational integral and estimate the ground-state energy in terms of c. Compare with the Numerov-method ground-state energy found in Prob. 4.31.

8.11 For a particle in a box of length l, use the variation function $\phi = x^k(l - x)^k$ for $0 \leq x \leq l$. You will need the integral

$$\int_0^l x^s(l - x)^t\,dx = l^{s+t+1}\frac{\Gamma(s + 1)\Gamma(t + 1)}{\Gamma(s + t + 2)}$$

where the gamma function obeys the relation $\Gamma(z + 1) = z\Gamma(z)$. The definition of the gamma function $\Gamma(z)$ need not concern you, since the gamma functions will ultimately cancel. (a) Show that the variational integral equals $(\hbar^2/ml^2)(4k^2 + k)/(2k - 1)$. (b) Find the optimum value of k and calculate the percent error in the ground-state energy for this k.

8.12 Consider a one-particle, one-dimensional system with $V = 0$ for $0 \leq x \leq l$ and $V = V_0$ elsewhere (Fig. 2.5). (a) Use the variation function $\phi = \sin[\pi(x + c)/(l + 2c)]$ for $-c \leq x \leq l + c$ and $\phi = 0$ elsewhere, where c is a positive variational parameter. Sketch ϕ and V on the same plot. Choose c to minimize the variational integral W and find the expression for W. (*Hint*: Use a simple substitution to put the integrals in the form of integrals in the Appendix.) (b) Find W for $V_0 = 20\hbar^2/ml^2$ and compare with the true ground-state energy $2.814\hbar^2/ml^2$ (Prob. 4.30c).

8.13 Prove that, for a system with a nondegenerate ground state, $\int \phi^* \hat{H} \phi \, d\tau > E_1$, if ϕ is any normalized, well-behaved function that is not equal to the true ground-state wave function. *Hint:* Let b be a positive constant such that $E_1 + b < E_2$. Turn (8.4) into an inequality by replacing all E_k's except E_1 with $E_1 + b$. (The notation $E_1, E_2, \ldots$, is as in Section 8.2.)

8.14 In 1971 a paper was published that applied the normalized variation function $N \exp(-br^2/a_0^2 - cr/a_0)$ to the hydrogen atom and stated that minimization of the variational integral with respect to the parameters b and c yielded an energy 0.7% above the true ground-state energy for infinite nuclear mass. Without doing any calculations, state why this result must be in error.

8.15 (a) Use the triangular function (7.35) as a variation function for the ground state of the particle in a box. Note that $f''(x)$ is infinite at $x = \frac{1}{2}l$ because of the discontinuity in $f'(x)$ at this point. Therefore, in evaluating $\int f^* \hat{H} f \, dx$, we run into difficulty in evaluating the integral of ff''. One way around this problem is to first show that

$$\int_0^l ff'' \, dx = -\int_0^l (f')^2 \, dx = -\int_0^{l/2} (f')^2 \, dx - \int_{l/2}^l (f')^2 \, dx \tag{8.97}$$

for any function obeying the boundary conditions. Then, using the expression on the right of (8.97), we can calculate the variational integral. Prove (8.97), and then calculate the percent error in the ground-state energy, using this triangular function. Note that there is no parameter in this trial function. [If you are ambitious, try this alternative procedure: Note that $f'(x)$ involves the Heaviside step function (Section 7.7), and therefore $f''(x)$ involves the Dirac delta function. Use the properties of the delta function to evaluate $\int_0^l ff'' \, dx$, and find the percent error using this triangular function.] (b) The variation function ϕ of the first example in Section 8.1 has discontinuities in ϕ' at $x = 0$ and $x = l$, so, strictly speaking, we should use one of the procedures of part (a) of this problem to evaluate $\langle \phi | \hat{H} | \phi \rangle$. Do this and show that the same value of $\langle \phi | \hat{H} | \phi \rangle$ is obtained.

8.16 (a) For the ground state of the hydrogen atom, use the Gaussian trial function $\phi = \exp(-cr^2/a_0^2)$. Find the optimum value of c and the percent error in the energy. (Gaussian variational functions are widely used in molecular quantum mechanics; see Section 15.4.) (b) Multiply the function in (a) by the spherical harmonic Y_2^0, and then minimize the variational integral. This yields an upper bound to the energy of which hydrogen-atom state?

8.17 For the system of Prob. 8.2, explain why the variation function $\phi = (2/l)^{1/2} \sin(2\pi x/l)$ gives an upper bound to the energy E_2 of the first excited state. (*Hint*: With the origin at the center of the box, V is an even function.) Use this ϕ to evaluate the variation integral for $V_0 = \hbar^2/ml^2$ and compare with the true value $E_2 = 20.23604\hbar^2/ml^2$.

8.18 Evaluate

$$\text{(a)} \quad \begin{vmatrix} 3 & 1 & i \\ -2 & 4 & 0 \\ 5 & 7 & \frac{1}{2} \end{vmatrix} \qquad \text{(b)} \quad \begin{vmatrix} 2 & 5 & 1 & 3 \\ 8 & 0 & 4 & -1 \\ 6 & 6 & 6 & 1 \\ 5 & -2 & -2 & 2 \end{vmatrix}$$

8.19 (a) Prove that the value of a determinant all of whose elements below the principal diagonal are zero is equal to the product of the diagonal elements. (b) How many terms are there in the expansion of an nth-order determinant?

8.20 Verify the block-diagonal determinant equation (8.34).

8.21 (a) Consider some permutation of the integers 1, 2, 3, ..., n. The permutation is an *even permutation* if an even number of interchanges of pairs of integers restores the permutation to the natural order 1, 2, 3, ..., n. An *odd permutation* requires an odd number of interchanges of pairs to reach the natural order. For example, the permutation 3124 is even, since two interchanges restore it to the natural order: 3124 → 1324 → 1234. Write down and classify (even or odd) all permutations of 123. (b) Verify that the definition (8.24) of the third-order determinant is equivalent to

$$\begin{vmatrix} a_{11} & a_{12} & a_{13} \\ a_{21} & a_{22} & a_{23} \\ a_{31} & a_{32} & a_{33} \end{vmatrix} = \sum (\pm 1) a_{1i} a_{2j} a_{3k}$$

where ijk is one of the permutations of the integers 123, the sum is over the 3! different permutations of these integers, and the sign of each term is plus or minus, depending on whether the permutation is even or odd. (c) How would we define the nth-order determinant using this type of definition?

8.22 Use Gaussian elimination with detached coefficients to solve

$$\begin{aligned} 2x_1 - x_2 + 4x_3 + 2x_4 &= 16 \\ 3x_1 \qquad - x_3 + 4x_4 &= -5 \\ 2x_1 + x_2 + x_3 - 2x_4 &= 8 \\ -4x_1 + 6x_2 + 2x_3 + x_4 &= 3 \end{aligned}$$

8.23 When Gaussian elimination is programmed for a computer (or done on an electronic calculator), one should use exchanges of equations to avoid dividing by a coefficient that is much smaller in absolute value than the other coefficients in the equations. Explain why division by an extremely small coefficient can lead to large errors in the results.

8.24 Gaussian elimination can be used to efficiently evaluate a determinant, as follows. Divide each element of row 1 of the determinant (8.20) by a_{11} and place the factor a_{11} in front of the determinant (Theorem IV of Section 8.3). Then subtract the appropriate multiples of the row 1 elements from row 2, row 3, ..., row n to make $a_{21}, a_{31}, \ldots, a_{n1}$ zero (Theorem V). Then divide the second-row elements by the current value of a_{22} and insert the factor a_{22} in front of the determinant, and so on. Ultimately, we get a determinant all of whose elements below the principal diagonal are zero. From Prob. 8.19, this determinant equals the product of its diagonal elements. Use this procedure to evaluate the determinant in Prob. 8.18b.

8.25 Solve each set of simultaneous equations, where $i = \sqrt{-1}$:

$$\text{(a)} \quad \begin{aligned} 8x - 15y &= 0 \\ -3x + 4y &= 0 \end{aligned} \qquad \text{(b)} \quad \begin{aligned} -4x + 3iy &= 0 \\ 5ix + \tfrac{15}{4} y &= 0 \end{aligned}$$

8.26 Solve each set of simultaneous equations using Gauss–Jordan elimination with detached coefficients or some other method:

$$\text{(a)}\quad \begin{aligned} x + 2y + 3z &= 0 \\ 3x + y + 2z &= 0 \\ 2x + 3y + z &= 0 \end{aligned} \qquad \text{(b)}\quad \begin{aligned} x + 2y + 3z &= 0 \\ x - y + z &= 0 \\ 7x - y + 11z &= 0 \end{aligned}$$

8.27 Write a computer program that uses Gaussian elimination to solve a system of n linear, simultaneous, inhomogeneous equations in n unknowns, where $n \le 10$. Test it on a couple of examples.

8.28 Consider the linear variation function $\phi = \sum_i c_i f_i$. State whether each of the following statements about the basis functions f_i is true or false. (a) Each f_i must be normalized. (b) Each f_i must be well behaved. (c) The functions $\{f_i\}$ must be orthogonal to one another. (d) Each f_i must be an eigenfunction of the system's Hamiltonian operator.

8.29 When the variation function $\phi = c_1 f_1 + c_2 f_2$ is applied to a certain quantum-mechanical problem, one finds $\langle f_1|\hat{H}|f_1\rangle = 4a$, $\langle f_1|\hat{H}|f_2\rangle = a$, $\langle f_2|\hat{H}|f_2\rangle = 6a$, $\langle f_1|f_1\rangle = 2b$, $\langle f_2|f_2\rangle = 3b$, $\langle f_1|f_2\rangle = b$, where a and b are known positive constants. Use this ϕ to find (in terms of a and b) upper bounds to the lowest two energies, and for each W, find c_1 and c_2 for the normalized ϕ.

8.30 Solve the second-order secular equation (8.57) for the special case where $H_{11} = H_{22}$ and $S_{11} = S_{22}$. (*Reminder:* f_1 and f_2 are real functions.) Then solve for c_1/c_2 for each of the two roots W_1 and W_2.

8.31 For the system of Prob. 8.2 with $V_0 = \hbar^2/ml^2$ (a particle in a box with a rectangular bump in the middle), consider the linear variation function

$$\phi = c_1 f_1 + c_2 f_2 = c_1(2/l)^{1/2}\sin(\pi x/l) + c_2(2/l)^{1/2}\sin(3\pi x/l)$$

for $0 \le x \le l$. (a) Explain why this variation function will give upper bounds to the energies E_1 and E_3 in (8.61). (b) Explain why f_1 and f_2 are orthonormal. (c) Note that $\langle f_i|\hat{H}|f_j\rangle = \langle f_i|\hat{T}|f_j\rangle + \langle f_i|V|f_j\rangle$. Explain why $\hat{T}f_j = \varepsilon_j f_j$ (for $0 \le x \le l$), and give the $\hat{T}$ eigenvalues ε_1 and ε_2. Show that $\langle f_i|\hat{T}|f_j\rangle = \delta_{ij}\varepsilon_j$. (d) Using the results of parts (b) and (c) to help evaluate the integrals, set up and solve the secular equation. Then find the variation functions ϕ_1 and ϕ_2 that correspond to W_1 and W_2. Compare W_1 and W_2 with the true energies $E_1 = 5.750345\hbar^2/ml^2$ and $E_3 = 44.808373\hbar^2/ml^2$. Compare W_1 with the value $W = 5.753112\hbar^2/ml^2$ found in Prob. 8.2 using $\phi = f_1$. (e) If we want to improve on the results of part (d) by using a three-term linear variation function, what would be a logical choice for f_1, f_2, and f_3?

8.32 Apply the linear variation function

$$\phi = c_1 x^2(l - x) + c_2 x(l - x)^2, \qquad 0 \le x \le l$$

to the particle in a one-dimensional box. (See Prob. 8.30.) Calculate the percent errors for the $n = 1$ and $n = 2$ energies. Sketch $x^2(l - x)$, $x(l - x)^2$ and the two approximate wave functions you find. (To help sketch the functions, find the nodes and the maxima and minima of each function.)

8.33 Show that if the change of variable $x' \equiv x - \frac{1}{2}l$ (corresponding to shifting the origin to the center of the box) is made in the functions (8.63), then f_1 and f_2 are even functions of x' and f_3 and f_4 are odd.

8.34 Verify the values given for H_{12}, H_{22}, S_{12}, and S_{22} in the example in Section 8.5.

8.35 Derive the roots (8.73) of the odd-function secular equation in the Section 8.5 example.

8.36 Use a spreadsheet or a program like Mathcad to plot the percent deviation of the variation function (8.75) from the true ground-state wave function versus x/l.

8.37 Derive the functions ϕ_2 and ϕ_3 in Eq. (8.76) of the Section 8.5 example.

8.38 Let the variation function of (8.40) be complex. Then $c_j = a_j + ib_j$, where a_j and b_j are real numbers. There are $2n$ parameters to be varied, namely the a_j's and b_j's. (a) Use the chain rule to show that the minimization conditions $\partial W/\partial a_i = 0, \partial W/\partial b_i = 0$ are equivalent to the conditions $\partial W/\partial c_i = 0, \partial W/\partial c_i^* = 0$. (b) Show that minimization of W leads to Eq. (8.54) and its complex conjugate, which may be discarded. Hence Eqs. (8.54) and (8.59) are valid for complex variation functions.

8.39 We wish to prove that the approximate wave functions obtained in the linear variation method are orthogonal and that the approximate energies obtained are upper bounds to the energies of the n lowest states. Let the approximate function ϕ_α have the value W_α for the variational integral and the coefficients $c_j^{(\alpha)}$ in (8.40). (We add α to distinguish the n different ϕ's.) We rewrite (8.54) as

$$\sum_k [(\langle f_i|\hat{H}|f_k\rangle - \langle f_i|f_k\rangle W_\alpha)c_k^{(\alpha)}] = 0, \quad i = 1, \dots, n \tag{8.99}$$

(a) Show that $\langle f_i|\hat{H} - W_\alpha|\phi_\alpha\rangle = 0$ by showing this integral to equal the left side of (8.99). (b) Use the result of (a) to show $\langle \phi_\beta|\hat{H} - W_\alpha|\phi_\alpha\rangle = 0$ and $\langle \phi_\alpha|\hat{H} - W_\beta|\phi_\beta\rangle^* = 0$ for all α and β. (c) Equate the two integrals in (b), and use the Hermitian property of $\hat{H}$ to show that $\langle \phi_\beta|\phi_\alpha\rangle(W_\beta - W_\alpha) = 0$. We conclude that, for $W_\alpha \neq W_\beta$, ϕ_α and ϕ_β are orthogonal. (For $W_\alpha = W_\beta$, we can form orthogonal linear combinations of ϕ_α and ϕ_β that will have the same value for the variational integral.) (d) Let $\phi_1, \phi_2, \dots, \phi_n$ be the normalized approximate wave functions found in the variation method, where the functions are listed in order of increasing value of the variational integral. Consider the function $g = \sum_{\alpha=1}^{m} b_\alpha \phi_\alpha$, where $m \le n$ and the coefficients b_α are chosen so that $\langle g|\psi_1\rangle = 0, \langle g|\psi_2\rangle = 0, \dots, \langle g|\psi_{m-1}\rangle = 0$ and so that g is normalized. Here $\psi_1, \psi_2, \dots,$ are the true wave functions of the lowest-energy states. Explain why $\langle g|\hat{H}|g\rangle \ge E_m$. (See Section 8.2.) (e) Use the results of parts (b) and (c) to show that $\langle \phi_\alpha|\hat{H}|\phi_\beta\rangle = 0$ for $\alpha \neq \beta$. (f) Use the result of (e) to show that $\langle g|\hat{H}|g\rangle = \sum_{\alpha=1}^{m} |b_\alpha|^2 W_\alpha$. (g) Use the orthonormality of the ϕ's to show that $\sum_{\alpha=1}^{m} |b_\alpha|^2 = 1$. (h) Use the results of (f) and (g) to show that $\langle g|\hat{H}|g\rangle \le W_m$. *Hint:* See Eq. (8.5). (i) Combine (h) and (d) to give the desired result: $W_m \ge E_m$ for $m = 1, 2, \dots, n$.

8.40 Find $\mathbf{A}^*$, $\mathbf{A}^{\mathrm{T}}$, and $\mathbf{A}^\dagger$ if

$$\mathbf{A} = \begin{pmatrix} 7 & 3 & 0 \\ 2 - i & 2i & i \\ 1 + i & 4 & 2 \end{pmatrix}$$

8.41 (a) Which of the following matrices are real? (b) Which are symmetric? (c) Which are Hermitian?

$$\mathbf{B} = \begin{pmatrix} i & 2 + i \\ 2 - i & 3 \end{pmatrix} \quad \mathbf{C} = \begin{pmatrix} i & 2 + i \\ 2 + i & -1 \end{pmatrix} \quad \mathbf{D} = \begin{pmatrix} 3 & 2 - i \\ 2 + i & -1 \end{pmatrix} \quad \mathbf{F} = \begin{pmatrix} 5 & 3 \\ 3 & -1 \end{pmatrix}$$

8.42 Verify the orthonormality equation (8.94) for the column vectors of a unitary matrix.

8.43 If the functions v and w are normalized and orthogonal, and if v and w are expanded in terms of the complete, orthonormal set $\{f_i\}$ as $v = \sum_i v_i f_i$ and $w = \sum_i w_i f_i$ (where the expansion coefficients v_i and w_i are constants), show that the column vectors $\mathbf{v}$ and $\mathbf{w}$ that consist of the expansion coefficients $v_1, v_2, \dots$ and $w_1, w_2, \dots$, respectively, are normalized and orthogonal, as defined by (8.84) and (8.85).

8.44 Without using a computer, find the eigenvalues and normalized eigenvectors of each of these matrices. Begin by solving the characteristic equation. Check that the sum of the eigenvalues equals the trace of the matrix.

$$\mathbf{A} = \begin{pmatrix} 0 & -1 \\ 3 & 2 \end{pmatrix} \quad \mathbf{B} = \begin{pmatrix} 2 & 0 \\ 9 & 2 \end{pmatrix} \quad \mathbf{C} = \begin{pmatrix} 4 & 0 \\ 0 & 4 \end{pmatrix}$$

8.45 If $\mathbf{A}$ is a diagonal square matrix of order three with unequal diagonal elements a_{11}, a_{22}, a_{33}, find the normalized eigenvectors of $\mathbf{A}$.

8.46 (a) Without using a computer, find the eigenvalues and normalized eigenvectors of

$$\mathbf{A} = \begin{pmatrix} 2 & 2 \\ 2 & -1 \end{pmatrix}$$

(b) Is $\mathbf{A}$ real and symmetric? Is $\mathbf{A}$ Hermitian? (c) Is the eigenvector matrix $\mathbf{C}$ orthogonal? Is the eigenvector matrix $\mathbf{C}$ unitary? (d) Write down $\mathbf{C}^{-1}$ without doing any calculations. (e) Verify that $\mathbf{C}^{-1}\mathbf{AC}$ equals the diagonal matrix of eigenvalues. (f) Verify Eq. (8.81) for each eigenvector.

8.47 For

$$\mathbf{A} = \begin{pmatrix} 2 & -2i \\ 2i & 2 \end{pmatrix}$$

find the eigenvalues and normalized eigenvectors and answer the questions of Prob. 8.46(b)–(e).

8.48 Find the eigenvector $\mathbf{c}^{(2)}$ for the eigenvalue λ_2 in the example in Section 8.6.

8.49 Find the eigenvalues and normalized eigenvectors of the matrix

$$\mathbf{A} = \begin{pmatrix} -1 & 0 & -2 \\ 0 & 5 & 0 \\ -2 & 4 & 2 \end{pmatrix}$$

You can avoid solving a cubic equation if you expand the characteristic determinant in the simplest way and do not fully multiply out the characteristic polynomial.

8.50 An efficient way to calculate the inverse of a square matrix $\mathbf{A}$ of order n is as follows: (a) Place the nth-order unit matrix $\mathbf{I}$ at the right of the matrix $\mathbf{A}$ to form an n-row, $2n$-column array, which we denote by $(\mathbf{A} \vdots \mathbf{I})$. (b) Perform Gauss–Jordan elimination on the rows of $(\mathbf{A} \vdots \mathbf{I})$ so as to reduce the $\mathbf{A}$ portion of $(\mathbf{A} \vdots \mathbf{I})$ to the unit matrix. At the end of this process, the array will have the form $(\mathbf{I} \vdots \mathbf{B})$. The matrix $\mathbf{B}$ is $\mathbf{A}^{-1}$. (If $\mathbf{A}^{-1}$ does not exist, it will be impossible to reduce the $\mathbf{A}$ portion of the array to $\mathbf{I}$.) Use this procedure to find the inverse of the matrix in Prob. 8.49.

8.51 Use a computer-algebra system to find the eigenvalues and normalized eigenvectors of the square matrix $\mathbf{B}$ of order six with elements $b_{jk} = (j^2 + k^2)/(j + k)$.

8.52 Excel can find eigenvalues and eigenvectors of real, symmetric matrices as follows. If the eigenvalues of the nth-order real, symmetric matrix $\mathbf{H}$ are arranged in increasing order: $\lambda_1 \le \lambda_2 \le \cdots \le \lambda_n$, an extension of a theorem due to Rayleigh and Ritz states that $\lambda_1 = \min(\mathbf{x}^{\mathrm{T}}\mathbf{Hx}/\mathbf{x}^{\mathrm{T}}\mathbf{x})$, where $\mathbf{x}$ is an nth-order nonzero column vector whose elements are varied to minimize the quantity in parentheses; also, $\lambda_2 = \min(\mathbf{y}^{\mathrm{T}}\mathbf{Hy}/\mathbf{y}^{\mathrm{T}}\mathbf{y})$ if $\mathbf{y}^{\mathrm{T}}\mathbf{c}_1 = 0$, where $\mathbf{c}_1$ is the eigenvector corresponding to λ_1; $\lambda_3 = \min(\mathbf{z}^{\mathrm{T}}\mathbf{Hz}/\mathbf{z}^{\mathrm{T}}\mathbf{z})$ if $\mathbf{z}^{\mathrm{T}}\mathbf{c}_1 = 0$ and $\mathbf{z}^{\mathrm{T}}\mathbf{c}_2 = 0$, where $\mathbf{c}_2$ is the eigenvector corresponding to λ_2; etc. (Note the close analogy with the results of Sections 8.1 and 8.2.) Use this theorem to have Excel find the eigenvalues and normalized eigenvectors of the matrix in Prob. 8.46. *Hints*: Assign names in Excel to the various matrices involved. To multiply matrices $\mathbf{A}$ and $\mathbf{B}$, select an appropriately sized rectangular array of cells where you want the matrix product to appear; then type =MMULT(A,B) and press the Control, Shift, and Enter keys simultaneously. The transpose of matrix **C** is found similarly using the formula =TRANSPOSE(C). To find λ_1, start with a guess for **x** and use the Solver to vary **x** so as to minimize $\mathbf{x}^T\mathbf{H}\mathbf{x}$ subject to the constraint that $\mathbf{x}^T\mathbf{x} = 1$. After you find the first eigenvalue and eigenvector, add the relevant orthogonality constraint to the Solver and find the next eigenvalue and eigenvector; and so on.

8.53 For a matrix of order 20 whose eigenvalues are $1, 2, 3, \ldots, 20$, the characteristic equation can be written as $\prod_{m=1}^{20}(\lambda - m) = 0$, where the product notation is defined by (17.22). Use a spreadsheet or computer-algebra system to graph the characteristic polynomial for λ in the range $\lambda = 0.9$ to 20.1. Choose the scale on the vertical axis so that the points where the curve crosses the horizontal axis are clearly visible. Now add the quantity $1 \times 10^{-8}\lambda^{19}$ to the characteristic polynomial, graph the altered characteristic polynomial, and notice what has happened to the roots of the characteristic polynomial. If you see less than 20 roots for the altered polynomial, explain where the missing roots are. In the original characteristic polynomial, the coefficient of λ^{19} is the sum of the integers 1 to 20, which is 210, so the change made was less than one part in 10^{10} in the λ^{19} coefficient, yet it caused a major change in the eigenvalues.

8.54 If $\mathbf{B} = \mathbf{M}^{-1}\mathbf{A}\mathbf{M}$, prove that the square matrices **A** and **B** have the same eigenvalues. Also, express the eigenvectors of **B** in terms of the eigenvectors of **A**. *Hints:* Start with the eigenvalue equation for **A**, replace **A** by its expression in terms of **B**, and multiply the resulting equation by $\mathbf{M}^{-1}$ on the left to obtain the eigenvalue equation for **B**.

8.55 If $\lambda_1, \lambda_2, \ldots, \lambda_n$ are the eigenvalues of **A**, find the eigenvalues and eigenvectors of $\mathbf{A}^2$.

8.56 This problem deals with the matrix solution of the linear-variation method when the basis functions are nonorthogonal. (a) If $\{f_i\}$ in $\phi = \sum_{i=1}^{n} c_i f_i$ is not an orthonormal set, we take linear combinations of the functions $\{f_i\}$ to get a new set of functions $\{g_m\}$ that are orthonormal. We have $g_m = \sum_k a_{km} f_k$, $m = 1, 2, \ldots, n$, where the coefficients a_{km} are constants and where $\langle g_j | g_m \rangle = \delta_{jm}$. [One procedure to choose the a_{km} coefficients is the Schmidt method (Section 7.2); another method is discussed in Prob. 8.57.] (a) In $\langle g_j | g_m \rangle = \delta_{jm}$, substitute the summation expression for each g and show that the resulting equation is equivalent to the matrix equation $\mathbf{A}^\dagger\mathbf{S}\mathbf{A} = \mathbf{I}$, where **I** is a unit matrix, **S** is the overlap matrix with elements S_{jk}, and **A** is the matrix of coefficients a_{km}. (b) Verify that the set of equations (8.54) can be written as $\mathbf{H}\mathbf{c} = W\mathbf{S}\mathbf{c}$, where **c** is the column vector of coefficients $c_1, c_2, \ldots, c_n$. As we did in going from (8.80c) to (8.96), we introduce the index i to label the various eigenvalues and eigenvectors and we write $\mathbf{H}\mathbf{c} = W\mathbf{S}\mathbf{c}$ as $\mathbf{H}\mathbf{c}^{(i)} = W_i\mathbf{S}\mathbf{c}^{(i)}$ for $i = 1, 2, \ldots, n$. Verify that $\mathbf{HC} = \mathbf{SCW}$, where **C** and **W** are the matrices in (8.87). (c) Since $\mathbf{A}\mathbf{A}^{-1} = \mathbf{I}$, we can write $\mathbf{HC} = \mathbf{SCW}$ as $\mathbf{H}\mathbf{A}\mathbf{A}^{-1}\mathbf{C} = \mathbf{S}\mathbf{A}\mathbf{A}^{-1}\mathbf{C}\mathbf{W}$. Multiply this last equation by $\mathbf{A}^\dagger$ on the left. Use the result of (a) to show that we get $\mathbf{H}'\mathbf{C}' = \mathbf{C}'\mathbf{W}$, where $\mathbf{C}' \equiv \mathbf{A}^{-1}\mathbf{C}$ and $\mathbf{H}' \equiv \mathbf{A}^\dagger\mathbf{H}\mathbf{A}$. Comparison with (8.88) shows that $\mathbf{H}'\mathbf{C}' = \mathbf{C}'\mathbf{W}$ is the eigenvalue equation for the $\mathbf{H}'$ matrix. The matrix procedure to solve the linear variation problem $\mathbf{HC} = \mathbf{SCW}$ with nonorthogonal basis functions is then as follows: (1) Compute the matrix elements of **H** and **S** using the nonorthogonal basis. (2) Use the overlap integrals and a procedure such as the Schmidt method to find a matrix **A** whose elements a_{km} transform the nonorthogonal functions $\{f_i\}$ to orthonormal functions $\{g_i\}$. (3) Calculate $\mathbf{H}'$ using $\mathbf{H}' \equiv \mathbf{A}^\dagger\mathbf{H}\mathbf{A}$. (4) Find the eigenvalues W_i and the eigenvectors $\mathbf{c}'^{(i)}$ of the $\mathbf{H}'$ matrix. (5) Use $\mathbf{C} = \mathbf{A}\mathbf{C}'$ to compute the coefficient matrix **C**. The eigenvalues W_i found in step 4 and the coefficients found in step 5 are the desired energy estimates and variation-function coefficients.

8.57 The **symmetric** (or **Löwdin**) **orthogonalization** procedure is often used to orthogonalize a basis set. Given the nonorthogonal basis set $\{f_i\}$, we form the set of functions $\{g_m\}$ as the linear combinations $g_m = \sum_k a_{km} f_k$. As shown in Prob. 8.56(a), for $\{g_m\}$ to be an orthogonal set, the matrix of transformation coefficients a_{km} must satisfy $\mathbf{A}^\dagger\mathbf{S}\mathbf{A} = \mathbf{I}$, where **S** is the overlap matrix with $S_{jk} = \langle f_j | f_k \rangle$. If the square matrix $\mathbf{B}$ satisfies $\mathbf{B}^2 = \mathbf{S}$, then $\mathbf{B}$ is a *square root* of $\mathbf{S}$, which is written as $\mathbf{B} = \mathbf{S}^{1/2}$. (A matrix can have more than one square root.) Thus $\mathbf{S}^{1/2}$ satisfies $\mathbf{S}^{1/2}\mathbf{S}^{1/2} = \mathbf{S}$. The inverse of $\mathbf{S}^{1/2}$ is written as $\mathbf{S}^{-1/2}$. By the definition of the inverse, $\mathbf{S}^{-1/2}\mathbf{S}^{1/2} = \mathbf{S}^{1/2}\mathbf{S}^{-1/2} = \mathbf{I}$. (a) Since $\mathbf{S}$ is a Hermitian matrix, it can be diagonalized by the unitary matrix $\mathbf{U}$ of its orthonormal eigenvectors, and we have $\mathbf{U}^\dagger\mathbf{S}\mathbf{U} = \mathbf{s}$, where $\mathbf{s}$ is the diagonal matrix of eigenvalues of $\mathbf{S}$. Show that $\mathbf{S} = \mathbf{U}\mathbf{s}\mathbf{U}^\dagger$. (b) Let s_i denote the eigenvalues of $\mathbf{S}$. Let $\mathbf{s}$ and $\mathbf{s}^{-1/2}$ denote the diagonal matrices whose diagonal elements are s_i and $s_i^{-1/2}$, respectively, where $s_i^{-1/2}$ denotes the reciprocals of the *positive* square roots of the $\mathbf{S}$ eigenvalues. [The eigenvalues of $\mathbf{S}$ can all be proven to be positive (see *Szabo and Ostlund*, p. 143), so no s_i is zero and all the numbers $s_i^{-1/2}$ exist.] Show that the matrix $\mathbf{M} \equiv \mathbf{U}\mathbf{s}^{1/2}\mathbf{U}^\dagger$ satisfies $\mathbf{M}^2 = \mathbf{U}\mathbf{s}\mathbf{U}^\dagger = \mathbf{S}$. Hence $\mathbf{M}$ is a square root of $\mathbf{S}$; $\mathbf{M} = \mathbf{S}^{1/2}$. (c) Show that the matrix $\mathbf{N} \equiv \mathbf{U}\mathbf{s}^{-1/2}\mathbf{U}^\dagger$ satisfies $\mathbf{M}\mathbf{N} = \mathbf{I}$. Since $\mathbf{M} = \mathbf{S}^{1/2}$, we have $\mathbf{S}^{1/2}\mathbf{N} = \mathbf{I}$, so $\mathbf{N} = \mathbf{S}^{-1/2}$ and $\mathbf{S}^{-1/2} = \mathbf{U}\mathbf{s}^{-1/2}\mathbf{U}^\dagger$. (d) The symmetric orthogonalization procedure takes the transformation matrix $\mathbf{A}$ to be $\mathbf{S}^{-1/2}$; that is, $\mathbf{A} = \mathbf{S}^{-1/2} = \mathbf{U}\mathbf{s}^{-1/2}\mathbf{U}^\dagger$. To show that this choice satisfies the requirement $\mathbf{A}^\dagger\mathbf{S}\mathbf{A} = \mathbf{I}$, we need to find $\mathbf{A}^\dagger$. Prove that $(\mathbf{BC})^\dagger = \mathbf{C}^\dagger\mathbf{B}^\dagger$ by finding the (i, j)th elements of $(\mathbf{BC})^\dagger$ and $\mathbf{C}^\dagger\mathbf{B}^\dagger$. Then set $\mathbf{C} = \mathbf{DE}$ and show that $(\mathbf{BDE})^\dagger = \mathbf{E}^\dagger\mathbf{D}^\dagger\mathbf{B}^\dagger$. Next, show that $\mathbf{A}^\dagger = \mathbf{U}\mathbf{s}^{-1/2}\mathbf{U}^\dagger$. Then show that $\mathbf{A}^\dagger\mathbf{S}\mathbf{A} = \mathbf{I}$. Hence, to use symmetric orthogonalization, we find the eigenvalues and orthonormal eigenvectors of $\mathbf{S}$, use the eigenvalues s_i to form the matrix $\mathbf{s}^{-1/2}$, use the eigenvectors to form $\mathbf{U}$, calculate the transformation matrix $\mathbf{A} = \mathbf{U}\mathbf{s}^{-1/2}\mathbf{U}^\dagger$, and take the orthonormal functions as $g_m = \sum_k a_{km} f_k$.

8.58 Suppose $\hat{A}g_n = a_n g_n$. Let the eigenfunctions g_n be expanded in terms of the complete orthonormal set $\{f_i\}$ according to $g_n = \sum_k c_k^{(n)} f_k$. Substitute this expansion into $\hat{A}g_n = a_n g_n$, multiply by f_i^*, integrate over all space, and show that the set of equations $\sum_k (A_{ik} - a_n \delta_{ik}) c_k^{(n)} = 0$ for $i = 1, 2, 3, \ldots$ is obtained, where $A_{ik} \equiv \langle f_i | \hat{A} | f_k \rangle$. This set of equations has the same form as the set (8.54) with $S_{ik} = \delta_{ik}$. Thus, just as the W values and the coefficients c_k in (8.54) with $S_{ik} = \delta_{ik}$ can be found by finding the eigenvalues and eigenvectors of the $\mathbf{H}$ matrix, the eigenvalues a_n of $\hat{A}$ and the expansion coefficients $c_k^{(n)}$ of the eigenfunctions g_n of $\hat{A}$ can be found by finding the eigenvalues a_n and eigenvectors $\mathbf{c}^{(n)}$ of the matrix $\mathbf{A}$, whose elements are A_{ik}.

8.59 Consider the double-well potential with $V = \infty$ for $x < 0$, $V = 0$ for $0 \le x \le \frac{1}{4}l$, $V = V_0$ for $\frac{1}{4}l < x < \frac{3}{4}l$, $V = 0$ for $\frac{3}{4}l \le x \le l$, and $V = \infty$ for $x > l$, where l and V_0 are positive constants. Use Mathcad or some other computer-algebra program to apply the linear variation method to this problem, taking the basis functions as the m lowest particle-in-a-box (pib) wave functions (2.23). Use dimensionless variables (Section 4.4 and Prob. 4.36). Set things up so that V_0 and m can be readily changed. Use the program's ability to find eigenvalues and eigenvectors to solve for the approximate energies and wave functions. Have the program graph the lowest four approximate wave functions. Show that $T_{jk} = k^2\pi^2\hbar^2/2ml^2$ but let the program evaluate the V_{jk} integrals. *Hint:* In Mathcad, you may need to adjust the value of the TOL variable. (a) For $V_0 = 100\hbar^2/ml^2$, find the lowest four energy levels using the following numbers of pib basis functions: 4; 8; 16; 32. Compare with the following exact reduced-energy values found from Eq. (4.92): 45.802165653, 46.107222914, 113.938076461, and 143.353993916. Which basis functions contribute substantially to the ground-state wave function? Which contribute substantially to the first-excited-state wave function? (b) With four basis functions, you will find that the variational energy of the one-node function that corresponds to the first excited state lies below the variational energy of the even function that corresponds to the ground state. Is this a violation of the variation theorem inequalities (8.60) to (8.62)?

8.60 Revise your solution to the double-well problem 8.59 so as to treat even and odd functions separately. Do this by introducing a parameter whose value is 1 or 0, depending on whether we are treating the even or the odd wave functions. Calculate the lowest two

Hamiltonian eigenvalues using the first 16 even pib functions. Repeat with the lowest 16 odd pib functions. Compare your results with those of Prob. 8.59.

8.61 Consider the one-particle, one-dimensional problem with $V = V_0(\hbar^2/ml^2)x$ for $0 \le x \le l$ and $V = \infty$ elsewhere. Modify your solution to Prob. 8.59 to estimate the lowest four energy eigenvalues for $V_0 = 200$ using the following numbers of particle-in-a-box basis functions: (a) 8; (b) 12. Note the appearance of the lowest four wave functions. Which particle-in-a-box states make substantial contributions to the ground state? to the first excited state?

8.62 Revise your solution to Prob. 8.59 to treat the one-dimensional harmonic oscillator using particle-in-a-box (pib) basis functions. Recall that in Section 4.4 we found that for $E_r \le 5$, the wave function can be taken as zero outside the region $-5 \le x_r \le 5$, where E_r and x_r are defined by (4.77) and (4.78). Therefore, we shall take the pib basis functions to extend from -5 to 5, with the center of the "box" at $x_r = 0$. Since the box has a length of 10 units in x_r, we have $f_j = (2/10)^{1/2} \sin[j\pi(x_r + 5)/10]$ for $|x_r| \le 5$ and $f_j = 0$ elsewhere. You will also need to revise the kinetic- and potential-energy matrix elements. Increase the number of pib basis functions until all energy values with $E_r < 5$ are accurate to three decimal places. Check the appearance of the lowest three variation functions. Which pib basis functions contribute most to the ground state? to the first excited state?

8.63 Revise your solution to Prob. 8.59 to apply the pib basis functions to the one-dimensional quartic oscillator with $V = cx^4$. See Prob. 8.62 for hints. Take the "box" to extend from $x_r = -3.5$ to 3.5, where x_r is as found in Prob. 4.31. Increase the number of pib basis functions until the lowest three energy values remain stable to three decimal places. Compare the lowest three energies with those found by the Numerov method in Prob. 4.31. Check the appearance of the lowest three variational functions. Now repeat for the box going from $x_r = -4.5$ to 4.5. For which box length do we get faster convergence to the true energies?

8.64 Apply the particle-in-a-box basis functions to the radial equation for the hydrogen atom for the $l = 0$ states. Recall that in Section 6.9, we expressed the radial factor in the H-atom wave function as $R(r) = r^{-1}F(r)$, where $F(r) = 0$ at $r = 0$. The variation function in this problem will have the form $\phi = r^{-1}F(r)Y_l^m(\theta, \phi)$; take the dimensionless function $F_r(r_r)$ to be a linear combination of 28 pib basis functions, where the box goes from $r_r = 0$ to 27, where r_r is defined in Section 6.9. Work out the proper forms for the integrals H_{jk} and S_{jk}. Find the estimates for the lowest three $l = 0$ energies. For the ground-state variation function, how many pib basis functions appear with a coefficient greater than 0.1?

8.65 True or false? (a) The transpose of a column vector is a row vector. (b) The matrix product **bc** of a row vector **b** with n elements times a column vector **c** with n elements is a scalar. (c) The diagonal elements of a Hermitian matrix must be real numbers. (d) The eigenvalues of a diagonal matrix are equal to the diagonal elements. (e) For a square matrix all of whose elements below the principal diagonal are zero, the eigenvalues are equal to the diagonal elements. (See Prob. 8.19.) (f) Every nonzero column vector with n elements is an eigenvector of the unit matrix of order n. (g) Every nonzero linear combination of two eigenvectors that correspond to the same eigenvalue of a matrix is an eigenvector of that matrix. (h) Zero is not allowed as an eigenvalue of a matrix. (i) A zero column vector is not allowed as an eigenvector of a matrix. (j) Every square matrix has an inverse. (k) If **AB** = **AC**, then the matrices **B** and **C** must be equal. (l) If **A** is a square matrix of order n and **c** is a column vector with n elements, then **Ac** is a column vector with n elements. (m) Every real symmetric matrix is a Hermitian matrix.

CHAPTER 9

Perturbation Theory

9.1 INTRODUCTION

We now discuss the second major quantum-mechanical approximation method, perturbation theory.

Suppose we have a system with a time-independent Hamiltonian operator $\hat{H}$ and we are unable to solve the Schrödinger equation

$$\hat{H}\psi_n = E_n\psi_n \tag{9.1}$$

for the eigenfunctions and eigenvalues of the bound stationary states. Suppose also that the Hamiltonian $\hat{H}$ is only slightly different from the Hamiltonian $\hat{H}^0$ of a system whose Schrödinger equation

$$\hat{H}^0\psi_n^{(0)} = E_n^{(0)}\psi_n^{(0)} \tag{9.2}$$

we can solve. An example is the one-dimensional anharmonic oscillator with

$$\hat{H} = -\frac{\hbar^2}{2m}\frac{d^2}{dx^2} + \frac{1}{2}kx^2 + cx^3 + dx^4 \tag{9.3}$$

The Hamiltonian (9.3) is closely related to the Hamiltonian

$$\hat{H}^0 = -\frac{\hbar^2}{2m}\frac{d^2}{dx^2} + \frac{1}{2}kx^2 \tag{9.4}$$

of the harmonic oscillator. If the constants c and d in (9.3) are small, we expect the eigenfunctions and eigenvalues of the anharmonic oscillator to be closely related to those of the harmonic oscillator.

We shall call the system with Hamiltonian $\hat{H}^0$ the **unperturbed system**. The system with Hamiltonian $\hat{H}$ is the **perturbed system**. The difference between the two Hamiltonians is the **perturbation**, $\hat{H}'$:

$$\hat{H}' \equiv \hat{H} - \hat{H}^0 \tag{9.5}$$

$$\boxed{\hat{H} = \hat{H}^0 + \hat{H}'} \tag{9.6}$$

(The prime does not refer to differentiation.) For the anharmonic oscillator with Hamiltonian (9.3), the perturbation on the related harmonic oscillator is $\hat{H}' = cx^3 + dx^4$.

In $\hat{H}^0\psi_n^{(0)} = E_n^{(0)}\psi_n^{(0)}$ [Eq. (9.2)], $E_n^{(0)}$ and $\psi_n^{(0)}$ are called the unperturbed energy and unperturbed wave function of state n. For $\hat{H}^0$ equal to the harmonic-oscillator

Hamiltonian (9.4), $E_n^{(0)}$ is $\left(n + \frac{1}{2}\right)h\nu$ [Eq. (4.47)], where n is a nonnegative integer. (n is used instead of v for consistency with the perturbation-theory notation.) Note that the superscript $^{(0)}$ does not mean the ground state. Perturbation theory can be applied to any state. The subscript n labels which state we are dealing with. The superscript $^{(0)}$ denotes the unperturbed system.

Our task is to relate the unknown eigenvalues and eigenfunctions of the perturbed system to the known eigenvalues and eigenfunctions of the unperturbed system. To aid in doing so, we shall imagine that the perturbation is applied gradually, giving a continuous change from the unperturbed to the perturbed system. Mathematically, this corresponds to inserting a parameter λ into the Hamiltonian, so that

$$\hat{H} = \hat{H}^0 + \lambda\hat{H}' \tag{9.7}$$

When λ is zero, we have the unperturbed system. As λ increases, the perturbation grows larger, and at $\lambda = 1$ the perturbation is fully "turned on." We inserted λ to help relate the perturbed and unperturbed eigenfunctions, and ultimately we shall set $\lambda = 1$, thereby eliminating it.

Sections 9.1 to 9.8 deal with time-independent Hamiltonians and stationary states. Section 9.9 deals with time-dependent perturbations.

9.2 NONDEGENERATE PERTURBATION THEORY

The perturbation treatments of degenerate and nondegenerate energy levels differ. This section examines the effect of a perturbation on a nondegenerate level. If some of the energy levels of the unperturbed system are degenerate while others are nondegenerate, the treatment in this section will apply to the nondegenerate levels only.

Nondegenerate Perturbation Theory. Let $\psi_n^{(0)}$ be the wave function of some particular unperturbed nondegenerate level with energy $E_n^{(0)}$. Let ψ_n be the perturbed wave function into which $\psi_n^{(0)}$ is converted when the perturbation is applied. From (9.1) and (9.7), the Schrödinger equation for the perturbed state is

$$\hat{H}\psi_n = (\hat{H}^0 + \lambda\hat{H}')\psi_n = E_n\psi_n \tag{9.8}$$

Since the Hamiltonian in (9.8) depends on the parameter λ, both the eigenfunction ψ_n and the eigenvalue E_n depend on λ:

$$\psi_n = \psi_n(\lambda, q) \quad \text{and} \quad E_n = E_n(\lambda)$$

where q denotes the system's coordinates. We now expand ψ_n and E_n as Taylor series (Prob. 4.1) in powers of λ:

$$\psi_n = \psi_n|_{\lambda=0} + \left.\frac{\partial\psi_n}{\partial\lambda}\right|_{\lambda=0}\lambda + \left.\frac{\partial^2\psi_n}{\partial\lambda^2}\right|_{\lambda=0}\frac{\lambda^2}{2!} + \cdots \tag{9.9}$$

$$E_n = E_n|_{\lambda=0} + \left.\frac{dE_n}{d\lambda}\right|_{\lambda=0}\lambda + \left.\frac{d^2E_n}{d\lambda^2}\right|_{\lambda=0}\frac{\lambda^2}{2!} + \cdots \tag{9.10}$$

By hypothesis, when λ goes to zero, ψ_n and E_n go to $\psi_n^{(0)}$ and $E_n^{(0)}$:

$$\psi_n|_{\lambda=0} = \psi_n^{(0)} \quad \text{and} \quad E_n|_{\lambda=0} = E_n^{(0)} \tag{9.11}$$

We introduce the following abbreviations:

$$\psi_n^{(k)} \equiv \frac{1}{k!}\frac{\partial^k \psi_n}{\partial \lambda^k}\bigg|_{\lambda=0}, \qquad E_n^{(k)} \equiv \frac{1}{k!}\frac{d^k E_n}{d\lambda^k}\bigg|_{\lambda=0}, \qquad k = 1, 2, \ldots \tag{9.12}$$

Equations (9.9) and (9.10) become

$$\psi_n = \psi_n^{(0)} + \lambda\psi_n^{(1)} + \lambda^2\psi_n^{(2)} + \cdots + \lambda^k\psi_n^{(k)} + \cdots \tag{9.13}$$

$$E_n = E_n^{(0)} + \lambda E_n^{(1)} + \lambda^2 E_n^{(2)} + \cdots + \lambda^k E_n^{(k)} + \cdots \tag{9.14}$$

For $k = 1, 2, 3, \ldots$, we call $\psi_n^{(k)}$ and $E_n^{(k)}$ the ***k*th-order corrections** to the wave function and energy. We shall assume that the series (9.13) and (9.14) converge for $\lambda = 1$, and we hope that for a small perturbation, taking just the first few terms of the series will give a good approximation to the true energy and wave function. (Quite often, perturbation-theory series do not converge, but even so, the first few terms of a nonconvergent series can often give a useful approximation.)

We shall take $\psi_n^{(0)}$ to be normalized: $\langle\psi_n^{(0)}|\psi_n^{(0)}\rangle = 1$. Instead of taking ψ_n as normalized, we shall require that ψ_n satisfy

$$\langle\psi_n^{(0)}|\psi_n\rangle = 1 \tag{9.15}$$

If ψ_n does not satisfy this equation, then multiplication of ψ_n by the constant $1/\langle\psi_n^{(0)}|\psi_n\rangle$ gives a perturbed wave function with the desired property. The condition $\langle\psi_n^{(0)}|\psi_n\rangle = 1$, called *intermediate normalization*, simplifies the derivation. Note that multiplication of ψ_n by a constant does not change the energy in the Schrödinger equation $\hat{H}\psi_n = E_n\psi_n$, so use of intermediate normalization does not affect the results for the energy corrections. If desired, at the end of the calculation, the intermediate-normalized ψ_n can be multiplied by a constant to normalize it in the usual sense.

Substitution of (9.13) into $1 = \langle\psi_n^{(0)}|\psi_n\rangle$ [Eq. (9.15)] gives

$$1 = \langle\psi_n^{(0)}|\psi_n^{(0)}\rangle + \lambda\langle\psi_n^{(0)}|\psi_n^{(1)}\rangle + \lambda^2\langle\psi_n^{(0)}|\psi_n^{(2)}\rangle + \cdots$$

Since this equation is true for all values of λ in the range 0 to 1, the coefficients of like powers of λ on each side of the equation must be equal, as proved after Eq. (4.11). Equating the λ^0 coefficients, we have $1 = \langle\psi_n^{(0)}|\psi_n^{(0)}\rangle$, which is satisfied since $\psi_n^{(0)}$ is normalized. Equating the coefficients of λ^1, of λ^2, and so on, we have

$$\langle\psi_n^{(0)}|\psi_n^{(1)}\rangle = 0, \qquad \langle\psi_n^{(0)}|\psi_n^{(2)}\rangle = 0, \qquad \text{etc.} \tag{9.16}$$

The corrections to the wave function are orthogonal to $\psi_n^{(0)}$ when intermediate normalization is used.

Substituting (9.13) and (9.14) into the Schrödinger equation (9.8), we have

$$(\hat{H}^0 + \lambda\hat{H}')(\psi_n^{(0)} + \lambda\psi_n^{(1)} + \lambda^2\psi_n^{(2)} + \cdots)$$
$$= (E_n^{(0)} + \lambda E_n^{(1)} + \lambda^2 E_n^{(2)} + \cdots)(\psi_n^{(0)} + \lambda\psi_n^{(1)} + \lambda^2\psi_n^{(2)} + \cdots)$$

Collecting like powers of λ, we have

$$\hat{H}^0\psi_n^{(0)} + \lambda(\hat{H}'\psi_n^{(0)} + \hat{H}^0\psi_n^{(1)}) + \lambda^2(\hat{H}^0\psi_n^{(2)} + \hat{H}'\psi_n^{(1)}) + \cdots$$
$$= E_n^{(0)}\psi_n^{(0)} + \lambda(E_n^{(1)}\psi_n^{(0)} + E_n^{(0)}\psi_n^{(1)}) + \lambda^2(E_n^{(2)}\psi_n^{(0)} + E_n^{(1)}\psi_n^{(1)} + E_n^{(0)}\psi_n^{(2)}) + \cdots \quad (9.17)$$

Now (assuming suitable convergence) for the two series on each side of (9.17) to be equal to each other for all values of λ, the coefficients of like powers of λ in the two series must be equal.

Equating the coefficients of the λ^0 terms, we have $\hat{H}^0\psi_n^{(0)} = E_n^{(0)}\psi_n^{(0)}$, which is the Schrödinger equation for the unperturbed problem, Eq. (9.2), and gives us no new information.

Equating the coefficients of the λ^1 terms, we have

$$\hat{H}'\psi_n^{(0)} + \hat{H}^0\psi_n^{(1)} = E_n^{(1)}\psi_n^{(0)} + E_n^{(0)}\psi_n^{(1)}$$
$$\hat{H}^0\psi_n^{(1)} - E_n^{(0)}\psi_n^{(1)} = E_n^{(1)}\psi_n^{(0)} - \hat{H}'\psi_n^{(0)} \quad (9.18)$$

The First-Order Energy Correction. To find $E_n^{(1)}$, we multiply (9.18) by $\psi_m^{(0)*}$ and integrate over all space, which gives

$$\langle\psi_m^{(0)}|\hat{H}^0|\psi_n^{(1)}\rangle - E_n^{(0)}\langle\psi_m^{(0)}|\psi_n^{(1)}\rangle = E_n^{(1)}\langle\psi_m^{(0)}|\psi_n^{(0)}\rangle - \langle\psi_m^{(0)}|\hat{H}'|\psi_n^{(0)}\rangle \quad (9.19)$$

where bracket notation [Eqs. (7.1) and (7.3)] is used. $\hat{H}^0$ is Hermitian, and use of the Hermitian property (7.12) gives for the first term on the left side of (9.19)

$$\langle\psi_m^{(0)}|\hat{H}^0|\psi_n^{(1)}\rangle = \langle\psi_n^{(1)}|\hat{H}^0|\psi_m^{(0)}\rangle^* = \langle\psi_n^{(1)}|\hat{H}^0\psi_m^{(0)}\rangle^*$$
$$= \langle\psi_n^{(1)}|E_m^{(0)}\psi_m^{(0)}\rangle^* = E_m^{(0)*}\langle\psi_n^{(1)}|\psi_m^{(0)}\rangle^* = E_m^{(0)}\langle\psi_m^{(0)}|\psi_n^{(1)}\rangle \quad (9.20)$$

where we used the unperturbed Schrödinger equation $\hat{H}^0\psi_m^{(0)} = E_m^{(0)}\psi_m^{(0)}$, the fact that $E_m^{(0)}$ is real, and (7.4). Substitution of (9.20) into (9.19) and use of the orthonormality equation $\langle\psi_m^{(0)}|\psi_n^{(0)}\rangle = \delta_{mn}$ for the unperturbed eigenfunctions gives

$$(E_m^{(0)} - E_n^{(0)})\langle\psi_m^{(0)}|\psi_n^{(1)}\rangle = E_n^{(1)}\delta_{mn} - \langle\psi_m^{(0)}|\hat{H}'|\psi_n^{(0)}\rangle \quad (9.21)$$

If $m = n$, the left side of (9.21) equals zero, and (9.21) becomes

$$\boxed{E_n^{(1)} = \langle\psi_n^{(0)}|\hat{H}'|\psi_n^{(0)}\rangle = \int \psi_n^{(0)*}\hat{H}'\psi_n^{(0)}\,d\tau} \quad \mathbf{(9.22)}$$

The first-order correction to the energy is found by averaging the perturbation $\hat{H}'$ over the appropriate unperturbed wave function.

Setting $\lambda = 1$ in (9.14), we have

$$E_n \approx E_n^{(0)} + E_n^{(1)} = E_n^{(0)} + \int \psi_n^{(0)*}\hat{H}'\psi_n^{(0)}\,d\tau \quad (9.23)$$

EXAMPLE For the anharmonic oscillator with Hamiltonian (9.3), evaluate $E^{(1)}$ for the ground state if the unperturbed system is taken as the harmonic oscillator.

The perturbation is given by Eqs. (9.3) to (9.5) as

$$\hat{H}' = \hat{H} - \hat{H}^0 = cx^3 + dx^4$$

and the first-order energy correction for the state with quantum number v is given by (9.22) as $E_v^{(1)} = \langle \psi_v^{(0)} | cx^3 + dx^4 | \psi_v^{(0)} \rangle$, where $\psi_v^{(0)}$ is the harmonic-oscillator wave function for state v. For the $v = 0$ ground state, use of $\psi_0^{(0)} = (\alpha/\pi)^{1/4} e^{-\alpha x^2/2}$ [Eq. (4.55)] gives

$$E_0^{(1)} = \langle \psi_0^{(0)} | cx^3 + dx^4 | \psi_0^{(0)} \rangle = \left(\frac{\alpha}{\pi}\right)^{1/2} \int_{-\infty}^{\infty} e^{-\alpha x^2}(cx^3 + dx^4)\, dx$$

The integral from $-\infty$ to ∞ of the odd function $cx^3 e^{-\alpha x^2}$ is zero. Use of the Appendix integral (A.10) with $n = 2$ and (4.33) for α gives

$$E_0^{(1)} = 2d\left(\frac{\alpha}{\pi}\right)^{1/2} \int_0^{\infty} e^{-\alpha x^2} x^4\, dx = \frac{3d}{4\alpha^2} = \frac{3dh^2}{64\pi^4 \nu^2 m^2}$$

The unperturbed ground-state energy is $E_0^{(0)} = \frac{1}{2}h\nu$ and $E_0^{(0)} + E_0^{(1)} = \frac{1}{2}h\nu + 3dh^2/64\pi^4\nu^2 m^2$.

EXERCISE Consider a one-particle, one-dimensional system with $V = \infty$ for $x < 0$ and for $x > l$, and $V = cx$ for $0 \le x \le l$, where c is a constant. (a) Sketch V for $c > 0$. (b) Treat the system as a perturbed particle in a box and find $E^{(1)}$ for the state with quantum number n. Then use Eq. (3.88) to state why the answer you got is to be expected. (*Partial Answer:* (b) $\frac{1}{2}cl$.)

The First-Order Wave-Function Correction. For $m \neq n$, Eq. (9.21) is

$$(E_m^{(0)} - E_n^{(0)}) \langle \psi_m^{(0)} | \psi_n^{(1)} \rangle = -\langle \psi_m^{(0)} | \hat{H}' | \psi_n^{(0)} \rangle, \quad m \neq n \tag{9.24}$$

To find $\psi_n^{(1)}$, we expand it in terms of the complete, orthonormal set of unperturbed eigenfunctions $\psi_m^{(0)}$ of the Hermitian operator $\hat{H}^0$:

$$\psi_n^{(1)} = \sum_m a_m \psi_m^{(0)}, \qquad \text{where} \quad a_m = \langle \psi_m^{(0)} | \psi_n^{(1)} \rangle \tag{9.25}$$

where Eq. (7.41) was used for the expansion coefficients a_m. Use of $a_m = \langle \psi_m^{(0)} | \psi_n^{(1)} \rangle$ in (9.24) gives

$$(E_m^{(0)} - E_n^{(0)}) a_m = -\langle \psi_m^{(0)} | \hat{H}' | \psi_n^{(0)} \rangle, \quad m \neq n$$

By hypothesis, the level $E_n^{(0)}$ is nondegenerate. Hence $E_m^{(0)} \neq E_n^{(0)}$ for $m \neq n$, and we may divide by $(E_m^{(0)} - E_n^{(0)})$ to get

$$a_m = \frac{\langle \psi_m^{(0)} | \hat{H}' | \psi_n^{(0)} \rangle}{E_n^{(0)} - E_m^{(0)}}, \quad m \neq n \tag{9.26}$$

The coefficients a_m in the expansion (9.25) of $\psi_n^{(1)}$ are given by (9.26) except for a_n, the coefficient of $\psi_n^{(0)}$. From the second equation in (9.25), $a_n = \langle \psi_n^{(0)} | \psi_n^{(1)} \rangle$. Recall that the choice of intermediate normalization for ψ_n makes $\langle \psi_n^{(0)} | \psi_n^{(1)} \rangle = 0$ [Eq. (9.16)]. Hence $a_n = \langle \psi_n^{(0)} | \psi_n^{(1)} \rangle = 0$, and Eqs. (9.25) and (9.26) give the first-order correction to the wave function as

$$\psi_n^{(1)} = \sum_{m \neq n} \frac{\langle \psi_m^{(0)} | \hat{H}' | \psi_n^{(0)} \rangle}{E_n^{(0)} - E_m^{(0)}} \psi_m^{(0)} \tag{9.27}$$

The symbol $\sum_{m \neq n}$ means we sum over all the unperturbed states except state n.

Setting $\lambda = 1$ in (9.13) and using just the first-order wave-function correction, we have as the approximation to the perturbed wave function

$$\psi_n \approx \psi_n^{(0)} + \sum_{m \neq n} \frac{\langle \psi_m^{(0)} | \hat{H}' | \psi_n^{(0)} \rangle}{E_n^{(0)} - E_m^{(0)}} \psi_m^{(0)} \tag{9.28}$$

[For $\psi_n^{(2)}$ and the normalization of ψ, see *Kemble*, Chapter XI.]

The Second-Order Energy Correction. Equating the coefficients of the λ^2 terms in (9.17), we get

$$\hat{H}^0 \psi_n^{(2)} - E_n^{(0)} \psi_n^{(2)} = E_n^{(2)} \psi_n^{(0)} + E_n^{(1)} \psi_n^{(1)} - \hat{H}' \psi_n^{(1)} \tag{9.29}$$

Multiplication by $\psi_m^{(0)*}$ followed by integration over all space gives

$$\langle \psi_m^{(0)} | \hat{H}^0 | \psi_n^{(2)} \rangle - E_n^{(0)} \langle \psi_m^{(0)} | \psi_n^{(2)} \rangle$$

$$= E_n^{(2)} \langle \psi_m^{(0)} | \psi_n^{(0)} \rangle + E_n^{(1)} \langle \psi_m^{(0)} | \psi_n^{(1)} \rangle - \langle \psi_m^{(0)} | \hat{H}' | \psi_n^{(1)} \rangle \tag{9.30}$$

The integral $\langle \psi_m^{(0)} | \hat{H}^0 | \psi_n^{(2)} \rangle$ in this equation is the same as the integral in (9.20), except that $\psi_n^{(1)}$ is replaced by $\psi_n^{(2)}$. Replacement of $\psi_n^{(1)}$ by $\psi_n^{(2)}$ in (9.20) gives

$$\langle \psi_m^{(0)} | \hat{H}^0 | \psi_n^{(2)} \rangle = E_m^{(0)} \langle \psi_m^{(0)} | \psi_n^{(2)} \rangle \tag{9.31}$$

Use of (9.31) and orthonormality of the unperturbed functions in (9.30) gives

$$(E_m^{(0)} - E_n^{(0)}) \langle \psi_m^{(0)} | \psi_n^{(2)} \rangle = E_n^{(2)} \delta_{mn} + E_n^{(1)} \langle \psi_m^{(0)} | \psi_n^{(1)} \rangle - \langle \psi_m^{(0)} | \hat{H}' | \psi_n^{(1)} \rangle \tag{9.32}$$

For $m = n$, the left side of (9.32) is zero and we get

$$E_n^{(2)} = -E_n^{(1)} \langle \psi_n^{(0)} | \psi_n^{(1)} \rangle + \langle \psi_n^{(0)} | \hat{H}' | \psi_n^{(1)} \rangle$$

$$E_n^{(2)} = \langle \psi_n^{(0)} | \hat{H}' | \psi_n^{(1)} \rangle \tag{9.33}$$

since $\langle \psi_n^{(0)} | \psi_n^{(1)} \rangle = 0$ [Eq. (9.16)]. Note from (9.33) that to find the *second*-order correction to the energy, we have to know only the *first*-order correction to the wave function. In fact, it can be shown that knowledge of $\psi_n^{(1)}$ suffices to determine $E_n^{(3)}$ also. In general, it can be shown that if we know the corrections to the wave function through the kth order, then we can compute the corrections to the energy through order $2k + 1$ (see *Bates*, Vol. I, p. 184).

Substitution of (9.27) for $\psi_n^{(1)}$ into (9.33) gives

$$E_n^{(2)} = \sum_{m \neq n} \frac{\langle \psi_m^{(0)} | \hat{H}' | \psi_n^{(0)} \rangle}{E_n^{(0)} - E_m^{(0)}} \langle \psi_n^{(0)} | \hat{H}' | \psi_m^{(0)} \rangle \tag{9.34}$$

since the expansion coefficients a_m [Eq. (9.26)] are constants that can be taken outside the integral. Since $\hat{H}'$ is Hermitian, we have

$$\langle \psi_m^{(0)} | \hat{H}' | \psi_n^{(0)} \rangle \langle \psi_n^{(0)} | \hat{H}' | \psi_m^{(0)} \rangle = \langle \psi_m^{(0)} | \hat{H}' | \psi_n^{(0)} \rangle \langle \psi_m^{(0)} | \hat{H}' | \psi_n^{(0)} \rangle^* = |\langle \psi_m^{(0)} | \hat{H}' | \psi_n^{(0)} \rangle|^2$$

and

$$E_n^{(2)} = \sum_{m \neq n} \frac{|\langle \psi_m^{(0)} | \hat{H}' | \psi_n^{(0)} \rangle|^2}{E_n^{(0)} - E_m^{(0)}} \tag{9.35}$$

which is the desired expression for $E_n^{(2)}$ in terms of the unperturbed wave functions and energies.

Inclusion of $E_n^{(2)}$ in (9.14) with $\lambda = 1$ gives the approximate energy of the perturbed state as

$$E_n \approx E_n^{(0)} + H'_{nn} + \sum_{m \neq n} \frac{|H'_{mn}|^2}{E_n^{(0)} - E_m^{(0)}} \tag{9.36}$$

where the integrals are over the unperturbed normalized wave functions.

For formulas for higher-order energy corrections, see *Bates,* Volume I, pages 181–185. The form of perturbation theory developed in this section is called *Rayleigh–Schrödinger perturbation theory*; other approaches exist.

Discussion. Equation (9.28) shows that the effect of the perturbation on the wave function $\psi_n^{(0)}$ is to "mix in" contributions from other states $\psi_m^{(0)}$, $m \neq n$. Because of the factor $1/(E_n^{(0)} - E_m^{(0)})$, the most important contributions (aside from $\psi_n^{(0)}$) to the perturbed wave function come from states nearest in energy to state n.

To evaluate the first-order correction to the energy, we must evaluate only the single integral H'_{nn}, whereas to evaluate the second-order energy correction, we must evaluate the matrix elements of $\hat{H}'$ between the nth state and all other states m, and then perform the infinite sum in (9.35). In many cases the second-order energy correction cannot be evaluated exactly. Third-order and higher-order energy corrections are even harder to deal with.

The sums in (9.28) and (9.36) are sums over different states rather than sums over different energy values. If some of the energy levels (other than the nth) are degenerate, we must include a term in the sums for each linearly independent wave function corresponding to the degenerate levels.

The reason we have a sum over states in (9.28) and (9.36) is that we require a complete set of functions for the expansion (9.25), and therefore we must include all linearly independent wave functions in the sum. If the unperturbed problem has continuum wave functions (for example, the hydrogen atom), we must also include an integration over the continuum functions, if we are to have a complete set. If $\psi_E^{(0)}$ denotes an unperturbed continuum wave function of energy $E^{(0)}$, then (9.27) and (9.35) become

$$\psi_n^{(1)} = \sum_{m \neq n} \frac{H'_{mn}}{E_n^{(0)} - E_m^{(0)}} \psi_m^{(0)} + \int \frac{H'_{E,n}}{E_n^{(0)} - E^{(0)}} \psi_E^{(0)}\, dE^{(0)}$$

$$E_n^{(2)} = \sum_{m \neq n} \frac{|H'_{mn}|^2}{E_n^{(0)} - E_m^{(0)}} + \int \frac{|H'_{E,n}|^2}{E_n^{(0)} - E^{(0)}} dE^{(0)} \tag{9.37}$$

where $H'_{E,n} \equiv \langle \psi_E^{(0)} | \hat{H}' | \psi_n^{(0)} \rangle$. The integrals in these equations are over the range of continuum-state energies (for example, from zero to infinity for the hydrogen atom). The existence of continuum states in the unperturbed problem makes evaluation of $E_n^{(2)}$ even harder.

The Variation–Perturbation Method. The variation–perturbation method allows one to accurately estimate $E^{(2)}$ and higher-order perturbation-theory energy corrections for the ground state of a system without evaluating the infinite sum in (9.36). The method is based on the inequality

$$\langle u|\hat{H}^0 - E_g^{(0)}|u\rangle + \langle u|\hat{H}' - E_g^{(1)}|\psi_g^{(0)}\rangle + \langle \psi_g^{(0)}|\hat{H}' - E_g^{(1)}|u\rangle \geq E_g^{(2)} \qquad (9.38)$$

where u is any well-behaved function that satisfies the boundary conditions, and where the subscript g refers to the ground state. For the proof of (9.38), see *Hameka*, Section 7-9. By taking u to be a trial function with parameters that we vary to minimize the left side of (9.38), we can estimate $E_g^{(2)}$. The function u turns out to be an approximation to $\psi_g^{(1)}$, the first-order correction to the ground-state wave function, and u can then be used to estimate $E_g^{(3)}$ also. Similar variational integrals can be used to find higher-order corrections to the ground-state energy and wave function.

9.3 PERTURBATION TREATMENT OF THE HELIUM-ATOM GROUND STATE

The helium atom has two electrons and a nucleus of charge $+2e$. We shall consider the nucleus to be at rest (Section 6.6) and place the origin of the coordinate system at the nucleus. The coordinates of electrons 1 and 2 are (x_1, y_1, z_1) and (x_2, y_2, z_2); see Fig. 9.1.

If we take the nuclear charge to be $+Ze$ instead of $+2e$, we can treat heliumlike ions such as H^-, Li^+, Be^{2+}. The Hamiltonian operator is

$$\hat{H} = -\frac{\hbar^2}{2m_e}\nabla_1^2 - \frac{\hbar^2}{2m_e}\nabla_2^2 - \frac{Ze'^2}{r_1} - \frac{Ze'^2}{r_2} + \frac{e'^2}{r_{12}} \qquad (9.39)$$

where m_e is the mass of the electron, r_1 and r_2 are the distances of electrons 1 and 2 from the nucleus, and r_{12} is the distance from electron 1 to 2. The first two terms are the operators for the electrons' kinetic energy [Eq. (3.48)]. The third and fourth terms are the potential energies of attraction between the electrons and the nucleus. The final term is the potential energy of interelectronic repulsion [Eq. (6.58)]. Note that the potential energy of a system of interacting particles cannot be written as the sum of potential energies of the individual particles; the potential energy is a property of the system as a whole.

The Schrödinger equation involves six independent variables, three coordinates for each electron. In spherical coordinates, $\psi = \psi(r_1, \theta_1, \phi_1, r_2, \theta_2, \phi_2)$.

The operator ∇_1^2 is given by Eq. (6.6) with r_1, θ_1, ϕ_1 replacing r, θ, ϕ. The variable r_{12} is $r_{12} = [(x_1 - x_2)^2 + (y_1 - y_2)^2 + (z_1 - z_2)^2]^{1/2}$, and by using the relations between Cartesian and spherical coordinates, we can express r_{12} in terms of $r_1, \theta_1, \phi_1, r_2, \theta_2, \phi_2$.

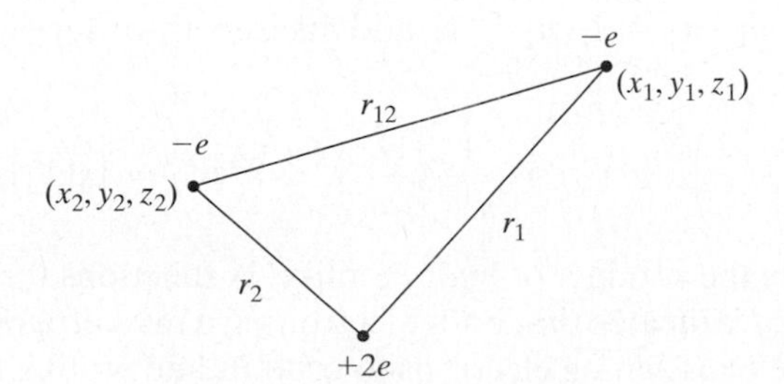

FIGURE 9.1 Interparticle distances in the helium atom.

Because of the e'^2/r_{12} term, the Schrödinger equation for helium cannot be separated in any coordinate system, and we must use approximation methods. The perturbation method separates the Hamiltonian (9.39) into two parts, $\hat{H}^0$ and $\hat{H}'$, where $\hat{H}^0$ is the Hamiltonian of an exactly solvable problem. If we choose

$$\hat{H}^0 = -\frac{\hbar^2}{2m_e}\nabla_1^2 - \frac{Ze'^2}{r_1} - \frac{\hbar^2}{2m_e}\nabla_2^2 - \frac{Ze'^2}{r_2} \tag{9.40}$$

$$\hat{H}' = \frac{e'^2}{r_{12}} \tag{9.41}$$

then $\hat{H}^0$ is the sum of two hydrogenlike Hamiltonians, one for each electron:

$$\hat{H}^0 = \hat{H}_1^0 + \hat{H}_2^0 \tag{9.42}$$

$$\hat{H}_1^0 \equiv -\frac{\hbar^2}{2m_e}\nabla_1^2 - \frac{Ze'^2}{r_1}, \quad \hat{H}_2^0 \equiv -\frac{\hbar^2}{2m_e}\nabla_2^2 - \frac{Ze'^2}{r_2} \tag{9.43}$$

The unperturbed system is a helium atom in which the two electrons exert no forces on each other. Although such a system does not exist, this does not prevent us from applying perturbation theory to this system.

Since the unperturbed Hamiltonian (9.42) is the sum of the Hamiltonians for two independent particles, we can use the separation-of-variables results of Eqs. (6.18) to (6.24) to conclude that the unperturbed wave functions have the form

$$\psi^{(0)}(r_1, \theta_1, \phi_1, r_2, \theta_2, \phi_2) = F_1(r_1, \theta_1, \phi_1)F_2(r_2, \theta_2, \phi_2) \tag{9.44}$$

and the unperturbed energies are

$$E^{(0)} = E_1 + E_2 \tag{9.45}$$

$$\hat{H}_1^0 F_1 = E_1 F_1, \qquad \hat{H}_2^0 F_2 = E_2 F_2 \tag{9.46}$$

Since $\hat{H}_1^0$ and $\hat{H}_2^0$ are hydrogenlike Hamiltonians, the solutions of (9.46) are the hydrogenlike eigenfunctions and eigenvalues. From Eq. (6.94), we have

$$E_1 = -\frac{Z^2}{n_1^2}\frac{e'^2}{2a_0}, \qquad E_2 = -\frac{Z^2}{n_2^2}\frac{e'^2}{2a_0} \tag{9.47}$$

$$E^{(0)} = -Z^2\left(\frac{1}{n_1^2} + \frac{1}{n_2^2}\right)\frac{e'^2}{2a_0}, \qquad \begin{array}{l} n_1 = 1, 2, 3, \ldots \\ n_2 = 1, 2, 3, \ldots \end{array} \tag{9.48}$$

where a_0 is the Bohr radius. Equation (9.48) gives the zeroth-order energies of states with both electrons bound to the nucleus. The He atom also has continuum states.

The lowest level has $n_1 = 1$, $n_2 = 1$, and its zeroth-order wave function is [Eq. (6.104)]

$$\psi_{1s^2}^{(0)} = \frac{1}{\pi^{1/2}}\left(\frac{Z}{a_0}\right)^{3/2} e^{-Zr_1/a_0} \cdot \frac{1}{\pi^{1/2}}\left(\frac{Z}{a_0}\right)^{3/2} e^{-Zr_2/a_0} = 1s(1)1s(2) \tag{9.49}$$

where $1s(1)1s(2)$ denotes the product of hydrogenlike $1s$ functions for electrons 1 and 2, and where the subscript indicates that both electrons are in hydrogenlike $1s$ orbitals. (Note that the procedure of assigning electrons to orbitals and writing the atomic wave function as the product of one-electron orbital functions is an *approximation*.) The energy of this unperturbed ground state is

$$E^{(0)}_{1s^2} = -Z^2(2)\frac{e'^2}{2a_0} \tag{9.50}$$

The quantity $-e'^2/2a_0$ is the ground-state energy of the hydrogen atom (taking the nucleus to be infinitely heavy) and equals -13.606 eV [Eqs. (6.105)–(6.108)]. If the electron mass m_e in a_0 is replaced by the reduced mass for ^{4}He, $-e'^2/2a_0$ is changed to -13.604 eV, and we shall use this number to (partly) correct for the nuclear motion in He. For helium, $Z = 2$ and (9.50) gives $-8(13.604 \text{ eV}) = -108.83$ eV:

$$E^{(0)}_{1s^2} = -108.83 \text{ eV} \tag{9.51}$$

How does this zeroth-order energy compare with the true ground-state energy of helium? The experimental first ionization energy of He is 24.59 eV. The second ionization energy of He is easily calculated theoretically, since it is the ionization energy of the hydrogenlike ion He^+ and is equal to $2^2(13.604 \text{ eV}) = 54.42$ eV. If we choose the zero of energy as the completely ionized atom [this choice is implicit in (9.39)], then the ground-state energy of the helium atom is $-(24.59 + 54.42)$ eV $= -79.01$ eV. The zeroth-order energy (9.51) is in error by 38%. We should have expected such a large error, since the perturbation term e'^2/r_{12} is not small.

The next step is to evaluate the first-order perturbation correction to the energy. The unperturbed ground state is nondegenerate, and

$$E^{(1)} = \langle \psi^{(0)} | \hat{H}' | \psi^{(0)} \rangle$$

$$E^{(1)} = \frac{Z^6 e'^2}{\pi^2 a_0^6} \int_0^{2\pi}\int_0^{2\pi}\int_0^{\pi}\int_0^{\pi}\int_0^{\infty}\int_0^{\infty} e^{-2Zr_1/a_0} e^{-2Zr_2/a_0} \frac{1}{r_{12}} r_1^2 \sin\theta_1$$
$$\times\, r_2^2 \sin\theta_2 \, dr_1\, dr_2\, d\theta_1\, d\theta_2\, d\phi_1\, d\phi_2 \tag{9.52}$$

The volume element for this two-electron problem contains the coordinates of both electrons; $d\tau = d\tau_1\, d\tau_2$. The integral in (9.52) can be evaluated by using an expansion of $1/r_{12}$ in terms of spherical harmonics, as outlined in Prob. 9.13. One finds

$$E^{(1)} = \frac{5Z}{8}\left(\frac{e'^2}{a_0}\right) \tag{9.53}$$

Recalling that $\frac{1}{2}e'^2/a_0$ equals 13.604 eV when the ^{4}He reduced mass is used, and putting $Z = 2$, we find for the first-order perturbation energy correction for the helium ground state:

$$E^{(1)} = \tfrac{10}{4}(13.604 \text{ eV}) = 34.01 \text{ eV}$$

Our approximation to the energy is now

$$E^{(0)} + E^{(1)} = -108.83 \text{ eV} + 34.01 \text{ eV} = -74.82 \text{ eV} \tag{9.54}$$

which, compared with the experimental value of -79.01 eV, is in error by 5.3%.

To evaluate the first-order correction to the wave function and higher-order corrections to the energy requires evaluating the matrix elements of $1/r_{12}$ between the ground unperturbed state and all excited states (including the continuum) and performing the appropriate summations and integrations. No one has yet figured out how to evaluate directly all the contributions to $E^{(2)}$. Note that the effect of $\psi^{(1)}$ is to mix into the wave-function contributions from other configurations besides $1s^2$. We call this *configuration interaction*. The largest contribution to the true ground-state wave function of helium comes from the $1s^2$ configuration, which is the unperturbed (zeroth-order) wave function.

$E^{(2)}$ for the helium ground state has been calculated using the variation–perturbation method, Eq. (9.38). Scherr and Knight used 100-term trial functions to get extremely accurate approximations to the wave-function corrections through sixth order and thus to the energy corrections through thirteenth order [C. W. Scherr and R. E. Knight, *Rev. Mod. Phys*., **35**, 436 (1963)]. For calculations of the energy corrections through order 401, see J. D. Baker et al., *Phys. Rev. A*, **41**, 1247 (1990). The second-order correction $E^{(2)}$ turns out to be -4.29 eV, and $E^{(3)}$ is $+0.12$ eV. Through third order, we have for the ground-state energy

$$E \approx -108.83 \text{ eV} + 34.01 \text{ eV} - 4.29 \text{ eV} + 0.12 \text{ eV} = -78.99 \text{ eV}$$

which is close to the experimental value -79.01 eV. Including corrections through thirteenth order, Scherr and Knight obtained a ground-state energy for helium of $-2.90372433(e'^2/a_0)$, which is nearly as good as the value $-2.90372438(e'^2/a_0)$ obtained from the purely variational calculations described in the next section.

The perturbation-theory series expansion for the He-atom energy can be proved to converge [R. Ahlrichs, *Phys. Rev. A*, **5**, 605 (1972)].

9.4 VARIATION TREATMENTS OF THE GROUND STATE OF HELIUM

In the last section, we wrote the helium-atom Hamiltonian as $\hat{H} = \hat{H}^0 + \hat{H}'$, where the ground-state eigenfunction $\psi_g^{(0)}$ of $\hat{H}^0$ is (9.49). What happens if we use the zeroth-order perturbation-theory ground-state wave function $\psi_g^{(0)}$ as the variation function ϕ in the variational integral? The variational integral $\langle\phi|\hat{H}|\phi\rangle = \langle\phi|\hat{H}\phi\rangle$ then becomes

$$\begin{aligned}\langle\phi|\hat{H}|\phi\rangle &= \langle\psi_g^{(0)}|(\hat{H}^0 + \hat{H}')\psi_g^{(0)}\rangle = \langle\psi_g^{(0)}|\hat{H}^0\psi_g^{(0)} + \hat{H}'\psi_g^{(0)}\rangle \\ &= \langle\psi_g^{(0)}|E_g^{(0)}\psi_g^{(0)}\rangle + \langle\psi_g^{(0)}|\hat{H}'\psi_g^{(0)}\rangle = E_g^{(0)} + E_g^{(1)}\end{aligned} \tag{9.55}$$

since $\hat{H}\psi_g^{(0)} = E_g^{(0)}\psi_g^{(0)}$, $\langle\psi_g^{(0)}|\psi_g^{(0)}\rangle = 1$, and $E_g^{(1)} = \langle\psi_g^{(0)}|\hat{H}'|\psi_g^{(0)}\rangle$ [Eq. (9.22)]. Use of $\psi_g^{(0)}$ as the variation function gives the same energy result as in first-order perturbation theory.

Now consider variation functions for the helium-atom ground state. If we used $\psi_g^{(0)}$ [Eq. (9.49)] as the trial function, we would get the first-order perturbation result, -74.82 eV. To improve on this result, we introduce a variational parameter into (9.49). We try the normalized function

$$\phi = \frac{1}{\pi}\left(\frac{\zeta}{a_0}\right)^3 e^{-\zeta r_1/a_0} e^{-\zeta r_2/a_0} \tag{9.56}$$

which is obtained from (9.49) by replacing the true atomic number Z by a variational parameter ζ (zeta). ζ has a simple physical interpretation. Since one electron tends to screen the other from the nucleus, each electron is subject to an effective nuclear charge somewhat less than the full nuclear charge Z. If one electron fully shielded the other from the nucleus, we would have an effective nuclear charge of $Z - 1$. Since both electrons are in the same orbital, they will be only partly effective in shielding each other. We thus expect ζ to lie between $Z - 1$ and Z.

We now evaluate the variational integral. To expedite things, we rewrite the helium Hamiltonian (9.39) as

$$\hat{H} = \left[-\frac{\hbar^2}{2m_e}\nabla_1^2 - \frac{\zeta e'^2}{r_1} - \frac{\hbar^2}{2m_e}\nabla_2^2 - \frac{\zeta e'^2}{r_2} \right] + (\zeta - Z)\frac{e'^2}{r_1} + (\zeta - Z)\frac{e'^2}{r_2} + \frac{e'^2}{r_{12}} \quad (9.57)$$

where the terms involving zeta were added and subtracted. The terms in brackets in (9.57) are the sum of two hydrogenlike Hamiltonians for nuclear charge ζ. Moreover, the trial function (9.56) is the product of two hydrogenlike 1s functions for nuclear charge ζ. Therefore, when these terms operate on ϕ, we have an eigenvalue equation, the eigenvalue being the sum of two hydrogenlike 1s energies for nuclear charge ζ:

$$\left[-\frac{\hbar^2}{2m_e}\nabla_1^2 - \frac{\zeta e'^2}{r_1} - \frac{\hbar^2}{2m_e}\nabla_2^2 - \frac{\zeta e'^2}{r_2} \right]\phi = -\zeta^2(2)\frac{e'^2}{2a_0}\phi \quad (9.58)$$

Using (9.57) and (9.58), we have

$$\int \phi^* \hat{H}\phi \, d\tau = -\zeta^2 \frac{e'^2}{a_0}\int \phi^*\phi \, d\tau + (\zeta - Z)e'^2 \int \frac{\phi^*\phi}{r_1} d\tau + (\zeta - Z)e'^2 \int \frac{\phi^*\phi}{r_2} d\tau + e'^2 \int \frac{\phi^*\phi}{r_{12}} d\tau \quad (9.59)$$

Let f_1 be a normalized 1s hydrogenlike orbital for nuclear charge ζ, occupied by electron 1. Let f_2 be the same function for electron 2:

$$f_1 = \frac{1}{\pi^{1/2}}\left(\frac{\zeta}{a_0}\right)^{3/2} e^{-\zeta r_1/a_0}, \qquad f_2 = \frac{1}{\pi^{1/2}}\left(\frac{\zeta}{a_0}\right)^{3/2} e^{-\zeta r_2/a_0} \quad (9.60)$$

Noting that $\phi = f_1 f_2$, we now evaluate the integrals in (9.59):

$$\int \phi^*\phi \, d\tau = \iint f_1^* f_2^* f_1 f_2 \, d\tau_1 \, d\tau_2 = \int f_1^* f_1 \, d\tau_1 \int f_2^* f_2 \, d\tau_2 = 1$$

$$\int \frac{\phi^*\phi}{r_1} d\tau = \int \frac{f_1^* f_1}{r_1} d\tau_1 \int f_2^* f_2 \, d\tau_2 = \int \frac{f_1^* f_1}{r_1} d\tau_1$$

$$\int \frac{\phi^*\phi}{r_1} d\tau = \frac{1}{\pi}\frac{\zeta^3}{a_0^3}\int_0^\infty e^{-2\zeta r_1/a_0}\frac{r_1^2}{r_1} dr_1 \int_0^\pi \sin\theta_1 \, d\theta_1 \int_0^{2\pi} d\phi_1 = \frac{\zeta}{a_0}$$

where the Appendix integral (A.8) was used. Also,

$$\int \frac{\phi^*\phi}{r_2} d\tau = \int \frac{f_2^* f_2}{r_2} d\tau_2 = \int \frac{f_1^* f_1}{r_1} d\tau_1 = \frac{\zeta}{a_0}$$

since it doesn't matter whether the label 1 or 2 is used on the dummy variables in the definite integral. Finally, we must evaluate $e'^2 \int (\phi^*\phi/r_{12}) \, d\tau$. This is the same as the integral (9.52) that occurred in the perturbation treatment, except that Z is replaced by ζ. Hence, from (9.53)

$$e'^2 \int \frac{\phi^*\phi}{r_{12}}\, d\tau = \frac{5\zeta e'^2}{8a_0} \tag{9.61}$$

The variational integral (9.59) thus has the value

$$\int \phi^*\hat{H}\phi\, d\tau = \left(\zeta^2 - 2Z\zeta + \tfrac{5}{8}\zeta\right)\frac{e'^2}{a_0} \tag{9.62}$$

As a check, if we set $\zeta = Z$ in (9.62), we get the first-order perturbation-theory result, (9.50) plus (9.53).

We now vary ζ to minimize the variational integral:

$$\frac{\partial}{\partial\zeta}\int \phi^*\hat{H}\phi\, d\tau = \left(2\zeta - 2Z + \tfrac{5}{8}\right)\frac{e'^2}{a_0} = 0$$

$$\zeta = Z - \tfrac{5}{16} \tag{9.63}$$

As anticipated, the effective nuclear charge lies between Z and $Z - 1$. Using (9.63) and (9.62), we get

$$\int \phi^*\hat{H}\phi\, d\tau = \left(-Z^2 + \tfrac{5}{8}Z - \tfrac{25}{256}\right)\frac{e'^2}{a_0} = -\left(Z - \tfrac{5}{16}\right)^2\frac{e'^2}{a_0} \tag{9.64}$$

Putting $Z = 2$, we get as our approximation to the helium ground-state energy $-(27/16)^2 e'^2/a_0 = -(729/256)2(13.604 \text{ eV}) = -77.48$ eV, as compared with the true value of -79.01 eV. Use of ζ instead of Z has reduced the error from 5.3% to 1.9%. In accord with the variation theorem, the true ground-state energy is less than the variational integral.

How can we improve our variational result? We might try a function that had the general form of (9.56), that is, a product of two functions, one for each electron:

$$\phi = u(1)u(2) \tag{9.65}$$

However, we could try a variety of functions u in (9.65), instead of the single exponential used in (9.56). A systematic procedure for finding the function u that gives the lowest value of the variational integral will be discussed in Section 11.1. This procedure shows that for the best possible choice of u in (9.65) the variational integral equals -77.86 eV, which is still in error by 1.5%. We might ask why (9.65) does not give the true ground-state energy, no matter what form we try for u. The answer is that, when we write the trial function as the product of separate functions for each electron, we are making an approximation. Because of the e'^2/r_{12} term in the Hamiltonian, the Schrödinger equation for helium is not separable, and the true ground-state wave function cannot be written as the product of separate functions for each electron. To reach the true ground-state energy, we must go beyond a function of the form (9.65).

The Bohr model gave the correct energies for the hydrogen atom but failed when applied to helium. Hence, in the early days of quantum mechanics, it was important to show that the new theory could give an accurate treatment of helium. The pioneering work on the helium ground state was done by Hylleraas in the years 1928–1930. To allow for the effect of one electron on the motion of the other, Hylleraas used variational functions that contained the interelectronic distance r_{12}. One function he used is

$$\phi = N[e^{-\zeta r_1/a_0}e^{-\zeta r_2/a_0}(1 + br_{12})] \tag{9.66}$$

where N is the normalization constant and ζ and b are variational parameters. Since

$$r_{12} = [(x_2 - x_1)^2 + (y_2 - y_1)^2 + (z_2 - z_1)^2]^{1/2} \tag{9.67}$$

the function (9.66) goes beyond the simple product form (9.65). Minimization of the variational integral with respect to the parameters gives $\zeta = 1.849$, $b = 0.364/a_0$, and a ground-state energy of -78.7 eV, in error by 0.3 eV. The $1 + br_{12}$ term makes the wave function larger for large values of r_{12}. This is as it should be, because the repulsion between the electrons makes it energetically more favorable for the electrons to avoid each other. Using a more complicated six-term trial function containing r_{12}, Hylleraas obtained an energy only 0.01 eV above the true ground-state energy.

Hylleraas's work has been extended by others. Using a 1078-term variational function, Pekeris found a ground-state energy of $-2.903724375(e'^2/a_0)$ [C. L. Pekeris, *Phys. Rev.*, **115**, 1216 (1959)]. With relativistic and nuclear-motion corrections added, this gave for E_i, the ionization energy of helium, $E_i/hc = 198310.69\text{ cm}^{-1}$, compared with the experimental value 198310.67 cm^{-1}. Using an improved variational function, Frankowski and Pekeris bettered Perkeris's result by obtaining $-2.90372437703(e'^2/a_0)$, a result believed to be within $10^{-11}(e'^2/a_0)$ of the true nonrelativistic ground-state energy [K. Frankowski and C. L. Pekeris, *Phys. Rev.*, **146**, 46 (1966)]. Drake and Yan used linear variational functions containing r_{12} to calculate the ground-state energy and many excited-state energies of He that are thought to be accurate to 1 part in 10^{14} or better [G. W. F. Drake and Z-C. Yan, *Chem. Phys. Lett.*, **229**, 486 (1994); *Phys. Rev. A*, **46**, 2378 (1992)]. These workers similarly calculated Li variational energies for the ground state and two excited states with 1 part in 10^9 accuracy or better [Z-C. Yan and G. W. F. Drake, *Phys. Rev. A*, **52**, 3711 (1995)]. Adding in relativistic and nuclear motion corrections, Drake and Yan found good agreement between theoretically calculated and experimental spectroscopic transition frequencies of He and Li. Variational calculations with functions containing r_{ij} become very hard for many-electron atoms because of the many terms and difficult integrals that occur.

9.5 PERTURBATION THEORY FOR A DEGENERATE ENERGY LEVEL

We now consider the perturbation treatment of an energy level whose degree of degeneracy is d. We have d linearly independent unperturbed wave functions corresponding to the degenerate level. We shall use the labels $1, 2, \ldots, d$ for the states of the degenerate level, without implying that these are necessarily the lowest-lying states. The unperturbed Schrödinger equation is

$$\hat{H}^0\psi_n^{(0)} = E_n^{(0)}\psi_n^{(0)} \tag{9.68}$$

with

$$E_1^{(0)} = E_2^{(0)} = \cdots = E_d^{(0)} \tag{9.69}$$

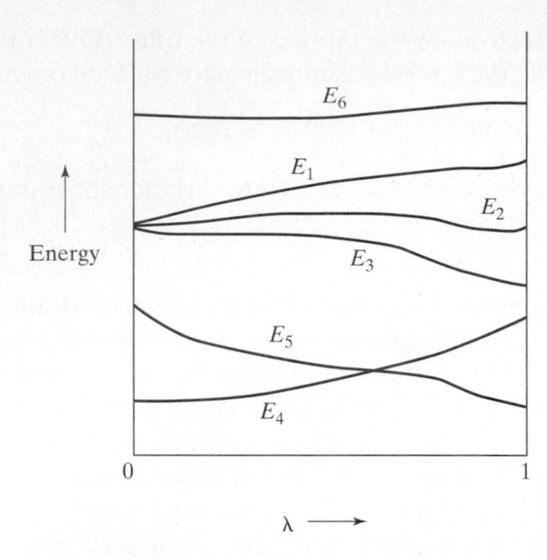

FIGURE 9.2 Effect of a perturbation on energy levels.

The perturbed problem is

$$\hat{H}\psi_n = E_n\psi_n \tag{9.70}$$

$$\hat{H} = \hat{H}^0 + \lambda\hat{H}' \tag{9.71}$$

As λ goes to zero, the eigenvalues of (9.70) go to the eigenvalues of (9.68); we have $\lim_{\lambda\to 0} E_n = E_n^{(0)}$. Figure 9.2 shows this for a hypothetical system with six states and a threefold-degenerate unperturbed level. Note that the perturbation splits the degenerate energy level. In some cases the perturbation may have no effect on the degeneracy or may only partly remove the degeneracy.

As $\lambda \to 0$, the eigenfunctions satisfying (9.70) approach eigenfunctions satisfying (9.68). Does this mean that $\lim_{\lambda\to 0}\psi_n = \psi_n^{(0)}$? Not necessarily. If $E_n^{(0)}$ is nondegenerate, there is a unique normalized eigenfunction $\psi_n^{(0)}$ of $\hat{H}^0$ with eigenvalue $E_n^{(0)}$, and we can be sure that $\lim_{\lambda\to 0}\psi_n = \psi_n^{(0)}$. However, if $E_n^{(0)}$ is the eigenvalue of the d-fold degenerate level, then (Section 3.6) any linear combination

$$c_1\psi_1^{(0)} + c_2\psi_2^{(0)} + \cdots + c_d\psi_d^{(0)} \tag{9.72}$$

is a solution of (9.68) with eigenvalue (9.69). The set of linearly independent normalized functions $\psi_1^{(0)}, \psi_2^{(0)}, \ldots, \psi_d^{(0)}$, which we use as eigenfunctions corresponding to the states of the degenerate level, is not unique. Using (9.72), we can construct an infinite number of sets of d linearly independent normalized eigenfunctions for the degenerate level. As far as the unperturbed problem is concerned, one such set is as good as another. For example, for the three degenerate $2p$ states of the hydrogen atom, we can use the $2p_1$, $2p_0$, and $2p_{-1}$ functions; the $2p_x$, $2p_y$, and $2p_z$, functions; or some other set of three linearly independent functions constructed as linear combinations of the members of one of the preceding sets. For the perturbed eigenfunctions that correspond to the d-fold degenerate unperturbed level, all we can say is that as λ approaches zero they each approach a linear combination of unperturbed eigenfunctions:

$$\lim_{\lambda\to 0}\psi_n = \sum_{i=1}^{d} c_i\psi_i^{(0)}, \qquad 1 \le n \le d \tag{9.73}$$

Our first task is thus to determine the **correct zeroth-order wave functions** (9.73) for the perturbation $\hat{H}'$. Calling these correct zeroth-order functions $\phi_n^{(0)}$, we have

$$\phi_n^{(0)} = \lim_{\lambda \to 0} \psi_n = \sum_{i=1}^{d} c_i \psi_i^{(0)}, \qquad 1 \le n \le d \tag{9.74}$$

Each different function $\phi_n^{(0)}$ has a different set of coefficients in (9.74). The correct set of zeroth-order functions depends on what the perturbation $\hat{H}'$ is.

The treatment of the d-fold degenerate level proceeds like the nondegenerate treatment of Section 9.2, except that instead of $\psi_n^{(0)}$ we use $\phi_n^{(0)}$. Instead of Eqs. (9.13) and (9.14), we have

$$\psi_n = \phi_n^{(0)} + \lambda\psi_n^{(1)} + \lambda^2\psi_n^{(2)} + \cdots, \qquad n = 1, 2, \ldots, d \tag{9.75}$$

$$E_n = E_d^{(0)} + \lambda E_n^{(1)} + \lambda^2 E_n^{(2)} + \cdots, \qquad n = 1, 2, \ldots, d \tag{9.76}$$

where (9.69) was used. Substitution into $\hat{H}\psi_n = E_n\psi_n$ gives

$$(\hat{H}^0 + \lambda\hat{H}')(\phi_n^{(0)} + \lambda\psi_n^{(1)} + \lambda^2\psi_n^{(2)} + \cdots)$$
$$= (E_d^{(0)} + \lambda E_n^{(1)} + \lambda^2 E_n^{(2)} + \cdots)(\phi_n^{(0)} + \lambda\psi_n^{(1)} + \lambda^2\psi_n^{(2)} + \cdots)$$

Equating the coefficients of λ^0 in this equation, we get $\hat{H}^0\phi_n^{(0)} = E_d^{(0)}\phi_n^{(0)}$. By the theorem of Section 3.6, each linear combination $\phi_n^{(0)}$ ($n = 1, 2, \ldots, d$) is an eigenfunction of $\hat{H}^0$ with eigenvalue $E_d^{(0)}$, and this equation gives no new information.

Equating the coefficients of the λ^1 terms, we get

$$\hat{H}^0\psi_n^{(1)} + \hat{H}'\phi_n^{(0)} = E_d^{(0)}\psi_n^{(1)} + E_n^{(1)}\phi_n^{(0)}$$

$$\hat{H}^0\psi_n^{(1)} - E_d^{(0)}\psi_n^{(1)} = E_n^{(1)}\phi_n^{(0)} - \hat{H}'\phi_n^{(0)}, \qquad n = 1, 2, \ldots, d \tag{9.77}$$

We now multiply (9.77) by $\psi_m^{(0)*}$ and integrate over all space, where m is one of the states corresponding to the d-fold degenerate unperturbed level under consideration; that is, $1 \le m \le d$. We get

$$\langle\psi_m^{(0)}|\hat{H}^0|\psi_n^{(1)}\rangle - E_d^{(0)}\langle\psi_m^{(0)}|\psi_n^{(1)}\rangle = E_n^{(1)}\langle\psi_m^{(0)}|\phi_n^{(0)}\rangle - \langle\psi_m^{(0)}|\hat{H}'|\phi_n^{(0)}\rangle,$$
$$1 \le m \le d \tag{9.78}$$

From Eq. (9.20), we have $\langle\psi_m^{(0)}|\hat{H}^0|\psi_n^{(1)}\rangle = E_m^{(0)}\langle\psi_m^{(0)}|\psi_n^{(1)}\rangle$. From (9.69), $E_m^{(0)} = E_d^{(0)}$ for $1 \le m \le d$, so $\langle\psi_m^{(0)}|\hat{H}^0|\psi_n^{(1)}\rangle = E_d^{(0)}\langle\psi_m^{(0)}|\psi_n^{(1)}\rangle$, and the left side of (9.78) equals zero. Equation (9.78) becomes

$$\langle\psi_m^{(0)}|\hat{H}'|\phi_n^{(0)}\rangle - E_n^{(1)}\langle\psi_m^{(0)}|\phi_n^{(0)}\rangle = 0, \qquad m = 1, 2, \ldots, d$$

Substitution of the linear combination (9.74) for $\phi_n^{(0)}$ gives

$$\sum_{i=1}^{d} c_i\langle\psi_m^{(0)}|\hat{H}'|\psi_i^{(0)}\rangle - E_n^{(1)}\sum_{i=1}^{d} c_i\langle\psi_m^{(0)}|\psi_i^{(0)}\rangle = 0 \tag{9.79}$$

The zeroth-order wave functions $\psi_i^{(0)}$ ($i = 1, 2, \ldots, d$) of the degenerate level can always be chosen to be orthonormal, and we shall assume this has been done:

$$\langle\psi_m^{(0)}|\psi_i^{(0)}\rangle = \delta_{mi} \tag{9.80}$$

for m and i in the range 1 to d. Equation (9.79) becomes

$$\sum_{i=1}^{d} [\langle\psi_m^{(0)}|\hat{H}'|\psi_i^{(0)}\rangle - E_n^{(1)}\delta_{mi}]c_i = 0, \qquad m = 1, 2, \ldots, d \tag{9.81}$$

This is a set of d linear, homogeneous equations in the d unknowns $c_1, c_2, \ldots, c_d$, which are the coefficients in the correct zeroth-order wave function $\phi_n^{(0)}$ in (9.74). Writing out (9.81), we have

$$\begin{aligned}
(H'_{11} - E_n^{(1)})c_1 + H'_{12}c_2 + \cdots + H'_{1d}c_d &= 0 \\
H'_{21}c_1 + (H'_{22} - E_n^{(1)})c_2 + \cdots + H'_{2d}c_d &= 0 \\
\cdots\cdots\cdots\cdots\cdots\cdots\cdots\cdots\cdots\cdots\cdots\cdots\cdots \\
H'_{d1}c_1 + H'_{d2}c_2 + \cdots + (H'_{dd} - E_n^{(1)})c_d &= 0 \\
H'_{mi} \equiv \langle \psi_m^{(0)} | \hat{H}' | \psi_i^{(0)} \rangle
\end{aligned} \tag{9.82}$$

For this set of linear homogeneous equations to have a nontrivial solution, the determinant of the coefficients must vanish (Section 8.4):

$$\boxed{\det[\langle \psi_m^{(0)} | \hat{H}' | \psi_i^{(0)} \rangle - E_n^{(1)} \delta_{mi}] = 0} \tag{9.83}$$

$$\begin{vmatrix}
H'_{11} - E_n^{(1)} & H'_{12} & \cdots & H'_{1d} \\
H'_{21} & H'_{22} - E_n^{(1)} & \cdots & H'_{2d} \\
\vdots & \vdots & \ddots & \vdots \\
H'_{d1} & H'_{d2} & \cdots & H'_{dd} - E_n^{(1)}
\end{vmatrix} = 0 \tag{9.84}$$

The **secular equation** (9.84) is an algebraic equation of degree d in $E_n^{(1)}$. It has d roots, $E_1^{(1)}, E_2^{(1)}, \ldots, E_d^{(1)}$, which are the first-order corrections to the energy of the d-fold degenerate unperturbed level. If the roots are all different, then the first-order perturbation correction has split the d-fold degenerate unperturbed level into d different perturbed levels of energies (correct through first order):

$$E_d^{(0)} + E_1^{(1)}, \qquad E_d^{(0)} + E_2^{(1)}, \qquad \ldots, \qquad E_d^{(0)} + E_d^{(1)}$$

If two or more roots of the secular equation are equal, the degeneracy is not completely removed in first order. In the rest of this section, we shall assume that all the roots of (9.84) are different.

Having found the d first-order energy corrections, we go back to the set of equations (9.82) to find the unknowns c_i, which determine the correct zeroth-order wave functions. To find the correct zeroth-order function

$$\phi_n^{(0)} = c_1\psi_1^{(0)} + c_2\psi_2^{(0)} + \cdots + c_d\psi_d^{(0)} \tag{9.85}$$

corresponding to the root $E_n^{(1)}$, we solve (9.82) for $c_2, c_3, \ldots, c_d$ in terms of c_1 and then find c_1 by normalization. Use of (9.80) in $\langle \phi_n^{(0)} | \phi_n^{(0)} \rangle = 1$ gives (Prob. 9.20)

$$\sum_{k=1}^{d} |c_k|^2 = 1 \tag{9.86}$$

For each root $E_n^{(1)}$, $n = 1, 2, \ldots, d$, we have a different set of coefficients $c_1, c_2, \ldots, c_d$, giving a different correct zeroth-order wave function.

In the next section, we shall show that

$$E_n^{(1)} = \langle \phi_n^{(0)} | \hat{H}' | \phi_n^{(0)} \rangle, \quad n = 1, 2, \ldots, d \tag{9.87}$$

which is similar to the nondegenerate-case formula (9.22), except that the correct zeroth-order functions have to be used.

Using procedures similar to those for the nondegenerate case, one can now find the first-order corrections to the correct zeroth-order wave functions and the second-order energy corrections. For the results, see *Bates*, Volume I, pages 197–198; *Hameka*, pages 230–231.

As an example, consider the effect of a perturbation $\hat{H}'$ on the lowest degenerate energy level of a particle in a cubic box. We have three states corresponding to this level: $\psi_{211}^{(0)}$, $\psi_{121}^{(0)}$, and $\psi_{112}^{(0)}$. These unperturbed wave functions are orthonormal, and the secular equation (9.84) is

$$\begin{vmatrix} \langle 211|\hat{H}'|211\rangle - E_n^{(1)} & \langle 211|\hat{H}'|121\rangle & \langle 211|\hat{H}'|112\rangle \\ \langle 121|\hat{H}'|211\rangle & \langle 121|\hat{H}'|121\rangle - E_n^{(1)} & \langle 121|\hat{H}'|112\rangle \\ \langle 112|\hat{H}'|211\rangle & \langle 112|\hat{H}'|121\rangle & \langle 112|\hat{H}'|112\rangle - E_n^{(1)} \end{vmatrix} = 0 \quad (9.88)$$

Solving this equation, we find the first-order energy corrections: $E_1^{(1)}, E_2^{(1)}, E_3^{(1)}$. The triply degenerate unperturbed level is split into three levels of energies (through first order): $(6h^2/8ma^2) + E_1^{(1)}$, $(6h^2/8ma^2) + E_2^{(1)}$, $(6h^2/8ma^2) + E_3^{(1)}$. Using each of the roots $E_1^{(1)}, E_2^{(1)}, E_3^{(1)}$, we get a different set of simultaneous equations (9.82). Solving each set, we find three sets of coefficients, which determine the three correct zeroth-order wave functions.

[If you are familiar with matrix algebra, note that solving (9.84) and (9.82) amounts to finding the eigenvalues and eigenvectors of the matrix whose elements are $\langle\psi_m^{(0)}|\hat{H}'|\psi_i^{(0)}\rangle$.]

9.6 SIMPLIFICATION OF THE SECULAR EQUATION

The secular equation (9.84) is easier to solve if some of the off-diagonal elements of the secular determinant are zero. In the most favorable case, all the off-diagonal elements are zero, and

$$\begin{vmatrix} H'_{11} - E_n^{(1)} & 0 & \cdots & 0 \\ 0 & H'_{22} - E_n^{(1)} & \cdots & 0 \\ \vdots & \vdots & \ddots & \vdots \\ 0 & 0 & \cdots & H'_{dd} - E_n^{(1)} \end{vmatrix} = 0 \quad (9.89)$$

$$(H'_{11} - E_n^{(1)})(H'_{22} - E_n^{(1)})\cdots(H'_{dd} - E_n^{(1)}) = 0$$

$$E_1^{(1)} = H'_{11}, \quad E_2^{(1)} = H'_{22}, \quad \ldots, \quad E_d^{(1)} = H'_{dd} \quad (9.90)$$

Now we want to find the correct zeroth-order wave functions. We shall assume that the roots (9.90) are all different. For the root $E_n^{(1)} = H'_{11}$, the system of equations (9.82) is

$$0 = 0$$
$$(H'_{22} - H'_{11})c_2 = 0$$
$$\cdots\cdots\cdots\cdots\cdots\cdots$$
$$(H'_{dd} - H'_{11})c_d = 0$$

Since we are assuming unequal roots, the quantities $H'_{22} - H'_{11}, \ldots, H'_{dd} - H'_{11}$ are all nonzero. Therefore, $c_2 = 0, c_3 = 0, \ldots, c_d = 0$. The normalization condition (9.86) gives $c_1 = 1$. The correct zeroth-order wave function corresponding to the first-order perturbation energy correction H'_{11} is then [Eq. (9.74)] $\phi_1^{(0)} = \psi_1^{(0)}$. For the root H'_{22}, the same reasoning gives $\phi_2^{(0)} = \psi_2^{(0)}$. Using each of the remaining roots, we find similarly: $\phi_3^{(0)} = \psi_3^{(0)}, \ldots, \phi_d^{(0)} = \psi_d^{(0)}$.

When the secular determinant is in diagonal form, the initially assumed wave functions $\psi_1^{(0)}, \psi_2^{(0)}, \ldots, \psi_n^{(0)}$ *are the correct zeroth-order wave functions for the perturbation* $\hat{H}'$.

The converse is also true. If the initially assumed functions are the correct zeroth-order functions, then the secular determinant is in diagonal form. This is seen as follows. From $\phi_1^{(0)} = \psi_1^{(0)}$ we know that the coefficients in the expansion $\phi_1^{(0)} = \sum_{i=1}^{d} c_i \psi_i^{(0)}$ are $c_1 = 1, c_2 = c_3 = \cdots = 0$, so for $n = 1$ the set of simultaneous equations (9.82) becomes

$$H'_{11} - E_1^{(1)} = 0, \qquad H'_{21} = 0, \qquad \ldots, \qquad H'_{d1} = 0$$

Applying the same reasoning to the remaining functions $\phi_n^{(0)}$, we conclude that $H'_{mi} = 0$ for $i \neq m$. Hence, use of the correct zeroth-order functions makes the secular determinant diagonal. Note also that the first-order corrections to the energy can be found by averaging the perturbation over the correct zeroth-order wave functions:

$$E_n^{(1)} = H'_{nn} = \langle \phi_n^{(0)} | \hat{H}' | \phi_n^{(0)} \rangle \tag{9.91}$$

a result mentioned in Eq. (9.87).

Often, instead of being in diagonal form, the secular determinant is in block-diagonal form. For example, we might have

$$\begin{vmatrix} H'_{11} - E_n^{(1)} & H'_{12} & 0 & 0 \\ H'_{21} & H'_{22} - E_n^{(1)} & 0 & 0 \\ 0 & 0 & H'_{33} - E_n^{(1)} & H'_{34} \\ 0 & 0 & H'_{43} & H'_{44} - E_n^{(1)} \end{vmatrix} = 0 \tag{9.92}$$

The secular determinant in (9.92) has the same form as the secular determinant in the linear-variation secular equation (8.66) with $S_{ij} = \delta_{ij}$. By the same reasoning used to show that two of the variation functions are linear combinations of f_1 and f_2 and two are linear combinations of f_3 and f_4 [Eq. (8.70)], it follows that two of the correct zeroth-order wave functions are linear combinations of $\psi_1^{(0)}$ and $\psi_2^{(0)}$ and two are linear combinations of $\psi_3^{(0)}$ and $\psi_4^{(0)}$:

$$\phi_1^{(0)} = c_1\psi_1^{(0)} + c_2\psi_2^{(0)}, \qquad \phi_2^{(0)} = c_1'\psi_1^{(0)} + c_2'\psi_2^{(0)}$$

$$\phi_3^{(0)} = c_3\psi_3^{(0)} + c_4\psi_4^{(0)}, \qquad \phi_4^{(0)} = c_3'\psi_3^{(0)} + c_4'\psi_4^{(0)}$$

where primes were used to distinguish different coefficients.

When the secular determinant of degenerate perturbation theory is in block-diagonal form, the secular equation breaks up into two or more smaller secular equations, and the set of simultaneous equations (9.82) for the coefficients c_i *breaks up into two or more smaller sets of simultaneous equations.*

Conversely, if we have, say, a fourfold-degenerate unperturbed level, and we happen to know that $\phi_1^{(0)}$ and $\phi_2^{(0)}$ are each linear combinations of $\psi_1^{(0)}$ and $\psi_2^{(0)}$ only, while $\phi_3^{(0)}$ and $\phi_4^{(0)}$ are each linear combinations of $\psi_3^{(0)}$ and $\psi_4^{(0)}$ only, we deal with two second-order secular determinants, rather than a fourth-order secular determinant.

How can we choose the right zeroth-order wave functions in advance and thereby simplify the secular equation? Suppose there is an operator $\hat{A}$ that commutes with both $\hat{H}^0$ and $\hat{H}'$. Then we can choose the unperturbed functions to be eigenfunctions of $\hat{A}$. Because $\hat{A}$ commutes with $\hat{H}'$, this choice of unperturbed functions will make the integrals H'_{ij} vanish if $\psi_i^{(0)}$ and $\psi_j^{(0)}$ belong to different eigenvalues of $\hat{A}$ [see Eq. (7.50)]. Thus, if the eigenvalues of $\hat{A}$ for $\psi_1^{(0)}, \psi_2^{(0)}, \ldots, \psi_d^{(0)}$ are all different, the secular determinant will be in diagonal form, and we will have the right zeroth-order wave functions. If some of the eigenvalues of $\hat{A}$ are the same, we get block-diagonal rather than diagonal form. In general, the correct zeroth-order functions will be linear combinations of those unperturbed functions that have the same eigenvalue of $\hat{A}$. (This is to be expected since $\hat{A}$ commutes with $\hat{H} = \hat{H}^0 + \hat{H}'$, so the perturbed eigenfunctions of $\hat{H}$ can be chosen to be eigenfunctions of $\hat{A}$.) For an example, see Prob. 9.22.

9.7 PERTURBATION TREATMENT OF THE FIRST EXCITED STATES OF HELIUM

Section 9.3 applied perturbation theory to the ground state of the helium atom. We now treat the lowest excited states of He. The unperturbed energies are given by (9.48). The lowest unperturbed excited states have $n_1 = 1, n_2 = 2$ or $n_1 = 2, n_2 = 1$, and substitution in (9.48) gives

$$E^{(0)} = -\frac{5Z^2}{8}\left(\frac{e'^2}{a_0}\right) = -\frac{20}{8}2\left(\frac{e'^2}{2a_0}\right) = -5(13.606 \text{ eV}) = -68.03 \text{ eV} \tag{9.93}$$

Recall that the $n = 2$ level of a hydrogenlike atom is fourfold degenerate, the 2s and three 2p states all having the same energy. The first excited unperturbed energy level of He is eightfold degenerate. The eight unperturbed wave functions are [Eq. (9.44)]

$$\begin{gathered}\psi_1^{(0)} = 1s(1)2s(2), \quad \psi_2^{(0)} = 2s(1)1s(2), \quad \psi_3^{(0)} = 1s(1)2p_x(2), \quad \psi_4^{(0)} = 2p_x(1)1s(2)\\ \psi_5^{(0)} = 1s(1)2p_y(2), \quad \psi_6^{(0)} = 2p_y(1)1s(2), \quad \psi_7^{(0)} = 1s(1)2p_z(2), \quad \psi_8^{(0)} = 2p_z(1)1s(2)\end{gathered} \tag{9.94}$$

where 1s(1)2s(2) signifies the product of a hydrogenlike 1s function for electron 1 and a hydrogenlike 2s function for electron 2. The explicit form of $\psi_8^{(0)}$, for example, is (Table 6.2)

$$\psi_8^{(0)} = \frac{1}{4(2\pi)^{1/2}}\left(\frac{Z}{a_0}\right)^{5/2} r_1 e^{-Zr_1/2a_0}\cos\theta_1 \cdot \frac{1}{\pi^{1/2}}\left(\frac{Z}{a_0}\right)^{3/2} e^{-Zr_2/a_0}$$

We chose to use the real 2p hydrogenlike orbitals, rather than the complex ones.

Since the unperturbed level is degenerate, we must solve a secular equation. The secular equation (9.84) assumes that the functions $\psi_1^{(0)}, \psi_2^{(0)}, \ldots, \psi_8^{(0)}$ are orthonormal. This condition is met. For example,

$$\begin{aligned}\int \psi_1^{(0)*}\psi_1^{(0)}\, d\tau &= \iint 1s(1)^*2s(2)^*1s(1)2s(2)\, d\tau_1\, d\tau_2\\ &= \int |1s(1)|^2\, d\tau_1 \int |2s(2)|^2\, d\tau_2 = 1\cdot 1 = 1\end{aligned}$$

$$\int \psi_3^{(0)*}\psi_5^{(0)}\, d\tau = \int |1s(1)|^2\, d\tau_1 \int 2p_x(2)^*2p_y(2)\, d\tau_2 = 1\cdot 0 = 0$$

where the orthonormality of the hydrogenlike orbitals has been used.

The secular determinant contains $8^2 = 64$ elements. The operator $\hat{H}'$ is Hermitian, and $H'_{ij} = (H'_{ji})^*$. Also, since $\hat{H}'$ and $\psi_1^{(0)}, \ldots, \psi_8^{(0)}$ are all real, we have $(H'_{ji})^* = H'_{ji}$, so $H'_{ij} = H'_{ji}$. The secular determinant is symmetric about the principal diagonal. This cuts the labor of evaluating integrals almost in half.

By using parity considerations, we can show that most of the integrals H'_{ij} are zero. First consider H'_{13}:

$$H'_{13} = \int_{-\infty}^{\infty}\int_{-\infty}^{\infty}\int_{-\infty}^{\infty}\int_{-\infty}^{\infty}\int_{-\infty}^{\infty}\int_{-\infty}^{\infty} 1s(1)2s(2)\frac{e'^2}{r_{12}}1s(1)2p_x(2)\,dx_1\,dy_1\,dz_1\,dx_2\,dy_2\,dz_2$$

An s hydrogenlike function depends only on $r = (x^2 + y^2 + z^2)^{1/2}$ and is therefore an even function. The $2p_x(2)$ function is an odd function of x_2 [Eq. (6.119)]. r_{12} is given by (9.67), and if we invert all six coordinates, r_{12} is unchanged:

$$r_{12} \rightarrow [(-x_1 + x_2)^2 + (-y_1 + y_2)^2 + (-z_1 + z_2)^2]^{1/2} = r_{12}$$

Hence, on inverting all six coordinates, the integrand of H'_{13} goes into minus itself. Therefore [Eq. (7.63)], $H'_{13} = 0$. The same reasoning gives $H'_{14} = H'_{15} = H'_{16} = H'_{17} = H'_{18} = 0$ and $H'_{23} = H'_{24} = H'_{25} = H'_{26} = H'_{27} = H'_{28} = 0$. Now consider H'_{35}:

$$H'_{35} = \int_{-\infty}^{\infty}\cdots\int_{-\infty}^{\infty} 1s(1)2p_x(2)\frac{e'^2}{r_{12}}1s(1)2p_y(2)\,dx_1\cdots dz_2$$

Suppose we invert the x coordinates: $x_1 \rightarrow -x_1$ and $x_2 \rightarrow -x_2$. This transformation will leave r_{12} unchanged. The functions $1s(1)$ and $2p_y(2)$ will be unaffected. However, $2p_x(2)$ will go over to minus itself, so the net effect will be to change the integrand of H'_{35} into minus itself. Hence (Prob. 7.29), $H'_{35} = 0$. Likewise, $H'_{36} = H'_{37} = H'_{38} = 0$ and $H'_{45} = H'_{46} = H'_{47} = H'_{48} = 0$. By considering the transformation $y_1 \rightarrow -y_1$, $y_2 \rightarrow -y_2$, we see that $H'_{57} = H'_{58} = H'_{67} = H'_{68} = 0$. The secular equation is thus

$$\begin{vmatrix} b_{11} & H'_{12} & 0 & 0 & 0 & 0 & 0 & 0 \\ H'_{12} & b_{22} & 0 & 0 & 0 & 0 & 0 & 0 \\ 0 & 0 & b_{33} & H'_{34} & 0 & 0 & 0 & 0 \\ 0 & 0 & H'_{34} & b_{44} & 0 & 0 & 0 & 0 \\ 0 & 0 & 0 & 0 & b_{55} & H'_{56} & 0 & 0 \\ 0 & 0 & 0 & 0 & H'_{56} & b_{66} & 0 & 0 \\ 0 & 0 & 0 & 0 & 0 & 0 & b_{77} & H'_{78} \\ 0 & 0 & 0 & 0 & 0 & 0 & H'_{78} & b_{88} \end{vmatrix} = 0$$

$$b_{ii} \equiv H'_{ii} - E^{(1)}, \qquad i = 1, 2, \ldots, 8$$

The secular determinant is in block-diagonal form and factors into four determinants, each of second order. We conclude that the correct zeroth-order functions have the form

$$\begin{aligned} \phi_1^{(0)} &= c_1\psi_1^{(0)} + c_2\psi_2^{(0)}, & \phi_2^{(0)} &= \bar{c}_1\psi_1^{(0)} + \bar{c}_2\psi_2^{(0)} \\ \phi_3^{(0)} &= c_3\psi_3^{(0)} + c_4\psi_4^{(0)}, & \phi_4^{(0)} &= \bar{c}_3\psi_3^{(0)} + \bar{c}_4\psi_4^{(0)} \\ \phi_5^{(0)} &= c_5\psi_5^{(0)} + c_6\psi_6^{(0)}, & \phi_6^{(0)} &= \bar{c}_5\psi_5^{(0)} + \bar{c}_6\psi_6^{(0)} \\ \phi_7^{(0)} &= c_7\psi_7^{(0)} + c_8\psi_8^{(0)}, & \phi_8^{(0)} &= \bar{c}_7\psi_7^{(0)} + \bar{c}_8\psi_8^{(0)} \end{aligned} \tag{9.95}$$

where the unbarred coefficients correspond to one root of each second-order determinant and the barred coefficients correspond to the second root.

The first determinant is

$$\begin{vmatrix} H'_{11} - E^{(1)} & H'_{12} \\ H'_{12} & H'_{22} - E^{(1)} \end{vmatrix} = 0 \tag{9.96}$$

We have

$$H'_{11} = \int_{-\infty}^{\infty} \cdots \int_{-\infty}^{\infty} 1s(1)2s(2)\frac{e'^2}{r_{12}} 1s(1)2s(2)\, dx_1 \cdots dz_2$$

$$H'_{11} = \iint [1s(1)]^2 [2s(2)]^2 \frac{e'^2}{r_{12}} d\tau_1\, d\tau_2$$

$$H'_{22} = \iint [1s(2)]^2 [2s(1)]^2 \frac{e'^2}{r_{12}} d\tau_1\, d\tau_2$$

The integration variables are dummy variables and may be given any symbols whatever. Let us relabel the integration variables in H'_{22} as follows: We interchange x_1 and x_2, interchange y_1 and y_2, and interchange z_1 and z_2. This relabeling leaves r_{12} [Eq. (9.67)] unchanged, so

$$H'_{22} = \iint [1s(1)]^2[2s(2)]^2 \frac{e'^2}{r_{12}} d\tau_2\, d\tau_1 = H'_{11} \tag{9.97}$$

The same argument shows that $H'_{33} = H'_{44}$, $H'_{55} = H'_{66}$, and $H'_{77} = H'_{88}$.

We denote H'_{11} by the symbol J_{1s2s}:

$$H'_{11} = J_{1s2s} = \iint [1s(1)]^2[2s(2)]^2 \frac{e'^2}{r_{12}} d\tau_1\, d\tau_2 \tag{9.98}$$

This is an example of a **Coulomb integral**, the name arising because J_{1s2s} is equal to the electrostatic energy of repulsion between an electron with probability density function $[1s]^2$ and an electron with probability density function $[2s]^2$. The integral H'_{12} is denoted by K_{1s2s}:

$$H'_{12} = K_{1s2s} = \iint 1s(1)2s(2) \frac{e'^2}{r_{12}} 2s(1)1s(2)\, d\tau_1\, d\tau_2 \tag{9.99}$$

This is an **exchange integral**: The functions on the left and right of e'^2/r_{12} differ from each other by an exchange of electrons 1 and 2. The general definitions of the Coulomb integral J_{mn} and the exchange integral K_{mn} are

$$J_{mn} \equiv \langle f_m(1)f_n(2)|e'^2/r_{12}|f_m(1)f_n(2)\rangle, \quad K_{mn} \equiv \langle f_m(1)f_n(2)|e'^2/r_{12}|f_n(1)f_m(2)\rangle \tag{9.100}$$

where the integrals go over the full range of the spatial coordinates of electrons 1 and 2, and f_m and f_n are spatial orbitals.

Substitution of (9.97) to (9.99) into (9.96) gives

$$\begin{vmatrix} J_{1s2s} - E^{(1)} & K_{1s2s} \\ K_{1s2s} & J_{1s2s} - E^{(1)} \end{vmatrix} = 0 \tag{9.101}$$

$$(J_{1s2s} - E^{(1)})^2 = (K_{1s2s})^2$$

$$E_1^{(1)} = J_{1s2s} - K_{1s2s}, \qquad E_2^{(1)} = J_{1s2s} + K_{1s2s} \tag{9.102}$$

We now find the coefficients of the correct zeroth-order wave functions that correspond to these two roots. Use of $E_1^{(1)}$ in (9.82) gives

$$K_{1s2s}c_1 + K_{1s2s}c_2 = 0$$
$$K_{1s2s}c_1 + K_{1s2s}c_2 = 0$$

Hence $c_2 = -c_1$. Normalization gives

$$\langle \phi_1^{(0)} | \phi_1^{(0)} \rangle = \langle c_1\psi_1^{(0)} - c_1\psi_2^{(0)} | c_1\psi_1^{(0)} - c_1\psi_2^{(0)} \rangle = |c_1|^2 + |c_1|^2 = 1$$
$$c_1 = 2^{-1/2}$$

where the orthonormality of $\psi_1^{(0)}$ and $\psi_2^{(0)}$ was used. The zeroth-order wave function corresponding to $E_1^{(1)}$ is then

$$\phi_1^{(0)} = 2^{-1/2}(\psi_1^{(0)} - \psi_2^{(0)}) = 2^{-1/2}[1s(1)2s(2) - 2s(1)1s(2)] \quad (9.103)$$

Similarly, one finds the function corresponding to $E_2^{(1)}$ to be

$$\phi_2^{(0)} = 2^{-1/2}(\psi_1^{(0)} + \psi_2^{(0)}) = 2^{-1/2}[1s(1)2s(2) + 2s(1)1s(2)] \quad (9.104)$$

We have three more second-order determinants to deal with:

$$\begin{vmatrix} H'_{33} - E^{(1)} & H'_{34} \\ H'_{34} & H'_{33} - E^{(1)} \end{vmatrix} = 0 \quad (9.105)$$

$$\begin{vmatrix} H'_{55} - E^{(1)} & H'_{56} \\ H'_{56} & H'_{55} - E^{(1)} \end{vmatrix} = 0 \quad (9.106)$$

$$\begin{vmatrix} H'_{77} - E^{(1)} & H'_{78} \\ H'_{78} & H'_{77} - E^{(1)} \end{vmatrix} = 0 \quad (9.107)$$

Consider H'_{33} and H'_{55}:

$$H'_{33} = \int_{-\infty}^{\infty} \cdots \int_{-\infty}^{\infty} 1s(1)2p_x(2) \frac{e'^2}{r_{12}} 1s(1)2p_x(2)\, dx_1 \cdots dz_2$$

$$H'_{55} = \int_{-\infty}^{\infty} \cdots \int_{-\infty}^{\infty} 1s(1)2p_y(2) \frac{e'^2}{r_{12}} 1s(1)2p_y(2)\, dx_1 \cdots dz_2$$

These two integrals are equal—the only difference between them involves replacement of $2p_x(2)$ by $2p_y(2)$—and these two orbitals differ only in their orientation in space. More formally, if we relabel the dummy integration variables in H'_{33} according to the scheme $x_2 \rightarrow y_2$, $y_2 \rightarrow x_2$, $x_1 \rightarrow y_1$, $y_1 \rightarrow x_1$, then r_{12} is unaffected and H'_{33} is transformed to H'_{55}. Similar reasoning shows $H'_{77} = H'_{33}$. Introducing the symbol J_{1s2p} for these Coulomb integrals, we have

$$H'_{33} = H'_{55} = H'_{77} = J_{1s2p} = \iint 1s(1)2p_z(2) \frac{e'^2}{r_{12}} 1s(1)2p_z(2)\, d\tau_1\, d\tau_2$$

Also, the exchange integrals involving the $2p$ orbitals are equal:

$$H'_{34} = H'_{56} = H'_{78} = K_{1s2p} = \iint 1s(1)2p_z(2) \frac{e'^2}{r_{12}} 2p_z(1)1s(2)\, d\tau_1\, d\tau_2$$

The three determinants (9.105) to (9.107) are thus identical and have the form

$$\begin{vmatrix} J_{1s2p} - E^{(1)} & K_{1s2p} \\ K_{1s2p} & J_{1s2p} - E^{(1)} \end{vmatrix} = 0$$

The determinant is similar to (9.101), and by analogy with (9.102)–(9.104), we get

$$E_3^{(1)} = E_5^{(1)} = E_7^{(1)} = J_{1s2p} - K_{1s2p} \tag{9.108}$$

$$E_4^{(1)} = E_6^{(1)} = E_8^{(1)} = J_{1s2p} + K_{1s2p} \tag{9.109}$$

$$\begin{aligned} \phi_3^{(0)} &= 2^{-1/2}[1s(1)2p_x(2) - 1s(2)2p_x(1)] \\ \phi_4^{(0)} &= 2^{-1/2}[1s(1)2p_x(2) + 1s(2)2p_x(1)] \\ \phi_5^{(0)} &= 2^{-1/2}[1s(1)2p_y(2) - 1s(2)2p_y(1)] \\ \phi_6^{(0)} &= 2^{-1/2}[1s(1)2p_y(2) + 1s(2)2p_y(1)] \\ \phi_7^{(0)} &= 2^{-1/2}[1s(1)2p_z(2) - 1s(2)2p_z(1)] \\ \phi_8^{(0)} &= 2^{-1/2}[1s(1)2p_z(2) + 1s(2)2p_z(1)] \end{aligned} \tag{9.110}$$

The electrostatic repulsion e'^2/r_{12} between the electrons has partly removed the degeneracy. The hypothetical eightfold-degenerate unperturbed level has been split into two nondegenerate levels associated with the configuration $1s2s$ and two triply degenerate levels associated with the configuration $1s2p$. It might be thought that higher-order energy corrections would further resolve the degeneracy. Actually, application of an external magnetic field is required to completely remove the degeneracy. Because the e'^2/r_{12} perturbation has not completely removed the degeneracy, any normalized linear combinations of $\phi_3^{(0)}$, $\psi_5^{(0)}$, and $\phi_7^{(0)}$ and of $\phi_4^{(0)}$, $\phi_6^{(0)}$, and $\phi_8^{(0)}$ can serve as correct zeroth-order wave functions.

To evaluate the Coulomb and exchange integrals in $E^{(1)}$ in (9.102) and (9.108), one uses the $1/r_{12}$ expansion given in Prob. 9.13. The results are

$$\begin{aligned} J_{1s2s} &= \frac{17}{81}\frac{Ze'^2}{a_0} = 11.42 \text{ eV}, \qquad J_{1s2p} = \frac{59}{243}\frac{Ze'^2}{a_0} = 13.21 \text{ eV} \\ K_{1s2s} &= \frac{16}{729}\frac{Ze'^2}{a_0} = 1.19 \text{ eV}, \qquad K_{1s2p} = \frac{112}{6561}\frac{Ze'^2}{a_0} = 0.93 \text{ eV} \end{aligned} \tag{9.111}$$

where we used $Z = 2$ and $e'^2/2a_0 = 13.606$ eV. Recalling that $E^{(0)} = -68.03$ eV [Eq. (9.93)], we get (Fig. 9.3)

$$\begin{aligned} E^{(0)} + E_1^{(1)} &= E^{(0)} + J_{1s2s} - K_{1s2s} = -57.8 \text{ eV} \\ E^{(0)} + E_2^{(1)} &= E^{(0)} + J_{1s2s} + K_{1s2s} = -55.4 \text{ eV} \\ E^{(0)} + E_3^{(1)} &= E^{(0)} + J_{1s2p} - K_{1s2p} = -55.7_5 \text{ eV} \\ E^{(0)} + E_4^{(1)} &= E^{(0)} + J_{1s2p} + K_{1s2p} = -53.9 \text{ eV} \end{aligned}$$

The first-order energy corrections seem to indicate that the lower of the two levels of the $1s2p$ configuration lies below the higher of the two levels of the $1s2s$ configuration. Study of the helium spectrum reveals that this is not so. The error is due to neglect of the higher-order perturbation-energy corrections.

−53.9 eV
1s2p
K_p
−55.4 eV
−55.7 eV
1s2s
K_s
−57.8 eV
J_p
J_s
$E^{(0)} + E^{(1)}$
−68.0 eV
$E^{(0)}$

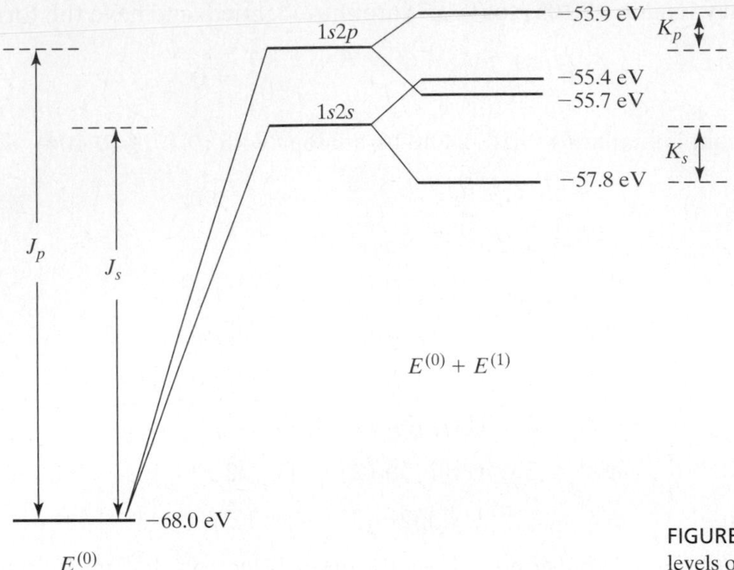

FIGURE 9.3 The first excited levels of the helium atom.

Using the variation–perturbation method (Section 9.2), Knight and Scherr calculated the second- and third-order corrections $E^{(2)}$ and $E^{(3)}$ for these four excited levels. [R. E. Knight and C. W. Scherr, *Rev. Mod. Phys.*, **35**, 431 (1963); for energy corrections through 17th order, see F. C. Sanders and C. W. Scherr, *Phys. Rev.*, **181**, 84 (1969).] Figure 9.4 shows their results (which are within 0.1 eV of the experimental energies). Figure 9.4 shows that Fig. 9.3 is quite inaccurate. Since the perturbation e'^2/r_{12} is not really very small, a perturbation treatment that includes only the $E^{(1)}$ correction does not give accurate results.

The first-order correction to the wave function, $\psi^{(1)}$, will include contributions from other configurations (configuration interaction). When we say that a level belongs to the configuration 1*s*2*s*, we are indicating the configuration that makes the largest contribution to the true wave function.

We started with the eight degenerate zeroth-order functions (9.94). These functions have three kinds of degeneracy. There is the degeneracy between hydrogenlike functions with the same *n*, but different *l*; the 2*s* and the 2*p* functions have the same energy. There is the degeneracy between hydrogenlike functions with the same *n* and *l*, but different *m*; the $2p_1$, $2p_0$, and $2p_{-1}$ functions have the same energy. (For convenience we used the real functions $2p_{x,y,z}$, but we could have started with the functions

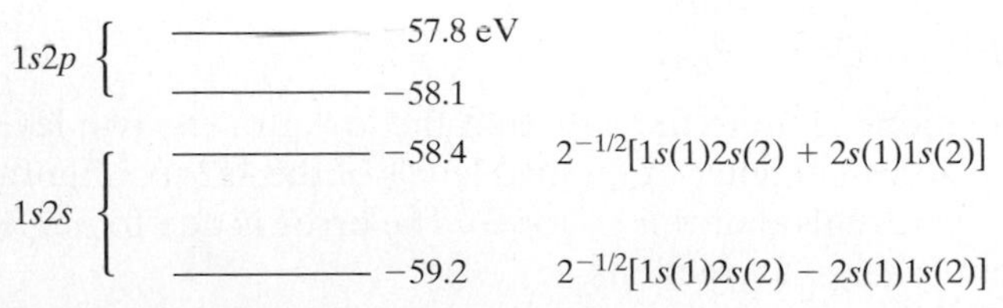

1s2p
57.8 eV
−58.1
−58.4
$2^{-1/2}[1s(1)2s(2) + 2s(1)1s(2)]$
1s2s
−59.2
$2^{-1/2}[1s(1)2s(2) - 2s(1)1s(2)]$

FIGURE 9.4 $E^{(0)} + E^{(1)} + E^{(2)} + E^{(3)}$ for the first excited levels of helium. Also shown are the correct zeroth-order wave functions for the 1*s*2*s* levels.

$2p_{1,0,-1}$.) Finally, there is the degeneracy between functions that differ only in the interchange of the two electrons between the orbitals; the functions $\psi_1^{(0)} = 1s(1)2s(2)$ and $\psi_2^{(0)} = 1s(2)2s(1)$ have the same energy. This last kind of degeneracy is called *exchange degeneracy*. When the interelectronic repulsion e'^2/r_{12} was introduced as a perturbation, the exchange degeneracy and the degeneracy associated with the quantum number l were removed. The degeneracy associated with m remained, however; each $1s2p$ helium level is triply degenerate, and we could just as well have used the $2p_1$, $2p_0$, and $2p_{-1}$ orbitals instead of the real orbitals in constructing the correct zeroth-order wave functions. Let us consider the reasons for the removal of the l degeneracy and the exchange degeneracy.

The interelectronic repulsion in helium makes the $2s$ orbital energy less than the $2p$ energy. Figures 6.9 and 6.8 show that a $2s$ electron has a greater probability than a $2p$ electron of being closer to the nucleus than the $1s$ electron(s). A $2s$ electron will not be as effectively shielded from the nucleus by the $1s$ electrons and will therefore have a lower energy than a $2p$ electron. [According to Eq. (6.94), the greater the nuclear charge, the lower the energy.] Mathematically, the difference between the $1s2s$ and the $1s2p$ energies results from the Coulomb integral J_{1s2s} being smaller than J_{1s2p}. These Coulomb integrals represent the electrostatic repulsion between the appropriate charge distributions. When the $2s$ electron penetrates the charge distribution of the $1s$ electron, it feels a repulsion from only the unpenetrated part of the $1s$ charge distribution. Hence the $1s$-$2s$ electrostatic repulsion is less than the $1s$-$2p$ repulsion, and the $1s2s$ levels lie below the $1s2p$ levels. The interelectronic repulsion in many-electron atoms lifts the l degeneracy, and the orbital energies for the same value of n increase with increasing l.

Now consider the removal of the exchange degeneracy. The functions (9.94) with which we began the perturbation treatment have each electron assigned to a definite orbital. For example, the function $\psi_1^{(0)} = 1s(1)2s(2)$ has electron 1 in the $1s$ orbital and electron 2 in the $2s$ orbital. For $\psi_2^{(0)}$ the opposite is true. The secular determinant was not diagonal, so the initial functions were not the correct zeroth-order wave functions. The correct zeroth-order functions do not assign each electron to a definite orbital. Thus the first two correct zeroth-order functions are

$$\phi_1^{(0)} = 2^{-1/2}[1s(1)2s(2) - 1s(2)2s(1)], \qquad \phi_2^{(0)} = 2^{-1/2}[1s(1)2s(2) + 1s(2)2s(1)]$$

We cannot say which orbital electron 1 is in for either $\phi_1^{(0)}$ or $\phi_2^{(0)}$. This property of the wave functions of systems containing more than one electron results from the indistinguishability of identical particles in quantum mechanics and will be discussed further in Chapter 10. Since the functions $\phi_1^{(0)}$ and $\phi_2^{(0)}$ have different energies, the exchange degeneracy is removed when the correct zeroth-order functions are used.

9.8 COMPARISON OF THE VARIATION AND PERTURBATION METHODS

The perturbation method applies to all the bound states of a system. Although the variation theorem stated in Section 8.1 applies only to the lowest state of a given symmetry, we can use the linear variation method to treat excited bound states.

Perturbation calculations are often hard to do because of the need to evaluate the infinite sums over discrete states and integrals over continuum states that occur in the second-order and higher-order energy corrections.

In the perturbation method, one can calculate the energy much more accurately (to order $2k + 1$) than the wave function (to order k). The same situation holds in the variation method, where one can get a rather good energy with a rather inaccurate wave function. If one calculates properties other than the energy, the results will generally not be as reliable as the calculated energy.

9.9 TIME-DEPENDENT PERTURBATION THEORY

In spectroscopy, we start with a system in some stationary state, expose it to electromagnetic radiation (light), and then observe whether the system has made a transition to another stationary state. The radiation produces a time-dependent potential-energy term in the Hamiltonian, so we must use the time-dependent Schrödinger equation. The most convenient approach here is an approximate one called time-dependent perturbation theory.

Let the system (atom or molecule) have the time-independent Hamiltonian $\hat{H}^0$ in the absence of the radiation (or other time-dependent perturbation), and let $\hat{H}'(t)$ be the time-dependent perturbation. The time-independent Schrödinger equation for the unperturbed problem is

$$\hat{H}^0\psi_k^0 = E_k^0\psi_k^0 \tag{9.112}$$

where E_k^0 and ψ_k^0 are the stationary-state energies and wave functions. The time-dependent Schrödinger equation (7.96) in the presence of the radiation is

$$-\frac{\hbar}{i}\frac{\partial\Psi}{\partial t} = (\hat{H}^0 + \hat{H}')\Psi \tag{9.113}$$

where the state function Ψ depends on the spatial and spin coordinates (symbolized by q) and on the time: $\Psi = \Psi(q, t)$. (See Chapter 10 for a discussion of spin coordinates.)

First suppose that $\hat{H}'(t)$ is absent. The unperturbed time-dependent Schrödinger equation is

$$-(\hbar/i)\partial\Psi^0/\partial t = \hat{H}^0\Psi^0 \tag{9.114}$$

The system's possible stationary-state state functions are given by (7.98) as $\Psi_k^0 = \exp(-iE_k^0 t/\hbar)\psi_k^0$, where the ψ_k^0 functions are the eigenfunctions of $\hat{H}^0$ [Eq. (9.112)]. Each Ψ_k^0 is a solution of (9.114). Moreover, the linear combination

$$\Psi^0 = \sum_k c_k\Psi_k^0 = \sum_k c_k \exp(-iE_k^0 t/\hbar)\psi_k^0 \tag{9.115}$$

with the c_k's being arbitrary time-independent constants, is a solution of the time-dependent Schrödinger equation (9.114), as proved in the discussion leading to Eq. (7.99). The functions Ψ_k^0 form a complete set (since they are the eigenfunctions of the Hermitian operator $\hat{H}^0$), so any solution of (9.114) can be expressed in the form (9.115). Hence (9.115) is the general solution of the time-dependent Schrödinger equation (9.114), where $\hat{H}^0$ is independent of time.

Now suppose that $\hat{H}'(t)$ is present. The function (9.115) is no longer a solution of the time-dependent Schrödinger equation. However, because the unperturbed functions Ψ_k^0 form a complete set, the true state function Ψ can at any instant of time be expanded as a linear combination of the Ψ_k^0 functions according to $\Psi = \sum_k b_k \Psi_k^0$. Because $\hat{H}$ is time-dependent, Ψ will change with time and the expansion coefficients b_k will change with time. Therefore,

$$\Psi = \sum_k b_k(t) \exp(-iE_k^0 t/\hbar)\psi_k^0 \tag{9.116}$$

In the limit $\hat{H}'(t) \rightarrow 0$, the expansion (9.116) reduces to (9.115).

Substitution of (9.116) into the time-dependent Schrödinger equation (9.113) and use of (9.112) gives

$$-\frac{\hbar}{i}\sum_k \frac{db_k}{dt}\exp(-iE_k^0 t/\hbar)\psi_k^0 + \sum_k E_k^0 b_k \exp(-iE_k^0 t/\hbar)\psi_k^0$$

$$= \sum_k b_k \exp(-iE_k^0 t/\hbar)E_k^0 \psi_k^0 + \sum_k b_k \exp(-iE_k^0 t/\hbar)\hat{H}'\psi_k^0$$

$$-\frac{\hbar}{i}\sum_k \frac{db_k}{dt}\exp(-iE_k^0 t/\hbar)\psi_k^0 = \sum_k b_k \exp(-iE_k^0 t/\hbar)\hat{H}'\psi_k^0$$

We now multiply by ψ_m^{0*} and integrate over the spatial and spin coordinates. Using the orthonormality equation $\langle \psi_m^0 | \psi_k^0 \rangle = \delta_{mk}$, we get

$$-\frac{\hbar}{i}\sum_k \frac{db_k}{dt}\exp(-iE_k^0 t/\hbar)\delta_{mk} = \sum_k b_k \exp(-iE_k^0 t/\hbar)\langle \psi_m^0 | \hat{H}' | \psi_k^0 \rangle$$

Because of the δ_{mk} factor, all terms but one in the sum on the left are zero, and the left side equals $-(\hbar/i)(db_m/dt)\exp(-iE_m^0 t/\hbar)$. We get

$$\frac{db_m}{dt} = -\frac{i}{\hbar}\sum_k b_k \exp[i(E_m^0 - E_k^0)t/\hbar]\langle \psi_m^0 | \hat{H}' | \psi_k^0 \rangle \tag{9.117}$$

Let us suppose that the perturbation $\hat{H}'(t)$ was applied at time $t = 0$ and that before the perturbation was applied the system was in stationary state n with energy $E_n^{(0)}$. The state function at $t = 0$ is therefore $\Psi = \exp(-iE_n^0 t/\hbar)\psi_n^0$ [Eq. (7.98)], and the $t = 0$ values of the expansion coefficients in (9.116) are thus $b_n(0) = 1$ and $b_k(0) = 0$ for $k \neq n$:

$$b_k(0) = \delta_{kn} \tag{9.118}$$

We shall assume that the perturbation $\hat{H}'$ is small and acts for only a short time. Under these conditions, the change in the expansion coefficients b_k from their initial values at the time the perturbation is applied will be small. To a good approximation, we can replace the expansion coefficients on the right side of (9.117) by their initial values (9.118). This gives

$$\frac{db_m}{dt} \approx -\frac{i}{\hbar}\exp[i(E_m^0 - E_n^0)t/\hbar]\langle \psi_m^0 | \hat{H}' | \psi_n^0 \rangle$$

Let the perturbation $\hat{H}'$ act from $t = 0$ to $t = t'$. Integrating from $t = 0$ to t' and using (9.118), we get

$$b_m(t') \approx \delta_{mn} - \frac{i}{\hbar}\int_0^{t'} \exp[i(E_m^0 - E_n^0)t/\hbar]\langle \psi_m^0 | \hat{H}' | \psi_n^0 \rangle \, dt \tag{9.119}$$

Use of the approximate result (9.119) for the expansion coefficients in (9.116) gives the desired approximation to the state function at time t' for the case that the time-dependent perturbation $\hat{H}'$ is applied at $t = 0$ to a system in stationary state n. [As with time-independent perturbation theory, one can go to higher-order approximations (see *Fong*, pp. 234–244).]

For times after t', the perturbation has ceased to act, and $\hat{H}' = 0$. Equation (9.117) gives $db_m/dt = 0$ for $t > t'$, so $b_m = b_m(t')$ for $t \geq t'$. Therefore, for times after exposure to the perturbation, the state function Ψ is [Eq. (9.116)]

$$\Psi = \sum_m b_m(t') \exp(-iE_m^0 t/\hbar)\psi_m^0 \qquad \text{for } t \geq t' \tag{9.120}$$

where $b_m(t')$ is given by (9.119). In (9.120), Ψ is a superposition of the eigenfunctions ψ_m^0 of the energy operator $\hat{H}^0$, the expansion coefficients being $b_m \exp(-iE_m^0 t/\hbar)$. [Compare Eqs. (9.120) and (7.65).] The work of Section 7.6 tells us that a measurement of the system's energy at a time after t' will give one of the eigenvalues E_m^0 of the energy operator $\hat{H}^0$, and the probability of getting E_m^0 equals the square of the absolute value of the expansion coefficient that multiplies ψ_m^0; that is, it equals $|b_m(t') \exp(-iE_m^0 t/\hbar)|^2 = |b_m(t')|^2$.

The time-dependent perturbation changes the system's state function from $\exp(-iE_n^0 t/\hbar)\psi_n^0$ to the superposition (9.120). Measurement of the energy then changes Ψ to one of the energy eigenfunctions $\exp(-iE_m^0 t/\hbar)\psi_m^0$ (reduction of the wave function, Section 7.9). The net result is a transition from stationary state n to stationary state m, the probability of such a transition being $|b_m(t')|^2$.

9.10 INTERACTION OF RADIATION AND MATTER

We now consider the interaction of an atom or molecule with electromagnetic radiation. A proper quantum-mechanical approach would treat both the atom and the radiation quantum mechanically, but we shall simplify things by using the classical picture of the light as an electromagnetic wave of oscillating electric and magnetic fields.

A detailed investigation, which we omit (see *Levine, Molecular Spectroscopy*, Section 3.2), shows that usually the interaction between the radiation's magnetic field and the atom's charges is much weaker than the interaction between the radiation's electric field and the charges, so we shall consider only the latter interaction. (In NMR spectroscopy the important interaction is between the magnetic dipole moments of the nuclei and the radiation's magnetic field. We shall not consider this case.)

Let the electric field $\mathscr{E}$ of the electromagnetic wave point in the x direction only. (This is plane-polarized radiation.) The electric field is defined as the force per unit charge, so the force on charge Q_i is $F = Q_i\mathscr{E}_x = -dV/dx$, where (4.26) was used. Integration gives the potential energy of interaction between the radiation's electric field and the charge as $V = -Q_i\mathscr{E}_x x$, where the arbitrary integration constant was taken as zero. For a system of several charges, $V = -\sum_i Q_i x_i \mathscr{E}_x$. This is the time-dependent perturbation $\hat{H}'(t)$. The space and time dependence of the electric field of an electromagnetic wave traveling in the z direction with wavelength λ and frequency

ν is given by (*Halliday and Resnick*, Section 41-8) $\mathscr{E}_x = \mathscr{E}_0 \sin(2\pi\nu t - 2\pi z/\lambda)$, where $\mathscr{E}_0$ is the maximum value of $\mathscr{E}_x$ (the amplitude). Therefore,

$$\hat{H}'(t) = -\mathscr{E}_0 \sum_i Q_i x_i \sin(2\pi\nu t - 2\pi z_i/\lambda)$$

where the sum goes over all the electrons and nuclei of the atom or molecule.

Defining ω and ω_{mn} as

$$\omega \equiv 2\pi\nu, \qquad \omega_{mn} \equiv (E_m^0 - E_n^0)/\hbar \tag{9.121}$$

and substituting $\hat{H}'(t)$ into (9.119), we get the coefficients in the expansion (9.116) of the state function Ψ as

$$b_m \approx \delta_{mn} + \frac{i\mathscr{E}_0}{\hbar}\int_0^{t'} \exp(i\omega_{mn}t)\left\langle \psi_m^0 \left| \sum_i Q_i x_i \sin(\omega t - 2\pi z_i/\lambda) \right| \psi_n^0 \right\rangle dt$$

The integral $\langle \psi_m^0 | \Sigma_i \cdots | \psi_n^0 \rangle$ in this equation is over all space, but significant contributions to its magnitude come only from regions where ψ_m^0 and ψ_n^0 are of significant magnitude. In regions well outside the atom or molecule, ψ_m^0 and ψ_n^0 are vanishingly small, and such regions can be ignored. Let the coordinate origin be chosen within the atom or molecule. Since regions well outside the atom can be ignored, the coordinate z_i can be considered to have a maximum magnitude of the order of one nm. For ultraviolet light, the wavelength λ is on the order of 10^2 nm. For visible, infrared, microwave, and radio-frequency radiation, λ is even larger. Hence $2\pi z_i/\lambda$ is very small and can be neglected, and this leaves $\Sigma_i Q_i x_i \sin \omega t$ in the integral.

Use of the identity (Prob. 1.27) $\sin \omega t = (e^{i\omega t} - e^{-i\omega t})/2i$ gives

$$b_m(t') \approx \delta_{mn} + \frac{\mathscr{E}_0}{2\hbar}\left\langle \psi_m^0 \left| \sum_i Q_i x_i \right| \psi_n^0 \right\rangle \int_0^{t'} [e^{i(\omega_{mn}+\omega)t} - e^{i(\omega_{mn}-\omega)t}]\, dt$$

Using $\int_0^{t'} e^{at}\, dt = a^{-1}(e^{at'} - 1)$, we get

$$b_m(t') \approx \delta_{mn} + \frac{\mathscr{E}_0}{2\hbar i}\left\langle \psi_m^0 \left| \sum_i Q_i x_i \right| \psi_n^0 \right\rangle \left[\frac{e^{i(\omega_{mn}+\omega)t'} - 1}{\omega_{mn} + \omega} - \frac{e^{i(\omega_{mn}-\omega)t'} - 1}{\omega_{mn} - \omega} \right] \tag{9.122}$$

For $m \neq n$, the δ_{mn} term equals zero.

As noted at the end of Section 9.9, $|b_m(t')|^2$ gives the probability of a transition to state m from state n. There are two cases where this probability becomes of significant magnitude. If $\omega_{mn} = \omega$, the denominator of the second fraction in brackets is zero and this fraction's absolute value is large (but not infinite; see Prob. 9.26). If $\omega_{mn} = -\omega$, the first fraction has a zero denominator and a large absolute value.

For $\omega_{mn} = \omega$, Eq. (9.121) gives $E_m^0 - E_n^0 = h\nu$. Exposure of the atom to radiation of frequency ν has produced a transition from stationary state n to stationary state m, where (since ν is positive) $E_m^0 > E_n^0$. We might suppose that the energy for this transition came from the absorption by the system of a photon of energy $h\nu$. This supposition is confirmed by a fully quantum-mechanical treatment (called *quantum field theory*) in which the radiation is treated quantum mechanically rather than classically. We have **absorption** of radiation with a consequent increase in the system's energy.

For $\omega_{mn} = -\omega$, we get $E_n^0 - E_m^0 = h\nu$. Exposure to radiation of frequency ν has induced a transition from stationary state n to stationary state m, where (since ν is positive) $E_n^0 > E_m^0$. The system has gone to a lower energy level, and a quantum-field-theory treatment shows that a photon of energy $h\nu$ is emitted in this process. This is **stimulated emission** of radiation. Stimulated emission occurs in lasers.

A defect of our treatment is that it does not predict **spontaneous emission,** the emission of a photon by a system not exposed to radiation, the system falling to a lower energy level in the process. Quantum field theory does predict spontaneous emission.

Note from (9.122) that the probability of absorption is proportional to $|\langle\psi_m^0|\sum_i Q_i x_i|\psi_n^0\rangle|^2$. The quantity $\sum_i Q_i x_i$ is the x component of the system's electric-dipole-moment operator $\hat{\boldsymbol{\mu}}$ (see Section 14.2 for details), which is [Eqs. (14.14) and (14.15)] $\hat{\boldsymbol{\mu}} = \mathbf{i}\sum_i Q_i x_i + \mathbf{j}\sum_i Q_i y_i + \mathbf{k}\sum_i Q_i z_i = \mathbf{i}\hat{\mu}_x + \mathbf{j}\hat{\mu}_y + \mathbf{k}\hat{\mu}_z$, where **i**, **j**, **k** are unit vectors along the axes and $\hat{\mu}_x, \hat{\mu}_y, \hat{\mu}_z$ are the components of $\hat{\boldsymbol{\mu}}$. We assumed polarized radiation with an electric field in the x direction only. If the radiation has electric-field components in the y and z directions also, then the probability of absorption will be proportional to

$$|\langle\psi_m^0|\hat{\mu}_x|\psi_n^0\rangle|^2 + |\langle\psi_m^0|\hat{\mu}_y|\psi_n^0\rangle|^2 + |\langle\psi_m^0|\hat{\mu}_z|\psi_n^0\rangle|^2 = |\langle\psi_m^0|\hat{\boldsymbol{\mu}}|\psi_n^0\rangle|^2$$

where Eq. (5.25) was used. The integral $\langle\psi_m^0|\hat{\boldsymbol{\mu}}|\psi_n^0\rangle = \boldsymbol{\mu}_{mn}$ is the **transition (dipole) moment**.

When $\boldsymbol{\mu}_{mn} = 0$, the transition between states m and n with absorption or emission of radiation is said to be **forbidden. Allowed** transitions have $\boldsymbol{\mu}_{mn} \neq 0$. Because of approximations made in the derivation of (9.122), forbidden transitions may have some small probability of occurring.

Consider, for example, the particle in a one-dimensional box (Section 2.2). The transition dipole moment is $\langle\psi_m^0|Qx|\psi_n^0\rangle$, where Q is the particle's charge and x is its coordinate and where $\psi_m^0 = (2/l)^{1/2}\sin(m\pi x/l)$ and $\psi_n^0 = (2/l)^{1/2}\sin(n\pi x/l)$. Evaluation of this integral (Prob. 9.27) shows it is nonzero only when $m - n = \pm 1, \pm 3, \pm 5, \ldots$ and is zero when $m - n = 0, \pm 2, \ldots$. The **selection rule** for a particle in a one-dimensional box is that the quantum number must change by an odd integer when radiation is absorbed or emitted.

Evaluation of the transition moment for the harmonic oscillator and for the two-particle rigid rotor gives the selection rules $\Delta v = \pm 1$ and $\Delta J = \pm 1$ stated in Sections 4.3 and 6.4.

The quantity $|b_m|^2$ in (9.122) is sharply peaked at $\omega = \omega_{mn}$ and $\omega = -\omega_{mn}$, but there is a nonzero probability that a transition will occur when ω is not precisely equal to $|\omega_{mn}|$, that is, when $h\nu$ is not precisely equal to $|E_m^0 - E_n^0|$. This fact is related to the energy–time uncertainty relation (5.14). States with a finite lifetime have an uncertainty in their energy.

Radiation is not the only time-dependent perturbation that produces transitions between states. When an atom or molecule comes close to another atom or molecule, it suffers a time-dependent perturbation that can change its state. Selection rules derived for radiative transitions need not apply to collision processes, since $\hat{H}'(t)$ differs for the two processes.

9.11 SUMMARY

For a system whose time-independent Schrödinger equation is $(\hat{H}^0 + \hat{H}')\psi_n = E_n\psi_n$, perturbation theory expresses the energies and wave functions of the nondegenerate levels as $E_n = E_n^{(0)} + E_n^{(1)} + E_n^{(2)} + \cdots$ and $\psi_n = \psi_n^{(0)} + \psi_n^{(1)} + \psi_n^{(2)} + \cdots$, where the unperturbed wave functions $\psi_n^{(0)}$ and energies $E_n^{(0)}$ satisfy $\hat{H}^0\psi_n^{(0)} = E_n^{(0)}\psi_n^{(0)}$. The first-order energy correction is $E_n^{(1)} = \int \psi_n^{(0)*}\hat{H}'\psi_n^{(0)}\,d\tau$. The second-order energy correction is given by (9.35). The first-order correction to the wave function is $\psi_n^{(1)} = \sum_{m\neq n} a_m\psi_m^{(0)}$, where the expansion coefficients a_m are given by (9.26). For a degenerate level with degree of degeneracy d, we have $\psi_n = \phi_n^{(0)} + \psi_n^{(1)} + \cdots$, where the correct zeroth-order wave functions are $\phi_n^{(0)} = \sum_{i=1}^{d} c_i\psi_i^{(0)}$ for $n = 1, \ldots, d$. The first-order energy corrections for the degenerate level are found from the secular equation $\det[\langle\psi_m^{(0)}|\hat{H}'|\psi_i^{(0)}\rangle - E_n^{(1)}\delta_{mi}] = 0$, and then the coefficients c_i are found by solving (9.82).

Perturbation theory was applied to the helium atom, with $\hat{H}'$ taken as e'^2/r_{12}. The unperturbed ground-state wave function is $\psi^{(0)} = 1s(1)1s(2)$. The first-order energy correction was calculated for the ground state. Degenerate perturbation theory was applied to the first group of helium excited states. We found that the $1s2s$ configuration gives rise to two nondegenerate energy levels with correct zeroth-order wave functions $[1s(1)2s(2) \pm 1s(2)2s(1)]/\sqrt{2}$; the $1s2p$ configuration gives rise to two triply degenerate levels with correct zeroth-order wave functions $[1s(1)2p(2) \pm 1s(2)2p(1)]/\sqrt{2}$, where $2p$ can be $2p_x$, $2p_y$, or $2p_z$. (These conclusions will be modified when electron spin is taken into account in Chapter 10.) The $1s2s$ levels lie below the $1s2p$ levels.

Time-dependent perturbation theory shows that, when an atom or molecule in a stationary state is exposed to electromagnetic radiation of frequency ν, the molecule may make a transition between two stationary states m and n whose energy difference is $h\nu$, provided the transition dipole moment $\langle\psi_m^0|\hat{\boldsymbol{\mu}}|\psi_n^0\rangle$ is nonzero for states m and n.

PROBLEMS

Sec.	9.2	9.3	9.4	9.5	9.6	9.7	9.10	general
Probs.	9.1–9.10	9.11–9.15	9.16	9.17–9.21	9.22	9.23–9.25	9.26–9.28	9.29–9.30

9.1 Consider a one-particle, one-dimensional system with $V = \infty$ for $x < 0$ and for $x > l$ and $V = C$ for $0 \le x \le l$, where C is a constant. (a) Sketch V for $C > 0$. (b) Treat the system as a perturbed particle in a box and find $E_n^{(1)}$ for the state with quantum number n.

9.2 (a) For the perturbed particle in a box of Prob. 9.1, find $E_n^{(2)}$ for the state with quantum number n. (b) Find $\psi_n^{(1)}$ for this system. (c) Use the results of Prob. 4.51 to explain why the results of Prob. 9.1 and 9.2 make sense.

9.3 For the anharmonic oscillator with Hamiltonian (9.3), evaluate $E^{(1)}$ for the first excited state, taking the unperturbed system as the harmonic oscillator. What is $E^{(0)}$?

9.4 Consider the one-particle, one-dimensional system with potential-energy

$$V = V_0 \text{ for } \tfrac{1}{4}l < x < \tfrac{3}{4}l, \qquad V = 0 \text{ for } 0 \le x \le \tfrac{1}{4}l \text{ and } \tfrac{3}{4}l \le x \le l$$

and $V = \infty$ elsewhere, where $V_0 = \hbar^2/ml^2$. Treat the system as a perturbed particle in a box. (a) Find the first-order energy correction for the general stationary state with quantum number n. (b) For the ground state and for the first excited state, compare $E^{(0)} + E^{(1)}$ with the true energies $5.750345\hbar^2/ml^2$ and $20.23604\hbar^2/ml^2$. Explain why $E^{(0)} + E^{(1)}$ for each of these two states is the same as obtained by the variational treatment of Probs. 8.2a and 8.17.

9.5 For the perturbed particle in a box of Prob. 9.4, explain (without doing any calculations) why we expect $E^{(1)}$ to be greatest for the $n = 1$ state.

9.6 For the perturbed particle in a box of Prob. 9.4, find the first-order correction to the wave function of the stationary state with quantum number n.

9.7 Consider the perturbed particle in a box of Prob. 9.4. (a) Explain why $\langle\psi_m^{(0)}|\hat{H}'|\psi_n^{(0)}\rangle = 0$ when $n = 1$ and m is an even integer. (b) Use a computer to evaluate $E^{(2)}$ for the ground state by summing over odd values of m in (9.35). Keep adding terms until the last added term is negligibly small in magnitude. Compare $E^{(0)} + E^{(1)} + E^{(2)}$ with the true ground-state energy $5.750345\hbar^2/ml^2$.

9.8 Assume that the charge of the proton is distributed uniformly throughout the volume of a sphere of radius 10^{-13} cm. Use perturbation theory to estimate the shift in the ground-state hydrogen-atom energy due to the finite proton size. The potential energy experienced by the electron when it has penetrated the nucleus and is at distance r from the nuclear center is $-eQ/4\pi\varepsilon_0 r$, where Q is the amount of proton charge within the sphere of radius r (*Halliday and Resnick*, Section 28-8). The evaluation of the integral is simplified by noting that the exponential factor in ψ is essentially equal to 1 within the nucleus.

9.9 True or false? The second-order correction $E^{(2)}$ to the ground-state energy is never positive.

9.10 (a) For an anharmonic oscillator with $\hat{H} = -(\hbar^2/2m)\,d^2/dx^2 + \frac{1}{2}kx^2 + cx^3$, take $\hat{H}'$ as cx^3. (a) Find $E^{(1)}$ for the state with quantum number v. (b) Find $E^{(2)}$ for the state with quantum number v. You will need the following integral (*Levine, Molecular Spectroscopy*, p. 154):

$$\begin{aligned}\langle\psi_{v'}^{(0)}|x^3|\psi_v^{(0)}\rangle &= [(v+1)(v+2)(v+3)/8\alpha^3]^{1/2}\delta_{v',v+3} \\ &\quad + 3[(v+1)/2\alpha]^{3/2}\delta_{v',v+1} + 3(v/2\alpha)^{3/2}\delta_{v',v-1} \\ &\quad + [v(v-1)(v-2)/8\alpha^3]^{1/2}\delta_{v',v-3}\end{aligned}$$

where the $\psi^{(0)}$'s are harmonic-oscillator wave functions and α is defined by (4.33). (c) Which unperturbed states contribute to $\psi_v^{(1)}$?

9.11 When Hylleraas began his calculations on helium, it was not known whether the isolated hydride ion H^- was a stable entity. Calculate the ground-state energy of H^- predicted by the trial function (9.56). Compare the result with the ground-state energy of the hydrogen atom, -13.60 eV, and show that this simple variation function (erroneously) indicates H^- is unstable with respect to ionization into a hydrogen atom and an electron. (More complicated variational functions give a ground-state energy of -14.35 eV.)

9.12 There is more than one way to divide a Hamiltonian $\hat{H}$ into an unperturbed part $\hat{H}^0$ and a perturbation $\hat{H}'$. Instead of the division (9.40) and (9.41), consider the following way of dividing up the helium-atom Hamiltonian:

$$\hat{H}^0 = -\frac{\hbar^2}{2m_e}\nabla_1^2 - \frac{\hbar^2}{2m_e}\nabla_2^2 - \left(Z - \frac{5}{16}\right)\frac{e'^2}{r_1} - \left(Z - \frac{5}{16}\right)\frac{e'^2}{r_2}$$

$$\hat{H}' = -\frac{5}{16}\frac{e'^2}{r_1} - \frac{5}{16}\frac{e'^2}{r_2} + \frac{e'^2}{r_{12}}$$

What are the unperturbed wave functions? Calculate $E^{(0)}$ and $E^{(1)}$ for the ground state. (See Section 9.4.)

9.13 One can show that (see *Eyring, Walter, and Kimball,* p. 369)

$$\frac{1}{r_{12}} = \sum_{l=0}^{\infty} \sum_{m=-l}^{l} \frac{4\pi}{2l+1} \frac{r_<^l}{r_>^{l+1}} [Y_l^m(\theta_1, \phi_1)]^* Y_l^m(\theta_2, \phi_2) \tag{9.123}$$

where $r_<$ means the smaller of r_1 and r_2, and $r_>$ is the larger of r_1 and r_2. Substitute this expansion into (9.52). Then multiply the right side by $Y_0^0(\theta_1, \phi_1)[Y_0^0(\theta_2, \phi_2)]^*(4\pi)$, which from (5.101) equals 1. Use the orthonormality of the spherical harmonics [Eq. (7.27)] to evaluate the angular integrals in terms of Kronecker deltas. Perform the sums to show that

$$E^{(1)} = \frac{16Z^6 e'^2}{a_0^6} \int_0^\infty \int_0^\infty e^{-2Zr_1/a_0} e^{-2Zr_2/a_0} \frac{1}{r_>} r_1^2 r_2^2 \, dr_1 \, dr_2$$

Next, integrate first over r_1 and write the r_1 integral as the sum of integrals from 0 to r_2 and from r_2 to ∞. In the range $0 \le r_1 \le r_2$, we have $r_> = r_2$; in the range $r_2 \le r_1 \le \infty$, we have $r_> = r_1$. Use indefinite integrals in the Appendix to do the r_1 integrals to obtain r_2 integrals, which are evaluated using an Appendix integral. Show that the result is (9.53).

9.14 Most (but not all) of the effect of nuclear motion in helium can be corrected for by replacing the electron's mass m_e by the reduced mass (6.59) in the expression for the energy. The energy of helium is proportional to what power of m_e? [See Eq. (9.64).] Use of μ instead of m_e multiplies the energies calculated on the basis of infinite nuclear mass by what factor?

9.15 (a) Use (9.53) to estimate $\langle 1/r_{12} \rangle^{-1}$ for the ground-state He atom. (b) Use the He experimental ground-state electronic energy of -79.01 eV to accurately find $\langle 1/r_{12} \rangle^{-1}$.

9.16 Calculate $\langle r_1 \rangle$ for the helium trial function (9.56). To save time, use the result of Prob. 6.21.

9.17 A certain unperturbed system has a doubly degenerate energy level for which the perturbation integrals have the values $H'_{11} = 4b$, $H'_{12} = 2b$, $H'_{22} = 6b$, where b is a positive constant, $H'_{jk} \equiv \langle \psi_j^{(0)} | \hat{H}' | \psi_k^{(0)} \rangle$, and $\langle \psi_j^{(0)} | \psi_k^{(0)} \rangle = \delta_{jk}$. (a) In terms of b, find the $E^{(1)}$ values for the perturbed system. (b) Find the normalized correct zeroth-order wave functions.

9.18 Explain why without solving for the $E^{(1)}$ values in Prob. 9.17, we can be certain that the sum of the $E^{(1)}$ values is $10b$.

9.19 Show that the secular equation (9.83) can be written as

$$\det[\langle \psi_m^{(0)} | \hat{H} | \psi_i^{(0)} \rangle - (E_n^{(0)} + E_n^{(1)})\delta_{mi}] = 0$$

9.20 Verify the normalization condition (9.86) for the coefficients in a correct zeroth-order wave function.

9.21 (a) For a particle in a square box of length l with origin at $x = 0$, $y = 0$, write down the wave functions and energy levels. (b) If the system of (a) is perturbed by

$$\hat{H}' = b \qquad \text{for } \tfrac{1}{4}l \le x \le \tfrac{3}{4}l \text{ and } \tfrac{1}{4}l \le y \le \tfrac{3}{4}l$$

where b is a constant and $\hat{H}' = 0$ elsewhere, find $E^{(1)}$ for the ground state. For the first excited energy level, find the $E^{(1)}$ values and the correct zeroth-order wave functions.

9.22 For a hydrogen atom perturbed by a uniform applied electric field in the z direction, the perturbation Hamiltonian is

$$\hat{H}' = e\mathscr{E}z = e\mathscr{E}r\cos\theta$$

where $\mathscr{E}$ is the magnitude of the electric field. Consider the effect of $\hat{H}'$ on the $n = 2$ energy level, which is fourfold degenerate. Since $\hat{H}'$ commutes with the angular-momentum operator $\hat{L}_z$, the ideas of Section 9.6 lead us to set up the secular determinant using the complex hydrogen-atom orbitals $2s$, $2p_1$, $2p_0$, and $2p_{-1}$, which are eigenfunctions of $\hat{L}_z$. Set up the secular determinant using the fact that matrix elements of $\hat{H}'$ between states with different values of the quantum number m will vanish. Also, use parity considerations to show that certain other integrals are zero (see Prob. 7.28). Evaluate the nonzero integrals, and find the first-order energy corrections and the correct zeroth-order wave functions. *Hint:* Choose the order of the orbitals so as to make the secular determinant block diagonal.

9.23 Consider the perturbation treatment of the helium configurations $1s3s$, $1s3p$, and $1s3d$. Without setting up the secular equation, but simply by analogy to the results of Section 9.7, write down the 18 correct zeroth-order wave functions. How many energy levels correspond to each of these three configurations, and what is the degeneracy of each energy level? The levels of which configuration lie lowest? highest?

9.24 We have considered helium configurations in which only one electron is excited. Get a rough estimate of the energy of the $2s^2$ configuration from Eq. (9.48). Compare this with the ground-state energy of the He^+ ion to show that the $2s^2$ helium configuration is unstable with respect to ionization to He^+ and an electron. If we had obtained a more accurate estimate of the $2s^2$ energy by including the first-order energy correction, would this increase or decrease our estimate of the $2s^2$ energy?

9.25 For helium the first-order perturbation energy correction is e'^2/r_{12} averaged over the correct unperturbed wave function. Show that if we evaluate $\langle e'^2/r_{12}\rangle$ using the incorrect zeroth-order functions $1s(1)2s(2)$ or $1s(2)2s(1)$, we get J_{1s2s} in each case. Now show that when we use the correct functions (9.103) and (9.104) to evaluate $\langle e'^2/r_{12}\rangle$, we get $J_{1s2s} \pm K_{1s2s}$ (as found from the secular equation). The exchange-integral contribution to the energy thus arises from the indistinguishability of electrons in the same atom.

9.26 Evaluate $\lim_{s\to 0}(e^{as} - 1)/s$. [$s$ corresponds to $\omega_{mn} \pm \omega$ in (9.122).]

9.27 Evaluate $\langle \psi_m^0|Qx|\psi_n^0\rangle$ for the particle in a one-dimensional box.

9.28 Find the selection rules for a charged particle in a three-dimensional box exposed to unpolarized radiation.

9.29 (a) Set f in (7.41) equal to $\hat{B}S$ and operate on each side of the resulting equation with $\hat{A}$. Then multiply by R^* and integrate over all space to obtain the sum rule

$$\sum_i \langle R|\hat{A}|g_i\rangle\langle g_i|\hat{B}|S\rangle = \langle R|\hat{A}\hat{B}|S\rangle$$

where the functions g_i form a complete, orthonormal set, the functions R and S are any two well-behaved functions, the operators $\hat{A}$ and $\hat{B}$ are linear, and the sum is over all members of the complete set. [See also Eq. (7.110). For other sum rules, see A. Dalgarno, *Rev. Mod. Phys.*, **35**, 522 (1963).] (b) An approximate way to evaluate $E_n^{(2)}$ in (9.35) is to replace $E_n^{(0)} - E_m^{(0)}$ by ΔE, where ΔE is some sort of average excitation energy for the problem, whose value can be roughly estimated from the spacings of the unperturbed levels. Use the sum rule of (a) to show that this replacement gives

$$E_n^{(2)} \approx \frac{1}{\Delta E}[\langle n|(\hat{H}')^2|n\rangle - \langle n|\hat{H}'|n\rangle^2]$$

where n stands for $\psi_n^{(0)}$.

9.30 True or false: (a) Every linear combination of solutions of the time-dependent Schrödinger equation is a solution of this equation. (b) Every linear combination of solutions of the time-independent Schrödinger equation is a solution of this equation. (c) The nondegenerate perturbation-theory formula $E^{(1)} = \langle\psi^{(0)}|\hat{H}'|\psi^{(0)}\rangle$ applies only to the ground state. (d) The exact ground-state He-atom wave function has the form $f(1)f(2)$, where $f(i)$, $i = 1, 2$, is a function of the coordinates of electron i.

CHAPTER 10

Electron Spin and the Spin–Statistics Theorem

10.1 ELECTRON SPIN

All chemists are familiar with the yellow color imparted to a flame by sodium atoms. Examining the spectrum of sodium, we find that the strongest yellow line (the D line) is actually two closely spaced lines. The sodium D line arises from a transition from the excited configuration $1s^2 2s^2 2p^6 3p$ to the ground state. The doublet nature of this and other lines in the Na spectrum indicates a doubling of the expected number of states available to the valence electron.

To explain this *fine structure* of atomic spectra, Uhlenbeck and Goudsmit proposed in 1925 that the electron has an *intrinsic* (built-in) angular momentum in addition to the orbital angular momentum due to its motion about the nucleus. If we picture the electron as a sphere of charge spinning about one of its diameters, we can see how such an intrinsic angular momentum can arise. Hence we have the term **spin angular momentum** or, more simply, **spin**. However, electron "spin" is not a classical effect, and the picture of an electron rotating about an axis has no physical reality. The intrinsic angular momentum is real, but no easily visualizable model can explain its origin properly. We cannot hope to understand microscopic particles based on models taken from our experience in the macroscopic world. Other elementary particles besides the electron have spin angular momentum.

In 1928, Dirac developed the relativistic quantum mechanics of an electron, and in his treatment electron spin arises naturally.

In the nonrelativistic quantum mechanics to which we are confining ourselves, electron spin must be introduced as an additional hypothesis. We have learned that each physical property has its corresponding linear Hermitian operator in quantum mechanics. For such properties as orbital angular momentum, we can construct the quantum-mechanical operator from the classical expression by replacing p_x, p_y, p_z by the appropriate operators. The inherent spin angular momentum of a microscopic particle has no analog in classical mechanics, so we cannot use this method to construct operators for spin. For our purposes, we shall simply use symbols for the spin operators, without giving an explicit form for them.

Analogous to the orbital angular-momentum operators $\hat{L}^2$, $\hat{L}_x$, $\hat{L}_y$, $\hat{L}_z$, we have the spin angular-momentum operators $\hat{S}^2$, $\hat{S}_x$, $\hat{S}_y$, $\hat{S}_z$, which are postulated to be linear and Hermitian. $\hat{S}^2$ is the operator for the square of the magnitude of the total spin angular momentum of a particle. $\hat{S}_z$ is the operator for the z component of the particle's spin angular momentum. We have

$$\hat{S}^2 = \hat{S}_x^2 + \hat{S}_y^2 + \hat{S}_z^2 \tag{10.1}$$

We postulate that the spin angular-momentum operators obey the same commutation relations as the orbital angular-momentum operators. Analogous to $[\hat{L}_x, \hat{L}_y] = i\hbar\hat{L}_z$, $[\hat{L}_y, \hat{L}_z] = i\hbar\hat{L}_x$, $[\hat{L}_z, \hat{L}_x] = i\hbar\hat{L}_y$ [Eqs. (5.46) and (5.48)], we have

$$[\hat{S}_x, \hat{S}_y] = i\hbar\hat{S}_z, \qquad [\hat{S}_y, \hat{S}_z] = i\hbar\hat{S}_x, \qquad [\hat{S}_z, \hat{S}_x] = i\hbar\hat{S}_y \tag{10.2}$$

From (10.1) and (10.2), it follows, by the same operator algebra used to obtain (5.49) and (5.50), that

$$[\hat{S}^2, \hat{S}_x] = [\hat{S}^2, \hat{S}_y] = [\hat{S}^2, \hat{S}_z] = 0 \tag{10.3}$$

Since Eqs. (10.1) and (10.2) are of the form of Eqs. (5.107) and (5.108), it follows from the work of Section 5.4 (which depended only on the commutation relations and not on the specific forms of the operators) that the eigenvalues of $\hat{S}^2$ are [Eq. (5.142)]

$$\boxed{s(s+1)\hbar^2, \qquad s = 0, \tfrac{1}{2}, 1, \tfrac{3}{2}, \ldots} \tag{10.4}$$

and the eigenvalues of $\hat{S}_z$ are [Eq. (5.141)]

$$\boxed{m_s\hbar, \qquad m_s = -s, -s+1, \ldots, s-1, s} \tag{10.5}$$

The quantum number s is called the **spin** of the particle. Although nothing in Section 5.4 restricts electrons to a single value for s, experiment shows that all electrons do have a single value for s, namely, $s = \frac{1}{2}$. Protons and neutrons also have $s = \frac{1}{2}$. Pions have $s = 0$. Photons have $s = 1$. However, Eq. (10.5) does not hold for photons. Photons travel at speed c in vacuum. Because of their relativistic nature, it turns out that photons can have either $m_s = +1$ or $m_s = -1$, but not $m_s = 0$ (see *Merzbacher*, Chapter 22). These two m_s values correspond to left circularly polarized and right circularly polarized light.

With $s = \frac{1}{2}$, the magnitude of the total spin angular momentum of an electron is given by the square root of (10.4) as

$$\left[\tfrac{1}{2}\left(\tfrac{3}{2}\right)\hbar^2\right]^{1/2} = \tfrac{1}{2}\sqrt{3}\hbar \tag{10.6}$$

For $s = \frac{1}{2}$, Eq. (10.5) gives the possible eigenvalues of $\hat{S}_z$ of an electron as $+\frac{1}{2}\hbar$ and $-\frac{1}{2}\hbar$. The electron spin eigenfunctions that correspond to these $\hat{S}_z$ eigenvalues are denoted by α and β:

$$\boxed{\hat{S}_z\alpha = +\tfrac{1}{2}\hbar\alpha} \tag{10.7}$$

$$\boxed{\hat{S}_z\beta = -\tfrac{1}{2}\hbar\beta} \tag{10.8}$$

Since $\hat{S}_z$ commutes with $\hat{S}^2$, we can take the eigenfunctions of $\hat{S}_z$ to be eigenfunctions of $\hat{S}^2$ also, with the eigenvalue given by (10.4) with $s = \frac{1}{2}$:

$$\hat{S}^2\alpha = \tfrac{3}{4}\hbar^2\alpha, \qquad \hat{S}^2\beta = \tfrac{3}{4}\hbar^2\beta \tag{10.9}$$

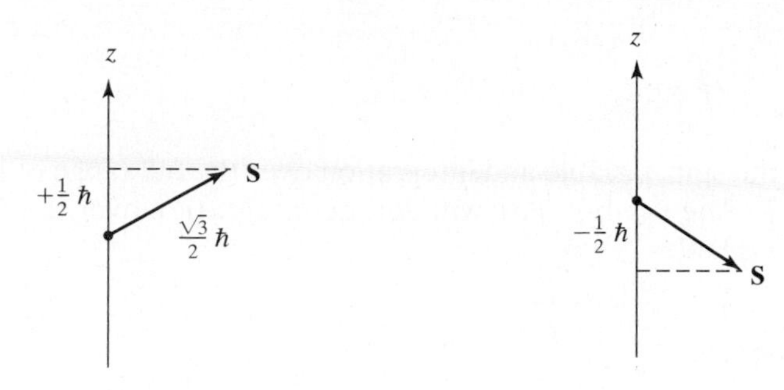

FIGURE 10.1 Possible orientations of the electron spin vector with respect to the z axis. In each case, **S** lies on the surface of a cone whose axis is the z axis.

$\hat{S}_z$ does not commute with $\hat{S}_x$ or $\hat{S}_y$, so α and β are not eigenfunctions of these operators. The terms *spin up* and *spin down* refer to $m_s = +\frac{1}{2}$ and $m_s = -\frac{1}{2}$, respectively. See Fig. 10.1. We shall later show that the two possibilities for the quantum number m_s give the doubling of lines in the spectra of the alkali metals.

The wave functions we have dealt with previously are functions of the spatial coordinates of the particle: $\psi = \psi(x, y, z)$. We might ask: What is the variable for the spin eigenfunctions α and β? Sometimes one talks of a spin coordinate ω, without really specifying what this coordinate is. Most often, one takes the spin quantum number m_s as being the variable on which the spin eigenfunctions depend. This procedure is quite unusual as compared with the spatial wave functions; but because we have only two possible electronic spin eigenfunctions and eigenvalues, this is a convenient choice. We have

$$\alpha = \alpha(m_s), \qquad \beta = \beta(m_s) \tag{10.10}$$

As usual, we want the eigenfunctions to be normalized. The three variables of a one-particle spatial wave function range continuously from $-\infty$ to $+\infty$, so normalization means

$$\int_{-\infty}^{\infty}\int_{-\infty}^{\infty}\int_{-\infty}^{\infty} |\psi(x, y, z)|^2\, dx\, dy\, dz = 1$$

The variable m_s of the electronic spin eigenfunctions takes on only the two discrete values $+\frac{1}{2}$ and $-\frac{1}{2}$ so normalization of the one-particle spin eigenfunctions means

$$\sum_{m_s=-1/2}^{1/2} |\alpha(m_s)|^2 = 1, \qquad \sum_{m_s=-1/2}^{1/2} |\beta(m_s)|^2 = 1 \tag{10.11}$$

Since the eigenfunctions α and β correspond to different eigenvalues of the Hermitian operator $\hat{S}_z$, they are orthogonal:

$$\sum_{m_s=-1/2}^{1/2} \alpha^*(m_s)\beta(m_s) = 0 \tag{10.12}$$

Taking $\alpha(m_s) = \delta_{m_s,1/2}$ and $\beta(m_s) = \delta_{m_s,-1/2}$, we can satisfy (10.11) and (10.12).

When we consider the complete wave function for an electron including both space and spin variables, we shall normalize it according to

$$\boxed{\sum_{m_s=-1/2}^{1/2} \int_{-\infty}^{\infty}\int_{-\infty}^{\infty}\int_{-\infty}^{\infty} |\psi(x, y, z, m_s)|^2\, dx\, dy\, dz = 1} \tag{10.13}$$

The notation

$$\int |\psi(x, y, z, m_s)|^2 \, d\tau$$

will denote summation over the spin variable and integration over the full range of the spatial variables, as in (10.13). The symbol $\int dv$ will denote integration over the full range of the system's spatial variables.

10.2 SPIN AND THE HYDROGEN ATOM

The wave function specifying the state of an electron depends not only on the coordinates x, y, and z but also on the spin state of the electron. What effect does this have on the wave functions and energy levels of the hydrogen atom?

To a very good approximation, the Hamiltonian operator for a system of electrons does not involve the spin variables but is a function only of spatial coordinates and derivatives with respect to spatial coordinates. As a result, we can separate the stationary-state wave function of a single electron into a product of space and spin parts:

$$\psi(x, y, z)g(m_s)$$

where $g(m_s)$ is either one of the functions α or β, depending on whether $m_s = \frac{1}{2}$ or $-\frac{1}{2}$. [More generally, $g(m_s)$ might be a linear combination of α and β; $g(m_s) = c_1\alpha + c_2\beta$.] Since the Hamiltonian operator has no effect on the spin function, we have

$$\hat{H}[\psi(x, y, z)g(m_s)] = g(m_s)\hat{H}\psi(x, y, z) = E[\psi(x, y, z)g(m_s)]$$

and we get the same energies as previously found without taking spin into account. The only difference spin makes is to double the possible number of states. Instead of the state $\psi(x, y, z)$, we have the two possible states $\psi(x, y, z)\alpha$ and $\psi(x, y, z)\beta$. When we take spin into account, the degeneracy of the hydrogen-atom energy levels is $2n^2$ rather than n^2.

10.3 THE SPIN–STATISTICS THEOREM

Suppose we have a system of several identical particles. In classical mechanics the identity of the particles leads to no special consequences. For example, consider identical billiard balls rolling on a billiard table. We can follow the motion of any individual ball, say by taking a motion picture of the system. We can say that ball number one is moving along a certain path, ball two is on another definite path, and so on, the paths being determined by Newton's laws of motion. Thus, although the balls are identical, we can distinguish among them by specifying the path each takes. The identity of the balls has no special effect on their motions.

In quantum mechanics the uncertainty principle tells us that we cannot follow the exact path taken by a microscopic "particle." If the microscopic particles of the system all have different masses or charges or spins, we can use one of these properties to distinguish the particles from one another. But if they are all identical, then the one way we had in classical mechanics of distinguishing them, namely by specifying their paths, is lost in quantum mechanics because of the uncertainty principle. Therefore, the wave function of a system of interacting identical particles must not distinguish among the particles. For example, in the perturbation treatment of the helium-atom excited states in Chapter 9, we saw that the function $1s(1)2s(2)$, which says that electron 1 is in the $1s$ orbital and electron 2 is in the $2s$ orbital, was not a correct zeroth-order wave function. Rather we had to use the functions $2^{-1/2}[1s(1)2s(2) \pm 1s(2)2s(1)]$, which do not specify which electron is in which orbital. (If the identical particles are well separated from one another so that their wave functions do not overlap, they may be regarded as distinguishable.)

We now derive the restrictions on the wave function due to the requirement of indistinguishability of identical particles in quantum mechanics. The wave function of a system of n identical microscopic particles depends on the space and spin variables of the particles. For particle 1, these variables are x_1, y_1, z_1, m_{s1}. Let q_1 stand for all four of these variables. Thus $\psi = \psi(q_1, q_2, \ldots, q_n)$.

We define the **exchange** or **permutation operator** $\hat{P}_{12}$ as the operator that interchanges all the coordinates of particles 1 and 2:

$$\boxed{\hat{P}_{12} f(q_1, q_2, q_3, \ldots, q_n) = f(q_2, q_1, q_3, \ldots, q_n)} \quad \textbf{(10.14)}$$

For example, the effect of $\hat{P}_{12}$ on the function that has electron 1 in a $1s$ orbital with spin up and electron 2 in a $3s$ orbital with spin down is

$$\hat{P}_{12}[1s(1)\alpha(1)3s(2)\beta(2)] = 1s(2)\alpha(2)3s(1)\beta(1) \quad (10.15)$$

What are the eigenvalues of $\hat{P}_{12}$? Applying $\hat{P}_{12}$ twice has no net effect:

$$\hat{P}_{12}\hat{P}_{12} f(q_1, q_2, \ldots, q_n) = \hat{P}_{12} f(q_2, q_1, \ldots, q_n) = f(q_1, q_2, \ldots, q_n)$$

Therefore, $\hat{P}_{12}^2 = \hat{1}$. Let w_i and c_i denote the eigenfunctions and eigenvalues of $\hat{P}_{12}$. We have $\hat{P}_{12} w_i = c_i w_i$. Application of $\hat{P}_{12}$ to this equation gives $\hat{P}_{12}^2 w_i = c_i \hat{P}_{12} w_i$. Substitution of $\hat{P}_{12}^2 = \hat{1}$ and $\hat{P}_{12} w_i = c_i w_i$ in $\hat{P}_{12}^2 w_i = c_i \hat{P}_{12} w_i$ gives $w_i = c_i^2 w_i$. Since zero is not allowed as an eigenfunction, we can divide by w_i to get $1 = c_i^2$ and $c_i = \pm 1$. The eigenvalues of $\hat{P}_{12}$ (and of any linear operator whose square is the unit operator) are $+1$ and -1.

If w_+ is an eigenfunction of $\hat{P}_{12}$ with eigenvalue $+1$, then

$$\hat{P}_{12} w_+(q_1, q_2, \ldots, q_n) = (+1) w_+(q_1, q_2, \ldots, q_n)$$

$$\boxed{w_+(q_2, q_1, \ldots, q_n) = w_+(q_1, q_2, \ldots, q_n)} \quad \textbf{(10.16)}$$

A function such as w_+ that has the property (10.16) of being unchanged when particles 1 and 2 are interchanged is said to be **symmetric** with respect to interchange of particles 1 and 2. For eigenvalue -1, we have

$$\boxed{w_-(q_2, q_1, \ldots, q_n) = -w_-(q_1, q_2, \ldots, q_n)} \quad \textbf{(10.17)}$$

The function w_- in (10.17) is **antisymmetric** with respect to interchange of particles 1 and 2, meaning that this interchange multiplies w_- by -1. There is no necessity for an arbitrary function $f(q_1, q_2, \ldots, q_n)$ to be either symmetric or antisymmetric with respect to interchange of 1 and 2.

Do not confuse the property of being symmetric or antisymmetric with respect to particle interchange with the property of being even or odd with respect to inversion in space. The function $x_1 + x_2$ is symmetric with respect to 1–2 interchange and is an odd function of x_1 and x_2. The function $x_1^2 + x_2^2$ is symmetric with respect to 1–2 interchange and is an even function of x_1 and x_2.

The operator $\hat{P}_{ik}$ is defined by

$$\hat{P}_{ik} f(q_1, \ldots, q_i, \ldots, q_k, \ldots, q_n) = f(q_1, \ldots, q_k, \ldots, q_i, \ldots, q_n) \tag{10.18}$$

The eigenvalues of $\hat{P}_{ik}$ are, like those of $\hat{P}_{12}$, $+1$ and -1.

We now consider the wave function of a system of n identical microscopic particles. Since the particles are indistinguishable, the way we label them cannot affect the state of the system. Thus the two wave functions

$$\psi(q_1, \ldots, q_i, \ldots, q_k, \ldots, q_n) \quad \text{and} \quad \psi(q_1, \ldots, q_k, \ldots, q_i, \ldots, q_n)$$

must correspond to the same state of the system. Two wave functions that correspond to the same state can differ at most by a multiplicative constant. Hence

$$\psi(q_1, \ldots, q_k, \ldots, q_i, \ldots, q_n) = c\psi(q_1, \ldots, q_i, \ldots, q_k, \ldots, q_n)$$

$$\hat{P}_{ik}\psi(q_1, \ldots, q_i, \ldots, q_k, \ldots, q_n) = c\psi(q_1, \ldots, q_i, \ldots, q_k, \ldots, q_n)$$

The last equation states that ψ is an eigenfunction of $\hat{P}_{ik}$. But we know that the only possible eigenvalues of $\hat{P}_{ik}$ are 1 and -1. We conclude that the wave function for a system of n identical particles must be symmetric or antisymmetric with respect to interchange of any two of the identical particles, i and k. Since the n particles are all identical, we could not have the wave function symmetric with respect to some interchanges and antisymmetric with respect to other interchanges. Thus the wave function of n identical particles must be either symmetric with respect to every possible interchange or antisymmetric with respect to every possible interchange of two particles. (The argument just given is not rigorous. The statement that the wave function of a system of identical particles must be either completely symmetric or completely antisymmetric with respect to interchange of two particles is called the *symmetrization postulate*.)

We have seen that there are two possible cases for the wave function of a system of identical particles, the symmetric and the antisymmetric cases. Experimental evidence (such as the periodic table of the elements to be discussed later) shows that for electrons only the antisymmetric case occurs. Thus we have an additional postulate of quantum mechanics, which states that *the wave function of a system of electrons must be antisymmetric with respect to interchange of any two electrons*.

In 1926, Dirac concluded (based on theoretical work and experimental data) that electrons require antisymmetric wave functions and photons require symmetric wave functions. However, Dirac and other physicists erroneously believed in 1926 that all material particles required antisymmetric wave functions. In 1930, experimental data indicated that α particles (which have $s = 0$) require symmetric wave functions, and physicists eventually realized that what determines whether a system of identical particles requires symmetric or antisymmetric wave functions is the spin of the particle. Particles with half-integral spin ($s = \frac{1}{2}, \frac{3}{2}$, and so on) require antisymmetric wave functions, while particles with integral spin ($s = 0, 1$, and so on) require symmetric wave functions. In 1940, the physicist Wolfgang Pauli used relativistic quantum field theory to prove this result. Particles requiring antisymmetric wave functions, such as electrons, are called **fermions** (after E. Fermi), whereas particles requiring symmetric wave functions, such as pions, are called **bosons** (after S. N. Bose). In nonrelativistic quantum mechanics, we must postulate that *the wave function of a system of identical particles must be antisymmetric with respect to interchange of any two particles if the particles have half-integral spin and must be symmetric with respect to interchange if the particles have integral spin.* This statement is called the **spin–statistics theorem** (since the statistical mechanics of a system of bosons differs from that of a system of fermions). [Many proofs of varying validity have been offered for the spin–statistics theorem; see I. Duck and E. C. G. Sudurshan, *Pauli and the Spin-Statistics Theorem*, World Scientific, 1997; *Am. J. Phys.*, **66**, 284 (1998); Sudurshan and Duck, *Pramana-J. Phys.*, **61**, 645 (2003). Several experiments have confirmed the validity of the spin–statistics theorem to extremely high accuracy; see G. M. Tino, *Fortschr. Phys.*, **48**, 537 (2000).]

The spin–statistics theorem has an important consequence for a system of identical fermions. The antisymmetry requirement means that

$$\psi(q_1, q_2, q_3, \ldots, q_n) = -\psi(q_2, q_1, q_3, \ldots, q_n) \tag{10.19}$$

Consider the value of ψ when electrons 1 and 2 have the same coordinates, that is, when $x_1 = x_2$, $y_1 = y_2$, $z_1 = z_2$, and $m_{s1} = m_{s2}$. Putting $q_2 = q_1$ in (10.19), we have

$$\begin{aligned} \psi(q_1, q_1, q_3, \ldots, q_n) &= -\psi(q_1, q_1, q_3, \ldots, q_n) \\ 2\psi &= 0 \\ \psi(q_1, q_1, q_3, \ldots, q_n) &= 0 \end{aligned} \tag{10.20}$$

Thus, two electrons with the same spin have zero probability of being found at the same point in three-dimensional space. (By "the same spin," we mean the same value of m_s.) Since ψ is a continuous function, Eq. (10.20) means that the probability of finding two electrons with the same spin close to each other in space is quite small. Thus the antisymmetry requirement forces electrons of like spin to keep apart from one another. To describe this, one often speaks of a **Pauli repulsion** between such electrons. This "repulsion" is not a real physical force, but a reflection of the fact that the electronic wave function must be antisymmetric with respect to exchange.

The requirement for symmetric or antisymmetric wave functions also applies to a system containing two or more identical composite particles. Consider, for example, an $^{16}O_2$ molecule. The ^{16}O nucleus has 8 protons and 8 neutrons. Each proton and each neutron has $s = \frac{1}{2}$ and is a fermion. Therefore, interchange of the two ^{16}O nuclei interchanges 16 fermions and must multiply the molecular wave function by $(-1)^{16} = 1$. Thus the $^{16}O_2$ molecular wave function must be symmetric with respect to interchange of the nuclear coordinates. The requirement for symmetry or antisymmetry with respect to interchange of identical nuclei affects the degeneracy of molecular wave functions and leads to the symmetry number in the rotational partition function [see *McQuarrie* (1976), pp. 104–105].

For interchange of two identical composite particles containing m identical bosons and n identical fermions, the wave function is multiplied by $(+1)^m(-1)^n = (-1)^n$.

A composite particle is thus a fermion if it contains an odd number of fermions and is a boson otherwise.

When the variational principle (Section 8.1) is used to get approximate electronic wave functions of atoms and molecules, the requirement that the trial variation function be well-behaved includes the requirement that it be antisymmetric.

10.4 THE HELIUM ATOM

We now reconsider the helium atom from the standpoint of electron spin and the antisymmetry requirement. In the perturbation treatment of helium in Section 9.3, we found the zeroth-order wave function for the ground state to be $1s(1)1s(2)$. To take spin into account, we must multiply this spatial function by a spin eigenfunction. We therefore consider the possible spin eigenfunctions for two electrons. We shall use the notation $\alpha(1)\alpha(2)$ to indicate a state where electron 1 has spin up and electron 2 has spin up; $\alpha(1)$ stands for $\alpha(m_{s1})$. Since each electron has two possible spin states, we have at first sight the four possible spin functions:

$$\alpha(1)\alpha(2), \qquad \beta(1)\beta(2), \qquad \alpha(1)\beta(2), \qquad \alpha(2)\beta(1)$$

There is nothing wrong with the first two functions, but the third and fourth functions violate the principle of indistinguishability of identical particles. For example, the third function says that electron 1 has spin up and electron 2 has spin down, which *does* distinguish between electrons 1 and 2. More formally, if we apply $\hat{P}_{12}$ to these functions, we find that the first two functions are symmetric with respect to interchange of the two electrons, but the third and fourth functions are neither symmetric nor antisymmetric and so are unacceptable.

What now? Recall that we ran into essentially the same situation in treating the helium excited states (Section 9.7), where we started with the functions $1s(1)2s(2)$ and $2s(1)1s(2)$. We found that these two functions, which distinguish between electrons 1 and 2, are not the correct zeroth-order functions and that the correct zeroth-order functions are $2^{-1/2}[1s(1)2s(2) \pm 2s(1)1s(2)]$. This result suggests pretty strongly that instead of $\alpha(1)\beta(2)$ and $\beta(1)\alpha(2)$, we use

$$2^{-1/2}[\alpha(1)\beta(2) \pm \beta(1)\alpha(2)] \tag{10.21}$$

These functions are the normalized linear combinations of $\alpha(1)\beta(2)$ and $\beta(1)\alpha(2)$ that are eigenfunctions of $\hat{P}_{12}$, that is, are symmetric or antisymmetric. When electrons 1 and 2 are interchanged, $2^{-1/2}[\alpha(1)\beta(2) + \beta(1)\alpha(2)]$ becomes $2^{-1/2}[\alpha(2)\beta(1) + \beta(2)\alpha(1)]$, which is the same as the original function. In contrast, $2^{-1/2}[\alpha(1)\beta(2) - \beta(1)\alpha(2)]$ becomes $2^{-1/2}[\alpha(2)\beta(1) - \beta(2)\alpha(1)]$, which is -1 times the original function. To show that the functions (10.21) are normalized, we have

$$\sum_{m_{s1}}\sum_{m_{s2}} \frac{1}{\sqrt{2}}[\alpha(1)\beta(2) \pm \beta(1)\alpha(2)]^* \frac{1}{\sqrt{2}}[\alpha(1)\beta(2) \pm \beta(1)\alpha(2)]$$

$$= \frac{1}{2}\sum_{m_{s1}}|\alpha(1)|^2 \sum_{m_{s2}}|\beta(2)|^2 \pm \frac{1}{2}\sum_{m_{s1}}\alpha^*(1)\beta(1)\sum_{m_{s2}}\beta^*(2)\alpha(2)$$

$$\pm \frac{1}{2}\sum_{m_{s1}}\beta^*(1)\alpha(1)\sum_{m_{s2}}\alpha^*(2)\beta(2) + \frac{1}{2}\sum_{m_{s1}}|\beta(1)|^2\sum_{m_{s2}}|\alpha(2)|^2 = 1$$

where we used the orthonormality relations (10.11) and (10.12).

Therefore, the four normalized two-electron spin eigenfunctions with the correct exchange properties are

$$\text{symmetric:} \begin{cases} \alpha(1)\alpha(2) & \textbf{(10.22)} \\ \beta(1)\beta(2) & \textbf{(10.23)} \\ [\alpha(1)\beta(2) + \beta(1)\alpha(2)]/\sqrt{2} & \textbf{(10.24)} \end{cases}$$

$$\text{antisymmetric: } [\alpha(1)\beta(2) - \beta(1)\alpha(2)]/\sqrt{2} \qquad \textbf{(10.25)}$$

We now include spin in the He zeroth-order ground-state wave function. The function $1s(1)1s(2)$ is symmetric with respect to exchange. The overall electronic wave function including spin must be antisymmetric with respect to interchange of the two electrons. Hence we must multiply the symmetric space function $1s(1)1s(2)$ by an antisymmetric spin function. There is only one antisymmetric two-electron spin function, so the ground-state zeroth-order wave function for the helium atom including spin is

$$\psi^{(0)} = 1s(1)1s(2) \cdot 2^{-1/2}[\alpha(1)\beta(2) - \beta(1)\alpha(2)] \qquad (10.26)$$

$\psi^{(0)}$ is an eigenfunction of $\hat{P}_{12}$ with eigenvalue -1.

To a very good approximation, the Hamiltonian does not contain spin terms, so the energy is unaffected by inclusion of the spin factor in the ground-state wave function. Also, the ground state of helium is still nondegenerate when spin is considered.

To further demonstrate that the spin factor does not affect the energy, we shall assume we are doing a variational calculation for the He ground state using the trial function $\phi = f(r_1, r_2, r_{12})2^{-1/2}[\alpha(1)\beta(2) - \beta(1)\alpha(2)]$, where f is a normalized function symmetric in the coordinates of the two electrons. The variational integral is

$$\int \phi^* \hat{H} \phi \, d\tau = \sum_{m_{s1}} \sum_{m_{s2}} \iint f^*(r_1, r_2, r_{12}) \frac{1}{\sqrt{2}} [\alpha(1)\beta(2) - \beta(1)\alpha(2)]^*$$

$$\times \hat{H} f(r_1, r_2, r_{12}) \frac{1}{\sqrt{2}} [\alpha(1)\beta(2) - \beta(1)\alpha(2)] \, dv_1 \, dv_2$$

Since $\hat{H}$ has no effect on the spin functions, the variational integral becomes

$$\iint f^* \hat{H} f \, dv_1 \, dv_2 \sum_{m_{s1}} \sum_{m_{s2}} \frac{1}{2} |\alpha(1)\beta(2) - \beta(1)\alpha(2)|^2$$

Since the spin function (10.25) is normalized, the variational integral reduces to $\iint f^* \hat{H} f \, dv_1 \, dv_2$, which is the expression we used before we introduced spin.

Now consider the excited states of helium. We found the lowest excited state to have the zeroth-order spatial wave function $2^{-1/2}[1s(1)2s(2) - 2s(1)1s(2)]$ [Eq. (9.103)]. Since this spatial function is antisymmetric, we must multiply it by a symmetric spin function. We can use any one of the three symmetric two-electron spin functions, so instead of the nondegenerate level previously found, we have a triply degenerate level with the three zeroth-order wave functions

$$2^{-1/2}[1s(1)2s(2) - 2s(1)1s(2)]\alpha(1)\alpha(2) \qquad (10.27)$$

$$2^{-1/2}[1s(1)2s(2) - 2s(1)1s(2)]\beta(1)\beta(2) \qquad (10.28)$$

$$2^{-1/2}[1s(1)2s(2) - 2s(1)1s(2)]2^{-1/2}[\alpha(1)\beta(2) + \beta(1)\alpha(2)] \qquad (10.29)$$

For the next excited state, the requirement of antisymmetry of the overall wave function leads to the zeroth-order wave function

$$2^{-1/2}[1s(1)2s(2) + 2s(1)1s(2)]2^{-1/2}[\alpha(1)\beta(2) - \beta(1)\alpha(2)] \quad (10.30)$$

The same considerations apply for the $1s2p$ states.

10.5 THE PAULI EXCLUSION PRINCIPLE

So far, we have not seen any very spectacular consequences of electron spin and the antisymmetry requirement. In the hydrogen and helium atoms, the spin factors in the wave functions and the antisymmetry requirement simply affect the degeneracy of the levels but do not (except for very small effects to be considered later) affect the previously obtained energies. For lithium, the story is quite different.

Suppose we take the interelectronic repulsions in the Li atom as a perturbation on the remaining terms in the Hamiltonian. By the same steps used in the treatment of helium, the unperturbed wave functions are products of three hydrogenlike functions. For the ground state,

$$\psi^{(0)} = 1s(1)1s(2)1s(3) \quad (10.31)$$

and the zeroth-order (unperturbed) energy is [Eq. (9.48)]

$$E^{(0)} = -\left(\frac{1}{1^2} + \frac{1}{1^2} + \frac{1}{1^2}\right)\left(\frac{Z^2e'^2}{2a_0}\right) = -27\left(\frac{e'^2}{2a_0}\right) = -27(13.606 \text{ eV}) = -367.4 \text{ eV}$$

The first-order energy correction is $E^{(1)} = \langle\psi^{(0)}|\hat{H}'|\psi^{(0)}\rangle$. The perturbation $\hat{H}'$ consists of the interelectronic repulsions, and

$$E^{(1)} = \int |1s(1)|^2|1s(2)|^2|1s(3)|^2\frac{e'^2}{r_{12}}\,dv + \int |1s(1)|^2|1s(2)|^2|1s(3)|^2\frac{e'^2}{r_{23}}\,dv + \int |1s(1)|^2|1s(2)|^2|1s(3)|^2\frac{e'^2}{r_{13}}\,dv$$

The way we label the dummy integration variables in these definite integrals cannot affect their value. If we interchange the labels 1 and 3 on the variables in the second integral, it is converted to the first integral. Hence these two integrals are equal. Interchange of the labels 2 and 3 in the third integral shows it to be equal to the first integral also. Hence

$$E^{(1)} = 3\iint |1s(1)|^2|1s(2)|^2\frac{e'^2}{r_{12}}\,dv_1\,dv_2 \int |1s(3)|^2\,dv_3$$

The integral over electron 3 gives 1 (normalization). The integral over electrons 1 and 2 was evaluated in the perturbation treatment of helium, and [Eqs. (9.52) and (9.53)]

$$E^{(1)} = 3\left(\frac{5Z}{4}\right)\left(\frac{e'^2}{2a_0}\right) = 153.1 \text{ eV}$$

$$E^{(0)} + E^{(1)} = -214.3 \text{ eV}$$

Since we can use the zeroth-order perturbation wave function as a trial variation function (recall the discussion at the beginning of Section 9.4), $E^{(0)} + E^{(1)}$ must be, according to the variation principle, equal to or greater than the true ground-state energy. The experimental value of the lithium ground-state energy is found by adding up the first, second, and third ionization energies, which gives [C. E. Moore, "Ionization Potentials and Ionization Limits," publication NSRDS-NBS 34 of the National Bureau of Standards (1970)]

$$-(5.39 + 75.64 + 122.45)\text{ eV} = -203.5\text{ eV}$$

We thus have $E^{(0)} + E^{(1)}$ as less than the true ground-state energy, which is a violation of the variation principle. Moreover, the supposed configuration $1s^3$ for the Li ground state disagrees with the low value of the first ionization potential and with all chemical evidence. If we continued on in this manner, we would have a $1s^Z$ ground-state configuration for the element of atomic number Z. We would not get the well-known periodic behavior of the elements.

Of course, our error is failure to consider spin and the antisymmetry requirement. The hypothetical zeroth-order wave function $1s(1)1s(2)1s(3)$ is symmetric with respect to interchange of any two electrons. If we are to have an antisymmetric $\psi^{(0)}$, we must multiply this symmetric space function by an antisymmetric spin function. It is easy to construct completely symmetric spin functions for three electrons, such as $\alpha(1)\alpha(2)\alpha(3)$. However, try as we may, it is impossible to construct a completely antisymmetric spin function for three electrons.

Let us consider how we can systematically construct an antisymmetric function for three electrons. We shall use f, g, and h to stand for three functions of electronic coordinates, without specifying whether we are considering space coordinates or spin coordinates or both. We start with the function

$$f(1)g(2)h(3) \tag{10.32}$$

which is certainly not antisymmetric. The antisymmetric function we desire must be converted into its negative by each of the exchange operators $\hat{P}_{12}$, $\hat{P}_{13}$, and $\hat{P}_{23}$. Applying each of these operators in turn to $f(1)g(2)h(3)$, we get

$$f(2)g(1)h(3), \qquad f(3)g(2)h(1), \qquad f(1)g(3)h(2) \tag{10.33}$$

We might try to construct the antisymmetric functions as a linear combination of the four functions (10.32) and (10.33), but this attempt would fail. Application of $\hat{P}_{12}$ to the last two functions in (10.33) gives

$$f(3)g(1)h(2) \quad \text{and} \quad f(2)g(3)h(1) \tag{10.34}$$

which are not included in (10.32) or (10.33). We must therefore include all six functions (10.32) to (10.34) in the desired antisymmetric linear combination. These six functions are the six $(3 \cdot 2 \cdot 1)$ possible permutations of the three electrons among the three functions f, g, and h. If $f(1)g(2)h(3)$ is a solution of the Schrödinger equation with eigenvalue E, then, because of the identity of the particles, each of the functions (10.32) to (10.34) is also a solution with the same eigenvalue E (exchange degeneracy), and any linear combination of these functions is an eigenfunction with eigenvalue E.

The antisymmetric linear combination will have the form

$$\begin{aligned} c_1f(1)g(2)h(3) + c_2f(2)g(1)h(3) + c_3f(3)g(2)h(1) + c_4f(1)g(3)h(2) \\ + c_5f(3)g(1)h(2) + c_6f(2)g(3)h(1) \end{aligned} \tag{10.35}$$

Since $f(2)g(1)h(3) = \hat{P}_{12} f(1)g(2)h(3)$, in order to have (10.35) be an eigenfunction of $\hat{P}_{12}$ with eigenvalue -1, we must have $c_2 = -c_1$. Likewise, $f(3)g(2)h(1) = \hat{P}_{13} f(1)g(2)h(3)$ and $f(1)g(3)h(2) = \hat{P}_{23} f(1)g(2)h(3)$, so $c_3 = -c_1$ and $c_4 = -c_1$. Since $f(3)g(1)h(2) = \hat{P}_{12} f(3)g(2)h(1)$, we must have $c_5 = -c_3 = c_1$. Similarly, we find $c_6 = c_1$. We thus arrive at the linear combination

$$c_1[f(1)g(2)h(3) - f(2)g(1)h(3) - f(3)g(2)h(1) - f(1)g(3)h(2) \\ + f(3)g(1)h(2) + f(2)g(3)h(1)] \tag{10.36}$$

which is easily verified to be antisymmetric with respect to 1–2, 1–3, and 2–3 interchange. [Taking all signs as plus in (10.36), we would get a completely symmetric function.]

Let us assume *f*, *g*, and *h* to be orthonormal and choose c_1 so that (10.36) is normalized. Multiplying (10.36) by its complex conjugate, we get many terms, but because of the assumed orthogonality the integrals of all products involving two different terms of (10.36) vanish. For example,

$$\int [f(1)g(2)h(3)]^* f(2)g(1)h(3)\, d\tau \\ = \int f^*(1)g(1)\, d\tau_1 \int g^*(2)f(2)\, d\tau_2 \int h^*(3)h(3)\, d\tau_3 = 0 \cdot 0 \cdot 1 = 0$$

Integrals involving the product of a term of (10.36) with its own complex conjugate are equal to 1, because *f*, *g*, and *h* are normalized. Therefore,

$$1 = \int |(10.36)|^2\, d\tau = |c_1|^2(1 + 1 + 1 + 1 + 1 + 1)$$

$$c_1 = 1/\sqrt{6}$$

We could work with (10.36) as it stands, but its properties are most easily found if we recognize it as the expansion [Eq. (8.24)] of the following third-order determinant:

$$\boxed{\frac{1}{\sqrt{6}} \begin{vmatrix} f(1) & g(1) & h(1) \\ f(2) & g(2) & h(2) \\ f(3) & g(3) & h(3) \end{vmatrix}} \tag{10.37}$$

(See also Prob. 8.21.) The antisymmetry property holds for (10.37) because interchange of two electrons amounts to interchanging two rows of the determinant, which multiplies it by -1.

We now use (10.37) to prove that it is impossible to construct an antisymmetric spin function for three electrons. The functions *f*, *g*, and *h* may each be either α or β. If we take $f = \alpha$, $g = \beta$, $h = \alpha$, then (10.37) becomes

$$\frac{1}{\sqrt{6}} \begin{vmatrix} \alpha(1) & \beta(1) & \alpha(1) \\ \alpha(2) & \beta(2) & \alpha(2) \\ \alpha(3) & \beta(3) & \alpha(3) \end{vmatrix} \tag{10.38}$$

Although (10.38) is antisymmetric, we must reject it because it is equal to zero. The first and third columns of the determinant are identical, so (Section 8.3) the determinant vanishes. No matter how we choose f, g, and h, at least two columns of the determinant must be equal, so we cannot construct a nonzero antisymmetric three-electron spin function.

We now use (10.37) to construct the zeroth-order ground-state wave function for lithium, including both space and spin variables. The functions f, g, and h will now involve both space and spin variables. We choose

$$f(1) = 1s(1)\alpha(1) \tag{10.39}$$

We call a function like (10.39) a **spin-orbital**. *A spin-orbital is the product of a one-electron spatial orbital and a one-electron spin function.*

If we were to take $g(1) = 1s(1)\alpha(1)$, this would make the first and second columns of (10.37) identical, and the wave function would vanish. This is a particular case of the **Pauli exclusion principle**: *No two electrons can occupy the same spin-orbital.* Another way of stating this is to say that no two electrons in an atom can have the same values for all their quantum numbers. The Pauli exclusion principle is a consequence of the more general antisymmetry requirement for the wave function of a system of identical spin-$\frac{1}{2}$ particles and is less satisfying than the antisymmetry statement, since the exclusion principle is based on approximate (zeroth-order) wave functions.

We therefore take $g(1) = 1s(1)\beta(1)$, which puts two electrons with opposite spin in the $1s$ orbital. For the spin-orbital h, we cannot use either $1s(1)\alpha(1)$ or $1s(1)\beta(1)$, since these choices make the determinant vanish. We take $h(1) = 2s(1)\alpha(1)$, which gives the familiar Li ground-state configuration $1s^2 2s$ and the zeroth-order wave function

$$\psi^{(0)} = \frac{1}{\sqrt{6}} \begin{vmatrix} 1s(1)\alpha(1) & 1s(1)\beta(1) & 2s(1)\alpha(1) \\ 1s(2)\alpha(2) & 1s(2)\beta(2) & 2s(2)\alpha(2) \\ 1s(3)\alpha(3) & 1s(3)\beta(3) & 2s(3)\alpha(3) \end{vmatrix} \tag{10.40}$$

Note especially that (10.40) is *not* simply a product of space and spin parts (as we found for H and He), but is a linear combination of terms, each of which is a product of space and spin parts.

Since we could just as well have taken $h(1) = 2s(1)\beta(1)$, the ground state of lithium is, like hydrogen, doubly degenerate, corresponding to the two possible orientations of the spin of the $2s$ electron. We might use the orbital diagrams

$$\underset{\overline{\uparrow\downarrow}}{1s} \quad \underset{\overline{\uparrow}}{2s} \qquad \text{and} \qquad \underset{\overline{\uparrow\downarrow}}{1s} \quad \underset{\overline{\downarrow}}{2s}$$

to indicate this. Each spatial orbital such as $1s$ or $2p_0$ can hold two electrons of opposite spin. A spin-orbital such as $2s\alpha$ can hold one electron.

Although the $1s^2 2p$ configuration will have the same unperturbed energy $E^{(0)}$ as the $1s^2 2s$ configuration, when we take electron repulsion into account by calculating $E^{(1)}$ and higher corrections, we find that the $1s^2 2s$ configuration lies lower for the same reason as in helium.

Consider some points about the Pauli exclusion principle, which we restate as follows: *In a system of identical fermions, no two particles can occupy the same state.* If we have a system of n interacting particles (for example, an atom), there is a single wave function (involving $4n$ variables) for the entire system. Because of the interactions between the particles, the wave function cannot be written as the product of wave functions of the individual particles. Hence, strictly speaking, we cannot talk of the states of individual particles, only the state of the whole system. If, however, the interactions between the particles are not too large, then as an initial approximation we can neglect them and write the zeroth-order wave function of the system as a product of wave functions of the individual particles. In this zeroth-order wave function, no two fermions can have the same wave function (state).

Since bosons require a wave function symmetric with respect to interchange, there is no restriction on the number of bosons in a given state.

In 1925, Einstein showed that in an ideal gas of noninteracting bosons, there is a very low temperature T_c (called the condensation temperature) above which the fraction f of bosons in the ground state is negligible but below which f becomes appreciable and goes to 1 as the absolute temperature T goes to 0. The equation for f for noninteracting bosons in a cubic box is $f = 1 - (T/T_c)^{3/2}$ for $T < T_c$ [*McQuarrie* (1976), Section 10-4]. The phenomenon of a significant fraction of bosons falling into the ground state is called *Bose–Einstein condensation*. Bose–Einstein condensation is important in determining the properties of superfluid liquid ^{4}He (whose atoms are bosons), but the interatomic interactions in the liquid make theoretical analysis difficult.

In 1995, physicists succeeded in producing Bose–Einstein condensation in a gas [*Physics Today*, August 1995, p. 17; C. E. Wieman, *Am. J. Phys.*, **64**, 847 (1996)]. They used a gas of $^{87}_{37}$Rb atoms. An ^{87}Rb atom has 87 nucleons and 37 electrons. With an even number (124) of fermions, ^{87}Rb is a boson. With a combination of laser light, an applied inhomogeneous magnetic field, and applied radiofrequency radiation, a sample of 10^4 ^{87}Rb atoms was cooled to 10^{-7} K, thereby condensing a substantial fraction of the atoms into the ground state. The radiofrequency radiation was then used to remove most of the atoms in excited states, leaving a condensate of 2000 atoms, nearly all of which were in the ground state. Each Rb atom in this experiment was subject to a potential-energy function $V(x, y, z)$ produced by the interaction of the atom's total spin magnetic moment with the applied magnetic field (Sections 6.8 and 10.9). The inhomogeneous applied magnetic field was such that the potential energy V was that of a three-dimensional harmonic oscillator (Prob. 4.19) plus a constant. The Rb atoms in the Bose–Einstein condensate are in the ground state of this harmonic-oscillator potential.

10.6 SLATER DETERMINANTS

Slater pointed out in 1929 that a determinant of the form (10.40) satisfies the antisymmetry requirement for a many-electron atom. A determinant like (10.40) is called a **Slater determinant**. All the elements in a given column of a Slater determinant involve the same spin-orbital, whereas elements in the same row all involve the same electron. (Since interchanging rows and columns does not affect the value of a determinant, we could write the Slater determinant in another, equivalent form.)

Consider how the zeroth-order helium wave functions that we found previously can be written as Slater determinants. For the ground-state configuration $1s^2$, we have the spin-orbitals $1s\alpha$ and $1s\beta$, giving the Slater determinant

$$\frac{1}{\sqrt{2}}\begin{vmatrix} 1s(1)\alpha(1) & 1s(1)\beta(1) \\ 1s(2)\alpha(2) & 1s(2)\beta(2) \end{vmatrix} = 1s(1)1s(2)\frac{1}{\sqrt{2}}[\alpha(1)\beta(2) - \beta(1)\alpha(2)] \quad (10.41)$$

which agrees with (10.26). For the states corresponding to the excited configuration $1s2s$, we have the possible spin-orbitals $1s\alpha$, $1s\beta$, $2s\alpha$, $2s\beta$, which give the four Slater determinants

$$D_1 = \frac{1}{\sqrt{2}}\begin{vmatrix} 1s(1)\alpha(1) & 2s(1)\alpha(1) \\ 1s(2)\alpha(2) & 2s(2)\alpha(2) \end{vmatrix} \qquad D_2 = \frac{1}{\sqrt{2}}\begin{vmatrix} 1s(1)\alpha(1) & 2s(1)\beta(1) \\ 1s(2)\alpha(2) & 2s(2)\beta(2) \end{vmatrix}$$

$$D_3 = \frac{1}{\sqrt{2}}\begin{vmatrix} 1s(1)\beta(1) & 2s(1)\alpha(1) \\ 1s(2)\beta(2) & 2s(2)\alpha(2) \end{vmatrix} \qquad D_4 = \frac{1}{\sqrt{2}}\begin{vmatrix} 1s(1)\beta(1) & 2s(1)\beta(1) \\ 1s(2)\beta(2) & 2s(2)\beta(2) \end{vmatrix}$$

Comparison with (10.27) to (10.30) shows that the $1s2s$ zeroth-order wave functions are related to these four Slater determinants as follows:

$$2^{-1/2}[1s(1)2s(2) - 2s(1)1s(2)]\alpha(1)\alpha(2) = D_1 \quad (10.42)$$

$$2^{-1/2}[1s(1)2s(2) - 2s(1)1s(2)]\beta(1)\beta(2) = D_4 \quad (10.43)$$

$$2^{-1/2}[1s(1)2s(2) - 2s(1)1s(2)]2^{-1/2}[\alpha(1)\beta(2) + \beta(1)\alpha(2)] = 2^{-1/2}(D_2 + D_3) \quad (10.44)$$

$$2^{-1/2}[1s(1)2s(2) + 2s(1)1s(2)]2^{-1/2}[\alpha(1)\beta(2) - \beta(1)\alpha(2)] = 2^{-1/2}(D_2 - D_3) \quad (10.45)$$

(To get a zeroth-order function that is an eigenfunction of the spin and orbital angular-momentum operators, we sometimes have to take a linear combination of the Slater determinants of a configuration; see Chapter 11.)

Next consider some notations used for Slater determinants. Instead of writing α and β for spin functions, one often puts a bar over the spatial function to indicate the spin function β, and a spatial function without a bar implies the spin factor α. With this notation, (10.40) is written as

$$\psi^{(0)} = \frac{1}{\sqrt{6}}\begin{vmatrix} 1s(1) & \overline{1s}(1) & 2s(1) \\ 1s(2) & \overline{1s}(2) & 2s(2) \\ 1s(3) & \overline{1s}(3) & 2s(3) \end{vmatrix} \quad (10.46)$$

Given the spin-orbitals occupied by the electrons, we can readily construct the Slater determinant. Therefore, a shorthand notation for Slater determinants that simply specifies the spin-orbitals is often used. In this notation, (10.46) is written as

$$\boxed{\psi^{(0)} = |1s\overline{1s}2s|} \quad \mathbf{(10.47)}$$

where the vertical lines indicate formation of the determinant and multiplication by $1/\sqrt{6}$.

We showed that the factor $1/\sqrt{6}$ normalizes a third-order Slater determinant constructed of orthonormal functions. The expansion of an nth-order determinant has $n!$ terms (Prob. 8.19). For an nth-order Slater determinant of orthonormal spin-orbitals, the same reasoning used in the third-order case shows that the normalization constant is $1/\sqrt{n!}$. We always include a factor $1/\sqrt{n!}$ in defining a Slater determinant of order n.

10.7 PERTURBATION TREATMENT OF THE LITHIUM GROUND STATE

Let us carry out a perturbation treatment of the ground state of the lithium atom. We take

$$\hat{H}^0 = -\frac{\hbar^2}{2m_e}\nabla_1^2 - \frac{\hbar^2}{2m_e}\nabla_2^2 - \frac{\hbar^2}{2m_e}\nabla_3^2 - \frac{Ze'^2}{r_1} - \frac{Ze'^2}{r_2} - \frac{Ze'^2}{r_3}, \quad \hat{H}' = \frac{e'^2}{r_{12}} + \frac{e'^2}{r_{23}} + \frac{e'^2}{r_{13}}$$

We found in Section 10.5 that to satisfy the antisymmetry requirement, the ground-state configuration must be $1s^22s$; the correct zeroth-order wave function is (10.40):

$$\begin{aligned}\psi^{(0)} = 6^{-1/2}[&1s(1)1s(2)2s(3)\alpha(1)\beta(2)\alpha(3) - 1s(1)2s(2)1s(3)\alpha(1)\alpha(2)\beta(3)\\ &- 1s(1)1s(2)2s(3)\beta(1)\alpha(2)\alpha(3) + 1s(1)2s(2)1s(3)\beta(1)\alpha(2)\alpha(3)\\ &+ 2s(1)1s(2)1s(3)\alpha(1)\alpha(2)\beta(3) - 2s(1)1s(2)1s(3)\alpha(1)\beta(2)\alpha(3)]\end{aligned} \tag{10.48}$$

What is $E^{(0)}$? Each term in $\psi^{(0)}$ contains the product of two $1s$ hydrogenlike functions and one $2s$ hydrogenlike function, multiplied by a spin factor. $\hat{H}^0$ is the sum of three hydrogenlike Hamiltonians, one for each electron, and does not involve spin. Thus $\psi^{(0)}$ is a linear combination of terms, each of which is an eigenfunction of $\hat{H}^0$ with eigenvalue $E_{1s}^{(0)} + E_{1s}^{(0)} + E_{2s}^{(0)}$, where these are hydrogenlike energies. Hence $\psi^{(0)}$ is an eigenfunction of $\hat{H}^0$ with eigenvalue $E_{1s}^{(0)} + E_{1s}^{(0)} + E_{2s}^{(0)}$. Therefore [Eq. (6.94)],

$$E^{(0)} = -\left(\frac{1}{1^2} + \frac{1}{1^2} + \frac{1}{2^2}\right)\left(\frac{Z^2e'^2}{2a_0}\right) = -\frac{81}{4}(13.606 \text{ eV}) = -275.5 \text{ eV} \tag{10.49}$$

The evaluation of $E^{(1)} = \langle\psi^{(0)}|\hat{H}'|\psi^{(0)}\rangle$ is outlined in Prob. 10.16. One finds

$$\begin{aligned}E^{(1)} = 2\iint 1s^2(1)2s^2(2)\frac{e'^2}{r_{12}}dv_1\,dv_2 + \iint 1s^2(1)1s^2(2)\frac{e'^2}{r_{12}}dv_1\,dv_2\\ - \iint 1s(1)2s(2)1s(2)2s(1)\frac{e'^2}{r_{12}}dv_1\,dv_2\end{aligned} \tag{10.50}$$

These integrals are Coulomb and exchange integrals:

$$E^{(1)} = 2J_{1s2s} + J_{1s1s} - K_{1s2s} \tag{10.51}$$

We have [Eqs. (9.52), (9.53), and (9.111)]

$$J_{1s1s} = \frac{5}{8}\frac{Ze'^2}{a_0}, \qquad J_{1s2s} = \frac{17}{81}\frac{Ze'^2}{a_0}, \qquad K_{1s2s} = \frac{16}{729}\frac{Ze'^2}{a_0}$$

$$E^{(1)} = \frac{5965}{972}\left(\frac{e'^2}{2a_0}\right) = 83.5 \text{ eV}$$

The energy through first order is -192.0 eV, as compared with the true ground-state energy of lithium, -203.5 eV. To improve on this result, we must calculate higher-order wave-function and energy corrections. This will mix into the wave function contributions from Slater determinants involving configurations besides $1s^22s$ (configuration interaction).

10.8 VARIATION TREATMENTS OF THE LITHIUM GROUND STATE

The zeroth-order perturbation wave function (10.40) uses the full nuclear charge ($Z = 3$) for both the 1s and 2s orbitals of lithium. We expect that the 2s electron, which is partially shielded from the nucleus by the two 1s electrons, will see an effective nuclear charge that is much less than 3. Even the 1s electrons partially shield each other (recall the treatment of the helium ground state). This reasoning suggests the introduction of two variational parameters b_1 and b_2 into (10.40).

Instead of using the $Z = 3$ 1s function in Table 6.2, we take

$$f \equiv \frac{1}{\pi^{1/2}}\left(\frac{b_1}{a_0}\right)^{3/2} e^{-b_1 r/a_0} \tag{10.52}$$

where b_1 is a variational parameter representing an effective nuclear charge for the 1s electrons. Instead of the $Z = 3$ 2s function in Table 6.2, we use

$$g = \frac{1}{4(2\pi)^{1/2}}\left(\frac{b_2}{a_0}\right)^{3/2}\left(2 - \frac{b_2 r}{a_0}\right) e^{-b_2 r/2a_0} \tag{10.53}$$

Our trial variation function is then

$$\phi = \frac{1}{\sqrt{6}}\begin{vmatrix} f(1)\alpha(1) & f(1)\beta(1) & g(1)\alpha(1) \\ f(2)\alpha(2) & f(2)\beta(2) & g(2)\alpha(2) \\ f(3)\alpha(3) & f(3)\beta(3) & g(3)\alpha(3) \end{vmatrix} \tag{10.54}$$

The use of different charges b_1 and b_2 for the 1s and 2s orbitals destroys their orthogonality, so (10.54) is not normalized. The best values of the variational parameters are found by setting $\partial W/\partial b_1 = 0$ and $\partial W/\partial b_2 = 0$, where the variational integral W is given by the left side of Eq. (8.9). The results are [F. B. Wilson, Jr., *J. Chem. Phys.*, **1**, 210 (1933)] $b_1 = 2.686$, $b_2 = 1.776$, and $W = -201.2$ eV. W is much closer to the true value -203.5 eV than the result -192.0 eV found in the last section. The value of b_2 shows substantial, but not complete, screening of the 2s electron by the 1s electrons.

We might try other forms for the orbitals besides (10.52) and (10.53) to improve the trial function. However, no matter what orbital functions we try, if we restrict ourselves to a trial function of the form of (10.54), we can never reach the true ground-state energy. To do this, we can introduce r_{12}, r_{23}, and r_{13} into the trial function or use a linear combination of several Slater determinants corresponding to various configurations (configuration interaction).

10.9 SPIN MAGNETIC MOMENT

Recall that the orbital angular momentum **L** of an electron has a magnetic moment $-(e/2m_e)\mathbf{L}$ associated with it [Eq. (6.128)]. It is natural to suppose that there is also a magnetic moment $\mathbf{m}_S$ associated with the electronic spin angular momentum **S**. We might guess that $\mathbf{m}_S$ would be $-(e/2m_e)$ times **S**. Spin is a relativistic phenomenon, however, and we cannot expect $\mathbf{m}_S$ to be related to **S** in exactly the same way that $\mathbf{m}_L$ is related to **L**. In fact, Dirac's relativistic treatment of the electron gave the result that (in SI units)

$$\mathbf{m}_S = -g_e \frac{e}{2m_e}\mathbf{S} \tag{10.55}$$

where Dirac's treatment gave $g_e = 2$ for the *electron g factor* g_e. Theoretical and experimental work subsequent to Dirac's treatment has shown that g_e is slightly greater than 2 [see P. Kusch, *Physics Today*, Feb. 1966, p. 23]: $g_e = 2(1 + \alpha/2\pi + \cdots) = 2.0023$, where the dots indicate terms involving higher powers of α and where the *fine-structure constant* α is

$$\alpha \equiv \frac{e^2}{4\pi\varepsilon_0 \hbar c} \equiv \frac{e'^2}{\hbar c} = 0.0072973525 \tag{10.56}$$

The magnitude of the spin magnetic moment of an electron is (in SI units)

$$|\mathbf{m}_S| = g_e \frac{e}{2m_e}|\mathbf{S}| = g_e \sqrt{\frac{3}{4}} \frac{e\hbar}{2m_e} \tag{10.57}$$

The ferromagnetism of iron is due to the electron's magnetic moment.

The two possible orientations of an electron's spin and its associated spin magnetic moment with respect to an axis produce two energy levels in an externally applied magnetic field. In electron-spin-resonance (ESR) spectroscopy, one observes transitions between these two levels. ESR spectroscopy is applicable to species such as free radicals and transition-metal ions that have one or more unpaired electron spins and hence have a nonzero total electron spin and spin magnetic moment.

NMR Spectroscopy. Many atomic nuclei have a nonzero spin angular momentum **I**. Similar to (10.4) and (10.5), the magnitude of **I** is $[I(I + 1)]^{1/2}\hbar$, where the nuclear-spin quantum number I can be 0, $\frac{1}{2}$, 1, etc., and the z component of **I** has the possible values $M_I\hbar$, where $M_I = -I, -I + 1, \ldots, I$. Some I values are 0 for every nucleus with an even number of protons and an even number of neutrons (for example, ${}^{16}_{8}\text{O}$ and ${}^{12}_{6}\text{C}$); $\frac{1}{2}$ for ${}^{1}_{1}\text{H}$, ${}^{13}_{6}\text{C}$, ${}^{19}_{9}\text{F}$, and ${}^{31}_{15}\text{P}$; 1 for ${}^{2}_{1}\text{H}$ and ${}^{14}_{7}\text{N}$; $\frac{3}{2}$ for ${}^{11}_{5}\text{B}$, ${}^{23}_{11}\text{Na}$, and ${}^{35}_{17}\text{Cl}$. If $I \neq 0$, the nucleus has a spin magnetic moment $\mathbf{m}_I$ given by an equation similar to (10.55):

$$\mathbf{m}_I = g_N(e/2m_p)\mathbf{I} \equiv \gamma\mathbf{I} \tag{10.58}$$

where m_p is the proton mass and the *nuclear g factor* g_N has a value characteristic of the nucleus. The quantity γ, called the **magnetogyric ratio** of the nucleus, is defined by $\gamma \equiv \mathbf{m}_I/\mathbf{I} = g_N e/2m_p$. Values of I, g_N, and γ for some nuclei are

nucleus	^{1}H	^{12}C	^{13}C	^{15}N	^{19}F	^{31}P
I	1/2	0	1/2	1/2	1/2	1/2
g_N	5.58569		1.40482	−0.56638	5.25773	2.2632
γ/(MHz/T)	267.522		67.283	−27.126	251.815	108.39

In nuclear-magnetic-resonance (NMR) spectroscopy, one observes transitions between nuclear-spin energy levels in an applied magnetic field. The nuclei most commonly studied are ^1H and ^{13}C. The sample (most commonly a dilute solution of the compound being studied) is placed between the poles of a strong magnet. The energy of interaction between an isolated nuclear spin magnetic moment $\mathbf{m}_I$ in an external magnetic field $\mathbf{B}$ is given by Eq. (6.131) as $E = -\mathbf{m}_I \cdot \mathbf{B}$. Using (10.58) for $\mathbf{m}_I$ and taking the z axis as coinciding with the direction of $\mathbf{B}$, we have

$$E = -\mathbf{m}_I \cdot \mathbf{B} = -\gamma(I_x\mathbf{i} + I_y\mathbf{j} + I_z\mathbf{k}) \cdot (B\mathbf{k}) = -\gamma B I_z$$

We convert this classical expression for the energy into a Hamiltonian operator by replacing the classical quantity I_z by the operator $\hat{I}_z$. Thus, $\hat{H} = -\gamma B \hat{I}_z$. Let $|M_I\rangle$ denote the function that is simultaneously an eigenfunction of the operators $\hat{I}^2$ (for the square of the magnitude of the nuclear-spin angular momentum) and $\hat{I}_z$. We have

$$\hat{H}|M_I\rangle = -\gamma B \hat{I}_z|M_I\rangle = -\gamma B M_I \hbar |M_I\rangle \tag{10.59}$$

Therefore, (10.59) gives the energy levels of the isolated nuclear spin in the applied magnetic field as

$$E = -\gamma \hbar B M_I, \qquad M_I = -I, -I + 1, \ldots, I$$

In NMR spectroscopy, the sample is exposed to electromagnetic radiation that induces transitions between nuclear-spin energy levels. The selection rule turns out to be $\Delta M_I = \pm 1$. The NMR transition frequency ν is found as follows:

$$h\nu = |\Delta E| = |\gamma| \hbar B |\Delta M_I| = |\gamma| \hbar B$$

$$\nu = (|\gamma|/2\pi)B = |g_N|(e/4\pi m_p)B \tag{10.60}$$

The value of γ differs greatly for different nuclei, and in any one experiment, one studies the NMR spectrum of one kind of nucleus. The most commonly studied nucleus is ^1H, the proton. The second most studied nucleus is ^{13}C. The ^{13}C isotope occurs in 1% abundance in carbon.

Equation (10.60) is for a nucleus isolated except for the presence of the external magnetic field $\mathbf{B}$. For a nucleus present in a molecule, we also have to consider the contribution of the molecular electrons to the magnetic field felt by each nucleus. In most ground-state molecules, the electron spins are all paired and there is no electronic orbital angular momentum. With no electronic spin or orbital angular momentum, the electrons do not contribute to the magnetic field experienced by each nucleus. However, the application of the external applied field $\mathbf{B}$ perturbs the molecular electronic wave function, thereby producing an electronic contribution to the magnetic field at each nucleus. This electronic contribution is proportional to the magnitude of the external field B and is usually in the opposite direction to B. Therefore, the magnetic field experienced by nucleus i in a molecule is $B - \sigma_i B = (1 - \sigma_i)B$, where the proportionality constant σ_i is called the **screening constant** or **shielding constant** for nucleus I, and is much less than 1. Equation (10.60) becomes for a nucleus in a molecule

$$\nu_i = (|\gamma|/2\pi)(1 - \sigma_i)B \tag{10.61}$$

The value of σ_i is the same for nuclei that are in the same electronic environment in the molecule. For example, in CH_3CH_2OH, the three CH_3 protons have the same σ_i, and the two CH_2 protons have the same σ_i. (A Newman projection of ethanol, which has a staggered conformation, shows two of the three CH_3 hydrogens closer to the OH group than is the third CH_3 hydrogen, but the low barrier to internal rotation in ethanol allows the three methyl hydrogens to be rapidly interchanged at room temperature, thereby making the electronic environment the same for these three hydrogens.)

We might thus expect the ^{1}H NMR spectrum of ethanol to show three peaks, one for the CH_3 protons, one for the CH_2 protons, and one for the OH proton, with the relative intensities of these peaks being 3:2:1. However, there is an additional effect, called spin–spin coupling, in which the nuclear spins of the protons on one carbon affect the magnetic field experienced by the protons on an adjacent carbon. Different possible orientations of the proton spins on one carbon produce different magnetic fields at the protons of the adjacent carbon, thereby splitting the NMR transition of the protons at the adjacent carbon. For example, the two CH_2 proton nuclei have the following four possible nuclear-spin orientations:

$$\uparrow\uparrow \quad \uparrow\downarrow \quad \downarrow\uparrow \quad \downarrow\downarrow \tag{10.62}$$

where the up and down arrows represent $M_I = \frac{1}{2}$ and $M_I = -\frac{1}{2}$, respectively. [Actually, because of the indistinguishability of identical particles, the middle two spin states in (10.62) must be replaced by linear combinations of these two states, to give nuclear spin states that are analogous to the electron spin functions (10.22) to (10.25).] The middle two spin states in (10.62) have the same effect on the magnetic field felt by the CH_3 protons, so the four spin states in (10.62) produce three different magnetic fields at the CH_3 protons, thereby splitting the CH_3 proton NMR absorption line into three closely spaced lines of relative intensities 1:2:1, corresponding to the number of CH_2 proton spin states that produce each magnetic field. One might expect the CH_2 protons to also split the OH proton NMR line. However, even a trace of water present in the ethanol will catalyze a rapid exchange of the OH proton between different ethanol molecules, thereby eliminating the splitting of the OH line.

A similar analysis of the possible CH_3 proton orientations (Prob. 10.23) shows that the CH_3 protons split the ethanol CH_2 proton NMR line into four lines of relative intensities 1:3:3:1. *The general rule is that a group of n equivalent protons on an atom splits the NMR line of protons on an adjacent atom into $n + 1$ lines.* The spin–spin splitting (which is transmitted through the chemical bonds) is too weak to affect the NMR lines of protons separated by more than three bonds from the protons doing the splitting. In the proton NMR spectrum of $CH_3CH_2C(O)H$, the CH_3 protons split the CH_2 proton line into 4 lines, and each of these is split into two lines by the C(O)H proton, so the CH_2 NMR line is split into 8 lines. (The intermolecular proton exchange in ethanol prevents the OH proton from splitting the CH_2 proton NMR absorption.)

Nuclei with $I = 0$ (for example, ^{12}C, ^{16}O) don't split proton NMR peaks. It turns out that nuclei with $I > \frac{1}{2}$ (for example, ^{35}Cl, ^{37}Cl, ^{14}N) generally don't split proton NMR peaks. The ^{19}F nucleus has $I = \frac{1}{2}$ and does split proton NMR peaks. Also, a quantum-mechanical analysis shows that the spin–spin interactions between equivalent protons don't affect the NMR spectrum.

The treatment just given (called a *first-order* analysis) is actually an approximation that is valid provided that the spin–spin splittings are much smaller than all the NMR frequency differences between chemically nonequivalent nuclei. In very large molecules, there will likely be chemically nonequivalent nuclei that are in only slightly different electronic environments, so the NMR frequency differences between them will be quite small and the first-order analysis will not hold. By increasing the strength of the applied magnetic field, one increases the NMR frequency differences between chemically nonequivalent nuclei, thereby tending to make the spectrum first-order, which is easier to analyze. Also, the signal strength is increased as the field is increased. Therefore, people try and use as high a field as is feasible. Current NMR research spectrometers have fields that correspond to proton NMR frequencies in the range 300 to 900 MHz.

NMR spectroscopy is the premier structural research tool in organic chemistry, and special NMR techniques allow the structures of small proteins to be determined with the aid of NMR.

10.10 LADDER OPERATORS FOR ELECTRON SPIN

The spin angular-momentum operators obey the general angular-momentum commutation relations of Section 5.4, and it is often helpful to use spin-angular-momentum ladder operators.

From (5.110) and (5.111), the raising and lowering operators for spin angular momentum are

$$\hat{S}_+ = \hat{S}_x + i\hat{S}_y \quad \text{and} \quad \hat{S}_- = \hat{S}_x - i\hat{S}_y \tag{10.63}$$

Equations (5.112) and (5.113) give

$$\hat{S}_+\hat{S}_- = \hat{S}^2 - \hat{S}_z^2 + \hbar\hat{S}_z \tag{10.64}$$

$$\hat{S}_-\hat{S}_+ = \hat{S}^2 - \hat{S}_z^2 - \hbar\hat{S}_z \tag{10.65}$$

The spin functions α and β are eigenfunctions of $\hat{S}_z$ with eigenvalues $+\frac{1}{2}\hbar$ and $-\frac{1}{2}\hbar$, respectively. Since $\hat{S}_+$ is the raising operator, the function $\hat{S}_+\beta$ is an eigenfunction of $\hat{S}_z$ with eigenvalue $+\frac{1}{2}\hbar$. The most general eigenfunction of $\hat{S}_z$ with this eigenvalue is an arbitrary constant times α. Hence

$$\hat{S}_+\beta = c\alpha \tag{10.66}$$

where c is some constant. To find c, we use normalization [Eq. (10.11)]:

$$1 = \sum_{m_s}[\alpha(m_s)]^*\alpha(m_s) = \sum(\hat{S}_+\beta/c)^*(\hat{S}_+\beta/c)$$

$$|c|^2 = \sum(\hat{S}_+\beta)^*\hat{S}_+\beta = \sum(\hat{S}_+\beta)^*(\hat{S}_x + i\hat{S}_y)\beta$$

$$|c|^2 = \sum(\hat{S}_+\beta)^*\hat{S}_x\beta + i\sum(\hat{S}_+\beta)^*\hat{S}_y\beta \tag{10.67}$$

We now use the Hermitian property of $\hat{S}_x$ and $\hat{S}_y$. For an operator $\hat{A}$ that acts on functions of the continuous variable x, the Hermitian property is

$$\int_{-\infty}^{\infty} f^*(x)\hat{A}g(x)\,dx = \int_{-\infty}^{\infty} g(x)[\hat{A}f(x)]^*\,dx$$

For an operator such as $\hat{S}_x$ that acts on functions of the variable m_s, which takes on discrete values, the Hermitian property is

$$\sum_{m_s} f^*(m_s)\hat{S}_x g(m_s) = \sum_{m_s} g(m_s)[\hat{S}_x f(m_s)]^* \qquad (10.68)$$

Taking $f = \hat{S}_+\beta$ and $g = \beta$, we can write (10.67) as

$$c^*c = \sum \beta(\hat{S}_x\hat{S}_+\beta)^* + i\sum \beta(\hat{S}_y\hat{S}_+\beta)^*$$

Taking the complex conjugate of this equation and using (10.63) and (10.65), we have

$$cc^* = \sum \beta^*\hat{S}_x\hat{S}_+\beta - i\sum \beta^*\hat{S}_y\hat{S}_+\beta$$

$$|c|^2 = \sum \beta^*(\hat{S}_x - i\hat{S}_y)\hat{S}_+\beta = \sum \beta^*\hat{S}_-\hat{S}_+\beta$$

$$|c|^2 = \sum \beta^*(\hat{S}^2 - \hat{S}_z^2 - \hbar\hat{S}_z)\beta$$

$$|c|^2 = \sum \beta^*(\tfrac{3}{4}\hbar^2 - \tfrac{1}{4}\hbar^2 + \tfrac{1}{2}\hbar^2)\beta = \hbar^2\sum \beta^*\beta = \hbar^2$$

$$|c| = \hbar$$

Choosing the phase of c as zero, we have $c = \hbar$, and (10.66) reads

$$\hat{S}_+\beta = \hbar\alpha \qquad (10.69)$$

A similar calculation gives

$$\hat{S}_-\alpha = \hbar\beta \qquad (10.70)$$

Since α is the eigenfunction with the highest possible value of m_s, the operator $\hat{S}_+$ acting on α must annihilate it [Eq. (5.135)]:

$$\hat{S}_+\alpha = 0$$

Likewise,

$$\hat{S}_-\beta = 0$$

From these last four equations, we get

$$(\hat{S}_+ + \hat{S}_-)\beta = \hbar\alpha, \qquad (\hat{S}_+ - \hat{S}_-)\beta = \hbar\alpha \qquad (10.71)$$

Use of (10.63) in (10.71) gives

$$\hat{S}_x\beta = \tfrac{1}{2}\hbar\alpha, \qquad \hat{S}_y\beta = -\tfrac{1}{2}i\hbar\alpha \qquad (10.72)$$

Similarly, we find

$$\hat{S}_x\alpha = \tfrac{1}{2}\hbar\beta, \qquad \hat{S}_y\alpha = \tfrac{1}{2}i\hbar\beta \qquad (10.73)$$

Matrix representatives of the spin operators are considered in Prob. 10.29.

10.11 SUMMARY

An elementary particle possesses a spin angular momentum of magnitude $[s(s+1)\hbar^2]^{1/2}$ and z component $m_s\hbar$, where $m_s = -s, -s+1, \ldots, s-1, s$. For an electron, $s = \frac{1}{2}$. The spin angular-momentum operators $\hat{S}_x$, $\hat{S}_y$, $\hat{S}_z$, and $\hat{S}^2$ obey relations analogous to those obeyed by the orbital angular-momentum operators. The electron spin eigenfunctions corresponding to the states $m_s = \frac{1}{2}$ and $m_s = -\frac{1}{2}$ are denoted by α and β. We have $\hat{S}_z\alpha = \frac{1}{2}\hbar\alpha$, $\hat{S}_z\beta = -\frac{1}{2}\hbar\beta$, $\hat{S}^2\alpha = \frac{1}{2}\left(\frac{3}{2}\right)\hbar^2\alpha$, and $\hat{S}^2\beta = \frac{1}{2}\left(\frac{3}{2}\right)\hbar^2\beta$. The spin functions α and β are orthonormal [Eqs. (10.11) and (10.12)]. For a one-electron system, the complete stationary-state wave function is the product of a spatial function $\psi(x, y, z)$ and a spin function (α or β or a linear combination of α and β).

According to the spin–statistics theorem, the complete wave function (including both space and spin coordinates) of a system of identical particles with half-integral spin must be antisymmetric with respect to interchange of any two such particles. The complete wave function of a system of identical particles with integral spin must be symmetric with respect to interchange of any two particles. Particles requiring antisymmetric wave functions are called fermions, and particles requiring symmetric wave functions are called bosons.

The two-electron spin eigenfunctions consist of the symmetric functions $\alpha(1)\alpha(2)$, $\beta(1)\beta(2)$, and $[\alpha(1)\beta(2) + \beta(1)\alpha(2)]/\sqrt{2}$ and the antisymmetric function $[\alpha(1)\beta(2) - \beta(1)\alpha(2)]/\sqrt{2}$. For the helium atom, each stationary state wave function is the product of a symmetric spatial function and an antisymmetric spin function or an antisymmetric spatial function and a symmetric spin function. Some approximate helium-atom wave functions are Eqs. (10.26) to (10.30).

A spin-orbital is the product of a one-electron spatial wave function and a one-electron spin function. An approximate wave function for a system of electrons can be written as a Slater determinant of spin-orbitals. Interchange of two electrons interchanges two rows in the Slater determinant, which multiplies the wave function by -1, ensuring antisymmetry. In such an approximate wave function, no two electrons can be assigned to the same spin-orbital. This is the Pauli exclusion principle and is a consequence of the antisymmetry requirement.

An electron has a spin magnetic moment $\mathbf{m}_S$ that is proportional to its spin angular momentum $\mathbf{S}$. Transitions between different nuclear-spin energy levels in an applied magnetic field give rise to the NMR spectrum of a molecule.

By using ladder operators, we found the effects of $\hat{S}_x$ and $\hat{S}_y$ on α and β.

PROBLEMS

Sec.	10.1	10.3	10.4	10.6	10.7
Probs.	10.1–10.5	10.6–10.8	10.9–10.11	10.12–10.15	10.16–10.17

Sec.	10.9	10.10	general
Probs.	10.18–10.24	10.25–10.27	10.28–10.30

10.1 Calculate the magnitude of the spin angular momentum of a proton. Give a numerical answer.

10.2 Calculate the angle that the spin vector **S** makes with the z axis for an electron with spin function α.

10.3 The most general spin function for an electron is $c_1\alpha + c_2\beta$, where c_1 and c_2 are constants. (a) Complete these equations: $\hat{S}^2(c_1\alpha + c_2\beta) = ?$, $\hat{S}_z(c_1\alpha + c_2\beta) = ?$, and $\hat{S}_z^2(c_1\alpha + c_2\beta) = ?$. (b) The requirement that $c_1\alpha + c_2\beta$ be normalized leads to what relation between c_1 and c_2?

10.4 Verify that taking the spin functions α and β as $\alpha(m_s) = \delta_{m_s,1/2}$ and $\beta(m_s) = \delta_{m_s,-1/2}$ gives functions that satisfy the orthonormality conditions (10.11) and (10.12).

10.5 (a) If S_z of a particle with spin quantum number s is measured, what are the possible outcomes? (b) If S_x of a particle with spin quantum number s is measured, what are the possible outcomes? (c) In the statistical-ensemble interpretation of quantum mechanics (Section 7.9), all properties of a particle have precise values at all times. Consider the relation $S^2 = S_x^2 + S_y^2 + S_z^2$ between a particle's spin magnitude and its spin components. For each of the spin quantum number values $s = \frac{1}{2}$, $s = 1$, and $s = \frac{3}{2}$, examine whether $S^2 = S_x^2 + S_y^2 + S_z^2$ can be obeyed with all of the four quantities S^2, S_x, S_y, and S_z simultaneously having experimentally observable values.

10.6 State whether each of the following is a boson or a fermion: (a) an electron; (b) a proton; (c) a neutron; (d) a photon; (e) a ^{12}C nucleus; (f) a ^{13}C nucleus; (g) a ^{12}C atom; (h) a ^{13}C atom; (i) an ^{14}N atom; (j) an ^{15}N atom.

10.7 (a) Show that $\hat{P}_{12}$ commutes with the Hamiltonian for the lithium atom. (b) Show that $\hat{P}_{12}$ and $\hat{P}_{23}$ do not commute with each other. (c) Show that $\hat{P}_{12}$ and $\hat{P}_{23}$ commute when they are applied to antisymmetric functions.

10.8 Show that $\hat{P}_{12}$ is Hermitian.

10.9 Classify each of these functions as symmetric, antisymmetric, or neither symmetric nor antisymmetric. (a) $f(1)g(2)\alpha(1)\alpha(2)$; (b) $f(1)f(2)[\alpha(1)\beta(2) - \beta(1)\alpha(2)]$; (c) $f(1)f(2)f(3)\beta(1)\beta(2)\beta(3)$; (d) $e^{-a(r_1-r_2)}$; (e) $[f(1)g(2) - g(1)f(2)][\alpha(1)\beta(2) - \alpha(2)\beta(1)]$; (f) $r_{12}^2 e^{-a(r_1+r_2)}$.

10.10 If electrons had a spin of zero, what would be the zeroth-order (interelectronic repulsions neglected) wave functions for the ground state and first excited state of lithium?

10.11 Explain why the function $Ne^{-cr_1}e^{-cr_2}(r_1 - r_2)$ should not be used as a trial variation function for the helium-atom ground state.

10.12 The antisymmetrization operator $\hat{A}$ is defined as the operator that antisymmetrizes a product of n one-electron functions and multiplies it by $(n!)^{-1/2}$. For $n = 2$,

$$\hat{A}f(1)g(2) = \frac{1}{\sqrt{2}}\begin{vmatrix} f(1) & g(1) \\ f(2) & g(2) \end{vmatrix}$$

(a) For $n = 2$, express $\hat{A}$ in terms of $\hat{P}_{12}$. (b) For $n = 3$, express $\hat{A}$ in terms of $\hat{P}_{12}$, $\hat{P}_{13}$, and $\hat{P}_{23}$.

10.13 Use theorems about determinants to show that taking the lithium spin-orbitals in a Slater determinant as $1s\alpha$, $1s\beta$, and $1s(c_1\alpha + c_2\beta)$, where c_1 and c_2 are constants, gives a wave function that equals zero.

10.14 A *permanent* is defined by the same expansion as a determinant except that all terms are given a plus sign. Thus the second-order permanent is

$$\overset{+}{\begin{vmatrix} a & b \\ c & d \end{vmatrix}}^{+} = ad + bc$$

Can you think of a use for permanents in quantum mechanics?

10.15 A muon has the same charge and spin as an electron, but a heavier mass. What would be the ground-state configuration of a lithium atom with two electrons and one muon?

10.16 Derive Eq. (10.50) for $E^{(1)}$ of lithium as follows. (a) Group together terms in $\psi^{(0)}$ that have the same spin factor, to get

$$\psi^{(0)} = a\beta(1)\alpha(2)\alpha(3) + b\alpha(1)\beta(2)\alpha(3) + c\alpha(1)\alpha(2)\beta(3) = A + B + C$$

where a, b, and c are certain spatial functions and $A \equiv a\beta(1)\alpha(2)\alpha(3)$, with similar definitions for B and C. Then verify that $E^{(1)} = \int |A|^2 H' d\tau + \int |B|^2 H' d\tau + \int |C|^2 H' d\tau + \cdots$, where the dots stand for six integrals that each contain two of the functions A, B, and C. Use orthogonality of spin functions [Eq. (10.12)] to show that each of the six integrals represented by dots is zero. (b) Use the normalization of spin functions to show that

$$E^{(1)} = \iiint a^2 H' dv_1\, dv_2\, dv_3 + \iiint b^2 H' dv_1\, dv_2\, dv_3 + \iiint c^2 H' dv_1\, dv_2\, dv_3$$

where spin is no longer involved. (c) Use relabeling of dummy integration variables to show that the three integrals in (b) are equal to one another. (d) Use orthonormality of the 1s and 2s orbitals and relabeling of integration variables to prove (10.50).

10.17 If we had incorrectly used as the zeroth-order Li ground-state wave function the nonantisymmetric function $1s(1)1s(2)2s(3)$, what would $E^{(1)}$ be calculated to be?

10.18 Calculate the magnitude of the spin magnetic moment of an electron.

10.19 (a) Use Eq. (6.131) to find the expression for the energy levels of the electron spin magnetic moment $\mathbf{m}_S$ in an applied magnetic field $\mathbf{B}$. (b) Calculate the ESR absorption frequency of an electron in a magnetic field of 1.00 T.

10.20 A ^{35}Cl nucleus has $I = 3/2$. (a) Find the magnitude of the spin angular momentum of a ^{35}Cl nucleus. (b) Find the possible values that can result if the z component of the spin angular momentum of a ^{35}Cl nucleus is measured. (c) The same as (b) for measurement of the y component.

10.21 (a) Verify the value of γ for ^{1}H given in the table after (10.58). (b) Calculate the NMR absorption frequency of a proton (^{1}H nucleus) in a magnetic field of 1.00 T.

10.22 (a) For a nucleus with $I = \frac{1}{2}$ and $g_N > 0$, sketch a graph of the nuclear-spin energy levels versus the applied field B. (b) Repeat (a) for a nucleus with $I = 1$ and $g_N > 0$.

10.23 For the three CH_3 protons in ethanol, draw diagrams similar to (10.62) showing the possible nuclear spin orientations. Deduce the number of lines into which the adjacent CH_2 proton NMR transition is split, and give the relative intensities of these lines.

10.24 The proton NMR spectrum of ethanol contains a triplet peak with relative intensities 1:2:1 for the CH_3 protons, a quartet with relative intensities 1:3:3:1 for the CH_2 protons, and a singlet peak for the OH proton. These peaks have total relative intensity ratios of 3:2:1. For each of the following molecules give a similar description of the proton NMR spectrum (a) $CH_3CH_2C(O)H$; (b) $CH_3CH_2OCH_2CH_3$; (c) benzene; (d) 1,4-dichlorobenzene; (e) 1,3-dichlorobenzene.

10.25 Verify the spin-operator equations (10.70) and (10.73).

10.26 Show that α and β are each eigenfunctions of $\hat{S}_x^2$ (but not of $\hat{S}_x$). Give a physical explanation of why these results make sense (see Prob. 10.27a).

10.27 (a) If the spin component S_x of an electron is measured, what possible values can result? (b) The functions α and β form a complete set, so any one-electron spin function can be written as a linear combination of them. Use Eqs. (10.72) and (10.73) to construct the two normalized eigenfunctions of $\hat{S}_x$ with eigenvalues $+\frac{1}{2}\hbar$ and $-\frac{1}{2}\hbar$. (c) Suppose a measurement of S_z for an electron gives the value $+\frac{1}{2}\hbar$; if a measurement of S_x is then carried out, give the probabilities for each possible outcome. (d) Do the same as in (b) for $\hat{S}_y$ instead of $\hat{S}_x$. In the

Stern–Gerlach experiment, a beam of particles is sent through an inhomogeneous magnetic field, which splits the beam into several beams each having particles with a different component of magnetic dipole moment in the field direction. For example, a beam of ground-state sodium atoms is split into two beams, corresponding to the two possible orientations of the valence electron's spin. (This neglects the effect of the nuclear spin; for a complete discussion, see H. Kopfermann, *Nuclear Moments*, Academic Press, 1958, pp. 42–51.) Prob. 10.27c corresponds to setting up a Stern–Gerlach apparatus with the field in the z direction and then allowing the $+\frac{1}{2}\hbar$ beam from this apparatus to enter a Stern–Gerlach apparatus that has the field in the x direction.

10.28 (a) Let Y_{jm} be the *normalized* eigenfunction of the generalized angular-momentum operators (Section 5.4) $\hat{M}^2$ and $\hat{M}_z$:

$$\hat{M}^2 Y_{jm} = j(j+1)\hbar^2 Y_{jm}, \qquad \hat{M}_z Y_{jm} = m\hbar Y_{jm}$$

From Section 5.4, the effect of $\hat{M}_+$ on Y_{jm} is to increase the $\hat{M}_z$ eigenvalue by $\hbar$:

$$\hat{M}_+ Y_{jm} = A Y_{j,m+1}$$

where A is a constant. Use the procedure that led to (10.69) and (10.70) to show that

$$\hat{M}_+ Y_{jm} = [j(j+1) - m(m+1)]^{1/2}\hbar Y_{j,m+1} \tag{10.74}$$

$$\hat{M}_- Y_{jm} = [j(j+1) - m(m-1)]^{1/2}\hbar Y_{j,m-1} \tag{10.75}$$

(b) Show that (10.74) and (10.75) are consistent with (10.69) and (10.70). (c) With $\mathbf{M} = \mathbf{L}$, the function Y_{jm} is the spherical harmonic $Y_l^m(\theta, \phi)$. Verify (10.74) directly for $l = 2$, $m = -1$. [Actually, for consistency with the phase choice of Eqs. (10.74) and (10.75), we must add the factor $(-i)^{m+|m|}$ to the definition (5.147) of the spherical harmonics; this introduces a minus sign for odd positive values of m.]

10.29 The eigenfunctions α and β of the Hermitian operator $\hat{S}_z$ form a complete, orthonormal set, and any one-electron spin function can be written as $c_1\alpha + c_2\beta$. We saw in Section 7.10 that functions can be represented by column vectors and operators by square matrices. For the representation that uses α and β as the basis functions, (a) write down the column vectors that correspond to the functions α, β, and $c_1\alpha + c_2\beta$; (b) use the results of Section 10.10 to show that the matrices that correspond to $\hat{S}_x$, $\hat{S}_y$, $\hat{S}_z$, and $\hat{S}^2$ are

$$\mathbf{S}_x = \frac{1}{2}\hbar\begin{pmatrix} 0 & 1 \\ 1 & 0 \end{pmatrix}, \quad \mathbf{S}_y = \frac{1}{2}\hbar\begin{pmatrix} 0 & -i \\ i & 0 \end{pmatrix}, \quad \mathbf{S}_z = \frac{1}{2}\hbar\begin{pmatrix} 1 & 0 \\ 0 & -1 \end{pmatrix}, \quad \mathbf{S}^2 = \frac{1}{4}\hbar^2\begin{pmatrix} 3 & 0 \\ 0 & 3 \end{pmatrix}$$

(c) Verify that the matrices in (b) obey $\mathbf{S}_x\mathbf{S}_y - \mathbf{S}_y\mathbf{S}_x = i\hbar\mathbf{S}_z$ [Eq. (10.2)]. (d) Find the eigenvalues and eigenvectors of the $\mathbf{S}_x$ matrix. Compare the results with those of Prob. 10.27.

10.30 True or false? (a) The allowed values of the quantum number s of an electron are $-\frac{1}{2}$ and $\frac{1}{2}$. (b) The magnitude of the z component S_z of the spin angular momentum of a particle with nonzero spin must always be less than the magnitude $|\mathbf{S}|$ of the spin angular momentum. (c) For every two-electron system, the spin factor in the wave function must be antisymmetric. (d) For every system of several fermions, interchange of the labels of two fermions in the wave function must multiply the wave function by -1. (e) An atom of ^{79}Br is a boson. (f) An atom of ^{3}He is a fermion. (g) The magnetic moment of a nucleus is much less than the magnetic moment of an electron. (h) For a proton in an external magnetic field that points in the positive z direction, the $M_I = -\frac{1}{2}$ state is higher in energy than the $M_I = \frac{1}{2}$ state.

CHAPTER 11

Many-Electron Atoms

11.1 THE HARTREE–FOCK SELF-CONSISTENT-FIELD METHOD

For hydrogen the exact wave function is known. For helium and lithium, very accurate wave functions have been calculated by including interelectronic distances in the variation functions. For atoms of higher atomic number, the best approach to finding a good wave function lies in first calculating an approximate wave function using the Hartree–Fock procedure, which we shall outline in this section. The Hartree–Fock method is the basis for the use of atomic and molecular orbitals in many-electron systems.

The Hamiltonian operator for an n-electron atom is

$$\hat{H} = -\frac{\hbar^2}{2m_e}\sum_{i=1}^{n}\nabla_i^2 - \sum_{i=1}^{n}\frac{Ze'^2}{r_i} + \sum_{i=1}^{n-1}\sum_{j=i+1}^{n}\frac{e'^2}{r_{ij}} \tag{11.1}$$

where an infinitely heavy point nucleus was assumed (Section 6.6). The first sum in (11.1) contains the kinetic-energy operators for the n electrons. The second sum is the potential energy (6.58) for the attractions between the electrons and the nucleus of charge Ze'. For a neutral atom, $Z = n$. The last sum is the potential energy of the interelectronic repulsions. The restriction $j > i$ avoids counting each interelectronic repulsion twice and avoids terms like e'^2/r_{ii}. The Hamiltonian (11.1) is incomplete, because it omits spin–orbit and other interactions. The omitted terms are usually small and will be considered in Sections 11.6 and 11.7.

The Hartree SCF Method. Because of the interelectronic repulsion terms e'^2/r_{ij}, the Schrödinger equation for an atom is not separable. Recalling the perturbation treatment of helium (Section 9.3), we can obtain a zeroth-order wave function by neglecting these repulsions. The Schrödinger equation would then separate into n one-electron hydrogenlike equations. The zeroth-order wave function would be a product of n hydrogenlike (one-electron) orbitals:

$$\psi^{(0)} = f_1(r_1, \theta_1, \phi_1)f_2(r_2, \theta_2, \phi_2)\cdots f_n(r_n, \theta_n, \phi_n) \tag{11.2}$$

where the hydrogenlike orbitals are

$$f = R_{nl}(r)Y_l^m(\theta, \phi) \tag{11.3}$$

For the ground state of the atom, we would feed two electrons with opposite spin into each of the lowest orbitals, in accord with the Pauli exclusion principle, giving the ground-state configuration. Although the approximate wave function (11.2) is qualitatively useful, it is gravely lacking in quantitative accuracy. For one thing, all the orbitals use the full nuclear charge Z. Recalling our variational treatments of helium and lithium, we know we can get a better approximation by using different effective atomic numbers for the different orbitals to account for screening of electrons. The use of effective atomic numbers gives considerable improvement, but we are still far from having an accurate wave function. The next step is to use a variation function that has the same form as (11.2) but is not restricted to hydrogenlike or any other particular form of orbitals. Thus we take

$$\phi = g_1(r_1, \theta_1, \phi_1)g_2(r_2, \theta_2, \phi_2)\cdots g_n(r_n, \theta_n, \phi_n) \tag{11.4}$$

and we look for the functions $g_1, g_2, \ldots, g_n$ that minimize the variational integral $\int \phi^* \hat{H} \phi\, dv / \int \phi^* \phi\, dv$. Our task is harder than in previous variational calculations, where we guessed a trial function that included some parameters and then varied the *parameters*. In (11.4) we must vary the *functions* g_i. [After we have found the best possible functions g_i, Eq. (11.4) will still be only an approximate wave function. The many-electron Schrödinger equation is not separable, so the true wave function cannot be written as the product of n one-electron functions.]

To simplify matters somewhat, we approximate the best possible atomic orbitals with orbitals that are the product of a radial factor and a spherical harmonic:

$$g_i = h_i(r_i)Y_{l_i}^{m_i}(\theta_i, \phi_i) \tag{11.5}$$

This approximation is generally made in atomic calculations.

The procedure for calculating the g_i's was introduced by Hartree in 1928 and is called the **Hartree self-consistent-field (SCF) method**. Hartree arrived at the SCF procedure by intuitive physical arguments. The proof that Hartree's procedure gives the best possible variation function of the form (11.4) was given by Slater and by Fock in 1930. [For the proof and a review of the SCF method, see S. M. Blinder, *Am. J. Phys.*, **33**, 431 (1965).]

Hartree's procedure is as follows. We first guess a product wave function

$$\phi_0 = s_1(r_1, \theta_1, \phi_1)s_2(r_2, \theta_2, \phi_2)\cdots s_n(r_n, \theta_n, \phi_n) \tag{11.6}$$

where each s_i is a normalized function of r multiplied by a spherical harmonic. A reasonable guess for ϕ_0 would be a product of hydrogenlike orbitals with effective atomic numbers. For the function (11.6), the probability density of electron i is $|s_i|^2$. We now focus attention on electron 1 and regard electrons 2, 3, ... , n as being smeared out to form a fixed distribution of electric charge through which electron 1 moves. We are thus averaging out the instantaneous interactions between electron 1 and the other electrons. The potential energy of interaction between point charges Q_1 and Q_2 is $V_{12} = Q_1'Q_2'/r_{12} = Q_1Q_2/4\pi\varepsilon_0 r_{12}$ [Eqs. (6.58) and (1.38)]. We now take Q_2 and smear it out into a continuous charge distribution such that ρ_2 is the charge density, the charge per unit volume. The infinitesimal charge in the infinitesimal volume dv_2 is $\rho_2\, dv_2$, and summing up the interactions between Q_1 and the infinitesimal elements of charge, we have

$$V_{12} = \frac{Q_1}{4\pi\varepsilon_0}\int \frac{\rho_2}{r_{12}}\, dv_2$$

For electron 2 (with charge $-e$), the charge density of the hypothetical charge cloud is $\rho_2 = -e|s_2|^2$, and for electron 1, $Q_1 = -e$. Hence

$$V_{12} = e'^2 \int \frac{|s_2|^2}{r_{12}}\, dv_2$$

where $e'^2 = e^2/4\pi\varepsilon_0$. Adding in the interactions with the other electrons, we have

$$V_{12} + V_{13} + \cdots + V_{1n} = \sum_{j=2}^{n} e'^2 \int \frac{|s_j|^2}{r_{1j}}\, dv_j$$

The potential energy of interaction between electron 1 and the other electrons and the nucleus is then

$$V_1(r_1, \theta_1, \phi_1) = \sum_{j=2}^{n} e'^2 \int \frac{|s_j|^2}{r_{1j}}\, dv_j - \frac{Ze'^2}{r_1} \tag{11.7}$$

We now make a further approximation beyond assuming the wave function to be a product of one-electron orbitals. We assume that the effective potential acting on an electron in an atom can be adequately approximated by a function of r only. This *central-field approximation* can be shown to be generally accurate. We therefore average $V_1(r_1, \theta_1, \phi_1)$ over the angles to arrive at a potential energy that depends only on r_1:

$$V_1(r_1) = \frac{\int_0^{2\pi} \int_0^{\pi} V_1(r_1, \theta_1, \phi_1) \sin\theta_1\, d\theta_1\, d\phi_1}{\int_0^{2\pi} \int_0^{\pi} \sin\theta\, d\theta\, d\phi} \tag{11.8}$$

We now use $V_1(r_1)$ as the potential energy in a one-electron Schrödinger equation,

$$\left[-\frac{\hbar^2}{2m_e} \nabla_1^2 + V_1(r_1) \right] t_1(1) = \varepsilon_1 t_1(1) \tag{11.9}$$

and solve for $t_1(1)$, which will be an improved orbital for electron 1. In (11.9), ε_1 is the energy of the orbital of electron 1 at this stage of the approximation. Since the potential energy in (11.9) is spherically symmetric, the angular factor in $t_1(1)$ is a spherical harmonic involving quantum numbers l_1 and m_1 (Section 6.1). The radial factor $R(r_1)$ in t_1 is the solution of a one-dimensional Schrödinger equation of the form (6.17). We get a set of solutions $R(r_1)$, where the number of nodes k interior to the boundary points ($r = 0$ and ∞) starts at zero for the lowest energy and increases by 1 for each higher energy (Section 4.2). We now *define* the quantum number n as $n \equiv l + 1 + k$, where $k = 0, 1, 2, \ldots$. We thus have 1s, 2s, 2p, and so on, orbitals (with orbital energy ε increasing with n) just as in hydrogenlike atoms, and the number of interior radial nodes $(n - l - 1)$ is the same as in hydrogenlike atoms (Section 6.6). However, since $V_1(r_1)$ is not a simple Coulomb potential, the radial factor $R(r_1)$ is not a hydrogenlike function. Of the set of solutions $R(r_1)$, we take the one that corresponds to the orbital we are improving. For example, if electron 1 is a 1s electron in the beryllium $1s^2 2s^2$ configuration, then $V_1(r_1)$ is calculated from the guessed orbitals of one 1s electron and two 2s electrons, and we use the radial solution of (11.9) with $k = 0$ to find an improved 1s orbital.

We now go to electron 2 and regard it as moving in a charge cloud of density

$$-e[|t_1(1)|^2 + |s_3(3)|^2 + |s_4(4)|^2 + \cdots + |s_n(n)|^2]$$

due to the other electrons. We calculate an effective potential energy $V_2(r_2)$ and solve a one-electron Schrödinger equation for electron 2 to get an improved orbital $t_2(2)$. We continue this process until we have a set of improved orbitals for all n electrons. Then we go back to electron 1 and repeat the process. We continue to calculate improved orbitals until there is no further change from one iteration to the next. The final set of orbitals gives the Hartree self-consistent-field wave function.

How do we get the energy of the atom in the SCF approximation? It seems natural to take the sum of the orbital energies of the electrons, $\varepsilon_1 + \varepsilon_2 + \cdots + \varepsilon_n$, but this is wrong. In calculating the orbital energy ε_1, we iteratively solved the one-electron Schrödinger equation (11.9). The potential energy in (11.9) includes, in an average way, the energy of the repulsions between electrons 1 and 2, 1 and 3, . . . , 1 and n. When we solve for ε_2, we solve a one-electron Schrödinger equation whose potential energy includes repulsions between electrons 2 and 1, 2 and 3, . . . , 2 and n. If we take $\sum_i \varepsilon_i$, we will count each interelectronic repulsion twice. To correctly obtain the total energy E of the atom, we must take

$$E = \sum_{i=1}^{n} \varepsilon_i - \sum_{i=1}^{n-1} \sum_{j=i+1}^{n} \iint \frac{e'^2 |g_i(i)|^2 |g_j(j)|^2}{r_{ij}} \, dv_i \, dv_j$$

$$E = \sum_i \varepsilon_i - \sum_i \sum_{j>i} J_{ij} \tag{11.10}$$

where the average repulsions of the electrons in the Hartree orbitals of (11.4) were subtracted from the sum of the orbital energies, and where the notation J_{ij} was used for Coulomb integrals [Eq. (9.100)].

The set of orbitals belonging to a given principal quantum number n constitutes a **shell**. The $n = 1, 2, 3, \ldots$ shells are the K, L, M, . . . shells, respectively. The orbitals belonging to a given n and a given l constitute a **subshell**. Consider the sum of the Hartree probability densities for the electrons in a filled subshell. Using (11.5), we have

$$2 \sum_{m=-l}^{l} |h_{n,l}(r)|^2 |Y_l^m(\theta, \phi)|^2 = 2|h_{n,l}(r)|^2 \sum_{m=-l}^{l} |Y_l^m(\theta, \phi)|^2 \tag{11.11}$$

where the factor 2 comes from the pair of electrons in each orbital. The spherical-harmonic addition theorem (*Merzbacher*, Section 9.7) shows that the sum on the right side of (11.11) equals $(2l + 1)/4\pi$. Hence the sum of the probability densities is $[(2l + 1)/2\pi]|h_{n,l}(r)|^2$, which is independent of the angles. A closed subshell gives a spherically symmetric probability density, a result called *Unsöld's theorem*. For a half-filled subshell, the factor 2 is omitted from (11.11), and here also we get a spherically symmetric probability density.

The Hartree–Fock SCF Method. The alert reader may have realized that there is something fundamentally wrong with the Hartree product wave function (11.4). Although we have paid some attention to spin and the Pauli exclusion principle by putting no more than two electrons in each spatial orbital, any approximation to the true wave function should include spin explicitly and should be antisymmetric to interchange of electrons (Chapter 10). Hence, instead of the spatial orbitals, we must use spin-orbitals and must take an antisymmetric linear combination of products of spin-orbitals. This was pointed out by Fock (and by Slater) in 1930, and an SCF calculation that uses antisymmetrized spin-orbitals is called a **Hartree–Fock calculation**. We have seen that a Slater determinant of spin-orbitals provides the proper antisymmetry. For example, to carry out a Hartree–Fock calculation for the lithium ground state, we start with the function (10.54), where f and g are guesses for the 1s and 2s orbitals. We then carry out the SCF iterative process until we get no further improvement in f and g. This gives the lithium ground-state Hartree–Fock wave function.

The differential equations for finding the Hartree–Fock orbitals have the same general form as (11.9):

$$\hat{F}u_i = \varepsilon_i u_i, \qquad i = 1, 2, \ldots, n \tag{11.12}$$

where u_i is the ith spin-orbital, the operator $\hat{F}$, called the **Fock** (or **Hartree–Fock**) **operator**, is the effective Hartree–Fock Hamiltonian, and the eigenvalue ε_i is the **orbital energy** of spin-orbital i. However, the Hartree–Fock operator $\hat{F}$ has extra terms as compared with the effective Hartree Hamiltonian given by the bracketed terms in (11.9). The Hartree–Fock expression for the total energy of the atom involves exchange integrals K_{ij} in addition to the Coulomb integrals that occur in the Hartree expression (11.10). See Section 14.3. [Actually, Eq. (11.12) applies only when the Hartree–Fock wave function can be written as a single Slater determinant, as it can for closed-subshell atoms and atoms with only one electron outside closed subshells. When the Hartree–Fock wave function contains more than one Slater determinant, the Hartree–Fock equations are more complicated than (11.12).]

The orbital energy ε_i in the Hartree–Fock equations (11.12) can be shown to be a good approximation to the negative of the energy needed to ionize a closed-subshell atom by removing an electron from spin-orbital i (Koopmans' theorem; Section 15.5).

Originally, Hartree–Fock atomic calculations were done by using numerical methods to solve the Hartree–Fock differential equations (11.12), and the resulting orbitals were given as tables of the radial functions for various values of r. [The Numerov method (Sections 4.4 and 6.9) can be used to solve the radial Hartree–Fock equations for the radial factors in the Hartree–Fock orbitals; the angular factors are spherical harmonics. See D. R. Hartree, *The Calculation of Atomic Structures*, Wiley, 1957; C. Froese Fischer, *The Hartree–Fock Method for Atoms*, Wiley, 1977.]

In 1951, Roothaan proposed representing the Hartree–Fock orbitals as linear combinations of a complete set of known functions, called **basis functions**. Thus for lithium we would write the Hartree–Fock 1s and 2s spatial orbitals as

$$f = \sum_i b_i \chi_i, \quad g = \sum_i c_i \chi_i \tag{11.13}$$

where the χ_i's are some complete set of functions, and where the b_i's and c_i's are expansion coefficients that are found by the SCF iterative procedure. Since the χ_i (chi i) functions form a complete set, these expansions are valid. The Roothaan expansion procedure allows one to find the Hartree–Fock wave function using matrix algebra (see Section 14.3 for details). The Roothaan procedure is readily implemented on a computer and is often used to find atomic Hartree–Fock wave functions and nearly always used to find molecular Hartree–Fock wave functions.

A commonly used set of basis functions for atomic Hartree–Fock calculations is the set of **Slater-type orbitals** (STOs) whose normalized form is

$$\frac{(2\zeta/a_0)^{n+1/2}}{[(2n)!]^{1/2}} r^{n-1} e^{-\zeta r/a_0} Y_l^m(\theta, \phi) \tag{11.14}$$

The set of all such functions with n, l, and m being integers obeying (6.96)–(6.98) but with ζ having all possible positive values forms a complete set. The parameter ζ is called the **orbital exponent**. To get a truly accurate representation of the Hartree–Fock orbitals, we would have to include an infinite number of Slater orbitals in the expansions. In practice, one can get very accurate results by using only a few judiciously chosen Slater orbitals. (Another possibility is to use Gaussian-type basis functions; see Section 15.4.)

Clementi and Roetti did Hartree–Fock calculations for the ground state and some excited states of the first 54 elements of the periodic table [E. Clementi and C. Roetti, *At. Data Nucl. Data Tables*, **14**, 177 (1974); Bunge and co-workers have recalculated these wave functions; C. F. Bunge et al., *At. Data Nucl. Data Tables*, **53**, 113 (1993); *Phys. Rev. A*, **46**, 3691 (1992); these atomic wave functions can be found at www.ccl.net/cca/data/atomic-RHF-wavefunctions/index.shtml]. For example, consider the Hartree–Fock ground-state wave function of helium, which has the form [see Eq. (10.41)]

$$f(1)f(2) \cdot 2^{-1/2}[\alpha(1)\beta(2) - \alpha(2)\beta(1)]$$

Clementi and Roetti expressed the 1s orbital function f as the following combination of five 1s Slater-type orbitals:

$$f = \pi^{-1/2} \sum_{i=1}^{5} c_i \left(\frac{\zeta_i}{a_0}\right)^{3/2} e^{-\zeta_i r/a_0}$$

where the expansion coefficients c_i are $c_1 = 0.76838$, $c_2 = 0.22346$, $c_3 = 0.04082$, $c_4 = -0.00994$, $c_5 = 0.00230$ and where the orbital exponents ζ_i are $\zeta_1 = 1.41714$, $\zeta_2 = 2.37682$, $\zeta_3 = 4.39628$, $\zeta_4 = 6.52699$, $\zeta_5 = 7.94252$. [The largest term in the expansion has an orbital exponent that is similar to the orbital exponent (9.63) for the simple trial function (9.56).] The Hartree–Fock energy is -77.9 eV, as compared with the true nonrelativistic energy, -79.0 eV. The 1s orbital energy corresponding to f was found to be -25.0 eV, as compared with the experimental helium ionization energy of 24.6 eV.

For the lithium ground state, Clementi and Roetti used a basis set consisting of two 1s STOs (with different orbital exponents) and four 2s STOs (with different orbital exponents). The lithium 1s and 2s Hartree–Fock orbitals were each expressed as a linear combination of all six of these basis functions. The Hartree–Fock energy is -202.3 eV, as compared with the true energy -203.5 eV.

Electron densities calculated from Hartree–Fock wave functions are quite accurate. Figure 11.1 compares the radial distribution function of argon (found by integrating the electron density over the angles θ and ϕ and multiplying the result by r^2) calculated by the Hartree–Fock method with the experimental radial distribution

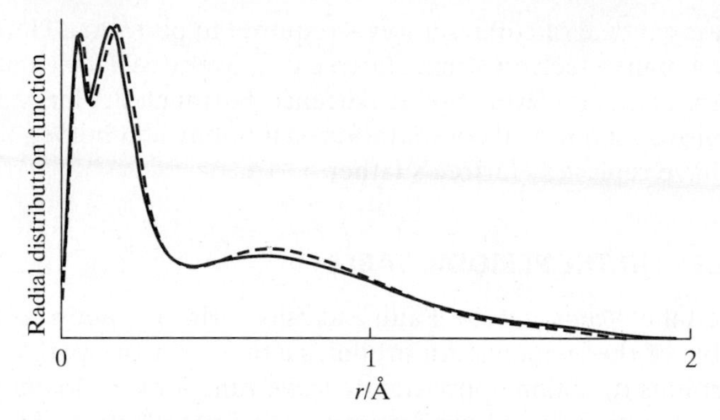

FIGURE 11.1 Radial distribution function in Ar as a function of r. The broken line is the result of a Hartree–Fock calculation. The solid line is the result of electron-diffraction data. [From L. S. Bartell and L. O. Brockway, *Phys. Rev.*, **90**, 833 (1953). Used by permission.]

function found by electron diffraction. (Recall from Section 6.6 that the radial distribution function is proportional to the probability of finding an electron in a thin spherical shell at a distance r from the nucleus.) Note the electronic shell structure in Fig. 11.1. The high nuclear charge in $_{18}Ar$ makes the average distance of the $1s$ electrons from the nucleus far less than in H or He. Thus there is only a moderate increase in atomic size as we go down a given group in the periodic table. Calculations show that the radius of a sphere containing 98% of the Hartree–Fock electron probability density gives an atomic radius in good agreement with the empirically determined van der Waals radius. [See C. W. Kammeyer and D. R. Whitman, *J. Chem. Phys.*, **56**, 4419 (1972).]

Although the radial distribution function of an atom shows the shell structure, the electron probability density integrated over the angles and plotted versus r does not oscillate. Rather, for ground-state atoms this probability density is a maximum at the nucleus (because of the s electrons) and continually decreases as r increases. Similarly, in molecules the maxima in electron probability density usually occur at the nuclei; see, for example, Fig. 13.7. [For further discussion, see H. Weinstein, P. Politzer, and S. Srebnik, *Theor. Chim. Acta*, **38**, 159 (1975).]

Accurate representation of a many-electron atomic orbital (AO) requires a linear combination of several Slater-type orbitals. For rough calculations, it is convenient to have simple approximations for AOs. We might use hydrogenlike orbitals with effective nuclear charges, but Slater suggested an even simpler method: to approximate an AO by a single function of the form (11.14) with the orbital exponent ζ taken as

$$\zeta = (Z - s)/n \tag{11.15}$$

where Z is the atomic number, n is the orbital's principal quantum number, and s is a screening constant calculated by a set of rules (see Prob. 15.54). A Slater orbital replaces the polynomial in r in a hydrogenlike orbital with a single power of r. Hence a single Slater orbital does not have the proper number of radial nodes and does not represent well the inner part of an orbital.

A great deal of computation is required to perform a Hartree–Fock SCF calculation for a many-electron atom. Hartree did several SCF calculations in the 1930s, when electronic computers were not in existence. Fortunately, Hartree's father, a retired engineer, enjoyed numerical computation as a hobby and helped his son. Nowadays computers have replaced Hartree's father.

11.2 ORBITALS AND THE PERIODIC TABLE

The orbital concept and the Pauli exclusion principle allow us to understand the periodic table of the elements. An orbital is a one-electron spatial wave function. We have used orbitals to obtain approximate wave functions for many-electron atoms, writing the wave function as a Slater determinant of one-electron spin-orbitals. In the crudest approximation, we neglect all interelectronic repulsions and obtain hydrogenlike orbitals. The best possible orbitals are the Hartree–Fock SCF functions. We build up the periodic table by feeding electrons into these orbitals, each of which can hold a pair of electrons with opposite spin.

Latter [R. Latter, *Phys. Rev.*, **99**, 510 (1955)] calculated approximate orbital energies for the atoms of the periodic table by replacing the complicated expression for the Hartree–Fock potential energy in the Hartree–Fock radial equations by a much simpler function obtained from the Thomas–Fermi–Dirac method, which uses ideas of statistical mechanics to get approximations to the effective potential-energy function for an electron and the electron-density function in an atom (*Bethe and Jackiw*, Chapter 5). Figure 11.2 shows Latter's resulting orbital energies for neutral ground-state atoms. These AO energies are in pretty good agreement with both Hartree–Fock and experimentally found orbital energies (see J. C. Slater, *Quantum Theory of Matter*, 2nd ed., McGraw-Hill, 1968, pp. 146, 147, 325, 326).

Orbital energies change with changing atomic number Z. As Z increases, the orbital energies decrease because of the increased attraction between the nucleus and the electrons. This decrease is most rapid for the inner orbitals, which are less well-shielded from the nucleus.

For $Z > 1$, orbitals with the same value of n but different l have different energies. For example, for the $n = 3$ orbital energies, we have $\varepsilon_{3s} < \varepsilon_{3p} < \varepsilon_{3d}$ for $Z > 1$. The splitting of these levels, which are degenerate in the hydrogen atom, arises from the interelectronic repulsions. (Recall the perturbation treatment of helium in Section 9.7.) In the limit $Z \to \infty$, orbitals with the same value of n are again degenerate, because the interelectronic repulsions become insignificant in comparison with the electron–nucleus attractions.

The relative positions of certain orbitals change with changing Z. Thus in hydrogen the $3d$ orbital lies below the $4s$ orbital, but for Z in the range from 7 through 20 the $4s$ is below the $3d$. For large values of Z, the $3d$ is again lower. At $Z = 19$, the $4s$ is lower; hence ${}_{19}K$ has the ground-state configuration $1s^2 2s^2 2p^6 3s^2 3p^6 4s$. Recall that s orbitals are more penetrating than p or d orbitals; this allows the $4s$ orbital to lie below the $3d$ orbital for some values of Z. Note the sudden drop in the $3d$ energy, which starts at $Z = 21$, when filling of the $3d$ orbital begins. The electrons of the $3d$ orbital do not shield each other very well; hence the sudden drop in $3d$ energy. Similar drops occur for other orbitals.

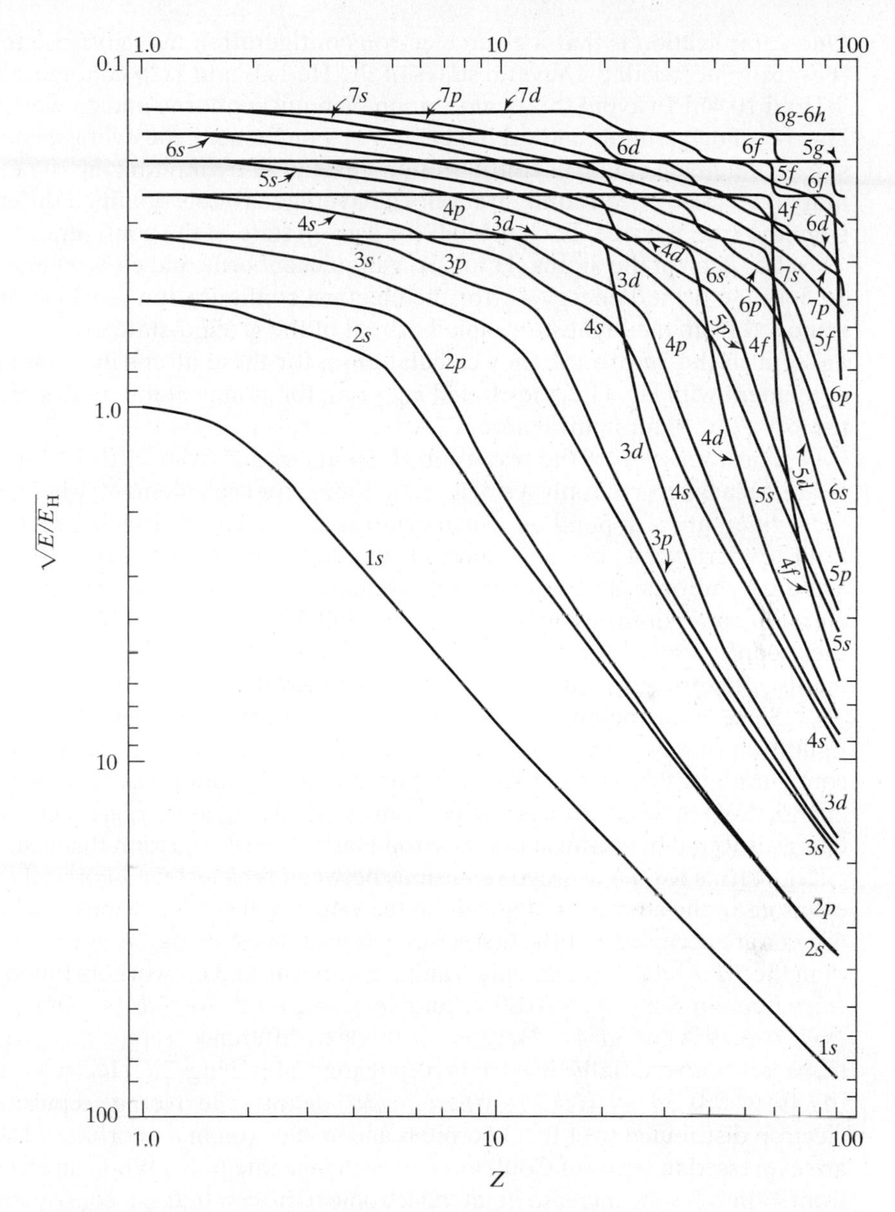

FIGURE 11.2 Atomic-orbital energies as a function of atomic number for neutral atoms, as calculated by Latter. [Figure redrawn by M. Kasha from R. Latter, *Phys. Rev.*, **99**, 510 (1955). Used by permission.] Note the logarithmic scales. E_H is the ground-state hydrogen-atom energy, –13.6 eV.

To help explain the observed electron configurations of the transition elements and their ions, Vanquickenborne and co-workers calculated Hartree–Fock $3d$ and $4s$ orbital energies for atoms and ions for $Z = 1$ to $Z = 29$ [L. G. Vanquickenborne, K. Pierloot, and D. Devoghel, *Inorg. Chem.*, **28**, 1805 (1989); *J. Chem. Educ.*, **71**, 469 (1994)].

One complication is that a given electron configuration may give rise to many states. [For example, recall the several states of the He $1s2s$ and $1s2p$ configurations (Sections 9.7 and 10.4).] To avoid this complication, Vanquickenborne and co-workers calculated Hartree–Fock orbitals and orbital energies by minimizing the average energy E_{av} of the states of a given electron configuration, instead of by minimizing the energy of each individual state of the configuration. The average orbitals obtained differ only slightly from the true Hartree–Fock orbitals for a given state of the configuration.

For each of the atoms $_1$H to $_{19}$K, Vanquickenborne and co-workers calculated the $3d$ average orbital energy ε_{3d} for the electron configuration in which one electron is removed from the highest-occupied orbital of the ground-state electron configuration and put in the $3d$ orbital; they calculated ε_{4s} for these atoms in a similar manner. In agreement with Fig. 11.2, they found $\varepsilon_{3d} < \varepsilon_{4s}$ for atomic numbers $Z < 6$ and $\varepsilon_{4s} < \varepsilon_{3d}$ for $Z = 7$ to 19 for neutral atoms.

For discussion of the transition elements with Z from 21 to 29, Fig. 11.2 is inadequate because it gives only a single value for ε_{3d} for each element, whereas ε_{3d} (and ε_{4s}) for a given atom depend on which orbitals are occupied. This is because the electric field experienced by an electron depends on which orbitals are occupied. Vanquickenborne and co-workers calculated ε_{3d} and ε_{4s} for each of the valence-electron configurations $3d^n4s^2$, $3d^{n+1}4s^1$, and $3d^{n+2}4s^0$ and found $\varepsilon_{3d} < \varepsilon_{4s}$ in each of these configurations of the neutral atoms and the +1 and +2 ions of the transition elements $_{21}$Sc through $_{29}$Cu (which is the order shown in Fig. 11.2).

Since $3d$ lies below $4s$ for Z above 20, one might wonder why the ground-state configuration of, say, $_{21}$Sc is $3d^14s^2$, rather than $3d^3$. Although $\varepsilon_{3d} < \varepsilon_{4s}$ for each of these configurations, this does not mean that the $3d^3$ configuration has the lower sum of orbital energies. When an electron is moved from $4s$ into $3d$, ε_{4s} and ε_{3d} are increased. An orbital energy is found by solving a one-electron Hartree–Fock equation that contains potential-energy terms for the average repulsions between the electron in orbital i and the other electrons in the atom, so ε_i depends on the values of these repulsions and hence on which orbitals are occupied. For the first series of transition elements, the $4s$ orbital is much larger than the $3d$ orbital. For example, Vanquickenborne and co-workers found the following $\langle r \rangle$ values in Sc: $\langle r \rangle_{3d} = 0.89$ Å and $\langle r \rangle_{4s} = 2.09$ Å for $3d^14s^2$; $\langle r \rangle_{3d} = 1.11$ Å and $\langle r \rangle_{4s} = 2.29$ Å for $3d^24s^1$. Because of this size difference, repulsions involving $4s$ electrons are substantially less than repulsions involving $3d$ electrons, and we have $(4s,4s) < (4s,3d) < (3d,3d)$, where $(4s,3d)$ denotes the average repulsion between an electron distributed over the $3d$ orbitals and an electron in a $4s$ orbital. (These repulsions are expressed in terms of Coulomb and exchange integrals.) When an electron is moved from $4s$ into $3d$, the increase in interelectronic repulsion that is a consequence of the preceding inequalities raises the orbital energies ε_{3d} and ε_{4s}. For example, for $_{21}$Sc, the $3d^14s^2$ configuration has $\varepsilon_{3d} = -9.35$ eV and $\varepsilon_{4s} = -5.72$ eV, whereas the $3d^24s^1$ configuration has $\varepsilon_{3d} = -5.23$ eV and $\varepsilon_{4s} = -5.06$ eV. For the $3d^14s^2$ configuration, the sum of valence-electron orbital energies is $-9.35 \text{ eV} + 2(-5.72 \text{ eV}) = -20.79$ eV, whereas for the $3d^24s^1$ configuration, this sum is $2(-5.23 \text{ eV}) - 5.06 \text{ eV} = -15.52$ eV. Thus, despite the fact that $\varepsilon_{3d} < \varepsilon_{4s}$ for each configuration, transfer of an electron from $4s$ to $3d$ raises the sum of valence-electron orbital energies in Sc. [As we saw in Eq. (11.10) for the Hartree method and will see in Section 14.3 for the Hartree–Fock method, the Hartree and Hartree–Fock expressions for the energy of an atom contain terms in addition to the sum of orbital energies, so we must look at more than the sum of orbital energies to see which configuration is most stable.]

For the +2 ions of the transition metals, the reduction in screening makes the valence 3*d* and 4*s* electrons feel a larger effective nuclear charge Z_{eff} than in the neutral atoms. By analogy to the H-atom equation $E = -(Z^2/n^2)(e'^2/2a)$ [Eq. (6.94)], the orbital energies ε_{3d} and ε_{4s} are each roughly proportional to Z_{eff}^2 and the energy difference $\varepsilon_{4s} - \varepsilon_{3d}$ is roughly proportional to Z_{eff}^2. The difference $\varepsilon_{4s} - \varepsilon_{3d}$ is thus much larger in the transition-metal ions than in the neutral atoms, the increase in valence-electron repulsion is no longer enough to make the 4*s* to 3*d* transfer energetically unfavorable, and the +2 ions have ground-state configurations with no 4*s* electrons.

Figure 11.2 shows that the separation between *ns* and *np* orbitals is much less than that between *np* and *nd* orbitals, giving the familiar ns^2np^6 stable octet.

The orbital concept is the basis for most qualitative discussions of the chemistry of atoms and molecules. The use of orbitals, however, is an approximation. To reach the true wave function, we must go beyond a Slater determinant of spin-orbitals.

11.3 ELECTRON CORRELATION

Energies calculated by the Hartree–Fock method are typically in error by about $\frac{1}{2}$% for light atoms. On an absolute basis this is not much, but for the chemist it is too large. For example, the total energy of the carbon atom is about −1000 eV, and $\frac{1}{2}$% of this is 5 eV. Chemical single-bond energies run about 5 eV. Calculating a bond energy by taking the difference between Hartree–Fock molecular and atomic energies, which are in error by several electronvolts for light atoms, is an unreliable procedure. We must seek a way to improve Hartree–Fock wave functions and energies. (Our discussion will apply to molecules as well as atoms.)

A Hartree–Fock SCF wave function takes into account the interactions between electrons only in an average way. Actually, we must consider the instantaneous interactions between electrons. Since electrons repel each other, they tend to keep out of each other's way. For example, in helium, if one electron is close to the nucleus at a given instant, it is energetically more favorable for the other electron to be far from the nucleus at that instant. One sometimes speaks of a *Coulomb hole* surrounding each electron in an atom. This is a region in which the probability of finding another electron is small. The motions of electrons are correlated with each other, and we speak of **electron correlation**. We must find a way to introduce the instantaneous electron correlation into the wave function.

Actually, a Hartree–Fock wave function does have some instantaneous electron correlation. A Hartree–Fock function satisfies the antisymmetry requirement. Therefore [Eq. (10.20)], it vanishes when two electrons with the same spin have the same spatial coordinates. For a Hartree–Fock function, there is little probability of finding electrons of the same spin in the same region of space, so a Hartree–Fock function has some correlation of the motions of electrons with the same spin. This makes the Hartree–Fock energy lower than the Hartree energy. One sometimes refers to a *Fermi hole* around each electron in a Hartree–Fock wave function, thereby indicating a region in which the probability of finding another electron with the same spin is small.

The **correlation energy** E_{corr} is the difference between the exact nonrelativistic energy E_{nonrel} and the (nonrelativistic) Hartree–Fock energy E_{HF}:

$$E_{corr} \equiv E_{nonrel} - E_{HF} \tag{11.16}$$

where E_{nonrel} and E_{HF} should both either include corrections for nuclear motion or omit these corrections. For the He atom, the (nonrelativistic) Hartree–Fock energy uncorrected for nuclear motion is $-2.86168\,(e'^2/a_0)$ [E. Clementi and C. Roetti, *At. Data Nucl. Data Tables*, **14**, 177 (1974)] and variational calculations (Section 9.4) give the exact nonrelativistic energy uncorrected for nuclear motion as $-2.90372\,(e'^2/a_0)$. Therefore, $E_{corr,He} = -2.90372(e'^2/a_0) + 2.86168(e'^2/a_0) = -0.04204(e'^2/a_0) = -1.14$ eV. For atoms and molecules where E_{nonrel} cannot be accurately calculated, one combines the experimental energy with estimates for relativistic and nuclear-motion corrections to get E_{nonrel}. For neutral atoms, $|E_{corr}|$ increases roughly linearly with the number n of electrons: $E_{corr} \approx -0.0170n^{1.31}(e'^2/a_0)$ [E. Clementi and G. Corongiu, *Int. J. Quantum Chem.*, **62**, 571 (1997)]. The percentage $(E_{corr}/E_{nonrel}) \times 100\%$ decreases with increasing atomic number. Some values are 0.6% for Li, 0.4% for C, 0.2% for Na, 0.1% for K.

We have already indicated two of the ways in which we may provide for instantaneous electron correlation. One method is to introduce the interelectronic distances r_{ij} into the wave function (Section 9.4). This method is only practicable for systems with a few electrons.

Another method is configuration interaction. We found (Sections 9.3 and 10.4) the zeroth-order wave function for the helium $1s^2$ ground state to be $1s(1)1s(2)[\alpha(1)\beta(2) - \beta(1)\alpha(2)]/\sqrt{2}$. We remarked that first- and higher-order corrections to the wave function will mix in contributions from excited configurations, producing **configuration interaction** (CI), also called **configuration mixing** (CM).

The most common way to do a configuration-interaction calculation on an atom or molecule uses the variation method. One starts by choosing a basis set of one-electron functions χ_i. In principle, this basis set should be complete. In practice, one is limited to a basis set of finite size. One hopes that a good choice of basis functions will give a good approximation to a complete set. For atomic calculations, STOs [Eq. (11.14)] are often chosen as the basis functions.

The SCF atomic (or molecular) orbitals ϕ_i are written as linear combinations of the basis-set members [see (11.13)], and the Hartree–Fock equations (11.12) are solved to give the coefficients in these linear combinations. The number of atomic (or molecular) orbitals obtained equals the number of basis functions used. The lowest-energy orbitals are the occupied orbitals for the ground state. The remaining unoccupied orbitals are called **virtual orbitals**.

Using the set of occupied and virtual spin-orbitals, one can form antisymmetric many-electron functions that have different orbital occupancies. For example, for helium, one can form functions that correspond to the electron configurations $1s^2$, $1s2s$, $1s2p$, $2s^2$, $2s2p$, $2p^2$, $1s3s$, and so on. Moreover, more than one function can correspond to a given electron configuration. Recall the functions (10.27) to (10.30) corresponding to the helium $1s2s$ configuration. Each such many-electron function Φ_i is a Slater determinant or a linear combination of a few Slater determinants. Use of more than one Slater determinant is required for certain open-shell functions such as (10.44) and (10.45). Each Φ_i is called a **configuration state function** or a **configuration**

function or simply a "configuration." (This last name is unfortunate, since it leads to confusion between an electron configuration such as $1s^2$ and a configuration function such as $|1s\overline{1s}|$.)

As we saw in perturbation theory, the true atomic (or molecular) wave function ψ contains contributions from configurations other than the one that makes the main contribution to ψ, so we express ψ as a linear combination of the configuration functions Φ_i:

$$\psi = \sum_i c_i \Phi_i \tag{11.17}$$

We then regard (11.17) as a linear variation function (Section 8.5). Variation of the coefficients c_i to minimize the variational integral leads to the equation

$$\det(H_{ij} - ES_{ij}) = 0 \tag{11.18}$$

where $H_{ij} \equiv \langle \Phi_i|\hat{H}|\Phi_j\rangle$ and $S_{ij} \equiv \langle \Phi_i|\Phi_j\rangle$. Commonly, the Φ_i functions are orthonormal, but if they are not orthogonal, they can be made so by the Schmidt method. [Only configuration functions whose angular-momentum eigenvalues are the same as those of the state ψ will contribute to the expansion (11.17); see Section 11.5.]

Because the many-electron configuration functions Φ_i are ultimately based on a one-electron basis set that is a complete set, the set of all possible configuration functions is a complete set for the many-electron problem: any antisymmetric many-electron function (including the exact wave function) can be expressed as a linear combination of the Φ_i functions. [For a proof of this, see *Szabo and Ostlund*, Section 2.2.7.] Therefore, if one starts with a complete one-electron basis set and includes all possible configuration functions, a CI calculation will give the exact atomic (or molecular) wave function ψ for the state under consideration. In practice, one is limited to a finite, incomplete basis set, rather than an infinite, complete basis set. Moreover, even with a modest-size basis set, the number of possible configuration functions is extremely large, and one usually does not include all possible configuration functions. Part of the "art" of the CI method is choosing those configurations that will contribute the most.

Because it generally takes very many configuration functions to give a truly accurate wave function, configuration-interaction calculations for systems with more than a few electrons are time-consuming, even on supercomputers. Other methods for allowing for electron correlation are discussed in Chapter 16.

In summary, to do a CI calculation, we choose a one-electron basis set χ_i, iteratively solve the Hartree–Fock equations (11.12) to determine one-electron atomic (or molecular) orbitals ϕ_i as linear combinations of the basis set, form many-electron configuration functions Φ_i using the orbitals ϕ_i, express the wave function ψ as a linear combination of these configuration functions, solve (11.18) for the energy, and solve the associated simultaneous linear equations for the coefficients c_i in (11.17). [In practice, (11.18) and its associated simultaneous equations are solved by matrix methods; see Section 8.6.]

As an example, consider the ground state of beryllium. The Hartree–Fock SCF method would find the best forms for the $1s$ and $2s$ orbitals in the Slater determinant $|1s\overline{1s}2s\overline{2s}|$ and use this for the ground-state wave function. [We are using the notation of Eq. (10.47).] Going beyond the Hartree–Fock method, we would include contributions from excited configuration functions (for example, $|1s\overline{1s}3s\overline{3s}|$) in a linear variation function for the ground state. Bunge did a CI calculation for the beryllium ground state using a linear combination of 650 configuration functions [C. F. Bunge, *Phys. Rev. A*, **14**, 1965 (1976)]. The Hartree–Fock energy is $-14.5730(e'^2/a_0)$, Bunge's CI result is $-14.6669(e'^2/a_0)$, and the exact nonrelativistic energy is $-14.6674(e'^2/a_0)$. Bunge was able to obtain 99.5% of the correlation energy.

A detailed CI calculation for the He atom is provided in Section 16.1.

11.4 ADDITION OF ANGULAR MOMENTA

For a many-electron atom, the operators for individual angular momenta of the electrons do not commute with the Hamiltonian operator, but their sum does. Hence we want to learn how to add angular momenta.

Suppose we have a system with two angular-momentum vectors $\mathbf{M}_1$ and $\mathbf{M}_2$. They might be the orbital angular-momentum vectors of two electrons in an atom, or they might be the spin angular-momentum vectors of two electrons, or one might be the spin and the other the orbital angular momentum of a single electron. The eigenvalues of $\hat{M}_1^2$, $\hat{M}_2^2$, $\hat{M}_{1z}$, and $\hat{M}_{2z}$ are $j_1(j_1+1)\hbar^2$, $j_2(j_2+1)\hbar^2$, $m_1\hbar$, and $m_2\hbar$, where the quantum numbers obey the usual restrictions. The components of $\hat{\mathbf{M}}_1$ and $\hat{\mathbf{M}}_2$ obey the angular-momentum commutation relations [Eqs. (5.46), (5.48), and (5.107)]

$$[\hat{M}_{1x}, \hat{M}_{1y}] = i\hbar\hat{M}_{1z}, \text{ etc.} \quad [\hat{M}_{2x}, \hat{M}_{2y}] = i\hbar\hat{M}_{2z}, \text{ etc.} \tag{11.19}$$

We define the **total angular momentum M** of the system as the vector sum

$$\mathbf{M} = \mathbf{M}_1 + \mathbf{M}_2 \tag{11.20}$$

M is a vector with three components:

$$\mathbf{M} = M_x\mathbf{i} + M_y\mathbf{j} + M_z\mathbf{k} \tag{11.21}$$

The vector equation (11.20) gives the three scalar equations

$$M_x = M_{1x} + M_{2x}, \qquad M_y = M_{1y} + M_{2y}, \qquad M_z = M_{1z} + M_{2z} \tag{11.22}$$

For the operator $\hat{M}^2$, we have

$$\hat{M}^2 = \hat{\mathbf{M}} \cdot \hat{\mathbf{M}} = \hat{M}_x^2 + \hat{M}_y^2 + \hat{M}_z^2 \tag{11.23}$$

$$\hat{M}^2 = (\hat{\mathbf{M}}_1 + \hat{\mathbf{M}}_2) \cdot (\hat{\mathbf{M}}_1 + \hat{\mathbf{M}}_2)$$

$$\hat{M}^2 = \hat{M}_1^2 + \hat{M}_2^2 + \hat{\mathbf{M}}_1 \cdot \hat{\mathbf{M}}_2 + \hat{\mathbf{M}}_2 \cdot \hat{\mathbf{M}}_1 \tag{11.24}$$

If $\hat{\mathbf{M}}_1$ and $\hat{\mathbf{M}}_2$ refer to different electrons, they will commute with each other, since each will affect only functions of the coordinates of one electron and not the other. Even if $\hat{\mathbf{M}}_1$ and $\hat{\mathbf{M}}_2$ are the orbital and spin angular momenta of the same electron, they will commute, as one will affect only functions of the spatial coordinates while the other will affect functions of the spin coordinates. Thus (11.24) becomes

$$\hat{M}^2 = \hat{M}_1^2 + \hat{M}_2^2 + 2\hat{\mathbf{M}}_1 \cdot \hat{\mathbf{M}}_2 \tag{11.25}$$

$$\hat{M}^2 = \hat{M}_1^2 + \hat{M}_2^2 + 2(\hat{M}_{1x}\hat{M}_{2x} + \hat{M}_{1y}\hat{M}_{2y} + \hat{M}_{1z}\hat{M}_{2z}) \tag{11.26}$$

We now show that the components of the total angular momentum obey the usual angular-momentum commutation relations. We have [Eq. (5.4)]

$$[\hat{M}_x, \hat{M}_y] = [\hat{M}_{1x} + \hat{M}_{2x}, \hat{M}_{1y} + \hat{M}_{2y}]$$
$$= [\hat{M}_{1x}, \hat{M}_{1y} + \hat{M}_{2y}] + [\hat{M}_{2x}, \hat{M}_{1y} + \hat{M}_{2y}]$$
$$= [\hat{M}_{1x}, \hat{M}_{1y}] + [\hat{M}_{1x}, \hat{M}_{2y}] + [\hat{M}_{2x}, \hat{M}_{1y}] + [\hat{M}_{2x}, \hat{M}_{2y}]$$

Since all components of $\hat{\mathbf{M}}_1$ commute with all components of $\hat{\mathbf{M}}_2$, we have

$$[\hat{M}_x, \hat{M}_y] = [\hat{M}_{1x}, \hat{M}_{1y}] + [\hat{M}_{2x}, \hat{M}_{2y}] = i\hbar\hat{M}_{1z} + i\hbar\hat{M}_{2z}$$
$$[\hat{M}_x, \hat{M}_y] = i\hbar\hat{M}_z \tag{11.27}$$

Cyclic permutation of x, y, and z gives

$$[\hat{M}_y, \hat{M}_z] = i\hbar\hat{M}_x, \qquad [\hat{M}_z, \hat{M}_x] = i\hbar\hat{M}_y \tag{11.28}$$

The same commutator algebra used to derive (5.109) gives

$$[\hat{M}^2, \hat{M}_x] = [\hat{M}^2, \hat{M}_y] = [\hat{M}^2, \hat{M}_z] = 0 \tag{11.29}$$

Thus we can simultaneously quantize M^2 and one of its components, say M_z. Since the components of the total angular momentum obey the angular-momentum commutation relations, the work of Section 5.4 shows that the eigenvalues of $\hat{M}^2$ are

$$\boxed{J(J+1)\hbar^2, \qquad J = 0, \tfrac{1}{2}, 1, \tfrac{3}{2}, 2, \ldots} \tag{11.30}$$

and the eigenvalues of $\hat{M}_z$ are

$$\boxed{M_J\hbar, \qquad M_J = -J, -J+1, \ldots, J-1, J} \tag{11.31}$$

We want to find out how the total-angular-momentum quantum numbers J and M_J are related to the quantum numbers j_1, j_2, m_1, m_2 of the two angular momenta we are adding in (11.20). We also want the eigenfunctions of $\hat{M}^2$ and $\hat{M}_z$. These eigenfunctions are characterized by the quantum numbers J and M_J, and, using ket notation (Section 7.3), we write them as $|JM_J\rangle$. Similarly, let $|j_1m_1\rangle$ denote the eigenfunctions of $\hat{M}_1^2$ and $\hat{M}_{1z}$ and $|j_2m_2\rangle$ denote the eigenfunctions of $\hat{M}_2^2$ and $\hat{M}_{2z}$. Now it is readily shown (Prob. 11.8) that

$$[\hat{M}_x, \hat{M}_1^2] = [\hat{M}_y, \hat{M}_1^2] = [\hat{M}_z, \hat{M}_1^2] = [\hat{M}^2, \hat{M}_1^2] = 0 \tag{11.32}$$

with similar equations with $\hat{M}_2^2$ replacing $\hat{M}_1^2$. Hence we can have simultaneous eigenfunctions of all four operators $\hat{M}_1^2$, $\hat{M}_2^2$, $\hat{M}^2$, $\hat{M}_z$, and the eigenfunctions $|JM_J\rangle$ can be more fully written as $|j_1j_2JM_J\rangle$. However, one finds that $\hat{M}^2$ does not commute with $\hat{M}_{1z}$ or $\hat{M}_{2z}$ (Prob. 11.10), so the eigenfunctions $|j_1j_2JM_J\rangle$ are not necessarily eigenfunctions of $\hat{M}_{1z}$ or $\hat{M}_{2z}$.

If we take the complete set of functions $|j_1m_1\rangle$ for particle 1 and the complete set $|j_2m_2\rangle$ for particle 2 and form all possible products of the form $|j_1m_1\rangle|j_2m_2\rangle$, we will have a complete set of functions for the two particles. Each unknown eigenfunction $|j_1j_2JM_J\rangle$ can then be expanded using this complete set:

$$|j_1j_2JM_J\rangle = \sum C(j_1j_2JM_j;\, m_1m_2)|j_1m_1\rangle|j_2m_2\rangle \qquad (11.33)$$

where the expansion coefficients are the $C(j_1 \cdots m_2)$'s. The functions $|j_1j_2JM_J\rangle$ are eigenfunctions of the commuting operators $\hat{M}_1^2$, $\hat{M}_2^2$, $\hat{M}^2$, and $\hat{M}_z$ with the following eigenvalues:

$\hat{M}_1^2$	$\hat{M}_2^2$	$\hat{M}^2$	$\hat{M}_z$
$j_1(j_1+1)\hbar^2$	$j_2(j_2+1)\hbar^2$	$J(J+1)\hbar^2$	$M_J\hbar$

The functions $|j_1m_1\rangle|j_2m_2\rangle$ are eigenfunctions of the commuting operators $\hat{M}_1^2$, $\hat{M}_{1z}$, $\hat{M}_2^2$, $\hat{M}_{2z}$ with the following eigenvalues:

$\hat{M}_1^2$	$\hat{M}_{1z}$	$\hat{M}_2^2$	$\hat{M}_{2z}$
$j_1(j_1+1)\hbar^2$	$m_1\hbar$	$j_2(j_2+1)\hbar^2$	$m_2\hbar$

Since the function $|j_1j_2JM_J\rangle$ being expanded in (11.33) is an eigenfunction of $\hat{M}_1^2$ with eigenvalue $j_1(j_1+1)\hbar^2$, we include in the sum only terms that have the same j_1 value as in the function $|j_1j_2JM_J\rangle$. (See Theorem 3 at the end of Section 7.3.) Likewise, only terms with the same j_2 value as in $|j_1j_2JM_J\rangle$ are included in the sum. Hence the sum goes over only the m_1 and m_2 values. Also, using $\hat{M}_z = \hat{M}_{1z} + \hat{M}_{2z}$, we can prove (Prob. 11.9) that the coefficient C vanishes unless

$$\boxed{m_1 + m_2 = M_J} \qquad \mathbf{(11.34)}$$

To find the total-angular-momentum eigenfunctions, one must evaluate the coefficients in (11.33). These are called *Clebsch–Gordan* or *Wigner* or *vector addition* coefficients. For their evaluation, see *Merzbacher*, Section 16.6.

Thus each total-angular-momentum eigenfunction $|j_1j_2JM_J\rangle$ is a linear combination of those product functions $|j_1m_1\rangle|j_2m_2\rangle$ whose m values satisfy $m_1 + m_2 = M_J$.

We now find the possible values of the total-angular-momentum quantum number J that arise from the addition of angular momenta with individual quantum numbers j_1 and j_2.

Before discussing the general case, we consider the case with $j_1 = 1$, $j_2 = 2$. The possible values of m_1 are $-1, 0, 1$, and the possible values of m_2 are $-2, -1, 0, 1, 2$. If we describe the system by the quantum numbers j_1, j_2, m_1, m_2, then the total number of possible states is fifteen, corresponding to three possibilities for m_1 and five for m_2. Instead, we can describe the system using the quantum numbers j_1, j_2, J, M_J, and we must have the same number of states in this description. Let us tabulate the fifteen possible values of M_J using (11.34):

$m_1 = -1$	0	1	
−3	−2	−1	$-2 = m_2$
−2	−1	0	−1
−1	0	1	0
0	1	2	1
1	2	3	2

where each M_J value in the table is the sum of the m_1 and m_2 values at the top and side. The number of times each value of M_J occurs is

value of M_J	3	2	1	0	−1	−2	−3
number of occurrences	1	2	3	3	3	2	1

The highest value of M_J is +3. Since M_J ranges from $-J$ to $+J$, the highest value of J must be 3. Corresponding to $J = 3$, there are seven values of M_J ranging from −3 to +3. Eliminating these seven values, we are left with

value of M_J	2	1	0	−1	−2
number of occurrences	1	2	2	2	1

The highest remaining value, $M_J = 2$, must correspond to $J = 2$. For $J = 2$, we have five values of M_J, which when eliminated leave

value of M_J	1	0	−1
number of occurrences	1	1	1

These remaining values of M_J clearly correspond to $J = 1$. Thus, for the individual angular-momentum quantum numbers $j_1 = 1$, $j_2 = 2$, the possible values of the total-angular-momentum quantum number J are 3, 2, and 1.

Now consider the general case. There are $2j_1 + 1$ values of m_1 (ranging from $-j_1$ to $+j_1$) and $2j_2 + 1$ values of m_2. Hence there are $(2j_1 + 1)(2j_2 + 1)$ possible states $|j_1m_1\rangle|j_2m_2\rangle$ with fixed j_1 and j_2 values. The highest possible values of m_1 and m_2 are j_1 and j_2, respectively. Therefore, the maximum possible value of $M_J = m_1 + m_2$ is $j_1 + j_2$ [Eq. (11.34)]. Since M_J ranges from $-J$ to $+J$, the maximum possible value of J must also be $j_1 + j_2$:

$$J_{max} = j_1 + j_2 \tag{11.35}$$

The second-highest value of M_J is $j_1 + j_2 - 1$, which arises in two ways: $m_1 = j_1 - 1$, $m_2 = j_2$ and $m_1 = j_1$, $m_2 = j_2 - 1$. Linear combinations of these two states must give one state with $J = j_1 + j_2$, $M_J = j_1 + j_2 - 1$ and one state with $J = j_1 + j_2 - 1$, $M_J = j_1 + j_2 - 1$. Continuing in this manner, we find that the possible values of J are

$$j_1 + j_2, \quad j_1 + j_2 - 1, \quad j_1 + j_2 - 2, \quad \ldots, \quad J_{min}$$

where J_{min} is the lowest possible value of J.

We determine J_{min} by the requirement that the total number of states be $(2j_1 + 1)(2j_2 + 1)$. For a particular value of J, there are $2J + 1$ M_J values, and so $2J + 1$ states correspond to each value of J. The total number of states $|j_1j_2JM_J\rangle$ for fixed j_1 and j_2 is found by summing the number of states $2J + 1$ for each J from J_{min} to J_{max}:

$$\text{number of states} = \sum_{J=J_{min}}^{J_{max}} (2J + 1) \tag{11.36}$$

This sum goes from $J_{\min}$ to $J_{\max}$. Let us now take the lower limit of the sum to be $J = 0$ instead of $J_{\min}$. This change adds to the sum terms with J values of $0, 1, 2, \ldots, J_{\min} - 1$. To compensate, we must subtract the corresponding sum that goes from $J = 0$ to $J = J_{\min} - 1$. Therefore, (11.36) becomes

$$\text{number of states} = \sum_{J=0}^{J_{\max}} (2J + 1) - \sum_{J=0}^{J_{\min}-1} (2J + 1)$$

Problem 6.14 gives $\sum_{l=0}^{n-1} (2l + 1) = n^2$. Replacing $n - 1$ with b, we have the result $\sum_{J=0}^{b} (2J + 1) = (b + 1)^2$. Therefore,

$$\text{number of states} = (J_{\max} + 1)^2 - J_{\min}^2 = J_{\max}^2 + 2J_{\max} + 1 - J_{\min}^2$$

Replacing $J_{\max}$ by $j_1 + j_2$ [Eq. (11.35)] and equating the number of states to $(2j_1 + 1)(2j_2 + 1) = 4j_1j_2 + 2j_1 + 2j_2 + 1$, we have

$$(j_1 + j_2)^2 + 2(j_1 + j_2) + 1 - J_{\min}^2 = 4j_1j_2 + 2j_1 + 2j_2 + 1$$

$$J_{\min}^2 = j_1^2 - 2j_1j_2 + j_2^2 = (j_1 - j_2)^2$$

$$J_{\min} = \pm(j_1 - j_2) \tag{11.37}$$

If $j_1 = j_2$, then $J_{\min} = 0$. If $j_1 \neq j_2$, then one of the values in (11.37) is negative and must be rejected [Eq. (11.30)]. Thus

$$J_{\min} = |j_2 - j_1| \tag{11.38}$$

To summarize, we have shown that *the addition of two angular momenta characterized by quantum numbers* j_1 *and* j_2 *results in a total angular momentum whose quantum number J has the possible values*

$$\boxed{J = j_1 + j_2, \quad j_1 + j_2 - 1, \quad \ldots, \quad |j_1 - j_2|} \tag{11.39}$$

EXAMPLE Find the possible values of the total-angular-momentum quantum number resulting from the addition of angular momenta with quantum numbers $j_1 = 2$ and $j_2 = 3$.

The maximum and minimum J values are given by (11.39) as $j_1 + j_2 = 2 + 3 = 5$ and $|j_1 - j_2| = |2 - 3| = 1$. The possible J values (11.39) are therefore $J = 5, 4, 3, 2, 1$.

EXERCISE Find the possible values of the total-angular-momentum quantum number resulting from the addition of angular momenta with quantum numbers $j_1 = 3$ and $j_2 = \frac{3}{2}$. (*Answer:* $J = \frac{9}{2}, \frac{7}{2}, \frac{5}{2}, \frac{3}{2}$.)

EXAMPLE Find the possible J values when angular momenta with quantum numbers $j_1 = 1$, $j_2 = 2$, and $j_3 = 3$ are added.

To add more than two angular momenta, we apply (11.39) repeatedly. Addition of $j_1 = 1$ and $j_2 = 2$ gives the possible quantum numbers 3, 2, and 1. Addition of j_3 to each of these values gives the following possibilities for the total-angular-momentum quantum number:

$$6, 5, 4, 3, 2, 1, 0; \qquad 5, 4, 3, 2, 1; \qquad 4, 3, 2 \tag{11.40}$$

We have one set of states with total-angular-momentum quantum number 6, two sets of states with $J = 5$, three sets with $J = 4$, and so on.

EXERCISE Find the possible J values when angular momenta with quantum numbers $j_1 = 1$, $j_2 = 1$, and $j_3 = 1$ are added. (*Answer:* $J = 3, 2, 2, 1, 1, 1, 0$.)

11.5 ANGULAR MOMENTUM IN MANY-ELECTRON ATOMS

Total Electronic Orbital and Spin Angular Momenta. The **total electronic orbital angular momentum** of an n-electron atom is defined as the vector sum of the orbital angular momenta of the individual electrons:

$$\mathbf{L} = \sum_{i=1}^{n} \mathbf{L}_i \tag{11.41}$$

Although the individual orbital-angular-momentum operators $\hat{\mathbf{L}}_i$ do not commute with the atomic Hamiltonian (11.1), one can show (*Bethe and Jackiw*, pp. 102–103) that $\hat{\mathbf{L}}$ does commute with the atomic Hamiltonian [provided spin–orbit interaction (Section 11.6) is neglected]. We can therefore characterize an atomic state by a quantum number L, where $L(L+1)\hbar^2$ is the square of the magnitude of the total electronic orbital angular momentum. The electronic wave function ψ of an atom satisfies $\hat{L}^2\psi = L(L+1)\hbar^2\psi$. The total-electronic-orbital-angular-momentum quantum number L of an atom is specified by a code letter, as follows:

L	0	1	2	3	4	5	6	7	8
letter	S	P	D	F	G	H	I	K	L

(11.42)

The total orbital angular momentum is designated by a capital letter, while lowercase letters are used for orbital angular momenta of individual electrons.

EXAMPLE Find the possible values of the quantum number L for states of the carbon atom that arise from the electron configuration $1s^22s^22p3d$.

The s electrons have zero orbital angular momentum and contribute nothing to the total orbital angular momentum. The $2p$ electron has $l = 1$ and the $3d$ electron has $l = 2$. From the angular-momentum addition rule (11.39), the total-orbital-angular-momentum quantum number ranges from $1 + 2 = 3$ to $|1 - 2| = 1$; the possible values of L are $L = 3, 2, 1$. The configuration $1s^22s^22p3d$ gives rise to P, D, and F states. [The Hartree–Fock central-field approximation has each electron moving in a central-field potential, $V = V(r)$. Hence, within this approximation, the individual electronic orbital angular momenta are constant, giving rise to a wave function composed of a single configuration that specifies the individual orbital angular momenta. When we go beyond the SCF central-field approximation, we mix in other configurations so we no longer specify precisely the individual orbital angular momenta. Even so, we can still use the rule (11.39) for finding the possible values of the total orbital angular momentum.]

EXERCISE Find the possible values of L for states that arise from the electron configuration $1s^22s^22p^63s^23p4p$. (*Answer:* $2, 1, 0$.)

The **total electronic spin angular momentum S** of an atom is defined as the vector sum of the spins of the individual electrons:

$$\mathbf{S} = \sum_{i=1}^{n} \mathbf{S}_i \tag{11.43}$$

The atomic Hamiltonian $\hat{H}$ of (11.1) (which omits spin–orbit interaction) does not involve spin and therefore commutes with the total-spin operators $\hat{S}^2$ and $\hat{S}_z$. The fact that $\hat{S}^2$ commutes with $\hat{H}$ is not enough to show that the atomic wave functions ψ are eigenfunctions of $\hat{S}^2$. The antisymmetry requirement means that each ψ must be an eigenfunction of the exchange operator $\hat{P}_{ik}$ with eigenvalue -1 (Section 10.3). Hence $\hat{S}^2$ must also commute with $\hat{P}_{ik}$ if we are to have simultaneous eigenfunctions of $\hat{H}$, $\hat{S}^2$, and $\hat{P}_{ik}$. Problem 11.17 shows that $[\hat{S}^2, \hat{P}_{ik}] = 0$, so the atomic wave functions are eigenfunctions of $\hat{S}^2$. We have $\hat{S}^2\psi = S(S+1)\hbar^2\psi$, and each atomic state can be characterized by a total-electronic-spin quantum number S.

EXAMPLE Find the possible values of the quantum number S for states that arise from the electron configuration $1s^2 2s^2 2p3d$.

Consider first the two 1*s* electrons. To satisfy the exclusion principle, one of these electrons must have $m_s = +\frac{1}{2}$ while the other has $m_s = -\frac{1}{2}$. If M_S is the quantum number that specifies the *z* component of the total spin of the 1*s* electrons, then the only possible value of M_S is $\frac{1}{2} - \frac{1}{2} = 0$ [Eq. (11.34)]. This single value of M_S clearly means that the total spin of the two 1*s* electrons is zero. Thus, although in general when we add the spins $s_1 = \frac{1}{2}$ and $s_2 = \frac{1}{2}$ of two electrons according to the rule (11.39), we get the two possibilities $S = 0$ and $S = 1$, the restriction imposed by the Pauli principle leaves $S = 0$ as the only possibility in this case. Likewise, the spins of the 2*s* electrons add up to zero. The exclusion principle does not restrict the m_s values of the 2*p* and 3*d* electrons. Application of the rule (11.39) to the spins $s_1 = \frac{1}{2}$ and $s_2 = \frac{1}{2}$ of the 2*p* and 3*d* electrons gives $S = 0$ and $S = 1$. These are the possible values of the total spin quantum number, since the 1*s* and 2*s* electrons do not contribute to S.

EXERCISE Find the possible values of S for states that arise from the electron configuration $1s^2 2s^2 2p^6 3s^2 3p4p$. (*Answer:* 0 and 1.)

Atomic Terms. A given electron configuration gives rise in general to several different atomic states, some having the same energy and some having different energies, depending on whether the interelectronic repulsions are the same or different for the states. For example, the 1*s*2*s* configuration of helium gives rise to four states: The three states with zeroth-order wave functions (10.27) to (10.29) all have the same energy; the single state (10.30) has a different energy. The 1*s*2*p* electron configuration gives rise to twelve states: The nine states obtained by replacing 2*s* in (10.27) to (10.29) by $2p_x$, $2p_y$, or $2p_z$ have the same energy; the three states obtained by replacing 2*s* in (10.30) by $2p_x$, $2p_y$, or $2p_z$ have the same energy, which differs from the energy of the other nine states.

Thus the atomic states that arise from a given electron configuration can be grouped into sets of states that have the same energy. One can show that states that arise from the same electron configuration and that have the same energy (with spin–orbit interaction neglected) will have the same value of L and the same value of S (see *Kemble*, Section 63a). A set of equal-energy atomic states that arise from the same electron configuration and that have the same L value and the same S value constitutes an atomic **term**. For a fixed L value, the quantum number M_L (where $M_L\hbar$ is the *z* component of the total electronic orbital angular momentum) takes on $2L + 1$ values ranging from $-L$ to $+L$. For a fixed S value, M_S takes on $2S + 1$ values. The atomic energy does not depend on M_L or M_S, and each term consists of $(2L + 1)(2S + 1)$ atomic states of equal energy. The degeneracy of an atomic term is $(2L + 1)(2S + 1)$ (spin–orbit interaction neglected).

Each term of an atom is designated by a **term symbol** formed by writing the numerical value of the quantity $2S + 1$ as a left superscript on the code letter (11.42) that gives the L value. For example, a term that has $L = 2$ and $S = 1$ has the term symbol ^3D, since $2S + 1 = 3$.

EXAMPLE Find the terms arising from each of the following electron configurations: (*a*) $1s2p$; (*b*) $1s^2 2s^2 2p3d$. Give the degeneracy of each term.

(a) The $1s$ electron has quantum number $l = 0$ and the $2p$ electron has $l = 1$. The addition rule (11.39) gives $L = 1$ as the only possibility. The code letter for $L = 1$ is P. Each electron has $s = \frac{1}{2}$, and (11.39) gives $S = 1, 0$ as the possible S values. The possible values of $2S + 1$ are 3 and 1. The possible terms are thus ^3P and ^1P. The ^3P term has quantum numbers $L = 1$ and $S = 1$, and its degeneracy is $(2L + 1)(2S + 1) = 3(3) = 9$. The ^1P term has $L = 1$ and $S = 0$, and its degeneracy is $(2L + 1)(2S + 1) = 3(1) = 3$. [The nine states of the ^3P term are obtained by replacing $2s$ in (10.27) to (10.29) by $2p_x$, $2p_y$, or $2p_z$. The three states of the ^1P term are obtained by replacing $2s$ in (10.30) by $2p$ functions.]

(b) In the two previous examples in this section, we found that the configuration $1s^2 2s^2 2p3d$ has the possible L values $L = 3, 2, 1$ and has $S = 1, 0$. The code letters for these L values are F, D, P, and the terms are

$$^1P, \quad ^3P, \quad ^1D, \quad ^3D, \quad ^1F, \quad ^3F \tag{11.44}$$

The degeneracies are found as in (a) and are 3, 9, 5, 15, 7, and 21, respectively.

Derivation of Atomic Terms. We now examine how to systematically derive the terms that arise from a given electron configuration.

First consider configurations that contain only completely filled subshells. In such configurations, for each electron with $m_s = +\frac{1}{2}$ there is an electron with $m_s = -\frac{1}{2}$. Let the quantum number specifying the z component of the total electronic spin angular momentum be M_S. The only possible value for M_S is zero ($M_S = \sum_i m_{si} = 0$); hence S must be zero. For each electron in a closed subshell with magnetic quantum number m, there is an electron with magnetic quantum number $-m$. For example, for a $2p^6$ configuration we have two electrons with $m = +1$, two with $m = -1$, and two with $m = 0$. Denoting the quantum number specifying the z component of the total electronic orbital angular momentum by M_L, we have $M_L = \sum_i m_i = 0$. We conclude that L must be zero. In summary, a configuration of closed subshells gives rise to only one term: ^1S. For configurations consisting of closed subshells and open subshells, the closed subshells make no contribution to L or S and may be ignored in finding the terms.

We now consider two electrons in different subshells; such electrons are called **nonequivalent**. Nonequivalent electrons have different values of n or l or both, and we need not worry about any restrictions imposed by the exclusion principle when we derive the terms. We simply find the possible values of L from l_1 and l_2 according to (11.39); combining s_1 and s_2 gives $S = 0, 1$. We previously worked out the pd case, which gives the terms in (11.44). If we have more than two nonequivalent electrons, we combine the individual l's to find the values of L, and we combine the individual s's to find the values of S. For example, consider a *pdf* configuration. The possible values of L are given by (11.40). Combining three spin angular momenta, each of which is $\frac{1}{2}$, gives $S = \frac{3}{2}, \frac{1}{2}, \frac{1}{2}$. Each of the three possibilities in (11.40) with $L = 3$ may be combined with each of the two possibilities for $S = \frac{1}{2}$, giving six ^2F terms. Continuing in this manner, we find that the following terms arise from a *pdf* configuration: $^2S(2)$, $^2P(4)$, $^2D(6)$, $^2F(6)$, $^2G(6)$, $^2H(4)$, $^2I(2)$, ^4S, $^4P(2)$, $^4D(3)$, $^4F(3)$, $^4G(3)$, $^4H(2)$, ^4I, where the number of times each type of term occurs is in parentheses.

Now consider two electrons in the same subshell (**equivalent** electrons). Equivalent electrons have the same value of n and the same value of l, and the situation is complicated by the necessity to avoid giving two electrons the same four quantum numbers. Hence not all the terms derived for nonequivalent electrons are possible. As an example, consider the terms arising from two equivalent p electrons, an np^2 configuration. (The carbon ground-state configuration is $1s^2 2s^2 2p^2$.) The possible values of m and m_s for the two electrons are listed in Table 11.1, which also gives M_L and M_S.

Note that certain combinations are missing from this table. For example, $m_1 = 1$, $m_{s1} = \frac{1}{2}$, $m_2 = 1$, $m_{s2} = \frac{1}{2}$ is missing, since it violates the exclusion principle. Another missing combination is $m_1 = 1$, $m_{s1} = -\frac{1}{2}$, $m_2 = 1$, $m_{s2} = \frac{1}{2}$. This combination differs from $m_1 = 1$, $m_{s1} = \frac{1}{2}$, $m_2 = 1$, $m_{s2} = -\frac{1}{2}$ (row 1) solely by interchange of electrons 1 and 2. Each row in Table 11.1 stands for a Slater determinant, which when expanded contains terms for all possible electron interchanges among the spin-orbitals. Two rows that differ from each other solely by interchange of two electrons correspond to the same Slater determinant, and we include only one of them in the table.

TABLE 11.1 Quantum Numbers for Two Equivalent p Electrons

m_1	m_{s1}	m_2	m_{s2}	$M_L = m_1 + m_2$	$M_S = m_{s1} + m_{s2}$
1	$\frac{1}{2}$	1	$-\frac{1}{2}$	2	0
1	$\frac{1}{2}$	0	$\frac{1}{2}$	1	1
1	$\frac{1}{2}$	0	$-\frac{1}{2}$	1	0
1	$-\frac{1}{2}$	0	$\frac{1}{2}$	1	0
1	$-\frac{1}{2}$	0	$-\frac{1}{2}$	1	-1
1	$\frac{1}{2}$	-1	$\frac{1}{2}$	0	1
1	$\frac{1}{2}$	-1	$-\frac{1}{2}$	0	0
1	$-\frac{1}{2}$	-1	$\frac{1}{2}$	0	0
1	$-\frac{1}{2}$	-1	$-\frac{1}{2}$	0	-1
0	$\frac{1}{2}$	0	$-\frac{1}{2}$	0	0
0	$\frac{1}{2}$	-1	$\frac{1}{2}$	-1	1
0	$\frac{1}{2}$	-1	$-\frac{1}{2}$	-1	0
0	$-\frac{1}{2}$	-1	$\frac{1}{2}$	-1	0
0	$-\frac{1}{2}$	-1	$-\frac{1}{2}$	-1	-1
-1	$\frac{1}{2}$	-1	$-\frac{1}{2}$	-2	0

The highest value of M_L in Table 11.1 is 2, which must correspond to a term with $L = 2$, a D term. The $M_L = 2$ value occurs in conjunction with $M_S = 0$, indicating that $S = 0$ for the D term. Thus we have a ^1D term corresponding to the five states

$$\begin{aligned} M_L &= 2 \quad 1 \quad 0 \quad -1 \quad -2 \\ M_S &= 0 \quad 0 \quad 0 \quad\;\; 0 \quad\;\; 0 \end{aligned} \tag{11.45}$$

The highest value of M_S in Table 11.1 is 1, indicating a term with $S = 1$. $M_S = 1$ occurs in conjunction with $M_L = 1, 0, -1$, which indicates a P term. Hence we have a ^3P term corresponding to the nine states

$$\begin{aligned} M_L &= 1 \quad 1 \quad\;\; 1 \quad 1 \quad 0 \quad\;\; 0 \quad -1 \quad -1 \quad -1 \\ M_S &= 1 \quad 0 \quad -1 \quad 1 \quad 0 \quad -1 \quad\;\; 1 \quad\;\; 0 \quad -1 \end{aligned} \tag{11.46}$$

Elimination of the states of (11.45) and (11.46) from Table 11.1 leaves only a single state, which has $M_L = 0$, $M_S = 0$, corresponding to a ^1S term. Thus a p^2 configuration gives rise to the terms ^1S, ^3P, ^1D. (In contrast, two nonequivalent p electrons give rise to six terms: ^1S, ^3S, ^1P, ^3P, ^1D, ^3D.)

Table 11.2a lists the terms arising from various configurations of equivalent electrons. These results may be derived in the same way that we found the p^2 terms, but this procedure can become quite involved. To derive the terms of the f^7 configuration would require a table with 3432 rows. More efficient methods exist [R. F. Curl and J. E. Kilpatrick, *Am. J. Phys.*, **28**, 357 (1960); K. E. Hyde, *J. Chem. Educ.*, **52**, 87 (1975)].

TABLE 11.2 Terms Arising from Various Electron Configurations

Configuration	Terms
(a) *Equivalent electrons*	
s^2; p^6; d^{10}	^1S
p; p^5	^2P
p^2; p^4	$^3P, {}^1D, {}^1S$
p^3	$^4S, {}^2D, {}^2P$
d; d^9	^2D
d^2; d^8	$^3F, {}^3P, {}^1G, {}^1D, {}^1S$
d^3; d^7	$^4F, {}^4P, {}^2H, {}^2G, {}^2F, {}^2D(2), {}^2P$
d^4; d^6	$^5D, {}^3H, {}^3G, {}^3F(2), {}^3D, {}^3P(2)$ $^1I, {}^1G(2), {}^1F, {}^1D(2), {}^1S(2)$
d^5	$^6S, {}^4G, {}^4F, {}^4D, {}^4P, {}^2I, {}^2H, {}^2G(2)$ $^2F(2), {}^2D(3), {}^2P, {}^2S$
(b) *Nonequivalent electrons*	
ss	$^1S, {}^3S$
sp	$^1P, {}^3P$
sd	$^1D, {}^3D$
pp	$^3D, {}^1D, {}^3P, {}^1P, {}^3S, {}^1S$

Note from Table 11.2a that the terms arising from a subshell containing N electrons are the same as the terms for a subshell that is N electrons short of being full. For example, the terms for p^2 and p^4 are the same. We can divide the electrons of a closed subshell into two groups and find the terms for each group; because a closed subshell gives only a ^1S term, the terms for each of these two groups must be the same. Table 11.2b gives the terms arising from some nonequivalent electron configurations.

To deal with a configuration containing both equivalent and nonequivalent electrons, we first find separately the terms from the nonequivalent electrons and the terms from the equivalent electrons. We then take all possible combinations of the L and S values of these two sets of terms. For example, consider an sp^3 configuration. From the s electron, we get a ^2S term. From the three equivalent p electrons, we get the terms ^2P, ^2D, and ^4S (Table 11.2a). Combining the L and S values of these terms, we have as the terms of an sp^3 configuration

$$^3P,\ ^1P,\ ^3D,\ ^1D,\ ^5S,\ ^3S \tag{11.47}$$

Hund's Rule. To decide which one of the terms arising from a given electron configuration is lowest in energy, we use the empirical **Hund's rule**: *For terms arising from the same electron configuration, the term with the largest value of S lies lowest; if there is more than one term with the largest S, then the term with the largest S and the largest L lies lowest.*

EXAMPLE Use Table 11.2 to predict the lowest term of (a) the carbon ground-state configuration $1s^22s^22p^2$; (b) the configuration $1s^22s^22p^63s^23p^63d^24s^2$.

(a) Table 11.2a gives the terms arising from a p^2 configuration as ^3P, ^1D, and ^1S. The term with the largest S will have the largest value of the left superscript $2S + 1$. Hund's rule predicts ^3P as the lowest term. (b) Table 11.2a gives the d^2 terms as ^3F, ^3P, ^1G, ^1D, ^1S. Of these terms, ^3F and ^3P have the highest S. ^3F has $L = 3$; ^3P has $L = 1$. Therefore, ^3F is predicted to be lowest.

EXERCISE Predict the lowest term of the $1s^22s2p^3$ configuration using (11.47). (*Answer:* ^5S.)

Hund's rule works very well for the ground-state configuration, but occasionally fails for an excited configuration.

Hund's rule gives only the lowest term of a configuration, and should not be used to decide the order of the remaining terms. For example, for the $1s^22s2p^3$ configuration of carbon, the observed order of the terms is

$$^5S < {}^3D < {}^3P < {}^1D < {}^3S < {}^1P$$

The ^3S term lies above the ^1D term, even though ^3S has the higher spin S.

It is not necessary to consult Table 11.2a to find the lowest term of a partly filled subshell configuration. We simply put the electrons in the orbitals so as to give the greatest number of parallel spins. Thus, for a d^3 configuration, we have

$$m\!: \quad \underset{+2}{\underline{\uparrow}} \quad \underset{+1}{\underline{\uparrow}} \quad \underset{0}{\underline{\uparrow}} \quad \underset{-1}{\underline{\quad}} \quad \underset{-2}{\underline{\quad}} \tag{11.48}$$

The lowest term thus has three parallel spins, so $S = \frac{3}{2}$, giving $2S + 1 = 4$. The maximum value of M_L is 3, corresponding to $L = 3$, an F term. Hund's rule thus predicts ^4F as the lowest term of a d^3 configuration.

The traditional explanation of Hund's rule is as follows: Electrons with the same spin tend to keep out of each other's way (recall the idea of Fermi holes), thereby minimizing the Coulombic repulsion between them. The term that has the greatest number of parallel spins (that is, the greatest value of S) will therefore be lowest in energy. For example, the ^3S term of the helium $1s2s$ configuration has an antisymmetric spatial function that vanishes when the spatial coordinates of the two electrons are equal. Hence the ^3S term is lower than the ^1S term.

This traditional explanation turns out to be wrong in most cases. It is true that the probability that the two electrons are very close together is smaller for the helium ^3S $1s2s$ term than for the ^1S $1s2s$ term. However, calculations with accurate wave functions show that the probability that the two electrons are very far apart is also less for the ^3S term. The net result is that the average distance between the two electrons is slightly less for the ^3S term than for the ^1S term, and the interelectronic repulsion is slightly greater for the ^3S term. The reason the ^3S term lies below the ^1S term is because of a substantially greater electron–nucleus attraction in the ^3S term as compared with the ^1S term. Similar results are found for terms of the atoms beryllium and carbon. [See J. Katriel and R. Pauncz, *Adv. Quantum Chem.*, **10**, 143 (1977).] The following explanation of these results has been proposed [I. Shim and J. P. Dahl, *Theor. Chim. Acta*, **48**, 165 (1978)]. The Pauli "repulsion" between electrons of like spin makes the average angle between the radius vectors of the two electrons larger for the ^3S term than for the ^1S term. This reduces the electronic screening of the nucleus and allows the electrons to get closer to the nucleus in the ^3S term, making the electron–nucleus attraction greater for the ^3S term. [See also R. E. Boyd, *Nature*, **310**, 480 (1984).]

Eigenvalues of Two-Electron Spin Functions. The helium $1s2s$ configuration gives rise to the term ^3S with degeneracy $(2L + 1)(2S + 1) = 1(3) = 3$ and to the term ^1S with degeneracy $1(1) = 1$. The three helium zeroth-order wave functions (10.27) to (10.29) must correspond to the triply degenerate ^3S term, and the single function (10.30) must correspond to the ^1S term. Since $S = 1$ and $M_S = 1, 0, -1$ for the ^3S term, the three spin functions in (10.27) to (10.29) should be eigenfunctions of $\hat{S}^2$ with eigenvalue $S(S + 1)\hbar^2 = 2\hbar^2$ and eigenfunctions of $\hat{S}_z$ with eigenvalues $M_S\hbar = \hbar, 0$, and $-\hbar$. The spin function in (10.30) should be an eigenfunction of $\hat{S}^2$ and $\hat{S}_z$ with eigenvalue zero in each case, since $S = 0$ and $M_S = 0$ here. We now verify these assertions.

From Eq. (11.43), the total-electron-spin operator is the sum of the spin operators for each electron:

$$\hat{\mathbf{S}} = \hat{\mathbf{S}}_1 + \hat{\mathbf{S}}_2 \tag{11.49}$$

Taking the z components of (11.49), we have

$$\hat{S}_z = \hat{S}_{1z} + \hat{S}_{2z} \tag{11.50}$$

$$\begin{aligned}\hat{S}_z\alpha(1)\alpha(2) &= \hat{S}_{1z}\alpha(1)\alpha(2) + \hat{S}_{2z}\alpha(1)\alpha(2)\\ &= \alpha(2)\hat{S}_{1z}\alpha(1) + \alpha(1)\hat{S}_{2z}\alpha(2)\\ &= \tfrac{1}{2}\hbar\alpha(1)\alpha(2) + \tfrac{1}{2}\hbar\alpha(1)\alpha(2)\\ \hat{S}_z\alpha(1)\alpha(2) &= \hbar\alpha(1)\alpha(2) \qquad (11.51)\end{aligned}$$

where Eq. (10.7) has been used. Similarly, we find

$$\hat{S}_z\beta(1)\beta(2) = -\hbar\beta(1)\beta(2) \qquad (11.52)$$

$$\hat{S}_z[\alpha(1)\beta(2) + \beta(1)\alpha(2)] = 0 \qquad (11.53)$$

$$\hat{S}_z[\alpha(1)\beta(2) - \beta(1)\alpha(2)] = 0 \qquad (11.54)$$

Consider now $\hat{S}^2$. We have [Eq. (11.26)]

$$\hat{S}^2 = (\hat{\mathbf{S}}_1 + \hat{\mathbf{S}}_2)\cdot(\hat{\mathbf{S}}_1 + \hat{\mathbf{S}}_2) = \hat{S}_1^2 + \hat{S}_2^2 + 2(\hat{S}_{1x}\hat{S}_{2x} + \hat{S}_{1y}\hat{S}_{2y} + \hat{S}_{1z}\hat{S}_{2z}) \qquad (11.55)$$

$$\begin{aligned}\hat{S}^2\alpha(1)\alpha(2) &= \alpha(2)\hat{S}_1^2\alpha(1) + \alpha(1)\hat{S}_2^2\alpha(2) + 2\hat{S}_{1x}\alpha(1)\hat{S}_{2x}\alpha(2)\\ &\quad + 2\hat{S}_{1y}\alpha(1)\hat{S}_{2y}\alpha(2) + 2\hat{S}_{1z}\alpha(1)\hat{S}_{2z}\alpha(2)\end{aligned}$$

Using Eqs. (10.7) to (10.9) and (10.72) and (10.73), we find

$$\hat{S}^2\alpha(1)\alpha(2) = 2\hbar^2\alpha(1)\alpha(2) \qquad (11.56)$$

Hence $\alpha(1)\alpha(2)$ is an eigenfunction of $\hat{S}^2$ corresponding to $S = 1$. Similarly, we find

$$\hat{S}^2\beta(1)\beta(2) = 2\hbar^2\beta(1)\beta(2)$$

$$\hat{S}^2[\alpha(1)\beta(2) + \beta(1)\alpha(2)] = 2\hbar^2[\alpha(1)\beta(2) + \beta(1)\alpha(2)]$$

$$\hat{S}^2[\alpha(1)\beta(2) - \beta(1)\alpha(2)] = 0$$

Thus the spin eigenfunctions in (10.27) to (10.30) correspond to the following values for the total spin quantum numbers:

		S	M_S	
triplet	$\alpha(1)\alpha(2)$	1	1	(11.57)
triplet	$2^{-1/2}[\alpha(1)\beta(2) + \beta(1)\alpha(2)]$	1	0	(11.58)
triplet	$\beta(1)\beta(2)$	1	−1	(11.59)
singlet	$2^{-1/2}[\alpha(1)\beta(2) - \beta(1)\alpha(2)]$	0	0	(11.60)

[In the notation of Section 11.4, we are dealing with the addition of two angular momenta with quantum numbers $j_1 = \frac{1}{2}$ and $j_2 = \frac{1}{2}$ to give eigenfunctions with total angular-momentum quantum numbers $J = 1$ and $J = 0$. The coefficients in (11.57) to (11.60) correspond to the coefficients C in (11.33) and are examples of Clebsch–Gordan coefficients.]

Figure 11.3 shows the vector addition of $\mathbf{S}_1$ and $\mathbf{S}_2$ to form $\mathbf{S}$. It might seem surprising that the spin function (11.58), which has the z components of the spins of the two electrons pointing in opposite directions, could have total spin quantum number $S = 1$. Figure 11.3 shows how this is possible.

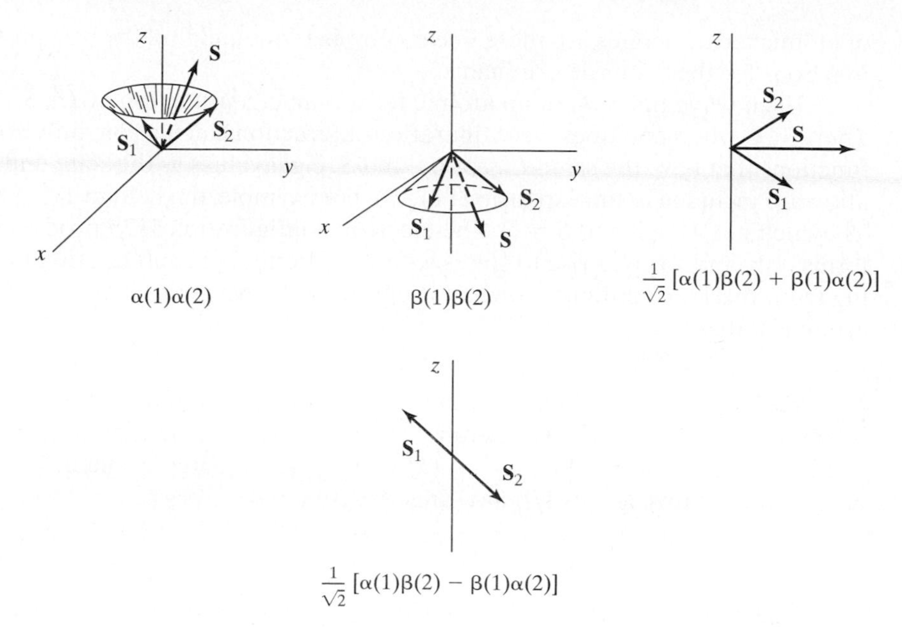

FIGURE 11.3 Vector addition of the spins of two electrons. For $\alpha(1)\alpha(2)$ and $\beta(1)\beta(2)$, the projections of $\mathbf{S}_1$ and $\mathbf{S}_2$ in the xy plane make an angle of 90° with each other (Prob. 11.16c).

Atomic Wave Functions. In Section 10.6, we showed that two of the four zeroth-order wave functions of the $1s2s$ helium configuration could be written as single Slater determinants, but the other two functions had to be expressed as linear combinations of two Slater determinants. Since $\hat{L}^2$ and $\hat{S}^2$ commute with the Hamiltonian (11.1) and with the exchange operator $\hat{P}_{ik}$, the zeroth-order functions should be eigenfunctions of $\hat{L}^2$ and $\hat{S}^2$. The Slater determinants D_2 and D_3 of Section 10.6 are not eigenfunctions of these operators and so are not suitable zeroth-order functions. We have just shown that the linear combinations (10.44) and (10.45) are eigenfunctions of $\hat{S}^2$, and they can also be shown to be eigenfunctions of $\hat{L}^2$.

For a configuration of closed subshells (for example, the helium ground state), we can write only a single Slater determinant. This determinant is an eigenfunction of $\hat{L}^2$ and $\hat{S}^2$ and is the correct zeroth-order function for the nondegenerate ^1S term. A configuration with one electron outside closed subshells (for example, the boron ground configuration) gives rise to only one term. The Slater determinants for such a configuration differ from one another only in the m and m_s values of this electron and are the correct zeroth-order functions for the states of the term. When all the electrons in singly occupied orbitals have the same spin (either all α or all β), the correct zeroth-order function is a single Slater determinant [for example, see (10.42)]. When this is not true, one has to take a linear combination of a few Slater determinants to obtain the correct zeroth-order functions, which are eigenfunctions of $\hat{L}^2$ and $\hat{S}^2$. The correct linear combinations can be found by solving the secular equation of degenerate perturbation theory or by operator techniques. Tabulations of the correct combinations for various configurations are available (*Slater, Atomic Structure*, Vol. II). Hartree–Fock calculations of atomic term energies use these linear combinations and find the best possible orbital functions for the Slater determinants.

Each wave function of an atomic term is an eigenfunction of $\hat{L}^2$, $\hat{S}^2$, $\hat{L}_z$, and $\hat{S}_z$. Therefore, when one does a configuration-interaction calculation, only configuration functions that have the same $\hat{L}^2$, $\hat{S}^2$, $\hat{L}_z$, and $\hat{S}_z$ eigenvalues as the state under consideration are included in the expansion (11.17). For example, the helium $1s^2$ ground term is ^1S, which has $L = 0$ and $S = 0$. The electron configuration $1s2p$ produces ^1P and ^3P terms only and so gives rise to states with $L = 1$ only. No configuration functions arising from the $1s2p$ configuration can occur in the CI wave function (11.17) for the He ground state.

Parity of Atomic States. Consider the atomic Hamiltonian (11.1). We showed in Section 7.5 that the parity operator $\hat{\Pi}$ commutes with the kinetic-energy operator. The quantity $1/r_i$ in (11.1) is $r_i^{-1} = (x_i^2 + y_i^2 + z_i^2)^{-1/2}$. Replacement of each coordinate by its negative leaves $1/r_i$ unchanged. Also

$$r_{ij}^{-1} = [(x_i - x_j)^2 + (y_i - y_j)^2 + (z_i - z_j)^2]^{-1/2}$$

and inversion has no effect on $1/r_{ij}$. Thus $\hat{\Pi}$ commutes with the atomic Hamiltonian, and we can choose atomic wave functions to have definite parity.

For a one-electron atom, the spatial wave function is $\psi = R(r)Y_l^m(\theta, \phi)$. The radial function is unchanged on inversion, and the parity is determined by the angular factor. In Prob. 7.28 we showed that Y_l^m is an even function when l is even and is an odd function when l is odd. Thus the states of one-electron atoms have even or odd parity according to whether l is even or odd.

Now consider an n-electron atom. In the Hartree–Fock central-field approximation, we write the wave function as a Slater determinant (or linear combination of Slater determinants) of spin-orbitals. The wave function is the sum of terms, the spatial factor in each term having the form

$$R_1(r_1) \cdots R_n(r_n) Y_{l_1}^{m_1}(\theta_1, \phi_1) \cdots Y_{l_n}^{m_n}(\theta_n, \phi_n)$$

The parity of this product is determined by the spherical-harmonic factors. We see that the product is an even or odd function according to whether $l_1 + l_2 + \cdots + l_n$ is an even or odd number. Therefore, the parity of an atomic state is found by adding the l values of the electrons in the electron configuration that gives rise to the state: if $\sum_i l_i$ is an even number, then ψ is an even function; if $\sum_i l_i$ is odd, ψ is odd. For example, the configuration $1s^2 2s 2p^3$ has $\sum_i l_i = 0 + 0 + 0 + 1 + 1 + 1 = 3$, and all states arising from this electron configuration have odd parity. (Our argument was based on the SCF approximation to ψ, but the conclusions are valid for the true ψ.)

Total Electronic Angular Momentum and Atomic Levels. The **total electronic angular momentum J** of an atom is the vector sum of the total electronic orbital and spin angular momenta:

$$\boxed{\mathbf{J} = \mathbf{L} + \mathbf{S}} \qquad \textbf{(11.61)}$$

The operator $\hat{\mathbf{J}}$ for the total electronic angular momentum commutes with the atomic Hamiltonian, and we may characterize an atomic state by a quantum number J, which has the possible values [Eq. (11.39)]

$$L + S,\quad L + S - 1,\quad \ldots,\quad |L - S| \tag{11.62}$$

We have $\hat{J}^2\psi = J(J + 1)\hbar^2\psi$.

For the atomic Hamiltonian (11.1), all states that belong to the same term have the same energy. However, when spin–orbit interaction (Section 11.6) is included in $\hat{H}$, one finds that the value of the quantum number J affects the energy slightly. Hence states that belong to the same term but that have different values of J will have slightly different energies. The set of states that belong to the same term and that have the same value of J constitutes an atomic **level**. The energies of different levels belonging to the same term are slightly different. (See Fig. 11.6 in Section 11.7.) To denote the level, one adds the J value as a right subscript on the term symbol. Each level is $(2J + 1)$-fold degenerate, corresponding to the $2J + 1$ values of M_J, where $M_J\hbar$ is the z component of the total electronic angular momentum $\mathbf{J}$. Each level consists of $2J + 1$ **states** of equal energy.

EXAMPLE Find the levels of a ^3P term and give the degeneracy of each level.

For a ^3P term, $2S + 1 = 3$ and $S = 1$; also, from (11.42), $L = 1$. With $L = 1$ and $S = 1$, (11.62) gives $J = 2, 1, 0$. The levels are $^3P_2$, $^3P_1$, and $^3P_0$.

The $^3P_2$ level has $J = 2$ and has $2J + 1 = 5$ values of M_J, namely, $-2, -1, 0, 1$, and 2. The $^3P_2$ level is 5-fold degenerate. The $^3P_1$ level is $2(1) + 1 = 3$-fold degenerate. The $^3P_0$ level has $2J + 1 = 1$ and is nondegenerate.

The total number of states for the three levels $^3P_2$, $^3P_1$, and $^3P_0$ is $5 + 3 + 1$. When spin–orbit interaction was neglected, we had a ^3P term that consisted of $(2L + 1)(2S + 1) = 3(3) = 9$ equal-energy states. With spin–orbit interaction, the 9-fold-degenerate term splits into three closely spaced levels: $^3P_2$ with 5 states, $^3P_1$ with 3 states, and $^3P_0$ with one state. The total number of states is the same for the term ^3P as for the three levels arising from this term.

EXERCISE Find the levels of a ^2D term and give the level degeneracies. (*Answer:* $^2D_{5/2}$, $^2D_{3/2}$; 6, 4.)

The quantity $2S + 1$ is called the **electron-spin multiplicity** (or the *multiplicity*) of the term. If $L \geq S$, the possible values of J in (11.62) range from $L + S$ to $L - S$ and are $2S + 1$ in number. If $L \geq S$, the spin multiplicity gives the number of levels that arise from a given term. For $L < S$, the values of J range from $S + L$ to $S - L$ and are $2L + 1$ in number. In this case, the spin multiplicity is greater than the number of levels. For example, if $L = 0$ and $S = 1$ (a ^3S term), the spin multiplicity is 3, but there is only one possible value for J, namely, $J = 1$. For $2S + 1 = 1, 2, 3, 4, 5, 6, \ldots$, the words **singlet**, **doublet**, **triplet**, **quartet**, **quintet**, **sextet**, . . . , are used to designate the spin multiplicity. The level symbol $^3P_1$ is read as "triplet P one."

For light atoms, the spin–orbit interaction is very small and the separation between levels of a term is very small. Note from Fig. 11.6 in Section 11.7 that the separations between the $^3P_0$, $^3P_1$, and $^3P_2$ levels of the helium $1s2p\ ^3P$ term are far, far less than the separation between the ^1P and ^3P terms of the $1s2p$ configuration.

Terms and Levels of Hydrogen and Helium. The hydrogen atom has one electron. Hence $L = l$ and $S = s = \frac{1}{2}$. The possible values of J are $L + \frac{1}{2}$ and $L - \frac{1}{2}$, except for $L = 0$, where $J = \frac{1}{2}$ is the only possibility. Each electron configuration gives rise to only one term, which is composed of one level if $L = 0$ and two levels if $L \neq 0$. The ground-state configuration $1s$ gives the term ^2S, which is composed of the single level $^2S_{1/2}$; the level is twofold degenerate ($M_J = -\frac{1}{2}, \frac{1}{2}$). The $2s$ configuration also gives a $^2S_{1/2}$ level. The $2p$ configuration gives rise to the levels $^2P_{3/2}$ and $^2P_{1/2}$; the $^2P_{3/2}$ level is fourfold degenerate, and the $^2P_{1/2}$ level is twofold degenerate. There are $2 + 4 + 2 = 8$ states with $n = 2$, in agreement with our previous work.

The helium ground-state configuration $1s^2$ is a closed subshell and gives rise to the single level $^1S_0$, which is nondegenerate ($M_J = 0$). The $1s2s$ excited configuration gives rise to the two terms ^1S and ^3S, each of which has one level. The $^1S_0$ level is nondegenerate; the $^3S_1$ level is threefold degenerate. The $1s2p$ configuration gives rise to the terms ^1P and ^3P. The levels of ^3P are $^3P_2$, $^3P_1$, and $^3P_0$; ^1P has the single level $^1P_1$.

Tables of Atomic Energy Levels. The spectroscopically determined energy levels for atoms with atomic number less than 90 are given in the tables of C. E. Moore and others: C. E. Moore, *Atomic Energy Levels*, National Bureau of Standards Circular 467, vols. I, II, and III, 1949, 1952, and 1958, Washington D.C.; these have been reprinted as Natl. Bur. Stand. Publ. NSRDS-NBS 35, 1971; W. C. Martin et al., *Atomic Energy Levels—The Rare-Earth Elements,* Natl. Bur. Stand. Publ. NSRDS-NBS 60, Washington, D.C., 1978. Revisions of Moore's tables for many elements are at physics.nist.gov/PhysRefData/ASD/index.html.

These tables also list the levels of many atomic ions. Spectroscopists use the symbol I to indicate a neutral atom, the symbol II to indicate a singly ionized atom, and so on. Thus C III refers to the C^{2+} ion.

The tables take the zero level of energy at the lowest energy level of the atom and list the level energies E_i as E_i/hc in cm^{-1}, where h and c are Planck's constant and the speed of light. The difference in E/hc values for two levels gives the wavenumber [Eq. (4.66)] of the spectral transition between the levels (provided the transition is allowed). An energy E of 1 eV corresponds to $E/hc = 8065.54$ cm^{-1} (Prob. 11.27).

Figure 11.4 shows some of the term energies of the carbon atom. The separations between levels of each term are too small to be visible in this figure.

11.6 SPIN–ORBIT INTERACTION

The atomic Hamiltonian (11.1) does not involve electron spin. In reality, the existence of spin adds an additional term, usually small, to the Hamiltonian. This term, called the **spin–orbit interaction**, splits an atomic term into levels. Spin–orbit interaction is a relativistic effect and is properly derived using Dirac's relativistic treatment of the electron. This section gives a qualitative discussion of the origin of spin–orbit interaction.

If we imagine ourselves riding on an electron in an atom, from our viewpoint, the nucleus is moving around the electron (as the sun appears to move around the earth). This apparent motion of the nucleus produces a magnetic field that interacts with the intrinsic (spin) magnetic moment of the electron, giving the spin–orbit interaction term in the Hamiltonian. The interaction energy of a magnetic moment **m** with a magnetic

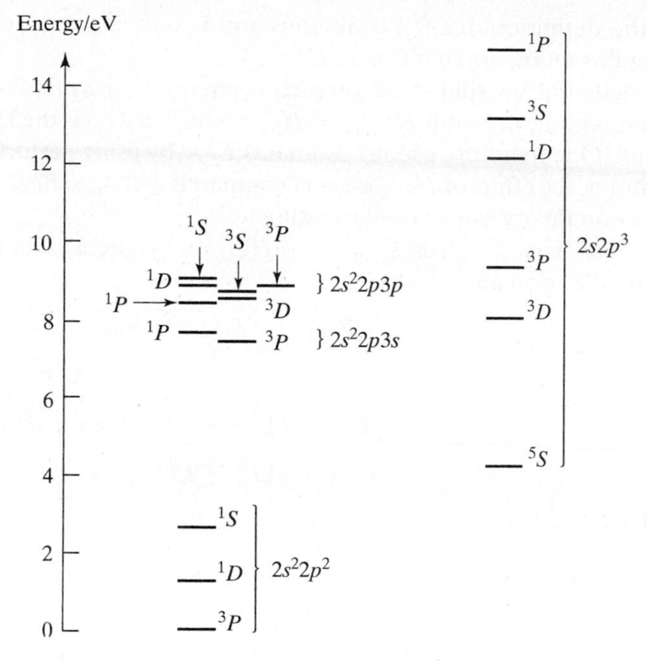

FIGURE 11.4 Some term energies of the carbon atom.

field **B** is given by (6.131) as $-\mathbf{m}\cdot\mathbf{B}$. The electron's spin magnetic moment $\mathbf{m}_S$ is proportional to its spin **S** [Eq. (10.57)], and the magnetic field arising from the apparent nuclear motion is proportional to the electron's orbital angular momentum **L**. Therefore, the spin–orbit interaction is proportional to $\mathbf{L}\cdot\mathbf{S}$. The dot product of **L** and **S** depends on the relative orientation of these two vectors. The total electronic angular momentum $\mathbf{J} = \mathbf{L} + \mathbf{S}$ also depends on the relative orientation of **L** and **S**, and so the spin–orbit interaction energy depends on J [Eq. (11.67)].

When a proper relativistic derivation of the spin–orbit-interaction term $\hat{H}_{\text{S.O.}}$ in the atomic Hamiltonian is carried out, one finds that for a one-electron atom (see *Bethe and Jackiw*, Chapters 8 and 23)

$$\hat{H}_{\text{S.O.}} = \frac{1}{2m_e^2c^2}\frac{1}{r}\frac{dV}{dr}\hat{\mathbf{L}}\cdot\hat{\mathbf{S}} \tag{11.63}$$

where V is the potential energy experienced by the electron in the atom and c is the speed of light. One way to get $\hat{H}_{\text{S.O.}}$ for a many-electron atom is to first neglect $\hat{H}_{\text{S.O.}}$ and carry out an SCF calculation (Section 11.1) using the central-field approximation to get an effective potential energy $V_i(r_i)$ for each electron i in the field of the nucleus and the other electrons viewed as charge clouds [Eqs. (11.7) and (11.8)]. One then sums (11.63) over the electrons to get

$$\hat{H}_{\text{S.O.}} \approx \frac{1}{2m_e^2c^2}\sum_i \frac{1}{r_i}\frac{dV_i(r_i)}{dr_i}\hat{\mathbf{L}}_i\cdot\hat{\mathbf{S}}_i = \sum_i \xi_i(r_i)\hat{\mathbf{L}}_i\cdot\hat{\mathbf{S}}_i \tag{11.64}$$

where the definition of $\xi_i(r_i)$ is obvious and $\hat{\mathbf{L}}_i$ and $\hat{\mathbf{S}}_i$ are the operators for orbital and spin angular momenta of electron i.

Calculating the spin–orbit interaction energy $E_{\text{S.O.}}$ by finding the eigenfunctions and eigenvalues of the operator $\hat{H}_{(11.1)} + \hat{H}_{\text{S.O.}}$, where $\hat{H}_{(11.1)}$ is the Hamiltonian of Eq. (11.1), is difficult. One therefore usually estimates $E_{\text{S.O.}}$ by using perturbation theory. Except for heavy atoms, the effect of $\hat{H}_{\text{S.O.}}$ is small compared with the effect of $\hat{H}_{(11.1)}$, and first-order perturbation theory can be used to estimate $E_{\text{S.O.}}$.

Equation (9.22) gives $E_{\text{S.O.}} \approx \langle\psi|\hat{H}_{\text{S.O.}}|\psi\rangle$, where ψ is an eigenfunction of $\hat{H}_{(11.1)}$. For a one-electron atom,

$$E_{\text{S.O.}} \approx \langle\psi|\xi(r)\hat{\mathbf{L}}\cdot\hat{\mathbf{S}}|\psi\rangle \tag{11.65}$$

We have

$$\mathbf{J}\cdot\mathbf{J} = (\mathbf{L} + \mathbf{S})\cdot(\mathbf{L} + \mathbf{S}) = L^2 + S^2 + 2\mathbf{L}\cdot\mathbf{S}$$

$$\mathbf{L}\cdot\mathbf{S} = \tfrac{1}{2}(J^2 - L^2 - S^2)$$

$$\hat{\mathbf{L}}\cdot\hat{\mathbf{S}}\psi = \tfrac{1}{2}(\hat{J}^2 - \hat{L}^2 - \hat{S}^2)\psi = \tfrac{1}{2}[J(J + 1) - L(L + 1) - S(S + 1)]\hbar^2\psi$$

since the unperturbed ψ is an eigenfunction of $\hat{L}^2$, $\hat{S}^2$, and $\hat{J}^2$. Therefore,

$$E_{\text{S.O.}} \approx \tfrac{1}{2}\langle\xi\rangle\hbar^2[J(J + 1) - L(L + 1) - S(S + 1)] \tag{11.66}$$

For a many-electron atom, it can be shown (*Bethe and Jackiw*, p. 164) that the spin–orbit interaction energy is

$$E_{\text{S.O.}} \approx \tfrac{1}{2}A\hbar^2[J(J + 1) - L(L + 1) - S(S + 1)] \tag{11.67}$$

where A is a constant for a given term; that is, A depends on L and S but not on J. Equation (11.67) shows that when we include the spin–orbit interaction, the energy of an atomic state depends on its total electronic angular momentum J. Thus each atomic term is split into levels, each level having a different value of J. For example, the $1s^22s^22p^63p$ configuration of sodium has the single term ^2P, which is composed of the two levels $^2P_{3/2}$ and $^2P_{1/2}$. The splitting of these levels gives the observed *fine structure* of the sodium D line (Fig. 11.5). The levels of a given term are said to form its *multiplet structure*.

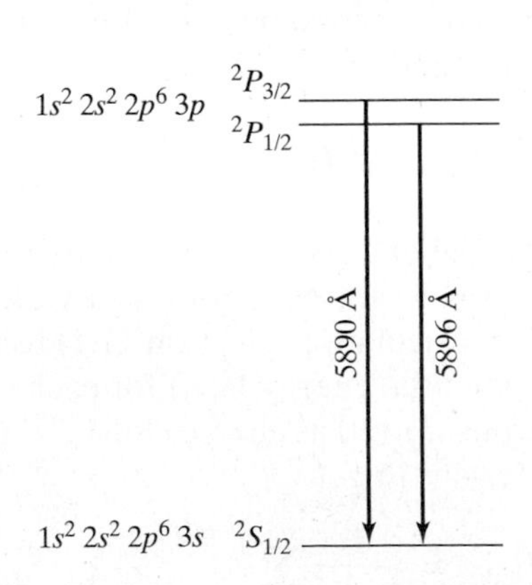

FIGURE 11.5 Fine structure of the sodium D line.

What about the order of levels within a given term? Since L and S are the same for such levels, their relative energies are determined, according to Eq. (11.67), by $AJ(J + 1)$. If A is positive, the level with the lowest value of J lies lowest, and the multiplet is said to be **regular**. If A is negative, the level with the highest value of J lies lowest, and the multiplet is said to be **inverted**. The following rule usually applies to a configuration with only one partly filled subshell: *If this subshell is less than half filled, the multiplet is regular; if this subshell is more than half filled, the multiplet is inverted.* (A few exceptions exist.) For the half-filled case, see Prob. 11.26.

EXAMPLE Find the ground level of the oxygen atom.

The ground electron configuration is $1s^2 2s^2 2p^4$. Table 11.2 gives ^1S, ^1D, and ^3P as the terms of this configuration. By Hund's rule, ^3P is the lowest term. [Alternatively, a diagram like (11.48) could be used to conclude that ^3P is the lowest term.] The ^3P term has $L = 1$ and $S = 1$, so the possible J values are 2, 1, and 0. The levels of ^3P are $^3P_2$, $^3P_1$, and $^3P_0$. The $2p$ subshell is more than half filled, so the rule just given predicts the multiplet is inverted, and the $^3P_2$ level lies lowest. This is the ground level of O.

EXERCISE Find the ground level of the Cl atm. (*Answer:* $^2P_{3/2}$.)

11.7 THE ATOMIC HAMILTONIAN

The Hamiltonian operator of an atom can be divided into three parts:

$$\hat{H} = \hat{H}^0 + \hat{H}_{\text{rep}} + \hat{H}_{\text{S.O.}} \tag{11.68}$$

where $\hat{H}^0$ is the sum of hydrogenlike Hamiltonians,

$$\hat{H}^0 = \sum_{i=1}^{n} \left(-\frac{\hbar^2}{2m_e}\nabla_i^2 - \frac{Ze'^2}{r_i} \right) \tag{11.69}$$

$\hat{H}_{\text{rep}}$ consists of the interelectronic repulsions,

$$\hat{H}_{\text{rep}} = \sum_{i} \sum_{j>i} \frac{e'^2}{r_{ij}} \tag{11.70}$$

and $\hat{H}_{\text{S.O.}}$ is the spin–orbit interaction (11.64):

$$\hat{H}_{\text{S.O.}} = \sum_{i=1}^{n} \xi_i \hat{\mathbf{L}}_i \cdot \hat{\mathbf{S}}_i \tag{11.71}$$

If we consider just $\hat{H}^0$, all atomic states corresponding to the same electronic *configuration* are degenerate. Adding in $\hat{H}_{\text{rep}}$, we lift the degeneracy between states with different L or S or both, thus splitting each configuration into *terms*. Next, we add in $\hat{H}_{\text{S.O.}}$, which splits each term into *levels*. Each level is composed of *states* with the same value of J and is $(2J + 1)$-fold degenerate, corresponding to the possible values of M_J.

We can remove the degeneracy of each level by applying an external magnetic field (the *Zeeman effect*). If $\mathbf{B}$ is the applied field, we have the additional term in the Hamiltonian [Eq. (6.131)]

$$\hat{H}_B = -\hat{\mathbf{m}} \cdot \mathbf{B} = -(\hat{\mathbf{m}}_L + \hat{\mathbf{m}}_S) \cdot \mathbf{B} \tag{11.72}$$

where both the orbital and spin magnetic moments have been included. Using $\mathbf{m}_L = -(e/2m_e)\mathbf{L}$, $\mathbf{m}_S = -(e/m_e)\mathbf{S}$, and $\beta_e \equiv e\hbar/2m_e$ [Eqs. (6.128), (10.55) with $g_e \approx 2$, and (6.130)], we have

$$\hat{H}_B = \beta_e\hbar^{-1}(\hat{\mathbf{L}} + 2\hat{\mathbf{S}})\cdot\mathbf{B} = \beta_e\hbar^{-1}(\hat{\mathbf{J}} + \hat{\mathbf{S}})\cdot\mathbf{B} = \beta_e B\hbar^{-1}(\hat{J}_z + \hat{S}_z) \qquad (11.73)$$

where β_e is the Bohr magneton and the z axis is taken along the direction of the field. If the external field is reasonably weak, its effect will be less than that of the spin–orbit interaction, and the effect of the field can be calculated by use of first-order perturbation theory. Since $\hat{J}_z\psi = M_J\hbar\psi$, the energy of interaction with the applied field is

$$E_B = \langle\psi|\hat{H}_B|\psi\rangle = \beta_e BM_J + \beta_e B\hbar^{-1}\langle S_z\rangle$$

Evaluation of $\langle S_z\rangle$ (*Bethe and Jackiw*, p. 169) gives as the final weak-field result:

$$E_B = \beta_e gBM_J \qquad (11.74)$$

where g (the *Landé g factor*) is given by

$$g = 1 + \frac{[J(J+1) - L(L+1) + S(S+1)]}{2J(J+1)} \qquad (11.75)$$

Thus the external field splits each level into $2J + 1$ states, each state having a different value of M_J.

Figure 11.6 shows what happens when we consider successive interactions in an atom, using the $1s2p$ configuration of helium as the example.

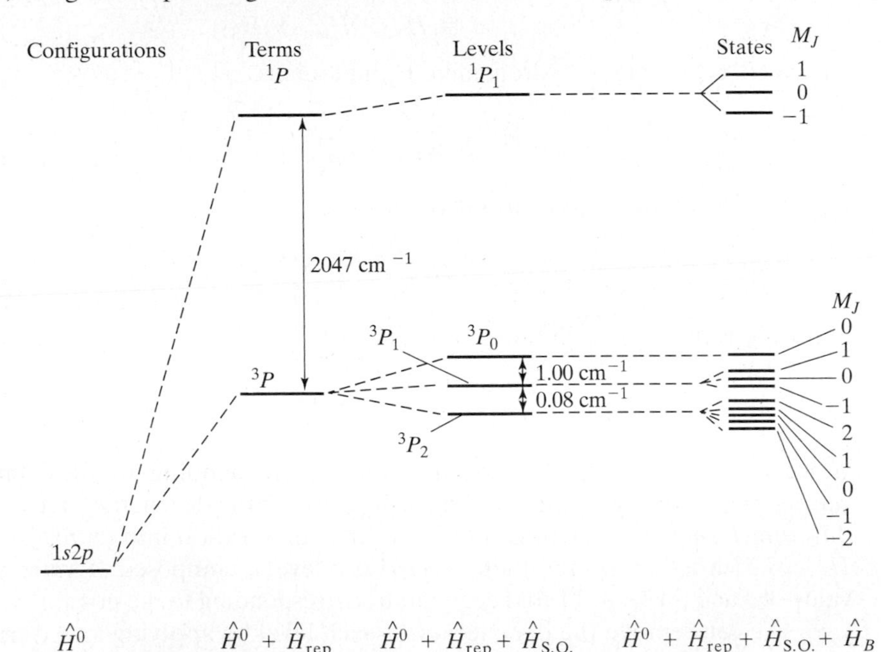

FIGURE 11.6 Effect of inclusion of successive terms in the atomic Hamiltonian for the $1s2p$ helium configuration. $\hat{H}_B$ is not part of the atomic Hamiltonian but is due to an applied magnetic field.

We have based the discussion on a scheme in which we first added the individual electronic orbital angular momenta to form a total-orbital-angular-momentum vector and did the same for the spins: $\mathbf{L} = \sum_i \mathbf{L}_i$ and $\mathbf{S} = \sum_i \mathbf{S}_i$. We then combined **L** and **S** to get **J**. This scheme is called **Russell–Saunders coupling** (or *L–S coupling*) and is appropriate where the spin–orbit interaction energy is small compared with the interelectronic repulsion energy. The operators $\hat{\mathbf{L}}$ and $\hat{\mathbf{S}}$ commute with $\hat{H}^0 + \hat{H}_{\text{rep}}$, but when $\hat{H}_{\text{S.O.}}$ is included in the Hamiltonian, $\hat{\mathbf{L}}$ and $\hat{\mathbf{S}}$ no longer commute with $\hat{H}$. ($\hat{\mathbf{J}}$ does commute with $\hat{H}^0 + \hat{H}_{\text{rep}} + \hat{H}_{\text{S.O.}}$.) If the spin–orbit interaction is small, then $\hat{\mathbf{L}}$ and $\hat{\mathbf{S}}$ "almost" commute with $\hat{H}$, and L–S coupling is valid.

As the atomic number increases, the average speed v of the electrons increases. As v/c increases, relativistic effects such as the spin–orbit interaction increase. For atoms with very high atomic number, the spin–orbit interaction exceeds the interelectronic repulsion energy, and we can no longer consider $\hat{\mathbf{L}}$ and $\hat{\mathbf{S}}$ to commute with $\hat{H}$; the operator $\hat{\mathbf{J}}$, however, still commutes with $\hat{H}$. In this case we first add in $\hat{H}_{\text{S.O.}}$ to $\hat{H}^0$ and then consider $\hat{H}_{\text{rep}}$. This corresponds to first combining the spin and orbital angular momenta of each electron to give a total angular momentum $\mathbf{j}_i$ for each electron: $\mathbf{j}_i = \mathbf{L}_i + \mathbf{S}_i$. We then add the $\mathbf{j}_i$'s to get the total electronic angular momentum: $\mathbf{J} = \sum_i \mathbf{j}_i$. This scheme is called *j–j coupling*. For most heavy atoms, the situation is intermediate between j–j and L–S coupling, and computations are difficult.

Several other effects should be included in the atomic Hamiltonian. The finite size of the nucleus and the effect of nuclear motion slightly change the energy (*Bethe and Salpeter*, pp. 102, 166, 351). There is a small relativistic term due to the interaction between the spin magnetic moments of the electrons (*spin–spin interaction*). We should also take into account the relativistic change of electronic mass with velocity. This effect is significant for inner-shell electrons of heavy atoms, where average electronic speeds are not negligible in comparison with the speed of light.

If the nucleus has a nonzero spin, the nuclear spin magnetic moment interacts with the electronic spin and orbital magnetic moments, giving rise to atomic *hyperfine structure*. The *nuclear spin angular momentum* **I** adds vectorially to the total electronic angular momentum **J** to give the *total angular momentum* **F** of the atom: $\mathbf{F} = \mathbf{I} + \mathbf{J}$. For example, consider the ground state of the hydrogen atom. The spin of a proton is $\frac{1}{2}$, so $I = \frac{1}{2}$; also, $J = \frac{1}{2}$. Hence the quantum number F can be 0 or 1, corresponding to the proton and electron spins being antiparallel or parallel. The transition $F = 1 \rightarrow 0$ gives a line at 1420 MHz, the 21-cm line emitted by hydrogen atoms in outer space. In 1951, Ewen and Purcell stuck a horn-shaped antenna out the window of a Harvard physics laboratory and detected this line. The frequency of the hyperfine splitting in the ground state of hydrogen is one of the most accurately measured physical constants: 1420.405751767 ± 0.000000001 MHz [L. Essen et al., *Nature*, **229**, 110 (1971)].

11.8 THE CONDON–SLATER RULES

In the Hartree–Fock approximation, the wave function of an atom (or molecule) is a Slater determinant or a linear combination of a few Slater determinants [for example, Eq. (10.44)]. A configuration-interaction wave function such as (11.17) is a linear combination of many Slater determinants. To evaluate the energy and other properties of atoms and molecules using Hartree–Fock or configuration-interaction wave functions, we must be able to evaluate integrals of the form $\langle D'|\hat{B}|D\rangle$, where D and D' are Slater determinants of orthonormal spin-orbitals and $\hat{B}$ is an operator.

Each spin-orbital u_i is a product of a spatial orbital θ_i and a spin function σ_i, where σ_i is either α or β. We have $u_i = \theta_i\sigma_i$ and $\langle u_i(1)|u_j(1)\rangle = \delta_{ij}$, where $\langle u_i(1)|u_j(1)\rangle$ involves a sum over the spin coordinate of electron 1 and an integration over its spatial coordinates. If u_i and u_j have different spin functions, then (10.12) ensures the orthogonality of u_i and u_j. If u_i and u_j have the same spin function, their orthogonality is due to the orthogonality of the spatial orbitals θ_i and θ_j.

For an n-electron system, D is

$$D = \frac{1}{\sqrt{n!}}\begin{vmatrix} u_1(1) & u_2(1) & \cdots & u_n(1) \\ u_1(2) & u_2(2) & \cdots & u_n(2) \\ \vdots & \vdots & \ddots & \vdots \\ u_1(n) & u_2(n) & \cdots & u_n(n) \end{vmatrix} \tag{11.76}$$

An example with $n = 3$ is Eq. (10.40). D' has the same form as D except that $u_1, u_2, \ldots, u_n$ are replaced by $u_1', u_2', \ldots, u_n'$.

We shall assume that the columns of D and D' are arranged so as to have as many as possible of their left-hand columns match. For example, if we were working with the Slater determinants $|1s\overline{1s}2s3p_0|$ and $|1s\overline{1s}3p_04s|$, we would interchange the third and fourth columns of the first determinant (thereby multiplying it by -1) and let $D = |1s\overline{1s}3p_02s|$ and $D' = |1s\overline{1s}3p_04s|$.

The operator $\hat{B}$ typically has the form

$$\hat{B} = \sum_{i=1}^{n}\hat{f}_i + \sum_{i=1}^{n-1}\sum_{j>i}\hat{g}_{ij} \tag{11.77}$$

where the *one-electron operator* $\hat{f}_i$ involves only coordinate and momentum operators of electron i and the *two-electron operator* $\hat{g}_{ij}$ involves electrons i and j. For example, if $\hat{B}$ is the atomic Hamiltonian operator (11.1), then $\hat{f}_i = -(\hbar^2/2m_e)\nabla_i^2 - Ze'^2/r_i$ and $\hat{g}_{ij} = e'^2/r_{ij}$.

Condon and Slater showed that the n-electron integral $\langle D'|\hat{B}|D\rangle$ can be reduced to sums of certain one- and two-electron integrals. The derivation of these Condon–Slater formulas uses the determinant expression of Prob. 8.21 together with the orthonormality of the spin-orbitals. (See *Parr*, pp. 23–27 for the derivation.) Table 11.3 gives the Condon–Slater formulas.

In Table 11.3, each matrix element of $\hat{g}_{12}$ involves summation over the spin coordinates of electrons 1 and 2 and integration over the full range of the spatial coordinates of electrons 1 and 2. Each matrix element of $\hat{f}_1$ involves summation over the spin coordinate of electron 1 and integration over its spatial coordinates. The variables in the sums and definite integrals are dummy variables.

If the operators $\hat{f}_i$ and $\hat{g}_{ij}$ do not involve spin, the expressions in Table 11.3 can be further simplified. We have $u_i = \theta_i\sigma_i$ and

$$\begin{aligned}\langle u_i(1)|\hat{f}_1|u_i(1)\rangle &= \int \theta_i^*(1)\hat{f}_1\theta_i(1)\,dv_1 \sum_{m_{s1}}\sigma_i^*(1)\sigma_i(1) \\ &= \int \theta_i^*(1)\hat{f}_1\theta_i(1)\,dv_1 = \langle\theta_i(1)|\hat{f}_1|\theta_i(1)\rangle\end{aligned}$$

TABLE 11.3 The Condon–Slater Rules

D and D' differ by	$\left\langle D' \middle\| \sum_{i=1}^{n} \hat{f}_i \middle\| D \right\rangle$	$\left\langle D' \middle\| \sum_{i=1}^{n-1} \sum_{j>i} \hat{g}_{ij} \middle\| D \right\rangle$
no spin-orbitals	$\sum_{i=1}^{n} \langle u_i(1)\|\hat{f}_1\|u_i(1)\rangle$	$\sum_{i=1}^{n-1}\sum_{j>i} [\langle u_i(1)u_j(2)\|\hat{g}_{12}\|u_i(1)u_j(2)\rangle$ $- \langle u_i(1)u_j(2)\|\hat{g}_{12}\|u_j(1)u_i(2)\rangle]$
one spin-orbital $u'_n \neq u_n$	$\langle u'_n(1)\|\hat{f}_1\|u_n(1)\rangle$	$\sum_{j=1}^{n-1} [\langle u'_n(1)u_j(2)\|\hat{g}_{12}\|u_n(1)u_j(2)\rangle$ $- \langle u'_n(1)u_j(2)\|\hat{g}_{12}\|u_j(1)u_n(2)\rangle]$
two spin-orbitals $u'_n \neq u_n,\ u'_{n-1} \neq u_{n-1}$	0	$\langle u'_n(1)u'_{n-1}(2)\|\hat{g}_{12}\|u_n(1)u_{n-1}(2)\rangle$ $- \langle u'_n(1)u'_{n-1}(2)\|\hat{g}_{12}\|u_{n-1}(1)u_n(2)\rangle$
three or more spin-orbitals	0	0

since σ_i is normalized. Using this result and the orthonormality of σ_i and σ_j, we get for the case $D = D'$ (Prob. 11.34)

$$\left\langle D \middle| \sum_{i=1}^{n} \hat{f}_i \middle| D \right\rangle = \sum_{i=1}^{n} \langle \theta_i(1)|\hat{f}_1|\theta_i(1)\rangle \tag{11.78}$$

$$\left\langle D \middle| \sum_{i=1}^{n-1} \sum_{j>i} \hat{g}_{ij} \middle| D \right\rangle = \sum_{i=1}^{n-1}\sum_{j>i} [\langle \theta_i(1)\theta_j(2)|\hat{g}_{12}|\theta_i(1)\theta_j(2)\rangle - \delta_{m_{s,i}m_{s,j}} \langle \theta_i(1)\theta_j(2)|\hat{g}_{12}|\theta_j(1)\theta_i(2)\rangle] \tag{11.79}$$

where $\delta_{m_{s,i}m_{s,j}}$ is 0 or 1 according to whether $m_{s,i} \neq m_{s,j}$ or $m_{s,i} = m_{s,j}$. Similar equations hold for the other integrals.

Let us apply these equations to evaluate $\langle D|\hat{H}|D\rangle$, where $\hat{H}$ is the Hamiltonian of an n-electron atom with spin–orbit interaction neglected and D is a Slater determinant of n spin-orbitals. We have $\hat{H} = \sum_i \hat{f}_i + \sum_i \sum_{j>i} \hat{g}_{ij}$, where $\hat{f}_i = -(\hbar^2/2m_e)\nabla_i^2 - Ze'^2/r_i$ and $\hat{g}_{ij} = e'^2/r_{ij}$. Introducing the Coulomb and exchange integrals of Eq. (9.100) and using (11.78) and (11.79), we have

$$\langle D|\hat{H}|D\rangle = \sum_{i=1}^{n} \langle \theta_i(1)|\hat{f}_1|\theta_i(1)\rangle + \sum_{i=1}^{n-1}\sum_{j>i} (J_{ij} - \delta_{m_{s,i}m_{s,j}} K_{ij}) \tag{11.80}$$

$$J_{ij} = \langle \theta_i(1)\theta_j(2)|e'^2/r_{12}|\theta_i(1)\theta_j(2)\rangle, \qquad K_{ij} = \langle \theta_i(1)\theta_j(2)|e'^2/r_{12}|\theta_j(1)\theta_i(2)\rangle \tag{11.81}$$

$$\hat{f}_1 = -(\hbar^2/2m_e)\nabla_1^2 - Ze'^2/r_1 \tag{11.82}$$

The Kronecker delta in (11.80) results from the orthonormality of the one-electron spin functions.

As an example, consider Li. The SCF approximation to the ground-state ψ is the Slater determinant $D = |1s\overline{1s}2s|$. The spin-orbitals are $u_1 = 1s\alpha$, $u_2 = 1s\beta$, and $u_3 = 2s\alpha$. The spatial orbitals are $\theta_1 = 1s$, $\theta_2 = 1s$, and $\theta_3 = 2s$. We have $J_{12} = J_{1s1s}$

and $J_{13} = J_{23} = J_{1s2s}$. Since $m_{s1} \neq m_{s2}$ and $m_{s2} \neq m_{s3}$, the only exchange integral that appears in the energy expression is $K_{13} = K_{1s2s}$. We get exchange integrals only between spin-orbitals with the same spin. Equation (11.80) gives the SCF energy as

$$E = \langle D|\hat{H}|D\rangle = 2\langle 1s(1)|\hat{f}_1|1s(1)\rangle + \langle 2s(1)|\hat{f}_1|2s(1)\rangle + J_{1s1s} + 2J_{1s2s} - K_{1s2s}$$

The terms involving $\hat{f}_1$ are hydrogenlike energies, and their sum equals $E^{(0)}$ in Eq. (10.49). The remaining terms equal $E^{(1)}$ in Eq. (10.51). As noted at the beginning of Section 9.4, $E^{(0)} + E^{(1)}$ equals the variational integral $\langle D|\hat{H}|D\rangle$, so the Condon–Slater rules have been checked in this case.

For an atom with closed subshells only (for example, ground-state Be with a $1s^2 2s^2$ configuration), the n electrons reside in $n/2$ different spatial orbitals, so $\theta_1 = \theta_2$, $\theta_3 = \theta_4$, and so on. Let $\phi_1 \equiv \theta_1 = \theta_2$, $\phi_2 \equiv \theta_3 = \theta_4, \ldots, \phi_{n/2} \equiv \theta_{n-1} = \theta_n$. If one rewrites (11.80) using the ϕ's instead of the θ's, one finds (Prob. 11.35) for the SCF energy of the ^1S term produced by a closed-subshell configuration

$$E = \langle D|\hat{H}|D\rangle = 2\sum_{i=1}^{n/2} \langle \phi_i(1)|\hat{f}_1|\phi_i(1)\rangle + \sum_{j=1}^{n/2}\sum_{i=1}^{n/2}(2J_{ij} - K_{ij}) \tag{11.83}$$

where $\hat{f}_1$ is given by (11.82) and where J_{ij} and K_{ij} have the forms in (11.81) but with θ_i and θ_j replaced by ϕ_i and ϕ_j. Each sum in (11.83) goes over all the $n/2$ different spatial orbitals.

For example, consider the $1s^2 2s^2$ electron configuration. We have $n = 4$ and the two different spatial orbitals are $1s$ and $2s$. The double sum in Eq. (11.83) is equal to $2J_{1s1s} - K_{1s1s} + 2J_{1s2s} - K_{1s2s} + 2J_{2s1s} - K_{2s1s} + 2J_{2s2s} - K_{2s2s}$. From the definition (11.81), it follows that $J_{ii} = K_{ii}$. The labels 1 and 2 in (11.81) are dummy variables, and interchanging them can have no effect on the integrals. Interchanging 1 and 2 in J_{ij} converts it to J_{ji}; therefore, $J_{ij} = J_{ji}$. The same reasoning gives $K_{ij} = K_{ji}$. Thus

$$J_{ii} = K_{ii}, \qquad J_{ij} = J_{ji}, \qquad K_{ij} = K_{ji} \tag{11.84}$$

Use of (11.84) gives the Coulomb- and exchange-integrals expression for the $1s^2 2s^2$ configuration as $J_{1s1s} + J_{2s2s} + 4J_{1s2s} - 2K_{1s2s}$. Between the two electrons in the $1s$ orbital, there is only one Coulombic interaction, and we get the term J_{1s1s}. Each $1s$ electron interacts with two $2s$ electrons, for a total of four $1s$–$2s$ interactions, and we get the term $4J_{1s2s}$. As noted earlier, exchange integrals occur only between spin-orbitals of the same spin. There is an exchange integral between the $1s\alpha$ and $2s\alpha$ spin-orbitals and an exchange integral between the $1s\beta$ and $2s\beta$ spin-orbitals, which gives the $-2K_{1s2s}$ term.

The magnitude of the exchange integrals is generally much less than the magnitude of the Coulomb integrals [for example, see Eq. (9.111)].

11.9 SUMMARY

The Hartree SCF method approximates the atomic wave function as a product of one-electron spatial orbitals [Eq. (11.2)] and finds the best possible forms for the orbitals by an iterative calculation in which each electron is assumed to move in the field produced by the nucleus and a hypothetical charge cloud due to the other electrons.

The more accurate Hartree–Fock method approximates the wave function as an antisymmetrized product (Slater determinant or determinants) of one-electron spin-orbitals and finds the best possible forms for the spatial orbitals in the spin-orbitals. Hartree–Fock calculations are usually done by expanding each orbital as a linear combination of basis functions and iteratively solving the Hartree–Fock equations (11.12). The Slater-type orbitals (11.14) are often used as the basis functions in atomic calculations. The difference between the exact nonrelativistic energy and the Hartree–Fock energy is the correlation energy of the atom (or molecule).

To go beyond the Hartree–Fock approximation and approach the true wave function and energy, we can use configuration interaction (CI), expressing ψ as a linear combination of configuration functions corresponding to various electron configurations [Eq. (11.17)].

Let $\mathbf{M}_1$ and $\mathbf{M}_2$ be two angular momenta with quantum numbers j_1, m_1 and j_2, m_2, and let $\mathbf{M}$ be their sum: $\mathbf{M} = \mathbf{M}_1 + \mathbf{M}_2$. We showed that, for the angular-momentum sum, $\hat{M}^2$ has eigenvalues $J(J + 1)\hbar^2$ and $\hat{M}_z$ has eigenvalues $M_J\hbar$, where the possible values of J and M_J are $J = j_1 + j_2, j_1 + j_2 - 1, \ldots, |j_1 - j_2|$ and $M_J = J, J - 1, \ldots, -J$.

For a many-electron atom where spin–orbit interaction is small, the individual electronic orbital angular momenta add to give a total electronic orbital angular momentum ($\mathbf{L} = \sum_i \mathbf{L}_i$), and similarly for the spin angular momenta ($\mathbf{S} = \sum_i \mathbf{S}_i$). The eigenvalues of the operators $\hat{L}^2$ and $\hat{S}^2$ for the squares of the magnitudes of the total electronic orbital and spin angular momenta are $L(L + 1)\hbar^2$ and $S(S + 1)\hbar^2$, respectively. For nonequivalent electrons (electrons in different subshells), the possible L values arising from a given electron configuration are readily found by using the angular-momentum addition rule (11.39), ignoring any closed subshells. For equivalent electrons, the possible L values can be found by consulting Table 11.2a. The possible S values are found using (11.39), ignoring all electrons that are paired.

With spin–orbit interaction neglected, atomic states that have the same energy have the same L value and the same S value. A set of equal-energy states with the same L and the same S constitutes a term. The term is denoted by the term symbol $^{2S+1}(L)$, where $2S + 1$ is called the spin multiplicity and (L) is a code letter that gives the L value [see (11.42)]. The degeneracy of an atomic term is $(2L + 1)(2S + 1)$.

When spin–orbit interaction is included, each term is split into a number of levels, each level having a different value of the total-electronic angular-momentum quantum number J. The total electronic angular momentum is $\mathbf{J} = \mathbf{L} + \mathbf{S}$, and the eigenvalues of $\hat{J}^2$ are $J(J + 1)\hbar^2$, where J ranges from $L + S$ to $|L - S|$ in integral steps. The symbol for a level is $^{2S+1}(L)_J$. Each level is $(2J + 1)$-fold degenerate, corresponding to the $2J + 1$ values of M_J, which range from J to $-J$. The degeneracy of an atomic level can be removed by application of an external magnetic field, which splits each level into $2J + 1$ states of slightly different energies.

In summary,

$$\text{Configurations} \xrightarrow[\text{repulsions}]{\text{interelectronic}} \text{Terms} \xrightarrow[\text{interaction}]{\text{spin-orbit}} \text{Levels} \xrightarrow[\text{field}]{\text{external}} \text{States}$$

The Condon–Slater rules give the values of one- and two-electron integrals involving Slater determinants and can be used to evaluate properties (such as the energy) for a wave function that is a linear combination of Slater determinants.

PROBLEMS

Sec.	11.1	11.2	11.4	11.5	11.6	11.7
Probs.	11.1–11.3	11.4–11.5	11.6–11.10	11.11–11.29	11.30–11.32	11.33

Sec.	11.8	general
Probs.	11.34–11.36	11.37–11.39

11.1 How many electrons can be put in each of the following: (a) a shell with principal quantum number n; (b) a subshell with quantum numbers n and l; (c) an orbital; (d) a spin-orbital?

11.2 If $R(r_1)$ is the radial factor in the function t_1 in the Hartree differential equation (11.9), write the differential equation satisfied by R.

11.3 Which STOs have the same form as hydrogenlike AOs?

11.4 Estimate the nonrelativistic $1s$ orbital energy in Ar. Check with Fig. 11.2.

11.5 At what atomic number does the second crossing of the $3d$ and $4s$ orbital energies occur in Fig. 11.2? Take account of the logarithmic scale. (Atomic spectral data show that the crossing actually occurs between 20 and 21.)

11.6 Give the possible values of the total-angular-momentum quantum number J that result from the addition of angular momenta with quantum numbers (a) $\frac{3}{2}$ and 4; (b) 2, 3, and $\frac{1}{2}$.

11.7 True or false? The angular-momentum addition rule (11.39) shows that the number of values of J obtained by adding j_1 and j_2 is always $2j_< + 1$, where $j_<$ is the smaller of j_1 and j_2 or is j_1 if $j_1 = j_2$.

11.8 Verify the angular-momentum commutation relations (11.32).

11.9 Prove that $m_1 + m_2 = M_J$ [Eq. (11.34)], as follows: Apply $\hat{M}_z = \hat{M}_{1z} + \hat{M}_{2z}$ to (11.33), substitute (11.33) in the resulting equation, combine terms, and use the linear independence of the functions involved to deduce (11.34).

11.10 Show that $[\hat{M}^2, \hat{M}_{1z}] = 2i\hbar(\hat{M}_{1x}\hat{M}_{2y} - \hat{M}_{1y}\hat{M}_{2x})$, where $\mathbf{M} = \mathbf{M}_1 + \mathbf{M}_2$.

11.11 True or false? (Answer without looking at Chapter 11 tables.) (a) The terms arising from the $1s^22s^22p3p$ configuration are the same as the terms arising from $1s^22s^22p^2$. (b) The terms arising from a $3d^3$ configuration are the same as those from a $3d^7$ configuration.

11.12 Verify the terms in Table 11.2b.

11.13 Find the terms that arise from each of the following electron configurations: (a) $1s^22s^22p^63s^23p5g$; (b) $1s^22s^22p3p3d$; (c) $1s^22s^22p^44d$. You may use Table 11.2a for part (c).

11.14 Which of the following electron configurations will contribute configuration functions to a CI calculation of the ground state of He? (a) $1s2s$; (b) $1s2p$; (c) $2s^2$; (d) $2s2p$; (e) $2p^2$; (f) $3d^2$.

11.15 Verify the spin-eigenfunction equations (11.52) to (11.54), (11.56), and the three equations following (11.56).

11.16 (a) Calculate the angle in Fig. 11.3 between the z axis and $\mathbf{S}$ for the spin function $\alpha(1)\alpha(2)$. (b) Calculate the angle between $\mathbf{S}_1$ and $\mathbf{S}_2$ for each of the functions (11.57) to (11.60). [*Hint*: Use $\mathbf{S}\cdot\mathbf{S} = (\mathbf{S}_1 + \mathbf{S}_2)\cdot(\mathbf{S}_1 + \mathbf{S}_2)$.] (c) If a vector $\mathbf{A}$ has components (A_x, A_y, A_z), what are the components of the projection of $\mathbf{A}$ in the xy plane? Use the answer to this question to find the angle between the projections of $\mathbf{S}_1$ and $\mathbf{S}_2$ in the xy plane for the function $\alpha(1)\alpha(2)$.

11.17 (a) If $\hat{S}^2 = (\hat{\mathbf{S}}_1 + \hat{\mathbf{S}}_2 + \cdots)\cdot(\hat{\mathbf{S}}_1 + \hat{\mathbf{S}}_2 + \cdots)$, show that $[\hat{S}^2, \hat{P}_{ik}] = 0$, where $\hat{P}_{ik}$ is the exchange operator. (b) Show that $[\hat{L}^2, \hat{P}_{ik}] = 0$, where $\mathbf{L}$ is the total electronic orbital angular momentum.

11.18 Of the atoms with $Z \leq 10$, which have ground states of odd parity?

11.19 Give the number of states that belong to each of the following terms: (a) ^4F; (b) ^1S; (c) ^3P; (d) ^2D.

11.20 How many states belong to each of the following carbon configurations? (a) $1s^22s^22p^2$; (b) $1s^22s^22p3p$.

11.21 Give the possible spin multiplicities of the terms that arise from each of the following electron configurations: (a) f; (b) f^2; (c) f^3; (d) f^7; (e) f^{12}; (f) f^{13}.

11.22 Give the levels that arise from each of the following terms, and give the degeneracy of each level: (a) ^1S; (b) ^2S; (c) ^3F; (d) ^4D.

11.23 For a state belonging to a $^3D_3$ level, give the magnitude of (a) the total electronic orbital angular momentum; (b) the total electronic spin angular momentum; (c) the total electronic angular momentum.

11.24 Give the symbol for the ground level of each of the atoms with $Z \leq 10$.

11.25 Give the symbol for the ground level of each of the atoms with $21 \leq Z \leq 30$. Which one of these atoms has the most degenerate ground level?

11.26 (a) Use a diagram like (11.48) to show that the lowest term of a single half-filled-subshell configuration has $L = 0$. (b) How many levels arise from a term with $L = 0$? Explain why no rule is needed to find the lowest level of the lowest term of a half-filled-subshell configuration.

11.27 Verify that if $E = 1$ eV, then $E/hc = 8065.54\ \text{cm}^{-1}$.

11.28 Consult Moore's table of atomic energy levels (Section 11.5) to find at least three electron configurations of the neutral carbon atom for which Hund's rule does not correctly predict the lowest term.

11.29 Selection rules for spectral transitions of atoms where Russell–Saunders coupling is valid are (*Bethe and Jackiw*, Chapter 11) $\Delta L = 0, \pm 1$; $\Delta S = 0$; $\Delta J = 0, \pm 1$ (but $J = 0$ to $J = 0$ is forbidden); $\Delta(\Sigma_i l_i) = \pm 1$, meaning that the change in configuration must change the sum of the l values of the electrons by ± 1. For most atomic spectral lines, only one electron changes its subshell; here the $\Delta(\Sigma_i l_i) = \pm 1$ rule becomes $\Delta l = \pm 1$ for the electron making the transition.

For the carbon atom, the levels that arise from the $1s^22s^22p^2$ configuration are

level	$^3P_0$	$^3P_1$	$^3P_2$	$^1D_2$	$^1S_0$
$(E/hc)/\text{cm}^{-1}$	0	16.4	43.4	10192.6	21648.0

and the energy levels of the $1s^22s2p^3$ configuration are

$^5S_2$	$^3D_3$	$^3D_1$	$^3D_2$	$^3P_1$	$^3P_2$	$^3P_0$
33735.2	64086.9	64089.8	64090.9	75254.0	75255.3	75256.1

$^1D_2$	$^3S_1$	$^1P_1$
97878	105798.7	119878

Use the above selection rules to find the wavenumbers of all the transitions that are allowed between pairs of these 15 levels.

11.30 Use Eq. (11.66) to calculate the separation between the $^2P_{3/2}$ and $^2P_{1/2}$ levels of the hydrogen-atom $2p$ configuration. (Because of other relativistic effects, the result will not agree accurately with experiment.)

11.31 Does Fig. 11.6 contain a violation of the rule given in Section 11.6 for determining whether a multiplet is regular or inverted?

11.32 Draw a diagram similar to Fig. 11.6 for the carbon $1s^22s^22p^2$ configuration. (The ^1S term is the highest.)

11.33 Use Eq. (11.74) to calculate the energy separation between the $M_J = \frac{1}{2}$ and $M_J = -\frac{1}{2}$ states of the $2p\ {}^2P_{1/2}$ hydrogen-atom level, if a magnetic field of 0.200 T is applied.

11.34 (a) Derive Eq. (11.78). (b) Derive (11.79).

11.35 For a closed-subshell configuration, (a) show that the double sum in (11.80) equals

$$\sum_{j>i}^{n/2}\sum_{i=1}^{n/2}(4J_{ij} - 2K_{ij}) + \sum_{i=1}^{n/2} J_{ii}$$

where the Coulomb and exchange integrals are defined in terms of the $n/2$ different spatial orbitals ϕ_i; (b) use (11.84) and the result of part (a) to derive (11.83).

11.36 Use the Condon–Slater rules to prove the orthonormality of two n-electron Slater determinants of orthonormal spin-orbitals.

11.37 Explain why it would be incorrect to calculate the experimental ground-state energy of lithium by taking $-(E_{2s} + 2E_{1s})$, where E_{2s} is the experimental energy needed to remove the 2s electron from lithium and E_{1s} is the experimental energy needed to remove a 1s electron from lithium.

11.38 The total magnetic moment **m** of an atom contains contributions from the magnetic moments $\mathbf{m}_L$, $\mathbf{m}_S$, and $\mathbf{m}_I$ associated with the total electronic orbital angular momentum **L**, the total electronic spin angular momentum **S**, and the nuclear spin angular momentum **I**. (a) For an atom with **L**, **S**, and **I** all nonzero, which one of these three contributions to **m** is by far the smallest? (b) For the ^{87}Rb ground electronic state (used in the Bose–Einstein condensation experiment of Section 10.5), which one of these three contributions is zero?

11.39 True or false? (a) The spin multiplicity of every term of an atom with an odd number of electrons must be an even number. (b) The spin multiplicity of every term of an atom with an even number of electrons must be an odd number. (c) The spin multiplicity of a term is always equal to the number of levels of that term. (d) In the Hartree SCF method, the energy of an atom equals the sum of the orbital energies of the electrons. (e) The Hartree–Fock method is capable of giving the exact nonrelativistic energy of a many-electron atom.

CHAPTER 12

Molecular Symmetry

12.1 SYMMETRY ELEMENTS AND OPERATIONS

Qualitative information about molecular wave functions and properties can often be obtained from the symmetry of the molecule. By the symmetry of a molecule, we mean the symmetry of the framework formed by the nuclei held fixed in their equilibrium positions. (Our starting point for molecular quantum mechanics will be the Born–Oppenheimer approximation, which regards the nuclei as fixed when solving for the electronic wave function; see Section 13.1.) The symmetry of a molecule can differ in different electronic states. For example, HCN is linear in its ground electronic state, but nonlinear in certain excited states. Unless otherwise specified, we shall be considering the symmetry of the ground electronic state.

Symmetry Elements and Operations. A **symmetry operation** is a transformation of a body such that the final position is physically indistinguishable from the initial position, and the distances between all pairs of points in the body are preserved. For example, consider the trigonal-planar molecule BF_3 (Fig. 12.1a), where for convenience we have numbered the fluorine nuclei. If we rotate the molecule counterclockwise by 120° about an axis through the boron nucleus and perpendicular to the plane of the molecule, the new position will be as in Fig. 12.1b. Since in reality the fluorine nuclei are physically indistinguishable from one another, we have carried out a symmetry operation. The axis about which we rotated the molecule is an example of a symmetry element. Symmetry elements and symmetry operations are related but different things, which are often confused. A **symmetry element** is a geometrical entity (point, line, or plane) with respect to which a symmetry operation is carried out.

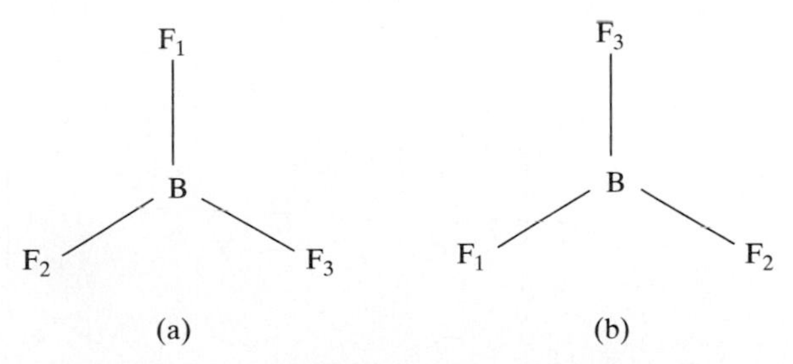

FIGURE 12.1 (a) The BF_3 molecule. (b) BF_3 after a 120° rotation about the axis through B and perpendicular to the molecular plane.

We say that a body has an ***n*-fold axis of symmetry** (also called an *n-fold proper axis* or an *n-fold rotation axis*) if rotation about this axis by $360/n$ degrees (where n is an integer) gives a configuration physically indistinguishable from the original position; n is called the *order* of the axis. For example, BF_3 has a threefold axis of symmetry perpendicular to the molecular plane. The symbol for an n-fold rotation axis is C_n. The threefold axis in BF_3 is a C_3 axis. To denote the operation of counterclockwise rotation by $(360/n)°$, we use the symbol $\hat{C}_n$. The "hat" distinguishes symmetry operations from symmetry elements. BF_3 has three more rotation axes; each B—F bond is a twofold symmetry axis (Fig. 12.2).

A second kind of symmetry element is a plane of symmetry. A molecule has a **plane of symmetry** if reflection of all the nuclei through that plane gives a configuration physically indistinguishable from the original one. The symbol for a symmetry plane is σ (lowercase sigma). (*Spiegel* is the German word for mirror.) The symbol for the operation of reflection is $\hat{\sigma}$. BF_3 has four symmetry planes. The plane of the molecule is a symmetry plane, since nuclei lying in a reflection plane do not move when a reflection is carried out. The plane passing through the B and F_1 nuclei and perpendicular to the plane of the molecule is a symmetry plane, since reflection in this plane merely interchanges F_2 and F_3. It might be thought that this reflection is the same symmetry operation as rotation by 180° about the C_2 axis passing through B and F_1, which also interchanges F_2 and F_3. This is not so. The reflection carries points lying above the plane of the molecule into points that also lie above the molecular plane, whereas the $\hat{C}_2$ rotation carries points lying above the molecular plane into points below the molecular plane. *Two symmetry operations are equal only when they represent the same transformation of three-dimensional space.* The remaining two symmetry planes in BF_3 pass through B—F_2 and B—F_3 and are perpendicular to the molecular plane.

The third kind of symmetry element is a **center of symmetry**, symbolized by i (no connection with $\sqrt{-1}$). A molecule has a center of symmetry if the operation of inverting all the nuclei through the center gives a configuration indistinguishable from the original one. If we set up a Cartesian coordinate system, the operation of **inversion** through the origin (symbolized by $\hat{i}$) carries a nucleus originally at (x, y, z) to $(-x, -y, -z)$. Does BF_3 have a center of symmetry? With the origin at the boron nucleus, inversion gives the result shown in Fig. 12.3. Since we get a configuration that is physically distinguishable from the original one, BF_3 does not have a center of sym-

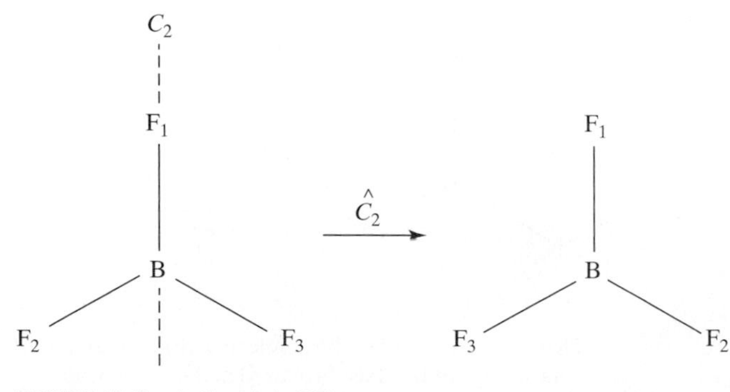

FIGURE 12.2 A C_2 axis in BF_3.

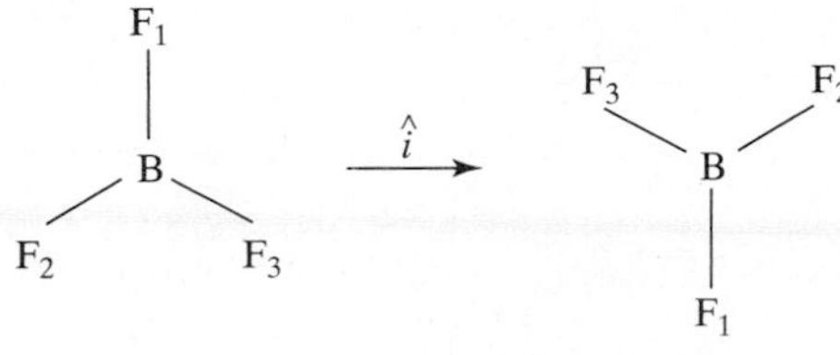

FIGURE 12.3 Effect of inversion in BF_3.

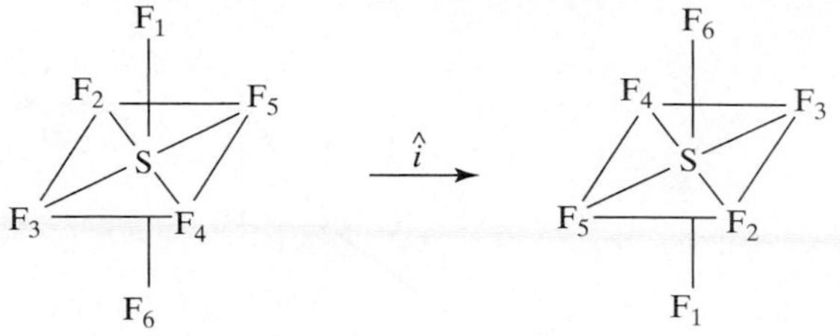

FIGURE 12.4 Effect of inversion in SF_6.

metry. For SF_6, inversion through the sulfur nucleus is shown in Fig. 12.4, and it is clear that SF_6 has a center of symmetry. (An operation such as $\hat{i}$ or $\hat{C}_n$ may or may not be a symmetry operation. Thus, $\hat{i}$ is a symmetry operation in SF_6 but not in BF_3.)

The fourth and final kind of symmetry element is an ***n*-fold rotation–reflection axis of symmetry** (also called an *improper axis* or an *alternating axis*), symbolized by S_n. A body has an S_n axis if rotation by $(360/n)°$ (n integral) about the axis, followed by reflection in a plane perpendicular to the axis, carries the body into a position physically indistinguishable from the original one. Clearly, if a body has a C_n axis and also has a plane of symmetry perpendicular to this axis, then the C_n axis is also an S_n axis. Thus the C_3 axis in BF_3 is also an S_3 axis. It is possible to have an S_n axis that is not a C_n axis. An example is CH_4. In Fig. 12.5 we have first carried out a 90° proper rotation ($\hat{C}_4$) about what we assert is an S_4 axis. As can be seen, this operation does not result in an equivalent configuration. When we follow the $\hat{C}_4$ operation by reflection in the plane perpendicular to the axis and passing through the carbon atom, we do get a configuration indistinguishable from the one existing before we performed the rotation and reflection. Hence CH_4 has an S_4 axis. The S_4 axis is not a C_4 axis, although it is a C_2 axis. There are two other S_4 axes in methane, each perpendicular to a pair of faces of the cube in which the tetrahedral molecule is inscribed.

The operation of counterclockwise rotation by $(360/n)°$ about an axis, followed by reflection in a plane perpendicular to the axis, is denoted by $\hat{S}_n$. An $\hat{S}_1$ operation is a 360° rotation about an axis, followed by a reflection in a plane perpendicular to the axis. Since a 360° rotation restores the body to its original position, an $\hat{S}_1$ operation is the same as reflection in a plane; $\hat{S}_1 = \hat{\sigma}$. Any plane of symmetry has an S_1 axis perpendicular to it.

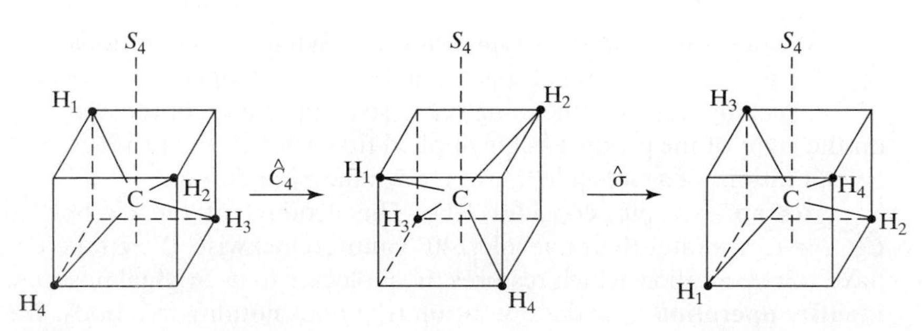

FIGURE 12.5 An S_4 axis in CH_4.

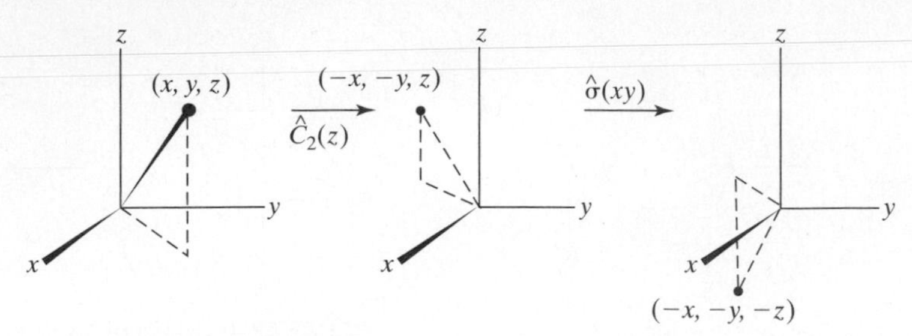

FIGURE 12.6 The $\hat{S}_2$ operation.

Consider now the $\hat{S}_2$ operation. Let the S_2 axis be the z axis (Fig. 12.6). Rotation by 180° about the S_2 axis changes the x and y coordinates of a point to $-x$ and $-y$, respectively, and leaves the z coordinate unaffected. Reflection in the xy plane then converts the z coordinate to $-z$. The net effect of the $\hat{S}_2$ operation is to bring a point originally at (x, y, z) to $(-x, -y, -z)$, which amounts to an inversion through the origin: $\hat{S}_2 = \hat{i}$. Any axis passing through a center of symmetry is an S_2 axis. Reflection in a plane and inversion are special cases of the $\hat{S}_n$ operation.

The $\hat{S}_n$ operation may seem an arbitrary kind of operation, but it must be included as one of the kinds of symmetry operations. For example, the transformation from the first to the third CH_4 configuration in Fig. 12.5 certainly meets the definition of a symmetry operation, but it is neither a proper rotation nor a reflection nor an inversion.

Performing a symmetry operation on a molecule gives a nuclear configuration physically indistinguishable from the original one. Hence the center of mass must have the same position in space before and after a symmetry operation. For the operation $\hat{C}_n$, the only points that do not move are those on the C_n axis. Therefore, a C_n symmetry axis must pass through the molecular center of mass. Similarly, a center of symmetry must coincide with the center of mass; a plane of symmetry and an S_n axis of symmetry must pass through the center of mass. The center of mass is the common intersection of all the molecular symmetry elements.

In discussing the symmetry of a molecule, we often place it in a Cartesian coordinate system with the molecular center of mass at the origin. The rotational axis of highest order is made the z axis. A plane of symmetry containing this axis is designated σ_v (for *vertical*); a plane of symmetry perpendicular to this axis is designated σ_h (for *horizontal*).

Products of Symmetry Operations. Symmetry operations are operators that transform three-dimensional space, and (as with any operators) we define the **product** of two such operators as meaning successive application of the operators, the operator on the right of the product being applied first. Clearly, the product of any two symmetry operations of a molecule must be a symmetry operation.

As an example, consider BF_3. The product of the $\hat{C}_3$ operator with itself, $\hat{C}_3\hat{C}_3 = \hat{C}_3^2$, rotates the molecule 240° counterclockwise. If we take $\hat{C}_3\hat{C}_3\hat{C}_3 = \hat{C}_3^3$, we have a 360° rotation, which restores the molecule to its original position. We define the **identity operation** $\hat{E}$ as the operation that does nothing to a body. We have $\hat{C}_3^3 = \hat{E}$. (The symbol comes from the German word *Einheit*, meaning unity.)

Now consider a molecule with a sixfold axis of symmetry, for example, C_6H_6. The operation $\hat{C}_6$ is a 60° rotation, and $\hat{C}_6^2$ is a 120° rotation; hence $\hat{C}_6^2 = \hat{C}_3$. Also $\hat{C}_6^3 = \hat{C}_2$. Therefore, a C_6 symmetry axis is also a C_3 and a C_2 axis.

Since two successive reflections in the same plane bring all nuclei back to their original positions, we have $\hat{\sigma}^2 = \hat{E}$. Also, $\hat{i}^2 = \hat{E}$. More generally, $\hat{\sigma}^n = \hat{E}$, $\hat{i}^n = \hat{E}$ for even n, while $\hat{\sigma}^n = \hat{\sigma}$, $\hat{i}^n = \hat{i}$ for odd n.

Do symmetry operators always commute? Consider SF_6. We examine the products of a $\hat{C}_4$ rotation about the z axis and a $\hat{C}_2$ rotation about the x axis. Figure 12.7 shows that $\hat{C}_4(z)\hat{C}_2(x) \neq \hat{C}_2(x)\hat{C}_4(z)$. Thus symmetry operations do not always commute. Note that we describe symmetry operations with respect to a fixed coordinate system that does not move with the molecule when we perform a symmetry operation. Thus the $C_2(x)$ axis does not move when we perform the $\hat{C}_4(z)$ operation.

Symmetry and Dipole Moments. As an application of symmetry, we consider molecular dipole moments. Since a symmetry operation produces a configuration physically indistinguishable from the original one, the direction of the dipole-moment vector must remain unchanged after a symmetry operation. (This is a nonrigorous, unsophisticated argument.) Hence, if we have a C_n axis of symmetry, the dipole moment must lie along this axis. If we have two or more noncoincident symmetry axes, the molecule cannot have a dipole moment, since the dipole moment cannot lie on two different axes. CH_4, which has four noncoincident C_3 axes, has no dipole moment. If there is a plane of symmetry, the dipole moment must lie in this plane. If there are

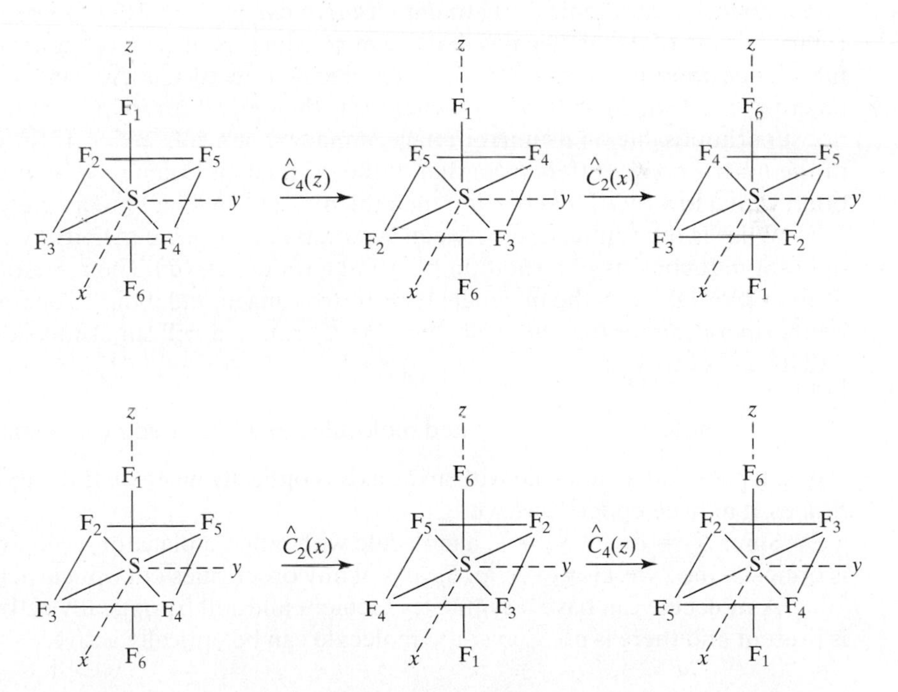

FIGURE 12.7 Products of two symmetry operations in SF_6. Top: $\hat{C}_2(x)\hat{C}_4(z)$. Bottom: $\hat{C}_4(z)\hat{C}_2(x)$.

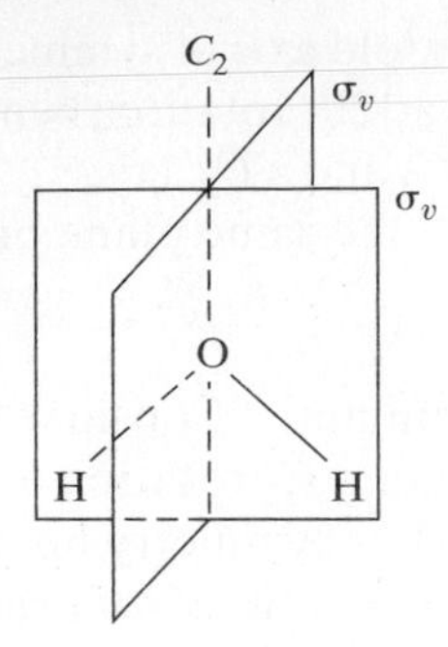

FIGURE 12.8 The symmetry elements of H_2O.

several symmetry planes, the dipole moment must lie along the line of intersection of these planes. In H_2O the dipole moment lies on the C_2 axis, which is also the intersection of the two symmetry planes (Fig. 12.8). A molecule with a center of symmetry cannot have a dipole moment, since inversion reverses the direction of a vector. A monatomic molecule has a center of symmetry. Hence atoms do not have dipole moments. (There is one exception to this statement; see Prob. 14.4.) Thus we can use symmetry to discover whether a molecule has a dipole moment. In many cases symmetry also tells us along what line the dipole moment lies.

Symmetry and Optical Activity. Certain molecules rotate the plane of polarization of plane-polarized light that is passed through them. Experimental evidence and a quantum-mechanical treatment (*Kauzmann*, pp. 703–713) show that the optical rotary powers of two molecules that are mirror images of each other are equal in magnitude but opposite in sign. Hence, if a molecule is its own mirror image, it is optically inactive: $\alpha = -\alpha$, $2\alpha = 0$, $\alpha = 0$, where α is the optical rotary power. If a molecule is not superimposable on its mirror image, it may be optically active. If the conformation of the mirror image differs from that of the original molecule only by rotation about a bond with a low rotational barrier, then the molecule will not be optically active.

What is the connection between symmetry and optical activity? Consider the $\hat{S}_n$ operation. It consists of a rotation ($\hat{C}_n$) and a reflection ($\hat{\sigma}$). The reflection part of the $\hat{S}_n$ operation converts the molecule to its mirror image, and if the $\hat{S}_n$ operation is a symmetry operation for the molecule, then the $\hat{C}_n$ rotation will superimpose the molecule and its mirror image:

$$\text{molecule} \xrightarrow{\hat{C}_n} \text{rotated molecule} \xrightarrow{\hat{\sigma}} \text{rotated mirror image}$$

We conclude that a molecule with an S_n axis is optically inactive. If the molecule has no S_n axis, it may be optically active.

Since $\hat{S}_1 = \hat{\sigma}$ and $\hat{S}_2 = \hat{\imath}$, a molecule with either a plane or a center of symmetry is optically inactive. However, an S_n axis of any order rules out optical activity.

A molecule can have a symmetry element and still be optically active. If a C_n axis is present and there is no S_n axis, the molecule can be optically active.

Symmetry Operations and Quantum Mechanics. What is the relation between the symmetry operations of a molecule and quantum mechanics? To classify the states of a quantum-mechanical system, we consider those operators that commute with the

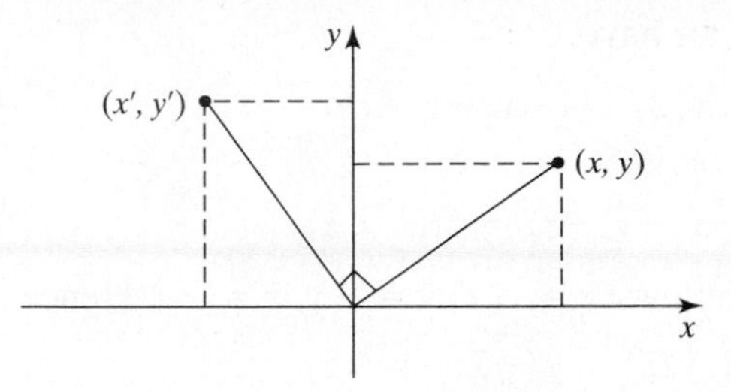

FIGURE 12.9 The effect of a $\hat{C}_4(z)$ rotation is to move the point at (x, y) to (x', y'). Use of trigonometry shows that $x' = -y$ and $y' = x$.

Hamiltonian operator and with each other. For example, we classified the states of many-electron atoms using the quantum numbers L, S, J, and M_J, which correspond to the operators $\hat{L}^2$, $\hat{S}^2$, $\hat{J}^2$, and $\hat{J}_z$, all of which commute with one other and with the Hamiltonian (omitting spin–orbit interaction). The symmetry operations discussed in this chapter act on *points* in three-dimensional space, transforming each point to a corresponding point. All the quantum-mechanical operators we have discussed act on *functions*, transforming each function to a corresponding function. Corresponding to each symmetry operation $\hat{R}$, we define an operator $\hat{O}_R$ that acts on functions in the following manner. Let $\hat{R}$ bring a point originally at (x, y, z) to the location (x', y', z'):

$$\hat{R}(x, y, z) \rightarrow (x', y', z') \tag{12.1}$$

The operator $\hat{O}_R$ is defined so that the function $\hat{O}_R f$ has the same value at (x', y', z') that the function f has at (x, y, z):

$$\hat{O}_R f(x', y', z') = f(x, y, z) \tag{12.2}$$

For example, let $\hat{R}$ be a counterclockwise 90° rotation about the z axis: $\hat{R} = \hat{C}_4(z)$; and let f be a $2p_x$ hydrogen orbital: $f = 2p_x = Nxe^{-k(x^2+y^2+z^2)^{1/2}}$. The shape of the $2p_x$ orbital is two distorted ellipsoids of revolution about the x axis (Section 6.7). Let these ellipsoids be "centered" about the points $(a, 0, 0)$ and $(-a, 0, 0)$, where $a > 0$ and $2p_x > 0$ on the right ellipsoid. The operator $\hat{C}_4(z)$ has the following effect (Fig. 12.9):

$$\hat{C}_4(z)(x, y, z) \rightarrow (-y, x, z) \tag{12.3}$$

For example, the point originally at $(a, 0, 0)$ is moved to $(0, a, 0)$, while the point at $(-a, 0, 0)$ is moved to $(0, -a, 0)$. From (12.2), the function $\hat{O}_{C_4(z)} 2p_x$ must have its contours centered about $(0, a, 0)$ and $(0, -a, 0)$, respectively. We conclude that (Fig. 12.10)

$$\hat{O}_{C_4(z)} 2p_x = 2p_y \tag{12.4}$$

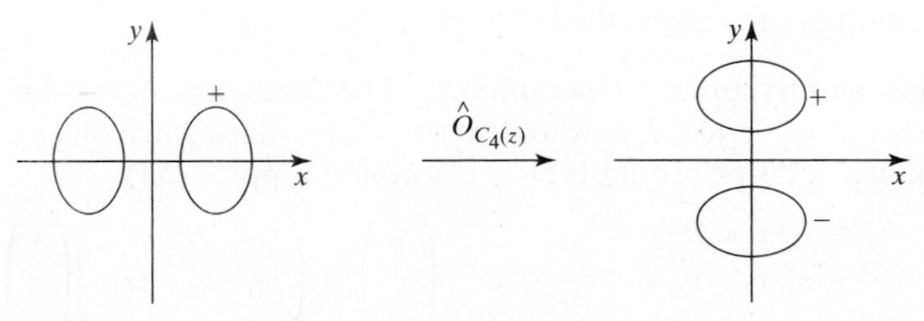

FIGURE 12.10 Effect of $\hat{O}_{C_4(z)}$ on a p_x orbital.

For the inversion operation, we have

$$\hat{i}(x, y, z) \rightarrow (-x, -y, -z) \tag{12.5}$$

and (12.2) reads

$$\hat{O}_i f(-x, -y, -z) = f(x, y, z)$$

We now rename the variables as follows: $\overline{x} = -x, \overline{y} = -y, \overline{z} = -z$. Hence

$$\hat{O}_i f(\overline{x}, \overline{y}, \overline{z}) = f(-\overline{x}, -\overline{y}, -\overline{z})$$

The point $(\overline{x}, \overline{y}, \overline{z})$ is a general point in space, and we can drop the bars to get

$$\hat{O}_i f(x, y, z) = f(-x, -y, -z)$$

We conclude that $\hat{O}_i$ is the parity operator (Section 7.5): $\hat{O}_i = \hat{\Pi}$.

The wave function of an n-particle system is a function of $4n$ variables, and we extend the definition (12.2) of $\hat{O}_R$ to read

$$\hat{O}_R f(x'_1, y'_1, z'_1, m_{s1}, \ldots, x'_n, y'_n, z'_n, m_{sn}) = f(x_1, y_1, z_1, m_{s1}, \ldots, x_n, y_n, z_n, m_{sn})$$

Note that $\hat{O}_R$ does not affect the spin coordinates. Thus, in looking at the parity of atomic states in Section 11.5, we looked at the spatial factors in each term of the expansion of the Slater determinant and omitted consideration of the spin factors, since these are unaffected by $\hat{\Pi}$.

When a system is characterized by the symmetry operations $\hat{R}_1, \hat{R}_2, \ldots,$ then the corresponding operators $\hat{O}_{R_1}, \hat{O}_{R_2}, \ldots$ commute with the Hamiltonian. (For a proof, see *Schonland*, Sections 7.1–7.3.) For example, if the nuclear framework of a molecule has a center of symmetry, then the parity operator $\hat{\Pi}$ commutes with the Hamiltonian for the electronic motion. We can then choose the electronic states (wave functions) as even or odd, according to the eigenvalue of $\hat{\Pi}$. Of course, all the symmetry operations may not commute among themselves (Fig. 12.7). Hence the wave functions cannot in general be chosen as eigenfunctions of all the symmetry operators $\hat{O}_R$. (Further discussion on the relation between symmetry operators and molecular wave functions is given in Section 15.2.)

There is a close connection between symmetry and the *constants of the motion* (these are properties whose operators commute with the Hamiltonian $\hat{H}$). For a system whose Hamiltonian is invariant (that is, doesn't change) under any translation of spatial coordinates, the linear-momentum operator $\hat{p}$ will commute with $\hat{H}$, and p can be assigned a definite value in a stationary state. An example is the free particle. For a system with $\hat{H}$ invariant under any rotation of coordinates, the operators for the angular-momentum components commute with $\hat{H}$, and the total angular momentum and one of its components are specifiable. An example is an atom. A linear molecule has axial symmetry, rather than the spherical symmetry of an atom; here only the axial component of angular momentum can be specified (Chapter 13).

Matrices and Symmetry Operations. The symmetry operation $\hat{R}$ moves the point originally at x, y, z to the new location x', y', z', where each of x', y', z' is a linear combination of x, y, z (for proof of this see *Schonland*, pp. 52–53):

$$\begin{aligned} x' &= r_{11}x + r_{12}y + r_{13}z \\ y' &= r_{21}x + r_{22}y + r_{23}z \\ z' &= r_{31}x + r_{32}y + r_{33}z \end{aligned} \quad \text{or} \quad \begin{pmatrix} x' \\ y' \\ z' \end{pmatrix} = \begin{pmatrix} r_{11} & r_{12} & r_{13} \\ r_{21} & r_{22} & r_{23} \\ r_{31} & r_{32} & r_{33} \end{pmatrix} \begin{pmatrix} x \\ y \\ z \end{pmatrix}$$

where $r_{11}, r_{12}, \ldots, r_{33}$ are constants whose values depend on the nature of $\hat{R}$. One says that the symmetry operation $\hat{R}$ is **represented** by the matrix **R** whose elements are $r_{11}, r_{12}, \ldots, r_{33}$. The set of functions x, y, z, whose transformations are described by **R**, is said to be the **basis** for this representation.

For example, from (12.3) and (12.5), for the $\hat{C}_4(z)$ operation, we have $x' = -y$, $y' = x$, $z' = z$; for $\hat{i}$, we have $x' = -x$, $y' = -y$, $z' = -z$. The matrices representing $\hat{C}_4(z)$ and $\hat{i}$ in the x, y, z basis are

$$\mathbf{C}_4(z) = \begin{pmatrix} 0 & -1 & 0 \\ 1 & 0 & 0 \\ 0 & 0 & 1 \end{pmatrix}, \qquad \mathbf{i} = \begin{pmatrix} -1 & 0 & 0 \\ 0 & -1 & 0 \\ 0 & 0 & -1 \end{pmatrix}$$

If the product $\hat{R}\hat{S}$ of two symmetry operations is $\hat{T}$, then the matrices representing these operations in the x, y, z basis multiply in the same way; that is, if $\hat{R}\hat{S} = \hat{T}$, then $\mathbf{RS} = \mathbf{T}$. (For proof, see *Schonland*, pp. 56–57.)

12.2 SYMMETRY POINT GROUPS

We now consider the possible combinations of symmetry elements. We cannot have arbitrary combinations of symmetry elements in a molecule. For example, suppose a molecule has one and only one C_3 axis. Any symmetry operation must send this axis into itself. The molecule cannot, therefore, have a plane of symmetry at an arbitrary angle to the C_3 axis; any plane of symmetry must either contain this axis or be perpendicular to it. (In BF_3 there are three σ_v planes and one σ_h plane.) The only possibility for a C_n axis noncoincident with the C_3 axis is a C_2 axis perpendicular to the C_3 axis. The corresponding $\hat{C}_2$ operation will send the C_3 axis into itself. Since $\hat{C}_3$ and $\hat{C}_3^2$ are symmetry operations, if we have one C_2 axis perpendicular to the C_3 axis, we must have a total of three such axes (as in BF_3).

The set of all the symmetry operations of a molecule forms a mathematical group. A **group** is a set of entities (called the *elements* or *members* of the group) and a rule for combining any two members of the group to form the product of these members, such that certain requirements are met. Let $A, B, C, D, \ldots$ (assumed to be all different from one another) be the members of the group and let $B * C$ denote the product of B and C. The product $B * C$ need not be the same as $C * B$. The requirements that must be met to have a group are as follows: (1) The product of any two elements (including the product of an element with itself) must be a member of the group (the *closure* requirement). (2) There is a single element I of the group, called the *identity element*, such that $K * I = K$ and $I * K = K$ for every element K of the group. (3) Every element K of the group has an *inverse* (symbolized by K^{-1}) that is a member of the group and that satisfies $K * K^{-1} = I$ and $K^{-1} * K = I$, where I is the identity element. (4) Group multiplication is *associative*, meaning that $(B * D) * G = B * (D * G)$ always holds for elements of the group.

The set of all symmetry operations of a three-dimensional body, with the rule of combination of $\hat{R}$ and $\hat{S}$ being successive performance of $\hat{R}$ and $\hat{S}$, forms a group. Closure is satisfied because the product of any two symmetry operations must be a symmetry operation. The identity element of the group is the identity operation $\hat{E}$, which does nothing. Associativity is satisfied [Eq. (3.6)]. The inverse of a symmetry operation

$\hat{R}$ is the symmetry operation that undoes the effect of $\hat{R}$. For example, the inverse of the inversion operation $\hat{i}$ is $\hat{i}$ itself, since $\hat{i}\hat{i} = \hat{E}$. The inverse of a $\hat{C}_3$ 120° counterclockwise rotation is a 120° clockwise rotation, which is the same as a 240° counterclockwise rotation: $\hat{C}_3\hat{C}_3^2 = \hat{E}$ and $\hat{C}_3^{-1} = \hat{C}_3^2$. *Note that it is the symmetry operations of a molecule (and not the symmetry elements) that are the members (elements) of the group.* We will make some use of group theory in Section 15.2, but a full development of group theory and its applications is omitted (see *Cotton* or *Schonland*).

For any symmetry operation of a molecule, the point that is the center of mass remains fixed. Hence the symmetry groups of isolated molecules are called **point groups**. For a crystal of infinite extent, we can have symmetry operations (for example, translations) that leave no point fixed, giving rise to *space groups.* We omit consideration of space groups.

Every molecule belongs to one of the symmetry point groups that we now list. For convenience the point groups have been classified into four divisions. Script letters denote point groups.

I. *Groups with no C_n axis:* $\mathscr{C}_1$, $\mathscr{C}_s$, $\mathscr{C}_i$

$\mathscr{C}_1$: If a molecule has no symmetry elements at all, it belongs to this group. The only symmetry operation is $\hat{E}$ (which is a $\hat{C}_1$ rotation). CHFClBr belongs to point group $\mathscr{C}_1$.

$\mathscr{C}_s$: A molecule whose only symmetry element is a plane of symmetry belongs to this group. The symmetry operations are $\hat{E}$ and $\hat{\sigma}$. An example is HOCl (Fig. 12.11).

$\mathscr{C}_i$: A molecule whose only symmetry element is a center of symmetry belongs to this group. The symmetry operations are $\hat{i}$ and $\hat{E}$.

II. *Groups with a single C_n axis:* $\mathscr{C}_n$, $\mathscr{C}_{nh}$, $\mathscr{C}_{nv}$, $\mathscr{S}_{2n}$

$\mathscr{C}_n$, $n = 2, 3, 4, \ldots$: A molecule whose only symmetry element is a C_n axis belongs to this group. The symmetry operations are $\hat{C}_n, \hat{C}_n^2, \ldots, \hat{C}_n^{n-1}, \hat{E}$. A molecule belonging to $\mathscr{C}_2$ is shown in Fig. 12.12.

$\mathscr{C}_{nh}$, $n = 2, 3, 4, \ldots$: If we add a plane of symmetry perpendicular to the C_n axis, we have a molecule belonging to this group. Since $\hat{\sigma}_h\hat{C}_n = \hat{S}_n$, the C_n axis is also an S_n axis. If n is even, the C_n axis is also a C_2 axis, and the molecule has the symmetry operation $\hat{\sigma}_h\hat{C}_2 = \hat{S}_2 = \hat{i}$. Thus, for n even, a molecule belonging to $\mathscr{C}_{nh}$ has a center of symmetry. (The group $\mathscr{C}_{1h}$ is the group $\mathscr{C}_s$ discussed previously.) Examples of molecules belonging to groups $\mathscr{C}_{2h}$ and $\mathscr{C}_{3h}$ are shown in Fig. 12.12.

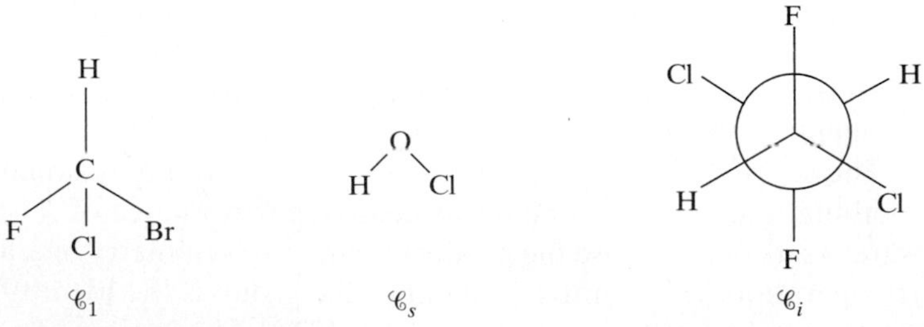

$\mathscr{C}_1$ $\mathscr{C}_s$ $\mathscr{C}_i$

FIGURE 12.11 Molecules with no C_n axis.

H C_2

H

H_2O_2; the O—O bond is perpendicular to the plane of the paper

$\mathscr{C}_2$

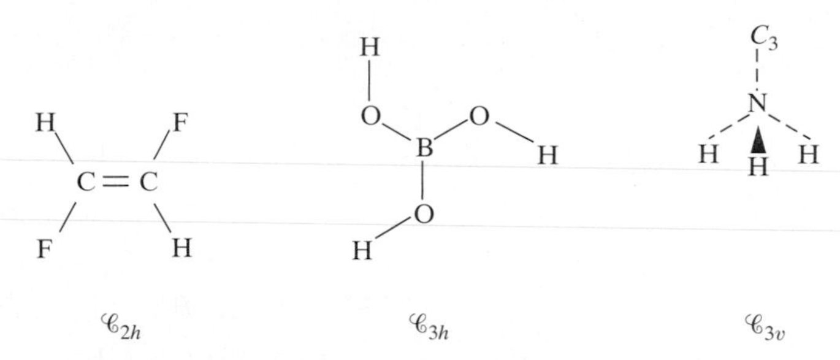

FIGURE 12.12 Molecules with a single C_n axis.

$\mathscr{C}_{nv}$, $n = 2, 3, 4, \ldots$: A molecule in this group has a C_n axis and n vertical symmetry planes passing through the C_n axis. (Group $\mathscr{C}_{1v}$ is the group $\mathscr{C}_s$.) H_2O with a C_2 axis and two vertical symmetry planes, belongs to $\mathscr{C}_{2v}$. NH_3 belongs to $\mathscr{C}_{3v}$. (See Fig. 12.12.)

$\mathscr{S}_n$, $n = 4, 6, 8, \ldots$: $\mathscr{S}_n$ is the group of symmetry operations associated with an S_n axis. First consider the case of odd n. We have $\hat{S}_n = \hat{\sigma}_h\hat{C}_n$. The operation $\hat{C}_n$ affects the x and y coordinates only, while the $\hat{\sigma}_h$ operation affects the z coordinate only. Hence these operations commute, and we have

$$\hat{S}_n^n = (\hat{\sigma}_h\hat{C}_n)^n = \hat{\sigma}_h\hat{C}_n\hat{\sigma}_h\hat{C}_n\cdots\hat{\sigma}_h\hat{C}_n = \hat{\sigma}_h^n\hat{C}_n^n$$

Now $\hat{C}_n^n = \hat{E}$, and, for odd n, $\hat{\sigma}_h^n = \hat{\sigma}_h$. Hence the symmetry operation $\hat{S}_n^n$ equals $\hat{\sigma}_h$ for odd n, and the group $\mathscr{S}_n$ has a horizontal symmetry plane if n is odd. Also,

$$\hat{S}_n^{n+1} = \hat{S}_n^n\hat{S}_n = \hat{\sigma}_h\hat{S}_n = \hat{\sigma}_h\hat{\sigma}_h\hat{C}_n = \hat{C}_n, \qquad n \text{ odd}$$

so the molecule has a C_n axis if n is odd. We conclude that the group $\mathscr{S}_n$ is identical to the group $\mathscr{C}_{nh}$ if n is odd. Now consider even values of n. Since $\hat{S}_2 = \hat{i}$, the group $\mathscr{S}_2$ is identical to $\mathscr{C}_i$. Thus it is only for $n = 4, 6, 8, \ldots$ that we get new groups. The S_{2n} axis is also a C_n axis: $\hat{S}_{2n}^2 = \hat{\sigma}_h^2\hat{C}_{2n}^2 = \hat{E}\hat{C}_n = \hat{C}_n$.

III. *Groups with one C_n axis and n C_2 axes:* $\mathscr{D}_n$, $\mathscr{D}_{nh}$, $\mathscr{D}_{nd}$

$\mathscr{D}_n$, $n = 2, 3, 4, \ldots$: A molecule with a C_n axis and n C_2 axes perpendicular to the C_n axis (and no symmetry planes) belongs to group $\mathscr{D}_n$. The angle between adjacent C_2 axes is π/n radians. The group $\mathscr{D}_2$ has three mutually perpendicular C_2 axes, and the symmetry operations are $\hat{E}, \hat{C}_2(x), \hat{C}_2(y), \hat{C}_2(z)$.

$\mathscr{D}_{nh}$, $n = 2, 3, 4, \ldots$: This is the group of a molecule with a C_n axis, n C_2 axes, and a σ_h symmetry plane perpendicular to the C_n axis. As in $\mathscr{C}_{nh}$, the C_n axis is also an S_n axis. If n is even, the C_n axis is a C_2 and an S_2 axis, and the molecule has a center of symmetry. Molecules in $\mathscr{D}_{nh}$ also have n vertical planes of symmetry, each such

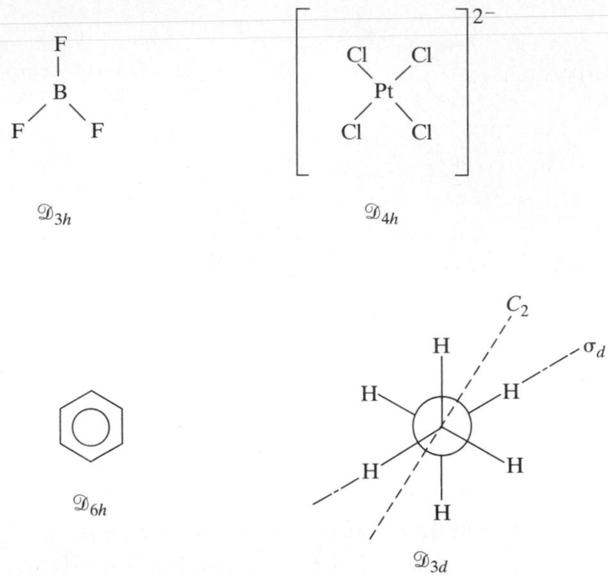

$\mathcal{D}_{3h}$ $\mathcal{D}_{4h}$ $\mathcal{D}_{6h}$ $\mathcal{D}_{3d}$

FIGURE 12.13 Molecules with a C_n axis and n C_2 axes.

plane passing through the C_n axis and a C_2 axis. (For the proof, see Prob. 12.27.) BF_3 belongs to $\mathcal{D}_{3h}$; $PtCl_4^{2-}$ belongs to $\mathcal{D}_{4h}$; benzene belongs to $\mathcal{D}_{6h}$ (Fig. 12.13).

$\mathcal{D}_{nd}$, $n = 2, 3, 4, \ldots$: A molecule with a C_n axis, n C_2 axes, and n vertical planes of symmetry, which pass through the C_n axis and bisect the angles between adjacent C_2 axes, belongs to this group. The n vertical planes are called **diagonal** planes and are symbolized by σ_d. The C_n axis can be shown to be an S_{2n} axis. The staggered conformation of ethane is an example of group $\mathcal{D}_{3d}$ (Fig. 12.13). [The symmetry of molecules with internal rotation (for example, ethane) actually requires special consideration; see H. C. Longuet-Higgins, *Mol. Phys.*, **6**, 445 (1963).]

IV. *Groups with more than one C_n axis, $n > 2$*: $\mathcal{T}_d, \mathcal{T}, \mathcal{T}_h, \mathcal{O}_h, \mathcal{O}, \mathcal{I}_h, \mathcal{I}, \mathcal{K}_h$

These groups are related to the symmetries of the Platonic solids, solids bounded by congruent regular polygons and having congruent polyhedral angles. There are five such solids: The tetrahedron has four triangular faces, the cube has six square faces, the octahedron has eight triangular faces, the pentagonal dodecahedron has twelve pentagonal faces, and the icosahedron has twenty triangular faces.

$\mathcal{T}_d$: The symmetry operations of a regular tetrahedron constitute this group. The prime example is CH_4. The symmetry elements of methane are four C_3 axes (each C—H bond), three S_4 axes, which are also C_2 axes (Fig. 12.5), and six symmetry planes, each such plane containing two C—H bonds. (The number of combinations of 4 things taken 2 at a time is $4!/2!2! = 6$.)

$\mathcal{O}_h$: The symmetry operations of a cube or a regular octahedron constitute this group. The cube and octahedron are said to be *dual* to each other; if we connect the midpoints of adjacent faces of a cube, we get an octahedron, and vice versa.

Hence the cube and octahedron have the same symmetry elements and operations. A cube has six faces, eight vertices, and twelve edges. Its symmetry elements are as follows: a center of symmetry, three C_4 axes passing through the centers of opposite faces of the cube (these are also S_4 and C_2 axes), four C_3 axes passing through opposite corners of the cube (these are also S_6 axes), six C_2 axes connecting the midpoints of pairs of opposite edges, three planes of symmetry parallel to pairs of opposite faces, and six planes of symmetry passing through pairs of opposite edges. Octahedral molecules such as SF_6 belong to $\mathcal{O}_h$.

$\mathcal{I}_h$: The symmetry operations of a regular pentagonal dodecahedron or icosahedron (which are dual to each other) constitute this group. The $B_{12}H_{12}^{2-}$ ion belongs to group $\mathcal{I}_h$. The twelve boron atoms lie at the vertices of a regular icosahedron (Fig. 12.14). The soccer-ball-shaped molecule C_{60} (buckminsterfullerene) belongs to $\mathcal{I}_h$. Its shape is a truncated icosahedron formed by slicing off each of the 12 vertices of a regular icosahedron (Fig. 12.14), thereby generating a figure with 12 pentagonal faces (5 faces meet at each vertex of the original icosahedron), 20 hexagonal faces (formed from the 20 triangular faces of the original icosahedron), and $12 \times 5 = 60$ vertices (5 new vertices are formed when one of the original vertices is sliced off).

$\mathcal{K}_h$: This is the group of symmetry operations of a sphere. (*Kugel* is the German word for sphere.) An atom belongs to this group.

For completeness, we mention the remaining groups related to the Platonic solids; these groups are chemically unimportant. The groups $\mathcal{T}$, $\mathcal{O}$, and $\mathcal{I}$ are the groups of symmetry proper rotations of a tetrahedron, cube, and icosahedron, respectively. These groups do not have the symmetry reflections and improper rotations of these solids or the inversion operation of the cube and icosahedron. The group $\mathcal{T}_h$ contains the symmetry rotations of a tetrahedron, the inversion operation, and certain reflections and improper rotations.

What groups do linear molecules belong to? A rotation by any angle about the internuclear axis of a linear molecule is a symmetry operation. A regular polygon of n sides has a C_n axis, and taking the limit as $n \to \infty$ we get a circle, which has a C_∞ axis. The internuclear axis of a linear molecule is a C_∞ axis. Any plane containing this axis is a symmetry plane. If the linear molecule does not have a center of symmetry (for example, CO, HCN), it belongs to the group $C_{\infty v}$. If the linear molecule has a center of symmetry (for example, H_2, C_2H_2), then it also has a σ_h symmetry plane and an infinite number of C_2 axes perpendicular to the molecular axis. Hence it belongs to $\mathcal{D}_{\infty h}$.

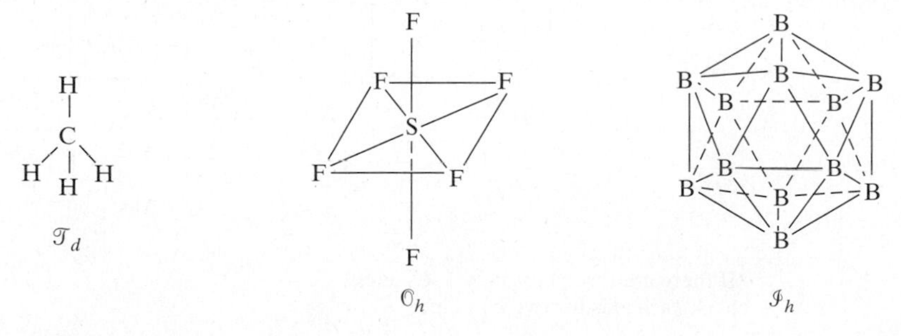

FIGURE 12.14 Molecules with more than one C_n axis, $n > 2$. (For $B_{12}H_{12}^{2-}$, the hydrogen atoms have been omitted for clarity.)

How do we find what point group a molecule belongs to? One way is to find all the symmetry elements and then compare with the above list of groups. A more systematic procedure is given in Fig. 12.15 [J. B. Calvert, *Am. J. Phys.*, **31**, 659 (1963)]. This procedure is based on the four divisions of point groups.

We begin by checking whether or not the molecule is linear. Linear molecules are classified in $\mathscr{D}_{\infty h}$ or $\mathscr{C}_{\infty v}$ according to whether or not there is a center of symmetry. If

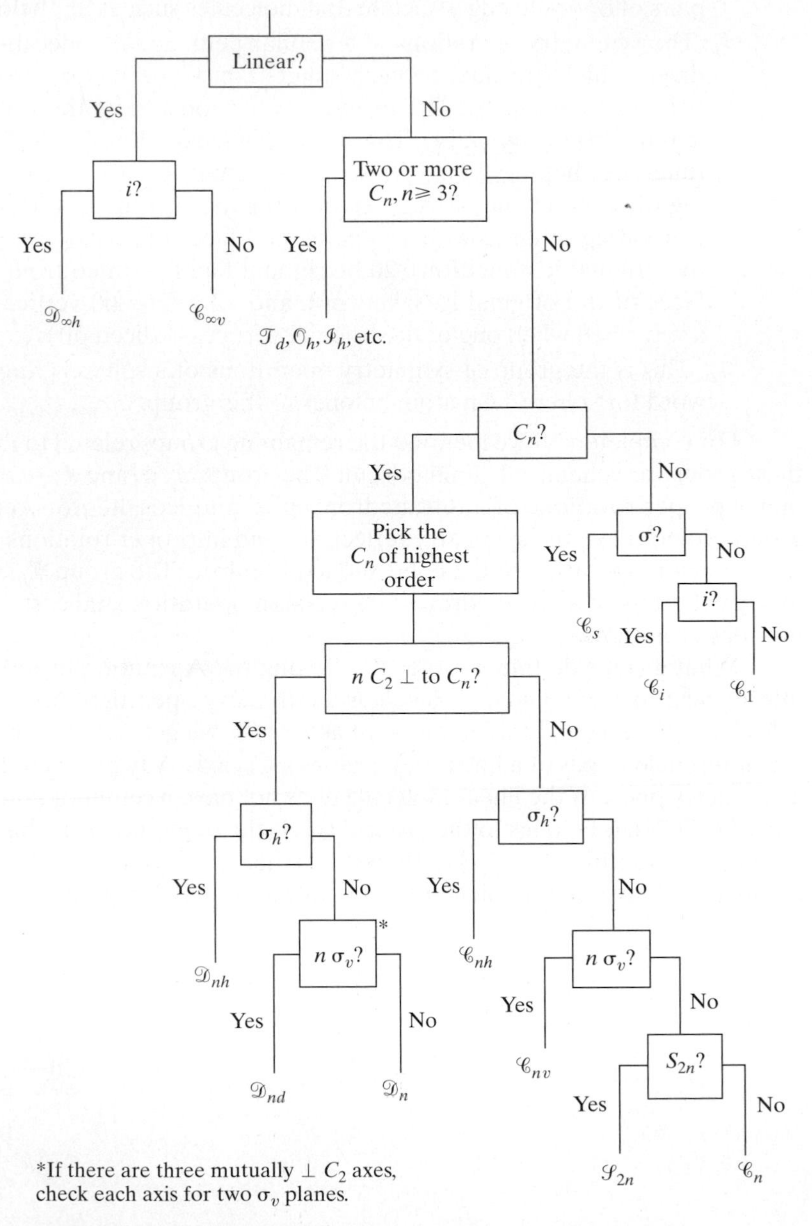

FIGURE 12.15 How to determine the point group of a molecule.

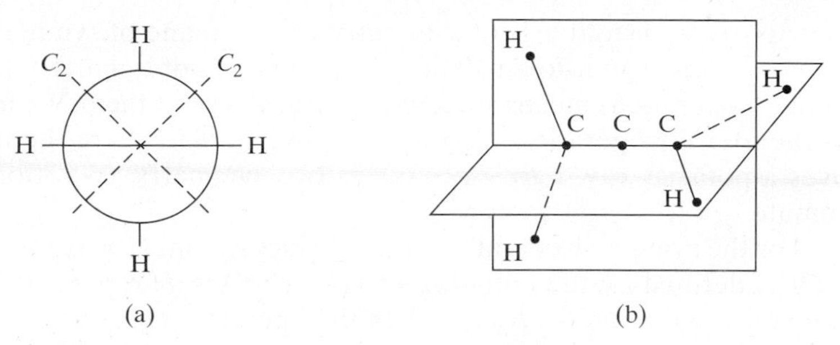

FIGURE 12.16 Two views of allene. The C=C=C axis is perpendicular to the plane of the paper in (a).

the molecule is nonlinear, we look for two or more rotational axes of threefold or higher order. If these are present, the molecule is classified in one of the groups related to the symmetry of the regular polyhedra (division IV). If these axes are not present, we look for any C_n axis at all. If there is no C_n axis, the molecule belongs to one of the groups $\mathscr{C}_s$, $\mathscr{C}_i$, $\mathscr{C}_1$ (division I). If there is at least one C_n axis, we pick the C_n axis of highest order as the main symmetry axis before proceeding to the next step. (If there are three mutually perpendicular C_2 axes, we may pick any one of these axes as the main axis.) We next check for n C_2 axes at right angles to the main C_n axis. If these are present, we have one of the division III groups. If these are absent, we have one of the division II groups. If we find the n C_2 axes, we look for a symmetry plane perpendicular to the main C_n axis. If it is present, the group is $\mathscr{D}_{nh}$. If it is absent, we check for n planes of symmetry containing the main C_n axis (if the molecule has three mutually perpendicular C_2 axes, we must try each axis as the main axis in looking for the two σ_v planes; the three C_2 axes are equivalent in the groups $\mathscr{D}_{nh}$ and $\mathscr{D}_n$, but not in $\mathscr{D}_{nd}$). If we find n σ_v planes, the group is $\mathscr{D}_{nd}$; otherwise it is $\mathscr{D}_n$. If the molecule does not have n C_2 axes perpendicular to the main C_n axis, we classify it in one of the groups $\mathscr{C}_{nh}$, $\mathscr{C}_{nv}$, $\mathscr{S}_{2n}$, or $\mathscr{C}_n$, by looking first for a σ_h plane, then for n σ_v planes, and, finally, if these are absent, checking whether or not the C_n axis is an S_{2n} axis. The procedure of Fig. 12.15 does not locate all symmetry elements. After classifying a molecule, check that all the required symmetry elements are indeed present. Although the above procedure might seem involved, it is really quite simple and is easily memorized.

The most common error students make in classifying a molecule is to miss the n C_2 axes perpendicular to the C_n axis of a molecule belonging to $\mathscr{D}_{nd}$. For example, it is easy to see that the C=C=C axis of allene is a C_2 axis, but the other two C_2 axes (Fig. 12.16) are often overlooked. Molecules with two equal halves "staggered" with respect to each other generally belong to $\mathscr{D}_{nd}$. Models may be helpful for those with visualization difficulties.

12.3 SUMMARY

A symmetry operation transforms an object into a position that is physically indistinguishable from the original position and preserves the distances between all pairs of points in the object. A symmetry element is a geometrical entity with respect to which a symmetry operation is performed. For molecules, the four kinds of symmetry elements are an n-fold axis of symmetry (C_n), a plane of symmetry (σ), a center of symmetry (i), and an n-fold rotation–reflection axis of symmetry (S_n). The product of symmetry operations means successive performance of them. We have $\hat{C}_n^n = \hat{E}$, where $\hat{E}$ is the identity operation; also, $\hat{S}_1 = \hat{\sigma}$, and $\hat{S}_2 = \hat{\imath}$, where the inversion operation moves a point at x, y, z to $-x$, $-y$, $-z$. Two symmetry operations may or may not commute.

For the symmetry operation $\hat{R}$ that brings a point at x, y, z to x', y', z', the operator $\hat{O}_R$ is defined by the equation $\hat{O}_R f(x', y', z') = f(x, y, z)$. If a molecule has the symmetry operations $\hat{R}_1, \hat{R}_2, \ldots$, then the operators $\hat{O}_{R_1}, \hat{O}_{R_2}, \ldots$, commute with the molecular Hamiltonian $\hat{H}$. If $\hat{R}_1, \hat{R}_2, \ldots$ all commute with one another, then the molecular wave functions can be taken as eigenfunctions of $\hat{O}_{R_1}, \hat{O}_{R_2}, \ldots$.

The set of all symmetry operations of a molecule constitutes a mathematical point group. The possible point groups of molecules are as follows. I. Groups with no C_n axis: $\mathscr{C}_1, \mathscr{C}_s, \mathscr{C}_i$. II. Groups with a single C_n axis: $\mathscr{C}_n, \mathscr{C}_{nh}, \mathscr{C}_{nv}, \mathscr{S}_{2n}$. III. Groups with one C_n axis and n C_2 axes: $\mathscr{D}_n, \mathscr{D}_{nh}, \mathscr{D}_{nd}$. IV. Groups with more than one C_n axis, $n > 2$: $\mathscr{T}_d, \mathscr{O}_h, \mathscr{I}_h$, and others.

PROBLEMS

Sec.	12.1	12.2	general
Probs.	12.1–12.15	12.16–12.32	12.33

12.1 True or false? (a) Symmetry operations always commute. (b) Symmetry operations never commute. (c) The product of two symmetry operations of a molecule must be a symmetry operation for that molecule.

12.2 Give all the symmetry elements of each of the following molecules: (a) H_2S; (b) NH_3; (c) CHF_3; (d) HOCl; (e) 1,3,5-trichlorobenzene; (f) CH_2F_2; (g) CHFClBr.

12.3 List all the symmetry operations of each of the molecules in Prob. 12.2.

12.4 Consider the square-planar ion $PtCl_4^{2-}$. Suppose we interchange two Cl atoms that are cis to each other. Does this interchange meet the definition of a symmetry operation (Section 12.1)? If so, express it in terms of some combination of the four kinds of symmetry operations discussed.

12.5 What symmetry operation is each of the following products of operations equal to? (a) $\hat{\sigma}^4$; (b) $\hat{\sigma}^7$; (c) $\hat{C}_4^2$; (d) $\hat{C}_4^6$; (e) $\hat{S}_4^2$; (f) $\hat{S}_6^3$; (g) $\hat{C}_{12}^3$; (h) $\hat{\imath}^3$.

12.6 Use Fig. 12.7 to state what symmetry operation in SF_6 each of the following products of symmetry operations is equal to. (a) $\hat{C}_2(x)\hat{C}_4(z)$; (b) $\hat{C}_4(z)\hat{C}_2(x)$.

12.7 For SF_6, which of the following pairs of operations commute? (a) $\hat{C}_4(z)$, $\hat{\sigma}(xy)$; (b) $\hat{C}_4(z)$, $\hat{\sigma}(yz)$; (c) $\hat{C}_2(z)$, $\hat{C}_2(x)$; (d) $\hat{\sigma}(xy)$, $\hat{\sigma}(yz)$; (e) $\hat{\imath}$, $\hat{\sigma}(xy)$.

12.8 What information does symmetry give about the dipole moment of each of the molecules in Prob. 12.2?

12.9 (a) Does H_2O_2 (Fig. 12.12) have an S_n axis? (b) Is it optically active? Explain.

12.10 For each of the following symmetry operations, find the matrix representative in the x, y, z basis. (a) $\hat{E}$; (b) $\hat{\sigma}(xy)$; (c) $\hat{\sigma}(yz)$; (d) $\hat{C}_2(x)$; (e) $\hat{S}_4(z)$; (f) $\hat{C}_3(z)$.

12.11 (a) Use SF_6 (Fig. 12.7) to verify that $\hat{C}_2(x)\hat{\sigma}(xy) = \hat{\sigma}(xz)$. (b) Write down the matrix representatives in the x, y, z basis of the three operations in part (a). Verify that these matrices multiply the same way the symmetry operations multiply.

12.12 (a) What are the eigenvalues of $\hat{O}_{C_4}$? (b) Is this operator Hermitian?

12.13 Do the same as in Prob. 12.12 for $\hat{O}_{C_2}$.

12.14 To what function is a $2p_z$ hydrogenlike orbital converted by (a) $\hat{O}_{C_4(z)}$; (b) $\hat{O}_{C_4(y)}$?

12.15 It is common to use rotation–inversion axes (rather than rotation–reflection axes) to classify the symmetry of crystals. Any S_n axis is equivalent to a rotation–inversion axis (symbolized by $\bar{p}$) whose order p may differ from n. A rotation–inversion operation consists of rotation by $2\pi/p$ radians followed by inversion. Show that

$$\hat{S}_n(z) = \hat{i}[\hat{C}_n(z)\hat{C}_2(z)]$$

Thus we have the following correspondence:

S_n	1	2	3	4
$\bar{p}$	$\bar{2}$	$\bar{1}$	$\bar{6}$	$\bar{4}$

Give the next three pairs of entries in this table.

12.16 State whether each of the following is a group. (a) All the integers (positive, negative, and zero) with the rule of combination for forming the product of two elements being addition. (b) All positive integers with the rule of combination being multiplication. (c) All real numbers except zero, with the rule of combination being multiplication.

12.17 Does the set of all square matrices of order four form a group if the rule for forming the "product" of two elements is matrix addition?

12.18 Give the point group of each of the following molecules. (a) CH_4; (b) CH_3F; (c) CH_2F_2; (d) CHF_3; (e) SF_6; (f) SF_5Br; (g) *trans*-SF_4Br_2; (h) CDH_3.

12.19 Give the point group of (a) $CH_2{=}CH_2$; (b) $CH_2{=}CHF$; (c) $CH_2{=}CF_2$; (d) *cis*-$CHF{=}CHF$; (e) *trans*-$CHF{=}CHF$.

12.20 Give the point group of (a) benzene; (b) fluorobenzene; (c) *o*-difluorobenzene; (d) *m*-difluorobenzene; (e) *p*-difluorobenzene; (f) 1,3,5-trifluorobenzene; (g) 1,4-difluoro-2,5-dibromobenzene; (h) naphthalene; (i) 2-chloronaphthalene.

12.21 Give the point group of (a) HCN; (b) H_2S; (c) CO_2; (d) CO; (e) C_2H_2;(f) CH_3OH; (g) ND_3; (h) OCS; (i) P_4; (j) PCl_3; (k) PCl_5; (l) $B_{12}Cl_{12}^{2-}$; (m) UF_6; (n) Ar.

12.22 Give the point group of (a) FeF_6^{3-}; (b) IF_5; (c) $CH_2{=}C{=}CH_2$; (d) C_8H_8, cubane; (e) $C_6H_6Cr(CO)_3$; (f) B_2H_6; (g) XeF_4; (h) F_2O; (i) spiropentane.

12.23 The number of members in a group is called the *order* of the group. Give the order of each of the following groups: (a) $\mathscr{C}_{3v}$; (b) $\mathscr{C}_s$; (c) $\mathscr{C}_{\infty v}$; (d) $\mathscr{D}_{3h}$.

12.24 The product of two members of a group must be a member of that group. (a) List the members (the symmetry operations) of the group $\mathscr{C}_{2v}$, using the x, y, and z axes to specify the axis or plane with respect to which each symmetry operation is performed. (b) For every possible product of two members of this group, state which symmetry operation it is equal to. Note that symmetry operations do not necessarily commute. (c) The *multiplication table* of a group shows all possible products of two members of the group. Write down the multiplication table for $\mathscr{C}_{2v}$ with each member of the group shown at the top and at the left of the table, and each entry in the table being the product of the member at the left of the row and the member at the top of the column of that entry.

12.25 The structure of ferrocene, $C_5H_5FeC_5H_5$, is an iron atom sandwiched midway between two parallel regular pentagons. For the eclipsed conformation, the vertices of the two pentagons are aligned; for the staggered conformation, one pentagon is rotated $2\pi/10$ radians with respect to the other. Electron diffraction results show that the gas-phase equilibrium conformation is the eclipsed one, with a quite low barrier to internal rotation of the rings. [See

S. Coriani et al., *ChemPhysChem*, **7,** 245 (2006).] What is the point group of (a) eclipsed ferrocene; (b) staggered ferrocene?

12.26 What is the point group of the tris(ethylenediamine)cobalt(III) complex ion? (Each $NH_2CH_2CH_2NH_2$ group occupies two adjacent positions of the octahedral coordination sphere.)

12.27 Consider a $\mathscr{D}_{nh}$ molecule with the z axis coinciding with the C_n axis and the x axis coinciding with one of the C_2 axes. Show that the product $\sigma(xy)\hat{C}_2(x)$ moves a point originally at (x, y, z) to $(x, -y, z)$. Therefore, $\hat{\sigma}(xy)\hat{C}_2(x) = \hat{\sigma}(xz)$. Since $\hat{C}_2(x)$ and $\hat{\sigma}(xy)$ are symmetry operations, their product must be a symmetry operation. Hence the xz plane is a symmetry plane. The same argument holds for any C_2 axis, so the molecule has n σ_v planes.

12.28 Give the point group of (a) a square-based pyramid; (b) a right circular cone; (c) a square lamina; (d) a square lamina with the top and bottom sides painted different colors; (e) a right circular cylinder; (f) a right circular cylinder with the two ends painted different colors; (g) a right circular cylinder with a stripe painted parallel to the axis; (h) a snowflake; (i) a doughnut; (j) a baseball (Fig. 12.17); (k) a $2p_z$ orbital; (l) a human being (ignore internal organs and slight external left–right asymmetries).

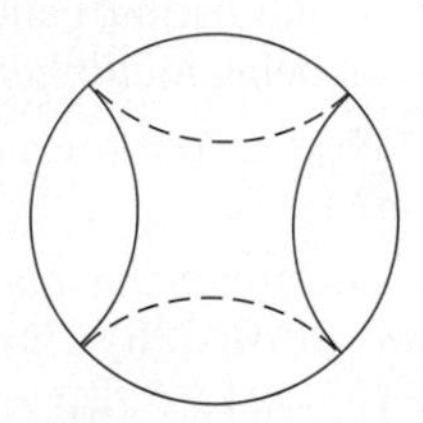

FIGURE 12.17 A baseball. The dashed and solid parts of the seam are in different hemispheres.

12.29 (a) What Platonic solid is dual to the regular tetrahedron? (b) How many vertices does a pentagonal dodecahedron have?

12.30 (a) For what values of n does the presence of an S_n axis imply the presence of a plane of symmetry? (b) For what values of n does the presence of an S_n axis imply the presence of a center of symmetry? (c) The group $\mathscr{D}_{nd}$ has an S_{2n} axis. For what values of n does it have a center of symmetry?

12.31 For which point groups can a molecule have a dipole moment?

12.32 For which point groups can a molecule be optically active?

12.33 Two people play the following game. Each in turn places a penny on the surface of a large chessboard. The pennies can be put anywhere on the board, as long as they do not overlap previously placed pennies. A penny may overlap more than one square. Once placed, a penny cannot be moved. When one of the players finds there is no room to place another penny on the board, she loses. With best play, will the person placing the first penny or her opponent win? Give the winning strategy.

CHAPTER 13

Electronic Structure of Diatomic Molecules

13.1 THE BORN–OPPENHEIMER APPROXIMATION

We now begin the study of molecular quantum mechanics. If we assume the nuclei and electrons to be point masses and neglect spin–orbit and other relativistic interactions (Sections 11.6 and 11.7), then the molecular Hamiltonian operator is

$$\hat{H} = -\frac{\hbar^2}{2}\sum_\alpha \frac{1}{m_\alpha}\nabla_\alpha^2 - \frac{\hbar^2}{2m_e}\sum_i \nabla_i^2 + \sum_\alpha\sum_{\beta>\alpha}\frac{Z_\alpha Z_\beta e'^2}{r_{\alpha\beta}} - \sum_\alpha\sum_i \frac{Z_\alpha e'^2}{r_{i\alpha}} + \sum_j\sum_{i>j}\frac{e'^2}{r_{ij}} \qquad (13.1)$$

where α and β refer to nuclei and i and j refer to electrons. The first term in (13.1) is the operator for the kinetic energy of the nuclei. The second term is the operator for the kinetic energy of the electrons. The third term is the potential energy of the repulsions between the nuclei, $r_{\alpha\beta}$ being the distance between nuclei α and β with atomic numbers Z_α and Z_β. The fourth term is the potential energy of the attractions between the electrons and the nuclei, $r_{i\alpha}$ being the distance between electron i and nucleus α. The last term is the potential energy of the repulsions between the electrons, r_{ij} being the distance between electrons i and j. The zero level of potential energy for (13.1) corresponds to having all the charges (electrons and nuclei) infinitely far from one another.

As an example, consider H_2. Let α and β be the two protons, 1 and 2 be the two electrons, and m_p be the proton mass. The H_2 molecular Hamiltonian operator is

$$\hat{H} = -\frac{\hbar^2}{2m_p}\nabla_\alpha^2 - \frac{\hbar^2}{2m_p}\nabla_\beta^2 - \frac{\hbar^2}{2m_e}\nabla_1^2 - \frac{\hbar^2}{2m_e}\nabla_2^2 + \frac{e'^2}{r_{\alpha\beta}} - \frac{e'^2}{r_{1\alpha}} - \frac{e'^2}{r_{1\beta}} - \frac{e'^2}{r_{2\alpha}} - \frac{e'^2}{r_{2\beta}} + \frac{e'^2}{r_{12}} \qquad (13.2)$$

The wave functions and energies of a molecule are found from the Schrödinger equation:

$$\hat{H}\psi(q_i, q_\alpha) = E\psi(q_i, q_\alpha) \qquad (13.3)$$

where q_i and q_α symbolize the electronic and nuclear coordinates, respectively. The molecular Hamiltonian (13.1) is formidable enough to terrify any quantum chemist. Fortunately, a very accurate, simplifying approximation exists. Since nuclei are much heavier than electrons ($m_\alpha \gg m_e$), the electrons move much faster than the nuclei. Hence, to a good approximation as far as the electrons are concerned, we can regard the nuclei as fixed while the electrons carry out their motions. Speaking classically, during the time of a cycle of electronic motion, the change in nuclear configuration is negligible. Thus, considering the nuclei as fixed, we omit the nuclear kinetic-energy terms from (13.1) to obtain the Schrödinger equation for electronic motion:

$$\boxed{(\hat{H}_{\text{el}} + V_{NN})\psi_{\text{el}} = U\psi_{\text{el}}} \quad \textbf{(13.4)}$$

where the **purely electronic Hamiltonian** $\hat{H}_{\text{el}}$ is

$$\hat{H}_{\text{el}} = -\frac{\hbar^2}{2m_e}\sum_i \nabla_i^2 - \sum_\alpha \sum_i \frac{Z_\alpha e'^2}{r_{i\alpha}} + \sum_j \sum_{i>j} \frac{e'^2}{r_{ij}} \quad (13.5)$$

The electronic Hamiltonian including nuclear repulsion is $\hat{H}_{\text{el}} + V_{NN}$. The nuclear-repulsion term V_{NN} is

$$V_{NN} = \sum_\alpha \sum_{\beta>\alpha} \frac{Z_\alpha Z_\beta e'^2}{r_{\alpha\beta}} \quad (13.6)$$

The energy U in (13.4) is the **electronic energy including internuclear repulsion**. The internuclear distances $r_{\alpha\beta}$ in (13.4) are not variables, but are each fixed at some constant value. Of course, there are an infinite number of possible nuclear configurations, and for each of these we may solve the electronic Schrödinger equation (13.4) to get a set of electronic wave functions and corresponding electronic energies. Each member of the set corresponds to a different molecular electronic state. The electronic wave functions and energies thus depend parametrically on the nuclear coordinates:

$$\psi_{\text{el}} = \psi_{\text{el},n}(q_i; q_\alpha) \quad \text{and} \quad U = U_n(q_\alpha)$$

where n symbolizes the electronic quantum numbers.

The variables in the electronic Schrödinger equation (13.4) are the electronic coordinates. The quantity V_{NN} is independent of these coordinates and is a constant for a given nuclear configuration. Now it is easily proved (Prob. 4.51) that the omission of a constant term C from the Hamiltonian does not affect the wave functions and simply decreases each energy eigenvalue by C. Hence, if V_{NN} is omitted from (13.4), we get

$$\hat{H}_{\text{el}}\psi_{\text{el}} = E_{\text{el}}\psi_{\text{el}} \quad (13.7)$$

where the **purely electronic energy** $E_{\text{el}}(q_\alpha)$ (which depends parametrically on the nuclear coordinates q_α) is related to the electronic energy including internuclear repulsion by

$$U = E_{\text{el}} + V_{NN} \quad (13.8)$$

We can therefore omit the internuclear repulsion from the electronic Schrödinger equation. After finding E_{el} for a particular configuration of the nuclei by solving (13.7), we calculate U using (13.8), where the constant V_{NN} is easily calculated from (13.6) using the assumed nuclear locations.

For H_2, with the two protons at a fixed distance $r_{\alpha\beta} = R$, the purely electronic Hamiltonian is given by (13.2) with the first, second, and fifth terms omitted. The nuclear repulsion V_{NN} equals e'^2/R. The purely electronic Hamiltonian involves the six

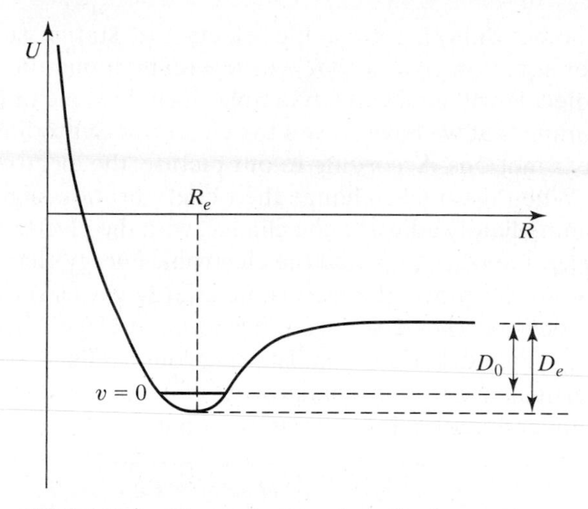

FIGURE 13.1 Electronic energy including internuclear repulsion as a function of the internuclear distance R for a diatomic-molecule bound electronic state.

electronic coordinates $x_1, y_1, z_1, x_2, y_2, z_2$ and involves the nuclear coordinates as parameters.

The electronic Schrödinger equation (13.4) can be dealt with by approximate methods to be discussed later. If we plot the electronic energy including nuclear repulsion for a bound state of a diatomic molecule against the internuclear distance R, we find a curve like Fig. 13.1. At $R = 0$, the internuclear repulsion causes U to go to infinity. The internuclear separation at the minimum in this curve is called the **equilibrium internuclear distance** R_e. The difference between the limiting value of U at infinite internuclear separation and its value at R_e is called the **equilibrium dissociation energy** (or the *dissociation energy from the potential-energy minimum*) D_e:

$$D_e \equiv U(\infty) - U(R_e) \tag{13.9}$$

When nuclear motion is considered (Section 13.2), one finds that the equilibrium dissociation energy D_e differs from the molecular ground-vibrational-state dissociation energy D_0. The lowest state of nuclear motion has zero rotational energy [as shown by Eq. (6.47)] but has a nonzero vibrational energy—the zero-point energy. If we use the harmonic-oscillator approximation for the vibration of a diatomic molecule (Section 4.3), then this zero-point energy is $\frac{1}{2}h\nu$. This zero-point energy raises the energy for the ground state of nuclear motion $\frac{1}{2}h\nu$ above the minimum in the $U(R)$ curve, so D_0 is less than D_e and $D_0 \approx D_e - \frac{1}{2}h\nu$. Different electronic states of the same molecule have different $U(R)$ curves (Figs. 13.5 and 13.19) and different values of R_e, D_e, D_0, and ν.

Consider an ideal gas composed of diatomic molecules AB. In the limit of absolute zero temperature, all the AB molecules are in their ground states of electronic and nuclear motion, so $D_0 N_A$ (where N_A is the Avogadro constant and D_0 is for the ground electronic state of AB) is the change in the thermodynamic internal energy U and enthalpy H for dissociation of 1 mole of ideal-gas diatomic molecules: $N_A D_0 = \Delta U_0^\circ = \Delta H_0^\circ$ for $AB(g) \rightarrow A(g) + B(g)$.

For some diatomic-molecule electronic states, solution of the electronic Schrödinger equation gives a $U(R)$ curve with no minimum. Such states are not bound and the molecule will dissociate. Examples include some of the states in Fig. 13.5.

Assuming that we have solved the electronic Schrödinger equation, we next consider nuclear motions. According to our picture, the electrons move much faster than the nuclei. When the nuclei change their configuration slightly, say from q'_α to q''_α, the electrons immediately adjust to the change, with the electronic wave function changing from $\psi_{\text{el}}(q_i; q'_\alpha)$ to $\psi_{\text{el}}(q_i; q''_\alpha)$ and the electronic energy changing from $U(q'_\alpha)$ to $U(q''_\alpha)$. Thus, as the nuclei move, the electronic energy varies smoothly as a function of the parameters defining the nuclear configuration, and $U(q_\alpha)$ becomes, in effect, the potential energy for the nuclear motion. The electrons act like a spring connecting the nuclei. As the internuclear distance changes, the energy stored in the spring changes. Hence the Schrödinger equation for nuclear motion is

$$\boxed{\hat{H}_N \psi_N = E \psi_N} \tag{13.10}$$

$$\hat{H}_N = -\frac{\hbar^2}{2} \sum_\alpha \frac{1}{m_\alpha} \nabla_\alpha^2 + U(q_\alpha) \tag{13.11}$$

The variables in the nuclear Schrödinger equation are the nuclear coordinates, symbolized by q_α. The energy eigenvalue E in (13.10) is the *total* energy of the molecule, since the Hamiltonian (13.11) includes operators for both nuclear energy and electronic energy. E is simply a number and does not depend on any coordinates. Note that for each electronic state of a molecule we must solve a different nuclear Schrödinger equation, since U differs from state to state. In this chapter we shall concentrate on the electronic Schrödinger equation (13.4).

In Section 13.2, we shall show that the total energy E for an electronic state of a diatomic molecule is approximately the sum of electronic, vibrational, rotational, and translational energies, $E \approx E_{\text{elec}} + E_{\text{vib}} + E_{\text{rot}} + E_{\text{tr}}$, where the constant E_{elec} [not to be confused with E_{el} in (13.7)] is given by $E_{\text{elec}} = U(R_e)$.

The approximation of separating electronic and nuclear motions is called the **Born–Oppenheimer approximation** and is basic to quantum chemistry. [The American physicist J. Robert Oppenheimer (1904–1967) was a graduate student of Born in 1927. During World War II, Oppenheimer directed the Los Alamos laboratory that developed the atomic bomb.] Born and Oppenheimer's mathematical treatment indicated that the true molecular wave function is adequately approximated as

$$\psi(q_i, q_\alpha) = \psi_{\text{el}}(q_i; q_\alpha)\psi_N(q_\alpha) \tag{13.12}$$

if $(m_e/m_\alpha)^{1/4} \ll 1$. The Born–Oppenheimer approximation introduces little error for the ground electronic states of diatomic molecules. Corrections for excited electronic states are larger than for the ground state, but still are usually small as compared with the errors introduced by the approximations used to solve the electronic Schrödinger equation of a many-electron molecule. Hence we shall not worry about corrections to the Born–Oppenheimer approximation. For further discussion of the Born–Oppenheimer approximation, see J. Goodisman, *Diatomic Interaction Potential Theory*, Academic Press, 1973, Volume 1, Chapter 1.

Born and Oppenheimer's 1927 paper justifying the Born–Oppenheimer approximation is seriously lacking in rigor. Subsequent work has better justified the Born–Oppenheimer approximation, but significant questions still remain; "the problem of the coupling of nuclear and electronic motions is, at the moment, without a sensible solution and ... is an area where much future work can and must be done" [B. T. Sutcliffe, *J. Chem. Soc. Faraday Trans.*, **89**, 2321 (1993); see also B. T. Sutcliffe and R. G. Woolley, *Phys. Chem. Chem. Phys.*, **7**, 3664 (2005)].

13.2 NUCLEAR MOTION IN DIATOMIC MOLECULES

Most of this chapter deals with the electronic Schrödinger equation for diatomic molecules, but this section examines nuclear motion in a bound electronic state of a diatomic molecule. From (13.10) and (13.11), the Schrödinger equation for nuclear motion in a diatomic-molecule bound electronic state is

$$\left[-\frac{\hbar^2}{2m_\alpha}\nabla_\alpha^2 - \frac{\hbar^2}{2m_\beta}\nabla_\beta^2 + U(R)\right]\psi_N = E\psi_N \tag{13.13}$$

where α and β are the nuclei, and the nuclear-motion wave function ψ_N is a function of the nuclear coordinates $x_\alpha, y_\alpha, z_\alpha, x_\beta, y_\beta, z_\beta$.

The potential energy $U(R)$ is a function of only the relative coordinates of the two nuclei, and the work of Section 6.3 shows that the two-particle Schrödinger equation (13.13) can be reduced to two separate one-particle Schrödinger equations, one for translational energy of the entire molecule, and one for internal motion of the nuclei relative to each other. We have

$$\psi_N = \psi_{N,\text{tr}}\psi_{N,\text{int}} \quad \text{and} \quad E = E_{\text{tr}} + E_{\text{int}} \tag{13.14}$$

The translational energy levels can be taken as the energy levels (3.72) of a particle in a three-dimensional box whose dimensions are those of the container holding the gas of diatomic molecules.

The Schrödinger equation for $\psi_{N,\text{int}}$ is [Eq. (6.43)]

$$\left[-\frac{\hbar^2}{2\mu}\nabla^2 + U(R)\right]\psi_{N,\text{int}} = E_{\text{int}}\psi_{N,\text{int}}, \qquad \mu \equiv m_\alpha m_\beta/(m_\alpha + m_\beta) \tag{13.15}$$

where $\psi_{N,\text{int}}$ is a function of the coordinates of one nucleus relative to the other. The best coordinates to use are the spherical coordinates of one nucleus relative to the other (Fig. 6.5 with m_N and m_e replaced by m_α and m_β). The radius r in relative spherical coordinates is the internuclear distance R, and we shall denote the relative angular coordinates by θ_N and ϕ_N. Since the potential energy in (13.15) depends on R only, this is a central-force problem, and the work of Section 6.1 shows that

$$\psi_{N,\text{int}} = P(R)Y_J^M(\theta_N, \phi_N), \qquad J = 0, 1, 2, \ldots, \quad M = -J, \ldots, J \tag{13.16}$$

where the Y_J^M functions are the spherical harmonic functions with quantum numbers J and M.

From (6.17), the radial function $P(R)$ is found by solving

$$-\frac{\hbar^2}{2\mu}\left[P''(R) + \frac{2}{R}P'(R)\right] + \frac{J(J+1)\hbar^2}{2\mu R^2}P(R) + U(R)P(R) = E_{\text{int}}P(R) \tag{13.17}$$

This differential equation is simplified by defining $F(R)$ as

$$F(R) \equiv RP(R) \tag{13.18}$$

Substitution of $P = F/R$ into (13.17) gives [Eq. (6.137)]

$$-\frac{\hbar^2}{2\mu}F''(R) + \left[U(R) + \frac{J(J+1)\hbar^2}{2\mu R^2}\right]F(R) = E_{\text{int}}F(R) \tag{13.19}$$

which looks like a one-dimensional Schrödinger equation with the effective potential energy $U(R) + J(J+1)\hbar^2/2\mu R^2$.

The most fundamental way to solve (13.19) is as follows: (a) Solve the electronic Schrödinger equation (13.7) at several values of R to obtain E_{el} of the particular molecular electronic state you are interested in; (b) add $Z_\alpha Z_\beta e'^2/R$ to each E_{el} value to obtain U at these R values; (c) devise a mathematical function $U(R)$ whose parameters are adjusted to give a good fit to the calculated U values; (d) insert the function $U(R)$ found in (c) into the nuclear-motion radial Schrödinger equation (13.19) and solve (13.19) by numerical methods.

A commonly used fitting procedure for step (c) is the method of *cubic splines*, for which computer programs exist (see *Press et al.*, Chapter 3; *Shoup*, Chapter 6).

As for step (d), numerical solution of the one-dimensional Schrödinger equation (13.19) is done using either the *Cooley–Numerov method* [see J. Tellinghuisen, *J. Chem. Educ.*, **66**, 51 (1989)], which is a modification of the Numerov method (Sections 4.4 and 6.9), or the *finite-element method* [see D. J. Searles and E. I. von Nagy-Felsobuki, *Am. J. Phys.*, **56**, 444 (1988)].

The solutions $F(R)$ of the radial equation (13.19) for a given J are characterized by a quantum number v, where v is the number of nodes in $F(R)$; $v = 0, 1, 2, \ldots$. The energy levels E_{int} [which are found from the condition that $P(R) = F(R)/R$ be quadratically integrable] depend on the quantum number J, which occurs in (13.19), and depend on v, which characterizes $F(R)$; $E_{\text{int}} = E_{v,J}$. The angular factor $Y_J^M(\theta_N, \phi_N)$ in (13.16) is a function of the angular coordinates. Changes in θ_N and ϕ_N with R held fixed correspond to changes in the spatial orientation of the diatomic molecule, which is rotational motion. The quantum numbers J and M are rotational quantum numbers. Note that Y_J^M is the wave function of a rigid two-particle rotor [Eq. (6.46)]. A change in the R coordinate is a change in the internuclear distance, which is a vibrational motion, and the quantum number v, which characterizes $F(R)$, is a vibrational quantum number.

Since accurate solution of the electronic Schrödinger equation [step (a)] is hard, one often uses simpler, less-accurate procedures than that of steps (a) to (d). The simplest approach is to expand $U(R)$ in a Taylor series about R_e (Prob. 4.1):

$$U(R) = U(R_e) + U'(R_e)(R - R_e) + \tfrac{1}{2}U''(R_e)(R - R_e)^2 + \tfrac{1}{6}U'''(R_e)(R - R_e)^3 + \cdots \tag{13.20}$$

At the equilibrium internuclear distance R_e, the slope of the $U(R)$ curve is zero (Fig. 13.1), so $U'(R_e) = 0$. We can anticipate that the molecule will vibrate about the equilibrium distance R_e. For R close to R_e, $(R - R_e)^3$ and higher powers will be small, and we shall neglect these terms. Defining the **equilibrium force constant** k_e as $k_e \equiv U''(R_e)$, we have

$$U(R) \approx U(R_e) + \tfrac{1}{2}k_e(R - R_e)^2 = U(R_e) + \tfrac{1}{2}k_e x^2 \tag{13.21}$$

$$k_e \equiv U''(R_e) \quad \text{and} \quad x \equiv R - R_e$$

We have approximated $U(R)$ by a parabola [Fig. 4.5 with $V \equiv U(R) - U(R_e)$ and $x \equiv R - R_e$].

With the change of independent variable $x \equiv R - R_e$, (13.19) becomes

$$-\frac{\hbar^2}{2\mu}S''(x) + \left[U(R_e) + \tfrac{1}{2}k_e x^2 + \frac{J(J+1)\hbar^2}{2\mu(x+R_e)^2}\right]S(x) \approx E_{\text{int}}S(x) \tag{13.22}$$

$$\text{where} \quad S(x) \equiv F(R) \tag{13.23}$$

Expanding $1/(x + R_e)^2$ in a Taylor series, we have (Prob. 13.6)

$$\frac{1}{(x+R_e)^2} = \frac{1}{R_e^2(1 + x/R_e)^2} = \frac{1}{R_e^2}\left(1 - 2\frac{x}{R_e} + 3\frac{x^2}{R_e^2} - \cdots\right) \approx \frac{1}{R_e^2} \tag{13.24}$$

We are assuming that $R - R_e = x$ is small, so all terms after the 1 have been neglected in (13.24). Substitution of (13.24) into (13.22) and rearrangement gives

$$-\frac{\hbar^2}{2\mu}S''(x) + \tfrac{1}{2}k_e x^2 S(x) \approx \left[E_{\text{int}} - U(R_e) - \frac{J(J+1)\hbar^2}{2\mu R_e^2}\right]S(x) \tag{13.25}$$

Equation (13.25) is the same as the Schrödinger equation for a one-dimensional harmonic oscillator with coordinate x, mass μ, potential energy $\tfrac{1}{2}k_e x^2$, and energy eigenvalues $E_{\text{int}} - U(R_e) - J(J+1)\hbar^2/2\mu R_e^2$. [The boundary conditions for (13.25) and (4.34) are not the same, but this difference is unimportant and can be ignored (*Levine, Molecular Spectroscopy*, p. 147).] We can therefore set the terms in brackets in (13.25) equal to the harmonic-oscillator eigenvalues, and we have

$$E_{\text{int}} - U(R_e) - J(J+1)\hbar^2/2\mu R_e^2 \approx (v + \tfrac{1}{2})h\nu_e$$

$$E_{\text{int}} \approx U(R_e) + (v + \tfrac{1}{2})h\nu_e + J(J+1)\hbar^2/2\mu R_e^2 \tag{13.26}$$

$$\nu_e = (k_e/\mu)^{1/2}/2\pi, \qquad v = 0, 1, 2, \ldots \tag{13.27}$$

where (4.25) was used for ν_e, the **equilibrium** (or **harmonic**) **vibrational frequency**. The molecular internal energy E_{int} is approximately the sum of the electronic energy $U(R_e) = E_{\text{elec}}$ (which differs for different electronic states of the same molecule), the vibrational energy $(v + \tfrac{1}{2})h\nu_e$, and the rotational energy $J(J+1)\hbar^2/2\mu R_e^2$. The approximations (13.21) and (13.24) correspond to a harmonic-oscillator, rigid-rotor approximation. From (13.26) and (13.14), the molecular energy $E = E_{\text{tr}} + E_{\text{int}}$ is approximately the sum of translational, rotational, vibrational, and electronic energies:

$$E \approx E_{\text{tr}} + E_{\text{rot}} + E_{\text{vib}} + E_{\text{elec}}$$

From (13.14), (13.16), (13.18), and (13.23), the nuclear-motion wave function is

$$\psi_N \approx \psi_{N,\text{tr}} S_v(R - R_e) R^{-1} Y_J^M(\theta_N, \phi_N) \tag{13.28}$$

where $S_v(R - R_e)$ is a harmonic-oscillator eigenfunction with quantum number v.

The approximation (13.26) gives rather poor agreement with experimentally observed vibration–rotation energy levels of diatomic molecules. The accuracy can be improved by the addition of the first- and second-order perturbation-theory energy corrections due to the terms neglected in (13.21) and (13.24). When this is done (see *Levine, Molecular Spectroscopy*, Section 4.2), the energy contains additional terms corresponding to vibrational anharmonicity [Eq. (4.62)], vibration–rotation interaction, and rotational centrifugal distortion of the molecule.

EXAMPLE An approximate representation of the potential-energy function of a diatomic molecule is the Morse function

$$U(R) = U(R_e) + D_e[1 - e^{-a(R-R_e)}]^2$$

Use of $U''(R_e) = k_e$ [Eq. (4.61)] and (13.27) gives $a = (k_e/2D_e)^{1/2} = 2\pi\nu_e(\mu/2D_e)^{1/2}$ (Prob. 4.28; the Morse functions in Prob. 4.28 and in this example differ because of different choices for the zero of energy). Use the Morse function and the Numerov method (Section 4.4) to (a) find the lowest six vibrational energy levels of the $^1H_2$ molecule in its ground electronic state, which has $D_e/hc = 38297\text{ cm}^{-1}$, $\nu_e/c = 4403.2\text{ cm}^{-1}$, and $R_e = 0.741$ Å, where h and c are Planck's constant and the speed of light; (b) find $\langle R\rangle$ for each of these vibrational states.

(a) The vibrational energy levels correspond to states with the rotational quantum number $J = 0$. Making the change of variables $x \equiv R - R_e$ and $S(x) \equiv F(R)$ [Eq. (13.23)] and substituting the Morse function into the nuclear-motion Schrödinger equation (13.19), we get for $J = 0$

$$-(\hbar^2/2\mu)S''(x) + D_e(1 - e^{-ax})^2S(x) = [E_{\text{int}} - U(R_e)]S(x) = E_{\text{vib}}S(x)$$

since for $J = 0$, $E_{\text{int}} = E_{\text{elec}} + E_{\text{vib}} = U(R_e) + E_{\text{vib}}$ [Eq. (13.26)]. As usual in the Numerov method, we switch to the dimensionless reduced variables $E_{\text{vib},r} \equiv E_{\text{vib}}/A$ and $x_r \equiv x/B$, where A and B are products of powers of the constants $\hbar$, μ, and a. The procedure of Section 4.4 gives (Prob. 13.7a) $A = \hbar^2a^2/\mu$ and $B = a^{-1}$, so

$$x_r \equiv x/B = ax, \qquad E_{\text{vib},r} \equiv E_{\text{vib}}/A = \mu E_{\text{vib}}/\hbar^2a^2 = E_{\text{vib}}(2D_e/h^2\nu_e^2)$$

where $a = 2\pi\nu_e(\mu/2D_e)^{1/2}$ was used. Substitution of $x = x_r/a$, $E_{\text{vib}} = \hbar^2a^2E_{\text{vib},r}/\mu$, $D_{e,r} = D_e/(\hbar^2a^2/\mu)$, $S(x) = S_r(x_r)B^{-1/2}$, and $S'' = B^{-5/2}S_r'' = B^{-1/2}B^{-2}S_r'' = B^{-1/2}a^2S_r''$ [Eqs. (4.80) and (4.81)] into the differential equation for $S(x)$ gives

$$S_r''(x_r) = [2D_{e,r}(1 - e^{-x_r})^2 - 2E_{\text{vib},r}]S_r(x_r) \equiv G_rS_r(x_r)$$

This last equation has the form of (4.84) with $G_r \equiv 2D_{e,r}(1 - e^{-x_r})^2 - 2E_{\text{vib},r}$, so we are now ready to apply the Numerov procedure of Section 4.4. For the H_2 ground electronic state, we find (Prob. 13.7b)

$$A = h^2\nu_e^2/2D_e = h^2c^2(4403.2\text{ cm}^{-1})^2/2hc(38297\text{ cm}^{-1}) = (253.12_9\text{ cm}^{-1})hc$$

$$B = 0.51412\text{ Å}, \qquad D_{e,r} = D_e/A = 151.29_4$$

We want to start and end the Numerov procedure in the classically forbidden regions. If we used the harmonic-oscillator approximation for the vibrational levels, the energies of the first six vibrational levels would be $(v + \frac{1}{2})h\nu_e$, $v = 0, 1, \ldots, 5$. The reduced energy of the sixth harmonic-oscillator vibrational level would be $5.5h\nu_e/A = 5.5h\nu_e/(253\text{ cm}^{-1})hc = 5.5(4403\text{ cm}^{-1})/(253\text{ cm}^{-1}) = 95.7$. Because of anharmonicity (Section 4.3), the sixth vibrational level will actually occur below 95.7, so we are safe in using 95.7 to find the limits of the classically allowed region. We have $D_{e,r}(1 - e^{-x_r})^2 = 95.7$, and with $D_{e,r} = 151.29$, we find $x_r = -0.58$ and $x_r = 1.58$ as the limits of the classically allowed region for a reduced energy of 95.7. Extending the range by 1.2 at each end, we would start the Numerov procedure at $x_r = -1.8$ and end at $x_r = 2.8$. However,

$x_r = (R - R_e)/B = (R - 0.741 \text{ Å})/(0.514 \text{ Å})$ and the minimum possible internuclear distance R is 0, so the minimum possible value of x_r is -1.44. We therefore start at $x_r = -1.44$ and end at 2.8. If we take an interval of $s_r = 0.04$, we will have about 106 points, which is adequate, but we will try for higher accuracy by taking $s_r = 0.02$ to give about 212 points. With these choices, we set up the Numerov spreadsheet in the usual manner (or use Mathcad or the computer program of Table 4.1); we find (Prob. 13.8) the following lowest six $E_{\text{vib},r}$ values: 8.572525, 24.967566, 40.362582, 54.757570, 68.152531, 80.547472. Using $E_{\text{vib},r} \equiv E_{\text{vib}}/A$, we find the lowest levels to be $E_{\text{vib}}/hc = 2169.95,\ 6320.01, 10216.94, 13860.73, 17251.38, 20388.90 \text{ cm}^{-1}$. Note the reduced spacings between levels as the vibrational quantum number increases. (For comparison, the harmonic-oscillator approximation gives the following values: 2201.6, 6604.8, 11008.0, 15411.2, 19814.4, 24217.6 cm^{-1}.)

It happens that the Schrödinger equation for the Morse function can be analytically solved virtually exactly, and the analytic solution (Prob. 13.10) gives the following lowest eigenvalues: 2169.96, 6320.03, 10216.97, 13860.78, 17251.47, 20389.02 cm^{-1}. Agreement between the Numerov Morse-function values and the analytic Morse-function values is very good. The experimentally observed lowest vibrational levels of H_2 are 2170.08, 6331.22, 10257.19, 13952.43, 17420.44, 20662.00 cm^{-1}. The deviations of the Morse-function values from the experimental values indicate that the Morse function is not a very accurate representation of the ground-state H_2 $U(R)$ function.

(b) We have $\langle x_r \rangle \approx \int_{-1.44}^{2.8} x_r |S_r|^2 dx_r \approx \sum_{x_r=1.44}^{2.8} x_r |S_r|^2 s_r$, where $s_r = 0.02$ is the interval spacing (not to be confused with the vibrational wave function S_r), and where the vibrational wave function S_r must be normalized. (See also Prob. 13.11.) We normalize S_r as described in Section 4.4, and then create a column of $x_r|S_r|^2 s_r$ values. Then we sum these values to find the following results for the six lowest vibrational states (Prob. 13.8b) $\langle x_r \rangle = 0.0440, 0.1365, 0.2360, 0.3435, 0.4605, 0.5884$. Using $x_r = (R - R_e)/B$, we find the following values: $\langle R \rangle = 0.763, 0.811, 0.862, 0.918, 0.978, 1.044$ Å. (To get accurate $\langle x_r \rangle$ values, $E_{\text{vib},r}$ must be found to many more decimal places than given in (a)—enough places to make the wave function close to zero at 2.8. If the spreadsheet does not allow you to enter enough decimal places to do this for $v = 0$, you can take the right-hand limit as 2.5 instead of 2.8.) Because of vibrational anharmonicity, the molecule gets larger as the vibrational quantum number increases. This effect is rather large for light atoms like hydrogen. The $v = 5$ Numerov–Morse vibrational wave function (Fig. 13.2) shows

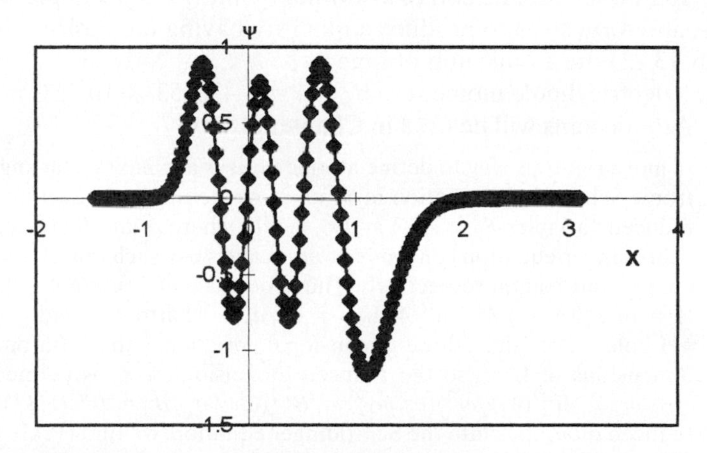

FIGURE 13.2 The $v = 5$ Morse vibrational wave function for H_2 as found by the Numerov method.

marked asymmetry about the origin ($x_r = 0$, which corresponds to $R = R_e$). For a spectacular example of the effect of anharmonicity on bond length, see the discussion of He_2 (the world's largest diatomic molecule) near the end of Section 13.7.

13.3 ATOMIC UNITS

Most quantum chemists report the results of their calculations using **atomic units**.

The hydrogen-atom Hamiltonian operator (assuming infinite nuclear mass) in SI units is $-(\hbar^2/2m_e)\nabla^2 - e^2/4\pi\varepsilon_0 r$. The system of atomic units is defined as follows. The units of mass, charge, and angular momentum are defined as the electron's mass m_e, the proton's charge e, and $\hbar$, respectively (rather than the kilogram, the coulomb, and the kg m^2/s); the unit of permittivity is $4\pi\varepsilon_0$, rather than the C^2 N^{-1} m^{-2}. (The atomic unit of mass used in quantum chemistry should not be confused with the quantity 1 amu, which is one-twelfth the mass of a ^{12}C atom.) When we switch to atomic units, $\hbar$, m_e, e, and $4\pi\varepsilon_0$ each have a numerical value of 1. Hence, to change a formula from SI units to atomic units, we simply set each of these quantities equal to 1. Thus, in SI atomic units, the H-atom Hamiltonian is $-\frac{1}{2}\nabla^2 - 1/r$, where r is now measured in atomic units of length rather than in meters. The ground-state energy of the hydrogen atom is given by (6.94) as $-\frac{1}{2}(e^2/4\pi\varepsilon_0 a_0)$. Since [Eq. (6.106)] $a_0 = 4\pi\varepsilon_0\hbar^2/m_e e^2$, the numerical value of a_0 (the Bohr radius) in atomic units is 1, and the ground-state energy of the hydrogen atom has the numerical value (neglecting nuclear motion) $-\frac{1}{2}$ in atomic units.

The atomic unit of energy, $e^2/4\pi\varepsilon_0 a_0$, is called the **hartree** (symbol E_h):

$$1 \text{ hartree} \equiv E_h \equiv e'^2/a_0 = e^2/4\pi\varepsilon_0 a_0 = m_e e^4/(4\pi\varepsilon_0)^2\hbar^2 = 27.21138 \text{ eV} \qquad (13.29)$$

The ground-state energy of the hydrogen atom is $-\frac{1}{2}$ hartree if nuclear motion is neglected. The atomic unit of length is called the **bohr**:

$$1 \text{ bohr} \equiv a_0 \equiv \hbar^2/m_e e'^2 = 4\pi\varepsilon_0\hbar^2/m_e e^2 = 0.5291772 \text{ Å} \qquad (13.30)$$

To find the atomic unit of any other quantity (for example, time) one combines $\hbar$, m_e, e, and $4\pi\varepsilon_0$ so as to produce a quantity having the desired dimensions. One finds (Prob. 13.13) the atomic unit of time to be $\hbar/E_h = 2.4188843 \times 10^{-17}$ s and the atomic unit of electric dipole moment to be $ea_0 = 8.478353 \times 10^{-30}$ C m.

Atomic units will be used in Chapters 13 to 17.

A more rigorous way to define atomic units is as follows. Starting with the H-atom electronic Schrödinger equation in SI units, we define (as in Section 4.4) the dimensionless reduced variables $E_r \equiv E/A$ and $r_r \equiv r/B$, where A and B are products of powers of the Schrödinger-equation constants $\hbar$, m_e, e, and $4\pi\varepsilon_0$ such that A and B have dimensions of energy and length, respectively. The procedure of Section 4.4 shows that (Prob. 13.12) $A = m_e e^4/(4\pi\varepsilon_0)^2\hbar^2 = e^2/4\pi\varepsilon_0 a_0 = e'^2/a_0 \equiv 1\,\text{hartree}$ and $B = \hbar^2 4\pi\varepsilon_0/m_e e^2 = a_0 \equiv 1$ bohr. For this three-dimensional problem, the H-atom wave function has dimensions of L$^{-3/2}$, so the reduced dimensionless ψ_r is defined as $\psi_r \equiv \psi B^{3/2}$. Also, $\partial^2\psi_r/\partial r_r^2 = B^{3/2}(\partial^2\psi/\partial r^2)(\partial r/\partial r_r)^2 = B^{3/2}(\partial^2\psi/\partial r^2)B^2 = B^{3/2}(\partial^2\psi/\partial r^2)a_0^2$. Introducing the reduced quantities into the Schrödinger equation, we find (Prob. 13.12) that the reduced H-atom Schrödinger equation is $-\frac{1}{2}\nabla_r^2\psi_r - (1/r_r)\psi_r = E_r\psi_r$, where ∇_r^2 is given by Eq. (6.6), with r replaced by r_r. In practice, people do not bother to include the r subscripts and instead write $-\frac{1}{2}\nabla^2\psi - (1/r)\psi = E\psi$.

13.4 THE HYDROGEN MOLECULE ION

We now begin the study of the electronic energies of molecules. We shall use the Born–Oppenheimer approximation, keeping the nuclei fixed while we solve, as best we can, the Schrödinger equation for the motion of the electrons. We shall usually be considering an isolated molecule, ignoring intermolecular interactions. Our results will be most applicable to molecules in the gas phase at low pressure. For inclusion of solvent effects, see Sections 15.17 and 17.6.

We start with diatomic molecules, the simplest of which is H_2^+, the hydrogen molecule ion, consisting of two protons and one electron. Just as the one-electron H atom serves as a starting point in the discussion of many-electron atoms, the one-electron H_2^+ ion furnishes many ideas useful for discussing many-electron diatomic molecules. The electronic Schrödinger equation for H_2^+ is separable, and we can get exact solutions for the eigenfunctions and eigenvalues.

Figure 13.3 shows H_2^+. The nuclei are at a and b; R is the internuclear distance; r_a and r_b are the distances from the electron to nuclei a and b. Since the nuclei are fixed, we have a one-particle problem whose purely electronic Hamiltonian is [Eq. (13.5)]

$$\hat{H}_{\text{el}} = -\frac{\hbar^2}{2m_e}\nabla^2 - \frac{e'^2}{r_a} - \frac{e'^2}{r_b} \tag{13.31}$$

The first term is the electronic kinetic-energy operator; the second and third terms are the attractions between the electron and the nuclei. In atomic units the purely electronic Hamiltonian for H_2^+ is

$$\hat{H}_{\text{el}} = -\tfrac{1}{2}\nabla^2 - \frac{1}{r_a} - \frac{1}{r_b} \tag{13.32}$$

In Fig. 13.3 the coordinate origin is on the internuclear axis, midway between the nuclei, with the z axis lying along the internuclear axis. The H_2^+ electronic Schrödinger equation is not separable in spherical coordinates. However, separation of variables is possible using *confocal elliptic coordinates* ξ, η, and ϕ. The coordinate ϕ is the angle of rotation of the electron about the internuclear (z) axis, the same as in spherical coordinates. The coordinates ξ (xi) and η (eta) are defined by

$$\xi \equiv \frac{r_a + r_b}{R}, \qquad \eta \equiv \frac{r_a - r_b}{R} \tag{13.33}$$

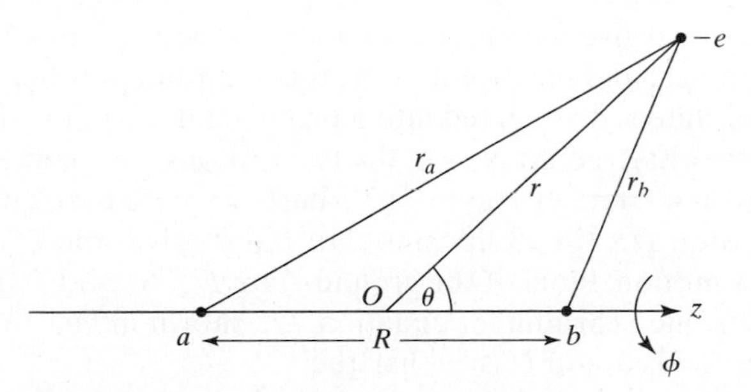

FIGURE 13.3 Interparticle distances in H_2^+.

The ranges of these coordinates are

$$0 \le \phi \le 2\pi, \qquad 1 \le \xi \le \infty, \qquad -1 \le \eta \le 1 \tag{13.34}$$

We must put the Hamiltonian (13.32) into these coordinates. We have

$$r_a = \tfrac{1}{2}R(\xi + \eta), \qquad r_b = \tfrac{1}{2}R(\xi - \eta) \tag{13.35}$$

We also need the expression for the Laplacian in confocal elliptic coordinates. One way to find this is to express ξ, η, and ϕ in terms of x, y, and z, the Cartesian coordinates of the electron, and then use the chain rule to find $\partial/\partial x$, $\partial/\partial y$, and $\partial/\partial z$ in terms of $\partial/\partial\xi$, $\partial/\partial\eta$, and $\partial/\partial\phi$. We then form $\nabla^2 \equiv \partial^2/\partial x^2 + \partial^2/\partial y^2 + \partial^2/\partial z^2$. The derivation of ∇^2 is omitted. (For a discussion, see *Margenau and Murphy*, Chapter 5.) Substitution of ∇^2 and (13.35) into (13.32) gives $\hat{H}_{el}$ of H_2^+ in confocal elliptic coordinates. The result is omitted.

For the hydrogen atom, whose Hamiltonian has spherical symmetry, the electronic angular-momentum operators $\hat{L}^2$ and $\hat{L}_z$ both commute with $\hat{H}$. The H_2^+ ion does not have spherical symmetry, and one finds that $[\hat{L}^2, \hat{H}_{el}] \neq 0$ for H_2^+. However, H_2^+ does have axial symmetry, and one can show that $\hat{L}_z$ commutes with $\hat{H}_{el}$ of H_2^+. Therefore, the electronic wave functions can be chosen to be eigenfunctions of $\hat{L}_z$. The eigenfunctions of $\hat{L}_z$ are [Eq. (5.81)]

$$\text{constant} \cdot (2\pi)^{-1/2} e^{im\phi}, \qquad \text{where } m = 0, \pm 1, \pm 2, \pm 3, \ldots \tag{13.36}$$

The z component of electronic orbital angular momentum in H_2^+ is $m\hbar$ or m in atomic units. The total electronic orbital angular momentum is not a constant for H_2^+.

The "constant" in (13.36) is a constant only as far as $\partial/\partial\phi$ is concerned, so the H_2^+ wave functions have the form $\psi_{el} = F(\xi, \eta)(2\pi)^{-1/2}e^{im\phi}$. One now tries a separation of variables:

$$\psi_{el} = L(\xi)M(\eta)(2\pi)^{-1/2}e^{im\phi} \tag{13.37}$$

Substitution of (13.37) into $\hat{H}_{el}\psi_{el} = E_{el}\psi_{el}$ gives an equation in which the variables are separable. One gets two ordinary differential equations, one for $L(\xi)$ and one for $M(\eta)$. Solving these equations, one finds that the condition that ψ_{el} be well-behaved requires that, for each fixed value of R, only certain values of E_{el} are allowed. This gives a set of different electronic states. There is no algebraic formula for E_{el}; it must be calculated numerically for each desired value of R for each state. In addition to the quantum number m, the H_2^+ electronic wave functions are characterized by the quantum numbers n_ξ and n_η, which give the number of nodes in the $L(\xi)$ and $M(\eta)$ factors in ψ_{el}.

For the ground electronic state, the quantum number m is zero. At $R = \infty$, the H_2^+ ground state is dissociated into a proton and a ground-state hydrogen atom; hence $E_{el}(\infty) = -\frac{1}{2}$ hartree. At $R = 0$, the two protons have come together to form the He^+ ion with ground-state energy: $-\frac{1}{2}(2)^2$ hartrees $= -2$ hartrees. Addition of the internuclear repulsion $1/R$ (in atomic units) to $E_{el}(R)$ gives the $U(R)$ potential-energy curve for nuclear motion. Plots of the ground-state $E_{el}(R)$ and $U(R)$, as found from solution of the electronic Schrödinger equation, are shown in Fig. 13.4. At $R = \infty$ the internuclear repulsion is 0, and U is $-\frac{1}{2}$ hartree.

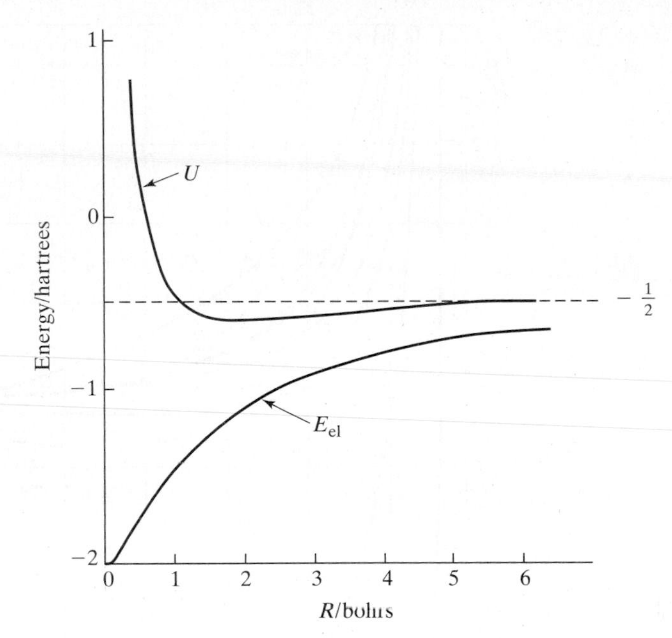

FIGURE 13.4 Electronic energy with (U) and without (E_{el}) internuclear repulsion for the H_2^+ ground electronic state.

The $U(R)$ curve is found to have a minimum at [see L. J. Schaad and W. V. Hicks, *J. Chem. Phys.*, **53**, 851 (1970)] $R_e = 1.9972$ bohrs $= 1.06$ Å, indicating that the H_2^+ ground electronic state is a stable bound state. The calculated value of E_{el} at 1.9972 bohrs is -1.1033 hartrees. Addition of the internuclear repulsion $1/R$ gives $U(R_e) = -0.6026$ hartree, compared with -0.5000 hartree at $R = \infty$. The ground-state binding energy is thus $D_e = 0.1026$ hartree $= 2.79$ eV. This corresponds to 64.4 kcal/mol = 269 kJ/mol. The binding energy is only 17% of the total energy at the equilibrium internuclear distance. Thus a small error in the total energy can correspond to a large error in the binding energy. For heavier molecules the situation is even worse, since chemical binding energies are of the same order of magnitude for most diatomic molecules, but the total electronic energy increases markedly for heavier molecules.

Note that the single electron in H_2^+ is sufficient to give a stable bound state.

Figure 13.5 shows the $U(R)$ curves for the first several electronic energy levels of H_2^+, as found by solving the electronic Schrödinger equation.

The angle ϕ occurs in $\hat{H}_{el}$ of H_2^+ only as $\partial^2/\partial\phi^2$. When ψ_{el} of (13.37) is substituted into $\hat{H}_{el}\psi_{el} = E_{el}\psi_{el}$, the $e^{im\phi}$ factor cancels, and we are led to differential equations for $L(\xi)$ and $M(\eta)$ in which the m quantum number occurs only as m^2. Since E_{el} is found from the $L(\xi)$ and $M(\eta)$ differential equations, E_{el} depends on m^2, and each electronic level with $m \neq 0$ is doubly degenerate, corresponding to states with quantum numbers $+|m|$ and $-|m|$. In the standard notation for diatomic molecules [F. A. Jenkins, *J. Opt. Soc. Am.*, **43**, 425 (1953)], the absolute value of m is called λ:

$$\lambda \equiv |m|$$

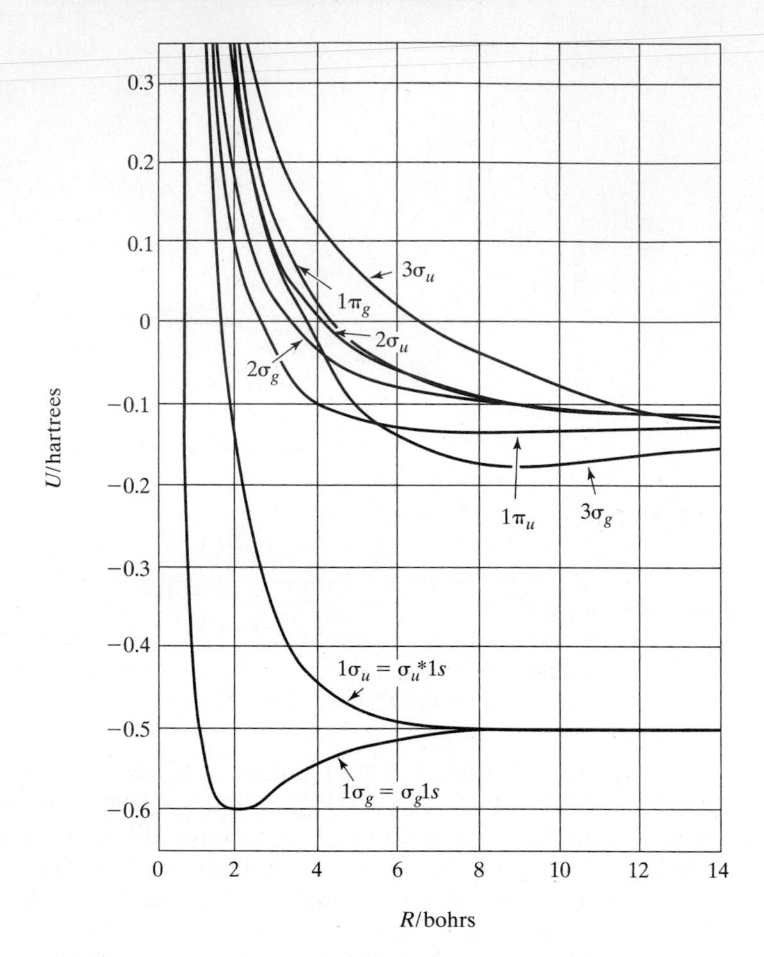

FIGURE 13.5 $U(R)$ curves for several H_2^+ electronic states. [Curves taken from J. C. Slater, *Quantum Theory of Molecules and Solids,* vol. 1, McGraw-Hill, 1963. Used by permission.]

(Some texts define λ as identical to m.) Similar to the s, p, d, f, g notation for hydrogen-atom states, a letter code is used to specify λ, the absolute value (in atomic units) of the component along the molecular axis of the electron's orbital angular momentum:

λ	0	1	2	3	4
letter	σ	π	δ	ϕ	γ

(13.38)

Thus the lowest H_2^+ electronic state is a σ state.

Besides classifying the states of H_2^+ according to λ, we can also classify them according to their parity (Section 7.5). From Fig. 13.12, inversion of the electron's coordinates through the origin O changes ϕ to $\phi + \pi$, r_a to r_b, and r_b to r_a. This leaves the potential-energy part of the electronic Hamiltonian (13.31) unchanged. We previously showed the kinetic-energy operator to be invariant under inversion. Hence the parity operator commutes with the Hamiltonian (13.31), and the H_2^+ electronic wave functions can be classified as either even or odd. For even electronic wave functions, we use the subscript g (from the German word *gerade*, meaning even); for odd wave functions, we use u (from *ungerade*).

The lowest σ_g energy level in Fig. 13.5 is labeled $1\sigma_g$, the next-lowest σ_g level at small R is labeled $2\sigma_g$, and so on. The lowest σ_u level is labeled $1\sigma_u$, and so on. The alternative notation $\sigma_g 1s$ indicates that this level dissociates to a $1s$ hydrogen atom. The meaning of the star in $\sigma_u^* 1s$ will be explained later.

For completeness we must take spin into account by multiplying each spatial H_2^+ electronic wave function by α or β, depending on whether the component of electron spin along the internuclear axis is $+\frac{1}{2}$ or $-\frac{1}{2}$ (in atomic units). Inclusion of spin doubles the degeneracy of all levels.

The H_2^+ ground electronic state has $R_e = 2.00$ bohrs $= 2.00\hbar^2/m_e e'^2$. The negative muon (symbol μ^-) is a short-lived (half-life 2×10^{-6} s) elementary particle whose charge is the same as that of an electron but whose mass m_μ is 207 times m_e. When a beam of negative muons (produced when ions accelerated to high speed collide with ordinary matter) enters H_2 gas, muomolecular ions that consist of two protons and one muon are formed. This species, symbolized by $(p\mu p)^+$, is an H_2^+ ion in which the electron has been replaced by a muon. Its R_e is $2.00\hbar^2/m_\mu e'^2 = 2.00\hbar^2/207 m_e e'^2 = 2.00/207$ bohr $= 0.0051$ Å. The two nuclei in this muoion are 207 times closer than in H_2^+. The magnitude of the vibrational-wave-function factor $S_v(R - R_e)$ in (13.28) is small but not entirely negligible for $R - R_e = -0.0051$ Å; so there is some probability for the nuclei in $(p\mu p)^+$ to come in contact, and nuclear fusion might occur. The isotopic nuclei ^{2}H (deuterium, D) and ^{3}H (tritium, T) undergo fusion much more readily than protons, so instead of H_2 gas, one uses a mixture of D_2 and T_2 gases. After fusion occurs, the muon is released and can then be recaptured to catalyze another fusion. Under the right conditions, one muon can catalyze 150 fusions on average before it decays. Unfortunately, at present, more energy is needed to produce the muon beam than is released by the fusion. (See J. Rafelski and S. E. Jones, *Scientific American*, July 1987, page 84.)

In the rest of this chapter, the subscript el will be dropped from the electronic wave function, Hamiltonian, and energy. It will be understood in Chapters 13 to 17 that ψ means ψ_{el}.

13.5 APPROXIMATE TREATMENTS OF THE H_2^+ GROUND ELECTRONIC STATE

For a many-electron atom, the self-consistent-field (SCF) method is used to construct an approximate wave function as a Slater determinant of (one-electron) spin-orbitals. The one-electron spatial part of a spin-orbital is an atomic orbital (AO). We took each AO as a product of a spherical harmonic and a radial factor. As an initial approximation to the radial factors, we can use hydrogenlike radial functions with effective nuclear charges.

For many-electron molecules, which (unlike H_2^+) cannot be solved exactly, we want to use many of the ideas of the SCF treatment of atoms. We shall write an approximate molecular electronic wave function as a Slater determinant of (one-electron) spin-orbitals. The one-electron spatial part of a molecular spin-orbital is a **molecular orbital** (MO). Because of the Pauli principle, each MO can hold no more than two electrons, just as for AOs. What kind of functions do we use for the MOs? Ideally, the analytic form of each MO is found by an SCF calculation (Section 14.3). In this section, we seek simple approximations for the MOs that will enable us to gain some qualitative understanding of chemical bonding. Just as we took the angular part of each AO to be the same kind of function (a spherical harmonic) as in the one-electron hydrogenlike atom, we shall take the angular part of each diatomic MO to be $(2\pi)^{-1/2}e^{im\phi}$, as in H_2^+. However, the ξ and η factors in the H_2^+ wave functions are complicated functions not readily usable in MO calculations. We therefore seek simpler functions that will provide reasonably accurate approximations to the H_2^+ wave functions and that can be used to construct molecular orbitals for many-electron diatomic molecules. With this discussion as motivation for looking at approximate solutions in a case where the Schrödinger equation is exactly solvable, we consider approximate treatments of H_2^+.

We shall use the variation method, writing down some function containing several parameters and varying them to minimize the variational integral. This will give an approximation to the ground-state wave function and an upper bound to the ground-state energy. By use of the factor $e^{im\phi}$ in the trial function, we can get an upper bound to the energy of the lowest H_2^+ level for any given value of m (see Section 8.2). By using linear variation functions, we can get approximations for excited states.

The H_2^+ ground state has $m = 0$, and the wave function depends only on ξ and η. We could try any well-behaved function of these coordinates as a trial variation function. We shall, however, use a more systematic approach based on the idea of a molecule as being formed from the interaction of atoms.

Consider what the H_2^+ wave function would look like for large values of the internuclear separation R. When the electron is near nucleus a, nucleus b is so far away that we essentially have a hydrogen atom with origin at a. Thus, when r_a is small, the ground-state H_2^+ electronic wave function should resemble the ground-state hydrogen-atom wave function of Eq. (6.104). We have $Z = 1$, and the Bohr radius a_0 has the numerical value 1 in atomic units; hence (6.104) becomes

$$\pi^{-1/2}e^{-r_a} \tag{13.39}$$

Similarly, we conclude that when the electron is near nucleus b, the H_2^+ ground-state wave function will be approximated by

$$\pi^{-1/2}e^{-r_b} \tag{13.40}$$

These considerations suggest that we try as a variation function

$$c_1\pi^{-1/2}e^{-r_a} + c_2\pi^{-1/2}e^{-r_b} \tag{13.41}$$

where c_1 and c_2 are variational parameters. When the electron is near nucleus a, the variable r_a is small and r_b is large, and the first term in (13.41) predominates, giving a function resembling (13.39). The function (13.41) is a linear variation function, and we are led to solve a secular equation, which has the form (8.57), where the subscripts 1 and 2 refer to the functions (13.39) and (13.40).

We can also approach the problem using perturbation theory. We take the unperturbed system as the H_2^+ molecule with $R = \infty$. For $R = \infty$, the electron can be bound to nucleus a with wave function (13.39), or it can be bound to nucleus b with wave function (13.40). In either case the energy is $-\frac{1}{2}$ hartree, and we have a doubly degenerate unperturbed energy level. Bringing the nuclei in from infinity gives rise to a perturbation that splits the doubly degenerate unperturbed level into two levels. This is illustrated by the $U(R)$ curves for the two lowest H_2^+ electronic states, which both dissociate to a ground-state hydrogen atom (see Fig. 13.5). The correct zeroth-order wave functions for the perturbed levels are linear combinations of the form (13.41), and we are led to a secular equation of the form (8.57), with W replaced by $E^{(0)} + E^{(1)}$ (see Prob. 9.19).

Before solving (8.57), let us improve the trial function (13.41). Consider the limiting behavior of the H_2^+ ground-state electronic wave function as R goes to zero. In this limit we get the He^+ ion, which has the ground-state wave function [put $Z = 2$ in (6.104)]

$$2^{3/2}\pi^{-1/2}e^{-2r} \tag{13.42}$$

From Fig. 13.3 we see that as R goes to zero, both r_a and r_b go to r. Hence as R goes to zero, the trial function (13.41) goes to $(c_1 + c_2)\pi^{-1/2}e^{-r}$. Comparing with (13.42), we see that our trial function has the wrong limiting behavior at $R = 0$; it should go to e^{-2r}, not e^{-r}. We can fix things by multiplying r_a and r_b in the exponentials by a variational parameter k, which will be some function of R; $k = k(R)$. For the correct limiting behavior at $R = 0$ and at $R = \infty$, we have $k(0) = 2$ and $k(\infty) = 1$ for the H_2^+ ground electronic state. Physically, k is some sort of effective nuclear charge, which increases as the nuclei come together. We thus take the trial function as

$$\phi = c_a 1s_a + c_b 1s_b \tag{13.43}$$

where the c's are variational parameters and

$$1s_a = k^{3/2}\pi^{-1/2}e^{-kr_a}, \qquad 1s_b = k^{3/2}\pi^{-1/2}e^{-kr_b} \tag{13.44}$$

The factor $k^{3/2}$ normalizes $1s_a$ and $1s_b$ [see Eq. (6.104)]. The molecular-orbital function (13.43) is a **linear combination of atomic orbitals**, an LCAO-MO. The trial function (13.43) was first used by Finkelstein and Horowitz in 1928.

For the function (13.43), the secular equation (8.57) is

$$\begin{vmatrix} H_{aa} - WS_{aa} & H_{ab} - WS_{ab} \\ H_{ba} - WS_{ba} & H_{bb} - WS_{bb} \end{vmatrix} = 0 \tag{13.45}$$

The integrals H_{aa} and H_{bb} are

$$H_{aa} = \int 1s_a^* \hat{H} 1s_a \, dv, \qquad H_{bb} = \int 1s_b^* \hat{H} 1s_b \, dv \tag{13.46}$$

where the H_2^+ electronic Hamiltonian operator $\hat{H}$ is given by (13.32). We can relabel the variables in a definite integral without affecting its value. Changing a to b and b to a changes $1s_a$ to $1s_b$ but leaves $\hat{H}$ unaffected (this would not be true for a heteronuclear diatomic molecule). Hence $H_{aa} = H_{bb}$. We have

$$H_{ab} = \int 1s_a^* \hat{H} 1s_b \, dv, \qquad H_{ba} = \int 1s_b^* \hat{H} 1s_a \, dv \tag{13.47}$$

Since $\hat{H}$ is Hermitian and the functions in these integrals are real, we conclude that $H_{ab} = H_{ba}$. The integral H_{ab} is called a *resonance* (or *bond*) *integral*. Since $1s_a$ and $1s_b$ are normalized and real, we have

$$S_{aa} = \int 1s_a^* 1s_a \, dv = 1 = S_{bb}$$

$$\boxed{S_{ab} = \int 1s_a^* 1s_b \, dv = S_{ba}} \qquad \textbf{(13.48)}$$

The **overlap integral** S_{ab} lies between 1 and 0, decreasing as the distance between the two nuclei increases.

The secular equation (13.45) becomes

$$\begin{vmatrix} H_{aa} - W & H_{ab} - S_{ab}W \\ H_{ab} - S_{ab}W & H_{aa} - W \end{vmatrix} = 0 \qquad (13.49)$$

$$H_{aa} - W = \pm(H_{ab} - S_{ab}W) \qquad (13.50)$$

$$W_1 = \frac{H_{aa} + H_{ab}}{1 + S_{ab}}, \qquad W_2 = \frac{H_{aa} - H_{ab}}{1 - S_{ab}} \qquad (13.51)$$

These two roots are upper bounds to the energies of the ground and first excited electronic states of H_2^+. We shall see that H_{ab} is negative, so W_1 is the lower-energy root.

We now find the coefficients in (13.43) for each of the roots of the secular equation. From Eq. (8.55), we have

$$(H_{aa} - W)c_a + (H_{ab} - S_{ab}W)c_b = 0 \qquad (13.52)$$

Substituting in W_1 from (13.51) [or using (13.50)], we get

$$c_a/c_b = 1 \qquad (13.53)$$

$$\phi_1 = c_a(1s_a + 1s_b) \qquad (13.54)$$

We fix c_a by normalization:

$$|c_a|^2 \int (1s_a^2 + 1s_b^2 + 2 \cdot 1s_a 1s_b) \, dv = 1 \qquad (13.55)$$

$$|c_a| = \frac{1}{(2 + 2S_{ab})^{1/2}} \qquad (13.56)$$

The normalized trial function corresponding to the energy W_1 is thus

$$\phi_1 = \frac{1s_a + 1s_b}{\sqrt{2}(1 + S_{ab})^{1/2}} \qquad (13.57)$$

For the root W_2, we find $c_b = -c_a$ and

$$\phi_2 = \frac{1s_a - 1s_b}{\sqrt{2}(1 - S_{ab})^{1/2}} \qquad (13.58)$$

Equations (13.57) and (13.58) come as no surprise. Since the nuclei are identical, we expect $|\phi|^2$ to remain unchanged on interchanging a and b; in other words, we expect no polarity in the bond.

We now consider evaluation of the integrals H_{aa}, H_{ab}, and S_{ab}. From (13.44) and (13.33), the integrand of S_{ab} is $1s_a 1s_b = k^3\pi^{-1}e^{-k(r_a+r_b)} = k^3\pi^{-1}e^{-kR\xi}$. The volume element in confocal elliptic coordinates is (*Eyring, Walter, and Kimball*, Appendix III)

$$dv = \tfrac{1}{8}R^3(\xi^2 - \eta^2)\, d\xi\, d\eta\, d\phi \tag{13.59}$$

Substitution of these expressions for $1s_a 1s_b$ and dv into S_{ab} and use of the Appendix integral (A.11) to do the ξ integral gives (Prob. 13.15)

$$S_{ab} = e^{-kR}(1 + kR + \tfrac{1}{3}k^2R^2) \tag{13.60}$$

The evaluation of H_{aa} and H_{ab} is considered in Prob. 13.16. The results are

$$H_{aa} = \tfrac{1}{2}k^2 - k - R^{-1} + e^{-2kR}(k + R^{-1}) \tag{13.61}$$

$$H_{ab} = -\tfrac{1}{2}k^2 S_{ab} - k(2 - k)(1 + kR)e^{-kR} \tag{13.62}$$

where $\hat{H}$ is given by (13.32) and so omits the internuclear repulsion.

Substituting the values for the integrals into (13.51), we get

$$W_{1,2} = -\tfrac{1}{2}k^2 + \frac{k^2 - k - R^{-1} + R^{-1}(1 + kR)e^{-2kR} \pm k(k - 2)(1 + kR)e^{-kR}}{1 \pm e^{-kR}(1 + kR + k^2R^2/3)} \tag{13.63}$$

where the upper signs are for W_1. Since $\hat{H}$ in (13.32) omits the internuclear repulsion $1/R$, W_1 and W_2 are approximations to the purely electronic energy E_{el}, and $1/R$ must be added to $W_{1,2}$ to get $U_{1,2}(R)$ [Eq. (13.8)].

The final task is to vary the parameter k at many fixed values of R so as to minimize first $U_1(R)$ and then $U_2(R)$. This can be done numerically using a computer (Prob. 13.18) or analytically. The results are that, for the $1s_a + 1s_b$ function (13.57), k increases almost monotonically from 1 to 2 as R decreases from ∞ to 0; for the $1s_a - 1s_b$ function (13.58), k decreases almost monotonically from 1 to 0.4 as R decreases from ∞ to 0. Since $0 < k \le 2$ and $S_{ab} > 0$, Eq. (13.62) shows that the integral H_{ab} is always negative. Therefore, W_1 in (13.51) corresponds to the ground electronic state $\sigma_g 1s$ of H_2^+. For the ground state, one finds $k(R_e) = 1.24$.

We might ask why the variational parameter k for the $\sigma_u^* 1s$ state goes to 0.4, rather than to 2, as R goes to zero. The answer is that this state of H_2^+ does not go to the ground state ($1s$) of He^+ as R goes to zero. The $\sigma_u^* 1s$ state has odd parity and must correlate with an odd state of He^+. The lowest odd states of He^+ are the $2p$ states (Section 11.5); since the $\sigma_u^* 1s$ state has zero electronic orbital angular momentum along the internuclear (z) axis, this state must go to an atomic $2p$ state with $m = 0$, that is, to the $2p_0 = 2p_z$ state.

Having found $k(R)$ for each root, one calculates W_1 and W_2 from (13.63) and adds $1/R$ to get the $U(R)$ curves. The calculated ground-state $U(R)$ curve has a minimum at 2.00 bohrs (Prob. 13.19), in agreement with the true R_e value 2.00 bohrs, and has $U(R_e) = -15.96$ eV, giving a predicted D_e of 2.36 eV, as compared with the true value 2.79 eV. (If we omit varying k but simply set it equal to 1, we get $R_e = 2.49$ bohrs and $D_e = 1.76$ eV.)

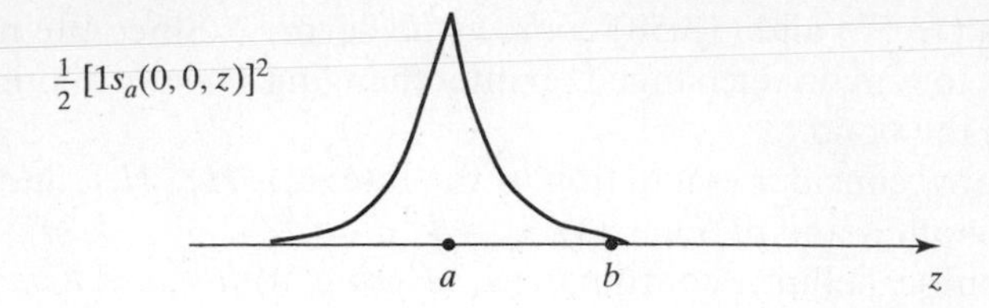

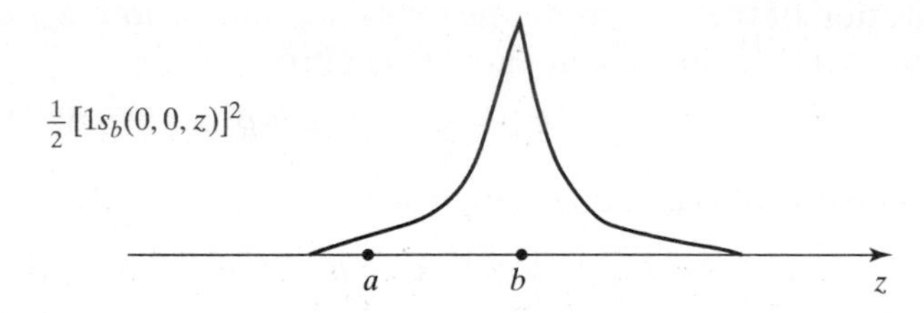

FIGURE 13.6 Atomic probability densities for H_2^+. Note the cusps at the nuclei.

Now consider the appearance of the trial functions for the $\sigma_g 1s$ and $\sigma_u^* 1s$ states at intermediate values of R. Figure 13.6 shows the values of the functions $(1s_a)^2$ and $(1s_b)^2$ at points on the internuclear axis (see also Fig. 6.7). For the $\sigma_g 1s$ function $1s_a + 1s_b$, we get a buildup of electronic probability density between the nuclei, as shown in Fig. 13.7. It is especially significant that the buildup of charge between the nuclei is greater than that obtained by simply taking the sum of the separate atomic charge densities. The probability density for an electron in a $1s_a$ atomic orbital is $(1s_a)^2$. If we add the probability density for half an electron in a $1s_a$ AO and half an electron in a $1s_b$ AO, we get

$$\tfrac{1}{2}(1s_a^2 + 1s_b^2) \tag{13.64}$$

However, in quantum mechanics, we do not add the separate atomic probability densities. Instead, we add the wave functions, as in (13.57). The H_2^+ ground-state probability density is then

$$\phi_1^2 = \frac{1}{2(1 + S_{ab})}[1s_a^2 + 1s_b^2 + 2(1s_a 1s_b)] \tag{13.65}$$

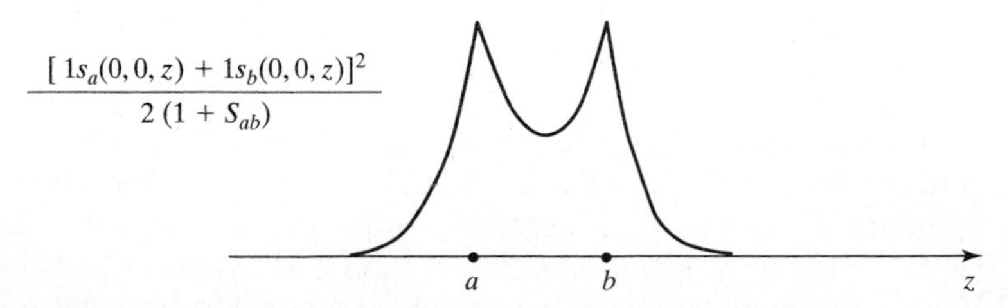

FIGURE 13.7 Probability density along the internuclear axis for the LCAO-MO function $N(1s_a + 1s_b)$.

The difference between (13.65) and (13.64) is

$$\phi_1^2 - \tfrac{1}{2}(1s_a^2 + 1s_b^2) = \frac{1}{2(1 + S_{ab})}[2(1s_a 1s_b) - S_{ab}(1s_a^2 + 1s_b^2)] \qquad (13.66)$$

Putting $R = 2.00$ and $k = 1.24$ in Eq. (13.60), we find that $S_{ab} = 0.46$ at R_e. (It might be thought that because of the orthogonality of different AOs, the overlap integral S_{ab} should be zero. However, the AOs $1s_a$ and $1s_b$ are eigenfunctions of *different* Hamiltonian operators, one for a hydrogen atom at a and one for a hydrogen atom at b. Hence the orthogonality theorem does not apply.)

Consider now the relative magnitudes of the two terms in brackets in (13.66) for points on the molecular axis. To the left of nucleus a, the function $1s_b$ is very small; to the right of nucleus b, the function $1s_a$ is very small. Hence outside the region between the nuclei, the product $1s_a 1s_b$ is small, and the second term in brackets in (13.66) is dominant. This gives a subtraction of electronic charge density outside the internuclear region, as compared with the sum of the densities of the individual atoms. Now consider the region between the nuclei. At the midpoint of the internuclear axis (and anywhere on the plane perpendicular to the axis and bisecting it), we have $1s_a = 1s_b$, and the bracketed terms in (13.66) become $2(1s_a)^2 - 0.92(1s_a)^2 \approx 1s_a^2$, which is positive. We thus get a buildup of charge probability density between the nuclei in the molecule, as compared with the sum of the densities of the individual atoms. This buildup of electronic charge between the nuclei allows the electron to feel the attractions of both nuclei at the same time, which lowers its potential energy. The greater the overlap in the internuclear region between the atomic orbitals forming the bond, the greater the charge buildup in this region.

The preceding discussion seems to attribute the bonding in H_2^+ mainly to the lowering in the average electronic potential energy that results from having the shared electron interact with two nuclei instead of one. This, however, is an incomplete picture. Calculations on H_2^+ by Feinberg and Ruedenberg show that the decrease in electronic potential energy due to the sharing is of the same order of magnitude as the nuclear repulsion energy $1/R$ and hence is insufficient by itself to give binding. Two other effects also contribute to the bonding. The increase in atomic orbital exponent ($k = 1.24$ at R_e versus 1.0 at ∞) causes charge to accumulate near the nuclei (as well as in the internuclear region), and this further lowers the electronic potential energy. Moreover, the buildup of charge in the internuclear region makes $\partial\psi/\partial z$ zero at the midpoint of the molecular axis and small in the region close to this point. Hence the z component of the average electronic kinetic energy [which can be expressed as $\frac{1}{2}\int |\partial\psi/\partial z|^2\, d\tau$; Prob. 7.7b] is lowered as compared with the atomic $\langle T_z \rangle$. (However, the *total* average electronic kinetic energy is raised; see Section 14.5.) For details, see M. J. Feinberg and K. Ruedenberg, *J. Chem. Phys.*, **54**, 1495 (1971); M. P. Melrose et al., *Theor. Chim. Acta*, **88**, 311 (1994).

Bader, however, has strongly criticized the views of Feinberg and Ruedenberg. Bader states (among other points) that H_2^+ and H_2 are atypical and that, in contrast to the increase of charge density in the immediate vicinity of the nuclei in H_2 and H_2^+, molecule formation that involves atoms other than H is usually accompanied by a substantial reduction in charge density in the immediate vicinity of the nuclei. See R. F. W. Bader in *The Force Concept in Chemistry*, B. M. Deb, ed., Van Nostrand Reinhold, 1981,

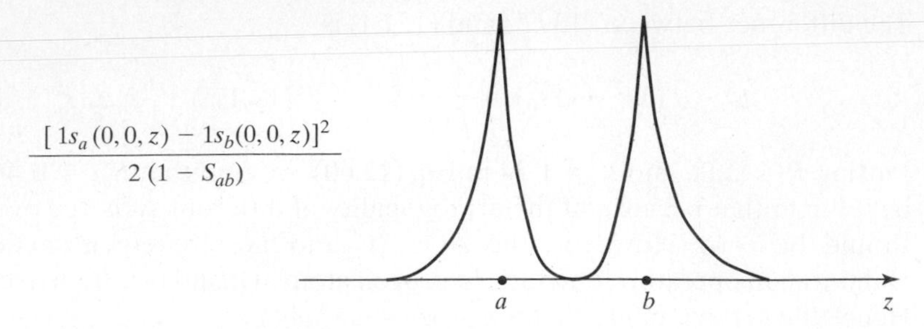

FIGURE 13.8 Probability density along the internuclear axis for the LCAO-MO function $N'(1s_a - 1s_b)$.

pp. 65–67, 71, 95–100, 113–115. Further study is needed before the origin of the covalent bond can be considered a settled question.

The σ_u^*1s trial function $1s_a - 1s_b$ is proportional to $e^{-r_a} - e^{-r_b}$. On a plane perpendicular to the internuclear axis and midway between the nuclei, we have $r_a = r_b$, so this plane is a nodal plane for the σ_u^*1s function. We do not get a buildup of charge between the nuclei for this state, and the $U(R)$ curve has no minimum. We say that the $\sigma_g 1s$ orbital is **bonding** and the σ_u^*1s orbital is **antibonding**. (See Fig. 13.8.)

Reflection of the electron's coordinates in the σ_h symmetry plane perpendicular to the molecular axis and midway between the nuclei converts r_a to r_b and r_b to r_a and leaves ϕ unchanged [Eq. (13.79)]. The operator $\hat{O}_{\sigma_h}$ (Section 12.1) commutes with the electronic Hamiltonian (13.32) and with the parity (inversion) operator. Hence we can choose the H_2^+ wave functions to be eigenfunctions of this reflection operator as well as of the parity operator. Since the square of this reflection operator is the unit operator, its eigenvalues must be $+1$ and -1 (Section 7.5). States of H_2^+ for which the wave function changes sign upon reflection in this plane (eigenvalue -1) are indicated by a star as a superscript to the letter that specifies λ. States whose wave functions are unchanged on reflection in this plane are left unstarred. Since orbitals with eigenvalue -1 for this reflection have a nodal plane between the nuclei, starred orbitals are antibonding.

Instead of using graphs, we can make contour diagrams of the orbitals (Section 6.7); see Fig. 13.9.

Sometimes the binding in H_2^+ is attributed to the resonance integral H_{ab}, since in the approximate treatment we have given, it provides most of the binding energy. This viewpoint is misleading. In the exact treatment of Section 13.4, there arose no such resonance integral. The resonance integral simply arises out of the nature of the LCAO approximation we used.

In summary, we have formed the two H_2^+ MOs (13.57) and (13.58), one bonding and one antibonding, from the AOs $1s_a$ and $1s_b$. The MO energies are given by Eq. (13.51) as

$$W_{1,2} = H_{aa} \pm \frac{H_{ab} - H_{aa}S_{ab}}{1 \pm S_{ab}} \tag{13.67}$$

where $H_{aa} = \langle 1s_a|\hat{H}|1s_a\rangle$, with $\hat{H}$ being the purely electronic Hamiltonian of H_2^+. The integral H_{aa} would be the molecule's purely electronic energy if the electron's wave function in the molecule were $1s_a$. In a sense, H_{aa} is the energy of the $1s_a$ orbital in the molecule.

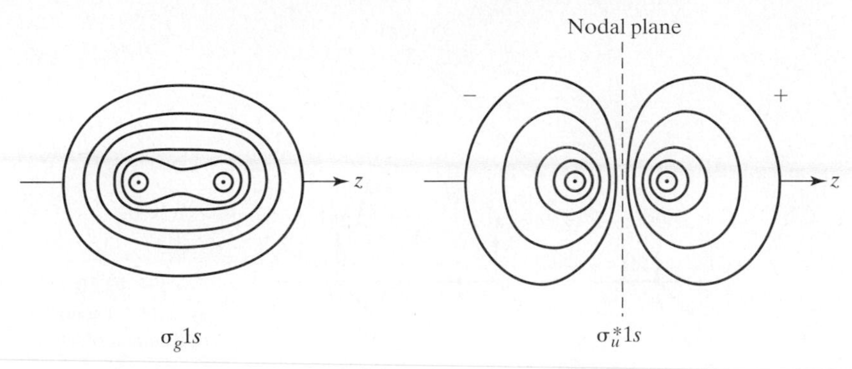

FIGURE 13.9 Contours of constant $|\psi|$ for the $\sigma_g 1s$ and $\sigma_u^* 1s$ MOs. The three-dimensional contour surfaces are generated by rotating these figures about the z axis. Note the resemblance of the antibonding-MO contours to those of a $2p_z$ AO.

In the limit $R = \infty$, H_{aa} becomes the $1s$ AO energy in the H atom. In the molecule, H_{aa} is substantially lower than the electronic energy of an H atom because the electron is attracted to both nuclei. A diagram of MO formation from AOs is given in Fig. 13.22. To get $U(R)$, the electronic energy including nuclear repulsion, we must add $1/R$ to (13.67).

Problem 13.20 outlines the use of Mathcad to create an animation showing how contour plots of the H_2^+ LCAO MOs ϕ_1 and ϕ_2 change as R changes.

We have described the lowest two H_2^+ electronic states according to the state of the hydrogen atom obtained on dissociation. This is a **separated-atoms description**. Alternatively, we can use the state of the atom formed as the internuclear distance goes to zero. This is a **united-atom description**. We saw that for the two lowest electronic states of H_2^+ the united-atom states are the $1s$ and $2p_0$ states of He^+. The united-atom designation is put on the left of the symbol for λ. The $\sigma_g 1s$ state thus has the united-atom designation $1s\sigma_g$. The $\sigma_u^* 1s$ state has the united-atom designation $2p\sigma_u^*$. It is not necessary to write this state as $2p_0\sigma_u^*$, because the fact that it is a σ state tells us that it correlates with the united-atom $2p_0$ state. For the united-atom description, the subscripts g and u are not needed, since molecular states correlating with $s, d, g, \ldots$ atomic states must be g, while states correlating with $p, f, h, \ldots$ atomic states must be u. From the separated-atoms states, we cannot tell whether the molecular wave function is g or u. Thus from the $1s$ separated-atoms state we formed both a g and a u function for H_2^+.

Before constructing approximate molecular orbitals for other H_2^+ states, we consider how the trial function (13.57) can be improved. From the viewpoint of perturbation theory, (13.57) is the correct zeroth-order wave function. We know that the perturbation of molecule formation will mix in other hydrogen-atom states besides $1s$. Dickinson in 1933 used a trial function with some $2p_0$ character mixed in (since the ground state of H_2^+ is a σ state, it would be wrong to mix in $2p_{\pm 1}$ functions); he took

$$\phi = [1s_a + c(2p_0)_a] + [1s_b + c(2p_0)_b] \tag{13.68}$$

where c is a variational parameter and where (Table 6.2)

$$1s_a = k^{3/2}\pi^{-1/2}e^{-kr_a}, \qquad (2p_0)_a = (2pz)_a = \frac{\beta^{5/2}}{4(2\pi)^{1/2}} r_a e^{-\beta r_a/2} \cos\theta_a$$

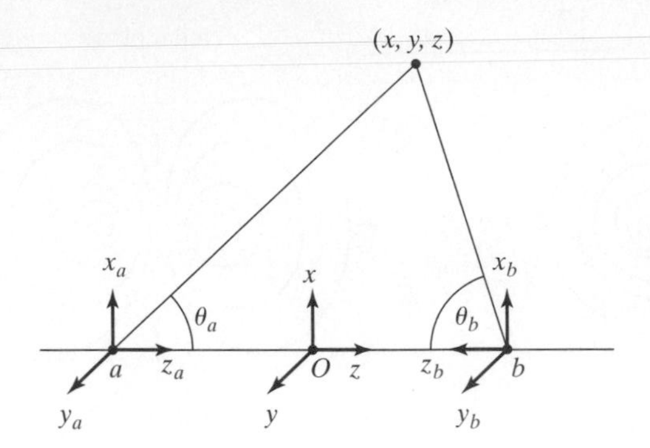

FIGURE 13.10 Coordinate systems for a homonuclear diatomic molecule.

with k and β being two other variational parameters. We have similar expressions for $1s_b$ and $(2p_0)_b$. The angles θ_a and θ_b refer to two sets of spherical coordinates, one set at each nucleus; see Fig. 13.10. The definitions of θ_a and θ_b correspond to using a right-handed coordinate system on atom a and a left-handed system on atom b. The coefficient c goes to zero as R goes to either zero or infinity.

The mixing together of two or more AOs on the same atom is called **hybridization**. The function $1s + c2p_0$ is a hybridized atomic orbital. Since the $2p_0$ function is positive in one lobe and negative in the other, the inclusion of $2p_0$ produces additional charge buildup between the nuclei, giving a greater binding energy. The hybridization allows for the polarization of the $1s_a$ and $1s_b$ atomic orbitals that occurs on molecule formation. The function (13.68) gives a $U(R)$ curve with a minimum at 2.01 bohrs. At this distance, the parameters have the values $k = 1.246$, $\beta = 2.965$, and $c = 0.138$ [F. Weinhold, *J. Chem. Phys.*, **54**, 530 (1971)]. The calculated D_e is 2.73 eV, close to the true value 2.79 eV.

The quantum mechanics and spectroscopy of H_2^+ are reviewed in C. A. Leach and R. E. Moss, *Ann. Rev. Phys. Chem.*, **46**, 55 (1995).

One final point. The approximate wave functions in this chapter are written in atomic units. When rewriting these functions in ordinary units, we must remember that wave functions are not dimensionless. A one-particle wave function ψ has units of length$^{-3/2}$ (Section 3.5). The AOs $1s_a$ and $1s_b$ that occur in the functions (13.57) and (13.58) are given by (13.44) in atomic units. In ordinary units, $1s_a = (k/a_0)^{3/2}\pi^{-1/2}e^{-kr_a/a_0}$.

13.6 MOLECULAR ORBITALS FOR H_2^+ EXCITED STATES

In the preceding section, we used the approximate functions (13.57) and (13.58) for the two lowest H_2^+ electronic states. Now we construct approximate functions for further excited states so as to build up a supply of H_2^+-like molecular orbitals. We shall then use these MOs to discuss many-electron diatomic molecules qualitatively, just as we used hydrogenlike AOs to discuss many-electron atoms.

To get approximations to higher H_2^+ MOs, we can use the linear-variation-function method. We saw that it was natural to take variation functions for H_2^+ as linear combinations of hydrogenlike atomic-orbital functions, giving LCAO-MOs. To get approximate MOs for higher states, we add in more AOs to the linear combination. Thus, to get approximate wave functions for the six lowest H_2^+ σ states, we use a linear combination of the three lowest $m = 0$ hydrogenlike functions on each atom:

$$\phi = c_1 1s_a + c_2 2s_a + c_3(2p_0)_a + c_4 1s_b + c_5 2s_b + c_6(2p_0)_b$$

As found in the preceding section for the function (13.43), the symmetry of the homonuclear diatomic molecule makes the coefficients of the atom-b orbitals equal to ± 1 times the corresponding atom-a orbital coefficients:

$$\phi = [c_1 1s_a + c_2 2s_a + c_3(2p_0)_a] \pm [c_1 1s_b + c_2 2s_b + c_3(2p_0)_b] \qquad (13.69)$$

where the upper sign goes with the even (g) states.

Consider the relative magnitudes of the coefficients in (13.69). For the two electronic states that dissociate into a $1s$ hydrogen atom, we expect that c_1 will be considerably greater than c_2 or c_3, since c_2 and c_3 vanish in the limit of R going to infinity. Thus the Dickinson function (13.68) has the $2p_0$ coefficient equal to one-seventh the $1s$ coefficient at R_e. (This function does not include a $2s$ term, but if it did, we would find its coefficient to be small compared with the $1s$ coefficient.) As a first approximation, we therefore set c_2 and c_3 equal to zero, taking

$$\phi = c_1(1s_a \pm 1s_b) \qquad (13.70)$$

as an approximation for the wave functions of these two states (as we already have done). From the viewpoint of perturbation theory, if we take the separated atoms as the unperturbed problem, the functions (13.70) are the correct zeroth-order wave functions.

The same argument for the two states that dissociate to a $2s$ hydrogen atom gives as approximate wave functions for them

$$\phi = c_2(2s_a \pm 2s_b) \qquad (13.71)$$

since c_1 and c_3 will be small for these states. The functions (13.71) are only an approximation to what we would find if we carried out the linear variation treatment. To find rigorous upper bounds to the energies of these two H_2^+ states, we must use the trial function (13.69) and solve the appropriate secular equation (8.59) (or use matrix algebra—Section 8.6).

In general, we have two H_2^+ states correlating with each separated-atoms state, and rough approximations to the wave functions of these two states will be the LCAO functions $f_a + f_b$ and $f_a - f_b$, where f is a hydrogenlike wave function. The functions (13.70) give the $\sigma_g 1s$ and $\sigma_u^* 1s$ states. Similarly, the functions (13.71) give the $\sigma_g 2s$ and $\sigma_u^* 2s$ molecular orbitals. The outer contour lines for these orbitals are like those for the corresponding MOs made from $1s$ AOs. However, since the $2s$ AO has a nodal sphere while the $1s$ AO does not, each of these MOs has one more nodal surface than the corresponding $\sigma_g 1s$ or $\sigma_u^* 1s$ MO.

Next we have the combinations

$$(2p_0)_a \pm (2p_0)_b = (2p_z)_a \pm (2p_z)_b \qquad (13.72)$$

giving the $\sigma_g 2p$ and $\sigma_u^* 2p$ MOs (Fig. 13.11). These are σ MOs even though they correlate with $2p$ separated AOs, since they have $m = 0$.

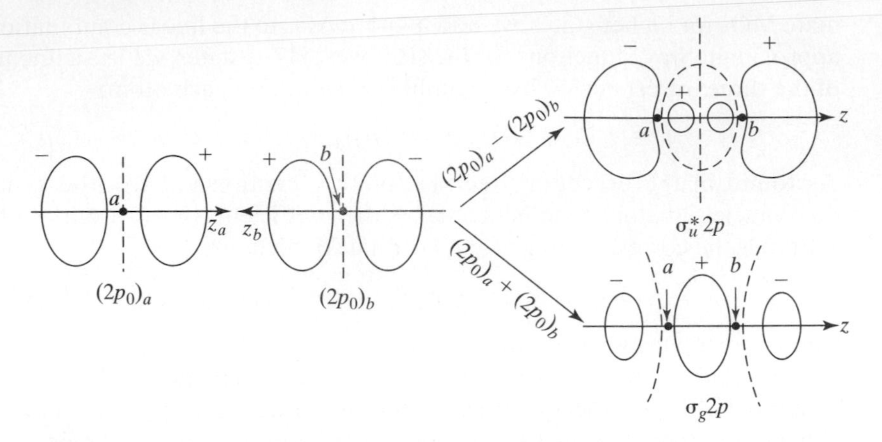

FIGURE 13.11 Formation of $\sigma_g 2p$ and $\sigma_u^* 2p$ MOs from $2p_z$ AOs. The dashed lines indicate nodal surfaces. The signs on the contours give the sign of the wave function. The contours are symmetric about the z axis. (Because of substantial $2s$–$2p$ hybridization, these contours are not accurate representations of true MO shapes. For accurate contours, see the reference of Fig. 13.20.)

The preceding discussion is oversimplified. For the hydrogen atom, the $2s$ and $2p$ AOs are degenerate, and so we can expect the correct zeroth-order functions for the $\sigma_g 2s$, $\sigma_u^* 2s$, $\sigma_g 2p$, and $\sigma_u^* 2p$ MOs of H_2^+ to each be mixtures of $2s$ and $2p$ AOs rather than containing only $2s$ or $2p$ character. [In the $R \to \infty$ limit, H_2^+ consists of an H atom perturbed by the essentially uniform electric field of a far-distant proton. Problem 9.22 showed that the correct zeroth-order functions for the $n = 2$ levels of an H atom in a uniform electric field in the z direction are $2^{-1/2}(2s + 2p_0)$, $2^{-1/2}(2s - 2p_0)$, $2p_1$, and $2p_{-1}$. Thus, for H_2^+, $2s$ and $2p_0$ in Eqs. (13.71) and (13.72) should be replaced by $2s + 2p_0$ and $2s - 2p_0$.] For molecules that dissociate into many-electron atoms, the separated-atoms $2s$ and $2p$ AOs are not degenerate but do lie close together in energy. Hence the first-order corrections to the wave functions will mix substantial $2s$ character into the $\sigma 2p$ MOs and substantial $2p$ character into the $\sigma 2s$ MOs. Thus the designation of an MO as $\sigma 2s$ or $\sigma 2p$ should not be taken too literally. For H_2^+ and H_2, the united-atom designations of the MOs are preferable to the separated-atoms designations, but we shall use mostly the latter.

For the other two $2p$ atomic orbitals, we can use either the $2p_{+1}$ and $2p_{-1}$ complex functions or the $2p_x$ and $2p_y$ real functions. If we want MOs that are eigenfunctions of $\hat{L}_z$, we will choose the complex p orbitals, giving the MOs

$$(2p_{+1})_a + (2p_{+1})_b \tag{13.73}$$

$$(2p_{+1})_a - (2p_{+1})_b \tag{13.74}$$

$$(2p_{-1})_a + (2p_{-1})_b \tag{13.75}$$

$$(2p_{-1})_a - (2p_{-1})_b \tag{13.76}$$

From Eq. (6.114) we have, since $\phi_a = \phi_b = \phi$,

$$(2p_{+1})_a + (2p_{+1})_b = \tfrac{1}{8}\pi^{-1/2}(r_a e^{-r_a/2} \sin\theta_a + r_b e^{-r_b/2} \sin\theta_b)e^{i\phi} \tag{13.77}$$

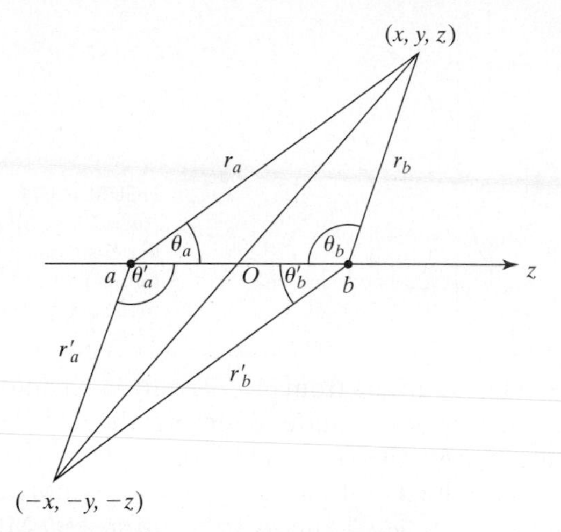

FIGURE 13.12 The effect of inversion of the electron's coordinates in H_2^+. We have $r'_a = r_b$, $r'_b = r_a$, and $\phi' = \phi + \pi$.

Since $\lambda = |m| = 1$, this is a π orbital. The inversion operation amounts to the coordinate transformation (Fig. 13.12)

$$r_a \rightarrow r_b, \qquad r_b \rightarrow r_a, \qquad \phi \rightarrow \phi + \pi \tag{13.78}$$

We have $e^{i(\phi+\pi)} = (\cos \pi + i \sin \pi)e^{i\phi} = -e^{i\phi}$. From Fig. 13.12 we see that inversion converts θ_a to θ_b and vice versa. Thus inversion converts (13.77) to its negative, meaning it is a u orbital. Reflection in the plane perpendicular to the axis and midway between the nuclei causes the following transformations (Prob. 13.23):

$$r_a \rightarrow r_b, \qquad r_b \rightarrow r_a, \qquad \phi \rightarrow \phi, \qquad \theta_a \rightarrow \theta_b, \qquad \theta_b \rightarrow \theta_a \tag{13.79}$$

This leaves (13.77) unchanged, so we have an unstarred (bonding) orbital. The designation of (13.77) is then $\pi_u 2p_{+1}$.

The function (13.77) is complex. Taking its absolute value, we can plot the orbital contours of constant probability density (Section 6.7). Since $|e^{i\phi}| = 1$, the probability density is independent of ϕ, giving a density that is symmetric about the z (internuclear) axis. Figure 13.13 shows a cross section of this orbital in a plane containing the nuclei. The three-dimensional shape is found by rotating this figure about the z axis, creating a sort of fat doughnut.

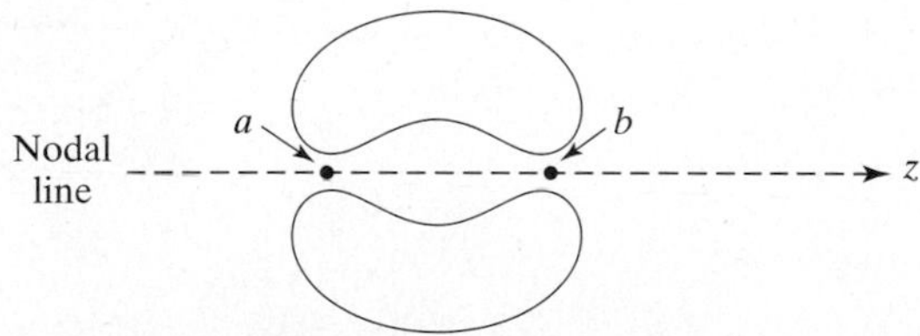

FIGURE 13.13 Cross section of the $\pi_u 2p_{+1}$ (or $\pi_u 2p_{-1}$) molecular orbital. To obtain the three-dimensional contour surface, rotate the figure about the z axis. The z axis is a nodal line for this MO (as it is for the $2p_{+1}$ AO.)

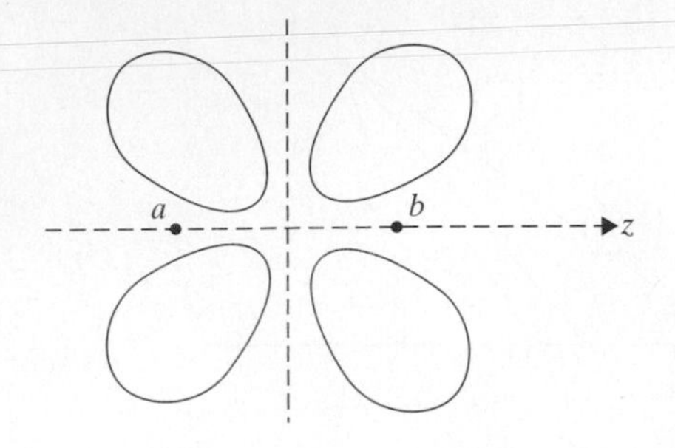

FIGURE 13.14 Cross section of the $\pi_g^*2p_{+1}$ (or $\pi_g^*2p_{-1}$) MO. To obtain the three-dimensional contour surface, rotate the figure about the z axis. The z axis and the xy plane are nodes.

The MO (13.75) differs from (13.77) only in having $e^{i\phi}$ replaced by $e^{-i\phi}$ and is designated $\pi_u 2p_{-1}$. The coordinate ϕ enters the H_2^+ Hamiltonian as $\partial^2/\partial\phi^2$. Since $\partial^2 e^{i\phi}/\partial\phi^2 = \partial^2 e^{-i\phi}/\partial\phi^2$, the states (13.73) and (13.75) have the same energy. Recall (Section 13.4) that the $\lambda = 1$ energy levels are doubly degenerate, corresponding to $m = \pm 1$. Since $|e^{i\phi}| = |e^{-i\phi}|$, the $\pi_u 2p_{+1}$ and $\pi_u 2p_{-1}$ MOs have the same shapes, just as the $2p_{+1}$ and $2p_{-1}$ AOs have the same shapes.

The functions (13.74) and (13.76) give the $\pi_g^*2p_{+1}$ and $\pi_g^*2p_{-1}$ MOs. These functions do not give charge buildup between the nuclei; see Fig. 13.14.

Now consider the more familiar alternative of using the $2p_x$ and $2p_y$ AOs to make the MOs. The linear combination

$$(2p_x)_a + (2p_x)_b \tag{13.80}$$

gives the $\pi_u 2p_x$ MO (Fig. 13.15). This MO is not symmetrical about the internuclear axis but builds up probability density in two lobes, one above and one below the yz

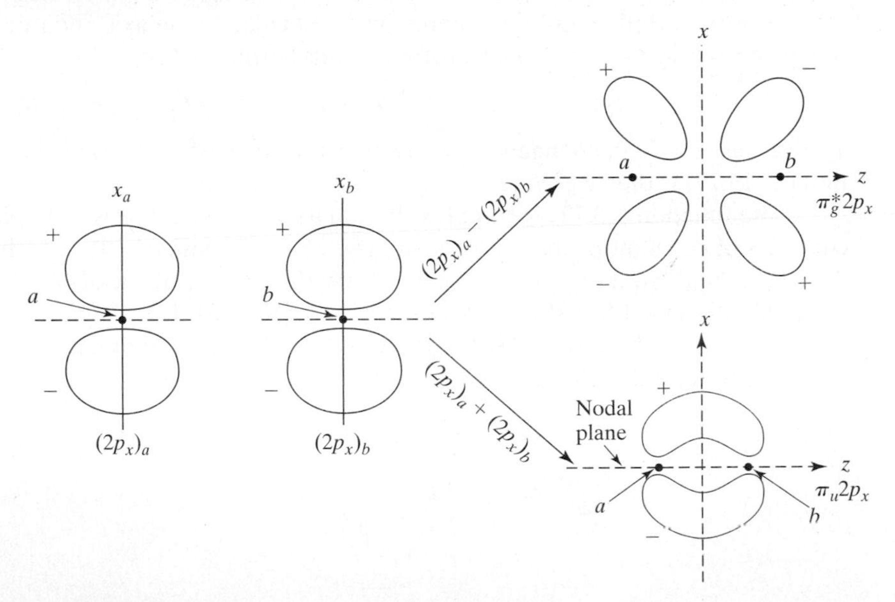

FIGURE 13.15 Formation of the $\pi_u 2p_x$ and $\pi_g^*2p_x$ MOs. Since $\phi = 0$ in the xz plane, the cross sections of these MOs in the xz plane are the same as for the corresponding $\pi_u 2p_{+1}$ and $\pi_g^*2p_{-1}$ MOs. However, the $\pi 2p_x$ MOs are not symmetrical about the z axis. Rather, they consist of blobs of probability density above and below the nodal yz plane.

plane, which is a nodal plane for this function. The wave function has opposite signs on each side of this plane. The linear combination

$$(2p_x)_a - (2p_x)_b \tag{13.81}$$

gives the $\pi_g^*2p_x$ MO (Fig. 13.15). Since the $2p_y$ functions differ from the $2p_x$ functions solely by a rotation of 90° about the internuclear axis, they give MOs differing from those of Fig. 13.15 by a 90° rotation about the z axis. The linear combinations

$$(2p_y)_a + (2p_y)_b \tag{13.82}$$

$$(2p_y)_a - (2p_y)_b \tag{13.83}$$

give the $\pi_u 2p_y$ and $\pi_g^*2p_y$ molecular orbitals. The MOs (13.80) and (13.82) have the same energy. The MOs (13.81) and (13.83) have the same energy. (Note that the g $\pi 2p$ MOs are antibonding, while the u $\pi 2p$ MOs are bonding.)

Just as the $2p_x$ and $2p_y$ AOs are linear combinations of the $2p_{+1}$ and $2p_{-1}$ AOs [Eqs. (6.118) and (6.120)], the $\pi_u 2p_x$ and $\pi_u 2p_y$ MOs are linear combinations of the $\pi_u 2p_{+1}$ and $\pi_u 2p_{-1}$ MOs. We can use any linear combination of the eigenfunctions of a degenerate energy level and still have an energy eigenfunction. Just as the $2p_{+1}$ and $2p_{-1}$ AOs are eigenfunctions of $\hat{L}_z$ and the $2p_x$ and $2p_y$ AOs are not, the $\pi_u 2p_{+1}$ and $\pi_u 2p_{-1}$ MOs are eigenfunctions of $\hat{L}_z$ and the $\pi_u 2p_x$ and $\pi_u 2p_y$ MOs are not. For the H_2^+ $\pi_u 2p$ energy level, we can use the pair of real MOs (13.80) and (13.82), or the pair of complex MOs (13.73) and (13.75), or any two linearly independent linear combinations of these functions.

We have shown the correlation of the H_2^+ MOs with the separated-atoms AOs. We can also show how they correlate with the united-atom AOs. As R goes to zero, the σ_u^*1s MO (Fig. 13.9) resembles more and more the $2p_z$ AO, with which it correlates. Similarly, the $\pi_u 2p$ MOs correlate with p united-atom states, while the π_g^*2p MOs correlate with d united-atom states.

13.7 MO CONFIGURATIONS OF HOMONUCLEAR DIATOMIC MOLECULES

We now use the H_2^+ MOs developed in the last section to discuss many-electron homonuclear diatomic molecules. (**Homonuclear** means the two nuclei are the same; **heteronuclear** means they are different.) If we ignore the interelectronic repulsions, the zeroth-order wave function is a Slater determinant of H_2^+-like one-electron spin-orbitals. We approximate the spatial part of the H_2^+ spin-orbitals by the LCAO-MOs of the last section. Treatments that go beyond this crude first approximation will be discussed later.

The sizes and energies of the MOs vary with varying internuclear distance for each molecule and vary as we go from one molecule to another. Thus we saw how the orbital exponent k in the H_2^+ trial function (13.54) varied with R. As we go to molecules with higher nuclear charge, the parameter k for the $\sigma_g 1s$ MO will increase, giving a more compact MO. We want to consider the order of the MO energies. Because of the variation of these energies with R and variations from molecule to molecule, numerous crossings occur, just as for atomic-orbital energies (Fig. 11.2). Hence we cannot give a

TABLE 13.1 Molecular-Orbital Nomenclature for Homonuclear Diatomic Molecules

Separated-Atoms Description	United-Atom Description	Numbering by Symmetry
$\sigma_g 1s$	$1s\sigma_g$	$1\sigma_g$
$\sigma_u^* 1s$	$2p\sigma_u^*$	$1\sigma_u$
$\sigma_g 2s$	$2s\sigma_g$	$2\sigma_g$
$\sigma_u^* 2s$	$3p\sigma_u^*$	$2\sigma_u$
$\pi_u 2p$	$2p\pi_u$	$1\pi_u$
$\sigma_g 2p$	$3s\sigma_g$	$3\sigma_g$
$\pi_g^* 2p$	$3d\pi_g^*$	$1\pi_g$
$\sigma_u^* 2p$	$4p\sigma_u^*$	$3\sigma_u$

definitive order. However, the following is the order in which the MOs fill as we go across the periodic table:

$$\sigma_g 1s < \sigma_u^* 1s < \sigma_g 2s < \sigma_u^* 2s < \pi_u 2p_x = \pi_u 2p_y < \sigma_g 2p < \pi_g^* 2p_x = \pi_g^* 2p_y < \sigma_u^* 2p$$

Each bonding orbital fills before the corresponding antibonding orbital. The $\pi_u 2p$ orbitals are close in energy to the $\sigma_g 2p$ orbital, and it was formerly believed that the $\sigma_g 2p$ MO filled first.

Besides the separated-atoms designation, there are other ways of referring to these MOs; see Table 13.1. The second column of this table gives the united-atom designations. The nomenclature of the third column uses $1\sigma_g$ for the lowest σ_g MO, $2\sigma_g$ for the second lowest σ_g MO, and so on.

Figure 13.16 shows how these MOs correlate with the separated-atoms and united-atom AOs. Because of the variation of MO energies from molecule to molecule, this diagram is not quantitative. (The word *correlation* is being used here to mean a correspondence; this is a different meaning than in the term *electron correlation*.)

Recall (Prob. 7.28 and Fig. 6.13) that $s, d, g, \ldots$ united-atom AOs are even functions and hence correlate with *gerade* (g) MOs, whereas $p, f, h, \ldots$ AOs are odd functions and hence correlate with *ungerade* (u) MOs.

A useful principle in drawing orbital correlation diagrams is the *noncrossing rule*, which states that for MO correlation diagrams of many-electron diatomic molecules, the energies of MOs with the same symmetry cannot cross. For diatomic MOs the word *symmetry* refers to whether the orbital is g or u and whether it is $\sigma, \pi, \delta \ldots$. For example, two σ_g MOs cannot cross on a correlation diagram. From the noncrossing rule, we conclude that the lowest MO of a given symmetry type must correlate with the lowest united-atom AO of that symmetry, and similarly for higher orbitals. [A similar noncrossing rule holds for potential-energy curves $U(R)$ for different electronic states of a many-electron diatomic molecule.] The proof of the noncrossing rule is a bit subtle; see C. A. Mead, *J. Chem. Phys.*, **70**, 2276 (1979) for a thorough discussion.

Just as we discussed atoms by filling in the AOs, giving rise to atomic configurations such as $1s^2 2s^2$, we shall discuss homonuclear diatomic molecules by filling in the

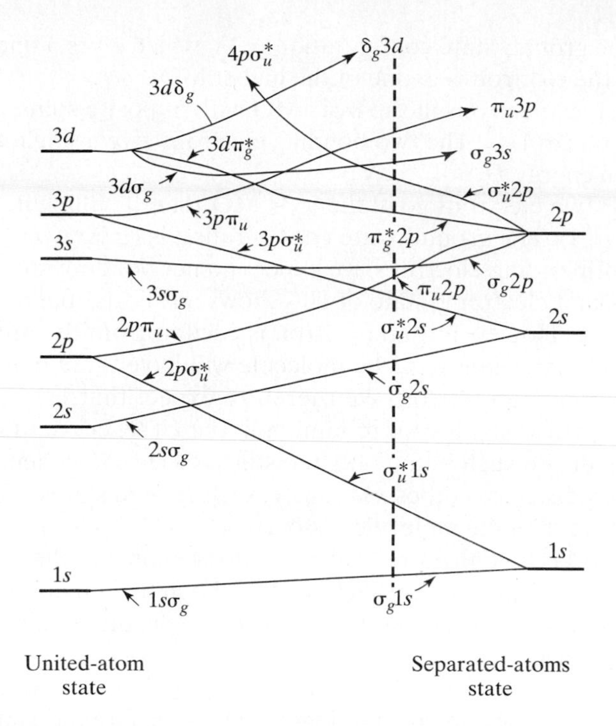

FIGURE 13.16 Correlation diagram for homonuclear diatomic MOs. (This diagram does not hold for H_2^+.) The dashed vertical line corresponds to the order in which the MOs fill.

MOs, giving rise to molecular electronic configurations such as $(\sigma_g 1s)^2(\sigma_u^* 1s)^2$. (Recall that with a single atomic configuration there is associated a hierarchy of terms, levels, and states; the same is true for a molecular configuration; see Section 13.8.)

Figure 13.17 shows the homonuclear diatomic MOs formed from the $1s$, $2s$, and $2p$ AOs.

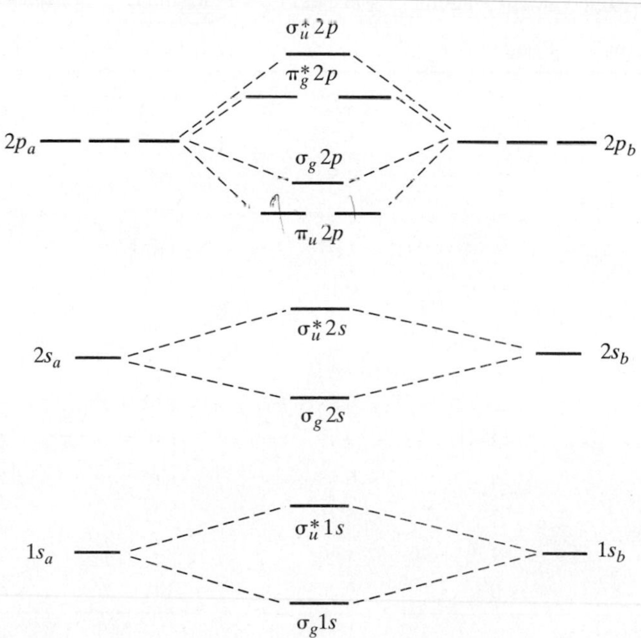

FIGURE 13.17 Homonuclear diatomic MOs formed from $1s$, $2s$, and $2p$ AOs.

For H_2^+ we have the ground-state configuration $\sigma_g 1s$, which gives a one-electron bond. For excited states the electron is in one of the higher MOs.

For H_2 we put the two electrons in the $\sigma_g 1s$ MO with opposite spins, giving the ground-state configuration $(\sigma_g 1s)^2$. The two bonding electrons give a single bond. The ground-state dissociation energy D_e is 4.75 eV.

Now consider He_2. Two electrons go in the $\sigma_g 1s$ MO, thereby filling it. The other two go in the next MO, $\sigma_u^* 1s$. The ground-state configuration is $(\sigma_g 1s)^2(\sigma_u^* 1s)^2$. With two bonding and two antibonding electrons, we expect no net bonding, in agreement with the fact that the ground electronic state of He_2 shows no substantial minimum in the potential-energy curve. However, if an electron is excited from the antibonding $\sigma_u^* 1s$ MO to a higher MO that is bonding, the molecule will have three bonding electrons and only one antibonding electron. We therefore expect that He_2 has bound excited electronic states, with a significant minimum in the $U(R)$ curve of each such state. Indeed, about two dozen such bound excited states of He_2 have been spectroscopically observed in gas discharge tubes. Of course, such excited states decay to the ground electronic state, and then the molecule dissociates.

The repulsion of two $1s^2$ helium atoms can be ascribed mainly to the Pauli repulsion between electrons with parallel spins (Section 10.3). Each helium atom has a pair of electrons with opposite spin, and each pair tends to exclude the other pair from occupying the same region of space.

Removal of an antibonding electron from He_2 gives the He_2^+ ion, with ground-state configuration $(\sigma_g 1s)^2(\sigma_u^* 1s)$ and one net bonding electron. Ground-state properties of this molecule are quite close to those for H_2^+; see Table 13.2 (which is discussed later in this section).

TABLE 13.2 Properties of Homonuclear Diatomic Molecules in Their Ground Electronic States

Molecule	Ground Term	Bond Order	D_e/eV	R_e/Å	$\tilde{\nu}_e$/cm^{-1}
H_2^+	$^2\Sigma_g^+$	$\frac{1}{2}$	2.79	1.06	2322
H_2	$^1\Sigma_g^+$	1	4.75	0.741	4403
He_2^+	$^2\Sigma_u^+$	$\frac{1}{2}$	2.5	1.08	1698
He_2	$^1\Sigma_g^+$	0	0.0009	3.0	
Li_2	$^1\Sigma_g^+$	1	1.07	2.67	351.4
Be_2	$^1\Sigma_g^+$	0	0.10	2.45	
B_2	$^3\Sigma_g^-$	1	3.1	1.59	1051
C_2	$^1\Sigma_g^+$	2	6.3	1.24	1855
N_2^+	$^2\Sigma_g^+$	$2\frac{1}{2}$	8.85	1.12	2207
N_2	$^1\Sigma_g^+$	3	9.91	1.10	2358
O_2^+	$^2\Pi_g$	$2\frac{1}{2}$	6.78	1.12	1905
O_2	$^3\Sigma_g^-$	2	5.21	1.21	1580
F_2	$^1\Sigma_g^+$	1	1.66	1.41	892
Ne_2	$^1\Sigma_g^+$	0	0.0036	3.1	14

Li_2 has the ground-state configuration $(\sigma_g 1s)^2(\sigma_u^* 1s)^2(\sigma_g 2s)^2$ with two net bonding electrons, leading to the description of the molecule as containing an Li—Li single bond. Experimentally, Li_2 is a stable species. In Li_2 the orbital exponent of the $1s$ AOs is considerably greater than in H_2^+ or H_2, because of the increase in the nuclear charges from 1 to 3. This shrinks the $1s_a$ and $1s_b$ AOs in closer to the corresponding nuclei. There is thus only very slight overlap between these two AOs, and the integrals S_{ab} and H_{ab} are very small for these AOs. As a result, the energies of the $\sigma_g 1s$ and $\sigma_u^* 1s$ MOs in Li_2 are nearly equal to each other and to the energy of a $1s$ Li AO. (For very small R, the $1s_a$ and $1s_b$ AOs do overlap appreciably and their energies then differ considerably.) The Li_2 ground-state configuration is often written as $KK(\sigma_g 2s)^2$ to indicate the negligible change in inner-shell orbital energies on molecule formation, which is in accord with the chemist's usual idea of bonding involving only the valence electrons. The orbital exponent of the $2s$ AOs in Li_2 is not much greater than 1, because these electrons are screened from the nucleus by the $1s$ electrons.

The Be_2 ground-state configuration $KK(\sigma_g 2s)^2(\sigma_u^* 2s)^2$ has no net bonding electrons.

The B_2 ground-state configuration $KK(\sigma_g 2s)^2(\sigma_u^* 2s)^2(\pi_u 2p)^2$ has two net bonding electrons, indicating a stable ground state, as is found experimentally. The bonding electrons are π electrons, which is at variance with the notion that single bonds are always σ bonds. We have two degenerate $\pi_u 2p$ MOs. Recall that when we had an atomic configuration such as $1s^2 2s^2 2p^2$ we obtained several terms, which because of interelectronic repulsions had different energies. We saw that the term with the highest total spin was generally the lowest (Hund's rule). With the molecular configuration of B_2 given above, we also have a number of terms. Since the lower (σ) MOs are all filled, their electrons must be paired and contribute nothing to the total spin. If the two $\pi_u 2p$ electrons are both in the same MO (for example, both in $\pi_u 2p_{+1}$), their spins must be paired (antiparallel), giving a total molecular electronic spin of zero. If, however, we have one electron in the $\pi_u 2p_{+1}$ MO and the other in the $\pi_u 2p_{-1}$ MO, their spins can be parallel, giving a net spin of 1; by Hund's rule, this term will be lowest, and the ground term of B_2 will have spin multiplicity $2S + 1 = 3$. Investigation of the electron-spin-resonance spectrum of B_2 trapped in solid neon at low temperature showed that the B_2 ground term is a triplet with $S = 1$ [L. B. Knight et al., *J. Am. Chem. Soc.*, **109**, 3521 (1987)].

C_2 has the ground-state configuration $KK(\sigma_g 2s)^2(\sigma_u^* 2s)^2(\pi_u 2p)^4$ with four net bonding electrons, indicating a stable ground state with a double bond. As mentioned, the $\pi_u 2p$ and $\sigma_g 2p$ MOs have nearly the same energy in many molecules. It was formerly believed that the ground-state configuration of C_2 was $KK(\sigma_g 2s)^2(\sigma_u^* 2s)^2(\pi_u 2p)^3(\sigma_g 2p)$. This configuration would allow two of the electrons to have their spins parallel, giving a triplet term. Because of the minimization of the repulsion between these two electrons, the triplet term of this configuration was thought to be lower than the $(\pi_u 2p)^4$ configuration, which can give only a singlet term. However, the electronic emission spectrum of C_2 shows the $(\pi_u 2p)^4$ singlet term to be the ground term by a small margin (0.09 eV) [E. A. Ballik and D. A. Ramsay, *J. Chem. Phys.*, **31**, 1128 (1959); *Astrophys. J.*, **137**, 61, 84 (1963)].

The N_2 ground-state configuration is $KK(\sigma_g 2s)^2(\sigma_u^* 2s)^2(\pi_u 2p)^4(\sigma_g 2p)^2$. The six net bonding electrons indicate a triple bond, in accord with the Lewis structure :N≡N:.

The O_2 ground-state configuration is

$$KK(\sigma_g 2s)^2(\sigma_u^* 2s)^2(\sigma_g 2p)^2(\pi_u 2p)^4(\pi_g^* 2p)^2$$

Spectroscopic evidence indicates that in O_2 (and in F_2) the $\sigma_g 2p$ MO is lower in energy than the $\pi_u 2p$ MO. The four net bonding electrons give a double bond. The $\pi_g^* 2p_x$ and $\pi_g^* 2p_y$ MOs have the same energy, and by putting one electron in each with parallel spins, we get a triplet term. By Hund's rule this is the ground term. This explanation of the paramagnetism of O_2 was one of the early triumphs of MO theory.

For F_2 the $\ldots(\pi_g^* 2p)^4$ ground-state configuration gives a single bond.

For Ne_2 the $\ldots(\pi_g^* 2p)^4(\sigma_u^* 2p)^2$ configuration gives no net bonding electrons and no chemical bond.

We can go on to describe homonuclear diatomic molecules formed from atoms of the next period. Thus the lowest electron configuration of Na_2 is $KKLL(\sigma_g 3s)^2$. However, there are some differences as compared with the corresponding molecules of the preceding period. For Al_2, the ground term is the triplet term of the $\ldots(\sigma_g 3p)(\pi_u 3p)$ configuration, which lies a mere 0.02 eV below the triplet term of the $\ldots(\pi_u 3p)^2$ configuration [C. W. Bauschlicher et al., *J. Chem. Phys.*, **86**, 7007 (1987)]. For Si_2, the ground term is the triplet term of the $\ldots(\sigma_g 3p)^2(\pi_u 3p)^2$ configuration, which lies 0.05 eV below the triplet term of the $\ldots(\sigma_g 3p)(\pi_u 3p)^3$ configuration [C. W. Bauschlicher and S. R. Langhoff, *J. Chem. Phys.*, **87**, 2919 (1987); M. R. Nimlos et al., *J. Chem. Phys.*, **87**, 5116 (1987)]. These results were established with massive CI calculations done on supercomputers.

Table 13.2 lists D_e, R_e, and $\widetilde{\nu}_e \equiv \nu_e/c$ for the ground electronic states of some homonuclear diatomic molecules, where ν_e is the harmonic vibrational frequency (13.27). The table also lists the *bond order*, which is one-half the difference between the number of bonding and antibonding electrons. [For a survey of the various methods that have been proposed to calculate bond orders, see J. J. Jules and J. R. Lombardi, THEOCHEM, **664–665**, 255 (2003).] As the bond order increases, D_e and ν_e tend to increase and R_e decreases. (The high ν_e of H_2 is due to its small reduced mass μ.) The term symbols in this table are explained in the next section.

Bonding MOs produce charge buildup between the nuclei, whereas antibonding MOs produce charge depletion between the nuclei. Hence removal of an electron from a bonding MO usually decreases D_e, whereas removal of an electron from an antibonding MO increases D_e. (Note that as R decreases in Fig. 13.16, the energies of bonding MOs decrease, while the energies of antibonding MOs increase.) For example, the highest filled MO in N_2 is bonding, and Table 13.2 shows that in going from the ground state of N_2 to that of N_2^+ the dissociation energy decreases (and the bond length increases). In contrast, the highest filled MO of O_2 is antibonding, and in going from O_2 to O_2^+ the dissociation energy increases (and R_e decreases). The designation of bonding or antibonding is not relevant to the effect of the electrons on the total energy of the molecule. Energy is always required to ionize a stable molecule, no matter which electron is removed. Hence both bonding and antibonding electrons in a stable molecule decrease the total molecular energy.

If the interaction between two ground-state He atoms were strictly repulsive (as predicted by MO theory), the atoms in He gas would not attract one another at all and the gas would never liquefy. Of course, helium gas can be liquefied. Configuration-interaction calculations and direct experimental evidence from scattering experiments show that as two He atoms approach each other there is an initial weak attraction, with the potential energy reaching a minimum at 3.0 Å of 0.0009 eV below the separated-atoms energy. At distances less than 3.0 Å, the force becomes increasingly repulsive because of overlap of the electron probability densities. The initial attraction (called a **London** or **dispersion force**) results from instantaneous correlation between the motions of the electrons in one atom and the motions of the electrons in the second atom. Therefore, a calculation that includes electron correlation is needed to deal with dispersion attractions.

The general term for all kinds of intermolecular forces is **van der Waals forces**. Except for highly polar molecules, the dispersion force is the largest contributor to intermolecular attractions. The dispersion force increases as the molecular size increases, so boiling points tend to increase as the molecular weight increases.

The slight minimum in the $U(R)$ curve at relatively large intermolecular separation produced by the dispersion force can be deep enough to allow the existence at low temperatures of molecules bound by the dispersion interaction. Such species are called **van der Waals molecules**. For example, argon gas at 100 K has a small concentration of Ar_2 van der Waals molecules. Ar_2 has $D_e = 0.012$ eV, $R_e = 3.76$ Å, and has seven bound vibrational levels ($v = 0, \ldots, 6$).

For the ground electronic state of He_2 [corresponding to the electron configuration $(\sigma_g 1s)^2(\sigma_u^* 1s)^2$], the zero-point vibrational energy is very slightly less than the dissociation energy D_e associated with the dispersion attraction, so the $v = 0, J = 0$ level is the only bound level. Because of the extremely weak binding, He_2 exists in significant amounts only at very low temperatures. He_2 was detected mass spectrometrically in a beam of helium gas cooled to 10^{-3} K by expansion [F. Luo et. al., *J. Chem. Phys.*, **98**, 3564 (1993); **100**, 4023 (1994)]. Accurate theoretical calculations on He_2 give $D_e = 0.000948$ eV, $D_0 = 0.00000014$ eV, $R_e = 2.97$ Å and give the average internuclear distance in He_2 as $\langle R \rangle \approx 46$ Å [J. B. Anderson, *J. Chem. Phys.*, **120**, 9886 (2004)]. $\langle R \rangle$ is huge because the $v = 0$ level lies so close to the dissociation limit.

Examples of diatomic van der Waals molecules and their R_e and D_e values include Ne_2, 3.1 Å, 0.0036 eV; HeNe, 3.2 Å, 0.0012 eV; Ca_2, 4.28 Å, 0.13 eV; Mg_2, 3.89 Å, 0.053 eV. Observed polyatomic van der Waals molecules include $(O_2)_2$, H_2–N_2, Ar–HCl, and $(Cl_2)_2$. For van der Waals bonding, R_e is significantly greater and D_e very substantially less than the corresponding values for chemically bound molecules. The Be_2 bond length of 2.45 Å is much shorter than is typical for van der Waals molecules, and the nature of the binding in Be_2 is not fully understood; see L. Füsti-Molnár and P. G. Szalay, *Chem. Phys. Lett.*, **258**, 400 (1996); J. Stärck and W. Meyer, *Chem. Phys. Lett.*, **258**, 421 (1996); R. L. Gdanitz, *Chem. Phys. Lett.*, **312**, 578 (1999). For more on van der Waals molecules, see *Chem. Rev.*, **88**, 813–988 (1988); **94**, 1721–2160 (1994).

13.8 ELECTRONIC TERMS OF DIATOMIC MOLECULES

We now consider the terms arising from a given diatomic molecule electron configuration.

For atoms, each set of degenerate atomic orbitals constitutes an atomic subshell. For example, the $2p_{+1}$, $2p_0$, and $2p_{-1}$ AOs constitute the $2p$ subshell. An atomic electronic configuration is defined by giving the number of electrons in each subshell; for example, $1s^2 2s^2 2p^4$. For molecules, each set of degenerate molecular orbitals constitutes a molecular **shell**. For example, the $\pi_u 2p_{+1}$ and $\pi_u 2p_{-1}$ MOs constitute the $\pi_u 2p$ shell. Each σ shell consists of one MO, while each π, δ, ϕ, ... shell consists of two MOs; σ shells are filled with two electrons, while non-σ shells hold up to four electrons. We define a molecular electronic **configuration** by giving the number of electrons in each shell, for example, $(\sigma_g 1s)^2(\sigma_u^* 1s)^2(\sigma_g 2s)^2(\sigma_u^* 2s)^2(\pi_u 2p)^3$.

For H_2^+, the operator $\hat{L}_z$ commutes with $\hat{H}$. For a many-electron diatomic molecule, one finds that the operator for the axial component of the total electronic orbital angular momentum commutes with $\hat{H}$. The component of electronic orbital angular momentum along the molecular axis has the possible values $M_L \hbar$, where $M_L = 0, \pm 1, \pm 2, \pm \ldots$. To calculate M_L, we simply add algebraically the m's of the individual electrons. Analogous to the symbol λ for a one-electron molecule, Λ is defined as

$$\Lambda \equiv |M_L| \tag{13.84}$$

(Some people define Λ as equal to M_L.) The following code specifies the value of Λ:

Λ	0	1	2	3	4
letter	Σ	Π	Δ	Φ	Γ

For $\Lambda \neq 0$, there are two possible values of M_L, namely, $+\Lambda$ and $-\Lambda$. As in H_2^+, the electronic energy depends on M_L^2, so there is a double degeneracy associated with the two values of M_L. Note that lowercase letters refer to individual electrons, while capital letters refer to the whole molecule.

Just as in atoms, the individual electron spins add vectorially to give a total electronic spin **S**, whose magnitude has the possible values $[S(S + 1)]^{1/2}\hbar$, with $S = 0, \frac{1}{2}, 1, \frac{3}{2}, \ldots$. The component of **S** along an axis has the possible values $M_S \hbar$, where $M_S = S, S - 1, \ldots, -S$. As in atoms, the quantity $2S + 1$ is called the **spin multiplicity** and is written as a left superscript to the code letter for Λ. Diatomic electronic states that arise from the same electron configuration and that have the same value for Λ and the same value for S are said to belong to the same electronic **term**. We now consider how the terms belonging to a given electron configuration are derived. (We are assuming Russell–Saunders coupling, which holds for molecules composed of atoms of not-too-high atomic number.)

A filled diatomic molecule shell consists of one or two filled molecular orbitals. The Pauli principle requires that, for two electrons in the same molecular orbital, one have $m_s = +\frac{1}{2}$ and the other have $m_s = -\frac{1}{2}$. Hence the quantum number M_S, which is the algebraic sum of the individual m_s values, must be zero for a filled-shell molecular configuration. Therefore, we must have $S = 0$ for a configuration containing only filled molecular shells. A filled σ shell has two electrons with $m = 0$, so M_L is zero. A filled π shell has two electrons with $m = +1$ and two electrons with $m = -1$, so M_L (which is the algebraic sum of the m's) is zero. The same situation holds for filled δ, ϕ, ... shells. Thus a closed-shell molecular configuration has both S and Λ equal to zero and gives rise to only a $^1\Sigma$ term. An example is the ground electronic configuration of H_2. (Recall that a filled-subshell atomic configuration gives only a ^1S term.) In deriving molecular terms, we need consider only electrons outside filled shells.

A single σ electron has $s = \frac{1}{2}$, so S must be $\frac{1}{2}$, and we get a $^2\Sigma$ term. An example is the ground electronic configuration of H_2^+. A single π electron gives a $^2\Pi$ term. And so on.

Now consider more than one electron. Electrons that are in different molecular shells are called *nonequivalent*. For such electrons we do not have to worry about giving two of them the same set of quantum numbers, and the terms are easily derived. Consider two nonequivalent σ electrons, a $\sigma\sigma$ configuration. Since both m's are zero, we have $M_L = 0$. Each s is $\frac{1}{2}$, so S can be 1 or 0. We thus have the terms $^1\Sigma$ and $^3\Sigma$. Similarly, a $\sigma\pi$ configuration gives $^1\Pi$ and $^3\Pi$ terms.

For a $\pi\delta$ configuration, we have singlet and triplet terms. The π electron can have $m = \pm 1$, and the δ electron can have $m = \pm 2$. The possible values for M_L are thus $+3$, -3, $+1$, and -1. This gives $\Lambda = 3$ or 1, and we have the terms $^1\Pi$, $^3\Pi$, $^1\Phi$, $^3\Phi$. (In atoms we add the *vectors* $\mathbf{L}_i$ to get the total $\mathbf{L}$; hence a *pd* atomic configuration gives P, D, and F terms. In molecules, however, we add the *z components* of the orbital angular momenta. This is an algebraic rather than a vectorial addition, so a $\pi\delta$ molecular configuration gives Π and Φ terms and no Δ terms.)

For a $\pi\pi$ configuration of two nonequivalent electrons, each electron has $m = \pm 1$, and we have the M_L values 2, -2, 0, 0. The values of Λ are 2, 0, and 0; the terms are $^1\Delta$, $^3\Delta$, $^1\Sigma$, $^3\Sigma$, $^1\Sigma$, and $^3\Sigma$. The values $+2$ and -2 correspond to the two degenerate states of the same Δ term. However, Σ terms are nondegenerate (apart from spin degeneracy), and the two values of M_L that are zero indicate two different Σ terms (which become four Σ terms when we consider spin).

Consider the forms of the wave functions for the $\pi\pi$ terms. We shall call the two π subshells π and π' and shall use a subscript to indicate the m value. For the Δ terms, both electrons have $m = +1$ or both have $m = -1$. For $M_L = +2$, we might write as the spatial factor in the wave function $\pi_{+1}(1)\pi'_{+1}(2)$ or $\pi_{+1}(2)\pi'_{+1}(1)$. However, these functions are neither symmetric nor antisymmetric with respect to exchange of the indistinguishable electrons and are unacceptable. Instead, we must take the linear combinations (we shall not bother with normalization constants)

$$^1\Delta{:}\quad \pi_{+1}(1)\pi'_{+1}(2) + \pi_{+1}(2)\pi'_{+1}(1) \tag{13.85}$$

$$^3\Delta{:}\quad \pi_{+1}(1)\pi'_{+1}(2) - \pi_{+1}(2)\pi'_{+1}(1) \tag{13.86}$$

Similarly, with both electrons having $m = -1$, we have the spatial factors

$$^1\Delta{:}\quad \pi_{-1}(1)\pi'_{-1}(2) + \pi_{-1}(2)\pi'_{-1}(1) \tag{13.87}$$

$$^3\Delta{:}\quad \pi_{-1}(1)\pi'_{-1}(2) - \pi_{-1}(2)\pi'_{-1}(1) \tag{13.88}$$

The functions (13.85) and (13.87) are symmetric with respect to exchange. They therefore go with the antisymmetric two-electron spin factor (11.60), which has $S = 0$. Thus (13.85) and (13.87) are the spatial factors in the wave functions for the two states of the doubly degenerate $^1\Delta$ term. The antisymmetric functions (13.86) and (13.88) must go with the symmetric two-electron spin functions (11.57), (11.58), and (11.59), giving the six states of the $^3\Delta$ term. These states all have the same energy (if we neglect spin–orbit interaction).

Now consider the wave functions of the Σ terms. These have one electron with $m = +1$ and one electron with $m = -1$. We start with the four functions

$$\pi_{+1}(1)\pi'_{-1}(2), \qquad \pi_{+1}(2)\pi'_{-1}(1), \qquad \pi_{-1}(1)\pi'_{+1}(2), \qquad \pi_{-1}(2)\pi'_{+1}(1)$$

Combining them to get symmetric and antisymmetric functions, we have

$$\begin{aligned}
&{}^1\Sigma^+\!: \quad \pi_{+1}(1)\pi'_{-1}(2) + \pi_{+1}(2)\pi'_{-1}(1) + \pi_{-1}(1)\pi'_{+1}(2) + \pi_{-1}(2)\pi'_{+1}(1) \\
&{}^1\Sigma^-\!: \quad \pi_{+1}(1)\pi'_{-1}(2) + \pi_{+1}(2)\pi'_{-1}(1) - \pi_{-1}(1)\pi'_{+1}(2) - \pi_{-1}(2)\pi'_{+1}(1) \\
&{}^3\Sigma^+\!: \quad \pi_{+1}(1)\pi'_{-1}(2) - \pi_{+1}(2)\pi'_{-1}(1) + \pi_{-1}(1)\pi'_{+1}(2) - \pi_{-1}(2)\pi'_{+1}(1) \\
&{}^3\Sigma^-\!: \quad \pi_{+1}(1)\pi'_{-1}(2) - \pi_{+1}(2)\pi'_{-1}(1) - \pi_{-1}(1)\pi'_{+1}(2) + \pi_{-1}(2)\pi'_{+1}(1)
\end{aligned} \tag{13.89}$$

The first two functions in (13.89) are symmetric. They therefore go with the antisymmetric singlet spin function (11.60). Clearly, these two spatial functions have different energies. The last two functions in (13.89) are antisymmetric and hence are the spatial factors in the wave functions of the two ${}^3\Sigma$ terms. The four functions in (13.89) are found to have eigenvalue +1 or −1 with respect to reflection of electronic coordinates in the xz σ_v symmetry plane containing the molecular (z) axis (Prob. 13.29). The superscripts + and − refer to this eigenvalue.

Examination of the Δ terms (13.85) to (13.88) shows that they are not eigenfunctions of the symmetry operator $\hat{O}_{\sigma_v}$. Since a twofold degeneracy (apart from spin degeneracy) is associated with these terms, there is no necessity that their wave functions be eigenfunctions of this operator. However, since $\hat{O}_{\sigma_v}$ commutes with the Hamiltonian, we can *choose* the eigenfunctions to be eigenfunctions of $\hat{O}_{\sigma_v}$. Thus we can combine the functions (13.85) and (13.87), which belong to a degenerate energy level, as follows:

$$(13.85) + (13.87) \qquad \text{and} \qquad (13.85) - (13.87)$$

These two linear combinations are eigenfunctions of $\hat{O}_{\sigma_v}$ with eigenvalues +1 and −1, and we could refer to them as ${}^1\Delta^+$ and ${}^1\Delta^-$ states. Since they have the same energy, there is no point in using the + and − superscripts. Thus the + and − designations are used only for Σ terms. However, when one considers the interaction between the molecular rotational angular momentum and the electronic orbital angular momentum, there is a very slight splitting (called Λ-*type doubling*) of the two states of a ${}^1\Delta$ term. It turns out that the correct zeroth-order wave functions for this perturbation are the linear combinations that are eigenfunctions of $\hat{O}_{\sigma_v}$, so in this case there is a point to distinguishing between Δ^+ and Δ^- states. The linear combinations (13.85) ± (13.87), which are eigenfunctions of $\hat{O}_{\sigma_v}$, are not eigenfunctions of $\hat{L}_z$ but are superpositions of $\hat{L}_z$ eigenfunctions with eigenvalues +2 and −2.

We can distinguish + and − terms for one-electron configurations. The wave function of a single σ electron has no phi factor and hence must correspond to a Σ^+ term. For a π electron, the MOs that are eigenfunctions of $\hat{L}_z$ are the π_{+1} and π_{-1} functions (whose probability densities are each symmetric about the z axis; Fig. 13.14). The π_{+1} and π_{-1} functions are not eigenfunctions of $\hat{O}_{\sigma_v}$, but the linear combinations $\pi_{+1} + \pi_{-1} = \pi_x$ and $\pi_{+1} - \pi_{-1} = \pi_y$ are. The π_x and π_y MOs (whose probability densities are not symmetric about the z axis; Fig. 13.15) are the correct zeroth-order functions if the perturbation of the electronic wave functions due to molecular rotation is considered. The π_x and π_y MOs have eigenvalues +1 and −1, respectively, for reflection in the xz plane, and eigenvalues −1 and +1, respectively, for reflection in the yz plane. (The operators $\hat{L}_z$ and $\hat{O}_{\sigma_v}$ do not commute; Prob. 13.30. Hence we cannot have all the eigenfunctions of $\hat{H}$ being eigenfunctions of both these operators as well. However, since each of these operators commutes with the electronic Hamiltonian and since there is no element of choice in the wave function of a nondegenerate level, all the σ MOs must be eigenfunctions of both $\hat{L}_z$ and $\hat{O}_{\sigma_v}$.)

Electrons in the same molecular shell are called *equivalent*. For equivalent electrons, there are fewer terms than for the corresponding nonequivalent electron configuration, because of the Pauli principle. Thus, for a π^2 configuration of two equivalent π electrons, four of the eight functions (13.85) to (13.89) vanish; the remaining functions give a $^1\Delta$ term, a $^1\Sigma^+$ term, and a $^3\Sigma^-$ term. Alternatively, we can make a table similar to Table 11.1 and use it to derive the terms for equivalent electrons.

Table 13.3 lists terms arising from various electron configurations. A filled shell always gives the single term $^1\Sigma^+$. A π^3 configuration gives the same result as a π configuration.

For homonuclear diatomic molecules, a g or u right subscript is added to the term symbol to show the parity of the electronic states belonging to the term. Terms arising from an electron configuration that has an odd number of electrons in molecular orbitals of odd parity are odd (u); all other terms are even (g). This is the same rule as for atoms.

The term symbols given in Table 13.2 are readily derived from the MO configurations. For example, O_2 has a π^2 configuration, which gives the three terms $^1\Sigma_g^+$, $^3\Sigma_g^-$, and $^1\Delta_g$. Hund's rule tells us that $^3\Sigma_g^-$ is the lowest term, as listed. The $v = 0$ levels of the $^1\Delta_g$ and $^1\Sigma_g^+$ O_2 terms lie 0.98 eV and 1.6 eV, respectively, above the $v = 0$ level of the ground $^3\Sigma_g^-$ term. Singlet O_2 is a reaction intermediate in many organic, biochemical, and inorganic reactions. [See A. A. Frimer, ed., *Singlet O_2*, vols. 1–4, CRC Press, 1985; C. S. Foote et al., eds., *Active Oxygen in Chemistry*, Springer, 1995; J. S. Valentine et al., eds., *Active Oxygen in Biochemistry*, Springer, 1995; C. Schweitzer and R. Schmidt, *Chem. Rev.*, **103**, 1685 (2003).]

TABLE 13.3 Electronic Terms of Diatomic Molecules

Configuration	Terms
$\sigma\sigma$	$^1\Sigma^+, {}^3\Sigma^+$
$\sigma\pi$; $\sigma\pi^3$	$^1\Pi, {}^3\Pi$
$\pi\pi$; $\pi\pi^3$	$^1\Sigma^+, {}^3\Sigma^+, {}^1\Sigma^-, {}^3\Sigma^-, {}^1\Delta, {}^3\Delta$
$\pi\delta$; $\pi^3\delta$; $\pi\delta^3$	$^1\Pi, {}^3\Pi, {}^1\Phi, {}^3\Phi$
σ	$^2\Sigma^+$
σ^2; π^4; δ^4	$^1\Sigma^+$
π; π^3	$^2\Pi$
π^2	$^1\Sigma^+, {}^3\Sigma^-, {}^1\Delta$
δ; δ^3	$^2\Delta$
δ^2	$^1\Sigma^+, {}^3\Sigma^-, {}^1\Gamma$

Most stable diatomic molecules have a $^1\Sigma^+$ ground term ($^1\Sigma_g^+$ for homonuclear diatomics). Exceptions include B_2, Al_2, Si_2, and O_2, and NO, which has a $^2\Pi$ ground term.

Spectroscopists prefix the ground term of a molecule by the symbol X. Excited terms of the same spin multiplicity as the ground term are designated as $A, B, C, \ldots$, while excited terms of different spin multiplicity from the ground term are designated as $a, b, c, \ldots$. Exceptions are C_2 and N_2, where the ground terms are $^1\Sigma_g^+$ but the letters $A, B, C, \ldots$ are used for excited triplet terms.

Just as for atoms, spin–orbit interaction can split a molecular term into closely spaced energy levels, giving a multiplet structure to the term. The projection of the total electronic spin **S** on the molecular axis is $M_S\hbar$. In molecules the quantum number M_S is called Σ (not to be confused with the symbol meaning $\Lambda = 0$):

$$\Sigma = S, S-1, \ldots, -S$$

The axial components of electronic orbital and spin angular momenta add, giving as the total axial component of electronic angular momentum $(\Lambda + \Sigma)\hbar$. (Recall that Λ is the absolute value of M_L. We consider Σ to be positive when it has the same direction as Λ, and negative when it has the opposite direction as Λ.) The possible values of $\Lambda + \Sigma$ are

$$\Lambda + S, \Lambda + S - 1, \ldots, \Lambda - S$$

The value of $\Lambda + \Sigma$ is written as a right subscript to the term symbol to distinguish the energy levels of the term. Thus a $^3\Delta$ term has $\Lambda = 2$ and $S = 1$ and gives rise to the levels $^3\Delta_3$, $^3\Delta_2$, and $^3\Delta_1$. In a sense, $\Lambda + \Sigma$ is the analog in molecules of the quantum number J in atoms. However, $\Lambda + \Sigma$ is the quantum number of the *z component* of total electronic angular momentum and therefore can take on negative values. Thus a $^4\Pi$ term has the four levels $^4\Pi_{5/2}$, $^4\Pi_{3/2}$, $^4\Pi_{1/2}$, and $^4\Pi_{-1/2}$. The absolute value of $\Lambda + \Sigma$ is called Ω:

$$\Omega \equiv |\Lambda + \Sigma| \tag{13.90}$$

The spin–orbit interaction energy in diatomic molecules can be shown to be well approximated by $A\Lambda\Sigma$, where A depends on Λ and on the internuclear distance R but not on Σ. The spacing between levels of the multiplet is thus constant. When A is positive, the level with the lowest value of $\Lambda + \Sigma$ lies lowest, and the multiplet is *regular*. When A is negative, the multiplet is *inverted*. Note that for $\Lambda \neq 0$ the spin multiplicity $2S + 1$ always equals the number of multiplet components. This is not always true for atoms.

Each energy level of a multiplet with $\Lambda \neq 0$ is doubly degenerate, corresponding to the two values for M_L. Thus a $^3\Delta$ term has six different wave functions [Eqs. (13.86), 13.88), (11.57) to (11.59)] and therefore six different molecular electronic states. in–orbit interaction splits the $^3\Delta$ term into three levels, each doubly degenerate. The ble degeneracy of the levels is removed by the Λ-type doubling mentioned usly.

r Σ terms ($\Lambda = 0$), the spin–orbit interaction is very small (zero in the first nation), and the quantum numbers Σ and Ω are not defined. term always corresponds to a single nondegenerate energy level.

13.9 THE HYDROGEN MOLECULE

The hydrogen molecule is the simplest molecule containing an electron-pair bond. The purely electronic Hamiltonian (13.5) for H_2 is

$$\hat{H} = -\tfrac{1}{2}\nabla_1^2 - \tfrac{1}{2}\nabla_2^2 - \frac{1}{r_{a1}} - \frac{1}{r_{a2}} - \frac{1}{r_{b1}} - \frac{1}{r_{b2}} + \frac{1}{r_{12}} \tag{13.91}$$

where 1 and 2 are the electrons and a and b are the nuclei (Fig. 13.18). Just as in the helium atom, the $1/r_{12}$ interelectronic-repulsion term prevents the Schrödinger equation from being separable. We therefore use approximation methods.

We start with the molecular-orbital approach. The ground-state electron configuration of H_2 is $(\sigma_g 1s)^2$, and we can write an approximate wave function as the Slater determinant

$$\frac{1}{\sqrt{2}}\begin{vmatrix} \sigma_g 1s(1)\alpha(1) & \sigma_g 1s(1)\beta(1) \\ \sigma_g 1s(2)\alpha(2) & \sigma_g 1s(2)\beta(2) \end{vmatrix} = \sigma_g 1s(1)\sigma_g 1s(2)\cdot 2^{-1/2}[\alpha(1)\beta(2) - \beta(1)\alpha(2)]$$

$$= f(1)f(2)\cdot 2^{-1/2}[\alpha(1)\beta(2) - \beta(1)\alpha(2)] \tag{13.92}$$

which is similar to (10.26) for the helium atom. To save time, we write f instead of $\sigma_g 1s$. As we saw in Section 10.4, omission of the spin factor does not affect the variational integral for a two-electron problem. Hence we want to choose f so as to minimize

$$\frac{\iint f^*(1)f^*(2)\hat{H}f(1)f(2)\,dv_1\,dv_2}{\iint |f(1)|^2|f(2)|^2 dv_1\,dv_2}$$

where the integration is over the spatial coordinates of the two electrons. Ideally, f should be found by an SCF calculation. For simplicity we can use an H_2^+-like MO. (The H_2 Hamiltonian becomes the sum of two H_2^+ Hamiltonians if we omit the $1/r_{12}$ term.) We saw in Section 13.5 that the function [Eq. (13.57)]

$$\frac{k^{3/2}}{(2\pi)^{1/2}(1 + S_{ab})^{1/2}}(e^{-kr_a} + e^{-kr_b})$$

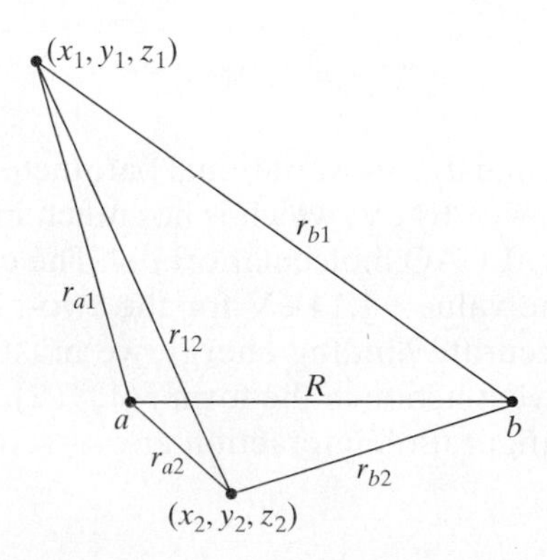

FIGURE 13.18 Interparticle distances in H_2.

gives a good approximation to the ground-state H_2^+ wave function. Hence we try as a variation function ϕ for H_2 the product of two such LCAO functions, one for each electron:

$$\phi = \frac{\zeta^3}{2\pi(1 + S_{ab})}(e^{-\zeta r_{a1}} + e^{-\zeta r_{b1}})(e^{-\zeta r_{a2}} + e^{-\zeta r_{b2}}) \tag{13.93}$$

$$\phi = \frac{1}{2(1 + S_{ab})}[1s_a(1) + 1s_b(1)][1s_a(2) + 1s_b(2)] \tag{13.94}$$

where the effective nuclear charge ζ will differ from k for H_2^+. Since

$$\hat{H} = \hat{H}_1^0 + \hat{H}_2^0 + 1/r_{12}$$

where $\hat{H}_1^0$ and $\hat{H}_2^0$ are H_2^+ Hamiltonians for each electron, we have

$$\iint \phi^* \hat{H} \phi \, dv_1 \, dv_2 = 2W_1 + \iint \frac{\phi^2}{r_{12}} dv_1 \, dv_2$$

where W_1 is given by (13.63) with k replaced by ζ. The evaluation of the $1/r_{12}$ integral is complicated and is omitted [see *Slater, Quantum Theory of Molecules and Solids*, Volume 1, page 65, and Appendix 6]. Coulson performed the variational calculation in 1937, using (13.93). [For the literature references of the H_2 calculations mentioned in this and later sections, see the bibliography in A. D. McLean, A. Weiss, and M. Yoshimine, *Rev. Mod. Phys.*, **32**, 211 (1960).] Coulson found $R_e = 0.732$ Å, which is close to the true value 0.741 Å; the minimum in the calculated $U(R)$ curve gave $D_e = 3.49$ eV, as compared with the true value 4.75 eV (Table 13.2). (Of course, the percent error in the total electronic energy is much less than the percent error in D_e, but D_e is the quantity of chemical interest.) The value of ζ at 0.732 Å is 1.197, which is less than k for H_2^+. We attribute this to the screening of the nuclei from each electron by the other electron.

How can we improve on the above simple MO result? We can look for the best possible MO function f in (13.92) to get the Hartree–Fock wave function for H_2. This was done by Kolos and Roothaan [W. Kolos and C. C. J. Roothaan, *Rev. Mod. Phys.*, **32**, 219 (1960)]. They expanded f in elliptic coordinates [Eq. (13.34)]. Since $m = 0$ for the ground state, the $e^{im\phi}$ factor in the SCF MO is equal to 1 and f is a function of ξ and η only. The expansion used is

$$f = e^{-\alpha\xi} \sum_{p,q} a_{pq} \xi^p \eta^q$$

where p and q are integers and α and a_{pq} are variational parameters. The Hartree–Fock results are $R_e = 0.733$ Å and $D_e = 3.64$ eV, which is not much improvement over the value 3.49 eV given by the simple LCAO molecular orbital. The correlation energy for H_2 is thus -1.11 eV, close to the value -1.14 eV for the two-electron helium atom (Section 11.3). To get a truly accurate binding energy, we must go beyond the SCF approximation of writing the wave function in the form $f(1)f(2)$. We can use the same methods we used for atoms: configuration interaction and introduction of r_{12} into the trial function.

First consider configuration interaction (CI). To reach the exact ground-state wave function, we include contributions from SCF (or other) functions for all the excited states with the same symmetry as the ground state. In the first approximation, only contributions from the lowest-lying excited states are included. The first excited configuration of H_2 is $(\sigma_g 1s)(\sigma_u^* 1s)$, which gives the terms ${}^1\Sigma_u^+$ and ${}^3\Sigma_u^+$. (We have one g and one u electron, so the terms are of odd parity.) The ground-state configuration $(\sigma_g 1s)^2$ is a ${}^1\Sigma_g^+$ state. Hence we do not get any contribution from the $(\sigma_g 1s)(\sigma_u^* 1s)$ states, since they have different parity from the ground state. Next consider the configuration $(\sigma_u^* 1s)^2$. This is a closed-shell configuration having the single state ${}^1\Sigma_g^+$. This is of the right symmetry to contribute to the ground-state wave function. As a simple CI trial function, we can take a linear combination of the MO wave functions for the $(\sigma_g 1s)^2$ and $(\sigma_u^* 1s)^2$ configurations. To simplify things, we will use the LCAO-MOs as approximations to the MOs. Thus we take

$$\phi = \sigma_g 1s(1)\sigma_g 1s(2) + c\sigma_u^* 1s(1)\sigma_u^* 1s(2) \tag{13.95}$$

where $\sigma_g 1s$ and $\sigma_u^* 1s$ are given by (13.57) and (13.58) with a variable orbital exponent and c is a variational parameter. This calculation was performed by Weinbaum in 1933. The result is a bond length of 0.757 Å and a dissociation energy of 4.03 eV, a considerable improvement over the Hartree–Fock result $D_e = 3.64$ eV. The orbital exponent has the optimum value 1.19. We can improve on this result by using a better form for the MOs of each configuration and by including more configuration functions. Hagstrom did a CI calculation in which the MOs were represented by expansions in elliptic coordinates. With 33 configuration functions, he found $D_e = 4.71$ eV, quite close to the true value 4.75 eV [S. Hagstrom and H. Shull, *Rev. Mod. Phys.*, **35**, 624 (1963)].

Now consider the use of r_{12} in H_2 trial functions. The first really accurate calculation of the hydrogen-molecule ground state was done by James and Coolidge in 1933. They used the trial function

$$\exp[-\delta(\xi_1 + \xi_2)] \sum c_{mnjkp}[\xi_1^m \xi_2^n \eta_1^j \eta_2^k + \xi_1^n \xi_2^m \eta_1^k \eta_2^j]r_{12}^p$$

where the summation is over integral values of m, n, j, k, and p. The variational parameters are δ and the c_{mnjkp} coefficients. The James and Coolidge function is symmetric with respect to interchange of electrons 1 and 2, as it should be, since we have an antisymmetric ground-state spin function. With 13 terms in the sum, James and Coolidge found $D_e = 4.72$ eV, only 0.03 eV in error. Their work has been extended by Kolos, Wolniewicz, and co-workers, who used as many as 279 terms in the sum. Since it is D_0 that is determined from the observed electronic spectrum, they used the Cooley–Numerov method (Section 13.2) to calculate the vibrational levels from their theoretical $U(R)$ curve and then calculated D_0. Including relativistic corrections and corrections to the Born–Oppenheimer approximation, they found $D_0/hc = 36118.1\ \text{cm}^{-1}$, in agreement with the spectroscopically determined value $36118.1 \pm 0.1\ \text{cm}^{-1}$ [W. Kolos et al., *J. Chem. Phys.*, **84**, 3278 (1986); L. Wolniewicz, *J. Chem. Phys.*, **99**, 1851 (1993)]. Their wave function is essentially indistinguishable from the true wave function, but its great complexity allows no simple physical interpretation.

13.10 THE VALENCE-BOND TREATMENT OF H_2

The first quantum-mechanical treatment of the hydrogen molecule was by Heitler and London in 1927. Their ideas have been extended to give a general theory of chemical bonding, known as the **valence-bond** (VB) theory. The valence-bond method is more closely related to the chemist's idea of molecules as consisting of atoms held together by localized bonds than is the molecular-orbital method. The VB method views molecules as composed of atomic cores (nuclei plus inner-shell electrons) and bonding valence electrons. For H_2, both electrons are valence electrons.

The first step in the Heitler–London treatment of the H_2 ground state is to approximate the molecule as two ground-state hydrogen atoms. The wave function for two such noninteracting atoms is

$$f_1 = 1s_a(1)1s_b(2)$$

where a and b refer to the nuclei and 1 and 2 refer to the electrons. Of course, the function

$$f_2 = 1s_a(2)1s_b(1)$$

is also a valid wave function. This then suggests the trial variation function

$$c_1f_1 + c_2f_2 = c_1 1s_a(1)1s_b(2) + c_2 1s_a(2)1s_b(1) \tag{13.96}$$

This linear variation function leads to the determinantal secular equation $\det(H_{ij} - S_{ij}W) = 0$ [Eq. (8.58)], where $H_{11} = \langle f_1|\hat{H}|f_1\rangle$, $S_{11} = \langle f_1|f_1\rangle, \ldots$.

We can also consider the problem using perturbation theory (as Heitler and London did). A ground-state hydrogen molecule dissociates to two neutral ground-state hydrogen atoms. We therefore take as the unperturbed problem two ground-state hydrogen atoms at infinite separation. One possible zeroth-order (unperturbed) wave function is $1s_a(1)1s_b(2)$. However, electron 2 could just as well be bound to nucleus a, giving the unperturbed wave function $1s_a(2)1s_b(1)$. These two unperturbed wave functions belong to a doubly degenerate energy level (exchange degeneracy). Under the perturbation of molecule formation, the doubly degenerate level is split into two levels, and the *correct* zeroth-order wave functions are linear combinations of the two unperturbed wave functions:

$$c_1 1s_a(1)1s_b(2) + c_2 1s_a(2)1s_b(1)$$

This leads to a 2×2 secular determinant that is the same as (8.57), except that W is replaced by $E^{(0)} + E^{(1)}$; see Prob. 9.19.

We now solve the secular equation. The Hamiltonian is Hermitian, all functions are real, and f_1 and f_2 are normalized; hence

$$H_{12} = H_{21}, \qquad S_{12} = S_{21}, \qquad S_{11} = S_{22} = 1$$

Consider H_{11} and H_{22}:

$$H_{11} = \langle 1s_a(1)1s_b(2)|\hat{H}|1s_a(1)1s_b(2)\rangle$$

$$H_{22} = \langle 1s_a(2)1s_b(1)|\hat{H}|1s_a(2)1s_b(1)\rangle$$

Interchange of the coordinate labels 1 and 2 in H_{22} converts H_{22} to H_{11}, since this relabeling leaves $\hat{H}$ unchanged. Hence $H_{11} = H_{22}$. The secular equation $\det(H_{ij} - S_{ij}W) = 0$ becomes

$$\begin{vmatrix} H_{11} - W & H_{12} - WS_{12} \\ H_{12} - WS_{12} & H_{11} - W \end{vmatrix} = 0 \tag{13.97}$$

This equation has the same form as Eq. (13.49), and by analogy to Eqs. (13.51), (13.57), and (13.58) the approximate energies and wave functions are

$$W_1 = \frac{H_{11} + H_{12}}{1 + S_{12}}, \qquad W_2 = \frac{H_{11} - H_{12}}{1 - S_{12}} \tag{13.98}$$

$$\phi_1 = \frac{f_1 + f_2}{\sqrt{2}(1 + S_{12})^{1/2}}, \qquad \phi_2 = \frac{f_1 - f_2}{\sqrt{2}(1 - S_{12})^{1/2}} \tag{13.99}$$

For S_{12} we have

$$S_{12} = \int f_1^* f_2 \, dv = \iint 1s_a(1)1s_b(2)1s_a(2)1s_b(1) \, dv_1 \, dv_2$$

$$S_{12} = \langle 1s_a(1)|1s_b(1)\rangle\langle 1s_a(2)|1s_b(2)\rangle = S_{ab}^2$$

where the overlap integral S_{ab} is defined by (13.48).

The numerators of (13.99) are

$$f_1 \pm f_2 = 1s_a(1)1s_b(2) \pm 1s_a(2)1s_b(1)$$

From our previous discussion, we know that the ground state of H_2 is a $^1\Sigma$ state with the antisymmetric spin factor (11.60) and a symmetric spatial factor. Hence ϕ_1 must be the ground state. The Heitler–London ground-state wave function is

$$\frac{1s_a(1)1s_b(2) + 1s_a(2)1s_b(1)}{\sqrt{2}(1 + S_{ab}^2)^{1/2}} \frac{1}{\sqrt{2}}[\alpha(1)\beta(2) - \alpha(2)\beta(1)] \tag{13.100}$$

The Heitler–London wave functions for the three states of the lowest $^3\Sigma$ term are

$$\frac{1s_a(1)1s_b(2) - 1s_a(2)1s_b(1)}{\sqrt{2}(1 - S_{ab}^2)^{1/2}} \begin{cases} \alpha(1)\alpha(2) \\ 2^{-1/2}[\alpha(1)\beta(2) + \beta(1)\alpha(2)] \\ \beta(1)\beta(2) \end{cases} \tag{13.101}$$

Now consider the ground-state energy expression. We write the molecular electronic Hamiltonian as the sum of two H-atom Hamiltonians plus perturbing terms:

$$\hat{H} = \hat{H}_a(1) + \hat{H}_b(2) + \hat{H}'$$

$$\hat{H}_a(1) = -\tfrac{1}{2}\nabla_1^2 - \frac{1}{r_{a1}}, \qquad \hat{H}_b(1) = -\tfrac{1}{2}\nabla_2^2 - \frac{1}{r_{b2}}, \qquad \hat{H}' = -\frac{1}{r_{b1}} - \frac{1}{r_{a2}} + \frac{1}{r_{12}}$$

We then have

$$H_{11} = \langle 1s_a(1)1s_b(2)|\hat{H}_a(1) + \hat{H}_b(2) + \hat{H}'|1s_a(1)1s_b(2)\rangle \tag{13.102}$$

For the integral involving $\hat{H}_a(1)$, we have

$$\langle 1s_a(1)1s_b(2)|\hat{H}_a(1)|1s_a(1)1s_b(2)\rangle = \langle 1s_a(1)|\hat{H}_a(1)|1s_a(1)\rangle\langle 1s_b(2)|1s_b(2)\rangle$$

The Heitler–London calculation does not introduce an effective nuclear charge into the $1s$ function. Hence $1s_a(1)$ is an eigenfunction of $\hat{H}_a(1)$ with eigenvalue $-\frac{1}{2}$ hartree, the hydrogen-atom ground-state energy. The $1s$ function is normalized, and we conclude that the $\hat{H}_a(1)$ integral equals $-\frac{1}{2}$ in atomic units. Similarly, the $\hat{H}_b(2)$ integral in (13.102) equals $-\frac{1}{2}$. Defining the *Coulomb integral Q* as

$$Q \equiv \langle 1s_a(1)1s_b(2)|\hat{H}'|1s_a(1)1s_b(2)\rangle \tag{13.103}$$

we have

$$H_{11} = Q - 1$$

For H_{12} we have

$$H_{12} = H_{21} = \langle 1s_a(2)1s_b(1)|\hat{H}_a(1) + \hat{H}_b(2) + \hat{H}'|1s_a(1)1s_b(2)\rangle$$

The $\hat{H}_a(1)$ integral is easily evaluated as

$$\langle 1s_a(2)|1s_b(2)\rangle\,\langle 1s_b(1)|\hat{H}_a(1)|1s_a(1)\rangle = -\tfrac{1}{2}S_{ab}^2$$

Defining the *exchange integral A* as

$$A \equiv \langle 1s_a(2)1s_b(1)|\hat{H}'|1s_a(1)1s_b(2)\rangle \tag{13.104}$$

we have

$$H_{12} = A - S_{ab}^2$$

Substitution into (13.98) gives

$$W_1 = -1 + \frac{Q + A}{1 + S_{ab}^2}, \qquad W_2 = -1 + \frac{Q - A}{1 - S_{ab}^2} \tag{13.105}$$

The quantity -1 hartree in these expressions is the energy of two ground-state hydrogen atoms. To obtain the $U(R)$ potential-energy curves, we add the internuclear repulsion $1/R$ to these expressions.

Many of the integrals needed to evaluate W_1 and W_2 have been evaluated in the treatment of H_2^+ in Section 13.5. The only new integrals are those involving $1/r_{12}$. The hardest one is the two-center, two-electron exchange integral:

$$\iint 1s_a(1)1s_b(2)\frac{1}{r_{12}}1s_a(2)1s_b(1)\,dv_1\,dv_2$$

Two-center means that the integrand contains functions centered on two different nuclei, a and b; *two-electron* means that the coordinates of two electrons occur in the integrand. This must be evaluated using an expansion for $1/r_{12}$ in confocal elliptic coordinates, similar to the expansion in Prob. 9.13 in spherical coordinates. Details of the integral evaluations are given in *Slater, Quantum Theory of Molecules and Solids*, Volume 1, Appendix 6. The results of the Heitler–London treatment are $D_e = 3.15$ eV, $R_e = 0.87$ Å. The agreement with the experimental values $D_e = 4.75$ eV, $R_e = 0.741$ Å is only fair. In this treatment, most of the binding energy is provided by the exchange integral A. The Heitler–London treatment of H_2 resembles the Heisenberg treatment of the helium $1s2s$ configuration given in Section 9.7. However, for H_2 the exchange integral (13.104) is negative (because of the contributions of the $-1/r_{b1} - 1/r_{a2}$ terms), whereas for He the exchange integral (9.99) contains only the interelectron-repulsion term and is positive. [In Hartree–Fock theory, the exchange integral (14.24) is defined to contain only the electron-repulsion term $1/r_{12}$.]

Consider some improvements on the Heitler–London function (13.100). One obvious step is the introduction of an orbital exponent ζ in the 1s function. This was done by Wang in 1928. The optimum value of ζ is 1.166 at R_e, and D_e and R_e are improved to 3.78 eV and 0.744 Å. Recall that Dickinson in 1933 improved the Finkelstein–Horowitz H_2^+ trial function by mixing in some $2p_z$ character into the atomic orbitals (hybridization). In 1931 Rosen used this idea to improve the Heitler–London–Wang function. He took the trial function

$$\phi = \phi_a(1)\phi_b(2) + \phi_a(2)\phi_b(1)$$

where the atomic orbital ϕ_a is given by $\phi_a = e^{-\zeta r_a}(1 + cz_a)$, with a similar expression for ϕ_b. This allows for the polarization of the AOs on molecule formation. The result is a binding energy of 4.04 eV. Another improvement, the use of ionic structures, will be considered in the next section. Still another improvement, the generalized valence-bond method, is discussed in Section 16.9.

13.11 COMPARISON OF THE MO AND VB THEORIES

Let us compare the molecular-orbital and valence-bond treatments of the H_2 ground state.

If ϕ_a symbolizes an atomic orbital centered on nucleus a, the spatial factor of the unnormalized LCAO-MO wave function for the H_2 ground state is

$$[\phi_a(1) + \phi_b(1)][\phi_a(2) + \phi_b(2)] \tag{13.106}$$

In the simplest treatment, ϕ is a 1s AO. The function (13.106) equals

$$\phi_a(1)\phi_a(2) + \phi_b(1)\phi_b(2) + \phi_a(1)\phi_b(2) + \phi_b(1)\phi_a(2) \tag{13.107}$$

What is the physical significance of the terms? The last two terms have each electron in an atomic orbital centered on a different nucleus. These are covalent terms, corresponding to equal sharing of the electrons between the atoms. The first two terms have both electrons in AOs centered on the same nucleus. These are ionic terms, corresponding to the chemical structures

$$H^- H^+ \quad \text{and} \quad H^+ H^-$$

The covalent and ionic terms occur with equal weight, so this simple MO function gives a 50–50 chance as to whether the H_2 ground state dissociates to two neutral hydrogen atoms or to a proton and a hydride ion. Actually, the H_2 ground state dissociates to two neutral H atoms. Thus the simple MO function gives the wrong limiting value of the energy as R goes to infinity.

How can we remedy this? Since H_2 is nonpolar, chemical intuition tells us that ionic terms should contribute substantially less to the wave function than covalent terms. The simplest procedure is to omit the ionic terms of the MO function (13.107). This gives

$$\phi_a(1)\phi_b(2) + \phi_b(1)\phi_a(2) \tag{13.108}$$

We recognize (13.108) as the Heitler–London function (13.100).

Although interelectronic repulsion causes the electrons to avoid each other, there is some probability of finding both electrons near the same nucleus, corresponding to an ionic structure. Therefore, instead of simply dropping the ionic terms from (13.107), we might try

$$\phi_{\text{VB,imp}} = \phi_a(1)\phi_b(2) + \phi_b(1)\phi_a(2) + \delta[\phi_a(1)\phi_a(2) + \phi_b(1)\phi_b(2)] \quad (13.109)$$

where $\delta(R)$ is a variational parameter and where the subscript indicates an improved VB function. In the language of valence-bond theory, this trial function represents **ionic–covalent resonance**. Of course, the ground-state wave function of H_2 does not undergo a time-dependent change back and forth from a covalent function corresponding to the structure H — H to ionic functions. Rather (in the approximation we are considering), the wave function is a time-independent mixture of covalent and ionic functions. Since H_2 dissociates to neutral atoms, we know that $\delta(\infty) = 0$. A variational calculation done by Weinbaum in 1933 using 1s AOs with an orbital exponent gave the result that at R_e the parameter δ has the value 0.26; the orbital exponent was found to be 1.19, and the dissociation energy was calculated as 4.03 eV, a modest improvement over the Heitler–London–Wang value of 3.78 eV. With δ equal to zero in (13.109), we get the VB function (13.108). With δ equal to 1, we get the LCAO-MO function (13.107). The optimum value of δ turns out to be closer to zero than to 1, and, in fact, the Heitler–London–Wang VB function gives a better dissociation energy than the LCAO-MO function.

Let us compare the improved valence-bond trial function (13.109) with the simple LCAO-MO function improved by configuration interaction. The LCAO-MO CI trial function (13.95) has the (unnormalized) form

$$\phi_{\text{MO,imp}} = [\phi_a(1) + \phi_b(1)][\phi_a(2) + \phi_b(2)] + \gamma[\phi_a(1) - \phi_b(1)][\phi_a(2) - \phi_b(2)]$$

Since we have not yet normalized this function, there is no harm in multiplying it by the constant $1/(1 - \gamma)$. Doing so and rearranging terms, we get

$$\phi_{\text{MO,imp}} = \phi_a(1)\phi_b(2) + \phi_b(1)\phi_a(2) + \frac{1 + \gamma}{1 - \gamma}[\phi_a(1)\phi_a(2) + \phi_b(1)\phi_b(2)]$$

There is also no harm done if we define a new constant δ as $\delta = (1 + \gamma)/(1 - \gamma)$. We see then that this improved MO function and the improved VB function (13.109) are *identical*. Weinbaum viewed his H_2 calculation as a valence-bond calculation with inclusion of ionic terms. We have shown that we can just as well view the Weinbaum calculation as an MO calculation with configuration interaction. (This was the viewpoint adopted in Section 13.9.)

The MO function (13.107) underestimates electron correlation, in that it says that structures with both electrons on the same atom are just as likely as structures with each electron on a different atom. The VB function (13.108) overestimates electron correlation, in that it has no contribution from structures with both electrons on the same atom. In MO theory, electron correlation can be introduced by configuration interaction. In VB theory, electron correlation is reduced by ionic–covalent resonance. The simple VB method is more reliable at large R than the simple MO method, since the latter predicts the wrong dissociation products.

To further fix the differences between the MO and VB approaches, consider how each method divides the H_2 electronic Hamiltonian into unperturbed and perturbation Hamiltonians. For the MO method, we write

$$\hat{H} = \left[\left(-\tfrac{1}{2}\nabla_1^2 - \frac{1}{r_{a1}} - \frac{1}{r_{b1}}\right) + \left(-\tfrac{1}{2}\nabla_2^2 - \frac{1}{r_{a2}} - \frac{1}{r_{b2}}\right)\right] + \frac{1}{r_{12}}$$

where the unperturbed Hamiltonian consists of the bracketed terms. In MO theory the unperturbed Hamiltonian for H_2 is the sum of two H_2^+ Hamiltonians, one for each electron. Accordingly, the zeroth-order MO wave function is a product of two H_2^+-like wave functions, one for each electron. Since the H_2^+ functions are complicated, we approximate the H_2^+-like MOs as LCAOs. The effect of the $1/r_{12}$ perturbation is taken into account in an average way through use of self-consistent-field molecular orbitals. To take instantaneous electron correlation into account, we can use configuration interaction.

For the valence-bond method, the terms in the Hamiltonian are grouped in either of two ways:

$$\hat{H} = \left[\left(-\tfrac{1}{2}\nabla_1^2 - \frac{1}{r_{a1}}\right) + \left(-\tfrac{1}{2}\nabla_2^2 - \frac{1}{r_{b2}}\right)\right] - \frac{1}{r_{a2}} - \frac{1}{r_{b1}} + \frac{1}{r_{12}}$$

$$\hat{H} = \left[\left(-\tfrac{1}{2}\nabla_1^2 - \frac{1}{r_{b1}}\right) + \left(-\tfrac{1}{2}\nabla_2^2 - \frac{1}{r_{a2}}\right)\right] - \frac{1}{r_{a1}} - \frac{1}{r_{b2}} + \frac{1}{r_{12}}$$

The unperturbed system is two hydrogen atoms. We have two zeroth-order functions consisting of products of hydrogen-atom wave functions, and these belong to a degenerate level. The correct ground-state zeroth-order function is the linear combination (13.100).

The MO method is used far more often than the VB method, because it is computationally much simpler than the VB method. The MO method was developed by Hund, Mulliken, and Lennard-Jones in the late 1920s. Originally, it was used largely for qualitative descriptions of molecules, but the electronic digital computer has made possible the calculation of accurate MO functions (Section 13.14). For a discussion of the relative merits of the MO and VB methods, see R. Hoffman et al., *Acc. Chem. Res.*, **36**, 750 (2003).

13.12 MO AND VB WAVE FUNCTIONS FOR HOMONUCLEAR DIATOMIC MOLECULES

The MO approximation puts the electrons of a molecule in molecular orbitals, which extend over the whole molecule. As an approximation to the molecular orbitals, we usually use linear combinations of atomic orbitals. The VB method puts the electrons of a molecule in atomic orbitals and constructs the molecular wave function by allowing for "exchange" of the valence electron pairs between the atomic orbitals of the bonding atoms. We compared the two methods for H_2. We now consider other homonuclear diatomic molecules.

We begin with the ground state of He_2. The separated helium atoms each have the ground-state configuration $1s^2$. This closed-subshell configuration does not have any unpaired electrons to form valence bonds, and the VB wave function is simply the antisymmetrized product of the atomic-orbital functions:

$$\frac{1}{\sqrt{24}}\begin{vmatrix} 1s_a(1) & \overline{1s_a}(1) & 1s_b(1) & \overline{1s_b}(1) \\ 1s_a(2) & \overline{1s_a}(2) & 1s_b(2) & \overline{1s_b}(2) \\ 1s_a(3) & \overline{1s_a}(3) & 1s_b(3) & \overline{1s_b}(3) \\ 1s_a(4) & \overline{1s_a}(4) & 1s_b(4) & \overline{1s_b}(4) \end{vmatrix} \qquad (13.110)$$

The subscripts a and b refer to the two atoms, and the bar indicates spin function β. The $1s$ function in this wave function is a helium-atom $1s$ function, which ideally is an SCF atomic function but can be approximated by a hydrogenlike function with an effective nuclear charge. In the shorthand notation of (10.47), the wave function (13.110) is

$$|1s_a\overline{1s_a}1s_b\overline{1s_b}| \qquad (13.111)$$

The VB wave function for He_2 has each electron paired with another electron in an orbital on the *same* atom and hence predicts no bonding.

In the MO approach, He_2 has the ground-state configuration $(\sigma_g 1s)^2(\sigma_u^* 1s)^2$. With no net bonding electrons, no bonding is predicted, in agreement with the VB method. The MO approximation to the wave function is

$$|\sigma_g 1s\,\overline{\sigma_g 1s}\,\sigma_u^* 1s\,\overline{\sigma_u^* 1s}| \qquad (13.112)$$

The simplest way to approximate the (unnormalized) MOs is to take them as linear combinations of the helium-atom AOs: $\sigma_g 1s = 1s_a + 1s_b$ and $\sigma_u^* 1s = 1s_a - 1s_b$. With this approximation, (13.112) becomes

$$|(1s_a + 1s_b)\overline{(1s_a + 1s_b)}(1s_a - 1s_b)\overline{(1s_a - 1s_b)}| \qquad (13.113)$$

We can add or subtract one column of the determinant from another column without changing the determinant's value. If we add column 1 to column 3 and column 2 to column 4, we simplify (13.113) to

$$4|(1s_a + 1s_b)\overline{(1s_a + 1s_b)}1s_a\overline{1s_a}|$$

We now subtract column 3 from column 1 and column 4 from column 2 to get

$$4|1s_b\overline{1s_b}1s_a\overline{1s_a}| \qquad (13.114)$$

The interchange of columns 1 and 3 and of columns 2 and 4 multiplies the determinant by $(-1)^2$, so (13.114) is equal to

$$4|1s_a\overline{1s_a}1s_b\overline{1s_b}|$$

which is identical (after normalization) to the VB function (13.111). This result is easily generalized to the statement that the simple VB and simple LCAO-MO methods give the same approximate wave functions for diatomic molecules formed from separated atoms with completely filled atomic subshells. We could now substitute the trial function (13.111) into the variational integral and calculate the repulsive curve for the interaction of two ground-state He atoms.

Before going on to Li_2, let us express the Heitler–London valence-bond functions for H_2 as Slater determinants. The ground-state Heitler–London function (13.100) can be written as

$$\tfrac{1}{2}(1 + S_{ab}^2)^{-1/2}\left\{\begin{vmatrix} 1s_a(1)\alpha(1) & 1s_b(1)\beta(1) \\ 1s_a(2)\alpha(2) & 1s_b(2)\beta(2) \end{vmatrix} - \begin{vmatrix} 1s_a(1)\beta(1) & 1s_b(1)\alpha(1) \\ 1s_a(2)\beta(2) & 1s_b(2)\alpha(2) \end{vmatrix}\right\}$$

$$= (2 + 2S_{ab}^2)^{-1/2}\{|1s_a\overline{1s_b}| - |\overline{1s_a}1s_b|\} \quad (13.115)$$

In each Slater determinant, the electron on atom a is paired with an electron of opposite spin on atom b, corresponding to the Lewis structure H—H. The Heitler–London functions (13.101) for the lowest H_2 triplet state can also be written as Slater determinants. Omitting normalization constants, we write the Heitler–London H_2 functions as

$$\text{Singlet:} \quad |1s_a\overline{1s_b}| - |\overline{1s_a}1s_b| \quad (13.116)$$

$$\text{Triplet:} \quad \begin{cases} |1s_a 1s_b| \\ |1s_a\overline{1s_b}| + |\overline{1s_a}1s_b| \\ |\overline{1s_a}\,\overline{1s_b}| \end{cases} \quad (13.117)$$

Now consider Li_2. The ground-state configuration of Li is $1s^2 2s$, and the Lewis structure of Li_2 is Li—Li, with the two $2s$ Li electrons paired and the $1s$ electrons remaining in the inner shell of each atom. The part of the valence-bond wave function involving the $1s$ electrons will be like the He_2 function (13.111), while the part of the VB wave function involving the $2s$ electrons (which form the bond) will be like the Heitler–London H_2 function (13.116). Of course, because of the indistinguishability of the electrons, there is complete electronic democracy, and we must allow every electron to be in every orbital. Hence we write the ground-state valence-bond function for Li_2 using 6×6 Slater determinants:

$$|1s_a\overline{1s_a}1s_b\overline{1s_b}2s_a\overline{2s_b}| - |1s_a\overline{1s_a}1s_b\overline{1s_b}\,\overline{2s_a}2s_b| \quad (13.118)$$

We have written down (13.118) simply by analogy to (13.111) and (13.116). For a fuller justification of it, we should show that it is an eigenfunction of the spin operators $\hat{S}^2$ and $\hat{S}_z$ with eigenvalue zero for each operator, which corresponds to a singlet state. This can be shown, but we omit doing so. To save space, (13.118) is sometimes written as

$$|1s_a\overline{1s_a}1s_b\overline{1s_b}\,\widehat{2s_a 2s_b}| \quad (13.119)$$

where the curved line indicates the pairing (bonding) of the $2s_a$ and $2s_b$ AOs.

The MO wave function for the Li_2 ground state is

$$|\sigma_g 1s\,\overline{\sigma_g 1s}\,\sigma_u^* 1s\,\overline{\sigma_u^* 1s}\,\sigma_g 2s\,\overline{\sigma_g 2s}| \quad (13.120)$$

If we approximate the two lowest MOs by $1s_a \pm 1s_b$ and carry out the same manipulations we did for the He_2 MO function, we can write (13.120) as

$$|1s_a\overline{1s_a}1s_b\overline{1s_b}\,\sigma_g 2s\,\overline{\sigma_g 2s}|$$

Recall the notation $KK(\sigma_g 2s)^2$ for the Li_2 ground-state configuration.

Now consider the VB treatment of the N_2 ground state. The lowest configuration of N is $1s^2 2s^2 2p^3$. Hund's rule gives the ground level as ${}^4S_{3/2}$, with one electron in each of the three $2p$ AOs. We can thus pair the two $2p_x$ electrons, the two $2p_y$ electrons, and the two $2p_z$ electrons to form a triple bond. The Lewis structure is :N≡N:. How is this

Lewis structure translated into the VB wave function? In the VB method, opposite spins are given to orbitals bonded together. We have three such pairs of orbitals and two ways to give opposite spins to the electrons of each bonding pair of AOs. Hence there are $2^3 = 8$ possible Slater determinants that we can write. We begin with

$$D_1 = |1s_a\overline{1s_a}2s_a\overline{2s_a}1s_b\overline{1s_b}2s_b\overline{2s_b}2p_{xa}\overline{2p_{xb}}2p_{ya}\overline{2p_{yb}}2p_{za}\overline{2p_{zb}}|$$

In all eight determinants, the first eight columns will remain unchanged, and to save space we write D_1 as

$$D_1 = |\cdots 2p_{xa}\overline{2p_{xb}}2p_{ya}\overline{2p_{yb}}2p_{za}\overline{2p_{zb}}| \tag{13.121}$$

Reversing the spins of the electrons in $2p_{xa}$ and $2p_{xb}$, we get

$$D_2 = |\cdots \overline{2p_{xa}}2p_{xb}2p_{ya}\overline{2p_{yb}}2p_{za}\overline{2p_{zb}}| \tag{13.122}$$

There are six other determinants formed by interchanges of spins within the three pairs of bonding orbitals, and the VB wave function is a linear combination of eight determinants (Prob. 13.33). The following rule (see *Kauzmann*, pages 421–422) gives a VB wave function that is an eigenfunction of $\hat{S}^2$ with eigenvalue 0 (as is desired for the ground state): The coefficient of each determinant is $+1$ or -1 according to whether the number of spin interchanges required to generate the determinant from D_1 is even or odd, respectively. Thus D_2 has coefficient -1. [Compare also (13.116).] Clearly, the single-determinant ground-state N_2 MO function is easier to handle than the eight-determinant VB function.

The VB method places great emphasis on pairing of electrons. In treating O_2, whose ground state is a triplet, the VB method runs into difficulties. It *is* possible to give a VB explanation of why O_2 has a triplet ground state, but the reasoning is involved [see B. J. Moss et al., *J. Chem. Phys.*, **63**, 4632 (1975)] in contrast to the simple MO explanation.

13.13 EXCITED STATES OF H_2

We have concentrated mostly on the ground electronic states of diatomic molecules. In this section we consider some of the excited states of H_2. Figure 13.19 gives the potential-energy curves for some of the H_2 electronic energy levels.

The lowest MO configuration is $(1\sigma_g)^2$, where the notation of the third column of Table 13.1 is used. This closed-shell configuration gives only a nondegenerate ${}^1\Sigma_g^+$ level, designated $X^1\Sigma_g^+$. The LCAO-MO function is (13.93).

The next-lowest MO configuration is $(1\sigma_g)(1\sigma_u)$, which gives rise to the terms ${}^1\Sigma_u^+$ and ${}^3\Sigma_u^+$ (Table 13.3). Since there is no axial electronic orbital angular momentum, each of these terms corresponds to one level. Spectroscopists have named these electronic levels $B^1\Sigma_u^+$ and $b^3\Sigma_u^+$. By Hund's rule, the b level lies below the B level. The LCAO-MO functions for these levels are [see Eqs. (10.27)–(10.30)]

$$b^3\Sigma_u^+:\quad 2^{-1/2}[1\sigma_g(1)1\sigma_u(2) - 1\sigma_g(2)1\sigma_u(1)]\begin{cases}\alpha(1)\alpha(2)\\ 2^{-1/2}[\alpha(1)\beta(2) + \alpha(2)\beta(1)]\\ \beta(1)\beta(2)\end{cases}$$

$$B^1\Sigma_u^+:\quad 2^{-1/2}[1\sigma_g(1)1\sigma_u(2) + 1\sigma_g(2)1\sigma_u(1)]2^{-1/2}[\alpha(1)\beta(2) - \alpha(2)\beta(1)]$$

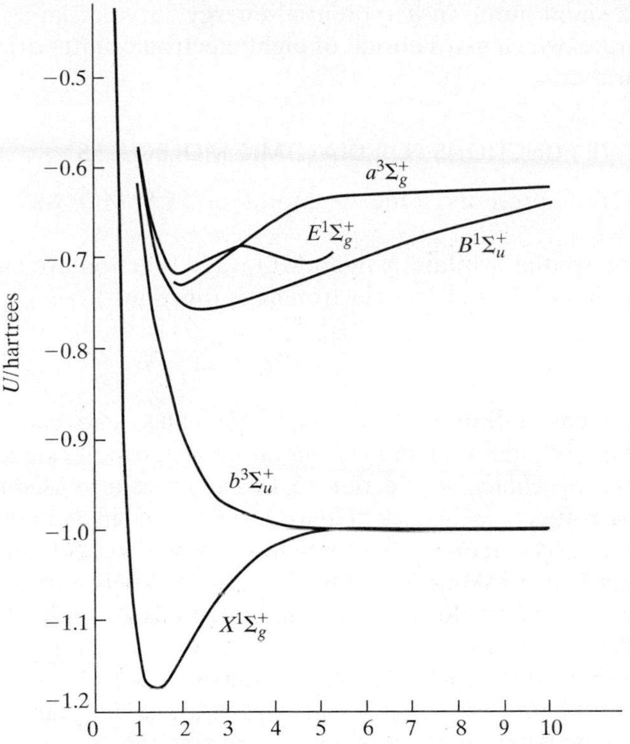

FIGURE 13.19 $U(R)$ curves for some electronic states of H_2. [See W. Kolos and L. Wolniewicz, *J. Chem. Phys.*, **43**, 2429 (1965); **45**, 509 (1966); J. Gerhauser and H. S. Taylor, *J. Chem. Phys.*, **42**, 3621 (1965).]

where $1\sigma_g \approx N(1s_a + 1s_b)$ and $1\sigma_u \approx N'(1s_a - 1s_b)$. The $b^3\Sigma_u^+$ level is triply degenerate. The $B^1\Sigma_u^+$ level is nondegenerate. The Heitler–London wave functions for the b level are given by (13.101). Both these levels have one bonding and one antibonding electron, and we would expect the potential-energy curves for both levels to be repulsive. Actually, the B level has a minimum in its $U(R)$ curve. The stability of this state should caution us against drawing too hasty conclusions from very approximate wave functions.

We expect the next-lowest configuration to be $(1\sigma_g)(2\sigma_g)$, giving rise to ${}^1\Sigma_g^+$ and ${}^3\Sigma_g^+$ levels. These levels of H_2 are designated $E^1\Sigma_g^+$ and $a^3\Sigma_g^+$. By Hund's rule, the triplet lies lower. The E state has two substantial minima in its $U(R)$ curve.

Although the $2\sigma_u$ MO fills before the two $1\pi_u$ MOs in going across the periodic table, the $1\pi_u$ MOs lie below the $2\sigma_u$ MO in H_2. The configuration $(1\sigma_g)(1\pi_u)$ gives rise to the terms ${}^1\Pi_u$ and ${}^3\Pi_u$, the triplet lying lower. These terms are designated $C^1\Pi_u$ and $c^3\Pi_u$. The c term gives rise to the levels $c^3\Pi_{2u}$, $c^3\Pi_{1u}$, and $c^3\Pi_{0u}$. These levels lie so close together that they are usually not resolved in spectroscopic work. The C level shows a slight hump in its potential-energy curve at large R. Each level is twofold degenerate, which gives a total of eight electronic states arising from the $(1\sigma_g)(1\pi_u)$ configuration.

13.14 SCF WAVE FUNCTIONS FOR DIATOMIC MOLECULES

This section presents some examples of SCF MO wave functions for diatomic molecules.

The spatial orbitals ϕ_i in an MO wave function are each expressed as a linear combination of a set of one-electron **basis functions** χ_s:

$$\phi_i = \sum_s c_{si}\chi_s \tag{13.123}$$

For SCF calculations on diatomic molecules, one can use Slater-type orbitals [Eq. (11.14)] centered on the various atoms of the molecule as the basis functions. (For an alternative choice, see Section 15.4.) The procedure used to find the coefficients c_{si} of the basis functions in each SCF MO is discussed in Section 14.3. To have a complete set of AO basis functions, an infinite number of Slater orbitals are needed, but the true molecular Hartree–Fock wave function can be closely approximated with a reasonably small number of carefully chosen Slater orbitals. A **minimal basis set** for a molecular SCF calculation consists of a single basis function for each inner-shell AO and each valence-shell AO of each atom. An **extended** basis set is a set that is larger than a minimal set. Minimal-basis-set SCF calculations are easier than extended-basis-set calculations, but the latter are much more accurate.

SCF wave functions using a minimal basis set were calculated by Ransil for several light diatomic molecules [B. J. Ransil, *Rev. Mod. Phys.*, **32**, 245 (1960)]. As an example, the SCF MOs for the ground state of Li_2 [MO configuration $(1\sigma_g)^2(1\sigma_u)^2(2\sigma_g)^2$] at $R = R_e$ are

$$\begin{aligned}
1\sigma_g &= 0.706(1s_a + 1s_b) + 0.009(2s_{\perp a} + 2s_{\perp b}) + 0.0003(2p\sigma_a + 2p\sigma_b) \\
1\sigma_u &= 0.709(1s_a - 1s_b) + 0.021(2s_{\perp a} - 2s_{\perp b}) + 0.003(2p\sigma_a - 2p\sigma_b) \\
2\sigma_g &= -0.059(1s_a + 1s_b) + 0.523(2s_{\perp a} + 2s_{\perp b}) + 0.114(2p\sigma_a + 2p\sigma_b)
\end{aligned} \tag{13.124}$$

The AO functions in these equations are STOs, except for $2s_\perp$. A Slater-type $2s$ AO has no radial nodes and is not orthogonal to a $1s$ STO. The Hartree–Fock $2s$ AO has one radial node $(n - l - 1 = 1)$ and is orthogonal to the $1s$ AO. We can form an orthogonalized $2s$ orbital with the proper number of nodes by taking the following normalized linear combination of $1s$ and $2s$ STOs of the same atom (Schmidt orthogonalization):

$$2s_\perp = (1 - S^2)^{-1/2}(2s - S \cdot 1s) \tag{13.125}$$

where S is the overlap integral $\langle 1s|2s\rangle$. Ransil expressed the Li_2 orbitals using the (nonorthogonal) $2s$ STO, but since the orthogonalized $2s_\perp$ function gives a better representation of the $2s$ AO, the orbitals have been rewritten using $2s_\perp$. This changes the $1s$ and $2s$ coefficients, but the actual orbital is, of course, unchanged; see Prob. 13.35. The notation $2p\sigma$ for an AO indicates that the p orbital points along the molecular (z) axis; that is, a $2p\sigma$ AO is a $2p_z$ AO. (The $2p_x$ and $2p_y$ AOs are called $2p\pi$ AOs.) The optimum orbital exponents for the orbitals in (13.124) are $\zeta_{1s} = 2.689, \zeta_{2s} = 0.634, \zeta_{2p\sigma} = 0.761$.

Our previous simple expressions for these MOs were

$$1\sigma_g = \sigma_g 1s = 2^{-1/2}(1s_a + 1s_b)$$
$$1\sigma_u = \sigma_u^* 1s = 2^{-1/2}(1s_a - 1s_b)$$
$$2\sigma_g = \sigma_g 2s = 2^{-1/2}(2s_a + 2s_b)$$

Comparison of these with (13.124) shows the simple LCAO functions to be reasonable first approximations to the minimal-basis-set SCF MOs. The approximation is best for the $1\sigma_g$ and $1\sigma_u$ MOs, whereas the $2\sigma_g$ MO has substantial $2p\sigma$ AO contributions in addition to the $2s$ AO contributions. For this reason the notation of the third column of Table 13.1 (Section 13.7) is preferable to the separated-atoms MO notation. The substantial amount of $2s$–$2p\sigma$ hybridization is to be expected, since the $2s$ and $2p$ AOs are close in energy [see Eq. (9.27)]. The hybridization allows for the polarization of the $2s$ AOs in forming the molecule.

Let us compare the $3\sigma_g$ MO of the F_2 ground state at R_e as calculated by Ransil using a minimal basis set with that calculated by Wahl using an extended basis set [A. C. Wahl, *J. Chem. Phys.*, **41**, 2600 (1964)]:

$$3\sigma_{g,\text{min}} = 0.038(1s_a + 1s_b) - 0.184(2s_a + 2s_b) + 0.648(2p\sigma_a + 2p\sigma_b)$$
$$\zeta_{1s} = 8.65, \qquad \zeta_{2s} = 2.58, \qquad \zeta_{2p\sigma} = 2.49$$
$$3\sigma_{g,\text{ext}} = 0.048(1s_a + 1s_b) + 0.003(1s_a' + 1s_b') - 0.257(2s_a + 2s_b)$$
$$+ 0.582(2p\sigma_a + 2p\sigma_b) + 0.307(2p\sigma_a' + 2p\sigma_b') + 0.085(2p\sigma_a'' + 2p\sigma_b'')$$
$$- 0.056(3s_a + 3s_b) + 0.046(3d\sigma_a + 3d\sigma_b) + 0.014(4f\sigma_a + 4f\sigma_b)$$
$$\zeta_{1s} = 8.27, \qquad \zeta_{1s'} = 13.17, \qquad \zeta_{2s} = 2.26$$
$$\zeta_{2p\sigma} = 1.85, \qquad \zeta_{2p\sigma'} = 3.27, \qquad \zeta_{2p\sigma''} = 5.86$$
$$\zeta_{3s} = 4.91, \qquad \zeta_{3d\sigma} = 2.44, \qquad \zeta_{4f\sigma} = 2.83$$

Just as several STOs are needed to give an accurate representation of Hartree–Fock AOs (Section 11.1), one needs more than one STO of a given n and l in the linear combination of STOs that is to accurately represent the Hartree–Fock MO. The primed and double-primed AOs in the extended-basis-set function are STOs with different orbital exponents. The $3d\sigma$ and $4f\sigma$ AOs are AOs with quantum number $m = 0$, that is, the $3d_0$ and $4f_0$ AOs. The total energies found are -197.877 and -198.768 hartrees for the minimal and extended calculations, respectively. The experimental energy of F_2 at R_e is -199.670 hartrees, so the error for the minimal calculation is twice that of the extended calculation. The extended-basis-set calculation is believed to give a wave function quite close to the true Hartree–Fock wave function. Therefore, the correlation energy in F_2 is about -0.90 hartrees $= -24._5$ eV.

In discussing H_2^+ and H_2, we saw how hybridization (the mixing of different AOs of the same atom) improves molecular wave functions. There is a tendency to think of hybridization as occurring only for certain molecular geometries. The SCF calculations make clear that all MOs are hybridized to some extent. Thus any diatomic-molecule σ MO is a linear combination of $1s$, $2s$, $2p_0$, $3s$, $3p_0$, $3d_0, \ldots$ AOs of the separated atoms.

To aid in deciding which AOs contribute to a given diatomic MO, we use two rules. First, only σ-type AOs (s, $p\sigma$, $d\sigma, \ldots$) can contribute to a σ MO; only π-type AOs ($p\pi$, $d\pi, \ldots$) can contribute to a π MO; and so on. Second, only AOs of reasonably

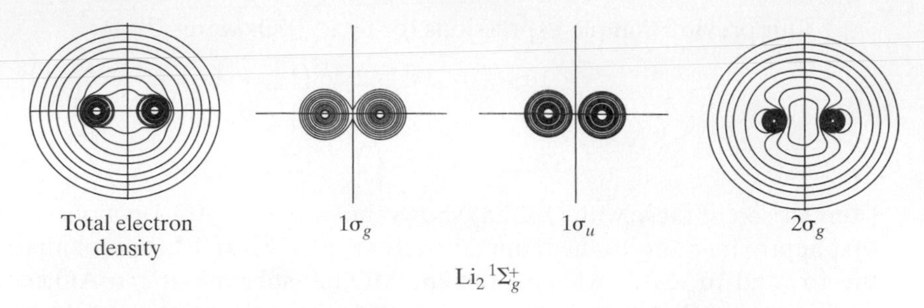

FIGURE 13.20 Hartree–Fock MO electron-density contours for the ground electronic state of Li_2 as calculated by Wahl. [A. C. Wahl, *Science,* **151**, 961 (1966); *Scientific American*, April 1970, p. 54; *Atomic and Molecular Structure*: 4 Wall Charts, McGraw-Hill, 1970.]

similar energy contribute substantially to a given MO. (For examples, see the minimal- and extended-basis-set MOs quoted above.)

Wahl plotted the contours of the near Hartree–Fock molecular orbitals of homonuclear diatomic molecules from H_2 through F_2. Figure 13.20 shows these plots for Li_2.

Of course, Hartree–Fock wave functions are only approximations to the true wave functions. It is possible to prove that a Hartree–Fock wave function gives a very good approximation to the electron probability density $\rho(x, y, z)$ for nuclear configurations in the region of the equilibrium configuration. A molecular property that involves only one-electron operators can be expressed as an integral involving ρ; see Eq. (14.8). Consequently, such properties are accurately calculated using Hartree–Fock wave functions. An example is the molecular dipole moment [Eq. (14.21)]. For example, the LiH dipole moment calculated with a near Hartree–Fock ψ is 6.00 D (debyes) [S. Green, *J. Chem. Phys.*, **54**, 827 (1971)], compared with the experimental value 5.83 D. (One debye = 3.33564×10^{-30} C m.) For NaCl, the calculated and experimental dipole moments are 9.18 D and 9.02 D [R. L. Matcha, *J. Chem. Phys.*, **48**, 335 (1968)]. An error of about 0.2 D is typical in such calculations, but where the dipole moment is small, the percent error can be large. An extreme example is CO, for which the experimental moment is 0.11 D with the polarity C^-O^+, but the near-Hartree–Fock moment is 0.27 D with the wrong polarity C^+O^-. However, a configuration-interaction wave function gives 0.12 D with the correct polarity [S. Green, *J. Chem. Phys.*, **54**, 827 (1971)].

A major weakness of the Hartree–Fock method is its failure to give accurate molecular dissociation energies. For example, an extended-basis-set calculation [P. E. Cade et al., *J. Chem. Phys.*, **44**, 1973 (1966)] gives $D_e = 5.3$ eV for N_2, as compared with the true value 9.9 eV. (To calculate the Hartree–Fock D_e, the molecular energy at the minimum in the $U(R)$ Hartree–Fock curve is subtracted from the sum of the Hartree–Fock energies of the separated atoms.) For F_2 the true D_e is 1.66 eV, while the Hartree–Fock D_e is -1.4 eV. In other words, the Hartree–Fock calculation predicts the separated atoms to be more stable than the fluorine molecule. A related defect of Hartree–Fock molecular wave functions is that the energy approaches the wrong limit as $R \rightarrow \infty$. Recall the MO discussion of H_2.

13.15 MO TREATMENT OF HETERONUCLEAR DIATOMIC MOLECULES

The treatment of heteronuclear diatomic molecules is similar to that for homonuclear diatomic molecules. We first consider the MO description.

Suppose the two atoms have atomic numbers that differ only slightly; an example is CO. We could consider CO as being formed from the isoelectronic molecule N_2 by a gradual transfer of charge from one nucleus to the other. During this hypothetical transfer, the original N_2 MOs would slowly vary to give finally the CO MOs. We therefore expect the CO molecular orbitals to resemble somewhat those of N_2. For a heteronuclear diatomic molecule such as CO, the symbols used for the MOs are similar to those for homonuclear diatomics. However, for a heteronuclear diatomic, the electronic Hamiltonian (13.5) is not invariant with respect to inversion of the electronic coordinates (that is, $\hat{H}_{\text{el}}$ does not commute with $\hat{\Pi}$), and the g, u property of the MOs disappears. The correlation between the N_2 and CO shell designations is

N_2	$1\sigma_g$	$1\sigma_u$	$2\sigma_g$	$2\sigma_u$	$1\pi_u$	$3\sigma_g$	$1\pi_g$	$3\sigma_u$
CO	1σ	2σ	3σ	4σ	1π	5σ	2π	6σ

MOs of the same symmetry are numbered in order of increasing energy. Because of the absence of the g, u property, the numbers of corresponding homonuclear and heteronuclear MOs differ. Figure 13.21 is a sketch of a contour of the CO $1\pi_{\pm1}$ MOs as determined by an extended-basis-set SCF calculation [W. M. Huo, *J. Chem. Phys.*, **43**, 624 (1965)]. Note its resemblance to the contour of Fig. 13.13, which is for the $1\pi_{u,\pm1}$ MOs of a homonuclear diatomic molecule.

The ground-state configuration of CO is $1\sigma^2 2\sigma^2 3\sigma^2 4\sigma^2 1\pi^4 5\sigma^2$, as compared with the N_2 configuration $(1\sigma_g)^2(1\sigma_u)^2(2\sigma_g)^2(2\sigma_u)^2(1\pi_u)^4(3\sigma_g)^2$.

As in homonuclear diatomics, the heteronuclear diatomic MOs are approximated as linear combinations of atomic orbitals. The coefficients are found by the procedure of Section 14.3. For example, a minimal-basis-set SCF calculation using Slater AOs (with nonoptimized exponents given by Slater's rules) gives for the CO 5σ, 1π, and 2π MOs at $R = R_e$ [B. J. Ransil, *Rev. Mod. Phys.*, **32**, 245 (1960)]:

$$5\sigma = 0.027(1s_{\text{C}}) + 0.011(1s_{\text{O}}) + 0.739(2s_{\perp\text{C}}) + 0.036(2s_{\perp\text{O}}) - 0.566(2p\sigma_{\text{C}}) - 0.438(2p\sigma_{\text{O}})$$

$$1\pi = 0.469(2p\pi_{\text{C}}) + 0.771(2p\pi_{\text{O}}), \qquad 2\pi = 0.922(2p\pi_{\text{C}}) - 0.690(2p\pi_{\text{O}})$$

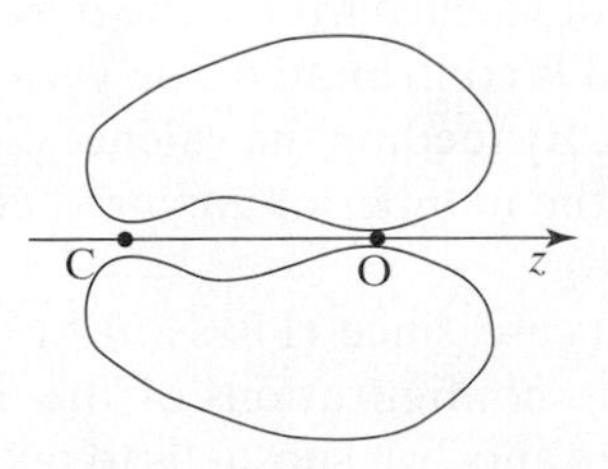

FIGURE 13.21 Cross section of a contour of the $1\pi_{\pm1}$ MOs in CO.

The expressions for the π MOs are simpler than those for the σ MOs because s and $p\sigma$ AOs cannot contribute to π MOs. For comparison, the corresponding MOs in N_2 at $R = R_e$ are (Ransil, op. cit.):

$$3\sigma_g = 0.030(1s_a + 1s_b) + 0.395(2s_{\perp a} + 2s_{\perp b}) - 0.603(2p\sigma_a + 2p\sigma_b)$$

$$1\pi_u = 0.624(2p\pi_a + 2p\pi_b), \qquad 1\pi_g = 0.835(2p\pi_a - 2p\pi_b)$$

The resemblance of CO and N_2 MOs is apparent. The 1σ MO in CO is found to be nearly the same as a $1s$ oxygen-atom AO; the 2σ MO in CO is essentially a carbon-atom $1s$ AO.

In general, for a heteronuclear diatomic molecule AB where the valence AOs of each atom are of s and p type and where the valence AOs of A do not differ greatly in energy from the valence AOs of B, we can expect the Figure 13.17 pattern of

$$\sigma s < \sigma^* s < \pi p < \sigma p < \pi^* p < \sigma^* p$$

valence-shell MOs formed from s and p valence-shell AOs to hold reasonably well. Figure 13.17 would be modified in that each valence AO of the more electronegative atom would lie below the corresponding valence AO of the other atom.

When the valence-shell AO energies of B lie very substantially below those of A, the s and $p\sigma$ valence AOs of B lie below the s valence-shell AO of A, and this affects which AOs contribute to each MO. Consider the molecule BF, for example. A minimal-basis-set calculation [Ransil, *Rev. Mod. Phys.*, **32**, 245 (1960)] gives the 1σ MO as essentially $1s_F$ and the 2σ MO as essentially $1s_B$. The 3σ MO is predominantly $2s_F$, with small amounts of $2s_B$, $2p\sigma_B$, and $2p\sigma_F$. The 4σ MO is predominantly $2p\sigma_F$, with significant amounts of $2s_B$ and $2s_F$ and a small amount of $2p\sigma_B$. This is quite different from N_2, where the corresponding MO is formed predominantly from the $2s$ AOs on each N. The 1π MO is a bonding combination of $2p\pi_B$ and $2p\pi_F$. The 5σ MO is predominantly $2s_B$, with a substantial contribution from $2p\sigma_B$ and a significant contribution from $2p\sigma_F$. This is unlike the corresponding MO in N_2, where the largest contributions are from $2p\sigma$ MOs on each atom. The 2π MO is an antibonding combination of $2p\pi_B$ and $2p\pi_F$. The 6σ MO has important contributions from $2p\sigma_B$, $2s_B$, $2s_F$, and $2p\sigma_F$.

We see from Fig. 11.2 that the $2p_F$ AO lies well below the $2s_B$ AO. This causes the $2p\sigma_F$ AO to contribute substantially to lower-lying MOs and the $2s_B$ AO to contribute substantially to higher-lying MOs, as compared with what happens in N_2. (This effect occurs in CO, although to a lesser extent. Note the very substantial contribution of $2s_C$ to the 5σ MO. Also, the 4σ MO in CO has a very substantial contribution from $2p\sigma_O$.)

For a diatomic molecule AB where each atom has s and p valence-shell AOs (this excludes H and transition elements) and where the A and B valence AOs differ widely in energy, we may expect the pattern of valence MOs to be $\sigma < \sigma < \pi < \sigma < \pi < \sigma$, but it is not so easy to guess which AOs contribute to the various MOs or the bonding or antibonding character of the MOs. By feeding the valence electrons into these MOs, we can make a plausible guess as to the number of unpaired electrons and the ground term of the AB molecule (Prob. 13.36).

Diatomic hydrides are a special case, since H has only a $1s$ valence AO. Consider HF as an example. The ground-state configurations of the atoms are $1s$ for H and $1s^2 2s^2 2p^5$ for F. We expect the filled $1s$ and $2s$ F subshells to take little part in the bonding. The four $2p\pi$ fluorine electrons are nonbonding (there are no π valence AOs on H). The hydrogen 1s AO and the fluorine 2$p\sigma$ AO have the same symmetry (σ) and have rather similar energies (Fig. 11.2), and a linear combination of these two AOs will form a σ MO for the bonding electron pair:

$$\phi = c_1(1s_{\text{H}}) + c_2(2p\sigma_{\text{F}})$$

where the contributions of $1s_{\text{F}}$ and $2s_{\text{F}}$ to this MO have been neglected. Since F is more electronegative than H, we expect that $c_2 > c_1$. (In addition, the $1s_{\text{H}}$ and $2p\sigma_{\text{F}}$ AOs form an antibonding MO, which is unoccupied in the ground state.)

The picture of HF just given is only a crude qualitative approximation. A minimal-basis-set SCF calculation using Slater orbitals with optimized exponents gives as the MOs of HF [B. J. Ransil, *Rev. Mod. Phys.*, **32**, 245 (1960)]

$$1\sigma = 1.000(1s_{\text{F}}) + 0.012(2s_{\perp\text{F}}) + 0.002(2p\sigma_{\text{F}}) - 0.003(1s_{\text{H}})$$

$$2\sigma = -0.018(1s_{\text{F}}) + 0.914(2s_{\perp\text{F}}) + 0.090(2p\sigma_{\text{F}}) + 0.154(1s_{\text{H}})$$

$$3\sigma = -0.023(1s_{\text{F}}) - 0.411(2s_{\perp\text{F}}) + 0.711(2p\sigma_{\text{F}}) + 0.516(1s_{\text{H}})$$

$$1\pi_{+1} = (2p\pi_{+1})_{\text{F}}, \qquad 1\pi_{-1} = (2p\pi_{-1})_{\text{F}}$$

The ground-state MO configuration of HF is $1\sigma^2 2\sigma^2 3\sigma^2 1\pi^4$. The 1σ MO is virtually identical with the 1s fluorine AO. The 2σ MO is pretty close to the 2s fluorine AO. The 1π MOs are required by symmetry to be the same as the corresponding fluorine π AOs. The bonding 3σ MO has its largest contribution from the 2$p\sigma$ fluorine and 1s hydrogen AOs, as would be expected from the discussion of the preceding paragraph. However, there is a substantial contribution to this MO from the 2s fluorine AO. (Since a single 2s function is only an approximation to the 2s AO of F, we cannot use this calculation to say exactly how much 2s AO character the 3σ HF molecular orbital has.)

For qualitative discussion (but not quantitative work), it is useful to have simple approximations for heteronuclear diatomic MOs. In the crudest approximation, we can take each valence MO of a heteronuclear diatomic molecule as a linear combination of two AOs ϕ_a and ϕ_b, one on each atom. (As the discussions of CO and BF show, this approximation is often quite inaccurate.) From the two AOs, we can form two MOs:

$$c_1\phi_a + c_2\phi_b \quad \text{and} \quad c_1'\phi_a + c_2'\phi_b$$

The lack of symmetry in the heteronuclear diatomic makes the coefficients unequal in magnitude. The coefficients are determined by solving the secular equation [see Eq. (13.45)]

$$\begin{vmatrix} H_{aa} - W & H_{ab} - WS_{ab} \\ H_{ab} - WS_{ab} & H_{bb} - W \end{vmatrix} = 0$$

$$(H_{aa} - W)(H_{bb} - W) - (H_{ab} - WS_{ab})^2 = 0 \tag{13.126}$$

where $\hat{H}$ is some sort of effective one-electron Hamiltonian. Suppose that $H_{aa} > H_{bb}$, and let $f(W)$ be defined as the left side of (13.126). The overlap integral S_{ab} is less than 1 (except at $R = 0$). [A rigorous proof of this follows from Eq. (3-114) in *Margenau and Murphy*.] The coefficient of W^2 in $f(W)$ is $(1 - S_{ab}^2) > 0$; hence $f(\infty) = f(-\infty) = +\infty > 0$. For $W = H_{aa}$ or H_{bb}, the first product in (13.126) vanishes. Hence $f(H_{aa}) < 0$ and $f(H_{bb}) < 0$. The roots of (13.126) occur where $f(W)$ equals 0. Hence, by continuity, one root must be between $+\infty$ and H_{aa} and the other between H_{bb} and $-\infty$. Therefore,

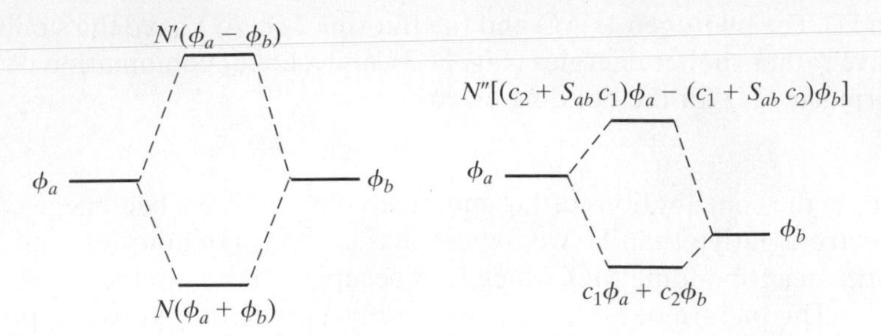

FIGURE 13.22 Formation of bonding and antibonding MOs from AOs in the homonuclear and heteronuclear cases. (See Prob. 13.37.)

the orbital energy of one MO is less than both H_{aa} and H_{bb} (the energies of the two AOs in the molecule; Section 13.5), and the energy of the other MO is greater than both H_{aa} and H_{bb}. One bonding and one antibonding MO are formed from the two AOs. Figure 13.22 shows the formation of bonding and antibonding MOs from two AOs, for the homonuclear and heteronuclear cases. These figures are gross oversimplifications, since a given MO has contributions from many AOs, not just two.

The coefficients c_1 and c_2 in the bonding heteronuclear MO in Fig. 13.22 are both positive, so as to build up charge between the nuclei. For the antibonding heteronuclear MO, the coefficients of ϕ_a and ϕ_b have opposite signs, causing charge depletion between the nuclei.

13.16 VB TREATMENT OF HETERONUCLEAR DIATOMIC MOLECULES

Consider the valence-bond ground state wave function of HF. We expect a single bond to be formed by the pairing of the hydrogen $1s$ electron and the unpaired fluorine $2p\sigma$ electron. The Heitler–London function corresponding to this pairing is [Eq. (13.118)]

$$\phi_{\text{cov}} = |\cdots 1s_{\text{H}}\overline{2p\sigma_{\text{F}}}| - |\cdots \overline{1s_{\text{H}}}2p\sigma_{\text{F}}| \qquad (13.127)$$

where the dots stand for $1s_{\text{F}}\overline{1s_{\text{F}}}2s_{\text{F}}\overline{2s_{\text{F}}}2p\pi_{x\text{F}}\overline{2p\pi_{x\text{F}}}2p\pi_{y\text{F}}\overline{2p\pi_{y\text{F}}}$. This function is essentially covalent, the electrons being shared by the two atoms. However, the high electronegativity of fluorine leads us to include a contribution from an ionic structure as well. An ionic valence-bond function has the form $\phi_a(1)\phi_a(2)$ [Eq. (13.109)]. Introduction of the required antisymmetric spin factor gives as the valence-bond function for an ionic structure in HF:

$$\phi_{\text{ion}} = |\cdots 2p\sigma_{\text{F}}\overline{2p\sigma_{\text{F}}}|$$

The VB wave function is then written as

$$\phi = c_1\phi_{\text{cov}} + c_2\phi_{\text{ion}} \qquad (13.128)$$

The optimum values of c_1 and c_2 are found by the variation method. This leads to the usual secular equation. We have ionic–covalent "resonance," involving the structures H—F and H^+F^-. The true molecular structure is intermediate between the covalent and ionic structures. A term $c_3|1s_{\text{H}}\overline{1s_{\text{H}}}|$ corresponding to the ionic structure H^-F^+ could also be included in the wave function, but this should contribute only slightly for HF. For molecules that are less ionic, both ionic structures might well be included.

For a highly ionic molecule such as NaCl, we expect the VB function to have $c_2 \gg c_1$. It might be thought that NaCl would dissociate to Na^+ and Cl^- ions, but this is not true. The ionization energy of Na is 5.1 eV, while the electron affinity of Cl is only 3.6 eV. Hence, in the gas phase the neutral separated ground-state atoms Na + Cl are more stable than the ground-state separated ions $Na^+ + Cl^-$. (In aqueous solution the ions are more stable because of the hydration energy, which makes the separated ions more stable than even the diatomic NaCl molecule.) If the nuclei are slowly pulled apart, a gas-phase NaCl molecule will dissociate to neutral atoms. Therefore, as R increases from R_e, the ratio c_2/c_1 in (13.128) must decrease, becoming zero at $R = \infty$. For intermediate values of R, the Coulombic attraction between the ions is greater than the 1.5-eV difference between the ionization potential and electron affinity, and the molecule is largely ionic. For very large R, the Coulombic attraction between the ions is less than 1.5 eV, and the molecule is largely covalent. However, if the nuclei in NaCl are pulled apart very rapidly, then the electrons will not have a chance to adjust their wave function from the ionic to the covalent wave function, and both bonding electrons will go with the chlorine nucleus, giving dissociation into ions.

Cesium has the lowest ionization energy, 3.9 eV. Chlorine has the highest electron affinity, 3.6 eV. (The electron affinity of fluorine is 3.45 eV.) Thus, even for CsCl and CsF, the separated ground-state neutral atoms are more stable than the separated ground-state ions. There are, however, cases of excited states of diatomic molecules that dissociate to ions.

13.17 THE VALENCE-ELECTRON APPROXIMATION

Suppose we want to treat Cs_2, which has 110 electrons. In the MO method, we would start by writing down a 110 × 110 Slater determinant of molecular orbitals. We would then approximate the MOs by functions containing variational parameters and go on to minimize the variational integral. Clearly, the large number of electrons makes this a formidable task. One way to simplify the problem is to divide the electrons into two groups: the 108 **core** electrons and the two 6*s* **valence** electrons, which provide bonding. We then try to treat the valence electrons separately from the core, taking the molecular energy as the sum of core- and valence-electron energies. This approach, introduced in the 1930s, is called the **valence-electron approximation**.

The simplest approach is to regard the core electrons as point charges coinciding with the nucleus. For Cs_2 this would give a Hamiltonian for the two valence electrons that is identical with the electronic Hamiltonian for H_2. If we then go ahead and minimize the variational integral for the valence electrons in Cs_2, with no restrictions on the valence-electron trial functions, we will clearly be in trouble. Such a procedure will cause the valence-electrons' MO to "collapse" to the $\sigma_g 1s$ MO, since the core electrons are considered absent. To avoid this collapse, one can impose the constraint that the variational functions used for the valence electrons be orthogonal to the orbitals of the core electrons. Of course, the task of keeping the valence orbitals orthogonal to the core orbitals means more work. A somewhat different approach is to drop the approximation of treating the core electrons as coinciding with the nucleus, and to treat them as a charge distribution that provides some sort of effective repulsive potential for the motion of the valence electrons. This leads to an effective Hamiltonian for the valence electrons, which is then used in the variational integral. The valence-electron approximation is widely used in approximate treatments of polyatomic molecules (Chapter 17).

13.18 SUMMARY

Because electrons are much lighter than nuclei, we can use the Born–Oppenheimer approximation to deal with molecules. This approximation takes the molecular wave function ψ as the product of wave functions for electronic motion and for nuclear motion: $\psi = \psi_{\text{el}}(q_i; q_\alpha)\psi_N(q_\alpha)$, where q_i and q_α are the electronic and nuclear coordinates, respectively. We solve the electronic Schrödinger equation for fixed positions of the nuclei. This equation is $(\hat{H}_{\text{el}} + V_{NN})\psi_{\text{el}} = U\psi_{\text{el}}$, where V_{NN} is the internuclear-repulsion term and $\hat{H}_{\text{el}}$ is the sum of operators for electronic kinetic energy, electron–nuclear attractions, and electron–electron repulsions. U is the electronic energy including internuclear repulsions. After finding $U(q_\alpha)$, we solve the nuclear-motion Schrödinger equation, which is $(\hat{T}_N + U)\psi_N = E\psi_N$, where $\hat{T}_N$ is the nuclear-kinetic-energy operator and E is the (total) molecular energy. Solution of the nuclear-motion Schrödinger equation shows that the molecular energy is approximately the sum of translational, rotational, vibrational, and electronic energies.

Quantum chemists use atomic units, in which energies are measured in hartrees and lengths in bohrs [Eqs. (13.29) and (13.30)].

The electronic Schrödinger equation for H_2^+ can be solved exactly to give wave functions that are eigenfunctions of $\hat{L}_z$, the operator for the component of electronic orbital angular momentum along the internuclear axis. The letters $\sigma, \pi, \delta, \phi, \ldots$ denote $\lambda \equiv |m|$ values of $0, 1, 2, 3, \ldots$, respectively, where $m\hbar$ is the $\hat{L}_z$ eigenvalue.

Approximate wave functions for the two lowest electronic states of H_2^+ are $N(1s_a + 1s_b)$ and $N'(1s_a - 1s_b)$, where $1s_a$ and $1s_b$ are $1s$ AOs centered on nuclei a and b, respectively.

Approximate MOs for homonuclear diatomic molecules were constructed as $N(1s_a \pm 1s_b)$, $N(2s_a \pm 2s_b), \ldots$, and in the separated-atoms notation, these were called $\sigma_g 1s$, $\sigma_u^* 1s$, $\sigma_g 2s$, $\sigma_u^* 2s, \ldots$, where σ means $\lambda = 0$, g or u denotes even or odd functions, and the star denotes an antibonding MO. An alternative notation, $1\sigma_g$, $1\sigma_u$, $2\sigma_g$, $2\sigma_u, \ldots$, numbers the MOs of each symmetry type in order of increasing energy. Using these approximate MOs, we examined the electron configurations, bond orders, and spin multiplicities of the ground terms of $H_2, He_2, Li_2, \ldots, Ne_2$.

A molecular electronic configuration gives rise to one or more electronic terms. Each diatomic-molecule term designation has the form ${}^{2S+1}(\Lambda)$, where S is the total electronic spin quantum number and (Λ) is a code letter $(\Sigma, \Pi, \Delta, \Phi, \ldots)$ that gives the $|M_L|$ value $(0, 1, 2, 3, \ldots)$, where $M_L\hbar$ is the component of electronic orbital angular momentum along the internuclear axis. Σ terms are designated Σ^+ or Σ^-, according to whether the eigenvalue of ψ_{el} for reflection in a plane containing the molecular axis is $+1$ or -1. For homonuclear diatomic molecules, a g or u subscript is added to the term symbol to show whether ψ_{el} is even or odd.

The simple LCAO-MO wave function for the H_2 ground electronic state is

$$\sigma_g 1s(1)\sigma_g 1s(2)2^{-1/2}[\alpha(1)\beta(2) - \beta(1)\alpha(2)]$$
$$\approx N[1s_a(1) + 1s_b(1)][1s_a(2) + 1s_b(2)]2^{-1/2}[\alpha(1)\beta(2) - \beta(1)\alpha(2)]$$

The simple covalent VB wave function for ground-state H_2 is constructed by interchange of the bonding electrons between the bonding atoms to give

$$N[1s_a(1)1s_b(2) + 1s_a(2)1s_b(1)]2^{-1/2}[\alpha(1)\beta(2) - \beta(1)\alpha(2)]$$

The H_2 MO function gives equal weight to covalent and ionic terms and thus underestimates electron correlation, which acts to keep electrons apart and makes the ionic terms less important than the covalent terms. The MO function can be improved by configuration interaction. The H_2 covalent VB wave function can be improved by the addition of a contribution from ionic terms.

The MO wave function for a $^1\Sigma$ diatomic-molecule state is a single Slater determinant of spin-orbitals. The VB wave function is a linear combination of Slater determinants that involve interchanges of spins within the pairs of bonding AOs.

A minimal basis set consists of one basis function for each inner shell and each valence AO. Examples of minimal- and extended-basis-set SCF wave functions for diatomic molecules were given in Sections 13.14 and 13.15. SCF wave functions give pretty accurate molecular geometries and dipole moments, but very inaccurate dissociation energies, due to improper behavior as $R \to \infty$.

PROBLEMS

Sec.	13.1	13.2	13.3	13.5	13.6	13.7
Probs.	13.1 13.5	13.6–13.11	13.12–13.14	13.15–13.22	13.23	13.24–13.25

Sec.	13.8	13.9	13.11	13.12	13.13	13.14	13.15	general
Probs.	13.26–13.30	13.31	13.32	13.33	13.34	13.35	13.36–13.37	13.38

13.1 True or false? (a) For a diatomic molecule, D_0 is greater than D_e. (b) The electronic wave function of a diatomic molecule changes when the internuclear distance changes. (c) The total energy of a molecule is the sum of the electronic energy (including nuclear repulsion) U and the energy E found by solving the nuclear Schrödinger equation $\hat{H}\psi_N = E\psi_N$.

13.2 For the H_2 ground electronic state, $D_0 = 4.4781$ eV. Find ΔH_0° for $H_2(g) \to 2H(g)$ in kJ/mol.

13.3 Use the D_0 value of H_2 (4.478 eV) and the D_0 value of H_2^+ (2.651 eV) to calculate the first ionization energy of H_2 (that is, the energy needed to remove an electron from H_2).

13.4 The infrared absorption spectrum of $^1H^{35}Cl$ has its strongest band at 8.65×10^{13} Hz. For this molecule, $D_0 = 4.43$ eV. (a) Find D_e for $^1H^{35}Cl$. (b) Find D_0 for $^2H^{35}Cl$.

13.5 (a) Verify that if anharmonicity is taken into account by inclusion of the $\nu_e x_e$ term in the vibrational energy, then $D_e = D_0 + \frac{1}{2}h\nu_e - \frac{1}{4}h\nu_e x_e$. (b) The $^7Li^1H$ ground electronic state has $D_0 = 2.4287$ eV, $\nu_e/c = 1405.65$ cm^{-1}, and $\nu_e x_e/c = 23.20$ cm^{-1}, where c is the speed of light. (These last two quantities are usually designated ω_e and $\omega_e x_e$ in the literature.) Calculate D_e for $^7Li^1H$.

13.6 Verify the Taylor-series expansion (13.24).

13.7 For the Morse-function H_2 example in Section 13.2: (a) Derive the expression for A and B in terms of $\hbar$, μ, and a. (*Hint*: Use the fact that the argument of an exponential function must be dimensionless.) (b) Calculate the numerical values of A, B, and $D_{e,r}$. (See Table A.3 in the Appendix.)

13.8 (a) Use the Numerov method with the endpoints and interval recommended in the Section 13.2 example to find the lowest six Morse-function vibrational levels of H_2. (b) Calculate $\langle x_r \rangle$ and $\langle R \rangle$ for these six levels. (c) Find the Morse-function $v = 6, 7, 8$, and 9 H_2 vibrational levels.

13.9 For the ground electronic state of $^{127}I_2$, $D_e/hc = 12550\,\text{cm}^{-1}$, $\nu_e/c = 214.5\,\text{cm}^{-1}$, and $R_e = 2.666$ Å. Use the Morse function and the Numerov method to calculate the six lowest vibrational energy levels of this electronic state. (The experimental E_{vib}/hc values for $v = 0, 2, 4$ are 107.19, 532.55, 953.01 cm^{-1}.)

13.10 Analytical solution of the Schrödinger equation for the Morse-function potential energy gives $E_{\text{vib}} = (v + \frac{1}{2})h\nu_e - (v + \frac{1}{2})^2 h^2 \nu_e^2 / 4D_e$. (The Morse-function U becomes infinite at $R = -\infty$. The expression given for E_{vib} corresponds to the boundary conditions that ψ becomes zero at $R = \infty$ and at $R = -\infty$, whereas for a diatomic molecule, we actually require that the vibrational wave function become zero at $R = 0$. This discrepancy is of no significance, since at $R = 0$, the Morse potential energy is very high and the Morse vibrational wave function is very close to zero.) For the $^1H_2$ ground electronic state, (a) calculate the six lowest analytical Morse vibrational energies to verify the values given in the Section 13.2 example; (b) find the maximum value of v predicted by the Morse function and compare with the true value $v = 14$.

13.11 If $\psi_{N,\text{int}}$ is the approximate nuclear wave function of (13.14) and (13.28), verify that $\langle R - R_e \rangle = \int |\psi_{N,\text{int}}|^2 (R - R_e)\, d\tau = \int_0^\infty |S_v|^2 (R - R_e)\, dR$.

13.12 (a) If Q stands for the dimension electric charge, show that $[4\pi\varepsilon_0] = \text{Q}^2\text{M}^{-1}\text{L}^{-3}\text{T}^2$. (b) Derive the expressions for A and B given in Section 13.3. (c) Verify the reduced form of the H-atom Schrödinger equation given in Section 13.3.

13.13 (a) Derive the expressions for the atomic units of time and electric dipole moment given in Section 13.3. (b) Find the expression for the atomic unit of electric field strength and calculate its value in volts per meter.

13.14 Give the numerical value in atomic units of each of the following quantities: (a) proton mass; (b) electron charge; (c) Planck's constant; (d) He^+ ground-state energy, assuming infinite nuclear mass; (e) one second; (f) c (speed of light); (g) hydrogen-atom ground-state energy, taking internal nuclear motion into account; (h) one debye [1 debye (D) $\approx 3.33564 \times 10^{-30}$ C m].

13.15 Derive (13.60) for the overlap integral S_{ab}.

13.16 (a) Derive (13.61) for the H_{aa} integral as follows. Add and subtract k/r_a in (13.32) to get $\hat{H} = \hat{H}_a + (k - 1)/r_a - 1/r_b$, where $\hat{H}_a$ is the Hamiltonian operator for a hydrogenlike atom of nuclear charge k with nucleus at a. Use this expression to write H_{aa} as the sum of three integrals. Evaluate the first integral using $\hat{H}_a 1s_a = -\frac{1}{2}k^2 1s_a$. Evaluate the $1/r_a$ integral using spherical coordinates with origin at a. Use confocal elliptic coordinates and Eqs. (13.59), (13.44), and (13.35) to evaluate the third integral. (b) Derive (13.62) for H_{ab}.

13.17 (a) Show that the variational integral W_1 for the ground state of H_2^+ [Eq. (13.63)] can be written as $W_1 = k^2 F(t) + kG(t)$, where $t \equiv kR$ and where F and G are certain functions of t. (b) Show that the minimization condition $\partial W_1 / \partial k = 0$ leads to

$$k = -\frac{G(t) + tG'(t)}{2F(t) + tF'(t)}$$

Using this equation, we can find k for a given value of t. We then use $R = t/k$ to find the value of R corresponding to our value of k.

13.18 Write a computer program that will calculate the optimum value of the orbital exponent k for the H_2^+ trial function (13.54) for a given R value. Have the program calculate $W_1 + 1/R$ [where W_1 is given by (13.63)] for k ranging from 0 to 3 in steps of 0.001 and have the program pick the k value that gives the smallest $W_1 + 1/R$. Write the program so that the value of R, the initial and final values of k, and the interval between k values are input from the keyboard for each run. Use the program to find k to the nearest 0.001 for R values of 0.5, 1.0, 2.0, 3.0, 4.0, and 6.0. Use the calculated energies to plot the $U(R)$ curve predicted by this trial function. (For more efficient ways to find the minimum of a function of one variable, see *Press et al.*, Chapter 10.)

13.19 Write a computer program that will simultaneously find the values of k and R that minimize $W_1 + 1/R$ in (13.63). On the first run, have R range from 0.1 to 6 in steps of 0.01 and have k range from 0 to 3 in steps of 0.01; compute $W_1 + 1/R$ for all possible pairs of k and R values in these ranges and have the computer find the pair of k and R values that gives the smallest $W_1 + 1/R$. Let these values be k' and R'. Rerun the program with k and R ranging from $k' - 0.01$ to $k' + 0.01$ and $R' - 0.01$ to $R' + 0.01$, respectively, each in steps of 0.0001 to find the optimum k and R, each accurate to 0.0001. (For much more efficient ways to find the minimum of a function of two variables, see *Press et al.*, Chapter 10.)

13.20 (a) Use Mathcad to create an animation showing how contours and three-dimensional plots of the H_2^+ LCAO MOs ϕ_1 and ϕ_2 [Eqs. (13.57) and (13.58)] for a plane containing the nuclei change as R changes from 3.8 to 0.1 bohr. Proceed as follows. Define the function $U(R, k)$ by adding the internuclear repulsion to (13.63). Include a parameter b in U, such that for $b = 1$ we get U for ϕ_1 and $b = -1$ we get U for ϕ_2. Specify the b value before defining U. Set R equal to 3.8 −FRAME/10, where the animation variable FRAME will later be defined to go from 0 to 37. To find the optimum k value at each R, use the Mathcad function root$(f(k), k)$, which finds the k value that makes $f(k) = 0$ provided we enter an initial guess for k. Take $f(k)$ as the derivative of $U(R, k)$ with respect to k. Do not find this derivative yourself but use the d/dx facility in Mathcad to have Mathcad take the derivative. Define an initial value for k before setting k equal to the root function. [In some versions of Mathcad, the root function will fail to find the solution for certain initial values of k. Use trial-and-error to find k values that work, and if a single initial k value does not work at all R values, use the if function (or nested if functions) to specify various k values at various R values. Different k values may be needed for the bonding and antibonding MOs.] To make the plot, define x_i and z_i to vary from −2.5 to 2.5 bohrs with increments of 1/6 bohr. Then use (13.57) and (13.58) to define phi(x, z) as the MO's value in the xz plane. Define the array (matrix) M by defining M_{ij} as phi(x_i, z_j). Create a contour plot, entering −2.5 and 2.5 as the limits for the axes. Also create a three-dimensional-surface plot taking appropriate values for the limits of y. Then create the animation. (b) By adding statements to the worksheet of part (a), use the solve block facility to find the predicted R_e for ϕ_1 and the optimum k at R_e. The conditions to be satisfied are that the derivative of U with respect to R must be zero and the derivative of U with respect to k must be zero. Also include conditions that R and k must be positive.

13.21 Use a modified form of the Mathcad worksheet for Prob. 13.20 to find the optimum k for each of the H_2^+ MOs (13.57) and (13.58) at each of these R/bohr values: (a) 10; (b) 6; (c) 4; (d) 2.5; (e) 2; (f) 1; (g) 0.1.

13.22 Use a spreadsheet to calculate the k values asked for in Prob. 13.21.

13.23 Verify Eq. (13.79) for a σ_h reflection.

13.24 Which species of each pair has the greater D_e? (a) Li_2 or Li_2^+; (b) C_2 or C_2^+; (c) O_2 or O_2^+; (d) F_2 or F_2^+.

13.25 Predict the bond order and the number of unpaired electrons for each of the following molecules: (a) S_2; (b) S_2^+; (c) S_2^-; (d) N_2^+; (e) N_2^-; (f) F_2^+; (g) F_2^-; (h) Ne_2^+; (i) Na_2^+; (j) Na_2^-; (k) H_2^-; (l) C_2^+; (m) C_2; (n) C_2^-.

13.26 Use the results of Prob. 13.25 and Table 13.3 to predict the ground term of each molecule in Prob. 13.25.

13.27 How many independent electronic wave functions correspond to each of the following diatomic-molecule terms: (a) $^1\Sigma^-$; (b) $^3\Sigma^+$; (c) $^3\Pi$; (d) $^1\Phi$; (e) $^6\Delta$?

13.28 Give the levels belonging to each of the terms in Prob. 13.27.

13.29 Show that the four functions of (13.89) have the indicated eigenvalues with respect to a $\sigma_v(xz)$ reflection of electronic coordinates. Start by showing that this reflection converts ϕ to $-\phi$ and leaves r_a and r_b unchanged.

13.30 Show that for a diatomic molecule $[\hat{L}_z, \hat{O}_{\sigma_v}] \neq 0$.

13.31 The ground state of H_2 has $^1\Sigma_g^+$ symmetry. What restriction does this impose on the values of m, n, j, and k in the James and Coolidge wave function (Section 13.9)?

13.32 In applying quantum chemistry to chemical reactions, which would be the more accurate approximation, the simple MO or the simple VB method?

13.33 Write down abbreviated expressions for the remaining six determinants of the N_2 VB function of Section 13.12. Use the rule given in that section to find the coefficient of each determinant in the wave function.

13.34 (a) Show that the simple MO wave function for the $b^3\Sigma_u^+$ level of H_2 is the same as the Heitler–London VB function for this level. (b) Show that the simple MO wave function for the $B^1\Sigma_u^+$ level of H_2 contains only ionic terms.

13.35 Verify that the $2s_\perp$ AO in (13.125) is orthogonal to the 1*s* AO and is normalized. (b) Let an MO ϕ have the form $\phi = a(1s) + b(2s) + \cdots$ when expressed using a nonorthogonal 2*s* STO and the form $\phi = c(1s) + d(2s_\perp) + \cdots$ when expressed using an orthogonalized 2*s* orbital. Show that $c = a + Sb$ and $d = b(1 - S^2)^{1/2}$, where $S = \langle 1s|2s \rangle$. (c) Let ζ_1 and ζ_2 be the orbital exponents of 1*s* and 2*s* STOs, respectively. Show that $S = 24\zeta_1^{3/2}\zeta_2^{5/2}/3^{1/2}(\zeta_1 + \zeta_2)^4$.

13.36 Use simple MO theory to predict the number of unpaired electrons and the ground term of each of the following: (a) BF; (b) BN; (c)BeS; (d) BO; (e) NO; (f) CF; (g) CP; (h) NBr; (i) ClO; (j) BrCl. Compare your results with the experimentally observed ground terms: (a) $^1\Sigma^+$; (b) $^3\Pi$; (c) $^1\Sigma^+$; (d) $^2\Sigma^+$; (e) $^2\Pi$; (f) $^2\Pi$; (g) $^2\Sigma^+$; (h) $^3\Sigma^-$; (i) $^2\Pi$; (j) $^1\Sigma^+$.

13.37 Use orthogonality (Section 8.5 and Prob. 8.39) to derive the expression given in Fig. 13.22 for the heteronuclear antibonding MO. Then do the same for the homonuclear antibonding MO. (b) Verify that the functions (13.57) and (13.58) are orthogonal.

13.38 True or false: (a) If $|\psi_1| = |\psi_2|$, then ψ_1 and ψ_2 must represent the same state. (b) The Hartree–Fock $U(R)$ curve for the H_2^+ ground electronic state shows the proper behavior as $R \rightarrow \infty$. (c) The Hartree–Fock $U(R)$ curve for the H_2 ground electronic state shows the proper behavior as $R \rightarrow \infty$.

CHAPTER 14

Theorems of Molecular Quantum Mechanics

This chapter discusses theorems that are used in molecular quantum mechanics. Section 14.1 expresses the electron probability density in terms of the wave function. Section 14.2 shows how the dipole moment of a molecule is calculated from the wave function. Section 14.3 gives the procedure for calculating the Hartree–Fock wave function of a molecule. Sections 14.4 to 14.7 discuss the virial theorem and the Hellmann–Feynman theorem, which are helpful in understanding chemical bonding.

14.1 ELECTRON PROBABILITY DENSITY

How is the wave function of a many-electron molecule related to the electron probability density? We want to find the probability of finding an electron in the rectangular volume element located at point (x, y, z) in space with edges dx, dy, dz. The electronic wave function ψ is a function of the spatial and spin coordinates of the n electrons. (For simplicity the parametric dependence on the nuclear configuration will not be explicitly indicated.) We know that

$$|\psi(x_1, \ldots, z_n, m_{s1}, \ldots, m_{sn})|^2 \, dx_1 \, dy_1 \, dz_1 \cdots dx_n \, dy_n \, dz_n \qquad (14.1)$$

is the probability of simultaneously finding electron 1 with spin m_{s1} in the volume element $dx_1 \, dy_1 \, dz_1$ at (x_1, y_1, z_1), electron 2 with spin m_{s2} in the volume element $dx_2 \, dy_2 \, dz_2$ at (x_2, y_2, z_2), and so on. Since we are not interested in what spin the electron we find at (x, y, z) has, we sum the probability (14.1) over all possible spin states of all electrons to give the probability of simultaneously finding each electron in the appropriate volume element with no regard for spin:

$$\sum_{m_{s1}} \cdots \sum_{m_{sn}} |\psi|^2 \, dx_1 \cdots dz_n \qquad (14.2)$$

Suppose we want the probability of finding electron 1 in the volume element $dx \, dy \, dz$ at (x, y, z). For this probability we do not care where electrons 2 through n are.

We therefore add the probabilities for all possible locations for these electrons. This amounts to integrating (14.2) over the coordinates of electrons $2, 3, \ldots, n$:

$$\left[\sum_{\text{all } m_s} \int \cdots \int |\psi(x, y, z, x_2, y_2, z_2, \ldots, x_n, y_n, z_n, m_{s1}, \ldots, m_{sn})|^2 \, dx_2 \cdots dz_n\right] dx\, dy\, dz \tag{14.3}$$

where there is a $(3n - 3)$-fold integration over x_2 through z_n.

Now suppose we ask for the probability of finding electron 2 in the volume element $dx\, dy\, dz$ at (x, y, z). By analogy to (14.3), this is

$$\left[\sum_{\text{all } m_s} \int \cdots \int |\psi(x_1, y_1, z_1, x, y, z, x_3, \ldots, z_n, m_{s1}, \ldots, m_{sn})|^2 \, dx_1\, dy_1\, dz_1\, dx_3 \cdots dz_n\right] dx\, dy\, dz \tag{14.4}$$

Of course, electrons do not come with labels, and this indistinguishability (Section 10.3) means that the probabilities (14.3) and (14.4) must be equal. This equality is readily proved. The wave function ψ is antisymmetric with respect to electron exchange, so $|\psi|^2$ is unchanged by an electron exchange. Interchanging the spatial and spin coordinates of electrons 1 and 2 in ψ in (14.4) and doing some relabeling of dummy variables, we see that (14.4) is equal to (14.3). Thus (14.3) gives the probability of finding any one particular electron in $dx\, dy\, dz$. Since the system has n electrons, the probability of finding *an* electron in $dx\, dy\, dz$ is n times (14.3). (In drawing this conclusion, we assume that the probability of finding more than one electron in the infinitesimal region $dx\, dy\, dz$ is negligible compared with the probability of finding one electron. This is certainly valid since the probability of finding two electrons will involve the product of six infinitesimal quantities as compared with the product of three infinitesimal quantities for the probability of finding one electron.)

Thus the probability density ρ for finding an electron in the neighborhood of point (x, y, z) is

$$\rho(x, y, z) = n \sum_{\text{all } m_s} \int \cdots \int |\psi(x, y, z, x_2, \ldots, z_n, m_{s1}, \ldots, m_{sn})|^2 \, dx_2 \cdots dz_n$$

$$\rho(\mathbf{r}) = n \sum_{\text{all } m_s} \int \cdots \int |\psi(\mathbf{r}, \mathbf{r}_2, \ldots, \mathbf{r}_n, m_{s1}, \ldots, m_{sn})|^2 \, d\mathbf{r}_2 \cdots d\mathbf{r}_n \tag{14.5}$$

where the vector notation for spatial variables (Section 5.2) is used. The atomic units of ρ are electrons/bohr^3.

ρ of a molecule is an experimentally observable quantity that can be found from measured x-ray diffraction intensities of molecular crystals or electron-diffraction intensities of gases. See P. Coppens and M. B. Hall (eds.), *Electron Distributions and the Chemical Bond*, Plenum, 1982; D. A. Kohl and L. S. Bartell, *J. Chem. Phys.*, **51**, 2891, 2896 (1969); P. Coppens, *J. Phys. Chem.*, **93**, 7979 (1989); P. Coppens, *Ann. Rev. Phys. Chem.*, **43**, 663 (1992).

To illustrate (14.5), consider the electron density for the simple VB and MO ground-state H_2 wave functions. The wave function is a product of a spatial factor and the spin function (11.60). (For more than two electrons, ψ cannot be factored into a product of a spatial part and a spin part; see Chapter 10.) Summation of (11.60) over m_{s1} and m_{s2} gives one (Section 10.4). Thus (14.5) becomes for H_2

$$\rho(x, y, z) = 2 \iiint |\phi(x, y, z, x_2, y_2, z_2)|^2 \, dx_2 \, dy_2 \, dz_2$$

where ϕ is the spatial factor. When ϕ is taken as the spatial factor in the VB function (13.100) or the MO function (13.94), we get (Prob. 14.1)

$$\rho_{\text{VB}} = \frac{1s_a^2 + 1s_b^2 + 2S_{ab}1s_a1s_b}{1 + S_{ab}^2}, \qquad \rho_{\text{MO}} = \frac{1s_a^2 + 1s_b^2 + 2(1s_a1s_b)}{1 + S_{ab}} \tag{14.6}$$

One finds (Prob. 14.2) that $\rho_{\text{MO}} > \rho_{\text{VB}}$ at the midpoint of the bond, so the MO function (which underestimates electron correlation) piles up more charge between the nuclei than the VB function.

The MO probability density in (14.6) is twice ρ for the H_2^+-like $1s_A + 1s_B$ MO [Eq. (13.65)]. One can prove that, for a many-electron MO wave function, ρ is found by multiplying the probability-density function of each MO by the number of electrons occupying it and summing the results:

$$\rho(x, y, z) = \sum_j n_j |\phi_j|^2 \tag{14.7}$$

where the sum is over the different orthogonal spatial MOs, and n_j (whose possible values are 0, 1, or 2) is the number of electrons in the MO ϕ_j. [We used (14.7) in Eq. (11.11).]

Calculations of ρ from SCF wave functions show that for nearly all molecules, local maxima in ρ occur only at the nuclei. An exception is the ground electronic state of Li_2, for which ρ has a small local maximum at the bond midpoint [*Bader*, Section E2.1; G. I. Bersuker et al., *J. Phys. Chem.*, **97**, 9323 (1993)].

Let $B(\mathbf{r}_i)$ be a function of the spatial coordinates x_i, y_i, z_i of electron i, where the notation of Section 5.2 is used. For an n-electron molecule, consider the average value

$$\left\langle \psi \left| \sum_{i=1}^{n} B(\mathbf{r}_i) \right| \psi \right\rangle = \int \psi^* \sum_{i=1}^{n} B(\mathbf{r}_i)\psi \, d\tau = \sum_{i=1}^{n} \int |\psi|^2 B(\mathbf{r}_i) \, d\tau$$

where ψ is the electronic wave function. Since the electrons are indistinguishable, each term in the sum $\sum_i \int |\psi|^2 B \, d\tau$ must have the same value. Therefore $\langle \psi | \sum_{i=1}^{n} B(\mathbf{r}_i) | \psi \rangle = \int n|\psi|^2 B(\mathbf{r}_1) \, d\tau$. Since $B(\mathbf{r}_1)$ depends only on x_1, y_1, z_1, before we integrate over x_1, y_1, z_1, we can integrate $n|\psi|^2$ over the spatial coordinates of electrons 2 to n and sum over all the spin coordinates. From Eq. (14.5), this produces the electron probability density $\rho(\mathbf{r}_1)$. Therefore, $\langle \psi | \sum_{i=1}^{n} B(\mathbf{r}_i) | \psi \rangle = \int \rho(\mathbf{r}_1) B(\mathbf{r}_1) \, d\mathbf{r}_1$. The subscript 1 on the integration variables is not needed, and the final result is

$$\int \psi^* \sum_{i=1}^{n} B(\mathbf{r}_i)\psi \, d\tau = \int \rho(\mathbf{r}) B(\mathbf{r}) \, d\mathbf{r} \tag{14.8}$$

where the integration is over the three spatial coordinates x, y, z. This result will be used later in this chapter and in Chapter 15.

14.2 DIPOLE MOMENTS

We now show how to calculate molecular dipole moments from wave functions.

The classical expression for the electric dipole moment $\boldsymbol{\mu}_{\text{cl}}$ of a set of discrete charges Q_i is

$$\boldsymbol{\mu}_{\text{cl}} = \sum_i Q_i \mathbf{r}_i \tag{14.9}$$

where $\mathbf{r}_i$ is the position vector from the origin to the ith charge [Eq. (5.33)]. The electric dipole moment is a vector whose x component is

$$\mu_{x,\text{cl}} = \sum_i Q_i x_i \tag{14.10}$$

with similar expressions for the other components. For a continuous charge distribution with charge density $\rho_Q(x, y, z)$, $\boldsymbol{\mu}_{\text{cl}}$ is found by summing over the infinitesimal elements of charge $dQ_i = \rho_Q(x, y, z)\, dx\, dy\, dz$:

$$\boldsymbol{\mu}_{\text{cl}} = \int \rho_Q(x, y, z)\mathbf{r}\, dx\, dy\, dz, \qquad \text{where} \quad \mathbf{r} = x\mathbf{i} + y\mathbf{j} + z\mathbf{k} \tag{14.11}$$

Now consider the quantum-mechanical definition of the electric dipole moment. Suppose we apply a uniform external electric field $\mathbf{E}$ to an atom or molecule and ask for the effect on the energy of the system. To form the Hamiltonian operator, we first need the classical expression for the energy. The electric field strength $\mathbf{E}$ is defined as $\mathbf{E} \equiv \mathbf{F}/Q$, where $\mathbf{F}$ is the force the field exerts on a charge Q. We take the z direction as the direction of the applied field: $\mathbf{E} = \mathscr{E}_z\mathbf{k}$. The potential energy V is [Eq. (4.26)]

$$dV/dz = -F_z = -Q\mathscr{E}_z \quad \text{and} \quad V = -Q\mathscr{E}_z z$$

This is the potential energy of a single charge in the field. For a system of charges,

$$V = -\mathscr{E}_z \sum_i Q_i z_i \tag{14.12}$$

where z_i is the z coordinate of charge Q_i. The extension of (14.12) to the case where the electric field points in an arbitrary direction follows from (4.26) and is

$$V = -\mathscr{E}_x \sum_i Q_i x_i - \mathscr{E}_y \sum_i Q_i y_i - \mathscr{E}_z \sum_i Q_i z_i = -\mathbf{E} \cdot \boldsymbol{\mu}_{\text{cl}} \tag{14.13}$$

This is the classical-mechanical expression for the energy of an electric dipole in a uniform applied electric field.

To calculate the quantum-mechanical expression, we use perturbation theory. The perturbation operator $\hat{H}'$ corresponding to (14.13) is $\hat{H}' = -\mathbf{E} \cdot \hat{\boldsymbol{\mu}}$, where the **electric dipole-moment operator** $\hat{\boldsymbol{\mu}}$ is

$$\hat{\boldsymbol{\mu}} = \sum_i Q_i \hat{\mathbf{r}}_i = \mathbf{i}\hat{\mu}_x + \mathbf{j}\hat{\mu}_y + \mathbf{k}\hat{\mu}_z \tag{14.14}$$

$$\hat{\mu}_x = \sum_i Q_i x_i, \quad \hat{\mu}_y = \sum_i Q_i y_i, \quad \hat{\mu}_z = \sum_i Q_i z_i \tag{14.15}$$

The first-order correction to the energy is [Eq. (9.22)]

$$E^{(1)} = -\mathbf{E} \cdot \int \psi^{(0)*} \hat{\boldsymbol{\mu}} \psi^{(0)} \, d\tau \tag{14.16}$$

where $\psi^{(0)}$ is the unperturbed wave function. Comparison of (14.16) and (14.13) shows that the quantum-mechanical quantity that corresponds to $\boldsymbol{\mu}_{\text{cl}}$ is the integral

$$\boldsymbol{\mu} = \int \psi^{(0)*} \hat{\boldsymbol{\mu}} \psi^{(0)} \, d\tau \tag{14.17}$$

$\boldsymbol{\mu}$ in (14.17) is the quantum-mechanical **electric dipole moment** of the system.

An objection to taking (14.17) as the dipole moment is that we considered only the first-order energy correction. If we had included $E^{(2)}$ in (14.16), the comparison with (14.13) would not have given (14.17) as the dipole moment. Actually, (14.17) is the dipole moment of the system in the absence of an applied electric field and is the *permanent* electric dipole moment. Application of the field distorts the wave function from $\psi^{(0)}$, giving rise to an *induced* electric dipole moment in addition to the permanent dipole moment. The induced dipole moment corresponds to the energy correction $E^{(2)}$. (For the details, see *Merzbacher*, Section 17.4.) The induced dipole moment $\boldsymbol{\mu}_{\text{ind}}$ is related to the applied electric field $\mathbf{E}$ by

$$\boldsymbol{\mu}_{\text{ind}} = \alpha \mathbf{E} \tag{14.18}$$

where α is the **polarizability** of the atom or molecule. The greater the polarizability of molecule B, the greater the London dispersion force (Section 13.7) between two B molecules.

The shift in the energy of a quantum-mechanical system caused by an applied electric field is called the *Stark effect*. The *first-order* (or *linear*) Stark effect is given by (14.16), and from (14.17) it vanishes for a system with no permanent electric dipole moment. The *second-order* (or *quadratic*) Stark effect is given by the energy correction $E^{(2)}$ and is proportional to the square of the applied field.

The electric dipole-moment operator (14.14) is an odd function of the coordinates. If the wave function in (14.17) is either even or odd, then the integrand in (14.17) is an odd function, and the integral over all space vanishes. We conclude that the permanent electric dipole moment $\boldsymbol{\mu}$ is zero for states of definite parity.

The permanent electric dipole moment of a molecule in electronic state ψ_{el} is

$$\boldsymbol{\mu} = \int \psi_{\text{el}}^* \hat{\boldsymbol{\mu}} \psi_{\text{el}} \, d\tau_{\text{el}} \tag{14.19}$$

The electronic wave functions of homonuclear diatomic molecules can be classified as g or u, according to their parity. Hence, a homonuclear diatomic molecule has a zero permanent electric dipole moment, a not too astonishing result. The same holds true for any molecule with a center of symmetry. The electric dipole-moment operator for a molecule includes summation over both the electronic and nuclear charges:

$$\hat{\boldsymbol{\mu}} = \sum_i (-e\mathbf{r}_i) + \sum_\alpha Z_\alpha e\mathbf{r}_\alpha \tag{14.20}$$

where $\mathbf{r}_\alpha$ is the vector from the origin to the nucleus of atomic number Z_α, and $\mathbf{r}_i$ is the vector to electron i. Since both the dipole-moment operator (14.20) and the electronic wave function depend on the parameters defining the nuclear configuration, the molecular electronic dipole moment $\boldsymbol{\mu}$ depends on the nuclear configuration. To indicate this, the quantity (14.19) can be called the dipole-moment *function* of the molecule. In writing (14.19), we ignored the nuclear motion. When the dipole moment of a molecule is experimentally determined, what is measured is the quantity (14.19) averaged over the zero-point vibrations (assuming the temperature is not high enough for there to be appreciable population of higher vibrational levels). We might use $\boldsymbol{\mu}_0$ and $\boldsymbol{\mu}_e$ to indicate the dipole moment averaged over zero-point vibrations and the dipole moment at the equilibrium nuclear configuration, respectively.

Since the second sum in (14.20) is independent of the electronic coordinates, we have

$$\begin{aligned}\boldsymbol{\mu} &= \int \psi_{\text{el}}^* \sum_i (-e\mathbf{r}_i)\psi_{\text{el}}\, d\tau_{\text{el}} + \sum_\alpha Z_\alpha e\mathbf{r}_\alpha \int \psi_{\text{el}}^*\psi_{\text{el}}\, d\tau_{\text{el}} \\ &= -e\int |\psi_{\text{el}}|^2 \sum_i \mathbf{r}_i\, d\tau_{\text{el}} + e\sum_\alpha Z_\alpha \mathbf{r}_\alpha\end{aligned}$$

Using (14.8), we have

$$\boldsymbol{\mu} = -e \iiint \rho(x, y, z)\mathbf{r}\, dx\, dy\, dz + e\sum_\alpha Z_\alpha \mathbf{r}_\alpha \tag{14.21}$$

where ρ is the electron probability density. Equation (14.21) is what would be obtained if we pretended that the electrons were smeared out into a continuous charge distribution whose charge density is given by (14.5) and we used the classical equation (14.11) to calculate $\boldsymbol{\mu}$.

14.3 THE HARTREE–FOCK METHOD FOR MOLECULES

A key development in quantum chemistry has been the computation of accurate self-consistent-field wave functions for many diatomic and polyatomic molecules. The principles of molecular SCF MO calculations are essentially the same as for atomic SCF calculations (Section 11.1). We shall restrict ourselves to closed-shell configurations. For open shells, the formulas are more complicated.

The molecular Hartree–Fock wave function is written as an antisymmetrized product (Slater determinant) of spin-orbitals, each spin-orbital being a product of a spatial orbital ϕ_i and a spin function (either α or β).

The expression for the Hartree–Fock molecular electronic energy E_{HF} is given by the variation theorem as $E_{\text{HF}} = \langle D|\hat{H}_{\text{el}} + V_{NN}|D\rangle$, where D is the Slater-determinant Hartree–Fock wave function and $\hat{H}_{\text{el}}$ and V_{NN} are given by (13.5) and (13.6). Since V_{NN} does not involve electronic coordinates and D is normalized, we have $\langle D|V_{NN}|D\rangle = V_{NN}\langle D|D\rangle = V_{NN}$. The operator $\hat{H}_{\text{el}}$ is the sum of one-electron operators $\hat{f}_i$ and two-electron operators $\hat{g}_{ij}$; we have $\hat{H}_{\text{el}} = \sum_i \hat{f}_i + \sum_j \sum_{i>j} \hat{g}_{ij}$, where $\hat{f}_i = -\frac{1}{2}\nabla_i^2 - \sum_\alpha Z_\alpha/r_{i\alpha}$ and $\hat{g}_{ij} = 1/r_{ij}$. The Hamiltonian $\hat{H}_{\text{el}}$ is the same as the Hamiltonian $\hat{H}$ for an atom except that $\sum_\alpha Z_\alpha/r_{i\alpha}$ replaces Z/r_i in $\hat{f}_i$. Hence Eq. (11.83) can be used to give $\langle D|\hat{H}_{\text{el}}|D\rangle$. Therefore, the Hartree–Fock energy of a diatomic or polyatomic molecule with only closed shells is

$$E_{\text{HF}} = 2\sum_{i=1}^{n/2} H_{ii}^{\text{core}} + \sum_{i=1}^{n/2}\sum_{j=1}^{n/2}(2J_{ij} - K_{ij}) + V_{NN} \tag{14.22}$$

$$H_{ii}^{\text{core}} \equiv \langle \phi_i(1)|\hat{H}^{\text{core}}(1)|\phi_i(1)\rangle \equiv \left\langle \phi_i(1)\middle| -\tfrac{1}{2}\nabla_1^2 - \sum_\alpha Z_\alpha/r_{1\alpha}\middle|\phi_i(1)\right\rangle \tag{14.23}$$

$$J_{ij} \equiv \langle \phi_i(1)\phi_j(2)|1/r_{12}|\phi_i(1)\phi_j(2)\rangle, \quad K_{ij} \equiv \langle \phi_i(1)\phi_j(2)|1/r_{12}|\phi_j(1)\phi_i(2)\rangle \tag{14.24}$$

where the one-electron-operator symbol was changed from $\hat{f}_1$ to $\hat{H}^{\text{core}}(1)$. The **one-electron core Hamiltonian**

$$\hat{H}^{\text{core}}(1) \equiv -\tfrac{1}{2}\nabla_1^2 - \sum_\alpha \frac{Z_\alpha}{r_{1\alpha}}$$

is the sum of the kinetic-energy operator for electron 1 and the potential-energy operators for the attractions between electron 1 and the nuclei. $\hat{H}^{\text{core}}(1)$ omits the interactions of electron 1 with the other electrons. The sums over i and j are over the $n/2$ occupied spatial orbitals ϕ_i of the n-electron molecule. In the **Coulomb integrals** J_{ij} and the **exchange integrals** K_{ij}, the integration goes over the spatial coordinates of electrons 1 and 2.

The Hartree–Fock method looks for those orbitals ϕ_i that minimize the variational integral E_{HF}. Each MO is taken to be normalized: $\langle \phi_i(1)|\phi_i(1)\rangle = 1$. Moreover, for computational convenience one takes the MOs to be orthogonal: $\langle \phi_i(1)|\phi_j(1)\rangle = 0$ for $i \neq j$. It might be thought that a lower energy could be obtained if the orthogonality restriction were omitted, but this is not so. A closed-shell antisymmetric wave function is a Slater determinant, and one can use the properties of determinants to show that a Slater determinant of nonorthogonal orbitals is equal to a Slater determinant in which the orbitals have been orthogonalized by the Schmidt or some other procedure; see Section 15.9 and F. W. Bobrowicz and W. A. Goddard, Chapter 4, Section 3.1 of *Schaefer, Methods of Electronic Structure Theory*. In effect, the Pauli antisymmetry requirement removes nonorthogonalities from the orbitals.

The derivation of the equation that determines the orthonormal ϕ_i's that minimize E_{HF} is complicated and is omitted. (For the derivation, see *Lowe and Peterson*, Appendix 7; *Szabo and Ostlund*, Sections 3.1 and 3.2; *Parr*, pages 21–23.) One finds that the closed-shell orthogonal Hartree–Fock MOs satisfy

$$\hat{F}(1)\phi_i(1) = \varepsilon_i\phi_i(1) \tag{14.25}$$

where ε_i is the orbital energy and where the **(Hartree–) Fock operator** $\hat{F}$ is (in atomic units)

$$\hat{F}(1) = \hat{H}^{\text{core}}(1) + \sum_{j=1}^{n/2}[2\hat{J}_j(1) - \hat{K}_j(1)] \tag{14.26}$$

$$\hat{H}^{\text{core}}(1) \equiv -\tfrac{1}{2}\nabla_1^2 - \sum_\alpha \frac{Z_\alpha}{r_{1\alpha}} \tag{14.27}$$

where the **Coulomb operator** $\hat{J}_j$ and the **exchange operator** $\hat{K}_j$ are defined by

$$\hat{J}_j(1)f(1) = f(1)\int |\phi_j(2)|^2 \frac{1}{r_{12}}\, dv_2 \tag{14.28}$$

$$\hat{K}_j(1)f(1) = \phi_j(1)\int \frac{\phi_j{}^*(2)f(2)}{r_{12}}\, dv_2 \tag{14.29}$$

where f is an arbitrary function and the integrals are definite integrals over all space.

The first term on the right of (14.27) is the operator for the kinetic energy of one electron. The second term is the potential-energy operators for the attractions between one electron and the nuclei. The Coulomb operator $\hat{J}_j(1)$ is the potential energy of interaction between electron 1 and a smeared-out electron with electronic density $-|\phi_j(2)|^2$. The factor 2 in (14.26) occurs because there are two electrons in each spatial orbital. The exchange operator has no simple physical interpretation but arises from the requirement that the wave function be antisymmetric with respect to electron exchange. The exchange operators are absent from the Hartree equations (11.9). The Hartree–Fock MOs ϕ_i in (14.25) are eigenfunctions of the same operator $\hat{F}$, the eigenvalues being the orbital energies ε_i.

The orthogonality of the MOs greatly simplifies MO calculations, causing many integrals to vanish. In contrast, the VB method uses atomic orbitals, and AOs centered on different atoms are not orthogonal. MO calculations are simpler than VB calculations, and the MO method is used far more often than the VB method.

The true Hamiltonian operator and wave function involve the coordinates of all n electrons. The Hartree–Fock Hamiltonian operator $\hat{F}$ is a one-electron operator (that is, it involves the coordinates of only one electron), and (14.25) is a one-electron differential equation. This has been indicated in (14.25) by writing $\hat{F}$ and ϕ_i as functions of the coordinates of electron 1. Of course, the coordinates of any electron could have been used. The operator $\hat{F}$ is peculiar in that it depends on its own eigenfunctions [see Eqs. (14.26) to (14.29)], which are not known initially. Hence the Hartree–Fock equations must be solved by an iterative process.

To obtain the expression for the orbital energies ε_i, we multiply (14.25) by $\phi_i^*(1)$ and integrate over all space. Using the fact that ϕ_i is normalized and using the result of Prob. 14.8, we obtain $\varepsilon_i = \int \phi_i^*(1)\hat{F}(1)\phi_i(1)\, dv_1$ and

$$\varepsilon_i = \langle \phi_i(1)|\hat{H}^{\text{core}}(1)|\phi_i(1)\rangle + \sum_j [2\langle \phi_i(1)|\hat{J}_j(1)|\phi_i(1)\rangle - \langle \phi_i(1)|\hat{K}_j(1)|\phi_i(1)\rangle]$$

$$\varepsilon_i = H_{ii}^{\text{core}} + \sum_{j=1}^{n/2}(2J_{ij} - K_{ij}) \tag{14.30}$$

where H_{ii}^{core}, J_{ij}, and K_{ij} are defined by (14.23) and (14.24).

Summation of (14.30) over the $n/2$ occupied orbitals gives

$$\sum_{i=1}^{n/2} \varepsilon_i = \sum_{i=1}^{n/2} H_{ii}^{\text{core}} + \sum_{i=1}^{n/2}\sum_{j=1}^{n/2}(2J_{ij} - K_{ij}) \tag{14.31}$$

Solving this equation for $\sum_i H_{ii}^{\text{core}}$ and substituting the result into (14.22), we obtain the Hartree–Fock energy as

$$E_{\mathrm{HF}} = 2\sum_{i=1}^{n/2} \varepsilon_i - \sum_{i=1}^{n/2}\sum_{j=1}^{n/2}(2J_{ij} - K_{ij}) + V_{NN} \tag{14.32}$$

Since there are two electrons per MO, the quantity $2\sum_i \varepsilon_i$ is the sum of the orbital energies. Subtraction of the double sum in (14.32) avoids counting each interelectronic repulsion twice, as discussed in Section 11.1.

A key development that made feasible the calculation of accurate molecular SCF wave functions was Roothaan's 1951 proposal to expand the spatial orbitals ϕ_i as linear combinations of a set of one-electron basis functions χ_s:

$$\phi_i = \sum_{s=1}^{b} c_{si}\chi_s \tag{14.33}$$

To exactly represent the MOs ϕ_i, the basis functions χ_s should form a complete set. This requires an infinite number of basis functions. In practice, one must use a finite number b of basis functions. If b is large enough and the functions χ_s well chosen, one can represent the MOs with negligible error.

To avoid confusion, we shall use the letters r, s, t, u to label the basis functions χ, and the letters i, j, k, l to label the MOs ϕ. (Often the Greek letters $\mu, \nu, \lambda, \sigma$ are used to label the basis functions.)

Substitution of the expansion (14.33) into the Hartree–Fock equations (14.25) gives

$$\sum_s c_{si}\hat{F}\chi_s = \varepsilon_i \sum_s c_{si}\chi_s$$

Multiplication by χ_r^* and integration gives

$$\sum_{s=1}^{b} c_{si}(F_{rs} - \varepsilon_i S_{rs}) = 0, \qquad r = 1, 2, \ldots, b \tag{14.34}$$

$$F_{rs} \equiv \langle \chi_r|\hat{F}|\chi_s\rangle, \qquad S_{rs} \equiv \langle \chi_r|\chi_s\rangle \tag{14.35}$$

The equations (14.34) form a set of b simultaneous linear homogeneous equations in the b unknowns c_{si}, $s = 1, 2, \ldots, b$, that describe the MO ϕ_i in (14.33). For a nontrivial solution, we must have

$$\det(F_{rs} - \varepsilon_i S_{rs}) = 0 \tag{14.36}$$

This is a secular equation whose roots give the orbital energies ε_i. The **(Hartree–Fock–) Roothaan equations** (14.34) must be solved by an iterative process, since the F_{rs} integrals depend on the orbitals ϕ_i (through the dependence of $\hat{F}$ on the ϕ_i's), which in turn depend on the unknown coefficients c_{si}.

One starts with guesses for the occupied-MO expressions as linear combinations of the basis functions, as in (14.33). This initial set of MOs is used to compute the Fock operator $\hat{F}$ from (14.26) to (14.29). The matrix elements (14.35) are computed, and the secular equation (14.36) is solved to give an initial set of ε_i's. These ε_i's are used to solve (14.34) for an improved set of coefficients, giving an improved set of MOs, which are then used to compute an improved $\hat{F}$, and so on. One continues until no further improvement in MO coefficients and energies occurs from one cycle to the next. The calculations are done using a computer. (The most efficient way to solve the Roothaan equations is to use matrix-algebra methods; see the last part of this section.)

We have used the terms SCF wave function and Hartree–Fock wave function interchangeably. In practice, the term *SCF wave function* is applied to any wave function obtained by iterative solution of the Roothaan equations, whether or not the basis set is large enough to give a really accurate approximation to the Hartree–Fock SCF wave function. There is only one true Hartree–Fock SCF wave function, which is the best possible wave function that can be written as a Slater determinant of spin-orbitals. Some of the extended-basis-set calculations approach the true Hartree–Fock wave function quite closely; such functions are called "near Hartree–Fock wave functions" or, less cautiously, "Hartree–Fock wave functions."

The Fock Matrix Elements. To solve the Roothaan equations (14.34), we first must express the *Fock matrix elements* (integrals) F_{rs} in terms of the basis functions χ. The Fock operator $\hat{F}$ is given by (14.26), and

$$F_{rs} = \langle \chi_r(1)|\hat{F}(1)|\chi_s(1)\rangle$$

$$F_{rs} = \langle \chi_r(1)|\hat{H}^{\text{core}}(1)|\chi_s(1)\rangle + \sum_{j=1}^{n/2} [2\langle \chi_r(1)|\hat{J}_j(1)\chi_s(1)\rangle - \langle \chi_r(1)|\hat{K}_j(1)\chi_s(1)\rangle] \quad (14.37)$$

Replacement of f by χ_s in (14.28), followed by use of the expansion (14.33), gives

$$\hat{J}_j(1)\chi_s(1) = \chi_s(1)\int \frac{\phi_j^*(2)\phi_j(2)}{r_{12}}\,dv_2 = \chi_s(1)\sum_t\sum_u c_{tj}^* c_{uj}\int \frac{\chi_t^*(2)\chi_u(2)}{r_{12}}\,dv_2$$

Multiplication by $\chi_r^*(1)$ and integration over the coordinates of electron 1 gives

$$\langle \chi_r(1)|\hat{J}_j(1)\chi_s(1)\rangle = \sum_t\sum_u c_{tj}^* c_{uj}\iint \frac{\chi_r^*(1)\chi_s(1)\chi_t^*(2)\chi_u(2)}{r_{12}}\,dv_1\,dv_2$$

$$\langle \chi_r(1)|\hat{J}_j(1)\chi_s(1)\rangle = \sum_{t=1}^{b}\sum_{u=1}^{b} c_{tj}^* c_{uj}(rs|tu) \quad (14.38)$$

where the **two-electron repulsion integral** is defined as

$$(rs|tu) \equiv \iint \frac{\chi_r^*(1)\chi_s(1)\chi_t^*(2)\chi_u(2)}{r_{12}}\,dv_1\,dv_2 \quad \textbf{(14.39)}$$

The widely used notation of (14.39) should not be misinterpreted as an overlap integral. Other notations, some mutually contradictory, are used for electron repulsion integrals, so it is always wise to check an author's definition.

Similarly, replacement of f by χ_s in (14.29) leads to (Prob. 14.9)

$$\langle \chi_r(1)|\hat{K}_j(1)\chi_s(1)\rangle = \sum_{t=1}^{b}\sum_{u=1}^{b} c_{tj}^* c_{uj}(ru|ts) \quad (14.40)$$

Substituting (14.40) and (14.38) into (14.37) and changing the order of summation, we get the desired expression for F_{rs} in terms of integrals over the basis functions χ:

$$F_{rs} = H_{rs}^{\text{core}} + \sum_{t=1}^{b}\sum_{u=1}^{b}\sum_{j=1}^{n/2} c_{tj}^* c_{uj}[2(rs|tu) - (ru|ts)]$$

$$F_{rs} = H_{rs}^{\text{core}} + \sum_{t=1}^{b}\sum_{u=1}^{b} P_{tu}[(rs|tu) - \tfrac{1}{2}(ru|ts)] \tag{14.41}$$

$$P_{tu} \equiv 2\sum_{j=1}^{n/2} c_{tj}^* c_{uj}, \qquad t = 1, 2, \ldots, b, \qquad u = 1, 2, \ldots, b \tag{14.42}$$

$$H_{rs}^{\text{core}} \equiv \langle \chi_r(1)|\hat{H}^{\text{core}}(1)|\chi_s(1)\rangle$$

The quantities P_{tu} are called **density matrix elements** or *charge, bond-order matrix elements*. [Some workers use the definition $P_{ut} \equiv 2\sum_j c_{tj}^* c_{uj}$.] Substitution of the expansion (14.33) into (14.7) for the electron probability density ρ gives for a closed-shell molecule:

$$\rho = 2\sum_{j=1}^{n/2} \phi_j^* \phi_j = 2\sum_{r=1}^{b}\sum_{s=1}^{b}\sum_{j=1}^{n/2} c_{rj}^* c_{sj}\chi_r^*\chi_s = \sum_{r=1}^{b}\sum_{s=1}^{b} P_{rs}\chi_r^*\chi_s \tag{14.43}$$

To express the Hartree–Fock energy in terms of integrals over the basis functions χ, we first solve (14.31) for $\sum_i\sum_j(2J_{ij} - K_{ij})$ and substitute the result into (14.32) to get

$$E_{\text{HF}} = \sum_{i=1}^{n/2} \varepsilon_i + \sum_{i=1}^{n/2} H_{ii}^{\text{core}} + V_{NN}$$

We have, using the expansion (14.33),

$$H_{ii}^{\text{core}} = \langle \phi_i|\hat{H}^{\text{core}}|\phi_i\rangle = \sum_r\sum_s c_{ri}^* c_{si}\langle \chi_r|\hat{H}^{\text{core}}|\chi_s\rangle = \sum_r\sum_s c_{ri}^* c_{si} H_{rs}^{\text{core}}$$

$$E_{\text{HF}} = \sum_{i=1}^{n/2} \varepsilon_i + \sum_r\sum_s\sum_{i=1}^{n/2} c_{ri}^* c_{si} H_{rs}^{\text{core}} + V_{NN}$$

$$E_{\text{HF}} = \sum_{i=1}^{n/2} \varepsilon_i + \frac{1}{2}\sum_{r=1}^{b}\sum_{s=1}^{b} P_{rs} H_{rs}^{\text{core}} + V_{NN} \tag{14.44}$$

An alternative expression for E_{HF} is useful. Multiplication of $\hat{F}\phi_i = \varepsilon_i\phi_i$ [Eq. (14.25)] by ϕ_i^* and integration gives $\varepsilon_i = \langle \phi_i|\hat{F}|\phi_i\rangle$. Substitution of $\phi_i = \sum_{s=1}^{b} c_{si}\chi_s$ [Eq. (14.33)] gives $\varepsilon_i = \sum_r\sum_s c_{ri}^* c_{si}\langle \chi_r|\hat{F}|\chi_s\rangle = \sum_r\sum_s c_{ri}^* c_{si} F_{rs}$. The first sum in (14.44) becomes $\sum_i \varepsilon_i = \sum_r\sum_s\sum_i c_{ri}^* c_{si} F_{rs} = \frac{1}{2}\sum_r\sum_s P_{rs}F_{rs}$, where the definition (14.42) of P_{rs} was used. Equation (14.44) becomes

$$E_{\text{HF}} = \frac{1}{2}\sum_{r=1}^{b}\sum_{s=1}^{b} P_{rs}(F_{rs} + H_{rs}^{\text{core}}) + V_{NN} \tag{14.45}$$

which expresses E_{HF} of a closed-shell molecule in terms of the density, Fock, and core-Hamiltonian matrix elements calculated with the basis functions χ_r.

EXAMPLE Do an SCF calculation for the helium-atom ground state using a basis set of two 1*s* STOs with orbital exponents $\zeta_1 = 1.45$ and $\zeta_2 = 2.91$. [By trial and error, these have been found to be the optimum ζ's to use for this basis set; see C. Roetti and E. Clementi, *J. Chem. Phys.*, **60**, 4725 (1974).]

From (11.14), the normalized basis functions are (in atomic units)

$$\chi_1 = 2\zeta_1^{3/2}e^{-\zeta_1 r}Y_0^0, \qquad \chi_2 = 2\zeta_2^{3/2}e^{-\zeta_2 r}Y_0^0, \qquad \zeta_1 = 1.45, \qquad \zeta_2 = 2.91 \tag{14.46}$$

To solve the Roothaan equations (14.34), we need the integrals F_{rs} and S_{rs}. The overlap integrals S_{rs} are

$$S_{11} = \langle\chi_1|\chi_1\rangle = 1, \qquad S_{22} = \langle\chi_2|\chi_2\rangle = 1$$

$$S_{12} = S_{21} = \langle\chi_1|\chi_2\rangle = 4\zeta_1^{3/2}\zeta_2^{3/2}\int_0^\infty e^{-(\zeta_1+\zeta_2)r}r^2\,dr = \frac{8\zeta_1^{3/2}\zeta_2^{3/2}}{(\zeta_1+\zeta_2)^3} = 0.8366$$

where the Appendix integral (A.8) was used.

The integrals F_{rs} are given by (14.41) and depend on H_{rs}^{core}, P_{tu}, and $(rs|tu)$. From (14.27), $\hat{H}^{\text{core}} = -\frac{1}{2}\nabla^2 - 2/r = -\frac{1}{2}\nabla^2 - \zeta/r + (\zeta - 2)/r$. The integrals H_{rs}^{core} are evaluated the same way that similar integrals were evaluated in the variation treatment of He in Section 9.4. We find (Prob. 14.12)

$$H_{11}^{\text{core}} = \langle\chi_1|\hat{H}^{\text{core}}|\chi_1\rangle = -\tfrac{1}{2}\zeta_1^2 + (\zeta_1 - 2)\zeta_1 = \tfrac{1}{2}\zeta_1^2 - 2\zeta_1 = -1.8488$$

$$H_{22}^{\text{core}} = \tfrac{1}{2}\zeta_2^2 - 2\zeta_2 = -1.5860$$

$$H_{12}^{\text{core}} = H_{21}^{\text{core}} = \langle\chi_1|\hat{H}^{\text{core}}|\chi_2\rangle = -\tfrac{1}{2}\zeta_2^2 S_{12} + \frac{4(\zeta_2 - 2)\zeta_1^{3/2}\zeta_2^{3/2}}{(\zeta_1+\zeta_2)^2}$$

$$H_{12}^{\text{core}} = H_{21}^{\text{core}} = \frac{\zeta_1^{3/2}\zeta_2^{3/2}(4\zeta_1\zeta_2 - 8\zeta_1 - 8\zeta_2)}{(\zeta_1+\zeta_2)^3} = -1.8826$$

Many of the electron-repulsion integrals $(rs|tu)$ are equal to one another. For real basis functions, one can show that (Prob. 14.13)

$$(rs|tu) = (sr|tu) = (rs|ut) = (sr|ut) = (tu|rs) = (ut|rs) = (tu|sr) = (ut|sr) \tag{14.47}$$

The electron-repulsion integrals are evaluated using the $1/r_{12}$ expansion (9.123) in Prob. 9.13. One finds [see Eq. (9.53) and Prob. 14.14]

$$(11|11) = \tfrac{5}{8}\zeta_1 = 0.9062, \qquad (22|22) = \tfrac{5}{8}\zeta_2 = 1.8188$$

$$(11|22) = (22|11) = (\zeta_1^4\zeta_2 + 4\zeta_1^3\zeta_2^2 + \zeta_1\zeta_2^4 + 4\zeta_1^2\zeta_2^3)/(\zeta_1+\zeta_2)^4 = 1.1826$$

$$(12|12) = (21|12) = (12|21) = (21|21) = 20\zeta_1^3\zeta_2^3/(\zeta_1+\zeta_2)^5 = 0.9536$$

$$(11|12) = (11|21) = (12|11) = (21|11) = \frac{16\zeta_1^{9/2}\zeta_2^{3/2}}{(3\zeta_1+\zeta_2)^4}\left[\frac{12\zeta_1 + 8\zeta_2}{(\zeta_1+\zeta_2)^2} + \frac{9\zeta_1+\zeta_2}{2\zeta_1^2}\right] = 0.9033$$

$$(12|22) = (22|12) = (21|22) = (22|21)$$
$$= \text{the } (11|12) \text{ expression with 1 and 2 interchanged} = 1.2980$$

To start the calculation, we need an initial guess for the ground-state AO expansion coefficients c_{si} in (14.33) so that we can get an initial estimate of the density matrix elements P_{tu} in (14.41). We saw in Section 9.4 that the optimum orbital exponent for a helium AO that consists of one 1s STO is $\frac{27}{16} = 1.6875$. Since the orbital exponent ζ_1 is much closer to 1.6875 than is ζ_2, we expect that the coefficient of χ_1 in $\phi_1 = c_{11}\chi_1 + c_{21}\chi_2$ will be substantially larger than the coefficient of χ_2. Let us take as an initial guess $c_{11}/c_{21} \approx 2$. [A more general method to get an initial guess for the c_{si} coefficients is to neglect the electron-repulsion integrals in (14.41) and approximate F_{rs} in the secular equation (14.36) by H_{rs}^{core} ($F_{rs} \approx H_{rs}^{\text{core}}$); we then solve (14.36) and (14.34).

This would give $c_{11}/c_{21} \approx 1.5$ (Prob 14.15).] The normalization condition $\int |\phi_1|^2\, d\tau = 1$ gives for real coefficients (Prob. 14.17)

$$c_{21} = (1 + k^2 + 2kS_{12})^{-1/2}, \qquad \text{where } k \equiv c_{11}/c_{21} \tag{14.48}$$

Substitution of $k = 2$ and $S_{12} = 0.8366$ gives $c_{21} \approx 0.3461$ and $c_{11} \approx 2c_{21} = 0.6922$.

With $n = 2$ and $b = 2$, Eq. (14.42) gives

$$P_{11} = 2c_{11}^*c_{11}, \quad P_{12} = 2c_{11}^*c_{21}, \quad P_{21} = P_{12}^*, \quad P_{22} = 2c_{21}^*c_{21} \tag{14.49}$$

The initial guess $c_{11} \approx 0.6922$, $c_{21} \approx 0.3461$ gives as the initial density matrix elements

$$P_{11} \approx 0.9583, \quad P_{12} = P_{21} \approx 0.4791, \quad P_{22} \approx 0.2396$$

The Fock matrix elements are found from (14.41) with $b = 2$. Using (14.47) and $P_{12} = P_{21}$ for real functions, we get (Prob. 14.16a)

$$F_{11} = H_{11}^{\text{core}} + \tfrac{1}{2}P_{11}(11|11) + P_{12}(11|12) + P_{22}[(11|22) - \tfrac{1}{2}(12|21)]$$
$$F_{12} = F_{21} = H_{12}^{\text{core}} + \tfrac{1}{2}P_{11}(12|11) + P_{12}[\tfrac{3}{2}(12|12) - \tfrac{1}{2}(11|22)] + \tfrac{1}{2}P_{22}(12|22)$$
$$F_{22} = H_{22}^{\text{core}} + P_{11}[(22|11) - \tfrac{1}{2}(21|12)] + P_{12}(22|12) + \tfrac{1}{2}P_{22}(22|22)$$

Substitution of the values of the H_{rs}^{core} and $(rs|tu)$ integrals listed previously gives (Prob. 14.16b)

$$F_{11} = -1.8488 + 0.4531P_{11} + 0.9033P_{12} + 0.7058P_{22} \tag{14.50}$$

$$F_{12} = F_{21} = -1.8826 + 0.4516_5P_{11} + 0.8391P_{12} + 0.6490P_{22} \tag{14.51}$$

$$F_{22} = -1.5860 + 0.7058P_{11} + 1.2980P_{12} + 0.9094P_{22} \tag{14.52}$$

Substitution of the initial guess for the P_{tu}'s into (14.50) to (14.52) gives as the initial estimates of the F_{rs}'s

$$F_{11} \approx -0.813, \quad F_{12} = F_{21} \approx -0.892, \quad F_{22} \approx -0.070$$

The initial estimate of the secular equation $\det(F_{rs} - S_{rs}\varepsilon_i) = 0$ is

$$\begin{vmatrix} -0.813 - \varepsilon_i & -0.892 - 0.8366\varepsilon_i \\ -0.892 - 0.8366\varepsilon_i & -0.070 - \varepsilon_i \end{vmatrix} \approx 0$$
$$0.3001\varepsilon_i^2 - 0.609_5\varepsilon_i - 0.739 \approx 0$$
$$\varepsilon_1 \approx -0.854, \quad \varepsilon_2 \approx 2.885$$

Substitution of the lower root ε_1 into the Roothaan equation (14.34) with $r = 2$ gives

$$c_{11}(F_{21} - \varepsilon_1 S_{21}) + c_{21}(F_{22} - \varepsilon_1 S_{22}) \approx 0$$
$$-0.177_5c_{11} + 0.784c_{21} \approx 0$$
$$c_{11}/c_{21} \approx 4.42$$

Substitution of $k = 4.42$ and $S_{12} = 0.8366$ in the normalization condition (14.48) gives

$$c_{21} \approx 0.189, \quad c_{11} = kc_{21} \approx 0.836$$

Substitution of these improved coefficients into (14.49) gives as the improved density matrix elements

$$P_{11} \approx 1.398, \quad P_{12} = P_{21} \approx 0.316, \quad P_{22} \approx 0.071$$

Substitution of these improved P_{tu}'s into (14.50) to (14.52) gives as the improved F_{rs} values

$$F_{11} \approx -0.880, \quad F_{12} = F_{21} \approx -0.940, \quad F_{22} \approx -0.124_6$$

The improved secular equation is

$$\begin{vmatrix} -0.880 - \varepsilon_i & -0.940 - 0.8366\varepsilon_i \\ -0.940 - 0.8366\varepsilon_i & -0.124_6 - \varepsilon_i \end{vmatrix} \approx 0$$

$$\varepsilon_1 \approx -0.918, \quad \varepsilon_2 \approx 2.810$$

The improved ε_1 value gives $c_{11}/c_{21} \approx 4.61$ and

$$c_{11} \approx 0.842, \quad c_{21} \approx 0.183$$

Another cycle of calculation yields (Prob. 14.18)

$$P_{11} = 1.418, \qquad P_{12} = P_{21} = 0.308, \qquad P_{22} = 0.067$$

$$F_{11} = -0.881, \qquad F_{12} = F_{21} = -0.940, \qquad F_{22} = -0.124_5 \tag{14.53}$$

$$\varepsilon_1 = -0.918, \qquad \varepsilon_2 = 2.809 \tag{14.54}$$

$$c_{11} = 0.842, \qquad c_{21} = 0.183$$

These last c's are the same as those for the previous cycle, so the calculation has converged and we are finished. The He ground-state SCF AO for this basis set is

$$\phi_1 = 0.842\chi_1 + 0.183\chi_2$$

The SCF energy is found from (14.44) with $n = 2$ and $b = 2$ as

$$E_{\text{HF}} = -0.918 + \tfrac{1}{2}[1.418(-1.8488) + 2(0.308)(-1.8826) + 0.067(-1.5860)] + 0$$

$$= -2.862 \text{ hartrees} = -77.9 \text{ eV}$$

A more precise calculation with $\zeta_1 = 1.45363$ and $\zeta_2 = 2.91093$ gives an SCF energy of -2.8616726 hartrees, as compared with the limiting Hartree–Fock energy -2.8616799 hartrees found with five basis functions [C. Roetti and E. Clementi, *J. Chem. Phys.*, **60**, 4725 (1974)].

Matrix Form of the Roothaan Equations. The Roothaan equations are most efficiently solved using matrix methods. The Roothaan equations (14.34) read

$$\sum_{s=1}^{b} F_{rs}c_{si} = \sum_{s=1}^{b} S_{rs}c_{si}\varepsilon_i, \quad r = 1, 2, \ldots, b$$

The coefficients c_{si} relate the MOs ϕ_i to the basis functions χ_s according to $\phi_i = \sum_s c_{si}\chi_s$. Let **C** be the square matrix of order b whose elements are the coefficients c_{si}. Let **F** be the square matrix of order b whose elements are $F_{rs} = \langle\chi_r|\hat{F}|\chi_s\rangle$. Let **S** be the square matrix whose elements are $S_{rs} = \langle\chi_r|\chi_s\rangle$. Let $\boldsymbol{\varepsilon}$ be the diagonal square matrix whose diagonal elements are the orbital energies $\varepsilon_1, \varepsilon_2, \ldots, \varepsilon_b$ so that the elements of $\boldsymbol{\varepsilon}$ are $\varepsilon_{mi} = \delta_{mi}\varepsilon_i$, where δ_{mi} is the Kronecker delta.

Use of the matrix multiplication rule (7.106) gives the (s, i)th element of the matrix product $\mathbf{C}\boldsymbol{\varepsilon}$ as $(\mathbf{C}\boldsymbol{\varepsilon})_{si} = \sum_m c_{sm}\varepsilon_{mi} = \sum_m c_{sm}\delta_{mi}\varepsilon_i = c_{si}\varepsilon_i$. Hence the Roothaan equations read

$$\sum_{s=1}^{b} F_{rs}c_{si} = \sum_{s=1}^{b} S_{rs}(\mathbf{C}\boldsymbol{\varepsilon})_{si} \tag{14.55}$$

From the matrix multiplication rule, the left side of (14.55) is the (r, i)th element of **FC**, and the right side is the (r, i)th element of $\mathbf{S}(\mathbf{C}\boldsymbol{\varepsilon})$. Since the general element of **FC** equals the general element of $\mathbf{SC}\boldsymbol{\varepsilon}$, these matrices are equal:

$$\mathbf{FC} = \mathbf{SC}\boldsymbol{\varepsilon} \tag{14.56}$$

This is the matrix form of the Roothaan equations.

The set of basis functions χ_s used to expand the MOs is not an orthogonal set. However, one can use the Schmidt or some other procedure to form orthogonal linear combinations of the basis functions to give a new set of basis functions χ'_s that is an orthonormal set: $\chi'_s = \sum_t a_{ts}\chi_t$ and $S'_{rs} = \langle\chi'_r|\chi'_s\rangle = \delta_{rs}$. (See Probs. 8.56 and 8.57 and *Szabo and Ostlund*, Section 3.4.5, for details of the orthogonalization procedure.) With this orthonormal basis set, the overlap matrix is a unit matrix, and the Roothaan equations (14.56) have the simpler form

$$\mathbf{F}'\mathbf{C}' = \mathbf{C}'\boldsymbol{\varepsilon} \tag{14.57}$$

where $F'_{rs} = \langle\chi'_r|\hat{F}|\chi'_s\rangle$ and $\mathbf{C}'$ is the matrix of the coefficients that relate the MOs ϕ_i to the orthonormal basis functions: $\phi_i = \sum_s c'_{si}\chi'_s$. It was shown in Prob. 8.56c that the $\mathbf{F}$ and $\mathbf{F}'$ matrices and the $\mathbf{C}$ and $\mathbf{C}'$ matrices are related by

$$\mathbf{F}' = \mathbf{A}^\dagger\mathbf{F}\mathbf{A} \quad \text{and} \quad \mathbf{C} = \mathbf{A}\mathbf{C}'$$

where $\mathbf{A}$ is the matrix of coefficients a_{ts} in $\chi'_s = \sum_t a_{ts}\chi_t$, so we can readily calculate $\mathbf{F}'$ from $\mathbf{F}$ and $\mathbf{C}$ from $\mathbf{C}'$. [$\mathbf{H}$ in Prob. 8.56 corresponds to $\mathbf{F}$ in (14.56).]

The matrix equation (14.57) has the same form as Eq. (8.88), which is $\mathbf{HC} = \mathbf{CW}$, where $\mathbf{C}$ and $\mathbf{W}$ [defined by (8.87)] are the eigenvector matrix and eigenvalue matrix, respectively, of $\mathbf{H}$. Thus, the orbital energies ε_i are the eigenvalues of the Fock matrix $\mathbf{F}'$ and each column of $\mathbf{C}'$ is an eigenvector of $\mathbf{F}'$. Because the Fock operator $\hat{F}$ is Hermitian, the Fock matrix $\mathbf{F}'$ is a Hermitian matrix. As noted in the paragraph preceding Eq. (8.95), the eigenvector matrix $\mathbf{C}'$ of the Hermitian matrix $\mathbf{F}'$ can be chosen to be unitary, meaning that its inverse equals its conjugate transpose [Eq. (8.93)] $\mathbf{C}'^{-1} = \mathbf{C}'^\dagger$. (With a unitary coefficient matrix $\mathbf{C}'$, the MOs ϕ_i are orthonormal; see Prob. 14.22.) Multiplication of (14.57) on the left by $\mathbf{C}'^{-1} = \mathbf{C}'^\dagger$ gives [see Eqs. (8.89) and (8.95)]

$$\mathbf{C}'^\dagger\mathbf{F}'\mathbf{C}' = \boldsymbol{\varepsilon} \tag{14.58}$$

which has the same form as Eq. (8.95).

The following procedure is commonly used to do an SCF MO calculation at a specified molecular geometry.

1. Choose a basis set χ_s.
2. Evaluate the H^{core}_{rs}, S_{rs}, and $(rs|tu)$ integrals.
3. Use the overlap integrals S_{rs} and an orthogonalization procedure to calculate the $\mathbf{A}$ matrix of coefficients a_{ts} that will produce orthonormal basis functions $\chi'_s = \sum_t a_{ts}\chi_t$.
4. Make an initial guess for the coefficients c_{si} in the MOs $\phi_i = \sum_s c_{si}\chi_s$. From the initial guess of coefficients, calculate the density matrix $\mathbf{P}$ in (14.42).
5. Use (14.41) to calculate an estimate of the Fock matrix elements F_{rs} from $\mathbf{P}$ and the $(rs|tu)$ and H^{core}_{rs} integrals.
6. Calculate the matrix $\mathbf{F}'$ using $\mathbf{F}' = \mathbf{A}^\dagger\mathbf{F}\mathbf{A}$.
7. Use a matrix-diagonalization method (Section 8.6) to find the eigenvalue and eigenvector matrices $\boldsymbol{\varepsilon}$ and $\mathbf{C}'$ of $\mathbf{F}'$.
8. Calculate the coefficient matrix $\mathbf{C} = \mathbf{A}\mathbf{C}'$.

9. Calculate an improved estimate of the density matrix from **C** using $\mathbf{P} = 2\mathbf{C}\mathbf{C}^\dagger$, which is the matrix form of (14.42) (Prob. 14.10c).
10. Compare the improved **P** with the preceding estimate of **P**. If all corresponding matrix elements differ by negligible amounts from each other, the calculation has converged and one uses the converged SCF wave function to calculate molecular properties. If the calculation has not converged, go back to step (5) to calculate an improved **F** matrix from the current **P** matrix and then do the succeeding steps.

One way to begin an SCF calculation is to initially estimate the Fock matrix elements by $F_{rs} \approx H_{rs}^{\text{core}}$, which amounts to neglecting the double sum in (14.41). This gives a very crude estimate. More commonly, SCF calculations get the initial estimate of the density matrix by doing a semiempirical calculation (Section 17.4) on the molecule. Semiempirical calculations are very fast. Still another possibility is to construct a guess for the **P** matrix by using the density matrices for the atoms composing the molecule. To find the equilibrium geometry of a molecule, one does a series of SCF calculations at many successive geometries (see Section 15.10). For the second and later SCF calculations of the series, one takes the initial guess of **P** as **P** for the SCF wave function of a nearby geometry.

14.4 THE VIRIAL THEOREM

We now derive the virial theorem. Let $\hat{H}$ be the time-independent Hamiltonian of a system in the bound stationary state ψ:

$$\hat{H}\psi = E\psi \tag{14.59}$$

Let $\hat{A}$ be a linear, time-independent operator. Consider the integral

$$\int \psi^*[\hat{H}, \hat{A}]\psi\, d\tau = \langle\psi|\hat{H}\hat{A} - \hat{A}\hat{H}|\psi\rangle = \langle\psi|\hat{H}|\hat{A}\psi\rangle - E\langle\psi|\hat{A}|\psi\rangle \tag{14.60}$$

where (14.59) was used. Since $\hat{H}$ is Hermitian, we have

$$\langle\psi|\hat{H}|\hat{A}\psi\rangle = \langle\hat{A}\psi|\hat{H}|\psi\rangle^* = E^*\langle\hat{A}\psi|\psi\rangle^* = E\langle\psi|\hat{A}\psi\rangle = E\langle\psi|\hat{A}|\psi\rangle$$

and Eq. (14.60) becomes

$$\int \psi^*[\hat{H}, \hat{A}]\psi\, d\tau = 0 \tag{14.61}$$

Equation (14.61) is the *hypervirial theorem.* [For some of its applications, see J. O. Hirschfelder, *J. Chem. Phys.*, **33**, 1462 (1960); J. H. Epstein and S. T. Epstein, *Am. J. Phys.*, **30**, 266 (1962).] In deriving (14.61), we used the Hermitian property of $\hat{H}$. The proof that $\hat{p}_x$ and $\hat{p}_x^2$ are Hermitian, and hence that $\hat{H}$ is Hermitian, requires that ψ vanish at $\pm\infty$ [see Eq. (7.17)]. Hence the hypervirial theorem does not apply to continuum stationary states, for which ψ does not vanish at ∞.

We now derive the virial theorem from (14.61). We choose $\hat{A}$ to be

$$\sum_i \hat{q}_i\hat{p}_i = -i\hbar\sum_i q_i\frac{\partial}{\partial q_i} \tag{14.62}$$

where the sum runs over the $3n$ Cartesian coordinates of the n particles. (Particle 1 has Cartesian coordinates q_1, q_2, q_3 and linear-momentum components p_1, p_2, p_3. In this chapter the symbol q will indicate a *Cartesian* coordinate.) To evaluate $[\hat{H}, \hat{A}]$, we use (5.4), (5.5), (5.8), and (5.9) to get

$$\left[\hat{H}, \sum_i \hat{q}_i\hat{p}_i\right] = \sum_i [\hat{H}, \hat{q}_i\hat{p}_i] = \sum_i \hat{q}_i[\hat{H}, \hat{p}_i] + \sum_i [\hat{H}, \hat{q}_i]\hat{p}_i$$
$$= i\hbar \sum_i q_i \frac{\partial V}{\partial q_i} - i\hbar \sum_i \frac{1}{m_i}\hat{p}_i^2 = i\hbar \sum_i q_i \frac{\partial V}{\partial q_i} - 2i\hbar\hat{T} \tag{14.63}$$

where $\hat{T}$ and $\hat{V}$ are the kinetic- and potential-energy operators for the system. Substitution of (14.63) into (14.61) gives

$$\left\langle \psi \left| \sum_i q_i \frac{\partial V}{\partial q_i} \right| \psi \right\rangle = 2\langle \psi | \hat{T} | \psi \rangle \tag{14.64}$$

Using $\langle B \rangle$ for the quantum-mechanical average of B, we write (14.64) as

$$\left\langle \sum_i q_i \frac{\partial V}{\partial q_i} \right\rangle = 2\langle T \rangle \tag{14.65}$$

Equation (14.65) is the quantum-mechanical **virial theorem**. Note that its validity is restricted to bound stationary states. (The word *vires* is Latin for "forces"; in classical mechanics, the derivatives of the potential energy give the negatives of the force components.)

For certain systems the virial theorem takes on a simple form. To discuss these systems, we introduce the idea of a homogeneous function. A function $f(x_1, x_2, \ldots, x_j)$ of several variables is **homogeneous of degree** n if it satisfies

$$\boxed{f(sx_1, sx_2, \ldots, sx_j) = s^n f(x_1, x_2, \ldots, x_j)} \tag{14.66}$$

where s is an arbitrary parameter. For example, the function $g = 1/y^3 + x/y^2z^2$ is homogeneous of degree -3, since $g(sx, sy, sz) = 1/s^3y^3 + sx/s^2y^2s^2z^2 = s^{-3}g(x, y, z)$.

Euler's theorem on homogeneous functions states that, if $f(x_1, \ldots, x_j)$ is homogeneous of degree n, then

$$\sum_{k=1}^{j} x_k \frac{\partial f}{\partial x_k} = nf \tag{14.67}$$

The theorem is proved as follows. Let

$$u_1 \equiv sx_1, \quad u_2 \equiv sx_2, \quad \ldots, \quad u_j = sx_j$$

Use of the chain rule gives for the partial derivative of the left side of (14.66) with respect to s

$$\frac{\partial f(u_1, \ldots, u_j)}{\partial s} = \frac{\partial f}{\partial u_1}\frac{\partial u_1}{\partial s} + \frac{\partial f}{\partial u_2}\frac{\partial u_2}{\partial s} + \cdots + \frac{\partial f}{\partial u_j}\frac{\partial u_j}{\partial s}$$
$$= x_1\frac{\partial f}{\partial u_1} + x_2\frac{\partial f}{\partial u_2} + \cdots + x_j\frac{\partial f}{\partial u_j} = \sum_{k=1}^{j} x_k \frac{\partial f}{\partial u_k}$$

The partial derivative of Eq. (14.66) with respect to s is thus

$$\sum_{k=1}^{j} x_k \frac{\partial f(u_1, \ldots, u_j)}{\partial u_k} = ns^{n-1} f(x_1, \ldots, x_j) \tag{14.68}$$

Let $s = 1$, so that $u_i = x_i$; Eq. (14.68) then gives (14.67). This completes the proof.

Now we return to the virial theorem (14.65). If the potential energy V is a homogeneous function of degree n when expressed in Cartesian coordinates, Euler's theorem gives

$$\sum_i q_i \frac{\partial V}{\partial q_i} = nV \tag{14.69}$$

and the virial theorem (14.65) simplifies to

$$\boxed{2\langle T \rangle = n\langle V \rangle} \tag{14.70}$$

for a bound stationary state. Since (Prob. 6.33)

$$\langle T \rangle + \langle V \rangle = E \tag{14.71}$$

we can write (14.70) in two other forms:

$$\langle V \rangle = \frac{2E}{n+2} \tag{14.72}$$

$$\langle T \rangle = \frac{nE}{n+2} \tag{14.73}$$

EXAMPLE Apply the virial theorem to (a) the one-dimensional harmonic oscillator; (b) the hydrogen atom; (c) a many-electron atom.

(a) For the one-dimensional harmonic oscillator, $V = \frac{1}{2}kx^2$, which is homogeneous of degree $n = 2$. Equations (14.70) and (14.72) give

$$\langle T \rangle = \langle V \rangle = \tfrac{1}{2}E = \tfrac{1}{2}h\nu\left(v + \tfrac{1}{2}\right) \tag{14.74}$$

This was verified for the ground state in Prob. 4.10.

(b) For the H atom, $V = -e'^2/(x^2 + y^2 + z^2)^{1/2}$ in Cartesian coordinates. V is a homogeneous function of degree -1. Hence

$$2\langle T \rangle = -\langle V \rangle \tag{14.75}$$

which was verified for the ground state in Prob. 6.34. For any hydrogen-atom bound stationary state,

$$\langle V \rangle = 2E \quad \text{and} \quad \langle T \rangle = -E \tag{14.76}$$

(c) For a many-electron atom with spin–orbit interaction neglected,

$$V = -Ze'^2 \sum_{i=1}^{n} \frac{1}{(x_i^2 + y_i^2 + z_i^2)^{1/2}} + \sum_i \sum_{j>i} \frac{e'^2}{[(x_i - x_j)^2 + (y_i - y_i)^2 + (z_i - z_j)^2]^{1/2}}$$

Replacing each of the $3n$ coordinates by s times the coordinate, we find that V is homogeneous of degree -1. Hence Eqs. (14.75) and (14.76) hold for any atom.

Now consider molecules. In the Born–Oppenheimer approximation, the molecular wave function is [Eq. (13.12)]

$$\psi = \psi_{\text{el}}(q_i; q_\alpha)\psi_N(q_\alpha)$$

where q_i and q_α symbolize the electronic and nuclear coordinates, respectively. ψ_{el} is found by solving the electronic Schrödinger equation (13.7):

$$\hat{H}_{\text{el}}\psi_{\text{el}}(q_i; q_\alpha) = E_{\text{el}}(q_\alpha)\psi_{\text{el}}(q_i; q_\alpha)$$

where E_{el} is the electronic energy and where

$$\hat{H}_{\text{el}} = \hat{T}_{\text{el}} + \hat{V}_{\text{el}} \tag{14.77}$$

$$\hat{T}_{\text{el}} = -\frac{\hbar^2}{2m_e}\sum_i\left(\frac{\partial^2}{\partial x_i^2} + \frac{\partial^2}{\partial y_i^2} + \frac{\partial^2}{\partial z_i^2}\right) \tag{14.78}$$

$$\hat{V}_{\text{el}} = -\sum_\alpha\sum_i \frac{Z_\alpha e'^2}{[(x_i - x_\alpha)^2 + (y_i - y_\alpha)^2 + (z_i - z_\alpha)^2]^{1/2}}$$

$$+ \sum_i\sum_{j>i} \frac{e'^2}{[(x_i - x_j)^2 + (y_i - y_j)^2 + (z_i - z_j)^2]^{1/2}} \tag{14.79}$$

Let the system be in the electronic stationary state ψ_{el}. If we put the subscript el on $\hat{H}$ and ψ in (14.59) and regard the variables of ψ_{el} to be the electronic coordinates q_i (with the nuclear coordinates q_α being parameters), then the derivation of the virial theorem (14.65) is seen to be valid for the electronic kinetic- and potential-energy operators, and we have

$$2\langle\psi_{\text{el}}|\hat{T}_{\text{el}}|\psi_{\text{el}}\rangle = \left\langle\psi_{\text{el}}\left|\sum_i q_i\frac{\partial V_{\text{el}}}{\partial q_i}\right|\psi_{\text{el}}\right\rangle \tag{14.80}$$

Viewed as a function of the electronic coordinates, V_{el} is *not* a homogeneous function, since

$$[(sx_i - x_\alpha)^2 + (sy_i - y_\alpha)^2 + (sz_i - z_\alpha)^2]^{-1/2}$$
$$\neq s^{-1}[(x_i - x_\alpha)^2 + (y_i - y_\alpha)^2 + (z_i - z_\alpha)^2]^{-1/2}$$

Thus the virial theorem for the average electronic kinetic and potential energies of a molecule will not have the simple form (14.75), which holds for atoms. We can, however, view V_{el} as a function of both the electronic and the nuclear Cartesian coordinates. From this viewpoint V_{el} *is* a homogeneous function of degree -1, since

$$[(sx_i - sx_\alpha)^2 + (sy_i - sy_\alpha)^2 + (sz_i - sz_\alpha)^2)]^{-1/2}$$
$$= s^{-1}[(x_i - x_\alpha)^2 + (y_i - y_\alpha)^2 + (z_i - z_\alpha)^2]^{-1/2}$$

Therefore, considering V_{el} as a function of both electronic and nuclear coordinates and applying Euler's theorem (14.67), we have

$$\sum_i q_i\frac{\partial V_{\text{el}}}{\partial q_i} + \sum_\alpha q_\alpha\frac{\partial V_{\text{el}}}{\partial q_\alpha} = -V_{\text{el}}$$

Using this equation in (14.80), we get

$$2\langle\psi_{\text{el}}|\hat{T}_{\text{el}}|\psi_{\text{el}}\rangle = -\langle\psi_{\text{el}}|\hat{V}_{\text{el}}|\psi_{\text{el}}\rangle - \left\langle\psi_{\text{el}}\middle|\sum_{\alpha} q_{\alpha}\frac{\partial V_{\text{el}}}{\partial q_{\alpha}}\middle|\psi_{\text{el}}\right\rangle \tag{14.81}$$

which contains an additional term as compared with the atomic virial theorem (14.75). Consider this extra term. We have

$$\left\langle\psi_{\text{el}}\middle|\sum_{\alpha} q_{\alpha}\frac{\partial V_{\text{el}}}{\partial q_{\alpha}}\middle|\psi_{\text{el}}\right\rangle = \sum_{\alpha} q_{\alpha}\int \psi_{\text{el}}^{*}\frac{\partial V_{\text{el}}}{\partial q_{\alpha}}\psi_{\text{el}}\,d\tau_{\text{el}}$$

where the nuclear coordinate q_a was taken outside the integral over electronic coordinates. In Section 14.7 we shall show that [see the bracketed sentence after Eq. (14.126)]

$$\int \psi_{\text{el}}^{*}\frac{\partial V_{\text{el}}}{\partial q_{\alpha}}\psi_{\text{el}}\,d\tau_{\text{el}} = \frac{\partial E_{\text{el}}}{\partial q_{\alpha}} \tag{14.82}$$

[Equation (14.82) is an example of the Hellmann–Feynman theorem.] Using these last two equations in the molecular electronic virial theorem (14.81), we get

$$2\langle\psi_{\text{el}}|\hat{T}_{\text{el}}|\psi_{\text{el}}\rangle = -\langle\psi_{\text{el}}|\hat{V}_{\text{el}}|\psi_{\text{el}}\rangle - \sum_{\alpha} q_{\alpha}\frac{\partial E_{\text{el}}}{\partial q_{\alpha}}$$

$$2\langle T_{\text{el}}\rangle = -\langle V_{\text{el}}\rangle - \sum_{\alpha} q_{\alpha}\frac{\partial E_{\text{el}}}{\partial q_{\alpha}} \tag{14.83}$$

where the q_{α}'s are the nuclear *Cartesian* coordinates. Using

$$\langle T_{\text{el}}\rangle + \langle V_{\text{el}}\rangle = E_{\text{el}} \tag{14.84}$$

we can eliminate either $\langle T_{\text{el}}\rangle$ or $\langle V_{\text{el}}\rangle$ from (14.83), which is the molecular form of the virial theorem.

Now consider a diatomic molecule. The electronic energy is a function of R, the internuclear distance: $E_{\text{el}} = E_{\text{el}}(R)$. The summation in (14.83) is over the nuclear Cartesian coordinates $x_a, y_a, z_a, x_b, y_b, z_b$. We have

$$\frac{\partial E_{\text{el}}}{\partial x_a} = \frac{dE_{\text{el}}}{dR}\frac{\partial R}{\partial x_a}, \quad \frac{\partial E_{\text{el}}}{\partial x_b} = \frac{dE_{\text{el}}}{dR}\frac{\partial R}{\partial x_b}$$

$$R = [(x_a - x_b)^2 + (y_a - y_b)^2 + (z_a - z_b)^2]^{1/2}$$

$$\frac{\partial R}{\partial x_a} = \frac{x_a - x_b}{R}, \quad \frac{\partial R}{\partial x_b} = \frac{x_b - x_a}{R} \tag{14.85}$$

with similar equations for the y and z coordinates. The sum in (14.83) becomes

$$\sum_{\alpha} q_{\alpha}\frac{\partial E_{\text{el}}}{\partial q_{\alpha}} = \frac{1}{R}\frac{dE_{\text{el}}}{dR}[x_a(x_a - x_b) + x_b(x_b - x_a) + y_a(y_a - y_b) + y_b(y_b - y_a) + z_a(z_a - z_b) + z_b(z_b - z_a)]$$

$$\sum_{\alpha} q_{\alpha}\frac{\partial E_{\text{el}}}{\partial q_{\alpha}} = R\frac{dE_{\text{el}}}{dR}$$

where $(x_a - x_b)^2 + (y_a - y_b)^2 + (z_a - z_b)^2 = R^2$ was used. The virial theorem (14.83) for a diatomic molecule becomes

$$2\langle T_{\text{el}}\rangle = -\langle V_{\text{el}}\rangle - R\frac{dE_{\text{el}}}{dR} \tag{14.86}$$

Using (14.84), we have the two alternative forms

$$\langle T_{\text{el}}\rangle = -E_{\text{el}} - R\frac{dE_{\text{el}}}{dR} \tag{14.87}$$

$$\langle V_{\text{el}}\rangle = 2E_{\text{el}} + R\frac{dE_{\text{el}}}{dR} \tag{14.88}$$

In deriving the molecular electronic virial theorem (14.83), we omitted the internuclear repulsion

$$V_{NN} = \sum_{\beta}\sum_{\alpha>\beta} \frac{Z_\alpha Z_\beta e'^2}{[(x_\alpha - x_\beta)^2 + (y_\alpha - y_\beta)^2 + (z_\alpha - z_\beta)^2]^{1/2}} \tag{14.89}$$

from the electronic Hamiltonian (14.77) to (14.79). Let

$$V = V_{\text{el}} + V_{NN}$$

where V_{el} is given by (14.79). We can rewrite the electronic Schrödinger equation $\hat{H}_{\text{el}}\psi_{\text{el}} = E_{\text{el}}\psi_{\text{el}}$ as [Eq. (13.4)]

$$(\hat{T}_{\text{el}} + \hat{V})\psi_{\text{el}} = U(q_\alpha)\psi_{\text{el}}$$

where

$$U(q_\alpha) = E_{\text{el}}(q_\alpha) + V_{NN}$$

$U(q_\alpha)$ is the potential-energy function for nuclear motion. Consider what happens to the right side of (14.83) when we add V_{NN} to V_{el} and E_{el}. We have

$$-\int \psi_{\text{el}}^*(\hat{V}_{\text{el}} + V_{NN})\psi_{\text{el}}\, d\tau_{\text{el}} - \sum_\alpha q_\alpha \frac{\partial U}{\partial q_\alpha}$$

$$= -\langle\psi_{\text{el}}|\hat{V}_{\text{el}}|\psi_{\text{el}}\rangle - V_{NN} - \sum_\alpha q_\alpha \frac{\partial E_{\text{el}}}{\partial q_\alpha} - \sum_\alpha q_\alpha \frac{\partial V_{NN}}{\partial q_\alpha} \tag{14.90}$$

Since V_{NN} is a homogeneous function of the nuclear Cartesian coordinates of degree -1, Euler's theorem gives

$$\sum_\alpha q_\alpha \frac{\partial V_{NN}}{\partial q_\alpha} = -V_{NN}$$

and (14.90) becomes

$$-\langle\psi_{\text{el}}|\hat{V}_{\text{el}} + V_{NN}|\psi_{\text{el}}\rangle - \sum_\alpha q_\alpha \frac{\partial U}{\partial q_\alpha} = -\langle\psi_{\text{el}}|\hat{V}_{\text{el}}|\psi_{\text{el}}\rangle - \sum_\alpha q_\alpha \frac{\partial E_{\text{el}}}{\partial q_\alpha} \tag{14.91}$$

Substitution of (14.91) into (14.83) gives

$$2\langle\psi_{\text{el}}|\hat{T}_{\text{el}}|\psi_{\text{el}}\rangle = -\langle\psi_{\text{el}}|\hat{V}_{\text{el}} + V_{NN}|\psi_{\text{el}}\rangle - \sum_{\alpha} q_{\alpha}\frac{\partial U}{\partial q_{\alpha}}$$

$$2\langle T_{\text{el}}\rangle = -\langle V\rangle - \sum_{\alpha} q_{\alpha}\frac{\partial U}{\partial q_{\alpha}} \tag{14.92}$$

which has the same form as (14.83), so the molecular electronic virial theorem holds whether or not we include the internuclear repulsion. Corresponding to Eqs. (14.86) to (14.88) for diatomic molecules, we have

$$2\langle T_{\text{el}}\rangle = -\langle V\rangle - R(dU/dR) \tag{14.93}$$

$$\langle T_{\text{el}}\rangle = -U - R(dU/dR) \tag{14.94}$$

$$\langle V\rangle = 2U + R(dU/dR) \tag{14.95}$$

The potential energy $V = V_{\text{el}} + V_{NN}$ takes the zero of energy with all particles (electrons and nuclei) at infinite separation from one another. Therefore, $U(R)$ in (14.93) to (14.95) does not go to zero at $R = \infty$ but goes to the sum of the energies of the separated atoms, which is negative.

The true wave functions for a system with V a homogeneous function of the coordinates must satisfy the form of the virial theorem (14.70). What determines whether an *approximate* wave function for such a system satisfies (14.70)? The answer is that, by inserting a variational parameter as a multiplier of each Cartesian coordinate and choosing this parameter to minimize the variational integral, we can make any trial variation function satisfy the virial theorem. (For the proof, see *Kauzmann*, page 229.) This process is called *scaling*, and the variational parameter multiplying each coordinate is called a *scale factor*. For a molecular trial function, the scaling parameter must be inserted in front of the nuclear Cartesian coordinates, as well as in front of the electronic coordinates.

Consider some examples. The zeroth-order perturbation wave function (9.49) for the heliumlike atom has no scale factor and so does not satisfy the virial theorem. If we were to calculate $\langle T\rangle$ and $\langle V\rangle$ for (9.49), we would find $2\langle T\rangle \neq -\langle V\rangle$; see Prob. 14.26. The Heitler–London trial function for H_2, Eq. (13.100), has no scale factor and does not satisfy the virial theorem. The Heitler–London–Wang function, which uses a variationally determined orbital exponent, satisfies the virial theorem. Hartree–Fock wave functions satisfy the virial theorem; note the scale factor in the Slater basis functions (11.14).

14.5 THE VIRIAL THEOREM AND CHEMICAL BONDING

We now use the virial theorem to examine the changes in electronic kinetic and potential energy that occur when a covalent chemical bond is formed in a diatomic molecule. For formation of a stable bond, the $U(R)$ curve must have a substantial minimum. At this minimum we have

$$\left.\frac{dU}{dR}\right|_{R_e} = 0 \tag{14.96}$$

and Eqs. (14.93) to (14.95) become

$$2\langle T_{\text{el}}\rangle|_{R_e} = -\langle V\rangle|_{R_e} \tag{14.97}$$

$$\langle T_{\text{el}}\rangle|_{R_e} = -U(R_e) \tag{14.98}$$

$$\langle V\rangle|_{R_e} = 2U(R_e) \tag{14.99}$$

These equations resemble those for atoms [Eqs. (14.75) and (14.76)]. At $R = \infty$ we have the separated atoms, and the atomic virial theorem gives

$$2\langle T_{\text{el}}\rangle|_{\infty} = -\langle V\rangle|_{\infty}, \qquad \langle T_{\text{el}}\rangle|_{\infty} = -U(\infty), \qquad \langle V\rangle|_{\infty} = 2U(\infty) \tag{14.100}$$

$U(\infty)$ is the sum of the energies of the two separated atoms. Equations (14.98)–(14.100) give

$$\langle T_{\text{el}}\rangle|_{R_e} - \langle T_{\text{el}}\rangle|_{\infty} = U(\infty) - U(R_e) \tag{14.101}$$

$$\langle V\rangle|_{R_e} - \langle V\rangle|_{\infty} = 2[U(R_e) - U(\infty)] \tag{14.102}$$

For bonding, we have $U(R_e) < U(\infty)$. Therefore, Eqs. (14.101) and (14.102) show that the average molecular potential energy at R_e is *less* than the sum of the potential energies of the separated atoms, whereas the average molecular kinetic energy is *greater* at R_e than at ∞. The decrease in potential energy is twice the increase in kinetic energy, and results from allowing the electrons to feel the attractions of both nuclei and perhaps from an increase in orbital exponents in the molecule (see Section 13.5). The equilibrium dissociation energy (13.9) is $D_e = \frac{1}{2}[\langle V\rangle|_{\infty} - \langle V\rangle|_{R_e}]$.

Consider the behavior of the average potential and kinetic energies for large R. The forces between uncharged atoms or molecules (other than those due to bond formation) are called *van der Waals forces*. For two neutral atoms, at least one of which is in an S state, quantum-mechanical perturbation theory shows that the van der Waals force of attraction is proportional to $1/R^7$, and the potential energy behaves like

$$U(R) \approx U(\infty) - \frac{A}{R^6}, \qquad R \text{ large} \tag{14.103}$$

where A is a positive constant. (See *Kauzmann*, Chapter 13.) This expression was first derived by London, and van der Waals forces between neutral atoms are called *London forces* or *dispersion forces*. (Recall the discussion near the end of Section 13.7.)

Substitution of (14.103) for U and dU/dR into (14.94) and (14.95), and use of (14.100) gives

$$\langle V\rangle \approx \langle V\rangle|_{\infty} + \frac{4A}{R^6}, \qquad \langle T_{\text{el}}\rangle \approx \langle T_{\text{el}}\rangle|_{\infty} - \frac{5A}{R^6}, \qquad R \text{ large} \tag{14.104}$$

Hence, as R decreases from infinity, the average potential energy at first increases, while the average kinetic energy at first decreases. The combination of these conclusions with our conclusions about $\langle V\rangle|_{R_e}$ and $\langle T_{\text{el}}\rangle|_{R_e}$ shows that $\langle V\rangle$ must go through a maximum somewhere between R_e and infinity and $\langle T_{\text{el}}\rangle$ must go through a minimum in this region.

Now consider small values of R. One can treat a diatomic molecule by applying perturbation theory to the united atom (UA) formed by merging the two atoms of the molecule. The perturbation is the difference between the molecular and the united-atom Hamiltonians: $H' = \hat{H}_{\text{mol}} - \hat{H}_{\text{UA}}$. One finds that the molecular purely electronic energy has the following form at small R [W. A. Bingel, *J. Chem. Phys.*, **30**, 1250 (1959); I. N. Levine, *J Chem. Phys.*, **40**, 3444 (1964); **41**, 2044 (1965); W. Byers Brown and E. Steiner, *J. Chem. Phys.* **44**, 3934 (1966)]:

$$E_{\text{el}} = E_{\text{UA}} + aR^2 + bR^3 + cR^4 + dR^5 + eR^5 \ln R + \cdots \tag{14.105}$$

where E_{UA} is the united-atom energy and a, b, c, d, e are constants. For $R \ll R_e$, we can use (14.105) and $U = E_{\text{el}} + V_{NN}$ [Eq. (13.8)] to write

$$U(R) \approx \frac{Z_a Z_b e'^2}{R} + E_{\text{UA}} + aR^2, \qquad R \text{ small} \tag{14.106}$$

The virial theorem then gives

$$\langle T_{\text{el}} \rangle \approx -E_{\text{UA}} - 3aR^2, \qquad R \text{ small}$$

$$\langle V \rangle \approx \frac{Z_a Z_b e'^2}{R} + 2E_{\text{UA}} + 4aR^2, \qquad R \text{ small}$$

The virial theorem (14.76) holds for the united atom: $\langle T_{\text{el}} \rangle|_0 = -E_{\text{UA}}$ and $\langle V_{\text{el}} \rangle|_0 = 2E_{\text{UA}}$. Therefore,

$$\langle T_{\text{el}} \rangle \approx \langle T_{\text{el}} \rangle|_0 - 3aR^2, \qquad R \text{ small} \tag{14.107}$$

$$\langle V \rangle \approx \frac{Z_a Z_b e'^2}{R} + \langle V_{\text{el}} \rangle|_0 + 4aR^2, \qquad R \text{ small} \tag{14.108}$$

$\langle V \rangle$ goes to infinity as R goes to zero, because of the internuclear repulsion.

Having found the general behavior of $\langle V \rangle$ and $\langle T_{\text{el}} \rangle$ as functions of R, we now draw Fig. 14.1. This figure is not for any particular molecule but resembles the known curves for H_2 and H_2^+ [W. Kolos and L. Wolniewicz, *J. Chem. Phys.*, **41**, 3663 (1964); *Slater, Quantum Theory of Molecules and Solids*, Volume 1, p. 36]. Similar curves hold for other diatomic molecules [see Fig. 1 in J. Hernandez-Trujillo et al., *Faraday Discuss.*, **135**, 79 (2007)].

How can we explain the changes in average kinetic and potential energy with R? Consider H_2^+. The electronic potential-energy function is

$$V_{\text{el}} = -\frac{e'^2}{r_a} - \frac{e'^2}{r_b} \tag{14.109}$$

If we plot V_{el} for points on the molecular axis for a large value of R, we get a curve like Fig. 14.2, which resembles two hydrogen-atom potential-energy curves (Fig. 6.6) placed side by side. We saw that the overlapping of the 1s AOs occurring in molecule formation increases the charge probability density between the nuclei for the ground state. However, Fig. 14.2 shows that the potential energy is relatively *high* in the region mid-way between the nuclei when R is large. Thus $\langle V \rangle$ initially increases as R decreases from infinity. Now consider the kinetic energy. The uncertainty principle (5.13) gives

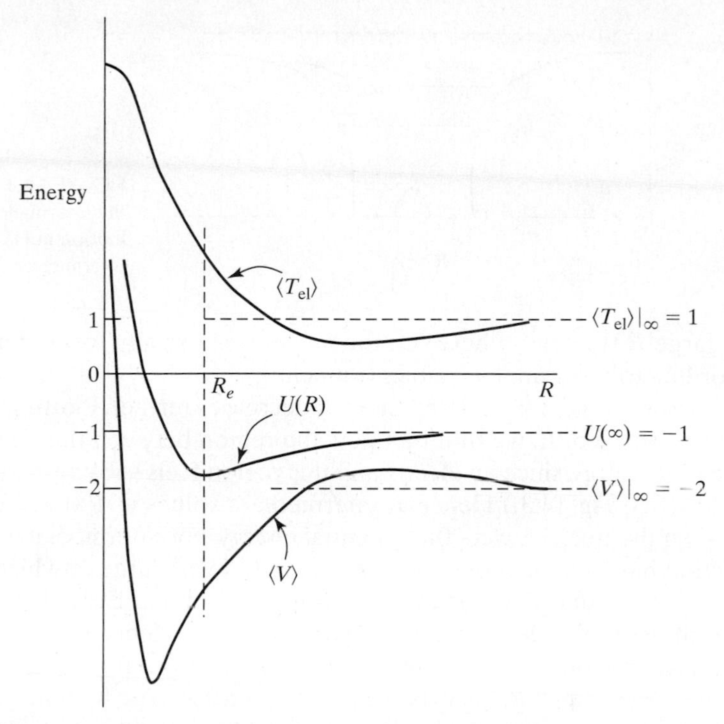

FIGURE 14.1 Variation of the average potential and kinetic energies of a diatomic molecule. The unit of energy is taken as the electronic kinetic energy of the separated atoms.

$(\Delta x)^2(\Delta p_x)^2 \geq \hbar^2/4$. For a stationary state, $\langle p_x \rangle$ is zero [see Eq. (3.92) and Prob. 14.28] and (5.11) gives $(\Delta p_x)^2 = \langle p_x^2 \rangle$. Hence a small value of $(\Delta x)^2$ means a large value of $\langle p_x^2 \rangle$ and a large value of the average kinetic energy (which equals $\langle p^2 \rangle/2m$). Thus a compact ψ_{el} corresponds to a large electronic kinetic energy. In the separated atoms, the wave function is concentrated in two rather small regions about each nucleus (Fig. 6.7). In the initial stages of molecule formation, the buildup of probability density between the nuclei corresponds to having a wave function that is less compact than it was in the separated atoms. Thus, as R decreases from infinity, the electronic kinetic energy initially decreases. The energies E_{el} of the two lowest H_2^+ states have been indicated in Fig. 14.2.

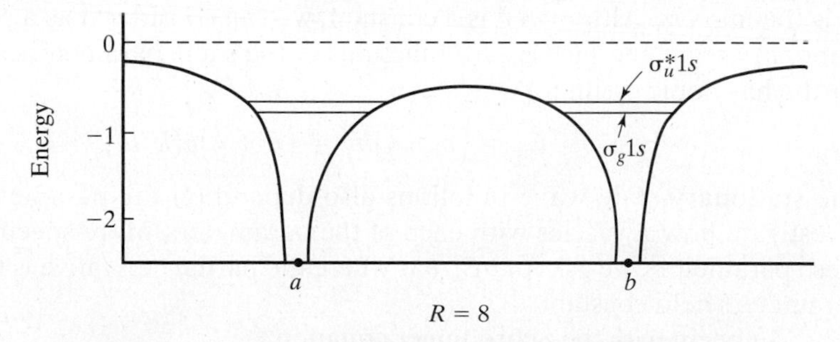

FIGURE 14.2 Potential energy along the internuclear axis for electronic motion in H_2^+ at a large internuclear separation. Atomic units are used.

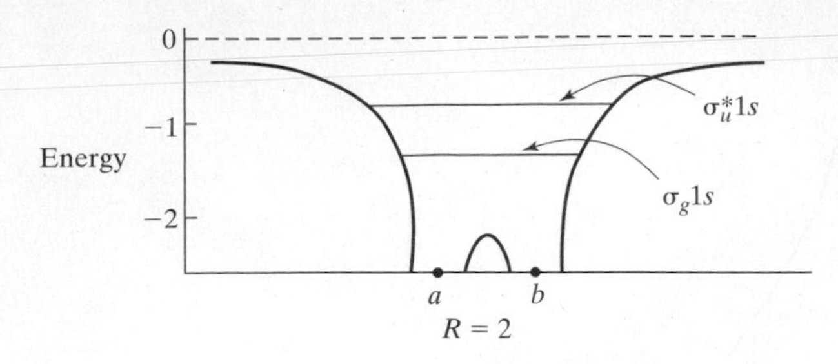

FIGURE 14.3 Potential energy along the internuclear axis for electronic motion in H_2^+ at an intermediate internuclear distance.

For large R the region between the nuclei is classically forbidden, but it is accessible according to quantum mechanics (tunneling).

Now consider what happens as R decreases further. Plotting (14.109) for an intermediate value of R, we find that now the region between the nuclei is a region of *low* potential energy, since an electron in this region feels substantial attractions from both nuclei. (See Fig. 14.3.) Hence at intermediate values of R the overlap charge buildup between the nuclei lowers the potential energy. For intermediate values of R, the wave function has become more compact compared with large R, which gives an increase in $\langle T_{\text{el}}\rangle$ as R is reduced. In fact, we see from Fig. 14.1 and Eq. (14.101) that $\langle T_{\text{el}}\rangle$ is *greater* at R_e in the molecule than in the separated atoms. Hence the molecular wave function at R_e is more compact than the separated-atoms wave functions.

For very small R, the average potential energy goes to infinity, because of the internuclear repulsion. *However*, for $R = R_e$, Fig. 14.1 shows that $\langle V\rangle$ is still decreasing sharply with decreasing R, and it is the increase in $\langle T_{\text{el}}\rangle$, and not the nuclear repulsion, that causes the $U(R)$ curve to turn up as R becomes less than R_e. The squeezing of the molecular wave function into a smaller region with the associated increase in $\langle T_{\text{el}}\rangle$ is more important than the internuclear repulsion in causing the initial repulsion between the atoms.

14.6 THE HELLMANN–FEYNMAN THEOREM

Consider a system with a time-independent Hamiltonian $\hat{H}$ that involves parameters. An obvious example is the molecular electronic Hamiltonian (13.5), which depends parametrically on the nuclear coordinates. However, the Hamiltonian of any system contains parameters. For example, in the one-dimensional harmonic-oscillator Hamiltonian operator $-(\hbar^2/2m)(d^2/dx^2) + \frac{1}{2}kx^2$, the force constant k is a parameter, as is the mass m. Although $\hbar$ is a constant, we can consider it as a parameter also. The stationary-state energies E_n are functions of the same parameters as $\hat{H}$. For example, for the harmonic oscillator

$$E_n = \left(v + \tfrac{1}{2}\right)h\nu = \left(v + \tfrac{1}{2}\right)\hbar(k/m)^{1/2} \tag{14.110}$$

The stationary-state wave functions also depend on the parameters in $\hat{H}$. We now investigate how E_n varies with each of the parameters. More specifically, if λ is one of these parameters, we ask for $\partial E_n/\partial\lambda$, where the partial derivative is taken with all other parameters held constant.

We begin with the Schrödinger equation

$$\hat{H}\psi_n = E_n\psi_n \tag{14.111}$$

where the ψ_n's are the normalized stationary-state eigenfunctions. Because of normalization, we have

$$E_n = \int \psi_n^* \hat{H} \psi_n \, d\tau \tag{14.112}$$

$$\frac{\partial E_n}{\partial \lambda} = \frac{\partial}{\partial \lambda} \int \psi_n^* \hat{H} \psi_n \, d\tau \tag{14.113}$$

The integral in (14.112) is a definite integral over all space, and its value depends parametrically on λ since $\hat{H}$ and ψ_n depend on λ. Provided the integrand is well behaved, we can find the integral's derivative with respect to a parameter by differentiating the integrand with respect to the parameter and then integrating. Thus

$$\frac{\partial E_n}{\partial \lambda} = \int \frac{\partial}{\partial \lambda}(\psi_n^* \hat{H} \psi_n) \, d\tau = \int \frac{\partial \psi_n^*}{\partial \lambda} \hat{H} \psi_n \, d\tau + \int \psi_n^* \frac{\partial}{\partial \lambda}(\hat{H} \psi_n) \, d\tau \tag{14.114}$$

We have

$$\frac{\partial}{\partial \lambda}(\hat{H}\psi_n) = \frac{\partial}{\partial \lambda}(\hat{T}\psi_n) + \frac{\partial}{\partial \lambda}(\hat{V}\psi_n) \tag{14.115}$$

The potential-energy operator is just multiplication by V, so

$$\frac{\partial}{\partial \lambda}(\hat{V}\psi_n) = \frac{\partial V}{\partial \lambda}\psi_n + V \frac{\partial \psi_n}{\partial \lambda} \tag{14.116}$$

The parameter λ will occur in the kinetic-energy operator as part of the factor multiplying one or more of the derivatives with respect to the coordinates. For example, taking λ as the mass of the particle, we have for a one-particle problem

$$\hat{T} = -\frac{\hbar^2}{2\lambda}\left(\frac{\partial^2}{\partial x^2} + \frac{\partial^2}{\partial y^2} + \frac{\partial^2}{\partial z^2}\right)$$

$$\frac{\partial}{\partial \lambda}(\hat{T}\psi) = -\frac{\hbar^2}{2}\frac{\partial}{\partial \lambda}\left[\frac{1}{\lambda}\left(\frac{\partial^2 \psi}{\partial x^2} + \frac{\partial^2 \psi}{\partial y^2} + \frac{\partial^2 \psi}{\partial z^2}\right)\right]$$

$$= \frac{\hbar^2}{2\lambda^2}\left(\frac{\partial^2 \psi}{\partial x^2} + \frac{\partial^2 \psi}{\partial y^2} + \frac{\partial^2 \psi}{\partial z^2}\right) - \frac{\hbar^2}{2\lambda}\left(\frac{\partial^2}{\partial x^2} + \frac{\partial^2}{\partial y^2} + \frac{\partial^2}{\partial z^2}\right)\left(\frac{\partial \psi}{\partial \lambda}\right)$$

since we can change the order of the partial differentiations without affecting the result. We can write this last equation as

$$\frac{\partial}{\partial \lambda}(\hat{T}\psi_n) = \left(\frac{\partial \hat{T}}{\partial \lambda}\right)\psi_n + \hat{T}\left(\frac{\partial \psi_n}{\partial \lambda}\right) \tag{14.117}$$

where $\partial \hat{T}/\partial \lambda$ is found by differentiating $\hat{T}$ with respect to λ just as if it were a function instead of an operator. Although we got (14.117) by considering a specific $\hat{T}$ and λ, the same arguments show it to be generally valid. Combining (14.116) and (14.117), we write

$$\frac{\partial}{\partial \lambda}(\hat{H}\psi_n) = \left(\frac{\partial \hat{H}}{\partial \lambda}\right)\psi_n + \hat{H}\left(\frac{\partial \psi_n}{\partial \lambda}\right) \tag{14.118}$$

Equation (14.114) becomes

$$\frac{\partial E_n}{\partial \lambda} = \int \frac{\partial \psi_n^*}{\partial \lambda} \hat{H}\psi_n \, d\tau + \int \psi_n^* \frac{\partial \hat{H}}{\partial \lambda}\psi_n \, d\tau + \int \psi_n^* \hat{H} \frac{\partial \psi_n}{\partial \lambda} \, d\tau \tag{14.119}$$

For the first integral in (14.119), we have

$$\int \frac{\partial \psi_n^*}{\partial \lambda} \hat{H}\psi_n \, d\tau = E_n \int \frac{\partial \psi_n^*}{\partial \lambda} \psi_n \, d\tau \tag{14.120}$$

The Hermitian property of $\hat{H}$ and (14.111) give for the last integral in (14.119)

$$\int \psi_n^* \hat{H} \frac{\partial \psi_n}{\partial \lambda} \, d\tau = \int \frac{\partial \psi_n}{\partial \lambda} (\hat{H}\psi_n)^* \, d\tau = E_n \int \psi_n^* \frac{\partial \psi_n}{\partial \lambda} \, d\tau$$

Therefore,

$$\frac{\partial E_n}{\partial \lambda} = \int \psi_n^* \frac{\partial \hat{H}}{\partial \lambda}\psi_n \, d\tau + E_n \int \frac{\partial \psi_n^*}{\partial \lambda} \psi_n \, d\tau + E_n \int \psi_n^* \frac{\partial \psi_n}{\partial \lambda} \, d\tau \tag{14.121}$$

The wave function is normalized, so

$$\int \psi_n^* \psi_n \, d\tau = 1, \qquad \frac{\partial}{\partial \lambda} \int \psi_n^* \psi_n \, d\tau = 0$$

$$\int \frac{\partial \psi_n^*}{\partial \lambda} \psi_n \, d\tau + \int \psi_n^* \frac{\partial \psi_n}{\partial \lambda} \, d\tau = 0 \tag{14.122}$$

Using (14.122) in (14.121), we obtain

$$\boxed{\frac{\partial E_n}{\partial \lambda} = \int \psi_n^* \frac{\partial \hat{H}}{\partial \lambda}\psi_n \, d\tau} \tag{14.123}$$

Equation (14.123) is the **(generalized) Hellmann–Feynman theorem**. [For a discussion of the origin of the Hellmann–Feynman and related theorems, see J. I. Musher, *Am. J. Phys.*, **34**, 267 (1966).]

EXAMPLE Apply the generalized Hellmann–Feynman theorem to the one-dimensional harmonic oscillator with λ taken as the force constant.

For the harmonic oscillator, $\hat{H} = -(\hbar^2/2m)(d^2/dx^2) + \frac{1}{2}kx^2$ and $\partial\hat{H}/\partial k = \frac{1}{2}x^2$. The energy levels are $E_v = (v + \frac{1}{2})h\nu = (v + \frac{1}{2})h(k/m)^{1/2}/2\pi$. We have $\partial E_v/\partial k = \frac{1}{2}(v + \frac{1}{2})hk^{-1/2}m^{-1/2}/2\pi = \frac{1}{2}(v + \frac{1}{2})h\nu/k$. Substitution in (14.123) gives

$$\int_{-\infty}^{\infty} \psi_v^* x^2 \psi_v \, dx = \left(v + \tfrac{1}{2}\right)h\nu/k \tag{14.124}$$

We have found $\langle x^2 \rangle$ for any harmonic-oscillator stationary state without evaluating any integrals. This result was also obtained from the virial theorem; see Eq. (14.74). For a third derivation, see *Eyring, Walter, and Kimball*, p. 79.

The derivation of the Hellmann–Feynman theorem assumes that $\partial\psi/\partial\lambda$ exists. For a state belonging to a degenerate energy level, this assumption may not be true and (14.123) need not hold. Changing the parameter's value from λ to $\lambda + d\lambda$ amounts to applying a perturbation $\hat{H}' \equiv \hat{H}(\lambda + d\lambda) - \hat{H}(\lambda) \approx (\partial\hat{H}/\partial\lambda)\,d\lambda$. This perturbation changes ψ from $\psi(\lambda)$ to $\psi(\lambda + d\lambda)$. If $\psi(\lambda)$ is not one of the correct zeroth-order wave functions (9.74) for the perturbation $\hat{H}'$, then $\psi(\lambda)$ need not equal $\lim_{d\lambda\to 0}\psi(\lambda + d\lambda)$, so ψ will make a discontinuous jump at $d\lambda = 0$ and $\partial\psi/\partial\lambda$ will not exist at this point. *The Hellmann–Feynman theorem (14.123) applies to the wave functions of a degenerate level only if we use the correct zeroth-order wave functions for the perturbation* $\hat{H}'$. These correct functions are found by solving (9.84) and (9.82) with $\hat{H}' \equiv (\partial\hat{H}/\partial\lambda)\,d\lambda$. [Since $E_n^{(1)}$ will be proportional to $d\lambda$, we can replace $\hat{H}'$ with $\partial\hat{H}/\partial\lambda$ and $E_n^{(1)}$ with $E_n^{(1)}/d\lambda$ in (9.84) and (9.82).] For further details, see references 4–7 in G. P. Zhang and T. F. George, *Phys. Rev. B*, **69**, 167102 (2004).

Application of the Hellmann–Feynman theorem to the hydrogenlike atom, with Z as the parameter, gives (Prob. 14.34a)

$$\int r^{-1}|\psi|^2\,d\tau = \left\langle\frac{1}{r}\right\rangle = \frac{Z}{n^2}\left(\frac{1}{a}\right) \qquad (14.125)$$

This result was also obtained from the virial theorem; see Eq. (14.76).

The Hellmann–Feynman theorem with E_n being the Hartree–Fock energy is obeyed by Hartree–Fock (as well as exact) wave functions. [See R. E. Stanton, *J. Chem. Phys.*, **36**, 1298 (1962).]

14.7 THE ELECTROSTATIC THEOREM

Hellmann and Feynman independently applied Eq. (14.123) to molecules, taking λ as a nuclear Cartesian coordinate. We now consider their results.

As usual, we are using the Born–Oppenheimer approximation, solving the electronic Schrödinger equation for a fixed nuclear configuration [Eqs. (13.4) to (13.6)]:

$$(\hat{T}_{\text{el}} + \hat{V})\psi_{\text{el}} = (\hat{T}_{\text{el}} + \hat{V}_{\text{el}} + \hat{V}_{NN})\psi_{\text{el}} = U\psi_{\text{el}}$$

where $\hat{T}_{\text{el}}, \hat{V}_{\text{el}}$, and $\hat{V}_{NN}$ are given by (14.78), (14.79), and (14.89). The Hamiltonian operator $\hat{T}_{\text{el}} + \hat{V}_{\text{el}} + \hat{V}_{NN}$ depends on the nuclear coordinates as parameters. If x_δ is the x coordinate of nucleus δ, the generalized Hellmann–Feynman theorem (14.123) gives

$$\frac{\partial U}{\partial x_\delta} = \int \psi_{\text{el}}^* \frac{\partial(\hat{T}_{\text{el}} + \hat{V}_{\text{el}} + \hat{V}_{NN})}{\partial x_\delta}\psi_{\text{el}}\,d\tau_{\text{el}} = \int \psi_{\text{el}}^*\left(\frac{\partial V_{\text{el}}}{\partial x_\delta} + \frac{\partial V_{NN}}{\partial x_\delta}\right)\psi_{\text{el}}\,d\tau_{\text{el}} \qquad (14.126)$$

since $\hat{T}_{\text{el}}$ is independent of the nuclear Cartesian coordinates, as can be seen from (14.78). [If we had omitted V_{NN} from V, we would have obtained Eq. (14.82), which was used in deriving the molecular electronic virial theorem.] From (14.79) we get

$$\frac{\partial V_{\text{el}}}{\partial x_\delta} = -\sum_i \frac{Z_\delta(x_i - x_\delta)e'^2}{r_{i\delta}^3} \qquad (14.127)$$

where $r_{i\delta}$ is the distance from electron i to nucleus δ. To find $\partial V_{NN}/\partial x_\delta$, we need to consider only internuclear repulsion terms that involve nucleus δ. Hence

$$\frac{\partial V_{NN}}{\partial x_\delta} = \frac{\partial}{\partial x_\delta}\sum_{\alpha\neq\delta}\frac{Z_\alpha Z_\delta e'^2}{[(x_\alpha - x_\delta)^2 + (y_\alpha - y_\delta)^2 + (z_\alpha - z_\delta)^2]^{1/2}} = \sum_{\alpha\neq\delta} Z_\alpha Z_\delta e'^2\frac{x_\alpha - x_\delta}{R_{\alpha\delta}^3}$$

where $R_{\alpha\delta}$ is the distance between nuclei α and δ. Since $\partial V_{NN}/\partial x_\delta$ does not involve the electronic coordinates and ψ_{el} is normalized, (14.126) becomes

$$\frac{\partial U}{\partial x_\delta} = -Z_\delta e'^2\int|\psi_{\text{el}}|^2\sum_i\frac{x_i - x_\delta}{r_{i\delta}^3}\,d\tau_{\text{el}} + \sum_{\alpha\neq\delta}Z_\alpha Z_\delta e'^2\frac{x_\alpha - x_\delta}{R_{\alpha\delta}^3} \tag{14.128}$$

Consider the integral in (14.128). Using Eq. (14.8) with $B(\mathbf{r}_i) = (x_i - x_\delta)/r_{i\delta}^3$, we get

$$\frac{\partial U}{\partial x_\delta} = -Z_\delta e'^2\iiint\rho(x, y, z)\frac{x - x_\delta}{r_\delta^3}\,dx\,dy\,dz + \sum_{\alpha\neq\delta}Z_\alpha Z_\delta e'^2\frac{x_\alpha - x_\delta}{R_{\alpha\delta}^3} \tag{14.129}$$

The variable r_δ is the distance between nucleus δ and point (x, y, z) in space:

$$r_\delta = [(x - x_\delta)^2 + (y - y_\delta)^2 + (z - z_\delta)^2]^{1/2}$$

What is the significance of (14.129)? In the Born–Oppenheimer approximation, $U(x_\alpha, y_\alpha, z_\alpha, x_\beta, \ldots)$ is the potential-energy function for nuclear motion, the nuclear Schrödinger equation being

$$\left(-\frac{\hbar^2}{2}\sum_\alpha\frac{1}{m_\alpha}\nabla_\alpha^2 + U\right)\psi_N = E\psi_N \tag{14.130}$$

The quantity $-\partial U/\partial x_\delta$ can thus be viewed [see Eq. (5.31)] as the x component of the effective force on nucleus δ due to the other nuclei and the electrons. In addition to (14.129), we have two corresponding equations for $\partial U/\partial y_\delta$ and $\partial U/\partial z_\delta$. If $\mathbf{F}_\delta$ is the effective force on nucleus δ, then

$$\mathbf{F}_\delta = -\mathbf{i}\frac{\partial U}{\partial x_\delta} - \mathbf{j}\frac{\partial U}{\partial y_\delta} - \mathbf{k}\frac{\partial U}{\partial z_\delta} \tag{14.131}$$

$$\mathbf{F}_\delta = -Z_\delta e'^2\iiint\rho(x, y, z)\,\frac{\mathbf{r}_\delta}{r_\delta^3}\,dx\,dy\,dz + e'^2\sum_{\alpha\neq\delta}Z_\alpha Z_\delta\frac{\mathbf{R}_{\alpha\delta}}{R_{\alpha\delta}^3} \tag{14.132}$$

where $\mathbf{r}_\delta$ is the vector from point (x, y, z) to nucleus δ,

$$\mathbf{r}_\delta = \mathbf{i}(x_\delta - x) + \mathbf{j}(y_\delta - y) + \mathbf{k}(z_\delta - z) \tag{14.133}$$

and where $\mathbf{R}_{\alpha\delta}$ is the vector from nucleus α to nucleus δ:

$$\mathbf{R}_{\alpha\delta} = \mathbf{i}(x_\delta - x_\alpha) + \mathbf{j}(y_\delta - y_\alpha) + \mathbf{k}(z_\delta - z_\alpha)$$

Equation (14.132) has a simple physical interpretation. Let us imagine the electrons smeared out into a charge distribution whose density is $-e\rho(x, y, z)$. The force on nucleus δ exerted by the infinitesimal element of electronic charge $-e\rho\,dx\,dy\,dz$ is [Eq. (6.56)]

$$-Z_\delta e'^2\frac{\mathbf{r}_\delta}{r_\delta^3}\rho\,dx\,dy\,dz \tag{14.134}$$

and integration of (14.134) shows that the total force exerted on δ by this hypothetical electron smear is given by the first term on the right of (14.132). The second term on the right of (14.132) is clearly the Coulomb's law force on nucleus δ due to the electrostatic repulsions of the other nuclei.

Thus *the effective force acting on a nucleus in a molecule can be calculated by simple electrostatics as the sum of the Coulombic forces exerted by the other nuclei and by a hypothetical electron cloud whose charge density* $-e\rho(x, y, z)$ *is found by solving the electronic Schrödinger equation*. This is the **Hellmann–Feynman electrostatic theorem.** The electron probability density depends on the parameters defining the nuclear configuration: $\rho = \rho(x, y, z; x_\alpha, y_\alpha, z_\alpha, x_\beta, \ldots)$.

It is quite reasonable that the electrostatic theorem follows from the Born–Oppenheimer approximation, since the rapid motion of the electrons allows the electronic wave function and probability density to adjust immediately to changes in nuclear configuration. The rapid motion of the electrons causes the sluggish nuclei to "see" the electrons as a charge cloud, rather than as discrete particles. The fact that the effective forces on the nuclei are electrostatic affirms that there are no "mysterious quantum-mechanical forces" acting in molecules.

Let us consider the implications of the electrostatic theorem for chemical bonding in diatomic molecules. We take the internuclear axis as the z axis (Fig. 14.4). By symmetry the x and y components of the effective forces on the two nuclei are zero. [Also, one can show that the z force components on nuclei a and b are related by $F_{z,a} = -F_{z,b}$ (Prob. 14.37). The effective forces on nuclei a and b are equal in magnitude and opposite in direction.]

From (14.134) and (14.133), the z component of the effective force on nucleus a due to the element of electronic charge in the infinitesimal region about (x, y, z) is

$$-Z_a e'^2 \rho[(z_a - z)/r_a^3]\, dx\, dy\, dz = Z_a e'^2 \rho(\cos\theta_a/r_a^2)\, dx\, dy\, dz \qquad (14.135)$$

since $\cos\theta_a = (-z_a + z)/r_a$. ($z_a$ is negative.) Similarly, the z component of force on nucleus b due to this charge is

$$-Z_b e'^2 \rho(\cos\theta_b/r_b^2)\, dx\, dy\, dz \qquad (14.136)$$

A positive value of (14.135) or (14.136) corresponds to a force in the $+z$ direction, that is, to the right in Fig. 14.4. When the force on nucleus a is algebraically greater than the force on nucleus b, then the element of electronic charge tends to draw a toward b. Hence electronic charge that is binding is located in the region where

$$Z_a e'^2 \rho(\cos\theta_a/r_a^2)\, dx\, dy\, dz > -Z_b e'^2 \rho(\cos\theta_b/r_b^2)\, dx\, dy\, dz \qquad (14.137)$$

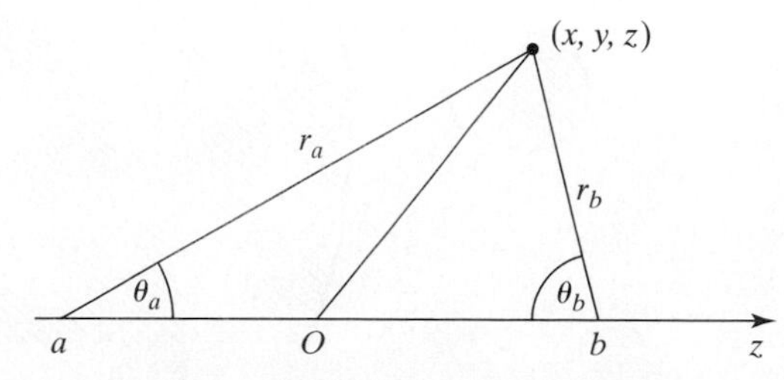

FIGURE 14.4 Coordinate system for a diatomic molecule. The origin is at O.

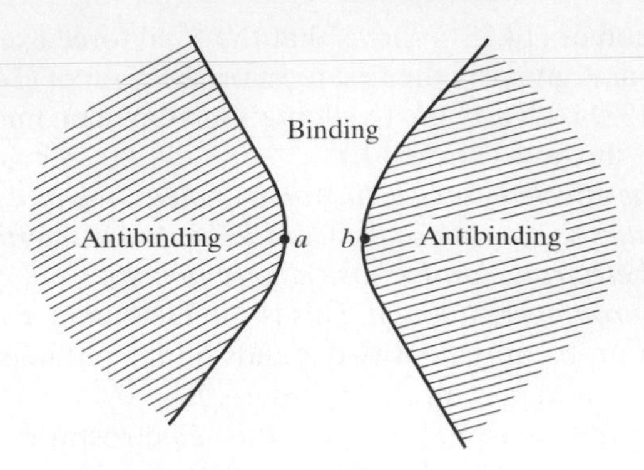

FIGURE 14.5 Cross section of binding and antibinding regions in a homonuclear diatomic molecule. To obtain the three-dimensional regions, rotate the figure about the internuclear axis.

Since the probability density ρ is nonnegative, division by ρ preserves the direction of the inequality sign, and the binding region of space is where

$$Z_a \frac{\cos \theta_a}{r_a^2} + Z_b \frac{\cos \theta_b}{r_b^2} > 0 \tag{14.138}$$

When the force on b is algebraically greater than that on a, the electronic charge element tends to draw b away from a. The antibinding region of space is thus characterized by a negative value for the left side of (14.138). The surfaces for which the left side of (14.138) equals zero divide space into the **binding** and **antibinding regions**. [T. Berlin, *J. Chem. Phys.*, **19**, 208 (1951); J. Hinze, *J. Chem. Phys.*, **101**, 6369 (1994); Berlin's ideas are extended to polyatomic molecules in T. Koga et al., *J. Am. Chem. Soc.*, **100**, 7522 (1978); X. Wang and Z. Peng, *Int. J. Quantum. Chem.*, **47**, 393 (1993).]

Figures 14.5 and 14.6 show the binding and antibinding regions for a homonuclear and a heteronuclear diatomic molecule. As might be expected, the binding region for a homonuclear diatomic molecule lies between the nuclei. Charge in this region tends to draw the nuclei together. Bonding leads to a transfer of charge probability density into the region between the nuclei because of the overlap between the bonding AOs.

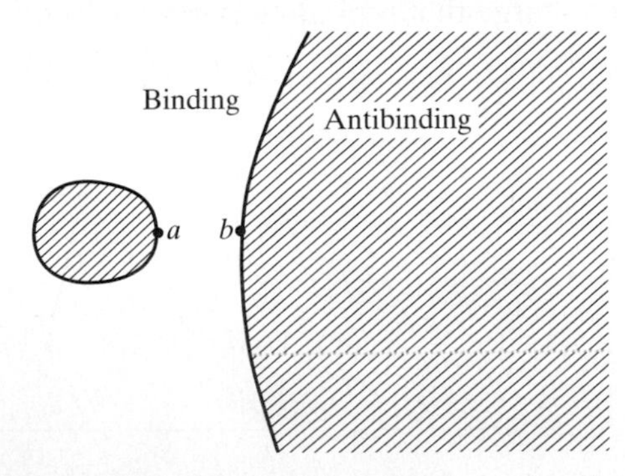

FIGURE 14.6 Binding and antibinding regions for a heteronuclear diatomic molecule with $Z_b > Z_a$.

Electronic charge that is "behind" the nuclei (to the left of nucleus a or to the right of nucleus b in Fig. 14.5) exerts a greater attraction on the nucleus that is nearer to it than on the other nucleus and thus tends to pull the nuclei apart.

Bader, Henneker, and Cade have taken the electron probability densities for homonuclear diatomic molecules at R_e as calculated from Hartree–Fock functions and subtracted off the probability densities for the corresponding separated atoms, as calculated from atomic Hartree–Fock wave functions [R. F. W. Bader, W. H. Henneker, and P. E. Cade, *J. Chem. Phys.*, **46**, 3341 (1967)]. They then plotted contours of this *density difference* $\Delta\rho$ for Li_2 through F_2. A contour with positive $\Delta\rho$ corresponds to a buildup in charge over that of the separated atoms. Not surprisingly they found a charge buildup in the central part of the binding region, between the nuclei. However, they also found that molecule formation is accompanied by a charge buildup in most of the volume of the antibinding regions. (The buildup is at the expense of charge probability density in the regions near the nuclei and containing the boundaries between binding and antibinding regions; see Fig. 3 of the paper.) Of course, the charge buildup between the nuclei is such that its contribution to the attractive force between the nuclei exceeds the Hellmann–Feynman repulsive force resulting from the charge buildup in the antibinding region. At $R = R_e$, the effective force on each nucleus is zero $(\partial U/\partial R|_{R_e} = 0)$, and the hypothetical electron smear must exert a net attractive Hellmann–Feynman force to counterbalance the internuclear repulsion. As mentioned (Section 13.14), Hartree–Fock wave functions provide accurate electron probability-density functions. However, the difference between the molecular and atomic ρ's is sensitive to small errors in the ρ's, and one might therefore question the accuracy of Hartree–Fock $\Delta\rho$ maps. Calculations on several diatomic molecules show that, when configuration-interaction wave functions are used in place of the Hartree–Fock wave functions, the $\Delta\rho$ contour values change by a few to 10% [M. E. Stephens and P. J. Becker, *Mol. Phys.*, **49**, 65 (1983); R. Moszynski and K. Szalewicz, *J. Phys. B*, **20**, 4347 (1987)]. This indicates that the Hartree–Fock $\Delta\rho$ maps are fairly accurate at R_e.

From the Hellmann–Feynman viewpoint, we seem to be considering chemical bonding solely in terms of potential energy, whereas the virial-theorem discussion involved both potential and kinetic energy. For the purposes of the Hellmann–Feynman discussion, we are imagining the electrons to be smeared out into a continuous charge distribution. Hence we make no reference to electronic kinetic energy. The use of the electrostatic theorem to explain chemical bonding has been criticized by some quantum chemists on the grounds that it hides the role of kinetic energy in bonding. [See the references cited after Eq. (13.66).]

In 1939, Feynman conjectured that the dispersion attraction between two molecules A and B at relatively large intermolecular distances is explainable as follows: The interactions between the two molecules cause the electron probability density of each molecule to be distorted and shifted somewhat toward the other molecule. The attractions of the nuclei in molecule A toward the distorted (polarized) electron density of molecule A and the attractions of the B nuclei toward the polarized B electron density then draw the two molecules together. In 1990, Hunt proved that the dispersion interaction between any two molecules in their ground electronic states results from the attractions of the nuclei in each molecule to the polarized electron density of the same molecule [K. L. C. Hunt, *J. Chem. Phys.*, **92**, 1180 (1990)].

For further applications of the Hellmann–Feynman electrostatic theorem, see B. M. Deb, ed., *The Force Concept in Chemistry*, Van Nostrand Reinhold, 1981.

14.8 SUMMARY

The electron probability density ρ of an n-electron molecule is found by summation of $|\psi|^2$ over spins, integration over the spatial coordinates of $n - 1$ electrons, and multiplication by n [Eq. (14.5)]. The dipole moment of a molecule is given by (14.21).

The best possible forms for the MOs ϕ_i of a molecule are the solutions of the Hartree–Fock equations $\hat{F}\phi_i = \varepsilon_i\phi_i$, where $\hat{F}$ is the Fock operator [Eqs. (14.26) to (14.29)] and ε_i is the orbital energy. To solve the Hartree–Fock equations, we expand ϕ_i using a set of basis functions: $\phi_i = \sum_s c_{si}\chi_s$; this leads to the Roothaan equations (14.34) for the coefficients c_{si} and orbital energies ε_i. Since the Fock operator $\hat{F}$ and its matrix elements F_{rs} [Eq. (14.41)] depend on the occupied MOs, which are unknown, the Roothaan equations are solved by an iterative process that starts with an initial guess for the occupied MOs.

For a bound stationary state, the quantum-mechanical virial theorem states that $2\langle T\rangle = \sum_i \langle q_i(\partial V/\partial q_i)\rangle$, where the sum is over the Cartesian coordinates of all the particles. If V is a homogeneous function of degree n, then $2\langle T\rangle = n\langle V\rangle$. For a diatomic molecule, the virial theorem becomes $\langle T_{\text{el}}\rangle = -U - R(dU/dR)$ and $\langle V\rangle = 2U + R(dU/dR)$, where $U(R)$ is the potential-energy function for nuclear motion. The virial theorem shows that at R_e, $\langle V\rangle$ of a diatomic molecule is less than the total $\langle V\rangle$ of the separated atoms, and $\langle T_{\text{el}}\rangle$ is greater than the total $\langle T_{\text{el}}\rangle$ of the separated atoms.

For a bound stationary state, the generalized Hellmann–Feynman theorem is $\partial E_n/\partial\lambda = \int \psi_n^*(\partial\hat{H}/\partial\lambda)\psi_n\, d\tau$, where λ is a parameter in the Hamiltonian. (In case of degeneracy, ψ_n must be a correct zeroth-order wave function for the perturbation of changing λ.) Taking λ as a nuclear coordinate, we are led to the Hellmann–Feynman electrostatic theorem, which states that the force on a nucleus in a molecule is the sum of the electrostatic forces exerted by the other nuclei and the electron charge density.

PROBLEMS

Sec.	14.1	14.2	14.3	14.4	14.5	14.6	14.7
Probs.	14.1–14.2	14.3–14.5	14.6–14.22	14.23–14.28	14.29–14.33	14.34–14.36	14.37–14.38

14.1 Derive (14.6) for ρ_{VB} and ρ_{MO}.

14.2 Show that ρ_{MO} of (14.6) is greater than ρ_{VB} of (14.6) at the midpoint of the line joining the nuclei.

14.3 Show that the dipole moment (14.9) of a system of charges is independent of the choice of the coordinate origin, *provided* the system has no net charge.

14.4 (a) Explain why the permanent dipole moment of a many-electron atom in a stationary state is always zero. (b) Explain why the permanent electric dipole moment of H can be nonzero for certain excited states. (c) Show qualitatively that two of the four correct zeroth-order functions of Prob. 9.22 give nonzero permanent dipole moments.

14.5 For NaCl, $R_e = 2.36$ Å. The ionization energy of Na is 5.14 eV, and the electron affinity of Cl is 3.61 eV. Use the simple model of NaCl as a pair of spherical ions in contact to estimate D_e and the dipole moment of NaCl. Compare with the experimental values $D_e = 4.25$ eV and $\mu = 9.0$ D. [One debye (D) is 3.33564×10^{-30} C m.]

14.6 Prove that the one-electron Hartree–Fock operator (14.26) is Hermitian.

14.7 Explain the origin of the extra terms in the molecular Hartree–Fock operator (14.26) as compared with the atomic Hartree operator of (11.9) and (11.7).

14.8 Verify that the Coulomb and exchange integrals J_{ij} and K_{ij} can be written in terms of the Coulomb and exchange operators of Section 14.3 as

$$J_{ij} = \langle\phi_i(1)|\hat{J}_j(1)|\phi_i(1)\rangle, \qquad K_{ij} = \langle\phi_i(1)|\hat{K}_j(1)|\phi_i(1)\rangle$$

14.9 Verify Eq. (14.40) for the $\hat{K}_j$ integral over basis functions.

14.10 (a) Use the definition (14.42) of P_{tu} to show that $P_{rs} = P_{sr}^*$, meaning that the density matrix $\mathbf{P}$ is a Hermitian matrix. (b) Show that Eq. (14.45) can be written as $E_{\text{HF}} = \frac{1}{2}\text{Tr}(\mathbf{P}^*\mathbf{F} + \mathbf{P}^*\mathbf{H}^{\text{core}}) + V_{NN}$, where Tr denotes the trace of a matrix (Section 7.10) and the $\mathbf{P}$, $\mathbf{F}$, and $\mathbf{H}^{\text{core}}$ matrices have elements P_{rs}, F_{rs}, and H_{rs}^{core}. (c) Verify that (14.42) can be written as $\mathbf{P}^* = 2\mathbf{C}\mathbf{C}^\dagger$, where $\mathbf{C}$ is the matrix of coefficients c_{uj}.

14.11 From (14.5), show that $\int_{-\infty}^{\infty}\int_{-\infty}^{\infty}\int_{-\infty}^{\infty}\rho\,dx\,dy\,dz = n$, where ρ is the electron probability density of an n-electron molecule. (b) Use the result of (a) and Eq. (14.43) to show that $n = \sum_r\sum_s P_{rs}S_{rs} = \sum_r\sum_s P_{rs}S_{sr}^*$. (c) Show that $n = \text{Tr}(\mathbf{PS}^*)$, which becomes $n = \text{Tr}(\mathbf{PS})$ for real basis functions. Here, $\mathbf{P}$ and $\mathbf{S}$ are the density and overlap matrices.

14.12 Verify the equations for H_{11}^{core} and H_{22}^{core} in the Section 14.3 example.

14.13 Verify the equalities (14.47) for electron-repulsion integrals.

14.14 Use Eq. (9.123) of Prob. 9.13 to verify the expressions for the integrals $(11|22)$ and $(22|22)$ in the Section 14.3 example.

14.15 For the He-atom SCF calculation in Section 14.3, find the initial estimate of c_{11}/c_{21} given by the approximation $F_{rs} \approx H_{rs}^{\text{core}}$.

14.16 (a) Verify the equations for F_{11}, F_{12}, and F_{22} that immediately precede Eq. (14.50). (b) Verify Eqs. (14.50) to (14.52) for F_{11}, F_{12}, and F_{22}.

14.17 Derive (14.48) from the normalization condition for ϕ_1.

14.18 Verify the numerical results for $P_{11}, P_{12}, P_{22}, F_{11}, F_{12}, F_{22}, \varepsilon_1, \varepsilon_2, c_{11}$, and c_{21} obtained on the last cycle of calculation in the Section 14.3 example [Eqs. (14.53), (14.54), and the preceding and following equations].

14.19 Repeat the He SCF calculation of Section 14.3 using the same basis functions but starting with the initial guess $c_{11} = c_{21}$ and c_{21} determined by the normalization condition (14.48).

14.20 (a) Write a computer program that will perform the helium-atom SCF calculation of Section 14.3. Have the input to the program be ζ_1, ζ_2, and the initial guess for c_{11}/c_{21}. Do not use the Section 14.3 values of the integrals, but have the program calculate all integrals from ζ_1 and ζ_2. Have the program print c_{11}, c_{21}, ε_1, and ε_2 for each cycle of calculation. Use the convergence criterion that c_{11} and c_{21} each differ from the c_{11} and c_{21} values of the previous cycle by less than 10^{-5} atomic units. (b) Use your program with $\zeta_1 = 1.45$ and $\zeta_2 = 2.91$ to find the number of iterations needed to reach convergence for each of these initial guesses for c_{11}/c_{21}: $100, 10, 1, 0, -1, -10, -100$. (c) Run the program with $\zeta_1 = 1.45363$ and $\zeta_2 = 2.91093$ to verify the SCF energy given for these ζ's at the end of the Section 14.3 example. (d) Run the program with ζ_1 changed by $+0.01$ and by -0.01 from the value in (c), and verify that each E_{HF} obtained is higher than that in (c). Repeat for ζ_2.

14.21 Calculate ρ for the He SCF wave function of the Section 14.3 example at $r = 0$ and at $r = 1$ bohr.

14.22 Given that $\phi_i = \sum_s c'_{si}\chi'_s$, where the χ'_s functions are orthonormal, show that the orbitals ϕ_i form an orthonormal set if the matrix $\mathbf{C}'$ of coefficients c'_{si} is unitary.

14.23 Which of the following functions are homogeneous? Give the degree of homogeneity. (a) $x + 3yz$; (b) 179; (c) x^2/yz^3; (d) $(ax^3 + bxy^2)^{1/2}$.

14.24 Let 1 and 2 be two bound stationary states of an atom, with $E_2 > E_1$. For which state is the average electronic kinetic energy larger?

14.25 Show that the hypervirial theorem follows from Eq. (7.112).

14.26 (a) Calculate $\langle T \rangle$ and $\langle V \rangle$ for the helium-atom trial function (9.56). All the needed integrals were evaluated in Chapter 9. (b) Verify that the virial theorem is satisfied for $\zeta = Z - 5/16$ but not for $\zeta = Z$.

14.27 A particle is subject to the potential energy $V = ax^4 + by^4 + cz^4$. If its ground-state energy is 10 eV, calculate $\langle T \rangle$ and $\langle V \rangle$ for the ground state.

14.28 Prove that for a bound stationary state: (a) $\langle p_x \rangle = 0$; (b) $\langle \partial V/\partial x \rangle = 0$. *Hint*: Use certain equations in Section 5.1.

14.29 The $U(R)$ curve for a diatomic-molecule repulsive electronic state can be roughly approximated by the function $ae^{-bR} - c$, where a, b, and c are positive constants with $a > c$. (This function omits the van der Waals minimum and fails to go to infinity at $R = 0$.) Sketch U, $\langle T_{\text{el}} \rangle$, and $\langle V \rangle$ as functions of R for this function.

14.30 Prove that $\partial \langle V \rangle/\partial R$ must be nonnegative at $R = R_e$; that is, $\langle V \rangle$ cannot be increasing with decreasing R as we go through the minimum in the $U(R)$ curve (Fig. 14.1). State and prove the corresponding theorem for $\langle T_{\text{el}} \rangle$.

14.31 Let ψ be the complete wave function for a molecule, with the Born–Oppenheimer approximation $\psi = \psi_{\text{el}}\psi_N$ not necessarily holding. Is it true that

$$2\langle \psi | \hat{T}_{\text{el}} + \hat{T}_N | \psi \rangle = -\langle \psi | \hat{V} | \psi \rangle$$

where $\hat{T}_{\text{el}}$ and $\hat{T}_N$ are the kinetic-energy operators for the electrons and nuclei and $\hat{V}$ is the complete potential-energy operator? Justify your answer.

14.32 Given that $D_e = 4.75$ eV and $R_e = 0.741$ Å for the ground electronic state of H_2, find $U(R_e)$, $\langle V \rangle|_{R_e}$, $\langle V_{\text{el}} \rangle|_{R_e}$, and $\langle T_{\text{el}} \rangle|_{R_e}$ for this state.

14.33 The Fues potential-energy function for nuclear vibration of a diatomic molecule is $U(R) = U(\infty) + D_e(-2R_e/R + R_e^2/R^2)$. Find the expressions for $\langle T_{\text{el}} \rangle$ and $\langle V \rangle$ predicted by this potential and comment on the results.

14.34 (a) Apply the generalized Hellmann–Feynman theorem with Z as the parameter to find $\langle 1/r \rangle$ for the hydrogenlike-atom bound states ψ_{nlm}. (b) Since hydrogenlike functions ψ_{nlm} with the same n but different l or m have the same energy, we must be sure that the functions ψ_{nlm} are the correct zeroth-order functions for the perturbation of varying Z. Use a theorem of Sec. 9.6 to verify this.

14.35 Use the generalized Hellmann–Feynman theorem to find $\langle p_x^2 \rangle$ for the one-dimensional harmonic-oscillator stationary states. Check that the result obtained agrees with the virial theorem.

14.36 Differentiate (9.7) and (9.14) with respect to λ, substitute the results into the generalized Hellmann–Feynman theorem, and let $\lambda = 0$ to derive the perturbation-theory equation $E_n^{(1)} = \langle \psi_n^{(0)} | \hat{H}' | \psi_n^{(0)} \rangle$.

14.37 Use $F_{z,a} = -\partial U/\partial z_a$ and (14.85) to show that in Fig. 14.4, $F_{z,a} = -(\partial U/\partial R) \times [(z_a - z_b)/R]$. Find a similar equation for $F_{z,b}$ and verify that $F_{z,a} = -F_{z,b}$.

14.38 The R_e values for the ground electronic states of HF, HCl, HBr, and HI are 0.9, 1.3, 1.4, and 1.6 Å. The surface enclosing the antibinding region "behind" the proton in these molecules intersects the internuclear axis at two points, one of which is the proton location. Calculate the distance between these two points of intersection for each of the hydrogen halides.

CHAPTER 15

Molecular Electronic Structure

15.1 AB INITIO, DENSITY-FUNCTIONAL, SEMIEMPIRICAL, AND MOLECULAR-MECHANICS METHODS

The electronic wave function of a polyatomic molecule depends on several parameters—the bond distances, bond angles, and dihedral angles of rotation about single bonds (these angles define the molecular conformation). A full theoretical treatment of a polyatomic molecule involves calculation of the electronic wave function for a range of each of these parameters. The equilibrium bond distances, bond angles, and dihedral angles are then found as those values that minimize the electronic energy including nuclear repulsion.

The four main approaches to calculating molecular properties are ab initio methods, semiempirical methods, the density-functional method, and the molecular-mechanics method.

Semiempirical molecular quantum-mechanical methods use a simpler Hamiltonian than the correct molecular Hamiltonian and use parameters whose values are adjusted to fit experimental data or the results of ab initio calculations. An example is the Hückel MO treatment of conjugated hydrocarbons (Section 17.2), which uses a one-electron Hamiltonian and takes the bond integrals as adjustable parameters rather than quantities to be calculated theoretically. In contrast, an **ab initio** (or **first principles**) calculation uses the correct Hamiltonian and does not use experimental data other than the values of the fundamental physical constants. A Hartree–Fock SCF calculation seeks the antisymmetrized product Φ of one-electron functions that minimizes $\int \Phi^* \hat{H} \Phi \, d\tau$, where $\hat{H}$ is the true Hamiltonian, and is thus an ab initio calculation. (Ab initio is Latin for "from the beginning" and indicates a calculation based on fundamental principles.) The term *ab initio* should not be interpreted to mean "100% correct." An ab initio SCF MO calculation uses the approximation of taking ψ as an antisymmetrized product of one-electron spin-orbitals and uses a finite (and hence incomplete) basis set.

The **density-functional method** (Section 16.4) does not attempt to calculate the molecular wave function but calculates the molecular electron probability density ρ and calculates the molecular electronic energy from ρ.

The **molecular-mechanics method** (Section 17.5) is not a quantum-mechanical method and does not use a molecular Hamiltonian operator or wave function. Instead, it views the molecule as a collection of atoms held together by bonds and expresses the molecular energy in terms of force constants for bond bending and stretching and other parameters.

This chapter discusses general principles of electronic structure calculations for polyatomic molecules and discusses the Hartree–Fock calculation method. As noted in Section 11.3, the Hartree–Fock method does not take account of electron correlation. Ab initio and density-functional methods that take account of electron correlation are discussed in Chapter 16. Chapter 17 discusses semiempirical methods and the molecular-mechanics method.

A free database of references to molecular ab initio and density-functional calculations is the Quantum Chemistry Literature Database at qcldb2.ims.ac.jp/.

The National Institute of Standards and Technology's Computational Chemistry Comparison and Benchmark Database (CCCBDB) at srdata.nist.gov/cccbdb/ tabulates the results of ab initio, density functional, and semiempirical calculations with various basis sets on 680 molecules. The EMSL Computational Results Database (CRDB) at www.emsl.pnl.gov/proj/crdb/ contains ab initio calculation results on 255 molecules using various basis sets. The program Spartan'06 (Section 15.14) gives access to a database of ab initio and density functional results for 140000 molecules, mostly for the 6-31G* basis set (Section 15.4).

15.2 ELECTRONIC TERMS OF POLYATOMIC MOLECULES

For polyatomic molecules, the operator $\hat{S}^2$ for the square of the total electronic spin angular momentum commutes with the electronic Hamiltonian operator, and, as for diatomic molecules, the electronic terms of polyatomic molecules are classified as singlets, doublets, triplets, and so on, according to the value of $2S + 1$. (The commutation of $\hat{S}^2$ and $\hat{H}_{\text{el}}$ holds provided spin–orbit interaction is omitted from the Hamiltonian. For molecules containing heavy atoms, spin–orbit interaction is considerable, and S is not a good quantum number.)

For linear polyatomic molecules, the operator $\hat{L}_z$ for the component of the total electronic orbital angular momentum along the molecular axis commutes with the electronic Hamiltonian, and the same term classifications are used as for diatomic molecules (Section 13.8), giving such possibilities as ${}^1\Sigma^+$, ${}^1\Sigma^-$, ${}^3\Sigma^+$, ${}^1\Pi$, and so on. For linear polyatomic molecules with a center of symmetry, the g, u classification is added.

For nonlinear polyatomic molecules, no orbital angular-momentum operator commutes with the electronic Hamiltonian, and the angular-momentum classification of electronic terms cannot be used. Operators that do commute with the electronic Hamiltonian are the symmetry operators $\hat{O}_R$ of the molecule (Section 12.1), and the electronic states of polyatomic molecules are classified according to the behavior of the electronic wave function on application of these operators. Consider H_2O as an example.

In its equilibrium configuration, the water molecule belongs to group $\mathscr{C}_{2v}$ with the symmetry operations

$$\hat{E} \qquad \hat{C}_2(z) \qquad \hat{\sigma}_v(xz) \qquad \hat{\sigma}_v(yz) \tag{15.1}$$

The standard convention [R.S. Mulliken, *J. Chem. Phys.*, **23**, 1997 (1955); **24**, 1118 (1956)] takes the molecular plane as the yz plane (Fig. 15.1). We readily find that each of the symmetry operations commutes with the other three. Therefore, the electronic wave functions can be chosen as simultaneous eigenfunctions of all four symmetry operators.

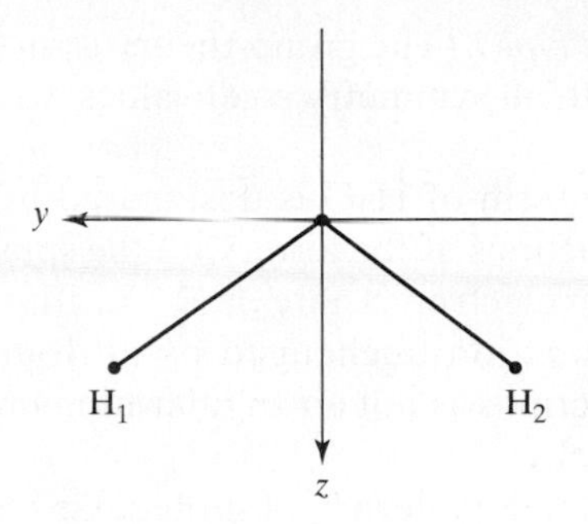

FIGURE 15.1 Coordinate axes for the H_2O molecule. The x axis is perpendicular to the molecular plane.

Since $\hat{O}_E$ is the unit operator, we have $\hat{O}_E\psi_{el} = \psi_{el}$. Each of the remaining symmetry operators satisfies $\hat{O}_R^2 = \hat{1}$, and so each has as its eigenvalues $+1$ and -1 [Eq. (7.55)]. Therefore, each electronic wave function of H_2O is an eigenfunction of $\hat{O}_E$ with eigenvalue $+1$ and an eigenfunction of each of the other three symmetry operators with the eigenvalue $+1$ or -1.

How many different combinations of these symmetry eigenvalues are there for H_2O? At first sight, we might think there are $(1)(2)(2)(2) = 8$ possible sets. However, certain sets can be ruled out, as we now show. Let the product of the symmetry operations $\hat{R}$ and $\hat{S}$ be the symmetry operation $\hat{T}$; $\hat{R}\hat{S} = \hat{T}$. Let ψ_{el} be an eigenfunction of $\hat{O}_R$, $\hat{O}_S$, and $\hat{O}_T$ with eigenvalues r, s, and t, respectively. Since the symmetry operators multiply the same way the symmetry operations do, we have

$$t\psi_{el} = \hat{O}_T\psi_{el} = \hat{O}_R(\hat{O}_S\psi_{el}) = s\hat{O}_R\psi_{el} = rs\psi_{el}$$

Dividing by ψ_{el}, we have $rs = t$ if $\hat{R}\hat{S} = \hat{T}$. Hence the eigenvalues of the symmetry operators must multiply in the same way the symmetry operations do. Now consider H_2O. Let us examine the set of symmetry eigenvalues

$$\begin{array}{cccc} \hat{E} & \hat{C}_2(z) & \hat{\sigma}_v(xz) & \hat{\sigma}_v(yz) \\ \hline 1 & -1 & -1 & -1 \end{array} \tag{15.2}$$

We find that $\hat{C}_2(z)\hat{\sigma}_v(xz) = \hat{\sigma}_v(yz)$. The set (15.2), however, has the product of the $\hat{O}_{C_2}$ and $\hat{O}_{\sigma_v(xz)}$ eigenvalues as $(-1)(-1) = 1$, which differs from the $\hat{O}_{\sigma_v(yz)}$ eigenvalue in (15.2). The set (15.2) must be discarded. Of the eight possible symmetry-eigenvalue combinations, only the following four sets are found to multiply properly (Prob. 15.1):

	$\hat{E}$	$\hat{C}_2(z)$	$\hat{\sigma}_v(xz)$	$\hat{\sigma}_v(yz)$
A_1	1	1	1	1
A_2	1	1	−1	−1
B_1	1	−1	1	−1
B_2	1	−1	−1	1

(15.3)

In this table the sets have been labeled A_1, A_2, B_1, and B_2. The letter A or B indicates whether the symmetry eigenvalue for the highest-order $\hat{C}_n$ or $\hat{S}_n$ operation of the molecule [$\hat{C}_2(z)$ for water] is $+1$ or -1, respectively. The subscripts 1 and 2 distinguish sets having the same letter label. Each possible set of symmetry eigenvalues in (15.3) is called a **symmetry species** (or *symmetry type*). (The group-theory term is *irreducible representation*.) The symmetry species with all symmetry eigenvalues +1 (A_1 for H_2O) is called the **totally symmetric species**.

Each molecular electronic term of H_2O is designated by giving the symmetry species of the electronic wave functions of the term, with the spin multiplicity $2S + 1$ as a left superscript. For example, an electronic state of H_2O with two electrons unpaired and with the electronic wave function unchanged by all four symmetry operators belongs to a $^3A_1$ term. (The subscript 1 is not an angular-momentum eigenvalue, but is part of the symmetry-species label.)

We now consider the **orbital degeneracy** of molecular electronic terms. This is degeneracy connected with the electrons' spatial (orbital) motion, as distinguished from spin degeneracy. Thus $^1\Pi$ and $^3\Pi$ terms of linear molecules are orbitally degenerate, while $^1\Sigma$ and $^3\Sigma$ terms are orbitally nondegenerate. Consider an operator $\hat{F}$ that commutes with the molecular electronic Hamiltonian and that does not involve spin. We have

$$\begin{aligned} \hat{F}\hat{H}_{\text{el}}\psi_{\text{el},i} &= \hat{F}E_{\text{el}}\psi_{\text{el},i} \\ \hat{H}_{\text{el}}(\hat{F}\psi_{\text{el},i}) &= E_{\text{el}}(\hat{F}\psi_{\text{el},i}) \end{aligned} \tag{15.4}$$

Thus $\hat{F}\psi_{\text{el},i}$ is an eigenfunction of $\hat{H}_{\text{el}}$ with eigenvalue E_{el}. Let the orbital degeneracy of the term to which $\psi_{\text{el},i}$ belongs be n. It follows from (15.4) that $\hat{F}\psi_{\text{el},i}$ must be some linear combination of the n orbitally degenerate wave functions of the term that have the same value of S_z as $\psi_{\text{el},i}$:

$$\hat{F}\psi_{\text{el},i} = \sum_{j=1}^{n} c_{ij,F}\psi_{\text{el},j} \tag{15.5}$$

where the c's are certain constants. (These arguments were previously given in Section 7.3.) For a level that is not orbitally degenerate ($n = 1$), Eq. (15.5) reduces to

$$\hat{F}\psi_{\text{el},i} = c_F\psi_{\text{el},i} \tag{15.6}$$

and here $\psi_{\text{el},i}$ *must* be an eigenfunction of $\hat{F}$. For $n > 1$, we can *choose* n linear combinations of the $\psi_{\text{el},j}$'s that are eigenfunctions of $\hat{F}$, but there is no *necessity* for the eigenfunctions of a degenerate level to be eigenfunctions of $\hat{F}$. If we have two operators $\hat{F}$ and $\hat{G}$ that commute with $\hat{H}_{\text{el}}$ but not with each other, we can pick the linear combinations to be eigenfunctions of $\hat{F}$ or of $\hat{G}$, but not in general of both $\hat{F}$ and $\hat{G}$ simultaneously.

Now consider symmetry operators. For H_2O all the symmetry operators commute among themselves, and so each electronic wave function is simultaneously an eigenfunction of all the symmetry operators. If all the electronic wave functions were orbitally nondegenerate, then it would automatically follow from (15.6) that they were simultaneously eigenfunctions of all the symmetry operators. We therefore suspect (but have not proved) that the electronic wave functions of H_2O are all orbitally nondegenerate. This statement is in fact correct. The letters A and B designate symmetry species of orbitally nondegenerate electronic terms. For any molecule all of whose symmetry operators commute with one another, the electronic wave functions are each simultaneous eigenfunctions of all symmetry operators, and the only symmetry species are nondegenerate A and B species.

For some point groups, the symmetry operators do not all commute. An example is $\mathcal{O}_h$ (Fig. 12.7). When the symmetry operators do not all commute, some of the electronic terms are orbitally degenerate. A symmetry operator applied to an electronic wave function of an orbitally degenerate term converts it to a linear combination of the wave functions of the term [Eq. (15.5)]. The effects of the symmetry operator $\hat{O}_R$ on the wave functions of an n-fold orbitally degenerate electronic term are specified by the n^2 numbers

$$c_{ij,R}, \qquad i = 1, \ldots, n, \qquad j = 1, \ldots, n \tag{15.7}$$

The n^2 numbers (15.7) when arranged in an $n \times n$ square array form a matrix (Section 7.10). If there are h symmetry operations in the molecular point group, then the h matrices of coefficients in (15.5) constitute the **symmetry species** of the degenerate-term wave functions in (15.5). The following letter labels are used for the symmetry species, according to the orbital degeneracy n:

n	1	2	3	4	5
Letter	A, B	E	T	G	H

(15.8)

Numerical subscripts distinguish different symmetry species having the same letter designation. For molecules with a center of symmetry, a g or u subscript is added, depending on whether the wave function has the eigenvalue $+1$ or -1 for inversion of all electronic spatial coordinates. The possible symmetry species can be found in a systematic way using group theory (see *Schonland*). As an example, group theory shows the possible symmetry species of a $\mathcal{D}_{6h}$ molecule to be

$$\begin{matrix} A_{1g} & A_{2g} & B_{1g} & B_{2g} & E_{1g} & E_{2g} \\ A_{1u} & A_{2u} & B_{1u} & B_{2u} & E_{1u} & E_{2u} \end{matrix} \tag{15.9}$$

We shall not give the numbers $c_{ij,R}$ that specify these symmetry species, except to note that A_{1g} is the totally symmetric symmetry species, with all c's equal to 1.

It is an empirical fact that for most molecules in their electronic ground states the electronic wave function belongs to the (nondegenerate) totally symmetric species. Also, the electronic spins are usually all paired in the ground state, and the ground state is a singlet. For H_2O the ground electronic state is $^1A_1$; for benzene it is $^1A_{1g}$.

We have been using the equilibrium-geometry point groups of H_2O and C_6H_6. For nonequilibrium nuclear configurations, the symmetry is in general less than that of $\mathcal{C}_{2v}$ or $\mathcal{D}_{6h}$. For reasonably small departures from equilibrium, however, the symmetry behavior should be given to a good approximation by the symmetry species of the equilibrium-geometry point group. Excited states sometimes differ in their equilibrium point group from the ground electronic state. For example, the point group of NH_3 is $\mathcal{D}_{3h}$ for several excited electronic states. (For properties of molecular electronic states, see G. Herzberg, *Electronic Spectra of Polyatomic Molecules*, Van Nostrand, 1966, Appendix VI.)

Corresponding to a given molecular electronic term, there are in general several electronic states. The interactions between electronic spin and electronic orbital motion and between electronic and nuclear motions split the energies of these states. These splittings are usually small.

15.3 THE SCF MO TREATMENT OF POLYATOMIC MOLECULES

The purely electronic nonrelativistic Hamiltonian for a polyatomic molecule is (in atomic units)

$$\hat{H}_{\text{el}} = -\frac{1}{2}\sum_i \nabla_i^2 - \sum_i \sum_\alpha \frac{Z_\alpha}{r_{i\alpha}} + \sum_i \sum_{j>i} \frac{1}{r_{ij}} \tag{15.10}$$

If interelectronic repulsions are neglected, the zeroth-order wave function is the product of one-electron spatial functions (molecular orbitals). Allowance for electron spin and antisymmetry gives a zeroth-order wave function that is an antisymmetrized product of molecular spin-orbitals, each spin-orbital being a product of a spatial MO and a spin function. The best possible variation function that has the form of an antisymmetrized product of spin-orbitals is the Hartree–Fock SCF function. [Improvements beyond the Hartree–Fock stage require some method (Chapter 16) to allow for electron correlation.] The MOs are usually expressed as linear combinations of basis functions, the coefficients being found by solution of the Roothaan equations (Section 14.3). If a large enough basis set is used, the MOs are accurate approximations to the Hartree–Fock MOs. If a minimal basis set is used, the MOs are only rough approximations to the Hartree–Fock MOs but are still referred to as SCF molecular orbitals. The SCF MO method is widely used in polyatomic-molecule electronic-structure calculations.

How are polyatomic MOs classified? As might be expected, the MOs of a polyatomic molecule show the same kinds of possible symmetry behavior as the overall electronic wave function does (Section 15.2); for the proof, see C. C. J. Roothaan, *Rev. Mod. Phys.*, **23**, 69 (1951). The MOs are therefore classified according to the symmetry species of the molecular point group. For example, the MOs of H_2O have the possible symmetry species a_1, a_2, b_1, and b_2 of (15.3). Lowercase letters are used for MO symmetry species. To distinguish MOs of the same symmetry species, we number them in order of increasing energy. Thus the lowest three a_1 MOs of water are called $1a_1$, $2a_1$, and $3a_1$. This nomenclature is similar to that of the third column in Table 13.1.

Each MO holds two electrons of opposite spin. MOs having the same energy constitute a **shell**. A shell that consists of a single MO of symmetry species a or b holds two electrons. A shell that consists of two e MOs having the same energy holds four electrons; and so on. For H_2O there are no degenerate symmetry species, and each shell holds two electrons. For C_6H_6 there are some doubly degenerate symmetry species [see (15.9)], so some of the benzene MOs occur in pairs having the same energy. Specification of the number of electrons in each shell specifies the molecular **electronic configuration**. Just as in atoms and diatomic molecules, a given electron configuration of a polyatomic molecule gives rise to one or more electronic terms. [For example, an $(e_{1g})^2$ configuration of a $\mathscr{D}_{6h}$ molecule gives the terms ${}^1A_{1g}$, ${}^1E_{2g}$, and ${}^3A_{2g}$.] The systematic method for finding the terms of a configuration uses group theory and will not be discussed here. For tables of the terms arising from various configurations, see G. Herzberg, *Electronic Spectra of Polyatomic Molecules*, Van Nostrand, 1966, pp. 330–334, 570–573. *A closed-shell configuration gives rise to a single nondegenerate term whose spin multiplicity is 1 and whose symmetry species is the totally symmetric one.* Most polyatomic molecules have a closed-shell ground state, for which the MO wave function is a single Slater determinant. For states arising from open-shell configurations, the MO wave function may require a linear combination of a few Slater determinants.

SCF MO Wave Functions for Open-Shell States. For SCF MO calculations on closed-shell states of molecules and atoms, electrons paired with each other are almost always given precisely the same spatial orbital function. A Hartree–Fock wave function in which electrons whose spins are paired occupy the same spatial orbital is called a **restricted Hartree–Fock** (RHF) wave function. (The unmodified term *Hartree–Fock wave function* is understood to mean the RHF wave function.)

Although the RHF wave function is generally used for closed-shell states, two different approaches are widely used for open-shell states. In the **restricted open-shell Hartree–Fock** (ROHF) method, electrons that are paired with each other are given the same spatial orbital function. For example, the ROHF wave function of the Li ground state is $|1s\overline{1s}2s|$, where the two 1s electrons occupy the same spatial MO. The 2s electron in this ROHF function has been given spin α. Since electrons with the same spin tend to keep away from each other (Pauli repulsion; Section 10.3), the interaction between the $2s\alpha$ and $1s\alpha$ electrons differs from the interaction between the $2s\alpha$ and $1s\beta$ electrons, and it seems reasonable to give the two 1s electrons slightly different spatial orbitals, which we shall call 1s and 1s′. This gives the **unrestricted Hartree–Fock** (UHF) wave function for the Li ground state as $|1s\overline{1s'}2s|$ where $1s \neq 1s'$. In a UHF wave function, the spatial orbitals of spin-α electrons are allowed to differ from those of spin-β electrons.

The UHF wave function gives a slightly lower energy than the ROHF wave function and is more useful in predicting electron-spin-resonance spectra (see *Szabo and Ostlund*, Section 3.8.6). The main problem with the UHF wave function is that it is not an eigenfunction of the spin operator $\hat{S}^2$ (nor can it be made an eigenfunction of $\hat{S}^2$ by taking a linear combination of a few UHF functions), whereas the true wave function and the ROHF wave function are eigenfunctions of $\hat{S}^2$. When a UHF wave function is found, one calculates $\langle S^2 \rangle$ for the UHF function. If the deviation of $\langle S^2 \rangle$ from $S(S + 1)\hbar^2$ is substantial, the UHF wave function should be viewed with suspicion.

15.4 BASIS FUNCTIONS

Most molecular quantum-mechanical methods, whether SCF, CI, perturbation theory (Section 16.2), coupled cluster (Section 16.3), or density functional (Sec. 16.4), begin the calculation with the choice of a set of basis functions χ_r, which are used to express the MOs ϕ_i as $\phi_i = \sum_i c_{ri}\chi_r$ [Eq. (14.33)]. (Density-functional theory uses orbitals called Kohn–Sham orbitals ϕ_i^{KS} that are expressed as $\phi_i^{\text{KS}} = \sum_i c_{ri}\chi_r$; see Sec. 16.4.) The use of an adequate basis set is an essential requirement for success of the calculation.

For diatomic molecules, the basis functions are usually taken as atomic orbitals, some centered on one atom, the remainder centered on the other atom. Each AO can be represented as a linear combination of one or more Slater-type orbitals (STOs). An STO centered on atom a has the form $Nr_a^{n-1}e^{-\zeta r_a}Y_l^m(\theta_a, \phi_a)$ [Eq. (11.14)]. For nonlinear molecules, the real form of the STOs, with Y_l^m replaced by $(Y_l^{m*} \pm Y_l^m)/2^{1/2}$ (Section 6.6), is used. Each MO ϕ_i is expressed as $\phi_i = \sum_r c_{ri}\chi_r$, where the χ_r's are the STO basis functions. We have LC-STO MOs.

For polyatomic molecules, the LC-STO method uses STOs centered on each of the atoms. The presence of more than two atoms causes difficulties in evaluating the needed integrals. For a triatomic molecule, one must deal with three-center, two-center, and one-center integrals. For a molecule with four or more atoms, one also has four-center integrals, but the number of centers in any one integral does not exceed four. Solution of the Roothaan equations requires evaluation of the electron-repulsion integrals $(rs|tu)$ and the H_{rs}^{core} integrals [Eq. (14.41)]. If the basis functions $\chi_r(1), \chi_s(1), \chi_t(2), \chi_u(2)$ are each centered on a different nucleus, then $(rs|tu)$ is a four-center integral. The H_{rs}^{core} integrals involve either one or two centers.

For the basis set $\chi_1, \chi_2, \ldots, \chi_b$, there are b different possibilities for each basis function in $(rs|tu)$, and use of the identities $(rs|tu) = (sr|tu) = \cdots$ [Eq. (14.47)] shows that there are about $b^4/8$ different electron-repulsion integrals to be evaluated. Accurate SCF molecular calculations on small- to medium-size molecules might use from 40 to 400 basis functions, producing from 300000 to 3×10^9 electron-repulsion integrals. Computer evaluation of three- and four-center integrals over STO basis functions is very time consuming.

To speed up molecular integral evaluation, Boys proposed in 1950 the use of **Gaussian-type functions** (GTFs) instead of STOs for the atomic orbitals in an LCAO wave function. A **Cartesian Gaussian** centered on atom b is defined as

$$g_{ijk} = N x_b^i y_b^j z_b^k e^{-\alpha r_b^2} \tag{15.11}$$

where i, j, and k are nonnegative integers, α is a positive **orbital exponent**, x_b, y_b, z_b are Cartesian coordinates with the origin at nucleus b, and r_b is the distance to nucleus b. The Cartesian-Gaussian normalization constant is

$$N = \left(\frac{2\alpha}{\pi}\right)^{3/4} \left[\frac{(8\alpha)^{i+j+k} i!\,j!\,k!}{(2i)!\,(2j)!\,(2k)!}\right]^{1/2}$$

When $i + j + k = 0$ (that is, $i = 0, j = 0, k = 0$), the GTF is called an ***s*-type** Gaussian. When $i + j + k = 1$, we have a ***p*-type** Gaussian, which contains the factor x_b, y_b, or z_b. When $i + j + k = 2$, we have a ***d*-type** Gaussian. There are six d-type Gaussians, with the factors x_b^2, y_b^2, z_b^2, $x_b y_b$, $x_b z_b$, and $y_b z_b$. If desired, five linear combinations (having the factors $x_b y_b$, $x_b z_b$, $y_b z_b$, $x_b^2 - y_b^2$, and $3z_b^2 - r_b^2$) can be formed to have the same angular behavior as the five real 3d AOs; a sixth combination with the factor $x_b^2 + y_b^2 + z_b^2 = r_b^2$ is like a 3s function. This sixth combination is often omitted from the basis set. Similarly, there are ten f-type Gaussians, and these could be combined to have the angular behavior of the seven real 4f AOs. [In general, linear combinations of Cartesian Gaussians can be formed to have the form $N r_b^l e^{-\alpha r_b^2}(Y_l^{m*} \pm Y_l^m)/2^{1/2}$.]

Note the absence of the principal quantum number n in (15.11). Any s AO (whether 1s or 2s or ...) is represented by a linear combination of several Gaussians with different orbital exponents, each Gaussian having the form $\exp(-\alpha r_b^2)$; any atomic p_x orbital is represented by a linear combination of Gaussians, each of the form $x_b \exp(-\alpha r_b^2)$; and so on. The Cartesian Gaussians form a complete set.

An occasionally used alternative to Cartesian Gaussians is *spherical Gaussians*, whose real form is $N r_b^{n-1} e^{-\alpha r_b^2}(Y_l^{m*} \pm Y_l^m)/2^{1/2}$.

The behavior of the Gaussian exponential factor is shown in Fig. 4.3a, where the origin is at nucleus b. A Gaussian function does not have the desired cusp at the nucleus and hence gives a poor representation of an AO for small values of r_b. To accurately represent an AO, we must use a linear combination of several Gaussians. Therefore, an

LC-GTF SCF MO calculation involves evaluation of very many more integrals than the corresponding LC-STO SCF MO calculation, since the number of two-electron integrals is proportional to the fourth power of the number of basis functions. However, Gaussian integral evaluation takes much less computer time than Slater integral evaluation. This is because the product of two Gaussian functions centered at two different points is equal to a single Gaussian centered at a third point. Thus all three- and four-center two-electron repulsion integrals are reduced to two-center integrals.

Let us discuss some of the terminology used to describe STO basis sets. A **minimal** (or **minimum**) **basis set** consists of one STO for each inner-shell and valence-shell AO of each atom (Section 14.3). For example, for C_2H_2 a minimal basis set consists of $1s, 2s, 2p_x, 2p_y$, and $2p_z$ AOs on each carbon and a $1s$ STO on each hydrogen. There are five STOs on each C and one on each H, for a total of 12 basis functions. This set contains two s-type STOs and one set of p-type STOs on each carbon and one s-type STO on each hydrogen. Such a set is denoted by $(2s1p)$ for the carbon functions and $(1s)$ for the hydrogen functions, a notation which is further abbreviated to $(2s1p/1s)$. The numbers of basis functions in a minimal STO set for the first part of the periodic table are

H, He	Li–Ne	Na–Ar	K, Ca	Sc–Kr
1	5	9	13	18

For each of the atoms Na to Ar, the minimal-basis AOs are $1s, 2s, 2p_{x,y,z}, 3s, 3p_{x,y,z}$ (but not $3d$).

A **double-zeta** (DZ) **basis set** is obtained by replacing each STO of a minimal basis set by two STOs that differ in their orbital exponents ζ (zeta). (Recall that a single STO is not an accurate representation of an AO. Use of two STOs gives substantial improvement.) For example, for C_2H_2 a double-zeta set consists of two $1s$ STOs on each H, two $1s$ STOs, two $2s$ STOs, two $2p_x$, two $2p_y$ and two $2p_z$ STOs on each carbon, for a total of 24 basis functions; this is a $(4s2p/2s)$ basis set. (Recall that we did a double-zeta SCF calculation on He in Section 14.3.) Since each basis function χ_r in $\phi_i = \sum_i c_{ri}\chi_r$ has its own independently determined variational coefficient c_{ri}, the number of variational parameters in a double-zeta-basis-set wave function is twice that in a minimal-basis-set wave function. A **triple-zeta** (TZ) **basis set** replaces each STO of a minimal basis set by three STOs that differ in their orbital exponents.

A **split-valence** (SV) **basis set** uses two (or more) STOs for each valence AO but only one STO for each inner-shell (core) AO. An SV basis set is minimal for inner-shell AOs and double zeta (or triple zeta or ...) for the valence AOs. Split-valence sets are called **valence double zeta** (VDZ), **valence triple-zeta** (VTZ), ... according to the number of STOs used for each valence AO.

AOs are distorted in shape and have their centers of charge shifted upon molecule formation. To allow for this **polarization**, one adds basis-function STOs whose l quantum numbers are greater than the maximum l of the valence shell of the ground-state atom. Any such basis set is a **polarized** (P) basis set. A common example is a **double-zeta plus polarization set** (DZ + P or DZP), which typically adds to a double-zeta set a set of five $3d$ functions on each "first-row" and each "second-row" atom and a set of three $2p$ functions ($2p_x, 2p_y, 2p_z$) on each hydrogen atom. In the quantum-chemistry literature, Li–Ne are called the **first-row** elements, even though they are actually the second row of the periodic table. This terminology will be used in this chapter and in chapters 16–18. A DZP STO basis set for $C_2H_5OSiH_3$ is designated as

($6s4p1d/4s2p1d/2s1p$), where the slashes separate the functions for atoms of different rows of the periodic table in decreasing order. For increased accuracy, higher-l polarization functions can be added.

Now consider Gaussian-basis-set terminology. Instead of using the individual Gaussian functions (15.11) as basis functions, the current practice is to take each basis function as a normalized linear combination of a few Gaussians, according to

$$\chi_r = \sum_u d_{ur} g_u \tag{15.12}$$

where the g_u's are normalized Cartesian Gaussians [Eq. (15.11)] centered on the same atom and having the same i, j, k values as one another, but different α's. The **contraction coefficients** d_{ur} are constants that are held fixed during the calculation. In (15.12), χ_r is called a **contracted Gaussian-type function** (CGTF) and the g_u's are called **primitive Gaussians**. By using contracted Gaussians instead of primitive Gaussians as the basis set, the number of variational coefficients to be determined is reduced, which gives large savings in computational time with little loss in accuracy if the contraction coefficients d_{ur} are well chosen.

The classifications given for STO basis sets also apply to CGTF basis sets if "STO" is replaced by "CGTF" in each definition. For example, a **minimal basis set** of contracted Gaussians consists of one contracted Gaussian function for each inner-shell AO and for each valence-shell AO. A DZ basis set has two CGTFs for each such AO, and a DZP set adds contracted Gaussians with higher l values to the DZ set, where $l \equiv i + j + k$ in (15.11).

In molecular calculations using CGTF basis functions, the orbital exponents and contraction coefficients of the basis functions are kept fixed at the predetermined values for the basis set used. Therefore, in a CGTF minimal-basis-set calculation, there is no way for the basis functions to adjust their sizes to differing molecular environments. By using a double-zeta basis set, we allow the AO sizes to vary from one molecule to another. For example, suppose we have two $1s$ CGTFs $1s'$ and $1s''$ centered on a certain H atom. Each function $1s'$ and $1s''$ is a linear combination of a few s-type primitive Gaussians [Eq. (15.12)]. Let the orbital exponents in the primitives of $1s'$ be much larger than those in $1s''$. Then $1s''$ is spread out over a much larger region of space than $1s'$. The expression for a given MO will contain the terms $c_1 1s' + c_2 1s''$, where the optimum c_1 and c_2 values are found by the SCF process. The size of the function $c_1 1s' + c_2 1s''$ will increase as the ratio c_2/c_1 increases (assuming c_1 and c_2 have the same signs).

By adding polarization functions, we allow the AO shapes to vary, thereby shifting charge density away from the nuclei and into the bonding regions in the molecule. For example, adding $2p$ functions to a $1s$ function on a hydrogen atom, we get the terms $c_1 1s + c_2 2p_x + c_3 2p_y + c_4 2p_z$ as part of an MO. This AO will be polarized in a direction determined by the values of c_2, c_3 and c_4. For example, suppose $c_3 = 0$, $c_4 = 0$, $c_1 > 0$, $c_2 > 0$. Then, since $2p_x$ has opposite signs on each side of the x axis, the term $c_2 2p_x$ will cancel some of the probability density of the $c_1 1s$ term on one side of the H atom and will augment it on the other side, thereby polarizing the $1s$ function in the positive x direction. (Recall the discussion of hybridization in Section 13.5.) Similarly, p-type AOs can be polarized by mixing in d-type AOs.

Several methods exist to form contracted Gaussian sets. Minimal CGTF sets are often formed by fitting STOs. One starts with a minimal basis set of one STO per AO, with the STO orbital exponents fixed at values found to work well in calculations on small molecules. Each STO is then approximated as a linear combination of N Gaussian functions, where the Gaussian orbital exponents and the coefficients in the linear combination are chosen to give the best least-squares fit to the STO. Most commonly, $N = 3$, giving a set of CGTFs called STO-3G; this basis set is defined for the atoms H through I. Since a linear combination of three Gaussians is only an approximation to an STO, the STO-3G basis set gives results not quite as good as a minimal-basis-set STO calculation. A minimal basis set of STOs for a compound containing only first-row elements and hydrogen is denoted by $(2s1p/1s)$. Since each STO is replaced by a linear combination of three primitive Gaussians (which is one contracted Gaussian), the STO-3G basis set for a compound of first-row atoms and H is denoted by $(6s3p/3s)$ contracted to $[2s1p/1s]$, where *parentheses indicate the primitive Gaussians and brackets indicate the contracted Gaussians.*

Suppose we want to fit a linear combination of three Cartesian GTFs to a $1s$ STO having $\zeta = 1$. In atomic units, this STO is [Eqs. (11.14) and (5.101)] $S(r; 1) = \pi^{-1/2}e^{-r}$, where the value of the parameter ζ is given in parentheses. From (15.11) and the following normalization-constant expression, the normalized s-type Gaussian is $(2\alpha/\pi)^{3/4}e^{-\alpha r^2}$. The desired normalized linear combination of three Gaussians has the form

$$G_{3N}(r; 1) = c_1(2\alpha_1/\pi)^{3/4}e^{-\alpha_1 r^2} + c_2(2\alpha_2/\pi)^{3/4}e^{-\alpha_2 r^2} + c_3(2\alpha_3/\pi)^{3/4}e^{-\alpha_3 r^2}$$

where the six parameters $c_1, c_2, c_3, \alpha_1, \alpha_2, \alpha_3$ must be adjusted to fit the function S, and N indicates a normalized function. Using a least-squares criterion of goodness of fit and the Solver in the Excel spreadsheet, we find (Prob. 15.9) $c_1 = 0.444615$, $c_2 = 0.535336$, $c_3 = 0.154340$, $\alpha_1 = 0.109814$, $\alpha_2 = 0.40575$, $\alpha_3 = 2.22746$.

Suppose we now want to fit a $1s$ STO with orbital exponent ζ with a linear combination of three GTFs. We can show (Prob. 15.10) that the correct fit is found by taking the function $G_{3N}(r; 1)$ and replacing each orbital exponent α_i by $\zeta^2\alpha_i$, while leaving the coefficients c_i unchanged. The quantity ζ is called a **scale factor**. The value $\zeta = 1.24$ is found to work well as the STO orbital exponent of an H-atom $1s$ basis function in SCF MO molecular calculations. Multiplication of the preceding orbital exponents by $(1.24)^2$ gives the orbital exponents for $\zeta = 1.24$ as 0.16885, 0.62388, and 3.42494. These orbital exponents, along with the coefficients listed above, define the STO-3G $1s$ orbital for H. To get the STO-3G $1s$ orbital for another atom, one uses the appropriate ζ for that atom and multiplies the $\zeta = 1$ orbital exponents by ζ^2.

Another way to form contracted Gaussians is to start with atomic GTF SCF calculations. Huzinaga used a $(9s5p)$ basis set of uncontracted Gaussians to do SCF calculations on the atoms Li–Ne. For example, for the ground state of the O atom, the optimized orbital exponents of the nine s-type basis GTFs were found to be [S. Huzinaga, *J. Chem. Phys.*, **42**, 1293 (1965)]

g_1	g_2	g_3	g_4	g_5	g_6	g_7	g_8	g_9
7817	1176	273.2	81.2	27.2	9.53	3.41	0.940	0.285

The expansion coefficients for the $1s$ SCF oxygen AO were found to be

g_1	g_2	g_3	g_4	g_5	g_6	g_7	g_8	g_9
0.0012	0.009	0.043	0.144	0.356	0.461	0.140	−0.0006	0.001

and the $2s$ SCF AO coefficients are

g_1	g_2	g_3	g_4	g_5	g_6	g_7	g_8	g_9
−0.0003	−0.002	−0.010	−0.036	−0.095	−0.196	−0.037	0.596	0.526

Suppose we want to form a valence-double-zeta (VDZ) $[3s2p]$ set of contracted GTFs for O. We see that the g_1, g_2, g_3, g_4, g_5, and g_7 coefficients are much larger for the $1s$ AO than for the $2s$ AO, the g_8 and g_9 coefficients are much larger for the $2s$ AO than for the $1s$ AO, and g_6 makes substantial contributions to both $1s$ and $2s$. We might therefore take the contracted 1s basis function as

$$1s = N(0.0012g_1 + 0.009g_2 + 0.043g_3 + 0.144g_4 + 0.356g_5 + 0.461g_6 + 0.140g_7)$$

where the normalization constant N is needed because g_8 and g_9 have been omitted. For a VDZ set, we need two basis functions for the $2s$ AO. These will be formed from g_6, g_8, and g_9, which are the main contributors to the $2s$ AO. Of these three, the function g_9 has the smallest orbital exponent and so falls off most slowly as r increases. (g_9 is called a **diffuse** function.) The outer region of an AO changes the most upon molecule formation, and to allow for this change we can take the diffuse function g_9 as one of the basis functions, giving as the 2s oxygen contracted basis functions

$$2s = N'(-0.196g_6 + 0.596g_8), \qquad 2s' = g_9$$

The $2p$ and $2p'$ CGTFs can be formed similarly.

Dunning's DZ $[4s2p]$ and Dunning and Hay's VDZ $[3s2p]$ contractions of Huzinaga's $(9s5p)$ first-row-atom AOs were often used in molecular calculations [T. H. Dunning, *J. Chem. Phys.*, **53**, 2823 (1970); T. H. Dunning and P. J. Hay in *Schaefer, Methods of Electronic Structure Theory*, pp. 1–27].

The 3-21G basis set (defined for the atoms H through Xe) and the 6-31G set (defined for H through Zn) are VDZ basis sets of CGTFs. In the 3-21G set, each inner-shell AO ($1s$ for Li–Ne; $1s, 2s, 2p_x, 2p_y, 2p_z$ for Na–Ar; and so on) is represented by a single CGTF that is a linear combination of three primitive Gaussians. For each valence-shell AO ($1s$ for H; $2s$ and the $2p$'s for Li–Ne; . . .; $4s$ and the $4p$'s for K, Ca, Ga–Kr; $4s$, the $4p$'s, and the five $3d$'s for Sc–Zn), there are two basis functions, one of which is a CGTF that is a linear combination of two Gaussian primitives and one which is a single diffuse Gaussian. The 6-31G set uses six primitives in each inner-shell CGTF and represents each valence-shell AO by one CGTF with three primitives and one Gaussian with one primitive. The orbital exponents and contraction coefficients d_{ur} in these basis sets were determined by using these basis sets to minimize the SCF energies of atoms. However, in the 3-21G basis set, the orbital exponents found for H in an atomic calculation are increased using a scale factor, and in the 6-31G set, the valence orbital exponents of H and of Li through O are scaled to be more appropriate for molecular calculations.

The 6-31G* basis set (defined for the atoms H through Zn) is a valence double-zeta polarized basis set that adds to the 6-31G set six d-type Cartesian–Gaussian polarization functions on each of the atoms Li through Ca and ten f-type Cartesian–Gaussian polarization functions on each of the atoms Sc through Zn. The 6-31G** basis set adds to the 6-31G* set a set of three p-type Gaussian polarization functions on each hydrogen and helium atom. The orbital exponents of the polarization functions in these two basis sets were determined as the average of the optimum values found in calculations on small molecules. Each polarization function in 6-31G* or 6-31G** consists of a single primitive Gaussian function. The 6-31G* and 6-31G** sets are sometimes denoted as 6-31G(d) and 6-31G(d, p), respectively. In the 6-31G* basis set, a phosphorus atom has 19 basis functions centered on it ($1s$, $2s$, $2p_x$, $2p_y$, $2p_z$, $3s$, $3s'$, $3p_x$, $3p_y$, $3p_z$, $3p'_x$, $3p'_y$, $3p'_z$, and six d's) and is [$4s3p1d$] for P.

For second-row atoms, d orbitals contribute significantly to the bonding. To allow for this, the 3-21G$^{(*)}$ basis set (defined for H through Ar) is constructed by the addition to the 3-21G set of a set of six d-type Gaussian functions on each second-row atom. For H–Ne, the 3-21G$^{(*)}$ set (which is sometimes called 3-21G*) is the same as the 3-21G set.

Anions, compounds with lone pairs, and hydrogen-bonded dimers have significant electron density at large distances from the nuclei. To improve the accuracy for such compounds, the 3-21+G and 6-31+G* basis sets are formed from the 3-21G and 6-31G* sets by the addition of four highly diffuse functions (s, p_x, p_y, p_z) on each nonhydrogen atom. A **highly diffuse function** is one with a very small orbital exponent (typically, 0.01 to 0.1). The 3-21++G and 6-31++G* sets also include a highly diffuse s function on each hydrogen atom.

Basis sets larger than the relatively small 6-31G* and 6-31G** basis sets are often used in calculations that include electron correlation. An example is the 6-311G** set defined for H–Ca, Ga–Kr and I, which is single zeta for the core AOs, triple zeta for the valence AOs, and contains five d-type Gaussian polarization functions on each first- and second-row atom and three p-type polarization functions on each hydrogen atom. The 6-311G** set cannot really be considered to be a large basis set. Large basis sets typically use more than one set of polarization functions on each atom. An example of a large basis set for correlation calculations is the 6-311++G($3df$, $3pd$) set, defined for H–Ar. The pluses indicate diffuse functions on all atoms; the letters in parentheses indicate that three sets of five d-type Gaussian polarization functions (each set having a different orbital exponent) are added to each nonhydrogen (giving fifteen d functions on nonhydrogens), one set of seven f-type Gaussians is added to each nonhydrogen, and three sets of three p-type Gaussians and one set of five d-type Gaussians are added to each hydrogen.

The basis sets STO-3G, 3-21G, 3-21G$^{(*)}$, 6-31G*, 6-31G**, etc., were developed by Pople and co-workers (see *Hehre et al.*, Section 4.3).

Dunning and co-workers have developed the CGTF basis sets cc-pVDZ, cc-pVTZ, cc-pVQZ, cc-pV5Z, and cc-pV6Z (collectively denoted cc-pVnZ, where n goes from 2 to 6), designed for use in calculation methods (such as CI) that include electron correlation. Here, cc-pVDZ stands for *correlation-consistent, polarized valence double-zeta*. Unlike the Pople-type functions, where the number of d functions used varies from set to set, the cc family of functions always uses five d functions, seven f functions, etc. These sets are defined for the elements H–Ar, Ca, and Ga–Kr for $n = 2$ and 3; for H–Ar and Ca for $n = 4$ and 5; and for H and B–Ne for $n = 6$. For first-row atoms, the contracted Gaussians present in the cc basis sets are

cc-pVDZ	cc-pVTZ	cc-pVQz	cc-pV5Z	cc-pV6Z
[$3s2p1d$]	[$4s3p2d1f$]	[$5s4p3d2f1g$]	[$6s5p4d3f2g1h$]	[$7s6p5d4f3g2h1i$]
14	30	55	91	140

where the last row gives the number of basis functions for a first-row atom [for example, 3(1) + 2(3) + 1(5) = 14 for cc-pVDZ]. Note that as we go from one basis set to the next, the number of sets of basis functions of each angular-momentum l value is increased by one and one set of functions with the next higher l value is added. The idea is that in going from one set to the next larger set, one adds basis functions that make similar contributions to the correlation energy. (Hence the name "correlation consistent.") For example, the first set of f functions has been found to lower the energy by a similar amount as the second set of added d functions, so in going from cc-pVDZ to cc-pVTZ, one adds a second set of d functions and the first set of f functions (and an additional s and a p set).

For an H atom, the cc-pVDZ set is $[2s1p]$ and the same pattern is followed as for first-row atoms. For example, in going to cc-pVTZ for H, one adds an s set, a p set, and a d set to get $[3s2p1d]$.

The cc-pVDZ set is roughly comparable to the 6-31G** set, since both are $[2s1p]$ for H and $[3s2p1d]$ for first-row atoms.

The addition of diffuse primitive nonpolarization and polarization functions to the cc-pVnZ basis sets gives the augmented sets aug-cc-pVDZ, aug-cc-pVTZ, etc., suitable for calculations on anions and hydrogen-bonded species. To form the set aug-cc-pVnZ from cc-pVnZ, the number of sets of basis functions of each angular-momentum l value is increased by one by the addition of diffuse primitives. Thus, aug-cc-pVTZ is $[5s4p3d2f]$ for a first-row atom. For calculation of the electric polarizability of molecules, the convergence rate with increase in basis-set size is greatly increased by using doubly augmented (d-aug-cc-pVnZ) sets. For a first-row atom, d-aug-cc-pVTZ is $[6s5p4d3f]$.

The addition of certain primitive Gaussians to the cc-pVnZ sets gives the cc-pCVDZ, cc-pCVTZ, . . . sets (where CV stands for core/valence), which are designed for calculations that include correlation effects involving the core electrons. Diffuse functions can be added to these CV sets to give the aug-cc-pCVnZ sets.

The correlation-consistent basis sets have the advantage that as the basis-set size is expanded, the calculated values of a molecular property usually converge smoothly to a limiting value, making it easy to find this value by extrapolation. Because of this feature, these basis sets are the most widely used sets in high-level ab initio calculations. The disadvantage of these basis sets is that they are rather large.

Commonly used Gaussian basis sets and literature references for them are available at the EMSL Gaussian Basis Set Exchange website (bse.pnl.gov/bse/portal) at the Environmental Molecular Sciences Laboratory of the Pacific Northwest Laboratory.

Because of the time savings given by Gaussians in multicenter-integral evaluation, nearly all current molecular ab initio calculations use contracted-Gaussian basis sets. Most semiempirical methods (which neglect large classes of integrals) use STOs. Density-functional calculations may use Gaussians, STOs, or other basis functions.

EXAMPLE Use the Gaussian Basis Set Exchange on the Internet to find the 3-21G basis functions for the oxygen atom.

On the basis-set form, choose the 3-21G basis set and click O on the periodic table. From the drop-down menu at the upper left choose All Electron. The website allows one to choose the basis-set output to correspond to the format used by one of several programs. If we choose

TABLE 15.1 The 3-21G Basis Set for the Oxygen Atom

O	0		
S	3	1.00	
	322.03700000	0.05923940	
	48.43080000	0.35150000	
	10.42060000	0.70765800	
SP	2	1.00	
	7.40294000	−0.40445300	0.24458600
	1.57620000	1.22156000	0.85395500
SP	1	1.00	
	0.37368400	1.00000000	1.00000000

Gaussian 94 and click on Get Basis Set, we get Table 15.1. The O in the first row indicates the oxygen atom. The zero after the O is required by the *Gaussian 94* program and can be ignored. On the second line, the S and 3 indicate that an s-type CGTF consisting of 3 primitive Gaussians follows; the 1.00 is a scale factor and since its value is 1, it can be ignored. The first and second columns in the next three rows give the orbital exponents and contraction coefficients, respectively. The large values of these orbital exponents show that this is the inner-shell (core) $1s$ AO. Thus the $1s$ CGTO is

$$1s = 0.0592394g_s(322.037) + 0.3515g_s(48.4308) + 0.707658g_s(10.4206)$$

where $g_s(322.037)$ denotes a normalized primitive s-type GTF with orbital exponent 322.037 [Eq. (15.11)]. The SP on the sixth and ninth lines indicates that orbital exponents and contraction coefficients for s-type and p-type CGTFs follow. These are for the valence $2s$ and $2p$ AOs. The 3-21G set uses the same orbital exponents for the $2s$ and $2p$ AOs, so as to speed up calculations. The first column of numbers gives the orbital exponents and the second and third columns give the contraction coefficients. Thus the valence CGTFs are $2s' = -0.404453g_s(7.40294) + 1.22156g_s(1.5762)$, $2p_x' = 0.244586g_{p_x}(7.40294) + 0.853955g_{p_x}(1.5762), \ldots, 2s'' = g_s(0.373684)$, $2p_x'' = g_{p_x}(0.373684), \ldots,$ where the dots indicate $2p_y$ and $2p_z$ CGTFs. [Sometimes people omit the normalization constant from each primitive Gaussian (15.11) and include it within the contraction coefficient for that Gaussian. The fact that the contraction coefficient for each outer $2s$ and $2p$ CGTF is 1.00 assures us that the primitive Gaussians are normalized.]

To describe a quantum-mechanical calculation, one specifies the method and the basis set. The letters HF (for Hartree–Fock) denote any ab initio SCF MO calculation, whether or not the basis set is large enough to come close to the Hartree–Fock limit. Thus the notation HF/3-21G denotes an ab initio SCF MO calculation that uses the 3-21G basis set.

15.5 THE SCF MO TREATMENT OF H_2O

For a minimal-basis-set MO treatment of H_2O (Fig. 15.1 in Section 15.2), we start with the $O1s$, $O2s$, $O2p_x$, $O2p_y$, and $O2p_z$, inner-shell and valence oxygen AOs and the H_11s and H_21s valence AOs of the hydrogen atoms. Linear combinations of these seven basis AOs give LCAO approximations to the seven lowest MOs of water. As stated in Section 15.3, the MOs can be chosen so that upon application of the molecular symmetry operators each MO transforms according to one of the symmetry species of the molecular point group. To aid in choosing the right linear combinations of AOs, we examine the symmetry behavior of the AOs.

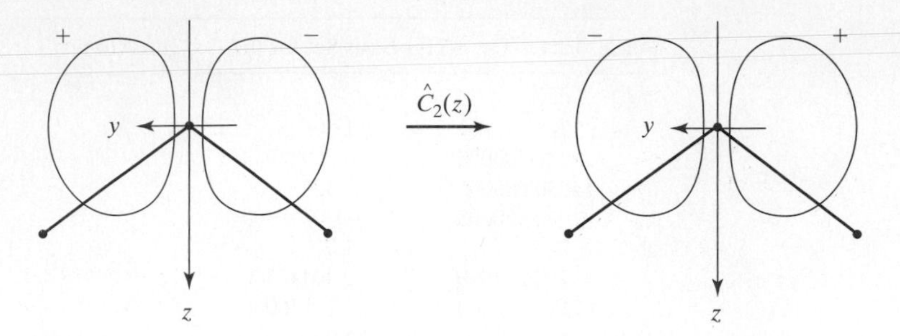

FIGURE 15.2 The effect of $\hat{C}_2(z)$ on the $2p_y$ oxygen AO in H_2O.

The $1s$ and $2s$ oxygen AOs are spherically symmetric, so rotation about the $C_2(z)$ axis and reflection in the xz or yz plane has no effect on them. They thus belong to the totally symmetric symmetry species a_1 of (15.3). The effect of a $\hat{C}_2(z)$ rotation on the oxygen $2p_y$ AO is shown in Fig. 15.2. This rotation sends $O2p_y$ into its negative (see also Fig. 12.10). Reflection in the xz plane sends $O2p_y$ into its negative, and reflection in the molecular yz plane sends $O2p_y$ into itself. The symmetry eigenvalues of $O2p_y$ are $1, -1, -1$, and 1. The symmetry species of this AO is b_2. Similarly, the symmetry species of $O2p_x$ and $O2p_z$ are found to be b_1 and a_1 respectively.

Now for the hydrogen $1s$ AOs. Reflection in the yz plane leaves each of them unchanged. However, rotation by 180° about the z axis sends H_11s into H_21s and vice versa (Fig. 15.3). The H_11s and H_21s functions are *not* eigenfunctions of $\hat{O}_{C_2(z)}$ and so do not transform according to any of the symmetry species of H_2O.

As a preliminary step in finding the MOs of a molecule, it is helpful (but not essential) to construct linear combinations of the original basis AOs such that each linear combination does transform according to one of the molecular symmetry species. Such linear combinations are called **symmetry orbitals** or **symmetry-adapted basis functions**. The symmetry orbitals are used as the basis functions χ_s in the expansions $\phi_i = \sum_s c_{si}\chi_s$ [Eq. (14.33)] of the MOs ϕ_i. *The use of basis functions that transform according to the molecular symmetry species simplifies the calculation by putting the secular determinant in block-diagonal form.* This will be illustrated below.

Each oxygen AO transforms according to one of the symmetry species of H_2O and can serve as a symmetry orbital. However, neither of the two hydrogen $1s$ AOs

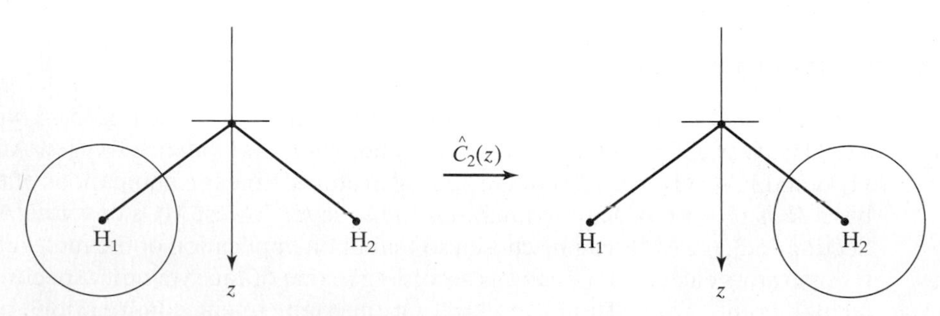

FIGURE 15.3 The effect of a $\hat{C}_2(z)$ rotation on the H_11s AO in H_2O.

belongs to a symmetry species of H_2O, and we must construct two symmetry orbitals from these AOs. Consider the linear combinations:

$$H_1 1s + H_2 1s \quad \text{and} \quad H_1 1s - H_2 1s \tag{15.13}$$

We have

$$\hat{O}_{C_2(z)}(H_1 1s + H_2 1s) = H_2 1s + H_1 1s$$
$$\hat{O}_{C_2(z)}(H_1 1s - H_2 1s) = H_2 1s - H_1 1s$$

Thus the first and second functions in (15.13) are eigenfunctions of $\hat{O}_{C_2(z)}$ with eigenvalues +1 and −1, respectively. Examination of the effects of the other three symmetry operators shows the functions (15.13) to belong to the symmetry species a_1 and b_2, respectively. We shall not bother to normalize the symmetry orbitals (15.13). The seven basis symmetry functions and their symmetry species are then

χ_1	χ_2	χ_3	χ_4	χ_5	χ_6	χ_7
$H_1 1s + H_2 1s$	$O1s$	$O2s$	$O2p_z$	$H_1 1s - H_2 1s$	$O2p_y$	$O2p_x$
a_1	a_1	a_1	a_1	b_2	b_2	b_1

(15.14)

Now consider the SCF secular determinant $\det(F_{rs} - \varepsilon_i S_{rs})$ [Eq. (14.36)]. We assert that

$$F_{rs} \equiv \langle \chi_r | \hat{F} | \chi_s \rangle = 0 \tag{15.15}$$

whenever χ_r and χ_s belong to different symmetry species. This result follows from the theorem [Eq. (7.50)] that $\langle g_j | \hat{B} | g_k \rangle = 0$ if g_j and g_k are eigenfunctions of a Hermitian operator $\hat{A}$ with different eigenvalues, where $\hat{A}$ commutes with $\hat{B}$. The symmetry orbitals of H_2O are eigenfunctions of the symmetry operators, each of which commutes with the electronic Hamiltonian and with the Fock operator $\hat{F}$. Symmetry orbitals χ_r and χ_s that belong to different symmetry species differ in at least one symmetry eigenvalue. Hence (15.15) follows. Moreover, since two eigenfunctions of a Hermitian operator that correspond to different eigenvalues are orthogonal, we have

$$S_{rs} \equiv \langle \chi_r | \chi_s \rangle = 0 \tag{15.16}$$

whenever χ_r and χ_s belong to different symmetry species. From (15.15) and (15.16), it follows that the use of symmetry orbitals puts the secular determinant of H_2O in block diagonal form, each block corresponding to a different symmetry species. The blocks are 4×4, 2×2, and 1×1. [For molecules with degenerate symmetry species (*E*, *T*, and so on), the symmetry orbitals of the degenerate species are not necessarily eigenfunctions of the symmetry operators. Nevertheless, the symmetry orbitals still put the secular determinant in block-diagonal form, as can be shown using group theory.]

The set of Roothaan simultaneous equations (14.34) then breaks up into one set of four simultaneous equations, one set of two simultaneous equations, and one set of one "simultaneous" equation (Section 9.6). The first set contains matrix elements involving only the four a_1 symmetry orbitals. Therefore, four of the lowest seven H_2O MOs are linear combinations of the four a_1 symmetry orbitals. These four MOs must have a_1 symmetry. Similarly, we have two MOs of b_2 symmetry and one MO of b_1 symmetry. The symmetry orbitals are *not* (in general) the MOs, but *each MO must be a linear combination of those symmetry orbitals having the same symmetry species as the MO.* The forms of the lowest MOs of H_2O are then

$$\begin{aligned} \phi_1 &= c_{11}\chi_1 + c_{21}\chi_2 + c_{31}\chi_3 + c_{41}\chi_4 & \phi_5 &= c_{55}\chi_5 + c_{65}\chi_6 \\ \phi_2 &= c_{12}\chi_1 + c_{22}\chi_2 + c_{32}\chi_3 + c_{42}\chi_4 & \phi_6 &= c_{56}\chi_5 + c_{66}\chi_6 \\ \phi_3 &= c_{13}\chi_1 + c_{23}\chi_2 + c_{33}\chi_3 + c_{43}\chi_4 & \phi_7 &= \chi_7 \\ \phi_4 &= c_{14}\chi_1 + c_{24}\chi_2 + c_{34}\chi_3 + c_{44}\chi_4 \end{aligned} \tag{15.17}$$

The next step in the SCF MO calculation is to choose explicit forms for the seven AOs. The orbital energies and the coefficients of the symmetry orbitals are then found using Roothaan's equations.

Pitzer and Merrifield did an H_2O minimal-basis-set calculation, representing each AO by a single STO [R. M. Pitzer and D. P. Merrifield, *J. Chem. Phys.*, **52**, 4782 (1970); S. Aung, R. M. Pitzer, and S. I. Chan, *J. Chem. Phys.*, **49**, 2071 (1968)]. They optimized the orbital exponents, finding 1.27 for H1s, 7.66 for O1s, 2.25 for O2s, and 2.21 for O2p. (To optimize the exponents, one must repeat the entire SCF iterative calculation for several different sets of orbital exponents to locate the set that gives the minimum energy. Since orbital-exponent optimization is time consuming, it is only feasible for small molecules.) The calculated orbital energies in hartrees are: $1a_1$, −20.56; $2a_1$, −1.28; $1b_2$, −0.62; $3a_1$, −0.47; $1b_1$, −0.40. The ground-state electron configuration of this ten-electron molecule is

$$(1a_1)^2(2a_1)^2(1b_2)^2(3a_1)^2(1b_1)^2 \tag{15.18}$$

The ground state has a closed-shell configuration and is a $^1A_1$ state.

The five lowest SCF MOs found by Pitzer and Merrifield at the experimental geometry are

$$\begin{aligned} 1a_1 &= 1.000(\text{O}1s) + 0.015(\text{O}2s_\perp) + 0.003(\text{O}2p_z) - 0.004(\text{H}_11s + \text{H}_21s) \\ 2a_1 &= -0.027(\text{O}1s) + 0.820(\text{O}2s_\perp) + 0.132(\text{O}2p_z) + 0.152(\text{H}_11s + \text{H}_21s) \\ 1b_2 &= 0.624(\text{O}2p_y) + 0.424(\text{H}_11s - \text{H}_21s) \\ 3a_1 &= -0.026(\text{O}1s) - 0.502(\text{O}2s_\perp) + 0.787(\text{O}2p_z) + 0.264(\text{H}_11s + \text{H}_21s) \\ 1b_1 &= \text{O}2p_x \end{aligned} \tag{15.19}$$

The O2$s_\perp$ orbital in (15.19) is an orthogonalized orbital [Eq. (13.125)]:

$$\text{O}2s_\perp = 1.028[\text{O}2s - 0.2313(\text{O}1s)] \tag{15.20}$$

where O2s is the ordinary 2s STO:

$$\text{O}2s = 2.25^{5/2}\pi^{-1/2}3^{-1/2}r_\text{O}\exp(-2.25r_\text{O})$$

where r_O is the distance to the oxygen nucleus. The MO approximation to the ground-state wave function of H_2O is the 10×10 Slater determinant

$$|1a_1\overline{1a_1}2a_1\overline{2a_1}1b_2\overline{1b_2}3a_1\overline{3a_1}1b_1\overline{1b_1}| \tag{15.21}$$

Consider the nature of the MOs. The lowest MO, $1a_1$, is essentially a pure nonbonding 1s oxygen AO, which is hardly surprising.

The oxygen part of the $2a_1$ MO is mostly O2s with some O2p_z mixed in. This mixing in (hybridization) of O2p_z adds to the value of the O2s AO along the positive z axis (which lies between the hydrogens; Fig. 15.1) and subtracts from O2s along the negative z axis. The combination of the hybridized O2s and O2p_z orbitals with the $H_1 1s$ and $H_2 1s$ orbitals in the $2a_1$ MO then gives electron probability-density buildup in the region enclosed by the three nuclei. Therefore, the $2a_1$ MO contributes to the bonding in water.

Consider the $1b_2$ MO. The $2p_y$ oxygen AO has its positive lobe on the H_1 side of the molecule, and so the positive lobe of O2p_y adds to $H_1 1s$ in the $1b_2$ MO, giving electron charge buildup between the H_1 and O nuclei. Similarly, the negative lobe of O2p_y adds to $-H_2 1s$, giving charge buildup between O and H_2 in this MO. Hence $1b_2$ is a bonding MO.

The hybridization of the $2s$ and $2p_z$ oxygen AOs in $3a_1$ builds up electron probability density in the region around the negative z axis, away from the hydrogens, giving this MO substantial lone-pair character. On the positive side of the z axis, the oxygen $2s$ and $2p_z$ AOs tend to cancel each other, and we get little bonding overlap between the oxygen hybrid and the hydrogen AOs. Alternatively, we can look separately at the overlap between O2$s_\perp$ and the hydrogens (which is negative or antibonding) and the overlap between O2p_z and the hydrogens (which is positive or bonding). Because of the approximate cancellation of these overlaps (see Section 15.6), the $3a_1$ MO has little bonding character and is best described as mainly a lone-pair MO.

The $1b_1$ MO is a nonbonding lone-pair $2p_x$ oxygen AO.

Figure 15.4 shows the shapes of the bonding MOs $2a_1$ and $1b_2$. Note that their symmetry eigenvalues are as given by (15.3). [For accurately plotted H_2O MO contours, see T. H. Dunning, R. M. Pitzer, and S. Aung, *J. Chem. Phys.*, **57**, 5044 (1972).]

Each of the SCF bonding MOs in (15.19) and Fig. 15.4 is delocalized over the entire molecule and does not resemble a chemical bond. The relation of these MOs to the picture presented in many chemistry books, where one bonding MO points along the O—H_1 bond and the other points along the O—H_2 bond, is discussed in Section 15.8.

The unoccupied $4a_1$ and $2b_2$ MOs of water calculated by Pitzer and Merrifield (for nonoptimized, Slater-rule exponents) are

$$\begin{aligned} 4a_1 &= 0.08(\text{O}1s) + 0.84(\text{O}2s_\perp) + 0.70(\text{O}2p_z) - 0.75(\text{H}_1 1s + \text{H}_2 1s) \\ 2b_2 &= 0.99(\text{O}2p_y) - 0.89(\text{H}_1 1s - \text{H}_2 1s) \end{aligned} \tag{15.22}$$

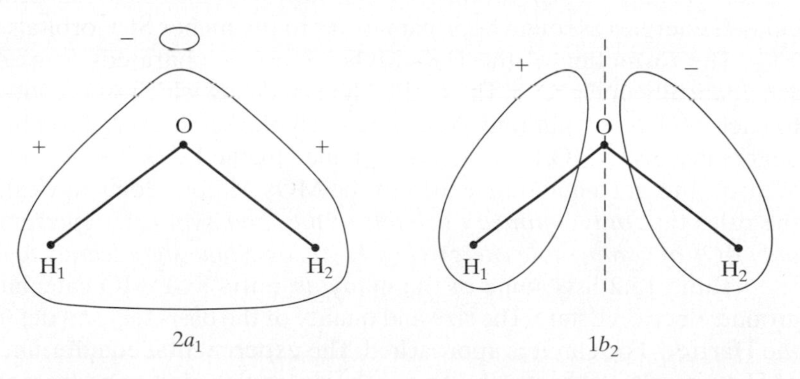

FIGURE 15.4 Sketches of the two main bonding MOs of H_2O.

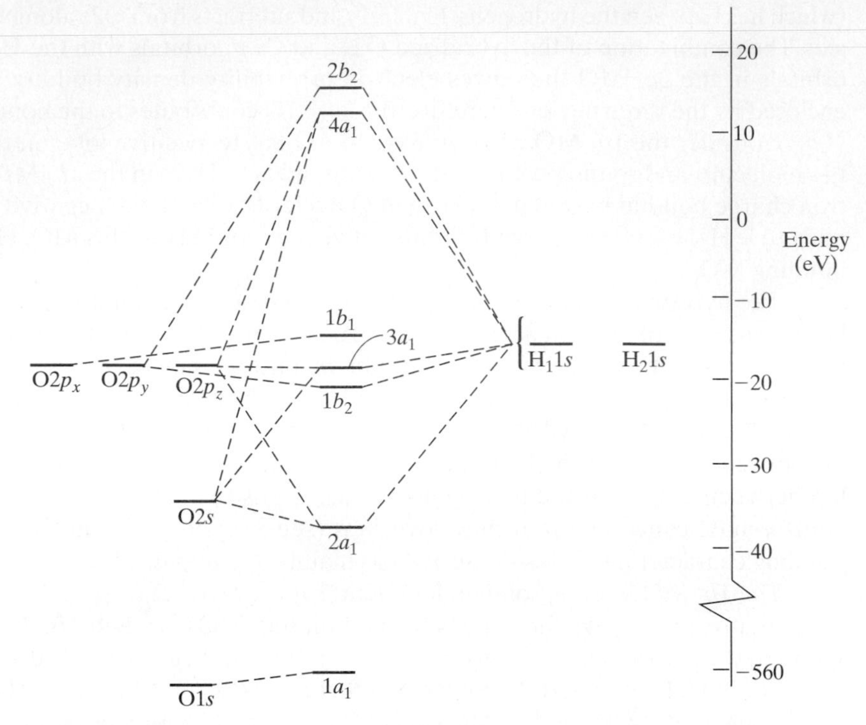

FIGURE 15.5 Formation of the H_2O. MOs from the minimal-basis AOs. The five lowest MOs are filled in the ground state. (Note the break in the scale.)

For these two MOs, the opposite signs of the oxygen and hydrogen AO coefficients give charge depletion between the nuclei. These MOs are antibonding.

The unoccupied orbitals (15.22) are called **virtual orbitals**. Because they were calculated for the electron configuration (15.18), they are not accurate representations of the SCF orbitals actually occupied in H_2O excited electronic states. When the electron configuration changes, the interorbital electronic interactions change, thereby changing the forms of all the SCF orbitals. We can, however, use the virtual orbitals and their calculated energies as rough approximations to the higher SCF orbitals and energies.

The formation of the H_2O MOs from the separated-atoms AOs is illustrated schematically in Fig. 15.5. The dashed lines indicate which AOs contribute significantly to each MO. Note that only AOs of roughly the same energy combine to a significant degree in a given MO. [This fact is explained by the $1/(E_n^{(0)} - E_m^{(0)})$ term in Eq. (9.28).] We can thus get a qualitative idea of the MOs without doing any calculations by using the rules that *only symmetry orbitals of the same symmetry species combine* and that *only AOs of comparable energy (Fig. 11.2) contribute significantly to a given MO*.

Table 15.2 lists some of the many ab initio SCF MO calculations on the H_2O ground electronic state. The size and quality of the basis set used determine how closely the Hartree–Fock limit is approached. The experimental equilibrium electronic energy of H_2O is −76.480 hartrees. (This is the energy needed to go from the molecule at its equilibrium geometry to nuclei and electrons at infinite separation from one another.)

TABLE 15.2 H_2O SCF MO Calculations[a]

Reference[b]	Basis Set[c]	Energy/E_h	μ/D	θ	R_{OH}/A
CCCBDB	STO-3G, 7	−74.966	1.71	100.0°	0.989
CCCBDB	3-21G, 13	−75.586	2.39	107.7°	0.967
Pitzer, Merrifield	Minimal STO, 7	−75.705	1.92	100.3°	0.990
CCCBDB	6-31G*, 19	−76.011	2.20	105.5°	0.947
CCCBDB	6-31G**, 25	−76.024	2.15	106.0°	0.943
CCCBDB	cc-pVDZ, 24	−76.027	2.04	104.6°	0.946
K. S. Kim et al.	6-311++G(2*d*,2*p*), 47	−76.057	2.02	106.3°	0.940
CCCBDB	cc-pVTZ, 58	−76.058	1.99	106.0°	0.941
J. Kim et al.	aug-cc-pVTZ, 92	−76.061	1.93	106.3°	0.941
Dunning et al.	[6*s*5*p*2*d*/3*s*1*p*], 43	−76.062	2.08	106.6°	0.941
Rosenberg et al.	$(5s4p2d/3s1p)_{STO}$, 39	−76.064	2.00	106.1°	0.940
Feller et al.	cc-pVQZ, 115	−76.0655		106.2°	0.940
Feller et al.	aug-cc-pVQZ, 172	−76.0667		106.3°	0.940
Amos	[8*s*6*p*4*d*2*f*/6*s*3*p*1*d*], 112	−76.0675		106.3°	0.940
K. S. Kim et al.	(13*s*8*p*4*d*2*f*/8*s*4*p*2*d*), 131	−76.0676	1.94	106.3°	0.940
Bakken et al.	cc-pV5Z, 201	−76.0678		106.33°	0.9396
Bakken et al.	aug-cc-pV5Z, 287	−76.0680		106.34°	0.9396
Pahl, Handy	special[c]	−76.06817		106.34°	0.9396
Estimated Hartree–Fock energy[d]		-76.068_3			
Nonrelativistic fixed-nuclei energy[e]		−76.438			
Experimental values		−76.480[e]	1.85	104.5°[f]	0.958[f]

[a]Energy/E_h is the total electronic energy including nuclear repulsion in hartrees at the calculated equilibrium geometry; μ, θ, and R_{OH} are the calculated electric dipole moment, equilibrium bond angle, and equilibrium bond length.

[b]CCCBDB is the Computational Chemistry Comparison and Benchmark DataBase (Sec. 15.1); R. M. Pitzer and D. P. Merrifield, *J. Chem. Phys.*, **52**, 4782 (1970); K. S. Kim et al., *J. Chem. Phys.*, **97**, 6649 (1992); J. Kim et al., *J. Chem. Phys.*, **102**, 310 (1995) [see also S. S. Xantheas and T. H. Dunning, *J. Chem. Phys.*, **99**, 8774 (1993)]; T. H. Dunning, R. M. Pitzer, and S. Aung, *J. Chem. Phys.*, **57**, 5044 (1972); B. J. Rosenberg and I. Shavitt, *J. Chem. Phys.*, **63**, 2162 (1975) and B. J. Rosenberg, W. C. Ermler, and I. Shavitt, *J. Chem. Phys.*, **65**, 4072 (1976); D. Feller et al., *J. Chem. Phys.*, **100**, 4981 (1994); R. D. Amos, *J. Chem. Soc. Faraday Trans. 2*, **83**, 1595 (1987); K. S. Kim et al., op. cit; V. Bakken et al., *Mol. Phys.*, **96**, 653 (1999); F. A. Pahl and N. C. Handy, *Mol. Phys.*, **100**, 3199 (2002).

[c]The number of basis functions is given. All the basis sets are GTFs except the Pitzer–Merrifield, Rosenberg et al., and Pahl–Handy sets. Brackets denote CGTF sets and parentheses denote an uncontracted set. A slash separates the oxygen and hydrogen basis functions. The Pahl–Handy set consists of plane waves and radial polynomials.

[d]This estimate is for the calculated Hartree–Fock equilibrium geometry. See D. Feller, C. M. Boyle, and E. R. Davidson, *J. Chem. Phys.*, **86**, 3424 (1987).

[e]A. Lüchow, J. B. Anderson, and D. Feller, *J. Chem. Phys.*, **106**, 7706 (1997).

[f]P. Jensen et al., *J. Mol. Spectrosc.*, **168**, 271 (1994).

All the calculations listed are nonrelativistic. The only significant relativistic correction will occur in the 1*s* inner-shell electrons of oxygen (see Section 11.7). Since this shell remains essentially unchanged on molecule formation, one can use the relativistic energy correction calculated for the O atom as the relativistic correction for H_2O. In addition, there is a small correction due to motion of the center of mass of the nuclei relative to the center of mass of the molecule. When these two corrections are subtracted from the experimental energy of −76.480 hartrees, we obtain the nonrelativistic fixed-nuclei H_2O energy as −76.438 hartrees.

Note that, except for the first three calculations in Table 15.2, the SCF MO calculations give a bond length that is somewhat shorter than the true length. This is a general characteristic of the HF method, as is the tendency of HF calculations to predict a dipole moment that is somewhat too large.

As mentioned in Section 15.4, the cc basis sets can be used to extrapolate to the complete-basis-set (CBS) limit for molecular properties. Energies in hartrees for SCF MO H_2O optimized-geometry calculations with the cc-pVnZ basis sets for $n = 2, 3, 4, 5$ are $-76.02705, -76.05777, -76.06552, -76.06778$, respectively. The following empirical formula has been found to work rather well for extrapolating cc-pVnZ SCF MO energies to the CBS limit:

$$E_{\text{SCF}}(n) = E_{\text{SCF}}(\infty) + Ae^{-Bn} \tag{15.23}$$

where $E_{\text{SCF}}(n)$ is the SCF energy found with the cc-pVnZ set, $E_{\text{SCF}}(\infty)$ is the CBS limit predicted by cc-pVnZ in the limit $n \to \infty$, and A and B are positive parameters whose values are found by a least-squares fit of the calculated energies. (This formula can be applied using either energies at optimized geometries or energies at a fixed geometry.) There are three unknown quantities in (15.23), A, B, and $E_{\text{SCF}}(\infty)$ and we have four calculated energies to do the least-squares fit. Although all four energies could be used, experience shows that the accuracy of the extrapolation is usually improved if the energy of the rather small cc-pVDZ set is not used in the fit. Use of the $n = 3, 4, 5$ energies gives (Prob. 15.21) $E_{\text{SCF}}(\infty) = -76.0687$ hartrees, close to the -76.068_3 value listed in Table 15.2.

Although the SCF MO calculations give good geometries and fairly good dipole moments, they give poor dissociation energies. For the near-Hartree–Fock calculations of Rosenberg et al., the difference between the energy of two isolated hydrogen atoms and one oxygen atom (as calculated with the oxygen basis set used in the H_2O calculation) and the calculated H_2O energy at the optimized geometry gives $D_e = 6.94$ eV, compared with the experimental value 10.09 eV. This large error in dissociation energy is typical of Hartree–Fock calculations, as we saw for diatomics. When H_2O is formed from $2H + O$, two new electron pairs are formed. Two electrons paired in the same MO move through the same region of space, and hence the correlation energy of such a pair is substantial. The Hartree–Fock calculation does not take into account this extra correlation in the molecule (as compared with the separated atoms) and hence gives too small a binding energy.

Hartree–Fock orbital energies have experimental as well as theoretical significance. In 1933, Koopmans gave arguments that indicate that the energy required to remove an electron from a closed-shell atom or molecule is reasonably well approximated by minus the orbital energy ε of the AO or MO from which the electron is removed, a result called **Koopmans' theorem**. A partial justification of this result is the fact that, if we neglect the change in the form of the MOs that occurs when the molecule is ionized, then the difference between the Hartree–Fock energies of the ion and the neutral closed-shell molecule can be shown to equal the orbital energy of the MO from which the electron was removed (see Prob. 15.18).

The energy needed to remove an electron from an MO of a molecule can be found experimentally using photoelectron spectroscopy. Here, one uses photons of known energy to knock electrons out of gas-phase molecules and measures the kinetic energies of the emitted electrons. A comparison of minus the near-Hartree–Fock orbital energies (Rosenberg and Shavitt, cited in Table 15.2) with the experimentally observed vertical ionization energies from the various H_2O MOs follows, where the energies are in electronvolts and the experimental values are in parentheses: $1a_1$, 559.5 (539.7); $2a_1$, 36.7 (32.2); $1b_2$, 19.5 (18.5); $3a_1$, 15.9 (14.7); $1b_1$, 13.8 (12.6). (The *vertical* ionization energy is the energy difference between the molecule M and the ion M^+ when M^+ is at the equilibrium geometry of M, whereas for the *adiabatic* ionization energy, M^+ is at the equilibrium geometry of M^+. In the unlikely event that the equilibrium geometries of M and M^+ are the same, these two ionization energies are equal. Otherwise, the vertical ionization energy is larger than the adiabatic ionization energy. Koopmans' theorem predicts the vertical ionization energy.) Koopmans'-theorem ionization energies are somewhat inaccurate because of (1) neglect of the change in the forms of the MOs that occurs on ionization, and (2) neglect of the change in correlation energy between the neutral molecule and the ion.

15.6 POPULATION ANALYSIS AND BOND ORDERS

A widely used (and widely criticized) method to analyze SCF MO wave functions is population analysis, introduced by Mulliken. He proposed a method that apportions the electrons of an n-electron molecule into net populations n_r in the basis functions χ_r and overlap populations n_{r-s} for all possible pairs of basis functions.

For the set of basis functions $\chi_1, \chi_2, \ldots, \chi_b$, each MO ϕ_i has the form

$$\phi_i = \sum_{s=1}^{b} c_{si}\chi_s = c_{1i}\chi_1 + c_{2i}\chi_2 + \cdots + c_{bi}\chi_b$$

For simplicity, we shall assume that the c_{si}'s and χ_s's are real. The probability density associated with one electron in ϕ_i is

$$|\phi_i|^2 = c_{1i}^2\chi_1^2 + c_{2i}^2\chi_2^2 + \cdots + 2c_{1i}c_{2i}\chi_1\chi_2 + 2c_{1i}c_{3i}\chi_1\chi_3 + 2c_{2i}c_{3i}\chi_2\chi_3 + \cdots$$

Integrating this equation over three-dimensional space and using the fact that ϕ_i and the χ_s's are normalized, we get

$$1 = c_{1i}^2 + c_{2i}^2 + \cdots + 2c_{1i}c_{2i}S_{12} + 2c_{1i}c_{3i}S_{13} + 2c_{2i}c_{3i}S_{23} + \cdots \tag{15.24}$$

where the S's are overlap integrals: $S_{12} = \int \chi_1\chi_2\, dv_1\, dv_2$, etc. Mulliken proposed that the terms in (15.24) be apportioned as follows. One electron in the MO ϕ_i contributes c_{1i}^2 to the net population in χ_1, c_{2i}^2 to the net population in χ_2, etc., and contributes $2c_{1i}c_{2i}S_{12}$ to the overlap population between χ_1 and χ_2, $2c_{1i}c_{3i}S_{13}$ to the overlap population between χ_1 and χ_3, etc.

Let there be n_i electrons in the MO ϕ_i ($n_i = 0, 1, 2$) and let $n_{r,i}$ and $n_{r-s,i}$ symbolize the contributions of electrons in the MO ϕ_i to the net population in χ_r and to the overlap population between χ_r and χ_s, respectively. We have

$$n_{r,i} = n_i c_{ri}^2, \qquad n_{r-s,i} = n_i(2c_{ri}c_{si}S_{rs}) \tag{15.25}$$

By summing over the occupied MOs ϕ_i, we obtain the Mulliken **net population** n_r in χ_r and the **overlap population** n_{r-s} for the pair χ_r and χ_s as

$$n_r = \sum_i n_{r,i} \quad \text{and} \quad n_{r-s} = \sum_i n_{r-s,i}$$

The sum of all the net and overlap populations equals the total number of electrons n in the molecule (Prob. 15.22): $\sum_r n_r + \sum_{r>s}\sum_s n_{r-s} = n$.

EXAMPLE For the H_2O MOs in (15.19), calculate the net and overlap population contributions from the $2a_1$ MO and find n_r for each basis function. Use H_11s and H_21s as basis functions, rather than the symmetry-adapted basis functions.

To find overlap populations, we need the overlap integrals. Since an orthogonalized 2*s* AO is used, all the basis functions centered on oxygen are mutually orthogonal, and the overlap populations are zero for all pairs of functions centered on oxygen. (For extended basis sets, this is not true. For example, a DZ basis set uses two 1*s*-type functions on oxygen and these functions are not orthogonal to each other.) Overlap integrals between basis STOs centered on different atoms can be found by interpolation in the tables of R. S. Mulliken et al., *J. Chem. Phys.*, **17**, 1248 (1949). One finds (Prob. 15.23)

$$\langle \mathrm{H_11}s|\mathrm{O1}s\rangle = \langle \mathrm{H_21}s|\mathrm{O1}s\rangle = 0.054, \quad \langle \mathrm{H_11}s|\mathrm{O2}s_\perp\rangle = \langle \mathrm{H_21}s|\mathrm{O2}s_\perp\rangle = 0.471$$
$$\langle \mathrm{H_11}s|\mathrm{O2}p_y\rangle = -\langle \mathrm{H_21}s|\mathrm{O2}p_y\rangle = 0.319, \quad \langle \mathrm{H_11}s|\mathrm{O2}p_z\rangle = \langle \mathrm{H_21}s|\mathrm{O2}p_z\rangle = 0.247$$
$$\langle \mathrm{H_11}s|\mathrm{H_21}s\rangle = 0.238$$

The contributions to the net populations in the basis functions from the two electrons in the $2a_1$ MO are given by (15.25) and (15.19) as $n_{\mathrm{O1}s,2a_1} = 2(-0.027)^2 = 0.0015$, $n_{\mathrm{O2}s_\perp,2a_1} = 2(0.820)^2 = 1.345$, $n_{\mathrm{O2}p_z,2a_1} = 0.035$, $n_{\mathrm{H_11}s,2a_1} = 2(0.152)^2 = 0.046$, $n_{\mathrm{H_21}s,2a_1} = 0.046$. The $2a_1$ contributions to the nonzero overlap populations are given by (15.25) and (15.19) as

$$n_{\mathrm{O1}s-\mathrm{H_11}s,2a_1} = 2(2)(-0.027)(0.152)(0.054) = -0.0009 = n_{\mathrm{O1}s-\mathrm{H_21}s,2a_1}$$
$$n_{\mathrm{O2}s_\perp-\mathrm{H_11}s,2a_1} = 0.235 = n_{\mathrm{O2}s_\perp-\mathrm{H_21}s,2a_1}$$
$$n_{\mathrm{O2}p_z-\mathrm{H_11}s,2a_1} = 0.020 = n_{\mathrm{O2}p_z-\mathrm{H_21}s,2a_1}, \quad n_{\mathrm{H_11}s-\mathrm{H_21}s,2a_1} = 0.022$$

The net population of O1*s* is found from (15.25) and (15.19) as the sum of contributions from each occupied MO:

$$n_{\mathrm{O1}s} = 2(1.000)^2 + 2(-0.027)^2 + 2(-0.026)^2 = 2.00$$

The net populations for the other basis functions are (Prob. 15.24a) $n_{\mathrm{O2}s_\perp} = 1.85$, $n_{\mathrm{O2}p_x} = 2.00$, $n_{\mathrm{O2}p_y} = 0.78$, $n_{\mathrm{O2}p_z} = 1.27$, $n_{\mathrm{H_11}s} = 0.54_5$, $n_{\mathrm{H_21}s} = 0.54_5$.

To decide whether the MO ϕ_i in a covalent molecule is bonding, we examine the sum of those overlap-population contributions $n_{r-s,i}$ for which the basis functions χ_r and χ_s lie on different atoms. If this interatomic overlap population contribution is substantially positive, the MO is bonding; if it is substantially negative, the MO is antibonding. If it is zero or near zero, the MO is nonbonding.

For example, for the $3a_1$ MO in (15.19), overlap of O1*s* with H_11s contributes $2(2)(-0.026)(0.264)(0.054) = -0.001_5$, overlap of O1*s* with H_21s contributes -0.001_5, overlap of $O2s_\perp$ with H_11s contributes $2(2)(-0.502)(0.264)(0.471) = -0.250$, overlap of $O2s_\perp$ with H_21s contributes -0.250, overlap of $O2p_z$ with H_11s contributes 0.205 and with H_21s contributes 0.205, and overlap of H_11s with H_21s contributes $2(2)(0.264)^2(0.238) = 0.066$. Summing, we get an interatomic overlap population of -0.03 for the $3a_1$ MO. This value is near zero, indicating a nonbonding (lone-pair) MO. The interatomic overlap population for $2a_1$ is found to be 0.53 and that for $1b_2$ is 0.50 (Prob. 15.24). These are bonding MOs. For the inner-shell $1a_1$ MO, we get 0.00.

Instead of apportioning the electrons into net populations in basis functions and overlap populations for pairs of basis functions, it is convenient for some purposes to apportion the electrons among the basis functions only, with no overlap populations. Mulliken proposed that this be done by splitting each overlap population n_{r-s} equally between the basis functions χ_r and χ_s. For each basis function χ_r, this gives a **gross population** N_r in χ_r that equals the net population n_r plus one-half the sum of the overlap populations between χ_r and all other basis functions:

$$N_r = n_r + \frac{1}{2}\sum_{s \neq r} n_{r-s}$$

The sum of all the gross populations equals the number of electrons in the molecule: $\sum_{r=1}^{b} N_r = n$

For example, the contribution to the gross population of $O2s_\perp$ from the $2a_1$ MO in (15.19) is

$$N_{O2s_\perp,2a_1} = 2[(0.820)^2 + (0.820)(0.152)(0.471) + (0.820)(0.152)(0.471)] = 1.58$$

One finds these other contributions to the gross population of $O2s_\perp$: 0.00 from $1a_1$, 0.25 from $3a_1$, zero from $1b_2$ and $1b_1$. Addition of these contributions gives a gross population (or occupation number) of 1.83 for $O2s_\perp$. Carrying out the calculation for the other basis functions, one finds (Prob. 15.24c) the following gross populations (where the contributions are listed in the order $1a_1, 2a_1, 1b_2, 3a_1, 1b_1$): $N_{O1s} = 2.00 + 0.00 + 0 + 0.00 + 0 = 2.00$; $N_{O2s_\perp} = 1.83$; $N_{O2p_x} = 0 + 0 + 0 + 0 + 2 = 2$; $N_{O2p_y} = 0 + 0 + 1.12 + 0 + 0 = 1.12$; $N_{O2p_z} = 0 + 0.05_5 + 0 + 1.44_5 + 0 = 1.50$; $N_{H_11s} = 0.00 + 0.18_4 + 0.44_2 + 0.150 + 0 = 0.77_6$; $N_{H_21s} = 0.77_6$.

Addition of the gross populations for all basis functions centered on atom B gives the **gross atomic population** N_B for atom B:

$$N_B = \sum_{r \in B} N_r$$

where the notation $r \in B$ denotes all basis functions centered on atom B. Provided all basis functions are atom centered (this is usually true), the sum of the gross atomic populations equals the number of electrons in the molecule. The Mulliken **net atomic charge** q_B on atom B with atomic number Z_B is defined as

$$q_B \equiv Z_B - N_B$$

For example, for the Pitzer–Merrifield H_2O wave function, the gross atomic populations, found by summing the gross populations of the basis functions on each atom, are $N_O = 2.00 + 1.83 + 2 + 1.12 + 1.50 = 8.45$, $N_{H_1} = 0.77_6 = N_{H_2}$; the net charges are $q_O = 8 - 8.45 = -0.45$, $q_{H_1} = 1 - 0.77_6 = 0.22_4 = q_{H_2}$. As expected, the oxygen is negatively charged.

One should not rely too much on numbers calculated by population analysis. Mulliken's assignment of half the overlap population to each basis function is arbitrary and sometimes leads to unphysical results (see *Mulliken and Ermler, Diatomic Molecules*, pages 36–38, 88–89). Moreover, a small change in basis set can produce a large change in the calculated net charges. For example, net atomic charges on each H atom in CH_4, NH_3, H_2O, and HF calculated from HF/STO-3G and HF/3-21G wave functions are (CCCBDB, Section 15.1)

	CH_4	NH_3	H_2O	HF
STO-3G	0.07	0.15	0.17	0.19
3-21G	0.20	0.29	0.37	0.45

Comparison of values calculated with the same basis set correctly shows the increasing charge on each H atom as the electronegativity increases from C to N to O to F, but comparison of values calculated with different basis sets could erroneously lead one to say that the C—H bond in CH_4 is more polar than the F—H bond in FH.

An improvement on Mulliken population analysis (MPA) is **natural population analysis** (NPA) [A. E. Reed, R. B. Weinstock, and F. Weinhold, *J. Chem. Phys.*, **83**, 735 (1985)], which uses ideas related to natural orbitals (Section 16.1). Here, one first calculates a set of orthonormal natural atomic orbitals (NAOs) from the AO basis set χ_r. The NAOs are then used to compute a set of orthonormal **natural bond orbitals** (NBOs), where each occupied NBO is classifiable as a core, lone pair, or bond orbital. Using these NBOs, one carries out a population analysis. NPA net atomic charges show less basis-set dependence than those from Mulliken population analysis. Other methods of assigning net atomic charges are discussed in the next section. Still another method of population analysis that yields net atomic charges is *Löwdin population analysis* (*Cramer*, Section 9.1.3.2).

A review of population analysis recommended that in view of the existence of improved methods, Mulliken population analysis should no longer be used [S. M. Bachrach in K. Lipkowitz and D. B. Boyd (eds.), *Reviews in Computational Chemistry*, vol. 5, VCH (1994), Chapter 3.]

Bond Orders. Many methods have been proposed to calculate bond orders from SCF MO wave functions. One useful definition is due to Mayer [I. Mayer, *Chem. Phys. Lett.*, **97**, 270 (1983)]. For a closed-shell molecule whose occupied MOs are $\phi_i = \sum_s c_{si}\chi_s$, the Mayer bond order B_{CD} between atoms C and D is defined as

$$B_{CD} \equiv \sum_{t\in C}\sum_{u\in D}\left(\sum_r P_{tr}S_{ru}\right)\left(\sum_r P_{ur}S_{rt}\right) \tag{15.26}$$

where the density matrix element P_{tr} is given by (14.42) as $P_{tr} \equiv \sum_j n_j c_{tj}^* c_{uj}$, the notation $t \in C$ denotes all basis functions centered on atom C, S_{rt} is the overlap integral between basis functions χ_r and χ_t, and n_j is the number of electrons in MO j. The Mayer bond orders usually have values close to those found from Lewis electron-dot formulas.

15.7 THE MOLECULAR ELECTROSTATIC POTENTIAL, MOLECULAR SURFACES, AND ATOMIC CHARGES

The Molecular Electrostatic Potential. The **electric potential** ϕ at a point P in space is defined as the reversible work per unit charge needed to move an infinitesimal test charge Q_t from infinity to P, which we write as $\phi_P \equiv w_{\infty\to P}/Q_t$. The SI unit of ϕ is the volt (V), where 1 V $\equiv$ 1 J/C. When we do reversible work w on the test charge, we change its potential energy V by w (just as reversibly raising or lowering a mass in the earth's gravitational field changes its potential energy). If we take the potential energy of Q_t as zero at infinity, we therefore have $V_{\mathrm{P}} = w_{\infty\to\mathrm{P}} = \phi_{\mathrm{P}} Q_t$. The electrical potential energy V of a charge at point P (where the electric potential is ϕ_P) is $\phi_{\mathrm{P}} Q_t$. From the definition of ϕ_{P}, it readily follows (Prob. 15.25) that in the space around a point charge Q, the electric potential (in SI units) is $\phi_{\mathrm{P}} = Q/4\pi\varepsilon_0 d$, where d is the distance between point P and the charge. The electric potential is a function of the location (x, y, z) of point P in space: $\phi = \phi(x, y, z)$.

If the system consists of a single point charge Q_A located at (x_A, y_A, z_A), then $\phi_1 = Q_A/4\pi\varepsilon_0 r_{1A}$, where r_{1A} is the distance between point A and point 1 with coordinates (x_1, y_1, z_1). If the system consists of several point charges, then each contributes to ϕ and

$$\phi_1 = \sum_i \frac{Q_i}{4\pi\varepsilon_0 r_{1i}} \tag{15.27}$$

If the system is a molecule, we view it as a collection of point-charge nuclei and electronic charge smeared out into a continuous distribution. Electrons are point charges and are not actually smeared out into a continuous charge distribution, but the electronic-charge-cloud picture is a reasonable approximation when considering interactions between two molecules that are not too close to each other. The probability of finding a molecular electron in a tiny volume $dv = dx\,dy\,dz$ is $\rho\,dv$, where ρ is the electron probability density [Eq. (14.5)]. Therefore, the amount of electronic charge in dv is $-e\rho\,dv$. Addition of the contributions of the molecular electronic charge and of the nuclei α gives the molecular electric potential as

$$\phi(x_1, y_1, z_1) = \sum_\alpha \frac{Z_\alpha e}{4\pi\varepsilon_0 r_{1\alpha}} - e \iiint \frac{\rho(x_2, y_2, z_2)}{4\pi\varepsilon_0 r_{12}}\,dx_2\,dy_2\,dz_2 \tag{15.28}$$

where r_{12} is the distance between points 1 and 2 and the integration is over all space. In atomic units, the e and the $4\pi\varepsilon_0$ disappear from (15.28).

Quantum chemists call ϕ the **molecular electrostatic potential** (MEP) or the **electrostatic potential** (ESP—no connection with parapsychology). Although the SI units of ϕ are volts, quantum chemists traditionally multiply (15.28) by the proton charge e (thereby converting its units to joules) and by the Avogadro constant N_{A} (thereby converting its units to J/mol). Thus a quantum chemist's MEP value at a point P is the molar electrical interaction energy between the molecule and a proton placed at point P, assuming that the molecule is not polarized by the proton.

To calculate $\phi(x, y, z)$ one calculates an approximate electronic wave function for the equilibrium geometry (Section 15.10), uses (14.5) to calculate $\rho(x, y, z)$ from the approximate wave function, and uses (15.28) to calculate ϕ. [In density-functional theory (DFT) (Section 16.4), ρ is calculated without first finding the molecular wave function.] The MEP is found not to be strongly affected by the choice of basis set or by the inclusion of electron correlation. The MEP is often calculated at the HF/6-31G* level of theory or using DFT. The MEP can be calculated from accurate experimental X-ray-diffraction crystallography data [N. Bouhmadia et al., *J. Chem. Phys.*, **116**, 6196 (2002)].

The electron probability density $\rho(x, y, z)$ tells us how the electronic charge is distributed in a molecule. The MEP $\phi(x, y, z)$ tells us the interaction energy between a nonpolarizing positive test charge at (x, y, z) and the nuclear charges and electronic charge distribution of the molecule. The MEP is commonly depicted by a contour map showing curves of constant ϕ in a particular plane through the molecule or by a surface in three-dimensional space on which ϕ is constant. MEPs are generally found to be positive throughout the spatial region within a molecule, due to the strong positive contributions from the nuclei. Outside a molecule, ϕ can be positive or negative. For each of the molecules H_2O and $HC(O)NH_2$, the MEP is negative in the region outside the oxygen atom and positive outside other atoms.

When two molecules approach each other, the MEP of each plays a key role in their interaction. An electrophilic (electron-loving) species will preferentially attack a molecule at sites where the MEP is most negative. (This is not invariably true, since the MEP ignores polarization of the molecule by the incoming species.) MEPs provide insight into molecular recognition processes such as enzyme–substrate and drug–receptor interactions.

If we model each atom in a molecule as a sphere of radius equal to the van der Waals radius of the atom (for bonded atoms, these spheres overlap, and the spheres of bonded atoms are truncated), the **van der Waals surface** of a molecule is defined by the outward-facing surfaces of these atomic spheres. The van der Waals surface is what one sees in the familiar space-filling CPK (R. B. Corey–Pauling–Koltun) molecular models. In discussing intermolecular interactions, the MEPs in the regions on and outside the van der Waals surface are most significant.

A more sophisticated surface than the van der Waals surface is an **isodensity surface**, defined as a surface on which the molecule's electron probability density ρ is constant. The molecular isodensity surface for which $\rho = 0.001$ electrons/bohr^3 = 0.006748 electrons/Å^3 was found to have about the same surface area as the van der Waals surface [C. K. Kim et al., *J. Comput. Chem.*, **25**, 2073 (2004)], and this isodensity surface is often used, although many workers prefer the 0.002 electrons/bohr^3 isodensity surface. For certain purposes, one wants values of the **molecular surface area** and the **molecular volume**. Of course, these quantities have no well-defined meaning, but these values are usually calculated as the area of and the volume enclosed by the van der Waals surface or by an isodensity surface with a specified ρ value.

As an alternative to showing surfaces of constant MEP, one often examines the MEP values on an isodensity surface of the molecule. These values constitute the **molecular surface electrostatic potential** (MSEP or MSESP). The MSEP is often depicted using a color scheme such as red denoting the most negative MEP values, blue the most positive, and other spectrum colors intermediate values. Negative regions of the MSEP arise from lone pairs on electronegative atoms, from π electrons (for example, the MSEP of benzene is negative within much of the hexagon above and below the molecular plane), and from strained carbon–carbon bonds (the MSEP of cyclopropane is negative in the three regions that lie near to and outside the three carbon–carbon bonds and near the plane of the C atoms).

Politzer and co-workers defined several statistical quantities (such as the average MSEP, the standard deviation of the MSEP) that are calculated from the MSEP and found empirical relations between condensed-phase organic-compound macroscopic properties that depend on intermolecular interactions (such as normal boiling point, heats of fusion and vaporization, surface tension) and the MSEP quantities. This allows prediction of these macroscopic properties from the MSEP with a fair degree of success [P. Politzer and J. S. Murray, *Fluid Phase Equilib.*, **185**, 129 (2001)].

For information on the role of the MEP in biochemical processes, see B. Honig and A. Nicholls, *Science*, **268**, 1144 (1995). For reviews of MEPs, see P. Politzer and J. S. Murray in K. B. Lipkowitz and D. B. Boyd (eds.), *Reviews in Computational Chemistry*, vol. 2., VCH, 1991, Chapter 7; P. Politzer and J. S. Murray, *Theor. Chem. Acc.*, **108**, 134 (2002).

Atomic Charges. Whereas the MEP is a well-defined, physically significant quantity that can be accurately calculated from a reasonably accurate wave function, there is no unique, well-defined answer to the question: What is the charge on a particular atom in a molecule? (The terms *atomic charge, net atomic charge,* and *partial atomic charge* are used synonymously.)

As noted in Section 15.6, Mulliken population analysis gives atomic charges that vary erratically as the basis set is improved. Better values are obtained from natural population analysis (NPA).

A popular way to get reasonable atomic charges Q_α is by fitting the MEP ϕ. One first uses a molecular wave function to calculate values of ϕ at a grid of many points in the region outside the molecule's van der Waals surface. One then places a charge Q_α at each nucleus α and calculates the quantity $\phi_i^{\text{approx}} \equiv \sum_\alpha Q_\alpha e/4\pi\varepsilon_0 r_{i\alpha}$ at each grid point. One then varies the Q_α values (subject to the constraint that they add to zero for a neutral molecule) so as to minimize the sum of the squares of the deviations $\phi_i^{\text{approx}} - \phi_i$ for the grid points. Various ways of choosing the grid points and of including other refinements give different schemes for finding the atomic charges Q_α, which are called **ESP** (electrostatic potential) **charges**. Three common schemes are the Merz–Singh–Kollman (MK or MSK) method, the CHELPG method (charges from electrostatic potentials, grid method), and the RESP (restrained ESP) method. The MEP depends most strongly on the charges assigned to atoms near the surface of the molecule and depends only quite weakly on the charges of atoms in the interior, so ESP charges of atoms in the interior cannot be accurately assigned.

Another procedure that yields atomic charges is Bader's **atoms-in-molecules** (AIM) theory [*Bader*; R. F. W. Bader, *Chem. Rev.*, **91**, 893 (1991)]. Here, one first calculates an approximate wave function for the molecule, and from this wave function one calculates the electron probability density $\rho(x, y, z)$. One now imagines drawing all the contour surfaces of constant ρ (isodensity surfaces). Then one draws lines (called *gradient paths*) throughout three-dimensional space such that at every point on a given line, that line is perpendicular to the isodensity surface that passes through the point. (A simple example is the ground-state H atom. Here the isodensity surfaces are spheres centered at the nucleus, and the gradient paths are radii emanating from the nucleus.) One finds that most such gradient paths start at infinity and end at one of the nuclei. For molecules containing rings (for example, benzene), many gradient paths will go from the point within the ring where ρ is a minimum to a nucleus. The three-dimensional region of space Ω_{A} belonging to atom A of the molecule is then defined as the region that contains all the gradient paths that end at the nucleus of atom A. The AIM charge on atom A is then defined by $Q_{\text{A}} = Z_{\text{A}} - \int_{\Omega_A} \rho \, dV$, where the integration goes over the region Ω_{A}.

The gradient vector $\nabla\rho$ at a point P can be shown to be perpendicular to the isodensity surface ρ = constant at P. There is only one perpendicular direction to a surface at a particular point, so a line drawn perpendicular to isodensity surfaces will have its direction at any point be in the direction of $\nabla\rho$. Hence such a line is called a *gradient path*. Because $\nabla\rho$ has a unique direction at each point in space (except for points where $\nabla\rho$ is zero or undefined), gradient paths from different nuclei cannot cross each other, and the gradient paths terminating at each nucleus divide space into nonoverlapping regions, one for each atom.

AIM charges have been criticized for being larger than seems chemically reasonable, and, according to Cramer, "are of little chemical utility" (*Cramer*, p. 317). For Bader's defense of AIM charges, see R. F. W. Bader and C. F. Matta, *J. Phys. Chem. A*, **108**, 8385 (2004). [See also C. F. Matta and R. F. W. Bader, *J. Phys. Chem. A*, **110**, 6365 (2006).]

Still another defined atomic charge is the generalized atomic polar tensor (GAPT) charge [J. Cioslowski, *J. Am. Chem. Soc.*, **111**, 8333 (1989); *Phys. Rev. Lett.*, **62**, 1469 (1989)], defined by $Q_A \equiv (\partial\mu_x/\partial x_A + \partial\mu_y/\partial y_A + \partial\mu_z/\partial z_A)/3$, where μ_x is the x component of the molecular dipole moment and x_A is the x coordinate of nucleus A. The GAPT charges of a neutral molecule can be shown to add to zero. GAPT charges show less basis-set dependence than Mulliken charges.

Cramer, Truhlar, and co-workers have devised the methods **Charge-Model 1** (CM1), CM2, and CM3 to calculate atomic charges [*Cramer*, Sec. 9.1.3.4 and C. P. Kelly et al., *Theor. Chem. Acc.*, **113**, 133 (2005)]. These methods start with a set of charges such as the Mulliken charges or the Löwdin charges and then improve these charges by using a specified formula that transfers charges from one atom to another to get the final CM charges. The formula used to transfer charges contains parameters whose values have been fixed at those values that allow CM-charge-calculated molecular dipole moments μ^{CM} of a large set of training molecules to give the best fit to the experimental dipole moments of these molecules, where the CM dipole moments are calculated from $\mu^{CM} = |\Sigma_B Q_B^{CM}\mathbf{r}_B|$, where Q_B^{CM} is the CM charge on atom B, and the sum goes over the atoms of the molecule. More specifically, CM2 and CM3 use the formula

$$Q_k^{CM} = Q_k^L + \sum_{k' \neq k} B_{kk'}^M (D_{kk'} + C_{kk'} B_{kk'}^M)$$

where Q_k^L is the charge given by Löwdin population analysis, $B_{kk'}^M$ is the Mayer bond order for atoms k and k', the sum goes over all atoms except atom k, and the values of the parameters $C_{kk'}$ and $D_{kk'}$ depend only on the atomic numbers of the two atoms. Different sets of C and D parameters are used depending which method (HF, DFT, etc.) and which basis set are used to find Q_k^L.

Still another method to find atomic charges gives the **Voronoi deformation density** (VDD) charges [C. F. Guerra et al., *J. Comp. Chem.*, **25**, 189 (2004)]. Here, one defines the region of space that belongs to atom A in a molecule as consisting of all those points that lie closer to the A nucleus than to any other nucleus. (This region of space is a polyhedron, called the *Voronoi polyhedron* or Voronoi cell, whose surfaces are planes that are perpendicular bisectors of the lines from A to its neighbors.) One defines the **deformation density** ρ_{def} as $\rho_{def} \equiv \rho - \Sigma_B \rho_B$, where ρ is the molecular electron probability density and $\Sigma_B \rho_B$ is the sum of spherically averaged electron densities calculated for isolated atoms that are located at the atomic positions in the molecule. The VDD charge of atom A in the molecule is then found by integrating ρ_{def} over the Voronoi cell of A and multiplying by -1 (electrons are negatively charged). The VDD charge attempts to measure how much charge is transferred to an atom from its neighbors during molecule formation. The VDD charges were found to show only small changes with change in basis set. The devisers of the VDD method found AIM and NPA charges to be too large and chemically unreasonable.

Charges on the O atom in H_2O using HF/cc-pV6Z calculations are -0.91 for NPA, -0.73 for RESP, and -1.23 for AIM, as compared with -0.27 for VDD (calculated with a density-functional method and a large basis set).

Calculated atomic charges are used as parameters in molecular-mechanics calculations (Section 17.5) to model the electrostatic interactions between nonbonded atoms. ESP-derived charges and CM1 charges are often used for this purpose.

For reviews of atomic charges, see M. M. Francl and L. E. Chirlian in K. B. Lipkowitz and D. B. Boyd (eds.), *Reviews in Computational Chemistry*, vol. 14, Wiley, 2000, Chapter 1; *Cramer*, Section 9.1.3.

15.8 LOCALIZED MOs

The idea of a chemical *bond* between a pair of atoms in a molecule is fundamental to chemistry. The experimental evidence supporting this concept is substantial. One can assign bond energies to various kinds of bonds and obtain good estimates for the heats of atomization of most molecules by adding the energies of the individual bonds. Other molecular properties (for example, magnetic susceptibility, dipole moment) can also be analyzed as the sum of contributions from individual bonds (and lone pairs). The infrared spectrum of a compound containing an OH group shows a characteristic band near $3600\,\text{cm}^{-1}$. This OH stretching vibrational band occurs at nearly the same frequency in HOH, HOCl, and CH_3OH. The length of an OH bond is nearly constant from molecule to molecule, about 0.96 Å.

The MO picture of H_2O presented in Section 15.5 appears to be gravely deficient, in that it is seemingly inconsistent with the existence of individual bonds in the molecule. Each of the bonding MOs in (15.19) is delocalized over the entire molecule. If one were to compare the bonding MOs for HOH and HOCl, one would find them to be quite different. Yet we know that the OH parts of these two molecules are similar. Actually, MO theory *can* explain the observed near invariance of a given kind of chemical bond, as we now show.

The MO approximation to the ground state of H_2O is a Slater determinant of the form

$$|\phi_1\overline{\phi_1}\phi_2\overline{\phi_2}\phi_3\overline{\phi_3}\phi_4\overline{\phi_4}\phi_5\overline{\phi_5}| \qquad (15.29)$$

Putting (15.29) into words, we say that two electrons are in each of the orthonormal MOs ϕ_1, ϕ_2, ϕ_3, ϕ_4, and ϕ_5. However, this MO description is not unique. The addition of a multiple of one column of a determinant to another column leaves the determinant unchanged in value. Adding column 7 to column 9 and column 8 to column 10 in (15.29), we get

$$|\phi_1\overline{\phi_1}\phi_2\overline{\phi_2}\phi_3\overline{\phi_3}\phi_4\overline{\phi_4}(\phi_4 + \phi_5)(\overline{\phi_4 + \phi_5})| \qquad (15.30)$$

This determinant leads to the description of the molecule as having the MOs $\phi_1, \phi_2, \phi_3, \phi_4$, and $\phi_4 + \phi_5$ each doubly occupied. Despite the different verbal descriptions, the wave functions (15.29) and (15.30) are identical.

[In discussing SCF calculations, we used an orthogonalized 2*s* AO of the form $a(1s) + b(2s)$, where 2*s* is a (nodeless) 2*s* Slater-type orbital. This procedure is justified by use of the freedom of adding a multiple of one column of a determinant to another. The determinantal wave function for the oxygen atom is the same whether it is the 2*s* or the orthogonalized $2s_\perp$ AO that is used.]

An objection that can be raised to (15.30) is that the MO $\phi_4 + \phi_5$ is neither normalized nor orthogonal to ϕ_4. Consider, however, the Slater determinant

$$|\phi_1\overline{\phi_1}\phi_2\overline{\phi_2}\phi_3\overline{\phi_3}(b\phi_4 + c\phi_5)(\overline{b\phi_4 + c\phi_5})(c\phi_4 - b\phi_5)(\overline{c\phi_4 - b\phi_5})| \quad (15.31)$$

where b and c are any two real constants such that

$$b^2 + c^2 = 1 \quad (15.32)$$

We now show (15.31) and (15.29) to be the same wave function. Multiplication of columns 7 and 8 of (15.29) by b and columns 9 and 10 by $-b^{-1}$ gives

$$|\ldots b\phi_4(\overline{b\phi_4})(-b^{-1}\phi_5)(\overline{-b^{-1}\phi_5})|$$

This step multiplies (15.29) by $b^2(-b^{-1})^2 = 1$. Next, $-bc$ times column 9 is added to column 7 and $-bc$ times column 10 is added to column 8 to give

$$|\cdots(b\phi_4 + c\phi_5)(\overline{b\phi_4 + c\phi_5})(-b^{-1}\phi_5)(\overline{-b^{-1}\phi_5})|$$

Finally, c/b times column 7 is added to column 9, c/b times column 8 is added to column 10, and (15.32) is used. This gives

$$|\cdots(b\phi_4 + c\phi_5)(\overline{b\phi_4 + c\phi_5})(c\phi_4 - b\phi_5)(\overline{c\phi_4 - b\phi_5})|$$

This completes the proof. The orthonormality of the MOs $c\phi_4 - b\phi_5$ and $b\phi_4 + c\phi_5$ readily follows from (15.32) and the orthonormality of ϕ_4 and ϕ_5. Thus we can describe the molecule as having the orthonormal MOs

$$\phi_1, \quad \phi_2, \quad \phi_3, \quad b\phi_4 + c\phi_5, \quad c_4\phi_4 - b\phi_5 \quad (15.33)$$

each doubly occupied.

There are an infinite number of other MO descriptions consistent with (15.29): Let d and e be any two real constants such that $d^2 + e^2 = 1$. If the procedure used to go from (15.29) to (15.31) is applied to columns 5, 6, 7, and 8 of (15.31), we end up with

$$|\phi_1\overline{\phi_1}\phi_2\overline{\phi_2}(d\phi_3 + be\phi_4 + ce\phi_5)(\overline{d\phi_3 + be\phi_4 + ce\phi_5})$$
$$\times (e\phi_3 - bd\phi_4 - cd\phi_5)(\overline{e\phi_3 - bd\phi_4 - cd\phi_5})(c\phi_4 - b\phi_5)(\overline{c\phi_4 - b\phi_5})|$$

We can thus describe the electronic configuration as having the orthonormal MOs

$$\phi_1, \quad \phi_2, \quad (d\phi_3 + be\phi_4 + ce\phi_5), \quad (e\phi_3 - bd\phi_4 - cd\phi_5), \quad (c\phi_4 - b\phi_5)$$

each doubly occupied. Continuing on in this manner, we can derive orthonormal MOs, each of which is a linear combination of all the original MOs $\phi_1, \phi_2, \phi_3, \phi_4, \phi_5$. For the general conditions that must be satisfied by the coefficients in these linear combinations, see *Szabo and Ostlund*, Section 3.2.3.

Thus for a given closed-shell electronic state of a molecule, there are many possible MO descriptions. The delocalized MOs (15.19) of H_2O are uniquely determined as the solutions of the SCF equations [Eq. (14.25)]

$$\hat{F}\phi_i = \varepsilon_i\phi_i, \quad i = 1, 2, \ldots \tag{15.34}$$

Hence the reader may be wondering how we can have other sets of MOs that also minimize the variational integral. Actually, Eq. (15.34) is not the most general equation satisfied by the MOs that minimize the variational integral. Instead, one finds in deriving the SCF equations that the Hartree–Fock MOs must satisfy

$$\hat{F}\phi_i = \sum_{j=1}^{N} \lambda_{ji}\phi_j, \qquad i = 1, \ldots, N \tag{15.35}$$

where the sum runs over the occupied MOs. The λ_{ji}'s are certain constants (Lagrangian multipliers), which can be chosen arbitrarily, subject to the MO orthonormality requirement. Different choices of the λ_{ji}'s give different sets of MOs, but each such set minimizes the variational integral. For a closed-shell configuration, one possible choice of the λ_{ji}'s can be shown to be

$$\lambda_{ji} = \varepsilon_i\delta_{ij} \tag{15.36}$$

This choice reduces (15.35) to (15.34). [For the proof of (15.35) and (15.36), see S. M. Blinder, *Am. J. Phys.*, **33**, 431 (1965); *Szabo and Ostlund*, Chapter 3.]

The delocalized MOs [such as (15.19) for H_2O] that satisfy (15.34) are called the **canonical MOs**. The canonical MOs are eigenfunctions of the Fock operator, which commutes with the molecular symmetry operators, and hence the canonical MOs each transform according to one of the possible molecular symmetry species. A set of MOs like (15.33) satisfies (15.35) but not (15.34) and does not necessarily transform according to the molecular symmetry species.

For an open-shell configuration, one can also find canonical delocalized MOs that satisfy (15.34) and that transform according to the molecular symmetry species. However, the form of $\hat{F}$ is more complicated than for the closed-shell case. [See C. C. J. Roothaan, *Rev. Mod. Phys.*, **32**, 179 (1960).]

Of the possible MO sets formed as linear combinations of the delocalized canonical MOs of water, a set that would appeal to a chemist is one for which the charge probability density of each bonding MO is localized in the region of one of the OH bonds. There are many ways of taking linear combinations of the delocalized MOs to get such **localized MOs (LMOs)**. A natural requirement that the two localized bonding MOs in water should satisfy is that they be equivalent to each other; that is, each localized bonding MO of water should be transformed into the other by the $\hat{C}_2(z)$ symmetry operation (Fig. 15.7). Localized MOs that are permuted among one another by a symmetry operation that permutes equivalent chemical bonds are called *equivalent orbitals*. As we shall see, localized MOs reconcile MO theory with the chemist's intuitive picture of chemical bonding.

For H_2O, the Lewis electron-dot formula suggests the localized MOs to be a pair of equivalent bonding orbitals $b(OH_1)$ and $b(OH_2)$, an inner-shell 1s oxygen orbital $i(O)$, and two lone-pair equivalent MOs $l_1(O)$ and $l_2(O)$ on oxygen.

The most "localized" MOs are those that are most separated from one another. By this, we mean that set of MOs for which the interelectronic repulsion between the different MOs viewed as "charge clouds" is a minimum. The energy of repulsion between the charge clouds of electron 1 in the MO ϕ_i and electron 2 in the MO ϕ_j is a Coulomb integral of the form (9.100). The total interorbital charge-cloud repulsion energy is then

$$4\sum_i \sum_{j>i} \int\int |\phi_i(1)|^2 |\phi_j(2)|^2 \frac{1}{r_{12}}\, dv_1\, dv_2 \tag{15.37}$$

where the sums run over the occupied MOs. (The factor 4 comes from the four interorbital repulsions of the electrons in each MO pair.) We define the localized MOs as those orthonormal MOs that minimize (15.37). [It turns out that minimization of (15.37) implies the maximization of the total *intra*orbital electron repulsion and the minimization of the magnitude of the (interorbital) exchange energy.] This definition was originally suggested by Lennard-Jones and Pople and has been applied to many molecules by Edmiston and Ruedenberg [C. Edmiston and K. Ruedenberg, *Rev. Mod. Phys.*, **35**, 457 (1963); *J. Chem. Phys.*, **43**, S97 (1965); P.-O. Löwdin, ed., *Quantum Theory of Atoms, Molecules, and the Solid State*, Academic Press, 1966, pages 263–280].

The MOs that minimize (15.37) are called the **energy-localized MOs**. For molecules with symmetry, we expect that the energy-localized MOs will also be equivalent orbitals, and this is borne out by the calculated energy-localized MOs. Energy-localized MOs are thus a generalization of equivalent MOs.

Liang and Taylor calculated the energy-localized MOs for H_2O, starting with the minimal-basis canonical MOs (15.19). In terms of the canonical (delocalized) MOs, Liang and Taylor found (J. H. Liang, Ph.D. thesis, Ohio State University, 1970; quoted in F. Franks, ed., *Water*, Vol. 1, Plenum, 1972, p. 42)

$$\begin{aligned}
i(\mathrm{O}) &= 0.99(1a_1) - 0.12(2a_1) \qquad\qquad\qquad + 0.06(3a_1) \\
b(\mathrm{OH}_1) &= 0.05(1a_1) + 0.57(2a_1) + 0.71(1b_2) + 0.42(3a_1) \\
b(\mathrm{OH}_2) &= 0.05(1a_1) + 0.57(2a_1) - 0.71(1b_2) + 0.42(3a_1) \\
l_1(\mathrm{O}) &= 0.08(1a_1) + 0.42(2a_1) \qquad\qquad\qquad - 0.57(3a_1) - 0.71(1b_1) \\
l_2(\mathrm{O}) &= 0.08(1a_1) + 0.42(2a_1) \qquad\qquad\qquad - 0.57(3a_1) + 0.71(1b_1)
\end{aligned} \tag{15.38}$$

The lone-pair $1b_1$ MO ($2p_{x\mathrm{O}}$) is equally divided between the equivalent lone-pair localized orbitals; note that $0.71 = (0.5)^{1/2}$. The inner-shell $1a_1$ MO contributes substantially to only the $i(\mathrm{O})$ MO. The bonding $1b_2$ MO is equally divided between the two bonding localized MOs. The bonding $2a_1$ MO makes substantial contributions to the two bonding localized MOs and smaller, but still substantial, contributions to the lone-pair localized MOs. The largely lone-pair $3a_1$ MO makes substantial contributions to the lone-pair localized MOs and lesser contributions to the bonding localized MOs.

In terms of the AOs, the energy-localized MOs of H_2O are

$$\begin{aligned}
i(\mathrm{O}) = -0.007(\mathrm{H}_1 1s) &- 0.007(\mathrm{H}_2 1s) + 0.99(\mathrm{O}1s) - 0.12(\mathrm{O}2s_\perp) \\
&+ 0.03(\mathrm{O}2p_z) \\
b(\mathrm{OH}_1) = 0.50(\mathrm{H}_1 1s) &- 0.10(\mathrm{H}_2 1s) + 0.02(\mathrm{O}1s) + 0.25(\mathrm{O}2s_\perp) \\
&+ 0.407(\mathrm{O}2p_z) + 0.441(\mathrm{O}2p_y)
\end{aligned}$$

$$b(\mathrm{OH}_2) = -0.10(\mathrm{H}_1 1s) + 0.50(\mathrm{H}_2 1s) + 0.02(\mathrm{O}1s) + 0.25(\mathrm{O}2s_\perp) + 0.407(\mathrm{O}2p_z) - 0.441(\mathrm{O}2p_y) \quad (15.39)$$

$$l_1(\mathrm{O}) = -0.09(\mathrm{H}_1 1s) - 0.09(\mathrm{H}_2 1s) + 0.09(\mathrm{O}1s) + 0.63(\mathrm{O}2s_\perp) - 0.39(\mathrm{O}2p_z) - 0.71(\mathrm{O}2p_x)$$

$$l_2(\mathrm{O}) = -0.09(\mathrm{H}_1 1s) - 0.09(\mathrm{H}_2 1s) + 0.09(\mathrm{O}1s) + 0.63(\mathrm{O}2s_\perp) - 0.39(\mathrm{O}2p_z) + 0.71(\mathrm{O}2p_x)$$

The MO wave function

$$|i(\mathrm{O})\overline{i(\mathrm{O})}b(\mathrm{OH}_1)\overline{b(\mathrm{OH}_1)}b(\mathrm{OH}_2)\overline{b(\mathrm{OH}_2)}l_1(\mathrm{O})\overline{l_1(\mathrm{O})}l_2(\mathrm{O})\overline{l_2(\mathrm{O})}|$$

is identical to (15.21).

Let us analyze these localized MOs for water. The $i(\mathrm{O})$ MO is nearly a pure $1s$ inner-shell oxygen AO.

To define the angle between the two localized bonding MOs, we draw the line from O to H_1 along which the electron probability density in $b(\mathrm{OH}_1)$ is a maximum, and we draw a similar line from O to H_2. The angle between these lines where they intersect at the O nucleus defines the angle between the localized bonding MOs. (If this angle differs significantly from the angle defined by straight lines between the nuclei, the bonds are said to be *bent*.) The angle between the localized bonding MOs is determined mainly by the oxygen $2p_y$ and $2p_z$ AO contributions (with a small influence by the hydrogen $1s$ AOs). For $b(\mathrm{OH}_1)$ the $\mathrm{O}2p_y\mathrm{O}2p_z$ contribution contains the terms $0.407z_\mathrm{O} + 0.441y_\mathrm{O}$. Let us rotate the coordinates in the zy plane by an angle $\alpha = \arctan(0.441/0.407) = 47\frac{1}{2}°$, as in Fig. 15.6. The relation between coordinates in the unrotated system and in the rotated $z'y'$ system is given by the well-known formulas

$$z' = z\cos\alpha + y\sin\alpha$$
$$y' = -z\sin\alpha + y\cos\alpha \quad (15.40)$$

From (15.40) and Fig. 15.6 we have $0.407z_\mathrm{O} + 0.441y_\mathrm{O} = 0.600z'_\mathrm{O}$. Multiplication by the exponential factor of the $2p$ oxygen AO then gives

$$0.407(\mathrm{O}2p_z) + 0.441(\mathrm{O}2p_y) = 0.60(\mathrm{O}2p_{z'}) \quad (15.41)$$

In other words, the hybridized $2p_z2p_y$ AO on the left of (15.41) is the same function as 0.60 times a $2p$ AO inclined at an angle of $47\frac{1}{2}°$ with the z axis. Thus the bonding $2p_z2p_y$ hybrid AOs of oxygen in $b(\mathrm{OH}_1)$ and $b(\mathrm{OH}_2)$ point in the general direction of the hydrogen atoms. The contribution from the hydrogen atoms to $b(\mathrm{OH}_1)$ is mostly from $\mathrm{H}_1 1s$, and the overlap between $\mathrm{H}_1 1s$ and the $2p_z2p_y$ oxygen hybrid then forms the $\mathrm{O}{-}\mathrm{H}_1$ chemical bond. The angle between the two hybrid oxygen AOs in $b(\mathrm{OH}_1)$ and $b(\mathrm{OH}_2)$ is 95°, and this is approximately the angle between the localized bonding MOs.

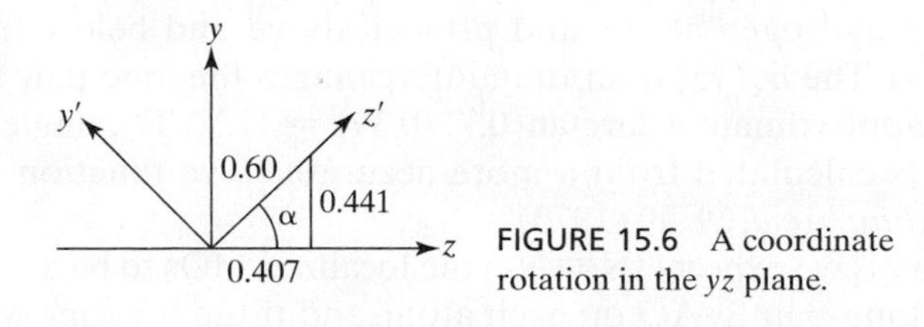

FIGURE 15.6 A coordinate rotation in the yz plane.

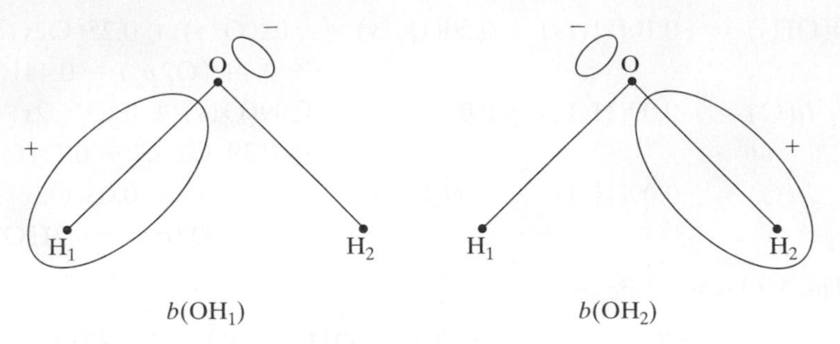

FIGURE 15.7 Rough sketches of the localized bonding MOs in H_2O.

Energy-localized H_2O bonding MOs calculated from an extended-basis-set SCF MO wave function were found to have an angle of 96° between the oxygen hybrids that contribute to these MOs and an angle of 103° between the localized MOs themselves, which is nearly the same as the $104\frac{1}{2}°$ molecular bond angle, indicating negligible bending for the bonds in H_2O. [W. von Niessen, *Theor. Chim. Acta*, **29**, 29 (1973).] In contrast, in the strained molecule cyclopropane, the angle between the carbon hybrids contributing to the carbon–carbon bonding energy-localized MOs deviates outward by 28° from the internuclear lines [M. D. Newton, E. Switkes, and W. N. Lipscomb, *J. Chem. Phys.*, **53**, 2645 (1970)].

The $b(\text{OH}_1)$ energy-localized MO is not completely confined to the region of the O—H_1 bond. We see from (15.39) that this MO has a small contribution from the H_21s AO. (The ratio of the contributions of two AOs to an MO is given essentially by the square of the ratio of their coefficients.) Consider the contributions of $O2s_\perp$, $O2p_y$, and $O2p_z$ to $b(\text{OH}_1)$. The $2p_y2p_z$ hybrid has a positive lobe in the region of the OH_1 bond, and the contribution of the $O2s_\perp$ AO reinforces this positive lobe. Overlap with H_11s then gives the O—H_1 bond. The $2p_y2p_z$ hybrid has a negative lobe on the side of the oxygen opposite the OH_1 bond. This negative lobe is partly, but not completely, canceled by the $O2s_\perp$ contribution. Thus the $b(\text{OH}_1)$ MO has a "tail" on the side of the O away from H_1, this tail being somewhat distorted toward H_2 by the $-0.10H_21s$ contribution (Fig. 15.7). (For accurately plotted contours, see F. Franks, ed., *Water*, Vol. 1, Plenum, New York, 1972, p. 46.) Despite this tail, this MO is far more localized than any of the bonding canonical MOs in (15.19). Because $b(\text{OH}_1)$ is mostly localized in the OH_1 region, we expect only a small change in its form on going from HOH to, say, HOCl. The observed near invariance of the O—H bond from molecule to molecule is explained by MO theory using localized bonding MOs. (Neglect of the contribution of H_21s to the $b(\text{OH}_1)$ localized MO gives a two-center orbital called a *bond orbital*.)

Finally, consider the two lone-pair MOs $l_1(\text{O})$ and $l_2(\text{O})$. These MOs are mainly localized on the oxygen atom and are equivalent to each other. They are directed away from the hydrogen atoms and project above and below the molecular (yz) plane (Fig. 15.8). The $\hat{\sigma}_v(yz)$ operation interchanges the lone-pair MOs. The angle between them is approximately $2\arctan(0.71/0.39) = 122°$. The angle between localized lone-pair MOs calculated from a more accurate wave function is 114°; W. von Niessen, *Theor. Chim. Acta*, **29**, 29 (1973).

For N_2 we expect (:N≡N:) the localized MOs to be an inner-shell 1s AO on each atom, a lone-pair 2s AO on each atom, and three bonding MOs spread over the two

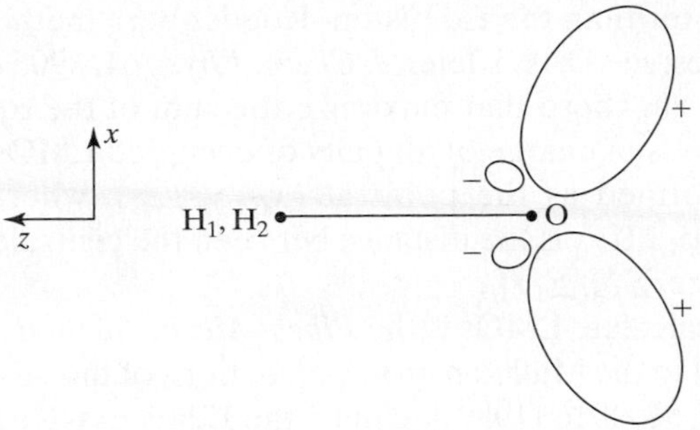

FIGURE 15.8 Rough sketches of the localized lone-pair MOs of H_2O.

atoms. The canonical-MO picture is that the triple bond consists of one σ bond and two π bonds, as in Section 13.7. The energy-localized bonding MOs were found by Edmiston and Ruedenberg to be three equivalent banana-shaped orbitals spaced 120° apart from one another. The AOs that contribute significantly to the localized bonding MOs are the $2s$, $2p\sigma$, $2p\pi_x$, and $2p\pi_y$ orbitals of each atom. The $i(N_a)$ and $i(N_b)$ localized MOs were found to be nearly pure $1s$ nitrogen AOs. Each of the $l(N_a)$ and $l(N_b)$ localized MOs is a hybrid of the $2s$ and $2p\sigma$ AOs of the relevant nitrogen atom, with the $2s$ AO making the larger contribution to the MO. Each lone-pair localized MO is directed away from the other nitrogen atom.

The concept of localized MOs is not as widely applicable as that of delocalized canonical MOs. Delocalized MOs are valid for any molecule. However (as noted in Section 11.5), the Hartree–Fock wave functions of nonclosed-shell electronic states are, in most cases, linear combinations of a few Slater determinants [for example, see (10.44) and (10.45)], and the above localization procedure does not apply to the open-shell orbitals in these wave functions. Thus, in a molecule in an excited electronic state with an open-shell configuration, the electrons in the incompletely filled MOs are delocalized over much of the molecule.

The great success of the concept of chemical bonds between pairs of atoms is a reflection of the fact that the electronic ground states of most molecules have closed-shell configurations, for which the localized MO description is just as valid as the delocalized MO description.

For a closed-shell ground-state molecule, those properties that involve only the ground-state wave function can be calculated just as well with either the localized or the delocalized MO description. Such properties include electron probability density, dipole moment, geometry, and heat of formation. Properties of a molecule that involve the wave function of the ground state and also the wave function of an open-shell excited state or the wave function of an open-shell ion cannot be calculated using a localized MO description. Such properties include the electronic absorption spectrum and molecular ionization energies.

Localized MOs are approximately transferable from molecule to molecule, which is not true for canonical MOs. The localized b(CH) MO in CH_4 is quite similar to the localized b(CH) MOs in C_2H_6 and in CH_3OH [S. Rothenberg, *J. Chem. Phys.*, **51**, 3389 (1969); *J. Am. Chem. Soc.*, **93**, 68 (1971)].

Calculation of the Edmiston–Ruedenberg energy-localized MOs is very time consuming. Boys (and Foster) proposed a method to find localized MOs that is computationally much faster than the Edmiston–Ruedenberg method and that gives similar results in most cases; see D. A. Kleier, *J. Chem. Phys.*, **61**, 3905 (1974). The Boys method defines the LMOs as those that maximize the sum of the squares of the distances between the centroids of charge of all pairs of occupied LMOs. The *centroid of charge* of orbital ϕ_i is defined as the point at (x_C, y_C, z_C), where $x_C \equiv \langle\phi_i|x|\phi_i\rangle$, $y_C \equiv \langle\phi_i|y|\phi_i\rangle$, $z_C \equiv \langle\phi_i|z|\phi_i\rangle$. If r_{ij} is the distance between the centroids of LMOs i and j, the Boys LMOs maximize $\sum_{j>i}\sum_i r_{ij}^2$.

Still another way to calculate LMOs is the *Pipek–Mezey method*, which maximizes a certain sum that is related to the Mulliken gross populations of the orbitals [J. Pipek and P. G. Mezey, *J. Chem. Phys.*, **90**, 4916 (1989)]. Unlike the Edmiston–Ruedenberg and the Boys LMOs, the Pipek–Mezey LMOs for a double bond consist of one σ and one π MO. (See Section 15.9.)

As well as providing insight into chemical bonds, LMOs are useful in speeding up correlation calculations on large molecules (see Section 16.2).

15.9 THE SCF MO TREATMENT OF METHANE, ETHANE, AND ETHYLENE

Methane. The AOs for a minimal-basis-set SCF MO calculation of CH_4 are the carbon $1s$, $2s$, $2p_x$, $2p_y$, and $2p_z$ AOs and a $1s$ AO on each hydrogen atom. The point group of CH_4 is $\mathcal{T}_d$. Group theory (see *Cotton* or *Schonland*) gives the possible symmetry species as A_1, A_2, E, T_1, and T_2. We shall not worry about specifying the symmetry behavior that corresponds to each symmetry species and shall use the species mainly as labels for the MOs. As usual, we set up the coordinate system with the z axis coinciding with the highest-order C_n or S_n axis. For methane this is an S_4 axis (Fig. 12.5). The coordinates of the hydrogen atoms H_1, H_2, H_3, H_4 are (q, q, q), $(q, -q, -q)$, $(-q, q, -q)$, and $(-q, -q, q)$, respectively, where $2q$ is the edge of the cube in which the molecule is inscribed (Fig. 15.9). Note the equivalence of the x, y, and z axes.

The carbon atom is at the center of the molecule, and the carbon $1s$ and $2s$ AOs are each sent into themselves by every symmetry operation. These AOs transform according to the totally symmetric species A_1. The carbon $2p_x$, $2p_y$, and $2p_z$ AOs are given by x, y, or z times a radial function. Their symmetry behavior is the same as that of the functions x, y, and z, respectively. From the formulas for rotation of coordinates [Eq. (15.52)], we see that any proper rotation sends each of the functions x, y, and z into some linear combination of x, y, and z. Any improper rotation is the product of some

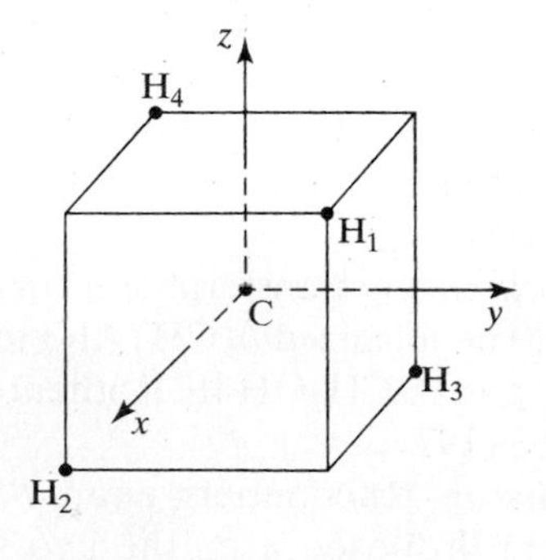

FIGURE 15.9 Coordinate axes for CH_4. The origin is at the center of the cube.

proper rotation and an inversion (Prob. 12.15); the inversion simply converts each coordinate to its negative. Hence the three carbon $2p$ orbitals are sent into linear combinations of one another by each symmetry operation. They must therefore transform according to one of the triply degenerate symmetry species. Further investigation (which is omitted) shows the symmetry species of the $2p$ AOs to be T_2.

Just as in H_2O, each $1s$ hydrogen AO in CH_4 does not transform according to any of the molecular symmetry species, and it is convenient to form symmetry-adapted basis functions by taking linear combinations of the $1s$ AOs. One obvious symmetry function is

$$\chi_1 = \mathrm{H}_1 1s + \mathrm{H}_2 1s + \mathrm{H}_3 1s + \mathrm{H}_4 1s \tag{15.42}$$

Since each methane symmetry operator permutes the hydrogen $1s$ orbitals among themselves, (15.42) is sent into itself by each symmetry operation and belongs to the totally symmetric species A_1. We need three more symmetry functions. The construction of these is not obvious without the use of group theory, and we shall simply write down the results. The remaining three orthogonal (unnormalized) symmetry-adapted basis functions can be taken as

$$\chi_2 = \mathrm{H}_1 1s + \mathrm{H}_2 1s - \mathrm{H}_3 1s - \mathrm{H}_4 1s \tag{15.43}$$

$$\chi_3 = \mathrm{H}_1 1s - \mathrm{H}_2 1s + \mathrm{H}_3 1s - \mathrm{H}_4 1s \tag{15.44}$$

$$\chi_4 = \mathrm{H}_1 1s - \mathrm{H}_2 1s - \mathrm{H}_3 1s + \mathrm{H}_4 1s \tag{15.45}$$

Each of these three functions is transformed into some linear combination of the three functions by each symmetry operation. For example, a $\hat{C}_3$ rotation about the CH_1 bond permutes the hydrogen atoms as follows: $1 \to 1, 2 \to 3, 3 \to 4, 4 \to 2$. The corresponding $\hat{O}_{C_3}$ operator transforms χ_2, χ_3, and χ_4 into χ_3, χ_4, and χ_2, respectively. These three symmetry functions therefore transform according to one of the triply degenerate symmetry species. The function χ_2 has positive signs for the hydrogen AOs with a positive x coordinate and negative signs for the hydrogen AOs with a negative x coordinate. We thus expect χ_2 to have the same symmetry behavior as the function x. Similarly, χ_3 and χ_4 behave as y and z, respectively. (As an example, note that a 120° counterclockwise rotation about the CH_1 bond has the following effect on the three unit vectors: $\mathbf{i} \to \mathbf{j}, \mathbf{j} \to \mathbf{k}, \mathbf{k} \to \mathbf{i}$; this is the same behavior shown by the functions χ_2, χ_3, and χ_4 for this rotation.) These three symmetry orbitals thus transform according to the same symmetry species as $C2p_x$, $C2p_y$, and $C2p_z$, that is, the species T_2.

The symmetry-adapted basis functions are

Symmetry function	χ_1	χ_2	χ_3	χ_4	$C1s$	$C2s$	$C2p_x$	$C2p_y$	$C2p_z$
Symmetry species	a_1	t_2	t_2	t_2	a_1	a_1	t_2	t_2	t_2

The nine lowest MOs therefore consist of three a_1 and six t_2 MOs. The six t_2 MOs belong to triply degenerate levels and thus fall into two different shells $1t_2$ and $2t_2$. Each such shell contains three MOs of equal orbital energy and each shell holds six electrons. SCF MO calculations give the three lowest shells as $1a_1$, $2a_1$, and $1t_2$, with orbital energies -11.20, -0.93, and -0.54 hartrees, respectively. The ground state of methane thus has the closed-shell electron configuration $(1a_1)^2(2a_1)^2(1t_2)^6$ and is a $^1A_1$ state.

Pitzer did SCF MO calculations on CH_4, using a minimal-basis set of STOs [R. M. Pitzer, *J. Chem. Phys.*, **46**, 4871 (1967)]. He found the optimized orbital exponents to be 1.17 for H1*s*, 5.68 for C1*s*, 1.76 for C2*s*, and 1.76 for C2p_z, which may be compared with the values 1.0, 5.7, 1.625, and 1.625 given by Slater's rules (Prob. 15.54). Pitzer found the minimum in energy to be at a bond distance of 1.089 Å, close to the experimental value 1.085 Å. The MOs at the experimental equilibrium bond length are

$$\begin{aligned}
1a_1 &= -0.005(\text{H}_1 1s + \text{H}_2 1s + \text{H}_3 1s + \text{H}_4 1s) + 1.001(\text{C}1s) + 0.025(\text{C}2s_\perp) \\
2a_1 &= 0.186(\text{H}_1 1s + \text{H}_2 1s + \text{H}_3 1s + \text{H}_4 1s) - 0.064(\text{C}1s) + 0.584(\text{C}2s_\perp) \\
1t_{2x} &= 0.318(\text{H}_1 1s + \text{H}_2 1s - \text{H}_3 1s - \text{H}_4 1s) + 0.554(\text{C}2p_x) \\
1t_{2y} &= 0.318(\text{H}_1 1s - \text{H}_2 1s + \text{H}_3 1s - \text{H}_4 1s) + 0.554(\text{C}2p_y) \\
1t_{2z} &= 0.318(\text{H}_1 1s - \text{H}_2 1s - \text{H}_3 1s + \text{H}_4 1s) + 0.554(\text{C}2p_z)
\end{aligned} \tag{15.46}$$

The $1a_1$ MO is essentially the carbon 1*s* AO. The $2a_1$ MO is a bonding combination of the carbon 2*s* AO and the symmetry orbital (15.42). This MO has charge buildup between the carbon atom and each of the four hydrogen atoms. The $1t_{2x}$ MO is a bonding MO; the function (15.43) is positive on the positive half of the *x* axis and negative on the negative half of the *x* axis, and its overlap with C2p_x gives charge buildup about the positive and negative sides of the *x* axis. Similarly, the bonding $1t_{2y}$ and $1t_{2z}$ MOs have charge buildup in the regions about the *y* and *z* axes, respectively.

We now consider the four localized (equivalent) bonding MOs of methane. Because of the tetrahedral symmetry, each of these orbitals must point along a CH bond, since otherwise they would not be equivalent to one another. (This is not true for water, where the equivalence requirement is satisfied by any two bonding MOs that make the same angle with the C_2 axis.) Each localized bonding MO is some linear combination of the five canonical occupied MOs:

$$b(\text{CH}_1) = a(1a_1) + b(2a_1) + d(1t_{2x}) + e(1t_{2y}) + f(1t_{2z}) \tag{15.47}$$

with similar expressions for $b(\text{CH}_2)$, $b(\text{CH}_3)$, and $b(\text{CH}_4)$. Since $1a_1$ is a nonbonding low-energy inner-shell orbital, we expect $|a| \ll |b|$. The $1a_1$ and $2a_1$ canonical MOs are directed equally to all four hydrogens, and so varying *a* or *b* in (15.47) does not direct $b(\text{CH}_1)$ preferentially to any one hydrogen. The $1t_{2x}$, $1t_{2y}$, and $1t_{2z}$ MOs are directed along the *x*, *y*, and *z* axes, respectively, so that by adjusting *d*, *e*, and *f* appropriately, we can get $b(\text{CH}_1)$ to be localized mainly in the region of the CH_1 bond. To fix *d*, *e*, and *f*, we use some properties of direction cosines.

If line *L* passes through the origin and makes the angles α, β, and γ with the *x*, *y*, and *z* axes, respectively, then the quantities

$$l \equiv \cos\alpha, \quad m \equiv \cos\beta, \quad n \equiv \cos\gamma \tag{15.48}$$

are the *direction cosines* of *L*. If (x_L, y_L, z_L) is a point on *L*, then clearly

$$x_L = r\cos\alpha, \qquad y_L = r\cos\beta, \qquad z_L = r\cos\gamma \tag{15.49}$$

where *r* is the distance from the origin. From $x^2 + y^2 + z^2 = r^2$ it follows that

$$l^2 + m^2 + n^2 = 1 \tag{15.50}$$

Let the lines L_1 and L_2 go from the origin to (x_1, y_1, z_1) and (x_2, y_2, z_2), respectively. If θ_{12} is the angle between L_1 and L_2, then [Eq. (5.20)]

$$\cos\theta_{12} = \frac{x_1x_2 + y_1y_2 + z_1z_2}{r_1r_2}$$

$$\cos\theta_{12} = l_1l_2 + m_1m_2 + n_1n_2 \tag{15.51}$$

Direction cosines are useful in discussing changes in coordinate axes. Let the $x'y'z'$ and the xyz Cartesian coordinate systems have a common origin, and let the $x'y'z'$ axes be obtained from the xyz axes by rotation, reflection, inversion, or some combination of these operations. Let the direction cosines of the x' axis with respect to the xyz system be l_1, m_1, n_1; let l_2, m_2, n_2 and l_3, m_3, n_3 be the direction cosines of the y' and z' axes, respectively. Let the vector $\mathbf{s}$ have coordinates (x, y, z) and (x', y', z') in the unrotated and rotated coordinate systems. We have $x' = \mathbf{s}\cdot\mathbf{i}'$, where $\mathbf{i}'$ is a unit vector along the x' axis. Since $\mathbf{i}'$ is of unit length, it follows from (15.49) and (15.48) that the coordinates of $\mathbf{i}'$ in the xyz system are l_1, m_1, and n_1. Hence

$$\begin{aligned} x' &= l_1x + m_1y + n_1z \\ y' &= l_2x + m_2y + n_2z \\ z' &= l_3x + m_3y + n_3z \end{aligned} \tag{15.52}$$

where the y' and z' equations follow from $y' = \mathbf{s}\cdot\mathbf{j}'$ and $z' = \mathbf{s}\cdot\mathbf{k}'$. [Equation (15.40) is a special case of (15.52).] Since the angle between any pair of the x', y', and z' axes is 90°, it follows from (15.51) that the direction cosines of these axes satisfy

$$\begin{aligned} l_1l_2 + m_1m_2 + n_1n_2 &= 0 \\ l_1l_3 + m_1m_3 + n_1n_3 &= 0 \\ l_2l_3 + m_2m_3 + n_2n_3 &= 0 \end{aligned} \tag{15.53}$$

Now we return to the determination of d, e, and f. The MOs t_{2x}, t_{2y}, and t_{2z} are directed along the x, y, and z axes, respectively, and the contributions of the carbon $2p_x$, $2p_y$, and $2p_z$ AOs to these MOs are

$$xe^{-\zeta r}, \quad ye^{-\zeta r}, \quad ze^{-\zeta r} \tag{15.54}$$

The contribution of the hydrogen AOs to the t_2 MOs has a more complicated form (the hydrogens are not at the coordinate origin), but we need not explicitly consider the hydrogen part of the MOs. This is because the hydrogen symmetry orbitals (15.43) to (15.45) have the same directional properties as the corresponding carbon $2p$ AOs (15.54) with which each is combined in the $1t_2$ MOs [Eq. (15.46)]. The linear combination (15.47) has as its carbon $2p$ contribution

$$(dx + ey + fz)e^{-\zeta r} \tag{15.55}$$

Let l_1, m_1, and n_1 be the direction cosines of the CH_1 line. We assert that if d, e, and f are chosen as proportional to these direction cosines, then $b(CH_1)$ will be directed toward H_1. To verify this, we set $d:e:f = l_1:m_1:n_1$ in (15.55) and use (15.52) to get

$$c(l_1x + m_1y + n_1z)e^{-\zeta r} = cx'e^{-\zeta r} \tag{15.56}$$

where c is some constant and the x' axis runs from C to H_1.

Similarly, by picking d, e, and f proportional to the direction cosines of the other CH lines, we form localized orbitals along these bonds. From (15.49) the direction cosines of the CH lines are

$$\begin{array}{llll} \mathrm{CH_1:} & 3^{-1/2}, 3^{-1/2}, 3^{-1/2} & \mathrm{CH_2:} & 3^{-1/2}, -3^{-1/2}, -3^{-1/2} \\ \mathrm{CH_3:} & -3^{-1/2}, 3^{-1/2}, -3^{-1/2} & \mathrm{CH_4:} & -3^{-1/2}, -3^{-1/2}, 3^{-1/2} \end{array} \tag{15.57}$$

To satisfy the equivalence requirement, the values of a and b in (15.47) must be the same for each bonding localized MO. The equivalent localized MOs for methane thus have the forms

$$\begin{aligned} b(\mathrm{CH_1}) &= a(1a_1) + b(2a_1) + 3^{-1/2}c(1t_{2x} + 1t_{2y} + 1t_{2z}) \\ b(\mathrm{CH_2}) &= a(1a_1) + b(2a_1) + 3^{-1/2}c(1t_{2x} - 1t_{2y} - 1t_{2z}) \\ b(\mathrm{CH_3}) &= a(1a_1) + b(2a_1) + 3^{-1/2}c(-1t_{2x} + 1t_{2y} - 1t_{2z}) \\ b(\mathrm{CH_4}) &= a(1a_1) + b(2a_1) + 3^{-1/2}c(-1t_{2x} - 1t_{2y} + 1t_{2z}) \end{aligned} \tag{15.58}$$

Orthonormality of the bonding localized MOs (15.58) requires that

$$a^2 + b^2 + c^2 = 1 \quad \text{and} \quad a^2 + b^2 - \tfrac{1}{3}c^2 = 0 \tag{15.59}$$

Hence

$$c = \tfrac{1}{2}\sqrt{3}, \quad (a^2 + b^2)^{1/2} = \tfrac{1}{2} \tag{15.60}$$

The equivalence, direction, and orthonormality requirements have fixed all but one parameter (the ratio a/b) in the localized bonding MOs of methane.

The $1t_{2x}$ MO points equally in the $+x$ and $-x$ directions. Similarly, the $1t_{2y}$ and $1t_{2z}$ MOs point equally on both sides of the carbon atom. This is not true of the bonding localized MOs: The $2a_1$ MO is positive in most of the bonding region between the carbon atom and the hydrogen atoms. [It is negative in the region very near the carbon atom, because of the $-0.06(\mathrm{C}1s)$ term and the negative portion of the orthogonalized C2s AO; see Eq. (15.20).] The linear combination $1t_{2x} + 1t_{2y} + 1t_{2z}$ points equally in the $(1, 1, 1)/\sqrt{3}$ and the $(-1, -1, -1)/\sqrt{3}$ directions. If b in (15.58) is taken as positive (as we have taken c), then the $2a_1$ MO adds to the positive half of this linear combination of the $1t_2$ MOs and cancels much of the negative half of this linear combination. With b and c having the same sign, the $b(\mathrm{CH_1})$ MO points mostly in the $(1, 1, 1)/\sqrt{3}$ direction, with only a small "tail" in the $(-1, -1, -1)/\sqrt{3}$ direction.

Pitzer's SCF calculation gives the energy-localized bonding and inner-shell methane MOs as

$$\begin{aligned} b(\mathrm{CH_1}) &= 0.055(1a_1) + 0.497(2a_1) + \tfrac{1}{2}(1t_{2x} + 1t_{2y} + 1t_{2z}) \\ b(\mathrm{CH_2}) &= 0.055(1a_1) + 0.497(2a_1) + \tfrac{1}{2}(1t_{2x} - 1t_{2y} - 1t_{2z}) \\ b(\mathrm{CH_3}) &= 0.055(1a_1) + 0.497(2a_1) + \tfrac{1}{2}(-1t_{2x} + 1t_{2y} - 1t_{2z}) \\ b(\mathrm{CH_4}) &= 0.055(1a_1) + 0.497(2a_1) + \tfrac{1}{2}(-1t_{2x} - 1t_{2y} + 1t_{2z}) \\ i(\mathrm{C}) &= 0.994(1a_1) - 0.111(2a_1) \end{aligned} \tag{15.61}$$

From (15.46) and (15.61), we have

$$i(\mathrm{C}) = 1.002(\mathrm{C}1s) - 0.040(\mathrm{C}2s_{\perp}) - 0.025(\mathrm{H_1}1s + \mathrm{H_2}1s + \mathrm{H_3}1s + \mathrm{H_4}1s)$$

$$b(\mathrm{CH}_1) = 0.024(\mathrm{C}1s) + 0.292(\mathrm{C}2s_\perp) + 0.569(\mathrm{H}_1 1s) - 0.066(\mathrm{H}_2 1s + \mathrm{H}_3 1s + \mathrm{H}_4 1s) + 0.277(\mathrm{C}2p_x + \mathrm{C}2p_y + \mathrm{C}2p_z) \quad (15.62)$$

From (15.54) to (15.56), the linear combination $dp_x + ep_y + fp_z$ is equal to a rotated p orbital pointing along the x' axis and containing the factor $dx + ey + fz = c(l_1x + m_1y + n_1z) = cx'$, where l_1, m_1, n_1 are the direction cosines of the x' axis relative to the x, y, z axes. Substitution of $l_1 = d/c, m_1 = e/c, n_1 = f/c$ into $l_1^2 + m_1^2 + n_1^2 = 1$ [Eq. (15.50)] gives $c = (d^2 + e^2 + f^2)^{1/2}$. Therefore, a localized MO that contains the terms $a\mathrm{C}2s_\perp + d\mathrm{C}2p_x + e\mathrm{C}2p_y + f\mathrm{C}2p_z$ can be viewed as containing the terms $a\mathrm{C}2s_\perp + (d^2 + e^2 + f^2)^{1/2}\mathrm{C}2p_{x'}$ and we say the hybridization of the carbon AO in this MO is

$$sp^{(d^2+e^2+f^2)/a^2} \quad (15.63)$$

For example, for the localized bonding MO (15.62) of CH_4, $a^2 = 0.0853$, $d^2 + e^2 + f^2 = 0.230$, and this MO is an $sp^{2.7}$ hybrid, which is close to the traditional sp^3 hybridization of VB theory (Section 16.8).

Ethane. The most fascinating property of C_2H_6 is the barrier to internal rotation about the carbon–carbon single bond. The staggered conformation of a C_2H_6 molecule is more stable than the eclipsed conformation by 0.125 eV = 0.00461 hartree, which is equivalent to 2.89 kcal/mol. In 1936, J. D. Kemp and Kenneth Pitzer discovered this fact in examining thermodynamic data for ethane. The correlation energy for an N-electron species is very roughly $-(0.04 \text{ hartree})(N - 1)$ [S. Kristyán, *Chem. Phys. Lett.*, **247**, 101 (1995)], so the Hartree–Fock energy of C_2H_6 will differ by about 0.7 hartree $\approx$ 20 eV from the true molecular energy. This SCF energy error is far greater than the barrier height. At first sight, it seems hopeless to expect an SCF MO calculation to give a meaningful result for the ethane barrier.

The minimal-basis-set AOs for ethane are the hydrogen $1s$ orbitals and the carbon $1s$, $2s$, and $2p$ orbitals, a total of $6(1) + 2(5) = 16$ basis AOs. To calculate the barrier in ethane, we must calculate the energy of the staggered and the eclipsed conformations, which requires two separate SCF calculations. One first forms appropriate linear combinations of the hydrogen AOs and of the carbon AOs to get symmetry-adapted basis functions. The Roothaan equations are then solved iteratively to give the basis-function coefficients and orbital energies, and the total molecular energy is found.

The pioneering ethane calculation is by Russell Pitzer (Kenneth Pitzer's son) and W. N. Lipscomb, who did an SCF MO calculation using a minimal STO basis set with orbital exponents chosen according to Slater's rules [R. M. Pitzer and W. N. Lipscomb, *J. Chem. Phys.*, **39**, 1995 (1963)]. They used the experimentally observed equilibrium geometry for staggered C_2H_6 and assumed the geometry of the eclipsed form to be that given by rigidly rotating one methyl group with respect to the other. Their calculated energies of -78.98593 and -78.99115 hartrees for the eclipsed and staggered forms, respectively, give a barrier of 3.3 kcal/mol, in reasonable agreement with experiment. Clementi and Popkie did a near-Hartree–Fock calculation with a basis set of 102 CGTFs [E. Clementi and H. Popkie, *J. Chem. Phys.*, **57**, 4870 (1972)]. They varied the bond angles and the CC bond length for each conformation to minimize the energy and found an increase of 0.02 Å in the CC length and a decrease of 0.3° in the HCH angle on going from the staggered to the eclipsed form. Their calculated barrier is 3.2 kcal/mol.

Why do SCF MO calculations give good values of the ethane rotational barrier? The answer lies in the correlation energy. Electrons paired in the same localized orbital move through the same region of space. Hence the intraorbital correlation for such a pair should be substantially greater in magnitude than the interorbital correlation energy for two electrons in different localized MOs. In line with this, it was formerly believed that the magnitude of the total interorbital molecular correlation energy was much less than the total intraorbital correlation energy. However, there are many more interorbital pair correlations than intraorbital correlations, and it is now recognized that the total interorbital correlation is not negligible and in some cases can be of comparable magnitude to the total intraorbital correlation. [See E. Steiner, *J. Chem. Phys.*, **54**, 1114 (1971).] Hence we must consider both kinds of correlation in ethane.

Rotation of a methyl group in ethane does not change any of the bonds, and thus intraorbital correlation should be essentially the same in the staggered and eclipsed forms. Moreover, we expect most of the interorbital correlations to be essentially unchanged in the two forms. In particular, correlations between the C—H bonds within each methyl group should remain virtually the same, and so should correlations between the C—C and C—H bonding pairs. It is only the correlations between C—H pairs of different methyl groups that should change, and these make the smallest contributions to interorbital correlation. Thus we expect the total correlation energy to be only slightly changed from staggered to eclipsed. Thus the energy error in an SCF MO calculation is about the same for the two forms, and the Hartree–Fock energy difference should yield a good estimate of the barrier. Recall that Hartree–Fock calculations give poor values for dissociation energies. This is because the number of electron pairs changes in forming a chemical bond from atoms, thereby changing the correlation energy substantially.

Some HF/6-31G* results for other rotational barriers calculated with geometry optimization of all structures are (values in kcal/mol): CH_3OH, 1.4 calculated versus 1.1 experimental; CH_3CHO, 1.0 calculated versus 1.2 experimental; CH_3NH_2, 2.4 calculated versus 2.0 experimental; CH_3SiH_3, 1.4 calculated versus 1.7 experimental (*Hehre et al.*, Section 6.4.1).

The physical origin of the ethane rotational barrier has been the subject of intense controversy. The steric-repulsion viewpoint attributes the barrier to the Pauli repulsion (Section 10.3) between the eclipsing localized C—H bonding electron pairs in eclipsed ethane [R. M. Pitzer, *Acc. Chem. Res.*, **16**, 207 (1983); F. M. Bickelhaupt and E. J. Baerends, *Angew. Chem. Int. Ed.*, **42**, 4183 (2003)]. The hyperconjugation viewpoint [A. E. Reed and F. Weinhold, *Isr. J. Chem.*, **31**, 277 (1991); V. Pophristic and L. Goodman, *Nature*, **411**, 565 (2001); F. Weinhold, *Nature*, **411**, 539 (2001); F. Weinhold, *Angew. Chem. Int. Ed.*, **42**, 4188 (2003)] attributes the barrier to greater stabilization of the staggered form by hyperconjugation. (In the language of the VB method, hyperconjugation stabilization in ethane is due to contributions of ionic resonance structures that have a double bond between the carbons and have an H^+ on one carbon and an H^- on the other carbon, with no bonds to these ionic H's. In the language of MO theory, hyperconjugation stabilization is due to interaction between a filled, localized bonding CH MO on one C and a vacant, localized antibonding CH MO on the second carbon. *Hyperconjugation* involves delocalization of localized σ electrons, whereas *conjugation* involves delocalization of localized π electrons. σ and π LMOs are defined at the end of Section 15.9.)

An analysis of ab initio VB wave functions for ethane and related molecules found that the main contributor to the barrier is steric repulsion, with hyperconjugation contributing only one-third of the ethane barrier [Y. Mo et al., *Angew. Chem. Int. Ed.*, **43**, 1986 (2004); L. Song et al., *J. Phys. Chem. A*, **109**, 2310 (2005); Y. Mo and J. Gao, *Acc. Chem. Res.*, **40**, 113 (2007)]. The experimentally observed structure of $CH_3CH_2CCCH_2CH_3$ has been interpreted as providing strong support for the steric-repulsion origin of the ethane barrier [R. K. Bohn, *J. Phys. Chem. A*, **108**, 6814 (2004)].

Ethylene. For C_2H_4, the ground-state equilibrium-geometry point group is $\mathcal{D}_{2h}$. The standard choice of coordinate axes is shown in Fig. 15.10.

There are eight symmetry operations for $\mathcal{D}_{2h}$. Each symmetry operation commutes with every other symmetry operation, so the electronic wave function must be an eigenfunction of all the symmetry operators, and we have only nondegenerate (A and B) symmetry species. Since the three rotations, the three reflections, and the inversion operation each have their squares equal to the identity operation, these symmetry operations must have the eigenvalues ± 1. All eight symmetry operations can be expressed as the product of one, two, or three of the reflections (each reflection simply converts one coordinate to its negative):

$$\hat{E} = [\hat{\sigma}(xy)]^2, \qquad \hat{i} = \hat{\sigma}(xy)\hat{\sigma}(xz)\hat{\sigma}(yz)$$

$$\hat{C}_2(x) = \hat{\sigma}(xy)\hat{\sigma}(xz), \qquad \hat{C}_2(y) = \hat{\sigma}(xy)\hat{\sigma}(yz), \qquad \hat{C}_2(z) = \hat{\sigma}(xz)\hat{\sigma}(yz)$$

Since the symmetry eigenvalues multiply the same way the symmetry operations do, specification of the eigenvalues of the three reflections is sufficient to fix all eight symmetry eigenvalues. There are thus $2^3 = 8$ possible symmetry species. The standard notation for these is given in Table 15.3. The g or u subscript corresponds to eigenvalue $+1$ or -1 for $\hat{i}$. For $\mathcal{D}_{2h}$ the designation A is used only for symmetry species with eigenvalue $+1$ for all three $\hat{C}_2$ rotations.

There are 14 minimal-basis-set AOs. It is easy to set up symmetry orbitals by trial and error. Unnormalized symmetry orbitals and their symmetry species are listed in Table 15.4.

The 14 lowest MOs include four a_g, four b_{1u}, two b_{2u}, two b_{3g}, one b_{3u}, and one b_{2g} MO. Eight of these MOs are occupied in the ground state. SCF MO calculations give as the electron configuration of the ^1A_g ground state [U. Kaldor and I. Shavitt, *J. Chem. Phys.*, **48**, 191 (1968)]

$$(1a_g)^2(1b_{1u})^2(2a_g)^2(2b_{1u})^2(1b_{2u})^2(3a_g)^2(1b_{3g})^2(1b_{3u})^2$$

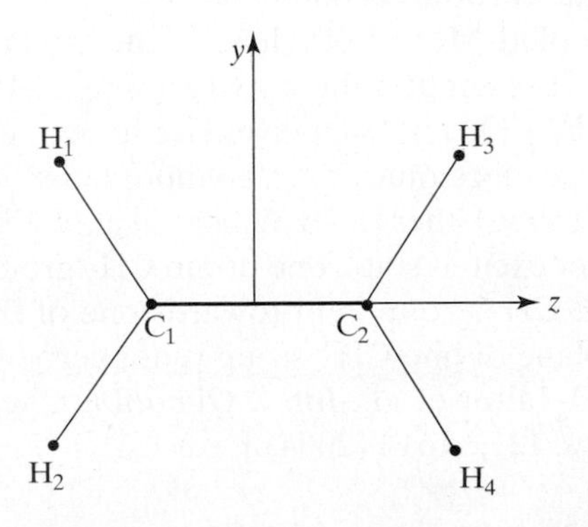

FIGURE 15.10 Coordinate axes for C_2H_4. The x axis is perpendicular to the molecular plane.

TABLE 15.3 Symmetry Species Eigenvalues for $\mathscr{D}_{2h}$

	$\hat{\sigma}(xy)$	$\hat{\sigma}(xz)$	$\hat{\sigma}(yz)$
A_g	+1	+1	+1
A_u	−1	−1	−1
B_{1g}	+1	−1	−1
B_{1u}	−1	+1	+1
B_{2g}	−1	+1	−1
B_{2u}	+1	−1	+1
B_{3g}	−1	−1	+1
B_{3u}	+1	+1	−1

TABLE 15.4 Symmetry-Adapted Basis Functions for C_2H_4

Symmetry Function	Symmetry Species
$H_11s + H_21s + H_31s + H_41s$	a_g
$C_11s + C_21s$	a_g
$C_12s + C_22s$	a_g
$C_12p_z - C_22p_z$	a_g
$H_11s + H_21s - H_31s - H_41s$	b_{1u}
$C_11s - C_21s$	b_{1u}
$C_12s - C_22s$	b_{1u}
$C_12p_z + C_22p_z$	b_{1u}
$H_11s - H_21s + H_31s - H_41s$	b_{2u}
$C_12p_y + C_22p_y$	b_{2u}
$H_11s - H_21s - H_31s + H_41s$	b_{3g}
$C_12p_y - C_22p_y$	b_{3g}
$C_12p_x + C_22p_x$	b_{3u}
$C_12p_x - C_22p_x$	b_{2g}

Each canonical MO of a planar molecule is classified as σ or π, according to whether its eigenvalue for reflection in the molecular plane is $+1$ or -1, respectively. (This usage is somewhat inconsistent with the $\sigma, \pi, \delta, \cdots$ classification of linear-molecule MOs. For linear molecules, the symbols σ and π signify an axial component of electronic orbital angular momentum of 0 and $\pm\hbar$, respectively. For nonlinear molecules we cannot specify the component of **L** along an internuclear line. For linear molecules, σ MOs are nondegenerate and π MOs are doubly degenerate. For nonlinear molecules, the σ–π classification is unrelated to the degeneracy.) For the ground state of water, all the occupied MOs are σ MOs except the lone-pair $1b_1$ MO, which is a π MO.

For ethylene, the only occupied π MO in the ground electronic state is the $1b_{3u}$ MO, the highest occupied MO. Since there is only one b_{3u} symmetry orbital in Table 15.4, the minimal-basis $1b_{3u}$ MO must be identical (apart from a normalization constant) to this symmetry orbital. One finds (using STOs) $1b_{3u} = 0.63(C_12p_x + C_22p_x)$. This bonding MO formed by overlap of the $2p_x$ AOs of carbon resembles the π_u MO of Fig. 13.15. The $1b_{3u}$ π MO accounts for the planarity of ethylene in its ground state. As one CH_2 group is rotated relative to the other, the overlap between the two carbon $2p_x$ AOs rapidly diminishes, and the energy of the $1b_{3u}$ MO increases; hence the molecule strongly resists torsion about the carbon–carbon bond.

The lowest-lying unoccupied MO of ethylene is the $1b_{2g}$ antibonding π MO: $1b_{2g} = 0.82(C_12p_x - C_22p_x)$. It resembles the π_g MO in Fig. 13.15. The excited ethylene configuration $(1a_g)^2 \cdots (1b_{3g})^2(1b_{3u})(1b_{2g})$ gives rise to two terms (a singlet and a triplet), and it is likely that these electronic states are nonplanar, with one CH_2 group rotated about 90° with respect to the other [K. B. Wiberg et al., *J. Phys. Chem.*, **96**, 10756 (1992)]. Moreover, in the singlet excited state, one of the CH_2 groups is pyramidalized, with the two CH bonds of one CH_2 group bent towards one of the hydrogens of the other CH_2 group, so that the plane of one CH_2 group is not perpendicular to the plane of the other CH_2 group [S. El-Taher et al., *Int. J. Quantum Chem.*, **82**, 242 (2001); M. Barbatti et al., *J. Chem. Phys.* **121**, 11614 (2004).]

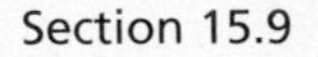

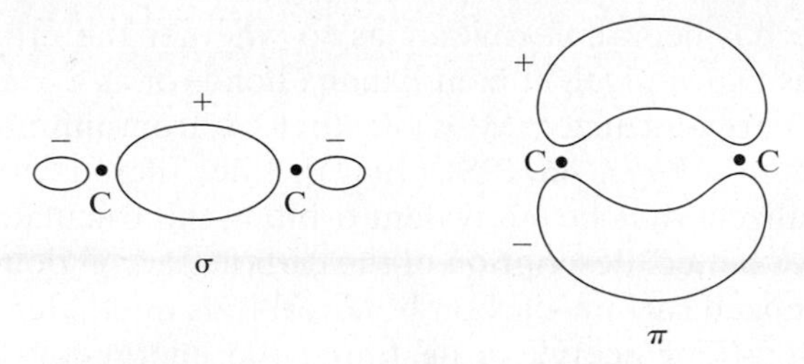

FIGURE 15.11 σ, π description of the ethylene double bond. The cross sections are taken in a plane perpendicular to the molecular plane, which is a nodal plane for the π MO.

The canonical σ MOs of ethylene each contain contributions from AOs of all six atoms and are delocalized over the whole molecule. By taking appropriate linear combinations of the canonical σ MOs, we can form localized σ MOs. We expect these MOs to consist of the following inner-shell and bonding MOs:

$$\begin{gathered} i(C_1), \qquad i(C_2) \\ b(C_1C_2), \quad b(C_1H_1), \quad b(C_1H_2), \quad b(C_2H_3), \quad b(C_2H_4) \end{gathered} \tag{15.64}$$

The MOs $b(C_1C_2)$ and $1b_{3u}$ give the familiar description of the carbon–carbon double bond as one σ and one π bond (Fig. 15.11). We still do not, however, have the equivalent MOs of ethylene, since the π MO $1b_{3u}$ is not equivalent to any of the σ MOs in (15.64), nor is it equivalent to itself; it goes into its negative upon a $\hat{C}_2(z)$ rotation. By adding and subtracting the $b(C_1C_2)$ and $1b_{3u}$ MOs, we can form two equivalent localized carbon–carbon bonding MOs:

$$b_1(C_1C_2) = 2^{-1/2}[b(C_1C_2) + 1b_{3u}] \tag{15.65}$$

$$b_2(C_1C_2) = 2^{-1/2}[b(C_1C_2) - 1b_{3u}] \tag{15.66}$$

The MOs (15.65) and (15.66) lead to the description of the carbon–carbon bond as composed of two "banana" bonds (Fig 15.12).

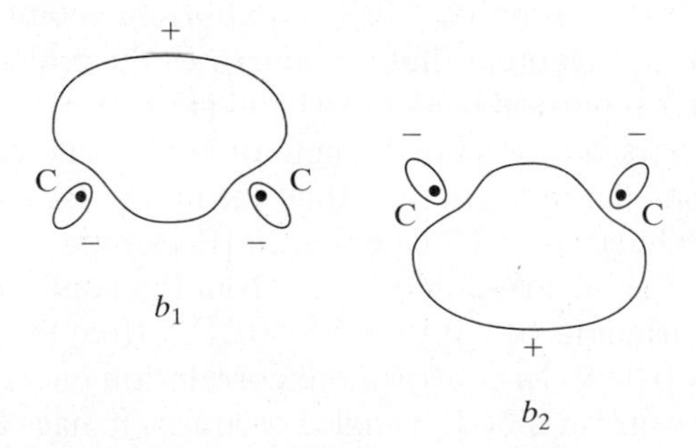

FIGURE 15.12 Banana-bond description of the ethylene double bond. The cross sections are the same as in Fig. 15.11.

There has been controversy as to whether the ethylene double bond is best described as two equivalent bent banana bonds or as a σ and a π bond. Kaldor calculated the energy-localized MOs of ethylene from minimal-basis-set SCF MOs [U. Kaldor, *J. Chem. Phys.*, **46**, 1981 (1967)]. Since there is no a priori necessity that the energy-localized MOs be equivalent orbitals, this calculation provides evidence as to which is the "better" description of the carbon–carbon double bond. Kaldor found the energy-localized carbon–carbon bond orbitals in ethylene to be the two equivalent banana bonds. For acetylene he found the energy-localized carbon–carbon bond orbitals to be three equivalent banana bonds and not one σ and two π bonds. Of course, although the electron densities in the individual MOs differ for the banana-bond versus σ–π descriptions, the total probability density for the four or six electrons in the double or triple bond is the same in either picture.

The energy-localized C_1H_1 bond MO in ethylene was found by Kaldor to be

$$0.3637(C_12s_\perp) + 0.4143(C_12p_y) - 0.2574(C_12p_z) + 0.4939(H_11s) + \cdots$$

where the dots indicate small contributions from other AOs. From (15.63), the hybridization of carbon in this localized MO is $sp^{1.8}$, which is close to the sp^2 of VB theory. The angle made by the carbon hybrid AO in $b(CH_1)$ with the C–C axis is $\pi - \arctan(1.61) = 122°$, essentially the same as the experimental bond angle. Hence the localized CH bond orbitals are not bent in planar ethylene.

We noted earlier in this section that each *canonical* MO of a planar molecule is classified as π or σ according to whether or not the molecular plane is a nodal plane for the MO. A *localized* bonding MO of a molecule (planar or nonplanar) is classified as σ or π or δ according to whether this MO has 0 or 1 or 2 nodal planes containing the line joining the nuclei of the two bonded atoms. The bonding localized MOs in water (Fig. 15.7) are σ MOs; the ethylene double-bond localized MOs in Fig. 15.11 consist of one σ and one π MO; the $b(CH_1)$ MO (15.62) in CH_4 is a σ MO. In certain transition-metal compounds (for example, $Re_2Cl_8^{2-}$), overlap of d AOs produces a δ localized bonding MO, and these compounds have a quadruple bond consisting of one σ, two π, and one δ MO; see F. A. Cotton et al. (eds.), *Multiple Bonds Between Metal Atoms*, 3rd ed., Springer, 2005.

15.10 MOLECULAR GEOMETRY

Equilibrium Geometry. The equilibrium geometry of a molecule corresponds to the nuclear arrangement that minimizes U, the molecular electronic energy including internuclear repulsion [Eqs. (13.4) and (13.8)].

The changes in U as a bond length or bond angle varies over moderate ranges are substantially smaller in magnitude than the molecular correlation energy. For example, for H_2O a large basis-set CI calculation [P. Hennig et al., *Theor. Chim. Acta*, **47**, 233 (1978)] found that a variation of $\pm 15°$ from the equilibrium 105° angle changed U by 0.008% and variations of -0.16 Å or $+0.21$ Å from the equilibrium OH bond lengths changed U by 0.07%. In contrast, the correlation energy of H_2O is 0.5% of U (Table 15.2). Also, changes in dihedral angles produce changes that are far less than the correlation energy (recall the discussion of ethane in Section 15.9).

Despite the smallness of these changes of molecular energy with nuclear locations as compared with the energy error (the correlation energy) inherent in the SCF MO method, ab initio SCF MO wave functions usually give good predictions (0 to 3% error) of equilibrium bond distances and angles in molecules not involving transition metals. Some HF/3-21G bond angles and lengths follow (*Hehre* et al., Section 6.2), where experimental values are in parentheses. NH_3: 112° (107°) and 1.00 Å (1.01 Å); H_2O 108° ($104\frac{1}{2}$°) and 0.97 Å (0.96 Å); C_2H_6: 108° (108°) for HCH, 1.08 Å (1.10 Å) for C—H, and 1.54 Å (1.53 Å) for C—C; C_2H_4: 116° (117°) for HCH and 1.32 Å (1.34 Å) for C═C; H_2CO: 115° ($116\frac{1}{2}$°) for HCH and 1.21 Å (1.20 Å) for C═O. (See also Section 18.1.)

Evidently, the correlation energy remains approximately constant for bond angle and length variations in the region of the equilibrium geometry.

The Potential-Energy Surface (PES). The geometry of a nonlinear molecule with N nuclei is defined by $3N - 6$ independent nuclear coordinates $q_1, q_2, \ldots, q_{3N-6}$ and its electronic energy U is a function of these coordinates. The number 6 is subtracted from the total number of nuclear coordinates because the three translational and three rotational degrees of freedom leave U unchanged. (A diatomic molecule has only two rotational degrees of freedom, the angles θ and ϕ in Fig. 6.3, and here U is a function of only one variable, the internuclear distance R.) The function U gives what is called the **potential-energy surface** (PES) for the molecule, so called because U is the potential energy in the Schrödinger equation (13.10) and (13.11) for nuclear motion. If U depended on two variables, then a plot of $U(q_1, q_2)$ in three dimensions would give a surface in ordinary three-dimensional space. Because of the large number of variables, U is a "surface" in an abstract "space" of $3N - 5$ dimensions. To find U, we must solve the electronic Schrödinger equation at many nuclear configurations, which is a formidable task for a large molecule. Calculation of U at one particular arrangement of the nuclei is called a **single-point calculation**, since it gives one point on the molecular PES.

A complicating fact is that a large molecule may have many minima on its PES. Figure 15.13 sketches the variation of the electronic energy U for butane, $CH_3CH_2CH_2CH_3$, as a function of the CCCC dihedral angle. At each point on this curve, all geometrical coordinates except the CCCC dihedral angle have been varied to yield the

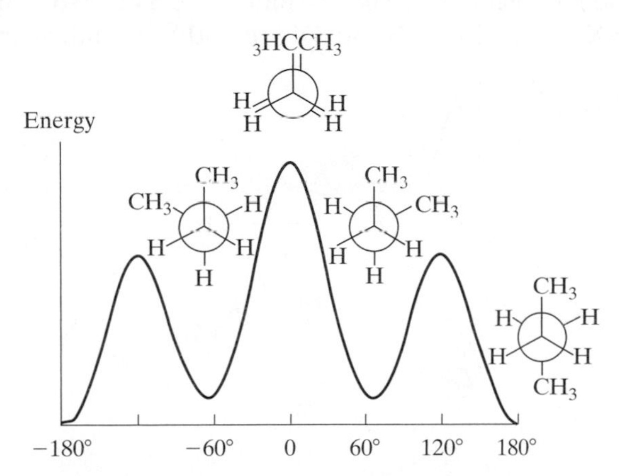

FIGURE 15.13 Electronic energy (including nuclear repulsion) for $CH_3CH_2CH_2CH_3$ versus the CCCC dihedral angle.

energy minimum for the particular fixed CCCC dihedral angle. The 0° dihedral angle gives the **syn** (or *cis*) conformation with the methyl groups eclipsing one another. This is an energy maximum with respect to variation of the dihedral angle. However, because the geometry has been optimized for all variables except the dihedral angle, the 0° point corresponds to an energy minimum for the $3N - 7$ remaining variables. The 0° point is a **first-order saddle point**, meaning that it is an energy maximum for one variable and an energy minimum for the remaining variables. (The point where the rider sits on a saddle is a point of maximum height on the curve that goes from one side of the saddle to the other and is a minimum point on the curve that goes from the back to the front of the saddle.) The energy minimum at about 60° corresponds to the **+gauche** conformation, and the minimum at 180° is the **anti** (or *trans*) conformation. The 180° minimum is the lowest-energy point on the butane PES and so is called the **global minimum**. The 60° minimum is a **local minimum**, meaning that it is lower in energy than all PES points in its immediate vicinity. The minimum near −60° is the −gauche conformation, which is a nonsuperimposable mirror image of the +gauche conformation.

The **conformation** of a molecule is specified by giving the values of all dihedral angles about single bonds. A conformation that corresponds to an energy minimum (local or global) is called a **conformer**.

Large molecules with many single bonds may have astronomical numbers of local minima, producing the **multiple-minima problem**. For example, consider the n-residue polypeptide $(\text{—NH—CHR—C(O)—})_n$. Even if one assumes that each OC—NH dihedral angle is 180° (its most common value) and ignores the conformations of the amino acid side chains R, the polypeptide has $2n$ adjustable dihedral angles (the torsion angles about the NH—CHR and CHR—C(O) bonds), each of which has three likely minima (60°, −60°, and 180°). This gives 3^{2n} possible conformations that are local energy minima. For $n = 40$, this is $3^{80} \approx 10^{38}$ conformations to be examined. Methods for searching for low-energy conformations of medium-size and large molecules are discussed in Section 15.11.

The **dihedral angle** $\omega = D(\text{RSTU})$ in Fig. 15.14a is defined as the angle ω between the half plane RST and the half plane STU. More precisely, the dihedral angle is angle RSX, where lines RS and SX are both perpendicular to line ST. In dealing with

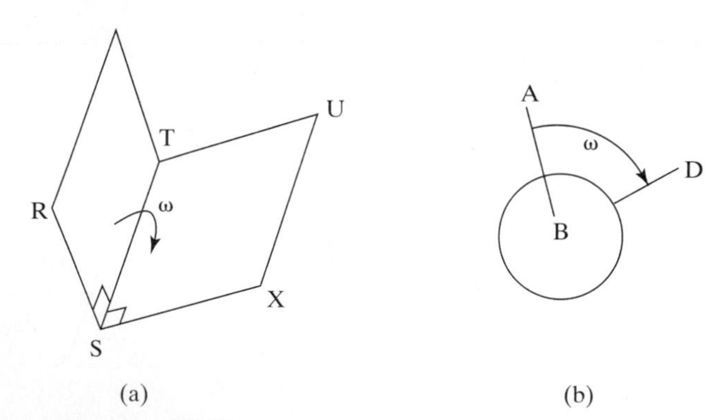

FIGURE 15.14 (a) The dihedral angle ω between the half planes RST and STU. (b) The dihedral angle $D(\text{ABCD})$. Atom C is behind B.

a molecular dihedral angle $D(\text{ABCD})$ involving four atoms A, B, C, D, the AB and CD bonds are not likely to be perpendicular to the BC bond, so one must project AB and CD into a plane perpendicular to line BC. This is done by drawing a Newman projection with line BC perpendicular to the plane of the paper (Fig. 15.14b). The range of dihedral angles is chosen to be either $0° \leq \omega < 360°$ or $-180° < \omega \leq 180°$. By definition, a clockwise rotation of the projected front bond AB to bring it to atom D corresponds to a positive dihedral angle $D(\text{ABCD})$.

Some references on geometry optimization are H. B. Schlegel, *Adv. Chem. Phys.*, **67**, 249 (1987); H. B. Schlegel in *Yarkony*, Part I, Chapter 8; *Leach*, Chapter 5; H. B. Schlegel, *J. Comput. Chem.*, **24**, 1514 (2003).

Geometry Optimization. Many systematic mathematical procedures (algorithms) exist to find a local minimum of a function of several variables. These procedures will find a local minimum in U in the neighborhood of the initially assumed molecular geometry. The process of finding such a minimum is called **geometry optimization** or **energy minimization**. For a molecule with several conformations, one must repeat the local-minimum search procedure for each possible conformation, so as to locate the global minimum. For large molecules, there may be too many conformations for all of them to be examined. Moreover, the true global-minimum equilibrium geometry might correspond to a highly unconventional structure that the researcher might not think to consider. For example, high-level calculations that use large basis sets and include electron correlation show that, for the vinyl cation, the classical structure (Fig. 15.15a) lies about 4 kcal/mol higher in energy than the true equilibrium structure (Fig. 15.15b), which has a three-center bond [C. Liang. T. P. Hamilton, and H. F. Schaefer, *J. Chem. Phys.*, **92**, 3653 (1990)]; the infrared spectrum of the vinyl cation also shows the three-center-bond structure to be more stable [M. W. Crofton et al., *J. Chem. Phys.*, **91**, 5139 (1989)]. The ethyl cation has a similar symmetrical bridged structure [B. Ruscic et al., *J. Chem. Phys.*, **91**, 114 (1989)].

Some procedures to find a local minimum in U require only repeated calculation of U at various values of its variables, but these procedures are very inefficient. More-efficient procedures require repeated calculation of both U and its derivatives. The set of $3N - 6$ first partial derivatives of U with respect to each of its variables constitutes a vector (in a "space" of $3N-6$ dimensions) called the **gradient** of U (Section 5.2). At a local minimum, the gradient must be zero, meaning that each of the $3N - 6$ first partial derivatives of U must be zero. Any point on the PES where the gradient is zero is called a **stationary** (or *critical*) **point**. A stationary point on a PES may be a minimum, a maximum, or a saddle point.

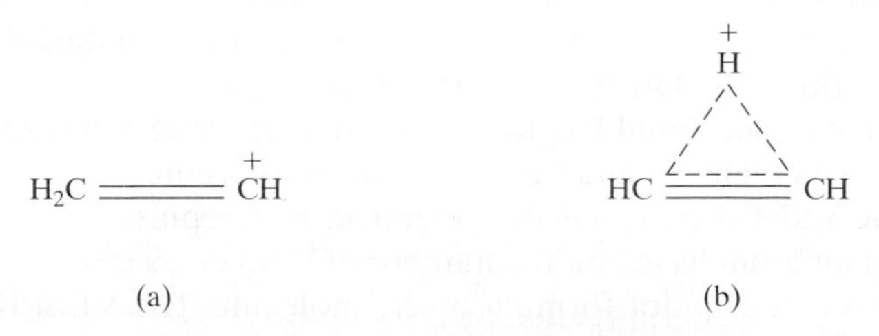

FIGURE 15.15 (a) Classical structure of the vinyl cation. (b) True equilibrium structure of the vinyl cation.

Analytical calculation of the gradient is the key to efficient geometry optimization. The SCF MO energy expression is Eq. (14.44), and its derivatives with respect to the nuclear coordinates would seem to involve the derivatives of the H_{rs}^{core} and $(rs|tu)$ integrals (which occur in ε_i), the derivatives of V_{NN}, and the derivatives of the SCF coefficients c_{si} (which occur in P_{rs}). However, the terms involving the derivatives of the c_{si}'s turn out to add up to zero, leaving only the derivatives of the integrals and of V_{NN}. The derivatives of the integrals are readily calculated, since the derivative of a Gaussian-type function with respect to a nuclear coordinate is another GTF. The derivatives of V_{NN} are trivial. Thus an analytic formula for the gradient of the ab initio SCF energy is known [see *Hehre* et al., Section 3.3.3; P. Pulay, *Adv. Chem. Phys.*, **69**, 241 (1987); P. Pulay in *Yarkony*, Part II, Chapter 9]. Once the SCF energy U and wave function have been found for some chosen geometry, the time needed to analytically calculate the energy gradient is roughly equal to the time needed to do the SCF wave function and energy calculation. (To calculate the gradient numerically requires varying the $3N - 6$ nuclear coordinates one at a time by a small amount, repeating the SCF calculation for each new geometry to get U for that geometry, and estimating each derivative as the ratio of the change in U to the change in the coordinate. Numerical evaluation of the gradient thus takes about $3N - 6$ times as long as analytic evaluation of the gradient.)

As well as using the energy gradient (the $3N - 6$ first partial derivatives $\partial U/\partial q_1, \partial U/\partial q_2, \ldots$), some energy minimization methods also use the second derivatives of U. The set of second derivatives $\partial^2 U/\partial q_1^2, \partial^2 U/\partial q_1 \partial q_2, \partial^2 U/\partial q_2 \partial q_1, \partial^2 U/\partial q_2^2, \ldots$, when arranged in a square array form a matrix called the **Hessian** or the **force-constant matrix** [since the second derivatives of U are force constants; see the equation after (13.21)]. An efficient way to find a local minimum of a function of several variables is the **Newton** (or **Newton–Raphson**) method, which approximates the function by a Taylor-series expansion that is terminated after the quadratic terms, and uses accurately evaluated first and second partial derivatives of the function (which occur in the linear and quadratic terms of the Taylor series). Because analytic calculation of the second derivatives in an ab initio SCF calculation is very costly in computer time, ab initio SCF geometry optimizations commonly use a modification of the Newton procedure, called the **quasi-Newton** (or *quasi-Newton–Raphson* or *variable metric*) method. In the quasi-Newton method, one does not calculate the Hessian directly, but instead starts with an estimate (or guess) for the Hessian and gradually improves (updates) this estimate using gradient information calculated at each step in the optimization cycle.

To optimize the geometry, one starts with a guess for the equilibrium structure. The guessed structure is based on typical values for bond lengths, bond angles estimated from a method such as the VSEPR method, and dihedral angles guessed based on experience with similar compounds. [One can use a model-builder computer program (Section 15.15) to get an initial structure guess.]

Some typical bond lengths are listed in Table 15.5, where Xn denotes an atom X bonded to n neighbors and C*ar* is an aromatic carbon.

The VSEPR (valence-shell electron-pair repulsion) method predicts the bond angles at an atom based on the number of valence electron pairs around that atom in the Lewis electron-dot formula of the molecule. The VESPR method predicts 180° angles for two pairs, 120° angles for three pairs, $109\frac{1}{2}°$ (tetrahedral) angles for four pairs, and 90° angles (octahedral arrangement of pairs) for six pairs. For example, the

TABLE 15.5 Some Typical Bond Lengths[a]

C4—H	C3—H	C2—H	C4—C4	C4—C3	C4—C2	C3—C3	C3—C2	C2—C2
1.09	1.08	1.06	1.54	1.52	1.46	1.46	1.45	1.38
C3=C3	C3=C2	C2=C2	C2≡C2	*CarCar*	O2—O2	O2—H	O2—C4	O2—C3
1.34	1.31	1.28	1.20	1.40	1.48	0.96	1.43	1.36
O2—C2	O1=C3	O1=C2	N3—H	N2—H	N3—C4	N2—C4	N3—N3	
1.36	1.22	1.16	1.01	0.99	1.47	1.47	1.45	

[a]Excerpted from J. A. Pople and M. Gordon, *J. Am. Chem. Soc.*, **89**, 4253 (1967). Lengths are in Å.

Lewis structure of H_2O shows four pairs around the O atom, which indicates a tetrahedral $109\frac{1}{2}°$ bond angle. Because lone pairs occupy a larger volume of space than bonded pairs, we expect the angle to be a little less than $109\frac{1}{2}°$. The observed angle is $104\frac{1}{2}°$. The VESPR method counts a double or triple bond as only one pair. Thus, 120° angles are predicted at each carbon in $H_2C{=}CH_2$, but because the double bond contains two pairs, we expect the HCH angle to be somewhat less than 120°. For further details on the VSEPR method, see any general chemistry textbook or R. J. Gillespie and I. Hargittai, *The VSEPR Model of Molecular Geometry*, Prentice Hall, 1991.

Some rules for predicting dihedral angles in acyclic organic compounds follow [adapted from J. A. Pople and M. Gordon, *J. Am. Chem. Soc.*, **89**, 4253 (1967)]. (1) The conformation about a bond connecting two atoms each with tetrahedral bond angles is usually staggered. (2) For an atom A with tetrahedral angles bonded to an atom B with trigonal (120°) bond angles, (a) one of the non-B atoms bonded to A lies in the same plane as B and the atoms bonded to B; (b) the lowest energy conformer usually has the double bond on B eclipsing a single bond to A. (3) When two bonded atoms A and B both have trigonal bond angles, all atoms bonded to A and B lie in the same plane.

For example, rule 2 tells us that one of the methyl hydrogens in $CH_3{-}CH{=}CH_2$ is in the same plane as the $CH{=}CH_2$ part of the molecule and eclipses the double bond, and further tells us that formic acid, HC(O)OH, is planar with the lowest-energy conformer having the OH hydrogen eclipsing the double bond.

After guessing the geometry, one searches for the minimum nearest the initially assumed geometry. One chooses a basis set and uses the SCF MO (or some other) method to approximately solve the electronic Schrödinger equation to find U and its gradient at the guessed initial geometry. Using the calculated values of U and ∇U (and perhaps information about the Hessian) the geometry optimization program changes the $3N - 6$ nuclear coordinates to a new set that is likely to be closer to a minimum than the initial set, and the SCF U and ∇U are calculated at this new structure. Using the results of the new calculation, a further improved set of nuclear coordinates is calculated, and the SCF calculation is repeated at the new geometry. The process is repeated until ∇U differs negligibly from zero, indicating that a minimum (at which ∇U is zero) may have been found. Typically, about $3N - 6$ to $2(3N - 6)$ repetitions of the SCF and gradient calculations are needed to find a minimum. The availability of analytical gradients makes possible the efficient determination of the ab initio equilibrium geometry of small- and medium-size molecules, and the introduction of analytical gradients into ab initio calculations (by Pulay in 1969) has been called a "revolution" [L. Schäfer, *J. Mol. Struct.*, **100**, 51 (1983)].

TABLE 15.6 HF/3-21G Geometry Optimization of a Formic Acid Conformer[a]

	R(CH)	R(C=O)	R(CO)	R(OH)	∠(HC=O)	∠(OCO)	∠(COH)
initial	1.080	1.220	1.360	0.960	120.0°	120.0°	109.5°
step 1	1.070	1.205	1.347	0.970	124.8	125.1	110.5
step 2	1.074	1.197	1.349	0.972	126.2	124.4	113.5
step 3	1.074	1.198	1.350	0.970	125.8	124.7	112.8
(step 4)	1.074	1.198	1.350	0.970	125.8	124.6	112.7

[a]In this conformer, the OH hydrogen eclipses the carbonyl oxygen. Bond lengths are in angstroms.

Some methods of geometry optimization can converge to a stationary point that is not a minimum, but is a saddle point. For example, if one does a HF/3-21G calculation on NH_3 starting with a planar geometry, the geometry optimizer in the program *Gaussian 03* converges to a planar geometry with 0.99 Å bond lengths and 120° angles. This geometry is a maximum with respect to motion of the nitrogen atom in the direction perpendicular to the molecular plane but is a minimum with respect to the remaining coordinates. Hence this stationary point is a first-order saddle point. *To be sure that one has found a minimum and not a saddle point, it is essential to test the nature of the stationary point found by the geometry optimization.* This is done by doing a vibrational-frequency calculation (Section 15.12) at the geometry found. For a true minimum, all the calculated frequencies will be real. For a first-order saddle point, one calculated frequency will be imaginary. Sad to say, there are optimized geometries reported in the literature that are saddle points rather than minima.

As an example of geometry optimization, a HF/3-21G geometry optimization of the formic acid molecule HC(O)OH using the *Gaussian 94* program converged in three steps as shown in Table 15.6. The initial geometry uses bond distances from Table 15.5 and bond angles given by VSEPR and assumes the molecule is planar, as predicted by the preceding dihedral-angle rules. A conformation with the OH hydrogen eclipsing the carbonyl oxygen was assumed. (For the other likely conformation, see Prob. 15.49.) The final optimized geometry is that at step 3, since the predicted coordinate changes from step 3 to step 4 are all less than the program's cutoff for convergence and the magnitudes of the gradient components at the step 3 geometry are all less than the cutoff value. The maximum bond-length and bond-angle changes from step 3 to step 4 would have been 0.0003 Å and 0.06°. A frequency calculation at the step 3 geometry gives all real vibrational frequencies, confirming that this geometry is a minimum and not a saddle point. The energy U and the magnitude $|\nabla U|$ of the gradient at the steps of this optimization are as follows:

	initial	step 1	step 2	step 3
U/hartrees	−187.694797	−187.699879	−187.700158	−187.700199
$\lvert\nabla U\rvert$/(hartrees/bohr)	0.0722	0.0217	0.0061	0.0007

Note the decreasing energy and the decreasing magnitude of the gradient as the minimum is approached.

The choice of coordinates used in the search affects the convergence rate of the optimization. One choice is the **Cartesian coordinates** of the nuclei. Another choice is to use bond distances, bond angles, and dihedral angles, which constitute the **primitive internal coordinates**. (Internal coordinates describe molecular vibrations but not translations or rotations.) For nonrigid molecules containing more than 30 atoms, certain linear combinations of the primitive internal coordinates, called *delocalized internal coordinates*, perform much better than Cartesian coordinates. For example, a semiempirical SCF optimization of the antibiotic jawsamycin ($C_{32}H_{13}N_3O_6$, $3N - 6 = 246$) required 409 cycles of coordinate changes using Cartesian coordinates but only 71 cycles using delocalized internal coordinates [J. Baker et al., *J. Chem. Phys.*, **105**, 192 (1996); **110**, 4986 (1999)]. Several other choices of internal coordinates are in use [see V. Bakken and T. Helgaker, *J. Chem. Phys.*, **117**, 9160 (2002); K. Németh and M. Challacombe, ibid., **121**, 2877 (2004)].

Ab initio SCF MO single-point calculations on molecules containing hundreds of atoms are now feasible with small basis sets. However, ab initio geometry optimization of such large molecules is not yet routinely possible, unless the molecule is highly symmetrical.

Notation. We shall use the ideas of an n-dimensional vector space—Section 5.2. Let $\mathbf{q}$ be the $(3N - 6)$-dimensional vector whose components are the nuclear coordinates that define the molecular geometry. [If Cartesian coordinates are used in the optimization, we deal with a $3N$-dimensional vector; if internal coordinates are used, we have a $(3N - 6)$-dimensional vector.] Thus the molecular geometry corresponds to a point in $(3N - 6)$-dimensional space. Let $\mathbf{q}_1$ be the initially assumed geometry, and let $\mathbf{q}_2, \mathbf{q}_3, \ldots, \mathbf{q}_k, \mathbf{q}_{k+1}, \ldots$ be the geometries generated by the optimization procedure. The geometry-optimization procedure produces a series of steps. The kth step is defined by the vector $\Delta\mathbf{q}_k$ that goes from $\mathbf{q}_k$ to $\mathbf{q}_{k+1}$; thus $\Delta\mathbf{q}_k = \mathbf{q}_{k+1} - \mathbf{q}_k$. The components of $\Delta\mathbf{q}_k$ are the changes in each nuclear coordinate for step k. The *length* and *direction* of the kth step are the length and direction of the vector $\Delta\mathbf{q}_k$.

The quasi-Newton Method. To illustrate the quasi-Newton method, we shall pretend that U is a function of only two variables X and Y. Let X_1 and Y_1 denote the initially assumed molecular geometry. If we neglect terms higher than quadratic, the Taylor series for a function of two variables is (*Sokolnikoff and Redheffer*, p. 336)

$$U(X,Y) \approx U(X_1,Y_1) + \left.\frac{\partial U}{\partial X}\right|_{X_1,Y_1}(X - X_1) + \left.\frac{\partial U}{\partial Y}\right|_{X_1,Y_1}(Y - Y_1) + \left.\frac{1}{2}\frac{\partial^2 U}{\partial X^2}\right|_{X_1,Y_1}(X - X_1)^2$$
$$+ \left.\frac{\partial^2 U}{\partial X\,\partial Y}\right|_{X_1,Y_1}(X - X_1)(Y - Y_1) + \left.\frac{1}{2}\frac{\partial^2 U}{\partial Y^2}\right|_{X_1,Y_1}(Y - Y_1)^2 \qquad (15.67)$$

Let $U_X \equiv \partial U/\partial X$, $U_Y \equiv \partial U/\partial Y$, $U_{XX} \equiv \partial^2 U/\partial X^2$, $U_{XY} \equiv \partial^2 U/\partial X\,\partial Y$, and $U_{YY} \equiv \partial^2 U/\partial Y^2$; let the subscript 1 denote evaluation at the point (X_1, Y_1). Then

$$U(X,Y) \approx U_1 + U_{X,1}(X - X_1) + U_{Y,1}(Y - Y_1) + \tfrac{1}{2}U_{XX,1}(X - X_1)^2$$
$$+ U_{XY,1}\,(X - X_1)(Y - Y_1) + \tfrac{1}{2}U_{YY,1}(Y - Y_1)^2 \qquad (15.68)$$

Just as the harmonic-oscillator approximation to a diatomic molecule's $U(R)$ function works well in the region near R_e (Fig 4.5), the quadratic approximation (15.68) works well in the region near a minimum.

If U were accurately a quadratic function of the coordinates in the region near (X_1, Y_1), then the second partial derivatives (the elements of the Hessian matrix) would be constants in this region, and the subscript 1 on the second partials would be unnecessary. Accurate ab initio SCF calculation of the second derivatives is very time-consuming, so one usually uses a quasi-Newton method, meaning that one starts with an approximation for the Hessian and improves this approximation as the geometry optimization proceeds. We therefore write

$$U(X, Y) \approx U_1 + U_{X,1}(X - X_1) + U_{Y,1}(Y - Y_1) + \tfrac{1}{2}U_{XX}^{(1)}(X - X_1)^2 + U_{XY}^{(1)}(X - X_1)(Y - Y_1) + \tfrac{1}{2}U_{YY}^{(1)}(Y - Y_1)^2 \tag{15.69}$$

where the superscript $^{(1)}$ denotes our first approximation to the Hessian matrix elements at (or near) the equilibrium geometry.

How do we get the initial guesses for the second derivatives of U? These derivatives are force constants, and force constants for stretching or bending a particular kind of bond length or angle are roughly constant from molecule to molecule. Thus if we are dealing with a compound containing an H—C=O group, we use known typical force constants for stretching the H—C and C=O bonds and for bending the HCO angle to help construct an initial estimate of the Hessian. [This approach is closely related to the molecular-mechanics method (Section 17.5).] A procedure for estimating the initial Hessian based on typical force constants is built into most programs that do geometry optimization. An alternative would be to get the initial Hessian from a semiempirical calculation (Chapter 17), which is much faster than an ab initio calculation.

Partial differentiation of (15.69) with respect to X and with respect to Y gives

$$\begin{aligned} U_X(X, Y) &\approx U_{X,1} + U_{XX}^{(1)}(X - X_1) + U_{XY}^{(1)}(Y - Y_1) \\ U_Y(X, Y) &\approx U_{Y,1} + U_{XY}^{(1)}(X - X_1) + U_{YY}^{(1)}(Y - Y_1) \end{aligned} \tag{15.70}$$

At a minimum, $U_X(X, Y)$ and $U_Y(X, Y)$ are zero. Let $(X_{2'}, Y_{2'})$ denote the point where the estimated first derivatives U_X and U_Y (the gradient components) on the left sides of (15.70) vanish. At $(X, Y) = (X_{2'}, Y_{2'})$, Eq. (15.70) becomes

$$\begin{aligned} 0 &= U_{X,1} + U_{XX}^{(1)}(X_{2'} - X_1) + U_{XY}^{(1)}(Y_{2'} - Y_1) \\ 0 &= U_{Y,1} + U_{XY}^{(1)}(X_{2'} - X_1) + U_{YY}^{(1)}(Y_{2'} - Y_1) \end{aligned} \tag{15.71}$$

Solving for $X_{2'}$ and $Y_{2'}$, we get

$$X_{2'} = X_1 + \frac{U_{XY}^{(1)}U_{Y,1} - U_{YY}^{(1)}U_{X,1}}{U_{XX}^{(1)}U_{YY}^{(1)} - (U_{XY}^{(1)})^2}, \quad Y_{2'} = Y_1 + \frac{U_{XY}^{(1)}U_{X,1} - U_{XX}^{(1)}U_{Y,1}}{U_{XX}^{(1)}U_{YY}^{(1)} - (U_{XY}^{(1)})^2} \tag{15.72}$$

Starting at the initially guessed geometry (X_1, Y_1), we have used the calculated gradient at point 1 and the initial guess for the Hessian to find point $2'$. The step from point 1 to point $2'$ calculated from (15.72) is called a *Newton (–Raphson) step*. If U were truly a quadratic function in the region we are working in and if we had accurate values for the Hessian matrix elements, the formula (15.72) would give us the minimum in U in a single step. In actuality, $(X_{2'}, Y_{2'})$ is only an approximation to the point that minimizes U.

We now use the ab initio SCF MO method (or some other method) to calculate U and its gradient at $(X_{2'}, Y_{2'})$.

We could now use $(X_{2'}, Y_{2'})$ as the new starting geometry for the next cycle of the geometry optimization, but faster convergence is obtained if instead we take the new starting point as $X_2 = X_1 + \alpha(X_{2'} - X_1)$, $Y_2 = Y_1 + \alpha(Y_{2'} - Y_1)$, where the value of α is found as follows. One expresses U as a polynomial (typically a cubic or quartic) whose coefficients are determined so that the U polynomial will have the values that were calculated for U at (X_1, Y_1) and at $(X_{2'}, Y_{2'})$, and the gradient of U will have the calculated gradient values at these two geometries. One then varies α to minimize the U polynomial, thereby giving the new predicted geometry (X_2, Y_2). This is an example of a **line search**. By varying α, we are searching along a line in $(3N - 6)$-dimensional space that joins (X_1, Y_1) and $(X_{2'}, Y_{2'})$.

Having obtained the new geometry (X_2, Y_2), we could now do an SCF calculation of U and its gradient at (X_2, Y_2), but it is accurate enough to just use the interpolated values of these quantities found from the U polynomial that was fitted to the data at points 1 and 2′.

One now uses the values of the gradient of U at points 1 and 2 to improve (update) the estimate of the Hessian by requiring that the improved Hessian satisfy Eq. (15.70) for points 1 and 2. Using a superscript $^{(2)}$ to denote the improved Hessian matrix elements, we require that

$$\begin{aligned} U_{X,2} &= U_{X,1} + U_{XX}^{(2)}(X_2 - X_1) + U_{XY}^{(2)}(Y_2 - Y_1) \\ U_{Y,2} &= U_{Y,1} + U_{XY}^{(2)}(X_2 - X_1) + U_{YY}^{(2)}(Y_2 - Y_1) \end{aligned} \tag{15.73}$$

There are three Hessian matrix elements to be solved for, but only two equations to be satisfied, so there is not a unique solution for the $U^{(2)}$'s, and several recipes have been proposed to find improved $U^{(2)}$'s that satisfy (15.73). One commonly used recipe is the Broyden, Fletcher, Goldfarb, Shanno (BFGS) procedure (*Leach*, Sec. 5.6).

Having improved the Hessian, we now use (15.72) with the $U^{(1)}$'s replaced by the $U^{(2)}$'s and with point 1 replaced by point 2 to calculate the new coordinates $(X_{3'}, Y_{3'})$. We then check for convergence by seeing if the absolute values of the predicted coordinate changes $X_{3'} - X_2$ and $Y_{3'} - Y_2$ are both less than some tiny fixed amount and if the gradient components $|U_{X,2}|$ and $|U_{Y,2}|$ are both less than some tiny amount. (The gradient must vanish at a minimum.) If all these conditions are met, the optimization is finished and the predicted geometry is point 2. If convergence has not been achieved, we calculate U and ∇U at point 3′, do a line search between points 2 and 3′ to locate point 3; and so on.

For a function of many variables, the best way to write the Taylor series and the geometry-search equations is using matrices.

Early in the quasi-Newton procedure when one is not very close to a minimum, the procedure may well predict large coordinate changes for which the quadratic approximation to the PES may be quite inaccurate and the predicted quasi-Newton step might make things worse, rather than better. To avoid this problem, one imposes a *trust radius*. When the length of a predicted step exceeds the trust radius, the coordinate changes are reduced by a scale factor; also, the direction of the step may be varied from the quasi-Newton prediction using some other search procedure.

In the quasi-Newton method, the next geometry is obtained from the Newton formula (15.72) plus a line search. A commonly used alternative to the quasi-Newton method is to calculate the next set of nuclear coordinates by a modified form of (15.72) in which the current coordinates X_1, Y_1 are replaced by linear combinations of the current coordinates and the coordinates in all the previous search steps, and the current gradient components are replaced by similar linear combinations of current and previous gradient components. The coefficients in the linear combinations are chosen so as to minimize the distance in the $3N - 6$ dimensional space from the point that is the linear combination of coordinates to the next predicted geometry point. No line search is needed. The Hessian may or may not be updated. This procedure is the **GDIIS** (geometry by direct inversion in the iterative subspace) **method** [P. Császár and P. Pulay, *J. Mol. Struct.*, **114**, 31 (1984); O. Farkas and H. B. Schlegel, *Phys. Chem. Chem. Phys.*, **4**, 11 (2002)].

In the molecular-mechanics method (Section 17.5), analytic evaluation of the second derivatives of U is rapid, so (provided the molecule is not very large) instead of the quasi-Newton method, one can use the Newton (–Raphson) method, in which the Hessian is accurately calculated instead of being estimated. The molecular-mechanics method allows geometry optimization for molecules containing thousands of atoms. For such large molecules, the Newton method is too computationally demanding, since one must deal with a large Hessian matrix. For very large molecules, molecular-mechanics geometry optimizations often use a modification of the Newton–Raphson method called the **block-diagonal Newton–Raphson method**. Here, one makes the approximation that $\partial^2 U/\partial q_i \partial q_j = 0$ whenever q_i and q_j are Cartesian coordinates of different atoms. This approximation puts the Hessian matrix in block-diagonal form, where each block is 3×3 and contains 9 second partial derivatives that involve only the coordinates of a particular atom. This allows us to deal with the atoms one at a time. For a 1000-atom molecule, instead of having to deal with a Hessian containing $3000^2 = 9 \times 10^6$ elements (or 2994^2 elements after vibrations and rotations are removed), we deal with 1000 matrices, each containing only $3^2 = 9$ elements.

An alternative when the size of the molecule prevents use of the quasi-Newton or Newton–Raphson methods is to use an optimization method that uses only the gradient and not the Hessian. Two such methods are the steepest-descent method and the conjugate-gradient method.

The Steepest-Descent and Conjugate-Gradient Methods. In the **steepest-descent method**, one begins by calculating U and ∇U at the initially assumed geometry. Let these quantities be U_1 and ∇U_1. Recall (Section 5.2) that the vector ∇U points in the direction of greatest rate of increase in U. In the steepest-descent method, each search step is taken in the direction for which U decreases the fastest, which means that the first step is in the direction of $-\nabla U_1$. (This direction is perpendicular to the contour surface of constant U that goes through point 1.) The size of the step is determined by a line search, as follows. One calculates U at several points along the $-\nabla U_1$ direction, fits a polynomial to the calculated U values on the line, and finds the minimum of the fitted polynomial, thereby giving point 2. One then calculates U_2 and ∇U_2 and does a line search in the direction of $-\nabla U_2$. One continues until the gradient and predicted step size have become negligibly small. The steepest-descent method can be very inefficient near the end of the search where ∇U is small, and so is used only at the beginning of the search when one is not close to a minimum point and ∇U is large. One then switches to another search method when ∇U becomes small.

An improvement on the steepest-descent method is the **conjugate-gradient method**. Here, the first step is the same as in the steepest-descent method, so

$\mathbf{q}_2 = \mathbf{q}_1 - \lambda_1 \nabla U_1$, where λ_1 is found from a line search. The direction of each subsequent step k is defined by a vector $\mathbf{d}_k$ (where $k = 2, 3, \ldots$) that is a linear combination of the negative gradient $-\nabla U_k$ and the preceding search direction. The explicit formulas for the conjugate-gradient method are

$$\mathbf{q}_{k+1} = \mathbf{q}_k + \lambda_k \mathbf{d}_k$$

$$\mathbf{d}_1 \equiv -\nabla U_1 \quad \text{and} \quad \mathbf{d}_k \equiv -\nabla U_k + \beta_k \mathbf{d}_{k-1} \quad \text{for } k > 1$$

The constant λ_k is found by a line search that minimizes U in the direction of $\mathbf{d}_k$. In the Fletcher–Reeves version of the conjugate gradient method, β_k is calculated from the formula

$$\beta_k \equiv (\nabla U_k \cdot \nabla U_k)/(\nabla U_{k-1} \cdot \nabla U_{k-1})$$

(An alternative formula for β_k is the Polak–Ribiere formula; see *Leach*, Section 5.4.4.) The idea of the conjugate-gradient method (which really should be called the conjugate-direction method) is to choose each new step in a direction that is conjugate to the directions used in the previous steps (where the word conjugate has a certain technical meaning that will not be defined here), so as to avoid undoing the minimization work done in previous steps.

The Truncated Newton Method. In the Newton and quasi-Newton methods, one solves a set of linear equations like (15.71) to find each Newton–Raphson step. For large molecules, repeated solution of these linear equations is time-consuming. Early in the search, when one is not very close to the minimum, the Newton step direction is not expected to be that accurate, and it is a waste of time to solve for this direction accurately. The **truncated Newton (TN) method** therefore solves these linear equations only approximately. The method is programmed so that the accuracy with which the linear equations are solved increases as the gradient decreases and one comes closer to the minimum. In the TN method the linear equations are often solved with a conjugate-gradient procedure, giving a procedure labeled TNCG. The TNCG method is often used in molecular-mechanics geometry optimizations.

15.11 CONFORMATIONAL SEARCHING

Large molecules may have huge numbers of conformations that are energy minima. One is usually interested in finding not only the lowest-energy conformation (the global minimum on the PES), but also all minima whose energies are low enough so that these conformations have significant populations at room temperature. The biologically active conformer of a biomolecule might not be the global-minimum conformation. Because of entropy effects, the global minimum might not be the most populated conformer at room temperature. Currently, no method exists that is guaranteed to find the global minimum and all low-lying minima of a large flexible molecule. Many methods of conformational searching exist. Because of the many possible conformers, the geometry optimization part of a conformational search for a large molecule is usually done using the molecular-mechanics method (Sections 15.1 and 17.5), rather than a quantum-mechanical calculation. (For a fuller discussion, see *Leach*, Chapter 9.)

In the **systematic** (or **grid**) **search method**, one uses a computer to systematically increment each dihedral angle involving rotation about a single bond by a fixed amount $\Delta\theta$ until all possible combinations of dihedral angles for the chosen $\Delta\theta$ have been generated. Typical values for $\Delta\theta$ are 30°, 60°, or 120°. The larger the value of $\Delta\theta$, the fewer the number of possibilities that have to be examined and the more likely it is that minima might be missed. When each new set of dihedral angles is generated, one first checks that the configuration produced does not have any nonbonded atoms too close to each other (as determined by the van der Waals radii of the atoms); this is called a **bump check**. If the conformation generated passes the bump check, one uses a geometry-optimization procedure to find the nearest energy minimum on the PES. When the nearest minimum has been found, one checks that it differs from any previously found minima. Special procedures are needed to apply a systematic search to ring compounds. Systematic searches are limited to molecules with no more than about 15 dihedral angles.

In the **random** (or **stochastic** or **Monte Carlo**) **search method**, one starts with a stable conformer and generates new initial configurations either by randomly changing the values of randomly chosen dihedral angles or by adding small random amounts to the Cartesian coordinates of each atom. This is followed by a bump check and energy minimization. (The term **Monte Carlo** denotes random sampling of points, rather than taking a regular grid of points. Monte Carlo is a European gambling resort where the laws of probability are continually tested.)

The **distance-geometry** search method describes the molecule by a **distance matrix** whose elements d_{ij} are the distances between all possible pairs of atoms i and j. One begins by assigning a minimum and maximum permitted value to each internuclear distance. For two atoms bonded to each other (1,2 atoms), the minimum and maximum permitted values are usually set equal to the typical bond length for the two atoms and the kind of bond (Table 15.5). The distance between two atoms A and C each bonded to the same atom B (1,3 atoms) is usually set equal to the value determined from the assigned A—B and B—C bond distances and the ABC bond angle, whose value can be taken as a standard value ($109\frac{1}{2}°$ for a tetrahedral angle, and so on). (Instead of using standard bond distances and bond angles to fix the 1,2 and 1,3 distances, one can use the bond distances and bond angles in a single conformer that has been energy minimized.) For two atoms A and D separated by three bonds (1,4 atoms, A—B—C—D) the minimum allowed A to D distance is the distance when the ABCD dihedral angle is 0 and the maximum allowed distance is the value corresponding to a 180° dihedral angle.

For a 1,n pair of atoms with $n > 4$, the minimum allowed distance is initially set equal to the sum of the van der Waals radii of the atoms, and the maximum allowed distance is temporarily set equal to some large number (say, 100 Å). These maximum allowed distances are then reduced using the **triangle inequality** $d_{AE} \leq d_{AG} + d_{GE}$, where A, E, and G are any three nuclei. (To see the validity of this relation, just join A, E, and G to form a triangle.) One repeatedly examines all possible sets of three nuclei and lowers each AE maximum-allowed distance that does not satisfy $u_{AE} \leq u_{AG} + u_{GE}$, where u denotes the upper bound (maximum allowed value). The preceding triangle inequality can be written as $d_{AG} \geq d_{AE} - d_{GE}$. One therefore repeatedly examines all trios of nuclei and raises any lower-bound distances l_{AG} that do not satisfy $l_{AG} \geq l_{AE} - u_{GE}$.

Once the final values of the minimum and maximum allowed distances have been arrived at, the distance-geometry method assigns each internuclear distance a random value that lies in its permitted range. These assigned internuclear distances may well not correspond to a possible arrangement of the atoms in three-dimensional space, so the distance-geometry method then uses mathematical procedures that produce a set of nuclear Cartesian coordinates for which the internuclear distances lie as close as possible to the randomly chosen values and that lie (as far as is possible) within the permitted ranges. Starting from these nuclear coordinates, one then does an energy minimization to find a conformer. (The equations of distance geometry, which find nuclear Cartesian coordinates from internuclear distances, aid in converting internuclear distances found by two-dimensional NMR spectroscopy of proteins into protein structures.)

The **genetic algorithm** (GA) method uses procedures analogous to mating, mutation, and survival of the fittest in living organisms. In a GA conformational search, each molecular dihedral angle is expressed as a string of n zeros and ones (*bits*—binary digits), where n is typically in the range 6 to 12. For example, with $n = 9$, the string 011010001 is the binary number $0(2^8) + 1(2^7) + 1(2^6) + 0(2^5) + 1(2^4) + 0(2^3) + 0(2^2) + 0(2^1) + 1(2^0) = 209$, and the dihedral angle is defined as $(209/2^9)360° = 147°$. (For technical reasons, the GA method often uses a different procedure, called Gray coding, to represent each angle as a binary string. In Gray coding, the binary representations of the successive decimal integers m and $m + 1$ differ by a single bit change. See R. Judson in K. B. Lipkowitz and D. B. Boyd (eds.), *Reviews in Computational Chemistry*, Vol. 10, Wiley, 1997, Chapter 1, which is a good review article on the GA method.) The set of single-bond dihedral angles in a molecule is encoded by stringing together the strings that encode each dihedral angle, thereby forming a "chromosome" containing nd bits, where d is the number of single-bond dihedral angles. A chromosome encodes the conformation of a molecule.

The GA method typically begins by generating a set of about 100 first-generation chromosomes, where each such chromosome is formed by setting each of its nd bits randomly equal to 0 or 1. For each chromosome, one does an energy calculation using the molecular-mechanics method to find the chromosome's molecular electronic energy U (really its steric energy; see Sec. 17.5). The lowest-energy chromosome is ranked as the fittest, the second lowest is the second fittest, and so on.

To form the next generation, one first takes the fittest 10% of chromosomes and moves them into the next generation; the fittest chromosome is moved unchanged, and the rest of the fittest 10% are each subjected to a small probability of a mutation (a random change in a bit). The remaining 90% of the second generation is formed by mating. Mating is done by first picking two chromosomes at random from the breeding pool, which typically consists of the fittest 40% of the first generation. One then forms two "children" of the chosen pair by replacing the first m bits of each chosen chromosome with the corresponding first m bits of the other chromosome, where m is a randomly chosen number in the range 1 to $nd - 1$. In addition to interchanging the first part of the two parent chromosomes, a small probability for mutation is included. Enough matings are done to make the size of the second generation equal that of the first. Energy evaluations and rankings are then done for the second-generation chromosomes, and the third generation is formed from the second the same way the second was formed from the first.

One continues for typically 100 generations. Then the members of the last generation are sorted into groups, where each group has a similar set of dihedral angles, and the lowest-energy member of each group is energy minimized (Section 15.10). The process can be repeated by choosing a new set of random first-generation chromosomes.

In the **molecular dynamics** search method, one begins with a conformation that is a minimum and then assigns to each atom a set of velocity components v_x, v_y, v_z that are randomly chosen from a Maxwell distribution at an elevated temperature (typically 500 K or 1000 K). The initial position and velocity of each atom are therefore known. One then applies Newton's second law of motion to each atom, where each component of the force on atom i is calculated as $F_{x,i} = -\partial U/\partial x_i$, where U is obtained from a molecular-mechanics force field (Section 17.5). One numerically integrates Newton's second law to get the position of each atom at the times $\Delta t, 2\Delta t, 3\Delta t, \ldots,$ where the time interval Δt is typically 10^{-15} s and one typically follows the atomic motions for 10^{-9} s. At equal time intervals, one samples configurations generated by the atomic motions (typically 10^3 to 10^4 configurations are taken) and each sampled configuration is subjected to energy minimization, so as to find a conformation. The elevated temperature used allows the molecule to climb over potential energy barriers to reach new regions of the PES that may contain lower-energy minima than the current region.

In the **Metropolis Monte Carlo** search method, one assigns the molecule an elevated temperature T (typically 1000 K). Starting with an initial conformation of the molecule, one randomly changes one (or a few) randomly chosen dihedral angles to give a new conformation whose energy is evaluated using a molecular-mechanics force field. If the new conformation has a lower energy than the initial one, the new conformation is accepted and becomes the starting conformation for the next random dihedral-angle change. If the new conformation has a higher energy, a random number r that lies between 0 and 1 is generated, and the new conformation is rejected unless $e^{-|\Delta E|/kT} > r$, where $|\Delta E|$ is the energy difference between the two successive conformations and k is Boltzmann's constant. This acceptance rule is the **Metropolis criterion**, named after the person who proposed it in conjunction with simulations designed to calculate the thermodynamic properties of fluids. One generates several thousand new conformations using this procedure, which produces a set of conformations whose energies are distributed according to the Boltzmann distribution law. From this set of conformations, one takes the nth, the $2n$th, the $3n$th, . . . , where n is typically 200, to give a few hundred conformations, each of which is then subjected to energy minimization.

Annealing consists of heating a solid to a high temperature, holding it at that temperature for a while, and then very slowly cooling it, thereby relieving strains and reaching the global minimum in the Gibbs free energy. Similarly, very slow cooling of a liquid favors formation of a single, highly ordered, low free-energy crystal. In contrast, rapid cooling of a liquid (quenching) leads to a polycrystalline or amorphous solid that is not the global free-energy minimum. (The minimization procedures of Section 15.10 find the nearest local minimum, rather than the global one, and are somewhat analogous to quenching.) **Simulated annealing** is a calculational procedure that uses a "cooling" process to obtain what is hoped will be the global energy minimum.

One can do simulated annealing with either the Metropolis Monte Carlo or the molecular dynamics search procedures. Using the Metropolis Monte Carlo procedure, one starts with a high-energy conformer that is a local minimum and assigns the molecule an elevated temperature T_{high} (say, 1000 K or 1500 K). One does several hundred random dihedral angle changes at T_{high}, accepting or rejecting each change using the Metropolis criterion. (The number of Monte Carlo steps needed at each temperature increases with an increase in the number of molecular dihedral angles and increases with an increase in the complexity of the PES.) Then the temperature in the acceptance/rejection inequality is reduced by a small amount; typically one multiplies it by 0.9. Several hundred Metropolis Monte Carlo steps are taken at the new temperature, which is then reduced by multiplying it by 0.9. The whole process is repeated until T reaches a very low temperature T_{low}, typically 50 K. (Also, when T is low enough so that a large percentage of the random dihedral changes lead to rejected conformations, one limits the size of the allowed dihedral-angle changes.) If the simulated annealing has been done appropriately for the molecule being studied, the conformation obtained at T_{low} will be the global minimum and no energy minimization will be needed. In practice, one often applies an energy-minimization procedure (Section 15.10) to the final T_{low} conformation. One then repeats the whole process several times, starting with a new initial conformation at T_{high}. If the same final conformation is found on the majority of the runs, one can have some confidence that the global minimum has been found.

The Metropolis Monte Carlo simulated annealing method gave excellent results when applied to a decapeptide containing 10 alanine residues. With 1000 steps taken at each T, 14 out of 20 runs reached the same minimum, which is therefore believed to be the global minimum [S. R. Wilson and W. Cui, *Biopolymers*, **29**, 225 (1990)]. However, this decapeptide with all its residues identical has a very symmetrical PES and is not a stringent test of simulated annealing. When Metropolis Monte Carlo simulated annealing was applied to the pentapeptide Met-enkephalin, it failed, reaching 24 different conformations on 24 runs [A. Nayeem, J. Vila, and H. A. Scheraga, *J. Comput. Chem.*, **12**, 594 (1991)]. (Met-enkephalin, with 24 single-bond dihedral angles, is a naturally occurring opiate neurotransmitter that has been widely used in conformational searches. Several methods have given the same global minimum for Met-enkephalin.)

Using the molecular dynamics approach for simulated annealing, after the atomic positions have evolved for several hundred time steps, one cools the system a bit by multiplying each velocity component by a scale factor that is slightly less than one. After another several hundred time steps, one again reduces the velocity components. This is repeated until the temperature is very low, typically 50 K.

Simulated annealing is widely used to refine biomolecular structures found from X-ray crystallography or NMR.

In the **diffusion-equation method** (DEM), one starts with a molecular-mechanics expression for the PES U as a function of the nuclear coordinates. One applies a certain mathematical operator $\hat{B}$ to U so as to smooth out its maxima and minima. $\hat{B}$ is a function of a parameter t that is analogous to time. As t increases, more and more minima in U disappear. One sets the value of t equal to a time t_0 at which one believes that only one minimum remains. The nature of $\hat{B}$ is such that the derivative $\partial U/\partial t$ obeys a differential equation that has the same form as the equation describing the process of diffusion of a solute in solution. One therefore solves this diffusion differential equation to find the smoothed PES U_{smoothed} at t_0. The smoothing process not only removes minima but also changes the locations of the minima that remain. One hopes and prays that U_{smoothed} has only one remaining minimum and that this minimum originated from the global minimum in the original PES. With only one minimum in $U_{smoothed}$, this minimum is easily located using one of the methods of Section 15.10. One then reverses the process and mathematically transforms $U_{smoothed}$ back to the original U in several steps, giving the series of functions $U_{smoothed}, U_{smoothed-1}, U_{smoothed-2}, \ldots, U$. One searches $U_{smoothed-1}$ for a minimum, starting the search at the location of the minimum in $U_{smoothed}$. Then the $U_{smoothed-2}$ surface is searched starting at the location of the minimum on $U_{smoothed-1}$. And so on. Thus the minimum on the $U_{smoothed}$ surface is traced back to where it occurred on the U surface. The DEM is quite fast. For example, in only 10 minutes of supercomputer time it found a structure close to the global minimum of Met-enkephalin on a certain molecular-mechanics PES [J. Kostrowicki and H. Scheraga, *J. Phys. Chem.*, **96**, 7442 (1992)]. A variation of the DEM procedure is to transform the PES to leave a few minima and then trace each of these back to the original PES. DEM is an example of a **potential-smoothing** (PS) search method. Other PS methods exist.

Unfortunately, the minimum found by the DEM is often not the global minimum, so DEM by itself is an unsatisfactory method. A two-step procedure has been proposed in which one first uses the DEM to locate a minimum and then searches the coordinate space in the region of this minimum. By confining the second search to the region near the DEM minimum, one greatly reduces the ranges of variables that must be searched [S. Nakamura et al., *J. Phys. Chem.*, **99**, 8374 (1995)]. Moreover, one can also do searching in the regions of the minima found during the various steps of the DEM reversal process.

In the **low-mode** conformational search method (LMOD or LMCS), one starts with a minimum-energy conformer and calculates the $3N - 6$ normal vibrational modes (Section 15.12). Vibrational modes that involve torsion about single bonds have low frequencies, so the method uses only modes having vibrational wavenumbers less than 250 cm^{-1}. One moves the atoms from their minimum-energy position along the paths they would follow in one of the low-frequency vibrational modes until a steep energy increase is seen. The high-energy structure that results is then subjected to energy minimization in the hope that a potential-energy barrier will be crossed in the minimization process, thereby leading to a new minimum-energy conformer. This procedure is repeated for each low-frequency mode of the original conformer, and then is repeated for random mixtures of the low-frequency modes. The low-mode search procedure is then applied to each new local-minimum conformer that has been found.

For any of the search methods used, all conformers found should be verified to be local minima, rather than saddle points, by calculating their vibrational frequencies and checking that these are all real.

The ring hydrocarbon cycloheptadecane ($C_{17}H_{34}$) was used to test several methods of conformational searching. At comparable search times, the following numbers of local-minimum conformers were found that have an energy within 3 kcal/mol of the global minimum on the PES of a certain molecular-mechanics potential-energy function, called the MM2 PES: random dihedral search—249; random Cartesian search—222; systematic search—211; distance geometry—176; molecular dynamics—169 [M. S. Saunders et al., *J. Am. Chem. Soc.*, **112**, 1419 (1990)]. Combining the search results gave a total of 262 minima in this energy range. Although this study found random searching to do better than a systematic search, a subsequent study of $C_{17}H_{34}$ using an improved systematic search method called SUMM (systematic unbounded multiple minima) found that SUMM was superior to all the methods used in the previous study. SUMM found all 262 minima [I. Kolossváry and W. C. Guida, *J. Comput. Chem.*, **14**, 691 (1993)]. In a comparison of SUMM with LMOD, SUMM took 52 hours of workstation time to find all 262 $C_{17}H_{34}$ low-energy conformers, whereas LMOD accomplished this task in only 28 hours [I. Kolossváry and W. C. Guida, *J. Am. Chem. Soc.*, **118**, 5011 (1996); *J. Comput. Chem.* **20**, 1671 (1999)].

Another study of $C_{17}H_{34}$ used a version of DEM plus local searching called PS-NMLS (potential-smoothing with normal-mode local searching) and found (in 13 days of workstation time) a total of 20469 $C_{17}H_{34}$ conformers on the MM2 PES [R.V. Pappu et al., *J. Phys. Chem. B*, **102**, 9725 (1998)]. The energy distribution of these conformers was roughly Gaussian-shaped, with a peak in the 8 to 9 kcal/mol range and maximum conformer energy about 25 kcal/mol above the global minimum.

Most medium-size flexible molecules exist as an equilibrium mixture of a great number of conformers whose shapes can differ considerably from one another. In contrast, a globular protein in its native, biologically active state has a well-defined three-dimensional shape that is determined by intramolecular interactions such as electrostatic attractions between charged amino and carboxylate groups, hydrogen bonding, and dispersion attractions and by interactions with the solvent. (There are significant exceptions to this statement; about 30% of human proteins contain large disordered, unstructured regions; see S. Everts, *Chem. Eng. News*, April 2, 2007, p. 58.) The conformation of a structured protein does fluctuate with time, but the fluctuations are not great and the overall shape is maintained.

A major problem is the protein-folding problem. Given the astronomical number of possible conformations of a protein molecule, how does a protein fold into its native state when it is synthesized, and how can we predict the three-dimensional structure of a protein knowing only its amino acid sequence? The native state of a protein is believed to be the Gibbs free-energy minimum of the protein in solution. If we neglect the entropy contribution to G and neglect the effect of the solvent, the native state would be the global energy minimum (GEM). Hence people are deeply interested in finding the GEM of a protein. Methods used to search for the GEM of a protein, include simulated annealing, DEM, genetic algorithms, and many others [M. Vàsquez, G. Némethy, and H. A. Scheraga, *Chem. Rev.*, **94**, 2183 (1994); H. A. Scheraga, *Biophys. Chem.*, **59**, 329 (1996); D. J. Wales and H. A. Scheraga, *Science*, **285**, 1368 (1999); C. Hardin et al., *Curr. Opin. Struct. Biol.*, **12**, 176 (2002); D. J. Wales, *Energy Landscapes*, Cambridge Univ. Press, 2003, Secs. 6.7 and 9.2; M. Nanais et al., *J. Comp. Chem.*, **26**, 1472 (2005); C. A. Floudas et al., *Chem. Eng. Sci.*, **61**, 966 (2006).]

15.12 MOLECULAR VIBRATIONAL FREQUENCIES

Geometry optimization (Section 15.10) yields a quantum-mechanical estimate of the molecular electronic energy U evaluated at a local minimum, and a conformational search (Section 15.11) yields an estimate of the global energy minimum. However, the nuclei in a molecule vibrate about their equilibrium positions, and it is essential to include the molecular vibrational zero-point energy E_{ZPE} if accurate quantum-mechanical estimates of energy differences are wanted. Calculation of E_{ZPE} requires knowing the molecular vibrational frequencies. Also, theoretical calculation of vibrational frequencies helps in analyzing infrared spectra; "it is virtually impossible to interpret and correctly assign the vibrational spectra of larger polyatomic molecules without quantum-mechanical calculations" (P. Pulay in *Yarkony*, Part II, Chapter 19). Finally, calculation of vibrational frequencies allows one to classify a stationary point on the PES found by a geometry-optimization method as a local minimum (all real vibrational frequencies) or an nth-order saddle-point (n imaginary frequencies).

The Schrödinger equation for nuclear motion in a molecule is $\hat{H}_N\psi_N = (\hat{T}_N + U)\psi_N = E\psi_N$, Eqs. (13.10) and (13.11). The nuclear Schrödinger equation for diatomic molecules was solved approximately in Section 13.2. For polyatomic molecules, derivations will be omitted. (For discussion of polyatomic-molecule vibrations, see *Wilson, Decius, and Cross.*) The total molecular energy E is approximately the sum of translational, rotational, vibrational, and electronic energies. In the harmonic-oscillator approximation, the vibrational energy of an N-atom molecule is the sum of $3N - 6$ normal-mode vibrational energies ($3N - 5$ for a linear molecule):

$$E_{\text{vib}} \approx \sum_{k=1}^{3N-6} \left(v_k + \tfrac{1}{2}\right) h\nu_k \tag{15.74}$$

where ν_k is the **harmonic** (or **equilibrium**) **vibrational frequency** for the kth normal mode and each vibrational quantum number v_k has the possible values 0, 1, 2, ... independent of the values of the other vibrational quantum numbers. For the ground vibrational state, each of the $3N - 6$ vibrational quantum numbers equals zero, and the zero-point energy in the harmonic-oscillator approximation is $E_{\text{ZPE}} \approx \frac{1}{2}\sum_{k=1}^{3N-6} h\nu_k$.

The harmonic vibrational frequencies of a molecule are calculated as follows. (1) Solve the electronic Schrödinger equation $(\hat{H}_{\text{el}} + V_{NN})\psi_{\text{el}} = U\psi_{\text{el}}$ for several molecular geometries to find the equilibrium geometry of the molecule (Section 15.10). (2) Calculate the set of second derivatives $(\partial^2 U/\partial X_i \partial X_j)_e$ of the molecular electronic energy U with respect to the $3N$ nuclear Cartesian coordinates of a coordinate system with origin at the center of mass, where these derivatives (the Hessian matrix elements—Section 15.10) are evaluated at the equilibrium geometry. These second derivatives can be evaluated analytically from ab initio SCF MO wave functions (and many other wave functions), although their ab initio calculation is time-consuming. (3) Form the **mass-weighted force-constant** (or mass-weighted Hessian) **matrix elements**

$$F_{ij} \equiv \frac{1}{(m_i m_j)^{1/2}} \left(\frac{\partial^2 U}{\partial X_i\, \partial X_j}\right)_e \tag{15.75}$$

where i and j each go from 1 to $3N$ and m_i is the mass of the atom corresponding to coordinate X_i. (4) Solve the following set of $3N$ linear equations in $3N$ unknowns:

$$\sum_{j=1}^{3N} (F_{ij} - \delta_{ij}\lambda_k) l_{jk} = 0, \qquad i = 1, 2, \ldots, 3N \tag{15.76}$$

In this set of equations, δ_{ij} is the Kronecker delta, and λ_k and the l_{jk}'s are as-yet unknown parameters whose significance will be seen shortly. In order that this set of homogenous equations have a nontrivial solution, the coefficient determinant must vanish:

$$\det(F_{ij} - \delta_{ij}\lambda_k) = 0 \tag{15.77}$$

This determinant is of order $3N$ and when expanded gives a polynomial whose highest power of λ_k is λ_k^{3N}, so the determinantal (secular) equation will yield $3N$ roots (some of which may be the same) for λ_k. The molecular harmonic vibrational frequencies are then calculated from

$$\nu_k = \lambda_k^{1/2}/2\pi$$

Six of the λ_k values found by solving (15.77) will be zero, yielding six frequencies with value zero, corresponding to the three translational and three rotational degrees of freedom of the molecule. (In practice, because the equilibrium geometry is never found with infinite accuracy, one may find six vibrational frequencies with values close to zero: $|\nu_k|/c < 50\text{ cm}^{-1}$.) The remaining $3N - 6$ vibrational frequencies are the molecular harmonic vibrational frequencies.

Note that *a vibrational-frequency calculation must be preceded by a geometry optimization using the same method and basis set as used for the frequency calculation.* "Frequencies" calculated at a point that is not a stationary point are not true vibrational frequencies.

Once the λ_k's have been found, we solve the set of equations (15.76) $3N - 6$ times, each time with a different one of the nonzero λ_k values, to yield the numbers l_{jk}. The quantity $m_s^{1/2} l_{jk}/m_j^{1/2} l_{sk}$ gives the ratio of the classical mechanical vibrational amplitude of the coordinate X_j to the amplitude of X_s for the kth normal mode. For example, for the diatomic molecule $^1H^{19}F$, there is only one normal mode ($k = 1$). Letting the coordinates $X_1, \ldots, X_6$ be $x_H, y_H, z_H, x_F, y_F, z_F$, respectively, with origin at the center of mass and z axis through the nuclei, one finds on solving (15.76) (*Levine, Molecular Spectroscopy*, Section 6.2) that $l_{31}/l_{61} = -(m_F/m_H)^{1/2}$. Hence $m_6^{1/2} l_{31}/m_3^{1/2} l_{61} = -m_F/m_H = -19$. Thus the vibrational amplitude (maximum displacement from equilibrium) of the H nucleus in HF is 19 times the amplitude of the F nucleus, with the vibrational displacements being in opposite directions. Figure 15.16 shows the forms of the normal modes of the water molecule.

In matrix notation, the equations (15.76) are $\mathbf{F}\mathbf{L}^{(k)} = \lambda_k \mathbf{L}^{(k)}$, where $\mathbf{F}$ has matrix elements F_{ij} and $\mathbf{L}^{(k)}$ is a column vector with elements l_{jk}. Thus the λ_k's are eigenvalues of the mass-weighted force-constant matrix, and are found by the usual matrix methods (Section 8.6).

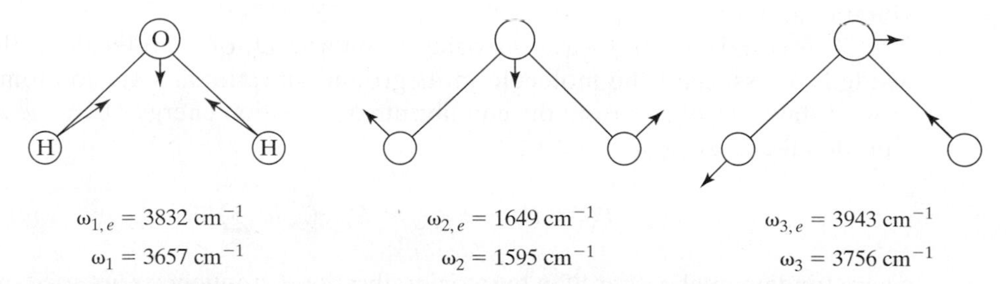

FIGURE 15.16 Normal modes of H_2O. The harmonic (ω_e) and fundamental (ω) wavenumbers of the modes are given.

The observed light-absorption frequency for the transition in which the vibrational quantum number v_k goes from 0 to 1 with no change in the other vibrational quantum numbers is called the **fundamental** (or **anharmonic**) **frequency** for the kth normal mode. Because of vibrational anharmonicity, a fundamental frequency is smaller than the corresponding harmonic frequency [recall Eq. (4.64) for diatomic molecules]. Vibrational frequencies are converted to wavenumbers by being divided by c [$\widetilde{\nu} \equiv 1/\lambda = \nu/c$, Eq. (4.66)] and wavenumbers are usually expressed in the units cm^{-1}. (Chemists have the bad habit of referring to the $\widetilde{\nu}$ values as "frequencies.") Fundamental vibrational wavenumbers are usually symbolized by ω. The harmonic and fundamental wavenumbers of gas-phase H_2O are shown in Fig. 15.16. Experimental values of harmonic vibrational frequencies are calculated from observed fundamental frequencies using anharmonicity constants found from analysis of vibrational infrared and Raman spectra. For medium and large molecules, often only the fundamental frequencies are known. For calculating E_{ZPE} from $\frac{1}{2}\sum_k h\nu_k$, more-accurate results are obtained by using the fundamental frequencies rather than the harmonic frequencies, since the fundamental frequencies incorporate anharmonicity corrections.

Harmonic vibrational frequencies calculated by the Hartree–Fock method are usually several percent higher than observed harmonic frequencies. Pretty good estimates of experimental *fundamental* frequencies can be found by multiplying ab initio SCF MO *harmonic* frequencies by an empirically found **scale factor**. For HF/6-31G* calculations, the vibrational scale factor is 0.895 [A. P. Scott and L. Radom, *J. Phys. Chem.*, **100**, 16502 (1996); this paper gives scale factors for many methods and basis sets, including separate scale factors for vibrational frequencies, zero-point energies, and thermodynamic properties]. The Scott–Radom study found that with the 0.895 scale factor, 84% of HF/6-31G* fundamental frequencies of a sample of 122 molecules were within 6% of the experimental value and recommended this basis set for HF vibrational-frequency calculations. For example, HF/6-31G* harmonic wavenumbers for H_2O are 4070, 1826, and 4188 cm^{-1}. With the 0.895 scale factor, the predicted HF/6-31G* fundamental wavenumbers of gas-phase H_2O are 3643, 1634, and 3748 cm^{-1}, in good agreement with the fundamentals in Fig. 15.16.

15.13 THERMODYNAMIC PROPERTIES

This section discusses how gas-phase thermodynamic properties can be calculated from molecular quantities such as the electronic energy, the equilibrium geometry, and the vibrational frequencies.

The (total) ground-state dissociation energy D_0 of a molecule is the energy needed to dissociate the molecule in its ground vibrational state to atoms in their ground states. D_0 differs from the equilibrium dissociation energy D_e by the zero-point vibrational energy E_{ZPE}:

$$D_0 = D_e - E_{\text{ZPE}} \approx D_e - \frac{1}{2}h\sum_{k=1}^{3N-6} \nu_k \qquad (15.78)$$

where fundamental, rather than harmonic, vibrational frequencies are used in E_{ZPE}. To calculate D_e, one calculates the molecular electronic energy U_e at the equilibrium geometry, calculates the ground-state energy of each atom in the molecule using the same method and basis set used in the molecular calculation, and takes the difference between the total atomic energies and the molecular energy. Even though the zero-point energy of a single vibrational mode is rather small, a medium-size or large molecule has many vibrational modes and its E_{ZPE} is substantial. For example, 1,3-butadiene, $CH_2CHCHCH_2$, with 24 normal modes, has $E_{ZPE} = 2.2$ eV, which corresponds to 50 kcal/mol.

Recall from Section 13.1 that $N_A D_0 = \Delta U_0^\circ = \Delta H_0^\circ$ for the process: gas-phase molecule $\rightarrow$ gas-phase atoms. This process is called *atomization* (at) and the 0 K gas-phase **atomization energy** is $\Delta U_{at,0}^\circ = N_A D_0$.

For example, for H_2O, a geometry-optimized HF/6-31G* calculation gives $U_e = -76.010746$ hartrees $= -76.010746 E_h$. In doing energy-difference calculations involving nonsinglet species (such as most ground-state atoms), one usually uses the UHF rather than the ROHF energy (Section 15.3). UHF/6-31G* ground-state atomic energies are $-0.498233 E_h$ for H and $-74.783931 E_h$ for O. The predicted D_e is then $D_e/E_h = 2(-0.498233) + (-74.783931) - (-76.010746) = 0.23035$ and Table A.2 gives $D_e = 6.27$ eV. The HF/6-31G* scaled fundamental vibrational wavenumbers of H_2O are (Section 15.12) 3643, 1634, and 3748 cm^{-1}, which give $E_{ZPE} \approx 0.56$ eV and $D_0 = 5.71$ eV, as compared with the experimental value $D_0 = 9.51$ eV. The $D_0 - 5.71$ eV value when multiplied by the Avogadro constant N_A gives a predicted atomization energy $\Delta U_{at,0}^\circ = 551$ kJ/mol $= 132$ kcal/mol. To compare this predicted value with experiment, one consults a table of thermodynamic data. Gas-phase enthalpies of formation $\Delta H_{f,0}^\circ = \Delta U_{f,0}^\circ$ at 0 K are 216.04 kJ/mol for H(g), 246.79 kJ/mol for O(g), and -238.92 kJmol for $H_2O(g)$ (*Chase et al.*). These data give the experimental atomization energy as $\Delta U_{at,0}^\circ = 917.79$ kJ/mol $= 219.4$ kcal/mol. Accurate calculation of dissociation energies requires inclusion of electron correlation (see Chapter 16).

Enthalpics of formation are more commonly used than atomization energies and are readily calculated from a predicted atomization energy and known thermodynamic data for the elements. For $H_2O(g)$, ΔH_f° is ΔH° for the reaction $H_2(g) + \frac{1}{2}O_2(g) \rightarrow H_2O(g)$. Using the scheme: reactants $\rightarrow$ atoms $\rightarrow$ products, we can calculate ΔH° for the formation reaction by subtracting the theoretically predicted atomization energy of the product from the experimental atomization energies of the reactants. Using the data in the preceding paragraph, we have as the HF/6-31G* prediction: $\Delta H_{f,0}^\circ(H_2O(g)) = [2(216.04) + 246.79 - 551]$ kJ/mol $= 128$ kJ/mol $=$ 30.6 kcal/mol. The experimental value is -238.9 kJ/mol $= -57$ kcal/mol. The error is, of course, the same as the error in the atomization energy.

So far, we have worked at 0 K. Values at 298 K are of more interest. Statistical mechanics (see any physical chemistry text) gives the translational contribution to the molar internal energy of an ideal gas as $\frac{3}{2}RT$, the rotational contribution as RT for linear molecules and $\frac{3}{2}RT$ for nonlinear molecules, and the vibrational contribution by a formula given in Prob. 15.37. (The electronic contribution is negligible at room temperature except for a few cases of molecules with low-lying excited electronic states.) One finds that only vibrational modes with wavenumbers below 900 cm^{-1} contribute significantly to the room-temperature vibrational internal energy (Prob. 15.37). For H_2, O_2, and H_2O, which have no low-frequency vibrations, the vibrational contributions to the internal energy are negligible at room temperature. For the formation reaction $H_2(g) + \frac{1}{2}O_2(g) \rightarrow H_2O(g)$, the molar internal energy of the product is

$\frac{3}{2}RT + \frac{3}{2}RT = 3RT$ higher at T than at 0 K, and the molar internal energy of the reactants is $\frac{3}{2}RT + RT + \frac{1}{2}\frac{3}{2}RT + \frac{1}{2}RT = 3.75RT$ higher at T than at 0. In addition, the formation reaction has a net change of $-\frac{1}{2}$ mole of gas, and the relations $H \equiv U + PV = U + nRT$ tell us that

$$\Delta H^\circ_{f,T} = \Delta U^\circ_{f,T} - \tfrac{1}{2}RT = \Delta U^\circ_{f,0} - 0.75RT - \tfrac{1}{2}RT = \Delta U^\circ_{f,0} - \tfrac{5}{4}RT = \Delta H^\circ_{f,0} - \tfrac{5}{4}RT$$

Thus the HF/6-31G* predicted value 30.6 kcal/mol for $\Delta H^\circ_{f,0}$ of $H_2O(g)$ corresponds to a predicted $\Delta H^\circ_{f,298}$ of 29.9 kcal/mol.

Statistical mechanics gives the molar entropy of an ideal gas as the sum of translational, rotational, vibrational, and electronic contributions. (See, for example, *Levine, Physical Chemistry*, Chapter 21.) The translational contribution depends only on the molar mass of the gas. The rotational contribution S_{rot} depends on the symmetry number and the principal moments of inertia. These quantities are readily found from the molecule's equilibrium geometry. The vibrational contribution S_{vib} depends on the molecular vibrational frequencies, which can usually be rather accurately calculated with the aid of a scale factor. The electronic contribution depends on the electronic degeneracy of the ground electronic state and in a few cases on the energies of any low-lying excited electronic states. Thus, ab initio quantum chemistry calculations can give accurate predictions of gas-phase room-temperature entropies for small molecules.

For example, East and Radom devised a procedure they call E1, which calculates S_{rot} from the MP2/6-31G* geometry (MP2 calculations are discussed in Section 16.2) and S_{vib} from HF/6-31G* scaled vibrational frequencies and the harmonic-oscillator approximation, except that internal rotations with barriers less than $1.4RT$ are treated as free rotations [A. L. L. East and L. Radom, *J. Chem. Phys.*, **106**, 6655 (1997)]. For 19 small molecules with no internal rotors, their E1 procedure gave gas-phase $S^\circ_{\text{m},298}$ values with a mean absolute deviation from experiment of only 0.2 J/mol-K and a maximum deviation of 0.6 J/mol-K. The E1 procedure was in error by up to $1\frac{1}{2}$ J/mol-K for molecules with one internal rotor and by up to 2 J/mol-K for molecules with two rotors. An improved procedure called E2 replaces the harmonic-oscillator potential for internal rotors by a cosine potential calculated using the MP2 method and a large basis set, and reduces the error to 1 J/mol-K for one-rotor molecules.

One is often interested in energy differences between two species B and C, such as different isomers or different conformers. The quantity $N_A(U_{e,\text{B}} - U_{e,\text{C}})$ gives the molar internal-energy difference at 0 K with zero-point energies neglected. The quantity $N_A[(U_{e,\text{B}} + E_{\text{ZPE,B}}) - (U_{e,\text{C}} + E_{\text{ZPE,C}})]$ gives the molar internal-energy difference at 0 K. Different conformers usually have similar zero-point vibrational energies, so the zero-point-energy contribution is often neglected here. Translational, rotational, and vibrational contributions are included to get energy differences at temperatures warmer than 0 K. As an example, a high-level ab initio calculation that included electron correlation found the following energy (and enthalpy) differences between the gauche and anti conformers of butane(g): 0.59 kcal/mol at 0 K with zero-point energy neglected, 0.70 kcal/mol at 0 K with zero-point energy included, and 0.64 kcal/mol at 298 K, as compared with an experimental value 0.67 kcal/mol averaged over the range 220 to 298 K [G. D. Smith and R. L. Jaffe, *J. Phys. Chem.*, **100**, 18718 (1996)].

15.14 AB INITIO QUANTUM CHEMISTRY PROGRAMS

This section surveys some of the available ab initio quantum chemistry program packages.

The program *Gaussian* (www.gaussian.com/), which exists in various versions (... *Gaussian 94, Gaussian 98, Gaussian 03*, ...) labeled by release year, is a widely used versatile program package that includes all common ab initio methods, such as Hartree–Fock, CI, MCSCF, MP (Section 16.2), and CC (Section 16.3), the density functional method, many semiempirical methods, and the molecular-mechanics method. *Gaussian* can optimize geometries, calculate vibrational frequencies, thermodynamic properties, and NMR shielding constants, search for transition states, calculate MEPs, and include the effects of a solvent. *Gaussian* is available in versions for supercomputers, workstations, PCs running Windows, and Macintosh computers.

GAMESS (General Atomic and Molecular Electronic Structure System) includes all the common ab initio, density functional, and semiempirical methods and has the advantage of being free. It runs on supercomputers, workstations, Macintoshes, and Windows PCs. For details, see www.msg.chem.iastate.edu/gamess/; M. W. Schmidt et al., *J. Comput. Chem.*, **14**, 1347 (1993); M. S. Gordon and M. W. Schmidt, "Advances in electronic structure theory: GAMESS a decade later," in *Theory and Applications of Computational Chemistry*, C. E. Dykstra et al. eds., Elsevier 2005 (GAMESS is sometimes called GAMESS-US to distinguish it from a different program GAMESS-UK.)

Q-Chem (www.q-chem.com/) is an ab initio package for computers using UNIX or LINUX and can also be run on Windows and Macintosh machines by using the Spartan program (see below) as a front end. Q-Chem allows calculations on large molecules and can do Hartree–Fock, MP, CC, and density-functional calculations [Y. Shao et al., *Phys. Chem. Chem. Phys.*, **8**, 3172 (2006)]. It incorporates methods such as CFMM and ONX to achieve linear scaling (Section 15.16) for large molecules.

Jaguar (www.schrodinger.com/) is a program for workstations or PCs that use Linux. It uses the pseudospectral method (Section 15.16) and can do HF, MP2, density functional, and GVB (Section 16.9) calculations [R. A. Friesner, et al., *J. Phys. Chem. A*, **103**, 1913 (1999)].

NWChem (www.emsl.pnl.gov/docs/nwchem/nwchem.html) is a free ab initio and density functional program designed to run on parallel supercomputers and workstation clusters [R. A. Kendall et al., *Computer Phys. Comm.*, **128**, 260 (2000)].

The ACES II program (www.qtp.ufl.edu/Aces2/) is designed for performing CC and MP (Sections 16.2 and 16.3) calculations.

Turbomole (www.cosmologic.de/QuantumChemistry/main_qChemistry.html) is an ab initio and density functional program that runs on workstations and Intel PCs using the LINUX operating system.

Molpro (www.molpro.net) is an ab initio and density functional program designed for highly accurate calculations on small- and medium-size molecules and includes many electron-correlation methods.

CADPAC (www-theor.ch.cam.ac.uk/software/cadpac.html) contains Hartree–Fock, density-functional, and commonly used correlation methods.

ORCA (ewww.mpi-muelheim.mpg.de/bac/logins/neese/description.php) is a UNIX and Windows ab initio, density, functional and semiempirical program that is free to academic users.

The following two programs have excellent graphical interfaces that allow easy setup of calculations and visualization of results.

Spartan (www.wavefun.com/) includes ab initio (Hartree–Fock, MP, CC, CI), density functional, semiempirical, and molecular mechanics methods, has several conformational searching methods and runs on workstations, PCs, and Macintoshes. Spartan Essential is a less expensive version that omits correlation methods, has fewer basis sets available, and runs on PCs and Macintoshes. Various student editions of Spartan also exist.

HyperChem (www.hyper.com/) includes Hartree–Fock, MP2, density-functional, semiempirical, and molecular mechanics methods and runs on PCs, Macs, and workstations.

Large-scale quantum chemistry calculations can be economically done on a cluster of networked off-the-shelf PCs running in parallel (*Chem. Eng. News*, Jan. 10, 2000, p. 27). Such a system is called a *Beowulf cluster*, after a NASA project that developed this concept. (In the epic poem *Beowulf*, the hero Beowulf slew the monster Grendel, just as a Beowulf cluster might be able to outperform a supercomputer, especially if the supercomputer is being shared by several users.) Beouwulf clusters (which typically contain from 8 to 128 PCs) are either put together by the researcher or can be bought.

15.15 PERFORMING AB INITIO CALCULATIONS

This section discusses some practical matters in using ab initio calculations.

Input. The input section of a calculation specifies the molecule, the calculational method to be used, the basis set, the kind of calculation (single-point, geometry optimization, frequency, and so on), and a molecular geometry (which will be used for a single-point calculation or optimized in an optimization calculation). For example, Table 15.7 shows the *Gaussian* input section for a geometry optimization of the H-eclipsing-O conformer of acetaldehyde, CH_3CHO (Fig. 15.17), using the Hartree–Fock method with the 3-21G basis set. The first line specifies the restricted Hartree–Fock method (the letters HF could have been used instead of RHF) and the 3-21G basis set. The keyword Opt requests a geometry optimization. If no keyword were present, a single-point calculation would be done. The second line is blank. The third line is a description of the calculation for the user's information and does not affect the computation. The fourth line is blank.

TABLE 15.7 Input for *Gaussian* CH_3CHO Geometry Optimization

```
# RHF/3-21G Opt

Acetaldehyde (H eclipsing O) HF/3-21G optimization

 0     1
 C1
 C2    1    1.52
 O3    1    1.22    2    120.0
 H4    2    1.09    1    109.5    3       0.0
 H5    2    1.09    1    109.5    3     120.0
 H6    2    1.09    1    109.5    3    -120.0
 H7    1    1.08    2    120.0    3     180.0
```

The first number in the fifth line specifies the molecular charge and the second number gives the spin multiplicity $2S + 1$. The next seven lines specify the initial guess for the geometry. The last line of the input file is blank.

The molecular-geometry guess in Table 15.7 is specified in internal coordinates (bond distances, bond angles, and dihedral angles) in a format called a **Z-matrix**. Each row in a Z-matrix specifies the location of an atom relative to previously specified atoms. The first column of the Z-matrix lists the atoms in the molecule. The numbers 1 through 7 are optional and were included for convenience. The order in which the atoms are listed in the first column is chosen by the user. In any given row (except the first), the third column of the Z-matrix specifies the distance (bond length) in angstroms between the atom in the first column and the atom whose row number the user listed in the second column. For example, the 1.52 in column 3 of row 2 of the Z-matrix specifies the C2 to C1 distance as 1.52 Å and the 1.09 in column 3 of row 5 specifies the H5 to C2 distance as 1.09 Å. (The bond distances and angles were chosen as standard values—Section 15.10 and Table 15.5.) In any given row of the Z-matrix (except the first and second), the fifth column specifies the angle in degrees for the bond formed by the atoms referred to in columns 1, 2, and 4 of that row, with the column-2 atom at the vertex of the angle. For example, the 120.0 in column 5 of row 3 specifies the angle O3C1C2 and the 109.5 in column 5 of row 4 specifies the angle H4C2C1. (*Gaussian* requires that all bond distances, angles, and dihedral angles have a decimal point.)

In any given row (except the first, second, and third), the entry in column 7 gives the dihedral angle for the four atoms listed in columns 1, 2, 4, and 6, in that order. For example, the 0.0 in column 7 of row 4 is the dihedral angle D(H4, C2, C1, O3). To determine this angle, we draw a Newman projection with the middle two atoms C2 and C1 perpendicular to the plane of the paper (Figure 15.17). This figure shows that H4 and O3 eclipse each other, which is a 0° dihedral angle. A 120° clockwise rotation is needed to turn the front atom H5 to reach O3, so row 5 has D(H5, C2, C1, O3) equal to 120°. Since a 120° counterclockwise rotation is needed to turn H6 to reach O3, row 6 has D(H6, C2, C1, O3) equal to −120°. Row 7 has D(H7, C1, C2, O3) = 180.0°. This is an unusual dihedral angle in that the end atoms H7 and O3 are both bonded to the same atom, C1. This situation could have been avoided by putting a 4 instead of a 3 in column 6 of row 7, but in a molecule such as $H_2C{=}O$, one cannot avoid a dihedral angle with the end atoms both bonded to the same atom.

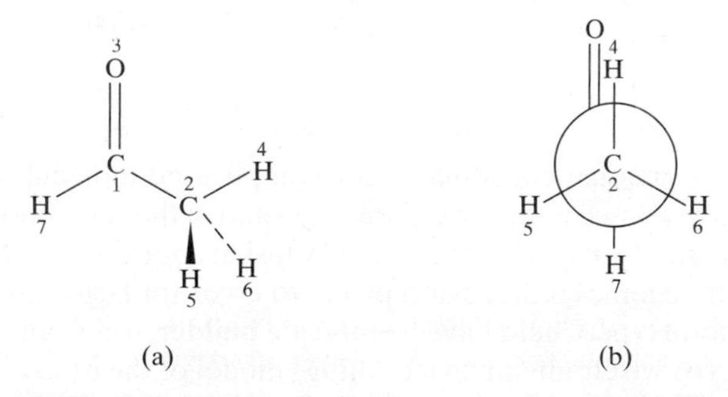

FIGURE 15.17 The H-eclipsing-O conformation of CH_3CHO.

A Z-matrix is constructed one row at a time from the assumed geometry. Row 1 contains only a single atom. Row 2 contains a bond length but no bond angle or dihedral angle. Row 3 contains a bond length and a bond angle, but no dihedral angle. Row 4 and all subsequent rows each contain a bond length, a bond angle, and a dihedral angle. The atom numbers used in a given row to specify which bond length, bond angle, or dihedral angle is being defined must all refer to atoms previously specified. For example, in row 4 of the Table 15.7 Z-matrix, we cannot put a 5 in column 4, since atom 5 has not yet been specified. It takes some practice to become good at constructing Z-matrices.

The hardest problem is getting the dihedral angles right. If row r of the Z-matrix contains the atoms r, i, k, d, in that order, then the last entry in row r is the dihedral angle D(r, i, k, d), whose value is found as follows: draw a Newman projection with the i to k bond perpendicular to the plane of the paper and with atom i in front of k. The dihedral angle D(r, i, k, d) is the angle needed to rotate the bond to r clockwise until it coincides with the bond to d. Chemists usually follow the convention that dihedral angles lie in the range −180° to 180°, so if a counterclockwise rotation is needed to make the bond to r coincide with the bond to d, then D(r, i, k, d) is in the range 0° to −180°.

In specifying a bond angle θ in a Z-matrix, the angle must be in the range $0 < \theta < 180°$. For a linear molecule such as CO_2, the forbidden angle of 180° is avoided as follows: One adds a dummy atom, symbolized by X. A convenient location for X is 1.0 Å from C with the fictitious C—X bond perpendicular to the molecular axis. The X atom is ignored in the calculation and simply serves as a reference point in the Z-matrix to define bond angles. When dealing with rings, it's a good idea to put a dummy atom in the center of the ring.

Sometimes one wants to hold one or more geometrical parameters constant during an optimization. For example, to find the electronic energy U as a function of the torsional angle in butane (Fig. 15.13), we would do several calculations, in each of which the dihedral angle D(CCCC) is held constant at a different value, while all other internal coordinates are optimized. A procedure for doing this in *Gaussian* is described in Prob. 15.45.

Instead of using bond lengths, angles, and dihedral angles in the Z-matrix, one can specify the nuclear positions using Cartesian coordinates for each atom. For example, for CO_2, if we take the z axis as the molecular axis and guess a bond length of 1.22 Å, the geometry specification is

```
C    0.0    0.0     0.0
O    0.0    0.0     1.22
O    0.0    0.0    −1.22
```

To obtain Cartesian coordinates for complicated molecules, one can use a molecule builder, as discussed in the next paragraph and in the subsection at the end of this section.

Instead of using a Z-matrix or Cartesian coordinates to input the guessed geometry (which becomes tedious and prone to error for large molecules), programs such as Spartan and HyperChem have a **molecule builder**, which allows one to construct a ball-and-stick (or wireframe or space-filling) model of the molecule on screen. The model is built from fragments selected by the user. Possible fragments are atoms, groups, rings, and so on. The builder uses standard bond lengths and bond angles and allows one to adjust dihedral angles to get a desired conformer, or the dihedral angle chosen by the builder can be used. One can interrogate the model to get the bond lengths, bond angles, and dihedral angles, and can obtain the Cartesian coordinates of the atoms in the model. Once the model has been constructed, one chooses the desired kind of calculation from a menu.

Several programs exist that have molecule builders that can provide input to programs like *Gaussian* and GAMESS that do not contain graphical interfaces and that can visualize results (such as displaying MO shapes and isodensity surfaces, animating normal vibrational modes, displaying predicted spectra, etc.) from the output of *Gaussian* and GAMESS. The program GaussView (www.gaussian.com) does this for *Gaussian* and the Windows program Chem3D Ultra (www.cambridgesoft.com) does this for *Gaussian*, GAMESS (which is included as part of Chem3D Ultra), and Jaguar. The workstation program Maestro (www.schrodinger.com/, free to academic users) provides a graphical interface for Jaguar. The Windows program Molden (www.cmbi.ru.nl/molden/), free to academic users, provides a graphical interface for *Gaussian* and GAMESS output. The free program Ghemical-GMS (www.uiowa.edu/~ghemical/ghemical.shtml) provides a graphical interface for GAMESS. The free Windows program ArgusLab (www.arguslab.com) provides an interface for *Gaussian* input and output. Q-Chem can use the Spartan program for graphical input and output. The free Windows and workstation program Molekel (www.cscs.ch/molekel/) will visualize output from *Gaussian* and GAMESS. The free program Ecce (ecce.emsl.pnl.gov/) provides a graphical interface to NWChem and *Gaussian*. The free Windows program Facio (www1.bbiq.jp/zzzfelis/Facio.html) is a graphical interface to GAMESS and *Gaussian*. Gabedit (gabedit.sourceforge.net/) is a free graphical interface to GAMESS, *Gaussian*, and other programs.

You can use a free online demonstration version of the program WebMO to build a molecule, view the corresponding Z-matrix, and even run a variety of computations (provided each computation does not exceed the time limit). Go to www.webmo.net/demo/ and follow the directions given in the tutorial at the bottom of the page.

Kinds of Calculations. So far in this chapter, single-point, geometry optimization, and frequency calculations have been mentioned (Sections 15.10 and 15.12). Other important kinds of calculations are discussed in later sections (transition-state searches, reaction-path following, PES scans in Section 16.10; inclusion of solvation effects in Section 15.17).

Although it is desirable that molecular geometry be optimized before molecular properties are calculated, ab initio geometry optimizations are impractical for very large molecules with little symmetry. For very large molecules we must be content with single-point ab initio calculations done at either an experimentally determined geometry or at a geometry optimized using a semiempirical or molecular-mechanics method. Geometry optimization with many of the methods that include electron correlation is limited to smaller molecules than can be optimized using the Hartree–Fock method. Since the Hartree–Fock method yields generally reliable geometries, a frequent procedure is to optimize the geometry with the Hartree–Fock method and then use a correlation method to do a single-point calculation of the molecular energy at the Hartree–Fock geometry.

Geometry optimization was discussed in Section 15.10, where it was noted that many of the optimization procedures used in quantum chemistry programs can converge to a stationary point that is a saddle point, rather than a minimum. To verify that a true minimum has been found, one can follow the geometry optimization with a frequency calculation. All $3N - 6$ calculated vibrational frequencies must be real to have a minimum. One way to avoid getting a saddle point instead of a minimum is to eliminate all symmetry in the starting geometry. For example, for NH_3, we would input different values for each bond length, different values for each bond angle, and would make sure no symmetry plane was present in the starting structure. (Recall that when all four atoms are in the same plane in the input geometry, *Gaussian 03* converges to a planar geometry.) Eliminating symmetry increases the time needed for the geometry optimization.

As noted in Section 15.12, a frequency calculation should be done at a minimum in the PES, so the frequency calculation should be preceded by a geometry optimization that uses the same method and basis set as the frequency calculation. Putting the two keywords Opt and Freq on the input line that begins with a # will cause *Gaussian* to follow the optimization with a frequency calculation.

Output. We now consider a few points about the output of a *Gaussian* calculation. Fuller details are given in *Foresman and Frisch* and in the online *Gaussian* manual (www.gaussian.com/). The CCCBDB (Section 15.1) contains the input and output files of many *Gaussian* calculations; these can be downloaded and viewed in a word processor.

The default procedure is for *Gaussian* not to print out the wave function. The keyword Pop=Reg will produce the coefficients of the MO expansions in terms of the basis functions for the five highest occupied and five lowest virtual MOs. The keyword Pop=Full produces all the MOs.

In a frequency calculation, *Gaussian* lists the harmonic frequencies (really the wavenumbers) in order of increasing value and in units of cm^{-1}. Any imaginary frequencies are listed first and are preceded by a minus sign as a signal that these are imaginary frequencies. Also listed are the symmetry species (Section 15.2) and relative coordinate displacements for the normal mode corresponding to each frequency.

Automatic Model Builders. An **automatic model builder** program is one that given the two-dimensional structure of a molecule (the atoms and their bonding linkages and specification of such stereochemical relations as cis or trans at each double bond) will attempt to construct a three-dimensional low-energy conformer without any user intervention. Comparison of the three-dimensional structures produced by several such programs with 639 accurately known organic structures found that CORINA (from the word *coordina*tes) gave the best overall results [J. Gasteiger et al., *J. Chem. Inf. Comput. Sci.*, **36**, 1030 (1996); J. Sadowski, J. Gasteiger, and G. Klebe, ibid., **34**, 1000 (1994); J. Sadowski and J. Gasteiger, *Chem. Rev.*, **93**, 2567 (1993)]. For 42% of the 639 structures, CORINA gave rms differences between the calculated and experimental atomic coordinates of less than 0.3 Å, and for 50% of the structures, the CORINA dihedral angles had an rms deviation from the true values of less than 15°. CORINA uses standard bond lengths and angles, assigns dihedral angles to minimize steric repulsions, and uses a simplified molecular-mechanics-like pseudo force field to do a geometry optimization for each ring in the molecule.

You can use the Internet to generate CORINA structures at no charge. Go to www.molecular-networks.com/online_demos/corina_demo.html, where you can either draw the molecule you are interested in or enter its structural formula as a SMILES string. SMILES is a way of representing a structural formula by a one-dimensional string. Some examples are CC=O for CH_3CHO (an equals sign denotes a double bond and hydrogens are omitted from organic compounds), CC(C)C for $(CH_3)_2CHCH_3$, and C1CCC1 for cyclobutane (the numeral ones indicate that the carbons preceding them are connected to each other). Further information on SMILES can be found at www.daylight.com/dayhtml/doc/theory/theory.smiles.html or in D. Weininger, *J. Chem. Inf. Comput. Sci.*, **28**, 31 (1988). After the SMILES string and the molecule's name are entered, click on Submit to produce a rotatable three-dimensional model of the CORINA structure (provided you have a molecule-viewer such as RASMOL or CHIME installed on your computer). To get the atomic Cartesian coordinates, use the option to download the structure to your computer. If you subsequently open the saved file in a word processor, you will see the Cartesian coordinates of the nuclei. The free online database ChemDB [cdb.ics.uci.edu/CHEM/Web/ and J. Chen et al., *Bioinformatics*, **21**, 4133 (2005)] contains over 4 million compounds whose three-dimensional structures have been estimated using CORINA. After entering the name or SMILES string of what you are looking for, the search may give you more than one result. Choose the molecule you want by clicking on the blue icon near its formula. You will get a rotatable three-dimensional model. In the drop-down list, choose the XYZ − Xmol XYZ format and click on Download Chemical. Save it to your computer. When you open the saved file with a word processor, you will see the Cartesian coordinates.

People doing drug design are keenly interested in finding structures whose shape fits a target receptor site and that have certain desired chemical properties that will enhance binding to the site. One way to find such structures is to search databases of structures. The Cambridge Structural Database [www.ccdc.cam.ac.uk/; F. H. Allen, *Acta Crystallogr. B*, **58**, 380 (2002)] contains over 400000 x-ray and neutron-diffraction-determined structures of small- and medium-size (typically 20 to 100 atoms) organic and organometallic compounds. Databases containing millions of three-dimensional structures produced by automatic model builders exist. ChemDB was mentioned above. Another such free database is ZINC (blaster.docking.org/zinc/) with over 4 million compounds, with compounds unlikely to be useful in screening for drugs eliminated. The ZINC structures were produced by the program OMEGA (www.eyesopen.com/products/applications/omega.html). Unlike CORINA, which produces a single conformer, OMEGA produces several low-energy conformers. For methods of searching structural databases for desired features, see *Leach*, Chapter 12; Y. C. Martin, M. G. Bures, and P. Willett, in K. Lipkowitz and D. B. Boyd (eds.)., *Reviews in Computational Chemistry*, Vol. 1, VCH, Chapter 6.

15.16 SPEEDING UP HARTREE–FOCK CALCULATIONS

Dealing with Electron-Repulsion Integrals. The calculation of the approximately $b^4/8$ **electron-repulsion integrals** (ERIs) $(rs|tu)$ [Eq. (14.39)] over the b basis functions consumes a major part of the time in an SCF MO calculation. Several methods are used to reduce the number of integrals evaluated.

Molecular symmetry is used to identify integrals that are equal, so that only one of them need be evaluated. For example, in H_2O, the integrals $(H_11s\ O2s|H_21s\ H_21s)$ and $(H_21s\ O2s|H_11s\ H_11s)$ are equal, provided the $O-H_1$ and $O-H_2$ bond distances are equal. Use of symmetry cuts the number of integrals to be evaluated in H_2O approximately in half.

In a large molecule, any one atom is far from most of the other atoms, and so a large fraction of the two-electron integrals are negligibly small for large molecules; $(rs|tu)$ will be very small if $\chi_r(1)$ and $\chi_s(1)$ or $\chi_t(2)$ and $\chi_u(2)$ are centered on widely separated nuclei. Hence, many programs test each $(rs|tu)$ integral to get its order of magnitude before it is calculated accurately. Integrals smaller than a certain threshold value can be neglected without affecting the accuracy of the overall calculation. Although the number of two-electron integrals increases as b^4, the number of such integrals whose value exceeds a fixed threshold increases only as b^2 for large molecules. Calculations on large molecules done with integrals of value less than 10^{-9} hartree neglected showed that the time to do an SCF calculation increased only as $b^{2.3}$ [R. Ahlrichs et al., *Chem. Phys. Lett.*, **162**, 165 (1989); see also D. L. Strout and G. E. Scuseria, *J. Chem. Phys.*, **102**, 8448 (1995)].

Many $(rs|tu)$ integrals involve basis functions representing inner-shell orbitals. These orbitals are little changed on molecule formation, and one can eliminate the need to explicitly represent them by using an **effective core potential** (ECP) or **pseudopotential** (Section 13.17). The ECP is a one-electron operator that replaces those two-electron Coulomb and exchange operators in the valence-electrons' Hartree–Fock equation $\hat{F}\phi_i = \varepsilon_i\phi_i$ that arise from interactions between the core electrons and the valence electrons. ECPs are derived from ab initio all-electron calculations on atoms. For compounds of main-group elements, calculations that use properly chosen ECPs give almost the same results as comparable all-electron ab initio calculations. For transition elements, obtaining accurate results with ECPs is harder. ECPs are reviewed in M. Krauss and W. J. Stevens, *Ann. Rev. Phys. Chem.*, **35**, 357 (1984); G. Frenking et al., in K. B. Lipkowitz and D. B. Boyd, eds., *Reviews in Computational Chemistry*, Vol. 8, Wiley, 1996, Chapter 2; T. R. Cundari et al. in *Reviews in Computational Chemistry*, Vol. 8, Chapter 3.

The integrals $(rs|tu)$ not only must be calculated but also must be stored and then recalled from memory as their values are needed in each SCF iteration (recall the SCF example in Section 14.3). Typically, 5 to 50 iterations are needed to achieve SCF convergence. For the large basis sets used in modern ab initio calculations, the number of $(rs|tu)$ values to be stored for a large-molecule calculation may exceed the internal (core) memory capacity of the computer, and the $(rs|tu)$ values must be stored on external memory disk drives. Locating and reading in the value of an integral from external memory is a relatively slow process.

To avoid the use of external storage memory, Almlöf developed the **direct SCF method** (not to be confused with the direct CI method of Section 16.1), in which no $(rs|tu)$ integrals are stored, but each two-electron integral is recomputed each time its value is needed. The direct SCF method allows ab initio calculations of large molecules and is very widely used.

When the direct SCF method was developed in the early 1980s, the internal memory of computers was rather limited. Nowadays, personal computers with internal memories of 1 to 4 gigabytes, hard-drive memories of hundreds of gigabytes, and very rapid speeds for reading and writing hard-drive data are common. This has made conventional calculations with all the integrals being stored more competitive with the direct SCF method, especially when several personal computers running in parallel are used [A. V. Mitin et al., *J. Comput. Chem.*, **24**, 154 (2003)].

A compromise between the direct method, which recomputes all integrals as needed, and the **conventional** stored-integral method, is the widely used **semidirect method**, which stores some integrals (those most time-consuming to calculate) and recomputes others.

The **pseudospectral** (PS) **method** for solving the Hartree–Fock equations (developed by Friesner and co-workers) uses both a basis-set expansion of each MO and a representation of each MO as a set of numerical values at chosen grid points in three-dimensional space. This allows the Fock matrix elements (14.37) to be evaluated without explicitly evaluating two-electron integrals. The first version of the PS method did not evaluate any multicenter two-electron integrals, but to achieve the required accuracy in the calculated energy, it was found necessary to evaluate a small fraction of two-electron integrals (those with large values) analytically. The PS method is a few times faster than conventional ab initio SCF calculations and may be useful for calculations on very large molecules [R. A. Friesner, *Ann. Rev. Phys. Chem.*, **42**, 341 (1991); H. G. Burnham et al., *J. Chem. Phys.*, **101**, 4028 (1994); D. T. Mainz et al., *J. Comput. Chem.*, **18**, 1863 (1997); Y. Cao et al., *J. Chem. Phys.*, **122**, 224116 (2005)].

Rapid Evaluation of Fock Matrix Elements. The expression (14.41) for the Fock matrix elements F_{rs} in Hartree–Fock theory for a closed-shell molecule can be written as

$$F_{rs} = \langle \chi_r | \hat{F} | \chi_s \rangle = H_{rs}^{\text{core}} + J_{rs} - \tfrac{1}{2} K_{rs}$$

$$J_{rs} \equiv \sum_{t=1}^{b} \sum_{u=1}^{b} P_{tu}(rs|tu), \qquad K_{rs} \equiv \sum_{t=1}^{b} \sum_{u=1}^{b} P_{tu}(ru|ts) \tag{15.79}$$

where the **Coulomb matrix element** J_{rs} and the **exchange matrix element** K_{rs} (which involve the basis functions χ_r and the density matrix elements P_{tu}) should not be confused with the Coulomb and exchange integrals J_{ij} and K_{ij} in (14.24) (which involve the MOs ϕ_i and ϕ_j). Computation of the F_{rs} values is time-consuming because of the huge number of electron-repulsion integrals $(rs|tu)$ that occur in J_{rs} and K_{rs}. To speed up calculation of the J_{rs} matrix elements, quantum chemists use ideas developed by the mathematicians Greengard and Rokhlin in their **fast multipole method** (FMM). The FMM speeds up the calculation of the potential energy $C\sum_i \sum_{j>i} Q_i Q_j / r_{ij}$ of a system of N point charges (or point masses), when N is very large. Here, Q_i and Q_j are charges (or masses) separated by a distance r_{ij} and C is a constant. The potential energy contains $\frac{1}{2}N(N-1)$ terms. The time required to calculate the potential energy is essentially proportional to N^2 when N is large, and we say that this is an $\mathrm{O}(N^2)$ calculation, where O stands for "order of."

In the FMM, the system of charges is imagined to be located in a box that is divided and subdivided repeatedly into various levels of smaller boxes. Interactions between charges close to each other (the *near field*) are calculated directly by the usual summation formula. Interactions between charges far from one another (the *far field*) are calculated using a multipole expansion. Given a group of charges close to one another, a **multipole expansion** expresses the contribution of the charge group to the potential energy at a point P outside the group as an infinite series. Each term in the series depends on the distance of the point P from the group of charges, a spherical harmonic function, and a multipole moment of the charge group. The multipole moments in successive terms of the series are the electric monopole moment (which equals the net charge of the group of charges), the electric dipole moment, the electric quadrupole moment, and so on. Accurate results can be obtained by including only a limited number of terms in the multipole expansions. If N is very large, the FMM method reduces the calculation from $O(N^2)$ to $O(N)$. A related method for dealing with this classical-physics problem is the **tree code** method of Barnes and Hut, which also uses boxes and multipole expansions but differs in calculational details from the FMM. For discussion of the FMM and tree-code methods, see L. Greengard, *Science*, **265**, 909 (1994).

Calculation of the Coulomb matrix element J_{rs} in (15.79) involves not point charges (as in the FMM method) but continuous distributions of charge defined by the basis functions. Therefore, quantum chemists modified the FMM method to deal with interactions involving continuous charge distributions. One such modification for rapid evaluation of the Coulomb matrix elements for large molecules is the **continuous fast multipole method** (CFMM) [C. A. White et al., *Chem. Phys. Lett.*, **253**, 268 (1996)]. Another is the Gaussian very fast multipole method (GvFMM) [M. C. Strain, G. E. Scuseria, and M. J. Frisch, *Science*, **271**, 51 (1996)].

The **quantum-chemical tree code** (QCTC) [M. Challacombe and E. Schwegler, *J. Chem. Phys.*, **106**, 5526 (1997)] is a modification of the classical tree-code method. The QCTC method allows calculation of the matrix elements J_{rs} of the Coulomb matrix **J** for large molecules in a time that is proportional to the number of basis functions b; this calculation is $O(b)$ and one says that the calculation exhibits **linear scaling** with size of the molecule. Challacombe and Schwegler used the QCTC method to do an ab initio SCF MO calculation on the 698-atom monomer of the P53 protein at a fixed geometry (obtained from a protein data bank) using the 3-21G basis set (3836 basis functions). They then calculated the molecular electrostatic potential (Section 15.7) of the P53 monomer. (The P53 protein is a tetramer and acts as a tumor suppressor. Mutations in the gene for this protein are found in half of human cancers.)

For small molecules, evaluation of **J** by conventional methods is faster than by a fast multipole method. The number of basis functions at which the FMM becomes faster is called the *crossover point* or the *onset point*.

When a fast-multipole method is used, short-range interelectronic interactions must still be computed by evaluating the relevant electron-repulsion integrals. These integrals are tested and those whose values are negligible are omitted, but this still leaves a lot of integrals to be evaluated for a large molecule. The electron-repulsion integrals $(rs|tu)$ are used to compute the Coulomb and exchange matrix elements J_{rs} and K_{rs}; the values of the individual $(rs|tu)$'s are not in themselves really needed. To speed up evaluation of the J_{rs} values, White and Head-Gordon devised what they call a *J-matrix engine*, which is an efficient procedure that calculates the J_{rs} values without explicitly forming all the $(rs|tu)$ integrals [C. A. White and M. Head-Gordon, *J. Chem. Phys.*, **104**, 2620 (1996)]. In the traditional approach, the $(rs|tu)$ integrals are calculated as intermediate quantities, which are then used to calculate the J_{rs} values. The J-matrix engine uses a different set of intermediate quantities to calculate the J_{rs}'s, so as to reduce the computational time without making any approximations.

An alternative to multipole methods to speed up the evaluation of **J** is the **resolution of the identity (RI) approximation** (also called the **density fitting** method). Here, one uses an **auxiliary basis set** g_k to expand the products $\chi_r(1)\chi_s(1)$ and $\chi_t(2)\chi_u(2)$ in the $(rs|tu)$ integral:

$$\chi_r(1)\chi_s(1) = \sum_k d_{rs,k}g_k(1), \qquad \chi_t(2)\chi_u(2) = \sum_k d_{tu,k}g_k(2) \tag{15.80}$$

The d_{rs} coefficients that give good fits to the $\chi_r(1)\chi_s(1)$ products are computed by the program. The g_k set usually is atom-centered Gaussians and has 2 to 4 times the number of basis functions as the χ_s set. Because the g_k basis set is incomplete, (15.80) is an approximation, and one must keep the errors introduced by the RI approximation small. The expansions (15.80) allow the very numerous four-center integrals $(rs|tu)$ to be expressed as sums containing less numerous three-center integrals involving the g_k functions and the d_{rs} coefficients. The RI method can speed up calculation of the exchange matrix elements K_{rs} as well as the J_{rs}'s. The RI method can give huge savings in time for calculations with very large basis sets χ_s [F. Weigend, *Phys. Chem. Chem. Phys.*, **4**, 4285 (2002); C. Hättig. *ibid.*, **7**, 59 (2005)].

The **multiplicative integral approximation** (MIA) is similar to the RI approximation in that it uses the expansions (15.80) to evaluate the $(rs|tu)$'s, but differs in details. Using the MIA method, Van Alsenoy and co-workers did an ab initio SCF MO geometry optimization (Section 15.10) of the 642-atom molecule crambin (a 46-residue protein) using the 4-21G basis set with starting coordinates taken as the structure found from X-ray crystallography [C. Van Alsenoy et al., *J. Phys. Chem. A*, **102**, 2246 (1998)]. At the time it was published, this was the largest ab initio geometry optimization ever done, although the energy minimum was not located as precisely as is done with smaller molecules. The calculation took 6300 hours (260 days) of CPU time on a workstation.

The *multipole accelerated resolution of the identity (MARI-J) approximation* finds the **J** matrix using a multipole method to evaluate the far-field ERIs and the RI method to calculate the near-field integrals [M. Sierka et al., *J. Chem. Phys.*, **118**, 9136 (2003)].

Still another method to speed up evaluation of **J** is the **Fourier-transform Coulomb (FTC) method**, which replaces the evaluation of the contributions of ERIs that involve three and four Gaussian basis functions with a numerical evaluation using a plane-wave basis set. A **plane-wave function** has the form $e^{i(kx+ly+mz)}$, where k, l, and m are constants. Combining the FTC, CFMM, and J-engine methods in one program gives a very rapid, accurate method for computing **J** for large molecules [L. Füsti-Molnar and J. Kong, *J. Chem. Phys.*, **122**, 074108 (2005)].

The ONX (order-N exchange) method and the LinK (linear exchange K) method use procedures to eliminate calculation of exchange matrix elements K_{rs} that are negligibly small and allow linear scaling to be achieved in the calculation of the exchange matrix **K** for large molecules that have a nonnegligible energy gap between the highest occupied and lowest vacant MO [E. Schwegler, M. Challacombe, and M. Head-Gordon, *J. Chem. Phys.*, **106**, 9703 (1997); C. Ochsenfeld, C. A. White, and M. Head-Gordon, ibid., **109**, 1663 (1998)].

15.17 SOLVENT EFFECTS

So far, we have treated the stationary-state quantum mechanics of an isolated molecule. The molecular properties so calculated are appropriate for gas-phase molecules not at high pressure. However, most of chemistry and biochemistry occurs in solution, and the solvent can have a major effect on the position of chemical equilibrium and on reaction rates. (For a survey of solvent effects on rates, equilibria, IR, UV, and NMR spectra, see C. Reichardt, *Solvents and Solvent Effects in Organic Chemistry*, VCH, 1988.) We now examine solvent effects on molecular and thermodynamic properties.

Consider a dilute solution of a polar solute molecule M in a polar solvent S; for example, a solution of CH_3Cl in water. The water molecules near the CCl side of a solute molecule will tend to be oriented with their positively charged hydrogen atoms toward the negatively charged Cl atom, while water molecules on the H_3C side of a solute molecule will tend to be oriented with their negatively charged O atoms toward the methyl group. In addition, the dipole moment of a solute molecule will induce a dipole moment in each nearby solvent molecule that adds to the permanent dipole moment. The net result of these orientation and induction effects is that the solvent acquires a bulk polarization in the region of each solute molecule. The polarized solvent generates an electric field, called the **reaction field**, at each solute molecule. The reaction field distorts the solute's molecular electronic wave function from what it was in the gas phase, thereby producing an induced dipole moment that adds to the permanent gas-phase dipole moment of M. The increased dipole moment of M further polarizes the solvent; and so on.

Because of the additional dipole moment induced by the solvent's reaction field, a polar molecule will have a larger dipole moment in a polar solvent than in the gas phase. Moreover, the dipole moment of a solute molecule will fluctuate with time, as the orientations of the nearby solvent molecules fluctuate. For example, for a solute water molecule in the solvent water (that is, for pure water), a molecular dynamics simulation in which the intermolecular interactions are modeled by placing fluctuating positive point charges on each H atom, and a fluctuating negative charge on each molecular C_2 axis near each O atom, and including a Lennard-Jones 6–12 potential interaction between each pair of O atoms (the TIP4P-FQ model, standing for transferable intermolecular potential with four interaction sites and fluctuating charges) gives the average electric dipole moment $\langle \mu \rangle$ at 25°C as 2.6 D (compared with 1.85 D in the gas phase) and gives a distribution of dipole moments with a full width at half maximum of 0.5 D [S. W. Rick, S. J. Stuart, and B. J. Berne, *J. Chem. Phys.*, **101**, 6141 (1994)]. $\langle \mu \rangle$ of liquid water cannot be directly measured (and its meaning is subject to some ambiguity, since the electron density between water molecules is not zero, giving rise to controversy about the best definition), but since this simulation gives good results for the dielectric constant and since other simulations with different models give similar results for $\langle \mu \rangle$, we can be reasonably confident in the results for $\langle \mu \rangle$. Several other theoretical studies gave values in the range 2.4 to 3.1 D; see A. V. Gubskaya and P. G. Kusalik, *J. Chem. Phys.*, **117**, 5290 (2002); B. Guillot, *J. Mol. Liq.*, **101**, 219 (2002).

The molecular electronic wave function and all molecular properties in solution will differ to some extent from their gas-phase counterparts.

The rigorous way to deal with solvent effects on molecular properties is to carry out quantum-mechanical calculations on a system consisting of a solute molecule surrounded by many solvent molecules; one repeats the calculations for various orientations of the solvent molecules and takes a suitable average over orientations to find average properties at a particular temperature and pressure. Such a calculation is usually impractical. Calculations that include a number of individual solvent molecules are called **explicit solvent** calculations and discussion of these is omitted (see *Cramer*, Chapter 12 for details).

Perhaps the most common way to calculate solvent effects is to use a **continuum solvent model.** Here, the molecular structure of the solvent is ignored and the solvent is modeled as a continuous dielectric of infinite extent that surrounds a cavity containing the solute molecule M. (A dielectric is a nonconductor of electricity.) The continuous dielectric is characterized by its *dielectric constant* (also called *relative permittivity*) ε_r, whose value is the experimental dielectric constant of the solvent at the temperature and pressure of the solution. The solute molecule can be treated classically as a collection of charges that interacts with the dielectric or it can be treated quantum mechanically. In a quantum-mechanical treatment, the interaction between a solute molecule M and the surrounding dielectric continuum is modeled by a term $\hat{V}_{\text{int}}$ that is added to the molecular electronic (fixed-nuclei) Hamiltonian $\hat{H}_{\text{M}}^{(0)}$, where $\hat{H}_{\text{M}}^{(0)}$ is for M in vacuum.

For reviews of continuum solvation models, see *Cramer*, Chap. 11; C. J. Cramer and D. G. Truhlar, *Chem. Rev.*, **99**, 2161 (1999); F. J. Luque et al., *Phys. Chem. Chem. Phys.*, **5**, 3827 (2003).

In the usual quantum-mechanical implementation of the continuum solvation model, the electronic wave function and electronic probability density of the solute molecule M are allowed to change on going from the gas phase to the solution phase, so as to achieve self-consistency between the M charge distribution and the solvent's reaction field. Any treatment in which such self-consistency is achieved is called a **self-consistent reaction-field** (SCRF) model. Many versions of SCRF models exist. These differ in how they choose the size and shape of the cavity that contains the solute molecule M and in how they calculate $\hat{V}_{\text{int}}$.

The Quantum-Onsager SCRF Method. In the **dipole-in-a-sphere** (or **quantum-Onsager** or Born–Kirkwood–Onsager) SCRF method, the molecular cavity is a sphere of radius a and the interaction between the molecular charge distribution and the reaction field is calculated by approximating the molecular charge distribution as an electric dipole located at the cavity center with electric dipole moment $\boldsymbol{\mu}$. In 1936, Onsager showed that the electric field in the cavity (the reaction field) produced by the polarization of the solvent by $\boldsymbol{\mu}$ is (in atomic units)

$$\mathbf{E}_R = \frac{2(\varepsilon_r - 1)}{(2\varepsilon_r + 1)a^3}\boldsymbol{\mu} \tag{15.81}$$

The potential energy of electrostatic interaction between $\boldsymbol{\mu}$ and the reaction field $\mathbf{E}_R$ is $\hat{V}_{\text{int}} = -\boldsymbol{\mu} \cdot \mathbf{E}_R$. The corresponding quantum-mechanical operator in atomic units is

$$\hat{V}_{\text{int}} = -\hat{\boldsymbol{\mu}} \cdot \mathbf{E}_R, \qquad \hat{\boldsymbol{\mu}} = -\sum_i \mathbf{r}_i + \sum_\alpha Z_\alpha \mathbf{R}_\alpha \tag{15.82}$$

where the electric dipole moment operator was taken from (14.20).

In an SCRF quantum-Onsager calculation, one starts by using a method such as HF, DFT (Sec. 16.4), MP2 (Sec. 16.2), or whatever, to calculate an electron probability density $\rho^{(0)}(\mathbf{r})$ for the isolated molecule, preferably at an optimized geometry. [$\rho^{(0)}$ is calculated from (14.5) in a wave-function-based method or from (16.45) in density-functional theory.] One then calculates the electric dipole moment in vacuum from (14.21) as $\boldsymbol{\mu}^{(0)} = -\int \rho^{(0)}(\mathbf{r})\mathbf{r}\,d\mathbf{r} + \sum_\alpha Z_\alpha \mathbf{r}_\alpha$. Then $\boldsymbol{\mu}^{(0)}$ is used in (15.81) to give an initial estimate $\mathbf{E}_R^{(0)}$ of the reaction field. From $\mathbf{E}_R^{(0)}$, one calculates an initial estimate of the operator $\hat{V}_{\text{int}}$ as $\hat{V}_{\text{int}}^{(0)} = -\hat{\boldsymbol{\mu}} \cdot \mathbf{E}_R^{(0)}$, where $\hat{\boldsymbol{\mu}}$ is given by (15.82). Using $\hat{V}_{\text{int}}^{(0)}$, one solves the equations of the quantum method being used and obtains an improved electron probability density $\rho^{(1)}$.

For example, suppose the Hartree–Fock method is being used. One finds that the part of $\hat{V}_{\text{int}}$ in (15.82) that involves electron 1 is added to the Fock operator $\hat{F}(1)$ of the Hartree–Fock equations (14.25) to give as the Hartree–Fock equations

$$[\hat{F}(1) + \mathbf{r}_1 \cdot \mathbf{E}_R]\phi_i(1) = \varepsilon_i \phi_i(1)$$

for the orbitals in the presence of the reaction field. Since the nuclei are fixed in solving the electronic Schrödinger equation, the interaction $-\sum_\alpha Z_\alpha \mathbf{R}_\alpha \cdot \mathbf{E}_R$ of the nuclear charges with the reaction field is a constant term, that, like V_{NN} in (13.6), appears in the electronic energy but not in the equations for the orbitals. [In density-functional theory (Section 16.4), the Kohn–Sham operator is similarly modified.]

From $\rho^{(1)}$, one calculates from (14.21) a value $\boldsymbol{\mu}^{(1)}$ for the molecular dipole moment that allows for the effect of $\mathbf{E}_R^{(0)}$. Using $\boldsymbol{\mu}^{(1)}$ in (15.79), one obtains an improved estimate $\mathbf{E}_R^{(1)}$ for the reaction field. Then an improved $\hat{V}_{\text{int}}^{(1)} = -\hat{\boldsymbol{\mu}} \cdot \mathbf{E}_R^{(1)}$ is used to solve for an improved electron density. And so on. Iterations are continued until there is no further change in ρ, $\boldsymbol{\mu}$, and $\mathbf{E}_R$. The molecular geometry should be reoptimized in the presence of the reaction field, but this step is sometimes omitted, since changes in geometry from the gas phase to the solution phase are usually small.

If the solute species has a net charge Q, the constant term $-(\varepsilon_r - 1)Q^2/\varepsilon_r a$ (first derived by Born), is added to V_{int}.

In the quantum-Onsager SCRF method, an uncharged solute molecule with no permanent dipole moment is unaffected by the solvent. In reality, the quadrupole and higher moments of the solute will interact with the solvent to give a reaction field.

With $\hat{V}_{\text{int}}$ included in the Hamiltonian operator, one obtains a molecular electronic energy (including nuclear repulsion) $U^{(f)}$ that is an eigenvalue of $\hat{H}_{\text{M}}^{(0)} + \hat{V}_{\text{int}}$. We have $U^{(f)} = \langle \psi^{(f)} | \hat{H}_{\text{M}}^{(0)} + \hat{V}_{\text{int}} | \psi^{(f)} \rangle$, where $\psi^{(f)}$ is the final electronic wave function of M, found when self-consistency is attained. The molecular electronic energy in vacuum is $U^{(0)} = \langle \psi^{(0)} | \hat{H}_{\text{M}}^{(0)} | \psi^{(0)} \rangle$. Thus

$$U^{(f)} = \langle \psi^{(f)} | \hat{H}_{\text{M}}^{(0)} + \hat{V}_{\text{int}} | \psi^{(f)} \rangle, \qquad U^{(0)} = \langle \psi^{(0)} | \hat{H}_{\text{M}}^{(0)} | \psi^{(0)} \rangle$$

The energy $U^{(f)}$ includes the energy of solute interaction with the solvent-produced reaction field, but there is another energy contribution that must also be included. This is the energy change in the solvent that results from its being polarized by the solute. The work required to polarize the solvent can be shown to be

$$E_{\text{pol}} = -\tfrac{1}{2}\langle \psi^{(f)} | \hat{V}_{\text{int}} | \psi^{(f)} \rangle \tag{15.83}$$

Gibbs Energy of Solvation. The standard Gibbs free energy change $\Delta G^\circ_{\text{solv}}$ for solvation of the solute species M in the solvent S at a particular temperature is defined as ΔG° for the process of gas-phase M plus liquid-phase S going to a solution of M in S. The standard states most commonly used in the definition of $\Delta G^\circ_{\text{solv}}$ are as follows. The solute standard state is the fictitious state with the solute at a 1 mole per liter concentration in the solvent but with the solute experiencing only interactions with solvent molecules and not with other solute molecules (as would be true in an ideally dilute solution). The gas-phase standard state of M is the fictitious state of pure gaseous solute at a 1 mole per liter concentration but with no intermolecular interactions (as would be true in an ideal gas). These standard states are not the ones most commonly used in thermodynamics. Confusion exists in the literature since people dealing with solvation energies often do not specify what standard states they are using.

Theoretical treatments of $\Delta G^\circ_{\text{solv}}$ separate it into several components. The **electrostatic contribution** $\Delta G^\circ_{\text{solv,el}}$ to $\Delta G^\circ_{\text{solv}}$ results from the electrostatic interactions between the solute and solvent. One can show that $\Delta G^\circ_{\text{solv,el}}$ can be found from an SCRF calculation as follows:

$$\begin{aligned}\Delta G^\circ_{\text{solv,el}} &= (U^{(f)} + E_{\text{pol}}) - U^{(0)} = U^{(f)} - \tfrac{1}{2}\langle\psi^{(f)}|\hat{V}_{\text{int}}|\psi^{(f)}\rangle - U^{(0)} \\ \Delta G^\circ_{\text{solv,el}} &= \langle\psi^{(f)}|\hat{H}^{(0)}_{\text{M}} + \tfrac{1}{2}\hat{V}_{\text{int}}|\psi^{(f)}\rangle - \langle\psi^{(0)}|\hat{H}^{(0)}_{\text{M}}|\psi^{(0)}\rangle\end{aligned} \qquad (15.84)$$

(Since $\Delta G^\circ_{\text{solv}}$ is expressed on a per-mole basis, the right sides of these equations should be multiplied by the Avogadro constant N_{A}.) Although energy is both a molecular and a macroscopic property, both entropy and free energy are macroscopic but not molecular properties. The dielectric continuum model uses the macroscopic property ε_r of the solvent, and the macroscopic treatment of the solvent allows us to find $\Delta G^\circ_{\text{solv,el}}$, which is a contribution to a macroscopic property. The dielectric continuum treatment of solvation is a combined quantum-mechanical and statistical-mechanical treatment.

In addition to the electrostatic contribution to $\Delta G^\circ_{\text{solv}}$, there are the following contributions. The **cavitation** contribution $\Delta G^\circ_{\text{solv,cav}}$ is the work needed to form the cavities in the solvent that are occupied by the solute molecules. The **dispersion** contribution $\Delta G^\circ_{\text{solv,dis}}$ results from London dispersion attractions between solute and solvent molecules. The **repulsion** contribution $\Delta G^\circ_{\text{solv,rep}}$ (often called the exchange-repulsion contribution) results from quantum-mechanical repulsions between solute and solvent molecules. The **molecular motion** (or **thermal**) contribution is $\Delta G^\circ_{\text{solv,mm}} = -RT\ln(z_{\text{M(sln)}}/z_{\text{M(g)}})$, where $z_{\text{M(sln)}}$ and $z_{\text{M(g)}}$ are the molecular partition functions of M in the solution and in the gas at the temperature T, and where 1 mol/L standard states are used in both the gas and in solution. This contribution results from changes in molecular motions on going from the gas to the solution phase. (Recall the statistical-mechanical formula $A - U_0 = -kT\ln Z$ for the Helmholtz free energy, where Z is the canonical partition function, and U_0 is the internal energy at absolute zero, with all molecules in the lowest energy level.) There is also a $P\,\Delta V$ contribution resulting from the difference between ΔA and ΔG, which is negligible. Some workers also include a contribution from specific structural interactions, such as hydrogen bonding, between solute and solvent molecules. Much of the effect of hydrogen bonding is included in the electrostatic contribution.

Several methods have been proposed to calculate $\Delta G^\circ_{\text{solv,cav}}$. The *scaled-particle-theory* formula (derived by Reiss, Frisch, Helfand, and Lebowitz and first applied to a solute in solution by Pierotti) calculates $\Delta G^\circ_{\text{solv,cav}}$ from the radii of the solute and solvent molecules (assumed spherical), the number of solvent molecules per unit volume, and the temperature and pressure [H. Reiss et al., *J. Chem. Phys.*, **32**, 119 (1960)]. For solvation methods that use a nonspherical molecular cavity, the scaled-particle theory spherical-cavity formula for $\Delta G^\circ_{\text{solv,cav}}$ was modified by Claverie to give what is often called the Pierotti–Claverie formula. The available methods give quite different results for $\Delta G^\circ_{\text{solv,cav}}$ and which method is best is unclear, since $\Delta G^\circ_{\text{solv,cav}}$ is not a measurable quantity. Monte Carlo simulations of liquid water indicate that the Pierotti–Claverie formula may be significantly in error [F. M. Floris et al., *J. Chem. Phys.*, **107**, 6353 (1997)]. For details, see J. Tomasi and M. Persico, *Chem. Rev.*, **94**, 2027 (1994), Section V.B.

The dispersion energy is commonly approximated as the sum of dispersion attractions between each atom of the solute molecule M and each atom of the surrounding solvent molecules, where each atom–atom dispersion energy has the form [Eq. (14.103)] $-d_{ij}/r_{ij}^6$, where r_{ij} is the distance between atoms i and j, and the coefficient d_{ij} is taken from experimental data or from theoretical calculations. The dispersion energy is proportional to the density of solvent molecules, and involves the distribution function that gives the probability density for atoms i and j being a distance r_{ij} apart. A commonly used approximation for this probability density is to take it as zero when the solvent atom j is in the molecular cavity of M and to take it as a constant otherwise. Once the solute–solvent dispersion energy is obtained, it must be converted to a dispersion free energy by a calculational procedure (see Tomasi and Persico, op. cit., Section V.C).

The repulsion contribution $\Delta G^\circ_{\text{solv,rep}}$ is calculated the same way as the dispersion contribution, except that $-d_{ij}/r^6$ is replaced by c_{ij}/r^{12} [recall the Lennard-Jones 6–12 potential—Eq. (17.91)]. The repulsion contribution is much smaller than the dispersion contribution and is often omitted.

In the molecular-motion contribution, the molecular partition function is the product of translational, rotational, vibrational, and electronic partition functions. If the molecule in solution is assumed to have the entire volume of the solution available to it, the ratio of gas-phase and solution-phase translational partition functions equals one. Likewise, the electronic partition function ratio will be one. It is unclear what one should use for the rotational partition function in solution, but if this is assumed to have the same form as that in the gas phase, the rotational partition function ratio (which involves the moments of inertia) will be very close to one, since structural changes from gas to solution are slight. Significant contributions to the vibrational partition function are made only by the low-frequency vibrational normal modes, and these modes sometimes show substantial changes in frequency on going from the gas phase to solution. If a vibrational calculation is done in the gas phase and in solution, one can calculate $\Delta G^\circ_{\text{solv,mm}}$, but the most common procedure is to omit it, assuming that its contribution is negligible.

The Multipole-Expansion Method. Two crude approximations in the quantum-Onsager SCRF method are the use of a spherical cavity for the solute molecular shape and the replacement of the actual distribution of solute molecular charge by a dipole. The true potential energy of interaction between a distribution of molecular charge and the surrounding continuum dielectric can be written as an infinite series (called a **multipole expansion**), in which the first term for a neutral molecule involves the molecular dipole moment [see Eq. (14.21)], the second term involves the molecular quadrupole moment, the third term involves the molecular octupole moment, and so on. Whereas the quantum-Onsager method includes only the first term, the multipole expansion method includes higher terms as well. The number of terms included is decided by the person doing the calculation. One finds that the terms after the Onsager dipole term make substantial contributions, and neglect of these terms is not justified. The fact that dipole-in-a-sphere SCRF calculations sometimes give good results has been ascribed to a partial cancellation of errors, with the neglect of the higher electrostatic terms partly canceling the error caused by using a spherical molecular shape [J. B. Foresman et al., *J. Phys. Chem.*, **100**, 16098 (1996)].

An improvement on a spherical molecular shape is an ellipsoidal molecular shape. Quantum-Onsager and multipole-expansion SCRF calculations using an ellipsoidal cavity give better results than spherical-cavity results, but the improvement is not great.

The PCM Method. Accurate ab initio calculation of solvent effects requires a molecular shape more realistic than a sphere or an ellipsoid. In the **polarizable-continuum model** (PCM) of Miertus, Scrocco, and Tomasi, each atomic nucleus in the solute molecule M is surrounded by a sphere of radius 1.2 times the van der Waals radius of that atom [see the references in V. Barone et al., *J. Comput. Chem.*, **19**, 404 (1998)]. (The volume occupied by these overlapping atomic spheres has sharp crevices where spheres intersect, so a smoothing procedure is sometimes used to eliminate the crevices.)

Since the PCM cavity has a complex shape, analytic expressions for the expansion coefficients in a multipole expansion cannot be found, and an analytic multipole expansion method is not feasible. Instead, a numerical method is used to find the solute–solvent interaction potential energy term $\hat{V}_{\text{int}}$. One can show from classical electrostatics that the electric potential ϕ_σ produced by the polarized dielectric continuum is equal to the electric potential produced by an **apparent surface charge** (ASC) distributed on the surface of the molecular cavity. The ASC is a continuous distribution of charge characterized by a surface charge density (charge per unit surface area) that varies from point to point of the cavity surface. In practice, one approximates this continuous ASC by replacing it with many point charges on the cavity surface. The cavity surface is divided into many tiny regions, and an apparent charge Q_k is placed in the kth region. If $\mathbf{r}_k$ is the point at which Q_k is located, then the electric potential $\phi_\sigma(\mathbf{r})$ due to the polarization of the dielectric is (in atomic units) [Eq. (15.27)]

$$\phi_\sigma(\mathbf{r}) = \sum_k \frac{Q_k}{|\mathbf{r} - \mathbf{r}_k|} \tag{15.85}$$

Classical electrostatics gives the following expression for the apparent charges:

$$Q_k = [(\varepsilon_r - 1)/4\pi\varepsilon_r]A_k \nabla\phi_{\text{in}}(\mathbf{r}_k) \cdot \mathbf{n}_k \tag{15.86}$$

where A_k is the area of the kth region, $\mathbf{r}_k$ is the point at which Q_k is located, $\nabla\phi_{\text{in}}(\mathbf{r}_k)$ is the gradient of the electric potential within the cavity evaluated in the limit as point $\mathbf{r}_k$ is approached, and $\mathbf{n}_k$ is a unit vector perpendicular to the cavity surface at $\mathbf{r}_k$ and pointing out of the cavity. [The gradient of ϕ is discontinuous at the cavity surface, so it is necessary to distinguish $\nabla\phi_{\text{in}}(\mathbf{r}_k)$ from $\nabla\phi_{\text{out}}(\mathbf{r}_k)$.] The electric potential within the cavity is the sum of the contribution $\phi_{\text{M,in}}$ from the charge distribution of the solute molecule M and the contribution $\phi_{\sigma,\text{in}}$ from the polarized dielectric: $\phi_{\text{in}} = \phi_{\text{M,in}} + \phi_{\sigma,\text{in}}$.

Since neither ϕ_{in} nor Q_k is known initially, one finds the apparent surface charges by the following iterative process. One initially neglects $\phi_{\sigma,\text{in}}$ and takes the initial estimate of ϕ_{in} as $\phi_{\text{in}}^{(00)} = \phi_{\text{M,in}}^{(0)}$ where $\phi_{\text{M,in}}^{(0)}$ is calculated from the electron density $\rho^{(0)}$ of an M molecule in vacuum. [$\rho^{(0)}$ is found from the wave function or, in DFT, the Kohn–Sham orbitals, of M in vacuum. Note that $\phi_{\text{M}}^{(0)}$ is the molecular electrostatic potential of Eq. (15.28).] Then (15.86) is used to find initial estimates $Q_k^{(00)}$ of the ASCs, which are used in (15.85) to find an initial estimate $\phi_{\sigma,\text{in}}^{(00)}$ of the electric potential produced by the polarized dielectric. The improved potential $\phi_{\text{in}}^{(01)} = \phi_{\text{M,in}}^{(0)} + \phi_{\sigma,\text{in}}^{(00)}$ is used in (15.86) to get improved charges $Q_k^{(01)}$, which are used in (15.85) to find $\phi_{\sigma,\text{in}}^{(01)}$, and so on. (The first superscript zero in $Q_k^{(00)}$ and $\phi_{\sigma,\text{in}}^{(01)}$ indicates the use of $\phi_{\text{M,in}}^{(0)}$.) One continues iterating until the charges converge to values $Q_k^{(0f)}$.

The converged charges are used to find an initial estimate of $\hat{V}_{\text{int}}$ as

$$\hat{V}_{\text{int}}^{(0)} = -\sum_i \phi_\sigma^{(0f)}(\mathbf{r}_i) + \sum_\alpha Z_\alpha \phi_\sigma^{(0f)}(\mathbf{r}_\alpha)$$

where the sums go over the electrons and nuclei and $\phi_\sigma^{(0f)}$ is found from (15.85) using the $Q_k^{(0f)}$'s. $\hat{V}_{\text{int}}^0$ is added to the molecular electronic Hamiltonian, which is used to get an improved electron density $\rho^{(1)}$ for M, which gives $\phi_{\text{M,in}}^{(1)}$, which gives the improved potential $\phi_{\text{in}}^{(10)} = \phi_{\text{M,in}}^{(1)} + \phi_{\sigma,\text{in}}^{(0f)}$, which is used in (15.86) to start a new cycle of charge and ϕ_σ iterations. One continues until everything has converged.

The original PCM method uses atomic spheres with radii 1.2 times the van der Waals radii to define the molecular cavity. The **isodensity polarizable continuum model** (IPCM) is a modification of the PCM that defines the surface of the molecular cavity as a contour surface of constant electron probability density of the solute molecule M [J. B. Foresman et al., *J. Phys. Chem.*, **100**, 16098 (1996)]. The isodensity value 0.0004 electrons/bohr^3 is commonly used, but other values have also been recommended [C.-G. Zhan and D. M. Chipman, *J. Chem. Phys.*, **109**, 10543 (1998)]. Since the solute's electronic wave function changes in each SCRF iteration, the size of the molecular cavity changes in each IPCM iteration. In the IPCM method, $\hat{V}_{\text{int}}$ is calculated from apparent surface charges. The **self-consistent isodensity PCM** (SCIPCM) method is a refinement of the IPCM method (*Foresman and Frisch*, Chapter 10), which allows geometry optimization and vibrational-frequency calculations to be done for the solute molecule in solution.

The united-atom Hartree–Fock (UAHF) PCM method uses atomic spheres to define the molecular cavity, but the assignment of sphere radii is more involved than in the original PCM method. In the UAHF PCM method [V. Barone, M. Cossi, and J. Tomasi, *J. Chem. Phys.*, **107**, 3210 (1997)], hydrogen atoms are not assigned spheres, but are included within the sphere of the atom they are bonded to (hence the name united atom). The sphere radius R_{X} of a nonhydrogen atom X is given by a formula that contains parameters. The method has 11 parameters, whose values were chosen so that the method gives good results for aqueous solvation free energies. (When the method is applied to a different solvent than water, different values must be used for the 11 parameters.) Using HF/6-31G* calculations, geometries optimized in vacuum but not reoptimized in solution, and including electrostatic, cavitation, dispersion, and repulsion contributions, the UAHF PCM method gave free energies of solvation of 43 uncharged organic molecules in water with a mean absolute error of 0.2 kcal/mol, and a maximum error of 0.6 kcal/mol, a very good result. (To give an idea of the order of magnitude of the contributions, for CH_3NH_2 in water, the electrostatic contribution was −4.4 kcal/mol, the cavitation contribution was 6.8 kcal/mol, and the dispersion plus repulsion contribution was −7.0 kcal/mol.)

The PCM method has been reformulated to eliminate the iterative calculation of the solute's wave function in solution. In this reformulation, the mutually consistent solute wave function in solution and the interaction operator $\hat{V}_{\text{int}}$ are found directly in a single SCF cycle, thereby speeding up the calculations [M. Cossi et al., *Chem. Phys. Lett.*, **255**, 327 (1996)].

The PCM method is also referred to as D-PCM (dielectric PCM).

The integral equation formulation PCM (IEF-PCM) method is a generalization of the PCM method that allows one to deal with anisotropic solvents such as liquid crystals, as well as with isotropic solvents [E. Cancès, B. Mennucci, and J. Tomasi, *J. Chem. Phys.*, **107**, 3032 (1997)]. IEF-PCM has given good results when applied to aqueous ionic solutions [M. Cossi et al., *Chem. Phys. Lett.*, **286**, 253 (1998)].

The COSMO (conductorlike solvation model) method of treating solvation resembles the PCM method in using a realistic solute-molecule shape and in using surface charges on the cavity surface around the solute molecule, but these charges are initially calculated using a condition suitable for a solvent medium that is an electrical conductor rather than a dielectric. The initial charges are then multiplied by the function $(\varepsilon_r - 1)/(\varepsilon_r + 0.5)$ to yield approximations for the charges suitable for the dielectric solvent (A. Klamt and G. Schüürmann, *J. Chem. Soc. Perkin Trans. 2*, **1993**, 799). The simplified procedure for finding the charges make COSMO computationally fast. A particular implementation of COSMO that allows efficient geometry optimization in solution is called C-PCM (conductor PCM) [V. Barone and M. Cossi, *J. Phys. Chem. A*, **102**, 1995 (1998)].

COSMO-RS (COSMO for real solvents) is an extension of COSMO beyond the dielectric-continuum approximation [A. Klamt et al., *J. Phys. Chem. A*, **102**, 5074 (1998); A. Klamt, *COSMO-RS*, Elsevier, 2005]. COSMO-RS begins with a quantum-mechanical COSMO calculation, and follows this with a statistical-mechanical procedure whose aim is to reduce the errors introduced by the continuum-solvent approximation. The ability of COSMO-RS to predict activity coefficients, vapor–liquid equilibria, and enthalpies of mixing was tested for many solutions [H. Grensemann and J. Gmehling, *Ind. Eng. Chem. Res.*, **44**, 1610 (2005)]. COSMO-RS predictions were found to be much less reliable than those of empirical group-contribution methods traditionally used by chemical engineers. Nevertheless, the authors recommended the use of COSMO-RS for situations where the parameters needed to use group-contribution methods are not available.

Chemical Equilibria in Solution. From gas-phase values of ΔU_r° and ΔS_r° for a reaction, calculated as described in Section 15.13, we can find a theoretical value of the gas-phase ΔG_r° for a reaction at the desired temperature. Then one calculates $\Delta G_{\text{solv}}^\circ$ for each reactant and product and combines these values with the gas-phase ΔG_r° value to find ΔG_r° for the reaction in solution, which then allows calculation of the equilibrium constant in solution.

This procedure has been applied to several isomerization reactions and also to calculate relative populations of different conformers in solution. For isomerizations or conformational changes, it's usually a good approximation to assume the two species have nearly the same cavitation and dispersion contributions, so these contributions will essentially cancel and need not be calculated.

For example, for the keto–enol equilibrium 2-pyridone $\rightleftharpoons$ 2-hydroxypyridine, dipole-in-a-sphere ab initio calculations with cavitation and dispersion contributions omitted gave the following values at 298 K in the gas phase and in the solvents cyclohexane ($\varepsilon_r = 2.0$) and acetonitrile ($\varepsilon_r = 36$): $\Delta G_r^\circ = -0.6, 0.4,$ and 2.3 kcal/mol, respectively, compared with the estimated experimental values −0.8, 0.3, and 3.0 kcal/mol [M. W. Wong, K. B. Wiberg, and M. J. Frisch, *J. Am. Chem. Soc.*, **114**, 1645 (1992)]. The more-polar keto form 2-pyridone (calculated gas-phase $\mu = 4.2$ D) was stabilized more than the enol form (calculated gas-phase $\mu = 1.5$ D) in solution.

Molecular Properties in Solution. The following data give examples of changes in calculated molecular properties from the gas phase to solution. For the gauche conformer of 1,2-dichloroethane, some SCIPCM density-functional calculated structural changes on going from the gas phase to a dilute solution in a solvent with $\varepsilon_r = 47$ are: r(CC) from 1.512 to 1.505 Å, r(CCl) from 1.810 to 1.820 Å, ∠CCCl from 112.9° to 112.8°, D(ClCCCl) from 69.9° to 68.6°, and the dipole moment changed from 2.92 to 3.82 D [K. B. Wiberg et al., *J. Phys. Chem.*, **99**, 9072 (1995)]. For formamide, $HCONH_2$, some PCM density-functional calculated vibrational wavenumber changes from the gas phase to the solvent water with $\varepsilon_r = 78$ are 162 to 366 cm^{-1} for NH_2 inversion, 650 to 682 cm^{-1} for NH_2 twisting, 1276 to 1302 cm^{-1} for CN stretching, and 1621 to 1588 cm^{-1} for NH_2 bending [V. Barone, M. Cossi, and J. Tomasi, *J. Comput. Chem.*, **19**, 404 (1998)].

Semiempirical Solvation Methods. See Section 17.6.

PROBLEMS

Sec.	15.2	15.4	15.5	15.6	15.7	15.9	15.10
Probs.	15.1–15.3	15.4–15.12	15.13–15.21	15.22–15.24	15.25–15.26	15.27–15.29	15.30–15.34

Sec.	15.11	15.13	15.15	15.16	general
Probs.	15.35–15.36	15.37–15.38	15.39–15.52	15.53	15.54

15.1 Verify that (15.3) are the possible symmetry species for $\mathscr{C}_{2v}$.

15.2 Work out the possible symmetry species for $\mathscr{D}_2$.

15.3 How many independent molecular wave functions correspond to (a) a ^3E term? (b) a ^1E term?

15.4 (a) Give the number of CGTFs used in a $[4s2p/2s]$ calculation of C_3H_7OH. (b) Give the number of CGTFs used in a DZ calculation of C_4H_9OH.

15.5 For C_4H_9OH, give the number of CGTFs used in a calculation with each of the following basis sets: (a) STO-3G; (b) 3-21G; (c) 6-31G*; (d) 6-31G**; (e) 6-31+G*; (f) cc-pVTZ; (g) cc-pVQZ; (h) aug-cc-pVDZ.

15.6 How many primitive and how many contracted GTFs are used in a calculation on $Si_{24}O_{60}H_{24}$ with each of these basis sets: (a) STO-3G; (b) 3-21G; (c) 6-31G*?

15.7 (a) Use the CCCBDB (Sec. 15.1) to find the HF/6-31G* electronic energy, dipole moment, and equilibrium geometry for the ground electronic state of NH_3 given by a HF/6-31G* calculation and by a HF/cc-pVDZ calculation. (b) Repeat (a) using the EMSL CRDB.

15.8 Use the Quantum Chemistry Literature Database (Section 15.1) to find references for HF/cc-pV5Z calculations on H_2O.

15.9 Use a spreadsheet to do a least-squares fit of the STO $S \equiv \pi^{-1/2}e^{-r}$ with a normalized linear combination G_{3N} of three normalized s-type Cartesian Gaussians. Suppose we divide space into infinitesimal volume elements $d\tau$, compute $(S - G_{3N})^2\,d\tau$ at a point in each volume element, sum these quantities to give the definite integral $I \equiv \int (S - G_{3N})^2\,d\tau$, and vary the six parameters in G_{3N} to minimize I. (a) Show that $I = 4\pi \int_0^\infty (S - G_{3N})^2 r^2\,dr$. (b) Set up a spreadsheet to compute I as $I \approx \sum_{i=1}^{300}[S(r_i) - G_{3N}(r_i)]^2 4\pi r_i^2\,\Delta r$, where r_i goes from 0 to 15 in steps of $\Delta r = 0.05$. (Beyond $r = 15$, the integrand in I is negligible.) Assign cells to the orbital exponents $\alpha_1, \alpha_2, \alpha_3$, the multiplicative coefficients d_1, d_2, d_3 of the individual normalized Gaussians in the unnormalized Gaussian (which we shall call G_3), and the coefficients $c_1 = Nd_1, c_2 = Nd_2, c_3 = Nd_3$ in the normalized Gaussian G_{3N}. Start with initial guesses for the six parameters (the α's and the d's). Put the r_i values in column A, the $S(r_i)$ values in column B, the unnormalized function G_3 in column C, the quantities $[G_3(r_i)]^2 r_i^2$ in column D, the normalized function $G_{3N} = NG_3$ in column E, and the quantities $4\pi\,\Delta r(S - G_{3N})^2 r_i^2$ in column F. Compute the normalization constant $N = \{4\pi\,\Delta r \sum_{i=1}^{300}[G_3(r_i)]^2 r_i^2\}^{-1/2}$ in some cell. Compute I in some cell. Graph S and G_{3N} on the same graph. Use the Solver in Excel to vary the 6 parameters in G_3 to minimize I. Rerun the Solver with different initial guesses for the parameters. (c) For which range of r is G_{3N} significantly less than S? Significantly greater than S?

15.10 The fit in Problem 15.9 gives $S(r) \approx \sum_{i=1}^{3} c_i G_i(r)$, where $S(r)$ is a 1s STO with orbital exponent 1 and G_i is a 1s GTO with orbital exponent α_i. Let $S(r, \zeta) \equiv \zeta^{3/2}\pi^{-1/2}e^{-\zeta r}$. Let $G_i(r, \zeta)$ be the function obtained by replacing α_i in $G_i(r)$ by $\alpha_i\zeta^2$. Show that $S(r, \zeta) \approx \sum_{i=1}^{3} c_i G_i(r, \zeta)$.

15.11 Use the EMSL Gaussian Basis Set Exchange on the Internet to find (a) the 6-31G** basis functions for H; (b) the 6-31G* basis functions for C.

15.12 Give an example of a molecule for which the HF/6-31G* and HF/6-31G** energies are the same.

15.13 For formaldehyde: (a) Work out the symmetry orbitals for a minimal-basis-set calculation; give the symmetry species of each symmetry orbital. (Choose the x axis perpendicular to the molecular plane.) (b) How many σ and how many π canonical MOs will result from a minimal-basis-set calculation? (See the Section 15.9 discussion of ethylene for the definition of σ and π MOs.) How many occupied σ and occupied π MOs are there for the ground state? (c) For each of the eight energy-localized MOs, state which AOs will make significant contributions. (d) What is the maximum-size secular determinant that occurs in finding the minimal-basis-set canonical MOs?

15.14 Work out the minimal-basis symmetry orbitals and their symmetry species for *cis*-1,2-difluoroethylene. Choose the x axis perpendicular to the molecular plane.

15.15 Sketch the $1a_1, 3a_1, 1b_1, 4a_1$, and $2b_2$ MOs of water.

15.16 Give the form of the normalization constants for the symmetry orbitals in Eq. (15.13).

15.17 Suppose a ground-state calculation gives us some virtual orbitals for the molecule M. In which one of the following species would an excited electron occupy an MO that was well approximated by a virtual orbital of ground-state M? (a) M; (b) M^+; (c) M^-. Explain.

15.18 Let E_{HF} be the Hartree–Fock energy of a closed-shell molecule; let $E^+_{k,\text{HF,approx}}$ be the approximate Hartree–Fock energy of the ion formed by removal of an electron from the kth

MO of this molecule, this energy being calculated using the MOs of the un-ionized molecule. For both the molecule and the ion, the Hartree–Fock wave function is a single determinant. Use Eqs. (11.80) to (11.82) [where (11.82) is modified to allow for the presence of several nuclei] and Eq. (14.30) to show that $E_{HF} - E^{+}_{k,\text{HF,approx}} = \varepsilon_k$, where ε_k is the orbital energy of MO k.

15.19 Write symmetry orbitals for a minimal-basis-set calculation of H_2.

15.20 Suppose an SCF calculation that includes $3d$ orbitals on oxygen is done on H_2O (Fig. 15.1). For each of the occupied ground-state H_2O MOs, use symmetry-species arguments to decide which of the following $3d$ oxygen AOs will contribute to that MO: $3d_{z^2}$, $3d_{xz}$, $3d_{yz}$, $3d_{xy}$, $3d_{x^2-y^2}$.

15.21 (a) Use (15.23) and the H_2O cc energies for $n = 3, 4$, and 5 given just before (15.23) to find the values of the three parameters in (15.23). There are three equations in three unknowns. Subtract the $n = 3$ equation from the $n = 4$ equation and from the $n = 5$ equation to obtain two equations in two unknowns. Then combine these equations to eliminate B. To solve the remaining equation for C, put all terms on one side of the equation and use the Excel Solver to find C. Then find the other two unknowns. (b) Instead of (15.23), the following equation has been proposed for extrapolation from aug-cc-pVQZ and aug-cc-pV5Z energies to the CBS limit [A. Karton and J. M. L. Martin, *Theor. Chem. Acc.*, **115**, 330 (2006)]:

$$E_{\text{SCF}}(\text{aug-}n) = E_{\text{SCF}}(\infty) + A(L + 1)e^{-9\sqrt{L}}$$

where L is the largest orbital-angular momentum l value that appears in the basis set. For H_2O, aug-cc-pVQZ and aug-cc-p5Z energies at optimized geometries are -76.066676 and -76.068009 hartrees. Find the CBS limit predicted by this extrapolation formula. Compare the two CBS values found in this problem to the estimate in Table 15.2.

15.22 (a) Use (15.24) to show that $1 = \sum_r c_{ri}^2 + 2\sum_{r>s}\sum_s c_{ri}c_{si}S_{rs}$. (b) Use the result of (a) to verify that $\sum_r n_r + \sum_{r>s}\sum_s n_{r-s} = n$.

15.23 Use the tables referred to in Problem 15.28c [or one of the other available overlap-integral tables; see the reference given after (17.71)] to verify the values of the H_2O overlap integrals given in Section 15.6.

15.24 (a) Verify the net populations given at the end of the Section 15.6 example. (b) Verify the H_2O $2a_1$ and $1b_2$ interatomic overlap populations given in Section 15.6. (c) Verify the gross populations given in Section 15.6 for the H_2O basis functions.

15.25 Use Coulomb's law to show that the electric potential ϕ_P at a point P a distance d from a point charge Q is $\phi_P = Q/4\pi\varepsilon_0 d$.

15.26 On Fig. 13.9 for ground-state H_2^+, sketch gradient paths that lie in the plane of the figure. Include the gradient path that does not end at a nucleus.

15.27 Frequently the x and z axes of ethylene are interchanged as compared with Fig. 15.10. What change does this cause in the MO labeling? [The symmetry-species designations in Table 15.3 are retained.]

15.28 (a) Does having the coefficient of C1s greater than 1.0 in the $1a_1$ methane MO in (15.46) violate the condition that the MO be normalized? Explain. (b) Use the results of Prob. 13.35 to express the $2a_1$ methane MO of (15.46) using a nonorthogonal $2s$ STO. (c) Verify that the $2a_1$ MO found in part (b) is normalized. [The needed overlap integrals can be found by interpolating in the tables of R. S. Mulliken, C. A. Rieke, D. Orloff, and H. Orloff, *J. Chem. Phys.*, **17**, 1248 (1949).]

15.29 Call the ethylene symmetry orbitals in Table 15.4 g_1 to g_{14}, in order. For each of the eight canonical ethylene MOs occupied in the ground state, decide, as best you can, which symmetry orbitals will make major contributions; give the sign of each such symmetry orbital in the MO expression. (*Hint*: Decide how many inner-shell and how many bonding canonical MOs there are; combine the symmetry orbitals so as to build up charge density between the nuclei for the bonding

MOs. *One further hint*: The third symmetry orbital in Table 15.4 makes no significant contribution to $3a_g$.) Sketch the canonical MOs. Assume an orthogonalized $2s$ orbital is used.

15.30 Find the components of the gradient and the elements of the Hessian for each of the following functions, where the c's are constants: (a) $U = c_1x^2 + c_2y^2 + c_3z^2$; (b) $U = c(x + y + z)^2$.

15.31 Give initial guesses for the geometry (bond lengths, bond angles, and dihedral angles) of the likely conformers of each of the following ground-state molecules. (a) CH_3OH; (b) C_2H_4; (c) CH_3NH_2; (d) $(CH_3)_2CO$.

15.32 For the function $U = 2x^2 - y^2$, (a) find the stationary points; (b) find the first-order saddle points.

15.33 Consider the function $U = 4(x - 1)^2 + 3(y - 2)^2$. (a) By inspection, find its minimum. (b) Start at an arbitrary point x_1, y_1 and show that the Newton–Raphson equation (15.72) with all partial derivatives evaluated exactly from U gives the minimum in one step.

15.34 For the function $U = 3x^2 + 6y^2$, (a) locate the minimum by inspection; (b) start at the point $x = 9, y = 9$ and apply the conjugate-gradient method to locate the minimum. Do the calculations by hand. Recall that the equation of the straight line that passes through point x_1, y_1 with slope m is $m = (y - y_1)/(x - x_1)$.

15.35 Write down the distance matrix for H_2O using the experimental geometry in Table 15.2. Take oxygen as atom 1.

15.36 True or false? (a) All triatomic molecules are planar. (b) The distance matrix is a symmetric matrix. (c) The diagonal elements of the distance matrix are zero. (d) The Hessian is a symmetric matrix.

15.37 The contribution of molecular vibrations to the molar internal energy U_m of a gas of nonlinear N-atom molecules is (zero-point vibrational energy not included) $U_{m,vib} = R\sum_{s=1}^{3N-6}\theta_s/(e^{\theta_s/T} - 1)$, where $\theta_s \equiv h\nu_s/k$ and ν_s is the vibrational frequency of normal mode s. Calculate the contribution to $U_{m,vib}$ at 25°C of a normal mode with wavenumber $\tilde{\nu}_s \equiv \nu_s/c$ of (a) 900 cm^{-1}; (b) 300 cm^{-1}; (c) 2000 cm^{-1}.

15.38 (a) For CO_2 a geometry-optimized HF/6-31G* calculation gives $U_e = -187.634176$ hartrees. UHF/6-31G* ground state energies are -37.680860 and -74.783931 hartrees for C and O, respectively. Calculate the predicted HF/6-31G* D_e of CO_2. (b) Unscaled HF/6-31G* vibrational wavenumbers for CO_2 are 745.8, 745.8, 1518.5, and 2585.0 cm^{-1}. Scale these by 0.89, find the HF/6-31G* prediction for D_0, and compare with the experimental value $D_0 = 16.56$ eV. Also calculate the predicted atomization energy in kcal/mol. (c) Thermodynamic tables give $\Delta H^\circ_{f,0}$ as 246.79 kJ/mol for O(g) and 711.2 kJ/mol for C(g) and give the molar enthalpy difference $H_{m,298} - H_{m,0}$ as 1.05 kJ/mol for graphite. Find the HF/6-31G* predictions for $\Delta H^\circ_{f,0}$ and $\Delta H^\circ_{f,298}$ of $CO_2(g)$. The experimental values are -393.2 kJ/mol and -393.5 kJ/mol.

15.39 Using standard bond lengths and bond angles, write Z-matrices using internal coordinates for each of the following: (a) CO_2; (b) CH_4; (c) H_2CO; (d) NH_3; (e) C_2H_4; (f) CH_3OH (do both staggered and eclipsed conformations); (g) CH_2ClCH_2Cl (anti and gauche conformations).

15.40 Given the following Z-matrix (where the semicolons are used to denote the end of a row), draw a Newman projection of the molecule. C1; C2 1 1.5; F3 1 1.4 2 109.5; Cl4 2 1.8 1 120.0 3 0.0; H5 1 1.1 2 109.5 4 120.0; H6 1 1.1 2 109.5 4 −120.0; O7 2 1.2 1 120.0 3 180.0

15.41 If the dihedral angle D(R, S, T, U) is 60°, what is the value of the dihedral angle D(U, T, S, R)?

15.42 For a nonlinear molecule containing N atoms, how many internal coordinates (bond lengths, bond angles, and dihedral angles) are specified in its Z-matrix? Justify your answer.

15.43 (a) For CH_2O, run HF/3-21G geometry optimization and vibrational-frequency calculations to obtain the predicted geometry, dipole moment, and vibrational frequencies.

Verify that all vibrational frequencies are real. (b) Repeat (a) using a HF/6-31G** calculation. Compare the calculated results to experimental results you find in the literature. (c) For each normal vibrational mode, *Gaussian* lists Cartesian-coordinate displacements of each atom using a so-called "standard orientation" coordinate system, for which the equilibrium geometry has been specified earlier in the calculation. Sketch each CH_2O normal mode using arrows to show atomic motions. (Certain programs can show animations of the normal modes; use this feature if you have access to such a program.)

15.44 Perform single-point HF/3-21G calculations on H_2O with the bond lengths held constant at 0.96Å and the bond angle taken as 100°, 102°, 104°, 106°, 108°, 110°, and 112°. Plot the energies versus bond angle and decide what order polynomial will give a good fit to the data. Use a graphing program or spreadsheet to fit the data with such a polynomial and find the 3-21G minimum-energy angle at this bond length.

15.45 In a Z-matrix for the *Gaussian* programs, letters can be used to specify some or all of the internal coordinates. For example, a Z-matrix for HOCl can be written as

```
O1
H2 1 R1
Cl3 1 R2 2 A1
 Variables:
R1 0.96
R2 1.6
A1 108.0
```

If one wanted to hold the bond angle constant at 108° during a geometry optimization, the last line of the preceding Z-matrix would be replaced by

```
 Constants:
A1 108.0
```

Do include a space before each of the words Variables and Constants. (In *Gaussian 98* and *03*, the keyword Opt must be replaced by Opt=Z-matrix.) Set up a Z-matrix for a partial geometry optimization for C_2H_6 in which the dihedral angles are held fixed at values corresponding to (a) the staggered form; (b) the eclipsed form. Then perform HF/6-31G** optimizations on these two conformations and take the energy difference to find the predicted barrier to internal rotation in kcal/mol.

15.46 Use CORINA on the Internet to build models and find Cartesian coordinates for the following molecules: (a) CH_3OH; (b) HC(O)OH.

15.47 Use ChemDB to answer Problem 15.46.

15.48 (a) Do a HF/3-21G geometry optimization and frequency calculation on NH_3 starting from a planar geometry. Is the final structure planar? Are all the vibrational frequencies real? (b) Repeat the calculation starting with a slightly nonplanar geometry. You might find it convenient to use a dummy atom in the Z-matrix. (c) What is the HF/3-21G predicted barrier to inversion of ammonia?

15.49 (a) Do HF/3-21G geometry optimizations on two conformers of HCOOH, one with OCOH dihedral angle of 0° and one with 180°. Compare the predicted geometries and the dipole moments of the two conformers. What is the predicted energy difference at 0 K omitting zero-point energy? (b) Do frequency calculations on these two conformers and use scaled vibrational frequencies to get the predicted 0 K energy difference including zero-point energy. Are all the vibrational frequencies real for each conformer?

15.50 (a) Do HF/6-31G* partial geometry optimizations of n-butane conformations with CCCC dihedral angles fixed at several values. Plot the energy versus dihedral angle. From the plot, estimate the barriers for the conversions gauche $\rightarrow$ anti and anti $\rightarrow$ gauche. (b) Start from a 60° dihedral angle and do a full geometry optimization to find the predicted dihedral angle for the gauche conformer. Calculate the predicted energy difference between gauche and anti conformers at 0 K, neglecting zero-point energy. See Prob. 15.45.

15.51 Do HF/6-31G* calculations to find the predicted geometries, dipole moments, and vibrational frequencies of the conformers of N_2H_4. What is the predicted energy difference (or differences) at 0 K?

15.52 Repeat Problem 15.51 for propene, CH_2CHCH_3.

15.53 For a minimal-basis STO calculation on H_2, how many different electron-repulsion integrals $(rs|tu)$ need to be calculated, taking into account (14.47) and the symmetry of the molecule?

15.54 Slater's rules for finding approximate orbital exponents of K-, L-, and M-shell Slater AOs are as follows. The orbital exponent ζ is taken as $(Z - s_{nl})/n$, where n is the principal quantum number and Z is the atomic number. The screening constant s_{nl} is calculated as follows: The AOs are divided into the following groups:

$$(1s) \quad (2s, 2p) \quad (3s, 3p) \quad (3d)$$

To find s_{nl}, the following contributions are summed: (a) 0 from electrons in groups to the right of the one being considered; (b) 0.35 from each other electron in the group considered, except that 0.30 is used in the 1s group; (c) for an s or p orbital, 0.85 from each electron whose quantum number n is one less than the orbital considered and 1.00 from each electron still further in; for a d orbital, 1.00 for each electron in a group to the left.

Calculate the orbital exponents according to Slater's rules for the atoms H, He, C, N, O, S, and Ar. The optimum values of ζ to use when approximating an AO as a single STO have been calculated and are given in E. Clementi and D. L. Raimondi, *J. Chem. Phys.*, **38**, 2686 (1963); E. Clementi et al., *J. Chem. Phys.*, **47**, 1300 (1967). Compare these optimum values with the above values found by Slater's rules. [For $n = 4$, Slater took ζ as $(Z - s_{nl})/3.7$; however, Slater's rules are generally unreliable for $n \geq 4$.]

CHAPTER 16

Electron-Correlation Methods

The four sources of error in ab initio molecular electronic calculations are (1) neglect of or incomplete treatment of electron correlation, (2) incompleteness of the basis set, (3) relativistic effects, and (4) deviations from the Born–Oppenheimer approximation. Deviations from the Born–Oppenheimer approximation are usually negligible for ground-state molecules. Relativistic effects will be discussed in Section 16.7. In calculations on molecules without heavy atoms, (1) and (2) are the main sources of error.

Almost all computational methods expand the MOs in a basis set of one-electron functions. The basis set has a finite number of members and hence is incomplete. The incompleteness of the basis set produces the **basis-set truncation error**.

The Hartree–Fock method, discussed in Chapter 15, neglects electron correlation. Chapter 16 discusses methods that include electron correlation.

A quantum-chemistry approximation method is **variational** if the energy calculated by the method is never less than the true energy of the state being calculated. Since an SCF MO wave function energy is equal to the variational integral (8.1), the SCF MO method is variational. Although being variational is a desirable property, we shall see that many of the calculation methods currently used are not variational.

Two other desirable properties are size extensivity and size consistency. These terms have been used by various people to mean somewhat different things. The definitions given here are from I. Shavitt, *Mol. Phys.*, **94**, 3 (1998). A method is **size extensive** if the computed energy of a system composed of n noninteracting identical systems equals n times the energy of one subsystem computed by the same method, and the computed energy of a uniform system (such as the uniform electron gas of Section 16.4) is proportional to the size of the system. Size extensivity is desirable so as to make the energy scale properly as the number of particles in the system increases. A method is **size consistent** if the computed energy of a molecule dissociated into two or more infinitely separated parts and treated as a single system equals the sum of the computed energies of each part. Size extensivity and size consistency are related but not equivalent properties, and a method can have one property but not the other. Size consistency applies only at infinite separation of the parts, but size extensivity applies at all geometries.

16.1 CONFIGURATION INTERACTION

To overcome the deficiencies of the Hartree–Fock wave function (for example, improper behavior as internuclear distances go to infinity and very inaccurate dissociation energies), one can introduce configuration interaction (CI), thus going beyond the Hartree–Fock approximation. Recall (Section 11.3) that in a molecular CI calculation one begins with a set of basis functions χ_i, does an SCF MO calculation to find SCF occupied and virtual (unoccupied) MOs, uses these MOs to form **configuration (state) functions** (CSFs) Φ_i, writes the molecular wave function ψ as a linear combination $\sum_i b_i\Phi_i$ of the CSFs, and uses the variation method to find the b_i's. The number of MOs produced equals the number of basis functions used. The type of MOs produced depends on the type of basis functions used. For example, if we include only *s* AOs in the basis set for a CI calculation on a linear molecule, we get only σ MOs, and no $\pi, \delta, \ldots$ MOs.

Each CSF is a linear combination of one to a few Slater determinants and is an eigenfunction of the spin operators $\hat{S}^2$ and $\hat{S}_z$ and satisfies the spatial symmetry requirements of the molecule. Alternatively, the CI wave function can be expressed as the equivalent linear combination of Slater determinants. When this is done, the number of Slater determinants is typically 4 or 5 times the number of CSFs.

The configuration functions in a CI calculation are classified as **singly excited, doubly excited, triply excited**, . . . , according to whether 1, 2, 3, . . . electrons are excited from occupied to unoccupied (virtual) orbitals. For example, the H_2 configuration function $2^{-1/2}|\sigma_u^*1s\,\overline{\sigma_u^*1s}|$ used in Eq. (13.95) is doubly excited (another term sometimes used is **doubly substituted**).

In the CI expansion $\psi = \sum_i b_i\Phi_i$, one includes only configuration functions that have the same symmetry properties (symmetry eigenvalues) as the state that is being approximated by the expansion. (This follows from Theorem 3 proved in Section 7.3.) For example, the ground state of H_2 is a $^1\Sigma_g^+$ state, and a CI calculation of the H_2 ground state would include only configuration functions that correspond to $^1\Sigma_g^+$ terms. A $\sigma_g\sigma_u$ electron configuration would have states of odd parity (u) and would not be included in ψ. A $\sigma_g\pi_g$ configuration would produce only Π terms (Table 13.3 in Section 13.8) and would not be included. For a π_g^2 or π_u^2 configuration, only the configuration function corresponding to the $^1\Sigma_g^+$ term (Table 13.3) would contribute to ψ.

The number of possible configuration functions with the proper symmetry increases extremely rapidly as the number of electrons and the number of basis functions increase. For n electrons and b basis functions, the number of configuration functions turns out to be roughly proportional to b^n. A CI calculation that includes all possible configuration functions with proper symmetry is called a **full CI** calculation. Because of the huge number of configuration functions, full CI calculations are out of the question except for small molecules (small n) and rather small basis sets (small b).

In most calculations (for example, molecular dissociation, excitation of valence-shell electrons to produce excited electronic states), one is looking at energy changes in processes affecting primarily the valence-shell electrons, so one expects the correlation energies involving the inner-shell electrons to change only slightly. Hence one usually makes the approximation of including only configuration functions that involve excitation of valence-shell electrons. The omission of excitations of inner-shell (core) electrons is called the **frozen-core** (FC) **approximation**. The contribution of core excitations is not always small, but their contribution changes only slightly with changes in environment. Use of the frozen-core approximation does not disqualify a calculation from being called full CI, provided all possible excitations of valence-shell electrons are included. The notation FCI(FC) denotes a full-CI calculation with the frozen-core approximation, but use of the FC approximation is so common, that people often omit the FC.

For a molecule with n electrons and with spin quantum number $S = 0$, the number of CSFs in a full CI calculation (with spatial symmetry restrictions ignored) is (*Wilson*, page 199)

$$\frac{b!(b+1)!}{\left(\frac{1}{2}n\right)!\left(\frac{1}{2}n+1\right)!\left(b-\frac{1}{2}n\right)!\left(b-\frac{1}{2}n+1\right)!} \tag{16.1}$$

where b is the number of one-electron basis functions used to express the MOs. For a 6-31G** full CI calculation of the small molecule CH_3OH, $b = 15 + 15 + 4(5) = 50$. For $n = 18$ and $b = 50$, the number of CSFs given by (16.1) is 7.6×10^{17}, so a full CI 6-31G** calculation on CH_3OH is not possible. One might therefore try a full CI calculation with the minimal STO-3G basis set, which has $b = 5 + 5 + 4(1) = 14$ for CH_3OH. This gives a mere 1.0×10^6 CSFs. This calculation is feasible, since an FCI(FC) calculation with 65×10^9 determinants has been done on the C_2 molecule [Z. Gan and R. J. Harrison, Proceedings of the ACM/IEEE SC 2005 Conference, p. 22], but would be a waste of time. Experience shows that in order to get a substantial portion of the correlation energy, one must use a large basis set. For example, for H_2O, full CI calculations with a (relatively small) DZ basis set and 256473 CSFs gives an energy of -76.158 hartrees, compared with the Hartree–Fock limit of -76.068 hartrees and the true nonrelativistic energy of -76.438 hartrees (Table 16.1), so only a relatively small portion of the -0.370-hartree correlation energy has been obtained. The SCF energy obtained with this DZ H_2O basis set is -76.010 hartrees, and the difference of -0.148 hartree between this SCF energy and the full-CI DZ energy is called the **basis-set correlation energy**. Even for a DZP basis set, a full CI calculation on H_2O gives only -76.257 hartrees, still quite far from -76.438 hartrees.

High-level correlation calculations may use basis sets that are triple zeta with two sets of polarization functions with different orbital exponents added on each atom; such a set is designated TZ2P. The correlation-consistent basis sets of Dunning and co-workers (Section 15.4) were especially designed for correlation calculations.

Since full CI calculations cannot be done for medium-size and large molecules, one must therefore decide which types of configuration functions are likely to make the largest contributions to ψ, and one includes only these. We expect the unexcited configuration function (the SCF MO wave function) to make the largest contribution. Which types of excited configurations make significant contributions to ψ? To answer this, we consider the effects of instantaneous electron correlation as a perturbation on the Hartree–Fock wave function. One finds that the first-order correction to the unperturbed (Hartree–Fock) wave function of a closed-shell state contains only doubly excited configuration functions (for justification, see Section 16.2). Thus we expect the most important correction to the Hartree–Fock wave function to come from doubly excited configuration functions. Although singly excited configuration functions are less important than double excitations in affecting the wave function, it turns out (see I. Shavitt, in *Schaefer, Methods of Electronic Structure Theory*, page 255) that single excitations have a significant effect on one-electron properties. [A *one-electron property* is one calculated as $\langle\psi|\hat{B}|\psi\rangle$, where the operator $\hat{B}$ is a sum of terms, each of which involves only a single particle; an example is the dipole moment; Eqs. (14.19) and (14.20).] Therefore, one usually includes single excitations in a CI calculation, and the most common type of CI calculation (designated CISD or CI-SD or SDCI) includes the singly and doubly excited configuration functions. CISD is an example of **limited CI**, which is anything less than a full CI calculation. (It turns out that the second-order correction to the Hartree–Fock function includes single, double, triple, and quadruple excitations.)

For calculations on a few 10-electron molecules, CISD gave about 94% of the basis-set correlation energy [R. J. Harrison and N. C. Handy, *Chem. Phys. Lett.*, **95**, 386 (1983)]. This result might seem to be a cause for optimism about CISD, but CISD has serious drawbacks. Calculations on a simple model system [F. Sasaki, *Int. J. Quantum Chem. Symp.*, **11**, 125 (1977)] indicate that the percentage of the basis-set correlation energy obtained by CISD decreases as the size of the molecule increases. For molecules that consist of first-row atoms, CISD is estimated to give 82% to 90% of the basis-set correlation energy of 20-electron molecules, 68% to 78% of the correlation energy of 50-electron molecules, and 55% to 67% for 100-electron molecules (Prob. 16.2).

A related defect is that CISD calculations are neither size extensive nor size consistent (see the beginning of Chapter 16). These properties are important whenever calculations on molecules of substantially different sizes are to be compared, as, for example, in calculation of the energy change in the dissociation reaction A $\rightarrow$ B + C.

To see that CISD is not size consistent, consider two infinitely separated helium atoms He_a and He_b. If we do a CISD calculation of He_a using a complete basis set, we get the exact energy E_a of He_a, since CISD is the same as full CI for this two-electron atom. Likewise for He_b. Now consider a complete-basis-set CISD calculation for the composite system of infinitely separated He_a and He_b. This composite system has four electrons, so a CISD calculation is not equivalent to a full CI calculation, and the CISD calculation will give an energy higher than the exact energy $E_a + E_b$ of the composite system. Therefore, CISD is not size consistent.

Full CI is size extensive and size consistent. SCF MO calculations are size extensive.

Because the CISD wave function is a variation function, the variation theorem assures us that the CISD energy cannot be less than the true energy. The CISD method is therefore variational.

Theory indicates that after double excitations, quadruple excitations are next in importance. Calculations on some 10-electron molecules found that CI with inclusion of single, double, triple, and quadruple excitations (CI-SDTQ) gave over 99% of the basis-set correlation energy (Harrison and Handy, op. cit.). For molecules containing only first-row atoms, CI-SDTQ is estimated to give the following percentages of the basis-set correlation energy: 98% to 99% for 20-electron molecules, 90% to 96% for 50-electron molecules, and 80% to 90% for 100-electron molecules (Sasaki, op. cit.). Provided we confine ourselves to molecules with no more than about 50 electrons, CI-SDTQ will come reasonably close to full CI and hence will be approximately size extensive. However, CI-SDTQ calculations with basis sets large enough to give good results for the correlation energy involve far too many CSFs to be practical.

An approximate formula due to Davidson [S. R. Langhoff and E. R. Davidson, *Int. J. Quantum Chem.*, **8**, 61 (1974)] is widely used to estimate the energy contribution ΔE_{Q} due to quadruple excitations:

$$\Delta E_{\text{Q}} \approx (1 - a_0^2)(E_{\text{CI-SD}} - E_{\text{SCF}}) \tag{16.2}$$

where a_0 is the coefficient of the SCF function Φ_0 in the normalized CI expansion $\psi = \sum_i a_i \Phi_i$, and $E_{\text{CI-SD}}$ and E_{SCF} are the CI-SD and SCF energies calculated with the basis set. For example, a DZ calculation on H_2O gave (in atomic units) at the equilibrium geometry $E_{\text{SCF}} = -76.009838$, $E_{\text{CI-SD}} = -76.150015$, and $a_0 = 0.97874$; from (16.2), we find $\Delta E_Q \approx -0.005897$, which, when added to $E_{\text{CI-SD}}$, gives -76.155912; this result is reasonably close to the CI-SDTQ DZ result of -76.157603 (the contribution of triple excitations is small).

Use of the Davidson correction reduces the size-consistency error. For example, CI-SD/DZP calculations on two H_2O molecules separated by 500 Å, a distance at which their interaction energy is utterly negligible, found that the CI-SD energy of this system exceeded twice the CI-SD energy of one H_2O molecule by 12.3 kcal/mol, which is the size-consistency error; the Davidson correction reduced this error to 3.8 kcal/mol [M. J. Frisch et al., *J. Chem. Phys.*, **84**, 2279 (1986)].

In the notation CI-SD/DZP, the method used precedes the slash and the basis-set designation follows the slash. For the SCF MO method, the abbreviation HF (for Hartree–Fock) is used, without implying that the Hartree–Fock limiting energy has been reached. Thus, HF/3-21G signifies an SCF MO calculation with the 3-21G basis set.

CISD calculations of molecular properties often do not give results of high accuracy. For example, a comparison of molecular geometries calculated by several correlation methods found that the CISD method gave the poorest results of those correlation methods studied [T. Helgaker et al., *J. Chem. Phys.*, **106**, 6430 (1997)]. Highly accurate CI results require a CISDTQ calculation, which is generally impractical.

Table 16.1 lists some calculations on H_2O that include electron correlation. The methods used in these calculations are discussed in Sections 16.1–16.6. All the calculations listed are nonrelativistic, fixed-nuclei calculations. The B3LYP/aug-cc-pVTZ and B3LYP/cc-pVTZ calculations give energies below the -76.438 true nonrelativistic energy. These are density-functional calculations and are not variational. The lowest-energy variational calculation is the MRCI calculation, which gives 97% of the correlation energy.

EXAMPLE The He SCF MO calculation in Section 14.3 used a basis set of two STOs χ_1 and χ_2. For the helium ground state treated with this basis set, (a) write down the configuration state functions (CSFs) that are present in the wave function in a full CI treatment, and (b) carry out a CI calculation that includes only doubly excited CSFs.

(a) Since we used two basis functions, the SCF calculation yielded two SCF orbitals ϕ_1 and ϕ_2. In the ground state of this two-electron atom, only ϕ_1 is occupied, and ϕ_2 is an unoccupied (virtual) orbital. We found $\phi_1 = 0.842\chi_1 + 0.183\chi_2$, but didn't bother to find ϕ_2. Use of the root $\varepsilon_2 = 2.809$ [Eq. (14.54)] and the final F_{rs}'s of (14.53) in the Roothaan equation (14.34) with $r = 1$ gives $-3.690c_{12} - 3.290c_{22} = 0$, and $c_{12}/c_{22} = -0.892$. The normalization condition (14.48) rewritten for c_{22} gives $c_{22} = 1.816$; hence $c_{12} = -1.620$. The SCF orbitals are

$$\phi_1 = c_{11}\chi_1 + c_{21}\chi_2 = 0.842\chi_1 + 0.183\chi_2$$
$$\phi_2 = c_{12}\chi_1 + c_{22}\chi_2 = -1.620\chi_1 + 1.816\chi_2$$

(Note that ϕ_2 has a node; ϕ_2 is an approximation to the 2s AO of helium.) The terms that arise from placing two electrons into these two s orbitals are a ^1S term with both electrons in ϕ_1, a ^1S term with both electrons in ϕ_2, a ^1S term with one electron in ϕ_1 and one in ϕ_2, and a ^3S term with one electron in ϕ_1 and one in ϕ_2. Since the ground state is ^1S, we include only the ^1S CSFs, which are [see Eqs. (10.41), (10.45), and (11.60)]:

$$\Phi_1 = |\phi_1\overline{\phi_1}|, \qquad \Phi_2 = |\phi_2\overline{\phi_2}|$$
$$\Phi_3 = \tfrac{1}{2}[\phi_1(1)\phi_2(2) + \phi_2(1)\phi_1(2)][\alpha(1)\beta(2) - \beta(1)\alpha(2)] \tag{16.3}$$

TABLE 16.1 H_2O Calculations That Include Correlation[a]

Reference[b]	Method[c]/Basis Set	Energy/E_h	μ/D	θ	R_{OH}/Å
Harrison, Handy	CISD/DZ	−76.150			
Harrison, Handy	FCI/DZ	−76.158			
CCCBDB	MP2(FC)/6-31G*	−76.197	2.24	104.0°	0.969
CCCBDB	CISD(FC)/6-31G*	−76.198	2.24	104.2°	0.966
CCCBDB	MP2(full)/6-31G*	−76.199	2.24	104.0°	0.969
CCCBDB	CCSD(FC)/6-31G*	−76.206	2.25	103.9°	0.969
CCCBDB	MP4(FC)/6-31G*	−76.207	2.25	103.8°	0.970
CCCBDB	CCSD(T)(FC)/6-31G*	−76.208	2.25	103.8°	0.971
CCCBDB	MP2(FC)/6-31G**	−76.220	2.20	103.8°	0.961
CCCBDB	CISD(FC)/6-31G**	−76.221	2.19	104.2°	0.958
CCCBDB	MP2(FC)/cc-pVDZ	−76.229	2.10	101.9°	0.965
CCCBDB	CCSD(T)(FC)/cc-pVDZ	−76.241	2.10	101.9°	0.966
Scuseria, Schaefer	CISDTQ/DZP	−76.270	2.13	104.5°	0.963
Schaefer et al.	CISD/TZ2P	−76.312	1.94	104.9°	0.952
CCCBDB	CISD(FC)/cc-pVTZ	−76.314	2.03	104.2°	0.953
CCCBDB	MP2(FC)/6-31+G(3df,2p)	−76.318	2.02	104.5°	0.959
CCCBDB	MP2(FC)/cc-pVTZ	−76.319	2.04	103.5°	0.959
CCCBDB	CCSD(T)(FC)/cc-pVTZ	−76.332	2.04	103.6°	0.959
Kim et al.	CISD(full)/($13s\ldots 2d$)[d]	−76.382		104.8°	0.952
Kim et al.	MP2(full)/($13s\ldots 2d$)	−76.391		104.2°	0.959
Kim et al.	CCSD(full)/($13s\ldots 2d$)	−76.396		104.4°	0.956
Kim et al.	CCSD(T)(full)/($13s\ldots 2d$)	−76.406		104.1°	0.959
Kim et al.	MP4(full)/($13s\ldots 2d$)	−76.407		104.1°	0.960
CCCBDB	B3LYP/6-31G*	−76.409	2.10	103.6°	0.969
CCCBDB	B3LYP/6-31G**	−76.420	2.04	103.7°	0.965
CCCBDB	B3LYP/cc-pVDZ	−76.421	1.94	102.7°	0.969
Lüchow et al.	FN-DQMC	−76.421[e]			
Lüchow et al.	MRCISD/aug-cc-pCV5Z	−76.427[e]			
CCCBDB	B3LYP/aug-cc-pVTZ	−76.445	1.85	104.8	0.965
CCCBDB	B3LYP/cc-pVTZ	−76.460	1.92	104.5	0.961
Nonrelativistic fixed-nuclei energy		−76.438			
Experimental values		−76.480	1.85	104.5°	0.958

[a]See footnote a to Table 15.2 in Section 15.6.

[b]R. J. Harrison and N. C. Handy, *Chem. Phys. Lett.*, **95**, 386 (1983); srdata.nist.gov/cccbdb/; G. E. Scuseria and H. F. Schaefer, III, *Chem. Phys. Lett.*, **146**, 23 (1988); H. F. Schaefer, III et al., in *Yarkony*, chapter 1; J. Kim et al., *J. Chem. Phys.*, **102**, 310 (1995); A. Lüchow, J. B. Anderson, and D. Feller, *J. Chem. Phys.*, **106**, 7706 (1997).

[c]FC denotes frozen-core calculations and full indicates that the frozen-core approximation is not used.

[d]($13s\ldots 2d$) is an uncontracted ($13s8p4d2f/8s4p2d$) set with 131 basis functions.

[e]Calculated at a geometry close to the experimental geometry.

(b) The CSF Φ_2 is doubly substituted, and Φ_3 is singly substituted, so we include only Φ_2 in addition to Φ_1. The variational function is $\psi = a_1\Phi_1 + a_2\Phi_2$, and the secular equation (11.18) is

$$\begin{vmatrix} \langle\Phi_1|\hat{H}|\Phi_1\rangle - E & \langle\Phi_1|\hat{H}|\Phi_2\rangle \\ \langle\Phi_2|\hat{H}|\Phi_1\rangle & \langle\Phi_2|\hat{H}|\Phi_2\rangle - E \end{vmatrix} = 0 \tag{16.4}$$

The orthogonality of the orbitals ϕ_1 and ϕ_2 (Section 14.3) ensures that $\langle\Phi_1|\Phi_2\rangle = 0$.

The function Φ_1 is the SCF wave function, so $\langle\Phi_1|\hat{H}|\Phi_1\rangle$ is equal to the SCF energy -2.862 hartrees calculated in the previous example: $\langle\Phi_1|\hat{H}|\Phi_1\rangle = -2.862$.

In evaluating the integrals in the secular determinant, we can omit the spin factors in the wave functions, since summation over these gives 1.

The He Hamiltonian (9.39) is $\hat{H} = \hat{H}^{\text{core}}(1) + \hat{H}^{\text{core}}(2) + 1/r_{12}$, and

$$\langle\Phi_2|\hat{H}|\Phi_2\rangle = \langle\phi_2(1)\phi_2(2)|\hat{H}^{\text{core}}(1) + \hat{H}^{\text{core}}(2) + 1/r_{12}|\phi_2(1)\phi_2(2)\rangle$$

$$\langle\Phi_2|\hat{H}|\Phi_2\rangle = \langle\phi_2(1)|\hat{H}^{\text{core}}(1)|\phi_2(1)\rangle\langle\phi_2(2)|\phi_2(2)\rangle + \langle\phi_2(2)|\hat{H}^{\text{core}}(2)|\phi_2(2)\rangle\langle\phi_2(1)|\phi_2(1)\rangle + \langle\phi_2(1)\phi_2(2)|1/r_{12}|\phi_2(1)\phi_2(2)\rangle$$

where $\langle\phi_2(2)|\phi_2(2)\rangle = 1 = \langle\phi_1(1)|\phi_1(1)\rangle$. To evaluate the integrals over the orbitals ϕ_i in terms of integrals over the basis functions χ_r, we substitute the expansion $\phi_i = \Sigma_r c_{ri}\chi_r$ [Eq. (14.33)] into the integrals. We get

$$\langle\phi_i(1)|\hat{H}^{\text{core}}(1)|\phi_j(1)\rangle = \sum_r\sum_s c_{ri}^*c_{sj}\langle\chi_r(1)|\hat{H}^{\text{core}}(1)|\chi_s(1)\rangle = \sum_{r=1}^{b}\sum_{s=1}^{b} c_{ri}^*c_{sj}H_{rs}^{\text{core}} \quad (16.5)$$

$$\langle\phi_i(1)\phi_j(2)|1/r_{12}|\phi_k(1)\phi_l(2)\rangle = \sum_r\sum_s\sum_t\sum_u c_{ri}^*c_{sj}^*c_{tk}c_{ul}\langle\chi_r(1)\chi_s(2)|1/r_{12}|\chi_t(1)\chi_u(2)\rangle$$

$$\langle\phi_i(1)\phi_j(2)|1/r_{12}|\phi_k(1)\phi_l(2)\rangle = \sum_{r=1}^{b}\sum_{t=1}^{b}\sum_{s=1}^{b}\sum_{u=1}^{b} c_{ri}^*c_{tk}c_{sj}^*c_{ul}(rt|su) \quad (16.6)$$

where the notation (14.39) is used. Using these equations, we find (Prob. 16.6)

$$\langle\Phi_2|\hat{H}|\Phi_2\rangle = 2(c_{12}^2H_{11}^{\text{core}} + 2c_{12}c_{22}H_{12}^{\text{core}} + c_{22}^2H_{22}^{\text{core}}) + c_{12}^4(11|11) + 4c_{12}^3c_{22}(11|12) + 2c_{12}^2c_{22}^2(11|22) + 4c_{12}^2c_{22}^2(12|12) + 4c_{12}c_{22}^3(12|22) + c_{22}^4(22|22) = 3.22_6$$

where the values of the integrals were taken from the example in Section 14.3. Similarly, we find (Prob. 16.6)

$$\langle\Phi_2|\hat{H}|\Phi_1\rangle = \langle\Phi_1|\hat{H}|\Phi_2\rangle = c_{11}^2c_{12}^2(11|11) + 2(c_{11}^2c_{12}c_{22} + c_{11}c_{12}^2c_{21})(11|12) + 2c_{11}c_{12}c_{21}c_{22}(11|22) + (c_{11}^2c_{22}^2 + 2c_{11}c_{21}c_{12}c_{22} + c_{12}^2c_{21}^2)(12|12) + 2(c_{11}c_{21}c_{22}^2 + c_{12}c_{22}c_{21}^2)(12|22) + c_{21}^2c_{22}^2(22|22) = 0.289_5$$

The $\hat{H}^{\text{core}}$ integrals vanish for $\langle\Phi_2|\hat{H}|\Phi_1\rangle$ because of the orthogonality of ϕ_1 and ϕ_2.

The secular equation (16.4) is

$$\begin{vmatrix} -2.862 - E & 0.289_5 \\ 0.289_5 & 3.22_6 - E \end{vmatrix} = 0$$

$$E = -2.876, 3.24$$

The lower root gives

$$E = -2.876 \text{ hartrees} = -78.25 \text{ eV}$$

as compared with the SCF energy of -77.87 eV and the true energy of -79.00 eV. This CI calculation has recovered 34% of the correlation energy. The CI ground-state ψ is found to be (Prob. 16.5) $\psi = 0.9989\Phi_1 - 0.0474\Phi_2$.

One finds that inclusion of the singly excited CSF Φ_3 gives only a very slight further improvement (see *Jørgensen and Oddershede*, page 38). Significant further improvement requires redoing the SCF calculation with a larger basis set, which will generate more virtual orbitals so that many more CSFs can be included in the CI calculation.

In a CI calculation that uses SCF MOs, there are two major computational tasks. One task is to transform the known integrals over the AO basis functions χ_r into integrals over the SCF MOs ϕ_i using Eqs. (16.5) and (16.6). Calculation of the integrals $\langle \phi_i \phi_j | 1/r_{12} | \phi_k \phi_l \rangle$ from the $(rs|tu)$ integrals is especially time consuming. For a basis set of b functions $\chi_1, \ldots, \chi_b$, there are b MOs ϕ_i and approximately $b^4/8$ different $\langle \phi_i \phi_j | 1/r_{12} | \phi_k \phi_l \rangle$ integrals to be computed [the factor 1/8 arises from equalities similar to (14.47)], and there are b^4 terms in the sums on the right side of (16.6). Hence the number of computations is apparently about $b^8/8$. For 100 basis functions, $b^8/8$ is about 10^{15}. Fortunately, use of a clever procedure can reduce the number of computations to about b^5 (see *Hehre et al.*, Section 3.3.4). The second task is to solve the CI secular equation to find the lowest-energy eigenvalue and the corresponding set of expansion coefficients. In an SCF calculation that uses 200 basis functions, one must solve for the eigenvalues and eigenvectors of a matrix whose order is 200. This is readily done using standard matrix diagonalization methods (Section 8.6). In an accurate CI calculation, one might use 10^8 CSFs to expand the wave function, and one must find the lowest eigenvalue and corresponding eigenvector of a matrix whose order is 10^8. Special techniques have been devised to find the lowest few eigenvalues and corresponding eigenvectors of a very large matrix (see I. Shavitt in *Schaefer, Methods of Electronic Structure Theory*, pp. 228–238); Davidson's method for doing this is widely used [see the references in J. H. van Lenthe and P. Pulay, *J. Comput. Chem.*, **11**, 1164 (1990)].

In a CI calculation with 10^8 CSFs, there are 10^{16} matrix elements H_{ij} between CSFs, which is too many to be stored in the internal memory of the computer. To avoid the problems of dealing with large matrices, Roos developed the **direct configuration interaction method** in 1972 (see B. O. Roos and P. E. M. Siegbahn, in *Schaefer, Methods of Electronic Structure Theory*, Chapter 7). The direct CI method avoids explicit calculation of the integrals H_{ij} between CSFs and avoids the need to solve the secular equation (11.18). Instead, the CI expansion coefficients in (11.16) and the energy are calculated directly from the one- and two-electron integrals over the basis functions. The direct CI method (not to be confused with the direct SCF method of Section 15.16) allows CI calculations with more than 10^{10} CSFs.

The CI procedure just discussed calculates SCF occupied and virtual orbitals from the basis functions and uses these SCF MOs to form configuration state functions. The convergence rate of this procedure is very slow, and huge numbers of CSFs must be included for accurate results. A major reason for the very slow convergence is that the excited (virtual) SCF orbitals have much of their probability density at large distances from the nuclei, whereas the ground-state wave function has most of its probability density reasonably near the nuclei.

Actually, there is no necessity to use SCF MOs in a CI calculation. Any set of MOs calculated from the basis set will produce the same final wave function provided a full CI calculation is carried out. Moreover, if the non-SCF MOs are well chosen, they can produce much faster convergence to the true wave function than is obtained with SCF

MOs, thereby allowing substantially fewer CSFs to be included in ψ. Some approaches that use this idea are the method of natural orbitals, the use of localized orbitals, and the multiconfiguration SCF method.

The INO Method. An alternative to the use of SCF MOs in CI calculations is provided by natural orbitals. For a CI wave function, which is a linear combination of Slater determinants, the electron probability density ρ can be shown to have the form [P.-O. Löwdin, *Rev. Mod. Phys.*, **32**, 328 (1960)]

$$\rho(x, y, z) = \sum_i \sum_j a_{ij}\phi_i^*(x, y, z)\phi_j(x, y, z)$$

where the ϕ's are all the MOs that appear in the Slater determinants of the CI wave function and the a_{ij}'s are a set of numbers. As we saw in Section 15.8, we can take a linear combination of MOs to form a new set of MOs without changing the overall wave function. It can be shown that there exists a set of MOs θ_i, called the **natural orbitals**, that have the property that when the CI wave function is expressed using the θ_i's, the probability density has the simple form

$$\rho(x, y, z) = \sum_i \lambda_i |\theta_i(x, y, z)|^2$$

where the *occupation numbers* λ_i lie between 0 and 2 and need not be integers. Note the resemblance of this expression to (14.7), which is for an SCF wave function.

It turns out that a CI calculation using natural orbitals converges much faster than one using SCF orbitals, so one needs to include far fewer CSFs to obtain a given level of accuracy. Unfortunately, the natural orbitals are defined in terms of the final CI wave function and cannot be calculated until a CI calculation using SCF orbitals has been completed. Hence several schemes have been devised to calculate approximate natural orbitals and use these in CI calculations.

For example, the *iterative natural-orbital* (INO) *method* starts by calculating a CI wave function using a manageable number of CSFs. From this CI wave function, one calculates approximate natural orbitals and uses these orbitals to construct an improved CI wave function; this process is then repeated to get further improvement. As an example, an INO calculation on the LiH ground state using only 45 CSFs obtained 87% of the correlation energy and gave a slightly lower energy than an ordinary CI wave function of 939 CSFs; see C. F. Bender and E. R. Davidson, *J. Phys. Chem.*, **70**, 2675 (1966).

Another way to speed up convergence is to use localized SCF MOs (Section 15.8) instead of canonical SCF MOs in the CSFs. CI calculations on 1,3-butadiene using this *localized correlation* method showed speedups by factors of 20 to 40, since fewer CSFs needed to be included [S. Saebø and P. Pulay, *Chem. Phys. Lett.*, **113**, 13 (1985)].

The MCSCF Method. In the **multiconfiguration SCF** (MCSCF) **method**, one writes the molecular wave function as a linear combination of CSFs Φ_i and varies not only the expansion coefficients b_i in $\psi = \sum_i b_i\Phi_i$, but also the forms of the molecular orbitals in the CSFs. The MOs are varied by varying the expansion coefficients c_{ri} that relate the MOs ϕ_i to the basis functions χ_r. For example, if we were to do an MCSCF calculation for the He ground state using the basis functions χ_1 and χ_2 of (14.46) and including only the CSFs Φ_1 and Φ_2 of (16.3), we would write as the MCSCF wave function

$$\psi = b_1\Phi_1 + b_2\Phi_2 = b_1|\phi_1\overline{\phi_1}| + b_2|\phi_2\overline{\phi_2}|$$
$$= b_1|(c_{11}\chi_1 + c_{21}\chi_2)(\overline{c_{11}\chi_1 + c_{21}\chi_2})| + b_2|(c_{12}\chi_1 + c_{22}\chi_2)(\overline{c_{12}\chi_1 + c_{22}\chi_2})|$$

and we would simultaneously vary the coefficients $b_1, b_2, c_{11}, c_{21}, c_{12}$, and c_{22} (subject to the conditions of orthonormality of ϕ_1 and ϕ_2 and normalization of ψ) to minimize the variational integral. The MCSCF values found for these six coefficients will differ from the corresponding values found in the SCF plus CI calculations we did, and the energy will be lower, since the c_{ri}'s will be optimal for the MCSCF ψ, rather than being optimal for the SCF ψ.

The optimum MCSCF orbitals can be found by an iterative process somewhat similar to the iterative process used to find SCF wave functions; see A. C. Wahl and G. Das, Chapter 3 in *Schaefer, Methods of Electronic Structure Theory*. By optimizing the orbitals, one can get good results with inclusion of relatively few CSFs. Because the orbitals are varied, the amount of calculation required in the MCSCF procedure is great, but advances in methods of computing MCSCF wave functions [R. Shepard, *Adv. Chem. Phys.*, **69**, 63 (1987)] have led to wide use of the MCSCF and related methods.

The most commonly used kind of MCSCF method is the **complete active space SCF** (CASSCF) method [B. O. Roos, *Adv. Chem. Phys.*, **69**, 399 (1987)]. Here, as usual, one writes the orbitals ϕ_i to be used in the CSFs as linear combinations of basis functions: $\phi_i = \sum_{r=1}^{b} c_{ri}\chi_r$. One divides the orbitals in the CSFs into **inactive** and **active orbitals**. The inactive orbitals are kept doubly occupied in all CSFs. The electrons not in the inactive orbitals are called **active electrons**. One writes the wave function as a linear combination of all CSFs Φ_i that can be formed by distributing the active electrons among the active orbitals in all possible ways and that have the same spin and symmetry eigenvalues as the state to be treated; $\psi = \sum_i b_i\Phi_i$. One then does an MCSCF calculation to find the optimum coefficients c_{ri} and b_i. A reasonable choice is to take the active orbitals as those MOs that arise from the valence orbitals of the atoms that form the molecule.

For example, the C_2 ground-state configuration is

$$(1\sigma_g)^2(1\sigma_u)^2(2\sigma_g)^2(2\sigma_u)^2(1\pi_u)^4$$

The $2s$ and $2p$ carbon AOs give rise to the $2\sigma_g, 2\sigma_u, 1\pi_{ux}, 1\pi_{uy}, 3\sigma_g, 1\pi_{gx}, 1\pi_{gy}$, and $3\sigma_u$ MOs, where the last four are unoccupied in the ground state. One might thus take the inactive orbitals as the $1\sigma_g$ and $1\sigma_u$ MOs (giving eight active electrons) and the active orbitals as the $2\sigma_g, 2\sigma_u, 1\pi_u, 3\sigma_g, 1\pi_g$, and $3\sigma_u$ MOs. A CASSCF calculation on the C_2 ground electronic state used a basis set of 82 functions and took the inactive orbitals as $1\sigma_g$ and $1\sigma_u$ and the active orbitals as $2\sigma_g, 2\sigma_u, 1\pi_u, 3\sigma_g, 1\pi_g, 3\sigma_u, 4\sigma_g$, and $4\sigma_u$ [W. P. Kraemer and B. O. Roos, *Chem. Phys.*, **118**, 345 (1987)]. Distribution of the eight active electrons among the ten active orbitals gave a wave function consisting of 1900 CSFs. The electronic energy was calculated for several points in the neighborhood of R_e, and the calculated $U(R)$ curve was substituted into the Schrödinger equation for nuclear motion, which was solved numerically (Section 13.2) to find vibrational–rotational energy levels, from which spectroscopic constants were found. The results were (experimental values in parentheses): $R_e/\text{Å} = 1.25$ (1.24), $D_e/\text{eV} = 6.06$ (6.3),

$\widetilde{\nu}_e/\text{cm}^{-1} = 1836$ (1855), and $\widetilde{\nu}_e x_e/\text{cm}^{-1} = 14.9$ (13.4), where $\widetilde{\nu}_e x_e \equiv \nu_e x_e/c$ [see Eq. (4.62)]. For comparison, the Hartree–Fock ψ gives $D_e = 0.8$ eV, $\widetilde{\nu}_e/\text{cm}^{-1} = 1905$, $\widetilde{\nu}_e x_e/\text{cm}^{-1} = 12.1$, and $R_e = 1.24$ Å.

With modern computational techniques, very large MCSCF wave functions can be calculated. A CASSCF calculation on FeO included 178916 CSFs and fully optimized the orbitals [P. J. Knowles and H.-J. Werner, *Chem. Phys. Lett.*, **115**, 259 (1985)]. For large molecules, use of all the valence orbitals as active gives rise to too many CSFs to be handled, and so the number of active orbitals must be limited in such cases. Typically, up to 14 active orbitals can be handled.

The MRCI Method. A widely used method that combines the MCSCF and conventional CI methods is the **multireference CI** (MRCI) **method**. In the conventional (or single-reference) CI method, one starts with the SCF wave function Φ_1 (which is called the **reference function**) and moves electrons out of occupied orbitals of Φ_1 into virtual SCF orbitals to produce CSFs $\Phi_2, \Phi_3, \ldots,$ and one writes the wave function as $\psi = \sum_i b_i \Phi_i$; one then varies the b_i's to minimize the variational integral. Recall that for H_2, although the Hartree–Fock function Φ_1 gives a reasonably good representation of the wave function in the region of R_e, in the limit as $R \to \infty$ we must take the variational function as a linear combination of two CSFs in order to get the proper dissociation behavior. For N_2, which has a triple bond, one finds that a linear combination of 10 CSFs is needed to give proper behavior at large R. In the MRCI method, one first does an MCSCF calculation to produce a wave function that is a linear combination of several CSFs $\Phi_1, \Phi_2, \ldots, \Phi_m$ with optimized orbitals and that has the proper behavior for all nuclear configurations. One then takes this MCSCF function and moves electrons out of occupied orbitals of the CSFs $\Phi_1, \Phi_2, \ldots, \Phi_m$ (called the *reference* CSFs) into virtual orbitals to produce further CSFs $\Phi_{m+1}, \ldots, \Phi_n$. Most commonly, one does a CISD calculation starting with the MCSCF function. One writes $\psi = \sum_{i=1}^{n} b_i \Phi_i$ and finds the optimum b_i's. Typically, the reference CSFs will contain singly and doubly excited CSFs and, in the final wave function, one will consider single and double excitations from the reference CSFs. Thus, the final MRCI wave function will include some triple and quadruple excitations.

CASSCF wave functions are often used as the starting point for MRCI calculations. The calculations that determined the ordering of the lowest states of Si_2 and Al_2 (Section 13.7) were CASSCF MRCI calculations.

When a CASSCF wave function is used for a MRCISD calculation, the number of CSFs produced may be too many to deal with, so various methods are used to reduce the amount of computation needed. One widely used procedure is **internally contracted** MRCI (icMRCI) [H.-J. Werner and P. J. Knowles, *J. Chem. Phys.*, **89**, 5803 (1988)]. Here, the optimized MCSCF function is treated as a single reference function (with fixed coefficients) from which one generates doubly excited functions. Each excited function is a linear combination of many ordinary CSFs with the coefficients within a given excited function being held fixed at the values for the MCSCF function. Thus, far fewer coefficients need to be calculated than in a conventional (uncontracted) MRCI calculation. (Singly excited functions are also included, but for technical reasons, these are not contracted but are treated as in uncontracted MRCI.) Experience has shown that the contracted MRCI wave function is almost as accurate as the uncontracted one.

MRCI methods that use localized orbitals have been developed; see D. Walter et al., *J. Chem. Phys.*, **118**, 8127 (2003); A. Venkatnathan et al., ibid., **120**, 1693 (2004).

Status of the CI Method. Although calculation of SCF MO wave functions for closed-shell, small- and medium-size molecules is essentially a routine procedure, molecular CI calculations often present special problems. To obtain reliable results, one must use sound judgment in choosing the basis set, the configuration functions to be included, and the procedure to be used.

Because of slow convergence, lack of size consistency, and disappointing results of CISD calculations, CI calculations have lost their former dominance in correlation calculations, and several other correlation methods have been developed (Sections 16.2–16.6). These correlation methods are much more efficient than CISD in giving accurate results for ground-state molecules at or near the equilibrium geometry. However, when it comes to dealing with geometries far from minima, as when dealing with complete potential-energy surfaces and chemical reactions, these methods do not perform so well, and multireference CI calculations (MRCI) are widely used to explore potential-energy surfaces and in studying chemical reactions (Section 16.10). "MRCI is perhaps the most widely applicable and commonly used method for modeling bond-breaking reactions and ground- and excited-state potential-energy surfaces in small molecules" [C. D. Sherrill and P. Piecuch, *J. Chem. Phys.*, **122**, 124104 (2005)].

The CI-Singles Method. The CI-singles (CIS) method (also called the Tamm–Dancoff approximation) is a computationally simple, often used procedure for treating excited states [J. B. Foresman, M. Head-Gordon, J. A. Pople, and M. Frisch, *J. Phys. Chem.*, **96**, 135 (1992)].

CIS wave functions for the lowest several excited states of a molecule are calculated as follows. A fixed molecular geometry is chosen. This is typically the optimized ground-state geometry found using a basis set and method known to give accurate ground-state geometries (or it might be the experimental ground-state geometry). A basis set that includes diffuse functions is used to calculate a single-determinant SCF MO ground-state wave function Φ_0 at the chosen geometry. This calculation also yields a set of unoccupied (virtual) orbitals, whose number depends on the size of the basis set used. Let ψ_i^a denote a singly excited Slater determinant in which the occupied spin-orbital i in Φ_0 is replaced by the virtual spin-orbital a. One forms the CIS linear variation function $\psi_{\text{CIS}} = \sum_a \sum_i c_{ia}\psi_i^a$, where the sums go over all the occupied and all the virtual spin-orbitals and c_{ia} is a variational coefficient. The equations of the linear variation method (Sections 8.5 and 8.6) are used to find the lowest several roots of the secular equation (where the molecular electronic Hamiltonian used to evaluate the matrix elements corresponds to the chosen ground-state geometry) and to find the coefficients that go with each root. Each of these roots is an approximation to the energy of an excited electronic state at the fixed molecular geometry that was chosen for the calculation.

When an electronic transition occurs from the ground state to an excited state, the much greater mass of the nuclei than that of the electrons means that the excited state has the greatest probability to be produced in a geometry that is close to the equilibrium geometry of the ground electronic state (the Franck–Condon principle), even though this geometry is not likely to be the equilibrium geometry of the excited state.

The excited state is thus produced in a high vibrational level. Hence the observed maximum-intensity frequency $\nu_{\max}$ in the electronic absorption spectrum corresponds to no change in geometry. The energy change $h\nu_{\max}$ is called the **vertical excitation energy**. The CIS prediction of the vertical excitation energy is found by taking the difference between the excited state energy found as a root of the secular equation and the ground state energy found from Φ_0. CIS predictions of vertical excitation energies are semiquantitatively correct, but not of high accuracy. Errors run about 1 eV in a quantity whose range is typically 3 to 10 eV.

The CIS wave function is found by solving for the coefficients c_{ia}. Since analytic gradients of CIS energies are available, one can then optimize the geometry of each excited state and also calculate its vibrational frequencies. CIS excited-state geometries and vibrational frequencies are more accurate than CIS vertical excitation energies [J. F. Stanton et al., *J. Chem. Phys.*, **103**, 4160 (1995)].

Note that the form of a CIS wave function differs from that of an ordinary CI wave function. In an ordinary CI wave function, the reference function (the SCF wave function for the state of interest) makes the largest contribution. In the CIS method for an excited state, the reference function is the SCF wave function for the ground state, and this reference function does not appear in the CIS wave function. (This makes the CIS wave function orthogonal to the ground-state wave function, which is desirable, so as to avoid having the variational calculation "collapse" to the ground state.) The CIS wave function includes only a modest amount of electron correlation.

16.2 MØLLER–PLESSET (MP) PERTURBATION THEORY

Physicists and chemists have developed various perturbation-theory methods to deal with systems of many interacting particles (nucleons in a nucleus, atoms in a solid, electrons in an atom or molecule), and these methods constitute **many-body perturbation theory** (MBPT). In 1934, Møller and Plesset proposed a perturbation treatment of atoms and molecules in which the unperturbed wave function is the Hartree–Fock function, and this form of MBPT is called **Møller–Plesset** (MP) **perturbation theory**. Actual molecular applications of MP perturbation theory began only in 1975 with the work of Pople and co-workers and Bartlett and co-workers [R. J. Bartlett, *Ann. Rev. Phys. Chem.*, **32**, 359 (1981); *Hehre et al.*].

The treatment of this section will be restricted to closed-shell, ground-state molecules. Also, the development will use spin-orbitals u_i, rather than spatial orbitals ϕ_i. For spin-orbitals, the Hartree–Fock equations (14.25) and (14.26) for electron m in an n-electron molecule have the forms (*Szabo and Ostlund*, Section 3.1)

$$\hat{f}(m)u_i(m) = \varepsilon_i u_i(m) \tag{16.7}$$

$$\hat{f}(m) \equiv -\tfrac{1}{2}\nabla_m^2 - \sum_\alpha \frac{Z_\alpha}{r_{m\alpha}} + \sum_{j=1}^{n} [\hat{j}_j(m) - \hat{k}_j(m)] \tag{16.8}$$

where $\hat{j}_j(m)$ and $\hat{k}_j(m)$ are defined by equations like (14.28) and (14.29) with spatial orbitals replaced by spin-orbitals and integrals over spatial coordinates of an electron replaced by integration over spatial coordinates and summation over the spin coordinate of that electron.

The MP unperturbed Hamiltonian is taken as the sum of the one-electron Fock operators $\hat{f}(m)$ in (16.7):

$$\hat{H}^0 \equiv \sum_{m=1}^{n} \hat{f}(m) \qquad (16.9)$$

The ground-state Hartree–Fock wave function Φ_0 is the Slater determinant $|u_1 u_2 \cdots u_n|$ of spin-orbitals. This Slater determinant is an antisymmetrized product of the spin-orbitals [for example, see Eq. (10.36)] and, when expanded, is the sum of $n!$ terms, where each term involves a different permutation of the electrons among the spin-orbitals. Each term in the expansion of Φ_0 is an eigenfunction of the MP $\hat{H}^0$; for example, for a four-electron system, application of $\hat{H}^0$ to a typical term in the Φ_0 expansion gives

$$[\hat{f}(1) + \hat{f}(2) + \hat{f}(3) + \hat{f}(4)]u_1(3)u_2(2)u_3(4)u_4(1)$$
$$= (\varepsilon_4 + \varepsilon_2 + \varepsilon_1 + \varepsilon_3)u_1(3)u_2(2)u_3(4)u_4(1)$$

where $\hat{f}(m)u_i(m) = \varepsilon_i u_i(m)$ [Eq. (16.7)] was used. Similarly, each other term in the expansion of $|u_1 u_2 u_3 u_4|$ is an eigenfunction of $\hat{H}^0$ with the same eigenvalue $\varepsilon_1 + \varepsilon_2 + \varepsilon_3 + \varepsilon_4$. Since Φ_0 is a linear combination of these $n!$ terms, Φ_0 is an eigenfunction of $\hat{H}^0$ with this eigenvalue:

$$\hat{H}^0 \Phi_0 = \left(\sum_{m=1}^{n} \varepsilon_m\right)\Phi_0 = E^{(0)}\Phi_0 \qquad (16.10)$$

The eigenfunctions of the unperturbed Hamiltonian $\hat{H}^0$ are the zeroth-order (unperturbed) wave functions [Eq. (9.2)], so the Hartree–Fock ground-state function Φ_0 is one of the zeroth-order wave functions. What are the other eigenfunctions of $\hat{H}^0$? The Hermitian operator $\hat{f}(m)$ has a complete set of eigenfunctions, these eigenfunctions being all the possible spin-orbitals of the molecule. The n lowest-energy spin-orbitals are occupied, and there are an infinite number of unoccupied (virtual) orbitals. The operator $\hat{H}^0 = \sum_{m=1}^{n} \hat{f}(m)$ is the sum of the operators $\hat{f}(m)$, and so the eigenfunctions of $\hat{H}^0$ are all possible products of any n of the spin-orbitals. However, the wave functions must be antisymmetric, so we must antisymmetrize these zeroth-order wave functions by forming Slater determinants. Thus, the zeroth-order wave functions are all possible Slater determinants formed using any n of the infinite number of possible spin-orbitals. (Of course, the n chosen spin-orbitals must all be different or the Slater determinant would vanish.)

The perturbation $\hat{H}'$ is the difference between the true molecular electronic Hamiltonian $\hat{H}$ and $\hat{H}^0$; $\hat{H}' = \hat{H} - \hat{H}^0$. Use of (15.10) for $\hat{H}$ and (16.9) and (16.8) for $\hat{H}^0$ gives (Prob. 16.8)

$$\hat{H}' = \hat{H} - \hat{H}^0 = \sum_{l}\sum_{m>l} \frac{1}{r_{lm}} - \sum_{m=1}^{n}\sum_{j=1}^{n} [\hat{j}_j(m) - \hat{k}_j(m)] \qquad (16.11)$$

The perturbation $\hat{H}'$ is the difference between the true interelectronic repulsions and the Hartree–Fock interelectronic potential (which is an average potential).

The MP first-order correction $E_0^{(1)}$ to the ground-state energy is [Eq. (9.22)] $E_0^{(1)} = \langle \psi_0^{(0)} | \hat{H}' | \psi_0^{(0)} \rangle = \langle \Phi_0 | \hat{H}' | \Phi_0 \rangle$, since $\psi_0^{(0)} = \Phi_0$. The subscript 0 denotes the ground state. We have

$$E_0^{(0)} + E_0^{(1)} = \langle \psi_0^{(0)} | \hat{H}^0 | \psi_0^{(0)} \rangle + \langle \Phi_0 | \hat{H}' | \Phi_0 \rangle = \langle \Phi_0 | \hat{H}^0 + \hat{H}' | \Phi_0 \rangle = \langle \Phi_0 | \hat{H} | \Phi_0 \rangle$$

But $\langle \Phi_0 | \hat{H} | \Phi_0 \rangle$ is the variational integral for the Hartree–Fock wave function Φ_0 and it therefore equals the Hartree–Fock energy E_{HF}. Hence (recall the beginning of Section 9.4)

$$E_0^{(0)} + E_0^{(1)} = E_{\text{HF}}$$

Note from (16.10) that the zeroth-order (unperturbed) eigenfunction Φ_0 of $\hat{H}^0$ has the eigenvalue $\sum_{m=1}^{n} \varepsilon_m$. Therefore, [Eq. (9.2)] $E_0^{(0)} = \sum_{m=1}^{n} \varepsilon_m$.

To improve on the Hartree–Fock energy, we must find the second-order energy correction $E_0^{(2)}$. From (9.35),

$$E_0^{(2)} = \sum_{s \neq 0} \frac{|\langle \psi_s^{(0)} | \hat{H}' | \Phi_0 \rangle|^2}{E_0^{(0)} - E_s^{(0)}} \tag{16.12}$$

We saw above that the unperturbed functions $\psi_s^{(0)}$ are all possible Slater determinants formed from n different spin-orbitals. Let $i, j, k, l, \ldots$ denote the occupied spin-orbitals in the ground-state Hartree–Fock function Φ_0, and let $a, b, c, d, \ldots$ denote the unoccupied (virtual) spin-orbitals. Each unperturbed wave function can be classified by the number of virtual spin-orbitals it contains; this number is called the **excitation level**. Let Φ_i^a denote the singly excited determinant that differs from Φ_0 solely by replacement of the occupied spin-orbital u_i by the virtual spin-orbital u_a. Let Φ_{ij}^{ab} denote the doubly excited determinant formed from Φ_0 by replacement of u_i by u_a and u_j by u_b; and so on.

Consider the matrix elements $\langle \psi_s^{(0)} | \hat{H}' | \Phi_0 \rangle$ in (16.12), where Φ_0 is a closed-shell single determinant. One finds (*Szabo and Ostlund*, Section 6.5) that this integral vanishes for all singly excited $\psi_s^{(0)}$'s; that is, $\langle \Phi_i^a | \hat{H}' | \Phi_0 \rangle = 0$ for all i and a. Also, $\langle \psi_s^{(0)} | \hat{H}' | \Phi_0 \rangle$ vanishes for all $\psi_s^{(0)}$'s whose excitation level is three or higher. This follows from the Condon–Slater rules (Table 11.3). Hence, we need consider only doubly excited $\psi_s^{(0)}$'s to find $E_0^{(2)}$. Also, the same reasoning applied to Eq. (9.27) shows that $\psi_0^{(1)}$, the first-order correction to the wave function, contains only doubly excited $\psi_s^{(0)}$'s.

The doubly excited function Φ_{ij}^{ab} is an eigenfunction of $\hat{H}^0 = \sum_m \hat{f}(m)$ with an eigenvalue that differs from the eigenvalue of Φ_0 solely by replacement of ε_i by ε_a and replacement of ε_j by ε_b. Hence in (16.12), $E_0^{(0)} - E_s^{(0)} = \varepsilon_i + \varepsilon_j - \varepsilon_a - \varepsilon_b$ for $\psi_s^{(0)} = \Phi_{ij}^{ab}$. Use of (16.11) for $\hat{H}'$ and of the Condon–Slater rules allows the integrals involving Φ_{ij}^{ab} to be evaluated; one finds (Prob. 16.9)

$$E_0^{(2)} = \sum_{b=a+1}^{\infty} \sum_{a=n+1}^{\infty} \sum_{i=j+1}^{n} \sum_{j=1}^{n-1} \frac{|\langle ab | r_{12}^{-1} | ij \rangle - \langle ab | r_{12}^{-1} | ji \rangle|^2}{\varepsilon_i + \varepsilon_j - \varepsilon_a - \varepsilon_b} \tag{16.13}$$

where n is the number of electrons and

$$\langle ab | r_{12}^{-1} | ij \rangle \equiv \iint u_a^*(1) u_b^*(2) r_{12}^{-1} u_i(1) u_j(2) \, d\tau_1 \, d\tau_2 \tag{16.14}$$

The integrals over spin-orbitals (which include a sum over spins) are readily evaluated in terms of electron-repulsion integrals. The sums over a, b, i, and j in (16.13) provide for the inclusion of all the doubly substituted $\psi_s^{(0)}$'s in (16.12).

Taking the molecular energy as $E^{(0)} + E^{(1)} + E^{(2)} = E_{\text{HF}} + E^{(2)}$ gives a calculation designated as MP2 or MBPT(2), where the 2 indicates inclusion of energy corrections through second order.

Formulas for the MP energy corrections $E^{(3)}$, $E^{(4)}$, and so on, have also been derived [for example, see R. Krishnan and J. A. Pople, *Int. J. Quantum Chem.*, **14**, 91 (1978)]. Since $\psi^{(1)}$, the first-order correction to the wave function, determines both $E^{(2)}$ and $E^{(3)}$ (Section 9.2), and since $\psi^{(1)}$, contains only doubly excited determinants, $E^{(3)}$ contains summations over only double substitutions. The MP $E^{(4)}$ involves summations over single, double, triple, and quadruple substitutions. MP calculations that include energy corrections through $E^{(3)}$ are designated MP3 or MBPT(3), and those that include corrections through $E^{(4)}$ are MP4 or MBPT(4).

To do an MP electron-correlation calculation, one first chooses a basis set and carries out an SCF MO calculation to obtain Φ_0, E_{HF}, and virtual orbitals. One then evaluates $E^{(2)}$ (and perhaps higher corrections) by evaluating the integrals over spin-orbitals in (16.13) in terms of integrals over the basis functions. One ought to use a complete set of basis functions to expand the spin-orbitals. The SCF calculation will then produce the exact Hartree–Fock energy and will yield an infinite number of virtual orbitals. The first two sums in (16.13) will then contain an infinite number of terms. Of course, one always uses a finite, incomplete basis set, which yields a finite number of virtual orbitals, and the sums in (16.13) contain only a finite number of terms. One thus has a basis-set truncation error in addition to the error due to truncation of the MP perturbation energy at $E^{(2)}$ or $E^{(3)}$ or whatever.

In MP4 calculations, evaluation of the terms that involve triply substituted determinants is very time consuming, so these terms are sometimes neglected [even though their contribution to $E^{(4)}$ is not small], giving an approximation to MP4 that is designated MP4(SDQ), where SDQ indicates inclusion of single, double, and quadruple excitations. Evaluation of $E^{(5)}$ is extremely time consuming and so is almost never done except by specialists investigating the convergence of the series.

To save time in MP2, MP3, and MP4 computations, the **frozen-core** (FC) approximation is usually used. Here, terms involving excitations out of core orbitals are omitted.

MP3 calculations take a lot longer than MP2 calculations but provide little improvement over MP2 molecular properties and so are rarely done (except as part of an MP4 calculation). By far, the most common MP level used is MP2; the next most common is MP4.

MP2 calculations are much faster than CI calculations, and most ab initio programs (Section 15.14) can perform MP calculations. Relative times for geometry-optimization calculations on CH_3OH using the 6-31G* basis set (38 basis functions) and the cc-pVTZ basis set (116 basis functions) are (srdata.nist.gov/cccbdb/) 1 for HF/6-31G*, 1.4 for MP2(FC)/6-31G*, 26 for MP4(FC)/6-31G*, 19 for HF/cc-pVTZ, 46 for MP2(FC)/cc-pVTZ, and 1250 for MP4(FC)/cc-pVTZ, where FC means frozen core.

In addition to their computational efficiency, MP calculations truncated at any order can be shown to be size extensive (*Szabo and Ostlund*, Section 6.7.4). However, MP calculations are not variational and can produce an energy below the true energy. Currently, size extensivity is viewed as more important than being variational.

A study of the convergence of the frozen-core MP perturbation series for small atoms and molecules [J. Olsen et al., *J. Chem. Phys.*, **105**, 5082 (1996)] found that for the cc-pVDZ basis set, the series usually converged, but when the basis set was augmented with diffuse functions, the MP series often diverged. For example, for the Ne atom, the MPn/aug-cc-pVDZ contributions for $n > 16$ increased in magnitude as n increased.

(The high-order MPn contributions were not calculated directly but were obtained as byproducts of FCI calculations.) For the F^- ion, the divergence of the perturbation series affected the reliability of the MP4/aug-cc-pVDZ results. Large-basis-set MP calculations on small diatomic molecules showed the results "to be far from converged at MP4," with MP4 errors in R_e and ν_e often larger than MP2 errors [T. H. Dunning and K. A. Peterson, *J. Chem. Phys.*, **108**, 4761 (1998)]. For further examples of MP series divergence, see M. L. Leininger et al., *J. Chem. Phys.*, **112**, 9213 (2000). These studies raise doubts about the use of MP perturbation theory beyond MP2 to calculate molecular properties. A mathematical analysis of when MP calculations converge or diverge is given in A. V. Sergeev and D. Z. Goodson, *J. Chem. Phys.*, **124**, 094111 (2006). Because of the convergence problems with the MP method, quantum chemists have largely stopped doing MP4 calculations (except for the use of MP4 calculations in the G3 method—Section 16.5), but the MP2 method is still widely used.

As with CI calculations, MP calculations with small basis sets are of little practical value, and MP calculations should use a 6-31G* or larger basis set for useful results. For DZP basis sets, MP2 calculations on closed-shell molecules typically yield 85% to 95% of the basis-set correlation energy [R. J. Bartlett, *Ann. Rev. Phys. Chem.*, **32**, 359 (1981)] and substantially improve the accuracy of equilibrium-geometry and vibrational-frequency predictions.

Experience indicates that in most electron-correlation calculations, the basis-set truncation error is larger than the error due to truncation of the correlation treatment. For example, when one goes from a 6-31G* basis set to a TZ2P basis set, the errors in MP2-predicted equilibrium single-bond lengths are reduced by a factor of 2 or 3 [E. D. Simandiras et al., *J. Chem. Phys.*, **88**, 3187 (1988)], but when one goes from MP2/TZ2P to MP3/TZ2P calculations, no improvement in geometry accuracy is obtained [I. L. Alberts and N. C. Handy, *J. Chem. Phys.*, **89**, 2107 (1988)].

The energy gradient in MP2 calculations is readily evaluated analytically [see P. Pulay, *Adv. Chem. Phys.*, **69**, 276 (1987)], and this allows MP2 geometry optimization to be done easily and allows calculation of MP2 vibrational frequencies.

The *direct* and *semidirect* MP2 methods, like the corresponding SCF methods (Section 15.16), speed up calculations on large molecules by recalculating all or some of the electron-repulsion integrals as needed, instead of storing them externally on disk and then retrieving them. A semidirect MP2(FC)/TZP geometry optimization of buckminsterfullerene, C_{60}, involved 1140 basis functions and found bond lengths of 1.446 and 1.406 Å [M. Häser, J. Almlöf, and G. E. Scuseria, *Chem. Phys. Lett.*, **181**, 497 (1991)], in agreement with the experimental values 1.45 and 1.40 Å.

A method to speed up MP2 calculations on large molecules is the **local MP2** (LMP2) method of Saebø and Pulay [S. Saebø and P. Pulay, *Ann. Rev. Phys. Chem.*, **44**, 213 (1993)]. Here, instead of using canonical SCF MOs in the Hartree–Fock reference determinant Φ_0, one transforms to localized MOs (Section 15.8). Also, instead of using the virtual orbitals found in the SCF calculation as the orbitals a and b in (16.13) to which electrons are excited, one uses atomic orbitals that are orthogonal to the localized occupied MOs. Also, in (16.13), one includes only unoccupied orbitals a and b that are in the neighborhood of the localized MOs i and j.

The LMP2 method has been combined with the pseudospectral (PS) method (Section 15.16) to further reduce computation times [R. B. Murphy et al., *J. Chem. Phys.*, **103**, 1481 (1995)].

As noted in Section 15.16, calculation of 2-electron repulsion integrals in SCF MO calculations on large molecules can be speeded up by expanding products like $\phi_i(1)\phi_j(1)$ using an auxiliary basis set—the RI approximation. This same approximation is also used in the MP2 method, giving the RI-MP2 method. [For a review of the RI approximation, see R. A. Kendall and H. A. Früchtl, *Theor. Chem. Acc.*, **97**, 158 (1997).] The RI-MP2 method is also called the DF-MP2 method, where DF stands for **density fitting**, since it fits the one-electron density function $\phi_i(1)\phi_j(1)$ with a series.

The DF-MP2 method has been combined with the LMP2 method to give the DF-LMP2 method [H. J. Werner et al., *J. Chem. Phys.*, **118**, 8149 (2003)]. The DF-LMP2 method is so efficient, that for large molecules, the preceding Hartree–Fock part of the calculation takes a lot longer than the MP2 part of the calculation.

For species involving open-shell ground states (for example, O_2, NO_2, and OH), one can base an MP calculation on the unrestricted SCF wave function (Section 15.3), giving calculations designated UMP2, UMP3, and so on. Unrestricted SCF wave functions are not eigenfunctions of $\hat{S}^2$, and this "spin contamination" can sometimes produce serious errors in UMP-calculated quantities [K. Wolinski and P. Pulay, *J. Chem. Phys.*, **90**, 3647 (1989)]. Alternatively, several versions of open-shell MP perturbation theory that are based on the ROHF wave function have been developed. It is not clear which of these ROHF MP methods is best [T. D. Crawford, H. F. Schaefer, and T. J. Lee, *J. Chem. Phys.*, **105**, 1060 (1996)].

Another limitation of MP calculations is that, although they work well near the equilibrium geometry, they do not work well at geometries far from equilibrium. For example, DZ calculations on H_2O showed that at the equilibrium geometry an MP2 calculation obtained 94% of the basis-set correlation energy, but at a geometry with the bonds at twice their equilibrium lengths an MP2 calculation obtained only 83% of the basis-set correlation energy [W. D. Laidig et al., *Chem. Phys. Lett.*, **113**, 151 (1985)]; also, the MP energy series at this stretched geometry converged erratically [N. C. Handy et al., *Theor. Chim. Acta*, **68**, 87 (1985)].

A third limitation is that MP calculations are not generally applicable to excited electronic states.

Because of these limitations, MP calculations have not made CI calculations obsolete, and multireference CI calculations are widely used for excited states and for geometries far from equilibrium.

Because of its computational efficiency and good results for molecular properties, the MP2 method is one of the two most commonly used methods for including correlation effects on molecular ground-state equilibrium properties. (The other widely used method is the density-functional method—Section 16.4.)

Table 16.1 lists results of some MP calculations on H_2O.

The MP2-R12 and MP2-F12 Methods. Because we always use a finite, and hence, incomplete, basis set, Eq. (16.13) gives only an approximation to the MP2 quantity $E_0^{(2)}$ The aim of the MP2-R12 method is to essentially eliminate this basis-set-truncation error and give a value of $E_0^{(2)}$, that is close to the value that would be obtained with a complete basis set. The MP2-R12 method starts with the variation–perturbation inequality (9.38), where $\psi_g^{(0)}$ is the SCF MO wave function Φ_0, and u (which can be any well-behaved function) is an approximation to the MP first-order ground-state wave-function correction $\psi^{(1)}$. In the MP2 method, the first-order correction to the wave function contains terms involving Slater determinants with two electrons excited from occupied to vacant orbitals. In the MP2-R12 method, the first-order correction to the wave function used as u in (9.38) contains additional terms of the form $c_{ij}r_{ij}\Phi_0$ for all pairs of electrons (this is bit of an oversimplification), where c_{ij} is a variational parameter, r_{ij} is the distance between electrons i and j, and Φ_0 is the unperturbed Hartree–Fock function. (When terms that involve r_{ij} are introduced into the wave function, one says the wave function is **explicitly correlated**.) Recall that inclusion of r_{12} in trial variation functions can give extremely accurate results for the He atom (Section 9.4). Whereas only two-electron integrals occur in the MP2 method, the inclusion of r_{ij} in the MP2-R12 method produces a very large number of three- and four-electron integrals containing factors like $r_{12}r_{34}/r_{23}$. These integrals are evaluated by using the resolution-of-the-identity method to express these integrals in terms of two-electron integrals and an auxiliary basis set. (Mathematically, this is not exactly the same as the use of the RI equation in density-fitting methods.) Since the auxiliary basis set is necessarily incomplete, the three- and four-electron integrals are being approximated, but with a proper basis set, the errors introduced are quite small. The variational parameters are evaluated so as to minimize the left side of (9.38), which gives the MP2-R12 value of $E_0^{(2)}$

The MP2-R12 method cannot give a better result than would be given by the MP2 method with a complete basis set, and the hope was that introduction of the r_{ij} terms into the MP2 wave function would yield essentially the same result as would be obtained by MP2 with a complete basis set. Although MP2-R12 results are significantly improved compared with MP2 results with the same medium-size basis set, it turns out that MP2-R12 results still differ substantially from the complete-basis-set limit. The reason for this is that the form $c_{ij}r_{ij}\Phi_0$ linear in r_{ij} is not the optimum form to use. When r_{ij} is replaced in the wave function by $e^{-\beta r_{ij}}$, where β is a constant, results very close to the MP2 complete-basis-set limit are obtained with only a moderate-size basis set [A. J. May et al., *Phys. Chem. Chem. Phys.*, **7**, 2710 (2005)]. The value used for the constant β has only a small effect on the results, and the value $\beta = 1.4$ has been found to work well. When any function other than the linear function r_{ij} is introduced into the MP2 wave function, the method is called **MP2-F12**, indicating that some function $f(r_{12})$ of interelectronic distances is being used. Tests of several functions found that the function $e^{-\beta r_{ij}}$ worked best [D. P. Tew and W. Klopper, *J. Chem. Phys.*, **123**, 074101 (2005)].

The MP2-R12 and MP2-F12 methods have been combined with the DF-MP2 and LMP2 methods to give the DF-LMP2-R12 method [H.-J. Werner and F. R. Manby, *J. Chem. Phys.*, **124**, 054114 (2006)] and the DF-LMP2-F12 method [F. R. Manby et al., *J. Chem. Phys.*, **124**, 094103 (2006)].

Since MP2-F12 is much better than MP2-R12, it will likely replace MP2-R12.

The SCS-MP2 and SOS-MP2 Methods. Because of the orthogonality of the spin functions α and β, the integrals in the MP2 energy expression (16.13) and (16.14) will be zero unless either (1) the virtual spin-orbital u_a has the same spin function as the occupied spin-orbital u_i and u_b has the same spin function as u_j, or (2) u_a has the same spin function as u_j and u_b has the same spin function as u_i. Thus, if u_i and u_j both have spin function α, then u_a and u_b must both have spin function α; if u_i and u_j both have spin function β, then u_a and u_b must both have spin function β; if u_i and u_j have different spin functions, then u_a and u_b must have different spin functions. We can therefore divide the MP2 correlation energy (16.13) into a *same-spin* ($\alpha\alpha$, $\beta\beta$) component (also called a "triplet" component) and an *opposite-spin* ($\alpha\beta$, $\beta\alpha$) component (also called a "singlet" component). We have $E^{(2)} = E^{(2)}_{\text{OS}} + E^{(2)}_{\text{SS}}$, where OS and SS indicate opposite spin and same spin. Calculations show that $E^{(2)}_{\text{OS}}$ is typically 3 or 4 times $E^{(2)}_{\text{SS}}$.

To improve the accuracy of MP2 calculations, Grimme proposed the **spin-component-scaled MP2 (SCS-MP2) method**, which takes $E^{(2)} = p_{\text{OS}}E^{(2)}_{\text{OS}} + p_{\text{SS}}E^{(2)}_{\text{SS}}$, where the empirical parameters p_{OS} and p_{SS} are given the values $p_{\text{OS}} = 1.2$ and $p_{\text{SS}} = 1/3$ [S. Grimme, *J. Chem. Phys.*, **118**, 9095 (2003)]. These values were found by a least-squares fit to a test set of reaction energies calculated by the QCISD(T) method (Section 16.3) with a large basis set. The SCS-MP2 method gives results for several properties that are significantly improved over MP2 results at no additional computational cost.

The **scaled-opposite-spin MP2 (SOS-MP2) method** omits the $E^{(2)}_{\text{SS}}$ contribution entirely and takes $E^{(2)} = 1.3E^{(2)}_{\text{OS}}$ [Y. Jung et al., *J. Chem. Phys.*, **121**, 9793 (2004); *J. Comput. Chem.*, **28**, 1953 (2007)]. The SOS-MP2 method gives results slightly less accurate than the SCS-MP2 method, but is faster and can be applied to larger molecules than SCS-MP2. Of course, the use of empirical parameters makes the SCS-MP2 and SOS-MP2 methods no longer strictly ab initio methods.

The CASPT2 Method. The MP method applies perturbation theory to a zeroth-order wave function (reference function) that is a single Slater determinant. Instead of starting with an SCF wave function as the zeroth-order wave function, one can start with an MCSCF wave function (Section 16.1) as the zeroth-order function and apply perturbation theory to get a generalization of MP theory. The MCSCF function most commonly used for this purpose is a CASSCF wave function (Section 16.1). The choice of zeroth-order Hamiltonian is not unique, and the $\hat{H}_0$ used is more complicated than (16.9). Inclusion of energy corrections through $E^{(2)}$ gives the CASPT2 (complete active space second-order perturbation theory) method [K. Andersson, P.-Å. Malmqvist, and B. O. Roos, *J. Chem. Phys.*, **96**, 1218 (1992); K. Andersson and B. O. Roos in *Yarkony*, Part I, Chapter 2]. CASPT2 results, although quite good, are typically of lower quality than MRCI results, but CASPT2 involves significantly less computational effort than MRCI [M. L. Abrams and C. D. Sherrill, *J. Phys. Chem. A*, **107**, 5611 (2003)]. One can carry the perturbation treatment to higher orders, giving CASPT3, etc. However, multireference MP calculations are found to diverge more often than single-reference MP calculations [J. Olsen and M. P. Fülscher, *Chem. Phys. Lett.*, **326**, 225 (2000)].

An alternative choice is to take the unperturbed MCSCF wave function as the generalized valence-bond (GVB) wave function (Section 16.9). When combined with a PS-LMP2 treatment, this gives the PS-GVB-LMP2 method [R. A. Friesner et al., *J. Phys. Chem. A*, **103**, 1913 (1999)].

16.3 THE COUPLED-CLUSTER METHOD

The **coupled-cluster** (CC) **method** for dealing with a system of interacting particles was introduced around 1958 by Coester and Kümmel in the context of studying the atomic nucleus. CC methods for molecular electronic calculations were developed by Čížek, Paldus, Sinanoglu, and Nesbet in the 1960s and by Pople and co-workers and Bartlett and co-workers in the 1970s. For reviews of the CC method, see R. J. Bartlett, *J. Phys. Chem.*, **93**, 1697 (1989); R. J. Bartlett in *Yarkony*, Part II, Chapter 16; T. J. Lee and G. E.

Scuseria, in S. R. Langhoff (ed.), *Quantum Mechanical Electronic Structure Calculations with Chemical Accuracy*, Kluwer, 1995, pp. 47–108; T. Helgaker et al., *J. Phys. Org. Chem.*, **17**, 913 (2004).

The fundamental equation in CC theory is

$$\psi = e^{\hat{T}}\Phi_0 \tag{16.15}$$

where ψ is the exact nonrelativistic ground-state molecular electronic wave function, Φ_0 is the normalized ground-state Hartree–Fock wave function, the operator $e^{\hat{T}}$ is defined by the Taylor-series expansion

$$e^{\hat{T}} \equiv 1 + \hat{T} + \frac{\hat{T}^2}{2!} + \frac{\hat{T}^3}{3!} + \cdots = \sum_{k=0}^{\infty} \frac{\hat{T}^k}{k!} \tag{16.16}$$

and the **cluster operator** $\hat{T}$ (no connection with kinetic energy) is

$$\hat{T} \equiv \hat{T}_1 + \hat{T}_2 + \cdots + \hat{T}_n \tag{16.17}$$

where n is the number of electrons in the molecule and the operators $\hat{T}_1, \hat{T}_2, \ldots$ are defined below. Proof of (16.15) is omitted [see references in R. F. Bishop and H. G. Kümmel, *Physics Today*, March 1987, p. 52], but its plausibility will be shown below. The wave function ψ in (16.15) is not normalized (Prob. 16.12) but can be normalized at the end of the calculation.

The **one-particle excitation operator** $\hat{T}_1$ and the **two-particle excitation operator** $\hat{T}_2$ are defined by

$$\hat{T}_1\Phi_0 \equiv \sum_{a=n+1}^{\infty}\sum_{i=1}^{n} t_i^a \Phi_i^a, \qquad \hat{T}_2\Phi_0 \equiv \sum_{b=a+1}^{\infty}\sum_{a=n+1}^{\infty}\sum_{j=i+1}^{n}\sum_{i=1}^{n-1} t_{ij}^{ab}\Phi_{ij}^{ab} \tag{16.18}$$

where Φ_i^a is a singly excited Slater determinant with the occupied spin-orbital u_i replaced by the virtual spin-orbital u_a, and t_i^a is a numerical coefficient whose value depends on i and a and will be determined by requiring that Eq. (16.15) be satisfied. The operator $\hat{T}_1$ converts the Slater determinant $|u_1 \cdots u_n| = \Phi_0$ into a linear combination of all possible singly excited Slater determinants. Φ_{ij}^{ab} is a Slater determinant with the occupied spin-orbitals u_i and u_j replaced by the virtual spin-orbitals u_a and u_b, respectively; t_{ij}^{ab} is a numerical coefficient. Similar definitions hold for $\hat{T}_3, \ldots, \hat{T}_n$. Since no more than n electrons can be excited from the n-electron function Φ_0, no operators beyond $\hat{T}_n$ appear in (16.17). The limits in (16.18) are chosen so as to include all possible single and double excitations without duplication of any excitation. By definition, when $\hat{T}_1$ or $\hat{T}_2$ or ... operates on a determinant containing both occupied and virtual spin-orbitals, the resulting sum contains only determinants with excitations from those spin-orbitals that are occupied in Φ_0 and not from virtual spin-orbitals. The function $\hat{T}_1^2\Phi_0 \equiv \hat{T}_1(\hat{T}_1\Phi_0)$ contains only doubly excited Slater determinants, and $\hat{T}_2^2\Phi_0$ contains only quadruply excited determinants. When $\hat{T}_1$ operates on a determinant containing only virtual spin-orbitals, the result is zero, by definition.

The effect of the $e^{\hat{T}}$ operator in (16.15) is to express ψ as a linear combination of Slater determinants that include Φ_0 and all possible excitations of electrons from occupied to virtual spin-orbitals. A full CI calculation also expresses ψ as a linear combination involving all possible excitations, and we know that a full CI calculation with a complete basis set gives the exact ψ. Hence, it is plausible that Eq. (16.15) is valid.

The mixing into the wave function of Slater determinants with electrons excited from occupied to virtual spin-orbitals allows electrons to keep away from one another and thereby provides for electron correlation.

In the CC method, one works with individual Slater determinants, rather than with CSFs, but each CSF is a linear combination of one or a few Slater determinants, and the CC and CI methods can each be formulated either in terms of individual Slater determinants or in terms of CSFs.

The aim of a CC calculation is to find the coefficients $t_i^a, t_{ij}^{ab}, t_{ijk}^{abc}, \ldots$ for all *i, j, k*, ..., and all *a, b, c*, ... Once these coefficients (called **amplitudes**) are found, the wave function ψ in (16.15) is known.

To apply the CC method, two approximations are made. First, instead of using a complete, and hence infinite, set of basis functions, one uses a finite basis set to express the spin-orbitals in the SCF MO wave function. One thus has available only a finite number of virtual orbitals to use in forming excited determinants. As usual, we have a basis-set truncation error. Second, instead of including all the operators $\hat{T}_1, \hat{T}_2, \ldots, \hat{T}_n$, one approximates the operator $\hat{T}$ by including only some of these operators. Theory shows (*Wilson*, p. 222) that the most important contribution to $\hat{T}$ is made by $\hat{T}_2$. The approximation $\hat{T} \approx \hat{T}_2$ gives

$$\psi_{\text{CCD}} = e^{\hat{T}_2}\Phi_0$$

Inclusion of only $\hat{T}_2$ gives an approximate CC approach called the **coupled-cluster doubles** (CCD) **method**. Since $e^{\hat{T}_2} = 1 + \hat{T}_2 + \frac{1}{2}\hat{T}_2^2 + \cdots$, the wave function ψ_{CCD} contains determinants with double substitutions, quadruple substitutions, hextuple substitutions, and so on. Recall (Section 16.1) that quadruple substitutions are next in importance after double substitutions in a CI wave function. The treatment of quadruple substitutions in the CCD method is only approximate. The CCD quadruple excitations are produced by the operator $\frac{1}{2}\hat{T}_2^2$, and so the coefficients of the quadruply substituted determinants are determined as products of the coefficients of the doubly substituted determinants [see Eq. (16.18)], rather than being determined independently, as in the CI-SDTQ method. The CCD approximation of the coefficients of the quadruply substituted determinants turns out to be pretty accurate.

We need equations to find the CCD amplitudes. Substitution of $\psi = e^{\hat{T}}\Phi_0$ [Eq. (16.15)] in the Schrödinger equation $\hat{H}\psi = E\psi$ gives

$$\hat{H}e^{\hat{T}}\Phi_0 = Ee^{\hat{T}}\Phi_0 \tag{16.19}$$

Multiplication by Φ_0^* and integration gives

$$\langle\Phi_0|\hat{H}|e^{\hat{T}}\Phi_0\rangle = E\langle\Phi_0|e^{\hat{T}}\Phi_0\rangle \tag{16.20}$$

We have $e^{\hat{T}}\Phi_0 = (1 + \hat{T} + \frac{1}{2}\hat{T}^2 + \cdots)\Phi_0 = \Phi_0 + \hat{T}\Phi_0 + \frac{1}{2}\hat{T}^2\Phi_0 + \cdots$. Since [Eq. (16.17)] $\hat{T} = \hat{T}_1 + \hat{T}_2 + \cdots + \hat{T}_n$, the functions $\hat{T}\Phi_0, \frac{1}{2}\hat{T}^2\Phi_0$, and so on, contain only Slater determinants with at least one occupied orbital replaced by a virtual orbital. Because of the orthogonality of the spin-orbitals, all such excited Slater determinants are orthogonal to Φ_0, as can be seen by replacing $\sum_{i=1}^n \hat{f}_i$ with 1 in Table 11.3. Therefore, $\langle\Phi_0|e^{\hat{T}}\Phi_0\rangle = \langle\Phi_0|\Phi_0\rangle = 1$, and (16.20) becomes

$$\langle\Phi_0|\hat{H}|e^{\hat{T}}\Phi_0\rangle = E \tag{16.21}$$

Multiplication of the Schrödinger equation (16.19) by Φ_{ij}^{ab*} and integration gives

$$\langle \Phi_{ij}^{ab} | \hat{H} | e^{\hat{T}} \Phi_0 \rangle = E \langle \Phi_{ij}^{ab} | e^{\hat{T}} \Phi_0 \rangle \tag{16.22}$$

Use of (16.21) to eliminate E from (16.22) gives

$$\langle \Phi_{ij}^{ab} | \hat{H} | e^{\hat{T}} \Phi_0 \rangle = \langle \Phi_0 | \hat{H} | e^{\hat{T}} \Phi_0 \rangle \langle \Phi_{ij}^{ab} | e^{\hat{T}} \Phi_0 \rangle \tag{16.23}$$

So far, the treatment is exact. We now invoke the CCD approximation $\hat{T} \approx \hat{T}_2$, and Eqs. (16.21) and (16.23) become

$$E_{\text{CCD}} = \langle \Phi_0 | \hat{H} | e^{\hat{T}_2} \Phi_0 \rangle \tag{16.24}$$

$$\langle \Phi_{ij}^{ab} | \hat{H} | e^{\hat{T}_2} \Phi_0 \rangle = \langle \Phi_0 | \hat{H} | e^{\hat{T}_2} \Phi_0 \rangle \langle \Phi_{ij}^{ab} | e^{\hat{T}_2} \Phi_0 \rangle \tag{16.25}$$

Since these equations are approximate, the exact energy E has been replaced by the CCD energy E_{CCD}. Also, the coefficients t_{ij}^{ab} (produced when $e^{\hat{T}_2}$ operates on Φ_0) in these equations are approximate. The first integral on the right side of (16.25) is

$$\langle \Phi_0 | \hat{H} | e^{\hat{T}_2} \Phi_0 \rangle = \langle \Phi_0 | \hat{H} | (1 + \hat{T}_2 + \tfrac{1}{2} \hat{T}_2^2 + \cdots) \Phi_0 \rangle$$

$$\langle \Phi_0 | \hat{H} | e^{\hat{T}_2} \Phi_0 \rangle = \langle \Phi_0 | \hat{H} | \Phi_0 \rangle + \langle \Phi_0 | \hat{H} | \hat{T}_2 \Phi_0 \rangle + 0 = E_{\text{HF}} + \langle \Phi_0 | \hat{H} | \hat{T}_2 \Phi_0 \rangle \tag{16.26}$$

where E_{HF} is the Hartree–Fock (or SCF) energy. The integral $\langle \Phi_0 | \hat{H} | \frac{1}{2} \hat{T}_2^2 \Phi_0 \rangle$ and similar integrals with higher powers of $\hat{T}_2$ vanish because $\hat{T}_2^2 \Phi_0$ contains only quadruply excited determinants; hence, $\hat{T}_2^2 \Phi_0$ differs from Φ_0 by four spin-orbitals, and the Condon–Slater rules (Table 11.3) show that the matrix elements of $\hat{H}$ between Slater determinants differing by four spin-orbitals are zero. Similar use of the Condon–Slater rules gives for the integral on the left of (16.25) (Prob. 16.13)

$$\langle \Phi_{ij}^{ab} | \hat{H} | e^{\hat{T}_2} \Phi_0 \rangle = \langle \Phi_{ij}^{ab} | \hat{H} | (1 + \hat{T}_2 + \tfrac{1}{2} \hat{T}_2^2) \Phi_0 \rangle \tag{16.27}$$

Also, use of the orthogonality of different Slater determinants gives (Prob. 16.13)

$$\langle \Phi_{ij}^{ab} | e^{\hat{T}_2} \Phi_0 \rangle = \langle \Phi_{ij}^{ab} | \hat{T}_2 \Phi_0 \rangle \tag{16.28}$$

Use of (16.26) to (16.28) in (16.25) gives

$$\langle \Phi_{ij}^{ab} | \hat{H} | (1 + \hat{T}_2 + \tfrac{1}{2} \hat{T}_2^2) \Phi_0 \rangle = (E_{\text{HF}} + \langle \Phi_0 | \hat{H} | \hat{T}_2 \Phi_0 \rangle) \langle \Phi_{ij}^{ab} | \hat{T}_2 \Phi_0 \rangle \tag{16.29}$$

$$i = 1, \ldots, n - 1; \quad j = i + 1, \ldots, n; \quad a = n + 1, \ldots; \quad b = a + 1, \ldots$$

Next, one uses the definition (16.18) of $\hat{T}_2$ to eliminate $\hat{T}_2$ from (16.29). $\hat{T}_2 \Phi_0$ is a multiple sum involving $t_{ij}^{ab} \Phi_{ij}^{ab}$, and $\hat{T}_2^2 \Phi_0 \equiv \hat{T}_2(\hat{T}_2 \Phi_0)$ is a multiple sum involving $t_{ij}^{ab} t_{kl}^{cd} \Phi_{ijkl}^{abcd}$. For each unknown amplitude t_{ij}^{ab}, there is one equation in (16.29), so the number of equations is equal to the number of unknowns. After replacing $\hat{T}_2 \Phi_0$ and $\hat{T}_2^2 \Phi_0$ by these multiple sums, one expresses the resulting integrals involving Slater determinants in terms of integrals over the spin-orbitals by using the Condon–Slater rules (Table 11.3); the integrals over spin-orbitals are then expressed in terms of integrals over the basis functions. The net result is a set of simultaneous nonlinear equations for the unknown amplitudes t_{ij}^{ab}, whose form is (see *Carsky and Urban*, pages 96–97)

$$\sum_{s=1}^{m} a_{rs} x_s + \sum_{t=2}^{m} \sum_{s=1}^{t-1} b_{rst} x_s x_t + c_r = 0, \qquad r = 1, 2, \ldots, m \tag{16.30}$$

where $x_1, x_2, \ldots, x_m$ are the unknowns t_{ij}^{ab}, the quantities a_{rs}, b_{rst}, and c_r are constants that involve orbital energies and electron-repulsion integrals over the basis functions, and m is the number of unknown amplitudes t_{ij}^{ab}. The set of equations (16.30) is solved iteratively, starting with an initial estimate for the x's found by neglecting many of the terms in (16.30). Once the x's (that is, the t_{ij}^{ab}'s) are known, the wave function is known from $\psi_{\text{CCD}} = e^{\hat{T}_2}\Phi_0$ [the displayed equation before (16.19)] and the energy is found from (16.24).

The next step in improving the CCD method is to include the operator $\hat{T}_1$ and take $\hat{T} = \hat{T}_1 + \hat{T}_2$ in $e^{\hat{T}}$; this gives the **CC singles and doubles** (CCSD) **method**. With $\hat{T} = \hat{T}_1 + \hat{T}_2 + \hat{T}_3$, one obtains the **CC singles, doubles, and triples** (CCSDT) **method** [J. Noga and R. J. Bartlett, *J. Chem. Phys.*, **86**, 7041 (1987)]. CCSDT calculations give very accurate results for correlation energies but are very demanding computationally and are only feasible for small molecules. Several approximate forms of CCSDT have been developed and bear such designations as CCSD(T), CCSDT-1, and CCSD[T] [formerly called CCSD + T(CCSD)]. The most widely used such approximation is CCSD(T), coupled cluster with inclusion of single and double excitations and perturbative inclusion of triple excitations.

Table 16.1 lists results of some CC calculations on H_2O.

The CCD, CCSD, CCSD(T), and CCSDT methods are size extensive but not variational. Analytic gradients are available for these methods. Usually, the frozen-core (FC) approximation is used in CC calculations. Here, excitations of inner-shell electrons are omitted.

For open-shell ground states (for example, OH, O_2), one can base a CC calculation on either a UHF or an ROHF wave function. Spin contamination is less serious for UHF-based CC wave functions than for UHF-MP2 wave functions.

A few versions of CC theory have been developed for treating excited states. One such version, the equation-of-motion (EOM) CCSD, has given very good results for vertical excitation energies. Analytic gradients for the EOM-CCSD method are available, allowing calculation of geometries and vibrational frequencies of excited states. For details see R. J. Bartlett in *Yarkony*, Part II, Chapter 16, Section 9.

Pople and co-workers developed the nonvariational **quadratic configuration-interaction** (QCI) **method**, which is intermediate between the CC and CI methods. The QCI method exists in the size-consistent forms QCISD, which is an approximation to CCSD, and QCISD(T), which is similar to CCSD[T]. QCISD(T) has given excellent results for correlation energies in many calculations [J. A. Pople, M. Head-Gordon, and K. Raghavachari, *J. Chem. Phys.*, **87**, 5968 (1987); **90**, 4635 (1989); K. Raghavachari and G. W. Trucks, *J. Chem. Phys.*, **91**, 1062, 2457 (1989); J. Paldus et al., *J. Chem. Phys.*, **90**, 4356 (1989)]. Unfortunately, QCISD(T) occasionally fails dramatically [L. A. Curtiss et al., *Chem. Phys. Lett*, **359**, 390 (2002)], so CCSD(T) calculations are currently preferred to QCISD(T) calculations.

The exact basis-set correlation energy (Section 16.1) is obtained by full CI, by CC calculations with $\hat{T}$ not truncated, and by MP perturbation-theory carried to infinite order (provided the series converges, which isn't always true). Full CI calculations with DZP basis sets were done for the H_2O, HF, and BH molecules at their equilibrium geometries and at geometries where the bond lengths were stretched to $1.5R_e$ and to $2R_e$. By comparing the DZP energies obtained by a partial correlation method with the full CI energies for these molecules, one can judge the accuracy of the correlation method for calculating molecular energies. The average absolute energy errors (deviations from FCI) in millihartrees (one millihartree corresponds to 0.627 kcal/mol) for various methods at the equilibrium geometry (At R_e) and at all three geometries (All R) are as follows (R. J. Bartlett in *Yarkony*, Part II, Chapter 16):

Method	MP2	CISD	MP3	CISDT	MP4-SDQ	CCD	CCSD	QCISD	MP4
At R_e	16.5	9.2	7.9	7.1	4.3	3.8	3.0	2.7	2.1
All R	27.8	22.0	22.9	16.7	11.3	12.8	7.1	6.4	5.9

Method	MP5	MP6	CCSD(T)	CCSDT-1	QCISD(T)	CCSDT	CISDTQ	CCSDTQ
At R_e	1.3	0.5	0.5	0.4	0.4	0.3	0.2	0.01
All R	5.1	1.4	1.2	0.9	0.8	0.8	1.1	0.03

The very accurate CCSDTQ, CCSDT, CISDTQ, CCSDT-1, and MP6 methods are much too computationally demanding to be used regularly. The CCSD(T) method "is widely considered to be the most accurate practicable method at hand" [A. Hesselmann et al., *J. Chem. Phys.*, **122**, 014103 (2005)]. A 1996 review article gave the maximum feasible molecular size for a CCSD(T) calculation requiring 10 energy and gradient evaluations as 8 to 12 first-row nonhydrogen atoms for a DZP basis set [M. Head-Gordon, *J. Phys. Chem.*, **100**, 13213 (1996)]. Advances since then have allowed larger molecules to be treated by CCSD(T).

Most of the methods used to improve the MP2 method (Section 16.2) have also been applied to the coupled-cluster method. Thus, the use of localized rather than canonical SCF MOs has given the LCCSD and LCCSD(T) methods (where the L is for local), which are much faster than the usual CC methods [M. Schütz and H.-J. Werner, *J. Chem. Phys.*, **114**, 661 (2001) and references cited therein]. An approximate version of LCCSD(T) called LCCSD(T0) has given "breathtaking" speedups of factors of 1000 to 10^6 for large molecules over conventional CCSD(T) and has allowed a single-point LCCSD(T0)/cc-pVDZ calculation on indinavir, a 92-atom molecule with 865 basis functions in this basis set [M. Schütz, *J. Chem. Phys.*, **113**, 9986 (2000)] and on a certain enzyme-catalyzed reaction with 49 atoms treated quantum mechanically using 1294 basis functions (mainly the cc-pVTZ set) [F. Claeyssens et al., *Angew. Chem. Int. Ed.*, **45**, 1 (2006)].

A disadvantage of such local correlation methods is that the approximation of neglecting interactions between distant orbitals introduces small discontinuities into the molecule's potential-energy surface, which might interfere with geometry optimization. A computationally efficient version of local CCSD that avoids such discontinuities has been developed [J. E. Subotnik et al., *J. Chem. Phys.*, **125**, 074116 (2006)].

As is done in the HF and MP methods, computation of electron-repulsion integrals in CC calculations can be speeded up by using the density-fitting (resolution-of-the-identity) method [M. Schütz and F. R. Manby, *Phys. Chem. Chem. Phys.*, **5**, 3349 (2003)].

As in the MP method, terms linear in interelectronic distances r_{ij} can be introduced into the CC wave function so as to reduce the basis-set truncation error, thereby giving the CCSD-R12 and CCSD(T)-R12 methods [J. Noga and P. Valiron, in *Computational Chemistry: Reviews of Current Trends*, Vol. 7, J. Leszczynski (Ed.), p. 131, World Scientific (2002)]. In these methods, the operator $\hat{T}$ in (16.15) is replaced by

$\hat{T} + \hat{R}$, where the operator $\hat{R}$ contains integrals that have the interelectronic distance r_{12} multiplied by various occupied and virtual orbitals.

Another way to reduce the basis-set truncation error is to use a formula that attempts to extrapolate incomplete-basis-set results to the complete basis set limit. One of many such approximate formulas is that due to Helgaker and co-workers [A. Halkier et al., *Chem. Phys. Lett.*, **286**, 243 (1998)]

$$E_n^{\text{corr}} = E_\infty^{\text{corr}} + A/n^3 \tag{16.31}$$

where E_n^{corr} is the correlation energy found from a CC calculation method [such as CCSD or CCSD(T)] using the cc-pVnZ basis set (or a related basis set such as aug-cc-pVnZ or cc-pCVnZ), A is a parameter, and E_∞^{corr} is the estimate of the correlation energy that would be found in the complete-basis-set (CBS) limit. (Note that $E_n^{\text{corr}} = E_n - E_n^{\text{HF}}$, where E_n and E_n^{HF} are the energies found by a CC method and by the Hartree–Fock method, respectively, with both calculations using the cc-pVnZ basis set.) Equation (16.31) needs to be combined with an extrapolation formula such as (15.23) for the estimated Hartree–Fock limit so as to obtain the estimated CBS limit for the total energy, according to $E_\infty = E_\infty^{\text{HF}} + E_\infty^{\text{corr}}$. Combining the results of cc-pVnZ and cc-pV(n − 1)Z CC calculations to eliminate A, we get from (16.31) (Prob. 16.14)

$$E_\infty^{\text{corr}} = \frac{n^3 E_n^{\text{corr}} - (n-1)^3 E_{n-1}^{\text{corr}}}{n^3 - (n-1)^3} \tag{16.32}$$

which allows E_∞^{corr} to be estimated from calculations with two correlation-consistent basis sets. Use of the cc-pVDZ and cc-pVTZ pair in (16.32) is found to give much lower accuracy than use of the cc-pVTZ and cc-pVQZ pair.

To overcome the failures of the CCSD(T) method in dealing with bond-breaking and open-shell species, several multireference CC theories have been developed, but these theories often show problem behavior. The reduced multireference CCSD(T) method [RMR CCSD(T)] has shown promising results in calculations on small molecules [X. Li and J. Paldus, *J. Chem. Phys.*, **124**, 174101 (2006)].

16.4 DENSITY-FUNCTIONAL THEORY

The electronic wave function of an n-electron molecule depends on $3n$ spatial and n spin coordinates. Since the Hamiltonian operator (15.10) contains only one-and two-electron spatial terms, one finds that the molecular energy can be written in terms of integrals involving only six spatial coordinates (*Pilar*, Section 10–5). In a sense, the wave function of a many-electron molecule contains more information than is needed and is lacking in direct physical significance. This has prompted the search for functions that involve fewer variables than the wave function and that can be used to calculate the energy and other properties.

The Hohenberg–Kohn Theorem. In 1964, Pierre Hohenberg and Walter Kohn proved that for molecules with a nondegenerate ground state, the ground-state molecular energy, wave function, and all other molecular electronic properties are uniquely determined by the ground-state electron probability density $\rho_0(x, y, z)$ (Section 14.1), a function of only three variables [P. Hohenberg and W. Kohn, *Phys. Rev.*, **136**, B864 (1964)].

(The zero subscript indicates the ground state.) One says that the ground-state electronic energy E_0 is a functional of ρ_0 and writes $E_0 = E_0[\rho_0]$, where the square brackets denote a functional relation. **Density-functional theory** (DFT) attempts to calculate E_0 and other ground-state molecular properties from the ground-state electron density ρ_0.

What is a functional? Recall that a *function* $f(x)$ is a rule that associates a number with each value of the variable x for which the function f is defined. For example, the function $f(x) = x^2 + 1$ associates the number 10 with the value 3 of x and associates a number with each other value of x. A **functional** $F[f]$ is a rule that associates a number with each function f. For example, the functional $F[f] = \int_{-\infty}^{\infty} f^*(x)f(x)\,dx$ associates a number, found by integration of $|f|^2$ over all space, with each quadratically integrable function $f(x)$. The variational integral $W[\phi] = \langle\phi|\hat{H}|\phi\rangle/\langle\phi|\phi\rangle$ is a functional of the variation function ϕ and gives a number for each well-behaved ϕ.

The proof of the Hohenberg–Kohn theorem is as follows. The ground-state electronic wave function ψ_0 of an n-electron molecule is an eigenfunction of the purely electronic Hamiltonian of Eq. (13.5), which, in atomic units, is

$$\hat{H} = -\frac{1}{2}\sum_{i=1}^{n} \nabla_i^2 + \sum_{i=1}^{n} v(\mathbf{r}_i) + \sum_{j}\sum_{i>j}\frac{1}{r_{ij}} \tag{16.33}$$

$$v(\mathbf{r}_i) = -\sum_{\alpha}\frac{Z_\alpha}{r_{i\alpha}} \tag{16.34}$$

The quantity $v(\mathbf{r}_i)$, the potential energy of interaction between electron i and the nuclei, depends on the coordinates x_i, y_i, z_i of electron i and on the nuclear coordinates. Since the electronic Schrödinger equation is solved for fixed locations of the nuclei, the nuclear coordinates are not variables for the electronic Schrödinger equation. Thus, $v(\mathbf{r}_i)$ in the electronic Schrödinger equation is a function of only x_i, y_i, z_i, which we indicate by using the vector notation of Section 5.2. In DFT, $v(\mathbf{r}_i)$ is called the **external potential** acting on electron i, since it is produced by charges external to the system of electrons.

Once the external potential $v(\mathbf{r}_i)$ and the number of electrons n are specified, the electronic wave functions and allowed energies of the molecule are determined as the solutions of the electronic Schrödinger equation. Hohenberg and Kohn proved that for systems with a nondegenerate ground state, the ground-state electron probability density $\rho_0(\mathbf{r})$ determines the external potential (except for an arbitrary additive constant) and determines the number of electrons. Hence, the ground-state wave function and energy (and, for that matter, all excited-state wave functions and energies) are determined by the ground-state electron density.

To see that $\rho_0(\mathbf{r})$ determines the number of electrons, we integrate (14.5) over all space and use the normalization of ψ to get $\int \rho_0(\mathbf{r})\,d\mathbf{r} = n$.

To see that $\rho_0(\mathbf{r})$ determines the external potential $v(\mathbf{r}_i)$, we suppose that this is false and that there are two external potentials v_a and v_b (differing by more than a constant) that each give rise to the same ground-state electron density ρ_0. Let $\hat{H}_a$ and $\hat{H}_b$ be the n-electron Hamiltonians (16.33) corresponding to $v_a(\mathbf{r}_i)$ and $v_b(\mathbf{r}_i)$, where v_a and v_b are not necessarily given by (16.34); they can be any external potential. Let $\psi_{0,a}$ and $\psi_{0,b}$ and $E_{0,a}$ and $E_{0,b}$ be the normalized ground-state wave functions and energies for these Hamiltonians. [Note that even if $\hat{H}_a$ is a molecular electronic Hamiltonian with v_a

given by (16.34), $v_b(\mathbf{r}_i)$ is not restricted to the form (16.34) but can be any function of $\mathbf{r}_i$.] $\psi_{0,a}$ and $\psi_{0,b}$ must be different functions, since they are eigenfunctions of Hamiltonians that differ by more than an additive constant (Prob. 16.16). If the ground state is nondegenerate, then there is only one normalized function, the exact ground-state wave function ψ_0, that gives the exact ground state energy E_0 when used as a trial variation function, and, according to the variation theorem, use of any normalized well-behaved function that differs from ψ_0 will make the variational integral greater than E_0 (Prob. 8.13): that is, $\langle\phi|\hat{H}|\phi\rangle > E_0$ if $\phi \neq \psi_0$ and the ground state is nondegenerate. Therefore, use of $\psi_{0,b}$ as a trial function with the Hamiltonian $\hat{H}_a$ gives

$$E_{0,a} < \langle\psi_{0,b}|\hat{H}_a|\psi_{0,b}\rangle = \langle\psi_{0,b}|\hat{H}_a + \hat{H}_b - \hat{H}_b|\psi_{0,b}\rangle = \langle\psi_{0,b}|\hat{H}_a - \hat{H}_b|\psi_{0,b}\rangle + \langle\psi_{0,b}|\hat{H}_b|\psi_{0,b}\rangle$$

The Hamiltonians $\hat{H}_a$ and $\hat{H}_b$ differ only in their external potentials v_a and v_b, so $\hat{H}_a - \hat{H}_b = \sum_{i=1}^{n}[v_a(\mathbf{r}_i) - v_b(\mathbf{r}_i)]$ and we have

$$E_{0,a} < \left\langle \psi_{0,b} \left| \sum_{i=1}^{n} [v_a(\mathbf{r}_i) - v_b(\mathbf{r}_i)] \right| \psi_{0,b} \right\rangle + E_{0,b}$$

The quantities $v_a(\mathbf{r}_i)$ and $v_b(\mathbf{r}_i)$ are one-electron operators, and using Eq. (14.8), we get

$$E_{0,a} < \int \rho_{0,b}(\mathbf{r})[v_a(\mathbf{r}) - v_b(\mathbf{r})]\,d\mathbf{r} + E_{0,b}$$

where, since the integration is over $\psi_{0,b}$, we get the electron density $\rho_{0,b}$ corresponding to $\psi_{0,b}$. If we go through the same reasoning with a and b interchanged, we get

$$E_{0,b} < \int \rho_{0,a}(\mathbf{r})[v_b(\mathbf{r}) - v_a(\mathbf{r})]\,d\mathbf{r} + E_{0,a}$$

By hypothesis, the two different wave functions give the same electron density: $\rho_{0,a} = \rho_{0,b}$. Putting $\rho_{0,a} = \rho_{0,b}$ and adding the last two inequalities, the two integrals cancel and we get $E_{0,a} + E_{0,b} < E_{0,b} + E_{0,a}$. This result is false, so our initial assumption that two different external potentials could produce the same ground-state electron density must be false. Hence, the ground-state electron probability density ρ_0 determines the external potential (to within an additive constant that simply affects the zero level of energy) and also determines the number of electrons. Hence ρ_0 determines the molecular electronic Hamiltonian and so determines the ground-state wave function, energy, and other properties.

The ground-state electronic energy E_0 is thus a functional of the function $\rho_0(\mathbf{r})$, which we write as $E_0 = E_v[\rho_0]$, where the v subscript emphasizes the dependence of E_0 on the external potential $v(\mathbf{r})$, which differs for different molecules.

The purely electronic Hamiltonian (13.5) is the sum of electronic kinetic-energy terms, electron–nuclear attractions, and electron–electron repulsions. Taking the average of (13.5) for the ground state, we have $E = \overline{T} + \overline{V}_{Ne} + \overline{V}_{ee}$, where, for notational convenience, overbars instead of angular brackets have been used to denote averages. Each of the average values in this equation is a molecular property determined by the ground-state electronic wave function, which, in turn, is determined by $\rho_0(\mathbf{r})$. Therefore, each of these averages is a functional of ρ_0:

$$E_0 = E_v[\rho_0] = \overline{T}[\rho_0] + \overline{V}_{Ne}[\rho_0] + \overline{V}_{ee}[\rho_0]$$

From (13.1), $\hat{V}_{Ne} = \sum_{i=1}^{n} v(\mathbf{r}_i)$, where $v(\mathbf{r}_i) = -\sum_\alpha Z_\alpha/\mathbf{r}_{i\alpha}$ in atomic units, so

$$\overline{V}_{Ne} = \left\langle \psi_0 \middle| \sum_{i=1}^{n} v(\mathbf{r}_i) \middle| \psi_0 \right\rangle = \int \rho_0(\mathbf{r})v(\mathbf{r})\, d\mathbf{r} \tag{16.35}$$

where (14.8) was used, and where $v(\mathbf{r})$ is the nuclear attraction potential-energy function (16.34) for an electron located at point $\mathbf{r}$. Thus $\overline{V}_{Ne}[\rho_0]$ is known, but the functionals $\overline{T}[\rho_0]$ and $\overline{V}_{ee}[\rho_0]$ are unknown. We have

$$E_0 = E_v[\rho_0] = \int \rho_0(\mathbf{r})v(\mathbf{r})\, d\mathbf{r} + \overline{T}[\rho_0] + \overline{V}_{ee}[\rho_0] = \int \rho_0(\mathbf{r})v(\mathbf{r})\, d\mathbf{r} + F[\rho_0] \tag{16.36}$$

where the functional $F[\rho_0]$ defined by $F[\rho_0] \equiv \overline{T}[\rho_0] + \overline{V}_{ee}[\rho_0]$, is independent of the external potential. Equation (16.36) does not provide a practical way to calculate E_0 from ρ_0, because the functional $F[\rho_0]$ is unknown.

The Hohenberg–Kohn Variational Theorem. To change (16.36) from a formal relation to a practical tool, we need a second theorem proven by Hohenberg and Kohn, and an approach developed by Kohn and Sham. Hohenberg and Kohn proved that for every trial density function $\rho_{tr}(\mathbf{r})$ that satisfies $\int \rho_{tr}(\mathbf{r})\, d\mathbf{r} = n$ and $\rho_{tr}(\mathbf{r}) \geq 0$ for all $\mathbf{r}$, the following inequality holds: $E_0 \leq E_v[\rho_{tr}]$, where E_v is the energy functional in (16.36). Since $E_0 = E_v[\rho_0]$, where ρ_0 is the true ground-state electron density, *the true ground-state electron density minimizes the energy functional* $E_v[\rho_{tr}]$ (just as the true normalized ground-state wave function minimizes the variational integral).

The proof of this Hohenberg–Kohn variational theorem is as follows. Let ρ_{tr} satisfy the above two conditions of integrating to n and being nonnegative. By the Hohenberg–Kohn theorem, ρ_{tr} determines the external potential v_{tr}, and this in turn determines the wave function ψ_{tr} that corresponds to the density ρ_{tr}. (Actually, this is only true if there exists an external potential v_{tr} that will give rise to an antisymmetric wave function that corresponds to ρ_{tr}. If this condition holds, ρ_{tr} is said to be *v-representable.* It turns out that not all ρ_{tr}'s are v-representable. This has not caused any practical difficulties in applications of DFT. Also, Levy has reformulated the Hohenberg–Kohn theorems in a way that eliminates the need for v-representability. See *Parr and Yang*, Sections 3.3, 3.4, and 7.3.) Let us use the wave function ψ_{tr} that corresponds to ρ_{tr} as a trial variation function for the molecule with Hamiltonian $\hat{H}$. The variation theorem gives

$$\langle \psi_{tr}|\hat{H}|\psi_{tr}\rangle = \left\langle \psi_{tr} \middle| \hat{T} + \hat{V}_{ee} + \sum_{i=1}^{n} v(\mathbf{r}_i) \middle| \psi_{tr} \right\rangle \geq E_0 = E_v[\rho_0]$$

Using the fact that the average kinetic and potential energies are functionals of the electron density, and using (16.35) with ψ_0 replaced by ψ_{tr}, this last equation becomes

$$\overline{T}[\rho_{tr}] + \overline{V}_{ee}[\rho_{tr}] + \int \rho_{tr}v(\mathbf{r})\, d\mathbf{r} \geq E_v[\rho_0] \tag{16.37}$$

The functionals $\overline{T}$ and $\overline{V}_{ee}$ are the same in (16.36) and (16.37) although the functions ρ_0 and ρ_{tr} differ. The left side of (16.37) differs from the corresponding expression in (16.36) only by having ρ_0 replaced by ρ_{tr}. Use of (16.36) with ρ_0 replaced by ρ_{tr} in (16.37) gives $E_v[\rho_{tr}] \geq E_v[\rho_0]$, which proves that no trial electron density can give a lower ground-state energy than the true ground-state electron density.

Hohenberg and Kohn proved their theorems only for nondegenerate ground states. Subsequently, Levy proved the theorems for degenerate ground states (see *Parr and Yang*, Section 3.4).

The Kohn–Sham (KS) Method. If we know the ground-state electron density $\rho_0(\mathbf{r})$, the Hohenberg–Kohn theorem tells us that it is possible in principle to calculate all the ground-state molecular properties from ρ_0, without having to find the molecular wave function. [In the traditional quantum-mechanical approach, one first finds the wave function and then finds ρ by integration; Eq. (14.5).] The Hohenberg–Kohn theorem does not tell us *how* to calculate E_0 from ρ_0 [since the functional F in (16.36) is unknown], nor does it tell us how to find ρ_0 without first finding the wave function. A key step toward these goals was taken in 1965 when Kohn and Sham devised a practical method for finding ρ_0 and for finding E_0 from ρ_0 [W. Kohn and L. J. Sham, *Phys. Rev.*, **140**, A1133 (1965)]. Their method is capable, in principle, of yielding exact results, but because the equations of the Kohn–Sham (KS) method contain an unknown functional that must be approximated, the KS formulation of DFT yields approximate results.

Kohn and Sham considered a fictitious reference system (denoted by the subscript s and often called *the noninteracting system*) of n noninteracting electrons that each experience the same external potential-energy function $v_s(\mathbf{r}_i)$, where $v_s(\mathbf{r}_i)$ is such as to make the ground-state electron probability density $\rho_s(\mathbf{r})$ of the reference system equal to the exact ground-state electron density $\rho_0(\mathbf{r})$ of the molecule we are interested in; $\rho_s(\mathbf{r}) = \rho_0(\mathbf{r})$. Since Hohenberg and Kohn proved that the ground-state probability-density function determines the external potential, once $\rho_s(\mathbf{r})$ is defined for the reference system, the external potential $v_s(\mathbf{r}_i)$ in the reference system is uniquely determined, although we might not know how to actually find it. The electrons do not interact with one another in the reference system, so the Hamiltonian of the reference system is

$$\hat{H}_s = \sum_{i=1}^{n}\left[-\tfrac{1}{2}\nabla_i^2 + v_s(\mathbf{r}_i)\right] \equiv \sum_{i=1}^{n}\hat{h}_i^{\mathrm{KS}}, \qquad \text{where} \quad \hat{h}_i^{\mathrm{KS}} \equiv -\tfrac{1}{2}\nabla_i^2 + v_s(\mathbf{r}_i) \tag{16.38}$$

$\hat{h}_i^{\mathrm{KS}}$ is the one-electron Kohn–Sham Hamiltonian. Use of a fictitious system of noninteracting electrons should not be too disturbing. Recall that we used a system of noninteracting electrons in the Section 9.3 perturbation treatment of the He atom. We can relate the fictitious Kohn–Sham reference system to the real molecule by writing the Hamiltonian $\hat{H}_\lambda = \hat{T} + \sum_i v_\lambda(\mathbf{r}_i) + \lambda\hat{V}_{ee}$, where the parameter λ ranges from 0 (no interelectronic repulsions, which is the reference system) to 1 (the real molecule), and v_λ is defined as the external potential that will make the ground-state electron density of the system with Hamiltonian $\hat{H}_\lambda$ equal to that of the real molecule's ground state.

Since the reference system s consists of noninteracting particles, the results of Section 6.2 and the antisymmetry requirement show that the ground-state wave function $\psi_{s,0}$ of the reference system is the antisymmetrized product (Slater determinant) of the lowest-energy Kohn–Sham spin-orbitals u_i^{KS} of the reference system, where the spatial part $\theta_i^{\mathrm{KS}}(\mathbf{r}_i)$ of each spin-orbital is an eigenfunction of the one-electron operator $\hat{h}_i^{\mathrm{KS}}$:

$$\psi_{s,0} = |u_1^{\mathrm{KS}}u_2^{\mathrm{KS}}\cdots u_n^{\mathrm{KS}}|, \qquad u_i^{\mathrm{KS}} = \theta_i^{\mathrm{KS}}(\mathbf{r}_i)\sigma_i \tag{16.39}$$

$$\hat{h}_i^{\mathrm{KS}}\theta_i^{\mathrm{KS}} = \varepsilon_i^{\mathrm{KS}}\theta_i^{\mathrm{KS}} \tag{16.40}$$

where σ_i is a spin function (either α or β) and the $\varepsilon_i^{\mathrm{KS}}$'s are Kohn–Sham orbital energies.

For a closed-shell ground state, the electrons are paired in the Kohn–Sham orbitals, with two electrons of opposite spin having the same spatial Kohn–Sham orbital (as in the RHF method).

Kohn and Sham rewrote the Hohenberg–Kohn equation (16.36) as follows. Let $\Delta\overline{T}$ be defined by

$$\Delta\overline{T}[\rho] \equiv \overline{T}[\rho] - \overline{T}_s[\rho] \tag{16.41}$$

where, for convenience, the zero subscript on ρ is omitted in this and many subsequent equations. $\Delta\overline{T}$ is the difference in the average ground-state electronic kinetic energy between the molecule and the reference system of noninteracting electrons with electron density equal to that in the molecule. Let

$$\Delta\overline{V}_{ee}[\rho] \equiv \overline{V}_{ee}[\rho] - \frac{1}{2}\iint \frac{\rho(\mathbf{r}_1)\rho(\mathbf{r}_2)}{r_{12}}\,d\mathbf{r}_1\,d\mathbf{r}_2 \tag{16.42}$$

where r_{12} is the distance between points x_1, y_1, z_1 and x_2, y_2, z_2. The quantity $\frac{1}{2}\iint \rho(\mathbf{r}_1)\rho(\mathbf{r}_2)r_{12}^{-1}\,d\mathbf{r}_1\,d\mathbf{r}_2$ is the classical expression (in atomic units) for the electrostatic interelectronic repulsion energy if the electrons were smeared out into a continuous distribution of charge with electron density ρ. The charge dQ_1 in a tiny volume element $d\mathbf{r}_1$ of such a distribution is $dQ_1 = -e\rho(\mathbf{r}_1)\,d\mathbf{r}_1$ and the potential energy of repulsion between dQ_1 and the charge in the volume element $d\mathbf{r}_2$ located at $\mathbf{r}_2$ is $e^2 r_{12}^{-1}\rho(\mathbf{r}_1)\rho(\mathbf{r}_2)\,d\mathbf{r}_1\,d\mathbf{r}_2$. Integration of this expression over $d\mathbf{r}_2$ gives the repulsion energy between dQ_1 and the charge distribution. Integration over $d\mathbf{r}_1$ and multiplication by $\frac{1}{2}$ then gives the total interelectronic repulsion energy for the continuous distribution, where the factor $\frac{1}{2}$ is needed to prevent counting each repulsion twice, once as the repulsion between dQ_1 and dQ_2 and once as the repulsion between dQ_2 and dQ_1.

With the definitions (16.41) and (16.42), Eq. (16.36) becomes

$$E_v[\rho] = \int \rho(\mathbf{r})v(\mathbf{r})\,d\mathbf{r} + \overline{T}_s[\rho] + \frac{1}{2}\iint \frac{\rho(\mathbf{r}_1)\rho(\mathbf{r}_2)}{r_{12}}\,d\mathbf{r}_1\,d\mathbf{r}_2 + \Delta\overline{T}[\rho] + \Delta\overline{V}_{ee}[\rho]$$

The functionals $\Delta\overline{T}$ and $\Delta\overline{V}_{ee}$ are unknown. Defining the **exchange–correlation energy functional** $E_{xc}[\rho]$ by

$$E_{xc}[\rho] \equiv \Delta\overline{T}[\rho] + \Delta\overline{V}_{ee}[\rho] \tag{16.43}$$

we have

$$E_0 = E_v[\rho] = \int \rho(\mathbf{r})v(\mathbf{r})\,d\mathbf{r} + \overline{T}_s[\rho] + \frac{1}{2}\iint \frac{\rho(\mathbf{r}_1)\rho(\mathbf{r}_2)}{r_{12}}\,d\mathbf{r}_1\,d\mathbf{r}_2 + E_{xc}[\rho] \tag{16.44}$$

The motivation for the definitions (16.41), (16.42), and (16.43) is to express $E_v[\rho]$ in terms of three quantities, the first three terms on the right side of (16.44), that are easy to evaluate from ρ and that include the main contributions to the ground-state energy, plus a fourth quantity E_{xc}, which, although not easy to evaluate accurately, will be a relatively small term. The key to accurate KS DFT calculation of molecular properties is to get a good approximation to E_{xc}.

Before we can evaluate the terms in (16.44), we need to find the ground-state electron density. Recall that the fictitious system of noninteracting electrons is defined to have the same electron density as that in the ground state of the molecule: $\rho_s = \rho_0$. It is readily proved (see Prob. 16.22) that the electron probability density of an n-particle system whose wave function [Eq. (16.39)] is a Slater determinant of the spin-orbitals $u_i^{\text{KS}} = \theta_i^{\text{KS}}\sigma_i$ is given by $\sum_{i=1}^{n} |\theta_i^{\text{KS}}|^2$. Therefore,

$$\rho = \rho_s = \sum_{i=1}^{n} |\theta_i^{\text{KS}}|^2 \tag{16.45}$$

How do we evaluate the terms in (16.44)? Using (16.34), we have $\int \rho(\mathbf{r})v(\mathbf{r})\,d\mathbf{r} = -\sum_\alpha Z_\alpha \int \rho(\mathbf{r}_1)r_{1\alpha}^{-1}\,d\mathbf{r}_1$, which is easily evaluated if $\rho(\mathbf{r})$ is known. The $\overline{T}_s$ term in (16.44) is the kinetic energy of the system of noninteracting electrons whose wave function ψ_s in (16.39) is equal to a Slater determinant of orthonormal Kohn–Sham spin-orbitals. We have $\overline{T}_s[\rho] = -\frac{1}{2}\langle\psi_s|\sum_i \nabla_i^2|\psi_s\rangle$. The Slater–Condon rules [Table 11.3 and Eq. (11.78)] give $\overline{T}_s[\rho] = -\frac{1}{2}\sum_i \langle\theta_i^{\text{KS}}(1)|\nabla_1^2|\theta_i^{\text{KS}}(1)\rangle$. Thus (16.44) becomes

$$E_0 = -\sum_\alpha Z_\alpha \int \frac{\rho(\mathbf{r}_1)}{r_{1\alpha}}\,d\mathbf{r}_1 - \frac{1}{2}\sum_{i=1}^{n} \langle\theta_i^{\text{KS}}(1)|\nabla_1^2|\theta_i^{\text{KS}}(1)\rangle + \frac{1}{2}\iint \frac{\rho(\mathbf{r}_1)\rho(\mathbf{r}_2)}{r_{12}}\,d\mathbf{r}_1\,d\mathbf{r}_2 + E_{xc}[\rho] \tag{16.46}$$

We can therefore find E_0 from ρ if we can find the KS orbitals θ_i^{KS} and if we know what the functional E_{xc} is. The electronic energy including nuclear repulsion is found by adding the internuclear repulsion V_{NN} to (16.46).

The Kohn–Sham orbitals are found as follows. The Hohenberg–Kohn variational theorem tells us that we can find the ground-state energy by varying ρ (subject to the constraint $\int \rho\,d\mathbf{r} = n$) so as to minimize the functional $E_v[\rho]$ Equivalently, instead of varying ρ, we can vary the KS orbitals θ_i^{KS}, which determine ρ by (16.45). (In doing so, we must constrain the θ_i^{KS}'s to be orthonormal, since orthonormality was assumed when we evaluated $\overline{T}_s$.) Just as one can show that the orthonormal orbitals that minimize the Hartree–Fock expression for the molecular energy satisfy the Fock equation (14.25), one can show that the Kohn–Sham orbitals that minimize the expression (16.46) for the molecular ground-state energy satisfy (for a proof, see *Parr and Yang*, Section 7.2):

$$\left[-\frac{1}{2}\nabla_1^2 - \sum_\alpha \frac{Z_\alpha}{r_{1\alpha}} + \int \frac{\rho(\mathbf{r}_2)}{r_{12}}\,d\mathbf{r}_2 + v_{xc}(1)\right]\theta_i^{\text{KS}}(1) = \varepsilon_i^{\text{KS}}\theta_i^{\text{KS}}(1) \tag{16.47}$$

where the function $v_{xc}(1)$ is defined by (16.50). From (16.40) and (16.38), alternative ways to write (16.47) are

$$[-\tfrac{1}{2}\nabla_1^2 + v_s(1)]\theta_i^{\text{KS}}(1) = \varepsilon_i^{\text{KS}}\theta_i^{\text{KS}}(1) \tag{16.48}$$

$$\hat{h}^{\text{KS}}(1)\theta_i^{\text{KS}}(1) = \varepsilon_i^{\text{KS}}\theta_i^{\text{KS}}(1) \tag{16.49}$$

The **exchange–correlation potential** v_{xc} is found as the functional derivative $\delta E_{xc}/\delta\rho$ of the exchange–correlation energy E_{xc}:

$$v_{xc}(\mathbf{r}) \equiv \frac{\delta E_{xc}[\rho(\mathbf{r})]}{\delta\rho(\mathbf{r})} \tag{16.50}$$

The precise definition of the functional derivative need not concern us (see *Parr and Yang*, Appendix A). The following formula allows one to find the **functional derivative** of most functionals that occur in DFT. For a functional defined by

$$F[\rho] = \int_e^f \int_c^d \int_a^b g(x, y, z, \rho, \rho_x, \rho_y, \rho_z)\, dx\, dy\, dz$$

where ρ is a function of x, y, and z that vanishes at the limits of the integral, and where $\rho_x \equiv (\partial\rho/\partial x)_{y,z}$ etc., the functional derivative can be shown to be given by

$$\frac{\delta F}{\delta \rho} = \frac{\partial g}{\partial \rho} - \frac{\partial}{\partial x}\frac{\partial g}{\partial \rho_x} - \frac{\partial}{\partial y}\frac{\partial g}{\partial \rho_y} - \frac{\partial}{\partial z}\frac{\partial g}{\partial \rho_z} \tag{16.51}$$

Note from (16.51) that the units of $\delta F/\delta\rho$ are not the same as the units of F/ρ, since F and g have different units. One can show that $\delta(c_1F_1 + c_2F_2)/\delta\rho = c_1\,\delta F_1/\delta\rho + c_2\,\delta F_2/\delta\rho$, where c_1 and c_2 are constants.

If $E_{xc}[\rho]$ is known, its functional derivative is readily found from (16.50) and (16.51), and so v_{xc} is known. The functional $E_{xc}[\rho]$ in (16.46) is a number. The functional derivative of $E_{xc}[\rho]$ is a function of ρ (see Prob. 16.17 for an example), and since ρ is a function of $\mathbf{r}$, v_{xc} is a function of $\mathbf{r}$, that is, of x, y, and z. Sometimes people write v_{xc} as $v_{xc}(\rho(\mathbf{r}))$. [In the Kohn–Sham eigenvalue equation (16.49), the variable is taken as $\mathbf{r}_1$ rather than as $\mathbf{r}$.] The functional derivative $\delta E_{xc}[\rho(\mathbf{r})]/\delta\rho(\mathbf{r})$ is a function of x, y, and z whose value at point (x, y, z) depends on how much $E_{xc}[\rho]$ changes when $\rho(x, y, z)$ changes by a tiny amount in a tiny region centered at (x, y, z).

The one-electron Kohn–Sham operator $\hat{h}^{\text{KS}}(1)$ in (16.49) and (16.47) is the same as the Fock operator (16.8) in the Hartree–Fock equations except that the exchange operators $-\sum_{j=1}^{n}\hat{k}_j$ in the Fock operator are replaced by v_{xc} (Prob. 16.18), which handles the effects of both exchange (antisymmetry) and electron correlation.

There is only one problem in using the Kohn–Sham method to find ρ and E_0. No one knows what the correct functional $E_{xc}[\rho]$ is. Therefore, both E_{xc} in the energy expression (16.46) and v_{xc} in (16.47) and (16.50) are unknown. Various approximations to E_{xc} will be discussed shortly.

The Kohn–Sham orbitals θ_i^{KS} are orbitals for the fictitious reference system of noninteracting electrons, so, strictly speaking, these orbitals have no physical significance other than in allowing the exact molecular ground-state ρ to be calculated from (16.45). The density-functional (DF) molecular wave function is not a Slater determinant of spin-orbitals. In fact, *there is no DF molecular wave function*. However, in practice, one finds that the occupied Kohn–Sham orbitals resemble molecular orbitals calculated by the Hartree–Fock method, and the Kohn–Sham orbitals can be used (just as Hartree–Fock MOs are used) in qualitative MO discussions of molecular properties and reactivities [see E. J. Baerends and O. V. Gritsenko, *J. Phys. Chem.*, **101**, 5383 (1997); Gritsenko et al., *J. Chem. Phys.*, **107**, 5007 (1997); R. Stowasser and R. Hoffmann, *J. Am. Chem. Soc.*, **121**, 3414 (1999)]. (Note that, strictly speaking, Hartree–Fock orbitals also have no physical reality, since they refer to a fictitious model system in which each electron experiences some sort of average field of the other electrons.)

For a closed-shell molecule, each Hartree–Fock occupied-orbital energy is a good approximation to the negative of the energy needed to remove an electron from that orbital (Koopmans' theorem). However, this is not true for Kohn–Sham orbital energies. The one exception is ε_i^{KS} for the highest-occupied KS orbital, which can be proved to be equal to minus the molecular ionization energy. (With the currently used approximations to E_{xc}, ionization energies calculated from KS highest-occupied-orbital energies agree poorly with experiment.)

Various approximate functionals $E_{xc}[\rho]$ are used in molecular DF calculations. To study the accuracy of an approximate $E_{xc}[\rho]$, one uses it in DF calculations and compares calculated molecular properties with experimental ones. The lack of a systematic procedure for improving $E_{xc}[\rho]$ and hence improving calculated molecular properties is the main drawback of the DF method.

In a "true" density-functional theory, one would deal with only the electron density (a function of three variables) and not with orbitals, and would search directly for the density that minimizes $E_v[\rho]$. Because the functional E_v is unknown, one instead uses the Kohn–Sham method, which calculates an orbital for each electron. Thus, the KS method represents something of a compromise with the original goals of DFT.

The exchange–correlation energy E_{xc} in (16.43) contains the following components: the *kinetic correlation energy* [the $\Delta\overline{T}$ term in (16.43), which is the difference in $\overline{T}$ for the real molecule and the reference system of noninteracting electrons—Eq. (16.41)], the *exchange energy* (which arises from the antisymmetry requirement), the *Coulombic correlation energy* (which is associated with interelectronic repulsions), and a *self-interaction correction* (SIC). The SIC arises from the fact that the classical charge-cloud electrostatic-repulsion expression $\frac{1}{2}\iint \rho(\mathbf{r}_1)\rho(\mathbf{r}_2)r_{12}^{-1}\,d\mathbf{r}_1\,d\mathbf{r}_2$ in (16.42) erroneously allows the portion of ρ in $d\mathbf{r}_1$ that comes from the smeared-out part of a particular electron to interact with the charge contributions of that same electron to ρ throughout space. In actuality, an electron does not interact with itself. (Note that for a one-electron molecule, there is no interelectronic repulsion, but the expression $\frac{1}{2}\iint \rho(\mathbf{r}_1)\rho(\mathbf{r}_2)r_{12}^{-1}\,d\mathbf{r}_1\,d\mathbf{r}_2$ erroneously gives an interelectronic repulsion.) The kinetic energy $\overline{T}_s$ of the reference system turns out to be close to $\overline{T}$ of the real molecule, and $\Delta\overline{T}/\overline{T}$ is small. However, the contribution of $\Delta\overline{T}$ to E_{xc} in (16.43) is not negligible.

The Local-Density Approximation (LDA). Hohenberg and Kohn showed that if ρ varies extremely slowly with position, then $E_{xc}[\rho]$ is accurately given by

$$E_{xc}^{\text{LDA}}[\rho] = \int \rho(\mathbf{r})\varepsilon_{xc}(\rho)\,d\mathbf{r} \tag{16.52}$$

where the integral is over all space, $d\mathbf{r}$ stands for $dx\,dy\,dz$, and $\varepsilon_{xc}(\rho)$ is the exchange plus correlation energy per electron in a homogeneous electron gas with electron density ρ. **Jellium** is a hypothetical electrically neutral, infinite-volume system consisting of an infinite number of interacting electrons moving in a space throughout which positive charge is continuously and uniformly distributed. The number of electrons per unit volume in the jellium has a nonzero constant value ρ. The electrons in the jellium constitute a **homogeneous** (or **uniform**) **electron gas**. Taking the functional derivative of E_{xc}^{LDA}, we find [Eqs. (16.50) and (16.51)]

$$v_{xc}^{\text{LDA}} = \frac{\delta E_{xc}^{\text{LDA}}}{\delta\rho} = \varepsilon_{xc}(\rho(\mathbf{r})) + \rho(\mathbf{r})\frac{\partial\varepsilon_{xc}(\rho)}{\partial\rho} \tag{16.53}$$

Kohn and Sham suggested the use of (16.52) and (16.53) as approximations to E_{xc} and v_{xc} in (16.46) and (16.47), a procedure that is called the **local density approximation** (LDA). One can show that ε_{xc} can be written as the sum of exchange and correlation parts:

$$\varepsilon_{xc}(\rho) = \varepsilon_x(\rho) + \varepsilon_c(\rho) \tag{16.54}$$

where

$$\varepsilon_x(\rho) = -\frac{3}{4}\left(\frac{3}{\pi}\right)^{1/3}(\rho(\mathbf{r}))^{1/3} \tag{16.55}$$

The correlation part $\varepsilon_c(\rho)$ has been calculated and the results have been expressed as a very complicated function $\varepsilon_c^{\text{VWN}}$ of ρ by Vosko, Wilk, and Nusair (VWN); see *Parr and Yang*, Appendix E; S. H. Vosko, L. Wilk, and M. Nusair, *Can. J. Phys.*, **58**, 1200 (1980). Thus

$$\varepsilon_c(\rho) - \varepsilon_c^{\text{VWN}}(\rho) \tag{16.56}$$

where $\varepsilon_c^{\text{VWN}}$ is a known function. From (16.50), (16.51), (16.52), (16.54), and (16.55), we get (Prob. 16.19)

$$v_{xc}^{\text{LDA}} = v_x^{\text{LDA}} + v_c^{\text{LDA}}, \quad v_x^{\text{LDA}} = -[(3/\pi)\rho(\mathbf{r})]^{1/3}, \quad v_c^{\text{LDA}} = v_c^{\text{VWN}} \tag{16.57}$$

$$E_x^{\text{LDA}} \equiv \int \rho\varepsilon_x \, d\mathbf{r} = -\frac{3}{4}\left(\frac{3}{\pi}\right)^{1/3}\int [\rho(\mathbf{r})]^{4/3}\, d\mathbf{r} \tag{16.58}$$

The method for finding the LDA quantities ε_x and ε_c is as follows (for fuller details, see *Parr and Yang*, Appendix E). Consider a uniform electron gas (UEG) with $\rho(\mathbf{r}) = k$, where k is some constant value. As noted after (16.51), $v_{xc} = v_{xc}(\rho(\mathbf{r}))$, and since $\rho(\mathbf{r})$ is a constant for a particular UEG, v_{xc} is a constant for a particular UEG. (Of course v_{xc} will have different values for two UEGs with different electron densities.) In the Kohn–Sham equations (16.47) for the reference system that corresponds to the UEG, the constant v_{xc} can be omitted without affecting the eigenfunctions (Prob. 4.51). Also, for a UEG, the second term in brackets in (16.47) (the external potential) must be replaced by the attraction between electron 1 and the uniformly distributed positive charge. Since the UEG is electrically neutral, the positive charge density equals the electron density, and the second and third terms in brackets in (16.47) cancel. Thus, the term $-\frac{1}{2}\nabla_1^2$ is the only surviving term in $\hat{h}^{\text{KS}}$ for the UEG. The UEG KS orbitals can thus be taken as three-dimensional free-particle wave functions $Ae^{i(k_x x + k_y y + k_z z)}$ [recall (2.30)], where the value of A is chosen to give the desired electron density in (16.45). Because the UEG is electrically neutral in each region of space, the sum of the electrostatic repulsions between the smeared-out electrons [the third term on the right side of (16.46)], the attractions between the smeared-out electrons and the continuous positive charge distribution [the first term on the right of (16.46) modified to correspond to continuous positive charges], and the repulsions between parts of the positive charge distribution [analogous to the internuclear repulsion term to be added to (16.46)] adds to zero. This leaves on the right side of the energy expression (16.46) only the E_{xc} term and the kinetic-energy term $\overline{T}_s$, which is readily evaluated from the known KS orbitals. Breaking E_{xc} into the sum of E_x and E_c [Eq. (16.59)], one evaluates E_x from (16.60) and the KS orbitals, with the result shown in (16.58). This leaves only E_c as unknown. One then does an accurate numerical solution of the UEG Schrödinger equation to find the energy for the particular density $\rho = k$. Combining this energy with the already calculated KS energy terms gives the unknown E_c for that ρ. Repetition of the entire procedure for many density values gives the UEG E_c as a function of ρ. From E_x and E_c, we find ε_x and ε_c.

The Functionals E_x and E_c. As an aid to developing approximate functionals for use in KS DFT, the functional E_{xc} is written as the sum of an **exchange-energy functional** E_x and a **correlation-energy functional** E_c:

$$E_{xc} = E_x + E_c \tag{16.59}$$

E_x is defined by the same formula used for the exchange energy in Hartree–Fock theory, except that the Hartree–Fock orbitals are replaced by the Kohn–Sham orbitals. The Hartree–Fock exchange energy of a closed-shell molecule is given by the terms in (14.22) that involve the exchange integrals K_{ij}. Replacing the Hartree–Fock orbitals by the Kohn–Sham orbitals in (14.24), we have for a closed-shell molecule

$$E_x = -\frac{1}{4}\sum_{i=1}^{n}\sum_{j=1}^{n} \langle \theta_i^{\text{KS}}(1)\theta_j^{\text{KS}}(2)|1/r_{12}|\theta_j^{\text{KS}}(1)\theta_i^{\text{KS}}(2)\rangle \tag{16.60}$$

where the factor of $\frac{1}{4}$ comes from the fact that in (14.22) we are summing over the orbitals, whereas in (16.60) we are summing over the electrons, which gives four times as many terms in the double sum in (16.60) (Prob. 16.20). Since, in practice, KS orbitals are found to rather closely resemble Hartree–Fock orbitals, the DFT exchange energy is close to the Hartree–Fock exchange energy. Now that E_x is defined, the correlation-energy functional E_c is defined as the difference between E_{xc} and E_x; that is $E_c \equiv E_{xc} - E_x$, and (16.59) follows. When E_x is evaluated using the definition (16.60) and E_c is evaluated by one of the currently available models (such as the LDA), one obtains poor results for molecular properties. Thus, in practice it is best to model both E_x and E_c, because this leads to cancellation of errors and better results. One therefore uses the LDA (or one of its improved versions discussed later) to find both E_x and E_c.

Both E_x and E_c are negative, with $|E_x|$ being much larger than $|E_c|$. The definition of E_c in DFT differs from the definition (11.16) of the correlation energy in Hartree–Fock theory, but analysis and calculations show that these two quantities are nearly equal [E. K. U. Gross et al. in B. B. Laird et al. (eds.) *Chemical Applications of Density Functional Theory*, American Chemical Society, 1996, Chapter 3]. Starting with accurate electron densities from MRCI wave functions for Li_2, N_2, and F_2, Gritsenko and co-workers used an iterative procedure to calculate KS orbitals and the KS quantities E_x and E_c for each of these molecules at three internuclear distances [O. V. Gritsenko et al., *J. Chem. Phys.*, **107**, 5007 (1997)]. The KS E_x and E_c values were found to be very close to the corresponding Hartree–Fock (HF) values E_x^{HF} and E_c^{HF} at the equilibrium internuclear distances, but agreement between these quantities decreased as the internuclear distances increased. For N_2 at R_e, values in hartrees for these quantities and for quantities in (16.41) are: $E_x = -13.114$, $E_x^{\text{HF}} = -13.008$; $E_c = -0.475$, $E_c^{\text{HF}} = -0.469$; $\Delta\overline{T} = 0.329$, $\overline{T} = 109.399$. The correlation kinetic energy $\Delta\overline{T}$ is a significant part of E_c.

The Xα Method. The following approximation for E_{xc} gives the **Xα method** (the X stands for exchange). Here, the correlation contribution to E_{xc} is omitted (it is substantially smaller in magnitude than the exchange contribution) and the exchange contribution is taken as

$$E_{xc} \approx E_x^{X\alpha} = -\frac{9}{8}\left(\frac{3}{\pi}\right)^{1/3} \alpha \int [\rho(\mathbf{r})]^{4/3}\, d\mathbf{r} \tag{16.61}$$

where α is an adjustable parameter; α values from $\frac{2}{3}$ to 1 have been used. Functional differentiation of (16.61) [Eqs. (16.50) and (16.51)] gives the Xα exchange potential as $v_x^{X\alpha} = -(3\alpha/2)(3\rho/\pi)^{1/3}$. Note that with $\alpha = \frac{2}{3}$, the Xα expression (16.61) for E_{xc} becomes equal to the exchange part (16.58) of the LDA E_{xc} expression. The Xα method gives rather erratic results in molecular calculations and has been superseded by better approximations to E_{xc}. The Xα method was developed by Slater prior to the work of Hohenberg, Kohn, and Sham, and was viewed by Slater as an approximation to the Hartree–Fock method. It is sometimes called the Hartree–Fock–Slater method. However, the Xα method is best viewed as a special case of DFT.

Performing Kohn–Sham Density-Functional Calculations. How does one do a molecular density-functional calculation with E_{xc}^{LDA} (or some other functional)? One starts with an initial guess for ρ, which is usually found by superposing calculated electron densities of the individual atoms at the chosen molecular geometry. From the initial guess for $\rho(\mathbf{r})$, an initial estimate of $v_{xc}(\mathbf{r})$ is found from (16.53) and (16.57) and this initial $v_{xc}(\mathbf{r})$ is used in the Kohn–Sham equations (16.47), which are solved for the initial estimate of the KS orbitals. In solving (16.47), the θ_i^{KS}'s are usually expanded in terms of a set of basis functions χ_r

$$\theta_i^{\text{KS}} = \sum_{r=1}^{b} c_{ri}\chi_r$$

to yield equations that resemble the Hartree–Fock–Roothaan equations (14.34) and (14.56), except that the Fock matrix elements $F_{rs} = \langle \chi_r|\hat{F}|\chi_s\rangle$ are replaced by the Kohn–Sham matrix elements $h_{rs}^{\text{KS}} = \langle \chi_r|\hat{h}^{\text{KS}}|\chi_s\rangle$, where $\hat{h}^{\text{KS}}$ is in (16.48) and (16.49). Thus, instead of (14.34), in KS DFT with a basis-set expansion of the orbitals, one solves the equations

$$\sum_{s=1}^{b} c_{si}(h_{rs}^{\text{KS}} - \varepsilon_i^{\text{KS}} S_{rs}) = 0, \qquad r = 1, 2, \ldots, b \tag{16.62}$$

The basis functions most commonly used in molecular KS DFT calculations are contracted Gaussians, but some DF programs use STOs or still other basis functions.

The KS equations can also be solved numerically, without using a basis-function expansion of the orbitals, but this choice is not often used.

The initially found θ_i^{KS}'s are used in (16.45) to get an improved electron density, which is then used to find an improved v_{xc}, which is then used in the KS equations (16.47) to find improved KS orbitals, and so on. The iterations continue until there is no further significant change in the density and the KS orbitals. KS DFT calculations involve iterations until self-consistency between the exchange–correlation potential v_{xc} and the KS orbitals θ_i^{KS} in (16.47) has been reached. They are thus a kind of SCF calculation.

Once the calculation has converged, the ground-state energy E_0 in (16.46) is found from the converged ρ and E_{xc}^{LDA} (or whatever functional is being used). The dipole moment can be calculated from ρ using (14.21), and other one-electron properties can be found from (14.8). Analytic gradients of the energy have been developed for KS DFT calculations, so the equilibrium geometry is readily found using one of the methods of Section 15.10. Analytic second derivatives of the KS DFT energy are available, and DF vibrational frequencies are readily calculated.

One significant difference between KS DFT calculations and Hartree–Fock calculations arises from the fact that $v_{xc}^{\text{LDA}}(\mathbf{r})$ and versions of v_{xc} more accurate than the LDA are very complicated functions of the coordinates. This makes it impossible to analytically evaluate the integrals $\langle \chi_r | v_{xc} | \chi_s \rangle$, which occur in h_{rs}^{KS}. Instead, $\langle \chi_r | v_{xc} | \chi_s \rangle$ is evaluated numerically by evaluating the integrand at each point of a grid of points in the molecule and performing a summation. [A less-used alternative approach is to expand $v_{xc}(\mathbf{r})$ using an auxiliary set of basis functions (not the same set used to expand the orbitals), where the expansion coefficients are chosen to give a good least-squares fit to values of $v_{xc}(\mathbf{r})$ evaluated at a grid of points.] The numerical evaluation of the $\langle \chi_r | v_{xc} | \chi_s \rangle$ integrals is the most time-consuming step in a DFT calculation with Gaussian basis functions. If too small a grid is used, inaccurate results will be obtained. If a very large grid is used, the calculation time can become prohibitive. Many DFT programs give the user the power to choose the grid size. [See J. M. L. Martin et al., *Comput. Phys. Commun.*, **133**, 189 (2001).]

DF calculations that use a basis-set expansion of the KS orbitals involve the same Coulomb matrix elements J_{rs} [Eq. (15.79)] that occur in Hartree–Fock calculations. Such DF calculations can be speeded up and linear scaling obtained for large molecules by the same techniques used to speed up Hartree–Fock calculations, namely, direct and semidirect methods, neglect of integrals smaller than a threshold value, the continuous fast multipole and quantum-chemical tree-code methods, the J-matrix engine. For details, see the references in Section 15.16. Also the conjugate-gradient density-matrix search method (Section 17.4) can be used to avoid diagonalizing the Kohn–Sham matrix.

Deviations of KS DF calculated results from the true values are due to use of approximate E_{xc} and v_{xc} expressions, and to basis-set inadequacies. See Chapter 18 for detailed comparisons.

The programs *Gaussian*, Q-Chem, Jaguar, ACES II, Turbomole, MOLPRO, NWChem, CADPAC, Spartan, ORCA, and HyperChem (Section 15.14) all contain KS DFT modules. Some programs that only do KS DF calculations are ADF (www.scm.com/) and deMon (www.demon-software.com/public_html/program.html). For a more complete list, see the article en.wikipedia.org/wiki/Density_functional_theory.

The Local-Spin-Density Approximation (LSDA). For open-shell molecules and molecular geometries near dissociation, the **local-spin-density approximation** (LSDA) gives better results than the LDA. Whereas in the LDA, electrons with opposite spins paired with each other have the same spatial KS orbital, the LSDA allows such electrons to have different spatial KS orbitals $\theta_{i\alpha}^{\text{KS}}$ and $\theta_{i\beta}^{\text{KS}}$. The LSDA is thus analogous to the UHF method (Section 15.3), which allows different spatial Hartree–Fock orbitals for electrons with different spins. The theorems of Hohenberg,

Kohn, and Sham do not require using different orbitals for electrons with different spins (unless an external magnetic field is present), and if the exact functional $E_{xc}[\rho]$ were known, one would not do so. With the approximate E_{xc} functionals that are used in KS DFT calculations, it is advantageous to allow the possibility of different orbitals for electrons with different spins, so as to improve calculated properties of open-shell species and species with geometries near dissociation.

The generalization of density-functional theory that allows different orbitals for electrons with different spins is called **spin-density-functional theory** (*Parr and Yang*, Chapter 8). In spin-DFT, one deals separately with the electron density $\rho^{\alpha}(\mathbf{r})$ due to the spin-α electrons and the density $\rho^{\beta}(\mathbf{r})$ of the spin-β electrons, and functionals such as E_{xc} become functionals of these two quantities: $E_{xc} = E_{xc}[\rho^{\alpha}, \rho^{\beta}]$. One deals with separate Kohn–Sham eigenvalue equations for the spin-α orbitals and the spin-β orbitals, where these equations have $v_{xc}^{\alpha} = \delta E_{xc}[\rho^{\alpha}, \rho^{\beta}]/\delta\rho^{\alpha}$ and similarly for v_{xc}^{β}. For species like CH_3 or the O_2 triplet ground state, the number n^{α} of α electrons differs from the number of β electrons, so here $\rho^{\alpha} \neq \rho^{\beta}$, and spin-DFT will give different orbitals for electrons with different spins.

Gunnarsson and Lundqvist did an LSDA spin-DFT calculation of the H_2 molecule, expanding the occupied KS orbitals using the $1s_a$ and $1s_b$ AOs as basis functions. For internuclear separations up to 3.2 bohrs, they found the lowest energy KS orbitals to be $\theta_{\alpha}^{\text{KS}} = \theta_{\beta}^{\text{KS}} = N(1s_a + 1s_b)$. However, for internuclear separations greater than 3.2 bohrs, they found the lowest-energy KS orbitals to be $\theta_{\alpha}^{\text{KS}} = N(1s_a + c1s_b)$ and $\theta_{\beta}^{\text{KS}} = N(c1s_a + 1s_b)$, where $c < 1$ and c decreased to zero as the internuclear separation increased to infinity. Having the spin-α electron's KS orbital differ from that of the spin-β electron allowed the H_2 molecular energy versus internuclear separation curve to show the proper dissociation behavior, corresponding to dissociation to one hydrogen atom with a spin-α electron and a second H atom with a spin-β electron. [Recall that the RHF wave function of H_2 dissociates improperly. The UHF H_2 wave function shows proper dissociation (*Szabo and Ostlund*, Section 3.8.7).]

As in the UHF method, allowing differing KS orbitals for electrons with different spins can produce a wave function for the reference system s that is not an eigenfunction of $\hat{S}^2$, but this spin-contamination is less of a problem in KS DFT than in the UHF method.

For species with all electrons paired and molecular geometries in the region of the equilibrium geometry, we can expect that $\rho^{\alpha} = \rho^{\beta}$, and spin-DFT will reduce to the ordinary form of DFT.

Despite the fact that ρ in a molecule is not a slowly varying function of position, the LSDA works surprisingly well for calculating molecular equilibrium geometries, vibrational frequencies, and dipole moments, even for transition-metal compounds, where Hartree–Fock calculations often give poor results. (For detailed results, see Chapter 18.) However, calculated LSDA molecular atomization energies are very inaccurate. Accurate dissociation energies require functionals that go beyond LSDA.

Gradient-Corrected (GGA) Functionals. The LDA and LSDA are based on the uniform-electron-gas model, which is appropriate for a system where ρ varies slowly with position. The integrand in the expression (16.52) for E_{xc}^{LDA} is a function of only ρ, and the integrand in E_{xc}^{LSDA} is a function of only ρ^{α} and ρ^{β}. Functionals that go beyond the LSDA aim to correct the LSDA for the variation of electron density with position. A common way to do this is by including the gradients [Eq. (5.30)] of ρ^α and ρ^β in the integrand. Thus

$$E_{xc}^{\text{GGA}}[\rho^\alpha, \rho^\beta] = \int f(\rho^\alpha(\mathbf{r}), \rho^\beta(\mathbf{r}), \nabla\rho^\alpha(\mathbf{r}), \nabla\rho^\beta(\mathbf{r}))\, d\mathbf{r} \tag{16.63}$$

where f is some function of the spin densities and their gradients. The letters GGA stand for **generalized-gradient approximation.** The term **gradient-corrected functional** is also used. (Gradient-corrected functionals have sometimes been called **"nonlocal"** functionals but, strictly speaking, this is a misuse of the mathematical meaning of nonlocal.) E_{xc}^{GGA} is usually split into exchange and correlation parts, which are modeled separately:

$$E_{xc}^{\text{GGA}} = E_x^{\text{GGA}} + E_c^{\text{GGA}} \tag{16.64}$$

Approximate gradient-corrected exchange and correlation energy functionals are developed using theoretical considerations such as the known behavior of the true (but unknown) functionals E_x and E_c in various limiting situations as a guide, with often some empiricism thrown in, by choosing the values of parameters in the functionals to give good performance for known values of various molecular properties.

Some commonly used GGA exchange functionals E_x are Perdew and Wang's 1986 functional (which contains no empirical parameters), designated PW86 or PWx86, Becke's 1988 functional, denoted B88, Bx88, Becke88, or B, and Perdew and Wang's 1991 exchange functional PWx91. The explicit form of the B88 exchange functional is

$$E_x^{\text{B88}} = E_x^{\text{LSDA}} - b \sum_{\sigma=\alpha,\beta} \int \frac{(\rho^\sigma)^{4/3} \chi_\sigma^2}{1 + 6b\chi_\sigma \ln[\chi_\sigma + (\chi_\sigma^2 + 1)^{1/2}]}\, d\mathbf{r}$$

where $\chi_\sigma \equiv |\nabla\rho^\sigma|/(\rho^\sigma)^{4/3}$, b is an empirical parameter whose value 0.0042 atomic units was determined by fitting known Hartree–Fock exchange energies (which are close to KS exchange energies) of several atoms, and [see (16.58) and Prob. 16.21]

$$E_x^{\text{LSDA}} = -\frac{3}{4}\left(\frac{6}{\pi}\right)^{1/3} \int [(\rho^\alpha)^{4/3} + (\rho^\beta)^{4/3}]\, d\mathbf{r} \tag{16.65}$$

The PWx86 functional (which has no empirical parameters) and the B88 exchange functional work about equally well in predicting molecular properties.

Commonly used GGA correlation functionals E_c include the Lee–Yang–Parr (LYP) functional, the Perdew 1986 correlation functional (P86 or Pc86), and the Perdew–Wang 1991 parameter-free correlation functional (PW91 or PWc91).

The Perdew–Burke–Ernzerhof (PBE) exchange and correlation functional has no empirical parameters [*Phys. Rev. Lett.*, **77**, 3865 (1996)].

Some E_x and E_c values in hartrees for the Ar atom are (A. D. Becke in *Yarkony*, Chapter 15): $E_x^{\text{HF}} = -30.19$, $E_x^{\text{LSDA}} = -27.86$, $E_x^{\text{B88}} = -30.15$; $E_c^{\text{HF}} = -0.722$, $E_c^{\text{LSDA}} = -1.431$, $E_c^{\text{PW91}} = -0.768$, where the Hartree–Fock (HF) values should be good estimates of the KS DFT values. These values may be compared with the total energy -527.54 hartrees for Ar, so the exchange and correlation energies are about 6% and 0.14%, respectively, of the total energy.

Any exchange functional can be combined with any correlation functional. For example, the notation BLYP/6-31G* denotes a DF calculation done with the Becke 1988 exchange functional and the Lee–Yang–Parr correlation functional, with the KS orbitals expanded in a 6-31G* basis set. The letter S (which acknowledges Slater's Xα method) denotes the LSDA exchange functional (16.65). VWN denotes the Vosko–Wilk–Nusair expression for the LSDA correlation functional (actually, these workers gave two different expressions for E_c^{LSDA}, which are sometimes referred to as VWN3 and VWN5). Thus an LSDA calculation can be denoted by the letters LSDA or by SVWN.

Meta-GGA Functionals. The GGA density functionals of the form (16.63) depend on the ground-state electron probability density ρ and its first derivatives. One way to improve on GGA functionals is to go to functionals that also depend on the second derivatives of ρ and/or a quantity called the kinetic-energy density (which is defined below). Such functionals are called **meta-GGA** (mGGA) functionals and have the form

$$E_{xc}^{\text{MGGA}}[\rho^\alpha, \rho^\beta] = \int f(\rho^\alpha, \rho^\beta, \nabla\rho^\alpha, \nabla\rho^\beta, \nabla^2\rho^\alpha, \nabla^2\rho^\beta, \tau_\alpha, \tau_\beta)\, d\mathbf{r} \qquad (16.66)$$

where the *Kohn–Sham kinetic-energy density* for the spin-α electrons is defined by

$$\tau_\alpha \equiv \frac{1}{2}\sum_i |\nabla\theta_{i\alpha}^{\text{KS}}|^2$$

where the $\theta_{i\alpha}^{\text{KS}}$'s are the Kohn–Sham orbitals for the spin-α electrons and the sum goes over the occupied orbitals. The reason for the name "kinetic-energy density" becomes clear if you look at Prob. 7.7(b). Most meta-GGA functionals omit $\nabla^2\rho^\alpha$ and $\nabla^2\rho^\beta$ from (16.66). The dependency of a meta-GGA functional on τ_α and τ_β can occur in the exchange part, or the correlation part, or in both parts of the functional.

Meta-GGA DFT calculations require a little more time than GGA calculations and can give better results than GGA calculations.

Becke's B95 meta-GGA correlation functional, containing two parameters whose values were fitted to atomic correlation energy data, is often used.

Hybrid Functionals. Hybrid exchange–correlation functionals are widely used. A hybrid functional mixes together the formula (16.60) for E_x with GGA (or meta-GGA) E_x and E_c formulas. For example, the popular B3LYP (or Becke3LYP) hybrid GGA functional (where the 3 indicates a three-parameter functional) is defined by

$$E_{xc}^{\text{B3LYP}} = (1 - a_0 - a_x)E_x^{\text{LSDA}} + a_0E_x^{\text{exact}} + a_xE_x^{\text{B88}} + (1 - a_c)E_c^{\text{VWN}} + a_cE_c^{\text{LYP}} \quad (16.67)$$

where E_x^{exact} (which is often denoted E_x^{HF}, since it uses a Hartree–Fock definition of E_x) is given by (16.60), and where the parameter values $a_0 = 0.20$, $a_x = 0.72$, and $a_c = 0.81$ were chosen to give good fits to experimental molecular atomization energies. The B3PW91 hybrid functional replaces E_c^{LYP} in (16.67) with E_c^{PW91}, and uses the same a values. Becke's one-parameter hybrid meta-GGA functional B1B95 (also called B1B96) is $E_{xc}^{\text{B1B95}} = (1 - a_0)E_x^{\text{B88}} + E_c^{\text{B95}} + a_0E_x^{\text{exact}}$, where the empirical-parameter value $a_0 = 0.28$ was found by fitting atomization energies. One says that B1B95 contains 28% exact exchange.

As an improvement on the B3LYP, B3PW91, and B1B95 hybrid functionals, Becke [A. D. Becke, *J. Chem. Phys.*, **107**, 8554 (1997); H. L. Schmider and Becke, *J. Chem. Phys.*, **108**, 9624 (1998)] proposed the B97 hybrid GGA functional

$$E_{xc} = E_x^{\text{GGA}} + c_x E_x^{\text{exact}} + E_c^{\text{GGA}} \tag{16.68}$$

where c_x is a parameter and E_x^{GGA} and E_c^{GGA} are certain GGA functionals that contain three and six parameters, respectively. The values of the 10 parameters in E_{xc} were determined as the set that gave the best fit to experimental energy data in the G2 test set (Section 16.5). Using a numerical method to solve the Kohn–Sham equations (so as to avoid basis-set truncation error), Becke found that the functional (16.68) gave a mean absolute error (MAE) of only 1.8 kcal/mol for 55 atomization energies, a significant improvement over the functional B3PW91, which had an MAE of 2.4 kcal/mol for these energies. Revised versions of B97, called B97-1, B97-2, and B97-3 exist. [For the references, see Table 2 in Y. Zhao et al., *J. Chem. Theory Comput.*, **2**, 364 (2006).] B98 is a hybrid GGA exchange–correlation functional that is a revision of B97.

Most density functionals give only fair accuracy for predicted activation energies (barrier heights) for chemical reactions. The BMK (Boese–Martin for kinetics) hybrid meta-GGA exchange–correlation functional uses 42% exact exchange and contains 16 parameters whose values were chosen to give good performance for barrier heights without significantly sacrificing performance on other properties [A. D. Boese and J. M. L. Martin, *J. Chem. Phys.*, **121**, 3405 (2004)].

To speed up density-functional calculations on large molecules, one can use the density fitting (RI) method (Section 16.2). Hybrid functionals have the disadvantage that when density fitting is used with hybrid functionals, the calculations are much slower than when a functional that does not include exact exchange is used.

GGA, meta-GGA, hybrid-GGA and hybrid-meta-GGA functionals give not only good equilibrium geometries, vibrational frequencies, and dipole moments, but also generally accurate molecular atomization energies. For example, B3LYP/6-311+G(2d,p) calculations on the G2 data set gave an MAE of 3.1 kcal/mol (*Foresman and Frisch*, Chapter 7). See Chapter 18 for details.

Use of Perturbation Theory. Grimme has proposed improving DFT energy results by using the MP2 second-order energy-correction formula, as follows [S. Grimme, *J. Chem. Phys.,* **124**, 034108 (2006)]. One first defines a hybrid-GGA DFT functional of the form

$$E_{xc}^{\text{hybrid-GGA}} = a_1 E_x^{\text{GGA}} + (1 - a_1)E_x^{\text{exact}} + a_2 E_c^{\text{GGA}}$$

and uses this functional to self-consistently solve for KS orbitals (both occupied and virtual) and orbital energies, by the standard DFT approach. One then calculates an improved value of E_{xc} as

$$E_{xc} = E_{xc}^{\text{hybrid-GGA}} + (1 - a_2)E_c^{\text{KS-MP2}}$$

where $E_c^{\text{KS-MP2}}$ is calculated from the MP2 equation (16.13) using the KS orbitals and orbital energies. This improved E_{xc} is then used in (16.46) to calculate the ground-state energy E_0. The two parameters a_1 and a_2 are chosen as values that give good fits to a chosen set of data. Grimme defined the B2-PLYP functional, also called B2PLYP (where the 2 indicates two parameters and the first P stands for perturbation), by taking E_x^{GGA} as the B88 exchange functional, taking E_c^{GGA} as the LYP correlation functional, and taking $a_1 = 0.47$, $a_2 = 0.73$. The mPW2-PLYP functional uses the modified Perdew–Wang functional for exchange and LYP for correlation, and has $a_1 = 0.45$, $a_2 = 0.75$. Since an MP2 energy calculation is required, this method requires significantly more time than a conventional DFT calculation, but the MP2 calculation can be speeded up by using the RI approximation (Section 16.2). Also, since the method is intermediate between DFT and MP2, to get good results a larger basis set is required than for DFT, but smaller than for MP2.

For the G3/05 test set (Section 16.5), mean absolute deviations from experiment are 2.5 kcal/mol for B2-PLYP and 2.1 kcal/mol for mPW2-PLYP, as compared with 4.4 kcal/mol for B3LYP [T. Schwabe and S. Grimme, *Phys. Chem. Chem. Phys.*, **8**, 4398 (2006)]. In these calculations, the geometry was found from a B3LYP calculation with a VTZ2P basis set and the single-point energy calculation was done using a polarized valence quadruple-zeta basis set.

Grimme calls B2PLYP and mPW2PLYP **double hybrid density functionals** (DHDF), since they mix in not only some exact exchange but also some correlation calculated by the MP2 method.

Evaluation of Functionals. Over 100 density functionals have been proposed. For listings and classifications of over 40 functionals, see N. E. Schultz et al., *J. Phys. Chem. A*, **109**, 11127 (2005); **109**, 4388 (2005); Y. Zhao and D. G. Truhlar, *J. Chem. Theory Comput.*, **1**, 415 (2005). There is no correct answer to the question "Which is the best DFT functional?", since one finds that functionals that give good results for organic compounds may give inferior results for inorganic compounds, and functionals that give good results for energy changes in reactions may give inferior results for activation energies of reactions. Thus the best functional to use depends on the kinds of compounds being studied and on which properties are being calculated. During the period 1995–2005, the B3LYP functional was the most widely used by chemists, since it gives good results for a number of commonly studied properties. (The success of B3LYP is something of an accident; see G. E. Scuseria and V. N. Staroverov, "Progress in the development of exchange-correlation functionals," Chapter 24 in *Theory and Applications of Computational Chemistry,* C. E. Dykstra et al., Elsevier, 2005.) Functionals that surpass the performance of B3LYP have been developed and will see increasing use. For example, the M06-L meta-GGA functional is claimed to give a better average performance than nearly all commonly used functionals [Y. Zhao and D. G. Truhlar, *J. Chem. Phys.*, **125**, 194101 (2006)]. This functional contains 37 optimized parameters. (M06 stands for the University of Minnesota 2006 and L stands for "local," indicating that no exact exchange is included.)

Zhao and Truhlar developed a set of meta-GGA exchange–correlation functionals that have the following percentages of exact (Hartree–Fock) exchange: 0% for M06-L, 27% for M06, 28% for M05, 54% for M06-2X, 56% for M05-2X, and 100% for M06-HF [Y. Zhao and D. G. Truhlar, *Theor. Chem. Acc.*, **120**, 215 (2008); *Acc. Chem. Res.*, **41**, 157 (2008)]. [For the functional $E_{xc} = (Y/100)E_x^{\text{exact}} + (1 - Y/100)E_x^{\text{DFT}} + E_c^{\text{DFT}}$, the percentage of exact exchange is Y.] These functionals each have about 35 parameters.

Zhao and Truhlar tested the performance of 15 functionals for a variety of properties and found that M06-2X, BMK, and M05-2X gave the best performance for main-group thermochemistry and kinetics. For example, for 177 energy changes, some mean absolute errors (MAEs) are 1.3 kcal/mol for M06-2X as compared with 2.6 kcal/mol for B98, and 3.6 kcal/mol for B3LYP; for 76 chemical-reaction barrier heights, the MAEs are 1.2 kcal/mol for M06-2X as compared with 3.6 kcal/mol for B98 and 4.5 kcal/mol for B3LYP. (These tests used polarized triple-zeta basis sets.) M06-L and M06 gave the best performance for transition-metal thermochemistry. M06-2X, M05-2X, M06-HF, M06, and M06-L gave the best performance for noncovalent interactions such as hydrogen bonding. The absence of exact exchange in M06-L speeds up calculations and allows larger systems to be treated.

Basis Sets. In the wave-function-theory (WFT) correlation methods of Sections 16.1–16.3, the adequacy of the basis set is crucial; use of too small a basis set gives unreliable results, and for the highest accuracy, one attempts to extrapolate to the complete basis-set limit. In DFT, which is based on the electron density ρ, rather than on the wave function, the basis set plays a smaller role, and good results can often be obtained with a rather small basis set, such as a DZP set. Moreover, in DFT most properties calculated with TZP basis sets have converged. However, the role of the basis set cannot be ignored in DFT. One DFT study found that the addition of diffuse functions to DZP basis sets provided much greater improvement in results than gotten by going to a TZP basis set, and this study recommended the use of diffuse functions on all nonhydrogen atoms in virtually all DFT calculations [B. J. Lynch et al., *J. Phys. Chem. A*, **107**, 1384 (2003)]; however, another study recommended against the use of diffuse functions in DFT calculations on uncharged organic compounds, since their use can produce less accurate results (especially for small basis sets) and leads to much longer computational times [S. Grimme et al., *J. Org. Chem.*, **72**, 2118 (2007)]. In DFT, one finds that a Pople basis set usually gives significantly better results than a correlation-consistent (cc) basis set of similar size [A. D. Boese et al., *J. Chem. Phys.*, **119**, 3005 (2003)]. Whereas in WFT, the cc basis sets usually allow a smooth extrapolation to the complete-basis-set limit, DFT/cc calculations sometimes show irregular convergence to a limit.

Excited States. DFT was originally developed as a ground-state theory, but several DFT theories that can calculate properties of excited electronic states have been developed. The most widely used such theory is **linear response time-dependent DFT (LR-TDDFT)**. Here, one applies time-dependent perturbation theory to the ground electronic state of a molecule perturbed by a weak oscillating time-dependent spatially uniform electric field. The effect of the applied field on the ground-state electronic density ρ is found to depend on a sum whose terms involve the energy differences $E_j - E_0$ between the various excited electronic states j and the ground electronic state, where these energies are calculated at the same molecular geometry. (Thus we are dealing with *vertical* excitation energies; see the CIS discussion in Section 16.1.) By using certain mathematical techniques, one can solve for the quantity $E_j - E_0$ for individual excited electronic states. Moreover, one can also calculate the strength of the electronic absorption to state j, so the electronic absorption spectrum can be predicted. By repeating the TDDFT calculation at various molecular geometries, one can find $E_j - E_0$ at each of these geometries for state j. Use of DFT-calculated E_0 values at these geometries then gives the potential-energy surface (Section 15.10) for the excited state j. Methods to find excited-state energy gradients analytically in LR-TDDFT have been developed, which allows efficient location of the equilibrium geometry of an excited state. Note that LR-TDDFT calculates properties of an excited state by examining the response of the ground-state electron density to the applied field.

The results of LR-TDDFT for vertical electronic excitation energies are quite good, with typical errors lying in the range 0.1 to 0.5 eV for situations where the theory is applicable, which is much better than for the CIS method (Section 16.1), which gives errors of 1 to 2 eV. Moreover, excited-state equilibrium geometries and vibrational frequencies are accurately calculated. However, LR-TDDFT works well only for excitations of a single valence electron to a low-lying excited state, and fails for molecules involving long-chain conjugated π–electron systems and for excited states involving charge transfer. Whether future functionals will increase the applicability of LR-TDDFT is unclear.

A review of single-reference methods for large-molecule excited-state calculations noted that LR-TDDFT can treat systems with up to 200 to 300 first-row atoms and stated that LR-TDDFT "is the most prominent method for the calculation of excited states of medium-sized and large molecules" [A. Dreuw and M. Head-Gordon, *Chem. Rev.*, **105**, 4009 (2005)].

The Past and Future of DFT. Hohenberg, Kohn, and Sham's work was published in 1964 and 1965. [For a personal account, see P. C. Hohenberg, W. Kohn, and L. J. Sham, *Adv. Quantum Chem.*, **21**, 7 (1990).] Quickly thereafter, physicists applied KS DFT using the LSDA to study solids with considerable success, and DFT became the dominant method of doing quantum-mechanical calculations on solids. Chemists were rather slow to apply DFT to molecules, because of numerical difficulties in doing reliable DFT molecular calculations, the lack of widely available molecular DFT computational programs, and perhaps partly in reaction to disappointment with the Xα method, which had been overpraised by some of its practitioners. The numerical difficulties were largely solved around 1980, and LSDA DFT molecular calculations in the 1980s achieved good results for molecular geometries but failed to give accurate dissociation energies.

The next major advance in DFT was the introduction of gradient-corrected functionals in the mid-1980s, which Becke found to give accurate dissociation energies. Also in 1988, analytic gradients were implemented in DFT, greatly facilitating calculation of equilibrium geometries. The 1989 DFT book by *Parr and Yang* helped draw the attention of quantum chemists to DFT. In 1993, provision for DF calculations was added to the popular program *Gaussian*. In the mid-1990s molecular DFT calculations experienced explosive growth.

DFT has the advantage of allowing for correlation effects to be included in a calculation that takes roughly the same time as a Hartree–Fock calculation, which does not include correlation. A 1996 review article [M. Head-Gordon, *J. Phys. Chem.*, **100**, 13220 (1996)] gave the following rough estimates of the maximum number of first-row nonhydrogen atoms in a molecule with no symmetry for which 10 energy and gradient evaluations could be done on a high-end workstation using a DZP basis set and various methods:

FCI	CCSD(T)	CCSD	MP2	HF	KS DFT
2	8 to 12	10 to 15	25 to 50	50 to 200	50 to 200

Whether KS DFT should be classified as an ab initio method is a matter of debate. If the true E_{xc} were known and used, then KS DFT would be an ab initio method. However, the true E_{xc} is unknown and must be replaced by a model E_{xc}, such as the LSDA or the LSDA with gradient corrections. Some people would consider that use of E_{xc}^{LSDA} disqualifies KS DFT as being an ab initio method, but others would not. Many of the gradient-corrected functionals contain empirical parameters, and in hybrid functionals, the mixing constant(s) are determined empirically. Use of functionals with empirically determined parameters clearly disqualifies a method as being ab initio, but the number of parameters used in these versions of DFT is far fewer than the number used in common semiempirical theories such as AM1 or PM3 (Section 17.4), which use different parameters for each kind of atom, which is not true in DFT. The KS DFT method is usually considered in a category by itself, distinct from ab initio methods such as HF, CI, MP, and CC.

Despite its successes, DFT is not a panacea. Some drawbacks and failings of DFT are the following.

The Hohenberg–Kohn–Sham theory is basically a ground-state theory. LR-TDDFT can only be applied to certain kinds of excited states. (One can use DFT to calculate the lowest state of each symmetry; for example, one can calculate the lowest singlet state and the lowest triplet state.)

Because approximate functionals are used, KS DFT is not variational and can yield an energy below the true ground-state energy. For example, a B3LYP/cc-pVTZ geometry optimization of H_2O gives an energy of -76.460 hartrees, compared with the true nonrelativistic energy of -76.438 hartrees (Table 16.1). Calculations with gradient-corrected functionals are size-consistent.

The true E_{xc} contains a self-interaction correction that exactly cancels the self-interaction energy in $\frac{1}{2}\iint \rho(\mathbf{r}_1)\rho(\mathbf{r}_2)r_{12}^{-1}\,d\mathbf{r}_1\,d\mathbf{r}_2$, but most currently used functionals are not completely free of self-interaction. Because of the self-interaction error, most currently available functionals give very incorrect $U(R)$ curves at large internuclear distances for symmetrical radical ions such as H_2^+, He_2^+, and F_2^+ and overestimate the intermolecular interaction in some charge-transfer complexes [Y. Zhang and W. Yang, *J. Chem. Phys.*, **109**, 2604 (1998)].

Although KS DFT yields good results for most molecular properties, with the presently available functionals, KS DFT cannot match the accuracy that the ab initio CCSD(T) method can achieve. Of course, CCSD(T) is limited to dealing with rather small molecules, whereas DFT can handle large molecules.

With methods such as CC, CI, and MP, the way to achieve more accurate results is clear. One uses larger basis sets and goes to higher orders of correlation (CCSD, CCSDT, . . . ; CISD, CISDT, . . . ; MP2, MP3, . . . , provided the perturbation series converges), although how far one can go for a given size molecule is limited by currently available computing power. In KS DFT, there is no clear way to construct more accurate E_{xc} functionals. One must try out new functionals one by one to see whether they will give improved results.

Many of the currently used E_{xc} functionals fail for van der Waals molecules. For example, the BLYP, B3LYP, and BPW91 functionals predict no binding in He_2 and Ne_2. However, the PBE functional works fairly well here [Y. Zhang et al., *J. Chem. Phys.*, **107**, 7921 (1997)]. Also, Adamo and Barone modified the parameters in the PW91 exchange functional to give the modified Perdew–Wang (mPW or MPW) exchange functional and found that the hybrid GGA one-parameter functional $E_{xc} = (1 - a)E_x^{\text{mPW91}} + aE_x^{\text{exact}} + E_c^{\text{PW91}}$ with $a = 0.25$ (called mPW1PW91) performs fairly well for He_2 and Ne_2 and works well for bond lengths, atomization energies, and vibrational frequencies of ordinary molecules [C. Adamo and V. Barone, *J. Chem. Phys.*, **108**, 664 (1998)]. A study of van der Waals dimers of rare-gas atoms and alkaline earth atoms using many functionals found the best performance for binding energies and geometries was given by the M05-2X functional [Y. Zhao and D. G. Truhlar, *J. Phys. Chem. A*, **110**, 5121 (2006)].

Noncovalent interactions are crucial in biological molecules. A test of 40 density functionals with the 6-31+G(d,p) basis set against a database of noncovalent interaction energies of biological significance, including both hydrogen-bonding systems and dispersion-dominated systems, found the M05-2X and the PWB6K (see the next paragraph) functionals gave the best results, with mean absolute errors around 1 kcal/mol [Y. Zhao and D. Truhlar, *J. Chem. Theory Comput.*, **3**, 289 (2007)].

Most of the currently available KS DFT functionals give only fair results for activation energies of reactions. However, a study of many methods of calculating barrier heights found good results from a few functionals, including PWB6K, BMK, BB1K, and MPW1K [J. Zheng et al., *J. Chem. Theory Comput.*, **3**, 5691 (2007)], with mean absolute errors running $1\frac{1}{4}$ to 2 kcal/mol with modest-size basis sets. The functional BB1K (where K stands for kinetics) is the same as B1B95 (see above) except that the parameter a_0 was fitted to kinetics data to give $a_0 = 0.42$. MPW1K is the mPW1PW91 functional with the parameter a changed to 0.428. PWB6K is a six-parameter hybrid meta-GGA functional based on the PW91 exchange functional and the B95 correlation functional; it has six parameters because the parameters within the exchange and correlation functionals (along with the fraction of exact exchange) were adjusted to fit kinetics data.

Although GGA, meta-GGA, and hybrid functionals usually give good results for molecular properties, currently available forms of these functionals are known to be significantly in error. DFT shows that the exact E_x, E_c, v_x, and v_c must satisfy certain conditions, and currently available functionals violate at least some of these conditions [see C. Filipi et al., in J. M. Seminario (ed.), *Recent Developments and Applications of Modern Density Functional Theory*, Elsevier, 1996, p. 295]. The Hohenberg–Kohn theorem applied to the reference system of noninteracting electrons assures us that the true ground-state electron density determines the external potential v_s in (16.48). Iterative methods have been devised that take a very accurate ground-state molecular electron density found from a high-level calculation (for example, CI) and use it to calculate v_s for the corresponding reference system. From v_s, we can use (16.47) and (16.48) to find $v_{xc}(\mathbf{r})$. The accurate v_{xc} found from an accurate ρ for a particular atom or molecule can then be compared with the v_{xc}'s found from currently used E_{xc}'s. The results show that many currently used v_{xc}'s are substantially in error [M. E. Mura, P. J. Knowles, and C. A. Reynolds, *J. Chem. Phys.*, **106**, 9659 (1997); E. J. Baerends et al., in B. B. Laird et al. (eds.) *Chemical Applications of Density-Functional Theory*, American Chemical Society, 1996, Chapter 2].

DFT is now the most common method used in quantum chemistry calculations. Also, DFT has been used to give quantitative definitions of such chemical concepts as electronegativity, hardness and softness, and reactivity; see *Parr and Yang*, Chapters 5 and 10; W. Kohn, A. D. Becke, and R. G. Parr, *J. Phys. Chem.*, **100**, 12974 (1996); P. Geerlings et al., *Chem. Rev.*, **103**, 1793 (2003).

16.5 COMPOSITE METHODS FOR ENERGY CALCULATIONS

A desirable goal is to compute a thermodynamic energy such as the molecular atomization energy or the enthalpy of formation, with chemical accuracy, which means an accuracy of 1 kcal/mol. Currently available functionals in DFT cannot do this. High-level methods such as CCSD(T), QCISD(T), and CISDTQ, with large basis sets, can do this but are much too costly to be feasible except for quite small molecules. The aim of the compound methods discussed in this section is to achieve 1 kcal/mol or better accuracy with a computational time that allows calculations on molecules containing several nonhydrogen atoms.

These compound methods use a series of ab initio calculations plus empirical corrections. The **Gaussian-3 (G3) method** (so called because it is an improvement on its predecessors, the G1 and G2 methods) is designed to give a result close to what would be obtained by a QCISD(T)/G3large calculation in much less computer time than required by such a calculation [L. A. Curtiss et al., *J. Chem. Phys.*, **109**, 7764 (1998); **114**, 9287 (2001)]; G3large is an improved version of the 6-311+G(3df,2p) basis set. In the G3 method, the zero-point energy is found by scaling the frequencies found in a HF/6-31G* frequency calculation with the factor 0.893. All subsequent calculations are done at the optimized geometry found in an MP2/6-31G* calculation. One then computes a base energy E^{base} from an MP4/6-31G* calculation. Various corrections to E^{base} are then found as differences between E^{base} and energies obtained from MP4/6-31+G(d), MP4/6-31G(2df,p), QCISD(T)/6-31G(d), and MP2/G3large single-point calculations. These corrections are added to E^{base}, and an empirical *higher-level correction* of $-An_{\text{pairs}} - Bn_{\text{unpaired}}$ hartrees is added to correct for basis-set incompleteness. Here, n_{pairs} is the number of valence electron pairs in the molecule and n_{unpaired} is the number of unpaired electrons; A and B are empirical parameters with $A = 0.0006386$, $B = 0.0002977$ for molecules and $A = 0.0006219$, $B = 0.0001185$ for atoms. The various corrections to E^{base} allow for the effects of including diffuse basis functions, polarization basis functions, and higher levels of electron correlation.

To test the accuracy of the G3 method, its developers compiled a set of accurately known thermochemical data (the **G3/05 test set**), consisting of 270 enthalpies of formation, 105 ionization energies, 63 electron affinities, 10 proton affinities, and 6 dimerization energies of H-bonded dimers [L. A. Curtiss et al., *J. Chem. Phys.*, **123**, 124107 (2005)]. Some medium-size molecules in the G3/05 set are naphthalene ($C_{10}H_8$), SF_6, and $C_6H_{13}Br$. The G3/05 set does not include any transition-metal compounds. (Accurate thermochemical data for such compounds is sparse and the G3 method has only been applied to nontransition-metal compounds.) The G3/05 set (which contains 454 energy changes) is an expansion of the earlier test sets G2 (125 energy changes), G2/97 (301 energy changes), and G3/99 (376 energy changes).

G3//B3LYP (also called G3B3 or G3B) is a modification of G3 that uses geometries and zero-point vibrational energies found from B3LYP/6-31G* calculations instead of from MP2/6-31G* and HF/6-31G* calculations [A. G. Baboul, *J. Chem. Phys.*, **110**, 7650 (1999)]. G3//B3LYP is faster than G3 and just as accurate as G3.

G3X (where the X stands for extension) is an improvement of the G3 method and uses B3LYP/6-31G(2df,p) geometries and zero-point energies and adds a *g* polarization function to the G3Large basis set for second- and third-row atoms [L. A. Curtiss et al.,

J. Chem. Phys., **114**, 108 (2001)]. For the G3/05 test set the mean absolute deviation of calculated values from experimental values is 1.01 kcal/mol for G3X and is 1.13 kcal/mol for G3 [L. A. Curtiss et al., *J. Chem. Phys.*, **123**, 124107 (2005)].

G3S and G3SX are modifications of G3 and G3X, respectively, in which instead of using the additive higher-level correction, one multiplies each of the various parts of the Hartree–Fock and the correlation energies by an empirically determined scale factor (six such factors are used). These methods give similar performance to G3 and G3X, and have the advantage of being applicable to determining the potential-energy surface of reactions where the products have a different number of unpaired electrons than the reactants [L. A. Curtiss et al., *J. Chem. Phys.*, **112**, 1125 (2000)].

The MP4 calculations are the most time-consuming steps in the G3 and G3X methods and prevent their applicability to large molecules. The G3(MP2) and G3X(MP2) methods [L. A. Curtiss et al., *J. Chem. Phys.*, **110**, 4703 (1999); **114**, 108 (2001)] are modifications of G3 and G3X that replace MP4 calculations with MP2 calculations, thereby speeding up the calculations and allowing somewhat larger molecules to be treated. For the G3/99 test set, mean absolute deviations from experiment are 1.31 kcal/mol for G3(MP2), 1.19 kcal/mol for G3X(MP2), 1.07 kcal/mol for G3, and 0.95 kcal/mol for G3X.

The **G4 method** is an improvement on G3 and differs from G3 as follows: G4 uses an extrapolation procedure [Eq. (15.23)] to estimate the Hartree–Fock energy in the complete-basis-set limit; uses a larger basis set; replaces the QCISD(T) calculation by a CCSD(T) calculation; uses B3LYP/6-31G(2df,p) to find the equilibrium geometry and the zero-point energy; and includes two additional empirical parameters in the higher-level correction [L. A. Curtiss et al., *J. Chem. Phys.*, **126**, 084108 (2007)]. For the G3/05 test set G4 has a mean absolute error of 0.83 kcal/mol, compared with 1.13 kcal/mol for G3. Relative computation times for benzene are 2 for G2, 1 for G3, and 3 for G4.

The **correlation-consistent composite approach** (ccCA) is based in part on the G3B method, but uses Dunning correlation-consistent basis sets instead of Pople basis sets, does extrapolations to the CBS limit, replaces the QCISD(T) calculation with a CCSD(T) calculation, includes a relativistic correction, and contains no empirical parameters [N. J. DeYonker et al., *J. Chem. Phys.*, **125**, 104111 (2006)]. For the G3/99 test set the mean absolute deviations from experiment are 0.96 and 0.97 kcal/mol for the versions ccCA-P and ccCA-S4 (which differ in the extrapolation formulas used), compared with 1.16 and 0.95 kcal/mol for G3 and G3X, respectively.

The CBS-Q, CBS-QB3, CBS-q, and CBS-4 methods are varieties of **CBS** (complete basis set) compound methods devised for calculations on molecules containing the atoms H to Ar. These methods use procedures designed to extrapolate calculated energies to the complete-basis-set limit. Like G3, the CBS methods include several correlated calculations done at a geometry optimized at a lower level of theory. The highest-level calculation used is QCISD(T)/6-31+G† (the 6-31+G† basis set is a slightly improved version of 6-31G*) in the CBS-Q and CBS-QB3 methods, is QCISD(T)/6-31G in the CBS-q method, and is MP4(SDQ)/6-31G in the CBS-4 method. Like G3, the CBS methods contain empirical corrections. For details see J. W. Ochterski, G. A. Petersson, and J. A. Montgomery, *J. Chem. Phys.*, **104**, 2598 (1996); J. A. Montgomery et al., *J. Chem. Phys.*, **110**, 2822 (1999).

The composite methods W1, W2, W3, and W4 (where the W stands for the Weizmann Institute, where the methods were developed) use high-level coupled-cluster calculations to achieve extraordinary accuracy in thermochemical quantities [A. Karton et al., *J. Chem. Phys.*, **125**, 144108 (2006) and references cited therein]. W1 and W2 use CCSD(T) and CCSD calculations with correlation-consistent basis sets, do extrapolations to the complete basis-set limit, and include relativistic corrections. W3 and W4 include CCSDT and CCSDTQ calculations, and W4 includes a CCSDTQ5 calculation with a small basis set. For various test sets of small molecules, the mean absolute deviation from experimental atomization energies or heats of formation is 0.6 kcal/mol for W1, 0.5 kcal/mol for W2, 0.2 kcal/mol for W3, and 0.1 kcal/mol for W4. These methods are limited to small molecules.

16.6 THE DIFFUSION QUANTUM MONTE CARLO METHOD

A **quantum Monte Carlo** (QMC) **method** uses a random process to solve the Schrödinger equation. Many QMC methods exist, but the **diffusion QMC** (DQMC) method is most commonly used for molecular calculations. Defining the imaginary time variable $\tau \equiv it/\hbar$, one finds (Prob. 16.25) that in the limit $\tau \to \infty$, the time-dependent state function Ψ (no matter what its form Ψ_0 is at $\tau = 0$, provided Ψ_0 includes some mixture of the ground state) becomes proportional to the ground-stationary-state wave function ψ_{gs}. Moreover, the equation for $\partial\Psi/\partial\tau$ has the same form as Fick's second law for the diffusion of a substance in a $3n$-dimensional space (where n is the number of electrons) with an added term corresponding to the substance undergoing a first-order reaction (with a rate constant that can take on positive or negative values) as well as diffusing. One simulates the diffusion and reaction processes on a computer by starting with a large number of particles (called *walkers*) distributed in the regions near the nuclei. One chooses a step size $\Delta\tau$ and at each increment in τ, the particles undergo random changes in coordinates, where the probability for a particular size change Δx in each coordinate is taken to be proportional to the probability that a diffusing particle will suffer a displacement by Δx in a given direction in time $\Delta\tau$; also, at each step $\Delta\tau$, each particle has a probability to either disappear or give birth to a second particle, corresponding to the first-order reaction term. Eventually, the distribution of walkers in $3n$-dimensional space will be proportional to the ground-state wave function, provided proper allowance is made for the antisymmetry requirement for the wave function.

Because of the antisymmetry requirement, the ground-state wave function has nodal surfaces in $3n$-dimensional space (see Prob. 16.26), and to ensure that the walkers converge to the ground-state wave function, one must know the locations of these nodes and must eliminate any walker that crosses a nodal surface in the simulation. In the **fixed-node** (FN) DQMC method, the nodes are fixed at the locations of the nodes in a known approximate wave function for the system, such as found from a large basis-set Hartree–Fock calculation. This approximation introduces some error, but FN-DQMC calculations are variational. In practice, the accuracy of FN-DQMC calculations is improved by a procedure called importance sampling. Here, instead of simulating the evolution of Ψ with τ, one simulates the evolution of f, where $f \equiv \Psi\psi_{tr}$, where ψ_{tr} is a known accurate trial variation function for the ground state. ψ_{tr} guides the random walkers to regions where $|\psi_{tr}|$ is large. ψ_{tr} typically has the form of a function of interelectronic coordinates r_{ij} multiplied by a linear combination of one or more Slater determinants of HF orbitals. (This ψ_{tr} can also be used to define the locations of the nodes.)

QMC calculations are especially well suited to be run on parallel computers, and the Quantum Monte Carlo at Home project using the home computers of tens of thousands of volunteers to run FN-DQMC calculations (qah.uni-muenster.de/index.php) has calculated accurate interaction energies for the DNA base pairs adenine–thymine and guanine–cytosine.

QMC methods have given very accurate results in some calculations on small systems, but the method can require very long calculation times, and no efficient method has yet been found to allow geometry optimization in a QMC calculation. (The FN-DQMC calculations on H_2O in Table 16.1 were done at the experimental geometry.) For a proposed FN-DQMC geometry optimization procedure, see S.-I. Lu, *J. Chem. Phys.*, **120**, 14 (2004). Currently, none of the widely used quantum chemistry program packages include the DQMC method, and its use is restricted to specialists. For more on DQMC, see Probs. 16.25 and 16.26, and J. B. Anderson, *Int. Rev. Phys. Chem.*, **14**, 85 (1995); K. Raghavachari and J. B. Anderson, *J. Phys. Chem.*, **100**, 12960 (1996); A. Lüchow and J. B. Anderson, *Ann. Rev. Phys. Chem.*, **51**, 501 (2000).

16.7 RELATIVISTIC EFFECTS

For a nonrelativistic hydrogenlike atom, the root-mean-square speed of a 1*s* electron is $Zc/137$, where Z is the nuclear charge and c is the speed of light (Prob. 16.27). Hence, for atoms of high atomic number, the average speed of inner-shell electrons is a significant fraction of c, and relativistic corrections to inner-shell orbitals and orbital energies are substantial for high-Z atoms. The valence electrons in an atom or molecule are well shielded from the nuclei, and their average speeds are far less than the speed of light, even in heavy atoms. Hence, it was formerly believed that one need not worry about relativistic corrections in molecules containing high-Z atoms. It is now realized that relativistic effects on properties of molecules with heavy atoms can be quite substantial.

The average radius of a hydrogenlike atom is proportional to the Bohr radius a_0, and a_0 is inversely proportional to the electron mass. Hence, the relativistic increase of mass with velocity shrinks the inner *s* orbitals in a heavy atom. To maintain their orthogonality to the inner *s* orbitals, the outer *s* orbitals are required to shrink also. The relativistic increase of mass also shrinks the *p* orbitals, but to a lesser extent than the *s* orbitals. Because of the relativistic shrinkage of the *s* and *p* orbitals, these orbitals screen the nucleus more effectively than in nonrelativistic atoms, and this produces an expansion of the *d* and *f* orbitals. Calculated relativistic contractions of the 6*s* orbital average radius in some atoms are 4% for $_{55}$Cs, 7% for $_{70}$Yb, 12% for $_{75}$Re, 18% for $_{79}$Au, and 12% for $_{86}$Rn [P. Pyykkö, *Chem. Rev.*, **88**, 563 (1988)]. Because of relativistic contraction, the atomic radius of Fr is less than that of Cs, which lies above Fr in the periodic table.

The relativistic form of the one-electron Schrödinger equation is the Dirac equation. One can do relativistic Hartree–Fock calculations using the Dirac equation to modify the Fock operator, giving a type of calculation called **Dirac–Fock** (or Dirac–Hartree–Fock). Likewise, one can use a relativistic form of the Kohn–Sham equations (16.49) to do relativistic density-functional calculations. (Relativistic Xα calculations are called Dirac–Slater or Dirac–Xα calculations.) Because of the complicated structure of the relativistic KS equations, relatively few all-electron fully relativistic KS molecular calculations that go beyond the Dirac–Slater approach have been done. [For relativistic DFT, see E. Engel and R. M. Dreizler, *Topics in Current Chemistry*, **181**, 1 (1996).]

All-electron Dirac–Fock relativistic calculations on molecules containing heavy atoms such as Au or U are very time-consuming. A commonly used approach is to do an all-electron atomic Dirac–Fock calculation on each type of atom in the molecule and use the result to derive a **relativistic effective core potential** (RECP) or pseudopotential (Section 15.16) for that atom. (Since the smallest parts of the relativistic effects are neglected in deriving RECPs, RECPs are sometimes called quasirelativistic ECPs.) One then does a molecular Hartree–Fock calculation in which only the valence electrons are treated explicitly. The valence electrons are treated nonrelativistically, and the effects of the core electrons are represented by adding the operator $\sum_\alpha \hat{U}_\alpha$ to the Fock operator $\hat{F}$, where $\hat{U}_\alpha$ is a relativistic ECP for atom α, and the sum goes over the atoms of the molecule. Here, it is assumed that the inner-shell AOs are not significantly changed on going from isolated atoms to the molecule. The results of the SCF calculation can be improved using CI or MP perturbation theory. MCSCF and MCSCF-CI calculations with RECPs are also done. RECPs can also be used in KS DFT calculations. RECPs for most elements of the periodic table are available at www.clarkson.edu/~pac/reps.html.

Another approach is to do a nonrelativistic calculation, using, for example, the Hartree–Fock method, and then use perturbation theory to correct for relativistic effects. For perturbation-theory formulations of relativistic Hartree–Fock calculations and relativistic KS DFT calculations, see W. Kutzelnigg, E. Ottschofski, and R. Franke, *J. Chem. Phys.*, **102**, 1740 (1995) and C. van Wüllen, *J. Chem. Phys.*, **103**, 3589 (1995); **105**, 5485 (1996).

Some examples of relativistic effects on molecular properties found from Hartree–Fock calculations with RECPs follow, where the numbers are listed in the order: nonrelativistic-ECP calculated value, relativistic-ECP calculated value, experimental value (see P. Pyykkö, op. cit. for references). (Also listed in parentheses are calculated nonrelativistic, calculated relativistic, and experimental values, where the calculated values are from nonrelativistic KS DFT and relativistic perturbation-theory KS DFT calculations using the B88P86 functional and a contracted Gaussian basis set; see C. van Wüllen, op. cit.) Equilibrium bond lengths: 2.80, 2.73, 2.48 Å in Ag_2; 1.73, 1.71, 1.70 Å in SnH_4; 1.76, 1.51, 1.52 Å (1.73, 1.56, 1.52 Å) in AuH; 2.83, 2.48, 2.47 Å (2.77, 2.58, 2.47 Å) in Au_2; 2.81, 2.53, 2.50 Å in Hg_2^{2+}; (2.10, 2.06, 2.06 Å) for the W–C length in $W(CO)_6$. Bond angle: 98.6°, 98.2°, 98° in $PbCl_2$. Dipole moments: 1.02, 0.92, 0.83 D for HBr, 0.71, 0.52, 0.45 D for HI. Harmonic vibrational wavenumber: 77, 163, 191 cm^{-1} (121, 165, 191 cm^{-1} for Au_2. Equilibrium dissociation energies: (52.3, 69.1, 77.4 kcal/mol) for AuH; (33.0, 47.1, 53.1 kcal/mol) for Au_2. Relativistic effects are substantial for these heavy-atom molecules.

Relativistic all-electron calculations on F_2, Cl_2, Br_2, I_2, and At_2 using various correlation methods found that the relativistic corrections were approximately independent of the level of theory used, except for At_2 [L. Visscher and K. G. Dyall, *J. Chem.*

Phys., **104**, 9040 (1996)]. For example, with a valence triple-zeta polarized basis set, the changes in D_e on going from a nonrelativistic to a relativistic calculation were −8, −7, −7, −7, −7 kcal/mol for Br_2 using the HF, MP2, CISD, CCSD, CCSD(T) methods, respectively; for I_2, these changes were −15, −13, −12.5, −13, −13 kcal/mol; for At_2, they were −30, −27, −24, −25, −24 kcal/mol.

Some reviews of relativistic quantum chemistry are P. Pyykkö, *Chem. Rev.*, **88**, 563 (1988); W. C. Ermler, R. B. Ross, and P. A. Christiansen, *Adv. Quantum. Chem.*, **19**, 139 (1988); K. Balasubramanian, *J. Phys. Chem.*, **93**, 6585 (1989); S. Wilson et al. (eds.), *The Effects of Relativity in Atoms, Molecules, and the Solid State*, Plenum, 1991; J. Almlöf and O. Gropen in K. B. Lipkowitz and D. B. Boyd (eds.), *Reviews in Computational Chemistry*, Vol. 8, Chapter 4, Wiley-VCH; N. Kaltsoyannis, *J. Chem. Soc., Dalton Trans.*, **1997**, 1; K. G. Dyall and K. Faegri, *Introduction to Relativistic Quantum Chemistry*, Oxford, 2007.

16.8 VALENCE-BOND TREATMENT OF POLYATOMIC MOLECULES

The valence-bond treatment of polyatomic molecules is closely tied to chemical ideas of structure. One begins with the atoms that form the molecule and pairs up the unpaired electrons to form chemical bonds. There are usually several ways of pairing up (coupling) the electrons. Each pairing scheme gives a VB **structure**. A Heitler–London-type function Φ_i (called a **bond eigenfunction**) is written for each structure i, and the molecular wave function ψ is taken as a linear combination $\sum_i c_i\Phi_i$ of the bond eigenfunctions. The variation principle is then applied to determine the coefficients c_i. The VB wave function is said to be a **resonance hybrid** of the various structures.

The weight of each resonance structure in the wave function is sometimes taken as proportional to the square of its coefficient in the wave function. Because the bond eigenfunctions are not mutually orthogonal, the electron probability density is not equal to the weighted sum of the probability densities of the various structures, and the $|c_i|^2$ quantities are somewhat lacking in direct physical significance. There are several other ways of assigning weights to VB structures. One procedure defines the *occupation number* n_i for the VB structure i as $n_i \equiv c_i^* \sum_j c_j S_{ij}$, where the sum goes over all the VB structures and $S_{ij} = \langle \Phi_i | \Phi_j \rangle$. The n_i's add to 1 (Prob. 16.28).

Water. The oxygen-atom ground-state electron configuration is $1s^2 2s^2 2p^4$ with an unpaired electron in each of the AOs $2p_y$ and $2p_z$. We thus assume that these AOs along with the hydrogen $1s$ AOs will form electron-pair bonds. The three possible ways of pairing these four AOs to get covalent structures are shown in Fig. 16.1.

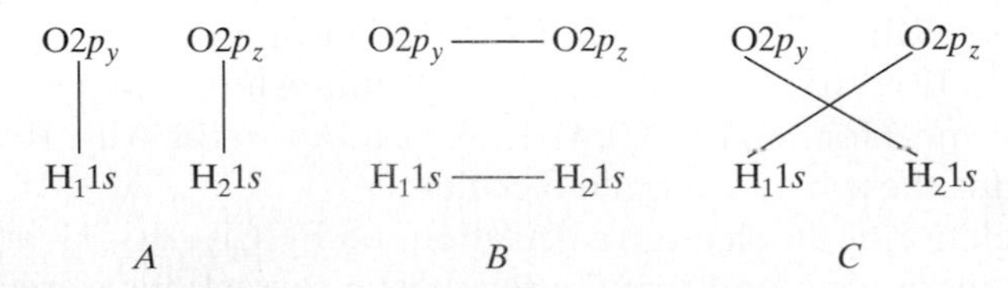

FIGURE 16.1 Ways of pairing valence AOs in H_2O.

Let

$$s_1 \equiv \mathrm{H}_1 1s, \qquad s_2 \equiv \mathrm{H}_2 1s \qquad p_y \equiv \mathrm{O}2p_y, \qquad p_z \equiv \mathrm{O}2p_z$$

The normalized VB function corresponding to structure A is [see Eq. (13.119)]

$$\Phi_A = N|\cdots \widehat{p_y s_1}\widehat{p_z s_2}| = N|\cdots p_y\bar{s}_1 p_z \bar{s}_2| - N|\cdots \bar{p}_y s_1 p_z \bar{s}_2|$$
$$- N|\cdots p_y \bar{s}_1 \bar{p}_z s_2| + N|\cdots \bar{p}_y s_1 \bar{p}_z s_2| \tag{16.69}$$

where the signs are determined by the rule in Section 13.12. The dots stand for $\mathrm{O}1s\overline{\mathrm{O}1s}\mathrm{O}2s\overline{\mathrm{O}2s}\mathrm{O}2p_x\overline{\mathrm{O}2p_x}$. Similarly,

$$\Phi_B = N|\cdots \widehat{p_y p_z}\widehat{s_1 s_2}|, \quad \Phi_C = N|\cdots \widehat{p_y s_2}\widehat{p_z s_1}| \tag{16.70}$$

We then take as a trial variation function ψ, a linear combination of the bond eigenfunctions of structures A, B, and C. However, the functions Φ_A, Φ_B, and Φ_C are not linearly independent; we have (Prob. 16.29)

$$\Phi_C = -(\Phi_A + \Phi_B) \tag{16.71}$$

It is wasted effort to include all three structures; we shall drop structure C, taking $\psi = c_A\Phi_A + c_B\Phi_B$.

For a molecule with n valence AOs to be paired, where n is even, Rumer showed in 1932 that the following procedure gives the linearly independent covalent structures for singlet states: The n AOs are arranged in a ring and lines are drawn between pairs of AOs; those structures where no lines cross are linearly independent. These are the VB **canonical covalent structures** of the molecule. Any structure with lines crossing is linearly dependent on the canonical structures and is omitted from the VB wave function. The number of ways of drawing $\frac{1}{2}n$ noncrossing lines between n points on a circle is

$$\frac{n!}{\left(\frac{1}{2}n\right)!\left(\frac{1}{2}n + 1\right)!} \tag{16.72}$$

[For a proof of (16.72), see J. Barriol, *Elements of Quantum Mechanics with Chemical Applications*, Barnes and Noble, 1971, pages 281–282.] H_2O has $4!/2!3! = 2$ canonical covalent structures, and these are A and B. (Actually, which structures are taken as the canonical ones is arbitrary, since the orbitals can be arranged in various ways on the ring.) To use Rumer's method when the number of orbitals to be paired is odd, we add a "phantom" orbital, whose contribution is subtracted at the end of the calculation.

Rumer's procedure is easily justified. Let $\Phi(\mid\;\mid)$, $\Phi(\sqcap)$, and $\Phi(\times)$ be three bond eigenfunctions that involve any number of AOs, but that differ only in the way they pair up a certain subset of four AOs. Each of these functions corresponds to one of the three different ways of pairing these four AOs (see Fig. 16.1). By a slight extension of (16.71), it follows that

$$\Phi(\times) = -\left[\Phi(\mid\;\mid) + \Phi(\sqcap)\right] \tag{16.73}$$

Any pairing scheme involving lines that cross can be shown by repeated application of (16.73) to be a linear combination of structures with no lines crossing. (See Prob. 16.30.)

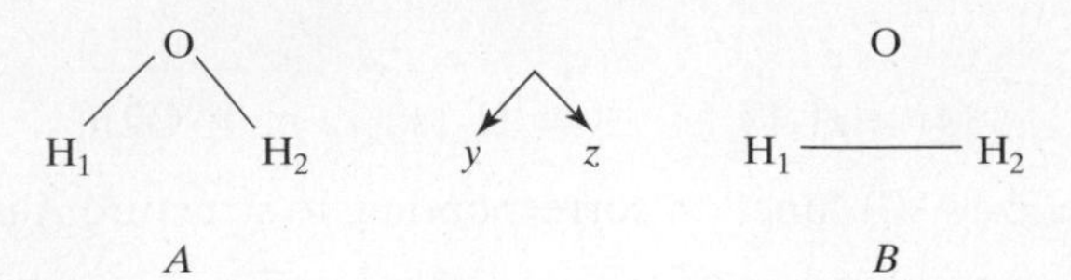

FIGURE 16.2 Structures corresponding to the pairing schemes A and B of Fig. 16.1.

The structures corresponding to the pairing schemes A and B for H_2O in Fig. 16.1 are shown in Fig. 16.2.

The separation between H_1 and H_2 is considerably greater than between O and each hydrogen. The small overlap between the hydrogen 1s AOs makes structure A far more significant than structure B. Since the $2p_y$ and $2p_z$ AOs are at 90° to each other, the VB structure A predicts a bond angle of 90°, since this allows maximum overlap between the bonding oxygen and hydrogen AOs. The observed angle of $104\frac{1}{2}°$ can be rationalized by considering electrostatic repulsions between the hydrogen atoms (ionic structures) and by allowing some mixing in (hybridization) of the 2s oxygen AO with the two bonding 2p AOs.

For H_2 the Heitler–London function was improved by inclusion of small contributions from ionic structures. Because of the considerable electronegativity difference between O and H, we expect ionic structures to be important in H_2O. Some ionic structures for H_2O are shown in Fig. 16.3. Because oxygen is much more electronegative than hydrogen, the contributions of G, H, and I are likely to be quite small. The point group $\mathscr{C}_{2v}$ has no degenerate symmetry species, and all the electronic wave functions of H_2O must be eigenfunctions of the $\hat{O}_{C_2}$ operator (see Section 15.2). The ground electronic state belongs to the totally symmetric symmetry species. Therefore, the coefficients of Φ_D and Φ_E in the wave function must be equal. In place of the terms $c_D\Phi_D + c_E\Phi_E$ in the VB wave function, we write $c_{DE}[N(\Phi_D + \Phi_E)]$, where the normalization constant is $N = (2 + 2\langle\Phi_D|\Phi_E\rangle)^{-1/2}$. The function $N(\Phi_D + \Phi_E)$ is called a VB **symmetry function** (or symmetry structure), whereas Φ_D and Φ_E are called **individual** VB functions. Thus, a reasonable variation function for H_2O is

$$c_A\Phi_A + c_B\Phi_B + c_{DE}[N(\Phi_D + \Phi_E)] + c_F\Phi_F \tag{16.74}$$

The ionic functions are written by analogy to the equation preceding (13.128); for example,

$$\Phi_D = |\cdots \widehat{p_y\bar{s}_1} p_z\bar{p}_z| = |\cdots p_y\bar{s}_1 p_z\bar{p}_z| - |\cdots \bar{p}_y s_1 p_z\bar{p}_z|$$

The linear variation function (16.74) leads to a secular equation

$$\det(H_{ij} - ES_{ij}) = 0 \tag{16.75}$$

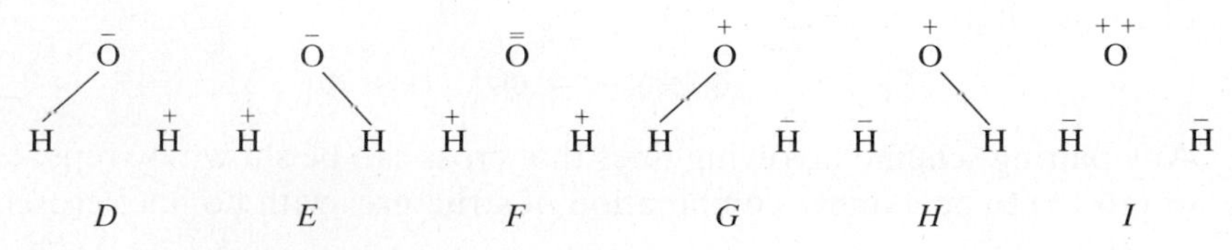

FIGURE 16.3 Some ionic structures for H_2O.

The AOs are expressed as STOs or as linear combinations of GTFs, and the matrix elements H_{ij} and overlap integrals between the functions $\Phi_A, \ldots, \Phi_F$ are calculated. The lowest root of the secular equation gives an approximation to the ground-state energy (higher roots correspond to excited singlet states). Evaluation of the corresponding coefficients gives the VB ground-state wave function. The calculation can be done semiempirically using experimental data to evaluate some of the integrals. Calculation of the matrix elements is involved, and approximations are often made; for example, overlap integrals between different AOs (but *not* between different structures) are neglected and exchange integrals involving interchange of the coordinates of more than two electrons are neglected. Systematic procedures for evaluating the matrix elements have been developed.

Peterson and Pfeiffer did an ab initio all-electron VB calculation on H_2O [C. Peterson and G. V. Pfeiffer, *Theor. Chim. Acta*, **26**, 321 (1972)]. They included 10 covalent and 39 ionic VB structures in their wave function. The large number of structures arises because they considered structures arising from such oxygen configurations as $1s^22s2p^5$ and $1s^22p^6$, as well as $1s^22s^22p^4$. The contribution of the ionic structures, as measured by the sum of their occupation numbers, was found to be 62%. The calculated energy and equilibrium geometry are -76.02 hartrees and 106.5°, 0.968 Å, which may be compared with the values in Table 15.2. They also did calculations on OH and O so as to calculate the first and second bond dissociation energies of H_2O. The calculated dissociation energies were not in good agreement with experiment.

Methane. The carbon-atom ground-state electron configuration $1s^22s^22p^2$ has two unpaired electrons and would seem to indicate a valence of 2. To get the well-known tetravalence of carbon, we assume that a $2s$ electron is promoted to the vacant $2p$ orbital, giving the configuration $1s^22s2p^3$. If we then assume one bond is formed with the $2s$ electron and three bonds are formed with the $2p$ electrons, the bonds are not all equivalent as they are known to be. Hence Pauling proposed that the $2s$ and $2p$ functions be linearly combined to form **hybridized** sp^3 atomic orbitals of the form

$$b_i(\mathrm{C}2s) + d_i(\mathrm{C}2p_x) + e_i(\mathrm{C}2p_y) + f_i(\mathrm{C}2p_z), \quad i = 1, \ldots, 4 \qquad (16.76)$$

For maximum overlap we want each function to point to a vertex of a tetrahedron. From the discussion following Eq. (15.47), the constants d_i, e_i, f_i are proportional to the direction cosines in (15.57). Also, each orbital should have the same value of b_i so that the four hybrid orbitals will be equivalent. Thus the orbitals (16.76) are of the form (15.58) with $a = 0$ and $2a_1, 1t_{2x}, 1t_{2y}, 1t_{2z}$, replaced by $\mathrm{C}2s, \mathrm{C}2p_x, \mathrm{C}2p_y, \mathrm{C}2p_z$. If we impose the requirement that the hybrid AOs be orthonormal, we have

$$b^2 + c^2 = 1, \quad b^2 - \tfrac{1}{3}c^2 = 0$$

$$c = \tfrac{1}{2}\sqrt{3}, \qquad b = \tfrac{1}{2}$$

The four equivalent sp^3 hybrid carbon AOs are then

$$\begin{aligned} \mathrm{te}_1 &= \tfrac{1}{2}[\mathrm{C}2s + \mathrm{C}2p_x + \mathrm{C}2p_y + \mathrm{C}2p_z] \\ \mathrm{te}_2 &= \tfrac{1}{2}[\mathrm{C}2s + \mathrm{C}2p_x - \mathrm{C}2p_y - \mathrm{C}2p_z] \\ \mathrm{te}_3 &= \tfrac{1}{2}[\mathrm{C}2s - \mathrm{C}2p_x + \mathrm{C}2p_y - \mathrm{C}2p_z] \\ \mathrm{te}_4 &= \tfrac{1}{2}[\mathrm{C}2s - \mathrm{C}2p_x - \mathrm{C}2p_y + \mathrm{C}2p_z] \end{aligned} \qquad (16.77)$$

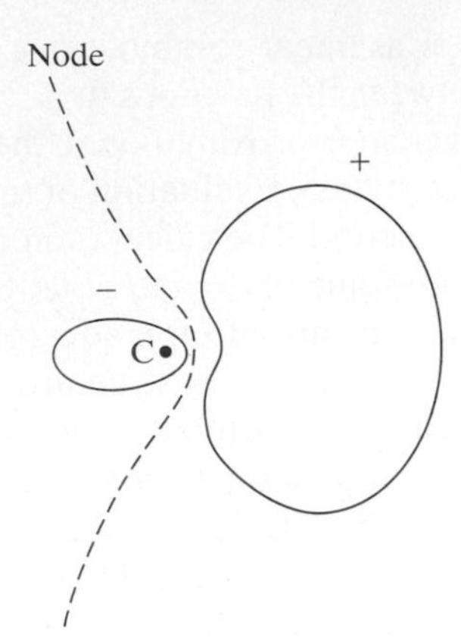

FIGURE 16.4 Carbon sp^3 hybrid orbital.

where "te" stands for tetrahedral. A typical sp^3 hybrid AO contour is shown in Fig. 16.4. (The three-dimensional shape is obtained by rotation about a horizontal axis through carbon.) Carbon sp^2 and sp hybrid AOs have similar shapes. [See I. Cohen and T. Bustard, *J. Chem. Educ.*, **43**, 187 (1966).]

There are many canonical covalent VB structures for methane, as well as ionic structures; however, chemical intuition suggests that the main contribution to the wave function is from the following covalent structure:

$$|\mathrm{C}1s\overline{\mathrm{C}1s}\ \widehat{s_1\mathrm{te}_1}\ \widehat{s_2\mathrm{te}_2}\ \widehat{s_3\mathrm{te}_3}\ \widehat{s_4\mathrm{te}_4}| \tag{16.78}$$

where $s_1 = \mathrm{H}_1 1s$, and so on. The structure (16.78) is a linear combination of $2^4 = 16$ determinants. Taking the CH_4 VB wave function as the single covalent structure (16.78) or the H_2O VB wave function as the function corresponding to structure A in Fig. 16.2 gives what is called the **perfect pairing** approximation.

Raimondi, Campion, and Karplus did an ab initio all-electron VB calculation on CH_4 using a minimal AO basis set of STOs [M. Raimondi, W. Campion, and M. Karplus, *Mol. Phys.*, **34**, 1483 (1977)]. Their wave function is a linear combination of 104 symmetry functions (and is a linear combination of a much larger number of individual functions) and contained 4900 Slater determinants. Very surprisingly, the function with the largest coefficient in the wave function is not the perfect-pairing function (16.78), but is an ionic symmetry function with two covalent bonds and two ionic bonds; in this function, one of the sp^3 hybrid AOs on C has two electrons and the H atom it points to has no electrons, giving a C^-H^+ ionic bond, and a second sp^3 hybrid has no electrons and the H atom it points to has two electrons, giving a C^+H^- ionic bond. The perfect-pairing function, with four covalent bonds, has the second-highest coefficient. Two symmetry functions with three covalent bonds and one ionic bond (one symmetry function with a $C^+\,H^-$ bond and one with a C^-H^+ bond) have coefficients almost as large as that of the perfect-pairing function. The perfect-pairing function has too little electron density in the overlap regions between atoms, and the great importance of the ionic structures is due to the fact that they increase the electron density in the overlap regions.

For ethylene the VB method uses sp^2-hybridized carbon AOs to form the σ bonds by overlap with the $1s$ hydrogen AOs. This leaves a p orbital on each carbon to form the

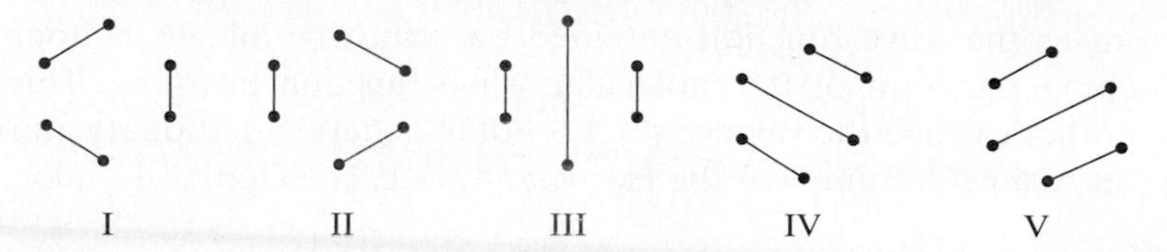

FIGURE 16.5 Canonical covalent structures for pairing the π orbitals in benzene.

bond. The HCH bond angle is predicted to be 120°, in reasonable agreement with the observed value of 117°. For acetylene each carbon is *sp* hybridized.

Conjugated Molecules. Consider benzene. The σ bonds are formed by sp^2 carbon hybrid AOs and 1*s* hydrogen orbitals. This leaves a *p* orbital on each carbon to form π bonds. There are $6!/3!4! = 5$ canonical covalent structures for pairing the π orbitals, and these are shown in Fig. 16.5. Structures I and II are the Kekulé structures, and III, IV, and V are the Dewar structures. The VB Dewar structures of benzene are formal ways of pairing up electrons in AOs. Each VB Dewar structure is based on a regular-hexagonal arrangement of carbon atoms. Likewise, structures I and II correspond to a regular hexagon of carbons and differ from the hypothetical molecule cyclohexatriene with alternating bond lengths. Each VB resonance structure is based on the same internuclear distances, but a different electron-pairing scheme.

The ground-state electronic wave function of C_6H_6 belongs to the totally symmetric symmetry species and is an eigenfunction of $\hat{O}_{C_6}$ with eigenvalue +1. The functions I and II combine with equal coefficients to form a single symmetry function, and the functions III, IV, and V combine to form a second symmetry function.

Benzene has six types of singly polar ionic structures. Four of the six types each have 12 associated individual structures and the other two types each have six individual structures, giving a total of 60 individual singly polar ionic structures (see Prob. 16.31). Doubly and triply polar ionic structures also occur.

Recall that increasing the number of functions f_i in a linear variation function $\sum_i c_i f_i$ improves the variation function, that is, lowers the value of the variational integral. If we were to consider structure I only, then the energy obtained would be considerably higher than when several VB structures are considered. The difference between the energy for the individual structure I and that found when all VB structures are included is the **resonance energy** of benzene. One says that benzene is "stabilized by resonance," but of course resonance is not a real phenomenon.

Such concepts as "configuration interaction," "resonance," "hybridization," and "exchange" are not real physical phenomena, but only artifacts of the approximations used in the calculations. Likewise, the concept of orbitals is but an approximation, and, strictly speaking, orbitals do not exist.

Atomic Valence States. The **valence state** of an atom for a given molecular electronic state is the state in which the atom exists in the molecule. Since individual atoms do not really exist in molecules, the valence-state concept is an approximate one. The VB approximation constructs molecular wave functions from wave functions of the individual atoms. We use the VB wave function to define the valence state of an atom as the wave function obtained on removing all other atoms to infinity, while keeping the form of the molecular wave function invariant. This process is purely hypothetical, and the valence state is not in general a stationary atomic state.

A simple example is the H_2 ground state. The Heitler–London function is

$$N[|1s_a\overline{1s_b}| - |\overline{1s_a}1s_b|]$$

Removal of hydrogen atom b leaves (at large internuclear separation, the normalization constant for each Slater determinant becomes $1/\sqrt{2}$)

$$2^{-1/2}1s_a - 2^{-1/2}\overline{1s_a}$$

The valence state is a hydrogen atom with a 1s spatial function and a 50% probability of having each of spin α or spin β. The valence-state ionization potential of hydrogen is thus 13.6 eV.

A less trivial example is H_2O. Although the H_2O ionic VB structures D and E are very important, it is traditional to ignore the ionic structures and find the oxygen valence state in H_2O from the perfect-pairing covalent function (16.69) corresponding to structure A. Removal of the H atoms from (16.69) gives

$$N[|\cdots 2p_y\overline{2p_z}| - |\cdots \overline{2p_y}2p_z| - |\cdots 2p_y\overline{2p_z}| + |\cdots \overline{2p_y}\overline{2p_z}|] \tag{16.79}$$

Each determinant in (16.79) belongs to the oxygen configuration $1s^22s^22p^4$. This configuration gives the terms ^3P, ^1D, and ^1S (Table 11.2). The first and last determinants are eigenfunctions of $\hat{S}_z$ with eigenvalues $+1\hbar$ and $-1\hbar$, respectively; hence these two determinants must correspond to states of the ^3P term (which has $S = 1$). Analysis (which we omit) of the other two determinants shows each of them to be an equal mixture (coefficients $1/\sqrt{2}$) of states belonging to the ^1D and ^3P terms. Thus the valence state is a mixture of states of the terms ^1D and ^3P and is not a stationary state of the atom. The valence-state wave function ψ_{vs} is not an eigenfunction of the atomic Hamiltonian. We can, however, calculate an average energy $\langle E_{\text{vs}}\rangle = \langle\psi_{\text{vs}}|\hat{H}|\psi_{\text{vs}}\rangle$. From $\psi_{\text{vs}} = N'[c_1\psi(^3P) + c_2\psi(^1D)]$, we have

$$\langle E_{\text{vs}}\rangle = |N'|^2[|c_1|^2E(^3P) + |c_2|^2E(^1D)]$$

The preceding discussion gives

$$|c_1|^2 = 1^2 + (2^{-1/2})^2 + (2^{-1/2})^2 + 1^2 = 3, \quad |c_2|^2 = (2^{-1/2})^2 + (2^{-1/2})^2 = 1$$

We have $|N'|^2 = (|c_1|^2 + |c_2|^2)^{-1} = \frac{1}{4}$ and

$$\langle E\rangle_{\text{vs}} = \tfrac{3}{4}E(^3P) + \tfrac{1}{4}E(^1D)$$

$\langle E\rangle_{\text{vs}}$ can then be calculated from tables of atomic energy levels (Section 11.5)

If 2s hybridization is included in the 2p bonding oxygen orbitals of water, the oxygen valence state is found to be a linear combination involving terms of the configurations $1s^22s^22p^4$, $1s^22s2p^5$, and $1s^22p^6$ (M. Kotani et al. in S. Flugge, ed., *Encyclopedia of Physics/Handbuch der Physik*, Springer, New York, 1961, Volume 37, pages 110–115). Hybridization gives a mixing of configurations in the valence state.

The valence state of carbon in CH_4 is important. Although the VB wave function of ground-state CH_4 is now known to have major contributions from ionic structures, the traditional way to find the C valence state is to start with the covalent perfect-pairing function (16.78). It is sometimes carelessly stated that the carbon sp^3 valence state corresponds to the ^5S term of the carbon atom $2s2p^3$ configuration. This is not correct. The $M_S = 2$ state of the $2s2p^3\ ^5S_2$ level has one $2s$ and three $2p$ electrons, with all four of these electrons having spin α. It is true that we can use the procedure of Section 15.8 to form linear combinations of the $2s$ and $2p$ orbitals and thereby put each of the four outer electrons of the $^5S_2\ M_S = 2$ state into an sp^3 hybrid AO without changing the wave function. However, each such hybrid AO would still have spin α. On the other hand, when we remove the hydrogen AOs from the CH_4 VB wave function (16.78), we are left with a linear combination of sixteen determinants in which each sp^3 hybrid AO has spin α in eight determinants and spin β in eight determinants. Thus the carbon valence state differs from the $2s2p^3\ ^5S_2\ M_S = 2$ state and in fact differs from the other states of this term. The valence state obtained on removal of the hydrogens from (16.78) turns out to be a mixture of states of the ^5S, ^3D, and ^1D terms of the $2s2p^3$ atomic configuration. [When other CH_4 VB structures besides (16.78) are included, we also get contributions from terms of the $2p^4$ and $2s^22p^2$ configurations to the C valence state.] The valence-state energy of carbon is well above the energy of the ground ^3P term of the $2s^22p^2$ configuration, but the energy gained by forming four bonds instead of two more than compensates for the energy needed to form the valence state.

Valence-state ionization potentials are used to estimate integrals in semiempirical calculations (Sections 17.2 and 17.3). The **valence-state ionization potential** for a $2p$ electron in an sp^3-hybridized carbon atom is the energy difference between the valence state of sp^3-hybridized C and the valence state of sp^2-hybridized C^+.

Status of the VB Method. The MO method puts electrons into orthogonal MOs. The VB method puts electrons into nonorthogonal AOs. This nonorthogonality makes VB calculations a huge computational task for a molecule with many electrons. To reduce the calculations, the σ electrons in a C_6H_6 ab initio VB calculation [J. M. Norbeck and G. A. Gallup, *J. Am. Chem. Soc.*, **96**, 3386 (1974)] were put into MOs, but even so this calculation did not optimize the molecular geometry, which is a simple task for an SCF MO calculation. The VB method allows for the change in the AOs that occurs on molecule formation by including ionic and other resonance structures, thereby complicating the wave function. One finds quite large contributions from ionic structures even for such compounds as CH_4 and C_6H_6. In contrast to the complicated VB wave function, a single-determinant SCF MO wave function usually furnishes a good description of the molecular ground electronic state in the region of the equilibrium geometry. The VB wave function does have the advantage of dissociating properly. For quantitative calculations, the MO method has greatly overshadowed the VB method, but there is still interest in generalizations of the VB method (see the next section).

Valence-bond theory has been used to describe the electronic structure of transition-metal complex ions, with such concepts as d^2sp^3 hybridization of the metal orbitals. However, the simple VB treatment of complex ions is not fully satisfactory and has been replaced by ligand-field theory, which is MO theory applied to species whose atoms have d (or f) electrons.

16.9 THE GVB, VBSCF, AND BOVB METHODS

The classical VB method described in the last section uses AOs that are optimized for individual atoms. Several modern ab initio VB methods have been developed that use orbitals optimized for the molecule. Three such methods are described in this section.

The **generalized valence-bond (GVB) method** was developed about 1970 by Goddard and co-workers [W. J. Hunt, P. J. Hay, and W. A. Goddard, *J. Chem. Phys.*, **57**, 738 (1972); P. J. Hay, W. J. Hunt, and W. A. Goddard, *J. Am. Chem. Soc.*, **94**, 8293 (1972); W. A. Goddard, T. H. Dunning, W. J. Hunt, and P. J. Hay, *Acc. Chem. Res.*, **6**, 368 (1973)].

The Heitler–London VB wave function for ground-state H_2 is [Eq. (13.100)] $1s_a(1)1s_b(2) + 1s_a(2)1s_b(1)$ multiplied by a normalization constant and a spin function. The GVB ground-state H_2 wave function replaces this spatial function by $f(1)g(2) + f(2)g(1)$, where the functions f and g are found by minimization of the variational integral. To find f and g, one expands each of them in terms of a basis set of AOs and finds the expansion coefficients by iteratively solving one-electron equations that resemble the equations of the SCF MO method.

Clearly, the GVB method will give a lower energy than the simple VB wave function. The GVB method allows for the change in the AOs that occurs on molecule formation by solving variationally for f and g. In the VB method, this change is allowed for by adding to the wave function terms that correspond to ionic and other resonance structures. The GVB wave function is thus much simpler than a VB wave function with resonance structures and the calculations are simpler.

The GVB method gives a D_e of 4.12 eV for ground-state H_2, as compared with 3.15 eV for the Heitler–London VB function, 3.78 eV for the Heitler–London–Wang function with an optimized orbital exponent, 4.03 eV for the Weinbaum function (13.109) that includes an ionic term, and 4.75 eV for the experimental value. At very large internuclear distance R, the GVB functions f and g approach the atomic orbitals $1s_a$ and $1s_b$. Thus, like the VB ψ (but unlike the MO ψ), the GVB wave function shows the correct behavior on dissociation. At intermediate distances, f is a linear combination of AOs that has its most important contribution from $1s_a$ but that has a significant contribution from $1s_b$ and lesser contributions from the other AOs (these contributions reflect the polarization of the $1s_a$ AO that occurs on molecule formation).

In the MO method, there is one orbital (an MO) for each electron pair. In the GVB method, there are two orbitals (f and g in the H_2 example) for each electron pair.

For CH_4 the VB wave function with resonance structures omitted is (16.78). For CH_4 the GVB wave function is

$$|\widehat{i_a i_b}\ \widehat{b_{1a} b_{1b}}\ \widehat{b_{2a} b_{2b}}\ \widehat{b_{3a} b_{3b}}\ \widehat{b_{4a} b_{4b}}| \tag{16.80}$$

The electrons are divided into pairs, and each pair is given two orbitals. In the VB method, the inner-shell orbitals i_a and i_b are each assumed to be $1s$ AOs on carbon, the bonding orbital b_{1a} is assumed to be an sp^3-hybridized carbon AO pointing toward H_1, and b_{1b} is assumed to be a $1s$ AO on hydrogen number 1. In the GVB method, no assumptions are made about the nature of the orbitals. One simply expands each of them in terms of the chosen basis set and solves the GVB equations until self-consistency is attained.

To simplify the calculation, the GVB method restricts orbitals in different pairs to be orthogonal to one another. (For example, b_{1a} and b_{2a} are assumed orthogonal, but b_{1a}

and b_{1b} are not assumed orthogonal.) This should be a good assumption since Pauli repulsion between pairs keeps them well separated spatially; moreover, a few test calculations have shown the orthogonality requirement to lead to little error.

For CH_4 one finds that i_a and i_b are essentially C1s AOs; b_{1a} is an orbital centered mainly on C with some contribution from H_1; b_{1a} points toward H_1 and has a carbon AO hybridization of $sp^{2.1}$ (using a minimal basis set). The hybridization differs from the sp^3 hybridization of the VB wave function as a result of the contribution of $H_1 1s$ to b_{1a}. The orbital b_{1b} is found to be mainly an $H_1 1s$ AO with some contribution of carbon AOs mixed in, thereby polarizing the orbital toward C.

Since inner-shell electrons are little changed on molecule formation, one can simplify the GVB calculation by assuming each of i_a and i_b to be a C1s AO, as is done in the VB wave function. This gives little loss in accuracy.

For C_2H_6 a GVB calculation with a minimal basis set gave a 3.1 kcal/mol rotational barrier, in good agreement with the 2.9 kcal/mol experimental value.

For C_2H_4 a GVB calculation gave a description of the double bond as composed of one σ and one π bond, in contrast to the energy-localized MOs (Section 15.9), which are two equivalent bent banana bonds.

For H_2O a GVB calculation produced an inner-shell pair on oxygen, two equivalent bonding pairs, and two equivalent lone pairs. With a DZP on oxygen basis set, an energy of -76.11 hartrees was obtained, which is below the Hartree–Fock limit of -76.07 hartrees (Table 15.2).

The GVB method (like the VB method) gives a description in terms of localized inner-shell, bonding, and lone pairs, whereas one must carry out a time-consuming special procedure to find localized MOs from canonical SCF MOs.

The GVB method is most applicable to molecules for which a single VB covalent structure is a good approximation.

The GVB method has been used to develop qualitative descriptions of chemical bonding; see W. A. Goddard and L. B. Harding, *Ann. Rev. Phys. Chem.*, **29**, 363 (1978).

The GVB method can be extended by writing the wave function as a linear combination of functions that correspond to different ways of coupling (combining) the electron spins to give a singlet state (just as is done in the VB method when the wave function is written as a linear combination of canonical covalent structures). This gives the *unrestricted* GVB wave function. A GVB wave function like (16.80) is sometimes called the GVB *perfect-pairing* (PP) wave function, to distinguish it from the unrestricted GVB wave function. The unmodified term GVB wave function usually refers to the GVB-PP wave function.

The **VBSCF** method [J. H. van Lenthe and G. G. Balint-Kurti, *J. Chem. Phys.*, **78**, 5699 (1983)] writes the molecular VB wave function as a linear combination of covalent and ionic VB structures and simultaneously optimizes the coefficients in the linear combination and the orbitals (which are expressed using a set of basis functions). The orbitals might be localized on individual atoms (as in the classical VB method of the last section) or they might be taken as semidelocalized (as in the GVB method).

The **breathing-orbital VB (BOVB)** method [P. C. Hiberty and S. Shaik, *Theor. Chem. Acc.*, **108**, 255 (2002) and references cited therein] differs in two ways from the VBSCF method. Each orbital in the BOVB method is always taken to be localized on an individual atom. The orbitals used in different structures are free to differ, so that each VB structure has its own set of optimized orbitals.

An excellent review of classical and modern valence-bond theory is S. Shaik and P. C. Hiberty in K. B. Lipkowitz et al. (eds.), *Reviews in Computational Chemistry*, vol. 20, Chapter 1, 2004, Wiley-VCH. Other reviews of modern VB theory are P. C. Hiberty and S. Shaik, *J. Comput. Chem.*, **28**, 137 (2007); A. Shurki, *Theor. Chem. Acc.*, **116**, 253 (2006).

16.10 CHEMICAL REACTIONS

The course of a chemical reaction is determined by the potential-energy function for nuclear motion $U(q_\alpha)$ (Section 13.1), where q_α indicates the coordinates of the N nuclei of the reactant molecules. To find the potential-energy surface (PES) $U(q_\alpha)$ (Section 15.10), we must solve the electronic Schrödinger equation at a very large number of nuclear configurations, which is a formidable task.

Once U is found, we look for the path of minimum potential energy on U connecting reactants and products. The point of maximum potential energy U on the minimum-energy path is called the **transition state**. This is a saddle point on the U surface, since it is a maximum point on a minimum-energy path. The transition state is not a stable molecule, and the transition from reactants to products is a smooth one. (However, in certain theories of reaction rates, it is convenient to ascribe properties such as entropy, free energy, and vibrational frequencies, to the transition state.) The energy difference between the transition state and the reactants (omitting zero-point vibrational energies) is called the **(classical) barrier height** for the forward reaction. For the reverse reaction, the surface U is the same as for the forward reaction.

When the U surface is known, it is possible (at least in principle) to calculate the reaction rate constant k as a function of temperature. Such calculations are extremely difficult. One must allow for quantum-mechanical tunneling through the barrier. Tunneling is important in reactions involving light species (e^-, H^+, H, H_2), including reactions that transfer such species between heavy molecules. Moreover, there is significant probability for molecules to traverse the reaction surface on paths that deviate somewhat from the minimum-energy path. Thus one must perform averaging over the various possible paths. (In other words, we must consider different approaches of the reactant molecules, calculate the probability of reaction occurring for each approach, and then suitably average these probabilities.) The rate constant thus depends not only on the barrier height, but on the whole shape of the reaction potential-energy surface. For qualitative discussion we can use the fact that a large barrier height means a small rate constant, and a low barrier means a fast reaction.

By measuring the experimental rate constant k as a function of temperature, one can determine an experimental **activation energy** E_a, where $k = A\exp(-E_a/RT)$. The experimental quantity E_a differs slightly from the barrier height on the U surface. [See I. Shavitt, *J. Chem. Phys.*, **49**, 4048 (1968); M. Menzinger and R. Wolfgang, *Angew. Chem. Int. Ed. Engl.*, **8**, 438 (1969); *Levine, Physical Chemistry*, Section 22.4.]

To determine a complete reaction surface $U(q_\alpha)$, one needs to solve the electronic Schrödinger equation at about 10 points on the surface for each of the $3N - 6$ variables, so one needs about 10^{3N-6} calculations. For three-, four-, and five-atom systems, one needs 10^3, 10^6, and 10^9 calculations. Moreover, since the Hartree–Fock method does not usually correctly describe the process of molecular dissociation, we must include electron correlation to calculate a PES accurately.

Except for systems with very few atoms, accurate ab initio calculation of the complete potential-energy surface is out of the question. Instead, one aims to find the most important features of the surface. One attempts to locate the points on the surface where all the first derivatives $\partial U/\partial q_\alpha$ (the components of the gradient) are zero. These are called stationary points. A stationary point may be a local minimum, a local maximum, or a saddle point. To determine the nature of a stationary point, one evaluates the $(3N - 6)^2$ second derivatives $\partial^2 U/\partial q_\alpha\, \partial q_\beta$ (the components of the Hessian) at the point and uses these second derivatives to find the vibrational frequencies (Section 15.12) at the stationary point. At a local minimum, all the vibrational frequencies are real numbers. A transition state is a first-order saddle point and has one and only one imaginary vibrational frequency. A higher-order saddle point has two or more imaginary vibrational frequencies and is not a transition state. (Equivalently, since the eigenvalues of the Hessian matrix are proportional to the squares of the vibrational frequencies, the Hessian has all positive eigenvalues at a local minimum and has one negative eigenvalue at a transition state.)

Local minima correspond to reactants, products, or reaction intermediates. A **reaction intermediate** (which should not be confused with a transition state) is a product in one elementary step and a reactant in a subsequent elementary step of a multistep mechanism. A reaction intermediate lies at a minimum in U for all nuclear displacements. Reaction intermediates are often too short lived to allow spectroscopic determination of their structure. Hence, a very significant application of quantum chemistry is the determination of the structures and relative energies of reaction intermediates; for ab initio SCF MO results, see *Hehre et al.*, Section 7.3.

Finding transition states on a PES is much harder than finding local energy minima (Section 15.10). Analytical gradients of U facilitate finding transition states. Techniques to find transition states are discussed in H. B. Schlegel, *Adv. Chem. Phys.*, **67**, 250 (1987); H. B. Schlegel in J. Bertran and I. G. Csizmadia (eds.) *New Theoretical Concepts for Understanding Organic Reactions*, Kluwer, 1989, pp. 33–53; H. B. Schlegel in *Yarkony*, Part I, Chapter 8; *Leach*, Section 5.9.

After finding a transition state (first-order saddle point) on a PES, one must determine the nature of the reactants and products for this transition state by following the downhill paths from the saddle point to the reactants and to the products. (See the discussion below of the IRC.) Moreover, a lower-energy transition state for these reactants and products might exist elsewhere on the surface, so one must explore the surface so as to find the lowest-energy transition state between given sets of reactants and products.

An aid to finding the reactants and products for a given transition state is the fact that the atomic displacements in the normal vibrational mode that corresponds to the imaginary frequency tend to be in the directions of the reactants and products for this transition state. In the Spartan program, one can produce an animation of this mode.

The PES of a reaction with more than a few atoms may have many first-order saddle points. For example, calculations on the PES of the $NH_2 + NO$ reaction found 3 possible sets of products, 9 reaction intermediates, and 23 first-order saddle points [E. W.-G. Diau and S. C. Smith, *J. Chem. Phys.*, **106**, 9236 (1997)].

Knowledge of the minima, transition states, and barrier heights on a surface gives a good idea of the reaction mechanism.

Once the structure and vibrational frequencies of the transition state and the barrier height have been calculated, a rough estimate of the rate constant can be found using Eyring's transition-state (activated complex) theory (see any physical chemistry text). For more precise results, one must locate the minimum-energy path between reactants and products.

Although the location of the transition state is independent of the nuclear coordinates used as the variables in $U(q_\alpha)$, one finds that the location of the minimum-energy path depends on the choice of coordinates used. (For example, for H_2O, one possible choice is the $O-H_1$ and $O-H_2$ distances and the HOH angle, and another choice is the $O-H_1$, $O-H_2$, and H_1-H_2 distances.) The minimum-energy path (MEP) usually used is the **intrinsic reaction coordinate** (IRC), which is defined as the path that would be taken by a classical particle sliding downhill with infinitesimal velocity from the transition state to each of the minima [see Schlegel, *Adv. Chem. Phys.*, **67**, 250 (1987)]; the IRC turns out to correspond to the minimum-energy path (the path of steepest descent from the transition state) on a surface whose coordinates are the mass-weighted Cartesian coordinates $m_\alpha^{1/2}x_\alpha$, $m_\alpha^{1/2}y_\alpha$, $m_\alpha^{1/2}z_\alpha$ of the nuclei, where m_α is the mass of nucleus α. The IRC is also called the **reaction path**. However, the reaction path is not the actual path taken by reacting molecules moving according to classical mechanics, since the molecules have translational, rotational, and vibrational kinetic energy. Techniques to determine the IRC are discussed in the preceding references for finding transition states.

Once the reaction path and the force-constant matrix at points along the reaction path have been found, one can use improved versions of transition-state theory (TST) called *generalized transition-state theory* to calculate rate constants more accurate than those given by TST [see references cited in D. G. Truhlar, R. Steckler, and M. S. Gordon, *Chem. Rev.*, **87**, 217 (1987)] and can construct a reaction-path Hamiltonian and use it to study such things as vibrational-energy transfer during the reaction [W. H. Miller et al., *J. Chem. Phys.*, **72**, 99 (1981); W. H. Miller, *J. Phys. Chem.*, **87**, 3811 (1983)].

The *Gaussian* program contains many features designed to find transition states, follow reaction paths, and explore PESs. To locate a transition state, one can input a guessed transition-state structure that is intermediate between the structures of reactants and products of the elementary reaction and use the keyword Opt=(TS,CalcFC). The TS tells *Gaussian* to use a search procedure that is appropriate to finding a first-order saddle point. The CalcFC tells *Gaussian* to accurately calculate the force constants (Hessian matrix elements) at the initial input geometry, rather than using the default procedure of estimating them (as is done in energy minimization—Section 15.10). Although using CalcFC increases the time needed for the calculation, it makes the transition-state search considerably more likely to succeed.

If a few guesses and Opt=(TS,CalcFC) don't succeed in locating a transition state, one can have *Gaussian* calculate an initial estimate of the transition-state structure from the structures of the reactants and products using a procedure called synchronous transit (ST) and then have *Gaussian* use this initial estimate as the starting point for the transition-state search. In this procedure, one uses the keyword Opt=QST2 and one inputs the structures of the reactants and of the products. For details, see *Foresman and Frisch*, p. 46, and the *Gaussian* manual, which is available at the *Gaussian* website (www.gaussian.com/).

The keyword IRC causes *Gaussian* to perform an IRC calculation. The input to such a job is the geometry of the transition state.

The keyword Scan allows one to use *Gaussian* to explore a PES. One specifies the variable(s) that will be scanned and the ranges of values they take on.

The most famous reaction surface is that for $H + H_2 \rightarrow H_2 + H$. The H_3 surface has been calculated to an extraordinary accuracy of 0.01 kcal/mol at 4067 points using a highly accurate special extrapolation procedure applied to results of multireference CI aug-cc-pVTZ and aug-cc-pVQZ wave functions, and an analytical potential function has been very accurately fitted to these points [S. L. Mielke, B. Garrett, and K. A. Peterson, *J. Chem. Phys.*, **116**, 4142 (2002)]. From this surface, rate constants in very good agreement with experiment have been calculated [S. L. Mielke et al., *Phys. Rev. Lett.*, **91**, 063201 (2003)]. This surface is called the CCI (complete configuration interaction) surface, where CCI indicates it is extremely close to what would be found from a full-CI calculation with a complete basis set. The effect on the H_3 barrier height of the main correction to the Born–Oppenheimer approximation is calculated in S. L. Mielke, *J. Chem. Phys.*, **122**, 224313 (2005)].

Comparisons of ab initio HF calculations with calculations that include electron correlation indicate that for many reactions ab initio HF transition-state structures are reasonably accurate, but for certain classes of reactions, HF transition-state geometries are unreliable [F. Bernardi and M. A. Robb, *Adv. Chem. Phys.*, **67**, 155 (1987)]. For further discussion, see Section 18.1. HF calculations do not generally give accurate energy differences between points on potential-energy surfaces.

The IMOMO, IMOMM, and ONIOM Methods. Accurate calculation of reaction barrier heights usually requires inclusion of electron correlation and use of a substantial-size basis set, and is impractical for medium and large molecules. To deal with this problem, the IMOMO (integrated molecular orbital plus molecular orbital) method of Morokuma and co-workers allows one to estimate energy changes involving large molecules by combining high-level quantum-mechanical calculations on a related smaller system (the *model* system) with low-level calculations on the actual system (the *real* system) [S. Humbel, S. Sieber, and K. Morokuma, *J. Chem. Phys.*, **105**, 1959 (1996)]. The IMOMO energy of a system is taken as

$$E_{\text{IMOMO}} = E(\text{real, low}) + [E(\text{model, high}) - E(\text{model, low})]$$

Roughly speaking, the quantity in brackets gives an estimate of $E(\text{real, high}) - E(\text{real, low})$ so that E_{IMOMO} is an approximation to $E(\text{real, high})$. The low level might be HF, DFT, MP2, or a semiempirical MO method (Section 17.4). The high level might be MP2, MP4, CCSD(T), CASSCF, and so on. Note that "MO" in the name IMOMO is not synonymous with Hartree–Fock but includes such MO-based methods as MP and CC; DFT does not use MOs but does use Kohn–Sham orbitals.

As an example, suppose one wants to calculate the barrier height for the reaction $Cl_a^- + H_aH_bC_a(Cl_b)C_bH_3 \rightarrow H_2C(Cl_a)CH_3 + Cl_b^-$, where the subscripts help distinguish identical atoms. The model system can be taken as $Cl^- + H_aH_bC_a(Cl)H_c$, in which an H atom (the *capping atom*) replaces the methyl group, so as to give a smaller system more easily treatable at a higher level. The methyl carbon in the real system that is replaced by the capping atom is called the *link atom*. The model system retains the atoms and bonds that are most directly involved in the reaction.

To optimize the reactants' geometries or the transition-state geometry, one varies the bond distances, angles, and dihedral angles in the real and the model systems (subject to certain constraints) so as to either minimize E_{IMOMO} or to locate a first-order saddle point. IMOMO constrains corresponding bond distances, bond angles, and dihedral angles in the real and model systems to be the same. For example, the $H_aC_aH_b$ angles in the real and the model system are equal and the $H_aC_aC_b$ angle in the real system equals the $H_aC_aH_c$ angle in the model system. Moreover, to avoid problems in geometry optimization, IMOMO takes the bond distance $C_a—H_c$ to the capping atom in the model system and the distance $C_a—C_b$ to the link atom in the real system to be frozen at standard values.

The method works rather well in finding barrier heights and produces large savings in computational time.

The IMOMM (integrated molecular orbital plus molecular mechanics) method resembles IMOMO, but the lower level used is the molecular-mechanics method (Section 17.5) and some details of the implementation differ from IMOMO [F. Maseras and K. Morokuma, *J. Comput. Chem.*, **16**, 1170 (1995)].

The IMOHC (integrated molecular orbital method with harmonic cap) method is a modification of IMOMO and IMOMM in which the bond distances to the link atom in the real system and the capping atom in the model system are not frozen but are allowed to vary in the optimization [J. C. Corchado and D. G. Truhlar, *J. Phys. Chem. A*, **102**, 1895 (1998)].

The ONIOM (our own n-layered integrated molecular orbital and molecular mechanics) method is an extension of IMOMO and IMOMM that performs calculations on n different systems using n levels of calculation. Each system and calculation level constitute a layer. The IMOMO and IMOMM methods are versions of ONIOM with $n = 2$. ONIOM3 is the three-layered version of ONIOM and uses a real system calculated at a low level, an intermediate model system calculated at a medium level, and a small model system calculated at a high level [M. Svensson et al., *J. Phys. Chem.*, **100**, 19357 (1996)].

PROBLEMS

Sec.	16.1	16.2	16.3	16.4	16.6	16.7
Probs.	16.1–16.7	16.8–16.11	16.12–16.14	16.15–16.24	16.25–16.26	16.27

Sec.	16.8	16.10
Probs.	16.28–16.35	16.36–16.37

16.1 Find the number of CSFs in a full CI calculation of CH_2SiHF using a 6-31G** basis set.

16.2 Let γ be the fraction of the basis-set correlation energy obtained by a CI-SD calculation on a molecule with n electrons. Sasaki showed that an approximate equation satisfied by γ is $\gamma^{-1} - 1 = \frac{1}{2}\beta\gamma n - \beta$, where β is a quantity whose value is typically 0.015 to 0.03 for molecules consisting of first-row atoms. For such molecules, use this equation to estimate the percent of the basis-set correlation energy obtained by CI-SD calculations for $n = 20, 50, 100$, and 200.

16.3 Frozen-core SCF/DZP and CI-SD/DZP calculations on H_2O at its equilibrium geometry gave energies of −76.040542 and −76.243772 hartrees. Application of the Davidson correction brought the energy to −76.254549 hartrees. Find the coefficient of Φ_0 in the normalized CI-SD wave function.

16.4 For a CI calculation of the H_2 ground electronic state, which of the following electron configurations will produce a CSF that will contribute to the wave function? (a) $(1\sigma_g)(2\sigma_g)$; (b) $(1\sigma_g)(2\sigma_u)$; (c) $(1\sigma_u)^2$; (d) $(1\pi_u)(1\pi_g)$; (e) $(1\pi_u)(3\sigma_u)$; (f) $(1\pi_u)^2$; (g) $(1\pi_u)(2\pi_u)$.

16.5 Verify the expression given for the CI ground-state wave function in the Section 16.1 example.

16.6 In the Section 16.1 CI example, verify the expressions given in terms of integrals over the basis functions for (a) $\langle\Phi_2|\hat{H}|\Phi_2\rangle$; (b) $\langle\Phi_2|\hat{H}|\Phi_1\rangle$.

16.7 For a CASSCF wave function in which the active orbitals are taken as those that arise from the 2*s* and 2*p* AOs, state the number of active electrons and the maximum number of electrons excited to virtual orbitals in the ground-state wave function for (a) C_2; (b) N_2; (c) O_2; (d) F_2.

16.8 Verify (16.11) for the MP perturbation $\hat{H}'$.

16.9 To derive the MP $E_0^{(2)}$, we set $\psi_s^{(0)} = \Phi_{ij}^{ab}$ in (16.12); the sum over excited states $s \neq 0$ in (16.12) is replaced by a quadruple sum over i, j, a, and b that produces all possible determinants that contain two excited spin-orbitals and that represent different states. (a) For this quadruple sum, explain why we want the limits for i and j to be as in (16.13); explain why $\Phi_{ij}^{ab} = 0$ if $a = b$, and explain why we want the limits for a and b to be as in (16.13). (b) Use the Condon–Slater rules to evaluate $\langle\Phi_{ij}^{ab}|\hat{H}'|\Phi_0\rangle$, and show that (16.13) follows.

16.10 True or false? A nonrelativistic MP2 calculation can give an energy that is less than the true nonrelativistic ground-state energy.

16.11 (a) Use MP2(FC)/6-31G* calculations to predict the geometry, vibrational frequencies, dipole moment, and atomization energy of CO_2. Use 0.95 as the scale factor for the vibrational frequencies. Compare with HF/6-31G* and experimental results (Prob. 15.38). (Note that frozen core is the default in *Gaussian* MP calculations, so the keyword MP produces an FC calculation.) (b) Repeat (a) for H_2O.

16.12 Show that ψ in (16.15) satisfies $\langle\psi|\Phi_0\rangle = \langle\Phi_0|\Phi_0\rangle = 1$, and show that $\langle\psi|\psi\rangle \neq 1$. In what earlier chapter in this book does an equation like $\langle\psi|\Phi_0\rangle = 1$ appear?

16.13 (a) Verify the CC equation (16.27). (b) Verify (16.28).

16.14 Verify the extrapolation equation (16.32).

16.15 Which of the following are functionals? (a) $\int f(x)\,dx$; (b) $\int_0^1 f(x)\,dx$; (c) $\int_0^2 [f(x) + 1]^2\,dx$; (d) $[f(x) + 1]^2$; (e) $df(x)/dx|_{x=0}$.

16.16 Let $\hat{H}_a$ and $\hat{H}_b$ be the operators obtained when $v(\mathbf{r}_i)$ in (16.33) is replaced by $v_a(\mathbf{r}_i)$ and $v_b(\mathbf{r}_i)$, respectively, where $v_a(\mathbf{r}_i)$ and $v_b(\mathbf{r}_i)$ differ by more than a constant. Prove that the ground-state wave functions $\psi_{0,a}$ and $\psi_{0,b}$ of these Hamiltonians must be different functions. *Hint:* Assume they are the same function, write the Schrödinger equations for $\hat{H}_a$ and $\hat{H}_b$, subtract one equation from the other, and show that this leads to $v_a(\mathbf{r}_i) - v_b(\mathbf{r}_i)$ equals a constant, thereby contradicting the given information and proving that the two wave functions cannot be equal.

16.17 (a) Find $\delta E_x^{X\alpha}/\delta\rho$, where $E_x^{X\alpha}$ is given by (16.61). (b) If $F[\rho] = \int \rho^{-1}\nabla\rho\cdot\nabla\rho\,dv$, where the integral is over all space, and ρ is a function of x, y, and z that vanishes at infinity, find $\delta F/\delta\rho$.

16.18 Verify that $\hat{h}^{KS}(1)$ in (16.49) is the same as the Fock operator (16.8) except that the exchange operators in the Fock operator are replaced by v_{xc}.

16.19 Verify Eqs. (16.57) and (16.58).

16.20 Verify (16.60) for E_x.

16.21 Verify that (16.65) for E_x^{LSDA} reduces to (16.58) for E_x^{LDA} if $\rho^\alpha = \rho^\beta$.

16.22 (a) For an n-electron molecule, show that the electron probability density is given by $\rho(\mathbf{r}) = n\langle\psi|\delta(\mathbf{r} - \mathbf{r}_1)|\psi\rangle = \langle\psi|\sum_{i=1}^{n}\delta(\mathbf{r} - \mathbf{r}_i)|\psi\rangle$, where $\delta(\mathbf{r} - \mathbf{r}_i) = \delta(x - x_i)\delta(y - y_i)\delta(z - z_i)$ and δ is the Dirac delta function (Section 7.7). (b) Use the Condon–Slater rules to show that if ψ is a single Slater determinant of spin-orbitals $u_i = \theta_i\sigma_i$, then $\rho(\mathbf{r}) = \sum_{i=1}^{n}|\theta_i(r)|^2$.

16.23 (a) For CO_2, use SVWN/6-31G*, BLYP/6-31G*, and B3LYP/6-31G* calculations to predict the equilibrium geometry, vibrational frequencies, and atomization energy. Do not scale the vibrational frequencies. Compare the results and the computational times with HF/6-31G* calculations (Prob. 15.38). (b) Repeat (a) for H_2O.

16.24 Repeat Prob. 15.49 for HCOOH using B3LYP/6-31G* calculations.

16.25 (a) Problem 7.46 showed that if $\hat{H}$ is independent of time, Ψ as a function of time is given by (7.100), where the c_n's are constants. Use (7.100) to explain why in the limit $\tau \to \infty$ (where $\tau \equiv it/\hbar$) we have $\Psi = c_{gs}e^{-\tau E_{gs}}\psi_{gs}$, where ψ_{gs} is the ground-state wave function. For convenience, we shift the zero level of energy by subtracting the constant V_{ref} from $\hat{H}$, where V_{ref} is chosen as the best estimate we have of the ground-state energy E_{gs}. Explain why this shift gives $\Psi = c_{gs}e^{-\tau(E_{gs}-V_{\text{ref}})}\psi_{gs}$. (Use of V_{ref} prevents Ψ from becoming very small or very large at large τ.) (b) Show that the time-dependent Schrödinger equation (7.96) for an n-electron molecule is $\partial\Psi/\partial\tau = \frac{1}{2}\sum_{i=1}^{n}\nabla_i^2\Psi - (V - V_{\text{ref}})\Psi$. This equation has the same form as the equation $\partial C/\partial t = D\,\nabla^2 C - kC$ for a species undergoing both diffusion (with diffusion coefficient D) and a first-order chemical reaction with rate constant k, where C is the concentration, provided we replace the three-dimensional space of the diffusion equation with a $3n$-dimensional "space" whose variables are the coordinates of the n electrons.

16.26 To properly apply the DQMC method, one must allow for the nodes produced by the antisymmetry requirement. (a) Consider a system of three electrons in a one-dimensional box, where we shall pretend that the interelectronic repulsions are small enough to be neglected. By analogy to the Li zeroth-order wave function (10.48), write down the ground-state wave function for this system. (b) Use orthogonality of the three different three-electron spin factors that multiply a, b, and c in Prob. 10.16 to show that each spatial factor that corresponds to a, b, or c is an eigenfunction of $\hat{H}$. Hence, in doing the DQMC computer simulation of the imaginary-time Schrödinger equation, we need deal with only one of the spatial factors, say, the one corresponding to a. For our system of electrons in a box, there are three spatial variables, the coordinates x_1, x_2, and x_3 of the three electrons, and the DQMC simulation is done in a three-dimensional space bounded by the sides of a cube, on which the wave function vanishes. Show that the spatial factor corresponding to a has a nodal surface defined by $x_2 = x_3$ and this is the only nodal surface within the cube. (*Hint:* Use a trigonometric identity for $\sin 2z$.) (c) Show that the nodal surface $x_2 = x_3$ is a plane that divides the cube into two regions of equal volume; that in one of these regions (the one with $x_3 > x_2$) the wave function is positive, and in the other region, the wave function is negative. Also show that for each point P in one region, there is a corresponding point (the one with the values of x_2 and x_3 interchanged) where the wave function has minus its value at P. In doing the DQMC simulation, one works entirely within one region and eliminates any walker that crosses the nodal surface. (Note that the concentration C in the diffusion equation must always be positive or zero.)

16.27 Use the viral-theorem result (14.76) to show that $\langle v^2\rangle^{1/2}/c = Ze'^2/\hbar c \approx Z/137$ for the ground-state H atom.

16.28 Prove that the sum of the VB occupation numbers n_i is 1.

16.29 Verify Eq. (16.71) for the H_2O covalent VB structures.

16.30 Consider the orbitals $1, 2, \ldots, 6$ arranged in order on a circle. Use (16.73) repeatedly to show that the scheme with the pairings 1–5, 2–4, 3–6 can be expressed as a linear combination of structures with no lines crossing.

16.31 Draw an example of each of the six types of singly polar ionic benzene structures. Work out how many individual structures of each type exist.

16.32 For 1,3-butadiene: (a) How many canonical covalent VB structures are there for the π electrons? (b) Draw these structures. (c) Draw the 12 individual singly polar ionic structures for the π electrons.

16.33 (a) How many canonical covalent structures are there for the naphthalene π electrons? (b) Of these, how many are Kekulé structures (no long bonds)? (c) Considering only the Kekulé structures, which bonds in naphthalene does the resonance method predict to be the shortest?

16.34 The three equivalent sp^2 hybrid AOs of carbon point to the corners of an equilateral triangle. Derive expressions for them, assuming orthonormality.

16.35 The two equivalent *sp* hybrid carbon AOs make an angle of 180° with each other. Derive them.

16.36 Consider the reaction HCN → CNH. (a) Find the HF/6-31G* equilibrium geometries of HCN and HNC. (Recall that 180° is not allowed as a Z-matrix bond angle.) (b) Find the HF/6-31G* transition-state structure for this reaction. *Hints*: We expect the transition-state (TS) structure to be roughly halfway between the reactant and product structures. Thus we expect a triangular TS with the CN distance somewhere between its values in the reactant and product, the HC distance somewhat longer than its values in the reactant, and the HN distance somewhat longer than its value in the product. Start with an initial guess for the TS structure. In *Gaussian*, one way to find a TS is to replace `Opt` by `Opt(CalcFC,TS)`.

16.37 (a) For formic acid, HCOOH, find the two stable conformers at the HF/6-31G* level; check that each conformer is a local minimum. Then find the structure of the HF/6-31G* transition state between these conformers. (See also Prob. 16.36.) What is the HF/6-31G* barrier to internal rotation about the CO single bond (omit zero-point energies)? (b) Repeat for vinyl alcohol.

CHAPTER 17

Semiempirical and Molecular-Mechanics Treatments of Molecules

Because of the difficulties in applying ab initio methods to medium and large molecules, many semiempirical methods have been developed to treat such molecules. The earliest semiempirical methods treated only the π electrons of conjugated molecules. This chapter begins with π-electron semiempirical methods (Sections 17.1 to 17.3) and then considers general semiempirical methods (Section 17.4).

The molecular-mechanics method (Section 17.5) is a nonquantum-mechanical method applicable to much larger molecules than semiempirical methods.

17.1 SEMIEMPIRICAL MO TREATMENTS OF PLANAR CONJUGATED MOLECULES

The canonical MOs of a planar unsaturated organic molecule can be divided into σ and π MOs according to whether the eigenvalue for reflection in the molecular plane is $+1$ or -1, respectively (Section 15.9). The earliest semiempirical methods for planar conjugated organic compounds (Sections 17.1–17.3) treated the π electrons separately from the σ electrons. Coulson stated that the justification for the σ–π separability approximation lies in the different symmetry of the σ and π orbitals and in the greater polarizability of the π electrons, which makes them more susceptible to perturbations such as those occurring in chemical reactions.

In the **π-electron approximation**, the n_π π electrons are treated separately by incorporating the effects of the σ electrons and the nuclei into some sort of effective π-electron Hamiltonian $\hat{H}_\pi$ (recall the similar valence-electron approximation; Sections 13.17 and 15.16):

$$\hat{H}_\pi = \sum_{i=1}^{n_\pi} \hat{H}_\pi^{\text{core}}(i) + \sum_{i=1}^{n_\pi} \sum_{j>i} \frac{1}{r_{ij}} \tag{17.1}$$

$$\hat{H}_\pi^{\text{core}}(i) = -\tfrac{1}{2}\nabla_i^2 + V(i) \tag{17.2}$$

where $V(i)$ is the potential energy of the ith π electron in the field produced by the nuclei and the σ electrons. (The *core* is everything except the π electrons.) The variational principle is then applied to find a π-electron wave function ψ_π that minimizes the variational integral $\int \psi_\pi^* \hat{H}_\pi \psi_\pi \, d\tau$ to give a π-electron energy E_π. The validity of the π-electron approximation has been discussed by Lykos and Parr (see *Parr*, pages 41–45, 211–218). Since (17.1) is not the true molecular electronic Hamiltonian, treatments that make the π-electron approximation are semiempirical. The main π-electron MO theories are the Hückel MO method (Section 17.2), and the Pariser–Parr–Pople method (Section 17.3).

17.2 THE HÜCKEL MO METHOD

The most celebrated semiempirical π-electron theory is the **Hückel molecular-orbital** (HMO) **method**, developed in the 1930s to treat planar conjugated hydrocarbons. Here the π-electron Hamiltonian (17.1) is approximated by the simpler form

$$\hat{H}_\pi = \sum_{i=1}^{n_\pi} \hat{H}^{\text{eff}}(i) \tag{17.3}$$

where $\hat{H}^{\text{eff}}(i)$ somehow incorporates the effects of the π-electron repulsions in an average way. This sounds rather vague, and in fact the Hückel method does not specify any explicit form for $\hat{H}^{\text{eff}}(i)$. Since the Hückel π-electron Hamiltonian is the sum of one-electron Hamiltonians, a separation of variables is possible (Section 6.2). We have

$$\hat{H}_\pi \psi_\pi = E_\pi \psi_\pi \tag{17.4}$$

$$\psi_\pi = \prod_{i=1}^{n_\pi} \phi_i \tag{17.5}$$

$$\hat{H}^{\text{eff}}(i)\phi_i = e_i \phi_i \tag{17.6}$$

$$E_\pi = \sum_{i=1}^{n_\pi} e_i \tag{17.7}$$

where the product notation in (17.5) is defined by Eq. (17.22). The wave function (17.5) takes no account of spin or the antisymmetry requirement. To do so, we must put each electron in a spin-orbital $u_i = \phi_i \sigma_i$, where σ_i is a spin function (either α or β). The wave function ψ_π is then written as a Slater determinant of spin-orbitals. Since $\hat{H}^{\text{eff}}(i)$ does not involve spin, we have $\hat{H}^{\text{eff}}(i)u_i = e_i u_i$. Each term in the antisymmetrized product function ψ_π has each electron in a different spin-orbital [see, for example, the equation preceding (10.48)]. When $\hat{H}_\pi$, which is being approximated as the sum of the $\hat{H}^{\text{eff}}(i)$'s, acts on each term in ψ_π, it gives the sum of the e_i's. Hence $\hat{H}_\pi \psi_\pi$ equals $\sum_i e_i \psi_\pi$, and (17.7) still holds when spin and antisymmetry are allowed for.

Since $\hat{H}^{\text{eff}}$ is not specified, there is no point in trying to solve (17.6) directly. Instead, the variational method is used.

The next assumption in the HMO method is to approximate the π MOs as LCAOs. In a minimal-basis-set calculation of a planar conjugated hydrocarbon, the only AOs of π symmetry are the carbon $2p\pi$ orbitals, where by $2p\pi$ we mean the real $2p$ AOs that are perpendicular to the molecular plane. We thus write

$$\phi_i = \sum_{r=1}^{n_C} c_{ri} f_r \tag{17.8}$$

where f_r is a $2p\pi$ AO on the rth carbon atom and n_C is the number of carbon atoms. Since (17.8) is a linear variation function, the optimum values of the coefficients for the n_C lowest π MOs satisfy Eq. (8.54):

$$\sum_{s=1}^{n_C} [(H_{rs}^{\text{eff}} - S_{rs}e_i)c_{si}] = 0, \quad r = 1, 2, \ldots, n_C \tag{17.9}$$

where the e_i's are the roots of the secular equation (8.58):

$$\det(H_{rs}^{\text{eff}} - S_{rs}e_i) = 0 \tag{17.10}$$

The key assumptions in the Hückel theory involve the integrals in (17.10). The integral H_{rr}^{eff} is assumed to have the same value for every carbon atom in the molecule. (For benzene the six carbons are equivalent, and this is no assumption. For 1,3-butadiene, $CH_2CHCHCH_2$, one would expect H_{rr}^{eff} for an end carbon and a middle carbon to differ slightly.) Moreover, H_{rr}^{eff} is assumed to be the same for carbon atoms in different planar hydrocarbons. The integral H_{rs}^{eff} is assumed to have the same value for any two carbon atoms bonded to each other and to vanish for two nonbonded carbons. The integral S_{rr} is equal to 1, since the AOs are normalized. The overlap integral S_{rs} is assumed to vanish for $r \neq s$. We have

$$\boxed{H_{rr}^{\text{eff}} = \int f_r^*(i)\hat{H}^{\text{eff}}(i)f_r(i)\, dv_i \equiv \alpha} \tag{17.11}$$

$$\boxed{H_{rs}^{\text{eff}} = \int f_r^*(i)\hat{H}^{\text{eff}}(i)f_s(i)\, dv_i \equiv \beta \quad \text{for } C_r \text{ and } C_s \text{ bonded}} \tag{17.12}$$

$$\boxed{H_{rs}^{\text{eff}} = 0 \quad \text{for } C_r \text{ and } C_s \text{ not bonded together}} \tag{17.13}$$

$$\boxed{S_{rs} = \int f_r^*(i)f_s(i)\, dv_i = \delta_{rs}} \tag{17.14}$$

$$f_r \equiv C_r 2p\pi \tag{17.15}$$

where δ_{rs} is the Kronecker delta. α is called the **Coulomb integral** and β is the **bond integral** (or resonance integral). Since carbons not bonded to each other are well separated in space, the assumption (17.13) is reasonable. However, taking the overlap integral as zero for carbons bonded to each other is a poor assumption. For Slater orbitals, S_{rs} for adjacent carbons ranges between 0.2 and 0.3, depending on the bond distance. Inclusion of overlap will be considered later.

We want each π MO to be normalized. Use of (17.8) and (17.14) gives $1 = \langle\phi_i|\phi_i\rangle = \langle\sum_r c_{ri}f_r|\sum_s c_{si}f_s\rangle = \sum_r\sum_s c_{ri}^*c_{si}\langle f_r|f_s\rangle = \sum_r\sum_s c_{ri}^*c_{si}\delta_{rs} = \sum_r c_{ri}^*c_{ri}$,

$$\sum_{r=1}^{n_C} |c_{ri}|^2 = 1 \tag{17.16}$$

This is the normalization condition for the ith Hückel MO.

In application of HMO theory to conjugated hydrocarbons, planarity is assumed. Occasionally this assumption does not hold. In gas-phase biphenyl the two rings are twisted at an angle of 44° with each other, because of steric interference between the *ortho* hydrogens.

From (17.9) it is clear that the order of the HMO secular determinant equals the number of conjugated atoms. Students sometimes make the error of assuming that this order always equals the number of π electrons.

Butadiene. To illustrate the HMO method, we consider 1,3-butadiene. The only thing of significance in the simple HMO treatment of a planar hydrocarbon is the topology of the carbon-atom framework; by this we mean which carbons are bonded together. No distinction is made between *s-cis* and *s-trans* butadiene. The terms *s-cis* and *s-trans* refer to the conformation about the single bond:

s-cis *s-trans*

The *s-trans* conformation is the dominant one observed for 1,3-butadiene. However, the UV, IR, and Raman spectra of 1,3-butadiene show the presence of small amounts of a second conformation. High-level ab initio calculations and analysis of spectra of gas-phase butadiene show that this second conformation is the nonplanar gauche form with a CCCC dihedral angle of 38° ± 3° [see the references in M. S. Deleuze and S. Knippenberg, *J. Chem. Phys.*, **125**, 104309 (2006)].

The numbering of the carbon atoms is

$$\underset{1}{CH_2}{=}\underset{2}{CH}{-}\underset{3}{CH}{=}\underset{4}{CH_2} \tag{17.17}$$

The Hückel assumptions (17.11) to (17.13) give $H_{11}^{\text{eff}} = H_{22}^{\text{eff}} = H_{33}^{\text{eff}} = H_{44}^{\text{eff}} = \alpha$, $H_{12}^{\text{eff}} = H_{23}^{\text{eff}} = H_{34}^{\text{eff}} = \beta$, and $H_{13}^{\text{eff}} = H_{14}^{\text{eff}} = H_{24}^{\text{eff}} = 0$. The secular equation (17.10) is

$$\begin{vmatrix} \alpha - e_k & \beta & 0 & 0 \\ \beta & \alpha - e_k & \beta & 0 \\ 0 & \beta & \alpha - e_k & \beta \\ 0 & 0 & \beta & \alpha - e_k \end{vmatrix} = 0 \tag{17.18}$$

We now divide each row of the determinant by β. This divides the determinant by β^4, and since $0/\beta^4 = 0$, we get

$$\begin{vmatrix} x & 1 & 0 & 0 \\ 1 & x & 1 & 0 \\ 0 & 1 & x & 1 \\ 0 & 0 & 1 & x \end{vmatrix} = 0 \tag{17.19}$$

where

$$x \equiv \frac{\alpha - e_k}{\beta}, \qquad e_k = \alpha - \beta x \tag{17.20}$$

A determinant in which all elements are zero except the elements of the principal diagonal and the elements immediately above and below this diagonal is called a *continuant*. When the elements immediately above the principal diagonal are all equal, those on the principal diagonal are all equal, and those immediately below the principal diagonal are all equal, the nth-order continuant can be shown to have the value (T. Muir, *The Theory of Determinants in the Historical Order of Development*, Dover, Volume 4, 1960, page 401)

$$\begin{vmatrix} a & b & 0 & 0 & \cdot & \cdot & \cdot & \cdot & \cdot & 0 \\ c & a & b & 0 & \cdot & \cdot & \cdot & \cdot & \cdot & 0 \\ 0 & c & a & b & \cdot & \cdot & \cdot & \cdot & \cdot & 0 \\ \cdot & \cdot & \cdot & \cdot & \cdot & \cdot & \cdot & \cdot & \cdot & \cdot \\ 0 & \cdot & \cdot & \cdot & \cdot & \cdot & 0 & c & a & b \\ 0 & \cdot & \cdot & \cdot & \cdot & \cdot & 0 & 0 & c & a \end{vmatrix} = \prod_{j=1}^{n} \left[a - 2(bc)^{1/2} \cos\left(\frac{j\pi}{n+1}\right) \right] \quad (17.21)$$

where n is the order of the determinant. The definition of the product notation used in (17.21) is

$$\prod_{j=1}^{n} b_j = b_1 b_2 b_3 \cdots b_n \quad (17.22)$$

Use of (17.21) in (17.19) gives

$$\prod_{j=1}^{4} \left(x - 2\cos\frac{j\pi}{5} \right) = 0$$

$$x = 2\cos(j\pi/5), \quad j = 1, 2, 3, 4 \quad (17.23)$$

$$x = -1.618, -0.618, 0.618, 1.618 \quad (17.24)$$

Alternatively, we can expand the secular determinant to yield the algebraic equation $x^4 - 3x^2 + 1 = 0$. The substitution $z = x^2$ yields $z^2 - 3z + 1 = 0$, and the quadratic formula gives $z = (3 \pm 5^{1/2})/2$. Hence $x = \pm z^{1/2} = (3 \pm 5^{1/2})^{1/2}/2^{1/2}$, $-(3 \pm 5^{1/2})^{1/2}/2^{1/2}$.

Corresponding to (17.19), the equations for the HMO coefficients of butadiene are

$$\begin{aligned} xc_{1j} + c_{2j} \qquad\qquad\qquad &= 0 \\ c_{1j} + xc_{2j} + c_{3j} \qquad\quad &= 0 \\ c_{2j} + xc_{3j} + c_{4j} &= 0 \\ c_{3j} + xc_{4j} &= 0 \end{aligned} \quad (17.25)$$

We must now substitute each of the roots (17.24) in turn into (17.25), as discussed in Section 8.5.

Consider first the root $x = -1.618$. The first equation of (17.25) reads $-1.618c_1 + c_2 = 0$ (where, for simplicity, the j subscript has been omitted). As discussed in Section 8.4, the solutions c_1, c_2, c_3, c_4 each contain an arbitrary multiplicative constant. Hence we shall solve for c_2, c_3, and c_4 in terms of c_1. We have $c_2 = 1.618c_1$. The second equation in (17.25) gives $c_3 = -c_1 - xc_2 = -c_1 + 1.618(1.618c_1) = 1.618c_1$. The fourth equation in (17.25) gives $c_4 = -c_3/x = 1.618c_1/1.618 = c_1$.

The normalization condition (17.16) is now used to fix c_1. Taking c_1 to be a real, positive number, we have $1 = c_1^2 + c_2^2 + c_3^2 + c_4^2 = c_1^2 + (1.618c_1)^2 + (1.618c_1)^2 + c_1^2 = 7.236c_1^2$, and $c_1 = 0.372$. Then $c_2 = 1.618c_1 = 0.602$, $c_3 = 1.618c_1 = 0.602$, and $c_4 = c_1 = 0.372$. The HMO corresponding to $x = -1.618$ is then $\phi = 0.372f_1 + 0.602f_2 + 0.602f_3 + 0.372f_4$. The energy of this HMO is given by (17.20) as $e = \alpha + 1.618\beta$.

Substitution of each of the three remaining roots into (17.25) yields three more HMOs. We find the following four normalized HMOs (Prob. 17.4):

$$\begin{aligned} \phi_1 &= 0.372f_1 + 0.602f_2 + 0.602f_3 + 0.372f_4 \\ \phi_2 &= 0.602f_1 + 0.372f_2 - 0.372f_3 - 0.602f_4 \\ \phi_3 &= 0.602f_1 - 0.372f_2 - 0.372f_3 + 0.602f_4 \\ \phi_4 &= 0.372f_1 - 0.602f_2 + 0.602f_3 - 0.372f_4 \end{aligned} \tag{17.26}$$

The HMO energies are

$$\begin{aligned} e_1 = \alpha + 1.618\beta, \quad e_2 = \alpha + 0.618\beta \\ e_3 = \alpha - 0.618\beta, \quad e_4 = \alpha - 1.618\beta \end{aligned} \tag{17.27}$$

The MO ϕ_1 in (17.26) has no nodes (other than the molecular plane) and leads to maximum charge buildup between the atoms. Clearly this must be the lowest-energy π MO. Its energy is $\alpha + 1.618\beta$ and, therefore, the bond integral β must be negative. [See also Eq. (13.62).] The HMOs and energy levels are sketched in Fig. 17.1. The number of vertical nodal planes is zero for the ground MO and increases by one for each higher MO.

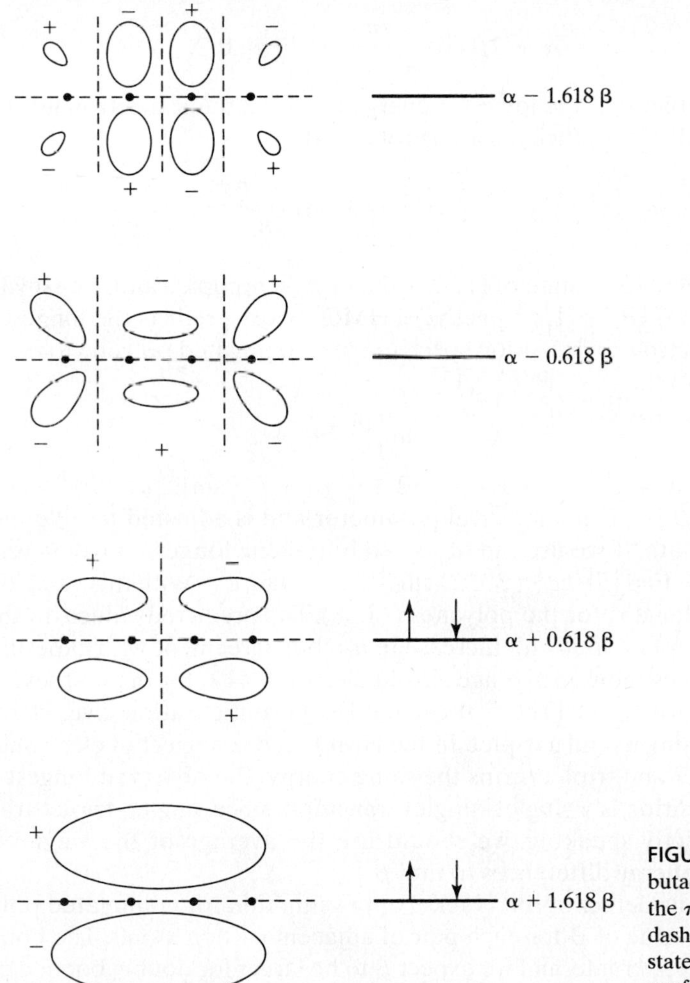

FIGURE 17.1 HMOs for butadiene. Nodal planes for the π MOs are indicated by dashed lines. The ground-state π-electron MO configuration is shown.

From the figure, it is clear that these MOs are orthogonal. We can approximate the energy of an electron in a carbon $2p\pi$ AO in the molecule by $\int f_i^* \hat{H}^{\text{eff}} f_i \, dv = \alpha$; an HMO is classified as bonding or antibonding according to whether its energy is less than or greater than α.

Conjugated Polyenes. Consider the conjugated polyene

$$\text{CH}_2{=}[\text{CH}{-}\text{CH}{=}]_s\text{CH}_2 \tag{17.28}$$

where s can be $0, 1, 2, \ldots$. Let n_C be the number of carbon atoms in the polyene. The HMO secular equation involves a continuant similar to (17.19) but of order n_C. Equation (17.21) with $b = c = 1$, $a = x$, and $n = n_\text{C}$ gives $x = 2\cos[j\pi/(n_\text{C} + 1)]$, where $j = 1, \ldots, n_\text{C}$. Since the values of x for $j = k$ and for $j = n_\text{C} + 1 - k$ are simply the negatives of each other, we can write

$$x = -2\cos\frac{j\pi}{n_\text{C} + 1}$$
$$e_j = \alpha + 2\beta\cos\frac{j\pi}{n_\text{C} + 1}, \qquad j = 1, 2, \ldots, n_\text{C} \tag{17.29}$$

Since β is negative, e_1 is the lowest π energy level. All the π-electron levels are nondegenerate. The HMO coefficients are (Prob. 17.5)

$$c_{rj} = \left(\frac{2}{n_\text{C} + 1}\right)^{1/2} \sin\frac{jr\pi}{n_\text{C} + 1} \tag{17.30}$$

In the ground electronic state of (17.28), the highest occupied and lowest vacant π MOs have $j = \frac{1}{2}n_\text{C}$ and $\frac{1}{2}n_\text{C} + 1$, respectively. HMO theory predicts the longest wavelength band of the electronic absorption spectrum of a conjugated polyene to occur at

$$\frac{1}{\lambda} = -\frac{4\beta}{hc}\sin\frac{\pi}{2n_\text{C} + 2} \tag{17.31}$$

where (17.29) and $\cos a - \cos b = -2\sin\left[\frac{1}{2}(a + b)\right]\sin\left[\frac{1}{2}(a - b)\right]$ were used. The bond integral β is a semiempirical parameter and is adjusted to give the best fit to experimental data. If we use the observed butadiene longest-wavelength absorption $\lambda = 217$ nm, we find $|\beta|/hc = 37300\ \text{cm}^{-1}$, $|\beta| = 4.62$ eV. With this value of β, we then calculate wavelengths for the polyenes (17.28). The predicted values do show the correct trend of increase in λ with increase in n_C, but agreement with experiment is poor; predicted λ values show an average absolute error of 44% for the first several members of the series (Prob. 17.3). [The first excited HMO configuration gives rise to two electronic terms, a singlet and a triplet. In the HMO model, neglect of electronic repulsions gives the singlet and triplet terms the same energy. The observed longest wavelength electronic transition is a singlet–singlet transition, since singlet–triplet transitions are forbidden. Strictly speaking, we should use the average of the singlet–singlet and singlet–triplet energy differences to find β.]

An obvious defect of the Hückel approximation for conjugated polyenes is the use of a single value of β for each pair of adjacent carbon atoms. The bond lengths in these molecules alternate, and we expect β to be larger for doubly bonded carbons than

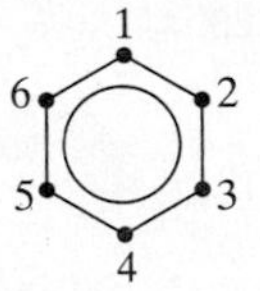

FIGURE 17.2 Numbering of carbons in benzene.

for singly bonded carbons. Lennard-Jones used two polyene bond integrals β_1 and β_2 for C—C and C=C bonds, respectively [J. E. Lennard-Jones, *Proc. Roy. Soc.*, **A158**, 280 (1937)]. With $\beta_1 = -3.32$ eV, $\beta_2 = -4.20$ eV, agreement with experiment is much improved over the single-β predictions; the average absolute error of predicted λ values is reduced to 9%.

Benzene. For benzene (Fig. 17.2), the HMO secular equation is

$$\begin{vmatrix} x & 1 & 0 & 0 & 0 & 1 \\ 1 & x & 1 & 0 & 0 & 0 \\ 0 & 1 & x & 1 & 0 & 0 \\ 0 & 0 & 1 & x & 1 & 0 \\ 0 & 0 & 0 & 1 & x & 1 \\ 1 & 0 & 0 & 0 & 1 & x \end{vmatrix} = 0 \qquad (17.32)$$

where x is given by (17.20).

The determinant in (17.32) is a special kind called a *circulant*. A circulant has only n independent elements; they appear in the first row, and succeeding rows are formed by successive cyclic permutations of these elements. The nth-order circulant $C(a_1, a_2, \ldots, a_n)$ is

$$C(a_1, a_2, \ldots, a_n) = \begin{vmatrix} a_1 & a_2 & a_3 & \cdots & a_n \\ a_n & a_1 & a_2 & \cdots & a_{n-1} \\ a_{n-1} & a_n & a_1 & \cdots & a_{n-2} \\ \cdot & \cdot & \cdot & \cdots & \cdot \\ a_2 & a_3 & a_4 & \cdots & a_1 \end{vmatrix}$$

The value of the nth-order circulant can be shown to be (T. Muir, *A Treatise on the Theory of Determinants*, Dover, 1960, pages 442–445)

$$C(a_1, a_2, \ldots, a_n) = \prod_{k=1}^{n} (a_1 + \omega_k a_2 + \omega_k^2 a_3 + \cdots + \omega_k^{n-1} a_n) \qquad (17.33)$$

where $\omega_1, \omega_2, \ldots, \omega_n$ are the n different nth roots of unity [Eq. (1.36)]:

$$\omega_k = e^{2\pi ik/n}, \qquad k = 1, 2, \ldots, n, \qquad i = \sqrt{-1} \qquad (17.34)$$

Substitution of (17.33) into (17.32), followed by the use of $\exp[2\pi ik(5/6)] = \exp(-2\pi ik/6)$ and $e^{i\theta} = \cos\theta + i\sin\theta$, gives

$$\prod_{k=1}^{6} (x + e^{2\pi ik/6} + e^{-2\pi ik/6}) = 0, \qquad i = \sqrt{-1}$$

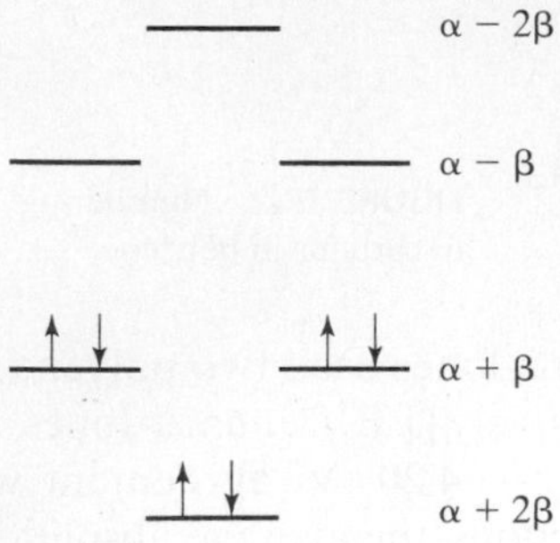

$\alpha - 2\beta$

$\alpha - \beta$

$\alpha + \beta$

$\alpha + 2\beta$

FIGURE 17.3 Hückel MO energies for benzene.

$$x = -2\cos\left(\frac{2\pi k}{6}\right), \qquad k = 1, \ldots, 6 \tag{17.35}$$

$$x = -1, +1, +2, +1, -1, -2$$

$$e_i = \alpha + 2\beta, \quad \alpha + \beta, \quad \alpha + \beta, \quad \alpha - \beta, \quad \alpha - \beta, \quad \alpha - 2\beta \tag{17.36}$$

There are two nondegenerate and two doubly degenerate Hückel levels. Figure 17.3 shows the HMO ground-state π-electron configuration.

The HMO coefficients can be found by solving the usual set of simultaneous equations, but it is simpler to use molecular symmetry. The $\hat{O}_{C_6}$ symmetry operator commutes with the π-electron Hamiltonian, so we can choose each MO to be an eigenfunction of this 60° rotation. Since $(\hat{O}_{C_6})^6 = \hat{1}$, the eigenvalues of $\hat{O}_{C_6}$ are the six sixth roots of unity (Prob. 7.24):

$$e^{2\pi ik/6}, \qquad k = 0, 1, \ldots, 5 \tag{17.37}$$

(Since $\hat{O}_{C_6}$ has some complex eigenvalues, it is not Hermitian. Only operators representing physical quantities need be Hermitian, and $\hat{O}_{C_6}$ does not correspond to any physical property of the molecule.) Substitution of $\phi_j = \sum_{r=1}^{n_C} c_{rj} f_r$ [Eq. (17.8)] into the eigenvalue equation $\hat{O}_{C_6}\phi_j = e^{2\pi ik/6}\phi_j$ gives

$$e^{2\pi ik/6}\phi_j = \hat{O}_{C_6}\phi_j$$

$$\sum_{r=1}^{6} c_{rj}e^{2\pi ik/6} f_r = \sum_{r=1}^{6} c_{rj}\hat{O}_{C_6} f_r = \sum_{r=1}^{6} c_{rj} f_{r-1} = \sum_{r=1}^{6} c_{r+1,\,j} f_r$$

where $f_0 \equiv f_6$ and $c_{7j} \equiv c_{1j}$. Equating the coefficients of corresponding AOs, we have

$$c_{r+1,\,j} = e^{2\pi ik/6}\, c_{rj} \tag{17.38}$$

The normalization condition (17.16) is $\sum_{r=1}^{6} |c_{rj}|^2 = 1$ and Eq. (17.38) shows that all the coefficients in the jth MO have the same absolute value. Hence

$$|c_{rj}| = 1/\sqrt{6}, \qquad r = 1, 2, \ldots, 6 \tag{17.39}$$

By setting $c_{1j} = 1/\sqrt{6}$ and using (17.38), we get the desired coefficients. To find which coefficients go with which energy, we evaluate the variational integral:

$$e_j = \int \phi_j^* \hat{H}^{\text{eff}} \phi_j \, dv = \sum_{r=1}^{6} \sum_{s=1}^{6} c_{rj}^* c_{sj} \int f_r^* \hat{H}^{\text{eff}} f_s \, dv$$

$$= \sum_{r=1}^{6} |c_{rj}|^2 \alpha + \sum_{r=1}^{6} c_{rj}^* c_{r+1,j} \beta + \sum_{r=1}^{6} c_{rj}^* c_{r-1,j} \beta$$

$$= \alpha + e^{2\pi i k/6} \sum_{r=1}^{6} |c_{rj}|^2 \beta + e^{-2\pi i k/6} \sum_{r=1}^{6} |c_{rj}|^2 \beta$$

$$= \alpha + 2\beta \cos(2\pi k/6), \qquad k = 0, \ldots, 5 \tag{17.40}$$

which agrees with (17.36). From (17.38), (17.39), and (17.40), the Hückel MOs and energies for benzene are

$$\begin{aligned}
\phi_1 &= 6^{-1/2}(f_1 + f_2 + f_3 + f_4 + f_5 + f_6) \\
\phi_2 &= 6^{-1/2}(f_1 + e^{\pi i/3} f_2 + e^{2\pi i/3} f_3 - f_4 + e^{4\pi i/3} f_5 + e^{5\pi i/3} f_6) \\
\phi_3 &= 6^{-1/2}(f_1 + e^{-\pi i/3} f_2 + e^{-2\pi i/3} f_3 - f_4 + e^{-4\pi i/3} f_5 + e^{-5\pi i/3} f_6) \\
\phi_4 &= 6^{-1/2}(f_1 + e^{2\pi i/3} f_2 + e^{4\pi i/3} f_3 + f_4 + e^{2\pi i/3} f_5 + e^{4\pi i/3} f_6) \\
\phi_5 &= 6^{-1/2}(f_1 + e^{-2\pi i/3} f_2 + e^{-4\pi i/3} f_3 + f_4 + e^{-2\pi i/3} f_5 + e^{-4\pi i/3} f_6) \\
\phi_6 &= 6^{-1/2}(f_1 - f_2 + f_3 - f_4 + f_5 - f_6)
\end{aligned} \tag{17.41}$$

$$e_1 = \alpha + 2\beta, \qquad e_2 = \alpha + \beta, \qquad e_3 = \alpha + \beta$$
$$e_4 = \alpha - \beta, \qquad e_5 = \alpha - \beta, \qquad e_6 = \alpha - 2\beta$$

The condition (17.38) determining the π-MO coefficients for benzene was derived solely from symmetry considerations, without use of the Hückel approximations. Thus the MOs (17.41) are (except for normalization constants) the correct minimal-basis-set SCF π-electron MOs for benzene. (The Hückel energies $e_1, \ldots, e_6$ are, however, not the true SCF orbital energies. The Hückel method ignores electron repulsions and takes the total π-electron energy as the sum of orbital energies. The SCF MO method takes electron repulsions into account in an average way, and the total SCF energy is not the sum of orbital energies.) A similar situation occurs for ethylene, where the minimal-basis-set π MOs (Section 15.9) are determined solely by symmetry. (An extended-basis-set benzene calculation would mix in $3p\pi, 3d\pi, \ldots,$ carbon AOs, and the contributions of these AOs must be determined by an explicit SCF calculation.)

The MOs in (17.41) for the degenerate π levels are complex. The two MOs of each degenerate level are complex conjugates of each other. By adding and subtracting these MOs, we get the commonly used real forms:

$$\begin{aligned}
\phi_{2,\text{real}} &= 2^{-1/2}(\phi_2 + \phi_3), \quad \phi_{3,\text{real}} = -2^{-1/2} i(\phi_2 - \phi_3) \\
\phi_{2,\text{real}} &= 12^{-1/2}(2f_1 + f_2 - f_3 - 2f_4 - f_5 + f_6) \\
\phi_{3,\text{real}} &= \tfrac{1}{2}(f_2 + f_3 - f_5 - f_6) \\
\phi_{4,\text{real}} &= 12^{-1/2}(2f_1 - f_2 - f_3 + 2f_4 - f_5 - f_6) \\
\phi_{5,\text{real}} &= \tfrac{1}{2}(f_2 - f_3 + f_5 - f_6)
\end{aligned} \tag{17.42}$$

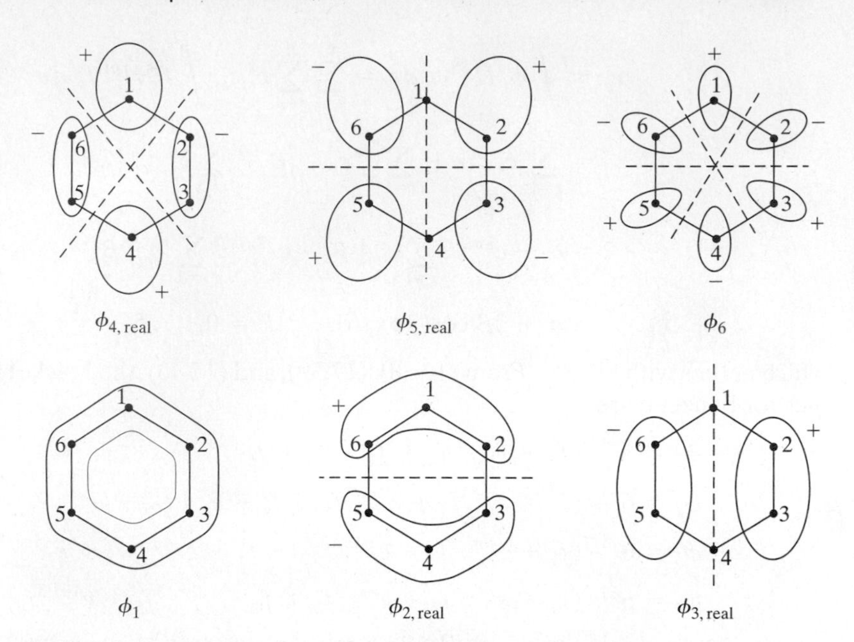

FIGURE 17.4 Benzene π MOs (real form). A top view is shown. The π MOs change sign on reflection in the molecular plane, which is a nodal plane for them. Dashed lines indicate nodal planes perpendicular to the molecular plane.

Figure 17.4 shows the real benzene π-electron MOs. Note the charge buildup between the nuclei for the bonding MOs.

The symmetry species of the benzene π MOs are (*Schonland*, p. 210)

MO	ϕ_1	ϕ_2	ϕ_3	ϕ_4	ϕ_5	ϕ_6
Symmetry species	a_{2u}	e_{1g}	e_{1g}	e_{2u}	e_{2u}	b_{2g}

The ground-state π-electron configuration is $(1a_{2u})^2(1e_{1g})^4$.

Monocyclic Conjugated Polyenes. For the monocyclic planar conjugated polyene C_nH_n, we can use the same treatment as for benzene, C_6H_6. Replacing 6 by n_C in (17.40), (17.38), and (17.39), we find as the HMO energies and coefficients

$$e_k = \alpha + 2\beta \cos\frac{2\pi k}{n_C}, \qquad k = 0, \ldots, n_C - 1 \tag{17.43}$$

$$c_{rk} = \frac{1}{\sqrt{n_C}} \exp\left[\frac{2\pi i(r-1)k}{n_C}\right], \qquad i = \sqrt{-1} \tag{17.44}$$

$$\phi_k = \frac{1}{\sqrt{n_C}} \sum_{r=1}^{n_C} \exp\left[\frac{2\pi i(r-1)k}{n_C}\right] f_r \tag{17.45}$$

where $f_r = C_r 2p\pi$. Note that the index k in these equations does not correspond to the actual order of the MOs: The lowest MO has $k = 0$; next are the MOs with $k = 1$ and $k = n_C - 1$; next are the MOs with $k = 2$ and $k = n_C - 2$; and so on.

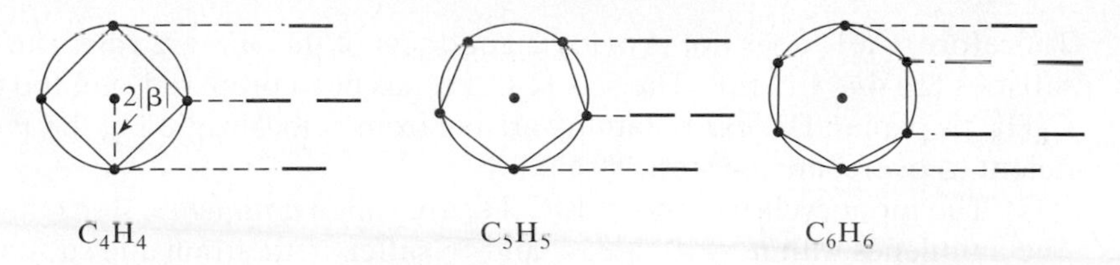

FIGURE 17.5 Mnemonic device for HMO energies of monocyclic planar hydrocarbons. The levels shown are for cyclobutadiene, the cyclopentadienyl radical, and benzene.

An amusing mnemonic device is available for the HMO energies (17.43) [A. A. Frost and B. Musulin, *J. Chem. Phys.*, **21**, 572 (1953)]. One inscribes a regular polygon of n_C sides in a circle of radius $2|\beta|$, putting an apex at the bottom of the circle. If a vertical scale of energy is set up with energy α coinciding with the center of the circle, then each polygon vertex is located at an HMO energy (Fig. 17.5). The method gives the correct degeneracies and spacings of the Hückel levels of the ring hydrocarbon C_nH_n (Prob. 17.8).

Consider the Hückel energy levels for the monocyclic polyene C_nH_n. The lowest shell consists of a nondegenerate level and holds two electrons. Each of the remaining low-lying shells consists of a doubly degenerate level and holds four electrons. (If n_C is even, the highest π energy level is nondegenerate, but this level is not occupied in the ground state.) To have a stable filled-shell π-electron configuration, we see that the number of π electrons must satisfy

$$n_\pi = 4m + 2, \qquad m = 0, 1, 2, \ldots \tag{17.46}$$

This is Hückel's famous $4m + 2$ rule, which ascribes extra stability to monocyclic conjugated systems that satisfy (17.46). With $4m + 1$ or $4m - 1$ π electrons, the compound is a free radical. With $4m$ π electrons, there are two electrons in a shell that can hold four electrons, and Hund's rule predicts a triplet (diradical) ground state.

Benzene satisfies the $4m + 2$ rule. The cyclopentadienyl radical $\cdot C_5H_5$ is one electron short of satisfying (17.46); the cyclopentadienyl anion $C_5H_5^-$ satisfies the $4m + 2$ rule; the cation $C_5H_5^+$ is predicted to have a triplet ground state and be highly reactive. These predictions are borne out; $C_5H_5^-$ is found to be considerably more stable than either $C_5H_5^+$ or $\cdot C_5H_5$. Similarly, $C_7H_7^+$ should be more stable than $\cdot C_7H_7$ or $C_7H_7^-$, as is verified experimentally; for example, the salt $C_7H_7^+Br^-$ is readily prepared.

C_4H_4 is predicted by the $4m + 2$ rule to have a triplet ground state. Cyclobutadiene, C_4H_4, first synthesized in 1965, is a highly reactive compound that dimerizes at temperatures above 35 K. Experimental observations and theoretical calculations agree that the ground electronic state is a singlet with a rectangular geometry and carbon–carbon bond lengths close to ordinary single- and double-bond lengths [see D. W. Whitman and B. K. Carpenter, *J. Am. Chem. Soc.*, **102**, 4272 (1980); W. T. Borden and E. R. Davidson, *Acc. Chem. Res.*, **14**, 69 (1981)]. The reason for this violation of Hund's rule is discussed in H. Kollmar and V. Staemmler, *J. Am. Chem. Soc.*, **99**, 3583 (1977).

HMO theory predicts that planar cyclooctatetraene, C_8H_8, would have a triplet ground state. Experimentally, C_8H_8 has a nonplanar "tub" structure with alternating single and double bonds. The nonplanarity results from steric strain in the planar geometry due to deviation from the 120° bond angle at the sp^2-hybridized carbons.

Therefore, C_8H_8 does not provide a good test of the $4m + 2$ rule. The dianion $C_8H_8^{2-}$ satisfies the $4m + 2$ rule. The salt $K_2C_8H_8$ has been prepared, and the evidence is that $C_8H_8^{2-}$ is planar. The extra stability arising from delocalization of the π electrons is sufficient to overcome the steric strain.

The monocyclic compounds C_nH_n are called *annulenes*. Benzene is [6]annulene. The annulenes with $n = 10, 12, 14$, and 16 suffer steric strain and substantial repulsions between nonbonded hydrogens, thereby preventing planarity and a clear-cut test of the $4m + 2$ rule. The compound [18]annulene has been much studied, but the results are conflicting. [18]annulene has a nearly planar 120°-angle structure with a cis configuration about every third double bond; this makes 6 of the 18 hydrogens lie in the interior of the ring. X-ray crystallography data of the solid have been interpreted as showing nearly equal carbon–carbon bond lengths, but this result is not conclusive. Various high-level ab initio calculations give differing results as to whether the bond lengths alternate. Theoretical calculations of the NMR proton shifts in [18]annulene found that good agreement with the experimental values was achieved only if the molecular geometry showed a carbon–carbon bond-length alternation of about 0.1 Å [C. S. Wannere et al., *Angew. Chem. Int. Ed.*, **43**, 4200 (2004)]. Despite this bond-length alternation, the authors of this paper view [18]annulene as an aromatic compound because of its considerable aromatic stabilization energy and the abnormal values of its proton NMR chemical shifts.

The stabilization gained by π-electron delocalization becomes negligible as n goes to infinity in the aromatic annulenes $C_{4m+2}H_{4m+2}$; see the discussion of delocalization energy later in this section.

In the preceding discussion, we used the same HMOs for a neutral compound and for its related ions. The Hückel method ignores electron repulsions, so the HMOs are unchanged when π electrons are added or removed.

The $4m + 2$ rule is sometimes applied to polycyclic systems; however, the pattern (17.43) of HMO levels holds only for a monocyclic system, and use of the $4m + 2$ rule for polycyclic systems is not justified.

The $4m + 2$ rule actually does not depend on the Hückel assumptions (17.11) to (17.14). The C_nH_n π MOs (17.45) were derived solely by symmetry considerations and are the correct SCF minimal-basis-set π MOs. With $k = 0$, we have an MO with all plus signs in front of the AOs. Clearly this MO has a lower energy than any of the others. For the remaining MOs, the pair with $k = j$ and $k = n_C - j$ are complex conjugates of each other and must have the same energy. [Since $\hat{H}$ is Hermitian, we have $\int \phi^* \hat{H} \phi \, dv = \int (\phi^*)^* \hat{H} \phi^* \, dv$.] Thus the excited MOs occur in pairs (except that when n_C is even, the MO with $k = \frac{1}{2} n_C$ is nondegenerate; with alternating plus and minus signs in front of the AOs, this is the highest-energy MO). The pattern of a nondegenerate lowest π level followed by doubly degenerate π levels thus holds for the minimal-basis-set SCF MOs. Of course, the energies (17.43) are not the correct SCF orbital energies.

Naphthalene. Now consider the HMO treatment of naphthalene. For butadiene and benzene, we set up the secular equation without bothering with the intermediate step of constructing symmetry orbitals from the $2p\pi$ AOs. For these molecules, the secular equation was easy enough to solve without the simplifications introduced by symmetry orbitals. For naphthalene the 10×10 secular determinant is difficult to deal with, and we first find symmetry orbitals. The point group of naphthalene (Fig. 17.6) is $\mathcal{D}_{2h}$.

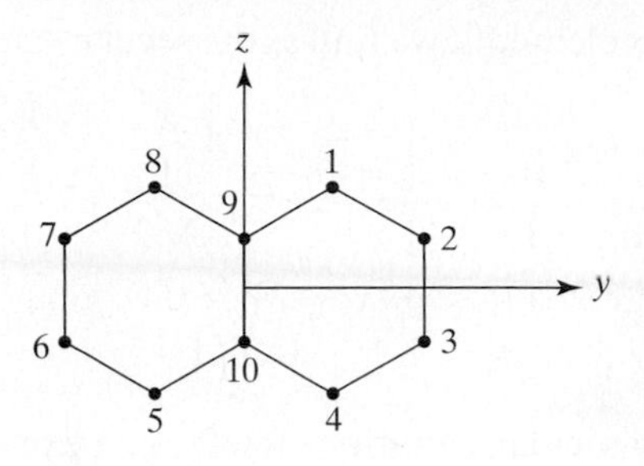

FIGURE 17.6 Axes for naphthalene. The x axis is perpendicular to the molecular plane.

The possible symmetry species are in Table 15.3. The $C2p\pi$ AOs all have eigenvalue -1 for reflection in the molecular (yz) plane. Each symmetry orbital will be some linear combination of AOs that are permuted among one another by the symmetry reflections (recall ethylene). Hence, to aid in finding the symmetry orbitals, we examine the effects of the $\hat{\sigma}(xy)$ and $\hat{\sigma}(xz)$ operations on the π AOs. We find the π AOs to fall into three sets:

$$1, 4, 5, 8 \qquad 2, 3, 6, 7 \qquad 9, 10$$

where the members of each set are permuted among one another by the symmetry reflections. (These are also the chemically equivalent carbons.) Each symmetry orbital must be some linear combination of the AOs of a given set. Naphthalene and ethylene have the same point group, and the pattern of the hydrogen symmetry orbitals in Table 15.4 gives us the symmetry orbitals of the first two of the above sets of naphthalene AOs. The symmetry orbitals and their readily verified symmetry species are then

$$b_{3u}\colon\quad g_1 = \tfrac{1}{2}(f_1 + f_4 + f_5 + f_8), \quad g_2 = \tfrac{1}{2}(f_2 + f_3 + f_6 + f_7), \quad g_3 = 2^{-1/2}(f_9 + f_{10})$$
$$a_u\colon\quad g_4 = \tfrac{1}{2}(f_1 - f_4 + f_5 - f_8), \quad g_5 = \tfrac{1}{2}(f_2 - f_3 + f_6 - f_7)$$
$$b_{2g}\colon\quad g_6 = \tfrac{1}{2}(f_1 - f_4 - f_5 + f_8), \quad g_7 = \tfrac{1}{2}(f_2 - f_3 - f_6 + f_7), \quad g_8 = 2^{-1/2}(f_9 - f_{10})$$
$$b_{1g}\colon\quad g_9 = \tfrac{1}{2}(f_1 + f_4 - f_5 - f_8), \quad g_{10} = \tfrac{1}{2}(f_2 + f_3 - f_6 - f_7)$$

The constants $\frac{1}{2}$ and $1/\sqrt{2}$ normalize the symmetry orbitals, provided the approximation $S_{rs} = \delta_{rs}$ [Eq. (17.14)] is used.

Instead of using the f_r's as basis functions, we set up the secular equation using the symmetry orbitals as basis functions:

$$\det[\langle g_p|\hat{H}^{\text{eff}}|g_q\rangle - \langle g_p|g_q\rangle e_k] = 0$$

The secular determinant is in block-diagonal form (Section 15.5), and we have two cubic and two quadratic equations to solve. Using $S_{rs} = \delta_{rs}$, we find that

$$\langle g_p|g_q\rangle = \delta_{pq}$$

when g_p and g_q belong to the same symmetry species. Also,

$$\langle g_1|\hat{H}^{\text{eff}}|g_1\rangle = \tfrac{1}{4}\langle f_1 + f_4 + f_5 + f_8|\hat{H}^{\text{eff}}|f_1 + f_4 + f_5 + f_8\rangle = \tfrac{1}{4}(\alpha + \alpha + \alpha + \alpha) = \alpha$$
$$\langle g_1|\hat{H}^{\text{eff}}|g_2\rangle = \tfrac{1}{4}\langle f_1 + f_4 + f_5 + f_8|\hat{H}^{\text{eff}}|f_2 + f_3 + f_6 + f_7\rangle = \tfrac{1}{4}(\beta + \beta + \beta + \beta) = \beta$$

Evaluating the remaining matrix elements, we find as the secular equation for the b_{3u} MOs

$$\begin{vmatrix} \alpha - e_k & \beta & \sqrt{2}\beta \\ \beta & \alpha + \beta - e_k & 0 \\ \sqrt{2}\beta & 0 & \alpha + \beta - e_k \end{vmatrix} = 0, \qquad \begin{vmatrix} x & 1 & \sqrt{2} \\ 1 & x + 1 & 0 \\ \sqrt{2} & 0 & x + 1 \end{vmatrix} = 0$$

$$(x + 1)(x^2 + x - 3) = 0$$

$$x = -1, \quad -\tfrac{1}{2} \pm \tfrac{1}{2}\sqrt{13}$$

Solution of the three remaining secular equations is left as an exercise. The naphthalene HMO energy levels in order of increasing energy are found to be (Prob. 17.15)

$$\alpha + 2.303\beta, \quad \alpha + 1.618\beta, \quad \alpha + 1.303\beta, \quad \alpha + \beta, \quad \alpha + 0.618\beta$$
$$\alpha - 0.618\beta, \quad \alpha - \beta, \quad \alpha - 1.303\beta, \quad \alpha - 1.618\beta, \quad \alpha - 2.303\beta$$

The levels are all nondegenerate ($\mathcal{D}_{2h}$ has only A and B symmetry species).

The HMO coefficients are found by solving the appropriate sets of simultaneous equations (Prob. 17.15).

Alternant Hydrocarbons. Note that the HMOs of naphthalene, like those of butadiene and benzene, are paired, meaning that for each HMO with the energy $\alpha - x\beta$ there is an HMO with energy $\alpha + x\beta$. This can be proved to be true for every alternant hydrocarbon (for the proof, see Prob. 17.20). An **alternant hydrocarbon** is a planar conjugated hydrocarbon in which the carbon atoms can be divided into a starred set and an unstarred set, with starred carbons bonded only to unstarred carbons, and vice versa (Fig. 17.7). All planar conjugated hydrocarbons are alternants except those containing a ring with an odd number of carbons.

Electronic Transitions. The predicted wavenumber for a transition between the highest occupied and lowest vacant HMOs of a conjugated hydrocarbon is

$$\frac{1}{\lambda} = \frac{|\beta|}{hc}\Delta x \tag{17.47}$$

where Δx is the difference in x values [Eq. (17.20)] for the two MOs. For naphthalene, $\Delta x = 1.236$, and the observed $1/\lambda$ is 34700 cm^{-1}. Choosing β to fit the observed wavelength for naphthalene (because of the orbital degeneracy of its first excited term, benzene is atypical), we find

$$|\beta|/hc = 28100 \text{ cm}^{-1}, \qquad |\beta| = 3.48 \text{ eV}$$

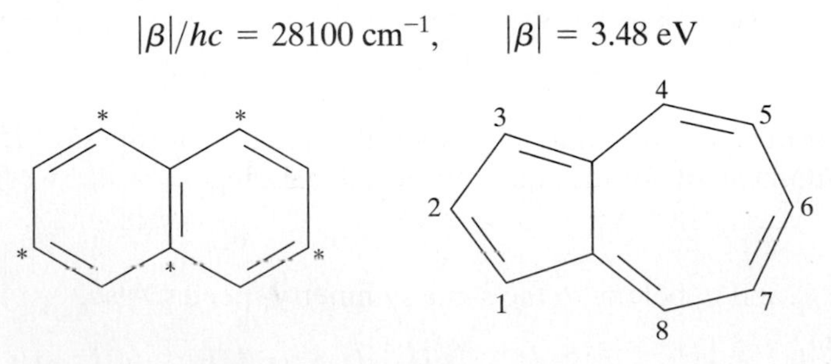

FIGURE 17.7 Naphthalene is an alternant hydrocarbon, and azulene is a nonalternant hydrocarbon.

Comparison of predicted and observed longest-wavelength absorptions for benzenoid hydrocarbons shows only fair agreement with experiment, with many deviations of a few thousand cm^{-1}.

Improved agreement can be obtained if we fit the frequencies to a straight line that does not pass through the origin; that is, we use

$$\frac{1}{\lambda} = \frac{1}{hc}|\beta|\,\Delta x + a \tag{17.48}$$

A least-squares fit of benzenoid hydrocarbon data gives [E. Heilbronner and J. N. Murrell, *J. Chem. Soc.*, **1962**, 2611]

$$a = 8200\ \text{cm}^{-1}, \qquad |\beta|/hc = 21900\ \text{cm}^{-1}, \qquad |\beta| = 2.72\ \text{eV} \tag{17.49}$$

These constants give a good fit to the data; the standard deviation is 600 cm^{-1}. Of course, with two parameters instead of one, the agreement is bound to be improved. The fact that a semiempirical theory with several adjustable parameters gives a good fit to experimental data cannot be taken as overwhelming proof of the validity of the theory.

We can partially justify the constant term in (17.48) as follows. The Hückel method neglects electron repulsions and therefore does not distinguish between singlet and triplet terms. Hence we should compare (17.47) with the energy difference between the ground state and the average of the energies of the singlet and triplet terms of the configuration with one electron excited to the lowest vacant HMO. The experimental frequencies are for singlet–singlet transitions. If we assume that the singlet–triplet splitting of the first excited configuration is reasonably constant for aromatic hydrocarbons, then $a = 8200\ \text{cm}^{-1} = 1.0$ eV can be interpreted as one-half this singlet–triplet splitting. Experimental values for half the singlet–triplet splitting in aromatic hydrocarbons are typically 0.7 to 0.8 eV, in reasonable agreement with $a = 1.0$ eV.

Delocalization Energy and Aromaticity. There are several properties of conjugated hydrocarbons that can be defined using HMO theory. We begin with **delocalization energy**. The corresponding VB term is *resonance energy*. The energy of the occupied ethylene π HMO, the $1b_{3u}$ MO in Section 15.9, is

$$\int \frac{1}{\sqrt{2}}(f_1 + f_2)^* \hat{H}^{\text{eff}} \frac{1}{\sqrt{2}}(f_1 + f_2)\, dv = \alpha + \beta$$

There are two π electrons in this MO, and the total Hückel π-electron energy for ethylene is $2\alpha + 2\beta$. For butadiene the total Hückel π-electron energy is

$$2(\alpha + 1.618\beta) + 2(\alpha + 0.618\beta) = 4\alpha + 4.472\beta \tag{17.50}$$

If butadiene had two isolated double bonds, its π-electron energy would be twice that of ethylene, namely $4\alpha + 4\beta$. The effect of delocalization is to change the butadiene π-electron energy by

$$4\alpha + 4.472\beta - (4\alpha + 4\beta) = 0.472\beta$$

Since β is negative, butadiene is stabilized by π-electron delocalization, and $0.472\,|\beta|$ is its HMO delocalization energy.

For benzene the π-electron energy is

$$2(\alpha + 2\beta) + 4(\alpha + \beta) = 6\alpha + 8\beta \tag{17.51}$$

as compared with $6\alpha + 6\beta$ for the π-electron energy of three isolated double bonds. The delocalization energy of benzene is $2\,|\beta|$. An "experimental" delocalization energy for benzene can be calculated as follows. The gas-phase enthalpy of hydrogenation of cyclohexene to cyclohexane is -28.6 kcal/mol. If benzene had three isolated double bonds, its gas-phase enthalpy of hydrogenation to cyclohexane would be three times -28.6 kcal/mol, which is -85.8 kcal/mol. The observed value is only -49.8 kcal/mol, indicating that benzene is more stable by 36 kcal/mol than it would be if its double bonds were isolated. (A similar delocalization energy is arrived at by adding up the bond energies of six C—H bonds, three C—C bonds, and three C=C bonds and comparing this to minus the enthalpy of formation of gas-phase benzene from its atoms; see Prob. 17.10.) Setting $2\,|\beta| = 36$ kcal/mol, we get $|\beta| = 18$ kcal/mol $= 0.8$ eV/molecule.

This $|\beta|$ is far less than the value 2.72 eV/molecule $= 63$ kcal/mol found by spectroscopic observations [Eq. (17.49)]. Part of the discrepancy is explainable as follows. In the hydrogenation of cyclohexene, the carbon–carbon double-bond length becomes a single-bond length. The figure of -85.8 kcal/mol applies to the hydrogenation of a hypothetical molecule with three isolated double bonds and three single bonds, with alternating bond lengths. We must therefore also consider the strain energy needed to compress three single bonds and stretch three double bonds to the benzene bond length. Even with this correction (Prob. 17.9), the thermochemical value of β differs sharply from the spectroscopic value. The difference is to be attributed to the crudity of the HMO method. One finds that a different value of β is required for each different physical property that is being considered. Moreover, the optimum values of β differ for chain and ring conjugated hydrocarbons.

The just-discussed traditional method of calculating Hückel delocalization energies of conjugated hydrocarbons by comparing a molecule's HMO π-electron energy E_π with $n_d(2\alpha + 2\beta)$, where n_d is the number of carbon–carbon double bonds, has serious shortcomings. This method predicts a substantial delocalization energy for the linear polyenes (17.28), whereas experiment shows the delocalization stabilization in these molecules to be small. For example, comparison of twice the enthalpy of hydrogenation of 1-butene with the enthalpy of hydrogenation of 1,3-butadiene gives the 1,3-butadiene delocalization energy as only 4 kcal/mol. Moreover, this method predicts substantial delocalization stabilization for certain cyclic polyenes that experiment reveals to be unstable, with no aromatic character. (A cyclic conjugated polyene is said to be **aromatic** when it shows substantially more stability than a hypothetical structure in which the double bonds do not interact with one another and when it undergoes substitution, rather than addition, when treated with electrophilic reagents like Br_2.)

To produce more reliable predictions of aromaticity, Hess and Schaad (following a suggestion of Dewar) calculated delocalization (resonance) energies of cyclic hydrocarbons by comparing the compounds' Hückel-theory E_π with a value calculated for a hypothetical acyclic conjugated polyene with the same number and kinds of bonds as in a localized structure of the cyclic hydrocarbon. [B. A. Hess and L. J. Schaad, *J. Am. Chem. Soc.*, **93**, 305, 2413 (1971); **94**, 3068 (1972); **95**, 3907 (1973); B. A. Hess, L. J. Schaad, and C. W. Holyoke, *Tetrahedron*, **28**, 3657, 5299 (1972); Schaad and Hess, *J. Chem. Educ.*,

51, 640 (1974); Hess and, Schaad, *Pure Appl. Chem.*, **52**, 1471 (1980); Schaad and Hess, *Chem. Rev.*, **101**, 1465 (2001).]

These workers found that the Hückel π-electron energies of noncyclic conjugated polyenes could be accurately calculated as $E_\pi \approx \sum_b n_b E_{\pi,b}$, where n_b is the number of bonds of a given type, $E_{\pi,b}$ is an empirical parameter, and the sum goes over all the types of carbon–carbon bonds. A conjugated hydrocarbon has three types of carbon–carbon single bonds and five types of double bonds, the types differing in the numbers of H atoms bonded to the carbons. Having found values of the parameters $E_{\pi,b}$ by fitting Hückel energies of noncyclic conjugated polyenes, Hess and Schaad then calculated the resonance energy of a cyclic polyene as the difference $\sum_b n_b E_{\pi,b} - E_\pi$ between the sum $\sum_b n_b E_{\pi,b}$ for a localized structure of the cyclic polyene and the cyclic polyene's Hückel π-electron energy E_π. For details, see Prob. 17.13.

Division of the Hess–Schaad resonance energy by the number of π electrons gives the resonance energy per π electron (REPE) of the compound. A compound with a substantially positive REPE value (greater than, say, 0.01 $|\beta|$) is predicted to be *aromatic*. A compound with a near-zero REPE is *nonaromatic*. A compound with a substantially negative REPE is predicted to be *antiaromatic*, being less stable than if its double bonds were isolated from one another. Some antiaromatic hydrocarbons are cyclobutadiene and fulvalene.

For the annulenes, C_nH_n, Hess–Schaad REPE versus n values are

n	4	6	8	10	12	14	16	18	20	22
REPE/$\|\beta\|$	−0.268	0.065	−0.061	0.026	−0.024	0.016	−0.011	0.012	−0.005	0.010

As n increases, the REPE becomes negligible, and the aromatic $4m + 2$ compounds and antiaromatic $4m$ compounds become nonaromatic.

Comparison with experiment shows that the Hess–Schaad method is quite successful in predicting aromaticity.

For further discussion of the controversial concept of aromaticity, see V. I. Minkin et al., *Aromaticity and Antiaromaticity*, Wiley, 1994; the May 2001 and October 2005 issues of *Chemical Reviews* are devoted to aromaticity and related topics.

π-Electron Charges and Bond Orders. Another defined quantity is the *π-electron charge*. The probability density for an electron in the HMO (17.8) is

$$|\phi_i|^2 = \sum_r \sum_s c_{ri}^* c_{si} f_r^* f_s \tag{17.52}$$

The HMO normalization condition (17.16) is $\sum_{r=1}^{n_C} |c_{ri}|^2 = 1$, and so it is natural to say that an electron in the MO ϕ_i has the probability $|c_{ri}|^2$ of being in the vicinity of the rth carbon atom. If n_i (= 0, 1, or 2) is the number of electrons in the MO ϕ_i, then the total **π-electron charge** q_r in the region of carbon-atom r is defined as

$$q_r \equiv \sum_i n_i |c_{ri}|^2 \tag{17.53}$$

where the sum is over the π MOs. q_r is often called the "π-electron density"; this name is misleading, since q_r is neither a charge density (which has dimensions of charge/volume) nor a probability density (which has dimensions of 1/volume). Rather, q_r is a pure number that gives the approximate number of π electrons in the vicinity of carbon atom r.

For 1,3-butadiene, $q_1 = 2|c_{11}|^2 + 2|c_{12}|^2 = 2(0.372)^2 + 2(0.602)^2 = 1.000$ and $q_2 = 2|c_{21}|^2 + 2|c_{22}|^2 = 1.000$.

For the ground state of a neutral alternant hydrocarbon, all the Hückel π-electron charges q_r are 1.

In the population-analysis discussion of Section 15.6, we saw that for a real MO ϕ_i expanded as the linear combination $\sum_r c_{ri} f_r$ of AOs f_r, the quantity $2c_{ri}c_{si}S_{rs}$ in Eq. (15.25) (where S_{rs} is the overlap integral) is a reasonable measure of the contribution of an electron in MO ϕ_i to the bonding overlap between AOs f_r and f_s. In the HMO method, overlap integrals are neglected, so we cannot use this population-analysis expression. The carbon–carbon bond distances in a conjugated compound are all reasonably similar, so we expect the $C2p\pi$–$C2p\pi$ overlap integral S_{rs} to have similar values for all pairs of bonded carbons. For nonbonded carbons, S_{rs} should be quite small. Since S_{rs} is approximately constant for bonded atoms, we can ignore the factor S_{rs} (and the constant factor 2) and take $c_{ri}c_{si}$ as the contribution of an electron in the real HMO ϕ_i to the π-electron bonding between bonded atoms r and s. If the MO ϕ_i is complex rather than real, one finds on integrating $|\phi_i|^2$ to produce the equation corresponding to (15.24) that the quantity $\frac{1}{2}(c_{ri}^* c_{si} + c_{si}^* c_{ri})$ occurs instead of $c_{ri}c_{si}$. Coulson therefore defined the **π-electron** (or **mobile**) **bond order** p_{rs} for the bond between bonded atoms r and s as

$$p_{rs} \equiv \frac{1}{2} \sum_i n_i (c_{ri}^* c_{si} + c_{si}^* c_{ri}) \tag{17.54}$$

where the sum is over the π MOs. When the coefficients are all real, (17.54) reduces to $p_{rs} = \sum_i n_i c_{ri} c_{si}$. That this definition is reasonable is indicated by the fact that it gives $p = 1$ for the π bond in ethylene.

Addition of the σ electrons' single bond gives the *total bond order* p_{rs}^{tot} as

$$p_{rs}^{\text{tot}} = 1 + p_{rs} \tag{17.55}$$

For butadiene we have $p_{12} = 2(0.372)(0.602) + 2(0.602)(0.372) = 0.894$ and $p_{23} = 2(0.602)(0.602) + 2(0.372)(-0.372) = 0.447$. The total bond orders are

$$CH_2 \overset{1.894}{—} CH \overset{1.447}{—} CH \overset{1.894}{—} CH_2 \tag{17.56}$$

The central bond has some double-bond character, and the end bonds have some single-bond character. (In VB theory, this is explained by contributions from such resonance structures as $\overset{-}{C}H_2—CH{=}CH—\overset{+}{C}H_2$.) The sum of the bond orders is 5.235, exceeding 5. [This is because the π-electron energy of butadiene exceeds that of two isolated double bonds; see Eq. (17.58).] For benzene each carbon–carbon bond order is found to be $5/3 = 1.667$.

Naturally, we expect a relation between the bond order and the bond length R_{rs} for carbon–carbon bonds. The simplest assumption is a linear relation: $R_{rs} = a + bp_{rs}^{\text{tot}}$, where a and b are constants. The π-electron MO coefficients for ethylene and for benzene are determined completely by symmetry and are independent of the Hückel approximations. We therefore use the bond orders and bond lengths of ethylene (2; 1.335 Å) and benzene (5/3; 1.397 Å) to find a and b; we get

$$R_{rs} = (1.707 - 0.186 p_{rs}^{\text{tot}})\ \text{Å} = (1.521 - 0.186 p_{rs})\ \text{Å} \tag{17.57}$$

Equation (17.57) works reasonably well. Thus the predicted naphthalene 1–2, 2–3, 1–9, and 9–10 bond lengths (in Å), as compared with the experimental gas-phase values (in parentheses) are 1.386 (1.381), 1.409 (1.417), 1.418 (1.422), and 1.425 (1.412), respectively. Note that the rings are not regular hexagons. For comparison, HF/6-31G** naphthalene 1–2, 2–3, 1–9, and 9–10 bond lengths are 1.358, 1.416, 1.420, and 1.409 Å, respectively [A. Hinchcliffe and H. J. Soscun Machado, *Chem. Phys. Lett.*, **214**, 64 (1993)].

A much-investigated question is whether the carbon–carbon bond lengths in very large conjugated linear and cyclic polyenes ($C_{2n}H_{2n+2}$ and $C_{4m+2}H_{4m+2}$ for m and n large) are equal or alternate in length. The HMO bond orders in a linear conjugated polyene become equal (except near the chain ends) in the limit of large n. The HMO bond orders in the cyclic polyene $C_{4m+2}H_{4m+2}$ are all equal. Hence it was formerly believed that in $C_{2n}H_{2n+2}$ and in $C_{4m+2}H_{4m+2}$ the bond lengths are all equal, except for those near the chain ends in $C_{2n}H_{2n+2}$. However, HMO calculations that take into account the strain energy involved in changing the σ bond lengths indicate strongly that for large n and m both linear and cyclic conjugated polyenes should have bonds that alternate in length (*Salem*, Section 8-4). Because of the crudity of the HMO method, these conclusions are not definitive.

The polymer *trans*-polyacetylene, *trans*-$(CH)_x$ contains a very long chain of alternating single and double carbon–carbon bonds. The term *trans* refers to the configurations of the two hydrogens bonded to each pair of doubly bonded carbons (Fig. 17.8). Analysis of the ^{13}C NMR spectrum of *trans*-$(CH)_x$ film gave 1.36 Å for the carbon–carbon double-bond length and 1.44 Å for the single-bond length, thereby proving the existence of bond-length alternation in long-chain polyenes [C. S. Yannoni and T. C. Clarke, *Phys. Rev. Lett.*, **51**, 1191 (1983)]. A density-functional B3LYP/6-31G* geometry optimization of the linear polyene *trans*-$C_{30}H_{32}$ gave 1.368 and 1.426 Å for the double- and single-bond lengths in the center of the molecule [C. Choi et al., *J. Chem. Phys.*, **107**, 6712 (1997)].

B3LYP/6-31G* calculations on the cyclic polyenes $C_{4m+2}H_{4m+2}$ up to $C_{66}H_{66}$ predict bond-length alternation for large m [C. H. Choi and M. Kertesz, *J. Chem. Phys.*, **108**, 6681 (1998)]. The fact that we can write two equivalent VB resonance structures for $C_{4m+2}H_{4m+2}$ such that a given bond is single in one structure and double in the other is no guarantee of equal bond lengths.

The Hückel π-electron energy E_π is related to the π bond orders p_{rs} and π-electron charges q_r; in fact (Prob. 17.16),

$$E_\pi = \alpha \sum_r q_r + 2\beta \sum_{s-r} p_{rs} \tag{17.58}$$

where the first sum is over the carbon atoms and the second sum is over the carbon–carbon bonds.

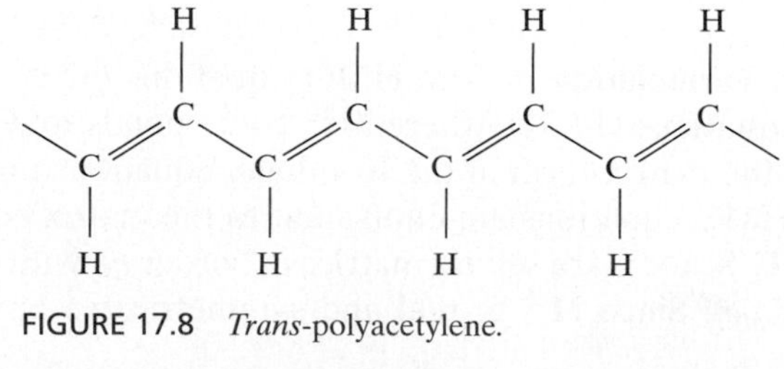

FIGURE 17.8 *Trans*-polyacetylene.

For conjugated species with partly filled degenerate MOs, there is an ambiguity in the q_r and p_{rs} values if the real forms of the MOs are used. For example, for $C_6H_6^-$, if we put the unpaired electron in $\phi_{4,\text{real}}$ of (17.42), then this electron contributes $\frac{1}{3}$ to q_1, $\frac{1}{12}$ to q_2, $\frac{1}{12}$ to q_3, $\frac{1}{3}$ to q_4, $\frac{1}{12}$ to q_5, and $\frac{1}{12}$ to q_6; but if we put the odd electron in $\phi_{5,\text{real}}$, then this electron contributes 0 to q_1 and q_4 and $\frac{1}{4}$ to each of q_2, q_3, q_5, and q_6. This ambiguity can be avoided by using the complex MOs, which have the symmetry of the molecule. Putting the odd electron in either ϕ_4 or ϕ_5 of (17.41), we get a contribution of $\frac{1}{6}$ to each q. Alternatively, we can average the contributions from ϕ_4 and ϕ_5 to give a $\frac{1}{6}$ contribution at each carbon.

The HMO quantities q_r and p_{rs} in (17.53) and (17.54) are closely related to the density matrix elements P_{rs} [Eq. (14.42)] of SCF theory. We see that the density matrix elements with $r = s$ (the diagonal elements) are equal to q_r; $P_{rr} = q_r$. Although HMO theory defines bond orders only for pairs of bonded atoms, if we formally define p_{rs} by (17.54) for nonbonded pairs of atoms also, then the definitions yield $p_{rs} = \frac{1}{2}(P_{rs} + P_{sr})$; for real HMOs, $p_{rs} = P_{rs}$.

Heteroatomic Conjugated Molecules. So far, we have applied the HMO method to hydrocarbons only. For planar conjugated molecules that involve π bonding to noncarbon atoms, the α and β integrals for the heteroatoms must be modified from the carbon values. For heteroatoms X and Y, we write

$$\alpha_X = \alpha_C + h_X\beta_{CC} \tag{17.59}$$

$$\beta_{XY} = k_{XY}\beta_{CC} \tag{17.60}$$

where h_X and k_{XY} are certain constants. The best values for these constants vary, depending on which molecular property is being considered, and the values used are based on a mixture of theory and guesswork. For details, see *Streitwieser*, Chapter 5.

Inclusion of Overlap. Aside from using the one-electron Hamiltonian (17.3), perhaps the most serious approximation of the simple HMO method is that of taking all overlap integrals equal to zero. Wheland proposed using a common nonzero value S for the overlap integral of carbons bonded to each other. This replaces each element β in the HMO secular equation (17.10) with $\beta - Se_i$. At the benzene carbon–carbon bond distance, S equals 0.25 for $2p\pi$ STOs with orbital exponent 1.625 (the value given by Slater's rules; Prob. 15.54). Inclusion of overlap in the HMO method is easy (see Prob. 17.23), but is rarely done. One finds that inclusion of overlap gives only slight changes in predicted transition frequencies of alternant hydrocarbons (Prob. 17.23d) and gives no change in the π-electron charges and bond orders when these are calculated with suitably modified definitions [B. H. Chirgwin and C. A. Coulson, *Proc. Roy. Soc.*, **A201**, 196 (1950)].

Matrix Formulation. The HMO equations (17.9) have the same form as the Roothaan equations (14.34), where H_{rs}^{eff} corresponds to F_{rs}, and e_i corresponds to ε_i. We showed the matrix form of the Roothaan equations to be $\mathbf{FC} = \mathbf{SC\varepsilon}$ [Eq. (14.56)]. Hence the HMO equations are equivalent to the matrix equation $\mathbf{H}^{\text{eff}}\mathbf{C} = \mathbf{SCe} = \mathbf{Ce}$, where $\mathbf{H}^{\text{eff}}$, $\mathbf{C}$, $\mathbf{S}$, and $\mathbf{e}$ are square matrices of order n_C with elements H_{rs}^{eff}, c_{si}, $S_{rs} = \delta_{rs}$, and $e_{mi} = \delta_{mi}e_i$. Since $\mathbf{H}^{\text{eff}}$ is real and symmetric, we have $\mathbf{C}^T\mathbf{H}^{\text{eff}}\mathbf{C} = \mathbf{e}$. Computer programs to do HMO calculations find the orthogonal matrix **C** that diagonalizes $\mathbf{H}^{\text{eff}}$. You can do online HMO calculations at www.chem.ucalgary.ca/SHMO/, www.stolaf.edu/depts/chemistry/courses/toolkits/247/js/huckel/, and www.chem.swin.edu.au/modules/mod3/interface.html.

Summary. Because of the simplicity of the method, carrying out HMO calculations became a favorite pastime of organic chemists, and the results of HMO calculations have been tabulated for hundreds of compounds. The HMO method was widely used to rationalize and predict the properties and reactivities of conjugated compounds.

The development of semiempirical theories more sophisticated than the HMO theory led some workers to argue that "Hückel theory has largely outlived its usefulness" (*Murrell and Harget*, page v). However, these more sophisticated theories have their failings (as we shall see), and the success of the Hess–Schaad use of Hückel theory to predict aromaticity indicates that HMO theory may still be useful, "especially on a qualitative level as a guide . . . in planning and interpreting experiments" (N. Trinajstic in *Segal*, Part A, Chapter 1).

17.3 THE PARISER–PARR–POPLE METHOD

Although the Hückel theory can be used to predict the longest-wavelength bands of aromatic hydrocarbons, it would be hopeless to try to use HMO theory to predict the complete electronic spectrum of an aromatic hydrocarbon. For example, Hückel theory, which neglects interelectronic repulsions, gives no separation between singlet and triplet electronic terms arising from the same configuration. Experimentally, separations of 1 or 2 eV are observed between such terms.

A semiempirical π-electron theory that takes electron repulsion into account and thereby improves on the Hückel method is the **Pariser–Parr–Pople** (PPP) **method**, developed in 1953. Here, the π-electron Hamiltonian (17.1) including electron repulsions is used, and the π-electron wave function is written as an antisymmetrized product of π-electron spin-orbitals. A minimal basis set of one $2p\pi$ STO on each conjugated atom is used, and the spatial π MOs ϕ_i are taken as linear combinations of these AOs: $\phi = \sum_{r=1}^{b} c_{ri} f_r$. The Roothaan equations are used to find SCF π MOs within the π-electron approximation.

The π-electron Hamiltonian $\hat{H}_\pi$ in (17.1) has the same form as the all-electron operator $\hat{H}_{\text{el}} = -\frac{1}{2}\sum_i \nabla_i^2 - \sum_i \sum_\alpha Z_\alpha / r_{i\alpha} + \sum_i \sum_{j>i} 1/r_{ij}$ for a molecule except that $\hat{H}_\pi^{\text{core}}(i)$ replaces $\hat{H}^{\text{core}}(i) = -\frac{1}{2}\nabla_i^2 - \sum_\alpha Z_\alpha / r_{i\alpha}$ [Eq. (14.27)], and the sums go over only the n_π π electrons rather than over all the electrons. Hence, similar to (14.25), the SCF π MOs satisfy $\hat{F}_\pi \phi_i = \varepsilon_i \phi_i$, where $\hat{F}_\pi$ is given by (14.26) with $\hat{H}^{\text{core}}(1)$ replaced by $\hat{H}_\pi^{\text{core}}$ and $n/2$ replaced by $n_\pi/2$. The SCF π-electron energy is given by (14.22) with H_{ii}^{core} replaced by $H_{\pi,ii}^{\text{core}}$ and $n/2$ replaced by $n_\pi/2$.

The Roothaan equations (14.34) and (14.41) become

$$\sum_s c_{si}(F_{\pi,rs} - \varepsilon_i S_{rs}) = 0, \qquad r = 1, \ldots, b \tag{17.61}$$

$$F_{\pi,rs} = H_{\pi,rs}^{\text{core}} + \sum_{t=1}^{b} \sum_{u=1}^{b} P_{tu}[(rs|tu) - \tfrac{1}{2}(ru|ts)] \tag{17.62}$$

In addition to assuming $\sigma-\pi$ separability, the PPP method makes further approximations. As in Hückel theory, overlap is neglected:

$$S_{rs} \equiv \langle f_r(1)|f_s(1)\rangle = \delta_{rs} \tag{17.63}$$

where δ_{rs} is the Kronecker delta. Consistent with the neglect of overlap integrals, when evaluating electron-repulsion integrals the PPP method makes the approximation of **zero differential overlap** (ZDO):

$$[f_r(1)]^* f_s(1)\, dv_1 = 0, \qquad \text{for } r \neq s \tag{17.64}$$

From (17.64) and $(rs|tu) \equiv \langle f_r(1)f_t(2)|1/r_{12}|\, f_s(1)f_u(2)\rangle$ [Eq. (14.39)], it follows that the electron-repulsion integrals are given by

$$(rs|tu) = \delta_{rs}\delta_{tu}(rr|tt) = \delta_{rs}\delta_{tu}\gamma_{rt} \tag{17.65}$$

where $\gamma_{rt} \equiv (rr|tt)$. Thus the method ignores many (but not all) of the electron-repulsion integrals, thereby greatly simplifying the calculation. In particular, all three- and four-center electron-repulsion integrals are ignored. The ZDO approximation is not used in the $H_{\pi,rs}^{\text{core}}$ integrals.

The ZDO approximation is at first sight rather drastic. However, a partial theoretical justification for it can be given by reinterpreting the AOs used to express the MOs as orthogonalized AOs (rather than ordinary AOs). Each orbital in a set of orthogonalized AOs is a linear combination of ordinary AOs, the coefficients being chosen so that the members of the set are mutually orthogonal. There are many ways to choose the linear combinations to produce an orthogonal set. One approach is to make the orthogonalized AOs (OAOs) resemble the ordinary AOs as much as possible by minimizing the sum of the squares of the deviations of the OAOs from the ordinary AOs; one minimizes the sum $\sum_i \int |\chi_{i,\text{OAO}} - \chi_i|^2\, dv$, where the sum goes over the set of AOs and where χ_i and $\chi_{i,\text{OAO}}$ are the ordinary and the orthogonalized AOs. This produces what are called *symmetrically orthogonalized* (or Löwdin) AOs (see Prob. 8.57; *Pilar*, Section 14-8). One finds that in each symmetrically orthogonalized AO the coefficient of one ordinary AO is substantially greater than the coefficients of the other ordinary AOs, so the symmetrically orthogonalized AOs are not drastically different from the ordinary AOs. One finds that with symmetrically orthogonalized AOs the integrals not neglected in the ZDO approximation undergo only small changes in value as compared with their values with ordinary AOs. These changes in value can be partly allowed for by the fact that many integrals are taken as empirical parameters. Moreover, with symmetrically orthogonalized AOs, the electron-repulsion integrals neglected in the ZDO approximation are generally found to be quite small (and all overlap integrals are, of course, zero). For details, see the references cited on page 27 of *Murrell and Harget*.

With the approximation (17.65) for $(rs|tu)$, the matrix elements $F_{\pi,rs}$ in (17.62) become (Prob. 17.24)

$$\begin{aligned} F_{\pi,rr} &= H_{\pi,rr}^{\text{core}} + \sum_t P_{tt}\gamma_{rt} - \tfrac{1}{2}P_{rr}\gamma_{rr} \\ F_{\pi,rs} &= H_{\pi,rs}^{\text{core}} - \tfrac{1}{2}P_{sr}\gamma_{rs}, \qquad r \neq s \end{aligned} \tag{17.66}$$

The PPP method does not attempt to explicitly specify $\hat{H}_{\pi}^{\text{core}}$ or to calculate the $H_{\pi,rs}^{\text{core}}$ integrals theoretically. Rather, the integrals $H_{\pi,rs}^{\text{core}}$ and γ_{rs} are calculated from approximate semiempirical formulas, some of which contain empirical parameters. For example, when the AOs f_r and f_s are on atoms R and S that are bonded to each other, $H^{\text{core}}_{\pi,rs}$ may be taken as $k\langle f_r|f_s\rangle$, where the value of the empirical parameter k is chosen so that the predictions of the theory give good agreement with experiment, and the overlap integral $\langle f_r|f_s\rangle$ is calculated from the STOs f_r and f_s, and not taken as zero as in (17.64). When the two different atoms R and S are not bonded to each other, $H^{\text{core}}_{\pi,rs}$ is taken as zero. (Several versions of the PPP theory exist, each of which uses a different set of semiempirical formulas to evaluate the integrals.) The two-center electron-repulsion integrals γ_{rs} are evaluated from a semiempirical formula that contains the one-center integrals γ_{rr} and γ_{ss} and the distance between atoms R and S. The one-center integral γ_{rr} is evaluated as the difference between the atomic valence-state ionization energy and electron affinity of atom R, where these two quantities are evaluated using spectroscopically determined atomic energy-level data. The one-center integral $H^{\text{core}}_{\pi,rr}$ is calculated from a semiempirical formula that involves the electron-repulsion integrals γ_{rs} for all the conjugated atoms $S \neq R$ and involves the orbital energy for a $2p\pi$ AO on atom R, this orbital energy being evaluated using atomic spectral data.

To do a PPP calculation, one starts with the HMO coefficients as an initial guess for the c_{si}'s, calculates the initial density matrix elements P_{rs}, calculates the initial $F_{\pi,rs}$ matrix elements, solves the equations (17.61) for π-electron orbital energies ε_i and an improved set of coefficients c_{si}, calculates improved P_{rs} values, and so on until convergence is reached. To improve the results, CI of the π electrons may be included.

The PPP method gives a good account of the electronic spectra of many, but not all, aromatic hydrocarbons. For more on the PPP method, see *Parr*, Chapter III; *Murrell and Harget*, Chapter 2; *Offenhartz*, Chapter 11.

The PPP method is little used nowadays and has been largely superseded by more-general semiempirical methods (Section 17.4). However, the PPP method is of historical importance, since many of the PPP approximations used to evaluate integrals are used in current semiempirical theories.

17.4 GENERAL SEMIEMPIRICAL MO AND DFT METHODS

The HMO and PPP methods apply only to planar conjugated molecules and treat only the π electrons. The semiempirical MO methods discussed in this section apply to all molecules and treat all the valence electrons.

Semiempirical MO theories fall into two categories: those using a Hamiltonian that is the sum of one-electron terms, and those using a Hamiltonian that includes two-electron repulsion terms, as well as one-electron terms. The Hückel method is a one-electron theory, whereas the Pariser–Parr–Pople method is a two-electron theory.

The Extended Hückel Method. The most important one-electron semiempirical MO method for nonplanar molecules is the **extended Hückel theory**. An early version was used by Wolfsberg and Helmholz in treating inorganic complex ions. The method was further developed and widely applied by Hoffmann [R. Hoffmann, *J. Chem. Phys.*, **39**, 1397 (1963); **40**, 2745, 2474, 2480 (1964); *Tetrahedron*, **22**, 521, 539 (1966); M. Wolfsberg and L. Helmholz, *J. Chem. Phys.*, **20**, 837 (1952)].

The extended Hückel (EH) method begins with the approximation of treating the valence electrons separately from the rest (Section 13.17). The valence-electron Hamiltonian is taken as the sum of one-electron Hamiltonians:

$$\hat{H}_{\text{val}} = \sum_i \hat{H}_{\text{eff}}(i) \tag{17.67}$$

where $\hat{H}_{\text{eff}}(i)$ is not specified explicitly. The MOs are approximated as linear combinations of the valence AOs f_r of the atoms:

$$\phi_i = \sum_{r=1}^{b} c_{ri} f_r \tag{17.68}$$

In the simple Hückel theory of planar hydrocarbons, each π MO contains contributions from one $2p\pi$ AO on each carbon atom. In the extended Hückel treatment of nonplanar hydrocarbons, each valence MO contains contributions from four AOs on each carbon atom (one $2s$ and three $2p$'s) and one $1s$ AO on each hydrogen atom. The AOs used are usually Slater-type orbitals with fixed orbital exponents determined from Slater's rules (Prob. 15.54). For the simplified Hamiltonian (17.67), the problem separates into several one-electron problems:

$$\hat{H}_{\text{eff}}(i)\phi_i = e_i \phi_i$$

$$E_{\text{val}} = \sum_i e_i \tag{17.69}$$

Application of the variation theorem to the linear trial function (17.68) gives as the secular equation and the equations for the MO coefficients

$$\det(H_{rs}^{\text{eff}} - e_i S_{rs}) = 0 \tag{17.70}$$

$$\sum_s [(H_{rs}^{\text{eff}} - e_i S_{rs}) c_{si}] = 0, \qquad r = 1, 2, \ldots, b \tag{17.71}$$

All this is similar to simple Hückel theory. However, the extended Hückel theory does not neglect overlap. Rather, *all* overlap integrals are explicitly evaluated using the forms chosen for the AOs and the internuclear distances at which the calculation is being done. Formulas for overlap integrals of STOs are readily available. [Footnotes 12 to 18 of D. M. Bishop et al., *J. Chem. Phys.*, **45**, 1880 (1966) give references to available tabulations of overlap integrals.] Since overlap is included, off-diagonal e_i's are present in the secular determinant.

Since $\hat{H}_{\text{eff}}(i)$ is not specified, there is the problem of what to use for the integrals H_{rs}^{eff}. For $r = s$, the one-electron integral $H_{rr}^{\text{eff}} \equiv \langle f_r | \hat{H}_{\text{eff}} | f_r \rangle$ looks like an average energy for an electron in the AO f_r centered on atom R in the molecule. Hence, the EH method takes H_{rr}^{eff} as equal to the orbital energy of the AO f_r for atom R in its valence state; the *valence state* is the hypothetical state of the atom in the molecule. The valence-state orbital energy can be found from atomic spectral data (see Section 16.8 for an example). By Koopmans' theorem (Section 15.5), the valence-state orbital energy is taken as equal to minus the valence-state ionization potential (VSIP) of f_r. For hydrogen and for carbon atoms in a molecule with only single bonds to carbon (sp^3-hybridized carbon), the VSIP parametrization gives

$$\langle \text{C}2s|\hat{H}_{\text{eff}}|\text{C}2s\rangle = -20.8 \text{ eV}, \qquad \langle \text{C}2p|\hat{H}_{\text{eff}}|\text{C}2p\rangle = -11.3 \text{ eV}$$
$$\langle \text{H}1s|\hat{H}_{\text{eff}}|\text{H}1s\rangle = -13.6 \text{ eV} \tag{17.72}$$

The carbon VSIPs for sp^2- and sp-hybridized carbon differ from those in (17.72), but the difference is often ignored and an average set of VSIPs is used for all carbons. VSIPs are tabulated in J. Hinze and H. H. Jaffé, *J. Am. Chem. Soc.*, **84**, 540 (1962); G. Pilcher and H. A. Skinner, *J. Inorg. Nuc. Chem.*, **24**, 937 (1962); L. C. Cusachs et al., *J. Chem. Phys.*, **44**, 835 (1966).

For the off-diagonal matrix elements $H_{rs}^{\text{eff}}, r \neq s$, Wolfsberg, Helmholz, and Hoffmann took

$$H_{rs}^{\text{eff}} = \tfrac{1}{2}K(H_{rr}^{\text{eff}} + H_{ss}^{\text{eff}})S_{rs} \tag{17.73}$$

where K is a numerical constant and the remaining quantities are evaluated as above. Since H_{rr}^{eff} and H_{ss}^{eff} are usually negative, (17.73) gives H_{rs}^{eff} as negative. The most commonly used value of K is 1.75 (although values between 1 and 3 have been suggested). In contrast to the simple Hückel theory, H_{rs}^{eff} is nonzero for all pairs of orbitals (unless S_{rs} vanishes for symmetry reasons).

Once the H_{rs}^{eff} and S_{rs} integrals have been evaluated, the secular equation is solved for the orbital energies and the MO coefficients are found. (Computer programs use matrix diagonalization to do this.) Since the H_{rs}^{eff} integrals do not depend on the MO coefficients, no iteration is needed. (The EH method is often used to provide an initial guess for the valence MO coefficients in ab initio SCF MO calculations.)

The total valence-electron energy is given by (17.69) as the sum of orbital energies. To predict molecular geometry using the EH method, one does a series of calculations over a range of bond distances and angles and looks for the nuclear configuration that minimizes (17.69). Note that (17.69) omits both electron–electron repulsions and nuclear–nuclear repulsions. It might be thought that such a theory would be useless for predicting molecular geometries, but this is not so.

The EH method was found to give rather accurate bond angles for molecules whose bonds are not highly polar, but to fail in bond-angle predictions for molecules with very polar bonds (for example, H_2O, which is predicted to be linear); see L. C. Allen and J. D. Russell, *J. Chem. Phys.*, **46**, 1029 (1967). The EH method is not reliable for predicting bond lengths, dipole moments, and barriers to internal rotation. The EH method is unreliable for predicting molecular conformations [B. Pullman, *Adv. Quantum Chem.*, **10**, 251 (1977)].

Because the EH method gives poor predictions of such molecular properties as bond lengths, dipole moments, energies, and rotational barriers, Jug concluded that this method "is obsolete" [K. Jug, *Theor. Chim. Acta*, **54**, 263 (1980)]. However, this judgment is too harsh, since the EH method has been used by Hoffmann and others to provide valuable qualitative insights into chemical bonding [for example, P. J. Hay, J. C. Thibeault, and R. Hoffmann, *J. Am. Chem. Soc.*, **97**, 4884 (1975)]. Gimarc noted that "the real value of the extended Hückel method is not in its quantitative results, which have never been impressive, but rather in the qualitative nature of the results and in the interpretations those results can provide" (B. M. Gimarc, *Molecular Structure and Bonding*, Academic Press, 1979, page 216).

The CNDO, INDO, and NDDO Methods. Several semiempirical two-electron MO generalizations of the PPP method were developed that are applicable to both planar and nonplanar molecules. The **complete neglect of differential overlap** (CNDO) **method** was proposed by Pople, Santry, and Segal in 1965. The **intermediate neglect of differential overlap** (INDO) **method** was proposed by Pople, Beveridge, and Dobosh in 1967. Both methods treat only the valence electrons explicitly. The valence-electron Hamiltonian has a similar form as (17.1):

$$\hat{H}_{\text{val}} = \sum_{i=1}^{n_{\text{val}}} [-\tfrac{1}{2}\nabla_i^2 + V(i)] + \sum_{i=1}^{n_{\text{val}}} \sum_{j>i} \frac{1}{r_{ij}} \equiv \sum_{i=1}^{n_{\text{val}}} \hat{H}_{\text{val}}^{\text{core}}(i) + \sum_{i=1}^{n_{\text{val}}} \sum_{j>i} \frac{1}{r_{ij}} \quad (17.74)$$

$$\hat{H}_{\text{val}}^{\text{core}}(i) \equiv -\tfrac{1}{2}\nabla_i^2 + V(i)$$

In (17.74), n_{val} is the number of valence electrons in the molecule, $V(i)$ is the potential energy of valence electron i in the field of the nuclei and the inner-shell (core) electrons, and $\hat{H}_{\text{val}}^{\text{core}}(i)$ is the one-electron part of $\hat{H}_{\text{val}}$. The CNDO and INDO methods are SCF MO methods that iteratively solve the Roothaan equations using approximations for the integrals in the Fock matrix elements.

The CNDO method uses a minimal basis set of valence Slater AOs f_r with orbital exponents fixed at values given by Slater's rules (Prob. 15.54), except that 1.2 is used for H1s. The valence MOs ϕ_i are written as $\phi_i = \sum_{r=1}^{b} c_{ri} f_r$. The molecular electronic energy is given by (14.22) with H_{ii}^{core} replaced by $H_{\text{val},ii}^{\text{core}}$, $n/2$ replaced by $n_{\text{val}}/2$, and V_{NN} replaced by the **core–core repulsion energy**

$$V_{cc} = \sum_{\alpha} \sum_{\beta>\alpha} \frac{C_\alpha C_\beta}{R_{\alpha\beta}}$$

where the **core charge** C_α on atom α equals the atomic number of atom α minus the number of core (inner-shell) electrons on α. The Roothaan equations are given by (17.61) with $F_{\pi,rs}$ replaced by $F_{\text{val},rs}$. The Fock matrix elements $F_{\text{val},rs}$ [Eq. (14.41)] are given by (17.62) with $H_{\pi,rs}^{\text{core}}$ replaced by $H_{\text{val},rs}^{\text{core}} \equiv \langle f_r | \hat{H}_{\text{val}}^{\text{core}} | f_s \rangle$:

$$F_{\text{val},rs} = H_{\text{val},rs}^{\text{core}} + \sum_{t=1}^{b} \sum_{u=1}^{b} P_{tu}[(rs|tu) - \tfrac{1}{2}(ru|ts)] \quad (17.75)$$

Henceforth in this section, the subscript val *will be omitted from* $F_{\text{val},rs}$, $H_{\text{val},rs}^{\text{core}}$, *and* $\hat{H}_{\text{val}}^{\text{core}}(i)$.

The CNDO method uses the ZDO approximation (17.64) for all pairs of AOs in overlap and electron-repulsion integrals. Thus, $S_{rs} = \delta_{rs}$ and $(rs|tu) = \delta_{rs}\delta_{tu}(rr|tt) \equiv \delta_{rs}\delta_{tu}\gamma_{rt}$ [Eq. (17.65)], where $(rs|tu)$ is given by (14.39). In the PPP method for conjugated hydrocarbons, there is only one basis AO per atom, the $2p\pi$ AO. In the CNDO method, there are several basis valence AOs on each atom (except hydrogens), and the ZDO approximation neglects electron-repulsion integrals containing the product $f_r(1)f_s(1)$ where f_r and f_s are different AOs centered on the same atom.

When the Roothaan equations (14.34) [or (14.56)] are solved exactly, the canonical MOs and the calculated values of molecular properties do not change if one changes the orientation of the coordinate axes; the calculated values are said to be *rotationally invariant*. Likewise, the results do not change if each basis AO on a particular atom is replaced by a linear combination of the basis AOs on that atom, and the results are *hybridizationally invariant*. When approximations are made in solving the Hartree–Fock–Roothaan equations, rotational and hybridizational invariance may not hold.

To maintain rotational and hybridizational invariance when the ZDO approximation is used, the CNDO method makes the additional approximation that the electron repulsion integral $\gamma_{rt} \equiv (rr|tt)$ depends only on which atoms the AOs f_r and f_t are centered on and does not depend on the nature of the orbitals f_r and f_t. Let the notation f_{r_A}, f_{t_B} indicate that the valence AOs f_r and f_t are centered on atoms A and B, respectively, where A and B might be the same atom or different atoms. The CNDO method takes $(r_A r_A|t_B t_B) \equiv \gamma_{r_A t_B} = \gamma_{AB}$ for all valence AOs f_r on A and all valence AOs f_t on B. In the CNDO method, all one-center valence-electron-repulsion integrals on atom A have the value γ_{AA}; all two-center valence electron repulsion integrals involving atoms A and B have the value γ_{AB}; all three- and four-center electron repulsion integrals are neglected (as a consequence of the ZDO approximation). The integrals γ_{AA} and γ_{AB} are evaluated using valence *s* STOs on A and B, and so depend on the orbital exponents, the principal quantum numbers of the valence electrons, and on the distance between atoms A and B.

CNDO takes the integral H_{rs}^{core} for $r \neq s$ as proportional to the overlap integral $S_{rs} = \langle f_r|f_s\rangle$ (recall a similar assumption in the PPP theory):

$$H_{r_A s_B}^{\text{core}} = \beta_{AB}^0 S_{r_A s_B} \qquad \text{for } r \neq s$$

where $S_{r_A s_B}$ is evaluated exactly [and not taken as δ_{rs}, even though it is taken as δ_{rs} in the Roothaan equations (17.61)] and the parameter β_{AB}^0 is taken as $\beta_{AB}^0 = \frac{1}{2}(\beta_A^0 + \beta_B^0)$, where the parameters β_A^0 and β_B^0 are chosen so that the coefficients in CNDO-calculated MOs of diatomic molecules resemble the coefficients in minimal-basis ab initio MOs. When A and B are the same atom, $S_{r_A s_A}$ is zero for $r \neq s$ by orthogonality of AOs on the same atom, and CNDO takes $H_{r_A s_A}^{\text{core}}$ as zero for $r \neq s$.

Now consider the integrals $H_{r_A r_A}^{\text{core}}$. We have $\hat{H}^{\text{core}}(1) = -\frac{1}{2}\nabla_1^2 + V(1)$, where $V(1)$ is the potential energy of valence electron 1 in the field of the core electrons and the nuclei. Breaking $V(1)$ into contributions from the individual atomic cores, we have

$$\hat{H}^{\text{core}}(1) = -\tfrac{1}{2}\nabla_1^2 + V_A(1) + \sum_{B \neq A} V_B(1) \tag{17.76}$$

where the basis AO f_{r_A} is centered on A. Then

$$H_{r_A r_A}^{\text{core}} = \langle f_{r_A}(1)|-\tfrac{1}{2}\nabla_1^2 + V_A(1)|f_{r_A}(1)\rangle + \sum_{B \neq A} \langle f_{r_A}(1)|V_B(1)|f_{r_A}(1)\rangle \tag{17.77}$$

Two versions of CNDO exist, called CNDO/1 and CNDO/2. The integral

$$U_{rr} \equiv \langle f_{r_A}(1)| - \tfrac{1}{2}\nabla_1^2 + V_A(1)|f_{r_A}(1)\rangle$$

looks like an average energy for an electron in the AO f_{r_A} in the molecule. Hence, CNDO/1 takes U_{rr} as the negative of the valence-state ionization energy from the AO f_{r_A}, where this ionization energy is found from atomic energy levels deduced from atomic spectral data. (This is an oversimplification; for details and for the CNDO/2 evaluation of U_{rr}, see *Murrell and Harget*, pp. 38–39.) To maintain rotational and hybridizational invariance, the integrals $\langle f_{r_A}|V_B|f_{r_A}\rangle \equiv V_{AB}$ in (17.77) are taken as equal for all valence AOs f_{r_A} on atom A and are calculated in CNDO/1 by using a valence s orbital for electron 1 and taking atomic core B as a point charge: $V_{AB} = -\langle s_A(1)|C_B/r_{1B}|s_A(1)\rangle$, where C_B is the core charge of atom B. With this choice for the electron–core attraction integral V_{AB} in CNDO/1, two neutral atoms or molecules separated by several angstroms experienced a substantial attraction to each other. To eliminate this spurious attraction, CNDO/2 uses the expression $V_{AB} = -C_B\gamma_{AB}$, where γ_{AB} is the electron-repulsion integral discussed earlier in this subsection.

With these approximations, the Fock matrix elements are evaluated and the Roothaan equations solved iteratively to find the CNDO orbitals and orbital energies.

The INDO method is an improvement on CNDO. In INDO, differential overlap between AOs on the same atom is not neglected in one-center electron-repulsion integrals $(rs|tu)$ where f_r, f_s, f_t, and f_u are all centered on the same atom, but is still neglected in two-center electron-repulsion integrals. Thus fewer two-electron integrals are neglected than in CNDO. Otherwise, the two methods are essentially the same. The INDO method gives an improvement on CNDO results, especially where electron spin distribution is important (for example, in calculating electron-spin-resonance spectra).

As to results, the CNDO and INDO methods give fairly good bond lengths and angles, somewhat erratic dipole moments, and poor dissociation energies. (For details on the CNDO and INDO methods, see *Pople and Beveridge*; *Leach*, Section 2.9; G. Klopman and R. C. Evans in *Segal*, Part A, page 29; *Murrell and Harget*, Chapter 3.)

Versions of CNDO and INDO parametrized to predict electronic spectra are called CNDO/S and INDO/S. These methods include some configuration interaction. Although the ground state of a closed-shell molecule is generally well represented by a single-determinant wave function, one typically requires CI for accurate representation of excited states. For details of these methods, see R. L. Ellis and H. H. Jaffé in *Segal*, Part B, page 49; J. Michl in *Segal*, Part B, page 99.

The CNDO and INDO methods are little used nowadays, since they have been made obsolete by the improved semiempirical methods discussed in the next subsection. The one exception to this statement is INDO/S, which is still widely used to calculate electronic spectra, since it gives good results for vertical excitation energies (Section 16.1) of large molecules, including transition-metal compounds [see M. C. Zerner in K. B. Lipkowitz and D. B. Boyd (eds.), *Reviews in Computational Chemistry*, vol. 2, VCH, 1991, pp. 333–335, 348–353]; INDO/S is often called ZINDO, after the ZINDO program of Zerner.

The **neglect of diatomic differential overlap** (NDDO) method (suggested by Pople, Santry, and Segal in 1965) is an improvement on INDO in which differential overlap is neglected only between AOs centered on different atoms: $f_r^*(1)f_s(1)\,dv_1 = 0$ only when AOs r and s are on different atoms. The degree of neglect of differential overlap in NDDO is more justifiable than in CNDO or INDO. The NDDO method satisfies the rotational and hybridizational invariance conditions without the need to use a common value for electron-repulsion integrals involving different valence AOs on a given atom.

A few initial attempts at parametrizing the NDDO method gave results that were rather disappointing (see G. Klopman and R. C. Evans in *Segal*, Part A, Chapter 2), and the method was little used until 1977, when Dewar and Thiel modified it to give the MNDO method, discussed later in this section.

The MNDO, AM1, PM3, PM5, PM6, RM1, and MNDO/d Methods. Pople's aim in the CNDO and INDO methods was to reproduce as well as possible the results of minimal-basis-set ab initio SCF MO calculations with theories requiring much less computer time than ab initio calculations. Since CNDO and INDO use approximations, we can expect their results to be similar to but less accurate than minimal-basis ab initio SCF MO results. Thus these methods do pretty well on molecular geometry but fail for binding energies. Dewar and co-workers devised several semiempirical SCF MO theories that closely resemble the INDO and NDDO methods. However, Dewar's aim was not to reproduce ab initio SCF wave functions and properties but to have a theory that would give molecular binding energies with chemical accuracy (within 1 kcal/mol) and that could be used for large molecules without a prohibitive amount of calculation. It might seem unlikely that one could devise an SCF MO theory that involves approximations to the ab initio Hartree–Fock method but that succeeds for binding energies, where the Hartree–Fock theory fails. However, by proper choice of the parameters in the semiempirical SCF theory, one can actually get better results than ab initio SCF calculations, because the choice of suitable parameters can compensate for the partial neglect of electron correlation in ab initio SCF theory.

The semiempirical theories of this subsection, which follow Dewar's approach to parametrization, will be called *Dewar-type theories*. These theories treat only the valence electrons, and most of these theories use a minimal basis set of valence Slater-type s and p AOs (with orbital exponents given values determined by parametrization) to expand the valence-electron MOs. (Extensions to include atoms with valence d orbitals are discussed later in this subsection.) The Fock–Roothaan equations (with the overlap integrals S_{rs} taken as δ_{rs}) are solved to find semiempirical SCF MOs. Some degree of neglect of differential overlap is used to eliminate many of the electron-repulsion integrals. In ab initio methods, the integrals occurring in the Fock matrix elements F_{rs} are evaluated accurately, but this is not the approach used in Dewar-type theories. Dewar-type theories take the one-center electron-repulsion integrals (ERIs) as parameters whose values are chosen to fit experimental atomic energy-level data and calculate the two-center ERIs from the values of the one-center ERIs and the internuclear distances using an approximate formula that may involve parameters. The remaining integrals are evaluated from approximate parameter-containing formulas that are designed not to give values that accurately reproduce ab initio values but to be consistent with the approximations used in the theory.

The Dewar-type theories are parametrized so as to yield good values of the 25°C *gas-phase* standard enthalpy of formation $\Delta H^\circ_{f,298}$. These theories calculate $\Delta H^\circ_{f,298}$ as follows. The molecular valence electronic energy U_{val} including nuclear repulsion is taken as the sum of the purely electronic energy $E_{\text{el,val}}$ of the valence electrons (which is found from the results of the semiempirical SCF calculation) and the **core–core repulsion energy** V_{cc} [compare Eq. (13.8)]:

$$U_{\text{val}} = E_{\text{el,val}} + V_{cc}$$

The Dewar-type theories treat the molecule as a collection of valence electrons and atomic cores, where each core consists of an atomic nucleus and the inner-shell (core) electrons. For example, the core of a carbon atom consists of the nucleus and the two

$1s$ electrons. The simplest approach would be to take $V_{cc} = \sum_{B>A} \sum_A V_{cc,AB} = \sum_{B>A} \sum_A C_A C_B / R_{AB}$, where C_A and C_B are the core charges of cores A and B. For example, for a carbon atom, $C_A = 6 - 2 = 4$, the number of valence electrons. Although this form of V_{cc} is used in CNDO and INDO, it is more consistent with the approximations used to evaluate the electron–core interaction integrals in Dewar-type theories to take

$$V_{cc} = \sum_{B>A} \sum_{A} [C_A C_B (s_A s_A | s_B s_B) + f_{AB}] \tag{17.78}$$

where the electron repulsion integral $(s_A s_A | s_B s_B)$ involves the valence s orbitals of atoms A and B (and is approximately proportional to $1/R_{AB}$), and f_{AB} is a small term whose form differs in the various theories. f_{AB} is an empirical function of R_{AB} that fine-tunes interatomic attractions and repulsions in the molecule, so as to improve agreement with experiment.

The equilibrium geometry is then found by minimizing U_{val} (Section 15.10) to give the equilibrium valence electronic energy including core repulsion: $U_{val,e} = E_{el,val,e} + V_{cc,e}$. Let molecule M be composed of the atoms $A_1, A_2, \ldots, A_n$. The molecular dissociation (or atomization) energy $D_{e,M}$ of M (Section 15.13) is calculated as $D_{e,M} = \sum_i E_{val,A_i} - U_{val,e,M}$, where E_{val,A_i} is the valence electronic energy of atom A_i calculated using the same Dewar-type method as used to calculate $U_{val,e,M}$, and the sum goes over all atoms of the molecule. For the dissociation reaction

$$\text{gas-phase molecules of M} \rightarrow \text{gas-phase atoms } A_1, A_2, \ldots A_n$$

we can write that at 25°C, $\Delta H^\circ_{298} = \sum_i \Delta H^\circ_{f,298,A_i(g)} - \Delta H^\circ_{f,298,M(g)}$, where $\Delta H^\circ_{f,298,M(g)}$ and $\Delta H^\circ_{f,298,A_i(g)}$ are the *gas-phase* 25°C standard enthalpies of formation of the molecular species M and of atom A_i. At 0 K, the enthalpy and energy changes for the dissociation reaction are equal, and if we ignore zero-point energy, we can write $\Delta H^\circ_0 = N_A D_e$, where the Avogadro constant N_A converts from a per molecule to a per mole basis. If we ignore the difference between 0 K and 298 K enthalpy changes and take $\Delta H^\circ_0 = \Delta H^\circ_{298}$, we have $N_A D_e = \sum_i \Delta H^\circ_{f,298,A_i(g)} - \Delta H^\circ_{f,298,M(g)}$. Using the preceding expression for D_e, we get the equation used to calculate a gas-phase $\Delta H^\circ_{f,298}$ value in a Dewar-type theory:

$$\Delta H^\circ_{f,298,M(g)} = N_A U_{val,e,M} - N_A \sum_i E_{val,A_i} + \sum_i \Delta H^\circ_{f,298,A_i(g)}$$

where $U_{val,e} = E_{el,val,e} + V_{cc,e}$ and the $\Delta H^\circ_{f,298,A_i(g)}$ values are taken from thermodynamics tables. This procedure ignores the zero-point vibrational energy (which should be subtracted from D_e to give D_0) and ignores the enthalpy changes from 0 K to 298 K (Section 15.13). The justification for ignoring these nonnegligible quantities is that the parameters of the theory are chosen to fit data that include experimental $\Delta H^\circ_{f,298}$ values of many compounds, so the parameter values include built-in corrections for the neglected quantities.

The parameter values in a Dewar-type theory are chosen as follows. One decides on a set of elements (for example, C, H, O, N, P, Si, S, F, Cl, Br, I) for which the theory is to be parametrized, and one chooses a set of molecules (containing only these elements) for which $\Delta H^\circ_{f,298}$ and the molecular geometry and dipole moment are known from experiment. One systematically varies the parameters of the theory so as to minimize the errors in the calculated heats of formation, geometries, and dipole moments.

This process resembles the process of optimizing a molecular geometry (Section 15.10). In geometry optimization, one varies the bond distances, angles, and dihedral angles so as to minimize the molecular electronic energy, including nuclear repulsion. In a Dewar-type-theory parametrization, one varies the parameters' values so as to minimize the weighted sums of the squares of the errors in calculated molecular properties Y_i, that is, to minimize $\sum_i W_i(Y_{i,\text{calc}} - Y_{i,\text{exper}})^2$, where the W_i's are weighting factors that determine the relative importance of the properties Y_i in the process. As in geometry optimization, it is hard to be sure the global minimum has been found (Section 15.11).

The first useful Dewar-type theory was the **MINDO/3** (third version of the modified INDO) method, published in 1975. For a sample of compounds containing no elements other than C, H, O, and N, the average absolute errors in MINDO/3 calculated properties are 11 kcal/mol in heats of formation, 0.022 Å in bond lengths, 5.6° in bond angles, 0.49 D in dipole moments, and 0.7 eV in ionization energies [M. J. S. Dewar and W. Thiel, *J. Am. Chem. Soc.*, **99**, 4907 (1977)]. Large errors in heats of formation occur for small-ring compounds, compounds with triple bonds, aromatic compounds, compact globular molecules, boron compounds, and molecules with lone-pair atoms. The errors in $\Delta H^\circ_{f,298}$ are larger than Dewar was aiming for, but MINDO/3 is still a significant achievement.

MINDO/3 is based on the INDO approximation. The remaining theories of this subsection are based on the far more justifiable NDDO approximation (see the preceding subsection). These NDDO-based Dewar-type theories give significantly better results than MINDO/3, and MINDO/3 is essentially obsolete, being little used nowadays.

Because MINDO/3 did not meet Dewar's aims, Dewar and Thiel developed the **MNDO** (modified neglect of diatomic overlap) method [M. J. S. Dewar and W. Thiel, *J. Am. Chem. Soc.*, **99**, 4899, 4907 (1977); Dewar and H. S. Rzepa, *J. Am. Chem. Soc.*, **100**, 58, 777, 784 (1978); Dewar and M. L. McKee, *J. Am. Chem. Soc.*, **99**, 5231 (1977)]. The MNDO method has been parametrized for compounds containing H, Li, Be, B, C, N, O, F, Al, Si, Ge, Sn, Pb, P, S, Cl, Br, I, Zn, and Hg. MNDO gives substantially improved results as compared with MINDO/3. For the same sample of C, H, O, N compounds used above for MINDO/3 errors, average absolute MNDO errors are 6.3 kcal/mol in heats of formation, 0.014 Å in bond lengths, 2.8° in bond angles, 0.30 D in dipole moments, and 0.5 eV in ionization energies [M. J. S. Dewar and W. Thiel, *J. Am. Chem. Soc.*, **99**, 4907 (1977)].

The MNDO valence-electron Hamiltonian $\hat{H}_{\text{val}}$ is given by (17.74) and the Fock matrix elements are given by (17.75). As in (17.76), the core Hamiltonian operator for valence electron 1 is written as $\hat{H}^{\text{core}}(1) = -\frac{1}{2}\nabla_1^2 + \sum_B V_B(1)$, where $V_B(1)$ is the part of the potential energy of electron 1 due to its interactions with the core (nucleus plus inner-shell electrons) of atom B. To evaluate the Fock matrix element $F_{\mu\nu}$ in (17.75), we need the core matrix elements $H^{\text{core}}_{\mu\nu}$, the electron-repulsion integrals $(\mu\nu|\lambda\sigma)$, and an initial guess for the density matrix elements $P_{\lambda\sigma}$, so as to start the SCF iterative process.

Here $\mu, \nu, \lambda, \sigma$ are used instead of r, s, t, u as subscripts to avoid confusion with the use of s to denote an s orbital.

The integrals occurring in the MNDO Fock matrix elements $F_{\mu\nu}$ are evaluated as follows.

The core matrix element $H^{\text{core}}_{\mu_A\nu_B} = \langle\mu_A(1)|\hat{H}^{\text{core}}(1)|\nu_B(1)\rangle$ (often called a core resonance integral) involving AOs centered on different atoms A and B is taken (as in CNDO) as proportional to the overlap integral (evaluated exactly) between these AOs:

$$H^{\text{core}}_{\mu_A\nu_B} = \tfrac{1}{2}(\beta_{\mu_A} + \beta_{\nu_B})S_{\mu_A\nu_B}, \qquad \text{A} \neq \text{B} \tag{17.79}$$

where the β's are parameters for each type of valence AO. For example, carbon has valence AOs of types 2s and 2p, and MNDO has parameters β_{C2s} and β_{C2p}.

The core matrix element $H^{\text{core}}_{\mu_A\nu_A} = \langle\mu_A(1)|\hat{H}^{\text{core}}(1)|\nu_A(1)\rangle$ involving different AOs centered on the same atom is evaluated using $\hat{H}^{\text{core}}(1) = -\frac{1}{2}\nabla_1^2 + V_A(1) + \sum_{B\neq A} V_B(1)$ [Eq. (17.76)] to write

$$H^{\text{core}}_{\mu_A\nu_A} = \langle\mu_A|-\tfrac{1}{2}\nabla^2 + V_A|\nu_A\rangle + \sum_{B\neq A}\langle\mu_A|V_B|\nu_A\rangle$$

The integral $\langle\mu_A| -\frac{1}{2}\nabla^2 + V_A|\nu_A\rangle$ can be shown to be zero using a theorem of group theory (*Offenhartz*, p. 325). In the crude approximation of taking electron 1 to interact with a pointlike core of charge C_B, we would have $V_B = -C_B/r_{1B}$ and $\langle\mu_A|V_B|\nu_A\rangle$ would equal $-C_B\langle\mu_A|1/r_{1B}|\nu_A\rangle$. Instead, similar to the CNDO/2 expression, MNDO takes $\langle\mu_A|V_B|\nu_A\rangle = -C_B(\mu_A\nu_A|s_Bs_B)$, where s_B is a valence s orbital on atom B and the ERI $(\mu_A\nu_A|s_Bs_B)$ is evaluated by an approximate procedure discussed briefly below. Thus

$$H^{\text{core}}_{\mu_A\nu_A} = -\sum_{B\neq A} C_B(\mu_A\nu_A|s_Bs_B) \quad \text{for } \mu_A \neq \nu_A \tag{17.80}$$

The core matrix element $H^{\text{core}}_{\mu_A\mu_A} = \langle\mu_A(1)|\hat{H}^{\text{core}}(1)|\mu_A(1)\rangle$ containing the same AO twice is evaluated using (17.77) to write

$$H^{\text{core}}_{\mu_A\mu_A} = \langle\mu_A|-\tfrac{1}{2}\nabla^2 + V_A|\mu_A\rangle + \sum_{B\neq A}\langle\mu_A|V_B|\mu_A\rangle \tag{17.81}$$

The integral $U_{\mu_A\mu_A} \equiv \langle\mu_A|-\frac{1}{2}\nabla^2 + V_A|\mu_A\rangle$ could be evaluated from atomic spectral data (as it is in CNDO), but instead, MNDO takes this integral as a parameter. For an atom with s and p valence AOs, MNDO thus has parameters denoted U_{ss} and U_{pp}. MNDO evaluates the remaining integrals in $H^{\text{core}}_{\mu_A\mu_A}$ in (17.81) with the approximation (similar to that used in CNDO/2) $\langle\mu_A|V_B|\mu_A\rangle = -C_B(\mu_A\mu_A|s_Bs_B)$, where the ERI is evaluated as discussed below. Thus

$$H^{\text{core}}_{\mu_A\mu_A} = U_{\mu_A\mu_A} - \sum_{B\neq A} C_B(\mu_A\mu_A|s_Bs_B) \tag{17.82}$$

The ERIs are evaluated as follows in MNDO. The ZDO approximation makes all three-center and four-center ERIs vanish. The one-center ERIs are not evaluated by integration but are assigned values that give a good fit to valence-state ionization energies (which are known from atomic spectral data). The one-center ERIs are either

Coulomb integrals of the form $g_{\mu\nu} \equiv (\mu_A\mu_A|\nu_A\nu_A)$ or exchange integrals of the form $h_{\mu\nu} \equiv (\mu_A\nu_A|\mu_A\nu_A)$. For an atom with only s and p valence AOs, there are six such one-center ERIs: $g_{ss}, g_{sp}, g_{pp}, g_{pp'}, h_{sp}, h_{pp'}$, where the prime denotes that the second p AO is along a different axis than the first. The values of these one-center ERIs found by fitting atomic data are less than the values given by direct integration, due to correlation effects that keep electrons apart.

The two-center ERIs are not evaluated directly by integration but are found from the values of the one-center ERIs and the internuclear distances using a complicated approximate procedure. [This procedure involves multipole expansions of the charge distributions; see M. J. S. Dewar and W. Thiel, *Theor. Chim. Acta*, **46**, 89 (1977).] The evaluation procedure gives values that are correct at the internuclear distances $R_{AB} = 0$, $R_{AB} = \infty$, and that are smaller in magnitude than the values found by accurate integration, so as to allow for electron correlation (and the use of a minimal basis set).

In MNDO, the core-core repulsion term is given by (17.78) with

$$f_{AB}^{MNDO} = C_A C_B (s_A s_A|s_B s_B)(e^{-\alpha_A R_{AB}} + e^{-\alpha_B R_{AB}}) \tag{17.83}$$

where α_A and α_B are parameters for atoms A and B. For the pairs of atoms O–H and N–H, a different function than (17.83) is used; namely, the first term $e^{-\alpha_A R_{AH}}$ in parentheses in (17.83) is replaced with $(R_{AH}/\text{Å})e^{-\alpha_A R_{AH}}$, where A is O or N.

In MNDO, there are thus six parameters to be optimized for each kind of atom: the one-center, one-electron integrals U_{ss} and U_{pp}, the STO orbital exponent ζ (MNDO takes $\zeta_s = \zeta_p$ for the valence AOs), β_s and β_p [Eq. (17.79)], and α [Eq. (17.83)]. For some atoms, MNDO assumes $\beta_s = \beta_p$ and these atoms have five parameters. For H, there are no valence p orbitals and there are four parameters for H.

In 1985 Dewar and co-workers published an improved version of MNDO called **AM1** (Austin model 1, named for the University of Texas at Austin); M. J. S. Dewar, E. G. Zoebisch, E. F. Healy, and J. J. P. Stewart, *J. Am. Chem. Soc.*, **107**, 3902 (1985). AM1 has been parametrized for H, B, Al, C, Si, Ge, Sn, N, P, As, Sb, O, S, Se, Te, F, Cl, Br, I, Zn, and Hg. The only differences between MNDO and AM1 are that the AM1 valence orbital exponents ζ_s and ζ_p on the same atom are allowed to differ and the core-repulsion function in AM1 is given by (17.78) with

$$f_{AB}^{AM1} = f_{AB}^{MNDO} + \frac{C_A C_B}{R_{AB}/\text{Å}}\left[\sum_k a_{kA}\exp[-b_{kA}(R_{AB} - c_{kA})^2] + \sum_k a_{kB}\exp[-b_{kB}(R_{AB} - c_{kB})^2]\right] \tag{17.84}$$

Here, f_{AB}^{MNDO} is given by (17.83), each sum in (17.84) contains two, three, or four Gaussian terms, depending which atom is involved, and the quantities a_{kA}, b_{kA}, and c_{kA} are parameters. [The expression given for f_{AB}^{AM1} in Dewar et al., *J. Am. Chem. Soc.*, **107**, 3902 (1985) is erroneous.] For example, for H there are three terms in the sum, giving nine additional parameters $a_{1H}, \ldots, c_{3H}$. With this modified core-repulsion function, new optimized values for the original MNDO parameters and optimized values for the parameters in the sums in (17.84) were found.

In 1989, Stewart re-parametrized AM1 to give the **PM3 method** (parametric method 3; methods 1 and 2 being MNDO and AM1) [J. J. P. Stewart, *J. Comput. Chem.*, **10**, 209, 221 (1989); **11**, 543 (1990); **12**, 320 (1991)]. PM3 differs from AM1 as follows. The one-center electron-repulsion integrals are taken as parameters to be optimized (rather than being found from atomic spectral data). The core-repulsion function has only two Gaussian terms per atom. A different method was used to optimize the PM3 parameters. PM3 has been parametrized for H, C, Si, Ge, Sn, Pb, N, P, As, Sb, Bi, O, S, Se, Te, F, Cl, Br, I, Al, Ga, In, Tl, Be, Mg, Zn, Cd, and Hg.

The **RM1 method** (Recife Model 1, so named because it was developed at the Federal University of Pernambuco in Recife, Brazil) has exactly the same structure as AM1, but all 191 parameters for the atoms C, H, O, N, S, P, F, Cl, Br, I were reevaluated using data from 1736 molecules (as compared with about 200 molecules used for AM1) [G. B. Rocha et al., *J. Comput. Chem.*, **27**, 1101 (2006); www.rm1.sparkle.pro.br/].

The **PDDG/PM3** and **PDDG/MNDO** methods are modifications of PM3 and MNDO that add a certain function, called the pairwise distance directed Gaussian (PDDG) function, containing additional parameters, to the core repulsion function, thereby significantly increasing the accuracy of these methods [M. P. Repasky et al., *J. Comput. Chem.*, **23**, 1601 (2002); I. Tubert-Brohman et al., *J. Comput. Chem.* **25**, 138 (2004); *J. Chem. Theory Comput.*, **1**, 817 (2005)].

A major limitation of the original versions of the MNDO, AM1, and PM3 methods is that they use a basis set of s and p valence AOs only, so they cannot be used with transition-metal compounds. (In Zn, Cd, and Hg, the d electrons are not valence electrons.) Moreover, for compounds containing such second-row elements as S, the contributions of d orbitals to MOs is significant, and these methods do not perform very well for such compounds.

Thiel and Voityuk [W. Thiel and A. A. Voityuk, *J. Am. Chem. Soc.*, **100**, 616 (1996)] extended MNDO to include d orbitals for many second-row and later elements, giving the **MNDO/d** method. MNDO/d does not add d orbitals for first-row elements, so for a compound containing only C, H, O, and N, MNDO/d is precisely the same as MNDO. Also, it was found that for Na, Mg, Zn, Cd, and Hg, inclusion of d orbitals made little difference, and MNDO/d uses only an sp basis for these elements (but the MNDO parameters for these five elements must be reoptimized in MNDO/d). MNDO/d parameters have been published for Al, Si, P, S, Cl, Br, and I, and the method has also been parametrized for several transition metals.

The PM3 method has been extended by the developers of the Spartan program (Section 15.14) to include d orbitals for many transition metals, giving the method PM3(tm), which is available in Spartan (www.wavefun.com).

Stewart revised the PM3 method to give the **PM5 method** (Stewart, MOPAC2002, Fujitsu Limited, Japan, 1999), which gives more-accurate $\Delta H^\circ_{f,298}$ values than PM3 and has been parametrized for 50 elements, including many transition elements.

In 2007, Stewart published the **PM6 method** [J. J. P. Stewart, *J. Mol. Model.*, **13**, 1173 (2007); openmopac.net/home.html), which has been parametrized for 70 elements. Data from 9000 compounds were used in the parametrization, including both experimental data and data from HF/6-31G* and B3LYP/6-31G* calculations. PM6 takes the core–repulsion term as (17.78) with

$$f_{AB} = C_A C_B (s_A s_A | s_B s_B) x_{AB} e^{-\alpha_{AB}(R_{AB} + 0.0003 R_{AB}^6)} + g_{AB}$$

where R_{AB} is in angstroms, $g_{AB} = 10^{-8}[(Z_A^{1/3} + Z_B^{1/3})/R_{AB}]^{12}$ (with Z_A and Z_B being the atomic numbers of A and B), and where x_{AB} and α_{AB} are two-atom parameters whose values depend on which atoms A and B are. The function g_{AB} is negligible at ordinary chemical distances, but becomes large in the rare situations where R_{AB} is very small; g_{AB} represents the repulsion between the cores of A and B. For the AB atom pairs OH and NH, the exponential $e^{-\alpha_{AB}(R_{AB}+0.0003R_{AB}^6)}$ in f_{AB} is replaced by $e^{-\alpha_{AB}R_{AB}^2}$, so as to give a better representation of hydrogen bonding. Also, for the atom pair CC, an additional term is included in f_{AB} so as to improve the accuracy for compounds with carbon–carbon triple bonds.

With 70 elements, PM6 has $70 \cdot 69/2 + 70 = 2485$ different possible pairs of atoms, for each of which pair a set of two-atom parameters is needed. Actually, PM6 includes parameters for about 450 pairs of atoms, and covers the situations most commonly encountered. For a molecule with an atom pair not parametrized in PM6, the contribution of that pair to the core–core interaction is negligible if the two atoms are separated by at least 4 Å.

PM6 uses the real form of Slater orbitals as the basis functions. Of course, for transition elements, *d* orbitals are required in addition to *s* and *p* valence orbitals. To achieve improved performance, PM6 also includes *d* orbitals for many main-group nonmetals such as Si, P, S, Cl, As, Se, Br, Sb, Te, and I.

PM6 gives a significant improvement in accuracy over its predecessors. For example, for 1157 compounds containing no elements other than C, H, N, and O, the average absolute error in gas-phase $\Delta H^\circ_{f,298}$ values in kcal/mol is 9.4 for AM1, 5.7 for PM3, 5.6 for PM5, 4.9 for RM1, and 4.6 for PM6; for 1774 compounds containing no elements other than H, C, N, O, F, P, S, Cl, Br, I, these errors are 12.6 for AM1, 8.0 for PM3, 6.8 for PM5, 6.6 for RM1, and 5.1 for PM6 (Stewart, op. cit.); PM6 gives a better account of hydrogen bonding than its predecessors.

At the time of publication of PM6, the semiempirical method PM4 had neither been published nor made available, and the PM5 method had not been published but was available for use in the program MOPAC2002.

Semiempirical methods are widely available in many programs. *Gaussian* (Section 15.14) includes the MNDO, AM1, PM3, MINDO/3, INDO, and CNDO methods. Spartan includes the MNDO, MNDO/d, AM1, PM3, PM3(tm), and RM1 methods. HyperChem has the MNDO, MNDO/d, AM1, PM3, RM1, MINDO/3, CNDO, INDO, INDO/S and extended Hückel methods. MOPAC2007 [J. J. P. Stewart, *J. Comput.-Aided Mol. Design*, **4**, 1 (1990); openmopac.net/home.html] has the MNDO, AM1, PM3, PM6, and RM1 methods and is available in Linux and Windows versions. MOPAC2002 (which is part of the program Scigress Explorer, www.fujitsu.com/us/services/solutions/lifesci/) has MNDO, MNDO/d, AM1, PM3, and PM5; AMPAC 8 (www.semichem.com/) has the SAM1, AM1, PM3, MNDO, and MNDO/d methods.

For assessment of the performance and deficiencies of these semiempirical methods, see Chapter 18.

The SCC-DFTB Method. The preceding methods discussed in this section are semiempirical MO methods. The SCC-DFTB method (self-consistent-charge density-functional tight-binding method) is a semiempirical DFT method [M. Elstner et al.,

Phys. Rev. B, **58**, 7260 (1998); M. Elstner, *Theor. Chem. Acc.*, **116**, 316 (2006)], somewhat similar to the semiempirical MO methods. The exchange-correlation energy functional used in SCC-DFTB is usually the PBE functional (Section 16.4). The SCC-DFTB method treats only the valence electrons explicitly, uses a minimal basis set of nonorthogonal AOs to expand the Kohn–Sham orbitals, neglects many integrals and uses approximations for many other integrals. The method contains two types of parameters: single-atom parameters (which are found from atomic DFT calculations), and parameters in interatomic repulsion-energy functions. The interatomic parameters are evaluated from bond-stretching energies calculated by DFT with the B3LYP functional with a DZP or TZP basis set. Thus the parameters are not found using experimental data, but using DFT calculations. The time for an SCC-DFTB calculation on a large molecule is similar to the time for an AM1 or PM3 calculation, but the results are usually more accurate than AM1 or PM3.

Semiempirical Calculations on Very Large Molecules. Methods that allow semiempirical calculations on molecules containing thousands of atoms are compared in A. D. Daniels and G. E. Scuseria, *J. Chem. Phys.*, **110**, 1321 (1999).

Yang and co-workers developed a "divide-and-conquer" method that combines calculations done on overlapping regions of the molecule [T.-S. Lee, D. M. York, and W. Yang, *J. Chem. Phys.*, **105**, 2744 (1996); S. L. Dixon and K. M. Merz, *J. Chem. Phys.*, **107**, 879 (1997)].

Stewart developed the MOZYME linear-scaling method that uses localized MOs to solve the SCF equations [J. J. P. Stewart, *Int. J. Quantum Chem.*, **58**, 133 (1996)]. The method is limited to closed-shell molecules that can be represented by a conventional Lewis structure.

The *conjugate-gradient density-matrix search method* speeds up large-molecule SCF calculations by avoiding diagonalization of the Fock matrix [step 7 after Eq. (14.58)]; instead, an improved estimate of the density matrix $\mathbf{P}$ is found by varying the P_{tu} matrix elements so as to minimize the energy calculated from $\mathbf{P}$ and the current estimate of the Fock matrix $\mathbf{F}'$. Then the improved $\mathbf{P}$ is used to calculate an improved $\mathbf{F}'$, which is used to calculate an improved $\mathbf{P}$, etc. This method allowed single-point AM1 energy calculations to be done on a 19995-atom polymer of glycine and a 6304-atom RNA molecule [A. D. Daniels, J. M. Millam, and G. E. Scuseria, *J. Chem. Phys.*, **107**, 425 (1997)].

The linear-scaling LocalSCF method [N. A. Anikin et al., *J. Chem. Phys.*, **121**, 1266 (2004)] uses weakly nonorthogonal localized MOs. A LocalSCF single-point AM1 calculation was done on a 119273-atom protein in 16000 s on a personal computer.

17.5 THE MOLECULAR-MECHANICS METHOD

The **molecular-mechanics** (MM) **method** is quite different from the semiempirical methods of the last section. Molecular mechanics is not a quantum-mechanical method, since it does not deal with an electronic Hamiltonian or wave function or an electron density. Instead, the method uses a model of a molecule as composed of atoms held together by bonds. Using such parameters as bond-stretching and bond-bending force constants, and allowing for interactions between nonbonded atoms, the method constructs a potential-energy expression that is a function of the atomic positions.

By minimizing this expression for various molecular conformers, the MM method predicts equilibrium geometries and relative energies. The method was developed by Westheimer, Hendrickson, Wiberg, Allinger, Warshel, and others, and is applicable to ground electronic states. Most applications have been to organic compounds, but applications to organometallic compounds and transition-metal coordination compounds are growing. Because MM calculations are much faster than quantum-mechanical calculations, geometry optimizations of molecules with 10^4 atoms and single-point energy calculations on systems with 10^6 atoms can be done.

Molecular mechanics deals with the changes in a molecule's electronic energy due to bond stretching (V_{str}), bond-angle bending (V_{bend}), out-of-plane bending (V_{oop}), internal rotation (torsion) about bonds (V_{tors}), interactions between these kinds of motion (which produce cross terms V_{cross}), van der Waals attractions and repulsions between nonbonded atoms (V_{vdW}), and electrostatic interactions between atoms (V_{es}). The sum of these contributions gives the molecular-mechanics potential energy V, called the **steric energy**, for motion of the atoms in the molecule (or molecules if the system being calculated has more than one molecule):

$$V = V_{str} + V_{bend} + V_{oop} + V_{tors} + V_{cross} + V_{vdW} + V_{es} \tag{17.85}$$

Some people use the term *strain energy* to designate V in (17.85) but other people use strain energy to denote a different quantity (see *Burkert and Allinger*, pp. 184–189).

The explicit expressions used for each of the terms in (17.85) define what is called a molecular-mechanics **force field**, since the derivatives of the potential-energy function determine the forces on the atoms. A force field contains analytical formulas for the terms in (17.85) and values for all the parameters that occur in these formulas. The MM method is sometimes called the *empirical-force-field method*. Empirical force fields are used not only for single-molecule molecular-mechanics calculations of energy differences, geometrics, and vibrational frequencies, but also for molecular-dynamics simulations of liquids and solutions, where Newton's second law is integrated to follow the motions of atoms with time in systems containing hundreds of molecules.

An MM force field assigns each atom in a molecule to one of a number of possible **atom types**, depending on the atom's atomic number and molecular environment. For example, some commonly used atom types in force fields for organic compounds are sp^3 (saturated) carbon, sp^2 (doubly bonded) carbon, sp (triply bonded) carbon, carbonyl carbon, aromatic carbon, and so on, H bonded to C, H bonded to O, H bonded to N, and so on. Different force fields contain somewhat different numbers and kinds of atom types, based on the decisions made by their constructors. A force field for organic compounds typically contains 50 to 75 atom types.

In ab initio and semiempirical molecular electronic single-point or geometry-optimization calculations, one inputs the atomic numbers of the atoms and a set of coordinates (Cartesian or internal) for each atom, and no specification is made as to which atoms are bonded to which atoms (Section 15.15). In a molecular-mechanics calculation, one must specify not only the initial atomic coordinates, but also which atoms are bonded to each atom, so that the V expression can be properly constructed. This specification is most conveniently done using a graphical computer interface to construct the molecule on screen (Section 15.15).

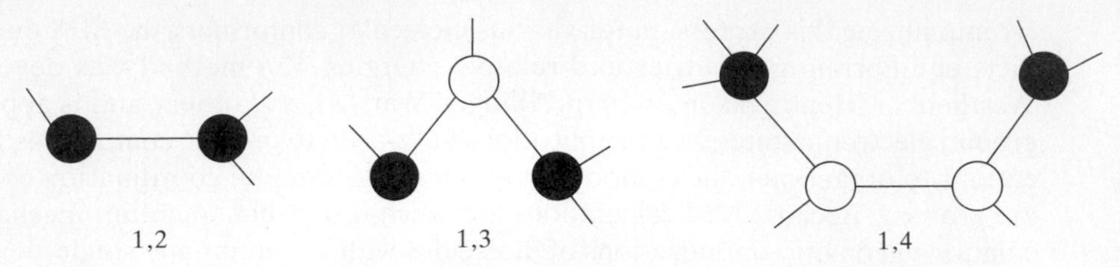

FIGURE 17.9 The shaded atoms are 1,2 atoms, 1,3 atoms, and 1,4 atoms.

The following terminology (Fig. 17.9) is used in molecular mechanics: 1,2 atoms are atoms bonded to each other; 1,3 atoms are separated by two bonds; 1,4 atoms are separated by three bonds; and so on.

Some commonly used force fields are the following. MM2 of Allinger and co-workers (*Burkert and Allinger*) is for small- to moderate-size organic compounds; MM3 is for small organic compounds, polypeptides, and proteins [N. L. Allinger and L. Yan, *J. Am. Chem. Soc.*, **115**, 11918 (1993) and references cited therein]; MM4 is an improved version [N. L. Allinger et al., *J. Comput. Chem.*, **24**, 1447 (2003); J. H. Lii et al., ibid., **24**, 1473, 1490, 1504 (2003); C. H. Langley et al., ibid., **22**, 1396, 1426, 1451, 1476 (2001)]. The following force fields are for polypeptides, proteins, and nucleic acids: AMBER (assisted model building with energy refinement) [W. D. Cornell et al., *J. Am. Chem. Soc.*, **117**, 5179 (1995); Y. Duan et al., *J. Comput. Chem.* **24,** 1999 (2003); amber.scripps.edu]; CHARMM (chemistry at Harvard molecular modeling) of Karplus and co-workers [B. R. Brooks et al., *J. Comput. Chem.*, **4**, 187 (1983); A. D. Mackerell, et al., *J. Am. Chem. Soc.*, **117**, 11946 (1995); *J. Phys. Chem. B*, **102**, 3586 (1998); N. Foloppe and A. D. Mackerell, *J. Comput. Chem.*, **21**, 86 (2000); yuri.harvard.edu/], OPLS (optimized potential for liquid simulations, sometimes called OPLS/AMBER) of Jorgensen and co-workers [W. L. Jorgensen and J. Tirado-Rives, *J. Am. Chem. Soc.*, **110**, 1657 (1988); J. Pranata et al., *J. Am. Chem. Soc.*, **113**, 2810 (1991); OPLS-AA version—W. L. Jorgensen et al., *J. Am. Chem. Soc.*, **118**, 11225 (1996); G. A. Kaminski et al., *J. Phys. Chem. B*, **105**, 6474 (2001)]; and GROMOS (Groningen molecular simulation) of van Gunsteren and Berendsen [X. Daura, A. E. Mark, and W. F. van Gunsteren, *J. Comput. Chem.*, **19**, 535 (1998); C. Oostenbrink et al., ibid., **25**, 1656 (2004); www.igc.ethz.ch/GROMOS/]. CHARMm is a commercially available version of CHARMM that is also applicable to smaller organic compounds [F. A. Momany and R. Rone, *J. Comput. Chem.*, **13**, 888 (1992)]. MMFF94 (Merck molecular force field 1994 version) of Halgren [T. A. Halgren, *J. Comput. Chem.*, **17**, 490, 520, 553, 616 (1996); T. A. Halgren and R. B. Nachbar, ibid., **17**, 587] and CFF93 and CFF95 [consistent force field 1993 and 1995 versions—M. J. Hwang, T. P. Stockfish, and A. T. Hagler, *J. Am. Chem. Soc.*, **116**, 2515 (1994); J. R. Maple et al., *J. Comput. Chem.*, **19**, 430 (1998)] are for calculations on small organic compounds, proteins, and nucleic acids. The Tripos force field (sometimes called SYBYL) [M. Clark et al., *J. Comput. Chem.*, **10**, 982 (1989)] is applicable to small organic molecules and proteins and is used for drug design.

The DREIDING force field [S. L. Mayo, B. D. Olafson, and W. A. Goddard, *J. Phys. Chem.*, **94**, 8897 (1990)] is a general force field applicable to organic and inorganic compounds of all the main-group elements. The UFF (universal force field) [A. K. Rappé et al., *J. Am. Chem. Soc.*, **114**, 10024 (1992)] is applicable to compounds of all elements in the periodic table.

Most of these force fields exist in several versions, and one should specify the version used in any calculation.

Some force fields for MM calculations on biomolecules (for example, OPLS, AMBER, CHARMM) exist in both united-atom (UA) and all-atom (AA) versions. A UA force field saves computational time by not explicitly including hydrogen atoms bonded to aliphatic carbon atoms. Instead, the field contains parameters for the CH_3, CH_2, and CH groups.

For reviews of force fields for calculations on biomolecules, see A. D. Mackerell, *J. Comput. Chem.*, **25**, 1584 (2004); J. W. Ponder and D. A. Case, *Adv. Protein Chem.*, **66**, 27 (2003); W. L. Jorgensen and J. Tirado-Rives, *Proc. Nat. Acad. Sci.*, **102**, 6665 (2005).

One should distinguish between a force field, which is defined by the expression for V in (17.85), and a molecular-mechanics program, which is a computer program that uses a force field to perform molecular-mechanics calculations. Sometimes MM programs and force fields have the same name. For example, CHARMM is also the name of a molecular-mechanics program. The CHARMM program has available in it, not only the CHARMM force field, but also MMFF94.

Some examples of MM computational times on a workstation are: geometry optimization of a 29-atom molecule: 0.3 s for MM2* (a modified version of MM2), 80 s for MNDO, 2×10^5 s for HF/6-31G*; geometry optimization of a 182-atom molecule: 100 s for MM2*, 1.7×10^4 s for AM1 [C. H. Reynolds, *J. Mol. Struc.* (*Theochem*), **401**, 267 (1997)].

We now consider the various terms in (17.85).

Stretching. The potential energy V_{str} of bond stretching is taken as the sum of potential energies $V_{str,ij}$ for stretching each bond of the molecule: $V_{str} = \Sigma_{1,2} V_{str,ij}$, where the sum is over all pairs of atoms bonded to each other, that is, over all 1,2 atom pairs. The simplest choice for $V_{str,ij}$ is to use the harmonic-oscillator approximation (Fig. 4.5) and take $V_{str,ij}$ as a quadratic function of the displacement of the bond length l_{ij} from its **reference** (or **natural**) **length** l_{IJ}^0; that is,

$$V_{str,ij} = \tfrac{1}{2} k_{IJ}(l_{ij} - l_{IJ}^0)^2 \tag{17.86}$$

similar to (13.21). Here, the capital letters I and J denote the atom types of atoms i and j in the molecule. The force constant k_{IJ} and the reference length l_{IJ}^0 depend on the atom types of the two atoms forming the bond. For example, in $CH_2{=}CH{-}CH_2{-}CH_3$, each carbon carbon single bond has a longer reference length and a smaller stretching force constant than the carbon–carbon double bond. Note that, because many terms in the steric energy (17.85) involve the atoms forming a particular bond, the reference value l_{IJ}^0 is not necessarily equal to the equilibrium bond length of the bond between atoms i and j in any particular molecule. However, it will likely be close to the equilibrium lengths of most bonds formed by atoms of types I and J. Many force fields omit the factor $\frac{1}{2}$ and write $V_{str,ij} = K_{IJ}(l_{ij} - l_{IJ}^0)^2$.

The force fields AMBER, CHARMM, and Tripos use the quadratic function (17.86) for $V_{str,ij}$. MM2 uses quadratic and cubic terms in $V_{str,ij}$. MM3, CFF93, and MMFF94 use quadratic, cubic, and quartic terms. For example, in MMFF94,

$$V_{str,ij} = (143.93\ \text{kcal/mol})\tfrac{1}{2}[k_{IJ}/(\text{mdyn/Å})](\Delta l_{ij}/\text{Å})^2[1 - (2\ \text{Å}^{-1})\Delta l_{ij} + \tfrac{7}{12}(2\ \text{Å}^{-1})^2\Delta l_{ij}^2] \tag{17.87}$$

where $\Delta l_{ij} \equiv l_{ij} - l_{IJ}^0$. It is traditional to use kcal/mol for molecular-mechanics energies, and the factor 143.93 converts from mdyn/Å to kcal/mol. The expression (17.87) has the form of the first three terms of a Taylor-series expansion of the Morse potential-energy function (Prob. 4.28) with the Morse exponential parameter equal to 2/Å; this value was found by trial and error to work well in MMFF94. The UFF and DREIDING force fields use a quadratic stretching function as the default, but can also use Morse functions.

A particular force field will contain values for the parameters k_{IJ} and l_{IJ}^0 for all the possible bonds formed by the atom types of that force field. Because different force fields use different expressions for at least some of the terms in the steric energy (17.85), the parameters of one force field are not really comparable to those of another field. (A mistake people sometimes make is to supply a missing parameter in one force field by using a parameter from another force field.) Some examples of l_{IJ}^0 values in Å are: for a bond between two sp^3 carbons, 1.508 in MMFF94, 1.54 in Tripos 5.2, 1.523 in MM2(87); for a bond between an sp^3 carbon and an sp^2 carbon, 1.482 in MMFF94, 1.501 in Tripos 5.2, 1.497 in MM2(87); for a double bond between two sp^2 carbons, 1.333 in MMFF94, 1.335 in Tripos 5.2, 1.337 in MM2(87).

Bending. The potential energy V_{bend} of bond bending is taken as the sum of potential energies $V_{\text{bend},ijk}$ for bending each bond angle of the molecule: $V_{\text{bend}} = \sum V_{\text{bend},ijk}$, where the sum is over all bond angles in the molecule. The simplest choice is to take $V_{\text{bend},ijk}$ as a quadratic function:

$$V_{\text{bend},ijk} = \tfrac{1}{2}k_{IJK}(\theta_{ijk} - \theta_{IJK}^0)^2 \quad (17.88)$$

where θ_{IJK}^0 is the reference value for the bond angle type IJK. This form is used in CHARMM, AMBER, and Tripos. MMFF94 uses quadratic and cubic terms in $V_{\text{bend},ijk}$ and uses a special form for bond angles near 180°. MM2 uses quadratic and sextic terms. CFF93 uses quadratic, cubic, and quartic terms.

Torsion. The term V_{tors} is taken as the sum of terms $V_{\text{tors},ijkl}$ over all 1,4 atom pairs: $V_{\text{tors}} = \Sigma_{1,4} V_{\text{tors},ijkl}$. For example, for ethane, H_3CCH_3, each hydrogen on the left carbon has a 1,4 relation to each of the three hydrogens on the right carbon, giving a total of nine terms in the V_{tors} sum. For H_2CCH_2, there are four torsion terms in the sum. A commonly used form (MMFF94, CFF93, MM2, MM3) for $V_{\text{tors},ijkl}$ is

$$V_{\text{tors},ijkl} = \tfrac{1}{2}[V_1(1 + \cos\phi) + V_2(1 - \cos 2\phi) + V_3(1 + \cos 3\phi)] \quad (17.89)$$

where ϕ is the dihedral angle $D(ijkl)$ (Fig. 15.14) and V_1, V_2, V_3 are parameters whose values depend on the atom types of i, j, k, and l. Tripos, CHARMm, and DREIDING use the following simpler form containing only one parameter: $\frac{1}{2}V_n[1 + \cos(n\phi - \phi_0)]$, where n gives the number of minima over 360° of the torsional potential and ϕ_0 determines the locations of the minima. For example, for $n = 3$ and $\phi_0 = 0$, we get a potential with minima at 60°, 180°, and 300°, as in C_2H_6; with $n = 2$ and $\phi_0 = \pi$, we get a potential with minima at 0° and 180°, as in C_2H_4.

The potential-energy change as one methyl group is rotated relative to the other in ethane can be determined experimentally by observation of transitions between vibrational energy levels produced by this potential. In an MM force field, V_{tors} in ethane is the sum of contributions $V_{\text{tors},ijkl}$ from nine dihedral angles, and an individual

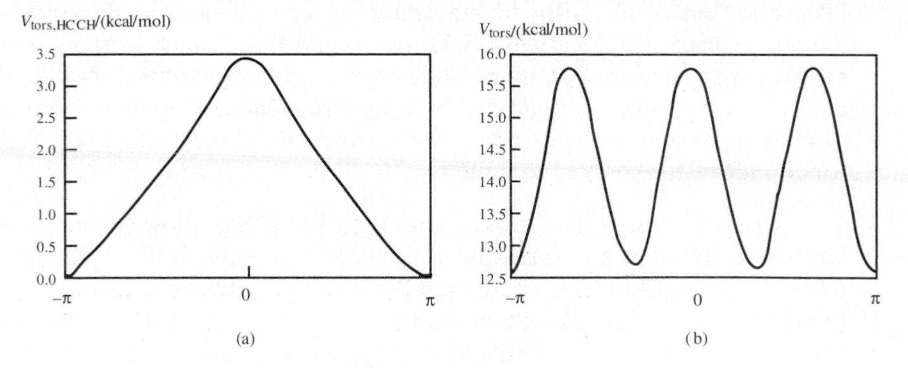

FIGURE 17.10 (a) HCCH torsional potential in ethane found from HF/6-31G* calculations with bond distances and bond angles held fixed during torsion [U. Dinur and A. T. Hagler, *J. Comput. Chem.*, **11**, 1234 (1990)]. (b) Sum of the nine HCCH torsional potentials of (a).

$V_{tors,ijkl}$ is not experimentally determinable. By taking the appropriate second derivative of the electronic energy of ethane and integrating the result, one can use ab initio calculations to estimate $V_{tors,ijkl}$. The result from a HF/6-31G* calculation is shown in Fig. 17.10a. When expressed as a series of terms involving cos $n\phi$, this curve turns out to have significant contributions from only the cos ϕ, cos 2ϕ, and cos 3ϕ terms. When the nine dihedral-angle contributions are added, the symmetry of C_2H_6 makes the contributions of the cos ϕ and cos 2ϕ terms add to zero, leaving only the cos 3ϕ term (Fig. 17.10b). Because of insufficient experimental data, MM2 and MM3 take V_1 and V_2 as zero for HCCH torsion (where C is a saturated carbon), but MMFF94 used results of ab initio conformational energy calculations to fit nonzero values for these quantities.

By using a large positive value for V_2 in (17.89), one can produce the high barrier to out-of-plane twisting in $H_2C{=}CH_2$.

Out-of-Plane Bending. The molecule cyclobutanone, $C_3H_6C{=}O$, contains an oxygen doubly bonded to one of the carbons of a four-membered ring. The carbonyl carbon and the three atoms bonded to it all lie in the same plane. The CCC bond angle at the carbonyl carbon is close to 90°, and the CCO bond angles are close to $\frac{1}{2}(360° - 90°) = 135°$. Because these 135° angles deviate considerably from the reference angles of approximately 120° at the carbonyl carbon, the bond-bending terms $V_{bend,ijk}$ at the carbonyl carbon will make the equilibrium position of the oxygen atom lie above the plane of the ring carbons, so as to produce near-120° CCO angles. To ensure planarity of the three atoms bonded to a carbonyl carbon (which is favored by pi bonding between C and O), one includes in V an out-of-plane (oop) bending term at each carbonyl carbon. This term can have the form $\frac{1}{2}k_{oop}\chi_{oop}^2$, where χ_{oop} is the angle between the CO bond and the plane of the carbonyl carbon and the two carbons bonded to it. (In actual practice, one either includes an out-of-plane bending term for each of the three atoms bonded to a carbonyl C or uses a single term where χ is taken as the average of the out-of-plane angles for each of the three atoms bonded to the carbonyl C.) Similar out-of-plane bending terms are used to enforce planarity at nitrogen atoms in amides.

[The question of planarity of the amide group is complicated; see T. A. Halgren, *J. Comput. Chem.*, **17**, 553 (1996); G. Forgarasi and P. G. Szalay, *J. Phys. Chem. A*, **101**, 1400 (1997).] Also, an out-of-plane bending term is often used at a carbon double bonded to another carbon, since this allows vibrational frequencies to be better reproduced.

Instead of out-of-plane bending terms, some force fields use so-called improper torsion terms to achieve the same result.

Cross Terms. The cross terms V_{cross} in (17.85) allow for interactions between stretching, bending, and torsional motions. For example, if the CO and OH bonds of a COH bond angle are stretched, then the distance between the end atoms of the COH bond angle is increased, which makes it easier to bend the COH angle. Likewise, reducing the COH angle tends to increase the OH and CO lengths. To allow for this interaction, one can add a stretch–bend cross term with the form $\frac{1}{2}k_{12}(\Delta l_1 + \Delta l_2)\Delta\theta$, where Δl_1, Δl_2, and $\Delta\theta$ are the deviations of the bond lengths and the bond angle from their reference values.

The most commonly used cross terms are stretch–bend, stretch–stretch for two bonds to the same atom, stretch–torsion, bend–torsion, and bend–bend for two angles with a common central atom. MM2 and MMFF94 include only stretch–bend interactions. The TRIPOS, AMBER, CHARMm, DREIDING, and UFF force fields have no cross terms. MM3 and MM4 include stretch–bend, bend–bend, and stretch–torsion terms.

Electrostatic Interactions. The electrostatic term V_{es} is commonly taken as the sum of electrostatic interactions involving all pairs of atoms except 1,2 and 1,3 pairs: $V_{\text{es}} = \sum_{1,\geq 4} V_{\text{es},ij}$, where atoms i and j have a 1,4 or greater relation. $V_{\text{es},ij}$ is calculated by assigning partial atomic charges Q_i to each atom and using the Coulombic potential energy expression

$$V_{\text{es},ij} = \frac{Q_i Q_j}{\varepsilon_r R_{ij}} \qquad (17.90)$$

where ε_r is a dielectric constant. For calculations modeling gas-phase molecules, ε_r is typically given a value in the range 1 to 1.5. An ε_r greater than 1 allows for screening interactions due to polarization of parts of the molecule lying between atoms i and j.

Various methods are used to assign the partial atomic charges. MMFF94 assumes the polarity of a given type of bond to be independent of its environment and assigns charges according to $Q_i = Q_{i,\text{formal}} + \sum \omega_{\text{KI}}$, where $Q_{i,\text{formal}}$ is the formal charge on atom i (found from the Lewis dot structure by dividing the electrons in each bond equally between the bonded atoms), ω_{KI} is a parameter representing the charge contribution to atom i from the bond between atoms i and k (of types I and K), and the sum goes over all bonds to atom i. Note that $\omega_{\text{KI}} = -\omega_{\text{IK}}$. Some values of ω_{KI} are 0 for a Csp^3–H bond, -0.138 for a Csp^3–Csp^2 bond. The ω_{KI} values were found by varying them to make MMFF94 give a good least-squares fit to HF/6-31G* dipole moments of several hundred molecules; the value zero for Csp^3–H bonds was assumed. [It should be noted that MMFF94 replaces R_{ij} in (17.90) with $R_{ij} + 0.05$ Å.]

In AMBER, the charges are found by using charges fit to HF/6-31G* electrostatic potentials (ESPs, Section 15.7) of related smaller molecules. For example, to get the charges on atoms of the amino acid residue —NHCH(R)C(O)— of a protein, one uses the charges of corresponding atoms of the molecule $CH_3C(O)NHCH(R)C(O)NH_2$. Similarly for the charges in DNA and RNA.

In OPLS, partial charges are assigned to the atoms in a protein based on the atom type. For example, all oxygen atoms in amide groups are given a partial charge of -0.50 and all carbons in amide groups are given a charge of $+0.50$.

Several other schemes based on electronegativity are used to assign charges; see *Leach*, Section 4.9.6; T. A. Halgren, *J. Comput. Chem.*, **17**, 520 (1996).

MM2 and MM3 do not use (17.90) but assign a dipole-moment vector to each bond and compute V_{es} as the sum of the potential energies of interaction between bond moments. Each bond dipole moment is located at the center of the bond and points along the bond. The values for the bond moment of given types of bond were chosen to fit known dipole moments of small molecules.

Some force fields reduce $V_{es,ij}$ of each 1,4 pair of atoms by multiplying it by a scaling factor. Some 1,4 electrostatic scaling factors are 0.75 in MMFF94 and 0.5 in AMBER.

Van der Waals Interactions. The van der Waals term in (17.85) is usually taken as the sum of interactions involving all possible 1,4, 1,5, 1,6, ... atom-pair interactions: $V_{vdW} = \sum_{1,\geq 4} V_{vdW,ij}$, where atoms i and j are in a 1,4 or greater relation. The 1,2 and 1,3 van der Waals interactions and electrostatic interactions are considered to be implicitly included in the bond-stretching and bond-bending parameters. Each van der Waals pair term $V_{vdW,ij}$ is the sum of an attraction due to London dispersion forces and a repulsion due mainly to Pauli repulsion. The AMBER, CHARMM, DREIDING, UFF, and TRIPOS force fields take $V_{vdW,ij}$ as a Lennard-Jones 12–6 potential (Fig. 17.11 and Prob. 17.36), which can be written in two equivalent forms:

$$V_{vdW,ij} = \varepsilon_{IJ}\left[\left(\frac{R_{IJ}^*}{R_{ij}}\right)^{12} - 2\left(\frac{R_{IJ}^*}{R_{ij}}\right)^{6}\right] = 4\varepsilon_{IJ}\left[\left(\frac{\sigma_{IJ}}{R_{ij}}\right)^{12} - \left(\frac{\sigma_{IJ}}{R_{ij}}\right)^{6}\right] \quad (17.91)$$

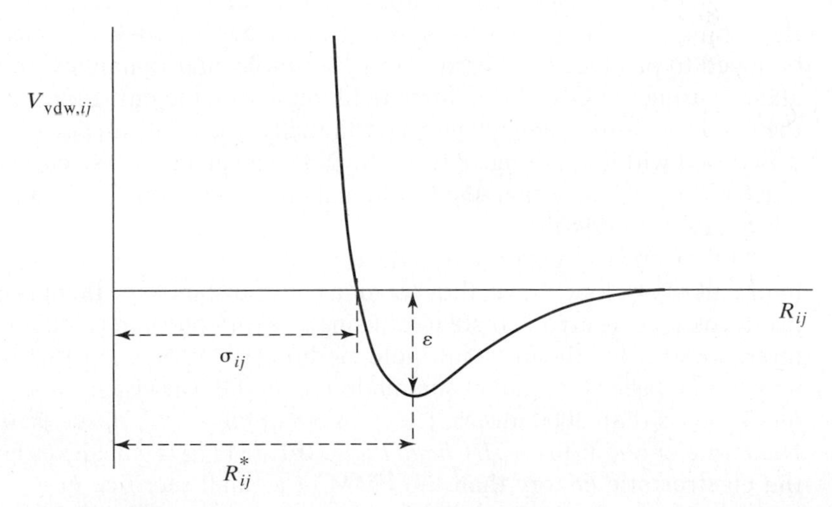

FIGURE 17.11 The Lennard-Jones potential as a function of interatomic distance.

where R_{ij} is the distance between atoms i and j, the well-depth parameter ε_{IJ} is the value of $V_{\text{vdW},ij}$ at the minimum in the interaction curve, the parameter R_{IJ}^* gives the value of R_{ij} at the minimum in $V_{\text{vdW},ij}$ and the alternative parameter σ_{IJ} is the R_{ij} value at which $V_{\text{vdW},ij}$ is zero. MM2 and MM3 use a function with an attractive term proportional to R_{ij}^{-6} and a repulsive term proportional to $e^{-aR_{ij}}$, where a is a parameter. MMFF94 uses a buffered 14–7 potential that was found to represent known van der Waals interactions between the rare gas atoms more accurately than other forms [T. A. Halgren, *J. Am. Chem. Soc.*, **114**, 7827 (1992)].

For 1,4 interactions, some force fields multiply $V_{\text{vdW},ij}$ by a scale factor that reduces its value.

As an example, for $CF_3CF_2CF_3$, a molecular-mechanics force field has 10 bond-stretching terms (8 for CF bonds and 2 for CC bonds), 18 bond-bending terms [there are $\frac{1}{2}(4)3 = 6$ bond angles at each C], 18 bond torsion terms (9 about each CC bond), 27 van der Waals terms (there are 18 1,4 interactions between the end atoms of the 18 dihedral angles plus 9 1,5 interactions between one set of CF_3 fluorines and the second set), 27 electrostatic terms, plus whatever cross terms the force field includes.

Cutoffs. The electrostatic and van der Waals interactions are called **nonbonded interactions** and consume the largest part of the time needed to calculate V_{steric} of a very large molecule. For a 3000-atom molecule, V_{es} and V_{vdW} are each the sum of about $\frac{1}{2}(3000)(2990) \approx 4 \times 10^6$ terms. (The last factor is reduced to acknowledge that 1,2 and 1,3 interactions are omitted.) To speed up MM calculations on large molecules (and molecular dynamics calculations on systems containing many molecules), many programs use a **cutoff**, meaning that $V_{\text{es},ij}$ and $V_{\text{vdW},ij}$ terms are omitted for atom pairs that are farther apart than some chosen distance.

If a cutoff is abruptly applied at a particular interatomic distance, this discontinuity can cause problems in energy minimization and molecular-dynamics calculations. To avoid this, one can use a cutoff that makes the nonbonded interactions go to zero gradually over a distance of, say, 1 Å. This is done using what is called a *switching function*.

Van der Waals interactions are proportional to $1/R_{ij}^6$ at large distances and so are short range. A van der Waals cutoff distance of 8 or 10 Å is typically used and was believed to produce little error. However, molecular-dynamics simulations of liquid alkanes using the GROMOS force field found that the enthalpies of vaporization and the vapor pressures changed very substantially when the van der Waals cutoff distance was varied within the range 8 to 14 Å, and so a van der Waals cutoff radius of 16 Å is much more justifiable than the traditional one of 8 or 10 Å [X. Daura et al., *J. Comput. Chem.*, **19**, 535 (1998)].

Often an electrostatic cutoff distance of 10 or 15 Å is used. However, electrostatic interactions are long range; thus use of a cutoff distance for them is not justifiable and produces serious errors. Instead of using a cutoff, one can calculate the electrostatic interactions using the fast-multipole method (FMM; Section 15.16), but this method may not be faster than direct summation of all the pairwise interactions when the system has less than 3000 atoms. The *structure-adapted multipole method* (SAMM) [C. Niedermeier and P. Tavan, *J. Chem. Phys.*, **101**, 734 (1994)] achieves faster evaluation of the electrostatic energy than the FMM at a small sacrifice in accuracy. In SAMM, instead of using the fixed subdivision of space that is used in the FMM method, space is divided into regions that correspond to various structural features of the biomolecule being studied, thereby allowing the multipole expansions to be truncated at lower orders than in the FMM method. For molecular-dynamics calculations on biomolecules, the SAMM method has been combined with a multiple-time-step method to yield FAMUSAMM (fast multiple-time-step structure-adapted multipole method); M. Eichinger et al., *J. Comput. Chem.*, **18**, 1729 (1997). In a multiple-time-step method, the forces between distant atoms are treated using a longer time step than is used for the forces between nearby atoms (which vary more rapidly).

Another method widely used to sum the electrostatic interactions in large systems is the *particle-mesh Ewald method*, which is a modification of a method originally used to sum electrostatic interactions in ionic crystals; here, the electrostatic-energy sum is replaced by the sum of sums that can each be calculated more rapidly than the original sum; see *Leach*, Section 6.8.1.

Hydrogen Bonding. For pairs of atoms that can hydrogen bond to each other, some force fields modify the van der Waals interaction to a form such as $A/R^{12} - C/R^{10}$, but many force fields contain no special terms for hydrogen bonding, and rely on the electrostatic and van der Waals terms to produce the hydrogen bond.

Conjugated Bonds. Conjugated bonds require special consideration in molecular mechanics. Consider, for example, $CH_2{=}CH{-}CH{=}CH_2$. Each of the four carbons is an sp^2 type. Hence, if the stretching force constant and reference length of each bond is assigned simply based on the types of the two atoms forming the bond, each carbon–carbon bond in 1,3-butadiene will have the same stretching constant and reference length. However, the central and end bonds actually have quite different lengths (1.46 and 1.34 Å). MMFF94 and Tripos make the reference bond lengths and stretching force constants depend not only on the atom types but also on the bond order in the Lewis structure. For two bonded sp^2 carbons, the MMFF94 reference length is 1.333 Å if the bond is a single bond and is 1.430 Å if it is a double bond, and the stretching constant for the double bond is about twice that for the single bond. Some force fields handle this problem in a slightly different way that is equivalent to the MMFF94 procedure. MM2, MM3, and MM4 use the more general procedure of performing a semiempirical π-electron calculation on the conjugated portion of the molecule to derive bond orders and use these orders to assign reference bond lengths and force constants to the conjugated bonds.

Parametrization. The general procedure for finding parameter values in a force field is to use experimental or ab initio theoretical information to choose an initial set of parameter values and then vary these parameters so as to minimize the deviations of force-field predicted molecular properties from experimental or ab initio calculated properties of a chosen set of molecules, called the **training set**. Molecular properties used in force-field parametrizations include molecular structures, conformational energy differences, vibrational frequencies, barriers to internal rotation, dipole moments, and intermolecular interactions. Earlier force fields relied mainly on experimental data in the parametrization process, but as ab initio calculations become increasingly able to accurately treat larger molecules, ab initio calculated properties are increasingly being used in parametrization (examples are MMFF94 and CFF93).

Initial estimates of the reference length l_{IJ}^0 and angle θ_{IJK}^0 can be taken as the typical experimental or ab initio calculated length of an IJ bond and an IJK angle. These initial values can be refined by adjusting them to minimize errors in predicted bond lengths and angles of the training-set molecules.

Bond bending and stretching force constants in polyatomic molecules are not experimentally observable quantities, since the observed vibrational frequencies are related in a complicated way to such force constants. However, one can write an approximate expression for the molecular vibrational potential energy as a sum of quadratic terms involving bond stretching, bending of independent bond angles, torsion involving independent dihedral angles, plus various cross terms. (Such a *valence force field* is not the same as a molecular-mechanics force field. In an MM force field, the bond-bending angles and the dihedral angles are not all independent of one another, and the valence force field omits nonbonded interactions.) By using a valence force field and combining vibrational-frequency data of many unstrained organic compounds, Schachtschneider and Snyder obtained bond stretching and bending force constants that are approximately transferable (*Burkert and Allinger*, pp. 20–21), and these can be used as initial estimates for these parameters. An alternative way to get initial estimates of the force constants is from analytical second derivatives of ab initio molecular electronic energies. The initial force constants can be improved by adjusting their values to minimize the errors in vibrational frequencies predicted by the force field.

An initial estimate of V_3 for HCCH (where C is saturated) can be found by fitting the rotational barrier in C_2H_6. Then V_3 for CCCH can be found by fitting the barrier in $CH_3CH_2CH_3$ and V_3 for CCCC from $CH_3CH_2CH_2CH_3$ (with V_1 and V_2 assumed to be zero). To refine V_3 and find V_1 and V_2 values, one can fit conformational energy differences (experimental or ab initio) for a set of training molecules. V_2 for HC=CH can be fit to the rotational barrier in C_2H_4.

For van der Waals parameters, one usually assumes that the parameters ε_{IJ} and R_{IJ}^* for interaction between unlike atom types can be calculated from the parameters $\varepsilon_{II}, \varepsilon_{JJ}, R_{II}^*, R_{JJ}^*$ for like interactions using mixing rules. The most common mixing rules are $\varepsilon_{IJ} = (\varepsilon_{II}\varepsilon_{JJ})^{1/2}$ and $R_{IJ}^* = \frac{1}{2}(R_{II}^* + R_{JJ}^*)$ or $R_{IJ}^* = (R_{II}^*R_{JJ}^*)^{1/2}$. MMFF94 uses more elaborate mixing rules, which were found by examining known rare-gas interaction data.

Sublimation energies of molecular compounds depend on van der Waals interactions, and such sublimation-energy data has been used to derive values of van der Waals parameters. Some force fields (for example, AMBER94, OPLS) adjust van der Waals parameters to give good values for properties such as densities and enthalpies of vaporization found in Monte Carlo simulations of liquid compounds.

Since all the parameters in a force field will influence all calculated molecular properties, parametrization is an iterative process. After adjusting a particular subset of parameters to fit some molecular property, the previously done fits of parameter subsets will be worsened, so one goes back and readjusts previously fitted parameters. One keeps iterating until no further significant improvement is obtained.

To use the molecular-mechanics method, one needs enough data to parametrize the force field. This makes it hard to apply molecular mechanics to novel types of compounds, but ab initio calculations may help in the parametrization.

An empirical force field may contain hundreds or thousands of parameters. MMFF94 has roughly the following numbers of parameters: 500 stretching force constants, 500 reference bond lengths, 2300 bending force constants, 2300 reference bond angles, 600 stretch–bend constants, 100 out-of-plane force constants, 2800 torsion parameters of which roughly half are zero, leaving about 1400 nonzero ones, 400 van der Waals parameters, and 600 electrostatic charge-increment parameters, for a total of about 9000 parameters. In contrast, UFF, which has 126 atom types and can treat compounds of all elements, contains only about 800 parameters. The number of parameters is kept relatively small in UFF by devices such as taking the reference bond angles θ^0_{IJK} to depend only on the atom type of the central atom J, calculating reference bond lengths l^0_{IJ} as the sum of bond radii of the atom types plus corrections for bond order and for electronegativity differences, calculating stretching force constants by an empirical rule that takes them as functions of l^0_{IJ} and effective-charge parameters for each atom type, and so on. Because of the relatively small number of parameters, UFF cannot achieve the accuracy of highly parametrized force fields such as MM3 or MMFF94, but the loss of accuracy is a tradeoff for the broad applicability of UFF.

Molecular Properties. What molecular properties can be calculated using molecular mechanics? A molecular-mechanics geometry optimization starts with the initially assumed geometry and finds the nearest local energy minimum by minimizing the steric energy V of (17.85) using one of the methods of Section 15.10. Because (17.85) provides an analytical expression for the energy, the first and second derivatives of V can be easily evaluated analytically, which facilitates energy minimization. Minimization of V yields the MM-predicted equilibrium geometry of a particular conformer. Many molecular-mechanics programs have built-in searching methods (Section 15.11) that locate many low-energy conformers.

It must be emphasized that the numerical value of the equilibrium steric energy of a conformer has no physical significance by itself. The zero level of V corresponds to a fictitious molecule in which all the bond lengths and angles have their reference values and torsional, van der Waals, and electrostatic interactions are absent. Different force fields will typically give very similar equilibrium geometries for a given conformer, but will give quite different steric energies for that conformer. Steric energies depend on how the force field was constructed and parametrized. In contrast, the electronic energy including nuclear repulsion found by an ab initio calculation has the physical significance of being the approximate energy relative to a well-defined system with all electrons and nuclei infinitely far apart from one another.

What does have physical significance in molecular mechanics is the steric-energy difference (calculated using the same force field) between two species having the same numbers and kinds of atoms and the same numbers and kinds of bonds. Thus, one can use differences in steric energies to meaningfully calculate energy differences between: (a) different conformers of the same molecule [for example, the anti and gauche conformers of butane (Section 15.10) or the chair and boat conformers of cyclohexane]; (b) different stereoisomers of the molecule (for example, cis and trans 1,2-dichloroethylene); (c) species differing by rotation about a bond (for example, eclipsed and staggered ethane); (d) different geometries of the same molecule (for example, planar NH_3

and pyramidal NH_3); (e) two molecules far apart from each other and the same pair forming a hydrogen bond. Intermolecular-interaction energies can be calculated.

The difference between the molecular-mechanics equilibrium steric energies of two species with the same numbers and kinds of bonds is an estimate of the difference between the equilibrium electronic energies of the two species. Corrections for zero-point vibrational energy differences and thermal energy differences should be added to the steric energy difference, but these are usually small and are often omitted.

A molecular-mechanics steric energy can be combined with empirical bond-energy parameters to calculate gas-phase heats of formation $\Delta H^\circ_{f,298}$, as discussed later in this section.

Although the empirical force fields discussed in this section provide a good representation of potential-energy surfaces in the regions of minima, one cannot use them to calculate the complete potential-energy surface for a chemical reaction, since these fields are incapable of describing bond breaking.

By taking $\sum_i Q_i r_i$ [Eq. (14.9)], where the sum goes over all the partial atomic charges, one can calculate a molecular dipole moment for a given conformation.

By evaluating the second partial derivatives of V at a local minimum, one can use the procedure of Section 15.12 to calculate molecular vibrational frequencies. Using the calculated structure and vibrational frequencies, one can estimate the gas-phase entropy S°_{298} (Section 15.13).

Heats of Formation. The molecular-mechanics zero level of energy has all bond lengths and angles at their reference values and no electrostatic, van der Waals, or torsional interactions. In such a hypothetical state, one can well approximate the molecular binding energy as the sum of empirical bond energies. Therefore, the equilibrium electronic energy U_{eq} of a molecule can be found by combining the molecular-mechanics-calculated equilibrium-geometry steric energy V_{steric} with bond energies. For example, for a saturated hydrocarbon with formula $C_{n_C}H_{n_H}$ (where n_C and n_H are the number of C atoms and H atoms), we have $U_{\text{eq}} = V_{\text{steric}} - n_{CH}b_{CH} - n_{CC}b_{CC}$, where n_{CH} and n_{CC} are the numbers of CH and CC single bonds in the molecule, b_{CH} and b_{CC} are CH and CC bond energies (which by convention are positive), and the zero level of energy is taken to correspond to separated (gas-phase) atoms $C(g)$ and $H(g)$. In the following discussion, all energies will be on a per mole basis.

For the formation reaction

$$n_C\text{C(graphite)} + \tfrac{1}{2}n_H H_2(g) \rightarrow C_{n_C}H_{n_H}(g)$$

the change in molar equilibrium electronic energy is $V_{\text{steric}} - n_{CH}b_{CH} - n_{CC}b_{CC} - n_C U_C - \frac{1}{2}n_H U_{H_2}$, where U_C and U_{H_2} are the molar equilibrium electronic energies of graphite and $H_2(g)$ with respect to the same zero level of energy as above. If we temporarily ignore zero-point vibrational energy, we can take this change in equilibrium electronic energy as equal to the change in standard-state thermodynamic internal energy for the formation reaction at absolute zero:

$$\Delta U^\circ_{f,0} = V_{\text{steric}} - n_{CH}b_{CH} - n_{CC}b_{CC} - n_C U_C - \tfrac{1}{2}n_H U_{H_2}$$

For a saturated hydrocarbon (acyclic or cyclic), it is not hard to see that the numbers of H and C atoms are related to the numbers of CH and CC bonds by

$$n_{\mathrm{H}} = n_{\mathrm{CH}} \quad \text{and} \quad n_{\mathrm{C}} = \tfrac{1}{4}n_{\mathrm{CH}} + \tfrac{1}{2}n_{\mathrm{CC}}$$

At temperature T, the change in translational energy for the above formation reaction is $\frac{3}{2}RT - \frac{1}{2}n_{\mathrm{H}}(\frac{3}{2}RT)$, and the change in rotational energy is $\frac{3}{2}RT - \frac{1}{2}n_{\mathrm{H}}RT$. The relation $\Delta H_f^\circ = \Delta U_f^\circ + \Delta(PV)^\circ = \Delta U_f^\circ + \Delta n_g RT$, where Δn_g is the change in number of moles of gas in the formation reaction, gives $\Delta H_f^\circ = \Delta U_f^\circ + (1 - \frac{1}{2}n_{\mathrm{H}})RT$. Combining all these relations and continuing to ignore the contribution from the change in vibrational energy, we have for the standard enthalpy of formation at temperature T

$$\Delta H_{f,T}^\circ = V_{\text{steric}} - n_{\mathrm{CH}}b_{\mathrm{CH}} - n_{\mathrm{CC}}b_{\mathrm{CC}} - (\tfrac{1}{4}n_{\mathrm{CH}} + \tfrac{1}{2}n_{\mathrm{CC}})U_{\mathrm{C}} - \tfrac{1}{2}n_{\mathrm{CH}}U_{\mathrm{H_2}} + 4RT - \tfrac{7}{4}n_{\mathrm{CH}}RT$$

$$\Delta H_{f,T}^\circ = V_{\text{steric}} - n_{\mathrm{CH}}(b_{\mathrm{CH}} + \tfrac{1}{4}U_{\mathrm{C}} + \tfrac{1}{2}U_{\mathrm{H_2}} + \tfrac{7}{4}RT) - n_{\mathrm{CC}}(b_{\mathrm{CC}} + \tfrac{1}{2}U_{\mathrm{C}}) + 4RT$$

Defining $a_{\mathrm{CH}} \equiv -(b_{\mathrm{CH}} + \frac{1}{4}U_{\mathrm{C}} + \frac{1}{2}U_{\mathrm{H_2}} + \frac{7}{4}RT)$ and $a_{\mathrm{CC}} \equiv -(b_{\mathrm{CC}} + \frac{1}{2}U_{\mathrm{C}})$, we have

$$\Delta H_{f,T}^\circ = V_{\text{steric}} + n_{\mathrm{CH}}a_{\mathrm{CH}} + n_{\mathrm{CC}}a_{\mathrm{CC}} + 4RT \tag{17.92}$$

One uses Eq. (17.92) to determine a_{CH} and a_{CC} by a least-squares fit to 25°C experimental ΔH_f° data for several gas-phase hydrocarbons. (Typically, $a_{\mathrm{CH}} \approx -4\frac{1}{2}$ kcal/mol and $a_{\mathrm{CC}} \approx 2\frac{1}{2}$ kcal/mol.) Then Eq. (17.92) can be used to find the gas-phase $\Delta H_{f,298}^\circ$ for any saturated hydrocarbon from its steric energy calculated by molecular mechanics. A similar derivation gives an analogous equation for other kinds of compounds. In arriving at (17.92), the contributions of vibrational energy were ignored. It is assumed that these contributions are allowed for when a_{CC} and a_{CH} are fit to 25°C ΔH_f° data. If more than one conformation is significantly populated at 25°C, one uses $\Delta H_f^\circ = \sum_i x_i \Delta H_{f,i}^\circ$, where x_i is the mole fraction of conformation i as calculated using the enthalpy and entropy differences between conformations, and where $\Delta H_{f,i}^\circ$ is calculated from (17.92) for each conformation. In practice, the accuracy of (17.92) is improved by including several correction terms; see Prob. 17.42.

Performance. How well does molecular mechanics perform? When a well-parametrized force field is applied to compounds similar to those used in the parametrization, one can get extremely good results. For monofunctional organic compounds, MM3 typically gives gas-phase ΔH_f° errors that run 0 to 1 kcal/mol (which is the same magnitude as the experimental errors in such data), gives bond lengths within 0.01 Å and bond angles within 2° of experimental values, and usually correctly predicts the most stable conformer.

However, one cannot expect such good performance when molecular mechanics is applied to broad classes of compounds. For a force field with 60 atom types, the number of possible types of IJKL dihedral angles is roughly $\frac{1}{2}(60)(45)(45)(60) \approx 3 \times 10^6$. The factor 45 is used instead of 60 because some atom types (for example, halogens, all H types, carbonyl oxygen) can only occur as end atoms of a molecular dihedral angle. For each of these 3 million types of dihedral angle, one needs V_1, V_2, and V_3 values. No available force field will have parameters for all the dihedral angles that might be encountered, so one is frequently faced with the problem of missing parameters. Most molecular-mechanics programs will use some procedure to estimate a value for each missing parameter. If torsion parameters for the IJKL dihedral angle are missing, the program might use torsion parameters for a dihedral angle of type IJKM, with one atom type replaced by a similar atom type (for example, hydrogen bonded to O might be replaced with hydrogen bonded to C). If this fails to produce the torsion parameters, the program might use a value that is typical for torsion about the JK bond, ignoring what types of atoms are bonded to J and K. When such estimated parameters are used in a molecular-mechanics calculation, the reliability of the result may be greatly diminished.

Currently available force fields for proteins contain significant deficiencies (see Section 18.5).

For further comparisons of results, see Chapter 18.

Programs. The Spartan program (Section 15.14) can do MM energy minimizations with the MMFF and Tripos force fields and does conformational searching. HyperChem (Section 15.14) has in it the force fields MM+ (an extension of MM2), AMBER, BIO+ (a version of CHARMM), and OPLS and can do energy minimization, molecular dynamics, and Monte Carlo calculations. Scigress Explorer (www.fujitsu.com/us/services/solutions/lifesci/) is a Windows molecular modeling program that includes molecular mechanics, can treat compounds of all elements, and has extended MM2 and MM3 force fields. PCModel (www.serenasoft.com) is a Windows and Macintosh program with the force fields MM3, MMFF94, Amber, and OPLS-AA. MacroModel [F. Mohamadi et al., *J. Comput. Chem.*, **11**, 440 (1990); www.schrodinger.com] is a workstation program with the force fields MM2, MM3, AMBER, OPLS-AA, AMBER94, and MMFF. Cerius2 (accelrys.com/products/cerius2/) is a workstation program with the DREIDING, UFF, and CFF95 force fields. SYBYL (www.tripos.com) is a suite of workstation molecular modeling programs with the Tripos, AMBER, MM2, and MMFF94 force fields. TINKER is a free molecular mechanics and molecular dynamics program for Windows, Macintosh, and Linux with the AMBER, CHARMM, and other force fields (dasher.wustl.edu/tinker/) and has a free graphical interface called Force Field Explorer (dasher.wustl.edu/ffe). CHARMM (yuri.harvard.edu/) is a workstation program with the CHARMM and MMFF94 force fields. ChemBio3D Ultra (www.camsoft.com/) is a Windows molecular-modeling program with MM2; a free trial version that lasts for two weeks is available (scistore.cambridgesoft.com/). ArgusLab (www.arguslab.com/) is a free program that includes the UFF field. BALLView (www.ballview.org) is a free molecular modeling Windows and Macintosh program with the AMBER and CHARMM force fields [A. Moll et al., *J. Comput.-Aided Mol. Des.*, **19**, 791 (2005); *Bioinformatics*, **22**, 365 (2006)]. *Gaussian 03* can do MM calculations with the AMBER, DREIDING, and UFF force fields.

References. The introduction of graphical user interfaces has led to widespread use and abuse of molecular-mechanics programs. For discussion of some errors people make in applying molecular mechanics, see K. Lipkowitz, *J. Chem. Educ.*, **72**, 1070 (1995). Some references on molecular mechanics are *Leach*, Chapter 4; *Cramer*, Chapter 2; *Burkert and Allinger*; A. K. Rappé and C. J. Casewit, *Molecular Mechanics Across Chemistry*, University Science Books, 1997. For applications of molecular mechanics to organometallic, bioinorganic and inorganic compounds, see C. R. Landis et al., in K. B. Lipkowitz and D. B. Boyd (eds.), *Reviews in Computational Chemistry*, vol. 6, VCH, 1995, Chapter 2; M. Zimmer, *Chem. Rev.*, **95**, 2629 (1995).

QM/MM Methods. Molecular mechanics can treat very large molecules, but is not well suited to treat chemical reactions. Quantum-mechanical methods (ab initio, density functional, and semiempirical) can treat chemical reactions, but are not well suited to treat very large molecules. To deal with chemical reactions in very large systems, one can use a combined quantum-mechanical/molecular-mechanics (QM/MM) approach, using quantum mechanics to treat the part of the system most affected by the reaction, and molecular mechanics to treat the rest. For a reaction in solution, one treats the reacting solute molecules using QM and the surrounding solvent molecules using MM. For an enzyme-catalyzed reaction, one treats the enzyme's active site and part or all of the substrate molecule(s) using QM and the rest of the enzyme and surrounding water molecules using MM. For enzyme-catalyzed reactions, one must deal with covalent bonds that join QM and MM regions. Several methods have been used to handle this problem. Although any quantum-mechanical method can be used for the QM region, as a practical matter semiempirical methods such as AM1, PM3, and SCC-DFTB are more often used. For enzyme reactions, Warshel and co-workers use an empirical valence-bond method for the QM region [J. Aqvist and A. Warshel, *Chem. Rev.*, **93**, 2523 (1993)].

The effective Hamiltonian for the QM/MM system is written as

$$\hat{H}_{\text{eff}} = \hat{H}_{\text{QM}} + \hat{H}_{\text{QM/MM}} + \hat{H}_{\text{MM}} \tag{17.93}$$

where $\hat{H}_{\text{QM}}$ is the Hamiltonian of the quantum-mechanical region in a vacuum, $\hat{H}_{\text{QM/MM}}$ is the Hamiltonian for the interaction between the QM and MM regions, and $\hat{H}_{\text{MM}}$ is taken as equal to the steric energy V_{MM} [Eq. (17.85)] of the MM region. If ψ_{QM} is the wave function of the QM region, the energy E of the system is taken as

$$E = \langle \psi_{\text{QM}} | \hat{H}_{\text{QM}} + \hat{H}_{\text{QM/MM}} | \psi_{\text{QM}} \rangle + V_{\text{MM}}$$

$\hat{H}_{\text{QM/MM}}$ is often approximated as the electrostatic interactions between the charges (electrons and nuclei) of the QM region and the partial atomic charges on the atoms in the MM region plus the van der Waals interactions between the atoms of the QM region and the atoms of the MM region:

$$\hat{H}_{\text{QM/MM}} = -\sum_i \sum_M \frac{Q_M}{r_{i,M}} + \sum_\alpha \sum_M \frac{Z_\alpha Q_M}{R_{\alpha M}} + \sum_\alpha \sum_M \varepsilon_{\alpha M} \left[\left(\frac{R^*_{\alpha M}}{R_{\alpha M}} \right)^{12} - 2 \left(\frac{R^*_{\alpha M}}{R_{\alpha M}} \right)^{6} \right]$$

where i and α denote respectively, the electrons and the nuclei in the QM region and M denotes an atom with partial atomic charge Q_M in the MM region. This approximation neglects polarization of solvent molecules by charges in the QM region. For a fixed configuration of nuclei (Born–Oppenheimer approximation), all terms in $\hat{H}_{\text{QM/MM}}$ are constants except the first double sum. In solving the electronic Schrödinger equation using the Hamiltonian $\hat{H}_{\text{eff}}$, the first double sum in $\hat{H}_{\text{QM/MM}}$ has the effect of extra nuclei with charges Q_{M}. This produces extra one-electron integrals in the Fock matrix. ψ_{QM} is a function of the electronic coordinates of the QM region and both ψ_{QM} and E are parametric functions of the nuclear coordinates of the QM region and the coordinates of the atoms in the MM region. Taking minus the partial derivatives of E with respect to the nuclear coordinates, we get the forces needed for a molecular dynamics simulation.

The *pseudobond* QM/MM method [Y. Zhang, T.-S. Lee, and W. Yang, *J. Chem. Phys.*, **110**, 46 (1999)] divides a large molecule into two parts by cutting a covalent bond X—Y between atoms X and Y and does the QM part of the calculation using an ab initio method. Let atom X be in the part of the molecule to be treated by QM, and suppose that Y is a carbon atom. The free valence created by cutting the bond is "capped" by singly bonding atom X to a one-free-valence pseudocarbon atom C_{ps} that has seven valence electrons (so as to be one electron short of an octet), a nuclear charge of seven, and no core electrons. The effect of the core electrons is handled by using an effective core potential (Section 15.16). The pseudobond $X-C_{ps}$ mimics the properties of the X—Y bond. Tests with the QM calculation done with the HF and DFT methods gave good results.

For reviews of the QM/MM method, see J. Gao in K. Lipkowitz and D. B. Boyd (eds.), *Reviews in Computational Chemistry*, vol. 7, VCH, 1996, Chapter 3; H. Lin and D. G. Truhlar, *Theor. Chem. Acc.*, **117**, 185 (2007).

17.6 EMPIRICAL AND SEMIEMPIRICAL TREATMENTS OF SOLVENT EFFECTS

Explicit-Solvent versus Continuum-Solvent Methods. Theoretical treatments of solvation can be categorized as either explicit-solvent approaches, in which many individual solvent molecules are explicitly included, or as continuum-solvent methods, where the solvent molecules are replaced by a continuous dielectric (Section 15.17).

In an explicit-solvent treatment, one applies the molecular dynamics or the Metropolis Monte Carlo method (Sections 15.11 and 17.5) to a system of a solute molecule (or molecules if a chemical reaction is being studied) surrounded by hundreds or thousands of solvent molecules, and by suitable averaging, one obtains thermodynamic or kinetic properties. The solvent molecules are treated using an empirical force field. Some of the models used to represent water molecules are discussed in *Leach*, Section 4.14. The solute molecule(s) can be modeled using molecular mechanics or semiempirical quantum mechanics (the QM/MM method of Section 17.5).

Continuum-solvent methods can be categorized as either classical or quantum-mechanical.

Classical Continuum-Solvent Methods. The simplest and crudest continuum-solvent method is the **solvent-accessible-surface area** (SASA) model, which assumes that the standard free-energy of solvation $\Delta G^\circ_{\text{solv}}$ can be expressed as

$$\Delta G^\circ_{\text{solv}} = \sum_i \sigma_i A_i \tag{17.94}$$

Recall (Section 15.7) that the van der Waals surface of a molecule is the outer surface formed by intersecting atomic spheres with van der Waals radii. The **solvent-accessible surface** is the surface traced out by the center of a spherical solute molecule as its rolls on the molecular van der Waals surface. For H_2O, a sphere of radius 1.4 Å is traditionally used in (17.94). A_i is the **solvent-accessible surface area** of atom i or group i, depending on whether an atom-based or group-based approach is used. The sum goes over all atoms (or groups) in the molecule. A_i is that portion of the surface area of a sphere centered at atom i and having radius $r_i + r_{\text{solvent}}$ (where these quantities are the van der Waals radius of atom i and the solvent molecular radius) that is not contained within any of the corresponding spheres centered at the other atoms in the molecule.

The σ_i's are empirical parameters called **atomic surface tensions** or **atomic solvation parameters.** They have the same dimensions as surface tension but are not macroscopic surface tensions. The σ_i's are found by fitting known $\Delta G^\circ_{\text{solv}}$ data for the solvent under study. The SASA method is especially useful for giving rapid estimates of $\Delta G^\circ_{\text{solv}}$ of biopolymers.

A related but more elaborate method than the SASA model is the SM5.0R method [G. D. Hawkins, C. J. Cramer, and D. G. Truhlar, *J. Phys. Chem. B*, **101**, 7147 (1997); Hawkins et al., *J. Org. Chem.* **63**, 4305 (1998)]. Here, (17.94) is used, but A_i is taken as the exposed van der Waals surface area of atom i (rather than the SASA of i) and σ_i, instead of being a constant for a given atom in a given solvent, has the form $\sigma_i = \sigma_i^{(0)} + \sum_{j \neq i} \sigma_{ij} f(R_{ij})$. Here, the sum goes over all atoms in the solute molecule except atom i, $\sigma_i^{(0)}$ and σ_{ij} are parameters whose values depend on the nature of atoms i and j, and $f(R_{ij})$ is a certain function of the distance R_{ij} between atoms i and j. The function f contains parameters and is defined to be zero for all distances greater than a certain cutoff distance. Inclusion of the sum allows σ_i to be affected by the environment of atom i in the molecule. The method contains about 40 parameters for water as the solvent and has been parametrized for compounds of H, C, N, O, S, F, Cl, Br, and I in water and in organic solvents.

Rather than develop a separate set of parameter values for each organic solvent, SM5.0R takes each surface tension coefficient $\sigma_i^{(0)}$ and σ_{ij} to depend on certain properties of the solvent such as its index of refraction, surface tension, and fraction of nonhydrogen atoms in the solvent molecule that are aromatic carbons, and to depend on parameters many of whose values depend on the nature of atom i or atoms i and j (whether they are H, C, N . . .) but whose values are the same for every organic solvent. The values of the parameters are fitted using a training set of 1836 $\Delta G^\circ_{\text{solv}}$ values for 227 uncharged solutes in 90 organic solvents.

The molecular geometry used to calculate R_{ij} is that of the gas-phase solute molecule. Test calculations on small uncharged solutes in water using AM1 geometries gave excellent results, but for charged solutes the performance was not so good. The SM5.0R method is useful for rapid estimations for large molecules. [The SM5 stands for solvation model version 5; the R indicates rigid (that is, gas-phase rather than solution-phase) geometries are used; the 0 indicates no use is made of atomic charges.]

Another classical method to estimate $\Delta G^\circ_{\text{solv}}$ is the **generalized Born/surface area** (GB/SA) method of Still and co-workers [W. C. Still et al., *J. Am. Chem. Soc.*, **112**, 6127 (1990); D. Qiu et al., *J. Phys. Chem. A*, **101**, 3005 (1997)]. In 1920, Born showed that when a charge Q distributed on the surface of a conducting sphere of radius a is transferred from vacuum to a dielectric medium with dielectric constant ε_r, the free-energy change is $\Delta G = -\frac{1}{2}(1 - 1/\varepsilon_r)Q^2/a$. Now imagine a system of several charged spheres having charges Q_i and radii a_i with R_{ij} being the distance between spheres i and j. If such a system is transferred from vacuum to a medium of dielectric constant ε_r, the free-energy change can be approximated by adding the Born expression $-\sum_i \frac{1}{2}(1 - 1/\varepsilon_r)Q_i^2/a_i$ to the intercharge electrostatic potential-energy change $\sum_{j>i}\sum_i Q_iQ_j/\varepsilon_r R_{ij} - \sum_{j>i}\sum_i Q_iQ_j/R_{ij}$ to get

$$\Delta G \approx -\left(1 - \frac{1}{\varepsilon_r}\right)\sum_{j>i}\sum_i \frac{Q_iQ_j}{R_{ij}} - \frac{1}{2}\left(1 - \frac{1}{\varepsilon_r}\right)\sum_i \frac{Q_i^2}{a_i} \qquad (17.95)$$

an equation sometimes called the *generalized Born equation*. If we model a solute molecule as a set of overlapping spheres with charges Q_i (where Q_i is a partial atomic charge) and radii a_i, then the generalized Born equation gives an approximation to the electrostatic contribution $\Delta G^\circ_{\text{solv,el}}$ (Section 15.17) to the solvation free energy. However, this method would overestimate the interactions between the solvent and charges buried within the solute molecule's interior. Instead, Still and co-workers proposed replacing (17.95) with

$$\Delta G^\circ_{\text{solv,el}} = -\frac{1}{2}\left(1 - \frac{1}{\varepsilon_r}\right)\sum_i \sum_j \frac{Q_i Q_j}{[R_{ij}^2 + a_i a_j e^{-R_{ij}^2/4a_i a_j}]^{1/2}} \tag{17.96}$$

where the sums go over all atoms in the molecule. For two charges, this expression reduces to the Born equation when $R_{ij} = 0$, gives a result close to the Onsager result for a dipole in a sphere (Section 15.17) when $R_{ij} \ll a_i a_j$, and gives a result close to the sum of the Born plus intercharge potential energy when $R_{ij} \gg a_i a_j$. The partial atomic charges Q_i to be used in (17.96) can be found by any of the usual methods used in molecular mechanics, but the best results are obtained using either charges fit to molecular electrostatic potentials or OPLS charges, which were derived from liquid simulations. The Born atomic radii a_i are calculated by a complicated procedure [see *Leach*, Chapter 11; D. Qiu et al., *J. Phys. Chem. A*, **101**, 3005 (1997)]. The term *Born radius* for a_i is misleading, in that a_i in (17.96) is some sort of average distance from the atomic charge Q_i to the dielectric continuum that surrounds the solute molecule.

Still and co-workers modeled the cavitation and van der Waals contributions to $\Delta G^\circ_{\text{solv}}$ by

$$\Delta G^\circ_{\text{cav}} + \Delta G^\circ_{\text{vdw}} = \sum_k \sigma_k A_k \tag{17.97}$$

where the sum goes over all atoms in the solute molecule, A_k is the solvent-accessible surface area for atom k, and the parameter σ_k is taken as 10 cal/(mol Å^2) for S and sp^3 C atoms, 7 cal/(mol Å^2) for P and sp^2 and sp C atoms, and 0 for O and N atoms. In (17.97), a united-atom approach (Section 17.5) is used for hydrogens, so there is no σ_k for H.

Addition of (17.96) and (17.97) gives the GB/SA expression for $\Delta G^\circ_{\text{solv}}$. The GB/SA expression is readily differentiated analytically, making it easy to use in molecular-mechanics energy minimizations (one finds the geometry that minimizes the sum of the MM steric energy and $\Delta G^\circ_{\text{solv}}$), and molecular-dynamics and Metropolis Monte Carlo simulations with inclusion of solvent effects. The GB/SA method is available in the MacroModel program (Section 17.5).

Quantum-Mechanical Continuum Solvation Models. Several ab initio continuum solvation models were discussed in Section 15.17. One can calculate $\Delta G^\circ_{\text{solv,el}}$ by such SCRF methods as the dipole-in-a-sphere, the multipole expansion, or the PCM methods using semiempirical methods such as AM1 or PM3 instead of an ab initio electronic-structure method. Thus the program MOPAC-93 implements the PCM calculation of solvent effects with semiempirical methods and the program AMPAC 6.0 implements the COSMO method.

Cramer, Truhlar, and co-workers developed a series of semiempirical continuum solvation models SMx, where SM stands for solvation model and $x = 1, 2, 3, 4, 5, 6, 8$ is the version number. The models SM1, SM2, SM3, and SM4 are obsolete. The SMx methods are quantum-mechanical versions of Still's GB/SA method. The earlier versions of SM5 (SM5.2 and SM5.4) were parametrized for use with the AM1 and PM3 semiempirical methods, but later versions of SM5 (SM5.42 and SM5.43) were also parametrized for use with the DFT and HF methods; SM6 is parametrized for use only with DFT calculations. In SMx, the solvation energy is written as

$$\Delta G^\circ_{\text{solv}} = \Delta E_{\text{EN}} + G_{\text{P}} + G^\circ_{\text{CDS}} \tag{17.98}$$

In this equation, the electronic and nuclear term ΔE_{EN} is the change in the electronic energy including nuclear repulsion of a solute molecule when it goes from the gas phase into solution; ΔE_{EN} corresponds to $\langle\psi^{(f)}|\hat{H}^0|\psi^{(f)}\rangle - \langle\psi^{(0)}|\hat{H}^0|\psi^{(0)}\rangle$ in Eq. (15.84). The polarization contribution G_{P} is the free-energy change due to electrostatic interactions between solute and solvent, including polarization of the solvent. G_P corresponds to $\langle\psi^{(f)}|V^{\text{int}}|\psi^{(f)}\rangle + E_{\text{pol}}$ in Section 15.17. The sum $\Delta E_{\text{EN}} + G_{\text{P}}$ equals $\Delta G^\circ_{\text{solv,el}}$ of (15.84). The cavitation–dispersion–solvent-disposition term G°_{CDS} is the free-energy change due to creation of the cavity around the solute molecule, dispersion interactions between solute and solvent molecules, and structural changes in the solvent, due to such things as solute–solvent hydrogen bonding. Other contributions are small and are neglected.

The SM5 series of models [G. D. Hawkins, C. J. Cramer, and D. G. Truhlar, *J. Phys. Chem.*, **100**, 19824 (1996) and references cited therein] has been parametrized for compounds containing H, C, N, O, F, S, Cl, Br, and I. In SM5, the polarization contribution G_{P} is taken as equal to the Still expression (17.96), except that the 4 in the exponential is replaced by 4.2 when the two atoms i and j are C and H. The Born radii a_i are calculated using Still's procedure or a faster procedure called pairwise descreening (PD) that is an approximation to Still's procedure [Hawkins et al., *J. Phys. Chem.*, **100**, 19824 (1996)].

In the version SM5.2PD, the atomic partial charges are taken as the Mulliken-population-analysis net charges (Section 15.6), which are given by $C_{\text{B}} - \Sigma_r P_{rr}$, where C_{B} is the charge on the core of atom B, the sum goes over all valence AOs centered on B, and P_{rr} is a density matrix element [Eq. (14.42)]. The SM5 model does an AM1 or PM3 or DFT or ab initio HF SCF calculation to minimize the sum of the molecular electronic energy including nuclear repulsion and G_{P}. Because G_{P} is included in the quantity to be minimized, one must modify the Fock matrix elements (14.41) by adding a term that is closely related to G_{P} and that contains the atomic partial charges Q_i that occur in G_{P} in (17.96). To do the SCF calculation, one starts with a guess for the density matrix elements P_{rs}, uses these to calculate the initial estimates of the atomic charges Q_i, and uses the initial Q_i's and P_{rs}'s to calculate the initial estimates of the Fock matrix elements. Solution of the Fock–Roothaan equations then allows improved density matrix elements to be found, which are used to start another cycle of the calculation. The calculation ends when self-consistency is reached.

The versions SM5.4 and SM5.4PD use improved charges (called class IV charges) instead of the Mulliken charges (called class II charges) used in SM5.2PD. Class IV charges are found as follows. For a training set of molecules with known dipole moments, one calculates the Mulliken charges Q_i using the AM1 or PM3 method and writes $Q_{i,\text{IV}} = g(Q_i)$, where g is a certain function that contains parameters and that maps the Mulliken charges into class-IV charges. One varies the parameters in g so that the molecular dipole moments calculated from the class-IV charges give a good fit to the known dipole moments. Once the parameters in g have been determined, one can use g to calculate class-IV charges from the Mulliken charges for any molecule. This particular method of getting class-IV charges is called CM1 (charge model 1).

The version SM5.42R [T. Zhu et al., *J. Chem. Phys.*, **109**, 9117 (1998)] uses the improved method CM2 of generating the class-IV charges (hence the 2 in 5.42). The version SM5.43R [J. D. Thompson et al., *J. Phys. Chem. A*, **108**, 6532 (2004)] uses the further-improved CM3 method to generate the class-IV charges. The version SM6 [C. P. Kelly et al., *J. Chem. Theory Comput.* **1**, 1133 (2005)] improves on SM5.43R by using charges that are based on the CM4 method and by using a larger and more varied data set for the parametrization.

The SM5 and SM6 models calculate G°_{CDS} using the SASA approach:

$$G^\circ_{\text{CDS}} = \sum_k \sigma_k A_k$$

where A_k is the solvent-accessible surface area of atom k, and σ_k has a form similar to that of σ_i in the SM5.0R model discussed near the beginning of this subsection.

The SM5 and SM6 models are available in the programs AMSOL and SMXGAUSS (comp.chem.umn.edu/solvation/). The program Spartan has the SM5.4 and SM5.0R models for aqueous solutions.

Performance of the SM5 and SM6 models is discussed in Section 18.3. For SM8, see A. V. Marenich et al., *J. Chem. Theory Comput.*, **3**, 2011 (2007).

COSMO-RS (COSMO for real solvents) is an extension of COSMO (Section 15.17) beyond the dielectric-continuum approximation [A. Klamt et al., *J. Phys. Chem. A*, **102**, 5074 (1998)]. COSMO-RS does not treat the solvent as a continuous dielectric, but treats solute and solvent molecules on equal footing and uses cavities and surface charges for both the solute and solvent molecules. COSMO-RS has eight general parameters plus two parameters for each different element, and does not use solvent-specific parameters. Thus for solutions containing compounds of H, C, N, O, and Cl, COSMO-RS has 18 parameters, whose values were chosen by fitting to experimental data.

17.7 CHEMICAL REACTIONS

Because of the difficulties involved in ab initio calculations of potential-energy surfaces for reactions, it would be highly desirable to have a semiempirical method that gives reliable reaction PES results. The MNDO, AM1, and PM3 methods have been widely applied to calculate relevant portions of PESs for chemical reactions (often with inclusion of solvent effects) so as to elucidate reaction mechanisms and transition-state structures.

An analysis of MNDO results for 24 simple organic reactions found that MNDO usually gave a fairly realistic transition-state structure [S. Schröder and W. Thiel, *J. Am. Chem. Soc.*, **107**, 4422 (1985)], with average absolute deviations from ab initio SCF MO results of 0.06 Å in bond lengths, 8° in bond angles, and $11\frac{1}{2}°$ in dihedral angles. However, the calculated MNDO barrier heights were not accurate, with an average absolute deviation of 22 kcal/mol from ab initio MP4 or MP3 results. MNDO was suggested as being suitable "for fast initial scans of potential surfaces to assess their qualitative features, but it has to be kept in mind that some of these characteristics may change at higher levels" (Schröder and Thiel, op. cit.).

Some shortcomings of AM1 for studying reactions are discussed by O. N. Ventura in S. Fraga (ed.), *Computational Chemistry*, Part B, Elsevier, 1992, pp. 605–607, where it is concluded that "there is no semiempirical method at present which can be used reliably in all situations [to study reactions]."

A method that improves the accuracy of a semiempirical PES for a particular reaction is NDDO-SRP of Truhlar and co-workers (NDDO with specific reaction parameters; recall that MNDO, AM1, and PM3 are all NDDO methods). Here, some of the parameters in the AM1 method are adjusted either to reproduce such experimental data as the reaction's energy change, barrier height, and rate constant or to reproduce a small number of points on the PES calculated with an ab initio method that includes correlation. See the references cited in W. Thiel, *Adv. Chem. Phys.*, **93**, 731 (1996).

Although the force fields mentioned in Section 17.5 do not apply to bond-breaking reactions, they can be applied to reactions involving only a change in molecular conformation. Some attempts have been made to develop force fields that will describe bond breaking; see F. Jensen and P.-O. Norrby, *Theor. Chem. Acc.*, **109**, 1 (2003); *Cramer*, Section 2.4.2.

MO theory has been used to draw qualitative conclusions about the course of chemical reactions. The most fruitful applications have come from the **Woodward–Hoffmann rules**, which predict the preferred path and stereochemistry for many important classes of organic reactions. As an example of the application of these rules, we consider the cyclization of a substituted *s-cis*-butadiene to a substituted cyclobutene. There are two possible steric courses the reaction can take, described as *conrotatory* or *disrotatory*, depending on whether the terminal groups rotate in the same or opposite senses as the reaction proceeds. Note the difference in products in Fig. 17.12.

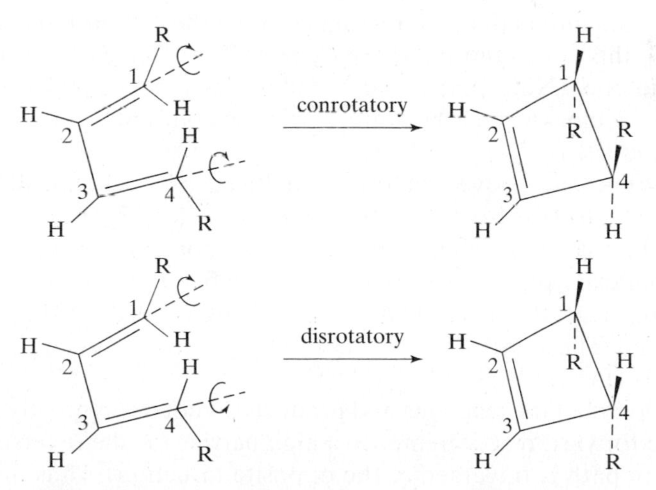

FIGURE 17.12 Conrotatory and disrotatory cyclizations. All atoms lie in the same plane except those shown with dashed or heavy bonds.

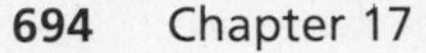

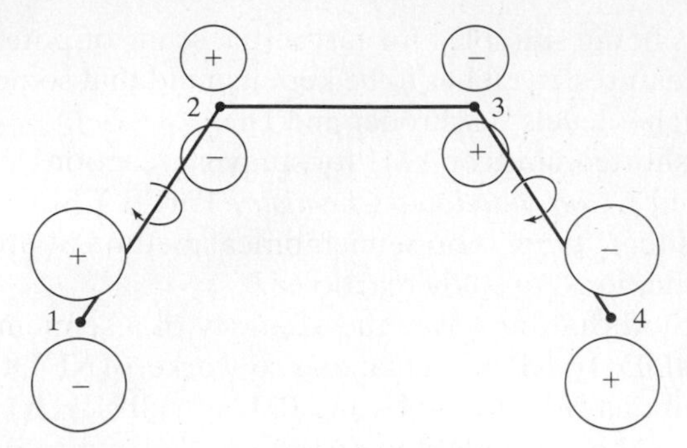

FIGURE 17.13 Relative phases of AOs in the highest occupied butadiene MO.

One finds that when the reaction is thermally induced the butadiene cyclization process is conrotatory, but when the reaction is photochemically induced, the process is disrotatory. The simplest approach that explains these facts starts with the assumption that the energy change during the reaction is determined primarily by the energy change in the highest-occupied MO (HOMO) of the substituted *s-cis*-butadiene molecule. This is the π MO ϕ_2 in Fig. 17.1. Figure 17.13 shows that for ϕ_2, a conrotatory motion of the π AOs on carbons 1 and 4 causes the positive lobe of $C_1 2p\pi$ to overlap the positive lobe of $C_4 2p\pi$; this overlap gives a bonding interaction of these AOs and leads to formation of the 1–4 σ bond of the cyclobutene. On the other hand, a disrotatory motion of these two AOs causes overlap of the positive lobe of one π AO with the negative lobe of the other π AO; this gives an antibonding interaction. Thus we expect the conrotatory process to have a lower activation energy than the disrotatory and to be preferred over the disrotatory. As a further check, we note that conrotatory motion leads to an antibonding interaction between the carbon 1 and 4 π AOs in the lowest π MO ϕ_1; this leaves the π AOs of carbons 2 and 3 in ϕ_1 to form the 2–3 π bond in the cyclobutene. (Note that if the butadiene is less symmetrically substituted than in Fig. 17.12, clockwise and counterclockwise conrotatory motions lead to a mixture of two products.)

When the above reaction is induced photochemically, an absorbed photon excites an electron from ϕ_2 to the butadiene π MO ϕ_3 (Fig. 17.1). The highest occupied MO is now ϕ_3, in which the $2p\pi$ AOs on carbons 1 and 4 have the same phases (rather than opposite phases as in ϕ_2) and in which a disrotatory motion produces positive bonding overlap between these AOs. Thus we predict disrotatory ring closure, as observed. The same reasoning correctly predicts the courses of other polyene ring closures (Prob. 17.43).

Provided the reactants and products do not differ greatly in energy, a high barrier for the forward reaction implies a high barrier for the reverse reaction (in which the reaction path is traversed in the opposite direction). Thus the above reasoning also applies in determining the course of the reverse, ring-opening, reactions.

Rather than considering only the energy change in the HOMO of butadiene, one can use a less approximate approach that looks at the energy changes in all MOs (occupied or unoccupied) that are involved in the bonds being broken or formed and that uses symmetry to correlate the MOs of the reactant(s) with the MOs of the product(s); see *Lowe and Peterson*, Section 14-9.

For further details of the Woodward–Hoffmann rules and their application to a wide variety of organic reactions, see R. B. Woodward and R. Hoffmann, *Angew. Chem. Intern. Ed.*, **8**, 781 (1969); *The Conservation of Orbital Symmetry*, Academic Press, 1970; R. E. Lehr and A. P. Marchand, *Orbital Symmetry*, Academic Press, 1972.

Pearson has applied orbital symmetry concepts to inorganic reactions. As two reactant molecules approach, electrons begin to flow from the HOMO of one to the lowest unoccupied MO (LUMO) of the other. These two MOs are called **frontier orbitals**. For a low activation energy, we require a positive overlap between these two MOs.

An example is the $H_2 + F_2 \rightarrow 2HF$ reaction. Consider a proposed mechanism in which the two molecules collide broadside to give a four-center transition state. The HOMO of H_2 is the $\sigma_g 1s$ MO; the LUMO of F_2 is the $\sigma_u^* 2p$ MO (Sections 13.6 and 13.7). Flow of electrons out of H_2 $\sigma_g 1s$ toward F_2 $\sigma_u^* 2p$ would lead to breaking of the H—H bond and formation of two H—F bonds. However, Fig. 17.14a shows that these two MOs do not have a positive overlap. Hence electron flow from H_2 to F_2 is forbidden by symmetry. Figure 17.14b shows the HOMO of F_2 and LUMO of H_2. Here there is a positive overlap, but flow of electrons out of the antibonding π_g^* MO would strengthen rather than weaken the F—F bond. We conclude that this bimolecular one-step mechanism has a high activation energy and is not favored. (The same reasoning applies to the famous $H_2 + I_2$ reaction.)

Further applications of orbital symmetry to inorganic reactions may be found in R. G. Pearson, *Chem. Eng. News*, Sept. 28, 1970, page 66; *Acc. Chem. Res.*, **4**, 152 (1971); *J. Am. Chem. Soc.*, **94**, 8287 (1972); *Symmetry Rules for Chemical Reactions*, Wiley, 1976.

The frontier-orbital approach sometimes fails. For a criticism of the frontier-orbital theory, see M. J. S. Dewar, *THEOCHEM*, **59**, 301 (1989).

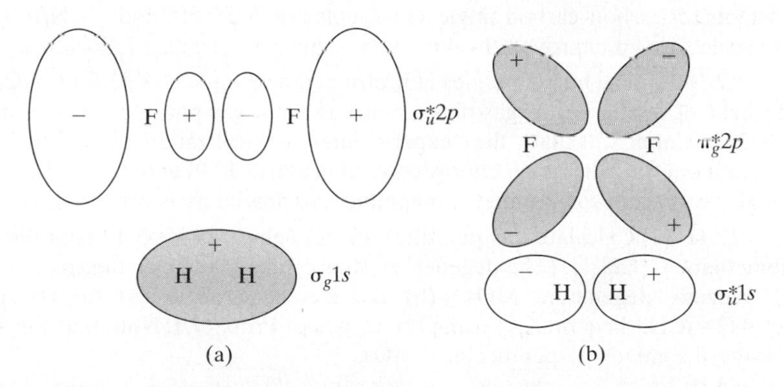

FIGURE 17.14 HOMOs and LUMOs for H_2 and F_2. Occupied MOs are shaded.

PROBLEMS

Sec.	17.2	17.3	17.4	17.5	17.7
Probs.	17.1–17.23	17.24	17.25–17.34	17.35–17.42	17.43–17.44

17.1 For the allyl radical $\cdot CH_2{-}CH{=}CH_2$, find (a) the HMOs and energies; (b) the mobile bond orders; (c) the π-electron charges; (d) the free valences (Prob. 17.19); (e) the delocalization energy.

17.2 Calculate the quantities (a) through (e) of Prob. 17.1 for the allyl cation and anion, $[CH_2CHCH_2]^+$ and $[CH_2CHCH_2]^-$. Which ion is predicted to be more stable?

17.3 For the polyenes (17.28), observed longest-wavelength electronic absorption bands are at 162.5, 217, 268, 304, 334, 364, 390, 410, and 447 nm for $s = 0, 1, 2, 3, 4, 5, 6, 7$, and 9, respectively. Compare these values with those given by the Hückel equation (17.31) and calculate the average absolute percent error.

17.4 Verify the butadiene HMO coefficients for ϕ_2, ϕ_3, and ϕ_4 in (17.26).

17.5 (a) Verify that the HMO coefficients (17.30) satisfy the HMO set of simultaneous equations. [*Hint*: Use the identity $\sin a + \sin b = 2\sin\frac{1}{2}(a+b)\cos\frac{1}{2}(a-b)$.] (b) Verify that the coefficients (17.30) give a normalized HMO. To evaluate the needed sum, express the sine function as exponentials and then use the formula for the sum of a geometric series.

17.6 Calculate the HMO total bond orders and the π-electron charges for the lowest excited state of butadiene.

17.7 Since only the topology of the carbon framework is of significance in the HMO method, the full symmetry of *s-trans*-butadiene need not be used to get the maximum simplification possible in the HMO method; instead, it is sufficient to use only the C_2 axis. (a) Write down the two possible symmetry species for the group $\mathscr{C}_2$. (b) Construct π-electron symmetry orbitals for butadiene, classifying them according to the symmetry species of $\mathscr{C}_2$. (c) Set up and solve the two Hückel secular equations for butadiene using the symmetry orbitals of (b) as basis functions.

17.8 Verify that the geometric construction of Fig. 17.5 gives the correct HMO energies of the cyclic polyene C_nH_n.

17.9 (a) Calculate the energy needed to compress three carbon–carbon single bonds and stretch three carbon–carbon double bonds to the benzene bond length 1.397 Å. Assume a harmonic-oscillator potential-energy function for bond stretching and compression. Typical carbon–carbon single- and double-bond lengths are 1.53 and 1.335 Å; typical stretching force constants for carbon–carbon single and double bonds are 500 and 950 N/m. (b) Use the result of (a) to calculate an improved β value for benzene from the data following Eq. (17.51).

17.10 Typical bond energies in kcal/mol are 99 for C—H, 83 for C—C, and 146 for C=C. The heat of formation of gas-phase benzene from gas-phase hydrogen and carbon atoms is −1323 kcal/mol. Calculate the "experimental" delocalization energy of benzene using these data, first omitting the strain-energy correction (Prob. 17.9) and then including it. (Because bond energies vary from compound to compound, this procedure is very rough.)

17.11 (a) Calculate the quantities (a) through (e) of Prob. 17.1 for the diradical trimethylenemethane $C(CH_2)_3$. (The degenerate MOs can be taken as either real or complex, similar to the benzene degenerate MOs.) (b) Do the same as in (a) for the propargyl diradical HC=C=CH. Save time by using the results of Prob. 17.1. Note that this linear molecule has two sets of spatially perpendicular π MOs.

17.12 The ionization energies of the first few polyacenes are 9.4 eV for benzene, 8.3 eV for naphthalene, 7.6 eV for anthracene, and 7.0 eV for tetracene. Use these data to calculate values of α and β in the HMO method. Compare the result with the β value in (17.49). Predict the ionization energy of pentacene. The HMO x values for the highest-occupied MOs of these polyacenes are -1.00, -0.618, -0.414, -0.295, and -0.220.

17.13 The terms involving α always cancel in calculations of the delocalization energy, so Hess and Schaad measured energies relative to α In this problem, α is therefore omitted. The Hess–Schaad $E_{\pi,b}$ values for the various conjugated-bond types are 2.0000β for $CH_2{=}CH$, 2.0699β for $CH{=}CH$, 2.0000β for $CH_2{=}C$, 2.1083β for $CH{=}C$, 2.1716β for $C{=}C$, 0.4660β for $CH{-}CH$, 0.4362β for $CH{-}C$, 0.4358β for $C{-}C$. (a) Verify that for 1,3-butadiene, $\sum_b n_b E_{\pi,b} = 4.466\beta$, which is very close to the Hückel E_π value 4.472β in (17.50). The Hess–Schaad method makes the delocalization energies of noncyclic conjugated polyenes essentially zero. (b) Verify that for benzene, $\sum_b n_b E_{\pi,b} = 7.6077\beta$ and that the Hess–Schaad REPE for benzene is $0.065|\beta|$. (c) Find the Hess–Schaad resonance energy and REPE for each of the following and predict whether the compound will be aromatic, nonaromatic, or antiaromatic: cyclobutadiene; planar [8]annulene; planar [18]annulene; azulene (Fig. 17.7), for which the x values in (17.20) of the occupied MOs are -2.3103, -1.6516, -1.3557, -0.8870, and -0.4773.

17.14 Estimate the carbon–carbon bond length in (a) $C_5H_5^-$; (b) $C_7H_7^+$; (c) $C_8H_8^{2-}$.

17.15 (a) Set up and solve the a_u, b_{2g}, and b_{1g} HMO secular equations for naphthalene. (b) Find the coefficients of the lowest naphthalene HMO.

17.16 Derive Eq. (17.58) for E_π. *Hint:* Start with E_π as the sum of orbital energies and use $e_i = \langle \phi_i | \hat{H}^{\text{eff}} | \phi_i \rangle$.

17.17 (a) For p_{rs}^{tot} equal to 1 and to 3, compare the bond-length predictions of Eq. (17.57) with experimental carbon–carbon single- and triple-bond lengths. (b) Look up or use a program to calculate the Hückel bond orders of azulene and compare the predicted bond lengths with the experimental values. [The experimental data can be found in R. J. Buenker and S. D. Peyerimhoff, *Chem. Phys. Lett.*, **3**, 37 (1969).]

17.18 Verify that the E_π expression (17.58) holds for 1,3-butadiene.

17.19 The *free-valence index* F_r for carbon atom r in a planar conjugated compound is defined as $F_r \equiv \sqrt{3} - \sum_s' p_{rs}$, where the sum is over the atoms bonded to atom r. [The quantity $\sqrt{3}$ is the value of $\sum_s' p_{rs}$ for the central atom of the diradical $C(CH_2)_3$, whose central C has the largest possible value for this sum of any trigonally bonded carbon.] F_r is a measure of the unused bonding power of atom r and is used to estimate the susceptibility of a given atom to attack by an uncharged free radical. Calculate F_r for each carbon in 1,3-butadiene, and state which carbon is predicted to be preferentially attacked by free radicals.

17.20 (a) Show that the equations satisfied by the HMO coefficients of a conjugated hydrocarbon are $x_i c_{ri} + \sum_{s \to r} c_{si} = 0$, $r = 1, \ldots, n_C$, where the sum is over carbons bonded to carbon r. (b) If the hydrocarbon is an alternant, then we can divide the carbons into two sets such that carbons in one set are bonded only to carbons in the other set. Verify that for an alternant hydrocarbon, if we replace x_i by $-x_i$ and multiply the coefficients of one set of carbons by -1 in each HMO equation of (a), we obtain equations that are satisfied. Hence (provided $x_i \neq 0$), for each HMO of an alternant hydrocarbon with $(\alpha - e_i)/\beta = x_i$, there is an HMO with $(\alpha - e_i)/\beta = -x_i$, whose coefficients are obtained by multiplying the coefficients of one set of carbons in the first HMO by -1.

17.21 (a) What is the HMO delocalization energy of cyclobutadiene? (b) Calculate the HMO total bond orders in cyclobutadiene.

17.22 Show that $\sum_r q_r = n_\pi$ in the HMO method.

17.23 (a) Show that, with inclusion of overlap between carbons bonded to each other (the Wheland method), the HMO equation for benzene is like (17.32) except that each x is replaced by w, where $w \equiv (\alpha - e_i)/(\beta - Se_i)$. Hence the values of w are the same as those found for x with overlap omitted. Provided a common value of S is used for any two bonded carbons in a molecule, the same situation holds for any planar conjugated hydrocarbon. (b) Verify that, with inclusion of overlap, $e_i = \alpha - w\gamma/(1 - Sw)$, where $\gamma \equiv \beta - S\alpha$. (c) With S taken as 0.25, find the Wheland e_i values for benzene in terms of α and γ. (d) Show that the predicted wavenumber of the transition between the highest-occupied (HO) and lowest-unoccupied (LU) Wheland π MOs is

$$\frac{1}{\lambda} = \frac{\gamma(w_{\text{HO}} - w_{\text{LU}})}{hc[1 + S^2 w_{\text{LU}} w_{\text{HO}} - S(w_{\text{LU}} + w_{\text{HO}})]}$$

Explain why $w_{\text{LU}} = -w_{\text{HO}}$ for an alternant hydrocarbon and use this relation to show that for an alternant hydrocarbon

$$1/\lambda = |\gamma|\Delta w/hc[1 - \tfrac{1}{4}S^2(\Delta w)^2]$$

where $\Delta w \equiv |w_{\text{HO}} - w_{\text{LU}}|$. Δw is identical to Δx calculated without overlap; also, $S = 0.25$ and Δw is typically about 1 or 2, so $\frac{1}{4}S^2(\Delta w)^2 \ll 1$. Hence the $1/\lambda$ value is nearly the same as calculated [Eq. (17.47)] without overlap, except that the empirical proportionality constant is interpreted as $|\gamma| = |\beta - S\alpha|$, rather than as $|\beta|$.

17.24 Verify (17.66) for the PPP F_π matrix elements.

17.25 Apply the extended Hückel method to H_2. Use $K = 1.75$; use the expression (13.60) of S as a function of R; plot the valence-electron energy as a function of R; compare the predicted value of R_e with the experimental value. (To avoid solving a secular equation, use symmetry orbitals.) Use 1.0 as the orbital exponent.

17.26 (a) Set up (but do not solve) the 8×8 extended Hückel secular determinant for methane at the equilibrium configuration; do not use symmetry orbitals. Take $K = 1.75$; use Slater's rules (Prob. 15.54) for the orbital exponents; evaluate the overlap integrals from the reference of Prob. 15.28. (b) Do the same as in (a), but now use symmetry orbitals.

17.27 Let the valence AOs r_A, s_A, t_B, u_B, v_C be centered on different atoms A, B, and C in a molecule, with r_A and s_A being different AOs. Which of the following integrals are neglected in the CNDO method? Which are neglected in the INDO method? Which are neglected in the MNDO method? Which are neglected in the AM1 method? (a) $(r_A r_A|s_A s_A)$; (b) $(r_A s_A|r_A s_A)$; (c) $(r_A r_A|t_B t_B)$; (d) $(r_A t_B|r_A t_B)$; (e) $(r_A r_A|r_A t_B)$; (f) $(r_A s_A|t_B u_B)$; (g) $(r_A r_A|t_B v_C)$.

17.28 The AM1 valence electronic energies of the atoms H and O are −11.396 eV and −316.100 eV, respectively. For H_2O at its AM1-calculated equilibrium geometry, the AM1 valence electronic energy (core–core repulsion omitted) is −493.358 eV and the AM1 core–core repulsion energy is 144.796 eV. For H(g) and O(g), $\Delta H^\circ_{f,298}$ values are 52.102 and 59.559 kcal/mol, respectively. Find the AM1 prediction of $\Delta H^\circ_{f,298}$ of $H_2O(g)$. The experimental value is −57.796 kcal/mol.

17.29 For each of the following molecules, do AM1 and PM3 calculations to find the predicted geometry, dipole moment, and gas-phase $\Delta H^\circ_{f,298}$ Compare with experimental $\Delta H^\circ_{f,298}$ values in thermodynamics tables. (a) $CH_3CH_2CH_3$; (b) H_2S; (c) benzene.

17.30 Calculate the barrier to internal rotation in ethane using AM1 and PM3. Compare the results to experiment and to the HF/6-31G** result (Prob. 15.45).

17.31 Find the AM1 geometry, dipole moment, and vibrational frequencies of formaldehyde. Compare the results with experiment and with the HF/3-21G and 6-31G** results (Prob. 15.43).

17.32 Find the AM1 geometries and dipole moments of the gauche and anti conformers of butane and compare with HF/6-31G* results (Prob. 15.50). Also find the AM1 difference in $\Delta H^\circ_{f,298}$ for these conformers.

17.33 Find the AM1 geometries and dipole moments of the two conformers of formic acid and compare with HF results (Prob. 15.49). Also find the AM1 difference in $\Delta H^\circ_{f,298}$ for these conformers.

17.34 Find the AM1 geometries of the reactant, product, and transition state in the reaction HCN → HNC and compare with the HF/6-31G* results (Prob. 16.36).

17.35 Give the number of bond-stretching terms, bond-bending terms, bond-torsion terms, van der Waals terms, and electrostatic terms in a force field for each of the following: (a) CF_3OH; (b) CCl_3CCl_2OH.

17.36 Equation (17.91) for the van der Waals interaction between two atoms has the form $V = a/R^{12} - b/R^6$. (a) If σ is the finite value of R at which V is zero, show that $V = b\sigma^6/R^{12} - b/R^6$. (b) If R^* is the value of R at the minimum in V, show that $R^* = 2^{1/6}\sigma$. (c) If $\varepsilon = V(\infty) - V(R^*)$ is the depth of the van der Waals well, show that $b = 4\sigma^6\varepsilon$. (d) Use these results to show that the van der Waals V can be written in each of the two forms in (17.91).

17.37 Use whatever molecular-mechanics program is available to you to calculate the geometries of the two conformers of HCOOH and find the energy difference between them. Omit zero-point vibrational energies and thermal energies in this and the following MM problems.

17.38 Use a molecular-mechanics program to calculate the barrier to internal rotation in ethane.

17.39 Use a molecular-mechanics program to find the geometries of the gauche and anti conformations of butane and the energy difference between them.

17.40 Use a molecular-mechanics program to find the geometries of the two conformers of vinyl alcohol (CH_2CHOH) and the energy difference between them.

17.41 Use a molecular-mechanics program to find the geometries and the energy difference between (a) cis and trans 1,2-difluoroethylene; (b) cis and trans 1,2-dichloroethylene; (c) cis and trans 1,2-diiodoethylene. Compare your results with available literature data as to which is the more stable isomer of each pair.

17.42 MM3 parameters in kcal/mol to be used with (17.92) to calculate $\Delta H^\circ_{f,298}$ values of saturated hydrocarbons are $a_{CH} = -4.590$ and $a_{CC} = 2.447$; in addition, the following corrections (in kcal/mol) are included: 1.045 for each CH_3 group, −2.627 for each carbon bonded to three other carbons, −6.641 for each carbon bonded to four other carbons, 0.42 for each bond (except $C-CH_3$ bonds) with a low rotational barrier, −1.780 for each four-membered ring, and −5.508 for each five-membered ring. Given the following MM3 steric energies, calculate the gas-phase $\Delta H^\circ_{f,298}$ of each compound and compare with the experimental values given in parentheses: (a) 2.05 kcal/mol for propane (−24.82 kcal/mol); (b) 3.18 kcal/mol for isobutane (−32.15 kcal/mol); (c) 32.63 kcal/mol for cyclobutane (6.78 kcal/mol). (d) How many conformations must be considered when calculating the MM $\Delta H^\circ_{f,298}$ of butane?

17.43 Consider the thermal reaction 1,3,5,-hexatriene → 1,3-cyclohexadiene. (a) Use the symmetry of the polyene HOMO to predict whether the reaction path is conrotatory or disrotatory. (The HMOs need not be found explicitly; all that is needed is the signs of the AOs in the HOMO and these can be found from the pattern of nodes; the number of vertical nodal planes is zero for the ground state and increases by 1 for each higher state.) (b) Do the same as in (a) when the reaction occurs photochemically. (c) State the general rules for the cyclization of the polyene (17.28) with n_π π electrons.

17.44 Examine the frontier orbitals and decide whether each of the following elementary reactions should have a high or low activation energy for a four-center broadside-collision reaction path. (a) $H_2 + D_2 \rightarrow 2HD$; (b) $N_2 + O_2 \rightarrow 2NO$; (c) $F_2 + Br_2 \rightarrow 2FBr$; (d) $H_2 + C_2H_4 \rightarrow C_2H_6$.

CHAPTER 18

Comparisons of Methods

This chapter examines the accuracies of ab initio, density-functional, semiempirical, and molecular-mechanics methods in calculating properties of ground-state closed-shell molecules. Unless otherwise stated, the comparisons are for molecules of first- and second-row atoms (H through Ar). The most commonly used measure of the accuracy of a method for calculation of property X is the **mean absolute error, MAE** (also called the *mean absolute deviation*, MAD, or the average unsigned error, AUE), which is the average of the absolute values of the errors of the calculated X values for a series of compounds: MAE $= (1/n)\sum_{i=1}^{n}|X_{i,\text{calc}} - X_{i,\text{true}}|$.

18.1 MOLECULAR GEOMETRY

Ab initio HF/STO-3G calculations give fairly good predictions of bond distances and good predictions of bond angles, but occasionally show very large bond-length errors; for example, an error of 0.72 Å in the Na_2 bond length and of 0.23 Å for NaH. Successive increases in basis-set size through the series STO-3G, 3-21G, 3-21G$^{(*)}$, and 6-31G* usually give improved bond-length accuracy. A study (*Hehre et al.*, Section 6.2) gave the MAEs listed in Table 18.1. (**Hypervalent** molecules have a Lewis dot structure with more than eight electrons around one or more atoms.) Hehre et al. concluded that for HF geometry predictions, the "relatively small 3-21G basis set (3-21G$^{(*)}$ for molecules incorporating second-row elements) appears to be the method of choice because of its wide applicability to molecules of moderate size."

Dihedral angles are usually calculated reasonably accurately by the ab initio HF method, but the number of comparisons with experiment is small (see J. A. Pople in

TABLE 18.1 Mean absolute errors in bond lengths (in Å) and angles[a]

	AH_n lengths	AB single bonds in H_mABH_n	AB multiple bonds in H_mABH_n	AB lengths, hypervalent species	Angles in H_mABH_n
HF/STO-3G	0.054	0.082	0.027		2.0°
HF/3-21G	0.016	0.067	0.017	0.125	1.7°
HF/3-21G$^{(*)}$	0.017	0.040	0.018	0.015	1.8°
HF/6-31G*	0.014	0.030	0.023	0.014	1.5°

[a]Data from *Hehre et al.*, Section 6.2. A and B denote nonhydrogen atoms.

Schaefer, Applications of Electronic Structure Theory, pages 11–12, 16–17). Exceptions include H_2O_2, where the 3-21G basis set predicts a 180° dihedral angle compared with the experimental angle of 120°, and cyclobutane and cyclopentane, where STO-3G substantially underestimates the nonplanarity. In these three molecules, HF/6-31G* calculations predict the conformational angles well.

MP2/6-31G* geometries are significantly more accurate than HF/6-31G* geometries. For example, MAEs in a sample of 73 bond lengths in H_mABH_n compounds are reduced from 0.021 Å in HF/6-31G* to 0.013 Å in MP2/6-31G* (*Hehre et al.*, pp. 156–161). A study of 184 small molecules examined the effect of various frozen-core correlation methods [D. Feller and K. A. Peterson, *J. Chem. Phys.*, **108**, 154 (1998)]; MAEs with the aug-cc-pVTZ basis set for AH lengths, AB lengths, and HAH angles were as follows:

	HF	MP2	MP4	CCSD	CCSD(T)
ΔAH/Å	0.014	0.011	0.007	0.009	0.009
ΔAB/Å	0.028	0.022	0.030	0.011	0.016
$\Delta\angle$HAH	1.6°	0.3°	0.3°	0.3°	0.4°

The HF errors increased with increase in the basis-set size for these three sets. MP4 results for AB lengths were less accurate than MP2 results.

Density-functional calculations give accurate molecular geometries with 6-31G* or larger basis sets. (As with other methods that include correlation, DFT should not be done with basis sets smaller than 6-31G*.) MAEs in bond lengths and bond angles for a sample of 108 molecules containing two to eight atoms are as follows [A. C. Scheiner, J. Baker, and J. W. Andzelm, *J. Comput. Chem.*, **18**, 775 (1997)]:

HF/6-31G**	MP2/6-31G**	BLYP/6-31G**	BPW91/6-31G**	B3PW91/6-31G**
0.021 Å	0.015 Å	0.021 Å	0.017 Å	0.011 Å
1.3°	1.1°	1.2°	1.2°	1.0°

The B3PW91 hybrid functional gave the best results of the functionals studied. These workers did DFT calculations with five different basis sets and found that as the basis set size increased, the errors in DFT geometries decreased significantly.

A study of 11 dihedral angles in organic compounds found the following MAEs [A. St.-Amant et al., *J. Comput. Chem.*, **16**, 1483 (1995)]: 3.8° for HF/6-31G*, 3.6° for MP2/6-31G*, and 3.4° for BP86 with a basis set that is TZP on nonhydrogens and DZP on hydrogens.

Semiempirical methods usually give satisfactory bond lengths and angles, but the results are not as accurate as ab initio or DFT results with a suitable-size basis set. For compounds containing no elements other than H, C, N, O, F, Al, Si, P, S, Cl, Br, and I, MAEs in 460 bond lengths and 196 bond angles are as follows [J. J. P. Stewart, *J. Comput. Chem.*, **12**, 320 (1991)]:

	MNDO	AM1	PM3
lengths	0.055 Å	0.051 Å	0.037 Å
angles	4.3°	3.8°	4.3°

For 712 bond lengths and 244 bond angles in compounds containing no elements except C, H, O, N, S, P, F, Cl, Br, and I, MAEs are as follows [J. J. P. Stewart, *J. Mol. Model.*, **13**, 1173 (2007)]:

	AM1	PM3	PM5	RM1	PM6
lengths	0.046 Å	0.037 Å	0.044 Å	0.036 Å	0.031 Å
angles	3.4°	3.8°	4.3°	4.0°	3.2°

For 900 bond angles in compounds of main-group elements, the MAEs are 8.8° for AM1, 8.5° for PM3, 8.6° for PM5, and 8.0° for PM6; for 2636 bond lengths in compounds of main-group elements, the MAEs are 0.131 Å for AM1, 0.104 Å for PM3, 0.121 Å for PM5, and 0.085 Å for PM6 (Stewart, op. cit.).

PDDG/PM3 gives geometries with accuracy similar to PM3 [M. P. Repasky et al., *J. Comput. Chem.*, **23**, 1601 (2002)].

MNDO, AM1, and PM3 do not include *d* orbitals and are not particularly accurate for geometries of molecules with elements from the second and later rows. MNDO/d includes *d* orbitals and gives improved performance for such compounds [W. Thiel and A. Voityuk, *J. Phys. Chem.*, **100**, 616 (1996)].

The performance of semiempirical methods for dihedral angles is quite mediocre. For 27 dihedral angles, MAEs are 11.7° for AM1, 16.6° for PM3, 16.7° for PM5, and 12.6° for PM6 and large errors occur for torsion angles in dipeptides [J. J. P. Stewart, *J. Mol. Model.*, **13**, 1173 (2007), supplementary material].

Molecular-mechanics force fields usually give good results for geometries for kinds of molecules for which the field has been properly parametrized. MMFF94 has root-mean-square (rms) deviations from MP2/6-31G* geometries (which are usually accurate) of 0.006 Å for 4205 bond lengths, 1.2° for 7021 bond angles, and 5.8° for 7974 dihedral angles. (For errors that follow a Gaussian distribution, the rms error is 1.25 times the mean absolute error.) For a sample of 30 organic compounds, rms errors in bond lengths and angles for several force fields are as follows [T. A. Halgren, *J. Comput. Chem.*, **17**, 553 (1996)]:

	MMFF94	MM3	UFF	CHARMm
lengths/Å	0.014 (95)	0.010 (94)	0.021 (51)	0.016 (44)
angles	1.2° (86)	1.2° (84)	2.5° (32)	3.1° (39)

where the number of lengths or angles is in parentheses.

For 76 medium-size organic compounds, DREIDING had rms errors of 0.035 Å in bond lengths, 3.2° in bond angles, and 8.9° in dihedral angles [S. L. Mayo et al., *J. Phys. Chem.*, **94**, 8897 (1990)] and Tripos had rms errors of 0.025 Å in bond lengths, 2.5° in bond angles, and 9.5° in dihedral angles [M. Clark et al., *J. Comput. Chem.*, **10**, 982 (1989)].

Transition-State Structures. Geometries of transition states are, as yet, experimentally unobservable. For transition states that contain a limited number of atoms, one can do a series of ab initio calculations in which the size of the basis set and the level of correlation are increased. If the geometries obtained by these calculations seem to be converging, one can be fairly confident that the highest-level ab initio calculation has given an accurate transition-state geometry. By comparing the high-level ab initio geometry with transition-state geometries found by other methods, one can judge the value of these methods for finding transition-state geometries. The number of such comparisons done is limited.

Although semiempirical, ab initio, and density-functional equilibrium geometries of a ground-state molecule are usually quite similar to one another (with bond-length differences from method to method usually no greater than 0.03 Å for first- and second-row atoms), calculated transition-state geometries vary considerably with choice of method and basis set, with bond-length differences of 0.1 or 0.2 Å being common. For example, some calculated chair-transition-state bond lengths in Å for the single bond being broken in the Cope rearrangement of 1,5-hexadiene are 1.66 for AM1, 2.02 for RHF/3-21G, 2.05 for RHF/6-31G*, 1.78 for MP2(FC)/6-31G*, 1.79 for MP4/6-31G*, 1.94 for CISD/6-31G*, 2.19 for CASSCF/6-31G*, 1.75 for SVWN/6-31G*, 2.03 for BLYP/6-31G*, 2.15 for BLYP/6-311+G**, 1.97 for B3LYP/6-31G*, and 2.04 for B3LYP/6-311+G** [O. Wiest, D. C. Montiel, and K. N. Houk, *J. Phys. Chem. A*, **101**, 8378 (1997)]. For this reaction, the methods that predict a longer bond length (greater than 1.95 Å) predict kinetic isotope effects in better agreement with experiment than those predicting a shorter bond length (less than 1.8 Å).

Although Hartree–Fock calculations with small basis sets such as 3-21G "sometimes fail to produce reasonable transition state geometries" (*Hehre*, Section 3.3), the Hartree–Fock method "with appropriate basis sets provides reasonable [transition-state] geometries in many cases" (Wiest, Montiel, and Houk, op. cit.).

LSDA density-functional calculations "are not adequate for the calculation of organic transition structures, " but B3LYP calculations "have been remarkably successful in predicting [transition-state] geometries . . . for a large number of organic reactions" (Wiest, Montiel, and Houk, op. cit.).

Hehre noted that current semiempirical methods "sometimes give very poor descriptions of [geometries of] reaction transition states" (*Hehre*, Chapter 3), and Houk and co-workers stated that "semiempirical methods . . . are not adequate for the calculation of organic transition structures" (Wiest, Montiel, and Houk, op. cit.). Despite this, it can be fruitful to use a fast semiempirical method in a preliminary investigation of a potential-energy surface to get initial estimates of transition-state structures. Each semiempirical structure is then used as the initial guess in an ab initio calculation of the transition-state geometry. Such a procedure can save considerable computational time (*Hehre*, Section 3.3).

18.2 ENERGY CHANGES

Atomization Energies, Heats of Formation, and Heats of Reaction. Gas-phase enthalpies of formation are readily calculated from atomization energies (Section 15.13), and gas-phase heats of reaction are readily found from gas-phase heats of formation of reactants and products. All three of these related quantities are considered in this subsection. *Throughout this section, only gas-phase reactions are considered* because this avoids the substantial contributions of intermolecular interactions in liquids and solids. To convert a calculated gas-phase reaction energy to one involving condensed phases, we use the experimental energies of condensation of substances.

For a data set of 55 atomization energies, 38 ionization potentials, 25 electron affinities, and 7 proton affinities of small molecules, the MAEs and maximum absolute errors in kcal/mol for several methods are as follows (calculations of Foresman, Frisch, Ochterski, and Frisch, reported in *Foresman and Frisch*, Chapter 7):

CBS-Q	G2	G2(MP2)	CBS-4	B3LYP/6-311+G(2d,p)
1.0	1.2	1.5	2.0	3.1
3.8	5.1	6.2	7.0	19.7

BLYP/6-311+G(2d,p)	B3LYP/6-31+G**	B3LYP/6-31G*	MP2/6-311+G(2d,p)	MP2/6-31+G**
3.9	3.9	7.9	8.9	11.4
15.9	33.8	54.2	39.2	44.0

PM3	AM1	HF/6-31+G**	HF/6-31G*	HF/STO-3G
17.2	18.8	46.7	51.0	93.3
69.9	95.5	179.8	184.2	313.9

The best results are given by the composite methods G2 and CBS, which cannot be applied to large molecules. Gradient-corrected DF methods do rather well. The Hartree–Fock method is useless for atomization energies, as discussed earlier.

A study of atomization energies of 66 small molecules with the aug-cc-pVnZ (n = D, T, Q) basis sets gave the following MAEs in kcal/mol [D. Feller and K. A. Peterson, *J. Chem. Phys.*, **108**, 154 (1998)], where the letters D, T, Q denote the basis set:

HF/D	HF/T	HF/Q	MP2/D	MP2/T	MP2/Q	MP4/D	MP4/T
85	66	62	15	5	5	16	4

MP4/Q	CCSD/D	CCSD/T	CCSD/Q	CCSD(T)/D	CCSD(T)/T	CCSD(T)/Q
2	21	11	7	18	$5\frac{1}{2}$	$2\frac{1}{2}$

MP2 is more accurate here than the more computationally expensive CCSD method.

A study of density-functional atomization energies of 44 small molecules (one to three nonhydrogens) gave the following MAEs in kcal/mol [J. M. Martell et al., *J. Phys. Chem. A*, **101**, 1927 (1997)]:

	BLYP	B3LYP	B3PW91
6-31G**	7.6	5.6	5.6
cc-pVDZ	7.3	8.5	8.4
cc-pVTZ	7.2	3.1	3.8

For a sample of 108 molecules with one to five nonhydrogen atoms, MAEs in kcal/mol in 108 atomization energies, 66 enthalpy changes for bond-dissociation reactions, 73 enthalpy changes for hydrogenation reactions, and 29 enthalpy changes for oxygenation (combustion) reactions were as shown in Table 18.2, where the TZ2P is a triple zeta set with two polarization functions per atom. The hybrid B3PW91 functional performed best. The Hartree–Fock method works rather well for hydrogenation reactions, which preserve the total number of electron-pair bonds.

TABLE 18.2 MAEs in kcal/mol for 108 atomization energies (atom.), 66 bond-dissociation energies (BD), 73 hydrogenation-reaction enthalpies (H_2), and 29 combustion energies (O_2)[a]

	atom.	BD	H_2	O_2
HF/6-31G**	119.2	58.8	8.5	44.5
MP2/6-31G**	22.0	8.8	7.0	11.2
SVWN/6-31G**	52.2	22.1	11.3	21.8
BPW91/6-31G**	7.4	5.9	10.1	27.6
BPW91/TZ2P	7.3	5.5	5.5	15.9
B3PW91/6-31G**	6.8	5.6	6.8	26.2
B3PW91/TZ2P	6.5	5.1	3.9	14.4

[a]A. C. Scheiner, J. Baker, and J. W. Andzelm, *J. Comput. Chem.*, **18**, 775 (1997).

For a set of 148 gas-phase $\Delta H^\circ_{f,298}$ values of molecules as large as C_6H_6 (the G2 neutral test set), MAEs and maximum errors in kcal/mol for density-functional methods using the 6-311+G(3df,2p) basis set and for composite methods are as follows [L. A. Curtiss et al., *J. Chem. Phys.*, **106**, 1063 (1997); **108**, 692 (1998)]:

G2	G2(MP2)	CBS-Q	CBS-q	CBS-4
1.6	2.0	1.6	2.1	3.1
8.2	10.1	11.2	10.3	14.4

SVWN	BLYP	BPW91	B3LYP	B3PW91
90.9	7.1	7.8	3.1	3.5
228.7	28.4	32.2	−20.1	−21.8

For the DFT calculations, the hybrid functionals perform best.

MAEs in kcal/mol for the 270 enthalpies of formation (second row of the table) and for all 454 energy differences (third row) in the G3/05 test set (Section 16.5) for some composite and DFT methods, where large basis sets were used for the DFT calculations, are as follows [L. A. Curtiss et al., *J. Chem. Phys.*, **126**, 084108 (2007); **123**, 124107 (2005); T. Schwabe and S. Grimme, *Phys. Chem. Chem. Phys.*, **8**, 4398 (2006)]:

G3	G3X	G4	B3LYP	B98	B2-PLYP	mPW2-PLYP
1.2	1.0	0.8	4.6	3.5		
1.1	1.0	0.8	4.1	3.3	2.5	2.1

The MP2-like perturbation correction produces substantially better performance for the B2-PLYP and mPW2-PLYP methods.

A database of 109 main-group molecular atomization energies was used to test over 30 density functionals using a quite large basis set and molecular geometries found from QCISD calculations with a quite large basis set [Y. Zhao et al., *J. Chem. Theory Comput.*, **2**, 364 (2006); Y. Zhao and D. G. Truhlar, *J. Chem. Phys.*, **125**, 194101 (2006)]. Some MAEs in kcal/mol are as follows: 1.9 for PW6B95, 2.2 for BMK, 2.3 for M05-2X, 2.6 for B1B95, 3.1 for B3PW91, 4.0 for M06-L, and 4.3 for B3LYP.

As to semiempirical methods, for a sample of 886 compounds of H, C, N, O, F, Al, Si, P, S, Cl, Br, and I, MAEs in kcal/mol for $\Delta H^\circ_{f,298}$ values are [J. J. P. Stewart, *J. Comput. Chem.*, **12**, 320 (1991)] 23.7 for MNDO, 14.2 for AM1, and 9.6 for PM3. For comparison, B3LYP/6-31G* calculations for this set give an average error of 5.2 kcal/mol. For 1774 compounds of C, H, O, N, P, S, F, Cl, Br and I, MAEs in $\Delta H^\circ_{f,298}$ in kcal/mol are as follows [J. J. P. Stewart, *J. Mol. Model.*, **13**, 1173 (2007)]:

AM1	PM3	PM5	RM1	PM6
12.6	8.1	6.8	6.6	5.1

MNDO/d includes *d* orbitals on many second-row elements and so gives improved performance for compounds of such elements. MAEs in kcal/mol for $\Delta H^\circ_{f,298}$ of 99 S-containing compounds are [W. Thiel and A. A. Voityuk, *J. Phys. Chem.*, **100**, 616 (1996)] 48.4 for MNDO, 10.3 for AM1, 7.5 for PM3, and 5.6 for MNDO/d.

For 1356 compounds containing no elements other than C, H, O, N, S, P, F, Cl, Br, and I, MAEs in $\Delta H^\circ_{f,298}$ in kcal/mol are as follows [K. W. Sattelmeyer et al., *J. Phys. Chem. A*, **110**, 13551 (2006)]:

AM1	PM3	RM1	PDDG/PM3
9.2	7.2	5.5	5.0

Although semiempirical methods yield fairly accurate heats of formation, their predictions of ΔH° of reactions are generally not satisfactory; see, for example, the bond-separation-energy data in the next subsection. "Errors in individual heats of formation of present-generation semi-empirical methods are random errors and large enough ... such that the overall error in a given reaction will be unacceptably large" (*Hehre*, p. 131). This conclusion was stated before the more-accurate semiempirical methods PM6, RM1, and PDDG/PM3 were developed, and so might not apply to them.

MM2 and MM3 usually give gas-phase heats of formation with 1 kcal/mol accuracy for compounds similar to those used in the parametrization. For example, the average absolute MM3 error in $\Delta H^\circ_{f,298}$ for a sample of 45 alcohols and ethers is 0.6 kcal/mol [N. L. Allinger et al., *J. Am. Chem. Soc.*, **112**, 8293 (1990)]. Most molecular-mechanics programs do not include provision for calculation of heats of formation.

Isodesmic Reactions and Bond-Separation Reactions. Although ab initio HF calculations fail in predicting molecular atomization energies, one can still use HF energies to estimate energy changes for certain types of reactions.

Recall that good barriers to internal rotation can usually be obtained from ab initio SCF MO calculations because of a near cancellation in correlation energies between different molecular conformations. As ethane goes from staggered to eclipsed, the number of chemical bonds of each type does not change. More generally, one might hope that a similar near cancellation of correlation energies might occur for an isodesmic chemical reaction; an **isodesmic reaction** (Greek *isos*, "equal"; *desm*, "bond") is one in which the number of bonds of each type does not change [W. J. Hehre et al., *J. Am. Chem. Soc.*, **92**, 4796 (1970)]. For example, the isodesmic reaction $CH_2{=}CHCH_2OH + CH_2{=}O \rightarrow CH_2{=}CHOH + CH_3CH{=}O$ has seven CH bonds, one CC double bond, one CC single bond, one CO double bond, one CO single bond, and one OH bond on each side.

A special kind of isodesmic reaction is a **bond-separation reaction**. Here, one starts with a molecule and converts it to products, each of which contains only one bond between nonhydrogen atoms. For example, starting with $CH_3{-}CH{=}C{=}O$, one would form the products $CH_3{-}CH_3$, $CH_2{=}CH_2$, and $CH_2{=}O$, in which the C—C, C=C, and C=O bonds are separated from one another. To balance the reaction, one adds an appropriate number of hydride molecules (for example, CH_4, NH_3, H_2O) to the left side. The bond-separation reaction for CH_3CHCO is then $CH_3{-}CH{=}C{=}O + 2CH_4 \rightarrow C_2H_6 + C_2H_4 + CH_2O$. The bond-separation reaction for benzene is $C_6H_6 + 6CH_4 \rightarrow 3C_2H_6 + 3C_2H_4$. (If the energy of a molecule could be represented as the sum of bond energies that were invariant from molecule to molecule, then the energy change for any isodesmic reaction would be zero. The energy change for a bond-separation reaction measures the interactions between the bonds in the molecule.)

If an SCF MO or DFT calculation could accurately predict the energy change for the bond-separation reaction of a large molecule, then we could use the known energies of the small product molecules like C_2H_6 to get a good estimate for the energy of the large molecule from an SCF calculation, without having to use expensive correlation methods.

A study of calculated bond-separation energies of 25 organic compounds found the following results in kcal/mol for the mean absolute errors, the maximum error, and the number of reactions where the error exceeded 5 kcal/mol (*Hehre*, pp. 134–136):

	AM1	HF/STO-3G	HF/3-21G$^{(*)}$	HF/6-31G*	SVWN/6-31G*	B3LYP/6-31G*	MP2/6-31G*
MAE	16.2	6.3	4.8	3.3	3.9	2.4	1.7
max.	43	23	17	16	9	8	7
no. > 5 kcal/mol	19	8	7	5	8	1	1

For bond separation reactions, AM1 is very unreliable; HF/STO-3G is unreliable; HF/3-21G$^{(*)}$ and SVWN/6-31G* are of marginal reliability; HF/6-31G* is usually reliable; and B3LYP/6-31G* and MP2/6-31G* are quite reliable. Despite the failure of HF/6-31G* in predicting atomization energies, it usually can reliably predict energy changes in bond-separation and other isodesmic reactions. Assuming the heats of formation of the small molecules in the bond-separation reaction are accurately known, the error in predicted heats of formation will be the same as the errors in the bond-separation reaction enthalpy changes.

Isomerization Energies. An important question is the relative energies of structural isomers. An isomerization is not necessarily an isodesmic reaction (an example is cyclopropane → propene). A study of about 40 Hartree–Fock calculated isomerization energies of organic compounds (*Hehre et al.*, Section 6.5.5) found that the STO-3G basis set gives unreliable results, with errors as high as 50 kcal/mol occurring. The 3-21G basis set [3-21G$^{(*)}$ for molecules with second-row atoms] did fairly well in some cases, but showed errors of up to 30 kcal/mol in other cases. The 6-31G*//3-21G$^{(*)}$ calculations were usually reliable; only one energy difference was in error by more than 10 kcal/mol, but a few were in error by 5 to 10 kcal/mol. (This notation denotes a 6-31G* single-point energy calculation done at the 3-21G$^{(*)}$ optimized geometry.) Hehre et al. note that the

Hartree–Fock 6-31G*//3-21G$^{(*)}$ "level of theory is perhaps the simplest available with which to calculate relative isomer energies for a wide variety of systems to reasonable accuracy."

As to semiempirical SCF MO methods, for a sample of 30 isomerization reactions [those in *Hehre*, Section 6.5.5, for which semiempirical results are given in J. J. P. Stewart, *J. Comput. Chem.*, **10**, 221 (1989)], MAEs are 9.1 kcal/mol for MNDO, 7.4 kcal/mol for AM1, and 5.8 kcal/mol for PM3. The largest individual error was 42 kcal/mol for MNDO, 24 kcal/mol for AM1, and 23 kcal/mol for PM3. For the 26 isomerizations for which 6-31G* and PM3 results are given, the MAEs are 2.4 kcal/mol for 6-31G*//3-21G$^{(*)}$ and 5.4 kcal/mol for PM3.

In summary, ab initio SCF MO calculations with a 6-31G* or larger basis set usually give isodesmic and isomerization reaction energies accurate to 5 kcal/mol. The STO-3G and 3-21G basis sets are not reliable for such calculations. MNDO, AM1, and PM3 are clearly inferior to HF/6-31G* calculations.

For a sample of 45 isomerizations, average absolute errors in kcal/mol are 2.9 for HF/6-31G*, 1.9 for MP2/6-31G*, 4.7 for SVWN/6-31G*, and 2.8 for B3LYP/6-31G* (*Hehre*, Table 2-14).

Calculations on 34 organic isomerization reactions, where a triple-zeta polarized basis set was used for ab initio and DFT methods (except where noted otherwise), gave the following MAEs in kcal/mol for reaction energies [S. Grimme et al., *J. Org. Chem.*, **72**, 2118 (2007); K. W. Sattelmeyer et al., *J. Phys. Chem. A*, **110**, 13551 (2006)]:

AM1	PM3	SCC-DFTB	RM1	B3LYP/6-31G*	PDDG/PM3	HF	B3LYP
7.2	5.0	5.	4.2	3.	2.7	2.7	2.3

MP2	B2-PLYP	BMK	SCS-MP2	CCSD(T)
1.45	1.3	1.3	1.0	0.7

Of course, CCSD(T) is too computationally demanding to be used except for small molecules. Note the improved performance of PDDG/PM3 as compared with the other semiempirical MO methods AM1, PM3, and RM1. Especially noteworthy is the excellent performance of the SCS-MP2 method (Section 16.2), which requires a similar computation time to MP2 and so can be widely applied. The rather poor performance of B3LYP in this and other references cited by Grimme at al., led these workers to "strongly recommend" that B3LYP not be used for thermochemical calculations.

Energy Differences between Conformers. A study of energy differences for eight pairs of conformers of organic compounds gave the following results for the average absolute errors in kcal/mol and the number of pairs where the error exceeded 0.5 kcal/mol (*Hehre*, pp. 175–185):

MNDO	AM1	PM3	HF/STO-3G	HF/3-21G$^{(*)}$	HF/6-31G*
1.4	1.3	1.8	0.9	1.0	0.7
7	4	6	6	6	3

SVWN/6-31G*	B3LYP/6-31G*	MP2/6-31G*	Blank
0.9	0.6	(0.5)	2.25
5	3	(2)	7

Parentheses indicate only seven energy differences were calculated. The blank entries are what would be obtained if all energy differences were taken as zero; that is, 2.25 kcal/mol is the average energy difference for the eight pairs. Thus the HF/STO-3G 0.9 kcal/mol average absolute error is substantial. The performance of the semiempirical methods is poor, and the HF/STO-3G, HF/3-21G$^{(*)}$, and SVWN/6-31G* methods do not do well. HF/6-31G*, B3LYP/6-31G*, and MP2/6-31G* do rather well. One study concluded that for conformational analysis, "results obtained at simple levels of theory are not always reliable. Indeed, for some problems, calculations must be performed at unexpectedly high levels before satisfactory agreement with experiment is achieved" [L. Radom et al., *J. Mol. Struct.*, **126**, 271 (1985)].

A study of 35 conformational energy differences found the following rms errors in kcal/mol [A. St.-Amant et al., *J. Comput. Chem.*, **16**, 1483 (1995)]:

HF/6-31+G**//HF/6-31G*	MP2/6-31+G**//MP2/6-31G*	SVWN/TZP	BP86/TZP
0.6	0.4	0.7	0.5

The average absolute conformational energy difference in this study was 1.6 kcal/mol.

For a sample of 15 conformational energy differences, MAEs in kcal/mol were as follows [K. W. Sattelmeyer et al., *J. Phys. Chem. A*, **110**, 13551 (2006)]:

PDDG/PM3	PM3	RM1	AM1	SCC-DFTB	B3LYP/6-31G*
1.8	1.8	1.5	1.4	1.2	0.4

Since the median value for the experimental energy differences in this study was 2 kcal/mol, only the B3LYP/6-31G* results can be considered accurate.

A study of 38 conformational energy differences found the following MAEs in kcal/mol for several MM force fields [K. Gundertofte et al., *J. Comput. Chem.*, **17**, 429 (1996); M. C. Nicklas, *J. Comput Chem.*, **18**, 1056 (1997)]: MM3—0.51, MM2—0.52, CHARMm—0.52, MMFF93—0.53, MMX—0.55, AMBER*—0.86, Tripos—1.11, DREIDING—1.20, UFF(no charges)—1.36, UFF(charges)—3.22. The average absolute energy difference for the 38 experimental values was 1.73 kcal/mol. Only the MM3, MM2, CHARMm, MMFF93, and MMX results are good. For more comparisons, see T. A. Halgren, ibid., **20**, 720 (1999).

Rotational Barriers. For 13 barriers to internal rotation about single bonds, the following average absolute percentage errors were obtained, where + denotes the 6-311+G(2d,p) basis set (*Hehre*, pp. 168–176): Tripos—64%, AM1—50%, HF/STO-3G —35%, HF/6-31G*—18%, HF/+—11%, MP2/6-31G*—21%, MP2/+—9%, SVWN/6-31G*—16%, SVWN/+—12%, B3LYP/6-31G*—17%, B3LYP/+—14%. The Tripos, AM1, and HF/STO-3G results are unsatisfactory. All the other methods give satisfactory results.

A study of 85 barriers about single bonds in 75 molecules gave the following average absolute percentage errors [H. F. Dos Santos and W. B. De Almeida, *J. Mol. Struct.* (*Theochem*), **335**, 129 (1995)]: MM2—17%, MNDO—63%, AM1—63%, PM3—61%. MM2 does well and the semiempirical methods do poorly. For 28 rotational barriers, MMFF94 had an average absolute percentage error of 14% [T. A. Halgren and R. B. Nachbar, *J. Comput. Chem.*, **17**, 587 (1996)].

Activation Energies. Hartree–Fock-calculated activation energies of reactions are unreliable, due to neglect of correlation, and are usually substantially larger than the corresponding experimental values. However, the *changes* in activation energy of a particular reaction due to different substituents can usually be calculated accurately with the Hartree–Fock method. For example, for the gas-phase Diels–Alder cycloaddition reactions of cyclopentadiene with ethylene substituted with from zero to four cyano groups, the mean absolute error for nine changes in activation energy is only 0.9 kcal/mol for HF/6-31G* calculations, and only one relative activation energy is in error by more than 1 kcal/mol (*Hehre*, Section 5.2). The HF/STO-3G method does poorly here, with a 4.1 kcal/mol average absolute error and eight relative activation energies in error by more than 1 kcal/mol. Likewise, the AM1 method fails here, with a 7.1 kcal/mol average absolute error and eight relative energies in error by more than 1 kcal/mol.

The activation energies for the reactions of species B with the species AX and AY with different substituents X and Y are the energy changes for the processes $B + AX \rightarrow [BAX]^{\ddagger}$ and $B + AY \rightarrow [BAY]^{\ddagger}$, where the double-dagger superscript denotes the transition state. The difference in activation energies for these two reactions corresponds to the hypothetical process $AX - AY \rightarrow [BAX]^{\ddagger} - [BAY]^{\ddagger}$ or $AX + [BAY]^{\ddagger} \rightarrow AY + [BAX]^{\ddagger}$. This is an isodesmic process, for which we can expect the correlation contribution to the energy change to be small and the Hartree–Fock method to work well.

As to correlated methods of predicting activation energies, MP2/6-31G* results are not reliable, but hybrid DFT calculations have given rather good results in many (but not all) cases. A study of 12 organic elementary gas-phase reactions (six of which had a radical species as a reactant) found the following average and maximum absolute errors in calculated barrier heights in kcal/mol [J. Baker, M. Muir, and J. Andzelm, *J. Chem. Phys.*, **102**, 2063 (1995)]:

MNDO	AM1	HF/3-21G	HF/6-31G*	MP2/6-31G*	BLYP/6-31G*	B3PW91/6-31G*
23.4	9.3	10.4	13.6	9.9	5.9	3.7
51.8	34.2	37.5	30.6	28.8	21.9	12.9

Only the hybrid DFT B3PW91/6-31G* results are reasonably satisfactory. High-level correlated methods such as G3, CCSD(T), and QCISD(T) give good results for activation energies but are too computationally expensive to be used routinely.

Semiempirical AM1 and PM3 activation energies are usually not reliable.

Truhlar and co-workers tested 205 combinations of methods and basis sets using a database of 24 reaction barrier heights [J. Zheng et al., *J. Chem. Theory Comput.*, **3**, 569 (2007)]. Some of their results are as follows, where the first number gives the MAE in kcal/mol and the number in parentheses is the relative computational cost of the method: G3X—0.6 (250); CCSD(T)/aug-cc-pVTZ—0.6 (13000); BMK/MG3S—1.4 (13); M05-2X/

MG3S—1.6 (13); B97-3/MG3S—2.0 (11); CCSD(T)/cc-pVTZ—2.1 (1660); M06-HF/6-31+G(d,p)—2.5 (2.7); M06-L/6-31+G(d,p)—4.6 (2.1); B3LYP/6-31+G(d,p)—4.9 (1.4); MP4/6-31G*—6.2 (460); MP2/6-31+G(d,p)—6.3 (1); HF/6-31G*—9.9 (0.15); PM3—12.8 (0.00005); AM1—12.9 (0.00005); PDDG/PM3—15.0 (0.00005); RM1—15.9 (0.00005); PM6—16.9 (0.00005); SCC-DFTB—19.3 (0.0004). Note the poor performance of all the semiempirical methods. [MG3S is a quite large basis set similar to but somewhat smaller than 6-311++G(3d2f,2df,2p).]

18.3 OTHER PROPERTIES

Dipole Moments. For a sample of 21 small molecules, average absolute errors in dipole moments were as follows: HF/STO-3G—0.65 D, HF/3-21G*—0.34 D, HF/6-31G*—0.30 D (*Hehre et al.*, Section 6.6.1); the STO-3G basis set is not reliable here. For a sample of 108 compounds, average absolute errors with the 6-31G** basis set were as follows: [A. C. Scheiner et al., *J. Comput. Chem.*, **18**, 775 (1997)]: HF—0.23 D, MP2—0.20 D, SVWN—0.23 D, BLYP—0.20 D, BPW91—0.19 D, B3PW91—0.16 D. These results are all quite good. Extremely accurate dipole moments were obtained with gradient-corrected functionals and a very large basis set (an uncontracted version of the aug-cc-pVTZ set); for BLYP the average absolute error was 0.06 D.

Semiempirical methods give dipole moments of fair accuracy. For 131 compounds containing no elements other than H, C, N, O, F, P, S, Cl, Br, and I, MAEs are as follows: AM1—0.38 D, PM3—0.36 D, PM5—0.50 D, RM1—0.33 D, PM6—0.37 D; for 313 compounds of main-group elements, MAEs are AM1—0.65 D, PM3—0.72 D, PM5—0.86 D, PM6—0.60 D [J. J. P. Stewart, *J. Mol. Model.*, **13**, 1173 (2007)]. For a small sample, PDDG/PM3 gave dipole moments slightly more accurate than AM1 and PM3 dipole moments [M. P. Repasky et al., *J. Comput. Chem.*, **23**, 1601 (2002)].

Vibrational Frequencies. As noted in Section 15.12, theoretically calculated vibrational frequencies are often multiplied by a scale factor to improve agreement with experiment. For a set of 122 molecules and 1066 vibrational frequencies, scaled theoretical harmonic vibrational frequencies showed the following rms deviations from experimental fundamental (anharmonic) vibrational frequencies; also given are the optimum scale factors and the percentages of scaled frequencies with less than 6% error and more than 20% error [A. P. Scott and L. Radom, *J. Phys. Chem.*, **100**, 16502 (1996)]:

	HF/3-21G	HF/6-31G*	MP2(FC)/6-31G*	BLYP/6-31G*
scale factor	0.908	0.895	0.943	0.994
rms error/cm^{-1}	87	50	63	45
% < 6% error	67	83	82	79
% > 20% error	9	2.5	3.5	1.9

	B3LYP/6-31G*	B3PW91/6-31G*	AM1	PM3
scale factor	0.961	0.957	0.953	0.976
rms error/cm^{-1}	34	34	126	159
% < 6% error	86	90	49	44
% > 20% error	0.9	1.6	15	17

The performance of AM1 and PM3 is mediocre at best. Scaled HF/6-31G* frequencies have good accuracy and this is an economical method to use for vibrational frequencies. Hybrid DFT scaled frequencies are even more accurate than HF/6-31G*.

The semiempirical SCC-DFTB method (Section 17.4) allows fairly accurate vibrational frequencies to be calculated very rapidly for large molecules. For a sample of 66 molecules and 1304 vibrational frequencies, MAEs in cm^{-1} for scaled calculated frequencies are 56 for SCC-DFTB, 69 for AM1, 74 for PM3, 30 for HF/cc-pVDZ, and 29 for B3LYP/cc-pVDZ [H. A. Witek and K. Morokuma, *J. Comput. Chem.*, **25**, 1858 (2004)].

A more elaborate scaling procedure is the modified SQM (scaled quantum-mechanical force field) method [J. Baker, A. A. Jarzecki, and P. Pulay, *J. Phys. Chem. A*, **102**, 142 (1998)]. Here, the force-constant matrix elements (15.75), rather than the vibrational frequencies, are scaled. The vibrations are expressed as stretches, bends, and torsions, and different scale factors (found from a training set of molecules) are used for the force constants of different kinds of vibrations. In all, 11 scale factors are used. With this procedure, B3LYP/6-31G* calculations after scaling gave an rms error of only 12 cm^{-1} for 843 fundamental frequencies of 30 test molecules. Only 8 out of 1506 frequencies of test and training molecules had errors exceeding 10%, and the average absolute percent error was 0.9%.

MM3 and MMFF94 give pretty accurate vibrational frequencies. For a sample of 157 frequencies of small molecules, the rms errors are 60 cm^{-1} for MMFF94 and 57 cm^{-1} for MM3 [T. A. Halgren, *J. Comput. Chem.*, **17**, 553 (1996)].

Entropies. Gas-phase entropies can be calculated from molecular properties as discussed in Section 15.13. Because semiempirical methods such as AM1 and PM3 give inaccurate results for low-frequency torsional vibrations, which contribute significantly to the entropy at room temperature, they do not give accurate entropies for compounds with internal rotation. Ab initio methods can be used to calculate entropies rather reliably, as discussed in Section 15.13.

Gibbs Energies of Solvation. The GB/SA method had an MAE of 0.9 kcal/mol for $\Delta G^\circ_{solv,298}$ of 30 uncharged compounds in water [D. Qiu et al., *J. Phys. Chem. A*, **101**, 3005 (1997)]. The more elaborately parametrized SM5 solvation models do better. For 13 versions of SM5 models, the MAE in $\Delta G^\circ_{solv,298}$ for 2084 values of uncharged solutes in aqueous and nonaqueous solvents ranged from 0.38 to 0.48 kcal/mol [G. D. Hawkins et al., in J. Gao and M. A. Thompson (eds.), *Combined Quantum Mechanical and Molecular Mechanical Methods*, Oxford, 1999]. Amazingly, the simple SM5.0R//AM1 model, which uses no charges (and thus needs no quantum-mechanical calculations except an AM1 calculation to get the gas-phase geometry), does as well as the more elaborate methods, having an MAE of 0.41 kcal/mol. The UAHF PCM model had an MAE of 0.2 kcal/mol for 43 uncharged organic molecules in water [V. Barone et al., *J. Chem. Phys.*, **107**, 3210 (1997)]. The COSMO-RS method had an MAE of 0.30 kcal/mol for 163 uncharged molecules in water [A. Klamt et al., *J. Phys. Chem. A*, **102**, 5074 (1998)].

The experimental $\Delta G^\circ_{solv,298}$ values in these studies are nearly all in the range +3 to −11 kcal/mol, for a range of 14 kcal/mol.

NMR Shielding Constants. To calculate NMR chemical shielding (screening) constants, the applied magnetic field **B** is treated as a perturbation, and one solves a set of equations called coupled perturbed equations. For details of the theory, see D. B. Chesnut in K. B. Lipkowitz and D. B. Boyd, *Reviews in Computational Chemistry*, vol. 8, VCH, 1996, Chapter 5. Just as the electric field is minus the gradient of the electric potential, the magnetic induction **B** is given by $\mathbf{B} = \nabla \times \mathbf{A}$, where **A** is the magnetic vector potential. Many different choices for **A** give the same **B**. A particular choice defines the "gauge" of the vector potential. Several methods have been proposed to give calculated quantum-mechanical results that are independent of the choice of gauge. The most widely used of these is the *gauge-including atomic orbital* (GIAO) method, in which each basis AO includes an exponential factor that contains **B**. Chemical shifts are reported in parts per million (ppm) and are the difference between screening constants in the molecule and that in a reference molecule such as tetramethylsilane for proton and ^{13}C NMR spectra. To calculate chemical shifts theoretically, one calculates shielding constants for the molecule of interest and the reference molecule and takes the differences. For 14 small molecules, MAEs in ^{13}C shifts in ppm using the GIAO method with a QZ2P basis set and MP2/TZ2P geometries were as follows: 9 for HF, 14 for LSDA, 6.5 for BPW91, 8 for BLYP, 7 for B3PW91, and 1.6 for MP2 [J. R. Cheeseman et al., *J. Chem. Phys.*, **104**, 5497 (1996)]. Since the observed shifts cover a range of about 200 ppm, all these results are good. The QZ2P basis set is too large for use with large molecules. For a series of molecules with two to eight nonhydrogen atoms, average absolute errors in ^{13}C shifts in ppm using the GIAO method with B3LYP/6-31G* geometries were 9 for HF/6-31G*, 9 for HF/6-311+G(2d,p), and 4 for B3LYP/6-311+G(2d,p) (Cheeseman et al., op. cit.). DFT calculations are most commonly used to predict NMR chemical shifts. MP2 calculations give better results than DFT calculations but are too time-consuming to use routinely on medium and large molecules.

Proton chemical shifts are harder to calculate accurately than ^{13}C shifts, since they span a smaller range (about 15 ppm) and are sensitive to solvent effects, but good results have been obtained with DFT.

Calculation of NMR spin–spin coupling constants is more challenging than calculation of shielding constants but can be done with fair accuracy with DFT, thereby allowing the complete prediction of the NMR spectrum. For example, the ^{1}H and ^{13}C NMR spectra of several medium-size natural-product molecules were calculated with a fair degree of accuracy using such methods as B3LYP/6-31G**, and agreement with experiment was improved when the effect of solvent was allowed for by using the PCM method [A. Bagno et al., *Chem. Eur. J.*, **12**, 5514 (2006)].

For more on calculating NMR chemical shifts and coupling constants, see E. R. Wilson, *Chem. Eng. News*, Sept. 28, 1998, p. 25; T. Helgaker et al., *Chem. Rev.*, **99**, 293 (1999); *Cramer*, Section 9.4; A. Wu et al., *J. Comput. Chem.*, **28**, 2431 (2007).

18.4 HYDROGEN BONDING

Hydrogen bonding is important throughout chemistry and biochemistry, and for a method to give reliable results when applied to large biological molecules, it must treat hydrogen bonding properly.

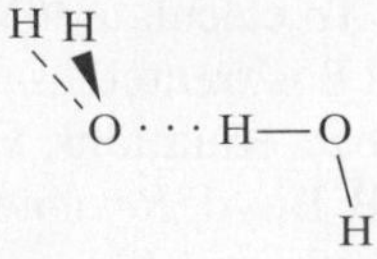

H H
O···H—O
H

FIGURE 18.1
The water dimer.

Many gas-phase hydrogen-bonded dimers have been characterized spectroscopically; examples include $(H_2O)_2$, $(HCl)_2$, $HF–H_2O$, and HF–HCN. For $(H_2O)_2$, the structure (determined by molecular-beam microwave spectroscopy) is shown in Fig. 18.1 [T. R. Dyke, et al., *J. Chem. Phys.*, **66**, 498 (1977)].

For H-bonded dimers, HF STO-3G, 3-21G, and 3-21G$^{(*)}$ calculations give equilibrium-geometry separations between the heavy atoms that are usually substantially in error (errors of 0.1 to 0.5 Å); 6-31G* calculations give heavy-atom separations in pretty good agreement with experiment (*Hehre et al.*, Table 6.32).

Suppose we want to calculate the dimerization energy of H_2O using the 3-21G basis set. The natural procedure would be to calculate the energy of the dimer $(H_2O)_2$ at its 3-21G equilibrium geometry using a 3-21G basis set on each of the six atoms of $(H_2O)_2$ and to calculate the energy of each H_2O monomer at its 3-21G equilibrium geometry using a 3-21G basis set on each of the three atoms of the monomer, and to take the dimerization energy as

$$\Delta\varepsilon = \varepsilon(AB)^{d}_{ab} - \varepsilon(A)^{m}_{a} - \varepsilon(B)^{m}_{b} \tag{18.1}$$

Here A and B stand for the monomer molecules and AB for the dimer. In this case, $A = H_2O$, $B = H_2O$, and $AB = (H_2O)_2$. The letters a, b, and ab symbolize the 3-21G basis set centered on the atoms of A, the 3-21G basis set centered on the atoms of B, and the 3-21G basis set centered on the atoms of AB, respectively. The letters d and m denote calculated dimer and monomer equilibrium geometries: $\varepsilon(A)^{m}_{a}$ is the energy of monomer A at the A equilibrium geometry calculated with the a basis set, with a similar meaning for $\varepsilon(B)^{m}_{b}$; the quantity $\varepsilon(AB)^{d}_{ab}$ is the energy of AB at the equilibrium geometry of AB calculated with the ab basis set. For $(H_2O)_2$, $\varepsilon(A)^{m}_{a} = \varepsilon(B)^{m}_{b}$, but this is not true for a mixed dimer such as $HF–H_2O$.

However, this procedure involves an inconsistency. When the monomer energy $\varepsilon(A)^{m}_{a}$ is calculated, the electrons of A have available to themselves only the 3-21G orbitals on the three atoms of A, whereas, when $\varepsilon(AB)^{d}_{ab}$ is calculated, the electrons of each H_2O molecule within the dimer have available not only the orbitals on their own nuclei, but also the orbitals on the nuclei of the other H_2O molecule. In effect, the dimer basis set is larger than that of each monomer, and this produces an artificial lowering of the dimer energy relative to that of the separated monomers. This artificial lowering is called the **basis-set superposition error** (BSSE). The BSSE would vanish in the limit of using a complete set for each monomer. The most common procedure to correct for the BSSE is to calculate the dimerization energy by adding the following correction to $\Delta\varepsilon$ of (18.1):

$$\varepsilon(A)^{d}_{a} - \varepsilon(A)^{d}_{ab} + \varepsilon(B)^{d}_{b} - \varepsilon(B)^{d}_{ab} \tag{18.2}$$

where d indicates that each of the four quantities in (18.2) is calculated with the A or B atoms located at their calculated locations in the dimer, and $\varepsilon(A)^{d}_{ab}$ is calculated with a basis set that consists of 3-21G orbitals on each nucleus of the monomer A and the appropriate 3-21G orbitals (called "ghost orbitals") centered at the three points in space that would correspond to the equilibrium positions of the other three nuclei in the dimer. This procedure, called the **counterpoise** (CP) **correction**, has been very controversial, but is now accepted by the majority of researchers as probably the best way to reduce the BSSE [G. Chalasinski and M. M. Szczesniak, *Chem. Rev.*, **94**, 1723 (1994); F. B. van Duijneveldt et al., *Chem. Rev.*, **94**, 1873; see also the references in A. Galano and J. R. Alvarez-Idaboy, *J. Comput. Chem.*, **27**, 1203 (2006)]. Because of the asymmetry of $(H_2O)_2$, $\varepsilon(A)_{ab}^{d} \neq \varepsilon(B)_{ab}^{d}$.

From the temperature and pressure dependences of the thermal conductivity of water vapor, ΔH_{373}° for $2H_2O(g) \rightarrow (H_2O)_2(g)$ has been found to be -3.6 ± 0.5 kcal/mol [L. A. Curtiss et al., *J. Chem. Phys.*, **71**, 2703 (1979)]; this ΔH_{373}° is found to correspond to an energy change involving nonvibrating, nonrotating, nontranslating species of $\Delta E_{el} = -5.0 \pm 0.7$ kcal/mol [E. M. Mas et al., *J. Chem. Phys.*, **113**, 6687 (2000)], where theoretically calculated vibrational frequencies of the dimer are used to help find the electronic energy change ΔE_{el}.

MP2 calculations with large aug-cc-pVnZ basis sets yielded an estimate of -5.0 ± 0.1 kcal/mol for the binding energy of the water dimer in the complete basis-set limit, and the effect of going to the MP4 level was found to be negligible [M. W. Feyereisen et al., *J. Phys. Chem.*, **100**, 2993 (1996)]; essentially the same result was found by a method called symmetry-adapted perturbation theory (SAPT), which treats the interactions between the monomers as a perturbation [E. M. Mas and K. Szalewicz, *J. Chem. Phys.*, **104**, 7606 (1996)]. CCSD(T) calculations extrapolated to the complete-basis-set limit gave -5.02 ± 0.05 kcal/mol [W. Klopper et al., *Phys. Chem. Chem. Phys.*, **22**, 2227 (2000)]. Thus theory and experiment are in agreement, and the ab initio calculations have a much smaller margin of error than the experimental result.

Density-functional studies of hydrogen-bonded dimers generally found good results with the B3LYP functional provided large enough basis sets (valence triple zeta with polarization functions) were used [B. Paizs and S. Suhai, *J. Comput. Chem.*, **19**, 575 (1998)].

The AM1 method applied to hydrogen bonding has serious failings. One study found AM1 heavy-atom separations in H-bonded species to be 0.1 to 0.9 Å too long, with an average error of 0.44 Å, and found AM1 hydrogen-bond energies to run one-fifth to two-thirds of the experimental values, being on average 57% too low [G. Buemi et al., *THEOCHEM*, **41**, 379 (1988)]. Another study concluded that "the AM1 method is an unreliable tool for studying systems with H-bonds" [A. A. Bliznyuk and A. A. Voityuk, *THEOCHEM*, **41**, 343 (1988)]. AM1 erroneously predicts $(H_2O)_2$ to have a structure with three hydrogen bonds, rather than the observed structure with one hydrogen bond [O. N. Ventura et al., *THEOCHEM*, **56**, 55 (1989)]. A study of AM1 and PM3 for 10 hydrogen-bonded species [J. Dannenberg, *J. Mol. Struct.* (*Theochem*), **401**, 287 (1997)], found that for species with OHO hydrogen bonds, both methods had serious problems.

PM6 gives improved performance in hydrogen-bond energies and in the geometries of hydrogen-bonded species as compared with AM1, PM3, and PM5 [J. J. P. Stewart, *J. Mol. Model.*, **13**, 1173 (2007)].

The original version of MM2 contained no special terms in the potential to provide for hydrogen bonding, but relied on the electrostatic interaction term V_{es} to produce hydrogen bonding. This procedure worked fairly well, but gave H-bond energies that were too small by 1 to 3 kcal/mol and gave distances between the two heavy atoms involved in the H bond that were several tenths of an angstrom too long. To overcome these errors, MM2 was modified by the addition of terms specific to hydrogen bonds, and MM2 now gives a good account of hydrogen bonding [N. L. Allinger, *J. Comput. Chem.*, **9**, 591 (1988)].

Since HF/6-31G* calculations give reasonably good descriptions of most H-bonded complexes [J. Pranata et al., *J. Am. Chem. Soc.*, **113**, 2810 (1991)], MMFF94 was parametrized to reproduce HF/6-31G* results for such complexes but with slightly larger dimerization energies and slightly smaller monomer–monomer distances. Both MMFF94 and OPLS give generally satisfactory descriptions of hydrogen bonding [T. A. Halgren, *J. Comput. Chem.*, **17**, 520 (1996)].

18.5 CONCLUSION

The overall reliability of the EH, CNDO, and INDO methods for calculating molecular properties is poor, and these methods were not considered in this chapter.

The ab initio SCF MO method is usually reliable for ground-state, closed-shell molecules, provided one uses a basis set of suitable size (at least 3-21G$^{(*)}$ for geometry and 6-31G* for energy differences) and uses appropriate procedures (such as bond-separation reactions) for finding energy differences. The STO-3G basis set is not generally reliable, and this basis set is little used nowadays.

MP2 perturbation theory usually substantially improves calculated properties as compared with HF results. DFT with gradient-corrected, hybrid, meta-GGA, and meta-hybrid functionals usually performs substantially better than the HF method.

The AM1, PM3, and other semiempirical methods are less reliable than HF calculations with basis sets of suitable size.

Molecular mechanics is usually reliable for those kinds of molecules for which the method has been properly parametrized, but some existing MM force fields are not very reliable. For small and medium organic compounds, MM2, MM3, MM4, and MMFF94 are generally reliable.

A comparison of the predictions of several force fields with ab initio local MP2 predictions of structure and energetics of several conformations of a tetrapeptide [M. D. Beachy et al., *J. Am. Chem. Soc.*, **119**, 5908 (1997)] found the MMFF94, OPLS, and AMBER force fields to give the best structure predictions and MMFF94 and OPLS fields to give the best energy predictions, but concluded that "a truly quantitative prediction of peptide and protein energetics via molecular mechanics is not yet available."

The comparisons of this section consider only compounds of H–Ar. For transition-metal (TM) compounds, ab initio SCF MO calculations often give poor results; see, for example, *Hehre et al.*, Section 6.2.7. DFT is currently by far the most widely used method to treat TM compounds. For a review of DFT applied to TM compounds, see J. N. Harvey, *Annu. Rep. Progr. Chem. Sec. C*, **102**, 203 (2006). A study of six density functionals with a large polarized quadruple zeta basis set found DFT to be "a useful semiquantitative tool" for TM chemistry [F. Furche and J. P. Perdew, *J. Chem. Phys.*, **124**, 044103 (2006)]. Geometries were found to be usually accurate but errors in energy differences were usually substantially larger than when dealing with main-group compounds. For example, for a set of 18 TM-compound reaction energies, the MAEs for the functionals were in the range 10 to 12 kcal/mol. Of the functionals tested, the BP86 and TPSS functionals showed the best price/performance ratio, and B3LYP behaved rather erratically.

This study did not include the M05, M06-L, or M06 functionals, and subsequent studies [Y. Zhao and D. G. Truhlar, *J. Chem. Phys.*, **124**, 224105 (2006); **125**, 194101 (2006); *Theor. Chem. Acc.*, **120**, 215 (2008); *Acc. Chem. Res.*, **41**, 157 (2008)] found these functionals to give superior performance, with MAEs of 8 (M05), 7 (M06-L), and 7 (M06) kcal/mol for the 18 reaction energies. These functionals were recommended for their wide applicability and accuracy.

The CCSD(T) method with an adequate basis set gives excellent results but can only be applied to rather small molecules. Some of the functionals developed subsequent to B3LYP provide much improved performance over B3LYP, but which functional provides the best results depends on what property is being calculated. The use of the RI method to speed up calculations is becoming widespread and allows larger molecules to be treated. The SCS-MP2 method provides a significant improvement over MP2 results and can give good results for fairly large molecules.

18.6 THE FUTURE OF QUANTUM CHEMISTRY

In the 1950s there was a general belief that meaningful ab initio calculation of molecular properties for all except very small molecules was out of the question. Quantum-chemistry books written in this period contain such statements as "we cannot hope ever to make satisfactory ab initio calculations [for organic compounds]" and "it is wise to renounce at the outset any attempt at obtaining precise solutions of the Schrödinger equation for systems more complicated than the hydrogen molecule ion." In 1959, Mulliken and Roothaan identified the "bottleneck" holding up accurate quantum-mechanical calculations on polyatomic molecules as the difficulty in evaluating multicenter integrals. This bottleneck has now been eliminated.

Ab initio HF calculations and geometry optimizations on moderate-size molecules have become routine, and computationally efficient methods (for example, DFT, MP2, and SCS-MP2) for inclusion of electron correlation are available. The degree of reliability of various quantum-mechanical methods and basis sets has been established by numerous calculations. The size of a molecule for which one can do an accurate ab initio or density-functional calculation is limited by the speed and storage capacity of the available electronic computers. As larger and faster computers are developed, it will become feasible to treat larger molecules.

The very substantial progress in quantum chemistry in recent years has made quantum-mechanical calculations a valuable tool to help decide a wide variety of questions of real chemical interest. Whereas years ago quantum-mechanical calculations on molecules were largely confined to journals read mainly by theoretical chemists, nowadays such calculations appear routinely in journals such as the *Journal of the American Chemical Society*, read by all kinds of chemists. Quantum chemistry is being applied to such problems as the hydration of ions in solution, surface catalysis, the structures and energies of reaction intermediates, the conformations of biological molecules, and the study of enzyme-catalyzed reactions. In many cases, theoretical calculations may not give definitive answers, but they are frequently good enough to allow for a very fruitful interaction of theory and experiment. Moreover, qualitative concepts such as the Woodward–Hoffmann rules have provided considerable insight into the course of chemical reactions and into chemical bonding.

The 1998 Nobel Prize in chemistry was shared by Walter Kohn, one of the developers of density-functional theory, and John A. Pople, one of the developers of the *Gaussian* series of programs and widely used Gaussian basis sets, the Pariser–Parr–Pople method, and the CNDO and INDO methods, and one of the first to apply the MP and CC methods to molecular calculations. The Nobel committee noted that computational quantum chemistry is "revolutionising the whole of chemistry."

In 1929, Dirac wrote, "The underlying physical laws necessary for the mathematical theory of ... the whole of chemistry are thus completely known, and the difficulty is only that the exact application of these laws leads to equations much too complicated to be soluble." Application of high-speed digital computers to quantum chemistry has overcome to a significant degree the difficulties referred to by Dirac. Of course, just a small fraction of chemically important problems have been successfully treated by quantum mechanics, but future prospects are bright.

Ab initio and density-functional calculations are now routinely used by many chemists as a valuable guide to experimental work and are revolutionizing the way chemistry is done. The future of quantum chemistry and the future of chemistry are inextricably linked.

PROBLEMS

18.1 (a) Write the bond-separation reaction for cyclopropene. (b) Calculated HF/6-31G* energies in hartrees are −152.91597 for CH_3CHO, −40.19517 for CH_4, −79.22875 for C_2H_6, and −113.86633 for H_2CO. Scaled HF/6-31G* zero-point vibrational energies in hartrees are 0.05404 for CH_3CHO, 0.04308 for CH_4, 0.07192 for C_2H_6, and 0.02633 for H_2CO. Calculate the ΔH_0° predicted by the HF/6-31G* method for the gas-phase CH_3CHO bond-separation reaction and compare with the experimental value of 11.5 kcal/mol.

Appendix

TABLE A.1 Physical Constants[a]

Constant	Symbol	SI Value
Speed of light in vacuum	c	2.99792458×10^{8} m/s
Proton charge	e	$1.6021765 \times 10^{-19}$ C
Permittivity of vacuum	ε_0	$8.8541878 \times 10^{-12}$ C^2 N^{-1} m^{-2}
Avogadro constant	N_A	6.022142×10^{23} mol^{-1}
Electron rest mass	m_e	9.109382×10^{-31} kg
Proton rest mass	m_p	1.672622×10^{-27} kg
Neutron rest mass	m_n	1.674927×10^{-27} kg
Planck constant	h	6.626069×10^{-34} J s
Faraday constant	$F = N_A e$	96485.34 C mol^{-1}
Permeability of vacuum	μ_0	$4\pi \times 10^{-7}$ N C^{-2} s^2
Bohr radius	$a_0 = \varepsilon_0 h^2/\pi m_e e^2$	$5.29177209 \times 10^{-11}$ m
Bohr magneton	$\beta_e = e\hbar/2m_e$	9.274009×10^{-24} J T^{-1}
Nuclear magneton	$\beta_N = e\hbar/2m_p$	5.050783×10^{-27} J T^{-1}
Electron g value	g_e	2.00231930436
Proton g value	g_p	5.5856947
Gas constant	R	8.3145 J mol^{-1} K^{-1}
Boltzmann constant	$k = R/N_A$	1.38065×10^{-23} J K^{-1}
Gravitational constant	G	6.674×10^{-11} m^3 kg^{-1} s^{-2}

[a]Adapted from P. J. Mohr, B. N. Taylor, and D. B. Newell (2007) "The 2006 CODATA Recommended Values of the Fundamental Physical Constants" at physics.nist.gov/constants and Mohr, Taylor, and Newell, *Rev. Mod. Phys.*, **80**, 633 (2008).

TABLE A.2 Conversion Factors[a]

1 erg $= 10^{-7}$ J
1 cal = 4.184 J
1 eV $= 1.6021765 \times 10^{-19}$ J $\hat{=}$ 23.06055 kcal/mol = 96.48534 kJ/mol
1 hartree $= 4.359744 \times 10^{-18}$ J = 27.21138 eV $\hat{=}$ 627.5095 kcal/mol = 2625.500 kJ/mol
1 debye $= 3.335641 \times 10^{-30}$ C m

[a]The symbol $\hat{=}$ means "corresponds to."

TABLE A.3 Relative Isotopic Masses

Isotope	Atomic Mass	Isotope	Atomic Mass
^{1}H	1.0078250	^{16}O	15.994915
^{2}H	2.014102	^{32}S	31.972071
^{12}C	12.000 . . .	^{35}Cl	34.968853
^{13}C	13.003355	^{37}Cl	36.965903
^{14}N	14.003074	^{127}I	126.90447

TABLE A.4 Greek Alphabet

Alpha	A	α	Iota	I	ι	Rho	P	ρ
Beta	B	β	Kappa	K	κ	Sigma	Σ	σ
Gamma	Γ	γ	Lambda	Λ	λ	Tau	T	τ
Delta	Δ	δ	Mu	M	μ	Upsilon	Υ	υ
Epsilon	E	ε	Nu	N	ν	Phi	Φ	ϕ
Zeta	Z	ζ	Xi	Ξ	ξ	Chi	X	χ
Eta	H	η	Omicron	O	o	Psi	Ψ	ψ
Theta	Θ	θ	Pi	Π	π	Omega	Ω	ω

TABLE A.5 Integrals[a]

$$\int x \sin bx \, dx = \frac{1}{b^2} \sin bx - \frac{x}{b} \cos bx \tag{A.1}$$

$$\int \sin^2 bx \, dx = \frac{x}{2} - \frac{1}{4b} \sin (2bx) \tag{A.2}$$

$$\int x \sin^2 bx \, dx = \frac{x^2}{4} - \frac{x}{4b} \sin (2bx) - \frac{1}{8b^2} \cos (2bx) \tag{A.3}$$

$$\int x^2 \sin^2 bx \, dx = \frac{x^3}{6} - \left(\frac{x^2}{4b} - \frac{1}{8b^3} \right) \sin (2bx) - \frac{x}{4b^2} \cos (2bx) \tag{A.4}$$

$$\int \sin ax \sin bx \, dx = \frac{\sin [(a-b)x]}{2(a-b)} - \frac{\sin[(a+b)x]}{2(a+b)}, \qquad a^2 \neq b^2 \tag{A.5}$$

$$\int x e^{bx} \, dx = \left(\frac{x}{b} - \frac{1}{b^2} \right) e^{bx} \tag{A.6}$$

$$\int x^2 e^{bx} \, dx = e^{bx} \left(\frac{x^2}{b} - \frac{2x}{b^2} + \frac{2}{b^3} \right) \tag{A.7}$$

$$\int_0^\infty x^n e^{-qx} \, dx = \frac{n!}{q^{n+1}}, \qquad n > -1, q > 0 \tag{A.8}$$

$$\int_0^\infty e^{-bx^2} \, dx = \frac{1}{2} \left(\frac{\pi}{b} \right)^{1/2}, \qquad b > 0 \tag{A.9}$$

$$\int_0^\infty x^{2n} e^{-bx^2} \, dx = \frac{(2n)!}{2^{2n+1} n!} \left(\frac{\pi}{b^{2n+1}} \right)^{1/2}, \qquad b > 0, n = 1, 2, 3, \ldots \tag{A.10}$$

$$\int_t^\infty z^n e^{-az} \, dz = \frac{n!}{a^{n+1}} e^{-at} \left(1 + at + \frac{a^2 t^2}{2!} + \cdots + \frac{a^n t^n}{n!} \right), \qquad n = 0, 1, 2, \ldots, a > 0 \tag{A.11}$$

[a]Indefinite integration of functions can be done at no charge at integrals.wolfram.com

Bibliography

Acton, F. S., *Numerical Methods That Work*, Harper & Row, 1970.

Atkins, P. W., and R. S. Friedman, *Molecular Quantum Mechanics*, 4th ed., Oxford University Press, 2005.

Bader, R. F. W., *Atoms in Molecules*, Oxford University Press, 1990.

Ballentine, L. E., *Quantum Mechanics: A Modern Development*, World Scientific, 1998.

Bates, D. R., ed., *Quantum Theory*, 3 vols., Academic Press, 1961.

Bethe, H. A., and R. W. Jackiw, *Intermediate Quantum Mechanics*, 3rd ed., Benjamin-Cummings, 1985.

Bethe, H. A., and E. E. Salpeter, *Quantum Mechanics of One- and Two-Electron Atoms*, Academic Press, 1957.

Blinder, S. M., *Introduction to Quantum Mechanics,* Elsevier Academic, 2004.

Burkert, U., and N. L. Allinger, *Molecular Mechanics* (ACS Monograph No. 177), American Chemical Society, 1982.

Carsky, P., and M. Urban, *Ab Initio Calculations*, Springer-Verlag, 1980.

Chase, M. W., et al., *JANAF Thermochemical Tables*, 3rd ed., American Chemical Society, 1985.

Christoffersen, R. E., *Basic Principles and Techniques of Molecular Quantum Mechanics*, Springer-Verlag, 1989.

Clark, T., *A Handbook of Computational Chemistry*, Wiley, 1985.

Cotton, F. A., *Chemical Applications of Group Theory*, 3rd ed., Wiley, 1990.

Cramer, C. J., *Essentials of Computational Chemistry*, 2nd ed., Wiley, 2005.

Dicke, R. H., and J. P. Wittke, *Introduction to Quantum Mechanics*, Addison-Wesley, 1960.

Dirac, P. A. M., *The Principles of Quantum Mechanics*, 4th ed., Oxford University Press, 1958.

Dykstra, C. E., *Introduction to Quantum Chemistry*, Prentice Hall, 1994.

Eyring, H., J. Walter, and G. E. Kimball, *Quantum Chemistry*, Wiley, 1944.

Fong, P., *Elementary Quantum Mechanics*, Addison-Wesley, 1962.

Foresman, J. B., and Æ. Frisch, *Exploring Chemistry with Electronic Structure Methods*, 2nd ed., Gaussian, 1996.

Griffiths, D. J., *Introduction to Quantum Mechanics*, Prentice Hall, 1995.

Halliday, D., and R. Resnick, *Physics*, 3rd ed., Wiley, 1978.

Hameka, H. F., *Quantum Mechanics*, Wiley, 1981.

Hehre, W. J., *Practical Strategies for Electronic Structure Calculations*, Wavefunction, 1995.

Hehre, W. J., L. Radom, P. v. R. Schleyer, and J. A. Pople, *Ab Initio Molecular Orbital Theory*, Wiley, 1986.

Helgaker, T., P. Jorgensen, and J. Olsen, *Molecular Electronic-Structure Theory,* Wiley, 2000.

Jammer, M., *The Conceptual Development of Quantum Mechanics*, McGraw-Hill, 1966.

Jensen, F., *Introduction to Computational Chemistry,* 2nd ed., Wiley, 2007.

Johnson, C. S., and L. G. Pedersen, *Problems and Solutions in Quantum Chemistry and Physics*, Addison-Wesley, 1974.

Jørgensen, P., and J. Oddershede, *Problems in Quantum Chemistry*, Addison-Wesley, 1983.

Karplus, M., and R. N. Porter, *Atoms and Molecules*, Benjamin, 1970.

Kauzmann, W., *Quantum Chemistry*, Academic Press, 1957.

Kemble, E. C., *The Fundamental Principles of Quantum Mechanics*, McGraw-Hill, 1937; Dover, 1958.

Leach, A. R., *Molecular Modelling*, 2nd ed., Pearson, 2001.

Levine, I. N., *Molecular Spectroscopy*, Wiley, 1975.

Levine, I. N., *Physical Chemistry*, 6th ed., McGraw-Hill, 2009.

Lewars, E., *Computational Chemistry*, Kluwer, 2003.

Lowe, J. P., and K. Peterson, *Quantum Chemistry*, 3rd ed., Elsevier Academic Press, 2006.

Margenau, H., and G. M. Murphy, *The Mathematics of Physics and Chemistry*, 2nd ed., Van Nostrand Reinhold, 1956.

McQuarrie, D. A., *Quantum Chemistry*, University Science, 1983.

McQuarrie, D. A., *Statistical Mechanics*, Addison Wesley, 1976.

Merzbacher, E., *Quantum Mechanics*, 3rd ed., Wiley, 1998.

Messiah, A., *Quantum Mechanics*, vols. 1 and 2, Halsted, 1963; Dover, 1999.

Mulliken, R. S., and W. C. Ermler, *Diatomic Molecules*, Academic Press, 1977.

Mulliken, R. S., and W. C. Ermler, *Polyatomic Molecules*, Academic Press, 1981.

Murrell, J. N., and A. J. Harget, *Semi-empirical Self-Consistent-Field Molecular Orbital Theories of Molecules*, Wiley-Interscience, 1971.

Murrell, J. N., S. F. A. Kettle, and J. M. Tedder, *Valence Theory*, 2nd ed., Wiley, 1970.

Offenhartz, P. O'D., *Atomic and Molecular Orbital Theory*, McGraw-Hill, 1970.

Park, D., *Introduction to the Quantum Theory*, 2nd ed., McGraw-Hill, 1974.

Parr, R. G., *Quantum Theory of Molecular Electronic Structure*, Benjamin, 1963.

Parr, R. G., and W. Yang, *Density-Functional Theory of Atoms and Molecules*, Oxford University Press, 1989.

Pauling, L., and E. B. Wilson, Jr., *Introduction to Quantum Mechanics*, McGraw-Hill, 1935; Dover, 1985.

Pilar, F. L., *Elementary Quantum Chemistry*, 2nd ed., McGraw-Hill, 1990.

Pople, J. A., and D. L. Beveridge, *Approximate Molecular Orbital Theory*, McGraw-Hill, 1970.

Press, W. H., S. A. Teukolsky, W. T. Vetterling, and B. P. Flannery, *Numerical Recipes*, 3rd ed., Cambridge University Press, 2007.

Ratner, M. A., and G. C. Schatz, *Introduction to Quantum Mechanics in Chemistry,* Prentice Hall, 2000.

Salem, L., *The Molecular Orbital Theory of Conjugated Systems*, Benjamin, 1966.

Schaefer, H. F., ed., *Applications of Electronic Structure Theory* (vol. 4 of *Modern Theoretical Chemistry*, W. Miller et al., eds.), Plenum, 1977.

Schaefer, H. F., *The Electronic Structure of Atoms and Molecules*, Addison-Wesley, 1972.

Schaefer, H. F., ed., *Methods of Electronic Structure Theory*, Plenum, 1977.

Schleyer, P. v. R., ed., *Encyclopedia of Computational Chemistry,* Wiley, 1998.

Schonland, D. S., *Molecular Symmetry*, Van Nostrand Reinhold, 1965.

Segal, G. A., ed., *Semiempirical Methods of Electronic Structure Calculation, Parts A and B*, Plenum, 1977.

Shoup, T. E., *Applied Numerical Methods for the Microcomputer*, Prentice-Hall, 1984.

Simons, J., *An Introduction to Theoretical Chemistry*, Cambridge University Press, 2003.

Slater, J. C., *Quantum Theory of Atomic Structure*, vols. I and II, McGraw-Hill, 1960.

Slater, J. C., *Quantum Theory of Molecules and Solids*, vol. I, *Electronic Structure of Molecules*, McGraw-Hill, 1963.

Sokolnikoff, I. S., and R. M. Redheffer, *Mathematics of Physics and Modern Engineering*, 2nd ed., McGraw-Hill, 1966.

Strang, G., *Linear Algebra and Its Applications*, 3rd ed., Harcourt, Brace, Jovanovich, 1988.

Streitwieser, A., Jr., *Molecular Orbital Theory for Organic Chemists*, Wiley, 1961.

Szabo, A., and N. S. Ostlund, *Modern Quantum Chemistry*, rev. ed., McGraw-Hill, 1989; Dover, 1996.

Taylor, A. E., and W. R. Mann, *Advanced Calculus*, 3rd ed., Wiley, 1983.

Whitaker, A., *Einstein, Bohr and the Quantum Dilemma*, Cambridge Univ. Press, 1996.

Wilson, E. B., Jr., J. C. Decius, and P. Cross, *Molecular Vibrations*, McGraw-Hill, 1955.

Wilson, S., *Electron Correlation in Molecules*, Oxford University Press, 1984.

Yarkony, D. R., ed., *Modern Electronic Structure Theory*, World Scientific, 1995.

Young, D., *Computational Chemistry*, Wiley, 2001.

Answers to Selected Problems

1.1 (a) F. (b) T. (c) T. (d) T. **1.2** (a) 1.867×10^{-19} J; (b) 5×10^{17}. **1.3** (a) 3.45 eV; (b) 451 nm. **1.4** 0.332 nm. **1.5** (a) 2.22×10^{-6} deg; (b) 2.22 nm. **1.7** $2mb^2x^2$. **1.8** (a) F. (b) F. **1.9** $3c\hbar^2/m = 6.67 \times 10^{-20}$ J. *Hint:* Use the time-independent Schrödinger equation. **1.10** (a) 3.29×10^{-6}; (b) 0.0753; (c) at $x = 0$. (d) *Hint:* Use the change of variable $z = -x$ in one of the integrals and use an integral in the Appendix. **1.11** 0.000216. **1.12** 4.978×10^{-6}. **1.13** None; the function in (c) is not normalized. **1.14** (a) 1/3; (b) 1/2. **1.15** 2.24, 0.0126. **1.16** $1 - 2(13)12/(26)25 = 13/25$. **1.17** (a) The Maxwell distribution of molecular speeds. **1.18** (b), (d), and (e) are imaginary. **1.21** (a) -1; (b) i; (c) 1; (d) 1; (e) $17 + 7i$; (f) $-0.1 - 0.7i$. **1.22** (a) -4; (b) $2i$; (c) $6 - 3i$; (d) $2e^{i\pi/5}$. **1.23** (a) 1, $\pi/2$; (b) 2, $\pi/3$; (c) 2, $4\pi/3$; (d) $5^{1/2}$, 296.6°. **1.25** (a) $e^{i\pi/2}$; (b) $e^{-i\pi}$; (c) $5^{1/2}e^{5.176i}$; (d) $2^{1/2}e^{i5\pi/4}$. **1.26** (a) 1, $-\frac{1}{2} + \frac{1}{2}\sqrt{3}i$, $-\frac{1}{2} - \frac{1}{2}\sqrt{3}i$. **1.28** (a) kg m s^{-2}; (b) kg m^2 s^{-2}. **1.29** 0.405 N. **1.30** (a) T. (b) F. (c) F. (d) T. (e) F. (f) T.

2.1 (a) $y = c_1e^{-3x} + c_2e^{2x}$; (b) $c_1 = -1/5$, $c_2 = 1/5$. **2.2** (b) $y = ae^x + bxe^x$. **2.3** (a) linear; (b) linear; (c) nonlinear. **2.4** (a) F. (b) F. (c) T. (d) F. (e) T. **2.6** (a) $\frac{1}{4} - (2n\pi)^{-1}\sin(n\pi/2)$; (b) 3; (c) $\frac{1}{4}$; (d) correspondence principle. **2.7** (a) 655; (b) 159. **2.9** (a) 1.8×10^{-17} J; (b) 11 nm; (c) UV. **2.10** 3.0×10^{26}. **2.11** 1.8 nm. **2.12** 4. **2.13** 1.0×10^{13} s^{-1}. **2.14** 3 and 2. **2.15** 323 nm. **2.16** The same energies and wave functions are obtained (although the mathematical expression for ψ looks different). **2.18** $e^{-iEt/\hbar}$ times (2.30). **2.21** 2. **2.22** 4.02 eV, 13.6 eV. **2.25** 0.264 nm. **2.26** 0.347 nm, 0.521 nm. **2.30** (a) F. (b) F. (c) T. (d) F. (e) T. (f) F.

3.1 (a) $-2x\sin(x^2 + 1)$; (b) $5\sin x$; (c) $\sin^2 x$; (d) x; (e) $-1/x^2$; (f) $36x^3 + 24x$. **3.2** (a) Operator; (b) function; (c) function; (d) operator; (e) operator; (f) function. **3.6** *Hint:* Read the definition of equality of operators. **3.8** (a) $20x^3$; (b) $6x^3$; (c) $x^2f'' + 4xf' + 2f$; (d) x^2f''. **3.10** $\hat{1}$. **3.12** (b) $\hat{A}$ and $\hat{B}$ linear and commute. **3.14** (a) $-\cos z$; (b) $2a + (4ax + 2b)d/dx$; (c) 0. **3.15** (a) Linear; (b) nonlinear; (c) linear; (d) nonlinear; (e) linear. **3.20** (a) Complex conjugation; (b) *Hint:* Consider an operator that is the product of three operators. **3.22** (a) Yes; (b) $1/p$ (c) $1/(p - a)$. **3.23** (a) Yes; (b) $1 - 2x$. **3.25** (a) Yes, 1; (b) no; (c) yes, -1; (d) yes, -1; (e) yes, -1. **3.26** The eigenfunctions are (2.30) with E replaced by the eigenvalues k, where $k \geq 0$. **3.29** (a) $i\hbar^3\partial^3/\partial y^3$; (b) $-i\hbar(x\partial/\partial y - y\partial/\partial x)$. **3.31** (a) $i\hbar$; (b) $2\hbar^2\partial/\partial x$; (c) 0; (d) 0; (e) $(\hbar^2/m)\partial/\partial x$; (f) $2yz\hbar^2\partial/\partial x$. **3.33** (a) $\int_0^2|\Psi(x,t)|^2\,dx$. (b) *Hint:* Try working Prob. 3.36. **3.34** (a) $\text{length}^{-1/2}$. **3.35** 7.58×10^{14} s^{-1}. **3.36** (a) 0.0108; (b) 0.306; (c) 0.306. **3.37** For (b), $n_x^2h^2/4a^2$. For (c), $n_z^2h^2/4c^2$. **3.42** (a) 17; (b) 6. **3.43** (a) Nondegenerate; (b) 6; (c) 4. **3.44** (a), (c), (d), (g). **3.45** (a) $a/2$; (b) $b/2$, $c/2$; (c) 0;

(d) $(1 - 3/2n_x^2\pi^2)a^2/3$, no, yes. **3.47** (a) No; (b) yes; (c) yes; (d) yes; (e) no. **3.49** (b) 12. **3.50** (a) T. (b) F. (c) F. (d) F. (e) F. (f) F. (g) F. (h) F. (i) T. (j) T. (k) T. (l) F. (m) T. (n) F. (o) F.

4.2 (a) $\Sigma_{n=0}^{\infty}(-1)^n x^{2n+1}/(2n+1)!$; (b) $\Sigma_{n=0}^{\infty}(-1)^n x^{2n}/(2n)!$. **4.3** (a) $\Sigma_{n=0}^{\infty} x^n/n!$. **4.4** (a) T. (b) T. (c) F. **4.6** (a) $c_{n+2} = (n^2 + n - 3)c_n/(n+1)(n+2)$; (b) $c_4 = -3c_0/8$, $c_5 = -3c_1/40$. **4.7** (a) Odd; (b) even; (c) odd; (d) neither; (e) even; (f) odd; (g) neither; (h) even. **4.9** (c) 0. **4.10** $h\nu/4, h\nu/4$. **4.12** $\pm(\alpha/9\pi)^{1/4}(2\alpha^{3/2}x^3 - 3\alpha^{1/2}x)e^{-\alpha x^2/2}$. **4.13** $e^{-\alpha x^2/2}(1 - 4\alpha x^2 + \frac{4}{3}\alpha^2 x^4)$. **4.14** $x = \pm\alpha^{-1/2}$. **4.18** $(2n + \frac{3}{2})h\nu, n = 0, 1, 2, \ldots$. **4.19** (a) $(v_x + \frac{1}{2})h\nu_x + (v_y + \frac{1}{2})h\nu_y + (v_z + \frac{1}{2})h\nu_z$; (b) 1, 3, 6, 10. **4.23** (a) 480 N/m; (b) 2.87×10^{-20} J; (c) 6.20×10^{13} Hz. **4.24** (a) 2989.96 cm^{-1}, 52.0 cm^{-1}; (b) 8346.0 cm^{-1}. **4.25** (a) 0.00142, 0.0160; (b) 0.159, 0.314. **4.29** With $s_r = 0.01$, $E/(\hbar^2/ml^2) = 4.93480218$, 19.73920752, 44.41320520. **4.30** (a) $s_r = 0.01$ and $x_{r,\max} - x_{r,\min} = 4$ give $E/(\hbar^2/ml^2) =$ 2.772516 and 10.605119. The method fails to find a third energy below 20. (b) $E_r = 3.356822$, 13.256836, 29.003101, 47.665198. (c) For (a), $E_r = 2.814429$, 10.751612, 19.991961. For (b), 3.413571, 13.475723, 29.452308, 48.143464. **4.31** For $s_r = 0.05$ and x_r ranging from -3.5 to 3.5, $E/m^{-2/3}\hbar^{4/3}c^{1/3} = 0.66798613, 2.39364258, 4.69678795$. **4.32** For $s_r = 0.02$ and x_r ranging from -3 to 3, $E/m^{-4/5}a^{1/5}\hbar^{8/5} = 0.7040487625, 2.7315324, 5.8841762$. **4.33** For $s_r = 0.05$ and $x_{r,\max} = 10$, we get $E/m^{-1/3}\hbar^{2/3}b^{2/3} = 1.855757, 3.244607, 4.381670$. **4.34** (c) For $s_r = 0.1$ and x_r from -6.5 to 6.5, we get $E/(\hbar^2/ma^2) = -6.125015, -3.125056, -1.125079, -0.120976$. **4.35** (c) $s_r = 0.05$ and x_r ranging from -6.5 to 6.5 gives 12 eigenvalues in this range. The lowest two are $E_r = 0.973365$ and 0.973395, and the highest is 9.794873. **4.36** (a) With $s_r = 0.01$, the lowest E_r is 5.740086; (c) With $s_r = 0.01$, the lowest two are $E_r = 63.869414269$ and 63.869414294. **4.41** 0.16; 0.12. **4.48** (b) *Hints:* Show that $\tanh ix = i \tan x$. $E_r = V_{0r}$ is not an eigenvalue. To help the Solver, you can add a constraint such as $E_r > 0.001$. For $V_{0r} = 1$, $E_r = 5.750345, 20.236043, 44.808373, 79.459210$. For $V_{0r} = 1000$, the lowest two are 66.399924233 and 66.399924251. **4.50** (b) V is an even function with 3 nodes and minima at $x = \pm(4a)^{-1/4}$; $V(\pm\infty) = \infty$. (c) Yes, since ψ has no interior nodes. **4.52** (a) F. (b) T. (c) T. (d) T. (e) F.

5.1 (a) No. (b) Yes. (c) Yes. (d) Yes. **5.3** $(3\hbar^3/i)\partial^2/\partial x^2$. **5.4** $\Delta x = (h/8\pi^2 m\nu)^{1/2}$, $\Delta p_x = (mh\nu/2)^{1/2}$, $\Delta x\, \Delta p_x = \hbar/2$. **5.5** $\Delta x = (5/192)^{1/2}l$, $\Delta p_x = (14)^{1/2}\hbar/l$. **5.8** 1, 0.707. **5.10** $|\mathbf{A}| = 7$, $|\mathbf{B}| = (33)^{1/2}$, $\mathbf{A}\cdot\mathbf{B} = 13$, $\mathbf{A}\times\mathbf{B} = -32\mathbf{i} - 18\mathbf{j} + 10\mathbf{k}$, $\theta = 71.1°$, $\mathbf{A}+\mathbf{B} = 2\mathbf{i} + 2\mathbf{j} + 10\mathbf{k}$, $\mathbf{A}-\mathbf{B} = 4\mathbf{i} - 6\mathbf{j} + 2\mathbf{k}$. **5.11** $\arccos(-1/3) = 109.47°$. **5.12** (b) 111.6°. **5.13** $\mathbf{grad}\, f = (4x - 5yz)\mathbf{i} - 5xz\mathbf{j} + (2z - 5xy)\mathbf{k}$; $\nabla^2 f = 6$. **5.14** (b) 3. **5.18** (a) $r = 5^{1/2}, \theta = \pi/2, \phi = 63.4°$. (b) $r = (10)^{1/2}, \theta = 184°, \phi = 180°$; (c) $r = (14)^{1/2}$, $\theta = 122.3°, \phi = 18.4°$; (d) $r = 3^{1/2}, \theta = 125.3°, \phi = 225°$. **5.19** (a) $x = -1, y = 0, z = 0$; (b) $x = 1.414, y = 0, z = 1.414$. **5.22** 35.3°, 65.9°, 90°, 114.1°, 144.7°. **5.34** (a) T. (b) F. (c) T. (d) T. (e) T. (f) F.

6.1 (a) T. (b) F. **6.5** (a) F. (b) T. **6.6** $0.31c, 0.64c, 0.91c, 1.20c, 1.24c$, and $1.80c$, where $c = 5.49 \times 10^{-19}$ J. **6.7** (a) T. (b) T. **6.9** (a) 1.1309 Å; (b) 230542 MHz and 345813 MHz; (c) 110189 MHz; (d) 2.945, 4.729. **6.10** Approximately 252.8 GHz. **6.13** 2×10^{39}. **6.15** (a) 10941.2 Å, 2.7400×10^{14} Hz; (b) 2735 Å, 1.096×10^{15} Hz. **6.16** 3971.2 Å, 3890.2 Å, 3647.1 Å. **6.17** (b) 4340.5 Å, 4101.8 Å. **6.20** -6.8 eV. **6.22** $5a/Z$. **6.23** $30a^2/Z^2$. **6.26** s states. **6.27** 14. **6.28** -108.8 eV. **6.29** a/Z. **6.30** At the origin (nucleus). **6.31** (a) 0.24. (b) 0.24. **6.34** (a) $-e'^2/a$; (b) $e'^2/2a$; (c) 1/137.0. **6.36** *Hint:* Use the fact that all directions of space are equivalent. **6.38** (a) All; (b) $\hat{H}, \hat{L}^2$; (c) all. **6.39** (a) A sphere centered at the origin; (b) m. **6.47** With $s_r = 0.05$ and r_r going from 1×10^{-12} to 6, we get $E/h\nu = 1.49999984, 3.4999985, 5.4999944, 7.499987$ for $l = 0$ and 2.499986, 4.499964, 6.499933, 8.499902 for $l = 1$. **6.49** With $s_r = 0.01$, $E/(\hbar^2/mb^2)$ is 4.93480 for the lowest $l = 0$ state and is 10.095357 for the lowest $l = 1$ state. **6.51** (a) 16×10^{-173}; (b) 1.6×10^{-51}; (c) 2.9×10^{-5}. **6.54** (a) F. (b) T. (c) F. (d) T. (e) F. (f) F.

7.1 (a) T. (b) T. (c) F. **7.8** $i(d/dx), 4\,d^2/dx^2$. **7.9** (c) and (d). **7.16** (b) $2p_x$, $2^{1/2}2p_1 - 2p_x$ or $2p_1, 2^{1/2}2p_x - 2p_1$; $\hat{H}, \hat{L}^2$. **7.17** (a) Use a table of integrals to help you. (b) $\pi^3/32 \simeq 1 - 1/3^3 + 1/5^3 - 1/7^3 + 1/9^3$; $\pi^3 \approx 31.021$; (c) -2.70%, 0.128%, -0.022%. **7.18** (b) $-1 = (4/\pi)(-1 + 1/3 - 1/5 + 1/7 - \cdots)$; -27.32%, -10.35%, -6.31%, -4.52%. **7.19** (a) F. (b) F. (c) T. **7.21** (c) Yes; no. **7.23** (b) and (c). **7.24** The n nth roots of 1. **7.30** (a) 0, 1, 0; (b) $\frac{1}{2}, 0, \frac{1}{2}$; (c) 0, 0, 1. **7.32** Probability 2/3 for $2\hbar^2$; probability 1/3 for $6\hbar^2$. **7.36** 1/4 for $h^2/8ml^2$, 3/4 for $h^2/2ml^2$. **7.37** Use a table of integrals. Outcomes are $n^2h^2/8ml^2$, where $n = 1, 2, 3, \ldots$. Probabilities are c_n^2, where $c_n/(210)^{1/2} = (6 - n^2\pi^2)(\sin n\pi)/n^4\pi^4 - (4\cos n\pi)/n^3\pi^3 - 2/n^3\pi^3$. **7.39** (b) $(2mE)^{1/2}, -(2mE)^{1/2}$; (c) probabilities $|c_1|^2$ for $(2mE)^{1/2}$ and $|c_2|^2$ for $-(2mE)^{1/2}$. **7.42** (a) 1; (b) 0; (c) 1; (d) 0. **7.44** $\frac{1}{2}f(0)$. **7.47** (a) 1.47×10^{-16} s. **7.49** (a) First row: 6, 2; second row: -12, -12. (b) First row: 2, 4; second row: 8, -8. (c) First row: 3, 0; second row: 4, 1; (d) First row: 6, 3; second row: 0, -9. (e) First row: -2, 5; second row: -16, -19. **7.50** First row of **CD**: $5i$, 10, 5; second row: 0, 0, 0; third row: $-i$, -2, -1. **DC** is a 1×1 matrix whose sole element is $5i - 1$. **7.56** (a) 0; (b) $\hbar$. **7.57** (b) $13h^2/32ml^2$; (c) Use product-to-sum trigonometric identities to evaluate the integral. $\langle x \rangle = \frac{1}{2}l + 8(3^{1/2}l/9\pi^2)\cos(3h^2t/8ml^2\hbar)$; $\langle x \rangle_{\max} = 0.656l$; $\langle x \rangle_{\min} = 0.344l$, since the cosine ranges from 1 to -1. **7.62** As a further hint, see the definition of a Hermitian operator. **7.63** (a) F. (b) T. (c) F. (d) F. (e) F. (f) T. (g) F. (h) F. (i) F. (j) F. (k) T. (l) F. (m) F. (n) T. (o) T. (p) F.

8.2 (a) $\langle \phi_1|\hat{H}|\phi_1\rangle = 5.753112\hbar^2/ml^2$; (b) $5.792969\hbar^2/ml^2$. **8.5** 1.3% error. **8.6** (a) $(5h^2/4\pi^2mb^2)$. **8.7** $0.72598\hbar^{3/2}a^{1/4}/m^{3/4}$. **8.9** 0% error. **8.11** (b) $k = 1.11237244$, 0.298%. **8.15** (a) $3h^2/2\pi^2ml^2$, 21.6%. **8.16** (a) $c = 8/9\pi$, 15% error; (b) $c = 8/729\pi$. **8.17** $20.23921\hbar^2/ml^2$. **8.18** (b) -84. **8.22** 1, 0, 4, -1. **8.24** -84. **8.25** (a) $x = 0$, $y = 0$. **8.26** (a) $x = 0$, $y = 0$, $z = 0$; (b) $x = -5k$, $y = -2k$, $z = 3k$. **8.28** (a) F. (b) T. (c) F. (d) F. **8.30** 1, -1. **8.31** $5.750518\hbar^2/ml^2$, $44.809711\hbar^2/ml^2$. **8.32** 1.3%, 6.4%. **8.44** For **A**: $\lambda = 1 + \sqrt{2}i$, $c_1 = -1/\sqrt{12} + i/\sqrt{6}$, $c_2 = \sqrt{3}/2$; $\lambda = 1 - \sqrt{2}i$, $c_1 = -1/\sqrt{12} - i/\sqrt{6}$, $c_2 = \sqrt{3}/2$. (Other answers are possible for the eigenvectors, depending on the choice of phase.) For **B**: $\lambda = 2$, $c_1 = 0$, $c_2 = 1$; $\lambda = 2$, $c_1 = 0$, $c_2 = 1$. For **C**: $\lambda = 4$, $c_1 = 1$, $c_2 = 0$; $\lambda = 4$, $c_1 = 0$, $c_2 = 1$. (Any linear combination of these eigenvectors is an eigenvector.) **8.46** (a) $\lambda = 3$, $c_1 = 2/5^{1/2}$, $c_2 = 1/5^{1/2}$; $\lambda = -2$, $c_1 = 1/5^{1/2}$, $c_2 = -2/5^{1/2}$. (b) Yes. Yes. (c) Yes. Yes. (d) $\mathbf{C}^{-1} = \mathbf{C}^{\mathrm{T}}$. **8.47** (a) $\lambda = 0$, $c_1 = i/2^{1/2}$, $c_2 = 1/2^{1/2}$; $\lambda = 4$, $c_1 = -i/2^{1/2}$, $c_2 = 1/2^{1/2}$. (b) No. Yes. (c) No. Yes. (d) $\mathbf{C}^{-1} = \mathbf{C}^{\dagger}$. **8.49** $\lambda = -2$, $c_1 = 2/5^{1/2} = 0.89443$, $c_2 = 0$, $c_3 = 1/5^{1/2} = 0.44721$; $\lambda = 3$, $c_1 = -0.44721$, $c_2 = 0$, $c_3 = 0.89443$; $\lambda = 5$, $c_1 = -0.27669$, $c_2 = 0.48420$, $c_3 = 0.83006$. **8.50** The rows of $\mathbf{A}^{-1}$ are $-1/3, 4/15, -1/3$; $0, 1/5, 0$; $-1/3, -2/15, 1/6$. **8.51** Eigenvalues: -3.366401, -0.185277, -0.004990, -0.000120, -0.0000011, 24.556790. Components of the eigenvector of the lowest eigenvalue: 0.6696, 0.4292, 0.1792, -0.0604, -0.2863, -0.4996. **8.53** The "missing" roots are imaginary numbers. **8.59** (a) With TOL $= 10^{-9}$ and 32 basis functions, we get 45.80785, 46.11184, 113.93885, 143.35815. **8.62** With TOL $= 10^{-8}$ and 13 basis functions, $E_r = 0.500000, 1.500001, 2.500002, 3.500215, 4.500204$ **8.64** With TOL $= 10^{-7}$ and 28 basis functions, $E_r = -0.4733, -0.1214, -0.0540$ **8.65** (a) T. (b) T. (c) T. (d) T. (e) T. (h) F. (j) F.

9.1 (b) C. **9.3** $15\,dh^2/64\pi^4\nu^2m^2$. **9.4** (a) $V_0/2 - (V_0/2n\pi)[\sin(\frac{3}{2}n\pi) - \sin(\frac{1}{2}n\pi)]$; (b) $5.753112\hbar^2/ml^2$, $20.23921\hbar^2/ml^2$. **9.6** $\sum_{k\neq n} H'_{kn}\psi_k^{(0)}/(E_n^{(0)} - E_k^{(0)})$, where $\psi_k^{(0)} = (2/l)^{1/2}\sin(k\pi x/l)$, $E_n^{(0)} - E_k^{(0)} = (n^2 - k^2)h^2/8ml^2$, $H'_{kn} = (V_0/\pi)[A_k/(n-k) - B_k/(n+k)]$ with $A_k = \sin[3(n-k)\pi/4] - \sin[(n-k)\pi/4]$, $B_k = \sin[3(n+k)\pi/4] - \sin[(n+k)\pi/4]$. **9.7** (b) $E^{(2)} = -0.0027338\hbar^2/ml^2$, $E^{(0)} + E^{(1)} + E^{(2)} = 5.750378\hbar^2/ml^2$. **9.8** 1.2×10^{-8} eV. **9.10** (a) $E^{(1)} = 0$ (parity); (b) $-(30v^2 + 30v + 11)c^2/8\alpha^3h\nu$. **9.11** -12.86 eV. **9.12** $E^{(0)} = -77.5$ eV; $E^{(1)} = 0$. **9.14** First power. **9.16** $3a/2\zeta$. **9.17** (a) $(5 \pm 5^{1/2})b$. **9.21** (b) $E^{(1)} = (1 + 2/\pi)^2 b/4$ for the ground state. $E^{(1)} = (1 + 2/\pi)b/4$, $(1 + 2/\pi)b/4$, for the states of the first excited level. **9.22** $0, 0, \pm 3e\mathscr{E}a_0$; $2p_1, 2p_{-1}, 2^{-1/2}(2s \mp 2p_0)$. **9.23** $1s3s$, two nondegenerate levels; $1s3p$, two triply degenerate levels; $1s3d$, two fivefold-degenerate levels. **9.24** -27.2 eV. **9.26** a. **9.30** (a) T. (b) F. (c) F. (d) F.

10.1 9.13×10^{-35} J s. **10.2** 54.7° (note that the component in the xy plane exceeds the z component). **10.5** (c) The equation cannot be obeyed for $s = 3/2$. **10.6** (a) Fermion; (b) fermion; (c) fermion; (d) boson. **10.9** (1) Neither. (2) Antisymmetric. (3) Symmetric. (4) Neither. (5) Symmetric. (6) Symmetric. **10.10** Ground state: $1s(1)1s(2)1s(3)$. **10.12** (a) $2^{-1/2}(1 - \hat{P}_{12})$. **10.14** One of my students gave the answer: "A permanent is used to put the wave into the function." **10.17** $2J_{1s2s} + J_{1s1s}$. **10.18** 1.61×10^{-23} J/T. **10.19** (b) 2.80×10^{10} Hz. **10.21** (b) 42.58 MHz. **10.27** (a) $\frac{1}{2}\hbar$ or $-\frac{1}{2}\hbar$; (b) $(\alpha + \beta)/2^{1/2}$, $(\alpha - \beta)/2^{1/2}$; (c) 0.5 for $\frac{1}{2}\hbar$, 0.5 for $-\frac{1}{2}\hbar$. **10.29** (d) eigenvalues $\frac{1}{2}\hbar, -\frac{1}{2}\hbar$. **10.30** (a) F. (b) T. (c) F. (d) F. (e) T. (f) T. (g) T.

11.1 (a) $2n^2$; (b) $4l + 2$; (c) 2; (d) 1. **11.5** 22. **11.6** 11/2, 9/2, 7/2, 5/2; (b) 11/2, 9/2, 9/2, 7/2, 7/2, 5/2, 5/2, 3/2, 3/2, 1/2. **11.11** (a) F. (b) T. **11.13** (a) $^1F, {}^1G, {}^1H, {}^3F, {}^3G, {}^3H$; (c) $^2D, {}^2S, {}^2P, {}^2D, {}^2F, {}^2G, {}^4P, {}^4D, {}^4F, {}^2P, {}^2D, {}^2F$. **11.14** (a), (c), (e), (f). **11.16** (a) 45°; (b) 70.53°, 70.53°, 70.53°, 180°. (c) 90°. **11.18** B, N, F. **11.19** (a) 28; (b) 1; (c) 9; (d) 10. **11.20** (a) 15; (b) 36. **11.21** (a) 2; (b) 1, 3; (c) 4, 2; (d) 8, 6, 4, 2; (e) 1, 3; (f) 2. **11.22** (a) $^1S_0$, 1; (b) $^2S_{1/2}$, 2; (c) $^3F_4$, 9; $^3F_3$, 7; $^3F_2$, 5; (d) $^4D_{7/2}$, 8; $^4D_{5/2}$, 6; $^4D_{3/2}$, 4; $^4D_{1/2}$, 2. **11.23** (a) $6^{1/2}\hbar$; (b) $2^{1/2}\hbar$; (c) $(12)^{1/2}\hbar$. **11.24** $^2S_{1/2}, {}^1S_0, {}^2S_{1/2}, {}^1S_0, {}^2P_{1/2}, {}^3P_0, {}^4S_{3/2}, {}^3P_2, {}^2P_{3/2}, {}^1S_0$. **11.25** $^2D_{3/2}, {}^3F_2, {}^4F_{3/2}, {}^7S_3, {}^6S_{5/2}, {}^5D_4, {}^4F_{9/2}, {}^3F_4, {}^2S_{1/2}, {}^1S_0$. Cobalt. **11.29** 64089.8, 75254.0, 105798.7, 64073.4, 64074.5, 75237.6, 75238.9, 75239.7, 105782.3, 64043.5, 64046.4, 64047.5, 75210.6, 75211.9, 105755.3, 87685, 109685, 98230 cm^{-1}. **11.30** 4.5×10^{-5} eV. **11.31** No. **11.33** 7.7×10^{-6} eV. **11.39** (a) T. (b) T. (c) F. (d) F. (e) F.

12.1 (a) F. (b) F. (c) T. **12.2** (a) $2\sigma_v, C_2$; (b) $3\sigma_v, C_3$; (c) $C_3, 3\sigma_v$; (d) σ; (g) none. **12.3** (a) $\hat{\sigma}_v, \hat{\sigma}'_v, \hat{C}_2, \hat{E}$; (b) $\hat{\sigma}_v, \hat{\sigma}'_v, \hat{\sigma}''_v, \hat{C}_3, \hat{C}_3^2, \hat{E}$; (c) same as (b); (d) $\hat{\sigma}, \hat{E}$; (g) $\hat{E}$. **12.5** (a) $\hat{E}$; (b) $\hat{\sigma}$; (c) $\hat{C}_2$; (d) $\hat{C}_2$; (h) $\hat{\imath}$. **12.6** (a) A $\hat{C}_2$ rotation about the line through F_3 and F_5. **12.7** (a) Yes. (b) No. (c) Yes. **12.8** (a) Lies along the C_2 axis; (b) lies along the C_3 axis; (e) is zero; (g) no information. **12.10** (a) The unit matrix of order 3. (b) A diagonal matrix with diagonal elements 1, 1, −1. (c) A diagonal matrix with diagonal elements −1, 1, 1. (d) A diagonal matrix with diagonal elements 1, −1, −1. **12.12** (b) No, since its eigenvalues are not all real. **12.15** $\overline{10}, \overline{3}, \overline{14}$. **12.16** (a) Yes. (b) No. **12.18** (a) $\mathcal{T}_d$; (b) $\mathscr{C}_{3v}$; (c) $\mathscr{C}_{2v}$; (d) $\mathscr{C}_{3v}$; (e) $\mathscr{O}_h$; (f) $\mathscr{C}_{4v}$; (g) $\mathscr{D}_{4h}$; (h) $\mathscr{C}_{3v}$. **12.19** (a) $\mathscr{D}_{2h}$. **12.20** (a) $\mathscr{D}_{6h}$; (b) $\mathscr{C}_{2v}$; (c) $\mathscr{C}_{2v}$; (d) $\mathscr{C}_{2v}$; (e) $\mathscr{D}_{2h}$. **12.21** (a) $\mathscr{C}_{\infty v}$. **12.22** (a) $\mathscr{O}_h$; (b) $\mathscr{C}_{4v}$. **12.23** (a) 6; (b) 2; (c) ∞. **12.26** The answer is not $\mathscr{D}_2$. **12.28** (a) $\mathscr{C}_{4v}$; (b) $\mathscr{C}_{\infty v}$; (c) $\mathscr{D}_{4h}$; (d) $\mathscr{C}_{4v}$; (e) $\mathscr{D}_{\infty h}$. **12.29** (a) The regular tetrahedron. **12.31** $\mathscr{C}_1, \mathscr{C}_s, \mathscr{C}_n, \mathscr{C}_{nv}$. **12.32** $\mathscr{C}_1, \mathscr{C}_n, \mathscr{D}_n, \mathcal{T}, \mathscr{O}, \mathscr{I}$. **12.33** The first player will win. The winning strategy is given in H. E. Dudeney, *Amusements in Mathematics*, Dover, 1958.

13.1 (a) F. (b) T. (c) F. **13.2** 432.07 kJ/mol. **13.3** 15.425 eV. **13.4** (a) 4.61 eV; (b) 4.48 eV. **13.5** (b) 2.5151 eV. **13.9** With $s_r = 0.01$ and x_r going from -0.70 to 0.80, we get 107.02, 319.68, 530.51, 739.51, 946.66, 1151.98 cm^{-1}. **13.14** (a) 1836.15; (b) -1; (c) 2π; (d) -2; (e) 4.134×10^{16} (one atomic unit of time $= \hbar a_0/e'^2 = 2.419 \times 10^{-17}$ s; (f) 137.036; (g) -0.49973; (h) 0.3934. **13.18** At $R = 0.5$, $k = 1.78$ and $U = 0.2682_4$. At $R = 1.0$, $k = 1.54$ and $U = -0.4410_0$. At $R = 2.0$, $k = 1.24$ and $U = -0.5865_1$. **13.19** $k = 1.238_0$ and $R = 2.003_3$. **13.21** (a) 0.998948, 1.001049; (b) 0.995439, 1.010694; (c) 1.028419, 1.015963; (g) 1.979895, 0.420001. **13.24** (a) Li_2; (b) C_2; (c) O_2^+; (d) F_2^+; (Actually, D_e of Li_2^+ is greater than that of Li_2). **13.25** (a) 2, 2; (b) 2.5, 1; (c) 1.5, 1; (l) $1\frac{1}{2}$, 1; (m) 2, 0; (n) $2\frac{1}{2}$, 1. **13.26** (a) $^3\Sigma_g^-$; (l) $^2\Pi_u$; (m) $^1\Sigma_g^+$; (n) $^2\Sigma_g^+$. **13.27** (a) 1; (b) 3; (c) 6; (d) 2. **13.28** (a) $^1\Sigma^-$; (b) $^3\Sigma^+$; (c) $^3\Pi_2, {}^3\Pi_1, {}^3\Pi_0$; (d) $^1\Phi_3$. **13.31** $j + k$ must be an even number. **13.32** VB. **13.36** (a) $0, {}^1\Sigma^+$; (b) $0, {}^1\Sigma^+$; (c) $0, {}^1\Sigma^+$; (d) $1, {}^2\Sigma^+$; (e) $1, {}^2\Pi$; (f) $1, {}^2\Pi$. **13.38** (a) F. (b) T. (c) F.

14.5 4.56 eV, 11.3 D. **14.19** The calculation converges to the same final result. **14.20** (b) 4 for each except 5 for -1. **14.23** (b) 0; (c) -2; (d) 3/2. **14.24** state 1. **14.26** (a) $\langle V \rangle = (5\zeta/8 - 2Z\zeta)e'^2/a_0$, $\langle T \rangle = \zeta^2 e'^2/a_0$. **14.27** $\langle V \rangle = 3\frac{1}{3}$ eV. **14.32** -31.95 eV, -63.90 eV, -83.33 eV, 31.95 eV. **14.34** (a) Z/n^2a. **14.35** $(v + \frac{1}{2})h\nu m$. **14.38** 0.45, 0.42, 0.28, and 0.25 Å. General formula: $R_e/[(Z_b/Z_a)^{1/2} - 1]$.

15.2 1, 1, 1, 1; 1, 1, -1, -1; 1, -1, 1, -1; 1, -1, -1, 1; where the symmetry eigenvalues are listed in the order $\hat{E}, \hat{C}_2(z), \hat{C}_2(y), \hat{C}_2(x)$. **15.3** (b) 2. **15.4** (a) 56. **15.5** (a) 35; (b) 65; (c) 95; (d) 125; (e) 115. **15.6** (a) 540; (b) 900; (c) 1404. **15.12** CO_2, for example. **15.13** (a) a_1: $H_1 1s + H_2 1s$, C1s, C2s, C2p_z, O1s, O2s, O_{2p_z}. b_1: C2p_x, O2p_x. b_2: C2p_y, O2p_y, $H_1 1s - H_2 1s$. (b) Ten σ, two π; seven σ, one π. (c) Partial answer: For i(C): C1s; for $b(CH_1)$: $H_1 1s$, C2s, C2p_y, C2p_z; for l_1(O): O2s, O2p_y, O2p_z. (d) 7×7. **15.14** There are nine a_1, nine b_2, two b_1, and two a_2 orbitals. **15.17** (c). **15.20** For the a_1 MOs, $3d_{z^2}$, and $3d_{x^2-y^2}$ contribute. For the $1b_2$ MO, $3d_{yz}$ contributes. For the $1b_1$ MO, $3d_{xz}$ contributes. **15.27** Interchanges the subscripts 1 and 3 for the b symmetry species. **15.28** (a) No. **15.29** Partial answer: $1a_g \approx g_2$; $1b_{1u} \approx g_6$; $2a_g \approx g_3 + g_1$; $2b_{1u} \approx g_5 + g_7 - g_8$, where the coefficients are omitted. **15.30** (a) Gradient: $2c_1 x\mathbf{i} + 2c_2 y\mathbf{j} + 2c_3 z\mathbf{k}$. Hessian rows: $2c_1$ 0 0; 0 $2c_2$ 0; 0 0 $2c_3$. **15.31** (a) $R_{CH} = 1.09$ Å, $R_{CO} = 1.43$ Å, $R_{OH} = 0.96$ Å, $\angle$HCH $= 109.5°$, $\angle$HCO $= 109.5°$, $\angle$COH a bit less than 109.5°, D(HCOH) $= 60°$. (b) $R_{CH} = 1.08$ Å, $R_{CC} = 1.34$ Å, $\angle$HCH a bit less than 120°, $\angle$HCC a bit more than 120°. **15.32** (a) One stationary point. (b) One saddle point. **15.33** (a) $x = 1$, $y = 2$. **15.35** Rows: 0 0.96 0.96; 0.96 0 1.51; 0.96 1.51 0 (where distances are in angstroms). **15.36** (a) T. (b) T. (c) T. (d) T. **15.37** (a) 0.14 kJ/mol; (b) 1.10 kJ/mol; (c) 0.0015 kJ/mol. **15.38** (a) 10.49 eV; (b) 10.18 eV, 234.8 kcal/mol; (c) 222.6 kJ/mol, 222.2 kJ/mol including vibrational contributions (Problem 15.37). **15.39** (a) Rows: C1; X2 1 1.0; O3 1 1.16 2 90.0; O4 1 1.16 2 90.0 3 180.0 (where the semicolons are not actually present in the Z-matrix). (b) C1; H2 1 1.09; H3 1 1.09 2 109.47; H4 1 1.09 2 109.47 3 120.0; H5 1 1.09 2 109.47 3 -120.0 (c) C1; O2 1 1.22; H3 1 1.08 2 120.0; H4 1 1.08 2 120.0 3 180.0 (d) N1; X2 1 1.0; H3 1 1.01 2 109.0; H4 1 1.01 2 109.0 3 120.0; H5 1 1.01 2 109.0 3 -120.0 **15.43** (a) $R_{CO} = 1.207$ Å, $R_{CH} = 1.083$ Å, $\angle$HCO $= 122.5°$, $\angle$HCH $= 114.9°$, planar; 2.66 D; 1337, 1378, 1693, 1916, 3163, 3234 cm^{-1}. (b) 2.66 D. **15.44** -75.5838626 hartrees at 100°; 108.0°. **15.45** 3.02 kcal/mol. **15.46** (a) Staggered conformation. (b) Planar with D(OCOH) $= 180°$. **15.48** (c) 1.6 kcal/mol (experimental value is 6 kcal/mol). **15.49** (a) 1.41 D, 4.55 D, 7.2 kcal/mol.

(b) 6.9 kcal/mol with a scale factor of 0.89. **15.50** (b) 0.9 kcal/mol. **15.51** 0 D, 2.24 D, 2.8 kcal/mol without zero-point energy, 2.5 kcal/mol with zero-point energy. **15.52** Only one conformer at a local minimum; 0.305 D. **15.53** 4. **15.54** 1.0 for H1*s*; 1.70 for He1*s*; 5.70 for C1*s*, 1.625 for C2*s* and C2*p*; 6.70 for N1*s*, 1.95 for N2*s* and N2*p*.

16.1 1.86×10^{28}. **16.2** 82% to 89% for $n = 20$; 44% to 55% for $n = 200$. **16.3** 0.9731. **16.7** (a) 8, 8; (b) 10, 6. **16.10** True. **16.11** (a) 1.180 Å, 16.43 eV, 378.9 kcal/mol. (b) 0.969 Å, 103.9°. **16.17** (a) $-(3\alpha/2)(3\rho/\pi)^{1/3}$. **16.23** (a) 20.47 eV, 17.04 eV, 16.38 eV. **16.32** (a) 2. **16.33** (a) 42; (b) 3. **16.36** (a) $R_{CH} = 1.059$ Å, $R_{CN} = 1.132$ Å; $R_{NH} = 0.985$ Å, $R_{CN} = 1.154$ Å. (b) $R_{CH} = 1.155$ Å, $R_{CN} = 1.169$ Å, $\angle$HCN = 77.5°. **16.37** (a) Both conformers at minima are planar. TS has D(HOCO) = 96.0°, and D(HOCH) = −86.4°. 13.5 and 7.4 kcal/mol from the low-energy and high-energy conformers, respectively.

17.1 (a) $\alpha + 2^{1/2}\beta, \alpha, \alpha - 2^{1/2}\beta$; $\phi_1 = \frac{1}{2}f_1 + 2^{-1/2}f_2 + \frac{1}{2}f_3$, $\phi_2 = 2^{-1/2}f_1 - 2^{-1/2}f_3$, $\phi_3 = \frac{1}{2}f_1 - 2^{-1/2}f_2 + \frac{1}{2}f_3$. (b) 0.707, 0.707; (c) 1, 1, 1; (d) 1.025, 0.318, 1.025; (e) 0.828β. **17.2** (a) Same as Problem 17.1; (b) 0.707, 0.707; 0.707; 0.707; (c) $\frac{1}{2}$, 1, $\frac{1}{2}$; 1.5, 1, 1.5; (d) 1.025, 0.318, 1.025; 1.025, 0.318, 1.025; (e) 0.828β, 0.828β. **17.6** $P_{12} = 1.448$, $P_{23} = 1.725$; $q_1 = 1.00$, $q_2 = 1.00$. **17.9** (a) 27 kcal/mol; (b) 1.4 eV. **17.10** 42 kcal/mol, 69 kcal/mol. **17.11** (a) $e_1 = \alpha + 3^{1/2}\beta$, $e_2 = \alpha$, $e_3 = \alpha$, $e_4 = \alpha - 3^{1/2}\beta$, $\phi_1 = 2^{-1/2}f_1 + 6^{-1/2}(f_2 + f_3 + f_4)$, $\phi_2 = 3^{-1/2}f_2 + 3^{-1/2}e^{2\pi i/3}f_3 + 3^{-1/2}e^{4\pi i/3}f_4$, $\phi_3 = 3^{-1/2}f_2 + 3^{-1/2}e^{-2\pi i/3}f_3 + 3^{-1/2}e^{-4\pi i/3}f_4$; $P_{12} = 0.577$; $q_1 = 1$, $q_2 = 1$, $q_3 = 1$, $q_4 = 1$; $F_1 = 0$, $F_2 = F_3 = F_4 = 1.155$, delocalization energy $= 1.464\beta$. (b) $P_{12} = P_{23} = 1.414$. *Hint:* Each set of spatially perpendicular MOs has three electrons. **17.12** $\alpha = -6.1$ eV, $\beta = -3.3$ eV; 6.9 eV. **17.13** (c) For azulene, $0.0231\,|\beta|$, aromatic. **17.14** (a) 1.40_1 Å; (b) 1.40_2 Å; (c) 1.40_9 Å. **17.15** (b) $\phi_1 = 0.301(f_1 + f_4 + f_5 + f_8) + 0.231(f_2 + f_3 + f_6 + f_7) + 0.461(f_9 + f_{10})$. **17.19** 0.84, 0.39, 0.39, 0.84; the end carbons. **17.23** (c) $\alpha + 1.33\gamma$, $\alpha + 0.80\gamma$, $\alpha + 0.80\gamma$, $\alpha - 1.33\gamma$, $\alpha - 1.33\gamma$, $\alpha - 4.0\gamma$. **17.25** Predicted $R_e = 0$. **17.27** CNDO: (b), (d), (e), (f), (g). INDO: (d), (e), (f), (g). MNDO: (d), (e), (g). **17.28** −59.24 kcal/mol. **17.29** (a) AM1: 0.004 D, −24.3 kcal/mol; PM3: 0.005 D, −23.7 kcal/mol (experimental dipole moment is 0.08 D). (b) AM1: 1.317 Å, 98.8°, 1.98 D, 4.0 kcal/mol; PM3: 1.290 Å, 93.5°, 1.77 D, −0.9 kcal/mol. (c) AM1: 22.0 kcal/mol; PM3: 23.5 kcal/mol. **17.30** 1.25 kcal/mol for AM1; 1.4 kcal/mol for PM3. **17.31** $R_{CO} = 1.227$ Å, $\angle$HCH = 115.5°, 2.32 D, lowest wavenumber is 1147 cm^{-1}. **17.32** D(CCCC) = 74.7° and $\mu = 0.01$ D for the gauche conformer. 0.7 kcal/mol. **17.33** 1.48 D, 4.02 D, 7.4 kcal/mol. **17.34** Transition state has $R_{CH} = 1.298$ Å, $R_{CN} = 1.216$ Å, $\angle$HCN = 67.5°. **17.35** (a) 5, 7, 3, 3, 3. **17.37** MM+ in HyperChem gives 3.9 kcal/mol. **17.38** MM+ in HyperChem gives 2.3 kcal/mol. **17.42** (a) −25.32 kcal/mol; (b) −32.50 kcal/mol; (c) 6.29 kcal/mol. **17.43** (a) Disrotatory; (b) conrotatory. **17.44** (a) High.

18.1 (b) 10.8 kcal/mol.

Index